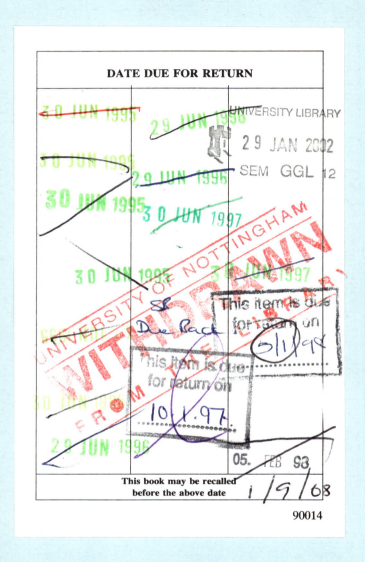

Polymers in Solution

Polymers in Solution

Their Modelling and Structure

Jacques des Cloizeaux and Gérard Jannink

Service de Physique Théorique du Centre d'Etudes Nucléaires de Saclay

Laboratoire Léon Brillouin au Centre d'Etudes Nucléaires de Saclay

Translated from the original French text
by J. des Cloizeaux
with the collaboration of G. Jannink

CLARENDON PRESS · OXFORD

1990

Oxford University Press, Walton Street, Oxford OX2 6DP
Oxford New York Toronto
Delhi Bombay Calcutta Madras Karachi
Petaling Jaya Singapore Hong Kong Tokyo
Nairobi Dar es Salaam Cape Town
Melbourne Auckland
and associated companies in
Berlin Ibadan

Published in the United States
by Oxford University Press, New York

British Library Cataloguing in Publication Data
des Cloizeaux, Jacques
Polymers in solution: their modelling and structure.
1. Polymers. Structure & properties
I. Title II. Jannink, Gerard
547.7
ISBN 0-19-852036-0

Library of Congress Cataloging-in-Publication Data
des Cloizeaux, Jacques.
[Polymères en Solution. English]
Polymers in solution: their modelling and structure/Jacques des Cloizeaux and Gérard Jannink translated from the original French text by J. des Cloizeaux, with the collaboration of G. Jannink.
Translation of: Polymères en solution.
1. Polymer solutions. I. Jannink, Gérard. II. Title.
QD381.9.S65D4713 1990 547.7'0454—dc20 89-70917
ISBN 0-19-852036-0

Typeset by Macmillan India Ltd, Bangalore-25

Printed in Great Britain by
Bookcraft (Bath) Ltd,
Midsomer Norton, Avon

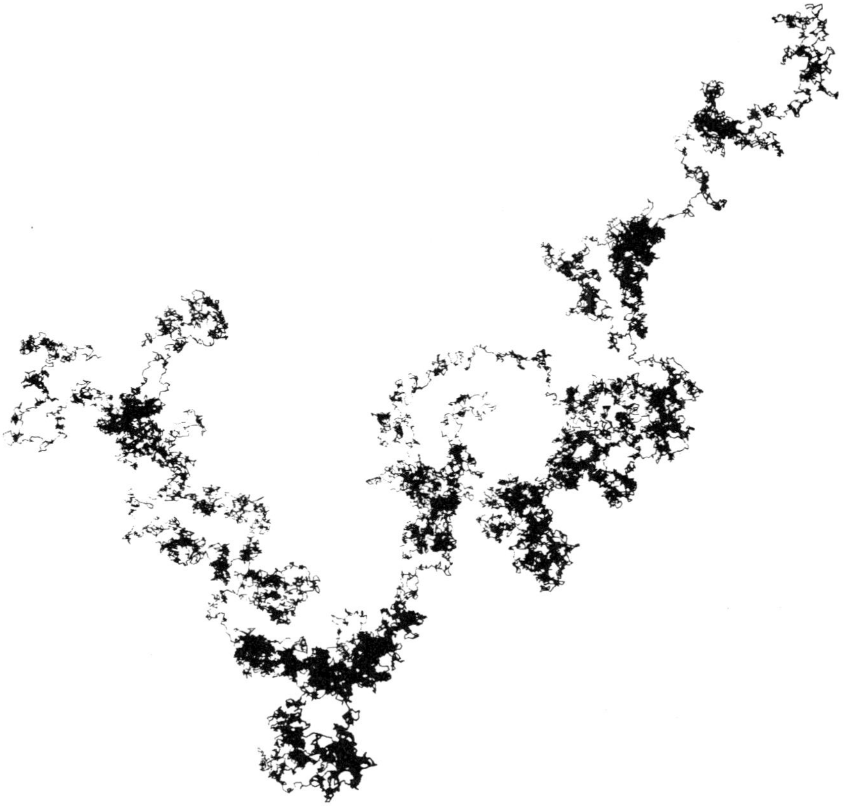

A Brownian chain (random walk) of 2×10^5 steps (from B. Duplantier and J.M. Luck, 1987)

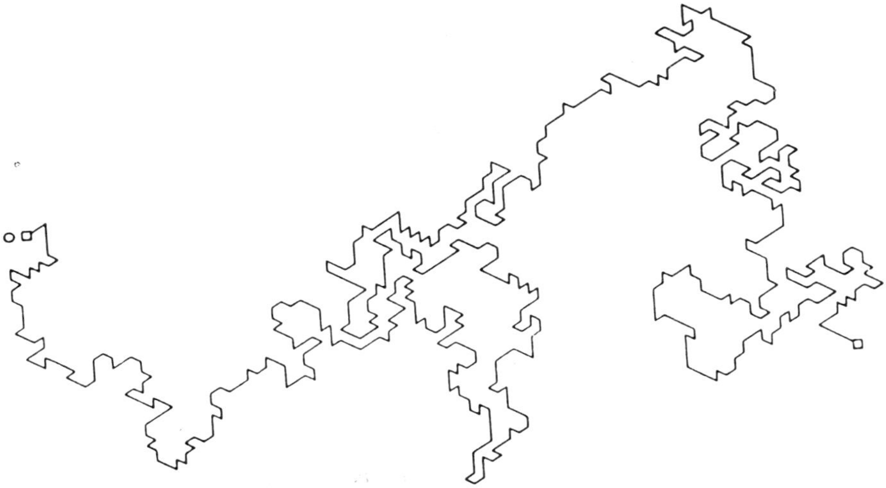

A self-avoiding chain (self-avoiding walk) drawn on a triangular lattice (from J.F. Renardy, 1971). The starting point of this chain is at O.

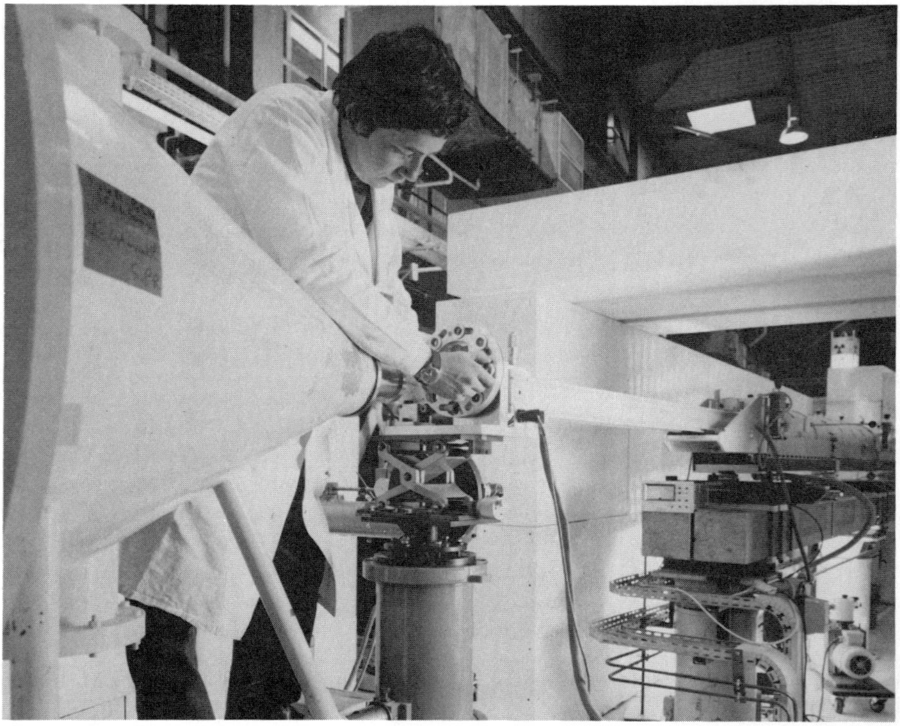

Preparation of a small-angle scattering experiment using neutrons: view of instruments at the Laboratoire Léon Brillouin in Saclay. Incident neutrons coming from the Orphéc source are guided in the tube seen on the right hand side of the figure. A sample holder stands at the tube end. The operator is setting a sample of a polymer solution in place. The cone seen on the left side is the front part of the void duct in which the scattered beam propagates; intensity is recorded on a detector placed at the rear of the duct.

PREFACE TO THE ENGLISH EDITION

The authors are pleased to present an English version of their book *Polymères en solution: leur modélisation et leur structure* to the scientific public. The text is practically the same, but a few recent results and references have been added; moreover, misprints, errors, and obscurities have been corrected. Since 1987, no great progress has been made in the field, except in two dimensions. The authors hope that the book will be useful to the community.

Saclay J.C.
23 December 1988 G.J.

PREFACE

This book is dedicated to the study of the static properties of flexible polymers in solution; its aim is to present the vast progress made by both theory and experiment over the last fifteen years. It might be asked why this question has aroused such lively interest and why great effort has been expended in this domain. What has progressively appeared is that long polymer chains, despite the variety in their chemical composition and physical properties, behave in a universal manner when in solution. Recognition of this fact is on a par with the formulation of the critical system concept of which the polymer in solution is a very good example.

At the same time, the use of new means of observation by diffraction of neutrons and photons has led to a better experimental knowledge of the critical behaviour. Finally, at the juncture between theory and experiment, the development of computers has, by means of network simulation, enabled interesting results to be obtained.

It should be noted, however, that many researchers in the field of polymers have been unwilling to accept the new principles, thereby minimizing their impact and interest. By stressing the diversity of the components of the different types of polymers, they have continued to use inadequate and superfluous parameters. Thus, the use of a poor notation has often masked the simplicity and beauty of the phenomena. Nevertheless, today there exists a substantial number of results which outstandingly show the validity of the universality hypothesis. Only appearing on a large scale, this universality is, moreover, compatible with the existence of specific physical properties which can be observed on more reduced scales.

As early as 1980, P.G. de Gennes brilliantly presented the question, in the light of the new principles, in his book *Scaling Concepts in Polymer Physics*, but the authors wanted to study thoroughly the concepts and develop the calculations which led them to treat the various aspects of a more restricted subject.

This work is the fruit of the collaboration of authors who have both worked for a good many years on polymer solutions, but these authors differ profoundly in character, education, and interests. One of them is a pure product of the French University system, the other carried out his studies in Switzerland and the United States: one is a theoretician, the other is more of an experimentalist. This disparity, which has not always facilitated the writing of the various chapters, despite the excellent relationship between the two, has certainly contributed to the value of the book. In this work, the authors have presented all the facets of the question, starting from first principles, and they have striven to extend their explanations as far as they could without entering into horrible complications. Moreover, the confrontation between theoretical and experimental points of view was inevitable; as G. Bachelard said of any scientific activity 'If it reasons, you must experiment; if it experiments, you must reason'. Difficulties nevertheless emerged. First, it is clear that theoretical propositions and experimental facts must be established in an autonomous way. The risk of 'seeing' what is predicted, or, inversely, of 'explaining' what is seen, without a true agreement, is great. Moreover, the fundamental quantities are obviously perceived in different ways by the theoretician and the experimentalist. Thus, to define the size of a polymer, the theoretician naturally uses the end-to-end distance, whereas the experimentalist, equally naturally, uses the radius of gyration. This is only an example; the reader will recognize others which are more serious. These difficulties, however, can be overcome and, on essential points, the views converge. Between theory and experiment, an agreement which is not only qualitative but also quantative will be sought and this is favoured by the universality of the behaviour of long polymers in solutions. Thus, and for more than one good reason, this book is a work of synthesis.

In particular, the authors endeavoured to present old and new results within the same scope thereby extending the work of their predecessors. This book is thus linked to *Polymer Chemistry* by Paul Flory, to *Modern Theory of Polymers* by Hiromi Yamakawa, and naturally to *Scaling Concepts in Polymer Physics* by Pierre Gilles de Gennes.

Despite writing the chapters quite independently, the authors wanted to give a true unity to the book. Thus, throughout the work, they aimed at using coherent notation and reasonable designations. Consequently, logic sometimes forced them to distance themselves somewhat from awkward traditions. Nevertheless, this problem of notation has not always been easy to solve, due to the large number of disciplines concerned by the study of polymers; namely, computer simulation, statistical mechanics and theory of liquids, description of the polymers themselves, field theory, study of the structure of a system by neutron or photon scattering, principles of renormalization theory and applications. The work thus contains very diverse elements.

The authors, however, had to restrict themselves and thus sacrificed the study of dynamic properties. No doubt it should not be regretted. In fact, these dynamic phenomena are very complex and not as well understood as the static phenomena, and for several reasons. First, we have to take into consideration the special hydrodynamic effects, a difficult task. In addition, it can be shown that the critical asymptotical limits are much more difficult to obtain in dynamics than in statics. In dynamics, we are always in an ill-defined crossover zone which is not easy to study with precision. For all these reasons, the authors considered only the static problem.

Within this framework, the book can be used as a basic manual by the student who wishes to instruct himself in the domain. It can also serve as a reference book, providing material for the experienced researcher. It is aimed, in the first instance, at the physicist, theoretician or experimentalist, as well as the physico-chemist. It may also interest certain mathematicians and chemists.

As with all the sciences, the study of polymers in solution developed in an anarchic fashion, drawing on very diverse disciplines. In order to help the reader to get his bearings, a good number of the chapters begin with a genesis which retraces the evolution of the ideas and techniques as the authors perceived them. These geneses are not historical accounts and do not attempt to be impartial; they simply serve as a guide to the reader. They might also interest the science historian, as the Memoirs of Saint-Simon interest the historian of the reign of Louis XIV.

Glancing through this preface, the reader will have understood that this book can be nothing other than imperfect, and for this, he or she has the authors' apologies, as do all the scientists whose works have been badly quoted or even unjustly forgotten.

The writing was nevertheless improved by the constructive criticisms which were made by certain physicists whom the authors would like to thank.

Saclay J.C.
18 May 1987 G.J.

ACKNOWLEDGEMENTS

Over many years, the authors benefited from numerous discussions with Pierre-Gilles de Gennes and Edouard Brézin. Moreover, the authors talked a lot about polymers with Henri Benoit, Paul Rempp, Claude Strazielle, Claude Picot, and Gilbert Weill from the Institut Charles Sadron in Strasbourg, and also with Sam Edwards and Robin Ball from the Cavendish Laboratory in Cambridge. Many experimental facts, observed at Saclay, have been discussed at length with Mireille Adam, Jean Pierre Cotton, Mohamed Daoud, Michel Delsanti, and Bernard Farnoux of the Laboratoire Léon Brillouin. Concerning theory and especially renormalization problems, the authors received valuable advice from their colleagues of the Service de Physique Théorique at Saclay, in particular from Michel Bergère, Mabrouk Benhamou, Claude Itzykson, Jean Zinn-Justin, and especially Bertrand Duplantier who provided assistance with Chapters 13, 14, and 16. K. van Holde, T. Witten, S. Doniach, J. Hodges, and Vendla Meyer helped us with the translation. To all, the authors express their gratitude.

The authors also thank Oxford University Press for accepting the translation provided and for the quality of the printed text.

CONTENTS

NOTATION AND MAIN SYMBOLS

UNITS

Atomic lengths are expressed in nanometers (1 nm = 10 Å).

CONVENTIONS

(a) As a rule, the symbols concerning the internal structure of polymer chains are printed in italic type; the symbols concerning a set of polymers are printed in bold italic type.
Example: N = number of links of a polymer
$\boldsymbol{N}$ = number of polymers in a set

(b) Sets are characterized by braces.
Example: $\{r\}$ = set of vectors $\vec{r}_j$ or $\vec{r}(s)$.

(c) A Fourier transform is indicated by a tilde.
Example: $\tilde{f}(k) = \int_{-\infty}^{+\infty} \mathrm{d}x\, \mathrm{e}^{ikx} f(x)$.

(d) When a Fourier transform $\tilde{f}(\vec{k}_1, \ldots, \vec{k}_n)$ is proportional to $\delta(\vec{k}_1 + \ldots + \vec{k}_n)$, we write:

$$\tilde{f}(\vec{k}_1, \ldots, \vec{k}_n) = (2\pi)^d \delta(\vec{k}_1 + \ldots + \vec{k}_n)\, \bar{f}(\vec{k}_1, \ldots, \vec{k}_{n-1}, -\vec{k}_1 - \ldots - \vec{k}_{n-1}).$$

(e) The average of an observable $\mathcal{O}$ over configurations is denoted by $\langle\!\langle \mathcal{O} \rangle\!\rangle$. On the contrary, the average of $\mathcal{O}$ over chain lengths is denoted by $\langle \mathcal{O} \rangle$; moreover, we use the notation

$$\mathcal{O}_{\mathrm{n}} = \langle \mathcal{O} \rangle \qquad \mathcal{O}_{\mathrm{w}} = \langle N\mathcal{O} \rangle / \langle N \rangle \qquad \mathcal{O}_{\mathrm{Z}} = \langle N^2 \mathcal{O} \rangle / \langle N^2 \rangle$$

where N is the number of links (or of monomers).

(f) When we put an index 'zero' on the upper left-hand side of a symbol, we refer to an unperturbed system (pure solvent, Brownian chain without interaction). Indices 'one' or 'two' in the same position refer to orders of perturbation

(g) In order to distinguish similar quantities pertaining to the solvent or to the solute, we use the following convention: an index 'zero' on the lower right-hand side of a symbol indicates that the corresponding quantity concerns the solvent; an index 'one' on the lower right-hand side of a symbol indicates that the corresponding quantity concerns the solute.

(h) The universal functions which appear in scaling laws are generally represented by symbols printed in upright type.

MAIN SYMBOLS

a	Constant of the Landau–Ginzburg model (coefficient of $(\vec{\nabla}\phi)^2$)
a_c	Critical value of the constant a for the Landau–Ginzburg model
a_B	Value of the two-leg vertex function for the Landau–Ginzburg model at $\vec{k}=0$ $(a_B = \boldsymbol{\Gamma}_B(\vec{0},\vec{0}))$
a_R	Value of the two-leg 'renormalized' vertex function for $\vec{k}=0$ $[a_R = \boldsymbol{\Gamma}_R(\vec{0},\vec{0})]$
$\boldsymbol{a}(\boldsymbol{b})$	The number of a chain ($N_{\boldsymbol{a}}$ = number of links of the chain $\boldsymbol{a}$)
$A(B)$	Number of links (A like N) or 'area' of a polymer chain (A like S)
A_2, A_3	Second and third virial coefficient (expansion in ρ) for a monodisperse solute $\left(\boldsymbol{\Pi}/\boldsymbol{RT} = \dfrac{\rho}{M} + A_2\rho^2 + A_3\rho^3 + \ldots\right)$
$A_{2,\Pi}$	Second virial coefficient (expansion in ρ) for a polydisperse solute $\left(\boldsymbol{\Pi}/\boldsymbol{RT} = \dfrac{\rho}{M_n} + A_{2,\Pi}\,\rho^2 + \ldots\right)$
$A_{2,\sim}$	Mean value of the second virial coefficient, obtained by measuring the osmotic compressibility in a scattering experiment
$\boldsymbol{A}_2, \boldsymbol{A}_3$	Second and third virial coefficient (expansion in $\boldsymbol{C}$) for a monodisperse solute $(\boldsymbol{\Pi\beta} = \boldsymbol{C} + \boldsymbol{A}_2\boldsymbol{C}^2 + \boldsymbol{A}_3\boldsymbol{C}^3)$
$A(x)$	Generating function for trees
$\mathfrak{T}$	Tree (tree diagram)
$\mathscr{A}(\mathscr{B})$	Subset of molecules ($\boldsymbol{N}_{\mathscr{A}}$ = number of polymer chains of the subset $\mathscr{A}$)
${}^0A_0(y), {}^0A_1(y)$	Numerical factors connected with the swelling ${}^0\mathfrak{X}_0$ in the vicinity of the Flory temperature.
$A_4(y)$	Numerical factor connected with the renormalization factor ${}^1\mathfrak{X}_4$
$\mathsf{A}(\mathsf{B})$	A represents a constant
Ⓐ	Avogadro number (Ⓐ $= 6.02249 \times 10^{23}$)
b	Two-body interaction
b_c	Two-body interaction at the top of the demixtion curve
$\mathfrak{b}$	Scattering length
$\mathfrak{b}^{\circ\!\to}$	Contrast length (with respect to the solvent)
$B(y)$	Generating function of vertex functions (associated with trees)
$\vec{B}(\vec{r}, t)$	Magnetic field

$B_j(\vec{r})$	Auxiliary field component used in the generating function of the Green's functions
$\mathsf{B}(u)$	Borel transform
c	Three-body interaction
c	Velocity of light
C	Monomer concentration or link concentration (number per unit volume)
$C_{\mathscr{A}\alpha}$	Monomer concentration
$C(\vec{r})$	Link concentration at the point of position vector $\vec{r}$
$C^{\mathrm{I}}, C^{\mathrm{II}}$	Link concentrations corresponding to the demixtion points of an isotherm ($C^{\mathrm{II}} > C^{\mathrm{I}}$)
$\boldsymbol{C}$	Chain concentration (Number of polymers per unit volume)
$\boldsymbol{C}(S)$	Concentration of chains of area S
$\boldsymbol{C}^*$	Chain concentration corresponding to the cross-over between dilute and semi-dilute solutions.
$\boldsymbol{C}^{**}$	Chain concentration corresponding, for semi-dilute solutions, to the crossover between good and poor solvents
$\mathscr{C}$	Area concentration (quantity analogous to C); $\mathscr{C}$ is a 'Brownian area' per unit volume: $\mathscr{C} = \boldsymbol{C}S$)
$\mathscr{C}_{\mathrm{k}}$	A concentration analogous to C and to $\mathscr{C}$, but for a Kuhnian chain ($\mathscr{C}_k = \boldsymbol{C}X^{1/\nu}$)
°C	Degrees Celsius
$\mathfrak{C}$	Chain
C	Integration path
$\mathbb{C}$	Specific heat
d	Dimension of space
$d_{\mathrm{a}1}, d_{\mathrm{a}2}$	Anomalous dimensions of fields
D	Hausdorff dimension of a fractal object
$\mathfrak{D}$	Diagram
$\mathscr{D}$	Diagram contribution to a partition function
$\mathscr{D}$	Diagram contribution before summation over some internal variables (areas or wave vectors)
e	Electric charge
E	Energy level
$\mathscr{E}$	Set (of molecules for instance)
$\mathbb{E}$	Average energy calculated over a large number of configurations

$\vec{E}$	Electric field
$\mathcal{E}$	Euler's constant
f	Fugacity
$\mathfrak{f}$	Another fugacity (renormalized)
$f(x)$	A function
$\mathrm{f}(\vec{x})$	Scaling function for the end-to-end distance of a long polymer in a good solvent
$\boldsymbol{f}$	Free energy per unit volume
$F_0(x, z)$, $F(x, g)$	Non-universal functions for the osmotic pressure of a polymer solution
$\mathrm{F}(x)$	Scaling function for the osmotic pressure in good solvent ($\boldsymbol{\Pi\beta} = \mathrm{F}(\boldsymbol{C}X^d)$)
F_∞	Universal constant concerning the osmotic pressure of semi-dilute solutions in good solvent (for $x \gg 1$, $\mathrm{F}(x) \simeq \mathrm{F}_\infty \mathrm{x}^{-1/(\nu d - 1)}$)
$\mathbb{F}$	Helmholtz free energy of an incompressible system or when the compressibility of the solvent is ignored ($\mathbb{F} = \mathbb{E} - \boldsymbol{\beta}^{-1}\mathbb{S}$)
$\mathscr{F}$	Helmholtz free energy of a compressible system
F	Force
$\mathfrak{g}$	Weight constant
g	Two-body osmotic parameter: a pure number which defines the second virial coefficient when the size of the isolated self-interacting polymer chain is chosen as the scaling length
$g(z)$	The osmotic parameter g as a function of the two-body interaction parameter z
g^*	The value of g in the Kuhnian limit ($g^* = g/(z \to \infty)$)
$g(\vec{r})$	Pair correlation function
$\mathbb{G}$	Gibbs free energy (free enthalpy)
$\mathfrak{G}$	Graph
$\mathrm{G}(x)$	Scaling function for the osmotic pressure ($\mathrm{F}(\boldsymbol{C}X^d) = \mathrm{G}(\rho/\rho^*_{\mathrm{G}})$)
$\boldsymbol{G}(\ldots)$	Green's function for spins on a lattice
$\mathscr{G}(\ldots)$	Connected Green's functions in continuous space
$\mathscr{G}_{\mathrm{G}}(\ldots)$	Non-connected Green's functions in continuous space (index G for general)
$\mathscr{G}(\vec{r}_1, \ldots, r_{2N}; a_1, \ldots, a_N)$	Connected Green's function with $2N$ legs in continuous space; this function associated with N fields, characterized by different values for the a_j, is the Laplace transform of the partition function $Z(\vec{r}_1, \ldots, \vec{r}_{2N}; a_1, \ldots, a_N)$

$\mathscr{G}(k_1, \ldots, k_{2N};\ a_1, \ldots, a_N)$ Green's function in reciprocal space; it is the Fourier transform of $\mathscr{G}(\vec{r}_1, \ldots, \vec{r}_{2N};\ a_1, \ldots, a_N)$

$\mathbf{G}_G\{B\}$	Generating function of the non-connected Green's functions
$\mathbf{G}\{B\}$	Generating function of the connected Green's functions
h	Column height
$\hbar$	Planck constant
h	Three-body osmotic parameter: a pure number which defines the third virial coefficient when one chooses as a scale, the size of the isolated polymer, in the same solvent and at the same temperature
$h(y)$	Three-body osmotic parameter as a function of the three-body interaction parameter y (for $z = 0$)
$\mathrm{h}(x)$	Scaling function related to the form function $H(\vec{q})$ of an isolated chain in good solvent
h_∞	Universal constant associated with the asymptotic behaviour of $\mathrm{h}(x)$ (for $x \gg 1$)
$H(\vec{q})$	Form function of a polymer ($\vec{q}$ is the transfer wave vector)
$\boldsymbol{H}(\vec{q})$	Structure function of the solute (polymer)
$\boldsymbol{H}^{\mathrm{I}}(\vec{q})$	Intramolecular structure function of the solute (intrachain)
$\boldsymbol{H}^{\mathrm{II}}(\vec{q})$	Intermolecular structure function of the solute (interchain)
$\mathfrak{H}$	Hamiltonian for chains
${}^{0}\mathfrak{H}_N$	Hamiltonian for N non-interacting Brownian chains
${}^{\mathrm{I}}\mathfrak{H}_N$	Contribution of the interaction to the Hamiltonian for N chains
$\mathscr{H}$	Hamiltonian for a spin system (or a field)
i	$= \sqrt{-1}$
$\|i\}\ (\|j\})$	Spin state
$I(q)$	Intensity of a scattered radiation: it is the number of particles that fall, during a period of time t, on a small detector of which the (angular) position corresponds to the deflection characterized by the transfer wave vector
I	Symbol representing an integral
$\mathfrak{I}$	Integration domain
$j(z)$	A function of the two-body interaction parameter: the product $zj(z)$ is a pure number which defines the second virial coefficient when, the size of the non-interacting polymer chain (Brownian chain) is chosen as the scaling length.

J	Contact energy (on a lattice)
J	Incident particle flux
$J(\vec{r})$	The flux of the particles that are scattered at the point of position vector $\vec{r}$
$\vec{J}$	$\vec{J} = \vec{S} + \vec{s}$: operator for the total spin
$\mathrm{J}(x)$	Scaling function associated with the structure function of a semi-dilute solution of polymers in good solvent
$\mathrm{J}_c(x)$	Scaling function associated with the structure function of a solution in poor solvent, in the vicinity of the critical demixtion point
$J(\vec{r})$	Source associated with $\mathscr{C}(\vec{r})$ (Edwards's method)
$\vec{k}$	Wave vector (in particular external wave vector on a diagram)
$\hat{k}$	Unit vector along $\vec{k}$ ($\hat{k} = \vec{k}/k$)
$\vec{k}_0$	Wave vector of an incoming radiation
$\vec{k}_f$	Wave vector of a scattered radiation (index f for final)
k_B	Conversion factor called Boltzmann constant $k_B = 1/\beta T = \boldsymbol{R}/$ Ⓐ $= 1.380 \times 10^{-23}$ joule K^{-1}
$\mathfrak{k}$	Connectivity of a lattice made of sites and bonds between these sites: number of bonds connecting each site to its neighbours
K	Degrees Kelvin (absolute degrees)
K_A	Experimental proportionality factor for converting the scattered intensity into a cross-section
K	Constant determined by the apparatus and the contrast: this constant makes it possible to convert the scattered intensity into the structure function of the solution
l	Characteristic length of a link ($\langle\langle u^2 \rangle\rangle = l^2 d$)
l_0	Length of a link
l_p	Persistence length
l_a	Atomic length
L	A length; in particular, the length of a chain of N links: $L = Nl_0$
$\mathscr{L}(\vec{r})$	Lagrangian associated with the Landau–Ginzburg model
m	Number of species constituting the solute
m	Molecular mass of a monomer ($m = 104$ for the monomer of polystyrene)
M	Molecular mass of a polymer chain
M_n, M_W, M_Z	Mean molecular masses ($M_n = \langle M \rangle, M_W = \langle MN \rangle / \langle N \rangle$, $M_Z = \langle MN^2 \rangle / \langle N^2 \rangle$
$\mathscr{M}$	A mass

M	Magnetization: variable appearing in the generating function of the vertex functions in field theory
$\boldsymbol{M}$	Number of sites of a lattice
M	Point in real space
n	Number of links of a chain (number of links of a sequence)
n	Index of refraction of a material (optical, neutron)
$\mathfrak{n}$	Number of components of the field (of the order parameter)
N	Number of links of a chain
$\boldsymbol{N}$	Number of polymers
N_{ξ}	Number of swollen sequences constituting a polymer chain in a semi-dilute solution for a good solvent; the size of a sequence is, by definition, the Kuhnian overlap length
$\mathcal{N}(N)$	Number of partitions of the integer N
$\mathcal{O}$	An observable; a function or a functional of the vectors defining the configurations of polymers; an operator
$\lvert\mathcal{O}\rangle)\}$	A quantum state associated with a neutron and its target (orbital state of the neutron, orbital state of the target and spin state for the total spin)
p	Number of contacts
p	A probability
p	The polydispersity of a polymer sample is $(1 + p)$ $\left(p = \left\langle\left(\frac{N}{\langle N\rangle} - 1\right)^2\right\rangle, \frac{M_\mathrm{W}}{M_\mathrm{n}} \simeq 1 + p\right)$
$\mathfrak{p}^+, \mathfrak{p}^-$	Projection operators on the states of total spin $S + 1/2$ and $S - 1/2$ (S is the spin of the nucleus and $1/2$ the spin of the neutron)
$\boldsymbol{P}$	Pressure
$\boldsymbol{P}^{\mathrm{I}}, \boldsymbol{P}^{\mathrm{II}}$	Pressure in the cells I and II
$\boldsymbol{P}_\mathrm{v}$	Vapour pressure
$\mathcal{P}$	Set of correlation points (on polymer chains)
P_μ	Stokes parameters ($\mu = 1, 2, 3, 4$): $\mathsf{P}_\mu = (\varepsilon/\hbar\omega)E^+\sigma_\mu E^-$
$P(\vec{r})$	Probability distribution associated with the vector joining the origin of a polymer to its extremity (end-to-end vector)
$\wp$	Permutation
$\mathfrak{P}$	A plane
$\vec{q}$	Transfer wave vector: difference between the scattered wave vector and the incident wave vector
$\vec{q}_j$	Internal wave vector in a diagram

Q	Partition function; in a more specific way, partition function of a discrete system
$\hat{\boldsymbol{Q}}_{\mathrm{G}}(V; N_0, \ldots, N_m)$	Partition function associated with the configurations of N_0 molecules of species zero, . . . , N_m molecules of species m, in a volume V; $\boldsymbol{Q}$ is printed in bold-faced type in order to express the fact that the quantities which are defined are the numbers N_j of molecules for each species; the index G (for general) indicates that the perturbation expansion of this partition function is represented by diagrams which are not always connected; the circumflex accent indicates that the solvent is explicitly taken into account
$Q_{\mathrm{G}}(V; \{f\})$	*General* grand partition function (discrete system); this quantity is expressed as a function of the set $\{f\}$ of fugacities
$Q(V; \{f\})$	*Connected* grand partition function (discrete system); the diagrams representing the various terms of the perturbation expansion of this function are connected diagrams
$\mathcal{Q}_{\mathrm{G}}(V; \{f\})$	*General* grand partition function (continuous system)
$\mathcal{Q}(V; \{f\})$	*Connected* grand partition function (continuous system)
$\mathcal{Q}(f)$	Function of the fugacity f, representing the connected grand function of a monodisperse ensemble of polymer chains in the limit where the volume V goes to infinity
$\vec{r}$	Position vector of a point
$\vec{r}_j$	Position vector of a point on a discrete chain
$\vec{r}(s)$	Position vector of a point on a continuous interacting Brownian chain; s represents the Brownian area measured along the chain, starting from its origin
r	Modulus of $\vec{r}(r = \lvert\vec{r}\rvert)$
$\hat{r}$	Unit vector along $\vec{r}\,(\hat{r} = \vec{r}/r)$
$\mathscr{R}$	Set of positions of correlation points
R	Square root of the mean square end-to-end distance
R_{G}	Square root of the mean square radius of gyration (in abbreviated form: radius of gyration)
R_{H}	Mean hydrodynamic radius
$\boldsymbol{R}$	Perfect gas constant ($\boldsymbol{R} = 8.317$ joule K^{-1})
$\mathfrak{R}$	Renormalization operator
s	Area of a segment of an interacting Brownian chain (Brownian 'extent')
s	'Extent' in a Kuhnian chain (quantity replacing the number of links which, for a continuous chain, is infinite)
s_0, s_1, s_2	Cut-off areas

$\vec{s}$	Spin of a neutron
S	Brownian area of a chain ($S = {}^0R^2/d$)
S	End-to-end 'extent' of a Kuhnian chain
S_t	Brownian area of a 'thermic sequence' (thermic blob)
${}^eS$	Area corresponding to S for a swollen chain (${}^eS = R^2/d$)
$\vec{\mathcal{S}}$	Spin of a nucleus
$\mathbb{S}$	Entropy
$\$$	Symmetry number
$\mathcal{S}_{\alpha,\beta}$	Matrix element playing a role in the calculation of diagrams (Fixman's method)
t	Non-universal reduced temperature $\left(t = \dfrac{T - T_c}{T_c}\right)$
t	Difference between a and its critical value ($t = a - a_c$)
$t(E)$	The collision operator 't' in potential theory; it is a function of the energy E
t	Time
T	Temperature
T_F	Flory temperature (Θ point)
T_c	Critical temperature for a magnetic system, or critical demixtion temperature for a polymer solution.
T_g	Glass transition temperature
T_f	Melting temperature
Tr	Trace (of a matrix)
$\mathcal{T}$	Transfer operator
$\vec{u}_j$	Vector defining the link $\vec{j}$ of a chain
u	'Renormalized coupling constant' in field theory: number proportional to the four-leg vertex function (for vanishing external wave vectors)
u	Pure number defining the two-body contact energy on a lattice
${}^0U_N$	Total number of closed chains of N links, drawn on a lattice, starting from the site at the origin
U_N	Total number of self-avoiding closed chains of N links drawn on a lattice, starting from the site at the origin; partition function of such chains
$U_{M,N}$	Number of Hamiltonian circuits on a lattice (M, N)
$\mathcal{U}$	Number representing the energy of a system in a given configuration
U	Energy density

v_e	'Excluded volume' corresponding to a two-body interaction
v_m	Volume of a monomer
$v_{\mathscr{A}}$	Partial molecular volume for a species $\mathscr{A}$
v	Partial volume of a monomer
v_r	Volume per site (for a lattice)
$v_m\{f\}$	Vertex function which m legs, for polymer diagrams; these vertex functions are represented by connected I-irreducible diagrams
V	Volume of the system
$V(\vec{r})$	Interaction potential
$w_{\mathscr{A}}$	Partial volume per unit mass associated with species $\mathscr{A}$
w	For a self-avoiding walk on a lattice, factor defining the attraction of adjacent points on the lattice [$w = \exp(-\boldsymbol{\beta}J)$]
w_c	Hypervolume representing a three-body interaction
$w(u)$	In field theory, quantity defining the variation of the 'effective interaction' with respect to the real interaction $\left(w(u) = -b\dfrac{\partial u}{\partial b}\right)$
$w(z)$	In polymer theory, quantity defining the variation of the osmotic pressure when the chain length increases $\left(w(z) = \varepsilon z\dfrac{\partial g}{\partial z}\right)$
$w[g]$	This function of g is equal to $w(z)$
$'w(y)$	In polymer theory, quantity defining the variation of the osmotic parameter h when the chain length increases $\left('w(y) = '\varepsilon y\dfrac{\partial h}{\partial y}\right)$
$'w[h]$	This function of h is equal to $'w(y)$
W	Gravitational weight of an object
$W, W(\ldots)$	Weight associated with a discrete configuration (on a lattice)
$\mathscr{W}, \mathscr{W}(\ldots)$	Weight associated with a configuration in continuous space
x_j	Component of $\vec{r}_j$
X	Length defining the size of an isolated chain interacting with itself in a solution ($X^2 = R^2/d$)
$\mathfrak{X}$	Swelling of a chain (e.g. in Flory's equation)
$\mathfrak{X}_0$	Swelling of a chain, associated with the mean square end-to-end distance ($\mathfrak{X}_0 = R^2/{}^0R^2$); this swelling is a renormalization factor

$\mathfrak{X}_N$ Renormalization factor associated with a star made of N chains starting from one point ($N = 1, 2, \ldots$); thus $\mathfrak{X}_1$ is associated with one end of a chain; we note that for simple reasons $\mathfrak{X}_2 = 1$

$\mathfrak{X}_G$ Gyration swelling ($\mathfrak{X}_G = R_G^2/{}^0R_G^2$); this swelling can be directly determined by radiation scattering

y Three-body interaction parameter ($y = cS^{3-d}(2\pi)^{-d}$)

$Y(N_1, \ldots, N_N)$ Ratio: $Y(N_1, \ldots, N_N) = \dfrac{Z(N_1, \ldots, N_N)}{Z(N_1) \ldots Z(N_N)}$ (discrete configurations)

$\mathscr{Y}(N_1, \ldots, N_N)$ Ratio $\mathscr{Y}(S_1, \ldots, S_N) = \dfrac{\mathscr{Z}(S_1, \ldots, S_N)}{\mathscr{Z}(S_1 \ldots \mathscr{Z}(S_N)}$ (continuous configurations)

z Two-body interaction parameter ($z = b\,S^{2-d/2}(2\pi)^{-d/2}$)

z_0 Value of z associated to the cut-off s_0 ($z_0 = b\,s_0^{2-d/2}(2\pi)^{-d/2}$)

Z Partition function

${}^0Z_N$ Total number of chains of N links, on a lattice, starting from the site at the origin

Z_N Total number of self-avoiding chains of N links on a lattice, starting from the site at the origin

$Z_N(\vec{r})$ Total number of self-avoiding chains of N links, starting from the site at the origin and arriving at the site of position vector $\vec{r}$

$Z(N_1, \ldots, N_N)$ Connected partition function of N interacting chains on a lattice; the chain of order j has N_j links

$\mathscr{Z}(S_1, \ldots, S_N)$ Connected partition function of N interacting chains in continuous space; the chain of order j has a Brownian area S_j; the chains are always assured to be distinguishable one from another (e.g. when $S_j = S$ for all j); in the same way, in all cases, initial and terminal points of a chain can be distinguished from each other

$\mathscr{Z}(N \times S)$ Connected partition function of N chains having the same Brownian area S

$\mathscr{Z}_G(\ldots)$ 'General' partition function (i.e. non-connected); the study of these functions can be reduced to the study of the connected partition functions $\mathscr{Z}(\ldots)$

$\mathscr{Z}(\vec{r}_1, \ldots, \vec{r}_E; j_1, \ldots, j_E; S_1, \ldots, S_N)$ Restricted partition function associated with the configurations of N interacting chains, bound to E correlation points (anchorage points); the vector $\vec{r}_\alpha$ defines the position in space of the point of abscissa S_α on the chain j_α

$\mathscr{Z}(\vec{r}_1, \ldots, \vec{r}_N; S_1, \ldots, S_N)$ Connected partition function of N chains, corresponding to the case where for each chain j, the position vector of the initial point is $\vec{r}_{2j-1}$ and the position vector of the terminal point is $\vec{r}_{2j}$

$^+\mathscr{Z}(\ldots)$ Regularized partition function of a continuous chain; the regularization is obtained by introducing a short distance cut-off; the quantity $\mathscr{Z}(\ldots)$ corresponding to $^+\mathscr{Z}(\ldots)$ is obtained by 'dimensional regularization', i.e. by analytic continuation with respect to the dimension d

$\mathscr{Z}_{\mathrm{I}}(\ldots)$ Partition function related to the I irreducible diagrams, i.e. the diagrams that cannot be separated into two disconnected parts by cutting one (and only one) interaction line

$\mathscr{Z}_N(z)$ Pure number giving $\mathscr{Z}(N \times S)$ (upto a dimensional factor) as a function of the interaction parameter z

$\mathfrak{Z}$ Renormalization factor of the field φ in field theory

α Critical exponent for closed chains; exponent associated with the return to the origin (in fact $\alpha = 2 - \nu d$); this exponent is the exponent α in the Landau–Ginzburg model for $\mathfrak{n} = 0$.

$\boldsymbol{\alpha}_{ij}$ Component of the polarization tensor

$\boldsymbol{\alpha}$ Polarizability

$\boldsymbol{\beta}$ The quantity $1/\boldsymbol{\beta}$ is an energy defining the temperature of the system ($\boldsymbol{\beta} = 1/k_{\mathrm{B}}T$ where k_{B} is Boltzmann's constant).

β Critical exponent related to the demixtion curve in the vicinity of the critical demixtion point (for an Ising model, exponent related to the magnetization curve in the vicinity of the Curie point); this exponent is the β exponent of the Landau–Ginzburg model for $\mathfrak{n} = 1$

γ Critical exponent for open chains; this exponent is the exponent γ of the Landau–Ginzburg model for $\mathfrak{n} = 0$

γ Critical exponent for the osmotic compressibility in the vicinity of the critical demixtion point; this exponent is the exponent γ of the Landau–Ginzburg model for $\mathfrak{n} = 1$

$\Gamma(x)$ Euler gamma function ($\Gamma(x+1) = x!$)

$\boldsymbol{\Gamma}$ Surface tension

$\boldsymbol{\Gamma}(r_1, \ldots, r_p; j_1, \ldots, j_p; a)$ In field theory, p-leg vertex function; the number j_α specifies which field component is associated with the (polymer) line ending in $\vec{r}_\alpha$; the vertex functions are only defined for even p; a is the constant of the Landau–Ginzburg model.

$\Gamma(M)$ Generating function of vertex functions; here M can be considered as a magnetization

δ	Critical exponent related to the long range decrease of the distribution function $P(\vec{r})$ ($\ln P(\vec{r}) \propto -r^{\delta}$; $\delta = 1/(1-\nu)$)
$\boldsymbol{\delta}$	Variation of a quantity
δ_{ij}	Kronecker symbol
$\delta(x)$	Dirac function (distribution)
Δ	Laplace operator
Δ_1	Critical exponent for deviation from scaling laws, used in magnetism ($\Delta_1 = \omega\nu$)
ε	given by $\varepsilon = 4 - d$
$'\varepsilon$	given by $'\varepsilon = 3 - d$
$\in$	Dielectric constant
$\zeta(x)$	Riemann function $\left(\zeta(x) = \sum_{n=1}^{\infty} \frac{1}{n^x}\right)$
ζ_N	Pure number giving $\mathscr{Z}_{\mathrm{R}}(N \times {}^{\mathrm{e}}S)$ (up to a dimensional factor) in the Kuhnian limit
η	Critical exponent ($2 - \eta = \gamma/\nu$)
$\boldsymbol{\eta}$	Very small number
η	Viscosity
$[\eta]$	'Intrinsic' viscosity of a solute in a solvent: $[\eta] = \lim_{\rho \to 0} (\eta - \eta_0)/\eta_0 \rho$ (ρ = solute concentration by mass, η_0 = solvent viscosity)
$\theta \equiv \theta_0$	Critical exponent for the small range behaviour of the distribution function $\mathrm{P}(\vec{r})$ ($\mathrm{P}(\vec{r}) \propto r^{\theta}$); in other words exponent related to the contact between the origin and the end of a Kuhnian chain
θ_1	Critical exponent related to the contact between the origin of a semi-infinite Kuhnian chain and an interior point of this chain
θ_2	Critical exponent related to the contact between two interior points of an infinite Kuhnian chain
$\mathrm{\theta}$	Diffraction angle
$\ominus(x)$	Step function: ($\ominus(x > 0) = 1$, $\ominus(0) = 1/2$, $\ominus(x < 0) = 0$)
κ	Subsidiary exponent for the long range behaviour of the distribution function $P(\vec{r})$.
$\mathrm{\kappa}$	Numerical factor involving the counting of partitions ($\mathrm{\kappa} = \exp \pi(2/3)^{1/2}$)
λ	Logarithm of the attrition coefficient: $\lambda = \ln \mu$

${}^0\Lambda$	Conversion factor for a Brownian chain: ${}^0R_G^2 = {}^0\Lambda\, M$
Λ	Conversion factor for a Kuhnian chain: $R_G^2 = \Lambda\, M^{2\nu}$
μ	Attrition coefficient
$\boldsymbol{\mu}$	Chemical potential
ν	Critical exponent for the swelling of a polymer chain in good solvent: this exponent is the inverse of the Hausdorff dimension of a Kuhnian chain; it is also the exponent ν of the Landau–Ginzburg model for $n = 0$
ν	Critical exponent for the correlation length in a polymer solution near the critical point of demixtion: this exponent is the exponent ν of the Landau–Ginzburg model for $n = 1$
ξ	Correlation length for a magnetic system in general
ξ	Correlation length (associated with concentrations) for a homogeneous solution in the vicinity of the demixtion point; correlation length of an Ising system ($n = 1$).
ξ_0	Brownian overlap length ($\xi_0 = \mathscr{C}^{-1/(d-2)}$)
ξ_k	Kuhnian overlap length ($\xi_k = \mathscr{C}_k^{-1/(d-1/\nu)}$)
ξ_e	Screening length for a polymer solution with chain overlap: quantity derived from $\boldsymbol{H}(\vec{q})$, where $\boldsymbol{H}(\vec{q})$ is considered as a function of the complex vector $\vec{q}$; ξ_e is given by the position of the singularity nearest to the real q space
ξ_t	Length of a 'thermal blob' or thermal sequence; these rather short sequences of a long swollen chain are nearly brownian and ξ_t represents the maximal length of these sequences.
$\Xi(q)$	Scattering cross-section per unit volume
Ξ	Universal constant relating the correlation length of a polymer solution near the critical point of demixtion, to the reduced temperature
π	Number pi ($\pi = 3.14\ldots$)
$\boldsymbol{\varpi}_\mu$	Polarization component ($\boldsymbol{\varpi}_\mu = P_\mu/P_4$ with $\boldsymbol{\mu} = 1, 2, 3$)
$\prod$	Symbol for product
$\boldsymbol{\Pi}$	Osmotic pressure
ρ	Solute mass per unit volume; ρ is a concentration by mass: $\rho = \boldsymbol{C}M/Ⓐ$
ρ_G^*	Concentration by mass corresponding to the cross-over from dilute to semi-dilute; $\rho_G^* = \boldsymbol{C}_G^*\, M/Ⓐ$ where $\boldsymbol{C}_G^* = (6R_G^2/d)^{-d/2}$
$\boldsymbol{\sigma}_1, \boldsymbol{\sigma}_2, \boldsymbol{\sigma}_3$	Pauli matrices
$\boldsymbol{\sigma}_4$	Unit matrix 2×2

σ_N	Critical exponent related to the asymptotic behaviour of the renormalization factor $\mathfrak{X}_N(z)$ as $z \to \infty$
$\sigma_N(z)$	Effective exponent considered as a function of z
$\sigma_N[g]$	Effective exponent considered as a function of g ($\sigma_N[g] = \sigma_N(z)$)
σ_a	Exponent for deviations from scaling laws ($\sigma_a = 2\Delta_1 = 2\omega\nu$)
$\vec{\sigma}(\vec{r})$	'Magnetic spin' located at a site of position vector $\vec{r}$
$\sum$	Symbol for sum (discrete sum)
$\mathbf{\Sigma}$	Diffraction differential cross-section
τ	Reduced temperature (universal) $$\tau = \frac{\beta_c - \beta}{\beta_c - \beta_F} = \left(\frac{T_F}{T_F - T_c}\right)\left(\frac{T - T_c}{T}\right)$$
τ	Ratio $\tau = S_t/S$
$\boldsymbol{\varphi}$	Phenyl group (C_6H_5—)
φ	Phase
$\vec{\varphi}$	n component field
φ	Solute volume/solution volume (i.e. volume fraction); number of occupied sites (monomer) per site, on a lattice
φ_c	Value of φ at the critical demixtion point.
$\boldsymbol{\varphi}$	Number of polymers per site, on a lattice ($\varphi = N\boldsymbol{\varphi}$)
χ	Flory's interaction parameter related to the polymer-solvent contact energy
χ	Magnetic susceptibility
$\chi(\vec{r})$	Source associated with $\mathscr{C}(\vec{r})$ [$\chi(\vec{r}) = iJ(\vec{r})$]
ψ	Solute mass/solution mass (i.e. mass fraction)
ω	Critical exponent for deviations from scaling laws (see σ_a and Δ_1)
ω	Pulsation
Ω	Solid angle
$\Omega(d)$	'Area' of a unit radius sphere in a d-dimensional space
$\mathbf{\Omega}$	Configuration
$\vec{\nabla}$	Partial differentiation operator
ℵ (aleph)	Universal ratio for the size of Kuhnian chains; $\aleph = 6\,R_G^2/R^2$
$\aleph(z)$	Ratio related to the size of interacting Brownian chains (and therefore partially swollen), as a function of z
⌈ (alif)	Universal ratio associated with strongly overlapping Kuhnian chains: ⌈ $= \xi_e/\xi_k$ (semi-dilute solutions of long polymers)

$\circlearrowleft$ Number of loops of a diagram made of polymer lines and interaction lines, each one carrying a wave vector, with flux conservation at the interaction points; the external wave vectors are fixed and the number of loops is equal to the number of independent wave vectors; in fact, to each independent wave vector, one assigns a circuit on the diagram (attention: do not confuse a loop and a closed polymer line)

1

POLYMERS AND POLYMER SOLUTIONS

GENESIS

The observation of polymers and the work devoted to their synthesis goes back to the beginning of the nineteenth century. However, the definitions given to them varied until 1920. Flory has shown that the many incidents which occurred during the formation of the basic notions are of great interest even now, and a few notable facts are mentioned below.

In 1805, Gough studied the elasticity of natural rubber and proposed two laws which were confirmed by Joule in 1859: (1) a piece of rubber, stretched under constant constraint, contracts in a reversible way when its temperature increases; (2) the stretching of rubber is a reversible exothermic transformation.

In 1826, Faraday established the chemical formula of the monomer unit of natural rubber or polyisoprene.

In 1860, Berthelot polymerized styrene $CH_2{=}CH\varphi$ (where φ represents the phenyl group C_6H_5) and obtained what he called metastyrene (nowadays known as polystyrene).

In 1863, Lourenço carried out a synthesis of polyethylene glycols of low molecular masses and isolated, one after the other, structures containing $N \leqslant 8$ monomer units.

In 1928, Mark and Meyer, measuring the osmotic pressure of rubber molecules in dilute solution, found molecular masses of the order of 4×10^5.

Along with these discoveries, of which only a small number are mentioned, the macromolecular hypothesis was developed, which in 1920 resulted in Staudinger's proposals. At this time, Staudinger claimed the existence of strongly bound structures with high molecular masses consisting essentially of linear chains which may either be independent or linked to one another in such a way as to form a network. It is always difficult to conceive that significant observations could have been made without a precise knowledge of the object under study, yet such was the situation. The research was centred on the colloidal properties of small molecules: clear conceptions of the chemical bond and reliable measurements of high molecular masses had been lacking until that time. Consequently, the macromolecular hypothesis, sometimes defended, sometimes criticized, was only gradually accepted.

During the 1920s, the linear chain model appeared as unacceptable because it seemed to imply the existence of free bonds at the end points. The polymer structure was therefore considered as a set of covalent cyclic molecules of low, molecular masses, held together by 'weak' valence bonds. Such a hypothesis led to the grouping of structures that were stable or unstable in solution, under the name of colloidal materials. But, in reality, a polymer molecule does not undergo dissociation in solution, which is a consequence of its completely covalent structure.

Structures of high molecular mass were conceivable, but their existence was denied. One of the arguments against this arose from erroneous interpretations concerning the variation of certain physical properties with the number N of links. For instance, a decrease in the melting temperature T_N had been observed for $N < 100$. The variation in T_N gradually becomes smaller when N increases and practically vanishes for N of the order of 100. This could mean that structures of degree larger than 100 do not exist; today the fact that T_N has a limiting value is explained by recognizing that the effects of the chain ends are gradually eliminated. A structure becomes a polymer when N is larger than 100. The fact that the physical properties depend on N is fundamental. However, at the present time (1988) certain types of behaviour are still not entirely understood; for instance, the variation of the volume of a chain with N.

After long controversies, the hypothesis of Staudinger was finally accepted around 1930. The image of a polymer that we now have, namely that of a long flexible chain, corresponds to a reversal in scientific attitude: the aim was to explain everything by starting from the chemical properties of the repetitive unit, but finally it was realized that the very existence of a chain structure was in itself fundamental.

Since 1930, the observation of linear structures in dilute solutions has revealed remarkable laws of behaviour. Each of these laws corresponded to non-classical behaviour. However, to unify and interpret the various observations, fundamental principles were lacking. The perturbation calculations developed in particular by Yamakawa and the Japanese school between 1960 and 1970 certainly constituted a good starting point for further developments, but the models remained ill-defined and it was not possible to use perturbation calculations to predict the behaviour of long chains. However, theoretical progress followed quickly when, in 1972, P.G. de Gennes showed that polymers in solution could be considered as a critical system of the Landau–Ginzburg type; in fact, as a consequence of this observation, the theory of polymers benefited from the great progress made in the domain of critical phenomena, following the work of K. Wilson.

At the same time, a new experimental technique, small-angle neutron scattering, was developed at the Centre d'Etudes Nucléaires de Saclay, and enabled numerous observations to be collected in areas which, until then, had not been accessible.

The 1970s correspond, in polymer research, to the introduction of new concepts and new methods: these have been continuously developing in the course of the following 15 years and they constitute the main subject of this book.

1. GENERALITIES AND DEFINITIONS

1.1 The chemical structure of a polymer

A polymer is a molecule composed of a succession of identical (or similar) polymer units:

$$- X - X - X - X - X -$$

The chain thus has a skeleton, side-groups, and two end-groups. Therefore, the formula of a linear polymer can be represented as follows:

$$A \left(X \right)_N B \tag{1.1.1}$$

The simplest species is polyethylene, whose skeleton consists of a series of carbon atoms each of which is linked together. Thus, its formula is

$$\left(CH_2 - CH_2 \right)_N = - CH_2 - CH_2 - CH_2 - \cdots \tag{1.1.2}$$

The polymer in solution that is the most currently used for configurational studies is polystyrene, which has the formula

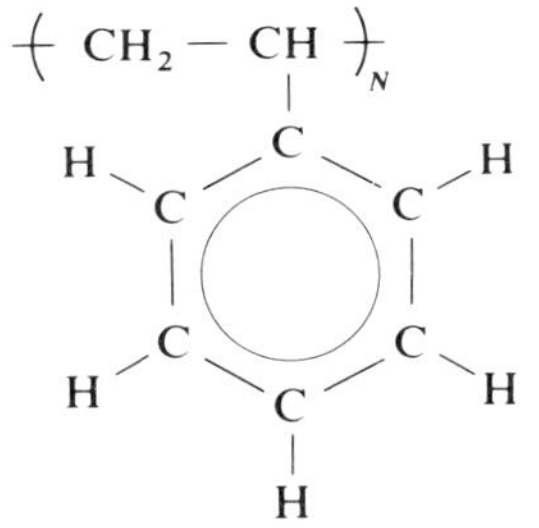

Sometimes other atoms like nitrogen or oxygen can periodically be found in the skeleton (polypeptides and polyoxyethylene). It may even happen that the skeleton does not contain any carbon atoms (polydimethylsiloxane): the carbon is then replaced by silicon.

A long molecule can also be obtained by binding (chemically), end-to-end, two or several polymer chains made with different units, for instance X and Y. In this case, the formula of the molecule can be written

$$A \left(X \right)_N \left(Y \right)_{N'} B \tag{1.1.3}$$

we say that this molecule is a block-copolymer.

Let us now examine in more detail the usual case where the skeleton is composed of a series of carbon atoms. The carbon has four valence bonds of the tetrahedral type (see Fig. 1.1). Two of these bonds can be used to find the chain

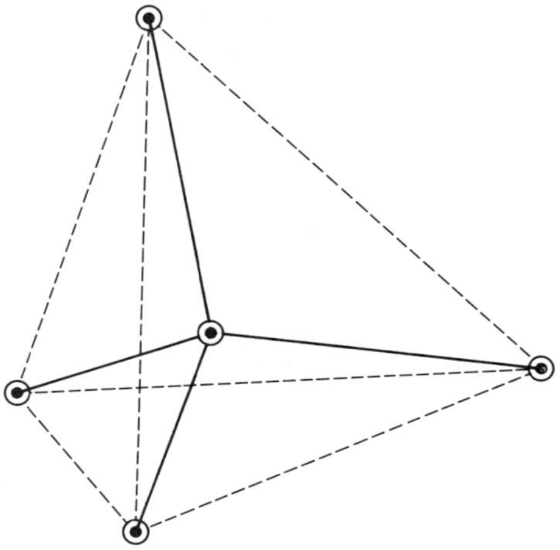

Fig. 1.1. A representation of the bonds of a carbon atom.

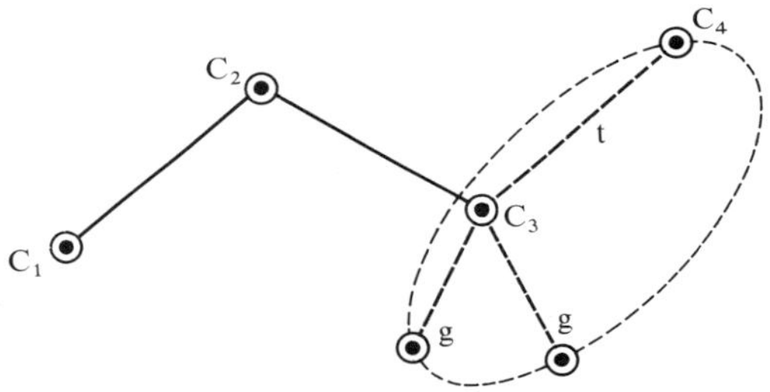

Fig. 1.2. Elementary configurations of four successive carbon atoms. The 'trans' position and the two 'gauche' positions are indicated.

units, the other two to find the side-groups. Four successive carbon atoms C_1, C_2, C_3, C_4 define an elementary configuration (see Fig. 1.2). Such a configuration is defined by the angle ψ of the two half-planes whose common edge is the straight line C_2C_3 and which pass respectively through C_1 and C_4. The energy $\mathcal{U}(\psi)$ has three minima. The absolute minimum corresponds to the value $\psi = \pi$. It is called the 'trans' configuration and is denoted by t. The other two minima correspond to the values $\psi = \pm\psi_0$, where ψ_0 is close to $\pi/3$; they are called 'gauche' configurations and are denoted by g^+ and g^-. According to Flory, the energy difference between gauche and trans for the alkanes is 500 ± 100 calories; in other words, we have

$$\mathcal{U}(\pi) - \mathcal{U}(\psi_0) \simeq 0.022 \text{ eV}.$$

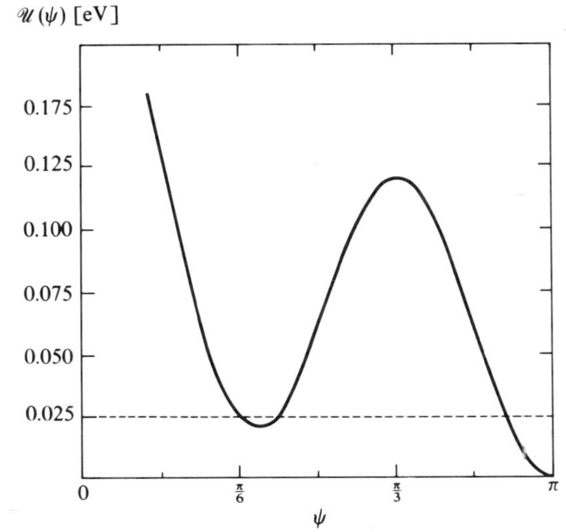

Fig. 1.3. The potential $\mathscr{U}(\psi)$ (in electronvolts) is plotted here against the angle ψ defining the configuration of the four carbon atoms. The minimum $\psi = \pi$ corresponds to the 'trans' position (after Flory).[1]

Moreover, $\mathscr{U}(\psi)$ is rather similar to the function represented in Fig. 1.3. As, at 17°C, we have $k_B T = 0.025$ eV, the three trans and gauche configurations are fairly stable at normal temperatures. Consequently, to a first approximation, we can set the linear chain of carbon atoms on a diamond lattice, and this fairly realistic simplification is useful for numerical simulations.

Polymers in solution or in the melt undergo thermal motions; the lifetime of a particular configuration is of the order of 10^{-9} seconds and the number of configurations is obviously very large.

1.2 Isomeries and tacticity

The notation (1.1.1) corresponds to an ideal chain. Actually, the isomerism of the monomer unit, when this unit is not symmetrical, generates periodicity defects. In fact, let us consider a unit of the vinyl type

$$\left(\underset{\displaystyle \mathrm{R}}{\underset{|}{\mathrm{CH}}} - \mathrm{CH_2} \right) \tag{1.1.4}$$

where R is the side group associated with the vinyl species under consideration ($R \rightarrow \varphi$ for polystyrene and $R \rightarrow CH_3$ for polypropylene). A normal connection is 'head–tail' (see Fig. 1.4).

$$- CH - CH_2 - CH - CH_2 -$$
$$\qquad |\qquad\qquad\qquad |$$
$$\qquad R\qquad\qquad\qquad R$$

$$- CH - CH_2 - CH_2 - CH -$$
$$\quad |\qquad\qquad\qquad\qquad\quad |$$
$$\quad R\qquad\qquad\qquad\qquad\quad R$$

$$- CH_2 - CH - CH - CH_2 -$$
$$\qquad\qquad | \qquad\quad |$$
$$\qquad\qquad R \qquad\quad R$$

(a)   (b)

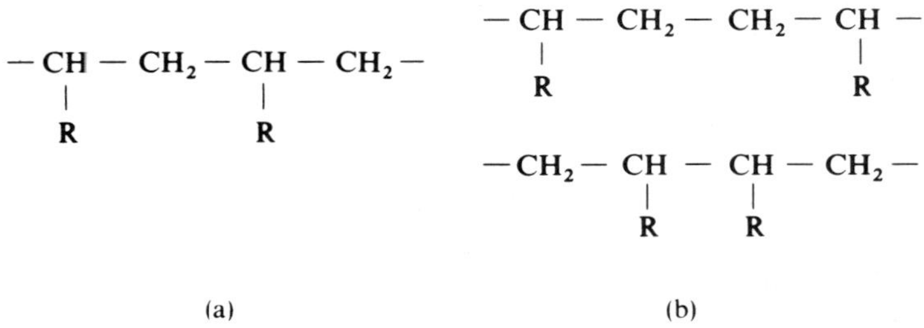

Fig. 1.4. (a) Normal 'head–tail' arrangement.
(b) 'Head–head' and 'tail–tail' arrangement.

However, the most important defects which have an influence on the physical properties of the amorphous polyvinyls, are connected with the chirality of the unit. In fact, in (1.1.4), the carbon bearing the radical is an asymmetric carbon. Let us consider the skeleton of the polyvinylic chain in its configuration t, t, t, and so on (see Fig. 1.5). The side-groups R may indifferently appear on either side of the skeleton plane. The setting of the Rs with respect to this plane is called tacticity. If the Rs are always in the same side, the chain is isotactic; if the Rs alternate, the chain is syndiotactic; if the Rs are disordered, the chain is atactic. The polyvinyl samples whose physical properties are studied in this book are atactic.

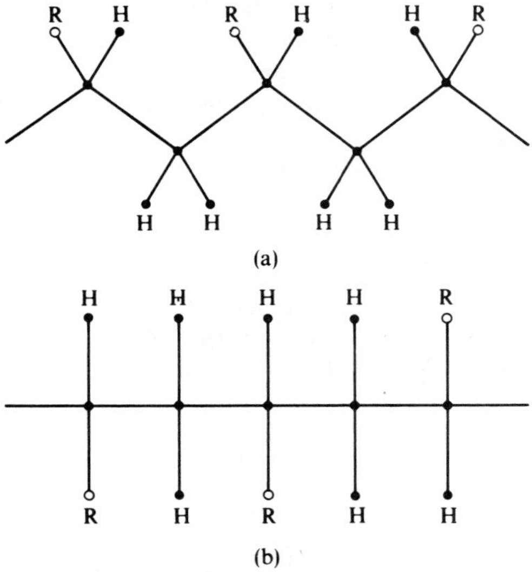

Fig. 1.5. Two views of the same polyvinylic chain in the configuration t, t, t . . . ; (a) side view; (b) view in which the plane containing the skeleton is perpendicular to the page.

However, let us note that, nowadays, more and more experiments are carried out with polydimethylsiloxane

$$\left(\begin{array}{c} CH_3 \\ | \\ Si - O \\ | \\ CH_3 \end{array} \right)_N$$

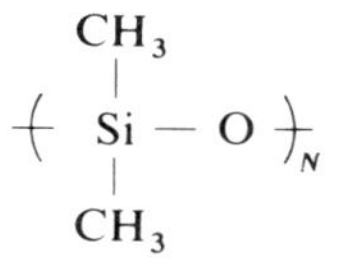

a molecule which has no connectivity and no tacticity defect.

1.3 Branching, non-linear polymers

The units which constitute the polymer may have, by accident or on purpose, a functionality larger than two. Thus, it is possible to obtain branched or reticulated structures whose elements are linear polymer segments. Two classes of structures can roughly be distinguished.

Branched polymers, star polymers, and comb polymers belong to the first class (see Fig. 1.6). These finite structures are obtained by binding several linear polymers together chemically; they are characterized by a certain regularity in the distribution of the branching points along the chain. In the second class, we find reticulated structures (networks) obtained by copolymerization of monomers whose functionality is on the average, larger than two (see Fig. 1.7). These monomers can be made of fairly long chains.

The reticulated structures are made up of clusters. If all the clusters have a finite size, the system is soluble and the solution is called sol. On the other hand, if the structure contains a cluster of infinite size, the system is a gel which is not soluble but which may swell in a solvent. The same reaction may lead either to sols or to gels according to the final branching rate. The sol–gel transition may be considered as a percolation transition. Note that an infinite cluster can be made either by chemical binding or (partially) by topological trapping [see Fig. 1.8]. From a mechanical point of view, a sol is viscous, a gel is elastic. Thus a piece of vulcanized rubber can be considered as a gel.

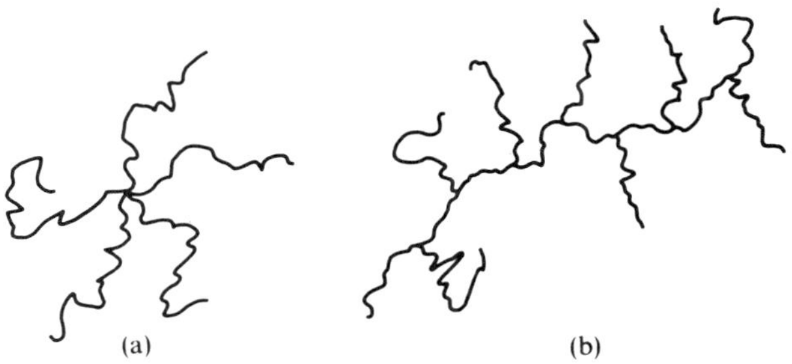

Fig. 1.6. Branched polymers: (a) star polymer, (b) comb polymer.

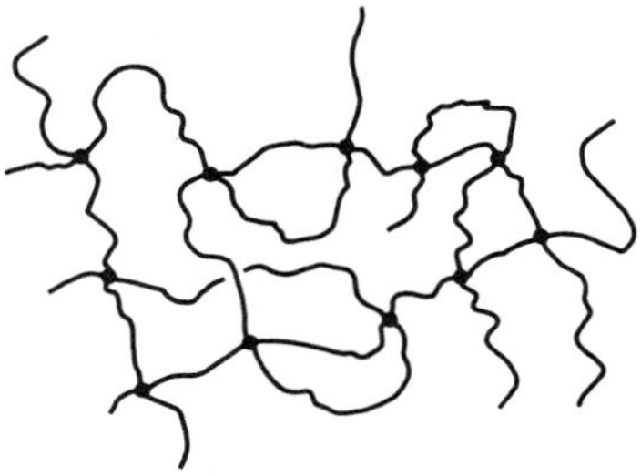

Fig. 1.7. A reticulated structure (network); here the functionality of the branching points is equal to four.

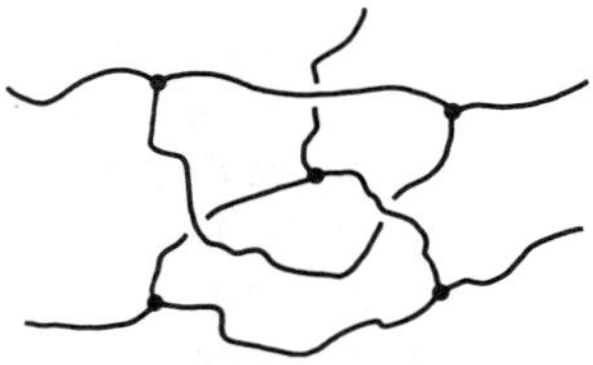

Fig. 1.8. Cluster consisting of two disconnected parts, bound by topological trapping.

1.4 Degree of polymerization and polydispersion

In a polymer sample, the degree N of polymerization varies from one polymer to another. We shall denote by P_N, the probability that a polymer taken at random in the sample be made of N monomer links. In other words, if the sample is put in solution and if $\boldsymbol{C}$ is the total number of polymers per unit volume, the number $\boldsymbol{C}_N$ of polymers with N links is given by

$$\boldsymbol{C}_N = P_N \boldsymbol{C} \qquad \sum_{N=1}^{\infty} P_N = 1. \tag{1.1.5}$$

In particular, if the polymers are linear the probabilities P_N define the *polydispersion* of the sample.

Let $\mathcal{O}(N)$ be a function of the number of links. For a sample, we define the average $\langle \mathcal{O} \rangle$ by the equality

$$\langle \mathcal{O} \rangle = \sum_N P_N \mathcal{O}(N).$$

In what follows, it will also be convenient to use the symbols $\mathcal{O}_\mathrm{n}$ (number average), $\mathcal{O}_\mathrm{W}$ (weight average), $\mathcal{O}_\mathrm{Z}$ (Z average) which are defined by the equalities

$$\begin{aligned} \mathcal{O}_\mathrm{n} &= \langle \mathcal{O} \rangle \\ \mathcal{O}_\mathrm{W} &= \frac{\langle N\mathcal{O} \rangle}{\langle N \rangle} \\ \mathcal{O}_\mathrm{Z} &= \frac{\langle N^2\mathcal{O} \rangle}{\langle N^2 \rangle} \end{aligned} \tag{1.1.6}$$

The *polydispersity* of the sample will be defined as the ratio

$$\frac{\langle N^2 \rangle}{\langle N \rangle^2} = 1 + p \qquad p \geqslant 0$$

where by definition, p is the polydispersity parameter $\left(p = \frac{\langle (N - \langle N \rangle)^2 \rangle}{\langle N \rangle^2} \right)$.

The molecular mass M of a polymer is a function of the number of links and, in this way, we define M_n, M_W and M_Z.

We note that we may generally write

$$M = mN + m_e$$

where m_e results from the end points, and therefore

$$M_n = m\langle N \rangle + m_e,$$

$$M_W = m\frac{\langle N^2 \rangle}{\langle N \rangle} + m_e, \tag{1.1.7}$$

and consequently

$$\frac{M_W}{M_n} = 1 + p - \frac{pm_e}{m\langle N \rangle + m_e}.$$

Thus, for $\langle N \rangle \gg 1$

$$\frac{M_W}{M_n} \simeq \frac{\langle N^2 \rangle}{\langle N \rangle^2} = 1 + p. \tag{1.1.8}$$

2. PHYSICAL PROPERTIES OF POLYMERS

Before studying the structure of polymers in solution, which is the subject of this book, we shall review a few general physical properties of polymers and solvents. This will enable us to understand why certain polymers are more adapted to a given experiment than others.

2.1 Physical properties of polymers in bulk

By definition, polymers in bulk are materials consisting exclusively of polymers, and they appear in liquid, solid, or viscous form depending on temperature.

The liquid polymer state is characterized by a viscosity which strongly increases with the molecular mass. On the other hand, the response of the polymer liquid to a mechanical constraint, at the beginning is of the elastic type, the viscous flow appearing only after a time delay which increases with the molecular mass. The behaviour of such a material is called viscoelastic.

For a large number of polymer species, the passage from the liquid state to the solid state by decrease of temperature is a continuous process. The structure of the solid is amorphous. For instance, this is what happens for atactic polystyrene and polydimethylsiloxane, materials whose structure in solution will be studied later on.

The transition between the viscous state and the solid state is characterized by a temperature T_g called glass transition temperature. This temperature corresponds to a discontinuity of physical parameters and in particular of the thermal expansion coefficient. The discontinuity observed when T decreases seems to result from the freezing, for $T < T_g$, of rotation and of large motions of chain elements.

We also note that certain polymer melts crystallize partially upon cooling. The transition occurs at a well-defined temperature T_f. Crystallization takes place only if the polymer has a perfect linear structure: for instance, it must not contain any asymmetric carbon. However, this tacticity condition is not sufficient, since polydimethylsiloxane, which is perfectly periodical, does not crystallize under normal conditions. On the contrary, polyethylene and isotactic polyethylene crystallize easily. In general, these polymers contain a large amorphous fraction. This is why they are called semi-crystalline. In certain conditions, it is possible to prepare polymer samples that are perfect crystals, in particular by polymerization *in situ* of a crystal made of monomers (polydiacetylene and polyoxyethylene).

2.2 Solubility of polymers in a liquid composed of small molecules

Among the liquids composed of small molecules, we may distinguish three kinds of solvents: non-polar solvents like benzene, polar solvents like nitrobenzene, and finally solvents with hydrogen bonds like water. Here, we are interested only in non-polar solvents: this is the simplest situation, because, in this case, the microstructure of the solvent has no long-range effect on the configuration of the polymers.

The amorphous polymers can easily be dissolved in non-polar solvents. The situation is different if the polymers are semi-crystalline, and in this case, to dissolve the polymer in the solvent, it is necessary to raise their temperature to a high level (for instance, 110 °C for polyethylene in xylene). Therefore, if a polymer is amorphous in its solid state, it is easier to study its solutions.

A solution is a monophasic state. It is also possible to obtain diphasic states, by cooling a solution. Below a certain temperature, demixtion occurs and two phases appear, one of them rich and the other one poor in polymer.

2.2.1 Reciprocal solubility of two liquids composed of small molecules

Attempts have been made to express the solubility of a material in a solvent, by starting from the thermodynamic properties of each constituent. This operation is possible only in definite situations, where it constitutes a kind of first approximation. To describe it, we shall first assume that the solvent and the other liquid are made of small molecules.

Let us then calculate the mixing energy of the two constituents (0 and 1) as a function

(1) of the interaction energies per unit volume;

$$U_{00},\ U_{11} \text{ and } U_{01} \equiv U_{10};$$

(2) of the volume fractions φ_0 and φ_1 with

$$\varphi_0 + \varphi_1 = 1. \tag{1.2.1}$$

To a first approximation, we may assume that the energy per unit volume of the mixture is

$$\frac{\mathbb{E}}{V} = \varphi_0^2 U_{00} + 2\varphi_0\varphi_1 U_{01} + \varphi_1^2 U_{11}, \tag{1.2.2}$$

and for the liquids when they are separated

$$\frac{{}^0\mathbb{E}}{V} = \varphi_0 U_{00} + \varphi_1 U_{11}. \tag{1.2.3}$$

From these quantities, we can deduce the vaporization energy $E_{\mathscr{A}}$ of the molecule $\mathscr{A}(\mathscr{A} = 0, 1)$, a quantity which should not be mistaken for the corresponding heat of vaporization. For a pure liquid, we have

$$E_{\mathscr{A}} = -v_{\mathscr{A}} U_{\mathscr{A}\mathscr{A}} \tag{1.2.4}$$

where $v_{\mathscr{A}}$ is the molecular volume.

The energy of mixing per unit volume is the difference

$$\frac{\mathbb{E} - {}^0\mathbb{E}}{V} = -\varphi_0\varphi_1(U_{00} + U_{11} - 2U_{01}).$$

This relation simplifies if we remark that for interactions of the van der Waals type, the interaction energies per unit volume are proportional to the product of the polarizabilities $\boldsymbol{\alpha}_{\mathscr{A}}$ of the molecules

$$U_{\mathscr{A}\mathscr{B}} \propto \boldsymbol{\alpha}_{\mathscr{A}}\boldsymbol{\alpha}_{\mathscr{B}}. \tag{1.2.5}$$

We shall assume that the proportionality coefficient is the same for all values of the indices $\mathscr{A} = 0, 1$; $\mathscr{B} = 0, 1$. This seems reasonable if the molecules are spherical and have the same volume. Therefore, we shall postulate the equality

$$U_{\mathscr{A}\mathscr{B}} = -\delta_{\mathscr{A}}\delta_{\mathscr{B}} \tag{1.2.6}$$

where by definition $\delta_{\mathscr{A}}$ is the 'solubility parameter' associated with the species $\mathscr{A}$. In this way

$$\frac{\mathbb{E} - {}^0\mathbb{E}}{V} = \varphi_0\varphi_1(\delta_0 - \delta_1)^2, \tag{1.2.7}$$

which is the result we were looking for.

The assumption (1.2.5) leads to an energy (1.2.7) which is positive. Thus, the mixing takes place with a loss of energy, which must be compensated by a gain of entropy

$$\mathbb{S} - {}^0\mathbb{S} = -\ln Z$$

where Z is the partition function of the mixture. Let us set the N_0, N_1 molecules on the sites of a lattice. In this case, the partition function Z is

$$Z = \frac{(N_0 + N_1)!}{N_0!N_1!}. \tag{1.2.8}$$

By using the approximate formula $N! \simeq (N/\mathrm{e})^N$, we obtain

$$\mathbb{S} - \mathbb{S}_0 = -\left(N_0 \ln \frac{N_0}{N_0 + N_1} + N_1 \ln \frac{N_1}{N_0 + N_1}\right). \tag{1.2.9}$$

The stability condition being

$$\mathbb{S} - {}^0\mathbb{S} > \beta(\mathbb{E} - {}^0\mathbb{E}), \tag{1.2.10}$$

we obtain for $N_0 = N_1$ and $v_0 = v_1 = v$

$$-\ln\frac{1}{2} > \frac{\beta v}{4}(\delta_1 - \delta_0)^2,$$

and therefore

$$|\delta_1 - \delta_0| < 1.66/\sqrt{\beta v}. \tag{1.2.11}$$

Traditionally, the solubility parameters are expressed in calorie$^{1/2}$ cm$^{-3/2}$. Let us take polystyrene as an example. The molecular volume of a monomer at 25 °C is

$$v = 0.153 \times 10^{-21}\ \mathrm{cm}^3$$

(see Fig. 5.15). Now, the condition (1.2.11) reads

$$|\delta_1 - \delta_0| < 4.2\ \mathrm{cal}^{1/2}\ \mathrm{cm}^{-3/2}. \tag{1.2.12}$$

Table 1.1

Species	Parameter δ [cal$^{1/2}$ cm$^{-3/2}$] at 25 °C
cyclohexane	8.14
deuterated cyclohexane	8.11
benzene	9.2
styrene	9.3
carbon disulphide	10.0

The values of the solubility parameter of a molecule can be determined experimentally[2] from measurements of the latent heat of vaporization. The values of these parameters for a few molecules are given in Table 1.1

Rarely do the binary mixtures of small non-polar molecules fail to satisfy the inequality (1.2.12)

2.2.2 Solubility of the amorphous polymers: good and poor solvents

Let us now consider the properties of amorphous polymers with respect to solvents. Is it possible to extrapolate the results obtained before for mixtures of small molecules? The answer is no, and we shall see in this book, that the energies and entropies of mixing are expressed in a manner which differs considerably from the one used in the preceding section.

Even in the case where the polymer is soluble *for all proportions* in the solvent, it is possible to distinguish two situations corresponding to differences in the swelling of the chain. In a 'poor solvent' the chain is only slightly swollen. In a 'good solvent', the chain is very swollen and has very different and peculiar properties. On the other hand, it is incorrect to assume that formulae (1.2.2) and (1.2.9) remain valid for chains.

However, one could think of representing the energy of mixing in a form analogous to (1.2.7)

$$\frac{\mathbb{E} - {}^{0}\mathbb{E}}{V} = \varphi_0 \varphi_1 (\delta_0 - \delta'_1)^2 \tag{1.2.13}$$

where δ_0 is the solubility parameter of the solvent and where δ'_1 is the solubility parameter of the chain: the latter should be close to the solubility parameter δ_1 of the monomer, because both $(\delta_1)^2$ and $(\delta'_1)^2$ are energies per unit volume, for molecules made of the same monomers. Unfortunately, δ'_1 cannot be measured because the vaporization energy which is proportional to the molecular volume (and to $(\delta'_1)^2$) is too large when the molecules are big. In practice, one replaces δ'_1 by δ_1.

On the other hand, the chain entropy (per link) is, of course, smaller than that of small molecules; nevertheless, the entropy of very flexible chains remains large. In brief, the chains are less soluble than the corresponding monomers and their solubility diminishes when their molecular mass increases.

2.2.3 Direct observation of a repulsion between polymer chains in dilute solutions

Long polymer chains are soluble in low molecular mass liquid solvent. Thus, in solution, polymer chains repel one another. This is remarkable, if we consider colloidal systems in general. Namely, (neutral) colloidal particles, dispersed in a liquid tend to aggregate as a result of the van der Waals forces. Such van der Waals forces exist also between polymer chains in solution, but they are weaker and counterbalanced by entropy contributions.

The fact that polymer chains in solution repel one another can now be observed directly, by mechanical measurement of the forces. The apparatus, developed by Israëlachvili and Tabor,[3] is made of two half-cylindrical mica shells, whose axes are perpendicular to each other (see Fig. 1.9).

The distance d between shells can be monitored with great accuracy (down to 0.1 nm) and the force $\mathscr{F}$ exerted between shells is measured as a function of $d(d \geqslant 0.1$ nm). When the apparatus is immersed in pure toluene, the observed[4] force–distance profile indicates the existence of attractive forces for $d \leqslant 10$ nm. However, when polystyrene chains ($M_W = 1.4 \times 10^5$) are terminally anchored on the half-shells, with a surface coverage $(3 \pm 0.5) \times 10^{-3}$ g/m^2, the result is very different. The force–distance profile indicates a strong repulsion (see Fig. 1.9). The onset of repulsion is at $d = 125 \pm 7$ nm, which is about twice the average end-to-end distance of the corresponding free polystyrene chain in toluene. The experiment thus shows very clearly the existence of repulsive forces, the study of which will be pursued throughout this book.

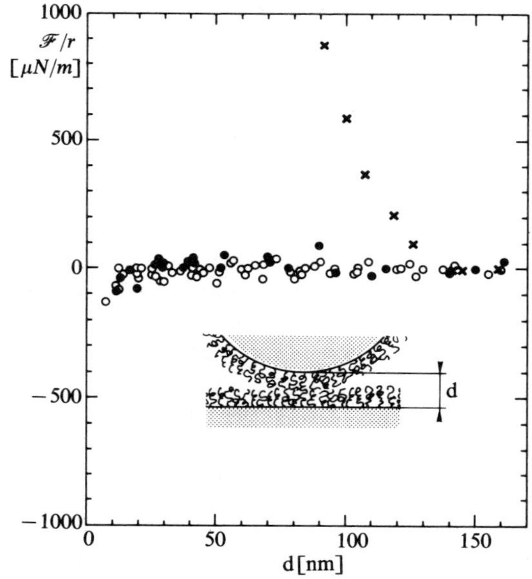

Fig. 1.9. (Top): Plot of the ratio force-to-radius of the half-shell against the separation distance. (Bottom): Schematic representation of the two half-cylindrical mica shells, on which polystyrene chains have been anchored.
●, ○ – measured values for bare mica shells in pure toluene
× – measured values for mica shells with end-grafted polystyrene ($M_W = 1.4 \times 10^5$). The surface coverage is $(3 \pm 0.5) \times 10^{-3}$ g/m^2 (from Taunton *et al.*[4]).

2.3 Gravity and solubility

Gravity acts on molecules, and, as in a gas, it has an effect on their concentration. This effect perturbs the (monophasic) state of a solution.

Using Boltzmann's law, we can calculate the concentration of the dissolved material in terms of the height h, measured from the bottom of the vessel containing the solution. The concentration $C(h)$ reads

$$C(h) = C(0)\exp(\boldsymbol{\beta}\mathscr{F}h) \tag{1.2.14}$$

where $\mathscr{F}$ is the gravity force, acting on the molecules of the dissolved material. In the case of a mixture consisting of a solute of volumic mass ρ_1 and of a solvent of volumic mass ρ_0, this force is equal to the Archimedes upward thrust

$$\mathscr{F} = \mathrm{g}v_1(\rho_1 - \rho_0) = \frac{M}{Ⓐ}\mathrm{g}\frac{(\rho_1 - \rho_0)}{\rho_1} \tag{1.2.15}$$

where v_1 is the molecular volume, M the molecular mass of the dissolved material, g the gravity intensity and Ⓐ Avogadro's number. Let us calculate the relative change of concentration for a molecular mass $M = 10^7$, a ratio $(\rho_1 - \rho_0)/\rho_1 = 0.66$, a height $h = 0.01$ m and a temperature $T = 0.066$ K. By introducing (1.2.15) in (1.2.14), we get

$$\frac{C(0) - C(h)}{C(0)} = 1 - \exp\left(\frac{M\mathrm{g}(\rho_1 - \rho_0)h\boldsymbol{\beta}}{Ⓐ\rho_1}\right) = 0.1.$$

In this example, the effect is perceptible, even experimentally. However, the experiments are generally made with samples of molecular mass smaller than $M = 10^7$ and, in this case, the concentration gradient resulting from gravity is negligible.

3. SYNTHESIS OF POLYMERS

3.1 Methods of preparation of polymers

The polymers which will be studied in this book are synthetic polymers and the industrial importance of such materials is well-known. However, we must recall that there also exist numerous natural polymers, and some of these have a great biological importance. Among the most common ones, we may cite natural rubber extracted from hevea latex and cellulose, with its derivatives, extracted from wood.

The synthesis of polymers[5] is the result of a series of chemical reactions starting from simple molecular elements. There are several well-defined types of reactions, with wide applications. Thus different means can be used to obtain polymers that are nominally identical but have very different 'qualities'. The qualities which are required for a sample are obtained by choosing an appropriate synthesis method and eventually by fractionating the product thus obtained. In particular, to study the structure of polymer solutions (by using methods described in this book), the experimentalists need samples that are as linear and monodisperse as possible.

There exist two main ways for the synthesis of polymers: polycondensation and polymerization. A polycondensation is a series of chemical reactions between difunctional compounds: thus the esterification of a diacid by a dialcohol produces a polyester. A polymerization is a chain reaction: the polymers have at one extremity an active site, and they grow step by step by adjunction of monomers containing a double bond; when a monomer clings to a chain, the double bond opens and the activity is transferred to the new extremity of the chain. Therefore, such a transformation requires an initiator able to create active sites. Industrially, the polymerization processes correspond to the largest tonnage and scientifically, for the preparation of good samples, they are also the most interesting.

It is possible to imagine other methods of preparation. Thus polypeptides and polynucleotides in live cells are made *in vivo* by replication processes. Consequently, the molecules produced in this way are (nearly) always strictly similar to the model, having the same molecular weight and the same tacticity with respect to the side-groups.

3.2 Polymerizations by chain reactions

3.2.1 General properties of polymerizations

A chain polymerization consists of three stages

(1) an initiating reaction creates an active site on a monomer;

(2) starting from the 'activated' monomer, the chain grows step by step (forming from 10^2 to 10^6 links) by repetitive addition of a monomer and regeneration of the active site at each step of the growth process;

(3) a termination reaction destroys the active site; the macromolecule is definitely constituted.

A chain polymerization can be radicalar, anionic, cationic, or stereospecific according to the type of initiator used. The reaction may occur in bulk, in suspension, in emulsion, as in solution. It is necessary to choose adequate temperatures for a good control of the processes.

(a) *In bulk*: the liquid monomer is placed in the presence of an initiator: the removal of heat may be difficult, especially when the monomer is a solvent of the polymer, because in this case the viscosity of the mixture increases in the course of the reaction.

(b) *In suspension*: the monomer is suspended in small droplets in a liquid (water, in general); the initiator is dissolved in the monomer and the reaction takes place in the droplets, the water being used for the removal of heat.

(c) *In emulsion*: the monomer is in aqueous emulsion and polymerization takes place in the micelles. However, the initiator is soluble in water (and not in the monomer). Each micelle can contain only one radical, and consequently very high molecular masses can be obtained.

(d) *In solution*: the monomer and the initiator are dissolved in a solvent; very often the polymer is also soluble in the solvent.

This last method, being more costly, is principally used in laboratories where by this means, better defined polymers can be prepared and the kinetics of the reaction can be studied.

3.2.2 Radicalar polymerizations

Radicalar polymerizations are very commonly used. In this case, the active site is an atom bearing a unpaired electron (free bond). The initiator is an organic material which can spontaneously split by homolytic breaking and in this way can produce radicals able to attack the monomer and thus to initiate the process. Actually, as initiators, one uses peroxides (R′–O–O–R″), hydroperoxides (R′–O–OH) (for instance hydrogen peroxide; H_2O_2, or cumene hydroperoxide) and aliphatic azoics (R′–N=R″). For each initiator, there is a domain of temperature often narrow (about 10 degrees) in which the homolytic decomposition occurs with an acceptable velocity.

One of the most common initiators is benzoic anhydride (benzoyl peroxide) with the formula

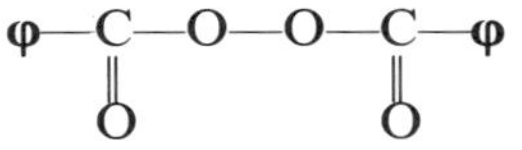

where φ represents a benzenic ring ($\varphi \rightarrow C_6H_5$).

For instance, this initiator is used for the preparation of polystyrene. Around 70 °C, benzoyl peroxide decomposes into 'primary' radicals.

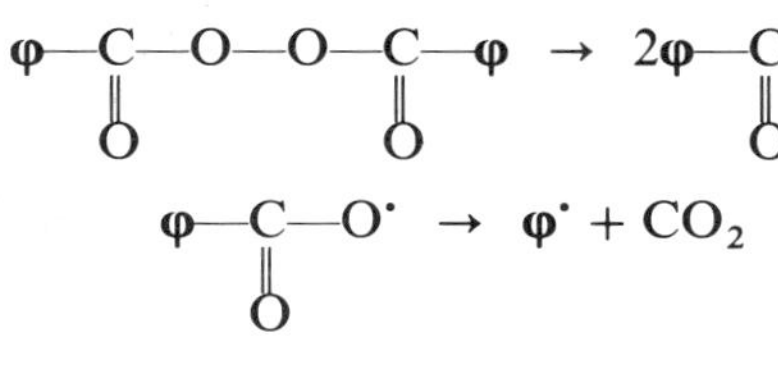

The radical φ can react on a monomer (styrene)

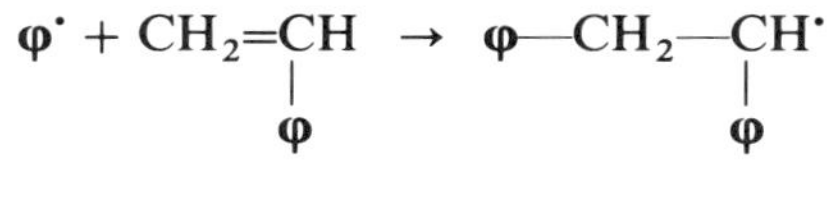

The molecule produced in this manner can now react on another monomer

$$\varphi\text{—}CH_2\text{—}\underset{\varphi}{\underset{|}{CH^\bullet}} + CH_2\text{=}\underset{\varphi}{\underset{|}{CH}} \rightarrow \varphi\text{—}CH_2\text{—}\underset{\varphi}{\underset{|}{H}}\text{—}CH_2\text{—}\underset{\varphi}{\underset{|}{CH^\bullet}}$$

and so on.

In a general way, the reaction can be written

$$R^\bullet + M \rightarrow R - M^\bullet$$

$$R - M_N^\bullet + M \rightarrow R - M_{N+1}^\bullet$$

where $R^\bullet$ is a primary radical and M is the monomer. The deactivation occurs by recombination of radicals with one another

$$R - M_N^\bullet + {}^\bullet M_{N'} - R \rightarrow R - M_{N+N'} - R$$

(and also $R^\bullet + {}^\bullet R \rightarrow R - R$) or by dismutation, i.e. by migration of an hydrogen from a radical to another radical. Transfer reactions of the radical may also occur with the monomer, with the solvent, with products added precisely for that purpose, or even with the polymer itself. If the active extremity of a chain $\mathfrak{C}'$ attacks the middle of another chain $\mathfrak{C}''$ as of the same chain, the free election may be transferred from $\mathfrak{C}'$ to $\mathfrak{C}''$ and simultaneously an hydrogen migrates from $\mathfrak{C}''$ to $\mathfrak{C}'$. In this way, the extremity of the chain is deactivated and terminated; but now a carbon of $\mathfrak{C}''$ carries a free electron which may serve as an interior for a new chain which gets grafted on the old one. In this way, one obtains parasitical branching, which may be a nuisance for applications.

The chain length depends on the relative quantities of polymer and initiator that are used but also on the transfer frequency. The mass polydispersion of the produced polymer is always large. The lifetime of a radical varies from 10^{-1} to 10 seconds. The polymerization reaction stops spontaneously by deactivation.

Finally, let us note that with this method, it is possible to prepare 'statistical copolymers' from mixtures of two species of monomers.

3.2.3 Anionic polymerizations

Anionic polymerizations have been developed from 1955 onwards especially by M. Swarc;[6] they have a great industrial and scientific importance. The active site is a carbon–metal bond (or sometimes an oxygen–metal or a sulphur–metal bond.) This bond is strongly polarized, $C^\ominus\text{–}Me^\oplus$, but weakly ionized in an organic medium.

An 'aprotic' solvent is used in order to prevent a proton from fixing on a carbanion C^- with transfer of ionization to a solvent molecule. Thus, for this reason, liquid ammonia which had been used for the first anionic polymerizations, is no longer chosen as a solvent.

Parasitical reaction: $- \overset{|}{\underset{|}{C}}{}^\ominus + NH_3 \rightarrow -\overset{|}{\underset{|}{C}}\text{—}H + NH_2^\ominus$

An 'aprotic solvent' like tetrahydrofuran does not contain labile hydrogen; it is also possible to use benzene, cyclohexane, dioxane, and dimethylformamide. The initiators are organometallics. Among them, we may cite butyl–lithium C_4H_9Li (Bu–Li or $Bu^{\ominus}–Li^{\oplus}$) and cumyl–potassium

$$\varphi - \overset{\overset{\displaystyle CH_3}{|}}{\underset{\underset{\displaystyle CH_3}{|}}{C}} - K$$

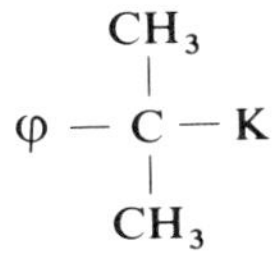

which act on the monomer by addition. Of course, any trace of water and air must be avoided and the experiment is performed in a current of argon or nitrogen. For instance, cumyl-K dissolved in tetrahydrofuran will react with styrene $CH_2–CH\varphi$ giving

$$\varphi - \overset{\overset{\displaystyle CH_3}{|}}{\underset{\underset{\displaystyle CH_3}{|}}{C}} - CH_2 - \overset{\overset{\displaystyle H}{|}}{\underset{\underset{\displaystyle \varphi}{|}}{C^{\ominus}}} K^{\oplus}$$

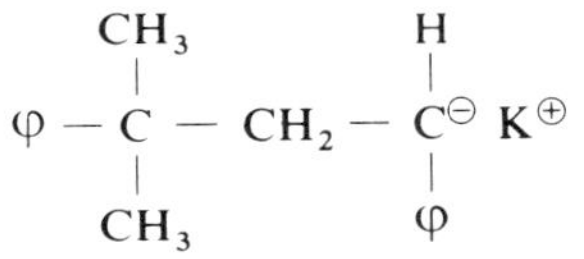

which itself reacts on another monomer, and so on. The reaction is very fast and exothermic: for instance, it is possible to operate with a solution containing 10% of monomer at $-70°C$ [in a bath of dry ice (solid CO_2)]. Each initiator molecule gives birth to one chain whose extremity remains active during weeks: this is what is called a living polymer: in general, solutions of living polymers have a beautiful colour (red, or blue, or green). On the contrary, butyl–lithium is colourless. The reaction may last a few seconds, a few minutes, or a few hours and stops by exhaustion of the monomer.

In order to terminate the reaction, one 'kills' the living polymers by adding a donor of protons; for instance, methanol. Thus

$$\underset{\underset{\displaystyle \varphi}{|}}{CH_2 - CH^{\ominus}} K^{\oplus} + CH_3OH \longrightarrow - CH_2 - \underset{\underset{\displaystyle \varphi}{|}}{CH_2} + CH_3OK$$

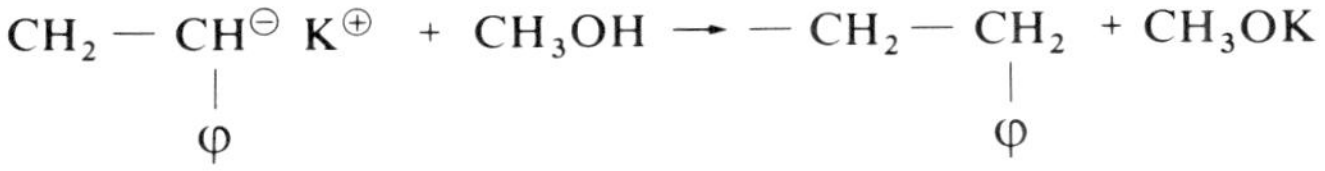

It is also possible to use living polymers in different manners; for instance, they may serve as initiators for the polymerization of a second type of polymers (or even a third one, and so on . . .) and, in this way, it is possible to make star polymers, or comb polymers, and gels.

In brief, this synthesis presents very interesting characteristics (especially from a scientific point of view).

(a) the rate of conversion is hundred per cent;

(b) the structural arrangement is regular, of the head-tail type
(example: $CH_2—CH\varphi—CH_2—CH\varphi—CH_2—CH\varphi—$);

(c) the availability of living polymers can be a source for various applications;
(d) the linear polymers that are synthesized in this manner have a well-defined, relatively-weak polydispersion.

Let us examine the last point in more detail. Let $P(t)$ be the probability distribution of the time needed to add a monomer to an existing chain. We may assume that P(t) is given by a Poisson law

$$P(t) = \tau^{-1} e^{-t/\tau}$$

with

$$\int_0^\infty dt\, P(t) = 1 \qquad \int_0^\infty dt\, t\, P(t) = \tau.$$

If a polymer chain is built monomer by monomer, from time zero onwards, the probability of obtaining a chain with N monomers after a time t is a function $P_N(t)$ given by the recurrence formulas

$$P_1(t) = \int_t^\infty dt'\, P(t')$$

$$P_N(t) = \int_0^t dt'\, P(t')\, P_{N-1}(t - t').$$

This leads to the result

$$P_N(t) = \frac{(t/\tau)^N}{N!} e^{-t/\tau},$$

with

$$\sum_N P_N(t) = 1.$$

Using this probability distribution, we get

$$\langle N \rangle = \sum_N N P_N(t) = t/\tau$$

$$\langle N^2 \rangle = \sum_N N^2 P_N(t) = t/\tau + t^2/\tau^2.$$

The polydispersity $(1 + p)$ is defined by the equality

$$1 + p = \frac{\langle N^2 \rangle}{\langle N \rangle^2}, \tag{1.3.1}$$

and for the process described here, we have

$$p = \frac{1}{\langle N \rangle}. \tag{1.3.2}$$

In practice, to obtain good samples, one has to start from very pure products. For instance, in 1985, the firm Toyo-Soda could provide a polystyrene sample of molecular mass $M = 7.75 \times 10^5$, i.e. $N = 7.45 \times 10^3$, with a (claimed) polydispersity equal to 1.007 which is the theoretical value given by (1.3.2). However, this value 1.01 is a lower bound. For large masses ($N > 10^4$), the polydispersity of the samples increases with the molecular mass, instead of decreasing as is suggested by (1.3.2). In general, this fact is explained by observing that the effect of the impurities must increase with the degree of polymerization.

3.2.4 Cationic polymerizations

Certain polymerizations which are now interpreted as cationic polymerizations have been known since the beginning of the twentieth century and even before. However, the mechanisms which come into play for this type of reaction have been understood only gradually, and their exact nature has been the subject of controversies. The active site is a strongly polarized bond $C^{\oplus}$–$A^{\ominus}$ where $A^{\ominus}$ is an anion. This anion often has a complex structure as $(SbF_6)^{\ominus}$. This structure must be quite symmetrical so as to ensure the stability of the ion (in SbF_6 the fluorines form an octahedron around Sb). Thus, the initiators are acids or substances such as SbF_5, BF_3, $TiCl_4$, $SnCl_4$ which can generate ions like $(SbF_6)^{\ominus}$ or $(BF_4)^{\ominus}$. During the reaction, the active site of the chain reacts on the monomer by incorporating this monomer and by regenerating the active site at the end of the chain, in the same way as in an anionic synthesis. However, the cationic reactivity is even stronger than the anionic one. Thus, isobutene with BF_3 at 100 °C polymerizes in a few minutes, forming polymers of high molecular masses. Secondary reactions occur, such as isomerization transfers, and they contribute towards stopping the chain reaction. In general, the polymers produced in this way have no stereospecificity.

3.2.5 Stereospecific polymerizations

Studies made by Ziegler around 1950 showed

(a) first, that aluminium hydride reacts around 60–90°C upon ethylene to give triethyl aluminium;

$$AlH_3 + 3CH_2{=}CH_2 \rightarrow Al{-}(CH_2{-}CH_3)_3;$$

(b) second, that such a metallic alkyl can initiate a polymerization with the cooperation of a catalyst consisting of the metallic halide of a transition metal. The typical combination

$$TiCl_4 + Al{-}(CH_2{-}CH_3)_3$$

allows one to obtain polyethylene.

Later, Natta (around 1955) was able to show that the monosubstituted polyethylene (propylene, butene, styrene) produced by means of this method is *stereoregular*, idio- or syndio-tactic according to the case. This important discovery opened up new possibilities and won a shared Nobel prize for Natta and Ziegler in 1963.

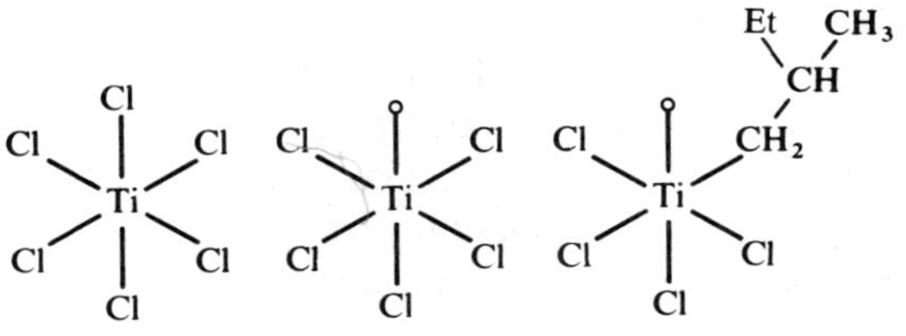

(a) (b) (c)

Fig. 1.10. A titanium atom in a crystal: (a) without empty site, (b) with an empty site, (c) with an empty site and an attached propylene monomer. The symbol Et represents an ethylene atom coming from the substitution of a chlorine atom to this radical in a triethyl aluminium molecule (according to P. Rempp).

Polymerization occurs in heterogeneous phase, and the catalytic activity strongly depends on the structure and the 'quality' of the microcrystals of titanium chloride which are used. In titanium chloride, the valence of titanium is six (see Fig. 1.10) and, on the edges of the crystal, there are empty chlorine sites which play the role of active centres. Titanium behaves like an electron attractor with respect to the double bond of the monomer. The polymerization process is rather complex: we shall sum it up by saying simply that the end of a growing chain is attached to an empty site of a titanium atom and in this way may react upon monomers attached to other empty sites, with formation of a longer chain.

3.3 Polycondensations

A condensation is a classical chemical reaction between functional groups, for instance

$$\text{acid} + \text{alcohol} \rightarrow \text{ester} + H_2O$$

$$\text{acid} + \text{amine} \rightarrow \text{amide} + H_2O.$$

A polycondensation is a series of such reactions, which takes place between bifunctional molecules. For example, the polycondensation of adipic acid with hexamethylene–diamine gives nylon 6-6. In fact, the chemical method of production is not so simple, but we may (naïvely) write the reaction in the form

$$HO{-}\underset{\displaystyle\overset{\|}{O}}{C}{-}(CH_2)_4{-}\underset{\displaystyle\overset{\|}{O}}{C}{-}OH \;+\; NH_2{-}(CH_2)_6{-}NH_2$$

$$\rightarrow HO{-}\underset{\displaystyle\overset{\|}{O}}{C}{-}(CH_2)_4{-}\underset{\displaystyle\overset{\|}{O}}{C}{-}NH{-}(CH_2)_6{-}NH_2 \;+\; H_2O$$

and so on. In the same way, nylon 9 is obtained by polycondensation of the amino-nanoic acid

$$NH_2{-}(CH_2)_8{-}\underset{\underset{O}{\|}}{C}{-}OH$$

which bears an acid function and an amine function.

Polycondensation may occur without elimination. Thus, the polycondensation of an alcohol and of a diisocyanate gives a polyurethane

$$(HO{-}R{-}OH) + (O{=}C{=}N{-}R'{-}N{=}C{=}O)$$

$$\rightarrow HO{-}R{-}O{-}\underset{\underset{O}{\|}}{C}{-}\underset{\underset{H}{|}}{N}{-}R'{-}N{=}C{=}O$$

and so on.

These polycondensations, which are of great industrial importance, obviously give very polydisperse products (the polydispersity is close to 2).

3.4 Fractionation

To obtain samples with small polydispersion, from a sample whose molecular mass distribution is relatively broad, fractionation may be useful. Two main methods can be used; precipitation fractionation and exclusion chromatography.

The first method is the simplest one. A polymer solution in a solvent A is contained in an ampulla the lowest part of which ends in a point (see Fig. 1.11). Then, to the solution is added a solvent B having the following properties: B and A are miscible but the mixture A + B is a 'poor solvent' for the polymer. By increasing the proportion of solvent B, one induces a precipitation of the longest chains, and this is the principle of this method. For instance, to a solution of polystyrene in toluene, we may add methanol or heptane.

The directions for operation are as follows. The solvent is added slowly, until the solution becomes turbid. The ampulla is then heated until all the polymer reverts into solution.

Then it should be left to cool and to separate slowly. There is phase separation: the phase containing the high mass polymers come together at the bottom of the ampulla and can be extracted by siphoning.

The size exclusion chromatography, which is generally used as an analysis technique, can also serve to prepare samples by fractionation. However, this method, which will be studied in detail in Section 6, calls for rather heavy equipment which is not to be found in most laboratories, and the operation in itself is rather costly

Other fractionation techniques are possible; in particular, the field-flow fractionation method seems interesting and will be briefly explained in Section 7.

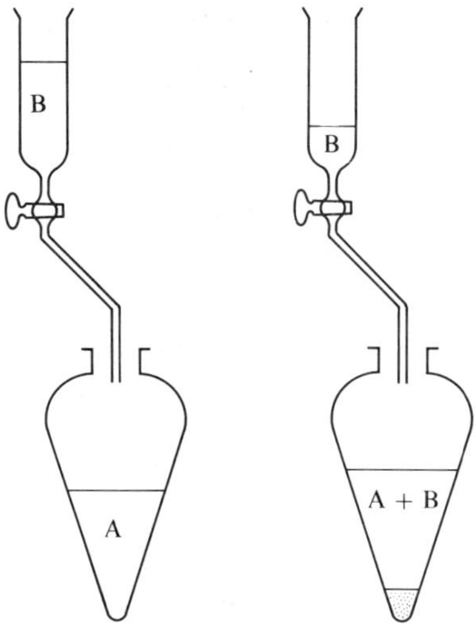

Fig. 1.11. Fractionation by precipitation: (a) the polymer is dissolved in solvent A; (b) the solvent B is mixed with solvent A. A non-soluble phase containing polymers with high molecular weights precipitates and is collected at the bottom of the ampulla.

4. DESCRIPTION OF A FEW POLYMERS

To perform the experiments described in this book, one needs a polymer with the following properties: a chain whose linear structure is perfect (no branching), which is quite flexible and soluble in non-polar solvents, and finally, which comprises a large number of links.

Common polymers like polyethylene, nylon, polyisoprene (natural rubber), and polystyrene have these properties to very diverse extents. From this point of view, polystyrene seems to be, by far, the best one. Another but less known polymer, polydimethylsiloxane, is perhaps even better, but its systematic use is more recent.

In this section, we shall describe the properties of the polymers just mentioned.

4.1 Polyethylene

Polyethylene is one of the simplest polymers from the point of view of its chemical structure. It was identified in 1900 by Bamberger and Tschirner.[7] Up to 1935, it was produced only in small quantities. In this year, for the first time, Fawcett et al. made a synthesis of this polymer by maintaining ethylene at 170°C for 4 hours in a steel bomb at a pressure of 2000 atmospheres. The molecular mass of the product thus obtained was of the order of 15 000. Since then, plants for the massive production of 'high pressure' polyethylene have been established, based on this principle, especially in England (Imperial Chemical Industry).

In 1955, Ziegler proposed a synthesis at atmospheric pressure, with the help of organo-metallic catalysts: the 'low pressure' polyethylene thus obtained is a better product than the 'high pressure' polyethylene: its density is larger, its branching rate is lower, and its crystallinity ratio is higher. The characteristics are indicated in Table 1.2.

The polyethylenes are so diverse that it is difficult to make a systematic study of their physical properties: it is not possible to find two samples that are identical from all points of view!

Polyethylene is used mainly for packing and for the isolation of electrical cables: polyethylene is one of the plastics which is least permeable to water vapour. Actually, numerous applications have been found for this product. Polyethylene has even been used to slow down neutrons in nuclear reactors!

At room temperature, the structure of polyethylene is semi-crystalline. Polyethylene samples whose origin is the same polymerization product may have various morphologies depending on the manner in which they are obtained.

By cooling down a dilute solution in xylene, one obtains lamellae with a high crystallinity. In 1957, A. Keller[8] showed[9] that these lamellae are made of folded chains (see Fig. 1.12): the chains are aligned and nearly perpendicular to the surface of the lamellae. A typical value of the thickness of these lamellae is 10 nm; the thickness is an increasing function of the temperature $T(T < T_F)$ of the solution during the crystallization process.

Solid polyethylene obtained directly from the molten state is also made of lamellae. These are set up radially around nucleation centres: they form so-called 'spherulitic' textures and the size of a spherulite is of the order of one micron. In this case, the amorphous fraction is larger (from 10 to 20 per cent).

Table 1.2.

	High pressure	Low pressure
Volumic mass [g cm^{-3}]	0.92	0.98
Branching rate = number of branchings per 100 carbon atoms	3	0
Extensibility	10%	400%
Demixtion temperature in solution in xylene for a 5% concentration	90 °C	70 °C
Crystalline structure	triclinic	orthorhombic
Melting temperature	110 °C	130 °C

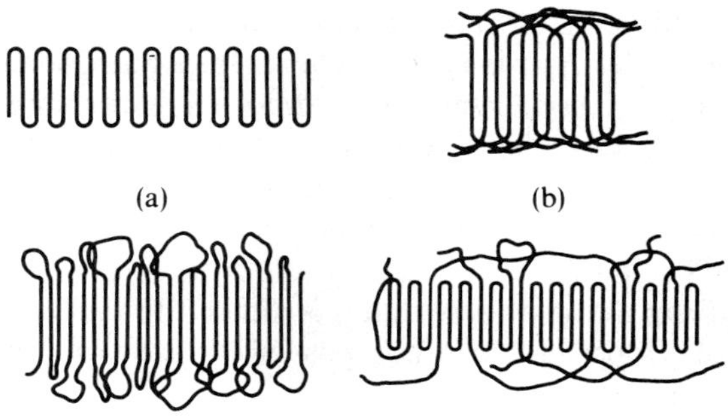

(a) (b) (c) (d)

Fig. 1.12. A two-dimensional representation of folded chains in a crystallite (according to Billmeyer)[9]; (a) ideal model, without amorphous fraction; (b), (c), and (d) models with amorphous fractions of various kinds.

When solidification occurs in a sample submitted to a longitudinal velocity gradient, other textures appear. One of them, the 'shish kebab'[10] characterized by alternate amorphous and crystalline layers is very well known. At present time, fibres made of completely aligned chains have also been obtained[10] but only in laboratories.

4.2 Nylon

Nylon is a polyamide, a condensation polymer (see Section 3.3), characterized by the repetitive bonding

$$\text{—NH—}\underset{\underset{\displaystyle O}{\|}}{C}\text{—}$$

and, as we have seen, nylon 6-6, which is the most common species, is obtained by condensation of adipic acid and hexamethylenediamine.

Nylons were discovered by W. Carothers in 1929, and they went on sale in 1938.

Nylon has a crystalline structure. The lattice has the property of being made essentially of 'dipolar' layers which form a 60° angle with the chain axis; the period, measured along the chain, is 1.74 nm. This arrangement results from the repetition of the acid–diamine hydrogen bond

$$(\text{—}\underset{|}{N}H\cdots O{=}\underset{|}{C}\text{—})$$

between adjacent chains. Compared with that of polyethylene, it is very different. Nylon is a synthetic textile which is frequently used with other fibres. Incidentally, it should be noted that silk is also made of chains of the amide type. A nylon fibre is made by hot extrusion through a small orifice (for instance, diameter $= 10^{-2}$ cm). The flow of the amorphous polymer liquid through this orifice creates a velocity gradient producing chain alignment. The crystallization occurs a few centimetres from the nozzle, by cooling down in the air. The chain orientation process is continued by cold stretching. The whole operation requires a precise control of temperature and flow.

The classical measurement unit of the linear density of a fibre is the 'denier', i.e. the mass in grams of 9000 m of fibre. The 'denier' of the extrusion product depends essentially on the nature of the polymer compound and of the extrusion flow. For nylon, the linear density varies from 30 to 300 denier. The ultimate stress is of the order of 500 newton/cm^2 (like common steel).

4.3 Polyisoprene: natural and synthetic rubber

Polyisoprene is a plant-polymer, extracted of hevea sap*. It can also be produced commercially.

Hevea sap is a suspension of 30% of latex in a serum. By adding an acid, globular clusters of polyisoprene linear chains are obtained, and the average number of links per chain is 300. The chemical structure of polyisoprene is

$$\left(\, CH_2 - \underset{\underset{CH_3}{|}}{C} = CH - CH_2 \,\right)_N$$

and we note that at every fourth carbon atom in the skeleton there is a double bond.

This double bond introduces a stereoisomerism; the unit may have two different forms:

the 'cis' form

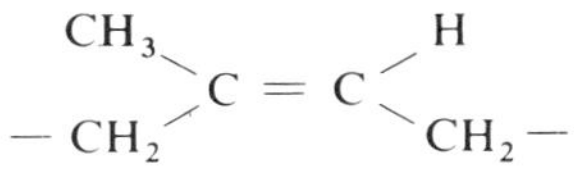

$$\begin{array}{c} CH_3 \diagdown \qquad\qquad H \diagup \\ \qquad\quad C = C \\ - CH_2 \diagup \qquad\quad \diagdown CH_2 - \end{array}$$

and the 'trans' form

$$\begin{array}{c} - CH_2 \diagdown \qquad\qquad H \diagup \\ \qquad\quad C = C \\ CH_3 \diagup \qquad\qquad \diagdown CH_2 - \end{array}$$

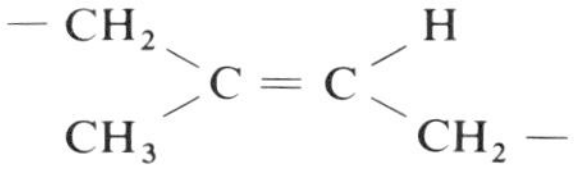

*The history of the hevea plantation is full of events. This tree was exclusively a product of the Brazilian forest until 1876. At that time, an Englishman, Wickman by name, smuggled hevea seeds from Amazonia, which were sown in Ceylon. Subsequently, rubber planting developed in Malaysia, Indonesia, and Indo-China. Regulating the level and the cost of production of the hevea latex has always been difficult in spite of international agreements such as the Stevenson plan (1922).

Actually, there are two varieties of polyisoprene:

(1) the 'hevea' variety, made of 'cis' units;

(2) the 'gutta-percha' variety, made of 'trans' units.

Cis-polyisoprene crytallizes partially between $0°$ and $35\,°C$ and the glass transition temperature T_g of the amorphous fraction corresponds to $-35\,°C$. On the other hand, trans-polyisoprene crystallizes in two forms, the α-form which melts at $65\,°C$ and the metastable β-form, obtained by quick cooling down, which melts at $56\,°C$.

The major discovery concerning polyisoprene is vulcanization, i.e. the creation of bridges between chains by means of sulphur atoms which saturate double bonds: actually, this goes back to 1839, when Goodyear observed the stability of this new material. Vulcanization is a series of reactions which can be shown schematically as follows

$$
\begin{array}{c}
CH3 \\
| \\
-CH_2-C=CH-CH_2- \\
\\
S \\
\\
-CH_2-C=CH-CH_2- \\
| \\
CH_3
\end{array}
$$

$$
\begin{array}{c}
CH_3 \\
| \\
-CH_2-C-CH=CH- \\
| \\
S \\
| \\
-CH_2-C-CH_2-CH_2- \\
| \\
CH_3
\end{array}
$$

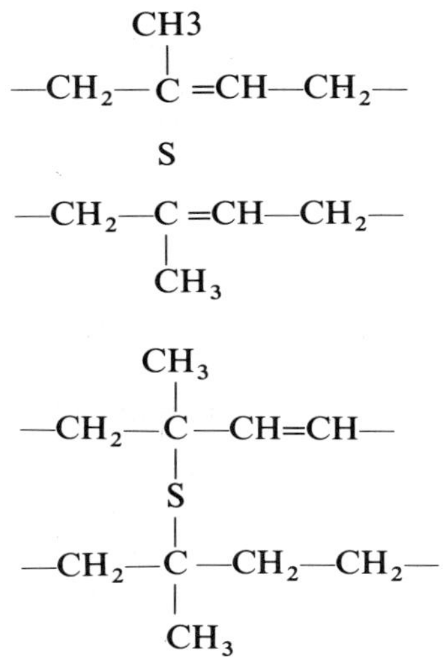

The reaction continues by addition of sulphur (and polyisoprene) and stops by exhaustion of sulphur.

The efficiency of a vulcanization is measured by the number of reticulations per atom of sulphur of the product. Up to 3 sulphur atoms per 100 carbon atoms, the efficiency of the vulcanization is of the order of 0.2. The molecular mass of the reticulated polyisoprene is infinite. In this state, it is a rubber.

Actually, real rubber is made by adding to the polyisoprene not only sulphur but also a large quantity of carbon black. Following this, reticulation is performed by applying pressure and by increasing the temperature. The elasticity and durability of the rubber are greatly improved thereby.

4.4 Polystyrene

Presently, polystyrene is a polymer material in common use, which is produced in large quantities (6.4×10^6 tons in 1980 in the Western world).

The first one to establish the chemical structure of the monomer was a pharmacist, E. Simon; he did this in Berlin around 1830 and gave the name 'styrol' to the monomer. He noted that the product solidifies after some time; J. Blyth and A. Hoffmann (in 1845) showed that during the process the chemical structure of the product does not change; they called it 'metastyrol'.

A long time before the macromolecular era, by adding a catalyst, J. Berthelot obtained a rapid formation of 'metastyrol'.

The polymer is a polyvinyl

$$\left(CH\varphi - CH_2 \right)_N$$

whose side-group φ is a benzenic ring. The presence of the group φ introduces a tacticity (see Section 1.2)

In the course of a radicalar or anionic synthesis, the benzene ring occurs nearly equally on both sides of the skeleton; however, successive pairs of benzene rings are slightly more numerous in syndiotactic position than in isotactic position.* However, this polystyrene is still called an atactic polymer. It has the property of being unable to crystallize even under large amplitude uniaxial stress.

The glass-transition temperature is about 110 °C. This polystyrene is used in light industry and as packing material. Its mechanical properties are good, provided that the residues of the synthesis are eliminated and that a plastifier is added.

In 1955, for the first time, G. Natta obtained definite amounts of stereoregular polystyrene by polymerization, in the presence of titanium chloride Cl_4Ti. This isotactic polystyrene has a crystallinity ratio which may reach 90 per cent. The melting temperature is 240 °C. The crystals belong to the rhombohedric system.

The atactic polystyrene is the polymer which has served the most for the study of the statistical behaviour of polymers in solution. The first significant viscosity measurements were made with rubbers and celluloses (1930), but it was very quickly recognized that polystyrene is a material which is especially suited for precise experiments. This polymer is perfectly linear and it is possible to obtain samples of molecular masses from 10^4 to 10^7 dalton, with an acceptable dispersion. Good solvents for polystyrene are benzene, carbon disulphide, and so on. Cyclohexane is also a solvent of polystyrene, but is not as good.

Light is strongly scattered by polystyrene in solution and the refractive index of the solution is very dependent of the mass concentration

$$\frac{dn}{d\rho} = 10^{-1}\,g^{-1}\,cm^3.$$

* Result obtained by nuclear magnetic resonance.

These two properties, which are related to each other (see Chapter 7), are the source of interesting experiments.

A peculiarity of polystyrene in bulk is the fact that it has a negative optical anisotropy. The optical anisotropy is revealed, for instance, by birefringence when the material is strained. Let us consider a uniaxial strain produced by a mechanical tension $\mathcal{T}$. For small deformations, the difference δn of the indices for directions parallel and perpendicular to the axis vary linearly with respect to $\mathcal{T}$

$$\delta n = c\mathcal{T}$$

where by definition c is the photo-elastic constant. The optical anisotropy is related to the link anisotropy, and, in general, the polarizability of the links is larger along its axis. However, a uniaxial deformation orients the links. Consequently, we can expect the refractive index to be higher along the strain axis: in this case the constant is positive. Thus, for polyisoprene $c = 2000$ brewster (1 brewster $= 10^{-8}$ cm^2 newton^{-1}). But, for polystyrene, the photoelastic constant is negative: $c = -5000$ brewster. Still, let us point out that the constant c for polystyrene becomes positive for large strains.

4.5 Polydimethylsiloxane

The structure of polydimethylsiloxane is

$$\left(\begin{array}{c} CH_3 \\ | \\ Si-O \\ | \\ CH_3 \end{array} \right)_N$$

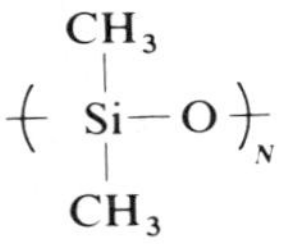

and its skeleton has the peculiarity of being uniquely constituted of oxygen and silicon atoms.

It is well-known that these chains are extremely flexible. Schulz and Haug[11] claim that this property is related to the following facts

1. First, the distance between neighbouring silicon atoms is relatively large: it is equal to 0.27 nm, whereas for a polystyrene chain

$$(CH\varphi-CH_2)_N$$

the distance between $CH\varphi$ groups is only 0.254 nm. On the other hand, the CH_3 group does not take up much room, and, particularly, less than a benzenic ring.

2. Second, the angle between the two bonds of the oxygen atom can easily be modified.

Thus, in this case, flexibility is not related to tacticity defects, as happens for polystyrene. The glass transition temperature is lower than $-100\,^\circ$C and the polymer in bulk is known to have good elastic properties.

Polydimethylsiloxane is soluble in toluene and also in styrene. It seems that for the study of polymers in solution, polydimethylsiloxane should be more interesting than polystyrene. However, it is only recently that it has been used for that purpose. This comes from the fact that, at the present time, synthesis of very long chains of this kind of polymer has still not been mastered.

5. MEASUREMENT OF MOLECULAR MASSES

Fundamentally, studying the behaviour of polymers in solution amounts to determining a statistical *geometrical* structure. For instance, we can interest ourselves in the size of a polymer in solution to see how it depends on the number of links (or on a quantity that is proportional to this number).

A priori, masses have nothing to do with the problem. However, to measure the number of links contained in a sample, the simplest way is to weigh the sample. So, one immediately obtains the mass concentration

$$\rho = \frac{\text{mass of polymer in solution}}{\text{volume of the solution}}.$$

Thus, we are led to express the size of a polymer as a function of its molecular mass, which can be considered as proportional to the number of links

Two types of average molecular masses can be determined experimentally: M_{n} and M_{W}.

The mass M_{n} can be determined by operating as follows. First, one determines the mass concentration ρ of the dissolved polymer and the corresponding chain concentration $\boldsymbol{C}$. The latter quantity is obtained, for small chain concentrations, by measuring the osmotic pressure (Van 't Hoff law, Chapter 5).

It is therefore possible to get an experimental value of the mass $\rho Ⓐ /\boldsymbol{C}$ where Ⓐ is the Avogadro number, and we shall verify that it is equal to M_{n} (see Section 1.4).

In fact,

$$\rho = \sum_N \rho_N \simeq \sum_N \frac{C_N M_N}{Ⓐ}$$

where N is the number of links of a chain. Therefore

$$\frac{\rho Ⓐ}{\boldsymbol{C}} = \frac{\sum_N C_N M_N}{\sum_N C_N} = \langle M \rangle = M_{\mathrm{n}},$$

which is the result we were looking for.

The average mass M_{W} is determined as follows. One measures the mass concentration ρ and the corresponding quantity

$$\sum_N N^2 C_N$$

which is obtained by radiation scattering. In fact, this sum is proportional to the intensity of radiation scattered by the solution, in the zero angle direction (see Chapter 7). Knowing the molecular mass m of a link, we deduce from it a value of the mass

$$\frac{m^2 \sum_N N^2 C_N}{\rho \text{Ⓐ}}$$

which is just equal to M_W, as we shall see. In fact,

$$m^2 \frac{\sum_N N^2 C_N}{\rho \text{Ⓐ}} \simeq m \frac{\sum_N N^2 C_N}{\sum_N N C_N} = m \frac{\langle N^2 \rangle}{\langle N \rangle} \simeq M_W.$$

6. SIZE-EXCLUSION CHROMATOGRAPHY

6.1 Object and principle of size-exclusion chromatography, nature of the gel

The chromatography which interests us in this book is a liquid state chromatography applied to polymer solutions. In practice, the available sample is polydisperse, and the question is the separation of polymers with different masses. The aim of this separation is either an analysis of the polydispersion of the sample, or a fractionation of the sample for further applications. In the first case, the required equipment is much lighter than in the second one.

The principle of the method is as follows: the polymer sample, dissolved in a proper solvent, is injected at time zero at the entrance of a column which, in general, is metallic and which is filled with a porous gel; then it is swept along the column by an eluent, i.e. pure solvent, which an adequate applied pressure forces to circulate through the column (see Fig. 1.13). The experiment shows that polymers with higher masses come out first.

After a time t, a certain amount of liquid has flowed through the column: this volume of liquid is measured and it is called the elution volume. Simultaneously, the instantaneous mass concentration of the solution issuing from the column is continuously measured. This measurement is made either by differential refraction or by light absorption.

If a polymer having a well-defined molecular mass is injected at the entrance of the column, a certain elution volume flows out and then a sharp signal indicates the passage of the polymer, provided that the presence of polymer in the solvent modifies the optical properties of the solution in an appreciable way. Thus, elution volume and molecular mass correspond to each other (see Figs 1.14 and 1.15)

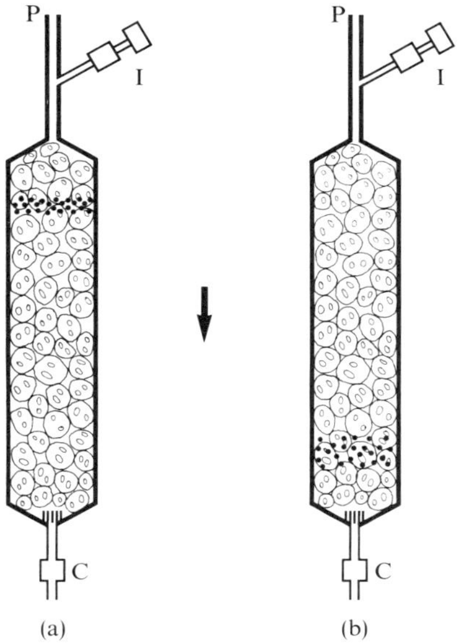

Fig. 1.13. Schematic representation of a chromatographic column (P = pump, I = injector, C = measurement cell). A monodisperse polymer sample, represented by dots, advances inside the column in the direction indicated by the arrow (successive configurations: (a), (b)).

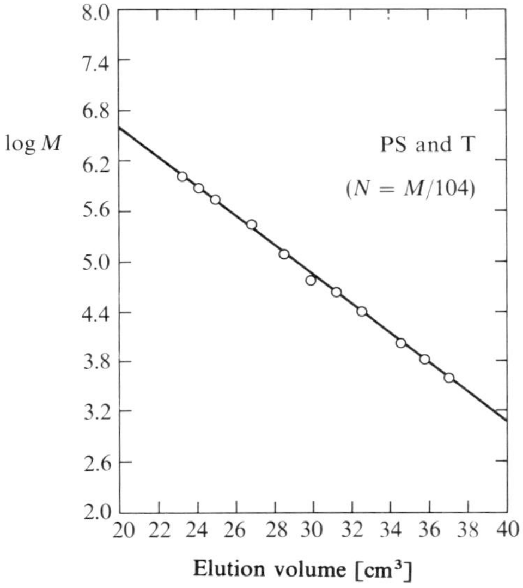

Fig. 1.14. Calibration curve of a chromatographic column used for measuring the polydispersion of polystyrene samples dissolved in toluene.

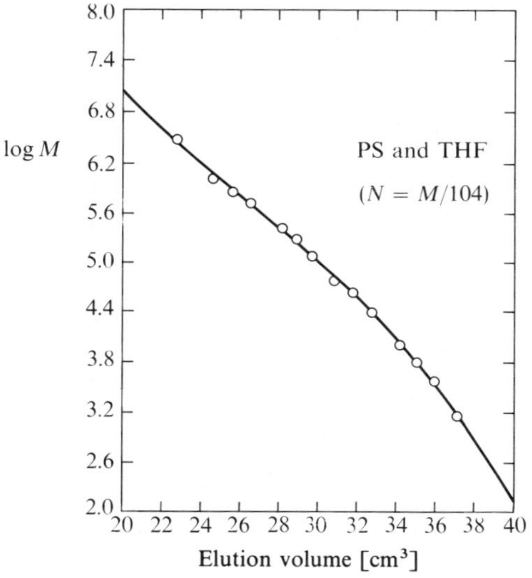

Fig. 1.15. Calibration curve of a chromatographic column used for measuring the polydispersion of polystyrene samples dissolved in tetrahydrofuran.

The method is obviously very general, but we must observe that there are two kinds of chromatographies: the interactive chromatography (also called enthalpic), and the size-exclusion chromatography (also called entropic). Interactive chromatography corresponds to the case where the molecules which are to be separated have some chemical affinity for the gel molecules. On the contrary, size-exclusion chromatography,[12,13] which is used for analysing or fractionating polymers, relies on size effects, i.e. geometrical factors, to separate the polymers from one another. Furthermore, in this case, one tries to avoid any attraction between the molecules in solution and the molecules of the gel; consequently, the gels whose polarity or polarizability is too large have to be discarded. Silica gels which are currently used in other domains must therefore be avoided in size-exclusion chromatography. Among the gels that can be used let us mention the microporous gels made by reticulation of polystyrene and divinylbenzene

$$CH_2 = CH - C \overset{CH-CH}{\underset{CH-CH}{\bigcirc}} C - CH = CH_2$$

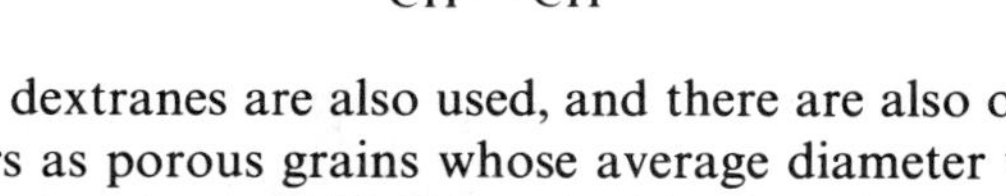

Porous glass and dextranes are also used, and there are also other possibilities.

The gel appears as porous grains whose average diameter varies from 10 to 100 microns. These grains are closely packed in the columns: the operation is rather delicate, and the methods used to fill the columns depend on the size of the grains.

Pumped through the column, the solvent flows slowly in the space between the grains. The very large molecules cannot enter the pores and come out first; their elution volume is equal to the interstitial volume between the grains. The smaller molecules can diffuse in the pores, where the solvent stagnates; consequently, they come out after the large molecules. It is generally assumed that, for molecules with very small sizes, the elution volume is equal to the total volume of solvent (interstitial volume + volume of the pores). For a molecule with an intermediary mass, the elution volume V_e might be given by

$$V_e = V_i + P V_p$$

where P is a partition coefficient, which is a function of the mass. This equality relies in the assumption that the molecules in the pores are in equilibrium with the molecules in the space between the grains. However, this interpretation of the experimental facts cannot be correct owing to a difficulty which will be explained in Section 6.4.

6.2 Analysis of the polydispersion of a sample

Only small quantities of product (of the order of 5 mg) are sufficient to analyse the mass polydispersion of a sample. Thus, a chromatographic column with a length of 60 cm and a diameter of 0.8 cm can do the job. For the gel, a mixture of constituents with different porosities has to be used; thus, the diameter of the

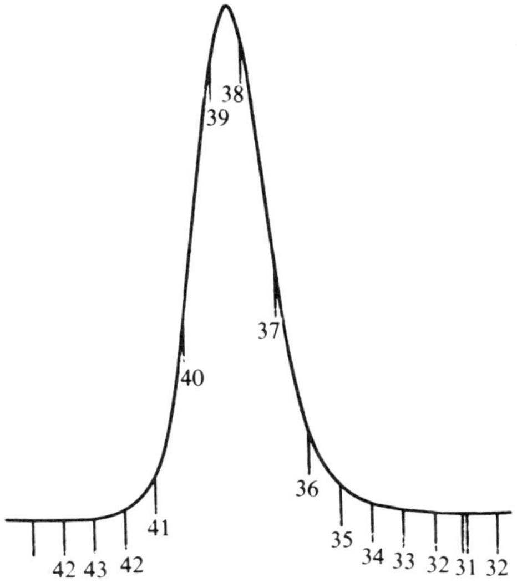

Fig. 1.16. Chromatogram of a polystyrene sample (according to Peyrouset and Panaris).[14] Here $M_n = 41\,000$ and $p = 0.01$. The numbers indicate the values of the masses in thousands of daltons.

pores may vary from 10 nm to 200 nm. The eluent flows through the column under pressure of the order of 200 to 2000 newton/cm^2 (20 to 200 bars). The calibration curve relates the molecular mass of the polymer to the elution volume. The shape of the curve depends essentially on the qualities of the column.

By adjusting the porosities of the constituents, it is possible to obtain a nearly linear relation between the elution volume and the logarithm of the molecular mass (see Figs 1.14 and 1.15). Of course, during the passage of the polymer through the column, there is also a small 'axial' diffusion.

At the exit, the (mass) polymer concentration is measured automatically, and simultaneously, the corresponding elution volume is continuously registered. In this way, one obtains a chromatogram (see Fig. 1.16) which defines the polydispersion of the sample under study. However, to determine the polydispersion curve with precision, from the chromatogram, it is necessary to take the axial diffusion into account.

6.3 Preparation of samples by size-exclusion chromatography

Preparing samples by chromatographic methods requires a heavier equipment. For instance, at Lacq,[14] in France, stainless steel columns with a diameter 6 cm and a height of 122 cm with cones at both ends were used in experiments: the gel used was made of silica microballs, with diameters in the range 100–200 microns, held back at the bottom of the column by a metallic sinter. With such an apparatus, it is possible to obtain narrow fractions of polystyrene, polyethylene (by operating at 150 °C), of polyvinylchloride, of polypropylene, of polyphenylsiloxane, and so on . . . For producing fractions of polydispersity close to 1.01 ($p = 0.01$), recycling is necessary and it is possible to fractionate 10 g per day, each fraction containing 0.5 g of polymer.

This method is good for samples of molecular mass smaller than one million, but beyond this, fractionation by precipitation seems more efficient.

6.4 Theoretical considerations on size-exclusion chromatography

At the present time (1988), it seems that there is no satisfactory theory of the size-exclusion chromatography, perhaps only because we lack a good model. We shall begin with a summary of the classical theory and later we shall run into a difficulty when trying to interpret this theory.

It is classically assumed that, in the column, the molecules lie either in the interstitial space, where they are swept along by the liquid, or in the pores where the liquid stagnates. It is assumed that at any moment, we have a concentration equilibrium. Then, we have to evaluate the partition coefficient P. We understand that, owing to size effects the molecule may occupy only a part of the volume of the pores and P may be defined as the ratio

$$P = \text{occupiable volume of the pores/total volume of the pores.}$$

For reasons of simplicity, the molecules will be represented by spheres of diameter $2r$ and the pores as cylinders of diameter $2r_0$ (see Fig. 1.17).

We see immediately that in this case

$$P = (1 - r/r_0)^2 \qquad r < r_0$$

$$P = 0 \qquad r > 0.$$

In Fig. 1.18, we represented $\ln(r/r_0)$ as a function of P. The number P varies linearly with respect to the elution volume (see Section 1.6.1) and $\ln(r/r_0)$ can be considered as a linear function of $\ln M$, where M is the molecular mass of the polymer. The curve of Fig. 1.18 has therefore to be compared to the calibration curves 14 and 15. Thus, it can be understood that by mixing three or four types of grains, with different porosities, it is possible to make columns which can separate polymers for very different values of the molecular mass. However, a question arises. In the model described above, what is the effective radius r of the polymer molecule? Is it the square root R of the mean square end-to-end distance or the square root R_G of the mean square radius of gyration, or another length? Let us now examine this problem.

It can be verified experimentally that the fractionation effect results from geometrical factors. In fact, Grubisic, Rempp, and Benoit (1967)[15] showed that for a given column and various polymers with different chemical compositions or different conformations, in solution in tetrahydrofuran, it is possible to get a

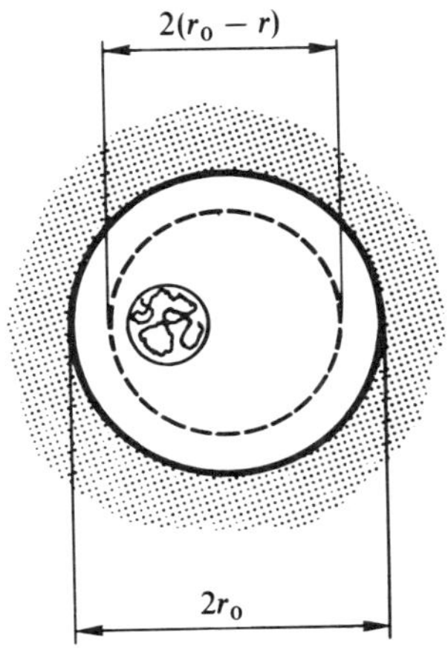

Fig. 1.17. Cylindrical pore of diameter $2r_0$ (front view). The pore contains a macromolecule of diameter $2r$. The dashed line defines the volume that can be occupied by the macromolecule.

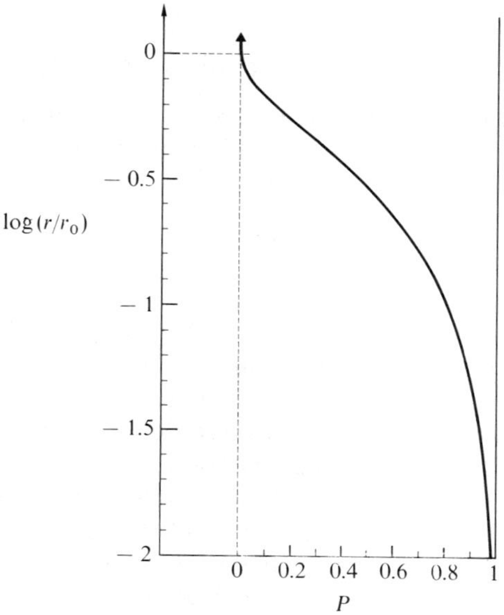

Fig. 1.18. Theoretical curve representing ln (r/r_0) as a function of the partition coefficient $P(r$ = radius of the macromolecule, $2r_0$ = diameter of a cylindrical pore, $P = (1 - r/r_0)^2)$.

unique calibration curve if, but only if, good variables are used (see Fig. 1.19). One of the variables is the elution volume; however, the other one is neither the logarithm of the mass, nor the logarithm of the radius of gyration (which is roughly proportional to it), but the logarithm of the hydrodynamical volume

$$[\boldsymbol{\eta}]M \equiv Ⓐ \frac{\boldsymbol{\eta} - \boldsymbol{\eta}_0}{\boldsymbol{\eta}_0 \boldsymbol{C}}$$

where $\boldsymbol{\eta}$ is the viscosity of a polymer solution containing $\boldsymbol{C}$ polymers per unit volume ($\boldsymbol{C}$ small), $\boldsymbol{\eta}_0$ the viscosity of the solvent, M the molecular mass of the polymer, and Ⓐ the Avogadro number.

On the other hand, it can be shown[16] that this volume is proportional to the product $R_G^2\ R_H$, where R_G is the radius of gyration of the chain and R_H the hydrodynamical radius (defined in Chapter 2, Section 2.4).

This interesting result is disturbing. In fact, we might believe that the polymers in solution stay for a while in the pores of the gel. The evidence showing that hydrodynamics plays a crucial role in chromatography seems to indicate that this idea is too simple. The authors quoted above suggested the following explanation: a molecule passing in front of a pore feels a velocity gradient, and the probability that it will enter the pore depends on this gradient;* however, the mechanism has to be analysed in detail and the question apparently remains unsolved!

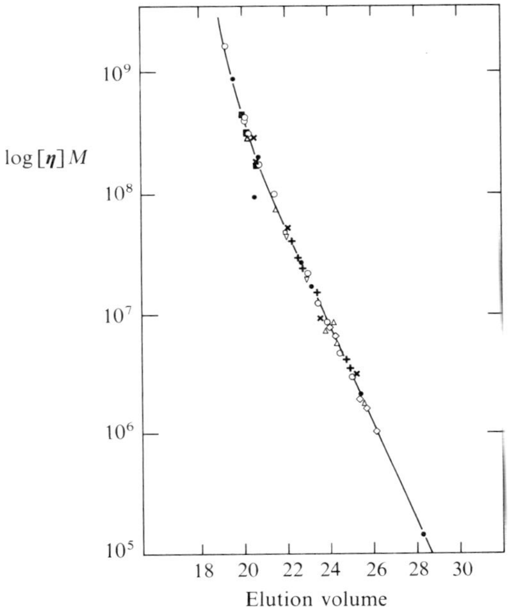

Fig. 1.19. Calibration curve for various polymers in tetrahydrofuran (according to Grubisic *et al.*[15]).
Polystyrene (PS); ● 'comb' polystyrene; + 'star' polystyrene;
△ heterogeneous grafted copolymer; × methyl-polymethacrylate;
◇ polyvinylchloride; ▽ grafted copolymer polystyrene–polymethyl methacrylate;
■ polyphenylsiloxane; □ polybutadiene

The quantity $[\eta]M = Ⓐ \dfrac{\eta - \eta_0}{\eta_0 C}$

whose logarithm is plotted against the elution volume is also a volume and it can be considered as proportional to $R_G^2\ R_H$ where R_G is the radius of gyration and R_H the hydrodynamical radius (see ref. 16).

* One might also assume as a few authors[17] do that the liquid flows in the pores. However, as was noticed by Janča,[13] this is hard to believe because, the 'useful' pores having a very small diameter, the velocity of the solution inside them must be very small. In fact, the flux through a cylindrical pore of radius r and length L is given by Poiseuille's law

$$\Phi = \frac{\pi r^4 \delta P}{8\eta L}$$

where δP is the pressure difference between the extremities.

7. FIELD-FLOW FRACTIONATION

Field-flow fractionation is a separation method which was introduced by Giddings[18] around 1960. The polymer solution flows in a flat ribbon-shaped duct. (see Fig. 1.20). A field perpendicular to the plane of the ribbon interacts with the polymers; this field may be a thermal gradient (or more simply the gravitation field).

Let $\boldsymbol{E}$ be the thickness of the ribbon and $h\,(0 < h < \boldsymbol{E})$ the coordinate of a point inside the ribbon along a direction perpendicular to this ribbon. The effect of the field is the creation of a concentration profile $C(h)$ of the form

$$C(h) = C(0)\,\mathrm{e}^{-h\mathfrak{g}}$$

where $\mathfrak{g}$ depends on the polymer mass. The velocity $v(h)$ which in the absence of a field is of the form

$$v(h) = v \sin(\Pi h/\mathrm{E})$$

may also depend on $\mathfrak{g}$ (particularly if the field is a thermal gradient)

In all cases, the mean velocity of a polymer which flows inside the ribbon is given by

$$\bar{v} = \frac{\displaystyle\int_0^E \mathrm{d}h\, C(h) v(h)}{\displaystyle\int_0^E \mathrm{d}h\, C(h)}.$$

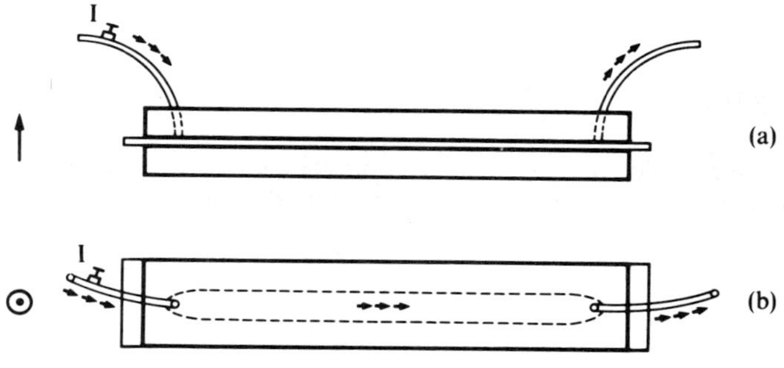

Fig. 1.20. Field-flow fractionation: schematic view of an apparatus: (a) side view, (b) top view. The duct is cut in a mylar sheet which is pressed between two blocks. The solvent flows in the direction indicated by the small arrows. The polymer is injected at the entrance (I = injector). An external field indicated by the solid arrow and by the point and circle, is applied to the apparatus. In an experiment described by Martin and Reynaud,[19] the field used is a thermal gradient. The temperature difference between the two (copper) blocks belongs to the range 10 °C–64 °C. The dimensions of the duct are: thickness = thickness of the mylar sheet = 0.1 mm, width = 2 cm, length = 40 cm. The pressure drop in the duct is of the order of a few millibars.

Therefore, $\bar{v}$ depends on $\mathfrak{g}$ and consequently on the mass of the polymer. This device can thus be used to separate polymers with different masses, and in this case, the light molecules come out first.

REFERENCES

1. Flory, P.J. (1969). *Statistical Mechanics of Chain Molecules.* Interscience-Wiley.
2. Hildebrand, J., Prausnitz, J., and Scott, R. (1970). *Regular and related solutions. The solubility of gases, liquids and solids.* Van Nostrand Reinhold.
3. Israëlachvili, J.N. and Tabor, D. (1972). *Proc. Roy. Soc.* (*London*) A **331**, 19.
4. Taunton, H.J., Toprakcioglu, C., Fetters, L.J. and Klein, J. (1988). *Nature* **332**, 712
5. More details can be found in the following books:
   Sigwalt, P. (1972). *Synthèse des composés macromoléculaires*, Vol. 1, 2nd part: *Chimie Macromoléculaire* (ed. G. Champetier). Hermann, Paris.
   Initiation à la chimie et la physicochimie macromoléculaires (1977). GFP (Groupe Français d'Etudes et d'Application des Polymères), 6 rue Boussingault, 67 Strasbourg, France.
   Elias, H.G. (1984). *Macromolecules* (transl. from German by J.W. Stafford), 2 vols. Plenum Press.
   Billmeyer, F.W. (1984). *Textbook of polymer science*, 3rd edn. John Wiley, New York.
6. Szwarc, M. (1964). *Proc. R. Soc.* (*London*) A **279**, 260.
7. Bamberger, E. and Tschirner, F. (1900). *Ber.* **33**, 955.
8. Keller, A. (1957). *Phil. Mag.* **2**, 1171.
9. Billmeyer, F. (1984). *Textbook of polymer science.* Wiley-Interscience, New York.
10. Pennings, A.J. (1977). *J. Polym. Sci. Polym. Symp.* **59**, 55
11. Schulz, G. and Haug, A. (1962). *Z. Phys. Chem.* (*Frankfurt*) **34**, 328.
12. Yau, W.W., Kirkland, J.J., and Bly, D.D. (1979). *Modern size exclusion liquid chromatography.* John Wiley.
13. Janča, J. (ed.) (1984). *Steric exclusion liquid chromatography.* Marcel Dekker, New York. See especially article 1: Principles of steric exclusion liquid chromatography, by J. Janča.
14. Peyrouset, A. and Panaris, R. (1974). *Bull. Soc. Chim. France* **11**, 2279.
15. Grubisic, Z., Rempp, P., and Benoit, H. (1967). *J. Polym. Sci.* (*Polym. Lett.*) **B5**, 753.
16. Weill, G. and des Cloizeaux, J. (1979). *J. Physique* **40**, 99.
17. Di Marzio, E.A. and Guttman, C.M. (1967). *J. Polym. Sci.* (*Polym. Lett.*) **B7**, 261; (1970) *Macromolecules* **3**, 681.
18. Giddings, J.C., Martin, M., and Myers, M.N. (1978). *J. Chromatogr.* **158**, 419.
19. Martin, M. and Reynaud, R. (1980). *Anal. Chem.* **52**, 2293.

2

GENERAL DESCRIPTION OF LONG CHAINS, UNIVERSALITY, CRITICAL PHENOMENA, AND SCALING LAWS

GENESIS

The application of universality principles and the use of scaling laws in physics, particularly in the field of polymers has been developing gradually, say from 1950 to 1980.

These notions have been introduced in order to describe random scale-invariant systems, whose importance had been recognized earlier. The manner in which this problem arose, appears clearly in the following text, taken from the introduction to the book *The Atoms* by Jean Perrin (1913).

'We all know, prior to rigorous definitions, how to get beginners to recognize that they are already acquainted with the idea of continuity. One plots in front of them a nice curve, plain enough, and applying a drawing rule to its contour, one says 'you see that there is a tangent at each point'. Or otherwise, in order to give the already more abstract idea of the true velocity of a projectile at a given point of its trajectory, one may say: 'Don't you feel that the average velocity between two neighbouring points of this trajectory finally ceases to vary appreciably when these points approach each other indefinitely?' And many minds, remembering indeed that this is the case for certain familiar motions, do not see that there are great difficulties here.

Mathematicians, however, have well understood the lack of rigor of these so-called geometrical considerations, and, for example, how childish it is to try to demonstrate, by drawing a curve, that any continuous function has a derivative. If it is true that differentiable functions are the simplest, and easiest to handle, they are nevertheless the exception; or, if geometrical language is preferred, curves which do not possess a tangent are the rule, and well-behaved curves such as the circle are most interesting cases, but very particular. At first, such restrictions seem only to be an intellectual exercise, doubtless clever but finally artificial and unproductive, when the desire for perfect rigor has been pushed to an obsession. And most often, those to whom one talks about tangentless curves or functions without derivatives believe at first that nature obviously does not show such complications and does not suggest this idea.

The contrary is true, however, and the logic of the mathematicians has maintained them closer to reality than the practical representations used by the physicists. This is what can be understood by considering without deliberate oversimplification certain entirely experimental facts. Such facts show up abundantly in the study of colloids. Let us observe, for instance, one of these: white flakes obtained by adding salt to water and soap. From a distance, the contour of this may appear sharp, but as soon as one gets nearer this distinctness vanishes. The eye no longer succeeds in fixing a tangent to a point: a straight line, which we would at first be inclined to consider as such, may under closer examination look to be equally as perpendicular as oblique to the contour. Using a magnifying glass, or a microscope, the uncertainty remains as great, because each time the magnification is increased one sees new sinuosities without ever feeling the pure and restful impression given, for instance, by a polished steel ball. In the same way that this ball can give a useful image of classical continuity, our flake can as logically suggest the more general concept of continuous functions without derivatives.

Also, we must note that the uncertainty as to the position of the tangential plane to a contour point is not exactly of the same order as the uncertainty concerning a tangent from a point on the coast of Britanny, which will be affected by the scale of the map that is used. The tangent here will differ in relation to the scale, but in every case it will be possible to fix it. The reason for this is that a map is a conventional drawing, where by construction any line has a tangent. On the contrary, it is the essential character of our flake (as it would be also of the coast if, instead of studying it on a map, one were to examine it on the spot, nearer or further away) that, at any scale, one can only guess at details without seeing them clearly, which absolutely inhibits the fixing of a tangent.

We are still in the realm of experimental reality, if, on applying the eye to the microscope, we observe the Brownian motion of any small particle tumbling about in a liquid suspension. In order to fix a tangent to its trajectory, we should find a limit, at least an approximate one, to the direction of the straight line joining the positions of this particle at two very closely succeeding instants. Now, as long as one is able to make the experiment, this direction varies wildly when the time interval between these two instants is decreased. So what is suggested to the unprejudiced observer by this study is once again the function without derivative, and by no means the curve with a tangent.

. . . The classical idea is certainly that one can break down any object into smaller parts which are practically homogeneous. In other words, it can be agreed that the *differentiation* of matter contained within a certain contour becomes weaker the more this contour is shrunk.

Now, far from being prescribed by experiment, I dare say that this conception rarely corresponds to it.

My eye looks in vain for a small 'practically homogeneous' region, on my hand, on the table where I write, on the trees, or on the ground which I can see from the window. And if, without being too demanding, I mark off a nearly homogeneous domain, on the trunk of a tree, for example, I need only to get closer in order to discern details on the bark which I just guessed at before, and to suspect the presence of still more. And then, when the naked eye becomes powerless, the magnifying glass, or the microscope, showing each part chosen successively at an ever-increasing scale, will reveal new details, and still newer ones, and when finally I shall have reached the actual limit of our power of resolution, the image that I shall retain will be far more differentiated than the one first perceived. It is well-known that a living cell is far from being homogeneous, that one grasps a complex organization of filaments and pellets immersed in an irregular plasma, where the eye guesses at things which it is futile to try to specify. Thus, the fragment of matter which at first we might hope to be somewhat homogeneous appears indefinitely spongy, and we have absolutely no presumptive evidence that by going further, one can finally reach 'homogeneity', or at least matter where properties would vary regularly from one point to another.

And it is not only living matter which is found to be indefinitely differentiated. The charcoal that would have been obtained by burning the bark observed a while ago, could also be shown to be indefinitely spongy. Vegetal earth, most of the rocks themselves, do not seem to be easily decomposable into smaller homogeneous parts. And we find, as the only examples of matter regularly continuous, crystals like diamond, liquids like water, or gases. So, the concept of continuity is derived on the whole by an arbitrary choice of mind from among experimental data.

It must also be recognized that often in spite of the profoundly irregular structure discovered with a little effort, the properties of this object can be very usefully represented approximately by continuous functions. Very simply, although wood is indefinitely spongy, it is still possible to speak meaningfully of the surface of a wooden beam to be painted, or of the volume displaced by a raft. In other words, at certain magnifications, for certain investigative processes, the regular continuum can represent phenomena, in the same manner as a tin sheet wrapped round a sponge, but not really following its delicate and intricate contour.'

One could not argue better. Thus, in this interesting text, ideas are introduced which we shall find again in the theory of polymers and which can be summarized as follows:

1. There are well-defined mathematical objects which cannot be represented by analytical functions, and these objects are unquestionably of interest.

2. These objects can appear inhomogeneous at any scale. It will be enough to postulate that these inhomogeneities are of the same nature at any scale in order to introduce the concept of critical object, i.e. of a random object which is scale-invariant.

3. Critical objects exist in nature. Jean Perrin mentions salted soapy water and Brownian trajectories of particles suspended in a fluid. There are many others, and some of them have aroused the interest of physicists: as, for example, the liquid vapour system at the critical point, the magnetic system at the Curie point, and turbulent systems in the inertial range.

1. CHAIN WITH INDEPENDENT LINKS AND THE BROWNIAN CHAIN

1.1 Definition

The theories whose aim is the study of the behaviour of polymers in solution, use idealistic models which become realistic in the limit of very long chains, but only in this limit. The chains of interacting points which schematize the polymers, have within this limit, very general properties which are fundamental both from a theoretical and from an experimental point of view.

The study of the Brownian chain, the simplest model, will enable us to introduce and to illustrate various basic notions which can easily be generalized to more realistic cases.

Let us consider a chain of points with position vectors $\vec{r}_j$, where j is an integer $j = 0, 1, \ldots, N$, made of independent links $\vec{u}_j$ in a d-dimension space

$$\vec{u}_j = \vec{r}_{j+1} - \vec{r}_j. \tag{2.1.1}$$

We shall postulate that the probability laws associated with each link are isotropic and identical. Let $p(\vec{u})$ be this function. Moreover, we shall assume that the mean square value of the length of a link is finite. Thus, we have

$$\int \mathrm{d}^d u\, p(\vec{u}) = 1$$

$$\langle\langle u_j^2 \rangle\rangle = \int \mathrm{d}^d u\, u^2\, p(\vec{u}) = l^2 d \tag{2.1.2}$$

where l is a characteristic length, and more generally

$$\langle\langle \vec{u}_j \cdot \vec{u}_l \rangle\rangle = \delta_{jl}\, l^2 d. \tag{2.1.3}$$

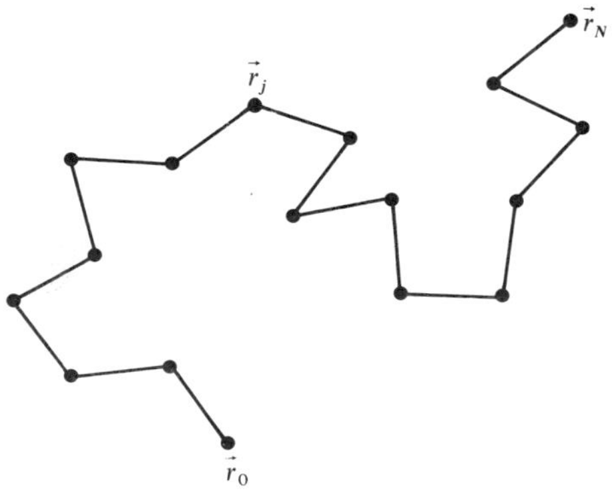

Fig. 2.1. A chain with N links (here $N = 16$).

This leads to the result

$$\langle\langle(\vec{r}_N - \vec{r}_0)^2\rangle\rangle = N\,l^2 d. \tag{2.1.4}$$

The vector $(\vec{r}_N - \vec{r}_0)$ is a random variable with a probability distribution $P_N(\vec{r})$

$$P_N(\vec{r}) = \langle\langle\delta(\vec{r} - \vec{r}_N + \vec{r}_0)\rangle\rangle.$$

Let $\tilde{P}_N(k)$ be its Fourier transform

$$\tilde{P}_N(\vec{k}) = \langle\langle e^{i\vec{k}\cdot(\vec{r}_N - \vec{r}_0)}\rangle\rangle = \int d^d r\, e^{i\vec{k}\cdot(\vec{r}_N - \vec{r}_0)}\, P_N(\vec{r}). \tag{2.1.5}$$

As the links of the chain are assumed to be independent, $\tilde{P}_N(k)$ can easily be expressed in terms of.the Fourier transform $\tilde{p}(\vec{k})$ of $p(\vec{u})$

$$\tilde{P}_N(k) = [\tilde{p}(k)]^N$$
$$\tilde{P}_N(\vec{0}) = \tilde{p}(0) = 1. \tag{2.1.6}$$

When N is large, $\tilde{P}_N(\vec{k})$ is appreciable only for small values of k. Accordingly, let us expand $\tilde{p}(\vec{k})$ with respect to k

$$\tilde{P}_N(\vec{k}) = \int d^d u\, e^{i\vec{k}\cdot\vec{u}}\, p(\vec{u}) \simeq 1 - \tfrac{1}{2}\langle\langle(\vec{k}\cdot\vec{u})^2\rangle\rangle + \ldots$$
$$= 1 - k^2 l^2/2 + \ldots. \tag{2.1.7}$$

Thus, for large values of N, we may write

$$\tilde{P}_N(\vec{k}) \simeq e^{-Nk^2 l^2/2} \tag{2.1.8}$$
$$P_N(\vec{r}) \simeq \frac{1}{(2\pi N l^2)^{d/2}} e^{-r^2/2Nl^2}.$$

This result is nothing other than the central limit theorem, since here $\vec{r}$ is the sum of independent random variables. We observe that the limiting law depends only on a single parameter (Nl^2). When $N \to \infty$, the effects related to the microstructure of the links disappear and the result depends on the function $p(\vec{u})$ only through the parameter l. Thus, in the limit $N \to \infty$, the behaviour of these chains with independent links, is universal, and the probability law

$$P_N(\vec{r}) = \frac{1}{(2\pi\langle\langle r^2\rangle\rangle/d)^{d/2}} \exp\left[-\frac{r^2 d}{2\langle\langle r^2\rangle\rangle}\right] \tag{2.1.9}$$

is a *universal law* which depends only on dimension d. Moreover it is a *scaling law* since the only argument in the law is $r^2/\langle\langle r^2\rangle\rangle$, the quantity $\langle\langle r^2\rangle\rangle$ being used to fix the scale.

We may now go to the continuous limit, which is called a Brownian chain. This limit is obtained by reducing the size of the links and by increasing their number so as to keep $\langle\langle r^2\rangle\rangle$ fixed.

We set

$$S = N\,l^2 \tag{2.1.10}$$

where S is a quantity which defines the size of the Brownian chain. Moreover, as each chain point can be labelled by its order j on the chain ($0 \leqslant j \leqslant N$), we may set

$$s = jl^2 \qquad 0 \leqslant s \leqslant S \tag{2.1.11}$$

and therefore we have

$$\frac{s}{S} = \frac{j}{N}.$$

The area s can thus be used as a curvilinear coordinate to locate a point on the continuous curve. Therefore, the configuration of a Brownian chain is defined by the vectorial function $\vec{r}(s)$ (where $\vec{r}(s)$ is the continuous limit of r_j).

Thus, for all s, we write

$$\langle\langle [\vec{r}(s) - \vec{r}(0)]^2 \rangle\rangle = sd, \tag{2.1.12}$$

and the distribution probability $P_N(\vec{r})$ is replaced by $P(\vec{r}, S)$, which is

$$P(\vec{r}, S) = \frac{1}{(2\pi S)^{d/2}} \exp(-r^2/2S). \tag{2.1.13}$$

Let us note that the length of a Brownian chain is infinite. In fact, let L be the length of a chain with independent links and l_0 the common length of the links. We can define L by the equality

$$L = N l_0 = N^{1/2}(N l_0^2)^{1/2} = N^{1/2} d^{1/2} S^{1/2}$$

(here $l_0^2 = l^2 d$). In the continuous limit $N \to \infty$ and S remains constant; therefore $L \to \infty$.

This is the reason why the curvilinear coordinates which is used to define the abscissa of a point on a Brownian chain is not a length but an area. This is connected with the fact that the Hausdorff dimensionality for a Brownian chain is not equal to one, as it is for a rectifiable curve, but is equal to two: thus in a sense, the Brownian chain has the characteristics of a surface (see Section 2.3).

Any portion of a Brownian chain, is in average and to a dilatation, absolutely identical to the whole chain. One says that there is *internal similarity*. The scale of a Brownian chain is determined only by one variable, namely S, which defines the mean square distance between the end points.

1.2 The general probability law associated with a Brownian chain

Let us now consider a chain made of N Gaussian independent links with the same probability law. The position of the point M_j of order $j (j = 0, \ldots, N)$ on the chain, can be represented by the vector $\vec{r}_j$.

With any configuration of the chain, we associate the weight

$$W(\vec{r}_0, \ldots, \vec{r}_N) = \exp\left[-\frac{1}{2l^2} \sum_{j=1}^{N} (\vec{r}_j - \vec{r}_{j-1})^2\right]. \tag{2.1.14}$$

Let us note that, here, we introduce a weight and not a probability distribution, in order to move more easily to the continuous limit. The weight defined above has a continuous limit; on the contrary the normalization constant has no limit when the number of variables becomes infinite. Thus, the probability distribution corresponding to the weight has no limit.

The weight (2.1.14) is used to calculate mean values, and obviously we have

$$\langle\langle [\vec{r}_j - \vec{r}_{j-1}]^2 \rangle\rangle = l^2 d$$
$$\langle\langle [\vec{r}_N - \vec{r}_0]^2 \rangle\rangle = N l^2 d. \tag{2.1.15}$$

Let us now consider the chain as a curve made of linear segments $M_{j-1} M_j$. The parameter s, with $0 \leqslant s < S = N l^2$ will serve to locate a point on this curve. The position vector of a point M corresponding to the coordinate s will be defined by the conditions

$$(j-1)l^2 \leqslant s \leqslant jl^2 \qquad j = 1, \ldots, N$$
$$\vec{r}(s) = \vec{r}_{j-1} + \left(\frac{s}{l^2} - j + 1\right)(\vec{r}_j - \vec{r}_{j-1}), \tag{2.1.16}$$

which, in particular, imply the relations

$$\vec{r}(jl^2) = \vec{r}_j.$$

Moreover, for $(j-1)l^2 < s < jl^2$, we have

$$\frac{\mathrm{d}\vec{r}(s)}{\mathrm{d}s} = \frac{1}{l^2}(\vec{r}_j - \vec{r}_{j-1}),$$

and therefore

$$\int_{(j-1)l^2}^{jl^2} \mathrm{d}s \left[\frac{\mathrm{d}\vec{r}(s)}{\mathrm{d}s}\right]^2 = \frac{1}{l^2}(\vec{r}_j - \vec{r}_{j-1})^2.$$

Thus the weight $W(\vec{r}_0, \ldots, \vec{r}_N)$ can be written in the form

$$W(\vec{r}_0, \ldots, \vec{r}_N) = \exp\left[-\frac{1}{2}\int_0^S \mathrm{d}s \left(\frac{\mathrm{d}\vec{r}(s)}{\mathrm{d}s}\right)^2\right]. \tag{2.1.17}$$

Now, let us increase the number of links N, while keeping S fixed. The weight $W(\vec{r}_0, \ldots, \vec{r}_N)$ becomes a functional $\mathscr{W}\{\vec{r}(s)\}$ of the vector function $\vec{r}(s)$; the right-hand side of (2.1.17) keeps the same form. In this way, we obtain

$$\mathscr{W}\{\vec{r}(s)\} = \exp\left[-\frac{1}{2}\int_0^S \mathrm{d}s \left(\frac{\mathrm{d}\vec{r}(s)}{\mathrm{d}s}\right)^2\right]. \tag{2.1.18}$$

The mean value of any functional $\mathcal{O}\{\vec{r}(s)\}$ of the function $\vec{r}(s)$, is defined by a functional integral

$$\langle\langle \mathcal{O}\{\vec{r}(s)\}\rangle\rangle = \frac{\int \mathrm{d}\{\vec{r}\}\,\mathcal{O}\{\vec{r}(s)\}\,\mathscr{W}\{\vec{r}(s)\}}{\int \mathrm{d}\{\vec{r}\}\,\mathscr{W}\{\vec{r}(s)\}} \tag{2.1.19}$$

where $\mathrm{d}\{\vec{r}\}$ is a volume element of the space of the functions $\vec{r}(s)$. We shall now explain how in practice such a value can be calculated. Let us consider, for example, an arbitrary vectorial function $\vec{a}(s)$ and let us suppose that we want to calculate the functional $\Phi\{\vec{a}(s)\}$ defined by

$$\Phi\{\vec{a}(s)\} = \left\langle\!\!\left\langle \exp\left[\int_0^S \mathrm{d}s\,\vec{a}(s)\cdot\frac{\mathrm{d}\vec{r}(s)}{\mathrm{d}s}\right]\right\rangle\!\!\right\rangle. \tag{2.1.20}$$

Applying the above-mentioned definitions, we find

$$\Phi\{\vec{a}(s)\} = \frac{\int \mathrm{d}\{\vec{r}\}\exp\left\{-\frac{1}{2}\int_0^S \mathrm{d}s\left[\left(\frac{\mathrm{d}\vec{r}(s)}{\mathrm{d}s}\right)^2 - 2\vec{a}(s)\frac{\mathrm{d}\vec{r}(s)}{\mathrm{d}s}\right]\right\}}{\int \mathrm{d}\{\vec{r}\}\exp\left\{-\frac{1}{2}\int_0^S \mathrm{d}s\left[\frac{\mathrm{d}\vec{r}(s)}{\mathrm{d}s}\right]^2\right\}}. \tag{2.1.21}$$

Let us introduce the functions

$$\vec{r}'(s) = \vec{r}(s) - \int_0^S \mathrm{d}s'\,\vec{a}(s') \tag{2.1.22}$$

The passage from $\vec{r}(s)$ to $\vec{r}'(s)$ in equivalent to a translation on the variables $\vec{r}(s)$; consequently

$$\mathrm{d}\{\vec{r}\} = \mathrm{d}\{\vec{r}'\}.$$

Now we use the fact that

$$\left[\frac{\mathrm{d}\vec{r}(s)}{\mathrm{d}s}\right]^2 - 2\vec{a}(s)\cdot\frac{\mathrm{d}\vec{r}(s)}{\mathrm{d}s} = \left[\frac{\mathrm{d}\vec{r}'(s)}{\mathrm{d}s^2}\right] - [\vec{a}(s)]^2. \tag{2.1.23}$$

We see that the integrals which appear in the numerator and in the denominator of the equation compensate each other; thus, we find the important relation:

$$\Phi\{\vec{a}(s)\} = \exp\left\{\frac{1}{2}\int_0^S \mathrm{d}s\,(\vec{a}(s))^2\right\}. \tag{2.1.24}$$

The expansion of $\Phi\{\vec{a}(s)\}$ to first- and second-order with respect to the components $a^i(s)$ of $\vec{a}(s)$ give

$$\int_0^S \mathrm{d}s\sum_i a^i(s)\left\langle\!\!\left\langle\frac{\mathrm{d}r^i(s)}{\mathrm{d}s}\right\rangle\!\!\right\rangle = 0$$

$$\int_0^S \mathrm{d}s\int_0^S \mathrm{d}s'\sum_{ij} a^i(s)\,a^j(s')\left\langle\!\!\left\langle\frac{\mathrm{d}r^i(s)}{\mathrm{d}s}\frac{\mathrm{d}r^j(s')}{\mathrm{d}s'}\right\rangle\!\!\right\rangle = \int_0^S \mathrm{d}s\sum_i [a^i(s)]^2$$

which implies

$$\left\langle\!\!\left\langle \frac{\mathrm{d}r^i(s)}{\mathrm{d}s} \right\rangle\!\!\right\rangle = 0$$

$$\left\langle\!\!\left\langle \frac{\mathrm{d}r^i(s)}{\mathrm{d}s}\frac{\mathrm{d}r^j(s')}{\mathrm{d}s'} \right\rangle\!\!\right\rangle = \delta(s - s')\delta_{ij} \tag{2.1.25}$$

We also note that the derivatives $\mathrm{d}r^j/\mathrm{d}s$ constitute a Gaussian random set (see Appendix A). Therefore the mean value of any product of these $\mathrm{d}r^j(s)/\mathrm{d}s$ can be calculated from (2.1.25) by applying Wick's theorem (see Appendix A).

1.3 Infinite Brownian chains

It is now possible to define infinite Brownian chains in the limit $S \to \infty$. The position of a point of the chain is given by the vector $\vec{r}(s)$ with $-\infty < s < +\infty$, and we can assume that the chain goes through the origin of coordinates. Thus, we postulate

$$\vec{r}(0) = 0$$

and we associate formally with a chain the (vanishing) weight

$$\mathscr{W}\{\vec{r}(s)\} = \exp\left\{-\frac{1}{2}\int_{-\infty}^{+\infty} \mathrm{d}s \left(\frac{\mathrm{d}\vec{r}(s)}{\mathrm{d}s}\right)^2\right\}. \tag{2.1.26}$$

This, we see that an infinite Brownian chain does not depend on any parameter. Moreover, we note that $\mathscr{W}\{\vec{r}(s)\}$ is invariant with respect to any transformation of the form $s \to \lambda^2 s$, $\vec{r}(s) \to \lambda \vec{r}(s)$. In other words, an infinite Brownian chain is statistically scale-invariant.

This scale-invariance is a characteristic property of critical phenomena. Thus, an infinite Brownian chain is a critical object.

Critical objects are very numerous in nature. In particular, they are to be found when we study the transitions which, for historical reasons, are called second-order phase transitions. For instance, in a fluid at the critical point, the density fluctuations remain correlated at distances extending to infinity. Thus, at the critical point, the fluid can be considered as a completely scale-invariant system, provided that the underlying microstructure is forgotten.

An infinite Brownian chain is a unique object, since such a chain does not depend on any parameter. This example shows that the appearance of scale-invariance, which goes along with the elimination of the microstructure details, is characterized by a drastic reduction of the parameters of the system. Thus, the final properties of a critical system present special features of universality and simplicity which the scientists consider to be highly interesting.

2 CHAIN WITH EXCLUDED VOLUME AND THE KUHNIAN CHAIN

2.1 Definition: exponent ν

A polymer chain is always made up of links which occupy a certain volume in space. The links cannot interpenetrate, and this exclusion leads to modifications of the chain statistics. Thus, we can try to generalize to self-avoiding chains, the results obtained for chains with independent links.

We shall admit that, when the number of links becomes very large, the influence of the microstructure is felt only globally. The behaviour of the chain becomes universal.

More precisely, let us consider a chain made of N links limited by the points of position vectors $\vec{r}_j$ ($j = 0, \ldots, N$) (see Fig. 2.2). For large values of N, we write

$$\langle\langle (\vec{r}_N - \vec{r}_0)^2 \rangle\rangle \simeq N^{2\nu} l_e^2 d \tag{2.2.1}$$

where ν is an exponent which belongs to the interval $1/2 < \nu < 1$. In fact, the value $\nu = 1/2$ corresponds to a chain with independent links and $\nu = 1$ to a rigid chain. We can admit that the effect of the excluded volume is a swelling of the chain. The behaviour of this chain is thus between that of a Brownian chain and that of a completely stretched chain (a rod). Of course, the exponent ν is a function of the space dimension d but does not depend on anything else, at least when the interactions between chain segments remain short-range. On the other hand, l_e is a complicated function of the chemical microstructure.

It may be generally admitted that the mean values of variables related to a chain can be calculated from the probability corresponding to each configuration, i.e. the set of all position vectors. This probability is given by a law '*a la Boltzmann*'

$$P(\vec{u}_1, \ldots, \vec{u}_N) = \frac{1}{Z}\exp[-{}^{c}\mathscr{U}\{\vec{u}\} - {}^{1}\mathscr{U}\{\vec{u}\}] \tag{2.2.2}$$

where ${}^{c}\mathscr{U}\{\vec{u}\}$ connects neighbouring points on the chain and where ${}^{1}\mathscr{U}\{u\}$ defines repulsive interactions.

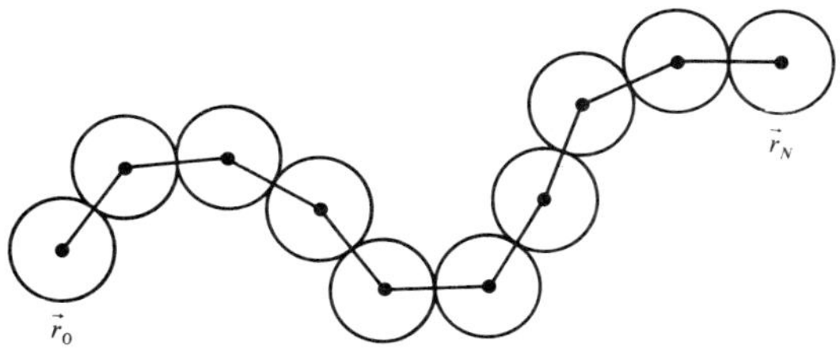

Fig. 2.2. A chain with excluded volume.

Actually, the quantities ${}^{c}\mathscr{U}$ and ${}^{I}\mathscr{U}$ are numbers, and are the products of certain free energies by $\boldsymbol{\beta} = 1/k_B T$. These free energies, which are also functions of $\boldsymbol{\beta}$, depend, in a complicated way, on the microstructure of the polymers in solution and on their solvents, in first approximation, ${}^{c}\mathscr{U}$ can be considered as independent of $\boldsymbol{\beta}$, since the length of the chain is practically constant; conversely, ${}^{I}\mathscr{U}$ would be proportional to $\boldsymbol{\beta}$ if the interactions were independent of temperature.

Typically, the interactions ${}^{c}\mathscr{U}$ and ${}^{I}\mathscr{U}$ for a realistic model can be expressed in the form

$$
\begin{aligned}
{}^{c}\mathscr{U}\{\vec{u}\} &= \sum_{j=1}^{N} V_c(\vec{u}_j) \\
{}^{I}\mathscr{U}\{\vec{u}\} &= \sum_{i,j} V(\vec{r}_i - \vec{r}_j)
\end{aligned}
\tag{2.2.3}
$$

where $V_c(\vec{r})$ and $V(\vec{r})$ are functions which depend only on $|\vec{r}|$ (and implicitly on temperature). Of course, the Brownian chain corresponds to the case where $V(\vec{r})$ and ${}^{I}\mathscr{U}\{\vec{u}\}$ vanish.

In spite of its merits, this model cannot be considered as really realistic since it excludes all correlations between successive links. It is not difficult, however, to convince oneself that correlations between successive links do not drastically modify the behaviour for large values of N. Still, it would be necessary to take them into account, if one wanted to make a real calculation of the length which appears in the asymptotic expression of $\langle\!\langle (\vec{r}_N - \vec{r}_0)^2 \rangle\!\rangle$ (see 2.2.1).

In a similar manner, for large values of N, the probability distribution associated with the vector $(\vec{r}_N - \vec{r}_0)$, reads

$$
\begin{aligned}
P_N(\vec{r}) &\equiv \langle\!\langle \delta(\vec{r} - \vec{r}_N + \vec{r}_0) \rangle\!\rangle \simeq \frac{1}{X^d}\mathrm{f}(r/X) \\
X &= \langle\!\langle (x_N - x_0)^2 \rangle\!\rangle^{1/2}
\end{aligned}
\tag{2.2.4}
$$

where x is a component of $\vec{r}$ and we expect $\mathrm{f}(y)$ to be a universal function that obeys the normalization conditions

$$
\begin{aligned}
&\int \mathrm{d}^d y\, \mathrm{f}(y) = 1, \\
&\int \mathrm{d}^d y\, y^2\, \mathrm{f}(y) = d.
\end{aligned}
\tag{2.2.5}
$$

As we did for the chain with independent links, we can eliminate the chemical microstructure by going to the continuous limit. The generalization is trivial. Let us set

$$
\begin{aligned}
\mathrm{S} &= N\, l_e^{1/\nu} \\
\mathrm{s} &= j\, l_e^{1/\nu}.
\end{aligned}
\tag{2.2.6}
$$

Keeping s fixed, we let N go to infinity, whereas the length l_e goes to zero. The quantity S remains proportional to the number of links. The position vector of a point on the chain is the vector $\vec{r}(\mathrm{s})$ with $0 \leqslant \mathrm{s} \leqslant \mathrm{S}$. The mean square distance between the end points remains invariant

$$R^2 \equiv \langle\!\langle (\vec{r}(\mathrm{S}) - \vec{r}(0))^2 \rangle\!\rangle = \mathrm{S}^{2\nu} d. \tag{2.2.7}$$

Thus, just as the chain with independent links has a continuous limit, which is the Brownian chain, the chain with excluded volume also has a continuous limit, which we call the Kuhnian chain.* Like the Brownian chain, the Kuhnian chain has an infinite length. In fact, the length L of a chain with excluded volume is (to a proportionality factor) equal to

$$L = N l_e = N^{1-\nu} \mathrm{S}^\nu, \tag{2.2.8}$$

which goes to infinity, when $N \to \infty$.

Only one length is associated with a Kuhnian chain and it defines its mean end-to-end distance. In fact, the size of a Kuhnian chain is defined by the 'course' S which has neither the dimension of a length (as for ordinary rectifiable curves), nor the dimension of an area (as for a Brownian chain), but an intermediate dimension, as can be seen from (2.2.7). This arises from the fact that the exponent ν of a Kuhnian chain is intermediate between 1/2 and 1, whereas the exponent of a 'normal curve' is $\nu = 1$ and the exponent of a Brownian curve is $\nu = 1/2$.

2.2 Probability laws associated with a Kuhnian chain

More generally, let us consider a Kuhnian chain of 'course' S and $(p+1)$ points on the chain (excluding the extremities). Let $\mathrm{s}_0, \ldots, \mathrm{s}_{p+1}$ be the courses of the chain segments connecting successive points (see Fig. 2.3). We have

$$0 \leqslant \mathrm{s}_j \leqslant \mathrm{S}$$

$$\mathrm{s}_0 + \ldots + \mathrm{s}_{p+1} = \mathrm{S}. \tag{2.2.9}$$

The set of vectors $\vec{u}_j = \vec{r}(\mathrm{s}_0 + \ldots + \mathrm{s}_j) - \vec{r}(\mathrm{s}_0 + \ldots + \mathrm{s}_{j-1})$ (where $j = 1, \ldots, p$) defines a configuration, and the following probability distribution is associated with it

$$P(\vec{u}_1, \ldots \vec{u}_p; \mathrm{s}_0, \ldots, \mathrm{s}_{p+1}) =$$

$$\left\langle\!\!\left\langle \prod_{j=1}^{p} \delta[\vec{u}_j - \vec{r}(\mathrm{s}_0 + \ldots + \mathrm{s}_j) + \vec{r}(\mathrm{s}_0 + \ldots + \mathrm{s}_{j-1})] \right\rangle\!\!\right\rangle \tag{2.2.10}$$

Here the vectors $\vec{u}_j$ are random variables and s_j the parameters. The mean value is calculated by giving to $\vec{r}(\mathrm{s})$ all possible forms.

* In honour of W. Kuhn who, as early as 1934, understood[1] that the excluded volume interactions involve the existence of an exponent ν different from 1/2, and who at that time had already calculated an approximate value ($\nu = 0.61$) for this exponent.

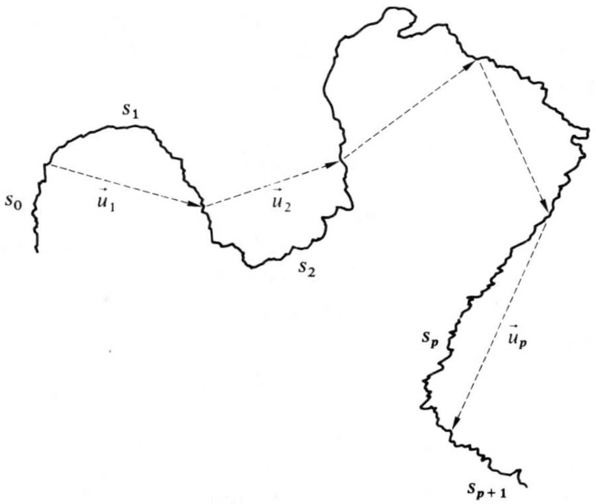

Fig. 2.3. A Kuhnian chain (schematic representation). The s_j are courses along the chain. The $\vec{u}_j$ are vectors defining the size and the orientation of segments.

By applying scaling laws, we may write

$$P(\vec{u}_1, \ldots, \vec{u}_p; s_0, \ldots, s_{p+1}) \simeq \frac{1}{(s_1 \ldots s_p)^{\nu d}} f\left(\frac{\vec{u}_1}{(s_1)^\nu}, \ldots, \frac{\vec{u}_p}{(s_p)^\nu}; \frac{s_0}{S}, \ldots, \frac{s_p}{S}\right) \quad (2.2.11)$$

where $S = s_0 + s_1 + \ldots + s_{p+1}$.

In particular, we can consider the limit where the chain is infinite. To reach this limit, we shall make s_0 and s_{p+1} become infinite, and therefore S will go to infinity. The new probability law reads*

$$P_\infty(\vec{u}_1, \ldots, \vec{u}_p; s_1, \ldots, s_p) = \frac{1}{(s_1 \ldots s_p)^{\nu d}} f_\infty\left(\frac{\vec{u}_1}{(s_1)^\nu}, \ldots, \frac{\vec{u}_p}{(s_p)^\nu}; \frac{s_1}{S}, \ldots, \frac{s_p}{S}\right) \quad (2.2.12)$$

In particular, for $p = 1$, we can write

$$P_\infty(\vec{r}; s) = \frac{1}{s^{\nu d}} f_\infty(r/s^\nu), \quad (2.2.13)$$

a formula analogous to eqn (2.2.4)

* The existence of an infinite chain is equivalent to the existence of the following limit

$$\lim_{\varepsilon \to 0} f(\vec{v}_1, \ldots, \vec{v}_p; m_0, \varepsilon m_1, \ldots, \varepsilon m_p) = f_\infty\left(\vec{v}_1, \ldots, v_p; \frac{m_1}{m_1 + \ldots + m_p}, \ldots, \frac{m_p}{m_1' + \ldots + m_p}\right)$$

In the Brownian case, the functions f(x) and $f_\infty(x)$ coincide. For a chain with exclusion, these functions will, of course, be different, but we may expect them to have very similar properties.

All the scaling relations which we described are reasonable assumptions that extrapolate eqn (2.2.13). We shall assume that they are exact, and they probably are exact, but this has yet to be proved mathematically.

2.3 Dimensionality of a random chain

A continuous random chain can be considered as a curve defined by a vector function $\vec{r}(s)$ with $0 < s < S$. However, it would not be right to conceive such a chain as a unidimensional object. In fact, a close relation exists between the dimensionality of a chain and the exponent ν.

In order to define the dimensionality of an object, we can proceed as follows. We introduce an arbitrarily small length $\varnothing$ and we look for the minimal number $N_\varnothing$ of balls of diameter $\varnothing$ that are necessary to cover the object completely (see Fig. 2.4). We discover that, when $\varnothing$ goes to zero, $N_\varnothing$ varies according to a law of the form

$$N_\varnothing = \mathscr{A} / \varnothing^D. \tag{2.2.14}$$

The number D defines the dimension* of the object in a Hausdorff sense,[2] and we shall call it *the dimensionality* of the object. Of course, this number has no connection with the dimension d of the space in which the object is embedded, in spite of the fact that the balls which cover the object have the dimension d. For usual objects, one easily verifies that the preceding definition of D corresponds to our intuition.

Let us now try to find out the dimensionality of a Kuhnian chain. For such a chain, the distance between end points is

$$|\vec{r}(S) - \vec{r}(0)| \sim S^\nu$$

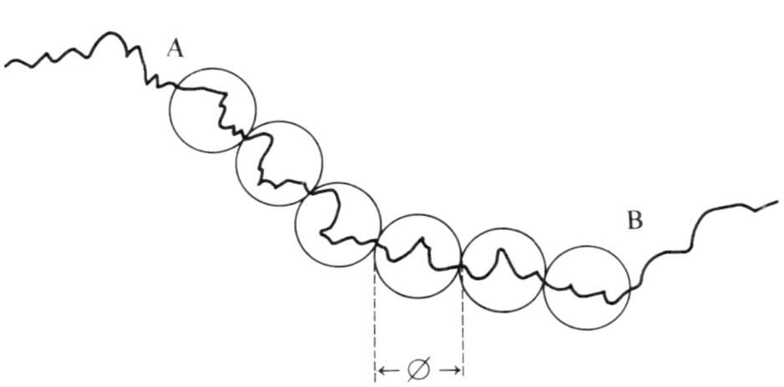

Fig. 2.4. Covering of a random chain with balls of diameter $\varnothing$.

* The dimension which is defined here is called 'capacity'; the true Hausdorff dimension is defined in a slightly more complicated way;[3] in the present case, these two notions coincide.

where the course S represents (in a way) the length of the chain. In a similar manner, the distance between two points of the chain is of the form $(0 < s < s' < S)$

$$|\vec{r}(s) - \vec{r}(s')| \sim |s - s'|^{\nu}$$

(the slowly varying factors have been omitted).

Let us consider the chain as made of N parts with equal causes S/N. The size of each chain segment is approximately

$$\varnothing \sim (S/N)^{\nu} \tag{2.2.15}$$

and therefore it is possible to cover each segment with a sphere of diameter $\varnothing$ (or nearly so). Now, the above equality can be written in the form

$$N \sim \frac{S}{\varnothing^{1/\nu}} \tag{2.2.16}$$

and, from a comparison with definition (2.2.14) we deduce the relation

$$D = 1/\nu. \tag{2.2.17}$$

The dimensionality of a continuous chain is the inverse of the exponent ν; thus, in general, it is not an integer, but when the chain is repulsive, its value is between 1 and 2.

In particular, a Brownian chain must be considered as an object of dimension two.

2.4 The radius of gyration and the hydrodynamic radius

Instead of choosing

$$R^2 = \frac{1}{2(N+1)^2} \langle\langle (\vec{r}_N - \vec{r}_0)^2 \rangle\rangle$$

to characterize the mean size of a polymer in solution, the experimentalists use the radius of gyration R_G, because this is a directly measurable quantity.

The radius of gyration of a chain with N links is defined by the average

$$R_G^2 = \frac{1}{2(N+1)^2} \left\langle\left\langle \sum_{ij} (\vec{r}_i - \vec{r}_j)^2 \right\rangle\right\rangle. \tag{2.2.18}$$

A simple geometric interpretation can be given for R_G^2. In fact, let $\vec{r}_G$ be the position vector of the centre of gravity of the chain

$$\vec{r}_G = \frac{1}{N+1} \sum_j \vec{r}_j.$$

Without taking into account any property of the chain under consideration, we can write

$$\begin{aligned}\frac{1}{2}\sum_{i=0}^{N}\sum_{j=0}^{N}(\vec{r}_i-\vec{r}_j)^2 &= \frac{1}{2}\sum_{ij}(\vec{r}_i-\vec{r}_G+\vec{r}_G-\vec{r}_j)^2\\ &= \sum_{ij}[(\vec{r}_i-\vec{r}_G)^2(\vec{r}_i-\vec{r}_G)(\vec{r}_j-\vec{r}_G)]\\ &= (N+1)\sum_{i}(\vec{r}_j-\vec{r}_G)^2. \qquad (2.2.19)\end{aligned}$$

In this way, we obtain the very general relation

$$R_G^2 = \frac{1}{N+1}\left\langle\!\!\left\langle \sum_{j=0}^{N}(\vec{r}_j-\vec{r}_G)^2 \right\rangle\!\!\right\rangle. \qquad (2.2.20)$$

The asymptotic behaviour of R_G is given by scaling laws and for large values of N, we have

$$R_G^2 = (\aleph/6)\langle\!\langle(\vec{r}_N-\vec{r}_0)^2\rangle\!\rangle \qquad (2.2.21)$$

where $\aleph$ is a constant which depends only on the type of chain under consideration and on the space dimension ($\aleph = 1$ for the Brownian chain).

On the other hand, the hydrodynamical radius R_H is defined by the expression

$$\frac{1}{R_H} = \frac{1}{2(N+1)^2}\left\langle\!\!\left\langle \sum_{ij}\frac{1}{|\vec{r}_i-\vec{r}_j|} \right\rangle\!\!\right\rangle. \qquad (2.2.22)$$

3. THE NUMBER OF CONFIGURATIONS AND THE ENTROPY OF A CHAIN WITH EXCLUDED VOLUME

3.1 Configurations of a chain drawn on a lattice: definition of γ

In the preceding sections, we have seen that the size of a large chain is characterized by an exponent ν which, for each type of chain, has a universal value. We shall now define another exponent γ whose real importance has been recognized only recently by the polymerists, but which plays an essential role in the Lagrangian theory of polymers. This exponent appears when one tries to characterize the asymptotic behaviour of the number of configurations.

First, we consider chains with N links, drawn on a d-dimensional lattice, from the site at the origin, and we assume that each site has $\mathfrak{f}$ neighbours ($\mathfrak{f}$ = connectivity, Fig. 2.5). If we deal with chains with independent links, the number of configurations is obviously equal to

$${}^{0}Z_N = \mathfrak{f}^N. \qquad (2.3.1)$$

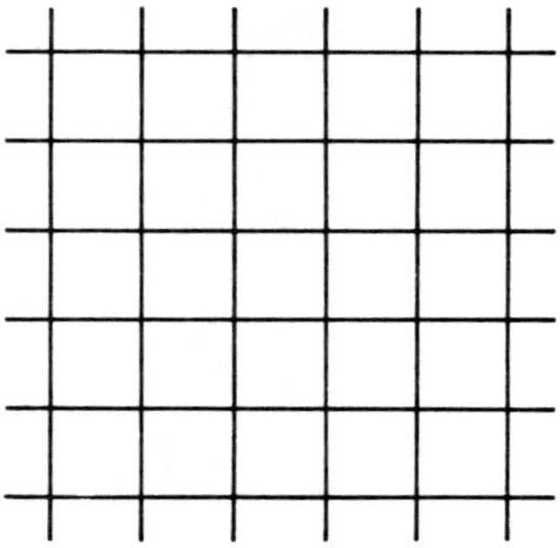

Fig. 2.5. Square lattice. Here the connectivity $\mathfrak{k}$ is equal to 4.

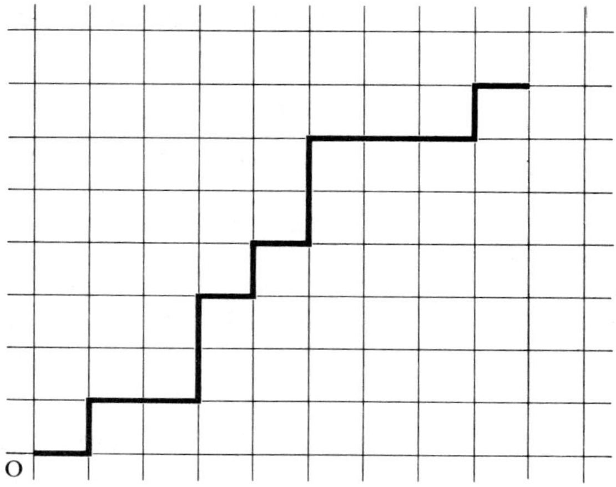

Fig. 2.6. Stair-chain on a lattice.

We shall now consider self-avoiding chains drawn on the lattice from the origin. Let Z_N be the number of configurations of these chains. This number is smaller than the number of chains for which only immediate returns are excluded, and therefore

$$Z_N < (\mathfrak{k} - 1)^N. \tag{2.3.2}$$

On the other hand, the number of self-avoiding walks is larger than the number of stair-chains on a lattice (Fig. 2.6). On a translation lattice,* the stair

* Here, a translation lattice is a lattice that is generated by transforming a point by application of the elements of a translation group.

chains are obtained by taking steps along any translation axis, always in the positive direction. We thus find the condition

$$Z_N > (t/2)^N. \tag{2.3.3}$$

These two conditions suggest the existence of a number μ defined by the limit

$$\lim_{N \to \infty} \frac{\ln Z_N}{N} = \ln \mu$$

and this result has been rigorously proved by Hammersley (see Chapter 3). This quantity is usually called the attrition coefficient.

Moreover, computer experiments show that the ratio Z_N/μ^N becomes infinite when $N \to \infty$. This fact leads to the assumption that for large values of N, Z_N is asymptotically given by a formula of the following type

$$Z_N \simeq Z_\square N^{\gamma - 1} \mu^N \tag{2.3.4}$$

where γ is an exponent, which, like ν, must be universal, i.e. for a given type of chain it must depend only on the space dimension. Of course, for a chain with independent links, $\gamma = 1$. On the other hand, $Z_\square$ is a constant which depends on the lattice.

The above representation of Z_N is generally accepted, and in particular by the physicist who do calculations with lattices, but no real proof of the correctness of this representation has yet been given.

In fact, one could a priori think that it would be more reasonable to believe that Z_N has a behaviour of the following type:

$$Z_N \simeq \mu^N \exp[\mathscr{A} N^\sigma], \tag{2.3.5}$$

and actually, as will be seen later (Chapter 4), the old Flory theory implicitly assumed such a behaviour (with $\sigma = 2\nu - 1$).

However, the new Lagrangian formalism, grounded on renormalization theory, implies the validity of eqn (2.3.4) (see Chapter 12), and precise calculations of the exponent γ have been performed. Consequently, it will be taken for granted that Z_N is given asymptotically by (2.3.4).

In a similar way, we can count the number of self-avoiding walks which start from the origin O and arrive at a point M defined by the position vector $\vec{r}$. Let $Z_N(\vec{r})$ be the number of these chains. Obviously, we have

$$Z_N(\vec{r}) = Z_N P_N(\vec{r}) \tag{2.3.6}$$

where $P_N(\vec{r})$ is the probability distribution associated with the chains with N links, which start from the origin and arrive at the point of position vector $\vec{r}$.

The asymptotic form of $Z_N(\vec{r})$ for large values of $\vec{r}$, has universal characteristics (for each type of chain) and eqns (2.3.4) and (2.3.6) can be summed up by postulating for Z_N the form

$$Z_N(\vec{r}) = \frac{Z_\square \mu^N N^{\gamma - 1}}{(l_c N^\nu)^d} f(r/l_c N^\nu). \tag{2.3.7}$$

The universal exponents γ and ν have to be determined, as well as the universal function $f(x)$ which is normalized by the conditions

$$\int d^d x \, f(x) = 1$$

$$\int d^d x \, x^2 f(x) = d \tag{2.3.8}$$

(d = dimension of space).

Conversely, $Z_\square$ and l_e are quantities which depend on all the peculiar features of the chain and not only on its general characteristics.

3.2 Configurations with return at the origin: definition of α

The number of chains which start from the origin and return to the origin after N links is given by $Z_N(\vec{0})$ (or more precisely by $\lim_{r \to 0} Z_N(\vec{r})$), and this is also an interesting quantity.

For large values of N, we might expect this quantity to be given by the expression

$$Z_N(\vec{0}) = Z_\square l_c^{-d} \mu^N N^{\gamma - 1 - \nu d} f(0). \tag{2.3.9}$$

However, this result is valid only if $f(x)$ does not vanish at the origin, an assumption which cannot be taken for granted; actually, we shall show later that for a self-avoiding chain, this function does vanish. In this case for large N, $Z_N(\vec{0})/Z_N$ varies as a power of N which is smaller than the power predicted by formula (2.3.9).

In general, we shall write

$$Z_N(\vec{0}) = \mathscr{B} l_c^{-d} \mu^N N^{\alpha - 2}. \tag{2.3.10}$$

In particular, let us assume that for small values of x

$$f(x) = f_0 x^\theta$$

We may admit that formula (2.3.10) is approximately valid for values of r of the order of l_e. On the other hand, we have approximately

$$Z_N(\vec{0}) \simeq Z_N(l_c) \simeq Z_\square l_c^{-d} N^{\gamma - 1 - \nu d}(f_0/N^{\theta\nu}), \tag{2.3.11}$$

and by comparing (2.3.10) to (2.3.11), we obtain the relation

$$\alpha = \gamma - \nu d + 1 - \theta\nu. \tag{2.3.12}$$

Knowing $Z_\square$, P_e, and $f(x)$ still does not enable us to determine the constant $\mathscr{B}$.

3.3 Entropy and continuous limit

Consider a random system. With each state $\boldsymbol{\Omega}_i$ of this system, we can associate a probability p_i. In statistical mechanics, the entropy of the system is defined by

$$\mathbb{S} = -\sum_i p_i \ln p_i \tag{2.3.13}$$

with

$$\sum_i p_i = 1.$$

In particular, if all the probabilities are equal, we have

$$p_i = p$$
$$\mathbb{S} = -\ln p. \tag{2.3.14}$$

If the system is a chain of origin O, on a lattice and if Z_N is the number of allowed configurations for the chain, we can write

$$p = \frac{1}{Z_N}, \tag{2.3.15}$$

and the corresponding entropy $\mathbb{S}_N$ is given by

$$\mathbb{S}_N = \ln Z_N. \tag{2.3.16}$$

Thus, for a chain with excluded volume and large values of N, we obtain from (2.3.4)

$$\mathbb{S}_N = N \ln \mu + (\gamma - 1) \ln N + \ln Z_{\square}. \tag{2.3.17}$$

In this formula, the quantities μ and $Z_{\square}$ depend on the lattice but γ is a universal exponent.

In a more general way, we may consider, on the lattice, chains whose points repel one another with short-range interactions. This condition will appear in the fact that with each configuration $\boldsymbol{\Omega}$ is associated a probability $p(\boldsymbol{\Omega})$

The entropy can be defined by the general formula

$$\mathbb{S}_N = -\sum_{\Omega} p(\boldsymbol{\Omega}) \ln p(\boldsymbol{\Omega}) \tag{2.3.18}$$

and the universality principle will lead us to admit that for large values of N, the expansion 2.3.17 remains valid.

Now, let us consider chains with N links, not on a lattice but in continuous space (see Fig. 2.3). Then, the number of configurations is infinite. This is not surprising since we are dealing with a classical system. This means that the entropy has to be renormalized by subtraction of an infinite constant.

Let $P(\vec{u}_1, \vec{u}_2, \ldots, \vec{u}_N)$ be the probability distribution associated with the links $\vec{u}_1, \ldots, \vec{u}_N$. A chain configuration can be represented by a point M in an Nd-dimensional phase space, the coordinates of M being the components of the vectors $\vec{u}_j$. Let us now pave this space with small cells of volume v. The position of a cell c_i can be defined by a point M_i of coordinates $\vec{u}^{\,i}_1, \ldots, \vec{u}^{\,i}_N$. Let p_i be the probability that a state of the chain be represented by a point belonging to the cell c_i. We have $p_i = v P(\vec{u}_2^i, \ldots, \vec{u}_N^i)$ and the entropy reads

$$\begin{aligned} \mathbb{S}(v) &= -\sum_i p_i \ln p_i \\ &= -\sum_i v P(\vec{u}_1^i, \ldots, \vec{u}_N^i) \ln P(\vec{u}_1^i, \ldots, \vec{u}_N^i) - \ln v \\ &= -\int \mathrm{d}^d u_1 \ldots \mathrm{d}^d u_N P(\vec{u}_1, \ldots, \vec{u}_N) \ln P(\vec{u}_1, \ldots, \vec{u}_N) - \ln v. \end{aligned} \tag{2.3.19}$$

When the size of the cells goes to zero, the last term becomes infinite, but this term is not at all dependent on the properties of the chain.

We can therefore subtract it without difficulty and we may define the 'classical' entropy by the equality

$$\mathbb{S}_N = -\int \mathrm{d}^d u_1 \ldots \mathrm{d}^d u_N P(\vec{u}_1, \ldots, \vec{u}_N) \ln P(\vec{u}_1, \ldots, \vec{u}_N). \tag{2.3.20}$$

In this case again, the expansion (2.3.17) remains valid. Now, however, the coefficient of N is not very significant; in fact, $\mathbb{S}_N$ is not invariant for scale transformations; thus, transformation $\vec{u}_j \to \lambda \vec{u}_j$ (for all j) leads to the result $\mathbb{S}_N \to \mathbb{S}_N - N \ln \lambda$.

Let us now go to the continuous limit by letting the number of links become infinite. In this case, again, $\mathbb{S}_N$ becomes infinite. However, by choosing a proper length scale (independent of N), it is always possible so to arrange that the coefficient of N in (2.3.17) vanishes; it is sufficient for that to put $\lambda = \mu$ in the above transformation. It is then possible to compare the entropy of two chains characterized respectively by the courses S and S_0. We find

$$\mathbb{S}(\mathrm{S}) - \mathbb{S}(\mathrm{S}_0) = (\gamma - 1) \ln (\mathrm{S}/\mathrm{S}_0). \tag{2.3.21}$$

The exponent γ thus remains a characteristic of continuous chains.

In other words, we can define entropy for continuous chains, but only after making adequate subtractions (renormalizations).

REFERENCES

1. Kuhn, W. (1934). *Kolloid Z.* **68**, 2.
2. Hausdorff, F. (1919). *Mathematische Annalen* **79**, 157.
   Hurewicz, W. and Wallman, H. (1941). *Dimension theory*, Ch. VII. Princeton University Press.
   Federer, H. (1969). *Geometric measure theory*. Springer, New York.
   Rogers, C.A. (1970). *Hausdorff measures*. Cambridge University Press, Cambridge.
   See also: Mandelbrot, B.B. (1977). *Fractals, form, chance and dimension*. Freeman, San Francisco.
3. Farmer, J.D., Ott, E. and Yorke, J.A. (1983). *Physica* D **7**, 153.

3

MATHEMATICAL MODELS OF CHAINS: STATIC PROPERTIES

1. NUMBER OF SELF-AVOIDING CHAINS ON A LATTICE

Though the static properties of polymers in solution are now quite well-known, only a small fraction of them has been proved in a perfectly rigorous manner. We owe the main results, concerning numbers of chains to the British mathematician Hammersley[1-4] who studied the problem around 1960, and some refinements are due to Kesten (1963).[5] These are summarized below in a modified form.

1.1 Hammersley theorems

These theorems have been demonstrated for self-avoiding chains drawn on two-dimensional square lattices. The proof could easily be extended to d-dimensional hypercubic lattices, and one could generalize to all kinds of lattices (see Fig. 3.1).

Theorem 1

Let Z_N be the number of self-avoiding chains with N links, drawn on a d-dimensional square lattice, from an origin O. The quantity $\lambda(N) = N^{-1} \ln Z_N$ tends to a limit λ when $N \to \infty$.

Proof (for d = 2)

Let Z_N be the number of self-avoiding chains starting from O. When we draw a self-avoiding chain, we must not return to the preceding site: therefore, the number of possible choices is not larger than three and $Z_N < 3^N$. On the other hand, the stair-chains (see Fig. 3.2) are self-avoiding chains and therefore $Z_N > 2^N$. Thus, the quantity

$$\lambda(N) = \frac{1}{N} \ln Z_N \tag{3.1.1}$$

is bounded

$$\ln 2 < \lambda(N) < \ln 3. \tag{3.1.2}$$

On the other hand, it is easy to see that

$$Z_{N+M} < Z_N Z_M. \tag{3.1.3}$$

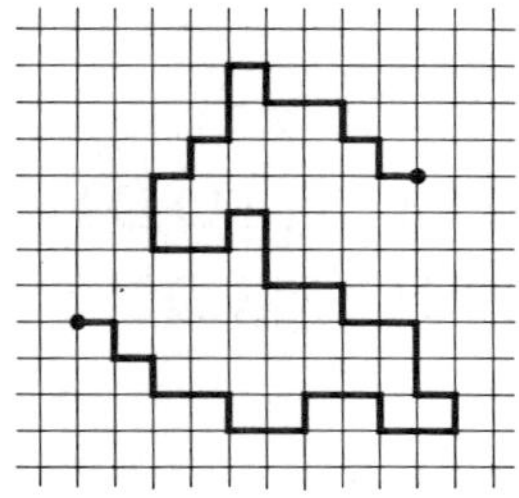

Fig. 3.1. Self-avoiding walk on a lattice.

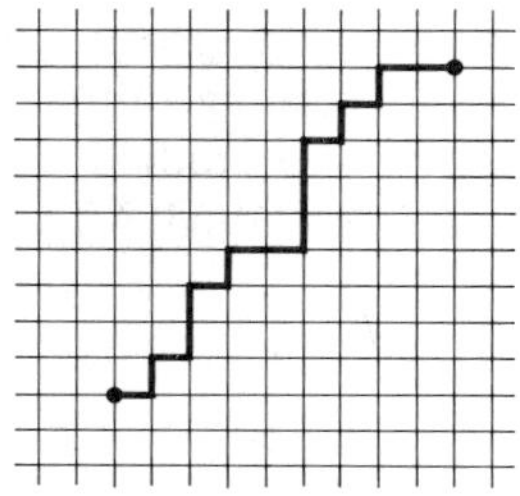

Fig. 3.2. A stair-chain.

Consequently, Z_N is a subadditive variable

$$\ln Z_{N+M} < \ln Z_N + \ln Z_N \tag{3.1.4}$$

and to obtain Hammersley's result, it is sufficient to apply a theorem by Hille[6] concerning such variables. Practically, one proceeds as follows. One starts from the equations

$$\lambda(N + M) < \frac{N\lambda(N)}{N + M} + \frac{M\lambda(M)}{N + M} \tag{3.1.5}$$

$$|\lambda(N)| < \mathscr{A}$$

and first, from them, one deduces by recurrence that

$$\lambda(MN) < \lambda(N). \tag{3.1.6}$$

Now, we set $\lambda = \inf \lambda(N)$. This lower bound exists since $\lambda(N)$ is bounded and it is never reached because of (3.1.6).

Let us show that, with any arbitrarily small positive number η, it is possible to associate a number $N(\eta)$ such that for any N larger than $N(\eta)$, we have

$$\lambda(N) - \lambda < \eta. \tag{3.1.7}$$

We observe that for any η, there exists an integer q such that

$$q > 4\mathscr{A}/\eta, \tag{3.1.8}$$

and an integer p such that

$$\lambda(p) - \lambda < \eta/2. \tag{3.1.9}$$

Then, let us set

$$N(\eta) = pq. \tag{3.1.10}$$

If $N > N(\eta) = pq$, we can write

$$N = mp + r \qquad m \geqslant q \qquad r < p$$

and therefore [See (3.1.5) and (3.1.6)]

$$\lambda(N) < \frac{mp\,\lambda(mp)}{mp + r} + \frac{r\lambda(r)}{mp + r} < \lambda(mp) + \frac{p|\lambda(r) - \lambda(mp)|}{mp}$$

$$< \lambda(p) + \frac{2\mathscr{A}}{m} < \lambda(p) + \frac{2\mathscr{A}}{q}.$$

Using (3.1.9) and (3.1.8), we get from the preceding inequality

$$\lambda(N) < \lambda + \eta$$

and therefore

$$0 < \lambda(N) - \lambda < \eta, \tag{3.1.11}$$

which is what we wanted to prove. Thus, μ, the attrition coefficient, is given by $\mu = e^{\lambda}$

Remarks

1. When the lattice is a Bravais lattice in a space of dimension d and when each site is connected to its nearest neighbours, inequality (3.1.2) is replaced by

$$\ln\left(\frac{\mathfrak{f}}{2}\right) < \lambda(N) < \ln(\mathfrak{f} - 1)$$

where $\mathfrak{f}$ is the lattice connectivity (the number of nearest neighbours of a site).

2. Hammersley's result indicates that $Z_N > \mu^N$; one might try to prove that actually, for N large, $Z_N \simeq Z_\square N^{\gamma-1}\mu^N$. However, Hammersley has been able only to find bounds of the form[3,4]

$$\mu^N < Z_N < \mu^N \exp[\mathscr{B}N^v] \qquad \text{with } v = \tfrac{1}{2}.$$

The modern theory, grounded on perturbation calculations, predicts the validity of the relation $Z_N \simeq Z_\square N^{\gamma-1}\mu^N$. However, if this formula is right, it will not be possible to prove it without using very sophisticated methods (see the criticism of Flory's theory in Chapter 8).

Theorem 2

Let U_N be the number of closed self-avoiding chains with N links, drawn on a d-dimensional square lattice, from an origin O (coinciding with the first, and the last point of the chain). The quantity $\lambda_0(N) = N^{-1} \ln U_N$ has a limit λ_0 when $N \to \infty$.

Proof (for d = 2)

Consider a closed chain (a circuit). The origin is given, the first point coincides with the last one, and the orientation is also given. Let U_N be the number of these circuits.

On these circuits, there are lower points, and among them there is a point which is the first one on the left. We shall call it the extremal lower point. In the same way, the extremal upper point will be the upper point on the left.

Let $2\, U_N^{|}$ be the number of circuits whose origin coincides with the extremal lower (upper) points. $U_N^{|}$ is the number of cycles (of various forms) that can be drawn on the lattice (there is no orientation on a cycle)

$$U_N^{|} = \frac{U_N}{2N}, \tag{3.1.12}$$

this number is obviously bounded

$$0 < U_N^{|} < 3^N. \tag{3.1.13}$$

Moreover, $\ln U_N^{|}$ is overadditive

$$\ln U_{N+M}^{|} > \ln U_N^{|} + \ln U_M^{|}.$$

In fact, let us consider a cycle with N links and a cycle with M links. Let us set these cycles on the lattice in such a way that the extremal lower point of the first one is exactly above the extremal upper point of the second one (see Fig. 3.3). We can open the cycles by removing, from each of them, the horizontal links which

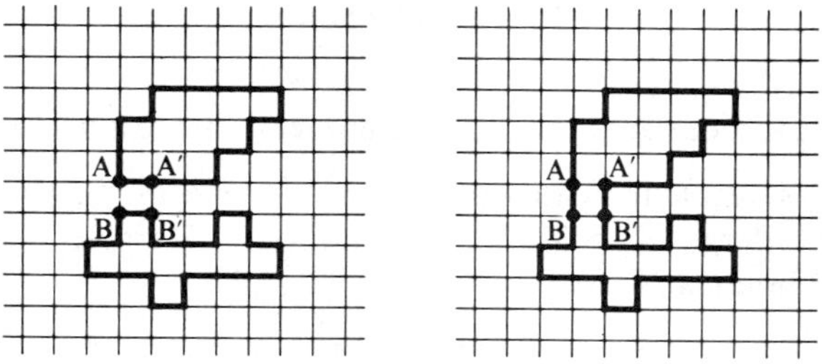

(a) (b)

Fig. 3.3. Constructing a cycle with two cycles. (a) The two cycles are set in proper position (A and B are extremal points according to the definition in the text). (b) The links AA′, BB′ are replaced by AB, A′B′.

start from the extremal points, and we can make one larger cycle with $(N + M)$ links by connecting the open cycles. Moreover, we see that for given values of N and M, the cycles with $N + M$ links thus obtained are distinct. Therefore

$$U^{|}_{M+N} > U^{|}_{M} U^{|}_{N}. \tag{3.1.14}$$

This property and the fact that $\lambda^{|}_{N}$ is bounded, implies the existence of the limit

$$\lambda_0 = \lim_{N \to \infty} \frac{1}{N} \ln U^{|}_{N}. \tag{3.1.15}$$

This can be easily established by proceeding as we did for the first theorem.

Consequently, the quantity

$$\lambda_0(N) = N^{-1} \ln U_N = N^{-1} \ln U^{|}_{N} - N^{-1} \ln(2N)$$

also has the same limit λ_0 when $N \to \infty$.

Theorem 3

For a chain with N links, drawn on a d-dimensional square lattice, let us denote by x_j the first coordinate of the point M_j of order j on the chain $(0 \leqslant j \leqslant N)$. Let $Z^{||}_{N}$ be the number of chains with N links, which satisfy the following requirements; the chains start from the origin O and all their points belong to the domain between the planes $x = x_0 = 0$ and $x = x_N$ (in other words, for all j with $0 \leqslant j \leqslant N$, we have, $x_N \geqslant x_j \geqslant x_0 = 0$). The quantity $\lambda^{||}_{N} = N^{-1} \ln Z^{||}_{N}$ tends to a limit $\lambda^{||}$ when $N \to \infty$ (see Fig. 3.4).

Proof

The proof is the same as for theorems 1 and 2. It uses the fact that $\lambda^{||}(N)$ is bounded and that the $Z^{||}_{N}$ obey the inequalities

$$Z^{||}_{N+M+1} > Z^{||}_{N} Z^{||}_{M} \tag{3.1.16}$$

which show that $\lambda^{||}(N)$ tends towards $\lambda^{||}$ from smaller values.

The presence of Z_{N+M+1} instead of Z_{N+M} in (3.1.16) comes from the fact that to construct one chain from two chains, one has to add a link in order to avoid

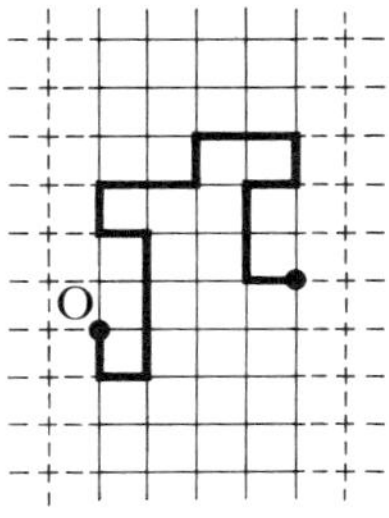

Fig. 3.4. Self-avoiding chain with 17 links and such that $0 \leqslant x_0 \leqslant x_j \leqslant x_{17} = 4$.

any overlap. The proof is not modified because it is always possible to redefine $Z_N^{||}$ by setting $W_N = Z_{N-1}^{||}$, and thus (3.1.16) reads $W_{N+M} > W_N W_M$.

Theorem 4

The limits $\lambda^{||}$ and λ are equal.

Proof

Let us consider a self-avoiding chain $\mathfrak{C}$ with N links on a hypercubic lattice. Let x_j be the first coordinate of the point M_j of order j on the chain $(0 \leqslant j \leqslant N)$. With this chain, we shall associate a chain $\mathfrak{C}^{|}$ having the property that the coordinate x_j of its points obey the inequalities $x_j^{|} \leqslant x_N^{\mathrm{I}}$ for any j. To construct $\mathfrak{C}_j$, in the case where we do not already have $x_j \leqslant x_N$ for any j, we proceed as follows.

We call chain extension the quantity $x_1 = \max x_j - \min x_j$. Then let σ_i be the largest integer for which $x_{\sigma_1} = \max x_j$. By cutting the chain in M_{σ_1}, we get two subchains. We can always assume that the subchain which contains the origin has the same extension as the chain itself. Otherwise, we would relabel the points on the chain by permuting the origin and the end point, and we would return to the preceding case. The subchain containing the end point M_N will be called the chain tail, its extension will be X_2 and we have in all cases $X_1 \geqslant X_2$.

Let us now transform this chain tail by symmetry with respect to the hyperplane $x = x_{\sigma_1}$. Thus, we generate a new self-avoiding chain with N links and the extension $X_1 + X_2$. For this new chain, we can define an index X_2 corresponding to a value x_{σ_2} defined as above.

Let us iterate the unfolding process of the chain, until we get, after p iterations, $\sigma_p = N$. The final chain is the chain $\mathfrak{C}^{|}$, which we were looking for.

In order to obtain it, we constructed a series of p chains with extensions X_1, $(X_1 + X_2), \ldots, (X_1 + \ldots + X_p)$. These quantities X obey the inequalities

$$X_1 \geqslant X_2 \geqslant \ldots \geqslant X_p > 0, \tag{3.1.17}$$

and constitute a partition of

$$X = X_1 + \ldots + X_p \leqslant N \tag{3.1.18}$$

where X is the extension of $\mathfrak{C}^{|}$.

Thus, to each chain $\mathfrak{C}$ corresponds one chain $\mathfrak{C}^{|}$ which is unique. On the contrary, several chains $\mathfrak{C}$ correspond in general to one chain $\mathfrak{C}^{|}$. However, this number is certainly smaller than $2\mathcal{N}(X)$ where $\mathcal{N}(X)$ is the number of partitions of the number X. In fact, let us start from a chain $\mathfrak{C}^{|}$; by folding a part of extension X_p, then a part of extension X_{p-1} and so on, in all possible ways compatible with the inequalities (3.1.17), it is possible to reconstruct all the chains $\mathfrak{C}$ from which we get $\mathfrak{C}^{|}$ (or the chains obtained by permutation of the origin and of the end point of $\mathfrak{C}$). Note that this process also gives a certain number of self-intersecting chains which are of no interest here.

Moreover, since $X \leqslant N$, we have $\mathcal{N}(X) \leqslant \mathcal{N}(N)$.

Then, let $Z_N^{|}$ be the number of self-avoiding chains $\mathfrak{C}_N^{|}$ that start from the origin O ($x_0 = 0$) in the half-space $x \geqslant 0$. We have

$$Z_N^{|} > \tfrac{1}{2}[\mathcal{N}(N)]^{-1} Z_N \tag{3.1.19}$$

(the factor 2 corresponds to the fact that for every partition of X, two different chains $\mathfrak{C}$ can be associated with $\mathfrak{C}^{|}$).

By applying, a second time, this chain unfolding technique to a chain $\mathfrak{C}^{|}$, we can construct a chain $\mathfrak{C}^{||}$ for which $x_0^{||} \leqslant x_j^{||} \leqslant x_N^{||}$ for any j. By proceeding as above, we can show that

$$Z_N^{||} > [\mathcal{N}(N)]^{-1} Z_N^{|} \tag{3.1.20}$$

(note that the factor 1/2 is absent), and finally, we obtain the result

$$Z_N > Z_N^{||} > \tfrac{1}{2}[\mathcal{N}(N)]^{-2} Z(N). \tag{3.1.21}$$

The asymptotic value of the number of partitions of N is given by the Hardy–Ramanujan formula (1917) (see Appendix B)

$$\ln \mathcal{N}(N) \sim \pi(2N/3)^{1/2} \qquad N \to \infty. \tag{3.1.22}$$

This means that for any N there exists a number A for which

$$\mathcal{N}(N) \leqslant e^{AN^{1/2}}. \tag{3.1.23}$$

Thus, setting

$$\lambda(N) = N^{-1} \ln Z_N \qquad \lambda^{||}(N) = N^{-1} \ln Z_N^{||} \tag{3.1.24}$$

we obtain from (3.1.21)

$$\lambda(N) > \lambda^{||}(N) > \lambda(N) - 2AN^{-1/2} - N^{-1} \ln 2.$$

Now, passing to the limit $N \to \infty$, we get the result

$$\lambda^{||} = \lambda.$$

Theorem 5

The limits λ_0 and λ are equal.*

Proof

By applying the unfolding technique used for the proof of the preceding theorem, but along another axis, we can start from $\mathfrak{C}^{||}$ chains and construct self-avoiding chains $\mathfrak{C}^{\square}$ so that for any j, we have simultaneously $x_0 \leqslant x_j \leqslant x_N$ and $y_0 \leqslant y_j \leqslant y_N$. (Here the first coordinate of $\mathbf{M}_j$ is x_j, the second one is y_j.)

Let $Z_N^{\square}$ be the number of chains $\mathfrak{C}^{\square}$ for which $x_0 = 0$, $y_0 = 0$. By arguing as above, we can easily show that [see (3.1.21)]

$$Z_N^{||} > Z_N^{\square} > \tfrac{1}{2}[\mathcal{N}(N)]^{-2} Z_N^{||}. \tag{3.1.25}$$

* The proof is not the original proof by Hammersley; the method used here is simpler and an error has been avoided.

Thus, setting

$$\lambda^{\square}(N) = N^{-1} \ln Z_N^{\square}$$

we can show as above that, for $N \to \infty$, $\lambda^{\square}(N)$ has a limit $\lambda^{\square}$ and that

$$\lambda^{\square} = \lambda^{||} = \lambda. \tag{3.1.26}$$

Let us now try to construct cycles on the lattice by using chains of the $\mathfrak{C}^{\square}$ type. Let us denote the point coordinate by x, the second one by y, and all the others by z.

Let x and y be two positive integers ($0 \leqslant x \leqslant N, 0 \leqslant y \leqslant N$) and let us mask two sites on the lattice, namely $A = (x, 0, 0)$ and $B \equiv (0, y, z)$. Now, we consider the set $\mathscr{E}$ of the chains with N links, which join A and B, in such a way that for any point M_j of a chain, we have $0 \leqslant x_j \leqslant x$, $0 \leqslant y_j \leqslant y$. Let $Z_N(A, B)$ be the number of these chains. We immediately see that

$$\sum_{A,B} Z_N(A, B) = Z_N^{\square}. \tag{3.1.27}$$

By symmetry with respect to the hyperplane $x = -1/2$, we generate from $\mathscr{E}$ a new set $\mathscr{E}_1$ of chains which join the points $A_1(-x-1, 0, 0)$ and $B_1(-1, y, z)$. In the same way, by symmetry with respect to the hyperplane $y = -1/2$, we generate from $\mathscr{E}$ a set $\mathscr{E}_2$ of chains which join the points $A_2(x, -1, 0)$ and $B_2(0, -y-1, z)$ and from $\mathscr{E}_1$ a set $\mathscr{E}_3$ of chains which join the points $A_3(-x-1, -1, 0)$ and $B_3(-1, -y-1, z)$. Now we have only to add the links AA_2, A_1A_3, BB_1 and B_2B_3 to form a set of self-avoiding cycles with $(4N + 4)$ links (see Fig. 3.5).

This leads to the following inequalities in which U_N is the number of circuits with N links.

$$U_{4N+4} > \sum_{A,B} [Z_N(A, B)]^4 > [\max Z_N(A, B)]^4. \tag{3.1.28}$$

Now, x and y can take at most N values, and the other coordinates can take at most $2N$ values. Consequently,

$$\max Z_N(A, B) \geqslant N^{-2}(2N)^{-(d-2)} \sum_{C,D} Z(C, D) = N^{-2}(2N)^{-(d-2)} Z_N^{\square}. \tag{3.1.29}$$

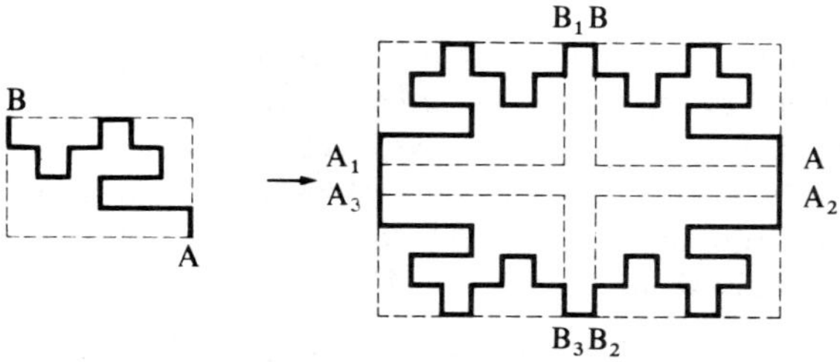

Fig. 3.5. A self-avoiding chain joining the opposite edges A and B of a rectangle and contained in this rectangle, is associated with symmetric chains, in order to construct a self-avoiding cycle.

Thus, we obtain the result

$$U_{4N+4} > (2N)^{-4d}(Z_N^{\square})^4, \tag{3.1.30}$$

and, on the other hand, we know that $U_N < Z_N$. Using the notation

$$\lambda_0(N) = N^{-1}\ln U_N$$
$$\lambda(N) = N^{-1}\ln Z_N$$
$$\lambda^{\square}(N) = N^{-1}\ln Z_N^{\square} \tag{3.1.31}$$

We get

$$\lambda(4N+4) > \lambda_0(4N+4) > \frac{N}{N+1}\lambda^{\square}(N) - \frac{d}{N+1}\ln(2N). \tag{3.1.32}$$

In the limit where N is infinite, we obtain

$$\lambda \geqslant \lambda_0 \geqslant \lambda^{\square}, \tag{3.1.33}$$

but as we have already proved the equality $\lambda^{\square} = \lambda$ [see (3.1.26)], from the preceding equation we deduce the result

$$\lambda_0 = \lambda. \tag{3.1.34}$$

1.2 Kesten bounds

Estimates concerning the convergence of the quantities $\lambda(N) = N^{-1}\ln Z_N$ and $\lambda_0(N) = N^{-1}\ln U_N$ have been given by Hammersley and by Kesten.

In particular, Kesten[5] has shown that, for a hypercubic lattice, there exist numbers A, B, C such that

$$|(Z_{N+2}/Z_N) - e^{2\lambda}| < AN^{-1/3}$$
$$-BN^{-1/3} < (U_{2N+3}/U_{2N+1}) - e^{2\lambda} < CN^{-1/4}.$$

Unfortunately, these bounds do not seem very good and do not tell us much, because the present theories indicate that the quantities $|(Z_{N+1}/Z_N) - e^{\lambda}|$ and $|(U_{N+1}/U_N) - e^{\lambda}|$ must be of order $N^{-\alpha}$.

Actually, the proofs given by Kesten are rather complicated and will be omitted here.

1.3 Self-avoiding spiral chains

Another type of self-avoiding chain, drawn on a square lattice, has been studied by Blöte and Hilhorst[7] who obtained exact results. These chains obey the following restriction: if one goes along the chain and if a site is a turning point, one turns always to the left (or always to the right). Consequently, in general, a chain is a double spiral (see Fig. 3.6). Thus, we are led to take an interest in chains that are self-avoiding simple spirals, since these chains form a subset of the double spirals (see Fig. 3.7).

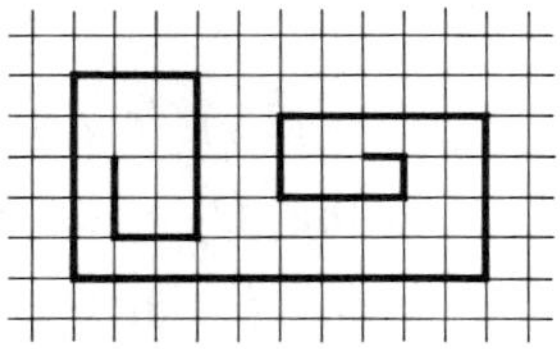

Fig. 3.6. Double spiral on a square lattice.

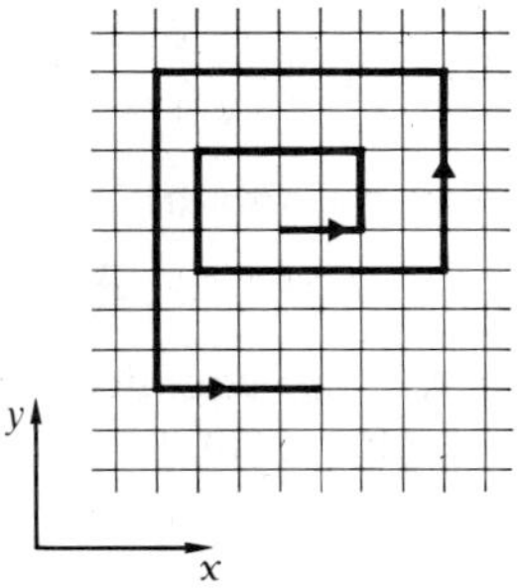

Fig. 3.7. Oriented simple spiral with two special properties: (1) there are only left turns; (2) the first link is directed along Ox and the last one is parallel to it.

For chains with N links and in the limit $N \to \infty$, Blöte and Hilhorst found the following results.

For chains spiralling towards the outside, the total number of chains starting from an origin O is given by

$$Z'_N \simeq \frac{1}{4\pi} 2^{1/2} N^{-1/2} \kappa^{N^{1/2}} \tag{3.1.35}$$

$$\kappa = \exp(\pi(2/3)^{1/2}) = 13.00195, \tag{3.1.36}$$

and the square root of the mean square end-to-end distance by

$$R'_N \simeq \frac{1}{2\pi} 3^{1/2} N^{1/2} \ln N.$$

In the same way, for all double-spiral chains, the total number of chains is given by

$$Z''_N \simeq 2^{-2} 3^{-5/4} \pi N^{-7/4} \kappa^{(2N)^{1/2}} \tag{3.1.37}$$

and the square root of the mean square end-to-end distance by

$$R''_N \simeq \frac{1}{2\pi} 3^{1/2} N^{1/2} \ln N. \tag{3.1.38}$$

These results are remarkable; in particular they show that the number of self-avoiding spirals with N links increases only as an exponential of $N^{1/2}$ whereas the number of self-avoiding chains increases as an exponential of N. Thus, we see from this example that the universality principles cannot be applied without discrimination and that one must be careful!

The number κ is the constant which appears in the famous Hardy–Ramanujan formula (1917) which gives the number of partitions of an integer N (see Appendix B). This fact is not fortuitous and our readers shall see why in what follows.

In order to illustrate the method used by Blöte and Hilhorst, we shall study only chains with N links, spiralling to the left. A spiral can be considered as made of segments of lengths x_j and y_j that are respectively parallel to the axis Ox and Oy.

To evaluate the number Z_N of configurations, one uses the generating function

$$Q(z) = \sum_{N=1}^{\infty} z^N Z_N. \tag{3.1.39}$$

In order to simplify the problem, we shall assume that the first segment has the same direction as Ox and that the last one is parallel to it (see Fig. 3.7). Then for a chain made of $(2L + 1)$ segments

$$x_{L+1} > 0$$

$$x_L > \ldots > x_1 > 0$$

$$y_L > \ldots > y_1 > 0$$

and $Q(z)$ is given by

$$Q(z) = \sum_L \sum_{\{x, y\}} z^{x_1 + \ldots + x_{L+1} + y_1 + \ldots + y_L}. \tag{3.1.40}$$

The sum over x_{L+1} is trivial and gives a factor $z/(1 - z)$; the sums over x and y are independent. Consequently,

$$Q(z) = \frac{z}{1 - z} \sum_L [g_L(z)]^2 \tag{3.1.41}$$

$$g_L(z) = \sum_{0 < x_1 < \ldots < x_L} z^{x_1 + \ldots + x_L}. \tag{3.1.42}$$

Now let us set

$$n_1 = x_1$$

$$n_j = x_j - x_{j-1} \qquad (j > 1).$$

We see that

$$x_1 + \ldots + x_L = n_L + 2n_{L-1} + \ldots + Ln_1. \tag{3.1.43}$$

By bringing (3.1.43) into (3.1.42) and by summing over the n_j, we find

$$g_L(z) = \prod_{p=1}^{L} \frac{z^p}{1 - z^p}.$$

The behaviour of Z_N for $N \gg 1$ is related to the singularity of $g_L(z)$ for $z = 1$ (with L large). Thus, let us assume $L \gg 1$ and let us set

$$z = 1 - \eta \qquad 0 < \eta \ll 1,$$

consequently

$$z^p \simeq \mathrm{e}^{-p\eta}$$

and

$$\ln g_L(z) = -\sum_{p=1}^{L} \ln(\mathrm{e}^{p\eta} - 1). \qquad (3.1.44)$$

We observe that for a given η, $g_L(z)$ is maximal with respect to L for values close to $L_0 = \ln 2/\pi$. Consequently, we shall assume in the following that L is close to L_0. Now, $\ln g_L(z)$ can be estimated by replacing the sum over L by an integral, but one must exercise caution. Thus, we shall write

$$\ln g_L(z) = -\sum_{p=1}^{L} [\ln(\mathrm{e}^{p\eta} - 1) - \ln(p\eta)] - \sum_{p=1}^{L} \ln(p\eta)$$

and only the first term will be replaced by an integral (by setting $t = p\eta$), the second one being left unchanged

$$\ln g_L(z) = -\frac{1}{\eta}\int_0^{L\eta + \eta/2} \mathrm{d}t[\ln(\mathrm{e}^t - 1) - \ln t] - L\ln\eta - \ln(L!).$$

Applying Stirling's formula $L! = (2L\pi)^{1/2}\,(L/e)^L$ and assuming that η is small, we find

$$\ln g_L(z) = -\frac{1}{\eta}\int_0^{L\eta} \mathrm{d}t\,\ln(\mathrm{e}^t - 1) - \tfrac{1}{2}\ln(2\pi/\eta).$$

It will be sufficient to evaluate $g_L(z)$ for such values of L as

$$L\eta = \ln 2 + O(\eta^{1/2}).$$

Remarking that

$$\int_0^{\ln 2} \mathrm{d}t\,\ln(\mathrm{e}^t - 1) = -\frac{\pi^2}{12}$$

and doing first-order corrections, we find

$$\ln g_L(z) = \frac{\pi^2}{12\eta} - \frac{1}{\eta}(L\eta - \ln 2)^2 - \tfrac{1}{2}\ln(2\pi/\eta).$$

Let us bring this expression in (3.1.41); we obtain

$$Q(z) \simeq \frac{1}{2\pi}\mathrm{e}^{\pi^2/6\eta}\sum_{L} \exp[-2(L\eta - \ln 2)^2/\eta]. \qquad (3.1.45)$$

Again, we may replace the sum over L by an integral and we get the result

$$Q(z) \simeq (8\pi\eta)^{-1/2} e^{\pi^2/6\eta} \qquad (\eta = 1 - z).$$

Then, Z_N is given by the contour integral [see (3.1.39)]

$$Z_N = \frac{1}{2\pi i} \oint dz\, z^{-N-1} Q(z) \tag{3.1.46}$$

which is performed in the complex z plane around the origin

The contour can be deformed and, in this way, we obtain for $N \gg 1$

$$Z_N = \frac{1}{(2\pi)^{3/2} i} \int_{z_c - i\infty}^{z_c + i\infty} \frac{dz}{(1-z)^{1/2}}\, z^{-N-1}\, e^{\pi^2/6(1-z)}$$

(with $0 < z_c < 1$).

This integral can be calculated by the steepest descent method and, we may choose for z_c the value which minimizes the function

$$\varphi(z) = -N \ln z + \frac{\pi^2}{6(1-z)}.$$

As

$$\varphi'(z) = -\frac{N}{z} + \frac{\pi^2}{6(1-z)^2}$$

it is sufficient to choose for z_c the value

$$z_c = 1 - \frac{\pi}{(6N)^{1/2}}$$

and, from it, we deduce

$$\varphi(z_c) \simeq \pi(2N/3)^{1/2}.$$

In this way, one obtains the dominant contribution

$$Z_N \sim \exp \varphi(z_c) = \exp\left[\pi(2N/3)^{1/2}\right].$$

The complete calculation can be made without difficulty, and we find

$$Z_N = \tfrac{1}{2} Z'_N$$

where Z'_N is given by (3.1.35).

On the other hand, we observe that the mean square end-to-end distance of a chain made of $(2L + 1)$ segments can be written

$$r_L^2\{x, y\} = \left[\sum_{j=1}^{L+1} (-)^j x_j\right]^2 + \left[\sum_{j=1}^{L} (-)^j y_j\right]^2.$$

Thus, the mean square end-to-end distance R_N^2 can be obtained from the function

$$Q_R(z) = \sum_L \sum_{\{x, y\}} r_L^2\{x, y\}\, z^{x_1 + \ldots + x_{L+1} + y_1 + \ldots + y_L} \tag{3.1.47}$$

which can be evaluated by applying the same kind of method as, for the calculation of $Q(z)$.

2. CHAINS COVERING A LATTICE

2.1 Hamiltonian walks and Hamiltonian circuits on a lattice

A Hamiltonian circuit is a closed chain drawn on a (non-oriented) lattice, so as to pass once and only once on each lattice site. A Hamiltonian walk is an open chain drawn on a (non-oriented) lattice, so as to pass once and only once on each lattice site.

First, we shall consider the case where every Hamiltonian circuit is drawn on a given rectangular lattice and we shall study how the number of circuits increases when the size of the lattice grows. The result is probably rather general and in this case the boundary conditions play a minor role. However, in certain cases, the boundary conditions may be important and this statement will be justified by considering certain hexagonal lattices and the Hamiltonian walks drawn on them. Finally, note that problems concerning Hamiltonian circuits are also studied, in Chapter 11, Section 6, in connection with 'field models'.

2.1.1 Hamiltonian circuits on a square lattice with a rectangular shape

Let us consider a square lattice with M columns and N lines, i.e. MN sites (see Fig. 3.8). Let $U_{M,N}$ be the number of Hamiltonian circuits that can be drawn on the lattice. If M and N are uneven $U_{M,N} = 0$ for the following reason. On the one hand, the number of segments on the circuit is equal to the number of sites, i.e. MN. On the other hand, the circuit is made up of an even number of vertical steps and of an even number of vertical steps; consequently, the total number of segments (elementary steps) is even. Thus MN must be even, and this property will be assumed in what follows. In this section, we show that $\ln U_{M,N}/MN$ has a limit when M and N become infinite (MN = even).

Let us start by showing that $\ln U_{M,N}/MN$ is bounded. First, we observe that two oriented circuits can be associated with each Hamiltonian circuit. Then, let us start from a corner; we see that at every step, we can choose between three directions at most. Thus we have

$$U_{M,N} < 3^{MN}. \tag{3.2.1}$$

Now, let us look for a lower bound of U_{MN}. We claim that, for $M > 2$, we have

$$U_{M,N+2} \geqslant 2U_{M,N}. \tag{3.2.2}$$

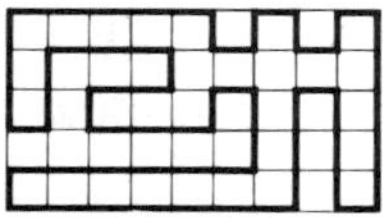

Fig. 3.8. Hamiltonian circuit on a square lattice made of M columns and N lines (here $M = 10$, $N = 6$).

To show this, we note that each circuit (M, N) where $M > 2$, can be used to generate two circuits $(M, N + 2)$ by proceeding as indicated in Fig. 3.9. We remark that the process is always possible and that all the circuits $(M, N + 2)$ obtained in this way are different.

In the same way, for $N > 2$, we have

$$U_{M+2,N} \geqslant 2U_{M,N}. \tag{3.2.3}$$

Let us apply the recurrence (3.2.3) by starting from $U_{3,4}$ and $U_{4,4}$ ($U_{3,4} = 2$, $U_{4,4} = 6$, see Fig. 3.10). We find

$$U_{2m+1,4} > 2^m \qquad (m \geqslant 2)$$
$$U_{2m+4,4} > 2^m. \tag{3.2.4}$$

Now, we claim that for all values of M, N and M', N', we have

$$U_{M,N+N'} \geqslant U_{M,N} U_{M,N'} \tag{3.2.5}$$

$$U_{M+M',N} \geqslant U_{M,N} U_{M',N}. \tag{3.2.6}$$

Let us prove (3.2.5). The case where MN and MN' are even is the only non-trivial one. Combining any circuit (M, N) and any circuit (M, N'), we can construct a circuit $(M, N + N')$ as shown on Fig. 3.11; moreover, all these circuits $(M, N + N')$ are different. These observations prove (3.2.5) and, in an analogous way (3.2.6).

Transforming eqns (3.2.4) by repeated application of (3.2.5), we obtain

$$U_{2m+1,4n} > 2^{mn} \qquad (m \geqslant 2, n \geqslant 1)$$
$$U_{2m,4n} > 2^{mn} \tag{3.2.7}$$

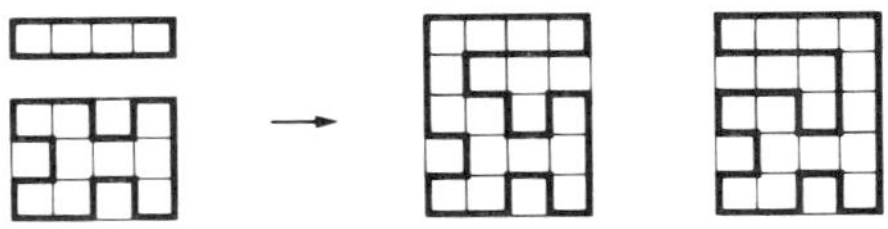

Fig. 3.9. By starting from one circuit (M, N), it is possible to make two Hamiltonian circuits $(M, N + 2)$.

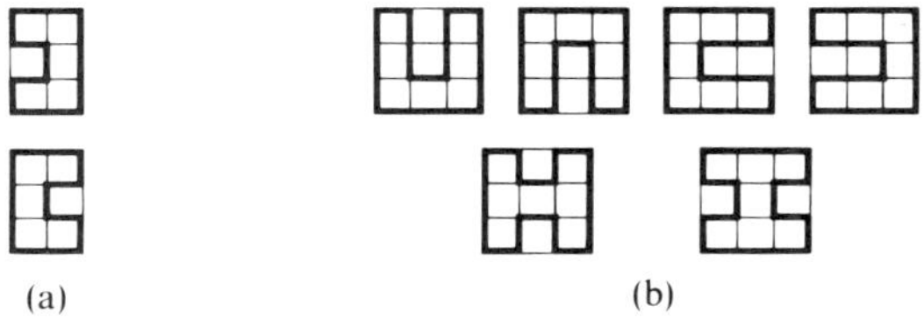

Fig. 3.10. (a) Circuits (3, 4); (b) circuits (4, 4).

Fig. 3.11. Construction of a Hamiltonian circuit $(M, N + N')$ from a circuit (M, N) and a circuit (M, N') (here $M = 6$, $N = 4$, $N' = 3$).

and using (3.2.2), we also obtain

$$U_{2m+1,4n+2} > 2^{mn}$$
$$U_{2m,4n+2} > 2^{mn}. \qquad (3.2.8)$$

The inequalities (3.2.7) and (3.2.8) show that for $M \geqslant 4$, $N \geqslant 4$

$$U_{M,N} > 2^{\frac{(M-1)}{2}\frac{(N-2)}{2}} > 2^{\frac{(M-2)(N-2)}{4}} > 2^{\frac{MN}{32}}. \qquad (3.2.9)$$

Thus, for even MN, (3.2.1) and (3.2.9) give

$$MN\frac{\ln 2}{32} < \ln U_{M,N} < MN \ln 3 \qquad (3.2.10)$$

(better bounds could easily be found).

The convergence of $\ln U_{MN}/MN$ when M and N become infinite, results from equations (3.2.5), (3.2.6), and (3.2.7) as can easily be shown by generalizing the method described in Section 1.1. Then, the quantity

$$\mu = \lim_{\substack{M\to\infty,\, N\to\infty \\ (MN\ \text{even})}} \frac{\ln U_{M,N}}{MN} \qquad (3.2.11)$$

can be considered as the entropy per site.

Let $Z_{M,N}$ be the number of non-oriented Hamiltonian walks that can be drawn on the lattice. We immediately see that

$$MN\, U_{M,N} < Z_{M,N} < MN\, 3^{MN}$$

(a circuit can generate MN distinguishable walks).

We may also think that

$$\lim_{M\to\infty\, N\to\infty} \frac{\lim Z_{M,N}}{MN} = \mu$$

but this statement remains to be proved.

The preceding results can be extended to d-dimensional lattices.

2.1.2 Hamiltonian circuit and Hamiltonian walks on a layered hexagonal lattice

By extrapolating the result proved in the preceding section, one might get the feeling that, for any lattice, it is possible to define a finite entropy per site. However, it is necessary to be careful: the lattice may become large, the number of points on the border may be negligible as compared to the total number of points of the lattice, and yet simultaneously, the structure of the border may retain a strong influence on the number of circuits or Hamiltonian walks drawn on the lattice.

A significant example has been given by Gordon, Kapadia, and Malakis (1976).[8] These authors considered hexagonal lattices made of $\mathbb{C}$ successive layers (see Fig. 3.12). The layer of rank $\mathbb{C}'$ contains $6\,(2\mathbb{C}'-1)$ sites and, consequently, the total number N of lattice sites with $\mathbb{C}$ layers is given by

$$N = \sum_{\mathbb{C}'=1}^{\mathbb{C}} 6(2\mathbb{C}'-1) = 6\mathbb{C}^2. \qquad (3.2.12)$$

At first, we observe that it is not possible to draw any Hamiltonian circuit on the lattice; to get this result, we have only to observe that such a circuit must pass through all the sites of the external layer and that this is not compatible with the exploration of the interior points.

Let us now consider the Hamiltonian walks drawn on the lattice and let $Z_N = Z(\mathbb{C})$ be their number. We see (again for the same kind of reason) that at least one extremity of the walk is on the external layer (see Fig. 3.13). Let $Z'(\mathbb{C})$ be the number of walks that have only one extremity on the external layer and $Z''(\mathbb{C})$ the number of walks that have both extremities on this layer. We have

$$Z(\mathbb{C}) = Z'(\mathbb{C}) + Z''(\mathbb{C}) \qquad (3.2.13)$$

with

$$Z'(1) = 0 \qquad Z''(1) = 6. \qquad (3.2.14)$$

Besides, it is not difficult to see that $Z'(\mathbb{C})$ and $Z''(\mathbb{C})$ obey the following recurrence relations

$$Z''(\mathbb{C}) = 2Z''(\mathbb{C}-1) \qquad (3.2.15)$$

$$Z'(\mathbb{C}) = 2Z'(\mathbb{C}-1) + 4Z''(\mathbb{C}-1). \qquad (3.2.16)$$

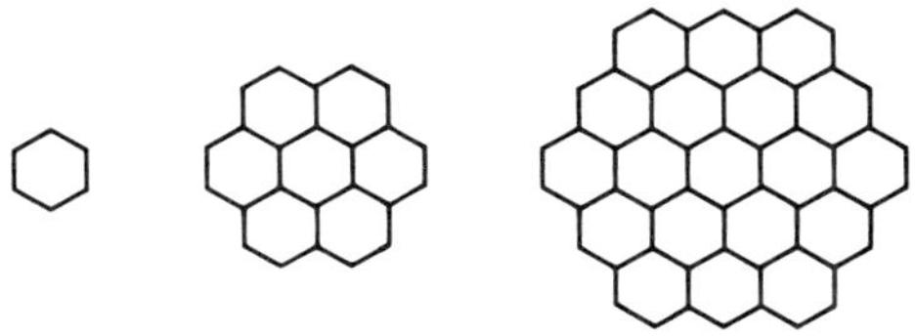

$\mathbb{C} = 1$ $\mathbb{C} = 2$ $\mathbb{C} = 3$

Fig. 3.12. Hexagonal lattices made of $\mathbb{C}$ layers ($\mathbb{C}$ = 1, 2, 3).

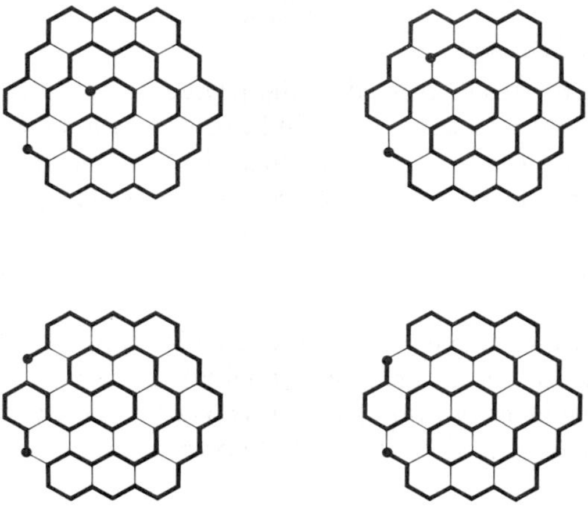

Fig. 3.13. Hamiltonian walk on a three-layer hexagonal lattice.

Relation (3.2.15) with (3.2.14) gives

$$Z''(\mathcal{C}) = 3.2^{\mathcal{C}}. \tag{3.2.17}$$

Let use this result in (3.2.18)

$$Z'(\mathcal{C}) = 2Z'(\mathcal{C} - 1) + 3.2^{\mathcal{C}-1}. \tag{3.2.18}$$

The solution of this equation, is of the form

$$Z'(\mathcal{C}) = (A\mathcal{C} + B)2^{\mathcal{C}}.$$

By using this equality in (3.2.18) and by taking (3.2.14) into account (this gives $A + B = 0$), we find

$$Z'(\mathcal{C}) = 6(\mathcal{C} - 1)2^{\mathcal{C}}. \tag{3.2.19}$$

By bringing (3.2.17) and (3.2.19) in (3.2.13), we get the result

$$Z(\mathcal{C}) = 3(2\mathcal{C} - 1)2^{\mathcal{C}}.$$

When $\mathcal{C} \to \infty$, $N \to \infty$, and consequently

$$\frac{\ln Z_N}{N} = \frac{\ln Z(\mathcal{C})}{6\mathcal{C}^2} \simeq \frac{\ln 2}{6\mathcal{C}}.$$

Thus, in the limit $N \to \infty$, the number of Hamiltonian walks on the lattice goes to infinity, but the entropy per site is zero, and this fact shows the (rather perverse) influence of the boundary conditions in the example under consideration

2.2 Oriented chain on a 'Manhattan' lattice: Kasteleyn's result

A polymer liquid can be represented by a set of self-avoiding chains, drawn on a lattice so that each lattice site is on one of the chains but only one. Again, we may try to evaluate the number Z of configurations of the system or its entropy $\mathbb{S} = \ln Z$.

The problem has not exactly been solved, but Kasteleyn[9] succeeded in handling a particularly interesting case. The lattice under consideration is an oriented lattice of the 'Manhattan' type (see Fig. 3.14). Kasteleyn considers the configurations of the circuits which follow the one way lanes on the lattice and which cover the whole lattice without any overlap.

Various results of graph theory enabled Kasteleyn to evaluate the number of configurations of the system in a rather simple manner (the origin of the chain is fixed once and for all in an arbitrary way). The calculation is simpler when the lattice is periodic.

Thus, let us consider a rectangular 'Manhattan' lattice consisting of M columns and N lines with cyclic boundary conditions (the lattice can be considered as drawn on a torus; consequently, as the lattice is oriented, M and N are even). The number $Z_{M,N}$ of configurations of the chain is given by the expression (Kasteleyn 1963).[9]

$$Z_{M,N} = \frac{1}{MN} 2^{1+(MN/2)} \sum_{m=1}^{M/2} \sum_{n=1}^{N/2}_{(m,\, n \neq M/2,\, N/2)} \left(\sin^2 \frac{2\pi m}{M} + \sin^2 \frac{2\pi n}{N} \right). \qquad (3.2.20)$$

The logarithm of this quantity is expressed as a sum. When M and N become infinite, this sum can be replaced by an integral, and it is not difficult to estimate the first correction terms. In this way, the following result is obtained

$$\ln Z_{M,N} \simeq MN \ln \mu + O(1) \qquad (3.2.21)$$

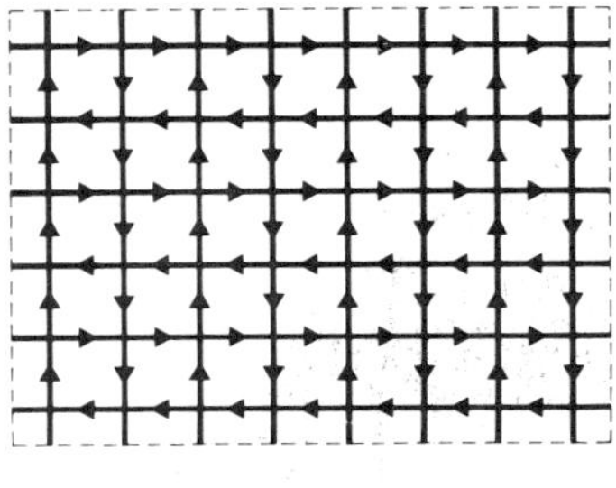

Fig. 3.14. 'Manhattan' lattice (with periodic boundary conditions). Every oriented circuit on the lattice must pass at each crossing and the orientation on the circuit must be compatible with the orientations on the lattice (traffic rules).

where MN is number of links of the chain and where μ is given by

$$\ln \mu = \frac{1}{2}\ln 2 + \frac{1}{4\pi^2}\int_0^\pi dx \int_0^\pi dy \int_0^\pi dy \ln(\sin^2 x + \sin^2 y), \qquad (3.2.22)$$

i.e. $\mu = 1.338$ (in agreement with the obvious inequality $1 < \mu < 2$)

Thus, the number of cycles that can be drawn on the lattice is given asymptotically by the simple formula

$$Z_{M,N} \simeq Z_\square \mu^{MN} \qquad (3.2.23)$$

($M \to \infty$, $N \to \infty$, $Z_\square$ = constant, as could be shown).

However, it seems clear that this simplicity results from the following facts:

(1) the system consists of only one closed chain;
(2) all border effects are eliminated by the cyclic boundary conditions which have been chosen.

More details concerning the partition function and the correlation functions of simple or multiple circuits on the Manhattan lattice can be found in articles by Duplantier and David (1987 and 1988).[10]

3. ON A FEW SIMPLE PROPERTIES OF CHAINS WITH REPULSIVE INTERACTIONS

One-dimensional chains with repulsive interactions have properties which can be studied rigorously, and we shall describe them below. In d-dimensional space, the effect of the excluded volume on the behaviour of the chains diminishes when d increases and simple arguments show that beyond $d = 4$, the chains with excluded volume behave like chains with independent links (quasi-Brownian chains).

Below four dimensions, the probability distribution of the vector joining the extremities is no longer Gaussian, and an interesting property of this distribution was established in a quasi-rigorous way by Fisher (1966).[11]

3.1 Chains in one-dimensional space

A one-dimensional lattice is made of regularly-spaced points on a straight line. Drawing a self-avoiding chain by starting from an origin O is a trivial operation (see Fig. 3.15). For any value N of the number of links, it is possible to draw two chains of length

$$|r_N - r_0| = Nl \qquad (3.3.1)$$

where l is the distance between two successive points on the lattice. Thus, in this case, we have

$$\mu = 1, \qquad \gamma = 1, \qquad \nu = 1.$$

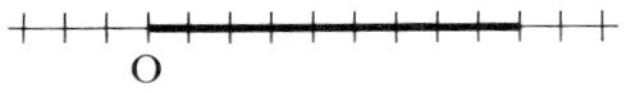

Fig. 3.15. Self avoiding chain on a one-dimensional lattice.

The passage to the continuous limit is also trivial and gives

$$|r(L) - r(0)| = L.$$

Consequently, the probability distribution between extremities is

$$P_L(r) = \frac{1}{2}\,\delta(r - L) = \frac{1}{2L}\,\delta\left(\frac{r}{L} - 1\right) \tag{3.3.2}$$

which obeys the scaling laws defined in Chapter 2.

Now, we would like to know the properties of a self-intersecting chain with repulsive interaction. The universality principle leads us to postulate that, for such a chain ($N \to \infty$), the following properties are valid:

1. The critical exponents γ and ν have the values

$$\gamma = 1, \qquad \nu = 1,$$

   and therefore for $N \gg 1$

$$Z_N \simeq Z_\square \mu^N$$

   where $Z_\square$ and μ are constants depending on the microscopic properties of the chain.

2. The length of the chain being defined, as always, by the equality $L = Nl$, the asymptotic law giving the end-to-end probability distribution can be written in the form

$$P_L(r) = \frac{1}{2\lambda L}\,\delta\left(\frac{r}{\lambda L} - 1\right) \tag{3.3.3}$$

   where λ is a number ($\lambda < 1$) which depends on the microscopic properties of the chain.

Though these properties have not been very rigorously proved, intuitive arguments show that they are satisfactory. Moreover, Balian and Toulouse[12] showed that in the continuous case the properties of a chain can be studied precisely by starting from a Lagrangian representation of the problem (see Chapter 11) and by using a transfer matrix method. In particular, these authors verified that the critical exponents are $\nu = 1$ and $\gamma = 1$ [see also the article by Thouless (1975)].[13]

This result can be understood as follows. Let us consider a chain on a one-dimensional lattice and let us admit that, when two points on the chain coincide, it costs an energy J. The probability associated with a chain is given by

Boltzmann's law

$$P = \frac{1}{Z} e^{-\beta p J} \tag{3.3.4}$$

where p is the number of contact points on the chain. Let us show that the extension $X = x_{max} - x_{min}$ of such a chain is certainly proportional to the number of links. The most probable configurations are obtained by minimizing the corresponding free energy (see Chapter 8, Section 1)

$$\mathbb{F} = \mathbb{E} - \beta^{-1}\mathbb{S} \tag{3.3.5}$$

(here $\mathbb{E} = \langle\!\langle p \rangle\!\rangle J$).

At equilibrium, we certainly have $\mathbb{F} < 0$. In fact, if we restricted ourselves to the only state for which the chain is completely stretched, we would find for the system, $\mathbb{F} = 0$, since in this case $\mathbb{E} = 0$ and $\mathbb{S} = 0$.

For the canonical ensemble, we have

$$\mathbb{E} = \langle\!\langle p \rangle\!\rangle J \quad \text{with} \quad p \geqslant \frac{N(N-X)}{2X}.$$

This inequality can be proved in two steps.

(a) First, it is assumed that N/X is an integer and then it is easy to show that in this case the lower bound of p is $N(N-X)/2X$.

(b) Second, it is shown that between two integer values of N/X the lower bound of p is a linear and increasing function of N, which implies that it is larger than $N(N-X)/2X$ inside this interval.

Thus

$$\mathbb{E} \geqslant \frac{NJ}{2}\left[N\left\langle\!\!\left\langle \frac{1}{X} \right\rangle\!\!\right\rangle - 1\right] \geqslant \frac{NJ}{2}\left[\frac{N}{\langle\!\langle X \rangle\!\rangle} - 1\right].$$

On the other hand, the entropy of this system of configurations is smaller than the logarithm of the total number 2^N of configurations of the chain

$$\mathbb{S} < N \ln 2.$$

Thus,

$$\mathbb{F} > \frac{N^2 J}{2\langle\!\langle X \rangle\!\rangle} - N\left(\beta^{-1} \ln 2 + \frac{J}{2}\right). \tag{3.3.6}$$

By combining this inequality with the inequality $\mathbb{F} < 0$, we obtain the result

$$\frac{N^2 J}{\langle\!\langle X \rangle\!\rangle} - N(2\beta^{-1} \ln 2 + J) < 0,$$

which can also be written as

$$\langle\!\langle X \rangle\!\rangle > N\beta J(\beta J + 2 \ln 2)^{-1}. \tag{3.3.7}$$

Thus, the average extension of the chain is proportional to N. In fact, we can find a lower bound of the sum

$$R_G^2 = \frac{1}{2(N+1)^2} \sum_{i,j} |x_i - x_j|^2.$$

For each chain

$$R_G^2(\mathfrak{C}) \geqslant \frac{1}{(N+1)^2} \sum_{n=1}^{X} (X - n + 1)n^2 = \frac{X(X+1)^2(X+2)}{12(N+1)^2} > \frac{X^4}{12N^2}.$$

Therefore,

$$R_G^2 > \frac{\langle\!\langle X^4 \rangle\!\rangle}{12N^2} > \frac{\langle\!\langle X \rangle\!\rangle^2}{12N^2} \tag{3.3.8}$$

and, from this inequality and from (3.3.7), one deduces the result

$$R_G > N 2^{-1/2} (\beta J)^4 (\beta J + 2 \ln 2)^{-4} \tag{3.3.9}$$

which shows that indeed $\nu = 1$.

Thus, a chain with repulsive interaction, whose origin is the site O, increases steadily along the lattice either in one direction (positive sense). or in the opposite direction (negative sense).

Now let us set $u_j = x_j - x_{j-1}$. For a chain growing in the positive direction, we have

$$\langle\!\langle u_j \rangle\!\rangle = \bar{u} \tag{3.3.10}$$

On the other hand, the variables $(u_j - \bar{u})$ and $(u_{j'} - \bar{u})$ are correlated only for finite values of $|j - j'|$. Thus, for chain growing in positive direction, the quantity $(x_N - x_0 - N\bar{u}) = \sum_{j=1}^{N} (u_j - \bar{u})$ can be considered as the sum of quasi-independent variables. Consequently, $(x_N - x_0 - N\bar{u})$ must be asymptotically Gaussian, and we may write

$$\langle\!\langle (x_N - x_0 - N\bar{u})^2 \rangle\!\rangle \simeq Nq \tag{3.3.11}$$

where q is a number $0 < q < 1$.

These results show that the asymptotic probability distribution for the one-dimensional vector joining the end points of the chain is given by

$$P_N(x) = \frac{1}{2\sqrt{2\pi Nq}} \{\exp[-(x - N\bar{u})^2/2Nq] + \exp[-(x + N\bar{u})^2/2Nq]\}. \tag{3.3.12}$$

This function is strongly peaked around two values, i.e. $x = N\bar{u}$ and $x = -N\bar{u}$, since the width of the Gaussian functions increases only like $N^{1/2}$ (see Fig. 3.16). Therefore, the law given by (3.3.12) is quite compatible with eqn (3.3.3) and the broadening associated with the Gaussian functions must simply be considered as a correction to scaling laws.

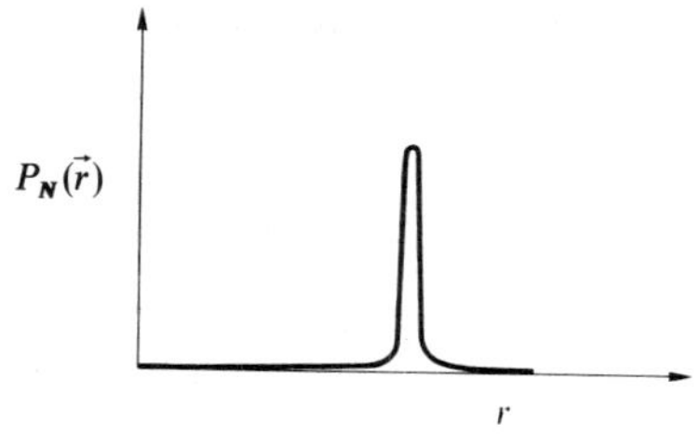

Fig. 3.16. Shape of $P_N(\vec{r})$ for a one-dimensional repulsive chain. Here, $r = |\vec{r}|$

It would not be very difficult to calculate $\bar{u}$ and q in terms of βJ; however, the question is rather academic: it seems that the one-dimensional chain is not of great physical interest.

3.2 Repulsive chains in d-dimensional space: dimensionality of a chain

The exclusion and repulsion effects* between links belonging to the same chain manifest themselves through a swelling, which may be characterized by a critical exponent ν defining the variation of the chain size with the number of links ($R_G \propto N^\nu$).

This exponent must be universal, but it depends on the dimension d of the space in which the chain is embedded. In particular, we have known for many years that the behaviour of chains with excluded volume is Brownian for $d > 4$. Actually, in this case, the repulsive potential can be treated by perturbation and the first corrective terms are convergent, whereas the same is not true for $d = 2$ and $d = 3$. (The case $d = 4$ appears as a limiting case; see Chapters 8 and 9.)

Thus, for $d > 4$, the nature of the asymptotic behaviour is not modified; on the contrary, for $d < 4$, summing divergencies may lead to the appearance of critical exponents. However, this argument is neither very simple nor very rigorous. It is, however, possible to reach the same conclusion in a much simpler and intuitive manner. In fact, let us consider, in a d-dimensional space domain, two simply connected objects with dimensions D and D' respectively, and let us assume that they have a random position. If $D + D' < d$, the probability that these objects have a common part has a zero measure. Actually, if they cut across each other, it is always possible to displace one object infinitesimally, so as to suppress the intersection. For instance, the statistics, of a set of segments ($D = 1$, $D' = 1$) in a three-dimensional space ($d = 3$) is quite trivial, whereas in two dimensions, intersection effects have to be taken into account. Thus, in general, two objects feel exclusion effects if and only if $D + D' \geqslant d$.

* Exclusion effects are commonly called 'excluded volume effects'.

In another connection, we showed (Chapter 2, Section 2.3) that the dimensionality of a chain with a size exponent ν is given by

$$D = \frac{1}{\nu}. \tag{3.3.13}$$

Exclusion effects will appear for two pieces of chain if but only if

$$2D = \frac{2}{\nu} \geqslant d,$$

i.e.

$$\nu \leqslant \frac{2}{d}. \tag{3.3.14}$$

For a Brownian chain $\nu = 1/2$ independently of the dimension of the space in which the chain is embedded; consequently for $d > 4$, the preceding condition is not realized.

For $d = 4$, exclusion effects appear, but we see that in this case, ν must remain equal to $1/2$. Thus, the case $d = 4$ is a marginal case.

On the contrary, for $d < 4$, the exponent ν must belong to the interval

$$\frac{1}{2} \leqslant \nu \leqslant \frac{2}{d}, \tag{3.3.15}$$

and these conditions are in full agreement with the experimental results (see Chapter 4) which say that $\nu \simeq 3/5$ for $d = 3$, $\nu \simeq 3/4$ for $d = 3$, and $\nu = 1$ for $d = 1$.

3.3 Asymptotic properties of correlations between chain ends: Fisher's result

Let us consider a d-dimensional hypercubic lattice and, on this lattice, the self-avoiding chains of origin O. Let $P_N(\vec{r})$ be the probability distribution of the end-to-end vector, $\vec{r}$. This vector $\vec{r}$ has coordinates $q_1, \ldots, q_d$ which are integers. The parity of $\vec{r}$ is by definition the parity of $\sum_j q_j$.

Thus, we see immediately that $P_N(\vec{r})$ differs from zero only if $\vec{r}$ and N have the same parity. In this case, and for large values of N, it is possible to postulate a scaling law of the form

$$P_N(\vec{r}) = \frac{1}{N^{\nu d}} f_{\square}(r/N^{\nu}) \tag{3.3.16}$$

(for $\vec{r}$ with the same parity as N; here $r = |\vec{r}|$ is a number), and as

$$\sum_r P_N(\vec{r}) = 1$$

we have

$$\frac{1}{2}\int \mathrm{f}_{\square}(x)\mathrm{d}^d x = 1.$$

The asymptotic properties of $\mathrm{f}_{\square}(x)$ are determined by Fisher's theorem. Note that the values $x \gg 1$ correspond, for the chain, to stretched configurations.

Fisher's theorem (stretched chain configurations)

Statement

The function $\mathrm{f}_{\square}(x)$ which, for a self-avoiding chain, describes the asymptotic behaviour of $P_N(\vec{r})$ when $N \to \infty$, decreases when $x \to \infty$ as $\exp[-D_{\square} x^{1/(1-\nu)}]$ ($D_{\square}$ = constant, ν = critical exponent)

Proof

We shall begin by showing that the function

$$P(t,\vec{r}) = \sum_{N=0}^{\infty} \mathrm{e}^{-Nt} P_N(\vec{r}) \tag{3.3.17}$$

decreases exponentially with r. More precisely, it will be established that

$$\text{(a)}\ \liminf\left[-\frac{1}{r}\ln P(t,\vec{r})\right] > t$$

$$\text{(b)}\ \limsup\left[-\frac{1}{r}\ln P(t,\vec{r})\right] < d[t + \ln(2d)]. \tag{3.3.18}$$

Condition (a) expresses the fact that $P(t,\vec{r})$ decreases at least exponentially with respect to r. Actually,

$$P_N(\vec{r}) = 0 \qquad r > N$$

$$P_N(\vec{r}) < 1 \qquad r \leqslant N,$$

therefore

$$P(t,\vec{r}) < \sum_{N \geqslant r} \mathrm{e}^{-Nt} = \frac{\mathrm{e}^{-rt}}{1-\mathrm{e}^{-t}}. \tag{3.3.19}$$

Condition (b) expresses the fact that, when r increases $P(t,\vec{r})$, does not decrease faster than exponentially. Actually, for $N > rd$

$$P_{N-1}(\vec{r}) + P_N(\vec{r}) > \frac{1}{(2d)^N}$$

since any site of the hypercube of edge r is the end point at least of one chain with N links or of one chain with $(N-1)$ links according to the site parity. Therefore,

$$P(t,\vec{r}) > \frac{1}{2}\sum_{N > rd} (2d)^{-N}\mathrm{e}^{-Nt} \tag{3.3.20}$$

$$= \frac{\exp\{-rd[t+\ln(2d)]\}}{2(1-\exp\{-[t+\ln(2d)]\}}.$$

In brief, $P(t,\vec{r})$ decreases exponentially with r. For large values of r and small values of t, we can replace $P_N(\vec{r})$, in the expression of $P(t,\vec{r})$, (3.3.17), by its asymptotic form (for large values of N)

$$P(t,\vec{r}) = \sum_{N=1}^{\infty} e^{-Nt} P_N(\vec{r})$$

$$\simeq \sum_{N=1}^{\infty} \frac{e^{-Nt}}{N^{\nu d}} f_{\square}(r/N^{\nu}) \tag{3.3.21}$$

and the sum can be replaced by an integral

$$P(t,\vec{r}) \simeq \int_0^{\infty} dN \frac{e^{-Nt}}{N^{\nu d}} f_{\square}(r/N^{\nu}). \tag{3.3.22}$$

The latter operation is allowable only if the integral converges, but if we admit with Fisher that for $x \gg 1$

$$f_{\square}(x) \propto x^k \exp(-D_{\square} x^{\delta}) \tag{3.3.23}$$

we verify that this condition holds.

The value of the integral is mainly determined by the product of two integrals and this product can be written in the form $e^{-\varphi}$ where

$$\varphi = Nt + D_{\square}(r/N^{\nu})^{\delta}.$$

In order to evaluate the integral, we apply the steepest descent method. At the saddle joint, we have

$$\partial\varphi/\partial N = 0,$$

which gives

$$N \propto r^{\frac{\delta}{1+\delta\nu}} t^{-\frac{1}{1+\delta\nu}}$$

$$\varphi \propto (rt^{\nu})^{\frac{\delta}{1+\delta\nu}}. \tag{3.3.24}$$

Moreover, as $P(t,\vec{r})$ must decrease exponentially with r, the function φ must be linear with respect to r and this condition implies the equality

$$\frac{\delta}{1+\delta\nu} = 1,$$

which leads immediately to the announced result

$$\delta = \frac{1}{1-\nu}. \tag{3.3.25}$$

Remark

The result proved by Fisher can easily be recovered by using a more direct method. This approach is less rigorous but it will enable us to understand better the origin of the result.

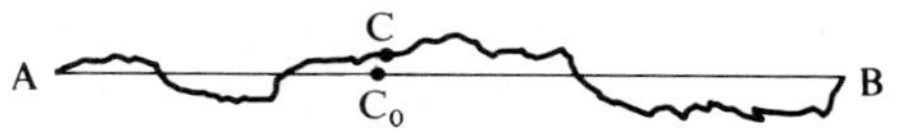

Fig. 3.17. A stretched chain. The point C remains always in the neighbourhood of C_0.

On a chain with N links ($N \gg 1$), let us consider the end points A and B and an intermediary point C; we can denote the numbers of links of AC and CB by N' and $(N - N')$ respectively (with $N' \gg 1$ and $N - N' \gg 1$). We remark that when the chain is strongly stretched, it becomes nearly straight (see Fig. 3.17) and that, in this case, the two subchains become practically independent. In such a configuration, the points A, C, B are nearly aligned and, for large values of N' and of $(N - N')$, the distance between C and its mean position C_0 is always small with respect to AB.

In this case, we may write

$$P_N(\vec{r}) \propto \int \mathrm{d}^d r'\, P_{N'}(\vec{r}\,')\, P_{N-N'}(\vec{r} - \vec{r}\,') \tag{3.3.26}$$

and, replacing $P_N(\vec{r})$ by its asymptotic form

$$P_N(\vec{r}) = \frac{1}{N^{\nu d}} \mathrm{f}_\square\,(r/N^\nu) \propto \exp[-D_\square (r/N^\nu)^\delta], \tag{3.3.27}$$

we obtain

$$\exp\left[-D_\square\left(\frac{r}{N^\nu}\right)^\delta\right] \propto \int \mathrm{d}^d\rho \exp\left\{-D_\square\left[\left(\frac{|\vec{r}\,'|}{(N')^\nu}\right)^\delta + \left(\frac{|\vec{r} - \vec{r}\,'|}{(N - N')^\nu}\right)^\delta\right]\right\}. \tag{3.3.28}$$

The integral on the right-hand side can be evaluated by using the steepest descent method. Let us set

$$\psi = \left[\frac{r'}{(N')^\nu}\right]^\delta + \left[\frac{r - r'}{(N - N')^\nu}\right]^\delta. \tag{3.3.29}$$

The position of C_0, which coincides which the saddle point, is determined by the equation

$$\partial\psi/\partial r' = 0$$

which reads

$$\frac{(r')^{\delta-1}}{(N')^{\nu\delta}} - \frac{(r - r')^{\delta-1}}{(N - N')^{\nu\delta}} = 0 \tag{3.3.30}$$

or

$$\frac{r'}{(N')^{\frac{\nu\delta}{\delta-1}}} = \frac{r - r'}{(N - N')^{\frac{\nu\delta}{\delta-1}}} = \frac{r}{(N')^{\frac{\nu\delta}{\delta-1}} + (N - N')^{\frac{\nu\delta}{\delta-1}}}. \tag{3.3.31}$$

On the other hand, in eqn (3.3.28) the saddle-point value of the integrant must coincide with the right-hand side, and this leads to the condition

$$\frac{(r')^{\delta}}{(N')^{\nu\delta}} + \frac{(r-r')^{\delta}}{(N-N')^{\nu\delta}} = \frac{r^{\delta}}{N^{\nu\delta}}. \tag{3.3.32}$$

Let us now eliminate r' and $(r - r')$ between (3.3.31) and (3.3.32): we obtain

$$\left[(N')^{\frac{\nu\delta}{\delta-1}} + (N-N')^{\frac{\nu\delta}{\delta-1}}\right]^{\delta-1} = N^{\nu\delta}. \tag{3.3.33}$$

It is remarkable that this condition can be satisfied, for all values of N and N', by setting

$$\delta = \frac{1}{1-\nu} \tag{3.3.34}$$

and, only for this value of δ.

Moreover, we observe that, in this case, eqn (3.3.31) reduces to the very simple form

$$\frac{r'}{N'} = \frac{r-r'}{N-N'} = \frac{r}{N}, \tag{3.3.35}$$

which corresponds exactly to the picture we have of a stretched chain. Moreover, a slightly more detailed calculation shows that applying the steepest descent method here is perfectly valid.

4. A STUDY OF A FEW SIMPLE MODELS

Polymers appear as chains with excluded volume, but the mathematical difficulties which are encountered when one tries to study these chains have induced many author to use simpler models. These simple chains are either chains with independent links or other asymptotically Brownian chains, and we shall now review these types of chain.

4.1 Asymptotically Brownian chain with independent links

This chain, of fundamental importance, was studied in the preceding chapter. Its properties can be summarized as follows. The various points of the chain have, vector positions $\vec{r}_j$ with $j = 0, \ldots, N$, and the vectors $\vec{u}_j$.

$$\vec{u}_j = \vec{r}_j - \vec{r}_{j-1}$$

which define the links are independent random variables. Their essential property is to have a finite mean square value

$$\langle\langle \vec{u}_j^{\,2} \rangle\rangle = l^2 d \tag{3.4.1}$$

and this fact implies that

$$\langle\!\langle (\vec{r}_N - \vec{r}_0)^2 \rangle\!\rangle = Nl^2 d. \tag{3.4.2}$$

One shows that, in this case, when N increases, the chain acquires a Brownian behaviour. This behaviour is characterized by the fact that the Fourier transform $\tilde{P}_N(\vec{k})$ of the probability distribution $P_N(\vec{r}) = \langle\!\langle \delta(\vec{r} - \vec{r}_N + \vec{r}_0) \rangle\!\rangle$ is asymptotically given by the simple expression

$$\tilde{P}_N(\vec{k}) = \langle\!\langle e^{i\vec{k}.(\vec{r}_N - \vec{r}_0)} \rangle\!\rangle \simeq e^{-Nk^2l2/2}. \tag{3.4.3}$$

The corresponding probability law

$$P_N(\vec{r}) = \frac{1}{(2\pi N l^2)^{d/2}} \exp\left[-\frac{r^2}{2Nl^2}\right] \tag{3.4.4}$$

can be used to calculate all the mean powers of $|\vec{r}_N - \vec{r}_0|$. Actually, we have

$$\begin{aligned}\frac{\langle\!\langle |\vec{r}_N - \vec{r}_0|^n \rangle\!\rangle}{[\langle\!\langle |\vec{r}_N - \vec{r}_0|^2 \rangle\!\rangle]^{n/2}} &= \frac{\int d^d r\, r^n e^{-r^2}}{\left[\int d^d r\, r^2 e^{-r^2}\right]^{n/2}} \left[\int d^d r\, e^{-r^2}\right]^{-1+n/2} \\ &= \frac{\int_0 dx\, x^{\frac{n}{2}+\frac{d}{2}-1}\, e^{-x}}{\left[\int_0^\infty dx\, x^{d/2}\, e^{-x}\right]^{n/2}} \left[\int dx\, x^{-1+d/2}\, e^{-x}\right]^{-1+n/2}\end{aligned}$$

whence the final result

$$\frac{\langle\!\langle |\vec{r}_N - \vec{r}_0|^n \rangle\!\rangle}{[\langle\!\langle |\vec{r}_N - \vec{r}_0|^2 \rangle\!\rangle]^{n/2}} = \left[\frac{2}{d}\right]^{n/2} \frac{\Gamma\left(\frac{n}{2} + \frac{d}{2}\right)}{\Gamma\left(\frac{d}{2}\right)} \tag{3.4.5}$$

Thus, for instance, for $n = -1$, $d = 3$

$$\left\langle\!\!\left\langle \frac{1}{|\vec{r}_N - \vec{r}_0|} \right\rangle\!\!\right\rangle = \left(\frac{6}{\pi}\right)^{1/2} \frac{1}{\{\langle\!\langle (\vec{r}_N - \vec{r}_0)^2 \rangle\!\rangle]^{1/2}}. \tag{3.4.6}$$

Besides, in order to interpret the diffraction experiments made with polymers, one needs the form function

$$H(\vec{q}) = \frac{1}{(N+1)^2} \left\langle\!\!\left\langle \sum_{jl} e^{i\vec{q}.(\vec{r}_j - \vec{r}_l)} \right\rangle\!\!\right\rangle \tag{3.4.7}$$

when the wavelength of the radiation, or of the incident particles, is large with respect to the length of the links ($ql \ll 1$), the Gaussian limit is reached and the calculation of $H(\vec{q})$ becomes very simple.

$$H(\vec{q}) \simeq \frac{1}{(N+1)^2} \sum_{ij} \exp[-|i-j| l^2 q^2/2]$$

$$= \frac{2}{(N+1)^2} \sum_{n=1}^{N} (N-n+1) \exp[-nl^2 q^2/2]$$

$$\simeq \frac{2}{N^2} \int_0^N \mathrm{d}n\, (N-n) \exp(-nl^2 q^2/2).$$

In this way, one obtains Debye's law

$$H(\vec{q}) = \frac{2}{x^2} [\mathrm{e}^{-x} - 1 + x] \tag{3.4.8}$$

where $x = Nl^2q^2/2 = R^2q^2/2d$, a law which has been especially useful to experimentalists.

If the wavelength is very small with respect to the size of the chain, the result becomes even simpler ($x \gg 1$)

$$H(\vec{q}) \simeq \frac{2}{x} = \frac{4}{Nl^2q^2}. \tag{3.4.9}$$

In the opposite case, when the wavelength is large with respect to the size of the chain, it is possible to expand $H(\vec{q})$ in powers of q

$$H(\vec{q}) = \left[1 - \frac{q^2}{2(N+1)^2 d} \left\langle\!\!\left\langle \sum_{ij} (\vec{r}_i - \vec{r}_j)^2 \right\rangle\!\!\right\rangle + \dots \right]. \tag{3.4.10}$$

This very general expression is known as the Guinier formula

$$H(\vec{q}) = 1 - \frac{q^2}{d} R_G^2 + \dots \tag{3.4.11}$$

where R_G is defined by

$$R_G^2 = \frac{1}{2(N+1)^2} \left\langle\!\!\left\langle \sum_{ij} (\vec{r}_i - \vec{r}_j)^2 \right\rangle\!\!\right\rangle \tag{3.4.12}$$

which is called the radius of gyration of the polymer (see Chapter 2, Section 2.4).

For an asymptotically Brownian chain, we can use the following expression

$$\left\langle\!\!\left\langle \frac{1}{2} \sum_{ij} (\vec{r}_i - \vec{r}_j)^2 \right\rangle\!\!\right\rangle = \frac{N(N+1)(N+2)}{6} l^2 \tag{3.4.13}$$

to derive the result

$$R_G^2 \simeq \frac{Nl^2 d}{6} = \frac{1}{6} \langle\!\langle (\vec{r}_N - \vec{r}_0)^2 \rangle\!\rangle \tag{3.4.14}$$

which could also be obtained by comparing (3.4.8) and (3.4.11).

Finally, we note that, for a Brownian chain, N and l are not independent. Setting $S = Nl^2$, we can express all the results in terms of S alone.

4.2 Non-asymptotically Brownian chains with independent links: chains with intermittence (Levy flights)

A chain with independent links is asymptotically Brownian if and only if the mean square length of a link is finite

$$\langle\langle u_j^2 \rangle\rangle < \infty .$$

However, there exists a class of chains with independent links which do not obey this condition and which nevertheless have a well-defined asymptotic limit.[14] These chains do not correspond to our current picture of a polymer, but it is better to be aware of their existence. They are interesting in themselves and exhibit the intermittence phenomenon, a concept which can be met in other branches of physics; for instance, hydrodynamics (theory of turbulence) and astrophysics.*

A Brownian chain is characterized by the fact that the characteristic function associated with a link $\vec{u}$ is of the form

$$\langle\langle e^{i\vec{k}.\vec{u}} \rangle\rangle = \exp\left[-k^2 l^2/2\right]. \tag{3.4.15}$$

We can generalize this definition. A ν-Brownian chain (a Lévy flight) is a chain with independent links, such, as

$$\langle\langle e^{i\vec{k}.\vec{u}} \rangle\rangle = \exp\left[-\nu (kl)^{1/\nu}\right]. \tag{3.4.16}$$

Consequently, the Fourier transform of the end-to-end probability distribution is

$$\tilde{P}_N(\vec{k}) = \langle\langle \exp[i\vec{k}\cdot(\vec{r}_N - \vec{r}_0)] \rangle\rangle = \exp[-N\nu (kl)^1/^\nu] \tag{3.4.17}$$

Of course, such chains, if they exist, are homogeneous, like the Brownian chain, and the function $P_N(\vec{r})$ must be of the form

$$P_N(\vec{r}) = \frac{1}{(N^\nu l)^d} f(r/N^\nu l). \tag{3.4.18}$$

Thus, the size of the chain must vary like N^ν.

In order to prove the existence of these chains, we have only to verify that the function $P_N(\vec{r})$ is really a probability distribution, i.e. that f(x), the Fourier transform of exp $(-k^{1/\nu})$ is positive. It is clean that this condition can be realized, and the simplest example of a ν-Brownian chain corresponds to the values $\nu = 1$, $d = 1$, in which case

$$\tilde{P}_N(\vec{k}) = e^{-N|k|l} \qquad P_N(x) = \frac{1}{\pi}\frac{Nl}{x^2 + (Nl)^2}. \tag{3.4.19}$$

* Let us also note that for a chain adsorbed on a surface, the successive contact points constitute an intermittent chain.[15]

However, as we shall see, ν-Brownian chains exist only for $\nu > 1/2$. In fact, the size of a ν-Brownian chain which is proportional to N^ν cannot be smaller than the size of a Brownian chain which is proportional to $N^{1/2}$.

In order to obtain this inequality, we observe that for small values of k, $\tilde{P}_N(\vec{k})$ can be written in the following formal manner.

$$\begin{aligned}\tilde{P}_N(\vec{k}) &= 1 - \frac{k^2}{2d}\langle\!\langle(\vec{r}_N - \vec{r}_0)^2\rangle\!\rangle + \dots \\ &= 1 - \frac{k^2}{2d} N \langle\!\langle u^2\rangle\!\rangle + \dots\end{aligned} \tag{3.4.20}$$

and, moreover, that, $\langle\!\langle u^2\rangle\!\rangle$ is assumed to be infinite. The fact that the coefficient in front of k^2 is infinite for a ν-Brownian chain shows that, in this case, the above expansion breaks down, but it clearly indicates that, in the vicinity of $k = 0$, the first term of the expansion of $\tilde{P}_N(\vec{k})$ in (fractional) powers of k must increase (in modulus) faster that k^2; therefore, since according to (3.4.17), we have

$$\tilde{P}_N(\vec{k}) \simeq 1 - N\nu(kl)^{1/\nu} + \dots$$

for small values of k, the condition $\nu > 1/2$ must hold.

For $\nu > 1/2$, the functions $P_N(\vec{r})$ corresponding to (3.4.17) are really positive, therefore they can be considered as probabilities, and they define chains.

We shall indirectly prove this by exhibiting a whole class of chains with independent links, and a ν-Brownian behaviour. Thus, let us associate with each link an independent probability distribution $P(\vec{u})$ whose asymptotic behaviour for large u is

$$P(\vec{u}) \simeq \frac{A}{u^{d+1/\nu}} \qquad \nu > 1/2. \tag{3.4.21}$$

Of course, $P(\vec{u})$ is integrable, and by definition

$$\int \mathrm{d}^d u P(\vec{u}) = 1,$$

but on the other hand

$$\langle\!\langle u^2\rangle\!\rangle = \int \mathrm{d}^d u\, u^2 P(\vec{u}) = \infty. \tag{3.4.22}$$

Let us now examine the behaviour of the function

$$\tilde{P}(\vec{k}) = \int \mathrm{d}^d u\, \mathrm{e}^{i\vec{k}.\vec{u}} P(\vec{u}), \tag{3.4.23}$$

for small values of k

$$\Delta_k \tilde{P}(\vec{k}) = -\int \mathrm{d}^d u\, u^2 \mathrm{e}^{i\vec{k}.\vec{u}} P(\vec{u}).$$

It is easy to verify that the integral converges for finite values of $\vec{k}$ but diverges when $k \to 0$. Consequently, for small values of k, we may replace $P(\vec{u})$ by its asymptotic form

$$\Delta_k \tilde{P}(\vec{k}) = -A \int \frac{d^d u}{u^{d-2+1/\nu}} e^{i\vec{k}.\vec{u}}. \tag{3.4.24}$$

Let us now use the identity (see Appendix M)

$$I_n(k) = \int d^d u \frac{1}{u^n} e^{i\vec{k}.\vec{u}} = \pi^{d/2} \frac{\Gamma\left(\frac{d}{2} - \frac{n}{2}\right)}{\Gamma\left(\frac{n}{2}\right)} (k/2)^{n-d} \tag{3.4.25}$$

which is valid for $0 < n < d$. We obtain

$$\Delta_k \tilde{P}(\vec{k}) = -\mathscr{B}\left(d + \frac{1}{\nu} - 2\right) k^{-2+1/\nu}$$

where

$$\mathscr{B} = A\, \pi^{d/2}\, 2^{1-1/\nu} \frac{\Gamma\left(1 - \frac{1}{2\nu}\right)}{\Gamma\left(\frac{d}{2} + \frac{1}{2\nu}\right)},$$

whence we get

$$\tilde{P}(\vec{k}) \simeq 1 - \mathscr{B}\nu k^{1/\nu}. \tag{3.4.26}$$

Thus, for a chain with N links

$$\tilde{P}_N(\vec{k}) = [\tilde{P}(\vec{k})]^N \simeq [1 - \mathscr{B}\nu k^{1/\nu}]^N \simeq \exp(-\mathscr{B}\nu k^{1/\nu}). \tag{3.4.27}$$

Accordingly, the chain is asymptotically ν-Brownian.

These ν-Brownian chains which are also called Lévy flights exhibit the intermittence property. This means that the global size of the chain does not result from an accumulation of links but from a very small number of very large links. When the number of links becomes larger, the probability that the chain contains a larger link increases simultaneously, and this is why the size of an intermittent chain is a rapidly growing function of the number of links

Remarks

Let us note that the projection of a ν-Brownian chain embedded in a d-dimensional space, on a d'-dimensional subspace is also a ν-Brownian chain, as can easily be seen by considering the definition (3.4.17).

Thus, the characteristic function of a ν-Brownian chain in three-dimensional space is

$$\tilde{P}_N(\vec{k}) = e^{-N\nu[(k_x^2 + k_y^2 + k_z^2)l^2]^{1/2\nu}} \tag{3.4.28}$$

and the characteristic function $\tilde{P}_N^{\perp}(\vec{k})$ of its projection, on the plane $z = 0$ is obtained by putting $k_z = 0$ in the preceding equation, which gives

$$\tilde{P}_N^{\perp}(\vec{k}) = e^{-N\nu[(k_x^2 + k_y^2)l^2]^{1/2\nu}}. \tag{3.4.29}$$

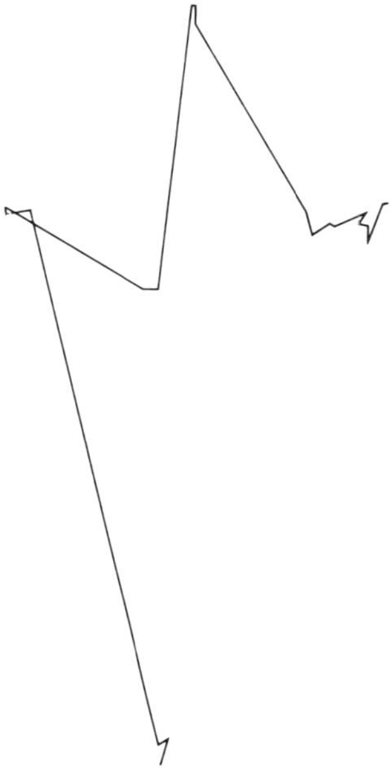

Fig. 3.18. An intermittent chain with independent links in two-dimensional space.

4.3 The Kratky–Porod chain

In a real polymer, the successive links are strongly correlated with one another. Thus, it is not very realistic to represent a polymer, even a short one, by a chain with independent links.

To remedy this deficiency, one must consider locally rigid chains. The simplest example of such chains is the Kratky–Porod chain (wormlike chain),[16] which is a continuous chain, and which, unlike the Brownian chain, has a finite length and, at each point, a well-defined tangent.

This chain can be defined by the following limiting process. We consider a chain in which the length of each link is a constant l_0 and in which the cosine of the angle of each link with the preceding one is also a constant [see Fig. 3.19]. Then, let u_j be the vector defining the link of order j. The links obey constraints which are expressed by the conditions

$$\begin{aligned} \langle\langle \vec{u}_j^2 \rangle\rangle &= l_0^2 \\ \langle\langle \vec{u}_{j+1} \cdot \vec{u}_j \rangle\rangle &= l_0^2 \cos\alpha. \end{aligned} \qquad (3.4.30)$$

The correlation existing between successive links induces correlations with a longer range on the chain. In fact, let us consider the link vector $\vec{u}_{j+n+1}$; we can

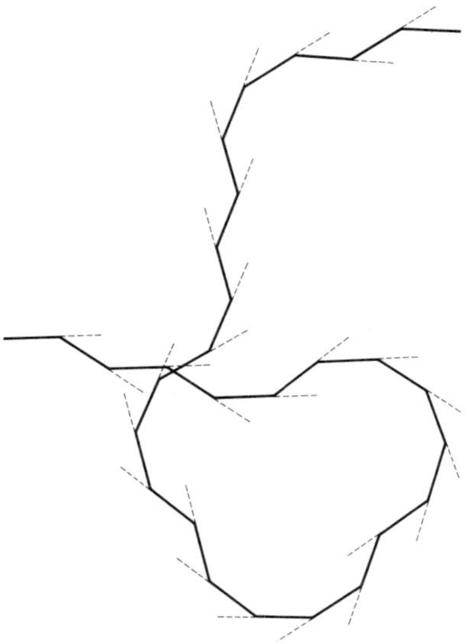

Fig. 3.19. A chain with links of definite length and such that the cosine of the angle between two successive links equals 4/5.

write it in the form

$$\vec{u}_{j+n+1} = \cos\alpha\, \vec{u}_{j+n} + \vec{v}_{j+n+1} \tag{3.4.31}$$

where $\vec{v}_{j+n+1}$ is a vector perpendicular to $\vec{u}_{j+n}$ and of length $l_0 \sin\alpha$. This vector, and the opposite one, always occurs with the same probability and therefore the average of the projection of $\vec{v}_{j+n+1}$ on the vector $\vec{u}_j$ is zero.

$$\langle\!\langle \vec{u}_j \cdot \vec{v}_{j+n+1} \rangle\!\rangle = 0$$

Consequently,

$$\langle\!\langle \vec{u}_j \cdot \vec{u}_{j+n+1} \rangle\!\rangle = \cos\alpha \langle\!\langle \vec{u}_j \cdot \vec{u}_{j+n} \rangle\!\rangle .$$

Thus, by recurrence, we obtain

$$\langle\!\langle \vec{u}_j \cdot \vec{u}_l \rangle\!\rangle = l_0^2 (\cos\alpha)^{|j-1|} . \tag{3.4.32}$$

Using this result, we could easily calculate the mean square value of the length of a chain segment, by taking into account the definition

$$\vec{r}_{N_0} - \vec{r}_0 = \sum_1^{N_0} \vec{u}_j .$$

However, it is more interesting to go directly to the continuous limit. Setting

$$L = N_0 l_0 \tag{3.4.33}$$

and keeping L fixed, we let l_0 decrease to zero and N_0 increase to infinity.

Simultaneously, we let α tend to zero in such a way that the ratio l_0/α^2 remains constant.

We set

$$\frac{2l_0}{\alpha^2} = l_p \qquad \text{and} \qquad N = \frac{N_0\alpha^2}{2} = L/l_p \tag{3.4.34}$$

where l_p is a length characterizing the rigidity of the chain. In the continuous limit, we write

$$\vec{r}(N) = r_{N_0}$$
$$\vec{t}(N) = \frac{\partial \vec{r}(N)}{l_p \partial N}. \tag{3.4.35}$$

The vector $\vec{t}(N)$ is unitary and tangent to the curve

$$[\vec{t}(N)]^2 = 1. \tag{3.4.36}$$

Thus, we have

$$\vec{t}(N) = \frac{\vec{u}_{N_0}}{l_0}.$$

Concurrently, in the limit ($N_0 \to \infty$, $\alpha \to 0$), we have

$$(\cos\alpha)^{N_0} \simeq \left(1 - \frac{\alpha^2}{2}\right)^{N_0} \simeq e^{-N_0\alpha^2/2} \simeq e^{-N}.$$

Consequently, using the preceding equation, we can express the correlations between the $\vec{t}(N)$, in the simple form

$$\langle\!\langle \vec{t}(N)\cdot\vec{t}(N')\rangle\!\rangle = e^{-|N-N'|} = e^{-|L-L'|/l_p}. \tag{3.4.37}$$

The length l_p appears here as the 'persistence length'. Thus, the Kratky–Porod chain is characterized by the two following properties

1. The unit vector $\vec{t}(N)$ is a Markovian variable, a property which can be defined as follows. Let us consider on the chain three points associated with the coordinates $N > N' > N''$ and the unit vectors $\vec{t}(N), \vec{t}(N')\, \vec{t}(N'')$, defining the tangents at these points. The variable $\vec{t}(N)$ is Markovian if the conditional probability that $\vec{t}(N) = \vec{t}$ when it is known that $\vec{t}(N') = \vec{t}'$ and $\vec{t}(N'') = t''$, depends only on $\vec{t}'$ and not at all on $\vec{t}''$.
2. The correlation $\langle\!\langle \vec{t}(N)\cdot\vec{t}(N')\rangle\!\rangle$ is given by eqn (3.4.37).

In principle, these two properties enable one to calculate the mean square distance between the extremities of a chain segment. To do this, it is sufficient to use the relation.

$$\vec{r}(N) = l_p \int_0^N dN'\,\vec{t}(N'). \tag{3.4.38}$$

Now, let us integrate eqn (3.4.37) with respect to N', between the values 0 and N. We obtain

$$\langle\langle [\vec{r}(N) - \vec{r}(0)]\cdot\vec{t}(N)\rangle\rangle = (1 - e^{-N})l_p. \tag{3.4.39}$$

Let us integrate the preceding equation with respect to N: we obtain

$$\langle\langle [\vec{r}(N) - \vec{r}(0)]^2\rangle\rangle = 2[N - 1 + e^{-N}]l_p^2, \tag{3.4.40}$$

For $N \ll 1$

$$\langle\langle [\vec{r}(N) - \vec{r}(0)]^2\rangle\rangle \simeq N^2 l_p^2, \tag{3.4.41}$$

and this result shows clearly that the chain is locally rigid. On the contrary, for $N \gg 1$

$$\langle\langle [\vec{r}(N) - \vec{r}(0)]^2\rangle\rangle = 2(N - 1)l_p^2 \tag{3.4.42}$$

and this equality is a typical characteristic of the Brownian behaviour. The Kratky–Porod chain is indeed asymptotically Brownian, because the correlations between links [see (3.4.32) and (3.4.37)] decrease exponentially with respect to the distance measured along the curve.

In principle, the other properties of the chain can be established without difficulty. Thus, various authors calculated the first moments[17] $\langle\langle [\vec{r}(N) - \vec{r}(0)]^{2n}\rangle\rangle$ (n = small integer), the asymptotic values of the probability distributions[18, 19] concerning the variables $\vec{r}(N) - \vec{r}(0)$ and $\vec{t}(N)$, and the curve that represents the form function

$$H(\vec{q}) = N^{-2}\left\langle\!\!\left\langle \int_0^N dN' \int_0^N dN'' \exp\{i\vec{q}\cdot[\vec{r}(N') - \vec{r}(N'')]\} \right\rangle\!\!\right\rangle \tag{3.4.43}$$

for an infinite chain[20] (see Fig. 3.20).

However, we have to note that the calculations are rather painful. On the other hand, other chains exhibit similar properties of local rigidity, in particular certain Gaussian chains, for which the calculations are much simpler (see below, same chapter).

Furthermore, the Kratky–Porod chain does not take into account any exclusion effects. For all these reasons, the Kratky–Porod chain does not seem very useful.

Remark

The Kratky–Porod chain can be defined and described in a slightly different way by studying the probability law corresponding to $\vec{t}(N)$. Let O be an origin and let us consider the vector $\overrightarrow{OM} = \vec{t}(N)$. The point M belongs to a sphere of unit radius and when N varies, the motion of M on the sphere is Brownian.

Let $P(\vec{t}, \vec{t}_0; N)$ be the probability that $\vec{t}(N) = \vec{t}$ when $\vec{t}(0) = 0$. It is possible to show that the function P is a solution of a diffusion equation on the sphere

$$\frac{\partial}{\partial N} P(\vec{t}, \vec{t}_0; N) = \frac{1}{2}\Delta_S P(\vec{t}, \vec{t}_0; N) + \frac{1}{4\pi}\delta(N)\delta_S(\vec{t} - \vec{t}_0). \tag{3.4.44}$$

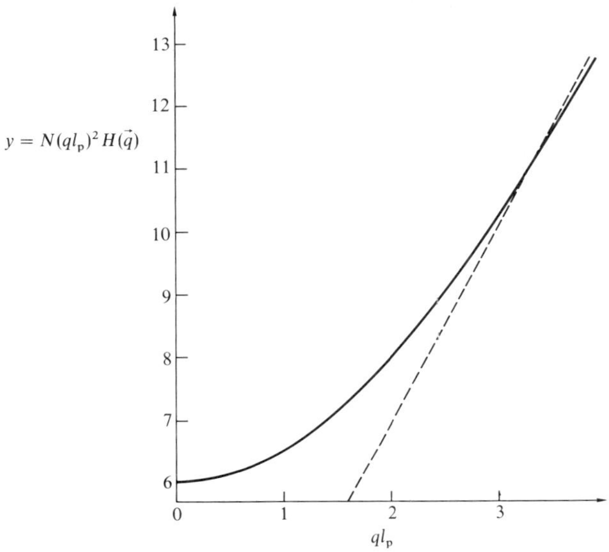

Fig. 3.20. Form function $H(\vec{q})$ of a Kratky–Porod chain in the infinite limit, according to ref. 20. Here l_p is the persistence length and Nl_p the length of the curve. The product $y = N(ql_p)^2 H(\vec{q})$ is a function of (ql_p) and the asymptot represented by a dashed line is given by the equation $y = \pi(ql_p) + 2/3$.

The operator Δ_S is the Laplacian on the sphere with unit radius, and can be written in the form

$$\Delta_S = \frac{1}{\sin\theta}\frac{\partial}{\partial\theta}\left(\sin\theta\frac{\partial}{\partial\theta}\right) + \frac{1}{\sin^2\theta}\frac{\partial^2}{\partial\varphi^2} \tag{3.4.45}$$

where θ and φ are the usual angular variables; here δ_S is a surface δ-function

The equation (3.4.45) can be solved by expanding $P(\vec{t}, \vec{t}_0; N)$ in spherical harmonics. In particular, we may choose $\vec{t}_0$ as the direction of the reference axis. In this case, the problem becomes simpler

$$\begin{aligned} \vec{t}.\vec{t}_0 &= \cos\theta \\ P(\vec{t}, \vec{t}_0; N) &\equiv P(\cos\theta; N) \\ \delta_S(\vec{t} - \vec{t}_0) &= \sum_{l=0}^{\infty} \frac{2l+1}{4\pi} \mathrm{P}_l(\cos\theta) \end{aligned} \tag{3.4.46}$$

where $\mathrm{P}_l(Z)$ is a Legendre polynomial.

We can expand $P(\cos\theta; N)$ in the form

$$P(\cos\theta; N) \equiv \sum_{t=1}^{\infty} a_l(N) \frac{2l+1}{4\pi} \mathrm{P}_l(\cos\theta). \tag{3.4.47}$$

By bringing this expression in (3.4.44), we obtain

$$\frac{\mathrm{d}}{\mathrm{d}N}a_l(N) = \frac{l(l+1)}{2}a_l(N) + \delta(N) \tag{3.4.48}$$

with $a_l(N) = 0$ for $N < 0$, which gives

$$a_l(N) = \mathrm{e}^{-l(l+1)N/2}. \tag{3.4.49}$$

Thus

$$P(\vec{t}, \vec{t}_0; N) = P(\cos\theta; N) = \sum_{l=0}^{\infty}\left(\frac{2l+1}{4\pi}\right)\mathrm{e}^{-l(l+1)N/2}\,\mathrm{P}_l(\cos\theta). \tag{3.4.50}$$

In particular, we have

$$\mathrm{P}_0(Z) = 1 \qquad \mathrm{P}_1(Z) = \mathrm{Z}, \tag{3.4.51}$$

and the polynomials $\mathrm{P}_1(Z)$ obey the relations

$$\int_{-1}^{+1}\mathrm{d}Z\,\mathrm{P}_l(Z)\mathrm{P}_m(Z) = \frac{2}{2l+1}\delta_{lm}. \tag{3.4.52}$$

Thus, the solution (3.4.52) can be used to calculate various quantities. For instance, we have

$$\begin{aligned}\langle\!\langle \vec{t}(N)\cdot\vec{t}(0)\rangle\!\rangle &= \langle\!\langle \cos\theta\rangle\!\rangle = 2\pi\int_0^{\pi}\mathrm{d}\theta\,\sin\theta\cos\theta\,P(\cos\theta; N)\\ &= 2\pi\int_{-1}^{+1}\mathrm{d}Z\,Z\,P(Z; N)\end{aligned} \tag{3.4.52}$$

we have

$$\int_{-1}^{+1}\mathrm{d}Z\,Z\,\mathrm{P}_l(Z) = \tfrac{2}{3}\delta_{1l};$$

we find in this way that

$$\langle\!\langle \vec{t}(N)\cdot\vec{t}(0)\rangle\!\rangle = \langle\!\langle \cos\theta\rangle\!\rangle = \mathrm{e}^{-N}, \tag{3.4.53}$$

an equality which coincides with eqn (3.4.37).

4.4 Chains with local correlations (asymptotically Brownian)

There is a whole class of chains for which the interactions between links involve only neighbours of finite order not larger than p. Thus, for a chain with nearest link interaction ($p = 1$), we have (N = number of links)

$$\mathscr{U} = \sum_1^N V_0(\vec{u}_j) + \sum_1^{N-1} V_1(\vec{u}_j, \vec{u}_{j+1}), \tag{3.4.54}$$

an expression which can be written in the form

$$\mathscr{U} = \sum_{q=1}^{N-1} \mathscr{U}_1(\vec{u}_q, \vec{u}_{q+1}) + \mathscr{U}_0(\vec{u}_1) \tag{3.4.55}$$

by setting

$$\mathscr{U}_0(\vec{u}) = V_0(\vec{u})$$
$$\mathscr{U}_1(\vec{u}, \vec{v}) = V_1(\vec{u}, \vec{v}) + V_0(\vec{v}).$$

When the interaction between links extends to neighbours of order p, formula (3.4.54) takes a more general form and we may set

$$\mathscr{U} = \sum_{q=1}^{N-p} \mathscr{U}_1(\vec{u}_q, \ldots, \vec{u}_{q+p}) + \mathscr{U}_0(\vec{u}_1, \ldots, \vec{u}_p). \tag{3.4.56}$$

Of course, the last term results from boundary effects; this does not mean, however, that there is any dissymmetry between the extremities of the chain (see above the case where $p = 1$).

Of course, the weight associated with a configuration is (see Chapter 2, Section 2.1)

$$W_N(\vec{u}_N, \ldots, \vec{u}_1) = \exp[-\mathscr{U}], \tag{3.4.57}$$

and the total number of configurations is represented by

$$Z_N = \int \mathrm{d}^d u_N \ldots \mathrm{d}^d u_1 \, W_N(\vec{u}_N, \ldots, \vec{u}_1)$$

where d is the dimension of the space in which the chain is embedded.

Let us now try to evaluate expressions of the form

$$\tilde{P}_N(\vec{k}_1, \ldots, \vec{k}_N) = \left\langle\!\!\left\langle \exp\left[i \sum_{j=1}^{N} \vec{k}_j \cdot \vec{u}_j \right] \right\rangle\!\!\right\rangle. \tag{3.4.58}$$

With this aim, let us introduce the transfer operator $\mathscr{T}(\vec{k})$ which acts in a dp-dimensional space

$$\langle\!\langle \vec{u}_p, \ldots, \vec{u}_1 | \mathscr{T}(\vec{k}) | \vec{v}_p, \ldots, \vec{v}_1 \rangle\!\rangle = \delta(\vec{v}_p - \vec{u}_{p-1}) \ldots \delta(\vec{v}_2 - \vec{u}_1)$$
$$\exp[-\mathscr{U}_1(\vec{u}_p, \ldots, u_1, v_1) + i\vec{k}\cdot\vec{u}_p]. \tag{3.4.59}$$

Thus, we see that

$$W_N(\vec{u}_N, \ldots, \vec{u}_1) = \int \mathrm{d}^d v_p \ldots \mathrm{d}^d v_2 \langle\!\langle \vec{u}_N, \ldots, \vec{u}_{N-p+1} | \mathscr{T}(\vec{0}) | \vec{v}_p, \ldots \vec{v}_2, \vec{u}_{N-p} \rangle\!\rangle$$
$$W_{N-1}(\vec{v}_p, \ldots, \vec{v}_2, \vec{u}_{N-p}, \ldots, \vec{u}_1)$$

from which we deduce

$$Z_N = \int \mathrm{d}^d u_p \ldots \mathrm{d}^d u_1 \int \mathrm{d}^d v_p \ldots \mathrm{d}^d v_1 \langle\!\langle \vec{u}_p, \ldots, \vec{u}_1 | [\mathscr{T}(\vec{0})]^{N-p} \, \vec{v}_p, \ldots, \vec{v}_1 \rangle\!\rangle$$
$$W_p(\vec{v}_p, \ldots, \vec{v}_1) \tag{3.4.60}$$

where

$$W_p(\vec{u}_p, \ldots, \vec{u}_1) = \exp[-\mathscr{U}_0]$$

and

$$\tilde{Z}_N(\vec{k}_1, \ldots, \vec{k}_N) = \frac{1}{Z_N}\int \mathrm{d}^d u_p \ldots \mathrm{d}^d u_1 \int \mathrm{d}^d v_p \ldots \mathrm{d}^d v_1$$

$$\langle \vec{u}_p, \ldots, \vec{u}_1 | \mathscr{T}(\vec{k}_N) \ldots \mathscr{T}(\vec{k}_{p+1}) | \vec{v}_p, \ldots, \vec{v}_1 \rangle$$

$$\exp\left[i \sum_{j=1}^{p} \vec{k}_j \cdot \vec{u}_j \right] Z_p(\vec{v}_p, \ldots, \vec{v}_1). \tag{3.4.61}$$

The operator $\mathscr{T}(\vec{k})$ can be diagonalized and, if the interactions are not signular, its spectrum is discrete

$$\mathscr{T}(\vec{k}) = \sum_{n=0}^{\infty} t_n(\vec{k}) | n(\vec{k}) \rangle \langle \bar{n}(\vec{k}) |. \tag{3.4.62}$$

Here the vectors $|n(\vec{k})\rangle$ and the vectors $\langle \bar{n}(\vec{k})|$ constitute a bi-orthogonal system

$$\langle \bar{n}(\vec{k}) | m(\vec{k}) \rangle = \delta_{nm}. \tag{3.4.63}$$

Using this representation, we immediately obtain

$$Z_N = \int \mathrm{d}^d u_p \ldots \mathrm{d}^d u_1 \int \mathrm{d}^d v_p \ldots \mathrm{d}^d v_1 \, W_p(\vec{v}_p, \ldots, \vec{v}_1)$$

$$\sum_n [t_n(\vec{0})]^{N-p} \langle \vec{u}_p, \ldots, \vec{u}_1 | n(0) \rangle \langle \bar{n}(0) | \vec{v}_p, \ldots, \vec{v}_1 \rangle. \tag{3.4.64}$$

In the same way, for $\vec{k}_1 = \vec{k}_2 = \ldots = \vec{k}_N = \vec{k}$, we have the simple representation

$$\tilde{P}_N(\vec{k}, \ldots, \vec{k}) = \frac{1}{Z_N}\int \mathrm{d}^d u_p \ldots \mathrm{d}^d u_1 \int \mathrm{d} v_p \ldots \mathrm{d} v_1 \, W_p(\vec{v}_p, \ldots, \vec{v}_1)$$

$$\sum_n [t_n(\vec{k})]^{N-p} \langle \vec{u}_p, \ldots, \vec{u}_1 | n(\vec{k}) \rangle \langle \bar{n}(\vec{k}) | \vec{v}_p, \ldots, \vec{v}_1 \rangle. \tag{3.4.65}$$

The matrix elements of $\mathscr{T}(\vec{0})$ are positive. Therefore the eigenvalue with the largest modulus is non-degenerate. The corresponding state which is associated with the eigenvector $|0(\vec{0})\rangle$ and the eigenvalue $t_0(\vec{0})$ is called the fundamental state. The quantities $t_0(\vec{0})$ and $\langle \vec{u}_p, \ldots, \vec{u}_1 | 0(\vec{0}) \rangle$ are positive.

For large values of N, the expression of Z_N becomes simpler and we have

$$Z_N = [t_0(\vec{0})]^{N-p} \left[\int \mathrm{d}^d u_p \ldots \mathrm{d}^d u_1 \, \langle \vec{u}_p, \ldots, \vec{u}_1 | 0(\vec{0}) \rangle \right.$$

$$\left. \int \mathrm{d}^d v_p \ldots \mathrm{d}^d v_1 \, Z_p(\vec{v}_p, \ldots, \vec{v}_1) \langle \bar{0}(\vec{0}) | \vec{v}_p, \ldots, \vec{v}_1 \rangle \right] \tag{3.4.66}$$

which is of the form

$$Z_N = \mathscr{A}\mu^N.$$

Thus, we verify that the critical exponent γ has the value $\gamma = 1$ which is typical for a Brownian chain. In the same way, for large values of N, we have ($|\vec{k}|$ remaining small)

$$\tilde{P}_N(\vec{k}, \ldots, \vec{k}) \simeq \left[\frac{t_0(\vec{k})}{t_0(0)}\right]^{N-p} \tag{3.4.67}$$

For small values of $\vec{k}$, $t_0(\vec{k})$ can be expanded as follows

$$t_0(\vec{k}) = t_0(\vec{0})\left[1 - \frac{k^2 l^2}{2} + \ldots\right] \tag{3.4.68}$$

(the first term in the expansion is proportional to k because $|0(\vec{0})\rangle$ is non-degenerate). Consequently, coming back to the definition of $\tilde{P}_N(\vec{k}_1, \ldots, \vec{k}_N)$ [see (3.4.58)] we find that

$$\langle\!\langle \exp[i\vec{k}\cdot(\vec{r}_N - \vec{r}_0)]\rangle\!\rangle \simeq \left[1 - \frac{k^2 l^2}{2}\right]^N \simeq \exp[-Nk^2 l^2/2]. \tag{3.4.69}$$

Thus, the probability law associated with the correlations between chain ends is Gaussian. In particular, the exponent ν has the value $\nu = 1/2$ and we find

$$\langle\!\langle (\vec{r}_N - \vec{r}_0)^2 \rangle\!\rangle = N\,dl^2. \tag{3.4.70}$$

The same property remains valid for the probability law associated with any points of the chain, if the points are far away on the chain. In this case, we have

$$\langle\!\langle \exp i\vec{k}\cdot(\vec{r}_{j+1})\rangle\!\rangle = \tilde{P}_N(\underbrace{\vec{0}, \ldots, \vec{0}}_{j}, \underbrace{\vec{k}, \ldots, \vec{k}}_{n}, \underbrace{\vec{0}, \ldots, 0}_{N-j-n})$$

$$\simeq \left[\frac{t_0(\vec{k})}{t_0(\vec{0})}\right]^n \simeq \exp(-nk^2 l^2/2). \tag{3.4.71}$$

Thus, the chain is asymptotically Brownian and this result is related to the property that the correlations between links decrease exponentially along the chain, as can easily be verified: to prove this, we have only to use the fact, that, as long as p is finite, the fundamental state of $\mathscr{T}(\vec{k})$ is non-degenerate, for small values of k.

4.5 Gaussian chains

A chain is Gaussian when the components of the set of vectors u_j associated with the chain links constitute a Gaussian set (see Appendix A). A Gaussian chain may consist of a finite or infinite number of discrete links, but it may also be continuous.

The study of the simple case where the chain is made of a finite number N of discrete links will be used as a basis for a general study of the Gaussian chains. In this case, the probability law of the chain is given by

$$P(\vec{u}_1, \ldots, \vec{u}_N) = \frac{1}{Z}\exp\left[-\frac{1}{2}\sum A_{jl}\vec{u}_j\cdot\vec{u}_l \right]$$

$$Z = [(2\pi)^N \det A]^{d/2} \tag{3.4.72}$$

(where d is the space dimension).

The Fourier transform of this probability law is

$$\tilde{P}(\vec{k}_1, \ldots, \vec{k}_N) = \left\langle\!\!\left\langle \exp\left(i\sum \vec{k}_j\cdot\vec{u}_j \right) \right\rangle\!\!\right\rangle$$

$$= \exp\left[-\frac{1}{2}\sum_{jl} B_{jl}\vec{k}_j\cdot\vec{k}_l \right] \tag{3.4.73}$$

where

$$\sum_m B_{jm}A_{ml} = \delta_{jl} \qquad B = A^{-1}.$$

The correlations between links are determined by [see eqn (A.21)]

$$\langle\!\langle \vec{u}_j\cdot\vec{u}_l \rangle\!\rangle = d\,B_{jl} \tag{3.4.74}$$

Giving these correlations defines the probability law completely and, consequently, all the properties of the chain. In particular, we can calculate the mean square distance between two chain points. As for $j > l$, we have

$$\vec{r}_j - \vec{r}_l = \sum_{m=l+1}^{j} \vec{u}_m$$

we find immediately that

$$\langle\!\langle (\vec{r}_j - \vec{r}_l)^2 \rangle\!\rangle = d \sum_{l<m\leqslant j}\ \sum_{l<n\leqslant j} B_{mn}. \tag{3.4.75}$$

Conversely, let us assume that we choose a priori the mean square distance

$$\langle\!\langle (\vec{r}_j - \vec{r}_l)^2 \rangle\!\rangle = d\,F_{jl}. \tag{3.4.76}$$

From it, we easily deduce B_{jl}. Indeed, the following identity

$$2\vec{u}_j\cdot\vec{u}_l = (\vec{r}_j - \vec{r}_{l-1})^2 + (\vec{r}_{j-1} - \vec{r}_l)^2 - (\vec{r}_j - \vec{r}_l)^2 - (\vec{r}_{j-1} - \vec{r}_{l-1})^2$$

can easily be verified and therefore

$$B_{jl} = \tfrac{1}{2}[F_{j,l-1} + F_{j-1,l} - F_{j,l} - F_{j-1,l-1}]. \tag{3.4.77}$$

In particular, by choosing a priori the values of the link correlations, we can construct chains corresponding to any value of the exponent ν. Thus, it is possible to assume that for large values of $|j - l|$

$$F_{jl} \simeq F|j - l|^{2\nu},$$

which implies

$$B_{jl} \simeq F\frac{\nu(2\nu - 1)}{|j - l|^{2-2\nu}}.$$

Thus, when all the B_{jl} are positive and decrease slowly when $|j - l|$ increases, the chain is swollen.

On the contrary, the quasi-Brownian chains (with $\nu = 1/2$) correspond to the case where the B_{jl} decrease rapidly when $|j - l|$ increases. A Gaussian chain is quasi-Brownian if

$$B_{jl} = b(|j - l)$$

with $\sum_1^{\infty} b(n) = b$ (finite),

and, in this case, for large values of $|j - l|$,

$$\langle\langle(\vec{r}_j - \vec{r}_l)^2\rangle\rangle \simeq d[b(0) + 2b]|j - l| \tag{3.4.78}$$

as can easily be verified by looking at (3.4.75).

Nevertheless, these quasi-Brownian chains can be rather rigid and the coefficients A_{ij} can be chosen so as to take stretching and bending elasticities simultaneously into account (Papadopoulos and Thomchick 1977)[21].

The generalization to the continuous case does not present any particular difficulty. A continuous chain is defined by a function $\vec{r}(\lambda)$, with $0 < \lambda < L$ and it possible to associate with it a Gaussian weight which is a functional of $\vec{r}(\lambda)$

$$W\{\vec{r}(\lambda)\} \propto \exp\left[-\int_0^L d\lambda \int_0^L d\lambda' A(\lambda, \lambda')\left[\frac{\partial\vec{r}(\lambda)}{\partial\lambda}\cdot\frac{\partial\vec{r}(\lambda')}{\partial\lambda'}\right]\right]. \tag{3.4.79}$$

Accordingly, we have [see (3.4.73)].

$$\left\langle\left\langle\frac{\partial\vec{r}(\lambda)}{\partial\lambda}\cdot\frac{\partial\vec{r}(\lambda')}{\partial\lambda'}\right\rangle\right\rangle = d\,B(\lambda, \lambda').$$

The kernels $A(\lambda, \lambda')$ and $B(\lambda, \lambda')$ obey the relation

$$\int d\lambda'' A(\lambda, \lambda'')B(\lambda'', \lambda') = \delta(\lambda - \lambda') \tag{3.4.80}$$

[see (3.4.61)] and, obviously, the Brownian case corresponds to the values

$$A(\lambda, \lambda') = B(\lambda, \lambda') = \delta(\lambda - \lambda').$$

We could also simulate a Kratky–Porod chain by setting

$$\left\langle\left\langle\frac{\partial\vec{r}(\lambda)}{\partial\lambda}\cdot\frac{\partial\vec{r}(\lambda')}{\partial\lambda'}\right\rangle\right\rangle = e^{-|\lambda - \lambda'|/l_p}.$$

Thus, it is possible, to a first approximation, to simulate any kind of chain by a Gaussian chain. However, the representation stops being faithful when one looks at the shape of the probability law associated with the vector joining the chain ends or when one considers non-quadratic mean values.

For a Gaussian chain, the probability distributions associated with the vectors joining two points on the chain are always Gaussian; moreover, if we know the B_{jl} we can calculate all mean values by direct (or indirect) application of Wick's theorem (see Appendix A). This theorem thus indicates that for a Gaussian chain, in all cases, we have [see eqn (A.25)]

$$\langle\!\langle |\vec{r}_j - \vec{r}_l|^4 \rangle\!\rangle = \left(1 + \frac{2}{d}\right) \langle\!\langle |\vec{r}_j - \vec{r}_l|^2 \rangle\!\rangle^2$$

which, in general, is not true for another kind of chain

In particular, it appears that the probability distribution of the vector joining the chain ends, for a chain with excluded volume, strongly differs from a Gaussian function, and this fact greatly diminishes the value of the Gaussian chains.

REFERENCES

1. Hammersley, J.M. (1961). *Quart. J. Math., Oxford* **12**, 250.
2. Hammersley, J.M. (1961). *Proc. Cambridge Phil. Soc.* **57**, 516 (apparently the proof given there applies only to the case $d = 2$, but it can be improved in order to apply for any d).
3. Hammersely, J.M. and Welsh, D.J.A. (1962). *Quart. J. Math.* **13**, 108.
4. Hammersley, J.M. (1963). *Sankhya: the Indian Journal of Statistics* A**25**, 29 and 269 (the proof given on p. 34 is not completely rigorous but can easily be corrected).
5. Kesten, H. (1963). *J. Math. Phys.* **4**, 960, and (1964). **5**, 1128.
6. Hille, E. (1972). *Methods in classical and functional analysis.* Addison-Wesley.
7. Blöte, H.W.J. and Hilhorst, J. (1984). *J. Phys.* A: *Math. Gen.* **17**, L111.
8. Gordon, M., Kapadia, P., and Malakis, A. (1976). *J. Phys.* A: *Math. Gen.* **9**, 751.
9. Kasteleyen, P.W. (1963). *Physics* **29**, 1329.
10. Duplantier, B. and David, F. (1988). *J. Stat. Phys.* **51**, 327. Duplantier, B. (1987). *J. Stat. Phys.* **49**, 411.
11. Fisher, M. E. (1966). *J. Chem. Phys.* **44**, 616.
12. Balian, R. and Toulouse, G. (1974). *Ann. Phys.* **83**, 28.
13. Thouless, D.J. (1975). *J. Phys.* C **8**, 1803.
14. Lévy, P. (1965). *Processus stochastiques et mouvement brownien.* Gauthier-Villars, Paris.
15. Bouchaud, E. and Daoud, M. (1987). *J. Phys.* A: *Math. Gen.* **20**, 1463.
16. Kratky, O. and Porod, G. (1949). *Recl. Trav. Chim.* **68**, 1106.
17. Nagai K. (1968). *J. Chem. Phys.* **48**, 5646.
18. Saito, N., Takahashi, K., and Yunoki, Y. (1967). *J. Phys. Soc. Japan* **22**, 219.
19. Gobush, W., Yamakawa, H., Stockmayer, W. H., and Magee, W. S. (1972) **57**, 2839.
20. des Cloizeaux, J. (1973). *Macromolecules* **6**, 403.
21. Papadopoulos, G.J. and Thomchick, J. (1977). *J. Phys.* A: *Math. Gen.* **10**, 1115.

4

COMPUTER EXPERIMENTS

1. GENERALITIES

Computer experiments have in the past been one of the best available sources of information concerning the experimental values of the critical exponents. Many of these computations were performed from 1955 to 1970 by English and American physicists. The polymers are simulated by chains constructed on lattices with the help of computers.

Two methods have been used. In one of these, rather short chains are studied but these chains are counted exactly and the values of the critical exponents are determined by extrapolation. In the other method, random processes are used to construct a certain number of rather long chains which give the value of the exponents more directly, provided that certain statistical corrections are made.

As we shall see, each method has advantages and disadvantages. The results are good and, in general the precision is (at best), of the order of a few per cent. However, the precision obtained is not really as good as it seems, because the calculated value of the exponents depends, to some extent, on the methods of analysis used for that purpose. All the results are in agreement with the scaling laws and the theoretical predictions. In fact, when contradictions appeared, it was always possible to solve them by modifying slightly the interpretation of the results.

In this chapter, we shall study the various methods, their results and their limitations, but for more details the reader may consult a review article by McKenzie (1976).[1]

2. SELF-AVOIDING CHAINS

To study the behaviour of self-avoiding chains, it is sufficient to consider the set of chains constructed in a lattice by starting from an origin O and by walking at random from site to site without passing the same site twice.

2.1 Method of exact counting

2.1.1 Description of the chains under study

The method consists in constructing all the chains with N links which starts from an origin site O. This method has been particularly developed in England and notably by Domb,[2] Sykes, and Martin. As the number of chains increases exponentially with the number of links, it is necessary to reduce the counting

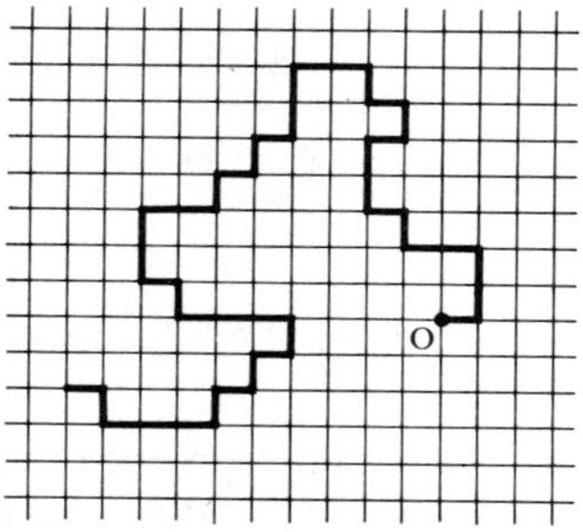

Fig. 4.1. (1) Self-avoiding chain drawn on a lattice by starting from an origin O.

work as much as possible, by taking into account all possible symmetries. The appropriate methods have been described by Sykes,[3] by Martin and Watts,[4] and by the McKenzies.[5]

Nevertheless, the chains for which the number of configurations can be counted exactly are always rather short. For the chains that have been studied, the maximum numbers N of links are as follows:

Triangular lattice ($d = 2$)

Counting of open or closed chains

$N = 17$ (Martin, Sykes, and Hioe 1967)[6]

Distance between end points

$N = 14$ (Martin and Watts 1971)[4]

Second virial coefficient (osmotic pressure)

$N = 7$ (McKenzie and Domb 1967).[7]

Square lattice ($d = 2$)

Counting of open or closed chains

$N = 14$ (Fisher and Hiley 1961)[8]

$N = 20$ (Martin and Watts 1971)[4]

Distance between end points

$N = 10$ (Fisher and Hiley 1961)[8]

$N = 20$ (Martin and Watts 1971)[4]

Hexagonal lattice ($d = 2$) (*honeycomb lattice*)

Counting of open or closed chains

$N = 34$ (Sykes, Guttmann, Watts, and Roberts 1972).[9]

Face-centred cubic lattice ($d = 3$)

Counting of open or closed chains

$N = 12$ (Martin, Sykes, and Hioe 1967)[6]

Distance between end points

$N = 10$ (McKenzie 1967, according to Martin and Watts 1971)[4]

Second virial coefficient

$N = 6$ (McKenzie and Domb 1967).[7]

Simple cubic lattice ($d = 3$)

Counting of open or closed chains

$N = 10$ (Fisher and Hiley 1961)[8]

$N = 15$ (Martin and Watts 1971)[4]

Distance between end points

$N = 8$ (Fisher and Hiley 1961)[8]

$N = 15$ (Martin and Watts 1971)[4]

Second virial coefficient

$N = 7$ (McKenzie and Domb 1967).[7]

Body-centred cubic lattice ($d = 3$)

Counting of open or closed chains

$N = 12$ (Martin and Watts 1971)[4]

Distance between chain ends

$N = 10$ (Watts 1974)[10]

Second virial coefficient

$N = 7$ (McKenzie and Domb 1967).[7]

Diamond lattice ($d = 3$)

Counting of chains

$N = 22$ (Sykes 1973, see Watts 1975)[11]

2.1.2 Calculation of the exponents γ and ν in two and three dimensions

The results obtained are extrapolated to large values of N. Various nearly equivalent methods have been used and they were reviewed by Gaunt and Guttmann[12] in 1974.

For instance, let us consider the total number Z_N of self-avoiding walks that can be drawn in a lattice by starting from the site at the origin. For large N, we shall try to represent Z_N by an expression of the form

$$Z_N \propto \mu^N N^{\gamma-1}. \tag{4.2.1}$$

Let us set

$$\mu_N = Z_{N+1}/Z_N.$$

To calculate μ, we can plot μ_N against N and extrapolate. On the other hand, for large values of N,

$$\ln \mu_N \simeq \ln \mu + \frac{\gamma - 1}{N}.$$

Thus γ can be estimated by plotting $\ln \mu_N$ against $1/N$. In general, the convergence is rather good. However, it must be realized that it would be equally reasonable to represent Z_N in slightly different forms. For instance, for N large, it would be valid to set

$$Z_N \sim \mu^N (N + a)^{\gamma-1} \qquad 0 < a < 1. \tag{4.2.2}$$

In both formulae, the exponent γ must be the same. However, when one tries to represent the experimental values by the second formula, it appears that the approximate value obtained for γ depends on the choice made a priori for a.

Of course, it is also possible to rely on more sophisticated methods and, in order to calculate the exponents γ and ν associated with chains drawn on various lattices, Watts[11] used Padé approximants. However, this method does not lead to spectacular improvement in the results. Actually, approximations of the Padé type are excellent when one tries to represent smooth analytic functions, but here this is not the case. The various functions of N which are to be interpolated or extrapolated, contain oscillating terms related to the lattice structure, and the presence of such terms, spoils the precision of the results.

Nevertheless, the calculated values of the critical exponents depend only slightly on the lattice type. The best values are obtained in two dimensions with the triangular lattice, and in three dimensions with the face-centred cubic lattice, for the following reasons. First, as each site has many neighbours, the angular fluctuations are reduced, and this is a favourable circumstance. Second, it is easy for the chain to come back close to itself and therefore, on these lattices, the chains very quickly *know* that they have to be self-avoiding.

The following results give an idea of the precision obtained with these types of chain:

For $d = 2$, triangular lattice

$\gamma = 1.330 \pm 0.003$ (Watts 1975)[11]

$\nu = 0.750 \pm 0.0025$ (Watts 1974)[10]

For $d = 3$, face-centred cubic lattice

$$\gamma = 1.1663 \pm 0.003 \text{ (Watts 1975)}^{11}$$
$$\nu = 0.60 \pm 0.02 \text{ (Watts 1974)}^{10} \tag{4.2.3}$$

Let us note that for $d = 3$, the results are close to those obtained by Le Guillou and Zinn-Justin[13] with renormalization group methods (see below Chapter 12, Section 3), namely*

$$\gamma = 1.1615 \pm 0.0011$$
$$\nu = 0.5880 \pm 0.0010 \tag{4.2.4}$$

For $d = 2$, they are close to the 'exact' results: $\nu = 43/32$ and $\nu = 3/4$ (see Chapter 12, Section 4).

Furthermore, certain authors like Domb and Hioe[14] measured simultaneously the mean square distance between chain ends $R_N^2 = \langle\langle(\vec{r}_N - \vec{r}_0)^2\rangle\rangle$ and the (mean) square radius of gyration. $R^2_{G,N}$, for the same chain: thus they could estimate the ratio

$$\aleph(N) = 6\, R^2_{G,N}/R^2_N. \tag{4.2.5}$$

When $N \to \infty$, this ratio has a limit $\aleph$ which must depend only on the type of chain under consideration.

For Brownian chains $\aleph = 1$ as we have seen in Chapter 3, Section 3.1.

For a chain with a size exponent ν, one could try to evaluate $\aleph$ by proceeding as follows. One might assume that the mean square distance between two points of the chain is given by

$$\langle\langle(\vec{r}_{j+n} - \vec{r}_j)^2\rangle\rangle = An^{2\nu}. \tag{4.2.6}$$

Then, the definition of the radius of gyration

$$R_G^2 = \frac{1}{2(N+1)^2} \sum_{i=0}^{N} \sum_{j=0}^{N} \langle\langle(\vec{r}_i - \vec{r}_j)^2\rangle\rangle \tag{4.2.7}$$

and the preceding assumption give,

$$\aleph = 6 \int_0^1 \mathrm{d}x\,(1-x)^{2\nu} = \frac{6}{(2\nu+1)(2\nu+2)}, \tag{4.2.8}$$

and we note that for a Brownian chain ($\nu = 1/2$), this formula gives the correct value $\aleph = 1$.

However, let us consider the case of self-avoiding chains.

For $d = 3$, $\nu = 0.59$, the preceding formula gives $\aleph = 0.87$ and the computer experiments[14] of Domb and Hioe give $\aleph = 6\,(0.155) = 0.93$. For $d = 2$, $\nu = 3/4$ the preceding formula gives $\aleph = 0.68$ and the computer experiments[14] $\aleph = 6(0.140) = 0.84$.

* However, it seems that at least one of the physicists who calculated γ has been too optimistic concerning the evaluation of his errors.

The reasons for this discrepancy are not difficult to discover. First, we observe that the radius of gyration does not reach its asymptotic behaviour as rapidly as the end-to-end distance, because important contributions to the radius of gyration, come from distances between rather close points. However, there exists a deeper reason for this discrepancy. In fact, formula (4.2.8) can be considered only a first approximation, because, for a chain with a finite size, the initial assumption given by (4.2.6) is erroneous.

Indeed, the scaling laws predict that $\langle\langle (\vec{r}_{j+n} - \vec{r}_j)^2 \rangle\rangle$ should be of the form

$$\langle\langle (\vec{r}_{j+n} - \vec{r}_j)^2 \rangle\rangle = A n^{2\nu} \mathrm{g}\left(\frac{n}{N}, \frac{j}{N}\right) \tag{4.2.9}$$

(with $\mathrm{g}(1, 0) = 1$), as was noticed by Domb and Hioe[14] who already in 1969 studied a few properties of this function. Thus, the exact formula giving $\aleph$ should read

$$\aleph = 6 \int_0^1 \mathrm{d}x \int_0^{1-x} \mathrm{d}y\, \mathrm{g}(x, y)\, x^{2\nu}, \tag{4.2.10}$$

but unfortunately, at the present time (1988), the function $\mathrm{g}(x, y)$ is only approximately known (see Chapter 13, Section 1.7).

In fact, by using renormalization methods, it was possible to calculate[15] the first two terms of the expansion of $\aleph$ in powers of $\varepsilon = 4 - d$. Using this result and the exact result $\aleph_{d=1} = 1/2$, we can represent $\aleph$ by the approximate expression (see Chapter 13, Section 1.4.2)

$$\aleph = 1 - 0.010417\,\varepsilon - 0.030628\,\varepsilon^2 - 0.007152\,\varepsilon^3. \tag{4.2.11}$$

In this way [see (13.1.63)], one finds

for $d = 3$, $\aleph = 0.952$
for $d = 2$, $\aleph = 0.799$.

The values thus obtained are in better agreement with the results of computer experiments.

Let us note that the effective values of $\aleph$ are larger than the values deduced from eqn (4.2.8). In fact, an internal segment of chain must be relatively more stretched than the chain itself, because the pieces of chain which are attached to each ends of the segment repel each other. Thus, the function $\mathrm{g}(x, y)$ defined by (4.2.9) must obey the inequality $\mathrm{g}(x, y) > 1$ ($x \neq 1$ and $y \neq 0$).

Certain authors[16] have tried to describe in a different way the fact that the central parts of a chain are relatively more swollen than the chain itself. They assumed that the mean square distance between end points $\langle\langle (\vec{r}_N - \vec{r}_0)^2 \rangle\rangle$ varies as $N^{2\nu}$ but that the mean square distance between internal points $(\vec{r}_i - \vec{r}_j)^2 \rangle\rangle$ varies as $|i - j|^{2\nu'}$ where ν' is smaller than ν. Thus they assume the equalities (with another notation).

$$\langle\langle (\vec{r}_i - \vec{r}_j)^2 \rangle\rangle = \left(\frac{|i - j|}{N}\right)^{2\nu'} \langle\langle (\vec{r}_N - \vec{r}_0)^2 \rangle\rangle \simeq A\,|i - j|^{2\nu'} N^{2(\nu - \nu')}. \tag{4.2.12}$$

As we have already seen, and as we shall also see later, such an assumption is unjustified from an experimental point of view. Moreover, it would lead to the conclusion that the infinite chain does not exist, since the preceding formula does not give any finite limit for $\langle\langle (r_i - r_j)^2 \rangle\rangle$ when $N \to \infty$, and this is difficult to believe.

On the other hand, as the infinite chain is a critical object, such behaviour would imply that it is not possible to construct a finite theory at the critical point. As we shall see later, this leads to a contradiction between the principles and the results of renormalization theory. Thus, the interpretation which has been described above has to be rejected.

2.1.3 Second virial coefficient

In the same way, it is possible to count the number of configurations obtained by placing on a lattice two self-avoiding chains without intersections. More precisely, let us draw a self-avoiding chain on the lattice starting from the origin O, and another self-avoiding chain starting from any other point O′. Let us now consider all the configurations for which the chains intersect each other (forbidden configurations). The number of forbidden configurations will be denoted by $Z_{N,M}$ (O is fixed but O′ may occupy any position). The value of $Z_{N,M}$ gives information concerning the osmotic pressure. In fact, the virial expansion for a set of chains with N links may be written

$$\Pi\beta = \boldsymbol{C} + \sum_{j=2}^{\infty} \boldsymbol{A}_j(N)\,\boldsymbol{C}^j \tag{4.2.13}$$

where $\boldsymbol{C}$ is the number of polymers per unit volume, and it is easy to prove the relation (see Chapter 9, Section 4.2)

$$\boldsymbol{A}_2(N) = \frac{Z_{N,N}}{2Z_N^2}\,v_r \tag{4.2.14}$$

where v_r is the volume per site.*

Thus, McKenzie and Domb[7] evaluated $Z_{N,N}$ and showed that the value of $\boldsymbol{A}_2(N)$ extrapolated to large values of N is proportional to a power of N

$$\boldsymbol{A}_2(N) \propto N^{2-\alpha'}$$

with

$$\alpha' = 0.51 \qquad d = 2$$

$$\alpha' = 0.28 \pm 0.02 \qquad d = 3.$$

* The correspondence with the notation used in Chapter 9 is given by the equalities

$$Z(N,N) = -v_r Z_{N,N}$$

$$Z(N) = Z_N$$

2.1.4 Tables and discussion

The shape of the distribution probability of the vector joining the chain ends has also been studied,[17] and, for large values of N, the validity of the scaling law

$$P_N(\vec{r}) = \frac{1}{(X_N)^d} \mathrm{f}(r/X_N) \tag{4.2.15}$$

(where $X_N^2 \equiv R_N^2/d \propto N^{2\nu}$), has been verified.

The chains that have been studied are not large enough to determine the behaviour of $\mathrm{f}(x)$ for small values of x. For rather short chains, Domb[17] observed that $P_N(\vec{r})$ has a depression at the origin, and this is compatible with fact that the function $\mathrm{f}(x)$ describing the asymptotic behaviour, vanishes at the origin, as will be shown later.

For large values of x, the behaviour of $\mathrm{f}(x)$ can be measured more easily, and one can verify that it is compatible with Fisher's predictions (see Chapter 3, Section 3.3).

A set of results appears in Tables 4.1 and 4.2. The exponents are given as usual in the form of fractions. In fact, many people believe that, in a general way, the critical exponents are fractional numbers; at the time of writing (1988), this belief

Table 4.1. Definition equations

Quantities	Symbols and equations
Space dimension	d
Number of chains	$Z_N \simeq Z_\square N^{\gamma-1}\mu^N$
Number of closed chains	$U_N \simeq U_\square N^{\alpha-2}\mu^N$
Mean square distance between chain ends	$R_N = (\langle\langle(\vec{r}_N - \vec{r}_0)^2\rangle\rangle)^{1/2} \propto N^\nu$
Mean size of a polymer	$X_N = (\langle\langle(\vec{r}_N - \vec{r}_0)^2\rangle\rangle/d)^{1/2}$
Distribution probability of the vector joining chain ends ($N \gg 1$)	$P_N(\vec{r}) = \frac{1}{(X_N)^d}\mathrm{f}(r/X_N)$
$r \gg R_N$	$\mathrm{f}(x) \propto x^k \exp[-\mathrm{D}x^\delta]$
$r \ll R_N$	$\mathrm{f}(x) \propto x^\theta \quad x \ll 1$
Number of polymer per unit volume of solution	C
Osmotic pressure	$\Pi\beta = C + \sum_{j>1} A_j(N)C^j$
Second virial coefficient	$A_2 \propto N^{2-\alpha'}$

Table 4.2. Exponents for self-avoiding chains.

(d = space dimension, $\mathfrak{k}$ = coordination number of the lattice; references are given in square brackets)

Lattice	d	$\mathfrak{k}$	μ[11]	γ	ν	α	α'	δ	κ
Linear	1	2	1	1	1	$-\infty$	1	—	—
Hexagonal (3)	2	3	1.8478	$\frac{4}{3}$[19]	—	$\frac{1}{2}$[19]	—	—	—
Square	2	4	2.6385	$\frac{4}{3}$[11]	$\frac{3}{4}$[10]	$\frac{1}{2}$[9]	—	4[17]	$\frac{2}{3}$[1]
Triangular	2	6	4.1520	$\frac{4}{3}$[11]	$\frac{3}{4}$[4]	$\frac{1}{2}$[6]	$\frac{1}{2}$[7]	4[18]	$\frac{2}{3}$[18]
Diamond	3	4	2.8792	$\frac{7}{6}$[11]	$\frac{3}{5}$[20]	—	—	—	—
Simple cubic	3	6	4.6838	$\frac{7}{6}$[11]	$\frac{3}{5}$[10]	$\frac{1}{4}$[9]	$\frac{1}{4}$[7]	$\frac{5}{2}$[19]	$\frac{1}{3}$[1]
Body-centred cubic	3	8	6.5295	$\frac{7}{6}$[11]	$\frac{3}{5}$[10]	$\frac{1}{4}$[9]	$\frac{1}{4}$[7]	$\frac{5}{2}$[1]	$\frac{1}{3}$[1]
Face-centred cubic	3	12	10.0355	$\frac{7}{6}$[11]	$\frac{3}{5}$[4]	$\frac{1}{4}$[6]	$\frac{1}{4}$[7]	$\frac{5}{2}$[18]	$\frac{1}{3}$[18]
Quasi-Brownian chain (2)	4			1	$\frac{1}{2}$	0		2	0
Brownian chain (1)	>4			1	$\frac{1}{2}$	0		2	0

(1) Brownian chains can be defined for any value of d.

(2) Quasi-Brownian chains correspond to a limiting case and differ from Brownian chains only by logarithmic factors.

(3) For the triangular and hexagonal lattices, which are dual, one verifies that $\mu_T + \mu_H = 5.9998$, but apparently, it has not been proved (1990) that $\mu_T + \mu_H = 6$.

is not supported by any serious argument, but this is true for the exactly soluble models (see, for instance, Chapter 12, Section 4).

The coefficient μ is not a universal quantity; obviously, it depends on the lattice and we see that for all self-avoiding chains

$$\mathfrak{k}/2 < \mu < \mathfrak{k} - 1,$$

as could be expected (see Chapter 2, Section 3.1).

On the contrary, we can verify that the critical exponents, γ, ν, α, α', δ, and κ depend only on the lattice dimensionality. Thus, the universality property of the critical exponents is well established.

Incidentally, let us observe that the same exponents can be obtained with partially repulsive chains provided that the repulsive part is dominant. Thus, Watson[21] showed that purely repulsive chains drawn on a honeycomb lattice can be transformed into chains drawn on the corresponding kagome lattice (see Fig. 4.2), in such a way that the interaction potentials for the new chains are

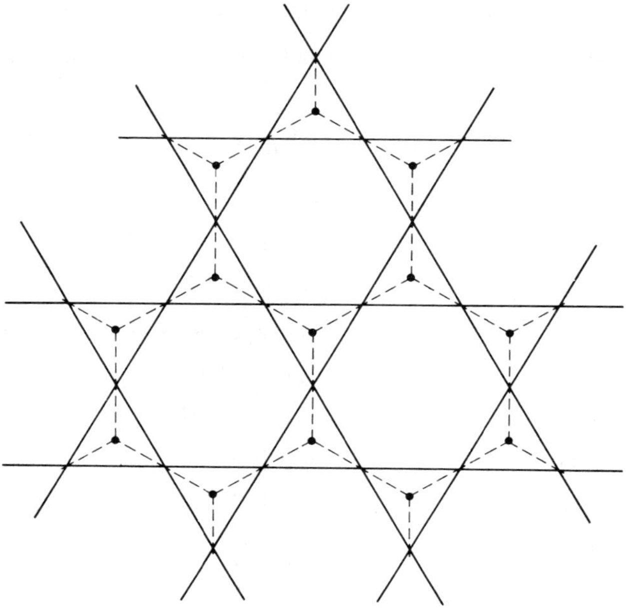

Fig. 4.2. Figure illustrating the correspondence between honeycomb lattice (dashed line) and kagome lattice (solid line).

partially attractive. Thus, this correspondence, which is exact, illustrates beautifully the universality principle since, by construction, the critical exponents coincide for both types of chain.

The exponents γ and ν can be considered as fundamental exponents from which all the other listed exponents can be deduced. Thus renormalization theory (see forthcoming chapters) shows that one should find

$$\alpha = 2 - \nu d, \tag{4.2.16}$$

and, if we consider the precision of the results, we may say that this relation is not incompatible with the values given for α and ν in Table 4.2 [the value of ν which is given for $d = 3$ seems slightly too large; see (4.2.4)].

Moreover, as we shall see later, the osmotic pressure obeys scaling laws which lead to the result

$$\alpha' = 2 - \nu d \tag{4.2.17}$$

and the theoretically predicted equality $\alpha = \alpha'$ is well verified with the values of α and α' given in Table 4.2.

The exponent θ which defines the behaviour of $P_N(\vec{r})$ for small values of r, can be obtained in terms of γ, ν, and α, as was previously shown [see (2.3.12)]

$$\theta\nu = \gamma + 1 - \nu d - \alpha, \tag{4.2.18}$$

or by using (4.2.7)

$$\theta = (\gamma - 1)/\nu, \tag{4.2.19}$$

which gives [see (4.2.3) and (4.2.4)]

$$d = 2 \qquad \theta = 0.44$$
$$d = 3 \qquad \theta = 0.275. \tag{4.2.20}$$

The exponent δ is directly related to ν by Fisher's theorem (see Chapter 3, Section 3.3)

$$\delta = 1/(1 - \nu), \tag{4.2.21}$$

a relation which is verified for the values of Table 4.2.

We shall also find (see Chapter 13) that κ can be expressed in terms of γ and ν

$$\kappa = (1 - d/2 + \nu d - \gamma)/(1 - \nu) \tag{4.2.22}$$

and this relation is also verified when each quantity is replaced by the experimental value given in Table 4.2 (for $d = 2$, $\kappa = 2/3$ and for $d = 3$, $\kappa = 1/3$).

2.2 The Monte Carlo method

2.2.1 The basis of the method

Originally, the Monte Carlo method was applied to polymers and developed by Wall *et al.*[22] This method is a process used to construct long random chains which afterwards serve to evaluate interesting quantities.

The Monte Carlo method and the exact-counting method are complementary, and each method has advantages and drawbacks.

The Monte Carlo method enables one to study much longer chains than the other method, and therefore, with it, it is easier to reach the asymptotic limit. Moreover, with the Monte Carlo method, it is possible not only to produce chains on a lattice, but also to construct chains of balls (necklaces) with excluded volume in three-dimensional space.

All these advantages are counterbalanced by a serious defect. In itself, the Monte Carlo method is not able to give all possible chains with the same probability, and compact chains appear more frequently than stretched chains (see Fig. 4.3). In fact, let us assume that, using this method, we want to construct a self-avoiding chain $\mathfrak{C}$ with N links on a lattice of coordination number $\mathfrak{k}$. The points on the chain will be denoted by $M_0, M_1, \ldots, M_N$. Arriving at point M_i, we find out that there are $\mathfrak{k}_i$ possible choices for M_{i+1} ($\mathfrak{k} - 1 \geqslant \mathfrak{k}_i \geqslant 0$). The integer $\mathfrak{k}_i$ is not constant but depends on the configuration of the first i links of the chain, and we shall assume that $\mathfrak{k}_i > 0$ (we are not at a dead end); otherwise, in principle, we have to start again from the beginning. Thus, choosing M_{i+1} at random among the $\mathfrak{k}_i$ possible sites, we take one step forward, and so on up to M_N. We verify that the probability of obtaining a chain $\mathfrak{C}$ with N links is

(a)

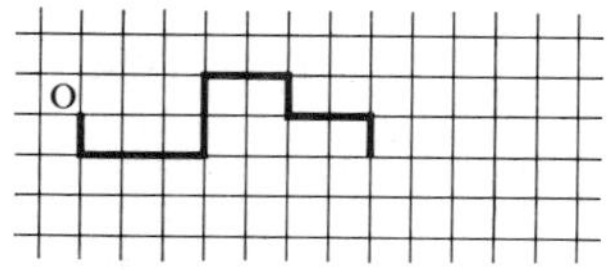

(b)

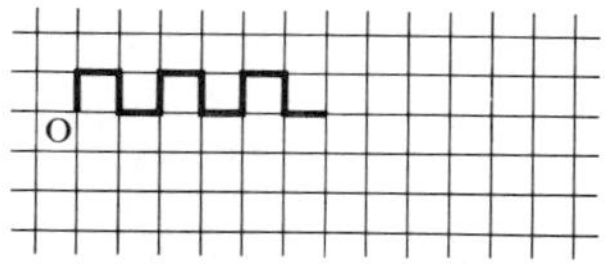

Fig. 4.3. Values of $W(\mathfrak{C})$ for chain with 12 links:
(a) for the less compact chain

$$1/W(\mathfrak{C}) = 4 \times 3^{11};$$

(b) for the more compact chain

$$1/W(\mathfrak{C}) = 4 \times 3^{6} \times 2^{5}.$$

proportional to the factor

$$W(\mathfrak{C}) = \prod_{0}^{N-1} (1/\mathfrak{k}_i) \tag{4.2.23}$$

($\mathfrak{k}_0 = \mathfrak{k}$) (see Fig. 4.3)

Thus, if a large number of chains is made, the number of chains obtained is proportional to $W(\mathfrak{C})$. But, when we calculate the statistical properties of chains, we must assign the same weight to each chain. Therefore, if we want to calculate the mean value of any function $\mathcal{O}(\mathfrak{C})$ of the configurations of a chain by using the Monte Carlo method, we must introduce statistical corrections. For this purpose, we have to assign the weight $[W(\mathfrak{C})]^{-1}$ to each chain $\mathfrak{C}$ produced with this method, and we note that the weight corresponding to each chain can easily be calculated. Thus, we may write

$$\langle\!\langle \mathcal{O} \rangle\!\rangle = \frac{\sum_{\mathfrak{C}} [W(\mathfrak{C})]^{-1} \mathcal{O}(\mathfrak{C})}{\sum_{\mathfrak{C}} [W(\mathfrak{C})]^{-1}}. \tag{4.2.24}$$

Omitting these statistical corrections would amount to overestimate the influence of compact chains.

It appears from Fig. 4.3 that the fluctuations of $W(\mathfrak{C})$ are large. However, this is not always the case; for instance, when the experiment is performed with a lattice which is not really compact (like the diamond lattice, or in a space of dimension superior or equal to three, or when the effective interaction is weak). In this case, one remains far from the asymptotic limit because the chain has not 'felt' the excluded volume constraints.

On the other hand, when the statistical corrections become large, the number of chains which have a non-negligible weight may be rather small, and, in this case, we have poor statistics, even if a large number of chains have been constructed. To obtain good results one must avoid extreme situations.

The method described above is the pure Monte Carlo method. With this method, it may happen that after getting i links, one cannot proceed further ($\mathfrak{k}_i = 0$); in this case, the whole process must be repeated. In order to save calculation time, Wall and Erpenbeck[23] invented a regeneration method which applies whenever it is impossible to go on with the construction of a chain. When applying this technique, one chooses once for all an integer p which defines the regeneration order, and with each integer i one associates the integer r_i, defined by

$$i = pq_i + r_i \qquad 0 < r_i \leqslant p \qquad (q_i = \text{integer})$$

$(r_i - 1)$ is the remainder of the division of $(i - 1)$ by p.

Then, one starts constructing the chain. If after i links, it is impossible to go on, one erases the last r_i links, and starts again by keeping the beginning of the chain. Thus, with this regeneration technique, it is possible to construct longer chains.

It is generally assumed that such a regeneration process does not perturb the statistics. This is not quite true. With this technique, it is easier to make compact chains, and therefore more statistical corrections are needed to obtain a correct result.

2.2.2 Results

In spite of a few drawbacks indicated above, the Monte Carlo method has furnished interesting results and a good number of them were reviewed in 1976 by McKenzie.[1]

Self-avoiding chains with a number of links of the order of one hundred are commonly made, but it has been possible to construct (non-trivial) chains with one thousand links.

First, these simulations lead to an evaluation of the coefficient μ (see Table 4.1 above) and the results thus obtained are in agreement with those given by the exact-counting method. Thus, for the square lattice, the value corresponding to the result of Wall and Erpenbeck[23] is $\mu = 2.64$, which is practically equal to the value $\mu = 2.6385$ obtained by exact counting.

In the same way, the values of ν obtained for chains on various lattices are close to those obtained by using other methods. For two-dimensional lattices,[24] they belong to the interval $0.59 \leqslant \nu \leqslant 0.61$; for three-dimensional lattices to the interval $0.59 \leqslant \nu \leqslant 0.61$.

The Monte Carlo method has been used to study fairly different situations, depending on several parameters. Calculations for lattices have been performed by adding a second-neighbour interaction.[25, 26, 27] Furthermore, various chain models with excluded volume in three-dimensional space and correlations between the directions of successive links have been studied.[28,29]

In many cases, it seems that the value of ν depends on the parameters, and in particular, when the angle between successive bonds is a constant of the model, the apparent value of ν is rather sensitive to the variation of this angle. Thus, a superficial view of certain results could lead to doubts concerning the validity of the universality principle. Nevertheless, this point of view does not seem to be correct and it has been criticized by various authors, for instance Mazur, McCrackin,[25] and Domb.[30]

Nowadays, it is generally recognized that as long as the chains remain globally repulsive, the exponent ν remains constant, independent of the interaction parameters. Still, the asymptotic limit might be approached only very slowly. In this case, moderately long chains may have a quasi-Brownian behaviour, whereas very long chains have an excluded volume behaviour, and the passage from one regime to the other one is continuous. Thus, for finite values of N, it is convenient to define an effective (apparent) exponent $\nu(N)$

$$\nu(N) = \frac{\delta(\ln R_N)}{\delta(\ln N)} \tag{4.2.25}$$

This effective exponent varies from 1/2 to ν when N goes from one to infinity, and very often this is the exponent that the computer experiments give rather than ν.

Actually, we must observe that this principle of stability and universality of the critical exponents is very general and applies to a great number of critical phenomena, as was shown by Griffiths.[31] As we shall see later, the critical exponents are related to the existence of fixed points for the Hamiltonians describing the systems under study. But, in general, these fixed points are isolated, and it is only in very special circumstances that there exist lines of fixed points, along which the critical exponents may vary in a continuous manner.

In reality, the results obtained with the Monte Carlo method are generally in good agreement with those given by exact counting. Thus, for instance, the values of $\aleph = 6 R^2_{G,N}/R^2_N$ which Wall and Erpenbeck[32] found by using the Monte Carlo method, are $\aleph_{WE} = 6(0.157)$ for $d = 3$, and $\aleph_{WE} = 6(0.145)$ for $d = 2$, whereas the values obtained later by Domb and Hioe[14] by exact counting are $\aleph_{DH} = 6(0.155)$ for $d = 3$ and $\aleph_{DH} = 6(0.140)$ for $d = 2$ (see Section 2.2).

Calculations of the end-to-end distribution have been performed by many authors but, in general, these calculations are not very useful, owing to a lack of precision in the results, or to a prejudice, for instance, when authors have represented their results in unacceptable analytic forms. We can, however, mention a quite comprehensive study of the subject by Mazur and McCrackin.[25] Among more recent work, we note articles by Havlin and Ben Avraham,[33] by Aragão de Carvalho and Caracciolo[34], and by Webman, Lebowitz, and Kalos.[35] These new studies confirm the previously known results and they are in good agreement with the predictions of renormalization theory concerning the values of the critical exponents ν and γ.

Very interesting results were found in 1971 by Renardy,[36] who studied the probability distribution of the end-to-end vector for a two-dimensional triangular lattice. He verified the validity of the scaling law

$$P_N(\vec{r}) = \frac{1}{\langle\langle r \rangle\rangle^2} f(r/\langle\langle r \rangle\rangle) \tag{4.2.26}$$

with $\langle\langle r \rangle\rangle \propto N^\nu$ and $\nu = 3/4$ (see Fig. 4.4), and for $f(x)$ he found the following results.

For large distances

$$f(x) \propto x^{-2.3 \pm 0.4} e^{-x^4}, \tag{4.2.27}$$

in agreement with Fisher's predictions $\left(\delta = \dfrac{1}{1-\nu} \simeq 4\right)$.

For short distances,

$$f(x) \propto x^{0.49 \pm 0.06}.$$

Thus, the calculation by Renardy indicated very clearly that $f(x)$ vanishes at the origin, a very important result from a theoretical point of view, as we shall see in following chapters. In fact, it was later shown[37] that the corresponding exponent θ is related to γ and ν by the relation $\theta = (\gamma - 1)/\nu$. The values $\gamma = 4/3$ and $\nu = 3/4$ which appear in Table 4.2 indicate that the value $\theta = 4/9 \simeq 0.444$ can be

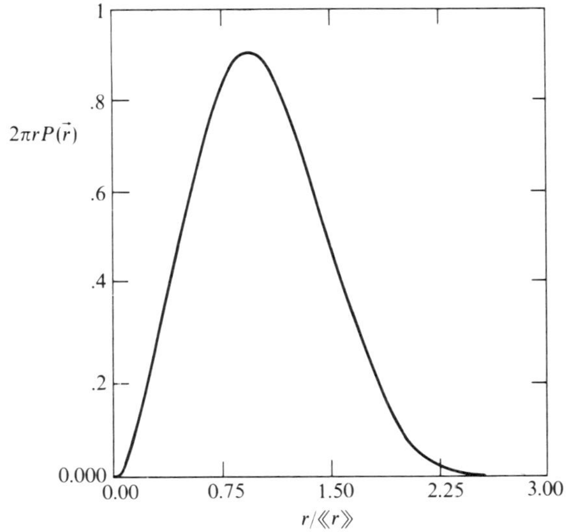

Fig. 4.4. Shape of $2\pi r P(\vec{r})$ for a triangular lattice (Monte Carlo method. Unpublished result by J.F. Renardy (1971)).

assigned to θ, in good agreement with Renardy's calculation, if one considers the precision of the results.

Finally, Bellemans and Janssens[38] used the Monte Carlo method to calculate the second virial coefficient for self-avoiding chains, on a simple cubic lattice. Their estimate of the critical exponent gives $\alpha' = 0.28$ in agreement with the exact counting method (the theory gives $\alpha' = 2 - \nu d \simeq 0.23$).

3. PARTIALLY ATTRACTIVE CHAINS

Random self-avoiding chains on a lattice simulate the behaviour of polymers in good solvent.

The only interaction in this model is a link–link repulsion; it is short-range and of the order of a lattice edge. Actually, this approximation is a rather imperfect representation of reality. The true interaction contains simultaneously, a short-range repulsive interaction, or hard core, and an attractive part whose range is a little longer and which results from van der Waals forces. Experimentally, the fact that the mixing of the polymer with the solvent is endothermic is a manifestation of these attractive forces. The shape of the true potential is indicated in Fig. 4.5.

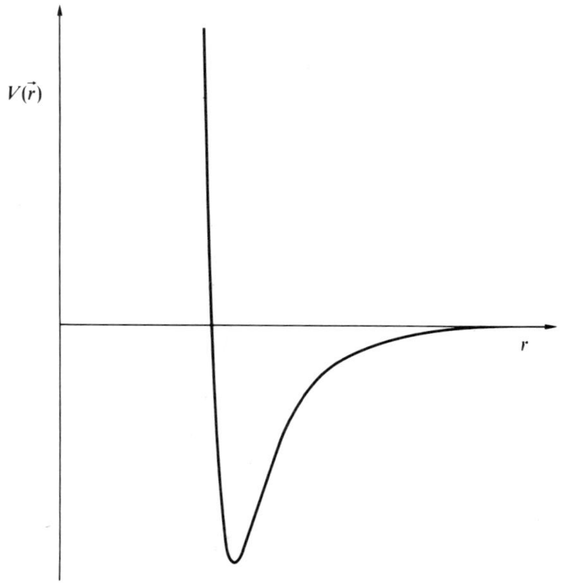

Fig. 4.5. Form of the potential $V(\vec{r})$ between links (hard core and van der Waals forces); the factor $\beta = 1/k_B T$ is incorporated in the potential.

When the temperature of a solution decreases, the solvent does not remain as good as it was and the importance of the attractive forces increases. An investigation of the temperature-dependence of the covolume (the excluded volume)

$$v_e = \int d^3r\{1 - \exp[-V(\vec{r})]\}, \tag{4.3.1}$$

where $V(\vec{r})$ may be considered as proportional to $\boldsymbol{\beta}$ (Boltzmann's law), shows the existence of a temperature T_F for which the covolume vanishes.* For $T > T_F$, we have $v_e > 0$: this is the good solubility region. For temperatures close to T_F, it is of course impossible to neglect the attractive part, and we guess that on average the configurations of the chains depend strongly on temperature in this domain. Orr[39] drew the attention to this problem as early as 1947, and to study configurations of chains on a lattice he proposed a method, the basic rules of which are as follows:

1. All chain intersections are forbidden.
2. An attractive interaction energy J is associated with each pair of points that are non-adjacent on the chain but nearest neighbours on the lattice. This estimation of the effect of the potential $V(\vec{r})$ is justified by the absence of long-range forces. According to Domb, the first Monte Carlo calculations of this kind were made by Wall and Mazur[40] in 1961. A study in greater depth was carried out by Mazur and McCrackin in 1968.[25] Exact-counting calculations started with W. Orr himself (1945). More detailed calculations were made later on by Rapaport (1974).[38]

3.1 The exact-counting method (partially attractive chains)

For a self-avoiding chain with $N + 1$ points (N links) on a square lattice or a simple cubic lattice, a contact can be defined as two points that are non-adjacent on the chain, and nearest neighbours on the lattice. Now let us consider the self-avoiding chains which have a given origin, N links, p contacts, and the same end-to-end vector $\vec{r}$; let $Z_{N,p}(\vec{r})$ be the number of configurations of these chains. This number has been counted by W. Orr[39] for $N \leqslant 8$ on the square lattice and for $N \leqslant 6$ on the simple cubic lattice.

Moreover, three types of results have been obtained by counting such chains, and these are reviewed below.

3.1.1 Correlations between chain ends

Fisher and Hiley[8] calculated mean square end-to-end distances $\langle\langle(\vec{r}_N - \vec{r}_0)^2\rangle\rangle$ in terms of the parameter $w = \exp(-\boldsymbol{\beta}J)$ by using data published by Orr. The

* A better definition of the Flory temperature T_F is given in Chapter 14.

basic quantities are the numbers $Z_{N,p}(\vec{r})$ defined above (N = number of links, p = number of contacts, $\vec{r}$ = end-to-end vector). By definition, we have

$$R_N^2 = \langle\langle(\vec{r}_N - \vec{r}_0)^2\rangle\rangle = \sum_{\vec{r}}\sum_{p} r^2 Z_{N,p}(\vec{r}) \Big/ \sum_{\vec{r}}\sum_{p} Z_{N,p}(\vec{r}). \tag{4.3.2}$$

The calculations have been made for values of w belonging to the interval 0 and 2.5. For large values of N, we expect to find for R_N a relation of the form

$$R_N \propto N^{\nu(w)} \tag{4.3.3}$$

Now $\nu_N(w)$ can be defined by the relation

$$2\nu_N(w) = N[R_{N+1}^2(w)/R_N^2(w) - 1] \tag{4.3.4}$$

and, obviously, we have

$$\lim_{N\to\infty} \nu_N(w) = \nu(w).$$

The chains being rather short, the exponents $\nu_N(w)$ fluctuate strongly with the parity of N. Fisher and Hiley[8] remedied this defect by using the values

$$\bar{\nu}_N(w) = \tfrac{1}{2}[\nu_N(w) + \nu_{N+1}(w)]$$

where dependence with respect to N is smoother. A graphical representation of $\bar{\nu}_N(w)$ in terms of $1/N$ leads to a simple determination of the extrapolated value

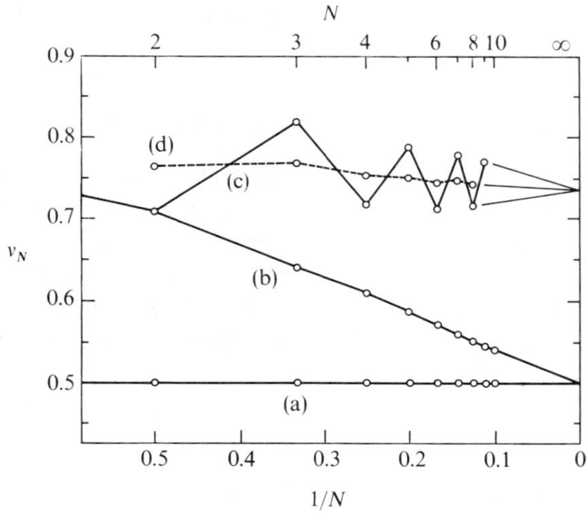

Fig. 4.6. Successive estimations of ν_N defined by $\langle\langle(\vec{r}_N - \vec{r}_0)^2\rangle\rangle = \mathscr{A} N^{2\nu_N}$ for chains drawn on a square lattice (from Fisher and Hiley.[8]): (a) without constraints; (b) without immediate returns; (c) without self-intersections. The dashed curve (d) corresponds to the average $\frac{1}{2}(\nu_N + \nu_{N+1})$ of case (c).

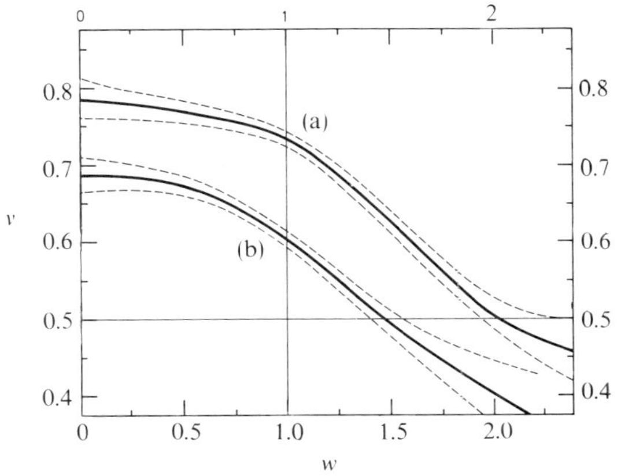

Fig. 4.7. Variation of the mean exponent with the interaction parameter: (a) for the square lattice ($d = 2$); (b) for the simple cubic lattice ($d = 3$). The dashed line gives the uncertainty range. (From Fisher and Hiley.[8])

(see, for instance, Fig. 4.6)

$$v(w) = \lim_{N\to\infty} \bar{v}_N(w).$$

It appears that, for a simple cubic lattice, there exists a value $w_F = 1.48 \pm 0.06$ for which $v(w) = 1/2$, which indicates that, in this case, the chain is Brownian (see Fig. 4.7). The corresponding temperature is given by the equality

$$k_B T_F / J = -0.26,$$

and for temperatures inferior to T_F, we have $v(w) < 1/2$. Besides, for the square lattice, Fig. 4.7 shows, in a rough way, that there exists also a temperature T_F, but larger than the preceding one.

3.1.2 Thermodynamical functions

(*a*) *Specific heat*

The number of self-avoiding chains with N links and p contacts is given by the sum

$$Z_{N,p} = \sum_{\vec{r}} Z_{N,p}(\vec{r}). \tag{4.3.5}$$

Therefore, the partition function of the chains is

$$Z_N(w) = \sum_{p=0}^{\infty} w^p Z_{N,p} \tag{4.3.6}$$

where $w = \exp(-\beta J)$

Rapaport[41] determined the specific heat of chains drawn on a face-centred cubic lattice (and also on a simple cubic lattice) for $N < 13$. The specific heat can be defined by the equality

$$\mathcal{C}_N(w) = \beta^2 \frac{\partial^2 \ln Z_N}{\partial \beta^2} = \beta^2 J^2 w \frac{\partial}{\partial w}\left(w \frac{\partial}{\partial w} \ln Z_N \right), \tag{4.3.7}$$

and it is convenient to introduce the reduced specific heat

$$\bar{\mathcal{C}}_N(w) = \mathcal{C}_N(w)/\beta^2 J^2 = \langle\langle p^2 \rangle\rangle - \langle\langle p \rangle\rangle^2. \tag{4.3.8}$$

$\bar{\mathcal{C}}_N(w)$ is a function of w and, for each N, this function has a maximum $\bar{\mathcal{C}}_{N,\max}$ for a value w_N of w (see Fig. 4.8), which decreases slowly with N. Thus, for $N = 10$, we have

$$k_B T_N/J = -(\ln w_N)^{-1} = -1.005.$$

The quantity, $\mathcal{C}_{N,\max} = \mathcal{C}_N(w_N)$ increases with N (see Fig. 4.9) and this is quite normal. The question which arises is whether $\mathcal{C}_{N,\max}/N$ has a finite limit when $N \to \infty$, or whether it diverges. Unfortunately, Rapaport's results do not provide any answer to this question.

(b) *Residual entropy of the condensed state*

When the attraction between links increases, the energy of the system made of N links, diminishes and reaches a minimal value. The configurations that are associated with this minimum have been discussed by W. Orr.[39] They are compact and are called globules (see Fig. 4.10). The energy differences result only from surface effects, and configurations with the same surface area are degenerate.

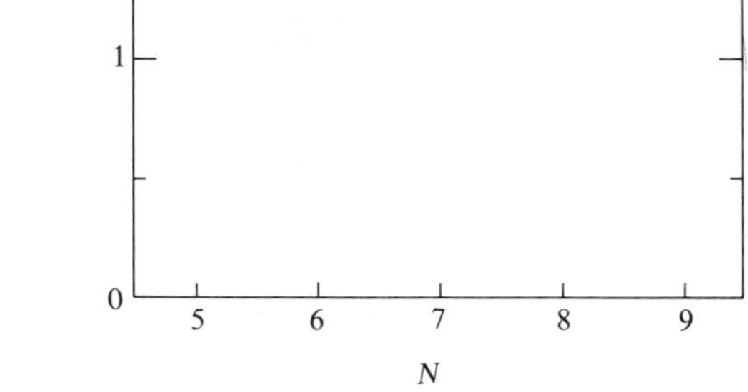

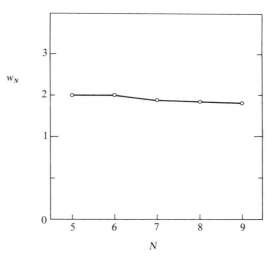

Fig. 4.8. Values of w_N corresponding, for each value of N, to the maximum of $C_N(w)$, in the case of a face-centred cubic lattice. Thus $C'_N(w_N) = 0$. (From Rapaport.[41])

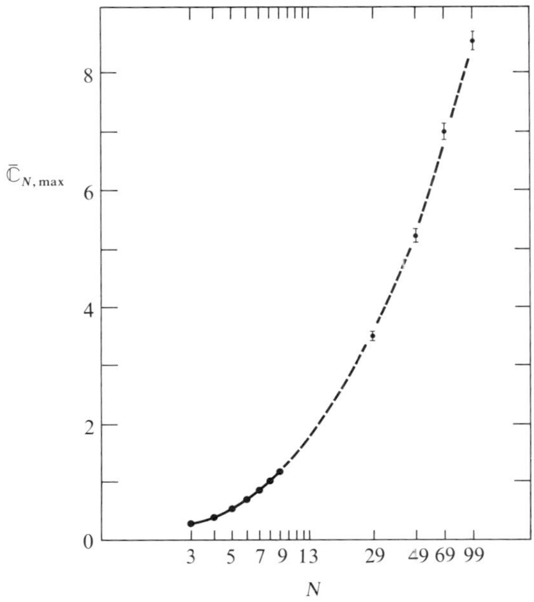

Fig. 4.9. Values of the maximum of the reduced specific heat for a face-centred cubic lattice. The reduced specific heat is defined by

$$\bar{C}_N(w) = C_N(w)/\beta^2 J^2 = C_N(w)/(\ln w)^2$$

and $\bar{C}_{N,\max}$ is the maximum of $\bar{C}_N(w)$. Thus $C_{N,\max} = C_N(w_N)$ ($N \leqslant 13$: calculation by Rapaport;[41] $N \geqslant 29$; calculations by Mazur and McCrackin[25])

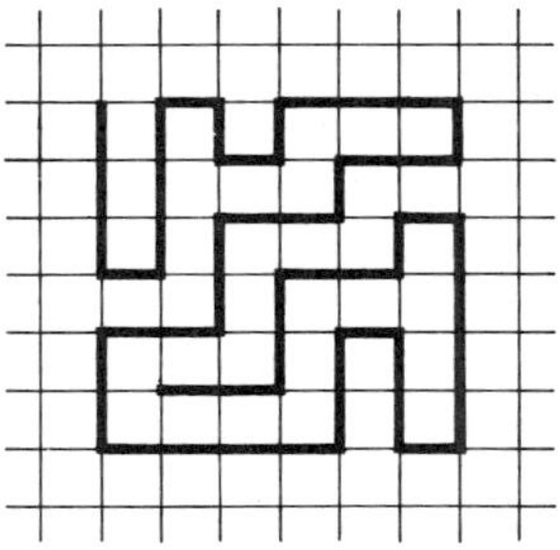

Fig. 4.10. A strongly attractive chain forming a globule.

In a general way, for a self-avoiding chain on a lattice, the average number of contacts per link can be defined by

$$\langle\!\langle p \rangle\!\rangle/N = N^{-1}\left(\sum_p p\, w^p Z_{N,p} \Big/ \sum_p w^p Z_{N,p}\right). \tag{4.3.9}$$

Obviously, this number is smaller than $\frac{1}{2}(\mathfrak{f} - 2)$ where $\mathfrak{f}$ is the lattice connectivity and has this limiting value when $w \to \infty$ and $N \to \infty$. When $w \to \infty$, i.e. when the chain is condensed and has a globular shape, the number of contacts reaches the maximum value $p(N)$ and

$$\lim_{N \to \infty} N^{-1} p(N) = \tfrac{1}{2}(\mathfrak{f} - 2). \tag{4.3.10}$$

In the same way, the residual entropy attains the limiting value (see Chapter 11, Section 6)

$$\mathbb{S} = \lim_{N \to \infty} N^{-1} [\ln Z_{N, p(N)}]. \tag{4.3.11}$$

3.2 Monte Carlo method (attractive chains)

Mazur and McCrackin used the Monte Carlo method to study attractive chains drawn on simple cubic or face-centred cubic lattices.

A self-avoiding chain $\mathfrak{C}$ constructed on a lattice by applying the Monte Carlo method, is produced with a probability which is proportional to the factor $W(\mathbf{\Omega})$ corresponding to the chain configuration $\mathbf{\Omega}$ (see Section 2.2). To calculate mean values, we must simultaneously correct this effect and take the attractive interactions into account. With each chain with $p(\mathbf{\Omega})$ contacts, is associated a weight $[W(\mathbf{\Omega})]^{-1}\, w^{p(\mathbf{\Omega})}$ where $w = \exp[-\beta J]$ (with $J < 0$).

The mean value of any function $\mathcal{O}(\mathbf{\Omega})$ of the chain configurations is given by

$$\langle\!\langle \mathcal{O} \rangle\!\rangle = \frac{\sum_{\mathbf{\Omega}} [W(\mathbf{\Omega})]^{-1}\, w^{p(\mathbf{\Omega})}\, \mathcal{O}(\mathbf{\Omega})}{\sum_{\mathbf{\Omega}} [W(\mathbf{\Omega})]^{-1}\, w^{p(\mathbf{\Omega})}}. \tag{4.3.12}$$

Thus, Mazur and McCrackin calculated a few quantities in terms of $\ln w$. The link numbers N of the chains which they considered belong to the range $0 < N < 100$ and differ by factors of 10. For each N, the number of constructed chains is of the order of 20 000.

Mazur and McCrackin calculated the mean square distance between end points and for the largest values of N, they represented it in the form

$$\langle\!\langle (\vec{r}_N - \vec{r}_0)^2 \rangle\!\rangle \simeq A N^{2\nu(w)}.$$

They observed that the effective exponent $\nu(w)$ decreases steadily when w increases. For $w = 1$, i.e. in the absence of attractive interaction $\nu(w) \simeq 0.60$. When w increases, $\nu(w)$ decreases and for $w = w_F$ it takes the value $\nu = 1/2$ which defines the Flory point. The corresponding values of w are given by

$\ln w_F = 0.30$ for the simple cubic lattice

$\ln w_F = 0.14$ for the face centred cubic lattice.

When w decreases beyond w_F, the exponent $\nu(w)$ quickly diminishes and becomes even smaller than the value $\nu = 2/3$ which corresponds to a compact globule, but this fact is not significant; it shows only that for the corresponding values of $\nu(w)$, one remains far from the asymptotic regime. Besides, it is expected that in the asymptotic regime $\nu(w)$ vary discontinuously and not continuously, with respect to w.

Mazur and McCrackin also calculated the distribution function of the distance between chain ends, but, unfortunately, their result does not allow us to get a very clear idea of this function in the asymptotic domain.

More recently, Kremer, Baumgärtner, and Binder[42] made a rather precise and complete study of the behaviour of self-avoiding chains on a diamond lattice with an interaction J (attractive: $J < 0$) between nearest neighbours. For the transition temperature, they found

$$k_B T_F / J = 1/\beta_F J = -\,2.25 \pm 0.05.$$

They confirmed that the specific heat $\mathbb{C}_N$ (where $\mathbb{C}_N \equiv \mathbb{C}_N(w)$, see (4.3.7) is a quantity which has a peak in the collapse zone (see Fig. 4.11). Unfortunately, the values of the ratio $\mathbb{C}_{N,\max}/N$ strongly fluctuate with the link number (see Fig. 4.12) and from the data it is not possible to draw conclusions concerning the limit of this quantity when $N \to \infty$. Apparently, one remains far from the asymptotic limit. Moreover, the theory predicts that $\mathbb{C}_{N,\max}/N$ is only slightly divergent as $\ln N$ (see Chapter 14, Section 6.8.8) and such a divergence is not easily detectable.

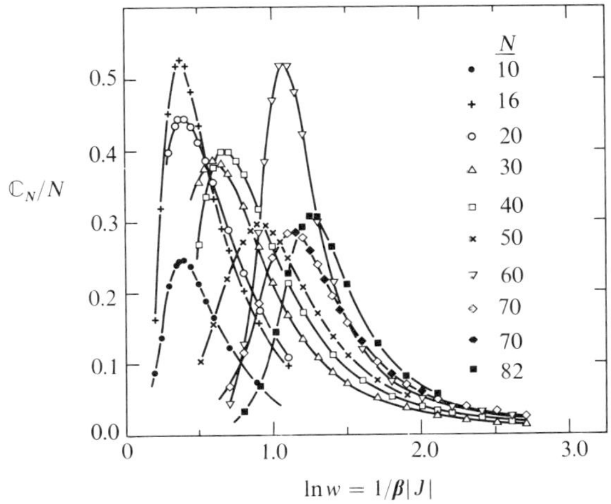

Fig. 4.11. Curve showing how $\mathbb{C}_N/N$ varies with temperature $T \propto 1/\beta J$, for various numbers of links. The peak clearly appears but the curves are very different from one another. (From Kremer *et al.* [43]).

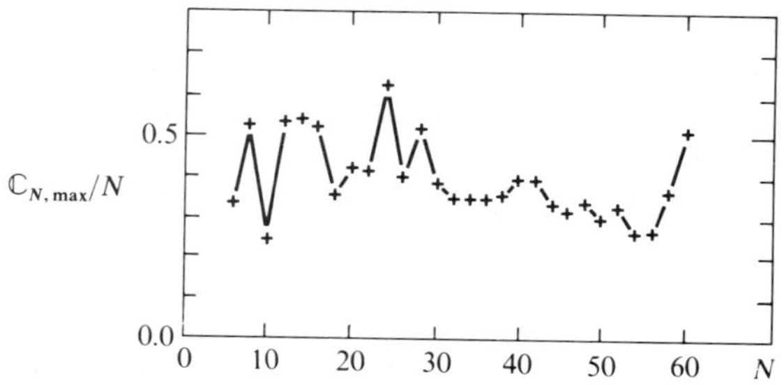

Fig. 4.12. Plot of $\mathbb{C}_{N,\max}/N$ against N. The results are erratic. (From Kremer, *et al.*[42])

REFERENCES

1. McKenzie, D.S. (1976). Polymers and scaling. *Phys. Rep.* **27C** (2), 36.
2. Domb, C. (1970). *Adv. Phys.* **19**, 339.
3. Sykes, M.F. (1961). *J. Math. Phys.* **2**, 52.
4. Martin, J.L. and Watts, M.G. (1971). *J. Phys.* A: *Gen. Phys.* **4**, 456.
5. McKenzie, D.S. and McKenzie, S. (1975). *J. Phys.* A: *Math. Gen.* **8**, 1755.
6. Martin, J.L., Sykes, M.F., and Hioe, F.T. (1967). *J. Chem. Phys.* **46**, 3478.
7. McKenzie, D.S. and Domb, C. (1967). *Proc. Phys. Soc.* **92**, 632. (Certain formulae in this article contain errors which do not affect the main results.)
8. Fisher, M.E. and Hiley, B.J. (1961). *J. Chem. Phys.* **34**, 1253.
9. Sykes, M.F., Guttmann, A. J., Watts, M.G., and Roberts, P.D. (1972). *J. Phys.* A: *Gen. Phys.* **5**, 653.
10. Watts, M.G. (1974). *J. Phys.* A: *Math. Nucl. Gen.* **7**, 489.
11. Watts, M.G. (1975). *J. Phys.* A: *Math. Gen.* **8**, 61.
12. Gaunt, D.S. and Guttmann, A.J. (1974). In *Phase transitions and critical phenomena*, Vol. 3 (eds. C. Domb and M.S. Green), p. 181. Academic Press.
13. Le Guillou, J.C. and Zinn-Justin, J. (1977). *Phys. Rev. Lett.* **39**, 95.
14. Domb, C. and Hioe, F.T. (1969). *J. Chem. Phys.* **51**, 1920.
15. Witten, T.A. and Schäfer L., *J. Phys.* A: *Math. Gen.* **11**, 1843.
    Benhamou, M. and Mahoux, G. (1985). *J. Physique Lett.* **46**, L. 689.
16. Mazur, J. and McIntyre, D. (1975). *Macromolecules* **8**, 464.
17. Domb, C., Gillis J., and Wilmers, G. (1965). *Proc. Phys. Soc.* **85**, 625; **86**, 426.
18. McKenzie, D.S. (1973). *J. Phys.* A: *Math. Nucl. Gen.* **6**, 338.
19. Hioe, F.T. (1967). Thesis, University of London.
20. Domb, C. (1963). *J. Chem. Phys.* **38**, 2957.
21. Watson, P.G. (1974). *Physica* **75**, 627.
22. Wall, F.T., Winder, S. and Gans, P.J. (1963). Monte Carlo methods applied to configurations of flexible polymer molecules. In *Methods in computational physics*, Vol. 1, *Statistical physics*. Academic Press, New York.
23. Wall, F.T. and Erpenbeck, J.J. (1959). *J. Chem. Phys.* **30**, 634.
24. Wall, F.T., Hiller, L.A. Jr. and Atchinson, W.F. (1955). *J. Chem. Phys.* **23**, 913.
25. Mazur, J. and McCrackin, F.L. (1968). *J. Chem. Phys.* **49**, 648.
26. Mark, P. and Windwer, S. (1967). *J. Chem. Phys.* **47**, 708.
    Mazur, J. (1964). *J. Chem. Phys.* **41**, 2256.

27. Wall, F.T., Windwer, S., and Gans, P.J. (1963). *J. Chem. Phys.* **38**, 2220.
28. Fleming, R.J. (1967). *Proc. Phys. Soc.* **90**, 1003.
Fleming, R.J. (1968). *J. Phys.* A. **1**, 404.
29. Loftus, E. and Gans, P.J. (1968). *J. Chem. Phys.* **49**, 3828.
Windwer, S. (1965). *J. Chem. Phys.* **43**, 115.
30. Domb, C. (1974). *Polymer* **15**, 259.
31. Griffiths, R.B. (1970). *Phys. Rev. Lett.* **24**, 1479.
32. Wall, F.T. and Erpenbeck, J.J. (1959). *J. Chem. Phys.* **30**, 637.
33. Havlin, S. and Ben Avraham, D. (1982). *J. Phys.* A: *Math. Gen.* **15**, L317; (1982). *Phys. Rev.* A. **26**, 1728.
34. Aragão de Carvalho, C. and Caracciolo, S. (1983). *J. Physique* **44**, 323.
35. Webman, I., Lebowitz, J.L., and Kalos, M.H. (1981). *Phys. Rev.* A. **23**, 316.
36. Renardy, J.F. (1971). Shape of self avoiding walks on a triangular lattice. Preprint, SPhT. C.E.N. Saclay, Nov. 1971
37. des Cloizeaux, J. (1974). *Phys. Rev.* A. **10**, 1665.
38. Bellemans, A. and Janssens, M. (1974). *Macromolecules* **7**, 809.
39. Orr, W.J.C. (1947). *Trans. Faraday Soc.* **43**, 12.
40. Wall, F.T., Mazur, J. (1961). *Ann. N.Y. Acad. Sci.* **89**, 608.
41. Rapaport, D.C. (1974). *Phys. Lett.* **48A**, 339.
42. Kremer, K., Baumgärtner, A., and Binder, K. (1982). *J. Phys.* A: *Math. Gen.* **15**, 2879.

5

OSMOTIC PRESSURE AND DENSITY

GENESIS

In the past, the basic experiment used to study the general properties of solutions was the measurement of the vapour pressure lowering of the solvent (Raoult 1890). But this method which is, in general, efficient is not really useful when the vapour pressure reduction is small. In such situations, it has been replaced by another method which is quite similar: it consists in measuring the osmotic pressure of the solution with respect to the solvent.

Apparently, the idea that it could be interesting to let a solvent and a solution communicate through a semi-permeable surface, came from biology, and more precisely from the study of natural membranes and from dialysis practices.

Osmotic pressure effects were first observed in cases where they are very strong, i.e. for charged systems (ionized proteins, Donnan effect 1911). However, before being able to study solutions of neutral polymers, the experimentalist had to wait for the manufacture of efficient semi-permeable membranes (collodion). Since that time (1920) up to the present (1988), this technique has been applied with great success to the study of polymers.

In fact, the decisive experimental proof of the macromolecular hypothesis has been given by osmotic pressure measurements in very dilute solutions. A second fundamental contribution is more recent. It deals with the study of systems of strongly overlapping chains. The observations made by Noda and his collaborators (1980) definitely showed that, in this case, the osmotic pressure dependence of the solute concentration follows a universal law; the crossover with the dilute state is also universal. These results gave an experimental proof of the principles which constitute the basis of the modern theory introduced by de Gennes and des Cloizeaux.

Very recently, this investigation method has been extended to two dimensions. Improvements have been made concerning the measurement technique of the surface osmotic pressure of polymers spread on a liquid–air interface. Vilanove and Rondelez thus measured, for the first time in 1980, the characteristic size exponent ν of chains in a two-dimensional space.

Let us note that these conclusions have been obtained after lengthy experimental work, in the course of which all kinds of uncertainties have been patiently eliminated. Thus, Flory reports that an exact determination of polymer molecular masses has been possible only quite recently, because, for a while, the experimentalists did not extrapolate the pressures obtained at finite concentrations carefully enough to zero concentration.

The interpretation which has recently been given concerning the osmotic pressure measurements and which aims at verifying the 'universality' principle, in solutions with strong overlap, did not immediately convince the polymerists because the same experimental data that had confirmed older theories were used to prove the modern theory! It is only when I. Noda obtained a complete set of observations and represented his results in an appropriated form, that the veracity of the modern theory became quite evident.

1. VOLUME AND SURFACE OSMOTIC PRESSURE

Let us consider a volume V occupied by a solution consisting of a solvent and of a solute. The number of solvent molecules is N_0, their concentration is $C_0 = N_0/V$. The number of solute molecules is N_1 and their concentration is $C_1 = N_1/V$ (in general $C_1 \ll C_0$).

Of course, these concentrations are uniform in the volume. Any concentration gradient has a tendency to disappear by molecular diffusion.

However, the volume can be divided into two cells I and II separated by a membrane which is permeable for the solvent but not for the solute, and we can assume that only cell I contains the solute. In this case, a pressure difference appears between both sides of the membrane. This is the 'volume' osmotic pressure.

$$\Pi = P^{\mathrm{I}} - P^{\mathrm{II}}. \tag{5.1.1}$$

Let us note that, at equilibrium, this pressure acts on the membrane. In fact, let us consider the resultant of the forces which are applied to the system (see Fig. 5.1). At equilibrium, this resultant vanishes; thus the reaction of the cylinder holder must be equal to $(P^{\mathrm{I}} - P^{\mathrm{II}}) \times$ (cylinder cross-section). The cylinder itself on which the membrane is fixed, is assumed to be at equilibrium. Thus, the force exerted on the membrane must be equal to the reaction of the holder, which is $\Pi \times$ (cylinder cross-section).

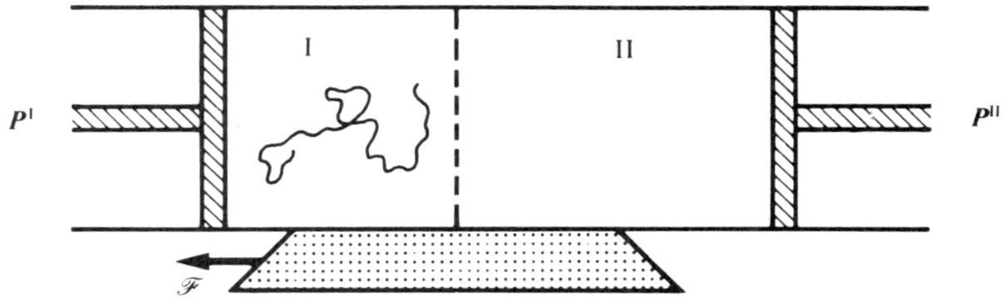

Fig. 5.1. Model system for the definition of the osmotic equilibrium between two cells I and II, with volumes V^{I} and V^{II}, separated by a semi-permeable membrane, which is bound to the bulk of the system. We have a cylinder in which two pistons can slide. The pressures P^{I} and P^{II} respectively are applied to these two pistons. The constituent which cannot permeate through the membrane is in cell I. The reaction of the cylinder holder is the force $\mathscr{F} = (P^{\mathrm{I}} - P^{\mathrm{II}}) \times$ (cylinder cross section)

In an analogous way, it is possible to consider adsorbed polymers on an interface (liquid–air, for instance). On a plane surface of area S, the N molecules have a uniform distribution, and the concentration is defined as the ratio $C = N/S$. We can again divide the surface into two parts I and II, separated by a barrier which the adsorbed polymers cannot cross; of course, the barrier should not be permeable to the solvent molecules since these molecules can always pass under it. In this case, one defines a 'surface' osmotic pressure which is exerted on the barrier.

1.1 Definition

Let us consider a solution containing $m + 1$ components, and a membrane which is not permeable to a few of them; let $N_0, N_1, \ldots, N_m$ be the number of molecules belonging to each species. The indices $\mathscr{A} = 0, \ldots, l$ $(l < m)$ are attributed to the components that are able to cross the membrane under consideration, and in particular the index 0 to the solvent; the indices $\mathscr{A} = l + 1, \ldots, m$ are attributed to the other components.

Let us define the osmotic equilibrium of the solution with respect to the species $\mathscr{A} = 0, \ldots, l$ with the help of the system represented in Fig. 6.1. The semi-permeable membrane is held in a cylinder, the volumes V^{I} and V^{II} are limited by pistons on which the pressures P^{II} and P^{II} are applied. The cell I is filled with the solution which consists of $m + 1$ components; it contains $N^{\mathrm{I}}_{\mathscr{A}}$ molecules belonging to the species $\mathscr{A} = 0, \ldots, m$. In cell II, there are only molecules able to cross the membrane; this cell contains $N^{\mathrm{II}}_{\mathscr{A}}$ molecules belonging to the species $\mathscr{A} = 0, \ldots l$.

The numbers

$$N_{\mathscr{A}} = N^{\mathrm{I}}_{\mathscr{A}} + N_{\mathscr{A}}{}^{\mathrm{II}} \tag{5.1.2}$$

being fixed ($\mathscr{A} = 0, \ldots, m$), the system represented in Fig. 5.1 is in osmotic equilibrium, if the volumes V^{I} and V^{II} are not zero, and if the numbers $N^{\mathrm{I}}_{\mathscr{A}}, N^{\mathrm{II}}_{\mathscr{A}}$ $(A = 0, \ldots, l)$ do not change with time for given pressures P^{I} and P^{II}. Conversely, we can fix the composition in each cell by writing

$$\begin{aligned} C^{\mathrm{I}}_{\mathscr{A}} &= N^{\mathrm{I}}_{\mathscr{A}}/V^{\mathrm{I}} > 0 \\ C^{\mathrm{II}}_{\mathscr{A}} &= N^{\mathrm{II}}_{\mathscr{A}}/V^{\mathrm{II}} = 0. \end{aligned} \tag{5.1.3}$$

These conditions (5.1.3) define a phase of the system in each cell.

The symbol $\{N^{\mathrm{I}}\}$ represents the set of the $N^{\mathrm{I}}_{\mathscr{A}}$ with $\mathscr{A} = 0, \ldots, m$, the symbol $\{N^{\mathrm{II}}\}$, the set of the $N^{\mathrm{II}}_{\mathscr{B}}$ with $\mathscr{B} = 0, \ldots, l$, and the symbols $\{C^{\mathrm{I}}\}$ and $\{C^{\mathrm{II}}\}$ are used in the same manner.

The osmotic pressure Π is the pressure difference $P^{\mathrm{I}} - P^{\mathrm{II}}$ which has to be exerted on both sides of the membrane, in order to ensure the coexistence[1] of phases I and II.

A very similar notation will be used in the case where polymers are adsorbed on a surface; thus the 'surface' osmotic pressure will also be denoted by Π but, in this case, the symbol Π represents a force per unit length.

1.2 Semi-permeable membranes and barriers for monolayers

The measurements of the 'volume' osmotic pressure (in three dimensions) are reliable if the semi-permeable membrane works well, and only in this case. In principle, this membrane is permeable to the 'solvent' species and impermeable to the 'solute' species. In the case of polymer solutes, this effect results from a size difference between the molecules under consideration. The membrane is made of a porous material and the diameter of the pores must be larger than the size of the solvent molecules and smaller than the size of the solvent molecules.

There are two types of semi-permeable membranes: swollen polymer gels (polymer networks) and organic materials (for example, cellulose esters). The interface must be sufficiently permeable to the solvent so as to ensure that equilibrium can be reached after a reasonable amount of time, but it must be quite impermeable to the solute. The equilibration time is sometimes reduced by application of a counter-pressure to the systems. The fact that the polymer chains diffuse rather slowly leads to another experimental difficulty. In general, the equilibrium is obtained by solvent transfer to the cell containing the solution. During this process, the solute concentration does not remain uniform in this cell. In particular, in the vicinity of the membrane, a more dilute region appears, called a stagnant zone. But the osmotic pressure is determined by the polymer concentration in the vicinity of the membrane. The asymptotic regime ($t \to \infty$) is attained only after a long time. It is thus necessary to stir the solution vigorously in order to achieve the right consistency in the solute concentration quickly.

In two dimensions, the measurement of the 'surface' osmotic pressure is made, in principle,* by setting a barrier on a liquid–air interface (Langmuir scales). The molecules of the liquid are free to pass from one side to the other side of the barrier: the polymer molecules are confined, and this confinement technique may seem easy to realize. Actually, some difficulty is experienced because an even spreading of the polymer molecules on the liquid–air interface must be achieved. Unfortunately, there are only few polymer–liquid couples for which the spreading is satisfactory. In general, water is used as the carrying liquid. But if a polymer chain is too hydrophobic like polystyrene, it has a tendency to avoid contact with water and to take a compact configuration. On the contrary, if a polymer chain is too hydrophilic like polymethylmetacrylate, it has a tendency to plunge partially under the interface. However, polyvinyl acetate seems to present balanced hydrophobic and hydrophilic tendencies. By means of this, we

* A more sensitive method consists in measuring the superficial tension lowering (see Section 2).

obtain a kind of two-dimensional solution though polymer overlaps remain possible.

1.3 Osmotic equilibrium, free energy, and chemical potential

Let us express the osmotic equilibrium condition by starting from the thermodynamic functions of the system. The number $N_{\mathscr{A}}$ of molecules being fixed, the system may be in various states i, to which the energy $\mathbb{E}_i$ and the probability p_i are attributed. The system is in equilibrium when the system entropy

$$\mathbb{S} = -\sum_i p_i \ln p_i \tag{5.1.4}$$

is maximum. Keeping the energy $\mathbb{E}$ fixed, we try to maximize $\mathbb{S}$ at constant volume V.

$$\mathbb{E} = \sum_i p_i \mathbb{E}_i. \tag{5.1.5}$$

Let us consider the function $\mathbb{S}(\mathbb{E}, V)$ which, for $\mathbb{E}$ and V fixed, is the maximum of $\mathbb{S}$. The temperature of the corresponding canonical ensemble is defined by introducing the Lagrange multiplier β and the Massieu function

$$\mathbb{S}(\mathbb{E}, V) - \beta \mathbb{E}. \tag{5.1.6}$$

The maximization of this expression with respect to $\mathbb{E}$ gives

$$\beta = \frac{\partial \mathbb{S}}{\partial \mathbb{E}}. \tag{5.1.7}$$

The Helmholtz free energy*

$$\mathscr{F} = \mathscr{F}(V, \beta) \tag{5.1.8}$$

can be associated with $\mathbb{S}$, by introducing the Legendre transformation

$$\begin{cases} \beta\mathscr{F} = \beta\mathbb{E} - \mathbb{S} \\ \mathbb{E} = \dfrac{\partial(\beta\mathscr{F})}{\partial\beta}. \end{cases} \tag{5.1.9}$$

The equilibrium, at V and β fixed, is determined by minimizing the function $\mathscr{F}$

It is also useful to consider the equilibrium of the system at fixed pressure P. For that, we associate the Gibbs free energy

$$\mathbb{G} = \mathbb{G}(P, \beta) \tag{5.1.10}$$

* In general, for solutions which are only slightly compressible, the Helmholtz free energy is not the interesting quantity; the interesting one is the free energy $\mathbb{F}$ which will be defined later.

with the Helmholtz free energy by introducing the Legendre transform

$$\begin{cases} \mathbb{G} = \mathscr{F} + \boldsymbol{P}\boldsymbol{V} \\ \boldsymbol{P} = -\dfrac{\partial \mathscr{F}}{\partial \boldsymbol{V}} \end{cases} \tag{5.1.11}$$

from which we immediately deduce

$$\boldsymbol{V} = \frac{\partial \mathbb{G}}{\partial \boldsymbol{P}}. \tag{5.1.12}$$

The equilibrium, at fixed pressure $\boldsymbol{P}$ is determined by minimizing the function $\mathbb{G}$. Let us now consider the numbers $N_{\mathscr{A}}$ as parameters of the system.

The chemical potential of species $\mathscr{A}$ is defined by the partial derivative

$$\mu_{\mathscr{A}} = \partial \mathbb{G}(\boldsymbol{P}, \{N\})/\partial N_{\mathscr{A}}. \tag{5.1.13}$$

From definition (5.1.11), we can extract the identity

$$\frac{\partial \mathbb{G}(\boldsymbol{P}, \{N\})}{\partial N_{\mathscr{A}}} = \frac{\partial \mathscr{F}(\boldsymbol{V}, \{N\})}{\partial N_{\mathscr{A}}} + \\ + \frac{\partial \mathscr{F}(\boldsymbol{V}, \{N\})}{\partial \boldsymbol{V}} \frac{\partial \boldsymbol{V}(\boldsymbol{P}, \{N\}}{\partial N_{\mathscr{A}}} + \boldsymbol{P} \frac{\partial \boldsymbol{V}(\boldsymbol{P}, \{N\})}{\partial N_{\mathscr{A}}}. \tag{5.1.14}$$

By replacing $\partial/\partial \boldsymbol{V}$ by its value $(-\boldsymbol{P})$, we get

$$\mu_{\mathscr{A}} = \frac{\partial \mathbb{G}(\boldsymbol{P}, \{N\})}{\partial N_{\mathscr{A}}} = \frac{\partial \mathscr{F}(\boldsymbol{V}, \{N\})}{\partial N_{\mathscr{A}}}. \tag{5.1.15}$$

Incidentally, we note that the phase stability implies a concavity of $\mathbb{G}$ directed upwards in the $N_{\mathscr{A}}$ space, i.e. the positivity of the matrix elements

$$\frac{\partial^2 \mathbb{G}(\boldsymbol{P}, \{N\}}{\partial N_{\mathscr{A}} \partial N_{\mathscr{A}}}.$$

Let us now consider phases I and II coexisting on both sides of the semi-permeable membrane. Let $\mathbb{G}^{\mathrm{I}}$ and $\mathbb{G}^{\mathrm{II}}$ be the corresponding Gibbs free energies. The free energy of the system is

$$\mathbb{G} = \mathbb{G}^{\mathrm{I}} + \mathbb{G}^{\mathrm{II}}. \tag{5.1.16}$$

The osmotic pressure and the equilibrium conditions can easily be expressed in terms of the function $\mathbb{G}$. The osmotic pressure is the pressure difference

$$\Pi = \boldsymbol{P}^{\mathrm{I}} - \boldsymbol{P}^{\mathrm{II}}. \tag{5.1.17}$$

For fixed concentrations, one of the pressures $\boldsymbol{P}^{\mathrm{I}}$ or $\boldsymbol{P}^{\mathrm{II}}$ can be considered as a parameter, and the other one is determined by the equilibrium conditions. However, in what follows, we shall adopt another point of view and it will be assumed that, for this problem, the data are, in one hand the number of molecules of each constituent, and, on the other hand, the osmotic pressures $\boldsymbol{P}^{\mathrm{I}}$ and $\boldsymbol{P}^{\mathrm{II}}$.

Thus, let us determine the osmotic equilibrium for given pressures $\boldsymbol{P}^{\mathrm{I}}$, $\boldsymbol{P}^{\mathrm{II}}$ and for given values of the $N_{\mathscr{A}}$. Minimizing the free energy, we obtain the relations

$$\frac{\partial G}{\partial N^{\mathrm{I}}_{\mathscr{A}}} = \frac{\partial G^{\mathrm{I}}}{\partial N^{\mathrm{I}}_{\mathscr{A}}} + \frac{\partial G^{\mathrm{II}}}{\partial N^{\mathrm{I}}_{\mathscr{A}}} = 0 \qquad \mathscr{A} = 0, \ldots, l. \tag{5.1.18}$$

By definition, we have

$$N^{\mathrm{I}}_{\mathscr{A}} + N^{\mathrm{II}}_{\mathscr{A}} = N_{\mathscr{A}} \qquad \mathscr{A} = 0, \ldots, l. \tag{5.1.19}$$

Thus, relation (5.1.19) may be written

$$\frac{\partial G^{\mathrm{I}}}{\partial N^{\mathrm{I}}_{\mathscr{A}}} - \frac{\partial G^{\mathrm{II}}}{\partial N^{\mathrm{I}}_{\mathscr{A}}} = 0, \tag{5.1.20}$$

or, by referring to (5.1.5),

$$\mu^{\mathrm{I}}_{\mathscr{A}}(\boldsymbol{P}^{\mathrm{I}}, \{N^{\mathrm{I}}\}) = \mu^{\mathrm{II}}_{\mathscr{A}}(\boldsymbol{P}^{\mathrm{II}}, \{N^{\mathrm{II}}\}) \tag{5.1.21}$$

for $\mathscr{A} = 0, \ldots, l$, but, conversely,

$$\mu^{\mathrm{I}}_{\mathscr{A}}(\boldsymbol{P}^{\mathrm{I}}, \{N^{\mathrm{I}}\}) \neq \mu^{\mathrm{II}}_{\mathscr{A}}(\boldsymbol{P}^{\mathrm{II}}, \{N^{\mathrm{II}}\}). \tag{5.1.22}$$

for $\mathscr{A} = l + 1, \ldots, m$.

The above conditions are sufficient to determine the equilibrium. In fact, the unknown quantities are, in each phase, the number of molecules of the components which are able to cross the membrane, i.e.

$$N^{\mathrm{I}}_{\mathscr{A}}, \ \mathscr{A} = 0, \ldots, l.$$

Hence, we get $N^{\mathrm{II}}_{\mathscr{A}}$ from (5.1.19)

$$N^{\mathrm{II}}_{\mathscr{A}} = N_{\mathscr{A}} - N^{\mathrm{I}}_{\mathscr{A}}. \tag{5.1.23}$$

In particular, these results determine the volumes of liquid in each cell.

1.4 Calculation of the osmotic pressure in terms of the chemical potentials at a given pressure

To measure osmotic pressure, an experiment is carried out in which two cells, one filled with solution and the other one filled with solvent, communicate through a semi-permeable membrane. Then, the pressure difference

$$\boldsymbol{\Pi} = \boldsymbol{P}^{\mathrm{I}} - \boldsymbol{P}^{\mathrm{II}} \tag{5.1.24}$$

is measured ($\boldsymbol{P}^{\mathrm{II}}$ is the pressure exerted on the solvent).

The chemical potentials of the solvent and of the species $\mathscr{A}$ that can cross the membrane are equal on both sides [see (5.1.21)]

$$\mu^{\mathrm{I}}_{\mathscr{A}}(\boldsymbol{P}^{\mathrm{I}}) = \mu^{\mathrm{II}}_{\mathscr{A}}(\boldsymbol{P}^{\mathrm{II}}) \qquad \mathscr{A} = 0, \ldots, l. \tag{5.1.25}$$

This condition leads us to compare the properties of two solutions submitted to pressures P^{I} and P^{II} whose difference is fixed by the osmotic equilibrium requirements.

However, it is easier to compare the properties of two isolated solutions when the same pressure is applied to them, and we would like to deduce from these properties the osmotic pressure that can be observed when solutions are brought into contact.

We shall now see that, if a few (generally justified) approximations hold, such an approach is quite possible.

Thus, let us consider two *isolated* solutions whose compositions coincide with those of the solutions in cells I and II at equilibrium (Fig. 5.2). The chemical potentials of a species $\mathscr{A}$, in these solutions, are respectively given by the functions $\mu^{\mathrm{I}}_{\mathscr{A}}(P)$ and $\mu^{\mathrm{II}}_{\mathscr{A}}(P)$.

It is especially interesting to evaluate the difference $\mu^{\mathrm{I}}_{\mathscr{A}}(P) - \mu^{\mathrm{II}}_{\mathscr{A}}(P)$ because it does not depend on pressure P and because, in certain cases, it is proportional to the osmotic pressure.

Actually, for any solution

$$\frac{\partial \mu_{\mathscr{A}}}{\partial P} = \frac{\partial^2 \mathbb{G}}{\partial P \partial N_{\mathscr{A}}} = \left.\frac{\partial V}{\partial N_{\mathscr{A}}}\right)_P \equiv v_{\mathscr{A}}(P). \tag{5.1.26}$$

The quantity $\partial V/\partial N_{\mathscr{A}}$ is the partial molecular volume of species $\mathscr{A}$ in the solution. In general, one may assume that this quantity is a function of pressure but does not depend on the composition. Therefore, we write

$$\left.\frac{\partial V^{\mathrm{I}}}{\partial N_{\mathscr{A}}}\right)_P = \left.\frac{\partial V^{\mathrm{II}}}{\partial N_{\mathscr{A}}}\right)_P = v_{\mathscr{A}}(P). \tag{5.1.27}$$

Thus, from relations

$$\mu^{\mathrm{I}}_{\mathscr{A}}(P^{\mathrm{I}}) = \mu^{\mathrm{II}}_{\mathscr{A}}(P^{\mathrm{II}})$$

$$\frac{\partial}{\partial P}\mu^{\mathrm{I}}_{\mathscr{A}}(P) = \frac{\partial}{\partial P}\mu^{\mathrm{II}}_{\mathscr{A}}(P) = v_{\mathscr{A}}(P) \tag{5.1.28}$$

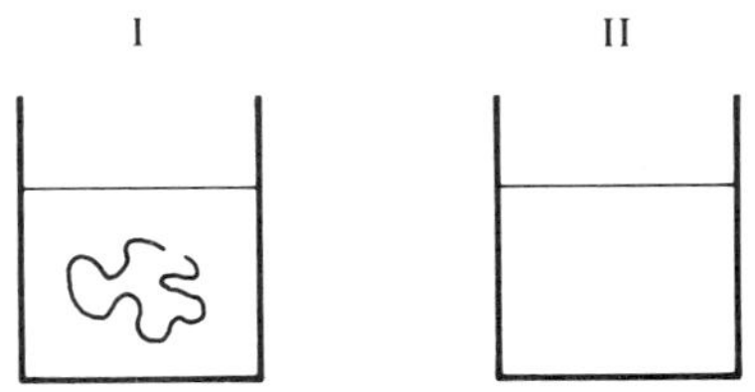

Fig. 5.2. Schematical representation of the experimental conditions for which the osmotic pressure is calculated. The same pressure is applied on both cells. Cell I contains the components which are not able to cross the membrane, in the model system of Fig. 5.1. The concentrations in each cell are the concentrations that would be obtained at osmotic equilibrium.

we immediately deduce

$$\mu^{\rm I}_{\mathscr{A}}(P) - \mu^{\rm II}_{\mathscr{A}}(P) = \int_{P^{\rm I}}^{P} dP' \frac{\partial \mu^{\rm I}_{\mathscr{A}}(P')}{\partial P'} - \int_{P^{\rm II}}^{P} dP'' \frac{\partial \mu^{\rm II}_{\mathscr{A}}(P'')}{\partial P''},$$

$$= \int_{P^{\rm II}}^{P^{\rm II}} dP v_{\mathscr{A}}(P), \tag{5.1.29}$$

and we observe that, in this case, the difference $\mu^{\rm I}_{\mathscr{A}}(P) - \mu^{\rm II}_{\mathscr{A}}(P)$ is independent of pressure. This relation simplifies again when $v_{\mathscr{A}}(P)$ is independent of P in the domain $(P^{\rm I}, P^{\rm II})$. We then have, for all P,

$$\Pi = P^{\rm I} - P^{\rm II} = -\frac{1}{v_{\mathscr{A}}}[\mu^{\rm I}_{\mathscr{A}}(P) - \mu^{\rm II}_{\mathscr{A}}(P)] \qquad \mathscr{A} = 0, \ldots, l. \tag{5.1.30}$$

Particularly when there is only one solvent ($l = 0$), the osmotic pressure is given immediately by the difference between the chemical potentials of the solvent in the solution and in the solvent, at the same pressure.

Differences of chemical potentials can thus be evaluated by measuring the osmotic pressure of a solution with respect to the solvent.

1.5 Osmotic pressure and vapour pressure of the solvent

The comparison made in the preceding section, between two isolated solutions suggests another experiment. Consider solutions I and II, whose compositions coincide with those of the solutions in Fig. 5.1. In the following setting (Fig. 1.3), these solutions are isolated from each other, and at equilibrium with their saturating vapour. These vapour pressures are respectively denoted by $P^{\rm I}_{\rm V}$ and $P^{\rm II}_{\rm V}$. Let us look for a relation between these pressures and the osmotic pressure which would appear between the two cells if they were in osmotic equilibrium.

Let us examine the simplest case ($m = 1$, $l = 0$), when the solution is made of one solvent and one solute. The solution is in cell I of Fig. 5.3, the solvent in cell II. Let us calculate the vapour pressure ratio

$$P^{\rm I}_{\rm V}/P^{\rm II}_{\rm V}.$$

It will be shown that if the solutes are non-volatile and if their concentrations are enough, this ratio is determined by the osmotic pressure Π

$$P^{\rm I}_{\rm V}/P^{\rm II}_{\rm V} = \exp(-\beta \Pi v_{\rm OL}). \tag{5.1.31}$$

where $v_{\rm OL}$ is the partial volume of the solvent in liquid state.

This approximate result can be established by comparing the chemical potentials of the solvent in the liquid and in the vapour state, in both cells. Let $\mu_{\rm OL}$ be the chemical potential of the liquid solvent. For a given osmotic equilibrium between the cells, we have

$$\mu^{\rm I}_{\rm OL}(P^{\rm I}) = \mu^{\rm II}_{\rm OL}(P^{\rm II})$$

$$\Pi = P^{\rm I} - P^{\rm II} \tag{5.1.32}$$

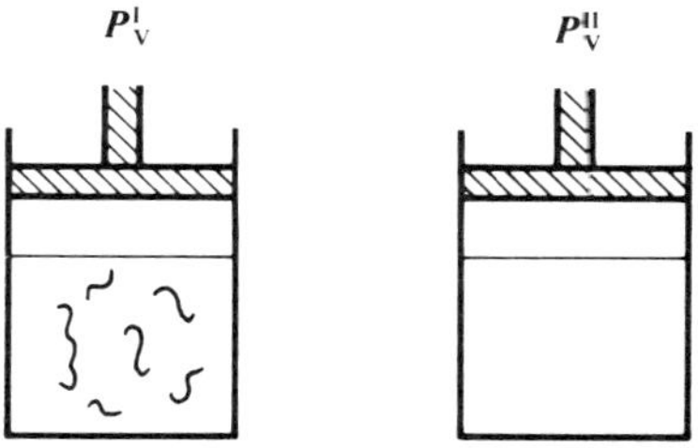

Fig. 5.3. Solution and pure solvent in equilibrium with their vapour.

where P^{I} and P^{II} are the pressures on both sides of the membrane. Let us now relate these pressures to the vapour pressures $P_{\mathrm{V}}^{\mathrm{I}}$ and $P_{\mathrm{V}}^{\mathrm{II}}$ corresponding to the situation described by Fig. 5.3. Let us use eqn (5.1.26)

$$\frac{\partial \mu_{\mathrm{OL}}}{\partial P} = v_{\mathrm{OL}}(P). \tag{5.1.33}$$

This equation applies to each cell and we shall assume that $v_{\mathrm{OL}}(P)$ is independent of pressure and composition

$$v_{\mathrm{OL}}(P) = v_{\mathrm{OL}}.$$

By integrating (5.1.33), successively in the intervals $(P^{\mathrm{I}}, P_{\mathrm{V}}^{\mathrm{I}})$ and $(P^{\mathrm{II}}, P_{\mathrm{V}}^{\mathrm{II}})$, we obtain

$$\mu_{\mathrm{OL}}^{\mathrm{I}}(P_{\mathrm{V}}^{\mathrm{I}}) - \mu_{\mathrm{OL}}^{\mathrm{I}}(P^{\mathrm{I}}) = (P_{\mathrm{V}}^{\mathrm{I}} - P^{\mathrm{I}})v_{\mathrm{OL}}$$
$$\mu_{\mathrm{OL}}^{\mathrm{II}}(P_{\mathrm{V}}^{\mathrm{II}}) - \mu_{\mathrm{OL}}^{\mathrm{II}}(P^{\mathrm{II}}) = (P_{\mathrm{V}}^{\mathrm{II}} - P^{\mathrm{II}})v_{\mathrm{OL}}.$$

Let us subtract, side by side, these two equalities and let us take (5.1.32) into account

$$\mu_{\mathrm{OL}}^{\mathrm{I}}(P_{\mathrm{V}}^{\mathrm{I}}) - \mu_{\mathrm{OL}}^{\mathrm{II}}(P_{\mathrm{V}}^{\mathrm{II}}) = (P_{\mathrm{V}}^{\mathrm{I}} - P_{\mathrm{V}}^{\mathrm{II}} - \Pi)v_{\mathrm{OL}}. \tag{5.1.34}$$

Let $\mu_{\mathrm{OV}}(P)$ be the chemical potential of the solvent vapour. We shall admit that the solute is not volatile and therefore that the chemical potential of the solvent is represented by the same function in both cells. Under these conditions, the liquid–vapour equilibrium implies the equalities

$$\mu_{\mathrm{OL}}^{\mathrm{I}}(P_{\mathrm{V}}^{\mathrm{I}}) = \mu_{\mathrm{OV}}(P_{\mathrm{V}}^{\mathrm{I}})$$
$$\mu_{\mathrm{OL}}^{\mathrm{II}}(P_{\mathrm{V}}^{\mathrm{II}}) = \mu_{\mathrm{OV}}(P_{\mathrm{V}}^{\mathrm{II}})$$

and eqn (5.1.34) can be rewritten in the form

$$\mu_{\mathrm{OL}}(P_{\mathrm{V}}^{\mathrm{I}}) - \mu_{\mathrm{OV}}(P_{\mathrm{V}}^{\mathrm{II}}) = (P_{\mathrm{V}}^{\mathrm{I}} - P_{\mathrm{V}}^{\mathrm{II}} - \Pi)v_{\mathrm{OL}}. \tag{5.1.35}$$

To calculate the difference of the solvent chemical potentials in the vapour state, we use the equation

$$\frac{\partial \mu_{\mathrm{OV}}}{\partial P} = v_{\mathrm{OV}}(P),$$

and we admit that the solvent vapour behaves like a perfect gas:

$$\beta P = [v_{0V}(P)]^{-1}. \tag{5.1.36}$$

Let us integrate the equation from P_V^{II} to P_V^{I}. We find

$$\mu_{0L}(P_V^{I}) - \mu_{0V}(P_V^{II}) = \beta^{-1} \ln(P_V^{I}/P_V^{II})$$

By substitution in eqn (5.1.35), we obtain

$$P_V^{I}/P_V^{II} = \exp[\beta(P_V^{I} - P_V^{II})v_{0L}]\exp[-\beta\Pi v_{0L}]. \tag{5.1.37}$$

This equation can be simplified. Equation (5.1.36) gives

$$\beta(P_V^{I} - P_V^{II})v_{0L} = v_{0L}\left(\frac{1}{v_{0V}(P_V^{I})} - \frac{1}{v_{0V}(P_V^{II})}\right) \ll 1.$$

Thus, the term which appears in the first exponential is very small. It can be neglected, and this leads to the predicted result

$$P_V^{I}/P_V^{II} \simeq \exp(-\beta\Pi v_{0L}). \tag{5.1.38}$$

However, let us note that this calculation is valid only in the case where $\beta\Pi v_{0L}$ is not very small.

1.6 Osmotic pressure for incompressible solutions

When solutions are incompressible, the partial molecular volumes

$$v_{\mathscr{A}} = \partial V/\partial N_{\mathscr{A}} \qquad \mathscr{A} = 0, \ldots, m \tag{5.1.39}$$

are pressure-independent, and we can also assume that they do not depend on the concentrations $C_{\mathscr{A}}$. The problem then becomes simpler.

The total volume of the system is

$$V = \sum_{\mathscr{A}} N_{\mathscr{A}} v_{\mathscr{A}}. \tag{5.1.40}$$

On the other hand, the volume V is given by eqn (5.1.12)

$$V = \frac{\partial \mathbb{G}}{\partial P}. \tag{5.1.41}$$

The solution of the equation

$$\frac{\partial \mathbb{G}}{\partial P} = \sum_{\mathscr{A}} N_{\mathscr{A}} v_{\mathscr{A}}$$

reads

$$\mathbb{G}(P,\{N\}) = \mathbb{F}\{N\} + P\sum_{\mathscr{A}} N_{\mathscr{A}} v_{\mathscr{A}}. \tag{5.1.42}$$

The function $\mathbb{F}\{N\}$ can easily be interpreted; it is related to the Helmholtz free energy $\mathscr{F}(V,\{N\})$ by the equality

$$\mathbb{F}\{N\} = \mathscr{F}\left(\sum_{\mathscr{A}} v_{\mathscr{A}} N_{\mathscr{A}}, \{N\}\right). \tag{5.1.43}$$

However, let us note that here $\mathscr{F}(V,\{N\})$ is a function which varies very rapidly with V and which is not of interest for us.

The function $\mathbb{F}\{N\}$ itself can be expressed as a function of the solute concentrations

$$C_{\mathscr{A}} = \frac{N_{\mathscr{A}}}{V} \qquad \mathscr{A} \neq 0 \tag{5.1.44}$$

under the form

$$\mathbb{F}\{N\} = V\Phi(\{C\}_{\mathscr{A} \neq 0}) \tag{5.1.45}$$

where volume and concentrations are considered as functions of the $N_{\mathscr{A}}$.

The chemical potentials are given by

$$\mu_0 = \frac{\partial \mathbb{G}}{\partial N_0} = \frac{\partial \mathbb{F}}{\partial N_0} + Pv_0 = v_0\left[\Phi - \sum_{\mathscr{B} \neq 0} C_{\mathscr{B}} \frac{\partial \Phi}{\partial C_{\mathscr{B}}}\right] \tag{5.1.46}$$

and for $\mathscr{A} = 0$

$$\mu_{\mathscr{A}} = \frac{\partial \mathbb{G}}{\partial N_{\mathscr{A}}} = \frac{\partial \mathbb{F}}{\partial N_{\mathscr{A}}} + Pv_{\mathscr{A}} = \left[\Phi - \sum_{\mathscr{B} \neq 0} C_{\mathscr{B}} \frac{\partial \Phi}{\partial C_{\mathscr{B}}} + P\right] + \frac{\partial \Phi}{\partial C_{\mathscr{A}}}. \tag{5.1.47}$$

Let us now calculate the osmotic pressure of a solution made of $(m + 1)$ constituents. The species with $\mathscr{A} = 0, \ldots, l$ can cross the membrane and are present in cells I and II. On the contrary, the species with $\mathscr{A} = l + 1, \ldots, m$ cannot cross the membrane and are found only in cell I. The osmotic equilibrium is defined by the conditions

$$\mu^{\mathrm{I}}_{\mathscr{A}} = \mu^{\mathrm{II}}_{\mathscr{A}} \qquad \mathscr{A} = 0, \ldots, l$$

which reads

$$\Phi^{\mathrm{I}} - \sum_{\mathscr{B} \neq 0} C^{\mathrm{I}}_{\mathscr{B}} \frac{\partial \Phi}{\partial C^{\mathrm{I}}_{\mathscr{B}}} + P^{\mathrm{I}} = \Phi^{\mathrm{II}} - \sum_{\mathscr{B} \neq 1} C^{\mathrm{II}}_{\mathscr{B}} \frac{\partial \Phi^{\mathrm{II}}}{\partial C^{\mathrm{II}}_{\mathscr{B}}} + P^{\mathrm{II}}, \tag{5.1.48}$$

and from these equalities, we deduce the osmotic pressure

$$\Pi = P^{\mathrm{I}} - P^{\mathrm{II}} = \left[\sum_{\mathscr{B}=1}^{m} C^{\mathrm{I}}_{\mathscr{B}} \frac{\partial \Phi^{\mathrm{I}}}{\partial C^{\mathrm{I}}_{\mathscr{B}}} - \Phi^{\mathrm{I}}\right] - \left[\sum_{\mathscr{B}=1}^{m} C^{\mathrm{II}}_{\mathscr{B}} \frac{\partial \Phi^{\mathrm{II}}}{\partial C^{\mathrm{II}}_{\mathscr{B}}} - \Phi^{\mathrm{II}}\right]. \tag{5.1.49}$$

Let us note that the first term on the right-hand side can be interpreted in a simple way by introducing the function of ${}^{0}\mathbb{F}(V,\{N\}_{\mathscr{A} \neq 0})$ which is defined by

$${}^{0}\mathbb{F}(V,\{N\}_{\mathscr{A} \neq 0}) = \mathbb{F}\left(\{N\}_{\mathscr{A} \neq 0}, v_1^{-1}\left(V - \sum_{\mathscr{B} \neq 0} v_{\mathscr{B}} N_{\mathscr{B}}\right)\right). \tag{5.1.50}$$

Equation (5.1.47) shows that

$$\frac{\partial}{\partial V}{}^{0}\mathbb{F}(V,\{N\}_{\mathscr{A}\neq 0}) = -\sum_{\mathscr{B}\neq 0}\left[C_{\mathscr{B}}\frac{\partial \Phi}{\partial C_{\mathscr{B}}} - \Phi\right].$$

Thus, in the case of $(m-1)$ solutes in one solvent, the osmotic pressure can be written as follows

$$\Pi = -\frac{\partial}{\partial V^{\mathrm{I}}}[{}^{0}\mathbb{F}(V^{\mathrm{I}},\{N^{\mathrm{I}}\}_{\mathscr{A}\neq 0}) - {}^{0}\mathbb{F}(V^{\mathrm{I}},\{0\}_{\mathscr{A}\neq 0})], \tag{5.1.51}$$

and we see that this relation is very analogous to the relation

$$P = -\frac{\partial \mathscr{F}(V,\{N\})}{\partial V}$$

which gives the pressure for compressible systems.

Let us now return to the general case. From eqn (5.1.49) we can deduce a very interesting relation which we shall use in this chapter, and which deals with small-angle scattering. We observe that Φ^{II} does not depend on the concentrations $C_{\mathscr{A}}$ with the indices $\mathscr{A} = l+1, \ldots, m$. Consequently, we have

$$\frac{\partial \Pi}{\partial C^{\mathrm{I}}_{\mathscr{A}}} = \sum_{\mathscr{B}=1}^{m} C^{\mathrm{I}}_{\mathscr{B}}\frac{\partial^2 \Phi^{\mathrm{I}}}{\partial C^{\mathrm{I}}_{\mathscr{B}}\,\partial C^{\mathrm{I}}_{\mathscr{A}}}, \tag{5.1.52}$$

and this relation is remarkable because it does not depend on the properties of the solution constituting phase I.

Using again the model shown on Fig. 5.1 let us now study the case where the solution is an incompressible liquid whose volume

$$V = V^{\mathrm{I}} + V^{\mathrm{II}}$$

is a function of the compositions $\{C^{\mathrm{I}}_{\mathscr{A}}\}$ and $\{C^{\mathrm{II}}_{\mathscr{A}}\}$ in each cell. For this purpose, let us consider the partial molecular volume which has been defined above

$$v_{\mathscr{A}} = \frac{\partial V(P,\{N\})}{\partial N_{\mathscr{A}}}.$$

As before, we assume that the partial volumes do not depend on pressure; however, we now admit that they may depend on the concentrations $C_{\mathscr{A}}$.

In particular, let us consider a solution made only of one solvent ($\mathscr{A} = 0$) and of one solute ($\mathscr{A} = 1$). The function $\Phi\{C^{\mathrm{I}}_{\mathscr{A}}\}$ can be written $\Phi\{C^{\mathrm{I}}_{1}\}$ and we can simplify the notation by setting $C^{\mathrm{I}}_{1} = C_1$. The partial volumes being constant, the osmotic pressure can be written in the form

$$\Pi = \Phi(0) - \Phi(C_1) + C_1\,\partial\Phi(C_1)/\partial C_1. \tag{5.1.53}$$

The expression of the osmotic pressure becomes even more complicated when concentration-dependent partial volumes $v_0(C_1)$ and $v_1(C_1)$ are introduced.

However, in this case, we may set

$$\mathbb{F}\{N\} = V\Phi(C_1). \tag{5.1.54}$$

The chemical potential of the solvent has the same expression as in (5.1.46)

$$\mu_0 = \frac{\partial \mathbb{G}}{\partial N_0} = \frac{\partial \mathbb{F}}{\partial N_0} + Pv_0 = v_0\left(\Phi - C_1\frac{\partial \Phi}{\partial C_1} + P\right) \tag{5.1.55}$$

with $v_0 \equiv v_0(C_1)$ defined by (5.1.39). The equilibrium condition is

$$\mu_0^{\mathrm{I}} = \mu_0^{\mathrm{II}}. \tag{5.1.56}$$

Now, the partial volume v_0 depends on the indices I, II. We have

$$v_0^{\mathrm{I}}\left(\Phi^{\mathrm{I}} - C_1^{\mathrm{I}}\frac{\partial \Phi^{\mathrm{I}}}{\partial C_1^{\mathrm{I}}} + P^{\mathrm{I}}\right) = v_0^{\mathrm{II}}(\Phi^{\mathrm{II}} + P^{\mathrm{II}}). \tag{5.1.57}$$

The osmotic pressure

$$\Pi = P^{\mathrm{I}} - P^{\mathrm{II}}$$

is now

$$\Pi = \Phi^{\mathrm{II}} - \Phi^{\mathrm{I}} + C_1 \partial\Phi^{\mathrm{I}}/\partial C_1 + (P^{\mathrm{II}} + \Phi^{\mathrm{II}})(v_0^{\mathrm{II}}/v_0^{\mathrm{I}} - 1), \tag{5.1.58}$$

which can also be written in the form

$$\Pi = \Phi(0) - \Phi(C_1) + C_1\frac{\partial \Phi(C_1)}{\partial C_1} + [P^{\mathrm{II}} + \Phi(0)]\left(\frac{v_0(0)}{v_0(C_1)} - 1\right). \tag{5.1.59}$$

We recover the preceding result (5.1.53) when the partial volumes v_0^{I}, v_0^{II} are strictly equal to v_0.

Let us differentiate the pressure with respect to C_1; we obtain

$$\frac{\partial \Pi}{\partial C_1} = C_1\frac{\partial^2 \Phi(C_1)}{(\partial C_1)^2} - (P^{\mathrm{II}} + \Phi(0))\frac{v_0(0)}{[v_0(C_1)]^2}\frac{\partial v_0(C_1)}{\partial C_1}. \tag{5.1.60}$$

This time, the pressure gradient depends on the properties of the solution in both phases [compare with (5.1.52)].

1.7 Examples

(a) *Incompressible solutions whose constituents have the same molecular volume*

A solution is incompressible if its volume V is the sum

$$V = \sum_{\mathscr{A}} N_{\mathscr{A}} v_{\mathscr{A}} \tag{5.1.61}$$

where the $v_{\mathscr{A}}$ are constant molecular volumes. The solution is ideal, if each configuration of the $N = \sum_{\mathscr{A}=0}^{m} N_{\mathscr{A}}$ molecules is associated with the same energy.

We shall assume that the molecular volumes are identical

$$v_{\mathscr{A}} \equiv v. \tag{5.1.62}$$

In this case, the free energy $\mathbb{F}\{N_{\mathscr{A}}\}$, which coincides with the Helmholtz free energy for $V = \sum_{\mathscr{A}} N_{\mathscr{A}} v_{\mathscr{A}}$ [see eqn (5.1.61)] is given by the expression (see Appendix C)

$$\mathbb{F}\{N\} = \mathbb{F}_0 + \beta^{-1} \sum_{\mathscr{A}=0}^{m} N_{\mathscr{A}} \ln N_{\mathscr{A}}/N \tag{5.1.63}$$

where $\mathbb{F}_0$ is the sum of the free energies that are associated to the pure constituents. By definition, the chemical potentials are

$$\mu_{\mathscr{A}} = \frac{\partial \mathbb{F}\{N\}}{\partial N_{\mathscr{A}}} + Pv \tag{5.1.64}$$

where P is the pressure. $\left(\text{But we note that we also have } \mu_{\mathscr{A}} = \frac{\partial \mathscr{F}}{\partial N_{\mathscr{A}}}!\right)$ We get

$$\mu_{\mathscr{A}} = {}^{0}\mu_{\mathscr{A}} + \beta^{-1} \ln (N_{\mathscr{A}}/N) + Pv \tag{5.1.65}$$

where ${}^{0}\mu_{\mathscr{A}}$ is the chemical potential associated with the constituent $\mathscr{A}$ in its pure state.

Let us examine the case where the membrane is permeable only to the main solvent $\mathscr{A} = 0$. The equilibrium between a solution (phase I) and a pure solvent (phase II), is given by the condition

$$\mu_0^{\mathrm{I}}(P^{\mathrm{I}}, \{N_{\mathscr{A}}^{\mathrm{I}}\}) = \mu_0^{\mathrm{II}}(P^{\mathrm{II}}, \{N_0^{\mathrm{II}}\}), \tag{5.1.66}$$

i.e. with 5.1.60)

$$\begin{aligned} \Pi = P^{\mathrm{I}} - P^{\mathrm{II}} &= -v^{-1}\beta^{-1} \ln(N_0^{\mathrm{I}}/N^{\mathrm{I}}) \\ &= -v^{-1}\beta^{-1} \ln\left[1 - \sum_{\mathscr{A}=1}^{m} (N_{\mathscr{A}}^{\mathrm{I}}/N^{\mathrm{I}})\right]. \end{aligned} \tag{5.1.67}$$

In particular, if the solution is dilute,

$$N_0^{\mathrm{I}} \gg \sum_{\mathscr{A}=1}^{m} N_{\mathscr{A}}^{\mathrm{I}} \tag{5.1.68}$$

we have

$$\Pi \simeq v^{-1}\beta^{-1} \sum_{\mathscr{A}=1}^{m} N_{\mathscr{A}}^{\mathrm{I}}/N^{\mathrm{I}}. \tag{5.1.69}$$

This is Van't Hoff's law.

For a membrane that is permeable to m constituents with $l < m$, we have the equilibrium conditions

$$\mu_{\mathscr{A}}^{\mathrm{I}}(P^{\mathrm{I}}, N_1^{\mathrm{I}}, \ldots, N_m^{\mathrm{I}}) = \mu_{\mathscr{A}}^{\mathrm{II}}(P^{\mathrm{II}}, N_1^{\mathrm{II}}, \ldots, N_l^{\mathrm{II}}) \qquad \mathscr{A} = 0, \ldots, l. \tag{5.1.70}$$

By introducing (5.1.65) in (5.1.70), we obtain

$$\Pi = P^{\mathrm{I}} - P^{\mathrm{II}} = -v^{-1}\beta^{-1}[\ln(N^{\mathrm{I}}_{\mathscr{A}}/N^{\mathrm{I}}) - \ln(N^{\mathrm{II}}_{\mathscr{A}}/N^{\mathrm{II}})]$$

for

$$\mathscr{A} = 0, \ldots, l. \tag{5.1.71}$$

Consequently,

$$N^{\mathrm{I}}_{\mathscr{B}}/N^{\mathrm{I}}_{\mathscr{A}} = N^{\mathrm{II}}_{\mathscr{B}}/N^{\mathrm{II}}_{\mathscr{A}}$$

$$\mathscr{A} = 0, \ldots, l \qquad \mathscr{B} = 0, \ldots, l. \tag{5.1.72}$$

Let 0 be the main solvent index. The ratio $N_{\mathscr{A}}/N_0$ is called the *molality* of species $\mathscr{A}$. For an ideal incompressible solution, the osmotic equilibrium requires that, on both sides of the membrane, the molalities are equal for all species that can cross it.

(*b*) *Ideal incompressible polymer solutions; the Flory–Huggins theory*

Polymers in solution form a dissymmetric system, in the sense that the molecular volume of the solute is much larger than the molecular volume of the solvent. Let us examine the following special case:

(a) the N_1 chains are made up of N monomers;

(b) each one of the N solvent molecules has a molecular volume v which is also equal to the volume of one monomer;

(c) the solution is incompressible.

The Flory–Huggins theory (see Chapter 8, Section 2.1) gives for the free energy [see (5.1.43)]

$$\mathbb{F}\{N\} = {}^0\mathbb{F} + \beta^{-1}\left[N_0 \ln \frac{N_0}{N_0 + NN_1} + N_1 \ln \frac{NN_1}{N_0 + NN_1}\right] \tag{5.1.73}$$

where ${}^0\mathbb{F}$ is the sum of the free energies of the pure constituents. This expression differs of the preceding one (5.1.63) in the sense that monomer rather than polymer number fractions are the arguments in the logarithmic functions. This is the dissymmetry effect, recognized by Flory and Huggins. Using the same approach as before, we obtain the osmotic pressure

$$\begin{aligned}\Pi &= -\beta^{-1}v^{-1}\left[\ln \frac{N_0}{N_0 + NN_1} + \left(1 - \frac{1}{N}\right)\frac{NN_1}{N_0 + NN_1}\right]\\ &= -\beta^{-1}v^{-1}\left[\ln\left(1 - \frac{NN_1}{N_0 + NN_1}\right) + \left(1 - \frac{1}{N}\right)\frac{NN_1}{N_0 + NN_1}\right]\end{aligned} \tag{5.1.74}$$

and in the limit of large dilutions, $N_0 \gg NN_1$, we recover Van't Hoff's law

$$\Pi = \beta^{-1}v^{-1}\frac{N_1}{N_0 + NN_1}. \tag{5.1.75}$$

Let us note that in the semi-dilute concentration domain

$$\frac{1}{N} \ll \frac{n\mathcal{N}_1}{\mathcal{N}_0 + N\mathcal{N}_1} \ll 1$$

we have, according to (5.1.74),

$$\Pi \simeq \frac{v^{-1}}{2} \beta^{-1} (N\mathcal{N})^2, \qquad (5.1.76)$$

but the dissymmetry effect is not expressed here in a correct manner. The results of the modern theory in this domain will be given in Chapter 8.

1.8 Transformations at constant chemical potentials: preferential adsorption

When a solution contains more than one solvent, these solvents do not, in general, play the same role. In particular, diluting a solution by adding a small quantity of solvent produces a variation of Gibbs free energy which depends on the nature of the added solvent.

For instance, when a small amount of precipitant is added to a dilute polymer solution, a large increase of the scattering cross-section is observed in the zero-angle limit.[2] At first, this phenomenon has been interpreted as resulting from a preflocculation or from the formation of a molecular complex. Actually, an analysis of this phenomenon shows that things are different: the simple explanation is that the distribution of the solvents does not remain uniform. The precipitant concentration is higher in the vicinity of the polymer chains than, on average, in the solution. The corresponding increase of cross-section is sometimes used to reinforce the signal measured in experiments when the polymer concentrations are very small.

Thus, the polymer chain has the remarkable property that its presence in the solution is able to modify the distribution of the various solvent molecules.

The study of the osmotic equilibrium between a polymer solution in two solvents and a mixture of these two solvents is quite instructive. In particular, we shall describe a limiting case in which we shall be able to define a coefficient characterizing the action of the polymers on the solvent distribution. First, let us attribute the index 0 to the 'main' solvent, the index 1 to the second solvent, and the index 2 to the polymer. We have $l = 1$, $m = 2$. The numbers of molecules are $\mathcal{N}^{\mathrm{I}}_{\mathscr{A}}$ ($\mathscr{A} = 0, 1, 2$) in cell I of Fig. 5.1 and $\mathcal{N}^{\mathrm{II}}_{\mathscr{A}}$ ($\mathscr{A} = 0, 1$) in cell II. The chemical potentials of the solvents are, respectively,

$$\begin{aligned} \boldsymbol{\mu}^{\mathrm{I}}_{\mathscr{A}} &\equiv \boldsymbol{\mu}^{\mathrm{I}}_{\mathscr{A}}(\mathcal{N}^{\mathrm{I}}_0, \mathcal{N}^{\mathrm{I}}_1, \mathcal{N}^{\mathrm{I}}_2; \boldsymbol{P}^{\mathrm{I}}), \\ \boldsymbol{\mu}^{\mathrm{II}}_{\mathscr{A}} &\equiv \boldsymbol{\mu}^{\mathrm{II}}_{\mathscr{A}}(\mathcal{N}^{\mathrm{II}}_0, \mathcal{N}^{\mathrm{II}}_1, 0; \boldsymbol{P}^{\mathrm{II}}), \end{aligned} \qquad (5.1.77)$$

where $\boldsymbol{P}^{\mathrm{I}}$ and $\boldsymbol{P}^{\mathrm{II}}$ are the pressures. Usually, these pressures $\boldsymbol{P}^{\mathrm{I}}$ and $\boldsymbol{P}^{\mathrm{II}}$ are fixed and also the quantities

$$N_{\mathscr{A}} = N^{\mathrm{I}}_{\mathscr{A}} + N^{\mathrm{II}}_{\mathscr{A}} \qquad \mathscr{A} = 0, 1$$

$$N_2 = N^{\mathrm{I}}_2 .$$

Then, the osmotic equilibrium conditions given by eqn (5.1.21)

$$\boldsymbol{\mu}^{\mathrm{I}}_{\mathscr{A}} = \boldsymbol{\mu}^{\mathrm{II}}_{\mathscr{A}} \qquad \mathscr{A} = 0, 1 \tag{5.1.78}$$

determine the amounts of solvents in each cell, i.e.

$$N^{\mathrm{I}}_{\mathscr{A}}, N^{\mathrm{II}}_{\mathscr{A}}, \qquad \mathscr{A} = 0, 1.$$

These quantities are thus functions of $\boldsymbol{P}^{\mathrm{I}}$, $\boldsymbol{P}^{\mathrm{II}}$ and of the $N_{\mathscr{A}}$ ($\mathscr{A} = 0, 1, 2$)

A study of the osmotic equilibrium is difficult to make in this case, but the problem simplifies in the following useful[3] limit. The pressures $\boldsymbol{P}^{\mathrm{I}}$ and $\boldsymbol{P}^{\mathrm{II}}$ remaining fixed, the number of molecules in cell II without polymers becomes infinite

$$N^{\mathrm{II}}_{\mathscr{A}} \to \infty, \qquad \mathscr{A} = 0, 1.$$

with $N^{\mathrm{II}}_1/N^{\mathrm{II}}_0 = a =$ constant. The osmotic equilibrium determines the amounts of solvents in cell I. The result becomes independent of the total quantities N_0, N_1 but depends on a.

In this way, one gets an osmotic equilibrium at fixed chemical potential. We call it dialysis equilibrium.

The equations defining the osmotic equilibrium are

$$\boldsymbol{\mu}^{\mathrm{I}}_{\mathscr{A}}(N^{\mathrm{I}}_0, N^{\mathrm{I}}_1, N^{\mathrm{I}}_2; \boldsymbol{P}^{\mathrm{I}}) = \boldsymbol{\mu}^{\mathrm{II}}_{\mathscr{A}}(a, \boldsymbol{P}^{\mathrm{II}}) = \boldsymbol{\mu}_{\mathscr{A}} \qquad \mathscr{A} = 0, 1 \tag{5.1.79}$$

where the potentials $\boldsymbol{\mu}_{\mathscr{A}}$ can be considered as parameters instead of a and $\boldsymbol{P}^{\mathrm{II}}$

We can now define the preferential adsorption coefficient λ

$$\lambda = \frac{\partial N^{\mathrm{I}}_1(N^{\mathrm{I}}_0, N^{\mathrm{I}}_2, \boldsymbol{\mu}_0, \boldsymbol{\mu}_1)}{\partial N^{\mathrm{I}}_2} \tag{5.1.80}$$

and we shall also be interested in the quantity

$$\frac{\partial \boldsymbol{\Pi}(N^{\mathrm{I}}_2, N^{\mathrm{I}}_0, \boldsymbol{\mu}_0, \boldsymbol{\mu}_1)}{\partial N^{\mathrm{I}}_2}$$

which, according to (5.1.11) is given by

$$\frac{\partial \boldsymbol{\Pi}(N^{\mathrm{I}}_2, N^{\mathrm{I}}_0, \boldsymbol{\mu}_0, \boldsymbol{\mu}_1)}{\partial N^{\mathrm{I}}_2} = \frac{\partial \boldsymbol{P}^{\mathrm{I}}(N^{\mathrm{I}}_2, N^{\mathrm{I}}_0, \boldsymbol{\mu}_0, \boldsymbol{\mu}_1)}{\partial N^{\mathrm{I}}_2} \tag{5.1.81}$$

Now, taking this relationship into account, we can determine the coupled equations which give λ and $\dfrac{\partial \boldsymbol{\Pi}(N^{\mathrm{I}}_2, N^{\mathrm{I}}_0, \boldsymbol{\mu}_0, \boldsymbol{\mu}_1)}{\partial N^{\mathrm{I}}_2}$.

We shall use the notation

$$a_{\mathscr{A}\mathscr{B}} = a_{\mathscr{B}\mathscr{A}} = N_1^{\mathrm{I}} \frac{\partial \mu_{\mathscr{A}}^{\mathrm{I}}(P^{\mathrm{I}},\{N\})}{\partial N_{\mathscr{B}}^{\mathrm{I}}}. \tag{5.1.82}$$

In addition, we know that the partial volume of species $\mathscr{A}$ is given by [See (5.1.26)]

$$v_{\mathscr{A}} = \frac{\partial V^{\mathrm{I}}(P^{\mathrm{I}},\{N\})}{\partial N_{\mathscr{A}}^{\mathrm{I}}} = \frac{\partial \mu_{\mathscr{A}}^{\mathrm{I}}(P^{\mathrm{I}},\{N\})}{\partial P^{\mathrm{I}}}.$$

Then, we get

$$\mathrm{d}\mu_{\mathscr{A}} = \sum_{\mathscr{B}=0}^{2} a_{\mathscr{A}\mathscr{B}}\,\mathrm{d}N_{\mathscr{B}}^{\mathrm{I}} + N_1^{\mathrm{I}} v_{\mathscr{A}}\,\mathrm{d}P^{\mathrm{I}}. \tag{5.1.83}$$

Let us express these relations by choosing N_0^{I}, N_2^{I}, μ_0, μ_1 as new parameters and, let us assume that N_2^{I} varies whereas N_0^{I}, μ_0, μ_1 remain constant. For $\mathscr{A} = 0$ in (5.1.83), we obtain

$$a_{\mathscr{A}1} \frac{\partial N_1^{\mathrm{I}}(N_0^{\mathrm{I}}, N_2^{\mathrm{I}}, \mu_0, \mu_1)}{\partial N_2^{\mathrm{I}}} + a_{\mathscr{A}2} + N_0^{\mathrm{I}} v_{\mathscr{A}} \frac{\partial P^{\mathrm{I}}(N_0^{\mathrm{I}}, N_2^{\mathrm{I}}, \mu_0, \mu_1)}{\partial N_2^{\mathrm{I}}} = 0. \tag{5.1.84}$$

Using the definition of λ (eqn (5.1.80)) and eqn (5.1.81), we obtain the coupled equations

$$a_{01}\lambda + N_0^{\mathrm{I}} v_0 \frac{\partial \Pi(N_0^{\mathrm{I}}, N_2^{\mathrm{I}}, \mu_0, \mu_1)}{\partial N_2^{\mathrm{I}}} = -\,a_{13}$$

$$a_{11}\lambda + N_0^{\mathrm{I}} v_1 \frac{\partial \Pi(N_0^{\mathrm{I}}, N_2^{\mathrm{I}}, \mu_0, \mu_1)}{\partial N_2^{\mathrm{I}}} = -\,a_{23}. \tag{5.1.85}$$

Thus, we find

$$\lambda = -\frac{a_{02}v_1 - a_{12}v_0}{a_{01}v_1 - a_{11}v_0}$$

$$\frac{\partial \Pi(N_0^{\mathrm{I}}, N_2^{\mathrm{I}}, \mu_0, \mu_1)}{\partial N_2^{\mathrm{I}}} = -\frac{1}{N_0^{\mathrm{I}}}\left(\frac{a_{01}a_{12}v_1 - a_{02}a_{11}v_0}{a_{01}v_1 - a_{11}v_0}\right). \tag{5.1.86}$$

It is convenient to eliminate the quantities which depend on the index 0; in this way, we shall get a simpler formula, in the limit $N_2^{\mathrm{I}} \to 0$. The volume V^{I} is a homogeneous function (with degree 1) of the quantities $N_{\mathscr{A}}^{\mathrm{I}}$. Consequently,

$$V^{\mathrm{I}} = \sum_{\mathscr{A}=0}^{2} N_{\mathscr{A}}^{\mathrm{I}} v_{\mathscr{A}}. \tag{5.1.87}$$

In the same way, the chemical potentials are homogeneous functions (with degree 0) of the $N_{\mathscr{A}}^{\mathrm{I}}$. Therefore

$$\sum_{\mathscr{B}=0}^{2} N_{\mathscr{B}}^{\mathrm{I}} a_{\mathscr{A}\mathscr{B}} = 0 \tag{5.1.88}$$

and the symmetry property of the $a_{\mathscr{A}\mathscr{B}}$ leads to the Gibbs–Duhem relations

$$\sum_{\mathscr{A}=0}^{2} N^{\mathrm{I}}_{\mathscr{A}} a_{\mathscr{A}\mathscr{B}} = 0. \tag{5.1.89}$$

By extracting v_0 and $a_{0\mathscr{B}}$ from the preceding relations and by introducing them in eqns (5.1.87), we obtain

$$\lambda = -\frac{a_{12}}{a_{11}}\left\{\frac{V^{\mathrm{I}} - N^{\mathrm{I}}_2[v_2 - (a_{22}/a_{12})v_1]}{V^{\mathrm{I}} - N^{\mathrm{I}}_2[v_2 - (a_{12}/a_{11})v_1]}\right\} \tag{5.1.90}$$

$$\left.\frac{\partial \Pi}{\partial N^{\mathrm{I}}_2}\right|_{N^{\mathrm{I}}_0, N^{\mathrm{I}}_2, \mu_0, \mu_1} = -\frac{a_{12}}{N^{\mathrm{I}}_0 v_1}\left\{1 - \frac{V - N^{\mathrm{I}}_2[v_2 - (a_{22}/a_{12})v_1]}{V - N^{\mathrm{I}}_2[v_2 - (a_{12}/a_{11})v_1]}\right\}. \tag{5.1.91}$$

On the other hand, the limiting preferential adsorption coefficient λ_0 of the second solvent with respect to the solute is defined by

$$\lambda_0 = \lim_{N^{\mathrm{I}}_2 \to 0} \lambda. \tag{5.1.92}$$

According to (5.1.91), we have

$$\lambda_0 = -\frac{a_{12}}{a_{11}}\left[1 + \lim_{N^{\mathrm{I}}_2 \to 0} \frac{N^{\mathrm{I}}_2}{V}(a_{22}/a_{12})v_1\right]. \tag{5.1.93}$$

We observe that the second term in brackets has a finite limit.

The coefficient

$$\lambda_0 = \lim_{N_2 \to 0} \left.\frac{\partial N_1}{\partial N_2}\right|_{\mu_0, \mu_1} \tag{5.1.94}$$

should not be mistaken for the coefficient which we shall denote by ζ_0

$$\zeta_0 = \lim_{N_2 \to 0} \left.\frac{\partial N_1}{\partial N_2}\right|_{P^{\mathrm{I}}, \mu_1} \tag{5.1.95}$$

and whose expression can be directly obtained from (5.1.81)

$$\zeta_0 = -\frac{a_{12}}{a_{11}}. \tag{5.1.96}$$

The following example shows how important this distinction is. In fact, let us consider an ideal incompressible solution of molecules with the same molecular volume. We have [see (5.1.60)]

$$\mu_{\mathscr{A}} = \mu^0_{\mathscr{A}} + \beta^{-1}\ln(N_{\mathscr{A}}/N) + Pv$$

where

$$N = N_0 + N_1 + N_2.$$

Then, we get

$$a_{\mathscr{A}\mathscr{B}} = N_0^{\mathrm{I}}(\partial \mu_{\mathscr{A}}^{\mathrm{I}}/\partial N_{\mathscr{B}})_{P, N_{\mathscr{C} \neq \mathscr{B}}} = N_0^{\mathrm{I}}\left(\frac{1}{N_{\mathscr{A}}^{\mathrm{I}}}\,\delta_{\mathscr{A}\mathscr{B}} - \frac{1}{N^{\mathrm{I}}}\right). \tag{5.1.97}$$

In particular,

$$a_{02} = a_{12} = a_{01} = -\, N_0^{\mathrm{I}}/N^{\mathrm{I}}$$

and

$$N_2^{\mathrm{I}} a_{22} = N_0^{\mathrm{I}}(1 - N_2^{\mathrm{I}}/N^{\mathrm{I}}) \neq 0. \tag{5.1.98}$$

In this case, according to eqns (5.1.87) and (5.1.96), on one hand we have

$$\lambda_0 = \lambda = 0$$

and, on the other hand,

$$\zeta_0 = N_1^{\mathrm{I}}/N_0^{\mathrm{I}}. \tag{5.1.99}$$

The result is still valid, when the chemical potentials of the solvents are given by the Flory–Huggins approximation [see (5.1.73)]

$$\mathbb{F}\{N\} = {}^0\mathbb{F} + \beta^{-1}\left[N_0 \ln \frac{N_0}{N_0 + N_1 + NN_2} + N_1 \ln \frac{N_1}{N_0 + N_1 + NN_2} \right.$$
$$\left. + N_2 \ln \frac{NN_2}{N_0 + N_1 + NN_2} \right]$$

where N is the polymerization degree.

In general, λ does not vanish when the solution is not ideal; for instance, when the polymer is dissolved in a mixture of a good and a poor solvent. Of course, in this case, it may be positive or negative.

1.9 Measurement of the 'volume' osmotic pressure

The osmotic pressure between two solutions is measured by means of an apparatus whose principle is indicated on Fig. 5.4. In the beginning, the cells I and II, separated by a semi-permeable membrane, are filled with pure solvent. Then, the levels are equal and the pressures P_0^{I} and P_0^{II} are equal.

At time $t = 0$, the solutes $\mathscr{A} = 0, \ldots, l$ are introduced in cell I and it is assumed that they cannot cross the membrane.

As a consequence, there is a flow of solvent from cell II to cell I which increases the dilution in cell I and goes on till the osmotic equilibrium is reached ($t \to \infty$). Then the levels are respectively $h^{\mathrm{I}} > h^{\mathrm{II}}$ and the pressures are $P^{\mathrm{I}} > P^{\mathrm{II}}$.

When the number of species $\mathscr{A}$ able to cross the membrane is larger than unity, the concentrations of these species $\mathscr{A} = 1, \ldots, l$ in cell II are especially important parameters, owing to preferential absorption (see Section 1.8). The concentrations can be *a priori fixed* by connecting cell II to a reservoir. In this case, we say that the osmotic pressure Π is measured at the dialysis equilibrium.

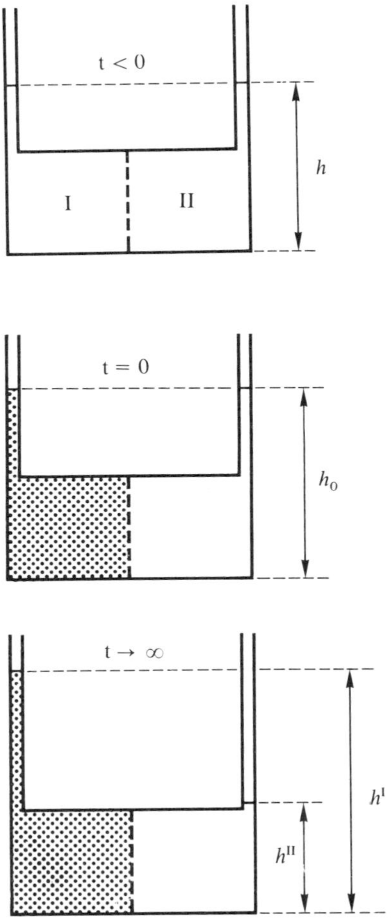

Fig. 5.4. Schematic representation of three successive situations met in the course of an osmotic pressure measurement. The white zones in the cells represent the solvents that are able to cross the semi-permeable membrane. The grey zones represent the solution containing species that are unable to cross the membrane.

Let ρ^{I} be the mass per unit volume of solution in I and ρ^{II} the mass per unit volume of solution in II. The balance of forces on the membrane gives

$$\boldsymbol{\Pi} = g(\rho^{\mathrm{I}} h^{\mathrm{I}} - \rho^{\mathrm{II}} h^{\mathrm{II}}) \qquad (5.1.101)$$

where g is the gravity intensity. Let us introduce the differences $\boldsymbol{\delta} h = h^{\mathrm{I}} - h^{\mathrm{II}}$ and $\boldsymbol{\delta}\rho = \rho^{\mathrm{I}} - \rho^{\mathrm{II}}$. We have

$$\boldsymbol{\Pi} = g(\rho^{\mathrm{I}} \boldsymbol{\delta} h + h^{\mathrm{II}} \boldsymbol{\delta}\rho). \qquad (5.1.102)$$

For polymer solutions containing heavy polymers, the second term on the right-hand side gives a contribution to the osmotic pressure which may be as large as 10 per cent.

Let us calculate a few characteristic values of a solution consisting of one solute and of one solvent. For this purpose, we shall use Van't Hoff's law [see eqn (5.1.78)]. For a dilute solution, we have

$$\Pi\beta = \frac{1}{v_0}(N_1/N) \tag{5.1.103}$$

where v_0 is the partial molecular volume of the solvent. This pressure may be huge. Let us take water as solvent, the molar volume is $V_0 = 18\ \text{cm}^3$ and the molecular volume $v_0 = V_0/Ⓐ \simeq 3 \times 10^{-23}\ \text{cm}^3$ where Ⓐ is Avogadro's number. For a number fraction of solute molecules $N_1/N = 10^{-2}$, formula (5.1.49) gives the osmotic pressure

$$\Pi \simeq 13 \times 10^5 \text{ pascal} \simeq 13 \text{ atmospheres.}$$

This is the pressure which would be exerted between a sugared cup of tea and pure water.

The molar fraction $N_1/N = 10^{-2}$ corresponds to a dilute solution when the molecules of solute and of solvent are about the same size. When the size of the solute molecules is one order of magnitude larger, it is only for smaller molar fractions, for instance $N_1/N = 10^{-5}$, that the solution can be considered as dilute. For polymer solutions, the osmotic pressures currently measured are of the order of 10^{-1} atmospheres.

1.10 Measurement of a 'surface' osmotic pressure.

At present time, the most sensitive technique for measuring 'surface' osmotic pressures is the Wilhelmy technique (1968);[4] it consists in measuring a lowering of surface tension. Let Γ_0 be the surface tension of a pure liquid, and Γ_{0+1} the surface tension of a liquid whose surface has been spread with a defined quantity of solute. It will be shown that the 'surface' osmotic pressure is simply

$$\Pi = \Gamma_0 - \Gamma_{0+1} > 0.$$

1.10.1 Surface tension and 'surface' osmotic pressure

Let us consider a mobile barrier of length L, vertically separating the thin layer from the pure liquid surface. It will be assumed that this barrier is wetted on both sides (see Fig. 5.5). The superficial tension Γ_0 of the bearing liquid exerts, on the barrier, a horizontal force

$$\mathscr{F}_0 = \Gamma_0 \times L \tag{5.1.104}$$

which is perpendicular to it.

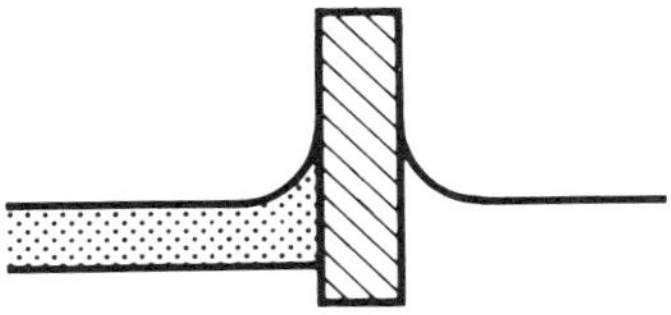

Fig. 5.5. Schematic representation of a mobile barrier bounding the thin layer.

In the same way, the surface tension Γ_{0+1} of the liquid covered with the thin layer exerts a horizontal force on the barrier

$$\mathscr{F}_{0+1} = \Gamma_{0+1} \times L$$

in the opposite direction.

By definition, the 'surface' osmotic pressure is the force which has to be exerted on the barrier to ensure equilibrium, and therefore

$$\Pi = -\mathscr{F}_{0+1}/L + \mathscr{F}_0/L = \Gamma_0 - \Gamma_{0+1},$$

which is the result we want.

The surface tensions, Γ_0 of the pure liquid, and Γ_{0+1} of the liquid covered with the adsorbed solute, are measured separately.

1.10.2 Effect of surface tension on a thin plaquette

Let us consider the surface of a liquid whose surface tension is to be measured and let us put a 'clean' platinum plaquette ($1 \times 1 \times 10^{-2}$ cm^3) vertically into contact with the liquid–air interface. The plaquette is suspended by a thread from a scale-pan of a balance, and one edge is in contact with the liquid (see Fig. 5.6). The liquid wets the platinum surface and a meniscus appears. The surface tension Γ exerts a downwards vertical force on the plaquette. This force is

$$\mathscr{F}_0 = 2\Gamma \times L$$

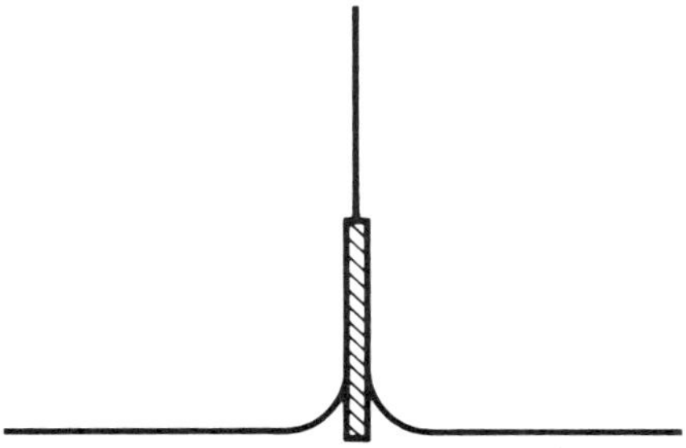

Fig. 5.6. Schematic view of a Wilhelmy plaquette, suspended by a thread, with its edge on the surface.

where L is the plaquette length. Its value can be obtained by measuring the difference δW between the weights of the free plaquette and of the plaquette in contact with the liquid. Thus

$$\Gamma = \delta W/2L.$$

By performing the operation successively for pure liquid and for the liquid with an adsorbed layer, one obtains the surface tensions Γ_0 and Γ_{0+1}.

The measurement sensitivity is of the order of 10^{-1} dyne cm^{-1} (10^{-4} newton m^{-1}).

1.11 Experimental results

The physicist who studies polymers in solution must determine how the osmotic pressure varies with concentration.

1.11.1 'Volume' osmotic pressure

The polymer concentration is defined by

$$\boldsymbol{C} = \boldsymbol{N}/\boldsymbol{V}$$

where $\boldsymbol{N}$ is the number of polymers and $\boldsymbol{V}$ the volume of solution. This quantity is convenient for theoretical studies of the osmotic pressure. However, it is not directly measurable. The measurable quantity corresponding to $\boldsymbol{C}$ is the mass concentration

$$\rho = \mathcal{M}/\boldsymbol{V}$$

where $\mathcal{M}$ is the mass (of solute).

We can thus take a mass $\mathcal{M}_1$ of solute and a given volume of solvent (the total volume). Let us put these quantities in the osmometer schematically represented on Fig. 5.4. From h^{I} we deduce the volume V^{I} of cell I. We shall assume that the solution is incompressible. The mass concentration of the solute is

$$\rho_1^{\mathrm{I}} = \frac{\mathcal{M}_1}{V^{\mathrm{I}}}$$

and the osmotic pressure, which is determined by h^{I}, h^{II}, is calculated with the help of formula (5.1.101)

$$\boldsymbol{\Pi} = \mathrm{g}(\rho^{\mathrm{I}} h^{\mathrm{I}} - \rho^{\mathrm{II}} h^{\mathrm{II}})$$

where g is the gravity intensity. In Fig. 5.7, we plotted osmotic pressure values measured by Noda, Kato, Kitano, and Nagasawa[5] for various values of the mass concentration ρ_2^{I} (denoted by ρ) of a poly(α-methyl)styrene sample, in toluene at 25°C

First, let us interpret these results by using Van't Hoff's law which is valid for very dilute solutions. The synthetically produced sample is polydisperse. We represent by $N_{\mathcal{A}}$ the number of molecules of molecular mass $\mathcal{M}_{\mathcal{A}}$ which it contains.

Then the Van't Hoff law reads

$$\Pi\beta = \sum_{\mathscr{A}} \frac{N_{\mathscr{A}}}{V} = \sum_{\mathscr{A}} C_{\mathscr{A}} = C.$$

The directly measurable mass concentration is

$$\rho = \frac{1}{\text{Ⓐ}} \sum_{\mathscr{A}} C_{\mathscr{A}} M_{\mathscr{A}}$$

where Ⓐ is the Avogadro number. Let us now express the osmotic pressure in terms of ρ. We get

$$\frac{\Pi}{RT} = \frac{\beta\Pi}{\text{Ⓐ}} = \rho/M_{\mathrm{n}} \tag{5.1.105}$$

where

$$M_{\mathrm{n}} = \sum_{\mathscr{A}} \frac{C_{\mathscr{A}} M_{\mathscr{A}}}{C}. \tag{5.1.106}$$

The quantity M_{n} is called the mean molecular mass. If the sample is made of identical particules with the same molecular mass M, then $M_{\mathrm{n}} = M$.

The value M_{n} corresponding to a given sample, is obtained from measurements of the osmotic pressure and of the mass concentration. In fact, we have

$$\lim_{\rho \to 0} \frac{\Pi}{RT\rho} = \frac{1}{M_{\mathrm{n}}}.$$

Thus, M_{n} is determined by extrapolating the measured values of Π and ρ. These values of ρ are as small as possible. However, they belong to a range in which the interchain interaction can be felt. Van't Hoff's law is not sufficient to reproduce the experimental result. In fact, Fig. 5.7 shows that the function $\Pi/(\rho g)$ has a non-negligible slope at the origin and this is an effect of the interaction. Actually, the osmotic pressure is given by a series expansion in powers of the concentration

$$\frac{\Pi}{RT} = \frac{\rho}{M_{\mathrm{n}}} + A_2\rho^2 + A_3\rho^3 + \ldots \tag{5.1.107}$$

where A_2 and A_3 are the second and the third virial coefficient.

With the help of this expression, let us extrapolate the measured values that appear in Fig. 5.7. By performing a 'mean square' adjustment of the pressures in the range $1\ \mathrm{g\,l^{-1}}$–$15\ \mathrm{g\,l^{-1}}$, we obtain $\dfrac{\Pi}{\rho g} = \dfrac{RT}{M_{\mathrm{n}} g} = 3.5$ m. Thus $M_{\mathrm{n}} g = RT/3.59$ newton.

For $T = 310$ K, $RT = 2.579 \times 10^3$ newton metre and $M_{\mathrm{n}} g = 7.184 \times 10^2$ newton.

Let us convert these results into molecular units. We have

$$M_{\mathrm{n}} = (7.184/9.81) \times 10^5 = 7.32 \times 10^4 \text{ dalton}.$$

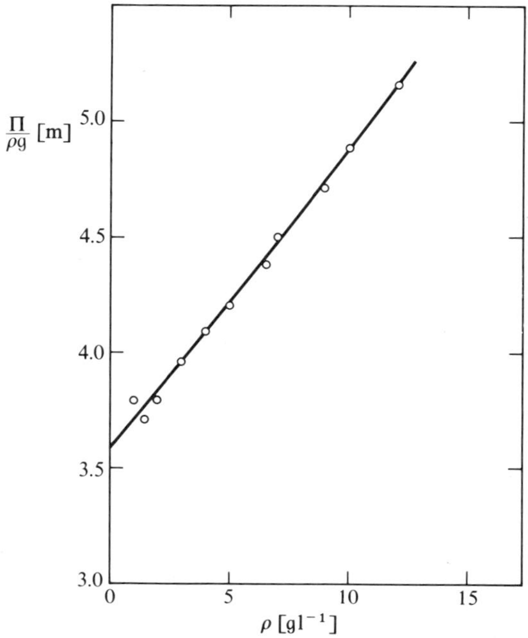

Fig. 5.7. The osmotic pressure Π of a poly(α-methyl)styrene sample in toluene at 25°C as a function of ρ^5. Here g is the gravity intensity. The measured values are represented by small circles. The curve is obtained in the interval 0–15 g/e by interpolation (polynomial of degree 3).

By interpolating the same experimental values, we also obtain

$$\boldsymbol{R}TA_2/\mathrm{g} = 0.116\ \mathrm{m}^4\,\mathrm{kg}^{-1}$$

$$\boldsymbol{R}TA_3/\mathrm{g} = 0.113 \times 10^{-2}\ \mathrm{m}^7\,\mathrm{kg}^{-2}$$

which leads to

$$A_2 = 4.4 \times 10^{-4}\ \mathrm{m}^3\,\mathrm{kg}^{-2} = 4.4 \times 10^{-4}\ \mathrm{cm}^3\,\mathrm{g}^{-2}$$

$$A_3 = 4.3 \times 10^{-6}\ \mathrm{m}^6\,\mathrm{kg}^{-3} = 4.3 \times 10^{-3}\ \mathrm{cm}^6\,\mathrm{g}^{-3}.$$

The fact that osmotic pressure is measured in a concentration range where the chain interactions are sizeable, compels us to examine the extrapolation process and its result $1/M_n$, in a critical way. In fact, let us note that the obtained value $1/M_n$ depends in a non-negligible way on the degree of the interpolating polynomial and on the concentration interval corresponding to the experiments.

The lower bound of this interval is related to the fact that there is a pressure threshold Π_s under which the values indicated by the osmometer loose any significance. This threshold defines a minimal concentration ρ_{min} which, according to Van't Hoff's approximation, is

$$\rho_{\min}(M_n) = \frac{\Pi_s M_n}{RT}.$$

Let us assume that $\Pi_s = 1$ pascal. In this case, $\rho_{\min} = 0.3\ \mathrm{g\,l^{-1}}$ for a molecular mass $M_n = 7 \times 10^4$.

The upper limit does not come from a technical limit, and if an exact and explicit extrapolation formula could be written, it would be possible to use measured values in a range as large as one would like. In fact, there is no such formula, and interpolation requires the use of limited expansions of type (5.1.107).

Each expansion of maximal degree q is valid in an interval $0 < \rho < \rho_{\min}$ in which one tries to represent the osmotic pressure, and the larger q is, the larger is $\rho_{\max}$. Actually, for a given q, one must to take care to choose for $\rho_{\min}$ a value which is not too large; otherwise, systematic errors occur and they lead to an underestimate of M_n when $q = 2$, and an overestimate of M_n when $q = 3$. According to Flory, this kind of error has prevented the macromolecular hypothesis from being accepted sooner.

On the other hand, for fixed ρ, the polymer overlap increases with M; the interval of ρ, corresponding to dilute solutions, becomes smaller and smaller. Consequently, measuring the osmotic pressure as a function of concentration cannot serve to determine the mass M_n of polymers with high molecular masses.

Actually, the dependence of the osmotic pressure with respect to polymer concentration, for polymers with high masses, reveals their spatial distribution. Let us present here two fundamental results. One of them shows an effect related to the quality of the solvent, the other an effect resulting from the size of the polymers. In both cases, the effects have to do with the polymer–polymer interaction.

Figure 5.8(a) shows how the osmotic pressure varies[6] with concentration for the same polyisobutylene sample, respectively in benzene and in cyclohexane solutions at 30 °C. We verify that the extrapolated quantities $\lim_{\rho \to 0} \frac{\Pi}{\rho g}$ do not depend on the nature of the solvent. On the contrary, we see that the contributions at finite concentrations do strongly depend on it: this is the first predicted effect. In particular, the coefficient A_2 corresponding to polyisobutylene–benzene solutions at 30 °C is ten times smaller than that of polybutylene–cyclohexane solutions. We then say that benzene is a poor solvent of polyisobutylene. Let us observe that the converse is true when the solute is polystyrene (Fig. 5.8(b)). Cyclohexane is a poor solvent of polystyrene but benzene is a good one for this polymer.

Figure 5.9 shows how the osmotic pressure varies with concentration for poly(α methyl)styrene with various molecular masses. The smallest molecular mass equals 7×10^4; the highest one equals 7×10^6 (this one was measured by

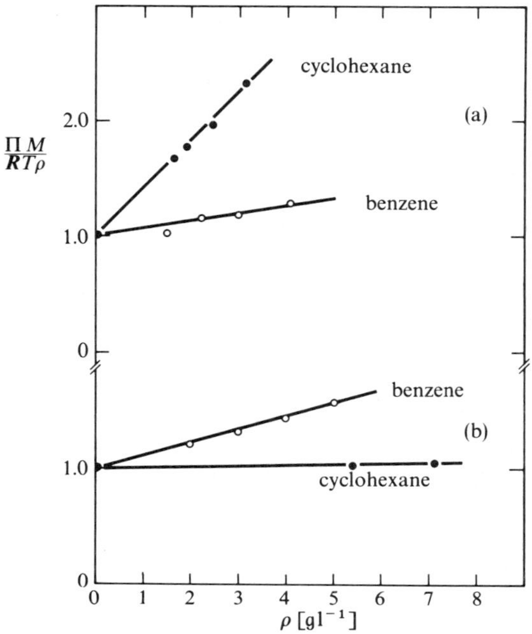

Fig. 5.8. (a) The osmotic pressure of a polyisobutylene sample in cyclohexane (•) and in benzene (○) at 37°C as a function of ρ.[6] (b) The osmotic pressure of a polystyrene sample in cyclohexane (•) and in benzene (○) as a function of ρ (ref. 7).

elastic light scattering; see Chapter 7). We verify that the extrapolated values

$$\lim_{\rho \to 0} \Pi/\rho g$$

are inversely proportional to the molecular mass. On the contrary, we see that beyond a certain concentration, the osmotic pressure only depends on concentration, and this is the second predicted effect.

From these two types of observations, which began to be made around 1950, the experimentalists have drawn interesting conclusions, which are valid for rather long chains in good solvents. They noted that the second virial coefficient A_2 values as the ratio R_G^3/M^2 where R_G is the radius of gyration of the chain [see (2.2.18)]. Thus, the volume $M^2 A_2$ is something like the volume occupied by the chain. On the other hand, a distinction was established betweeen a dilute regime in which the chains which repel one another remain far apart, and a semi-dilute regime in which the chains are obliged to interpenetrate deeply.

We shall give more details in Chapter 15, where these experiments will be interpreted and compared with results of the modern theory of polymer solutions. However, we have already described a fundamental property of these

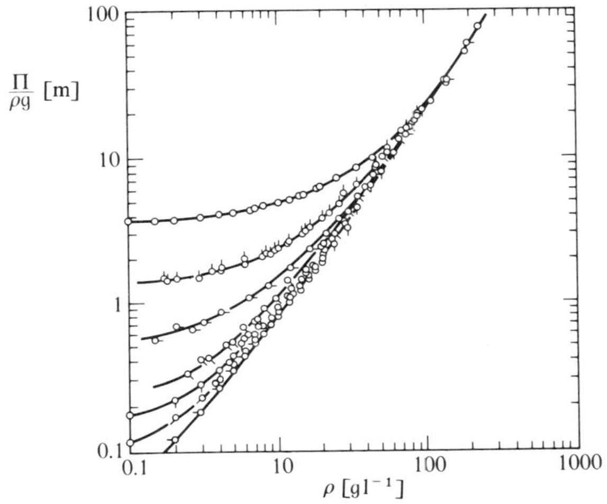

Fig. 5.9. The osmotic pressure of a sample of poly(α-methyl)styrene in toluene at 25 °C plotted against ρ for various molecular masses:[5]

○ $M_n = 7 \times 10^4$
$M_n = 20 \times 10^4$
○- $M_n = 50.6 \times 10^4$
$M_W = 119 \times 10^4$
$M_W = 182 \times 10^4$
$M_W = 330 \times 10^4$
-○ $M_W = 747 \times 10^4$

The average masses M_n M_W are defined respectively by eqns (1.1.6) and (1.1.7). M_n is measured by osmotic pressure and M_W by light scattering (see Chapter 7).

systems, which has only been recognized quite recently but which can be deduced from the experimental results mentioned above, by means of a simple renormalization operation. In fact, we may wonder whether a modification of the quality of the solvent or of the molecular mass of the solute is not equivalent to an effective concentration change. In order to answer this question, we shall consider two different solutions: The first one is a solution of polystyrene of molecular mass $M_{n1} = 7 \times 10^4$ in benzene at 37 °C and the second one a solution of polystyrene of molecular mass $M_{n2} = 15.5 \times 10^4$ in benzene at 30 °C. These samples have very different characteristics: the molecular masses of the solutes are very different and the solute–solvent interactions do not have quite the same strength (see Figs 5.8 and 5.9). Let us examine the values of the osmotic pressure, measured by Strazielle,[7] and let us use the representation

$$\Pi M_n / RT\rho = F(\rho),$$

which is convenient because here $F(\rho)$ is a pure number and $F(0) = 1$ as was shown before.

In Fig. 5.10, we plotted values $F_1(\rho_i)$ and $F_2(\rho_j)$ measured for various concentrations ρ_i and ρ_j and associated with the samples of molecular masses

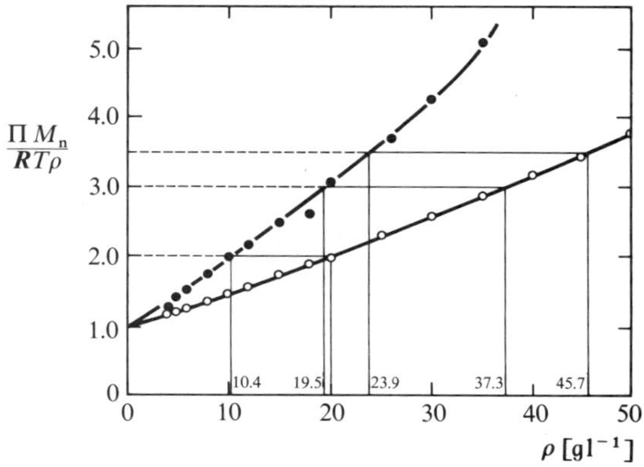

Fig. 5.10. The reduced osmotic pressures of two polystyrene samples, each one in a different solvent, plotted against ρ:
• $M_{n1} = 72\,000$ in toluene at 37°C; ○ $M_{n2} = 155\,000$ in benzene at 30°C (curves obtained by graphical interpolation).[7] The horizontal lines start from an arbitrary point on one curve. The corresponding abscissae, corresponding to the intersection of one line with the two curves, have a ratio which is independent of the height of the line under consideration.

M_{n1} and M_{n2} respectively. By graphical interpolation, we obtained the curves which appear in Fig. 5.10.

The renormalization operation for the concentrations is performed as follows. Let us arbitrarily choose an ordinate F. From the curves 1 and 2 and from the equality

$$F_1(\rho_1) = F_2(\rho_2) = F$$

we extract two values ρ_1 and ρ_2, and their ratio. We see that for three ordinates $F = 3.5$, $F = 3$, and $F = 2$, the ratio ρ_1/ρ_2 has the same value

$$\rho_1/\rho_2 = 1.92.$$

This shows that the curves are affine. Thus, if we renormalize, by a factor 1.92, the concentrations corresponding to the osmotic pressures of the first sample, we can transform the curve $F_1(\rho)$ into a curve which coincides with $F_2(\rho)$.

What is the meaning of this effective change of concentrations? Could differences in molecular masses and interactions induce variations of a more fundamental quantity? The interpretation of Fig. 5.9, briefly given above, tells us what this quantity can be. In fact, we have already observed that the product MA_2 varies with the quality of the solvent, like the mean volume $(R_G)^3$ occupied by a chain. But, in the limit $\rho \to 0$, the mass concentration of solute in this volume equals $\dfrac{M_n}{(R_G)^3}$ $\left(\text{for a given sample } i\text{, we write } \dfrac{M_{n,i}}{(R_{G,i})^3}\right)$. Now, let us come back to the polystyrene solutions 1 and 2 of Fig. 5.10. The scale transformation of concentrations which enabled us to superimpose the two osmotic pressure curves may come from a change of the volume 'occupied' by a chain; more precisely, we could postulate the equality

$$\frac{\rho_1}{M_{n,1} R_{G,1}^{-3}} = \frac{\rho_2}{M_{n,2} R_{G,2}^{-3}}. \quad (5.1.108)$$

Let us verify this assumption. If we admit that the volumes $R_{G,i}^3$ are related to the molecular masses by an equality of the form

$$R_{G,i}^2 = \Lambda_i (M_{n,i})^{2\nu} \quad (5.1.109)$$

where Λ_i is a constant which depends on the solvent and where ν is the size-critical exponent (excluded volume effect), we obtain, for $\nu = 0.59$,

$$\rho_1/\rho_2 = 1.73(\Lambda_2/\Lambda_1)^{3/2}.$$

we have just seen that

$$\rho_1/\rho_2 = 1.92$$

which gives

$$\Lambda_2/\Lambda_1 = (1.92/1.73)^{2/3} = 1.072.$$

Actually, it is known[8] that polystyrene swells a little more in benzene than in toluene, and this fact agrees with our interpretation.

In our evaluation, we did not take into account an eventual polydispersion difference concerning the molecular masses of the two samples. Such a difference would introduce a correction factor in (5.1.107).

Finally, we may say that polymers in solution interact with one another and that this interaction which is observed by osmotic pressure in the dilute concentration domain, in good solvents, is related to the chain size. We shall later examine this relation in detail, and we shall also see that it extends to the semi-dilute regime (see Chapter 13).

1.11.2 Osmotic pressure for polyelectrolytes: singular behaviour

Though the study of linear polyelectrolytes is beyond the scope of this book, their behaviour is solution differs so much from the behaviour of neutral polymers, that it is worthwhile to mention a few striking properties of these solutions.

A polyelectrolyte is a polymer which, dissolved in water, ionizes and gives one poly-ion and small-size counter-ions. Thus, sodium polystyrene sulphonate is a typical poly-electrolyte.

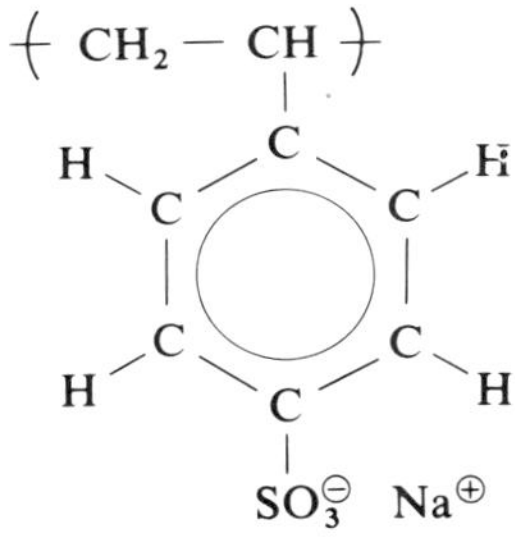

The number of free charges carried by the chain is called the ionization degree Z and the number of $Na^{\oplus}$ counter-ions simultaneously produced also equals Z (where $Z \leqslant N$).

It can be shown that an isolated polymer, in solution in pure water is always completely ionized* ($Z = N$); then the polyion has the shape of a nearly rigid rod. This entails that the solutions that are currently used are nearly always semi-dilute, i.e. in these solutions the chains overlap one another.

The membranes used to measure the osmotic pressure are impermeable to polyions but permeable to counter-ions. In spite of this, the osmotic pressure of polyelectrolytes in pure water and for rather small concentrations (but in the semi-dilute regime), is huge. Let us assume that the polyelectrolyte has been put in cell I (see Fig. 5.1). It partially ionizes but both cells must remain practically neutral. The counter-ions which, theoretically, can cross the membrane, are retained in cell I and a contact-potential difference appears at the boundary between the cells. Actually, it looks as if the counter-ions contributed like polyions to the osmotic pressure. Let $\boldsymbol{C}$ be the polyion concentration. We may write approximately

$$\boldsymbol{\Pi\beta} = \boldsymbol{C}(1 + Z) \tag{5.1.110}$$

where Z is smaller than N but of the same order ($N \gg 1$). In fact, we note that, in the case under consideration, the counter-ions form a dilute solution and, therefore, their contribution to the osmotic pressure, obeys Van't Hoff's law: this is what the preceding equation says.

Osmotic pressure measurements of sodium polystyrene–sulphonate, with a molecular mass $M = 4.5 \times 10^5$ have been made by Takahashi and Nagasawa,[10] and the result is represented in Fig. 5.11. The abscissa is the mass-concentration

$$\rho = \boldsymbol{C}M/Ⓐ\ ,$$

the ordinate is the reduced osmotic pressure

$$\boldsymbol{\Pi}M/\boldsymbol{R}T\rho = \boldsymbol{\Pi\beta}/\boldsymbol{C}.$$

Thus, eqn (5.1.108) can be written

$$\boldsymbol{\Pi}M/\boldsymbol{R}T\rho = \mathrm{Z}$$

and we can see from Fig. 5.11 that for a solution in pure water and a mass-concentration $\rho > \rho^* = 5 \times 10^{-3}$ g/l, we have effectively a behaviour of this type with Z = constant. The ρ^* concentration is the overlap concentration for 150 nm long rods. On the other hand, we find

$$Z/N = 0.2,$$

which shows that the chains are only partially ionized.

* This is true only for a finite chain; if the chain is infinite, counter-ions may condense on the polyion: this is Manning's condensation.[9]

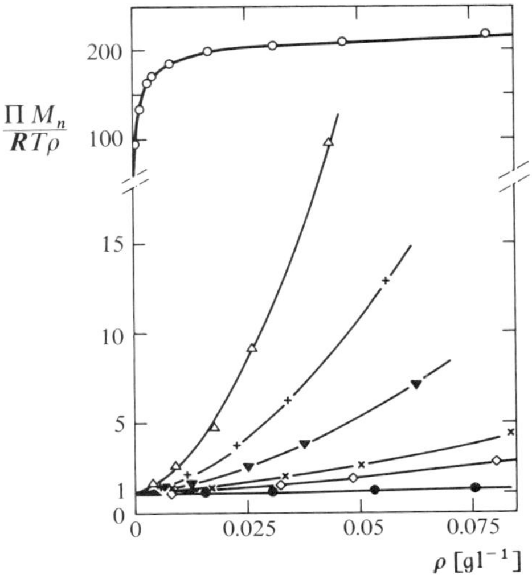

Fig. 5.11. Measured values of the osmotic pressure of an aqueous solution of sodium polystyrene-sulphonate for various values of the mass concentration $\rho[g/l]$ of polyelectrolyte. Molecular mass $M = 3.5 \times 10^5$.[10]
o Polyelectrolyte in pure water. The other points correspond to sodium polystyrene-sulphonate in water with added salt (ClNa).
The concentrations in added salt are

△ $\rho_s = 0.005$ g/l	× $\rho_s = 0.05$ g/l
+ $\rho_s = 0.01$ g/l	◇ $\rho_s = 0.1$ g/l
▼ $\rho_s = 0.02$ g/l	● $\rho_s = 0.5$ g/l

The situation changes in a dramatic way if salt is added to the polyelectrolyte solution. For instance, let us assume that a small amount of ClNa or of any similar salt is added in cell I (or in cell II). Each ClNa gives a $Cl^{\ominus}$ ion and a $Na^{\oplus}$ ion that is able to cross the membrane. The entropy of the system is larger if all the $Na^{\oplus}$ counter-ions are distributed as uniformly as possible, over both cells, but electrical neutrality has to be preserved. Consequently, the co-ions $Cl^{\ominus}$ are attracted into cell II. This is the *Donnan effect*. Simultaneously, the osmotic pressure diminishes in a spectacular way, and the same is true for the contact potential between the two cells. If a lot of salt was added, one would find again

$$\Pi\beta = C$$

(see Fig. 5.11).

Salt pumps, in aqueous solution, work according to this principle; polyelectrolyte is added to a cell and the salt is driven out into the other one. After this, it is easy to get rid of the polyelectrolyte.

1.11.3 'Surface' osmotic pressure

Vilanove and Rondelez[11] measured the 'surface' osmotic pressure of a polymer as a function of concentration. They used poly-vinyl acetate (PVAc) spread on an air–liquid interface. They determined the osmotic pressure by using the Wilhelmy method (see Section 1.10).

Two samples of PVAc with molecular masses $M = 1.92 \times 10^5$ and $M = 4.5 \times 10^5$ have been studied successively. The osmotic pressures have been obtained for 20 different mass concentrations ρ, between 1×10^{-3} and 2×10^{-3} g/m^2. The results are shown in Fig. 5.12. We see that the pressures associated with both PVAc samples superimpose.* The above mentioned experi-

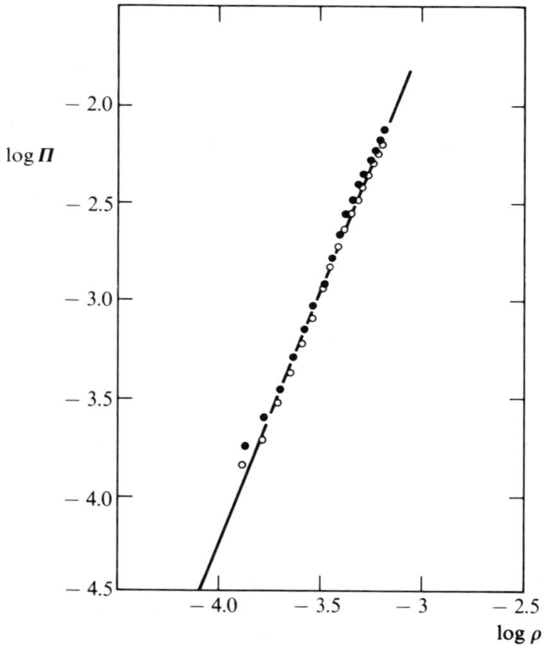

Fig. 5.12. The osmotic surface pressures of two polyvinyl-acetate samples plotted against concentration (ref. 11).

o $M = 192\,000$ ● $M = 450\,000$
Surface pressure in $N\,m^{-1}$.
Concentration in g m^{-2}.

* The curves superimpose because here we are dealing with the same polymer–solvent couple and not because of a more general universality principle

mentalists interpreted the observed variation with the help of the formula[12]

$$\Pi\beta \propto \rho^{\frac{\nu d}{\nu d - 1}}$$

where, here, $d = 2$ and where ν is the corresponding size exponent of the polymer. The value extracted from the results is $\nu = 0.79 \pm 0.01$. It is to be compared with the 'exact' value $\nu = 0.75$ (see Chapter 12, Section 4).

The air–water interface must be considered as a good solvent for the polyvinyl acetate used in this experiment. Vilanove and Rondelez[11] also used another polymer namely methyl-polymethacrylate, but in this case, the air–water interface must be considered as a poor solvent, and a different exponent ν is obtained. This exponent is interpreted as a tricritical exponent (see Chapter 14, Section 7.1). The value which they found for the exponent is

$$\nu_t = 0.56 \pm 0.01.$$

2. DENSITOMETRY

Densitometry is the measurement of masses per unit volume. The densitometry technique has reached a very high degree of accuracy owing to the mechanical resonance technique, developed in 1967 by Kratky, Leopold, and Stabinger.[13]

At that time, it seemed to be particularly interesting to know precisely the mass per unit volume of a solution.

Let us first note that the masses per unit volume for the solvent and for the solution are parameters that are needed in order to exploit osmotic pressure measurements. We may thus recall relation (5.1.101)

$$\Pi\beta = g(\rho^{\mathrm{I}}h^{\mathrm{I}} - \rho^{\mathrm{II}}h^{\mathrm{II}})$$

where ρ^{I} and ρ^{II} are masses per unit volume for the solution and for the solvent, h^{I} and h^{II} the level heights in the corresponding cells, and g the gravity intensity.

We shall see, in next chapter, that the interpretation of scattering experiments made with solutions, also requires a precise knowledge of a quantity related to the mass per unit volume; namely, the partial mass per unit volume.

Still, densitometry should not be considered only as an auxiliary technique whose data are used as parameters. In fact, this technique developed with the study of the volume changes which occur when a material is dissolved in a solvent. Typically, if one compares the volume of two equal quantities of solvent and solute to the volume of their solution, one finds a relative volume change of the order of 10^{-3}.

The fact that a solution is not a constant volume system, but that its volume depends on its composition in a non-trivial way, is of great importance. Let us point out typical effects the appearance of which, in solutions, is related to this volume change:

1. Solvatation[14] is the formation of solvent–solute complexes. This phenomenon entails a local modification of the solvent structure and consequently a volume change. Thus, hydration of ions in aqueous solutions is a special instance of solvatation, and it can be observed by measuring the mobility of the ions as a function of the ratio (electric charge of an ion)/(size of an ion).
2. When a polymer solution is heated, a second demixtion point occurs at a temperature T_2 and for a concentration C_2. In 1967, Flory, Orwoll, and Vrij[15] showed that the existence of this second critical point is related to variations of the volume of the solution, depending on its composition. Of course, the point (T_2, C_2) differs from the point (T_1, C_1) which is found by lowering the temperature and which has a completely different origin.
3. Another interesting effect has been mentioned by Francois, Candau, and Benoit,[16] who observed an anomalous decrease of the partial volume per unit mass of solute when the polymerization degree increases.

In connection with the preceding phenomena, another effect is related to the dissolution of a polymer in a mixture of two solvents. The relevant quantity in this case is preferential adsorption. Let N_0, N_1, and N_2 be respectively the number of molecules of the main solvent, of the second solvent, and of the polymer. Preferential adsorption is by definition [see (5.1.85) and (5.1.92)]

$$\lambda_0 = \lim_{N_2 \to 0} \left.\frac{\partial N_1}{\partial N_2}\right)_{\mu_0, \mu_1}$$

for fixed values μ_0 and μ_1 of the chemical potentials of the solvents. Hert, Strazielle, and Benoit[17] studied systematically how λ_0 varies, for linear and branched polymers, with respect to the composition of the solvent mixture, the molecular mass M of the polymer chains, and the temperature T of the solution. They noted that $\lambda_0(M, T)$ varies in the same way as $v(M, T)$, the polystyrene partial volume in the solvent.

According to the above-mentioned authors, this behaviour similarity can be understood in simple terms; preferential adsorption occurs when a non-adsorbed molecule is replaced by an adsorbed molecule; the partial volume varies when a solvent molecule is replaced by a 'free' volume. In some way, the free volume plays the role of a second solvent.

The volume changes which occur during dissolution can be directly obtained, by subtraction, from measurements of the volumes of the constituents before mixing and after mixing. However, these changes of volume can be observed in a more indirect manner, which turns out to be more precise. This method consists in measuring the mass per volume of solution as a function of composition. We shall describe this measurement method in the following sections, but first we must establish the relations which exist between masses per volume, partial volumes per mass for the solvent and the solute, and volume of solution.

2.1 Partial volume per mass

Let V be the volume of a solution, $\mathcal{M}_0$ the mass of solvent, $\mathcal{M}_1$ the mass of solute. The partial volumes per mass for the solvent and the solute are the quantities

$$w_{\mathcal{A}} = \left.\frac{\partial V}{\partial \mathcal{M}_{\mathcal{A}}}\right)_{\mathcal{M}_{\mathcal{B}}(\mathcal{B} \neq \mathcal{A})} \qquad \mathcal{A} = 0, 1 \tag{5.2.1}$$

where $\mathcal{A} = 0$ for the solvent and $\mathcal{A} = 1$ for the solute.

The partial volumes per mass are proportional to the partial molecular volumes defined by (5.1.26)

$$v_{\mathcal{A}} = \left.\frac{\partial V}{\partial N_{\mathcal{A}}}\right)_{N_{\mathcal{B}}(\mathcal{B} \neq \mathcal{A})} \tag{5.2.2}$$

where $N_{\mathcal{A}}$ is the number of molecules of species $\mathcal{A}$ in the solution. The relation is

$$w_{\mathcal{A}} = v_{\mathcal{A}} Ⓐ / M_{\mathcal{A}}$$

where $M_{\mathcal{A}}$ is the molecular mass of species $\mathcal{A}$ and Ⓐ the Avogadro number. We have

$$V = w_0 \mathcal{M}_0 + w_1 \mathcal{M}_1 \tag{5.2.3}$$

where V is a homogeneous function of $\mathcal{M}_0$ and $\mathcal{M}_1$ (V_0 and V_1 are functions or the ratio $\mathcal{M}_1/\mathcal{M}_0$).

Let us divide this equation by $(\mathcal{M}_0 + \mathcal{M}_1)$. Thus, the volume per mass of solution

$$w_s = V/(\mathcal{M}_0 + \mathcal{M}_1)$$

and the mass fraction of solute

$$\psi = \mathcal{M}_1/(\mathcal{M}_0 + \mathcal{M}_1)$$

immediately appear. We get

$$w_s = \psi w_1 + (1 - \psi) w_0.$$

The directly measurable quantities are w_s, ψ and ${}^0w_0$ which is the partial volume per mass for pure solvent. Starting from these quantities, we can define the *apparent* volume per mass of solute $\bar{w}_1$ by writing

$$w_s = (1 - \psi)\,{}^0w_0 + \psi \bar{w}_1 \tag{5.2.4}$$

The relation between the partial volumes per mass w_1 and $\bar{w}_2$ is determined as follows. First, let us write (5.2.4) in the form $V = {}^0w_0 \mathcal{M}_0 + \bar{w}_1 \mathcal{M}_1$ and let us differentiate V with respect to $\mathcal{M}_1$. We get

$$\begin{aligned} w_1 = \left.\frac{\partial V}{\partial \mathcal{M}_1}\right)_{\mathcal{M}_0} &= \bar{w}_1 + \mathcal{M}_1 \left.\frac{\partial \bar{w}_1}{\partial \mathcal{M}_1}\right)_{\mathcal{M}_0} \\ &= \bar{w}_1 + \mathcal{M}_1 \left.\frac{\partial \bar{w}_1}{\partial \psi}\frac{\partial \psi}{\partial \mathcal{M}_1}\right)_{\mathcal{M}_0} \\ &= \bar{w}_1 + \psi(1-\psi)\frac{\partial \bar{w}_1}{\partial \psi}. \end{aligned} \tag{5.2.5}$$

In the same way, for the solvent, we have

$$w_0 = \left.\frac{\partial V}{\partial \mathcal{M}_0}\right)_{\mathcal{M}_1} = {}^0w_0 - \psi^2 \frac{\partial \bar{w}_1}{\partial \psi}$$

It is thus possible to obtain the partial volume w_1 from the apparent volume per mass $\bar{w}_1$ and from the mass fraction ψ.

2.2 Measurement of the apparent partial volume per mass

The quantity $\bar{w}_1$ can be expressed in terms of (measured) masses per volume

$$\rho_s = w_s^{-1}$$

$${}^0\rho_0 = ({}^0w_0)^{-1}$$

In fact, according to (5.2.4), we have

$$\bar{w}_1 = {}^0w_0 + \frac{1}{\psi}(w_s - {}^0w_0)$$

which gives

$$\bar{w}_1 = \frac{1}{{}^0\rho_0}\left[1 - \frac{(\rho_s - {}^0\rho_0)}{\psi \rho_s}\right]. \tag{5.2.6}$$

The difference $\rho_s - {}^0\rho_0$ can be measured with the help of immersion scales (see Fig. 5.13). Let V_c be the volume of the cylinders, and $\mathcal{M}$ the mass which has to be added in one side to maintain equilibrium. We then have $\rho_s - {}^0\rho_0 = M/V_c$.

However, the apparatus that gives the most precise measurement of $\rho_s - {}^0\rho_0$ works in a completely different manner. This apparatus was devised by Kratky *et al.*[13] It consists of a tube, one end of which is set in a vertical wall, and of a system which excites transversal vibrations of this tube (see Fig. 5.14). The tube is filled with solution. The tube-plus-solution device constitutes a system which has its own vibration pulsation ω. The mass per volume of solution ρ_s is obtained from the observed value of ω. The whole thing can be compared to a vibrating beam and the elastic forces are practically independent of the solution.

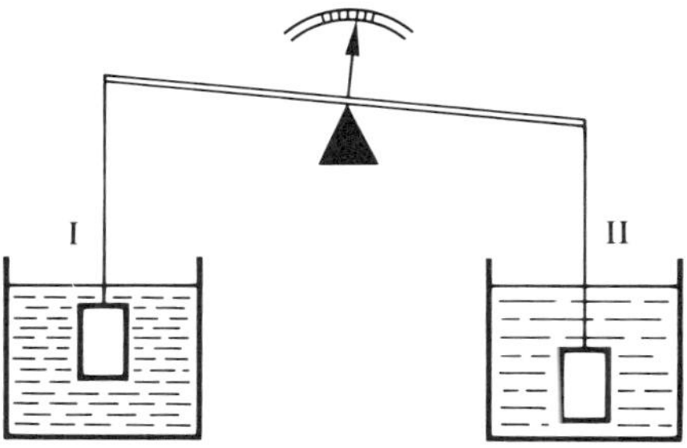

Fig. 5.13. Immersion scales used for measuring the difference of the masses per volume of solution (I) and of pure solvent (II).
Both immersed cylinders have the same volume V.

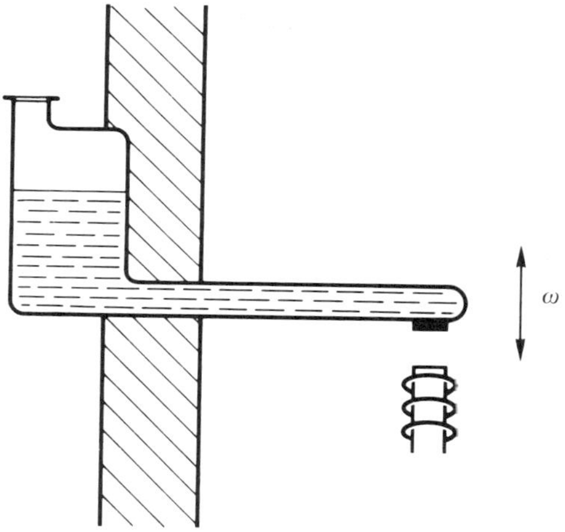

Fig. 5.14. Schematic view of the Kratky, Leopold, and Stabinger apparatus for the measurement of the mass per volume of solution. The solution is contained in a tube set in a wall. The protruding end of the tube is fitted with a magnet. An electromagnetic system produces transversal vibrations of the tube and enables the experimentalist to determine the resonance pulsation ω.

Thus, pulsation ω is a simple function of the mass $\mathscr{M}$ of solution in the tube

$$\omega^2 = \frac{E}{\mathscr{M} + \mathscr{M}_0} \tag{5.2.7}$$

where E and $\mathscr{M}_0$ are phenomenological quantities characterizing the tube; E has the dimension of an elastic constant, $\mathscr{M}_0$ of a mass. Thus, we have

$$\mathscr{M} = \frac{E}{\omega^2} - \mathscr{M}_0 .$$

We can write

$$\rho^{s} = A\left[\frac{1}{\omega^2} - \frac{1}{\omega_0^2}\right]$$

where A and ω_0 are apparatus parameters which can be obtained by performing calibration experiments with air and water. A typical value of the resonance pulsation is $6 \times 10^2\ \mathrm{s}^{-1}$ and the absolute precision for the measurement of the vibration period is 10^{-5} s. This precision is revealed by the great sensitivity of the measurement to temperature variations; these variations should be inferior to a millidegree ($\delta T \geqslant 10^{-3}$ K) in order to avoid fluctuations in the recording of pulsation ω.

2.3 Partial volume of a polymer

With the help of the above-described apparatus, Francois *et al.*[16] discovered a characteristic effect in linear polymer solutions. A preliminary study had already shown to these authors that the partial volume per mass and the branching ratio are related. A measurement of partial volumes per mass should in principle show a chain-end effect. This effect was observed; however, another unexpected effect appeared.

Problems concerning end-effects for long chains have already been mentioned in Chapter 1. The chemical structure of a chain with polymerization degree N can be represented by

$$\mathrm{A}\!\left(X\right)_N\mathrm{B}$$

where X is a monomer and A, B the chain ends. For a given polymer, A and B depend on the fabrication process. In general, the densities of these monomers are different: for relatively low values of the polymerization degree N ($N < 10^2$), the volume v_1 of the chain appreciably depends on the chemical nature of the chain ends. Let us set

$$v_1 = (N + 2)v_{\mathrm{m}} + v' \tag{5.2.9}$$

where v_{m} is the monomer volume and where v' is the error which is made by attributing to the chain ends a volume equal to the monomer volume. This volume v' is positive or negative according to the nature of A and B. The molecular volume per unit monomer is

$$v = v_1/(N + 2) = v_{\mathrm{m}} + v'/(N + 2). \tag{5.2.10}$$

In the following, $(N + 2)$ will be replaced by N, but before discussing this formula, let us consider the process which leads to the determination of v.

The quantities measured in the experiment are the mass per volume of solution ρ_{s} and the mass per volume of pure solvent ${}^{0}\rho_0$. The apparent partial volume per mass of polymer is deduced from these (measured) quantities by applying eqn (5.2.6). The partial volume per mass w_1 is itself a function of $\bar{w}_1$ and of ψ given by (5.2.5). The chain partial volume is thus given by

$$v_1 = w_1 M / Ⓐ$$

(where M is the molecular mass) and the partial volume per monomer is

$$v = v_1 / N.$$

The authors quoted above determined the *apparent* partial volume per mass of polymer $\bar{w}_1$. However, they did not obtain the real volume per mass w_1. Therefore, we shall only consider the apparent volume of the chain

$$\bar{v}_1 = \bar{w}_1 M / Ⓐ \tag{5.2.11}$$

and the apparent volume of the monomer

$$\bar{v} = \bar{v}_1 / N. \tag{5.2.12}$$

A systematic study of $\bar{v}_1(N)$ considered as a function of the nature of the chain-ends and of the molecular mass has established the following facts

1. For masses M smaller than 10^4 dalton ($N < 100$), the partial volume is very dependent on the nature of the chain-ends. In most cases, $\bar{v}(N)$ decrease with N, and this behaviour corresponds to the fact that the volume of the end-groups is larger than the volume of a monomer. However, there are polymerization initiators for which the volume error v' is negative; in this case, $\bar{v}(N)$ increases with N.

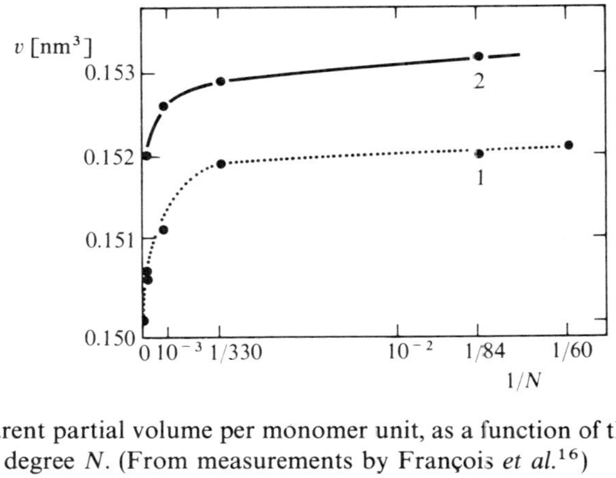

Fig. 5.15. Apparent partial volume per monomer unit, as a function of the inverse of the polymerization degree N. (From measurements by François *et al.*[16])

(1) in good solvent (benzene at room temperature);

(2) in poor solvent (cyclohexane at 35 °C).

The chain-end effect is observed for low polymerization degrees ($N < 10^2$). We note that, for $N > 10^2$, the partial volume varies less in poor solvent than in good solvent.

2. This end-effect is negligible for $N > 100$, in the sense that the curves associated with different values of v' converge towards a unique curve when $N > 100$.

3. However, the partial volume is not independent of N, even for $N \sim 10^3$, as can be seen from Fig. 5.15, on which values of $\bar{v}(N)$, measured by Francois, Candau and Benoit[16] for linear polystyrene initiated with cumyl–potassium, are plotted against $1/N$. We see that (5.2.9) is not realistic and must be replaced by a formula of the form

$$v_1 = (N + 2)v_m + v' + f(N)$$

where $f(N)$ is a slowly varying function such that $f(N) \simeq Nv''$ for $N < 100$ and $f(+\infty) = f$. This formula could lead us to believe that the chain is able to polarize the medium at distances larger than one could a priori imagine.

4. We observe that, for a given polymerization degree, the partial volume is higher in cyclohexane than in benzene. The origin of this difference is not known exactly. It may be related to two different facts: (a) the molar volume of cyclohexane is larger than the molar volume of benzene; (b) benzene is a better solvent of polystyrene than cyclohexane. Incidentally, we may also note, in connection with these facts, that the partial volume of a chain attains its highest value when it is in the molten state.

5. The same effect has been observed for molten linear polymers[18] (see Fig. 5.16) and also for branched polymers in solution (see Fig. 5.17). The study of

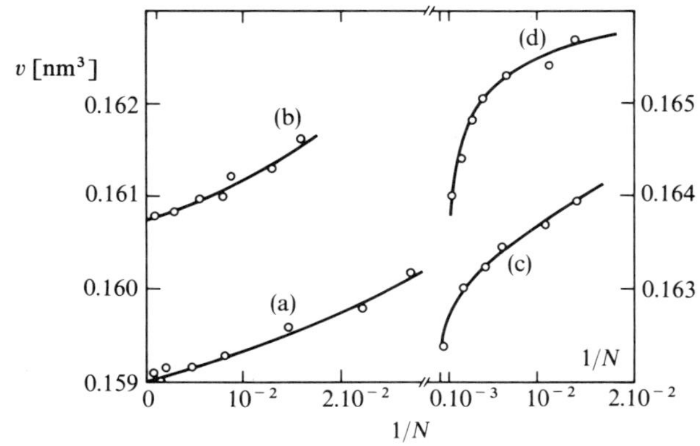

Fig. 5.16. Volume per monomer unit of molten polystyrene, as a function of the inverse of the polymerization degree, at various temperatures (from measurements by Fox and Flory)[18]:

(a) 140 °C; (b) 170 °C; (c) 204 °C; (d) 237 °C.

We note that an 'anomalous' behaviour appears for high values of N and temperatures larger than 200 °C.

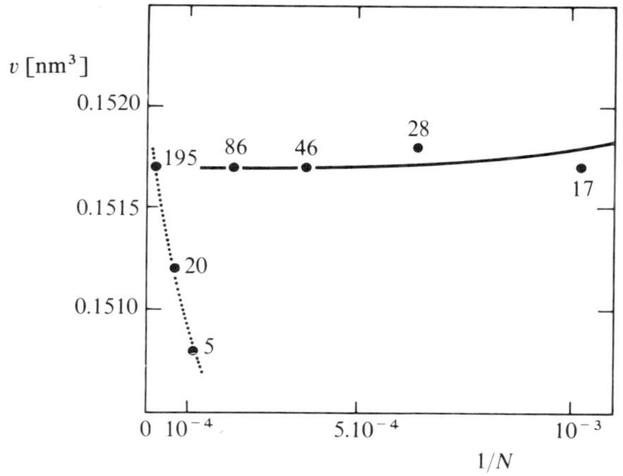

Fig. 5.17. Partial volume of comb-shaped polystyrene chains, in solution in benzene, as a function of the inverse of the total number of links (skeleton and teeth).

The number of comb teeth is indicated on the figure near each measurement point. Curve on the left: the number of links constituting the skeleton is constant (equal to 7×10^3). Curve in the upper part: the number of links in each tooth is constant (equal to 45).

molten polymers is of particular interest because it shows that this anomalous effect is temperature-dependent. Thus, it appears much more clearly at 237 °C than at 170 °C.

A coherent explanation of all these facts concerning the behaviour of $v_1/N)$ is still lacking, in spite of a few attempts.[19]

REFERENCES

1. Guggenheim, E.A. (1967). *Thermodynamics*, Chapter 1. North-Holland, Amsterdam.
2. Ewart, E.H., Roe, C.P., Debye, P., and McCartney, J.R. (1946). *J. Chem. Phys.* **14**, 687.
3. Chu, S.G. and Munk, P. (1978). *Macromolecules* **11**, 101, 879.
4. Adamson, A.W. (1976). Physical Chemistry of Surfaces, Chapter III. John Wiley, New York.
5. Noda, I., Kato, N., Kitano, T., and Nagasawa, M. (1981). *Macromolecules* **14**, 668
6. Krigbaum, W.R. and Flory, P.J. (1953). *J. Am. Chem. Soc.* **75**, 1775.
7. Values communicated to the authors by C. Strazielle Centre de Recherche des Macromolécules (Institut Charles Sadron), Strasbourg.
8. Orofino, T.A. and Flory, P.J. (1957). *J. Chem. Phys.* **26**, 1067.
9. Manning, G.S. (1969). *J. Chem. Phys.* **51**, 924.
10. Takahashi, A., Kato, N., and Nagasawa, M. (1970) *J. Phys. Chem.* **74**, 944.
11. Vilanove, R. and Rondelez, F. (1980). *Phys. Rev. Lett.* **45**, 1502.
12. des Cloizeaux, J. (1975). *J. Physique* **36**, 281.

13. Kratky, O., Leopold, H., and Stabinger, H. (1973). In *Methods in enzymology*, Vol. XXVII, p. 98 (presented by C. Hirs and S. Timascheff).
14. Bernal, J.D. and Fowler, R.H. (1933). *J. Chem. Phys.* **1**, 515.
15. Flory, P.J., Orwoll, R.A., and Vrij, A. (1964). *J. Am. Chem. Soc.* **86**, 3507, 3515.
16. Francois, J., Candau, F., and Benoit, H. (1974). *Polymer* **15**, 618.
17. Hert, M., Strazielle, C., and Benoit, H. (1973). *Makromol. Chem.* **172**, 169.
18. Fox, T.G. and Flory, P.J. (1950). *J. Appl. Phys.* **21**, 581.
19. Khokhlov, A.R. (1978). *Polymer* **19**, 1387.

6

RADIATION SCATTERING

GENESIS

The interaction between radiation and condensed matter produces multiple effects, which have been the subject of many observations and of very different studies. Scattering has been one of the first observed effects.

Originally, scattering was studied in order to know more about the nature of light and the interaction of light with matter. The discovery of the Tyndall effect (1863) is a typical example. When white light encounters a material, the scattered light often appears as coloured, the reason being that for shorter wavelengths the scattering is stronger.

Later, interference effects appeared as especially important. In fact, it is possible to derive from them correlations between scattering centres. In 1912, von Laue obtained the first diffraction plates for X-rays, and since 1912 scattering has commonly been used to investigate microscopic structures in condensed matter.

For about thirty years, neutron radiation sources have been available and in 1947, Fermi predicted that this radiation would be very useful for the study of complex organic structures. To-day, this prediction has been realized and numerous observations of correlations in such media are the exclusive result of neutron experiments.

1. ELECTROMAGNETIC RADIATION AND NEUTRON RADIATION

Scattering experiments can be performed by using very different kinds of radiation. The choice made by the experimentalist depends on a number of conditions which will be defined in the present chapter and in the next.

1.1 Electromagnetic radiation

An electromagnetic wave of wave vector $\vec{k}_0$ and of pulsation ω is described, classically,[1] by an electric field* $\vec{E}(\vec{r}, t)$ and a magnetic field† $\vec{B}(\vec{r}, t)$ at position vector $\vec{r}$ and time t. The plane wave propagating in direction OZ of the reference

* Formerly called induction.

system OX, OY, OZ is defined by the equalities.

$$\vec{E}(z, t) = \vec{E}^{-}e^{i(k_0 z - \omega_0 t)} + \vec{E}^{+}e^{-i(k_0 z - \omega_0 t)}$$
$$\vec{B}(z, t) = \vec{B}^{-}e^{i(k_0 z - \omega_0 t)} + \vec{B}^{+}e^{-i(k_0 z - \omega_0 t)} \tag{6.1.1}$$

where the $\vec{E}^{\pm}$, $\vec{B}^{\pm}$ are vectors whose components are complex numbers such that $\vec{E}^{-} = (\vec{E}^{+})^*$. More generally, a coherent radiation can be decomposed into two parts

$$\vec{E}(\vec{r}, t) = \vec{E}^{-}(\vec{r}, t) + \vec{E}^{+}(\vec{r}, t)$$
$$\vec{B}(\vec{r}, t) = \vec{B}^{-}(\vec{r}, t) + \vec{B}^{+}(\vec{r}, t) \tag{6.1.2}$$

where the symbol $-$ is associated with terms of type $e^{-i\omega t}$ and the symbol $+$ with terms of type $e^{i\omega t}$. Consequently, as the fields are real

$$\vec{E}^{-}(\vec{r}, t) = [\vec{E}^{+}(\vec{r}, t)]^* \qquad \text{and} \qquad \vec{B}^{-}(\vec{r}, t) = [\vec{B}^{+}(\vec{r}, t)]^*.$$

The fields obey Maxwell's equations

$$\begin{cases} \partial_t \vec{B}(\vec{r}, t) = -\vec{\nabla} \wedge \vec{E}(\vec{r}, t) \\ \vec{\nabla}\cdot\vec{B}(\vec{r}, t) = 0 \end{cases}$$
$$\begin{cases} \epsilon\partial_t \vec{E}(\vec{r}, t) = \mu^{-1}\vec{\nabla} \wedge \vec{B}(\vec{r}, t) \\ \vec{\nabla}\cdot\vec{E}(\vec{r}, t) = 0 \end{cases} \tag{6.1.3}$$

For a plane wave (6.1.1) propagating in direction OZ, we deduce from (6.1.3), (6.1.4);

$$B_Z^{\pm} = E_Z^{\pm} = 0$$

$$B_Z^{\pm} = -\frac{k_0}{\omega_0} E_Y^{\pm} \qquad \text{and} \qquad B_Y^{\pm} = \frac{k_0}{\omega_0} E_X^{\pm}.$$

where ϵ is the dielectric constant and μ the magnetic permeability. These two quantities are related to the velocity c of light

$$1/c = (\epsilon\mu)^{1/2}. \tag{6.1.4}$$

The energy density associated with the radiation

$$U_{\text{ph}}(\vec{r}, t) = \epsilon\vec{E}^{+}(\vec{r}, t)\cdot\vec{E}^{-}(\vec{r}, t) + \mu^{-1}\vec{B}^{+}(\vec{r}, t)\cdot\vec{B}^{-}(\vec{r}, t)$$

can be written

$$U_{\text{ph}} = \epsilon(|E_X^{+}|^2 + |E_Y^{+}|^2). \tag{6.1.5}$$

Here, the axes OX, OY are called polarization axes. The wave polarization is determined by the complex numbers E_X^-, E_Y^-. If E_X^-/E_Y^- is real, this polarization is linear. In the special case where $E_X^-/E_Y^- = \pm i$, this polarization is circular. In general, it is elliptic.

Each number E_X^-, E_Y^- is defined by two-real numbers. However, polarization is defined by only two observables:

the amplitude ratio $|E_X^-|/|E_Y^+|$

the phase difference δ

$$e^{i\delta} = \frac{E_Y^-}{|E_Y^-|} \bigg/ \frac{E_X^-}{|E_X^-|}. \tag{6.1.6}$$

To-day, lasers are the sources used to light-scattering experiments. The incident radiation is coherent and can be simply described with the help of plane waves (6.1.1). The measurable quantities associated with this type of radiation are:

(a) The wave vector $\vec{k}_0$. A typical value is

$$k_0 = 1.5 \times 10^{-4}\,\text{nm}^{-1}$$

For comparison, k_0 is about 0.3 nm^{-1} for X-rays.

(b) The photon flux J. The flux is related to the energy density U_{ph} of the radiation (6.1.5) by the equation

$$J = c\,U_{\text{ph}}/\hbar\omega_0. \tag{6.1.7}$$

(c) The polarization [see (6.1.6)].

In general, these elements are not sufficient to characterize a scattered radiation. For instance, the anisotropy of the scattering molecules may introduce orientation and polarization fluctuations. The electric field $\vec{E}^{\pm}$, associated with a given propagation direction, has in this case random features. Let p_α be the probability associated with the value ${}^\alpha E_I^\pm$ of the I-component of the field (the index $I = X, Y$ refers to the two polarization axes).

Then, it is convenient to define a density matrix D, of second-order, with elements $D_{L,J}$

$$D_{IJ} = \frac{\epsilon}{\hbar\omega_0}\sum_\alpha p_\alpha\, {}^\alpha E_I^-\, {}^\alpha E_J^+ .$$

The matrix D is definite positive. Let us introduce the basis σ_μ ($\mu = 1, 2, 3, 4$) made of Pauli matrices of components $\sigma_{\mu,IJ}$ with $I, J = X, Y$

$$\sigma_1 = \begin{vmatrix} 0 & 1 \\ 1 & 0 \end{vmatrix} \quad \sigma_2 = \begin{vmatrix} 0 & -i \\ i & 0 \end{vmatrix} \quad \sigma_3 = \begin{vmatrix} 1 & 0 \\ 0 & -1 \end{vmatrix} \tag{6.1.8}$$

and of the unit matrix

$$\sigma_4 = \begin{vmatrix} 1 & 0 \\ 0 & 1 \end{vmatrix}.$$

The decomposition

$$D = \frac{1}{2}\sum_{\mu=1}^{4} P_\mu \boldsymbol{\sigma}_\mu \tag{6.1.9}$$

defines parameters P_μ which are called Stokes parameters. They are given by

$$P_\mu = \sum_{I,J=X,Y} D_{IJ}\boldsymbol{\sigma}_{\mu,IJ} = \mathrm{Tr}(D\boldsymbol{\sigma}_\mu). \tag{6.1.10}$$

The vector whose components are the parameters P_μ is called the Stokes vector. A radiation is characterized by these P_μ ($\mu = 1, \ldots, 4$). In particular, the polarization vector[2] ϖ is defined by its three components

$$\varpi_\mu = P_\mu / P_4 \quad (\mu = 1, 2, 3)$$

The number $\varpi = |\vec{\varpi}|$ is called polarization factor.[2] As will be seen, its value varies from zero to one according to the state of the beam. Two cases can be distinguished.

1. *The pure case.* The radiation is a plane wave [see (6.1.1)]. Equation (6.1.10) can be written in the form

$$P_\mu = \frac{\epsilon}{\hbar\omega_0}\sum_{IJ} E_I^+ \,\boldsymbol{\sigma}_{\mu,IJ}\, E_J^- = \frac{\epsilon}{\hbar\omega} E^+ \boldsymbol{\sigma}_\mu E^- . \tag{6.1.11}$$

More explicitly, we have

$$P_1 = \frac{2\epsilon}{\hbar\omega_0}|E_X^+||E_Y^+|\cos\delta$$

$$P_2 = \frac{2\epsilon}{\hbar\omega_0}|E_X^+||E_Y^+|\sin\delta$$

where δ is the previously defined phase [see (6.1.6)]

$$P_3 = \frac{\epsilon}{\hbar\omega_0}(|E_X^+|^2 - |E_Y^+|^2).$$

Parameter P_4 is proportional to the energy density U_{ph}

$$P_4 = \frac{\epsilon}{\hbar\omega_0}[|E_X^+|^2 + |E_Y^+|^2] = \frac{U_{\mathrm{ph}}}{\hbar\omega_0}.$$

From these expressions, we deduce the relation

$$P_4^2 = P_1^2 + P_2^2 + P_3^2$$

and consequently $\varpi = 1$

Thus, three Stokes parameters are sufficient to characterize a plane electromagnetic wave.

A correspondence between the parameter P_μ and the polarizations of the plane waves is given in Table 6.1.

Table 6.1.

Polarization	Parameters	$\frac{\hbar\omega_0}{\epsilon}P_1$	$\frac{\hbar\omega_0}{\epsilon}P_2$	$\frac{\hbar\omega_0}{\epsilon}P_3$	$\frac{\hbar\omega_0}{\epsilon}P_4$
	$E_Y = 0$	0	0	$\lvert E_X^+\rvert^2$	$\lvert E_X^+\rvert^2$
Linear	$E_X = 0$	0	0	$-\lvert E_Y^+\rvert^2$	$\lvert E_Y^+\rvert^2$
	$E_X = E_Y$	$2\lvert E_X^+\rvert^2$	0	0	$2\lvert E_X^+\rvert^2$
Circular	$\lvert E_X = \lvert E_Y\rvert$, $\delta = \pm\pi/2$	0	$\pm 2\lvert E_X^+\rvert^2$	0	$2\lvert E_X^+\rvert^2$

Finally, let us note that P_2 and P_4 are invariant for rotations of the polarization axes with respect to the propagation axis OZ. The quantity ϖ_2 is called the wave helicity. The polarization ellipticity is associated with ϖ_2, the obliquity with ϖ_1.

2. *The mixed case.* The components (I, $J = X$, Y) of the electric field are of random nature and relation (6.1.10) can be written

$$P_\mu = \frac{\epsilon}{\hbar\omega_0} \sum_{IJ=X,Y} \sum_\alpha p_A\, {}^\alpha E_I^+\, \boldsymbol{\sigma}_{\mu,IJ}\, {}^\alpha E_J^- \tag{6.1.12}$$

or else

$$P_\mu = \frac{\epsilon}{\hbar\omega_0} \langle\!\langle E^+ \boldsymbol{\sigma}_\mu E^- \rangle\!\rangle. \tag{6.1.13}$$

Here, the four Stokes parameters are needed to describe the radiation state. The parameters P_μ for $\mu = 1, 2, 3$ have a special meaning connected to the fact that the trace of the Pauli matrices is zero. Then, let us compare

$$\sum_{\mu=1}^{3} P_\mu^2$$

to P_4^2. From (6.1.12) are get

$$P_4^2 - P_1^2 - P_2^2 - P_3^2 =$$
$$\frac{2}{\hbar\omega_0^2} \sum_{\alpha,\beta} p_\alpha p_\beta ({}^\alpha E_X^+\, {}^\beta E_Y^- - {}^\alpha E_Y^+\, {}^\beta E_X^-)({}^\alpha E_X^-\, {}^\beta E_Y^+ - {}^\alpha E_Y^-\, {}^\beta E_Y^+).$$

The two factors on the right-hand side being complex conjugate, the product is positive or zero

$$P_4^2 \geqslant P_1^2 + P_2^2 + P_3^2. \tag{6.1.14}$$

Inequality (6.1.14) gives

$$\varpi = |\vec{\varpi}| \leqslant 1. \tag{6.1.15}$$

The polarization factor ϖ varies from zero (completely mixed case) to one (pure case).

1.2 Neutron radiation

The neutron is a particle of mass $m_n = 1.0098$ dalton, with a zero electrical dipole (or smaller than 10^{-26} esu), a spin 1/2 and a magnetic moment which is equal to -1.91 nuclear magneton, where

$$1 \text{ nuclear magneton} = 5.05 \times 10^{-27} \text{ joule/tesla}$$
$$= 5.05 \times 10^{-24} \text{ erg/gauss.}$$

The neutron spin is given by the operator

$$\vec{s} = \frac{\hbar}{2}\vec{\sigma}$$

where the components of $\vec{\sigma}$ are the Pauli matrices σ_μ ($\mu = 1, 2, 3$) [See (6.1.8)]. The polarization vector is defined by the equality

$$\vec{\varpi} = \langle\!\langle \vec{\sigma} \rangle\!\rangle \tag{6.1.16}$$

where the angular brackets $\langle\!\langle\ \rangle\!\rangle$ indicate than an average has been made.* The polarization factors of a neutron i

$$\vec{\varpi} = |\vec{\varpi}_i| \tag{6.1.17}$$

always equals 1.

We may now introduce the polarization factor associated with a flux of N neutrons. Here, the polarization vector is[3]

$$\vec{\varpi} = \frac{1}{N}\sum_i \vec{\varpi}_i \tag{6.1.18}$$

and $\varpi = |\vec{\varpi}|$ is the polarization factor. A beam is depolarized when the spins of the N neutrons point in non-correlated random directions. A beam is polarized if all the spins point in the same direction ($\varpi = 1$).

A beam of monocinetic neutrons, propagating along the incident direction OZ is characterized by

(a) a wave vector with modulus k_0 (typically 3 nm);

(b) a flux $\vec{J}$ equal to the number of incident neutrons per unit surface (perpendicular to OZ) and per unit time;

(c) a polarization factor and a polarization direction.

1.2.1 Neutron sources

The production and the control of a neutron beam requires a lot of equipment of monumental size. Accordingly, the number of installations providing proper

* We can express the components $\bar{\omega}_\omega$ ($\mu = 1, 2, 3$) of the vector $\vec{\varpi}$ as functions of the density matrix D, as we did in the case of an electromeagnetic wave [see (6.1.10) and (6.1.15) and ref. 5].

beams is small.* In most facilities (in 1983), the neutrons are produced by fission chain reactions in uranium. These reactions occur in a reactor, in which the neutron flux remains constant. The neutrons which are very energetic just after the fission are thermalized by successive collisions in a moderator, and finally they diffuse outside the reactor through various channels towards the samples which are to be studied by scattering.

The heat produced during the fission process is removed by circulation of a cooler. Even if the heat outflow (which equals 6×10^7 at the ILL at Grenoble) is maximized, the neutron flux must be maintained below $10^{16}\ \text{cm}^{-2}\ \text{s}^{-1}$, in order to ensure that the temperature of the reactor elements remains under a critical technological threshold. The flux is low as compared to the photon flux obtained with a light source ($10^{19}\ \text{cm}^{-2}\,\text{s}^{-1}$ for a laser with a cross-section of $10^{-2}\ \text{cm}^{-2}$ and a power of 10^{-2} watt).† However, neutrons can be used to perform experiments that could not be carried out with photons (lack of contrast, for instance).

In facilities built after 1980, as for instance at the High Energy Laboratory at Tsukuba in Japan, the neutron sources are of the pulsed type: the instantaneous fluxes can be huge, though they are small on average. Several methods can be used to produce such fluxes: let us mention, for instance, the stripping of heavy nuclei by protons. The protons are accelerated up to an energy of 800 MeV and are sent on a series of tungsten plaquettes. They cross these plaquettes virtually without deviation but break the nuclei which stay on their trajectory. Fast neutrons are parts of the fragments of nuclei. These neutrons, after slowing down, are used in scattering experiments. They are collimated in beams which are, in general, perpendicular to the direction of the protons.

On the contrary, the results of experiments presented here have been produced with continuous neutron sources (controlled chain reactions). The beams thus produced, are more suitable to study long polymers than the pulsed beams: in fact, such studies require neutrons with a large wavelength, i.e. slow neutrons.

1.2.2 Flux and incident wave vector

Let us examine the characteristics of a neutron radiation adapted to the determination of molecular sizes (polymer) of the order of 10 nm. In this case, it is proper to use a neutron beam of wave vector $k_0 \simeq 5\ \text{nm}^{-1}$.

However, the thermalized flux extracted from the source has a very broad distribution of wave vectors. To fulfil the required conditions, one has to select a flux fraction δJ with wave vectors centred on $\vec{k}_0$ in the distribution $J_0(\vec{k})$

$$\delta J = \delta\Omega\, \delta k\, k^2 J_0(\vec{k})$$

where $\delta\Omega$ is a solid angle which is determined by geometrical conditions

* At present (1988) there are three of them in France (a) two at Grenoble: Laue–Langevin Institute and Centre d'Etudes Nucleaires de Grenoble; (b) one at Saclay: Centre d'Etudes Nucleuires de Saclay.

† In this case, the useful power is limited by the heating of the irradiated polymer solution.

(collimators and so on) and where δk is a fixed width which the experimentalist chooses as the best. Then, let us calculate $J_0(\vec{k})$ by assuming that the thermalized flux has a Maxwellian distribution

$$J_0(\vec{k}) = J_\square \frac{(2\pi)^{-3/2}}{k_M^3} \exp\left(-\frac{k^2}{2k_M^2}\right) \tag{6.1.19}$$

(index M for Maxwell)
where $k_M = \hbar^{-1}(m_n/\beta)^{1/2}$ is such that $\langle k^2 \rangle = 3\,k_M^2$. At k_0 fixed, $J_0(\vec{k}_0)$ is maximum when

$$k_M = k_0/\sqrt{3} = 0.577\,k.$$

However, at room temperature, $k_M = 30\ \text{nm}^{-1}$. Thus the selected part of the flux can be increased by cooling down the neutrons. This operation is performed by sending them through a flask of liquid hydrogen or liquid deuterium, with a temperature of about 25 K. Then, the spectrum is centred on a wave vector with modulus $k_0 = 12\ \text{nm}^{-1}$. The flask contains 1 litre of hydrogen or 10 litres of deuterium. It is set in the moderator where the neutron flux is maximum. In fact, if the flask of hydrogen (or deuterium) is set at some distance from the core, the number of neutrons that can be cooled down, is diminished by solid-angle effects. However, for security reasons, it is preferable to put this vessel at a distance from the core of the reactor. Risks have not always been assessed objectively. Let us note that the American Security regulations have been opposed to the implantation, in the United States, of cold sources in the moderators, whereas at the same time, such facilities have been planned and put into service in Germany and in France. The fact that, during the 1970s, there was absolutely no American contribution to the observation of small-angle neutron scattering has sometimes been attributed to this interdiction; however, it would be fairer to acknowledge the perspicacity and the boldness of those who were in charge of the European nuclear centres. The neutron flux for an average wave vector $k_0 = 12\ \text{nm}^{-1}$, obtained as indicated above, is of the order of $10^7\ \text{cm}^{-2}\,\text{s}^{-1}$.

2. THE SCATTERING EXPERIMENT

Let us set the sample under study at the origin of the laboratory system. An incident radiation propagates along OZ and falls on this sample. In certain conditions, which will be examined later, this radiation is scattered in all space directions, with an orientation distribution which depends on the structure of the sample.

In principle, for the incident radiation, it is sufficient to determine the following parameters

(a) the modulus k_0 of the wave vector;

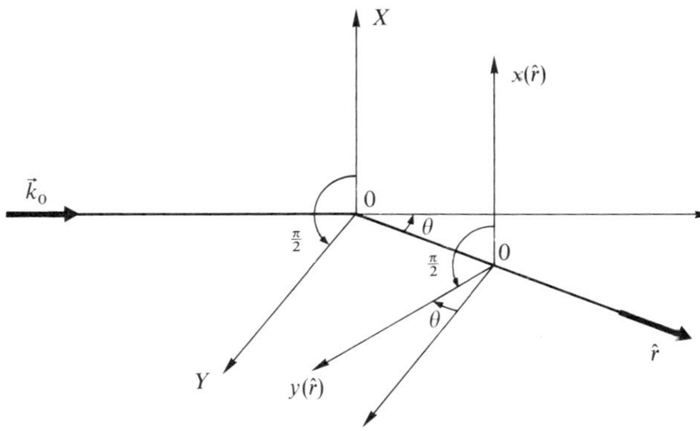

Fig. 6.1. Geometry of a scattering experiment. The propagation vector of the incident beam is $\vec{k}_0$. The direction of propagation of the scattered beam is given by the unit vector $\hat{r}$. The angle θ between $\hat{k}_0$ and $\vec{r}$ is the scattering angle. The plane (OZ, $\hat{r}$) is the scattering plane. The polarization axes are: OX, OZ for the incident radiation; O$x(\hat{r})$ (which coincides with OX), O$y(\hat{r})$ for the scattered radiation.

The angle between OY and O$y(\hat{r})$ equals θ. Scattering is studied in the plane which is orthogonal to OX.

(b) the neutron flux or the photon flux J_0 impinging on the sample;

(c) eventually, the polarization.

For the scattered radiation, it is sufficient to determine

(a) the direction* $\hat{k} = \vec{r}/|\vec{r}|$, corresponding to the scattering angle θ (angle between OZ and ($\hat{r}$);

(b) the neutron or photon flux $J(\vec{r})$;

(c) eventually, the polarization.

The scattering experiment consists in measuring the data J_0, k_0 and the fluxes $J(\vec{r})$ and in changing the orientation $\hat{k}$ and/or the value of k_0 (see Fig. 6.1).

When the polarization is a parameter in the experiment, it is convenient to choose polarization axes in the reference system of the laboratory. For the incident radiation, these axes OX and OY are orthogonal to the direction OZ. For the scattered radiation, the polarization axes which are orthogonal to $\hat{r}$, are defined with respect to the scattering plane (OZ, $\vec{r}$). These axes are

O$x(\hat{r})$, perpendicular to the scattering plane.

O$y(\hat{r})$, in the scattering plane.

The axes OX and O$x(\vec{r})$ coincide.

* We use the symbol $\hat{k}$ for the ratio $\vec{r}/|\vec{r}|$ because the direction in which the neutron is scattered corresponds to a wave with wave vector $\vec{k} = k\hat{k}$.

Actually, it will be necessary to specify other characteristics of the radiation, such as the divergence of the incident flux with respect to the axis OZ, the distribution of the wave number k_0 and the random noise. The values of these parameters are imposed on the experimentalist by requirements related to the experimental process and by technological constraints

3. RADIATION SCATTERING BY AN ELEMENTARY CENTRE

We shall now study the interaction of a neutron or of an electromagnetic radiation with an elementary scatterer. This scatterer is a nucleus or an electron according to the type of radiation. In fact, the problem corresponding to scattering by a unique centre is not as academic as it might seem. The results which will be obtained in this section will be used later (Section 4), when dealing with scattering by a set of centres.

3.1 Neutron scattering

3.1.1 Interaction potentials

The neutron interactions are nuclear or magnetic. Here, we shall be interested in the nuclear collision of a neutron with a a nucleus.

Let $\vec{r}$ be the vector joining the nucleus to the neutron. Let $\vec{s}$ be the spin operator for the neutron and $\vec{S}$ the spin operator for the nucleus. The interaction potential is (to a first approximation) a central potential which can be written

$$V(\vec{r}) \simeq V_0(\vec{r}) + (\vec{s}\cdot\vec{S})V_1(\vec{r}). \tag{6.3.1}$$

This potential is spin-dependent. When the radiation is scattered by a set of elementary centres, this dependence is the origin of the so-called coherent and incoherent scattering. The potential thus defined produces an isotropic collision around the incidence axis.

The spin–orbit coupling will be neglected, and this approximation will enable us to separate the spin and the space variables.

3.1.2 Stationary collision-state

The collision of a neutron with an elementary centre can be described by a stationary state. Let us define this state for the Schrödinger equation of the neutron–nucleus system.

Actually, the scattering nucleus (whose position vector is $\vec{r}_A$) is bound to an infinite structure. The position vector of the neutron will be represented by $\vec{r}_n$. An orbital state of the neutron will be denoted by the symbol $|\,\rangle$, an orbital state of the scattering nucleus by $|)$, and a spin state of the neutron–nucleus system by $|\}$. Thus a state of the system will be identified by the symbol $|\rangle)\}$.

The state $|\Psi(t)\rangle)\}$ obeys the Schrödinger equation[4]

$$i\hbar\partial_t|\Psi(t)\rangle)\} = \mathscr{H}|\Psi(t)\rangle)\}$$

where $\mathscr{H}$ is the Hamiltonian.

This Hamiltonian is the sum of two terms

$$\mathscr{H} = {}^0\mathscr{H} + V_{An}$$

where ${}^0\mathscr{H}$ is the Hamiltonian of the system in the absence of interaction and where V_{An} is the operator associated with the potential $V(\vec{r}_A - \vec{r}_n)$ [see (6.3.1) for the definition of $V(\vec{r})$.]

The stationary collision-state is an eigenstate $|\Psi\rangle)\}$ which satisfies the boundary conditions corresponding to the experiment. Let us define these conditions.

First, let us consider the states $|\Phi\rangle)\}$ associated with a neutron plane wave and a nucleus state

$${}^0\mathscr{H}|\Phi\rangle)\} = E_I|\Phi\rangle)\}$$

(I for initial).

The stationary collision state is a state $|\Psi\rangle)\}$ which is associated with $|\Phi\rangle)\}$ with the same energy, and which is a solution of the equation

$$\mathscr{H}|\Psi\rangle)\} = E_I|\Psi\rangle)\}. \tag{6.3.2}$$

it reads

$$|\Psi\rangle)\} = |\Phi\rangle)\} + |\zeta\rangle)\}$$

where $|\zeta\rangle)\}$ a neutron outgoing wave. It can be proved that[5]

$$|\Psi\rangle)\} = |\Phi\rangle)\} + T(E_I)|\Phi\rangle)\} \tag{6.3.3}$$

where the operator $T(E)$ is

$$T(E) = V_{An} + V_{An}\frac{I}{E - \mathscr{H} + i0}V_{An}.$$

We may write the initial state in the form

$$|\Phi\rangle)\} = |\vec{k}_0\rangle|\alpha)|i\}.$$

Here $|\vec{k}_0\rangle$ is the state corresponding to the wave vector $\vec{k}_0$. The state $|\alpha)$ is the orbital state of the scattering nucleus, with energy E_α (the spin–orbit coupling is neglected). Finally, the state $|i\}$ is the spin state of the neutron *and* of the nucleus. This state belongs to the subspace, the basic vectors of which are

$$|s_z, S_z\}$$

and their number is

$$(2s + 1)(2S + 1), \tag{6.3.4}$$

i.e. $2(2S + 1)$ for $s = 1/2$.

Let $|\vec{r}\rangle$ be the state associated with the position vector $\vec{r}$. The (partial) projection of the state $|\Phi\rangle)\}$ in real space is

$$\langle\vec{r}|\Phi\rangle)\} = (2\pi)^{-3/2} e^{i\vec{k}_0\cdot\vec{r}}|\alpha)|i\}. \tag{6.3.5}$$

On the other hand, the corresponding energy $E_{\rm I}$ is

$$E_{\rm I} = E_\alpha + E_{k_0} \qquad \text{where} \qquad E_{k_0} = \hbar^2 k_0^2/2m_{\rm n}. \tag{6.3.6}$$

3.1.3 Collision amplitude

When an incident neutron with wave vector $\vec{k}_0$ encounters a nucleus with state $|\alpha)$, the neutron is scattered in a direction $\hat{r} = \vec{r}/|\vec{r}|$ and, at the same time, the collision modifies the orbital state $|\alpha)$ of the nucleus and the spin state $|i\}$ of the neutron–nucleus system.

We shall represent by $|\gamma)$ the final orbital state of the nucleus and by $|j\}$ the final spin state of the neutron–nucleus system. The energy of the system is the same before and after the collision. Consequently, it will be convenient to define $k_{\alpha\to\gamma}$ by the relation

$$k_{\alpha\to\gamma}^2 = k_0^2 - 2\frac{m_{\rm n}}{\hbar^2}(E_\gamma - E_\alpha) \tag{6.3.7}$$

where E_α and E_γ are the non-perturbed energies of the scattering nucleus, corresponding respectively to the states $|\alpha)$ and $|\gamma)$.

By using these notation, it is not difficult to show with the help of (6.3.3) that for large values of r, the projection $\langle\vec{r}|\Psi\rangle)\}$ can be written in the form[5]

$$(2\pi)^{3/2}\langle\vec{r}|\Psi\rangle)\} \simeq |\alpha)|i\}\, e^{i\vec{k}_0\cdot\vec{r}} + \sum_{\gamma,\, j}|\gamma)|j\}\, A(\vec{k}_0, \hat{r}; \alpha, \gamma; i, j)\frac{e^{ik_{\alpha\to\gamma}r}}{r} \tag{6.3.8}$$

where $A(\vec{k}_0, \hat{r}; \alpha, \gamma; i, j)$ is given by

$$A(\vec{k}_0, \hat{k}; \alpha, \gamma; i, j) = -\frac{4\pi^2 m_{\rm n}}{\hbar^2}\{j|(\gamma|\langle k_{\alpha\to\gamma}\hat{k}|T(E_{\rm I})|\vec{k}_0\rangle|\alpha)|i\}. \tag{6.3.9}$$

Thus, $A(\vec{k}_0, \hat{r}; \alpha, \gamma; i, j)$ characterizes the scattering process; this quantity is called the 'collision amplitude'.

3.1.4 Factorizations

Let us consider the matrix element

$$\{j|(\gamma|\langle k_{\alpha\to\gamma}\hat{k}|T(E_{\rm I})|\vec{k}_0\rangle|\alpha)|i\} \tag{6.3.10}$$

which determines the collision amplitude (6.3.9). Here the operator $T(E)$ acts in a product space. The subspaces are associated respectively with the neutron orbit, with the spin of the neutron–nucleus system, and with the orbit of the nucleus which itself is bound to an infinite structure.

In these subspaces, $T(E)$ induces transformations between states which, in principle, are coupled to one another. The coupling is given by the Hamiltonian $\mathscr{H}$ whose explicit expression is

$$\mathscr{H} = \mathscr{H}_A - (\hbar^2/2m_n)\Delta_n + V(\vec{r}_n - \vec{r}_A) \tag{6.3.11}$$

with

$$\mathscr{H}_A = -(\hbar^2/2m_A)\Delta_A + V_A(\vec{r}_A)$$

(the operator V_A differs from the operator V_{An} where A and n refer respectively to the neutron and to the nucleus). The potential $V_A(\vec{r})$ corresponds to the binding of the nucleus to the infinite structure, $V(\vec{r})$ is the neutron–nucleus interaction potential (corresponding to V_{An}). The neutron–nucleus coupling depends on the difference $(\vec{r}_n - \vec{r}_A)$

To calculate the matrix elements, it would be convenient to operate separately in each of the three subspaces. We already admitted that the spin–orbit couplage is negligible: this entails an immediate factorization.

Let us now try to eliminate the coupling introduced by the term $V(\vec{r}_A - \vec{r}_0)$ in the Hamiltonian (6.3.11). For this purpose, let us introduce the unitary transformation

$$U = \exp[(\vec{\nabla}_n - i\vec{k}_0)\cdot\vec{r}_A]. \tag{6.3.12}$$

The operators which appear in $\mathscr{H}$ are transformed as follows. First, we have

$$UV(\vec{r}_n - \vec{r}_A)U^{-1} = V(\vec{r}_n), \tag{6.3.13}$$

and this result is exactly what we are aiming at. We can also write this relation in the form

$$UV_{An}U^{-1} \simeq V_n$$

which defines V_n. On the other hand, the operators V_A and Δ_n remain invariant in the transformation U. The same is not true for the operator Δ_A. In fact, we have

$$\mathrm{U}\Delta_A U^{-1} = \Delta_A + \mathscr{K}(m_A, \vec{k}_0)$$

where

$$\mathscr{K}(m_A, \vec{k}_0) = (\hbar^2/2m_A)(2\vec{\nabla}_A - \vec{\nabla}_n + i\vec{k}_0)\cdot(\vec{\nabla}_n - i\vec{k}_0)$$

is a neutron–nucleus coupling term which has been produced by transformation U. However, the result remains interesting: in fact, the new coupling term is negligible when the nucleus mass m_A is high (with respect to m_n) and also when the modulus of the outgoing wave vector has a value which is close to k_0. Let us assume that we are in this case: we put $\mathscr{K}(m_A, k_0) = 0$, and consequently we can write

$$U\mathscr{H}U^{-1} = \mathscr{H}_A + \mathscr{H}_n \tag{6.3.14}$$

where $\mathscr{H}_A$ is given by (6.3.11) and

$$\mathscr{H}_n = -(\hbar^2/2m_n)\Delta_n + V(\vec{r}_n). \tag{6.3.15}$$

Thus, the transformation uncouples the Hamiltonian.

Let us now examine how this transformation acts on the states of the system. We have

$$(\vec{r}_A|\langle\vec{r}_n|U|\vec{k}\rangle|\gamma) = e^{i(\vec{k}-\vec{k}_0)\cdot\vec{r}_A}\langle\vec{r}_n|(\vec{r}_A|\gamma)|\vec{k}\rangle.$$

The matrix element (6.3.10) can be written in the form

$$\{j|(\gamma|\langle k_{\alpha\to\gamma}\hat{k}|T(E_I)|\vec{k}_0\rangle|\alpha)|i\} = \{j|(\gamma|\langle k_{\alpha\to\gamma}\hat{k}|U^{-1}[UT(E_I)U^{-1}]U|\vec{k}_0\rangle|\alpha)|i\}.$$

The operator $UT(E)U^{-1}$ can easily be calculated by using the results (6.3.13), (6.3.14), and (6.3.15). Let us introduce the operator

$$t(E) = V_n + V_n\frac{I}{E-\mathscr{H}_n+i0}V_n \tag{6.3.16}$$

where $\mathscr{H}_n$ and $t(E)$ act only on the orbital states of the neutron and on the neutron–nucleus spin states. It can be shown that

$$\{j|(\gamma|\langle k_{\alpha\to\gamma}\hat{k}|T(E_I)|\vec{k}_0\rangle|\alpha)|i\} = \{j|\langle k_{\alpha\to\gamma}\hat{k}|t(E_{k_0})|\vec{k}_0\rangle|i\}(\gamma|e^{-i\vec{q}_{\alpha\to\gamma}\cdot\vec{r}_A}|\alpha) \tag{6.3.17}$$

where $q_{\alpha\to\gamma} = \hat{k}\,k_{\alpha\to\gamma} - \vec{k}_0$ is the wave vector transfer, and

$$E(k_0) = \hbar^2k_0^2/2m_n.$$

The expression (6.3.17) is the result we were looking for.

Thus, the collision amplitude $A(\vec{k}_0, \hat{k};\ \alpha, \gamma;\ i, j)$ which appears in (6.3.8) and which is given by (6.3.9) can be expressed as follows

$$A(\vec{k}_0, \hat{k};\ \alpha, \gamma;\ i, j) = -\frac{4\pi^2 m_n}{\hbar^2}\{j|\langle k_{\alpha\to\gamma}\hat{k}|t(E_{k_0})|\vec{k}_0\rangle|i\}(\gamma|e^{-i\vec{q}_{\alpha\to\gamma}\cdot\vec{r}_A}|\alpha).$$

3.1.5 Static approximation

Let us note that the matrix element

$$\{j|\langle k_{\alpha\to\gamma}\hat{k}|t(E_{k_0})|\vec{k}_0\rangle|i\}$$

has been taken between states which do not belong to the energy shell. The only matrix elements

$$(\gamma|e^{-i\vec{q}_{\alpha\to\gamma}\cdot\vec{r}_A}|\alpha)$$

that are large are those for which the energy difference $E_\alpha - E_\gamma$ is small with respect to the energy E_{k_0} of the incident neutron, and we shall neglect all the others. This is the static approximation.

In this case, eqn (6.3.17) shows that

$$k_{\alpha\to\gamma} \simeq k_0.$$

Thus, the wave vector transfer is

$$\vec{q}_{\alpha\to\gamma} \simeq \vec{q} = k_0\hat{k} - \vec{k}_0.$$

It does not depend on states α, γ; this is the elastic wave vector transfer

$$q = 2k_0 \sin(\theta/2)$$

where

$$\cos\theta = \hat{k}\cdot\hat{k}_0. \tag{6.3.18}$$

Thus, every non-negligible collision matrix element can be expressed in the form

$$\{j|\langle k_0\hat{k}|t(E_{k_0})|\vec{k}_0\rangle|i\}(\gamma|e^{-i\vec{q}\cdot\vec{r}_A}|\alpha). \tag{6.3.19}$$

The first matrix element in this product is calculated on the energy shell. It could be directly obtained from a study of scattering by the potential (6.3.1). It will be admitted, in the static approximation, that the nucleus is at the origin of coordinates ($\vec{r}_A = 0$). Then the second matrix element in the product reduces to the quantity $(\gamma|\alpha) = \delta_{\gamma\alpha}$. The collision amplitude which appears in (6.3.8) can be written

$$A(\vec{k}_0, \hat{k};\, \alpha, \gamma;\, i, j) = \delta_{\alpha\gamma} A(k_0, \hat{k};\, i, j)$$

where

$$A(k_0, \hat{k};\, i, j) = -\frac{4\pi^2 m_n}{\hbar^2}\{j|\langle k_0\hat{k}|t(E_I)|\vec{k}_0\rangle|i\}, \tag{6.3.20}$$

and eqn (6.3.8) becomes

$$(2\pi)^{3/2}\langle\vec{r}|\Psi\rangle)\} = |\alpha)|i\}e^{i\vec{k}_0\cdot\vec{r}} + \sum_j |\alpha)|j\}A(\vec{k}_0, \hat{k};\, i, j)\frac{e^{ik_0 r}}{r}. \tag{6.3.21}$$

3.1.6 Collision lengths

The operator $t(E)$ which appears in eqn (6.3.20) is defined by (6.3.16)

$$t(E) = V_n + V_n \frac{I}{E - \mathscr{H}_n + i0} V_n$$

where V_n is the central potential V defined by (6.3.1).

Let us examine the special case where the potential V is independent of the spin of the nucleus. Then, the amplitude does not depend on the spin states i, j and we have

$$A(\vec{k}_0, \hat{k};\, i, j) \equiv A(\vec{k}_0, \hat{k}) = -\frac{4\pi^2 m_n}{\hbar^2}\langle k_0\hat{k}|t(E_I)|\vec{k}_0\rangle. \tag{6.3.22}$$

The amplitude is a complex number, whose absolute value has the dimension of a length. In the limit $k_0 \to 0$, this number becomes real. It is called the collision length.

With a nucleus of spin zero, we can associate a potential V_0. If this nucleus is bound, the collision length is denoted by $\mathfrak{b}$ and we have

$$\mathfrak{b} = -\lim_{k_0 \to 0} A(\vec{k}_0, \hat{k}) \tag{6.3.23}$$

If the nucleus is free, the modulus of the collision length is smaller than the modulus of $\mathfrak{b}$, but the difference is appreciable only for nuclei whose mass is not large with respect to the neutron mass. Here, we shall only consider scattering by bound nuclei.

The length $\mathfrak{b}$ is independent of the scattering angle; the collision with a bound nucleus is *isotropic* in the limit $k_0 \to 0$. In practice, it remains so for $k_0 \simeq 10^{-1}$ nm; this is the Born approximation.[4]

For carbon[6]

$$\mathfrak{b} = (0.6648 \pm 0.0013) \times 10^{-5}\ \text{nm}.$$

3.1.7 Scattering with spin-flip

The contribution of protons (spin 1/2) to neutron diffraction by organic molecules is important and complex. To study it, we consider the scattering produced by the potential

$$V = V_0 + (\vec{s}\cdot\vec{S})\,V_1$$

where the spin operators, $\vec{s}$ for the neutron and $\vec{S}$ for the nucleus, act on the states $|i\}$ of the neutron–nucleus system.

3.1.7.1 *Spin-dependent collision operators*

When there is a neutron–nucleus spin interaction, several collision channels are open. It is then possible to introduce operators $t(E)$ which depend in a characteristic way on the operator $\vec{s}\cdot\vec{S}$ and on the spin state $|i\}$. In this way, we can consider all the possible channels together.

Let $\vec{J}$ be the total spin operator

$$\vec{J} = \vec{S} + \vec{s}.$$

The operator $\vec{S}\cdot\vec{s}$ can easily be expressed in terms of $\vec{J}$. In fact,

$$2\vec{s}\cdot\vec{S} = (\vec{s} + \vec{S})^2 - s^2 - S^2. \qquad (6.3.24)$$

When this operator is applied to a state with total spin $J = S \pm 1/2$, it is diagonal and we may simply write

$$2\vec{s}\cdot\vec{S} = J(J+1) - 3/4 - S(S+1).$$

Thus, for $J = S + 1/2$,

$$2\vec{s}\cdot\vec{S} = S$$

and, for $J = S - 1/2$,

$$2\vec{s}\cdot\vec{S} = -(S+1).$$

The projectors on the spin states $J = S \pm 1/2$ are, respectively

$$\mathfrak{p}^+ = (S + 1 + 2\vec{s}\cdot\vec{S})/(2S+1) = (J - S + 1/2)(J + S + 1/2)/(2S+1)$$
$$\mathfrak{p}^- = (S - 2\vec{s}\cdot\vec{S})/(2S+1) = -(J - S - 1/2)(J + S + 3/2)/(2S+1). \qquad (6.3.25)$$

Then the interaction potential can be written

$$V = V^{+}\mathfrak{p}^{+} + V^{-}\mathfrak{p}^{-}$$

where

$$V^{+} = V_0 + V_1 S/2$$

and

$$V^{-} = V_0 - V_1(S+1)/2.$$

With the potential $V^{\pm}$, we can associate operators $t^{\pm}$ for which

$$t(E) = t^{+}\mathfrak{p}^{+} + t^{-}\mathfrak{p}^{-}$$

These are the operators that were looked for.

Consequently, the collision amplitude of the neutron with initial spin state $|i\}$ and final spin state $|j\}$ is

$$A(\vec{k}_0, \hat{k}; i, j) = -\frac{4\pi^2 m_{\mathrm{n}}}{\hbar^2}\{j| < k_0, \hat{k}|t^{+}\mathfrak{p}^{+} + t^{-}\mathfrak{p}^{-}|\vec{k}_0\rangle|i\}. \qquad (6.3.26)$$

This amplitude is a complex number, but when $k_0 \to 0$, it becomes real

$$\lim_{k_0 \to 0} A(\vec{k}_0, \hat{k}; i, j) = -\{j|\mathfrak{b}^{+}\mathfrak{p}^{+} + \mathfrak{b}^{-}\mathfrak{p}^{-}|i\} \qquad (6.3.27)$$

where $\mathfrak{b}^{+}$ and $\mathfrak{b}^{-}$ are collision lengths associated respectively with the channels $S + 1/2$ and $S - 1/2$.

For the proton (spin 1/2), we have[6]

$$\mathfrak{b}^{+} = 1.085 \times 10^{-5}\ \mathrm{nm},$$

$$\mathfrak{b}^{-} = -4.74 \times 10^{-5}\ \mathrm{nm}.$$

The fact that the collision length $\mathfrak{b}^{-}$ associated with one channel is strongly negative, explains why the neutron scattering experiments with polymers have been so successful. This collision length (for the proton) is very different from the carbon collision length ($\mathfrak{b} = (0.66535 \pm 0.00014) \times 10^{-5}$ nm) and we shall see, in Chapter 7, all the possibilities that such a situation offers. However, to use this channel ($S - 1/2$) in the best way, it would be necessary to use polarized beams and targets and to carry out the experiment for two different polarization directions. In principle, this can be done. However, until 1986 the experiments were performed without polarizing the target. In such conditions, the benefit resulting from the fact that $\mathfrak{b}^{-}$ is very negative, is smaller, and, moreover, a large random noise appears.

It is convenient to replace $\mathfrak{p}^{+}$ and $\mathfrak{p}^{-}$ in (6.3.27) by their values (6.3.25)

$$\lim_{k_0 \to 0} A(\vec{k}_0, \hat{k}; i, j) = -\left\{j\left|\left(\frac{(S+1)\mathfrak{b}^{+} + S\mathfrak{b}^{-}}{2S+1} + 2(\vec{s}\cdot\vec{S})\frac{(\mathfrak{b}^{+} - \mathfrak{b}^{-})}{2S+1}\right)\right|i\right\}.$$

Let us set

$$B^{+} = \frac{(S+1)\mathfrak{b}^{+} + S\mathfrak{b}^{-}}{2S+1} \qquad \text{and} \qquad B^{-} = \frac{2(\mathfrak{b}^{+} - \mathfrak{b}^{-})}{2S+1}.$$

We obtain

$$\lim_{k_0 \to 0} A(\vec{k}_0, \hat{k};\, i, j) = -\{j|(B^+ + B^-(\vec{s}\cdot\vec{S}))|i\}.$$

For finite k_0, we shall replace $A(\vec{k}_0, \hat{k};\, i, j)$ by this limit: this is the Born approximation[4]

$$A(\vec{k}_0, \hat{k};\, i, j) = -\{j|(B^+ + B^-(\vec{s}\cdot\vec{S}))|i\}. \tag{6.3.28}$$

3.1.7.2 *Collision with spin-flip*

Let us consider the special case where the incident beam is polarized: the spin of each neutron points in the same direction (for instance, OX). We denote by $|\uparrow\}_{\mathrm{n}}$ this neutron spin state. Thus, the initial total spin state of the neutron–nucleus system is

$$|i\} = |i'\}_{\mathrm{N}}|\uparrow\}_{\mathrm{n}} \tag{6.3.29}$$

where $|i'\}_{\mathrm{N}}$ is the initial spin state of the nucleus. When a beam is prepared in this way, it is interesting to choose the collision channel for which the neutron spin flips, so that its final direction is opposite to the initial one. In fact, such an experiment enables the experimentalist to directly evaluate the random noise which occurs in other experiments. The final spin state is denoted by $|\downarrow\}_{\mathrm{n}}$ and, after collision, the total spin state is

$$|j\} = |j'\}_{\mathrm{N}}|\downarrow\}_{\mathrm{n}} \tag{6.3.30}$$

where $|j'\}_{\mathrm{N}}$ is the nucleus spin state after collision

Let us calculate the amplitude associated with such a collision. Taking into account (6.3.28), (6.3.29) and (6.3.30), we have by definition

$$A(\vec{k}_0, \hat{k};\, i', j';\, \uparrow, \downarrow) \simeq -{}_{\mathrm{N}}\{j'|_{\mathrm{n}}\{\downarrow|(B^+ + B^-(\vec{s}\cdot\vec{S}))|\uparrow\}_{\mathrm{n}}|i'\}_{\mathrm{N}}. \tag{6.3.31}$$

The first term on the right does not depend on the spin of the nucleus; it cannot contribute to the amplitude associated with the neutron spin-flip. The second term depends on the spin of the nucleus but only $s_x S_x$ and $s_y S_y$ contribute to the spin-flip.

Thus, we obtain

$$A(\vec{k}_0, \hat{k};\, i', j';\, \uparrow, \downarrow) = -{}_{\mathrm{N}}\left\{j'\left|\frac{B^-}{2}(S_x + iS_y)\right|i'\right\}_{\mathrm{N}} \tag{6.3.32}$$

where

$$B^- = 2(b^+ - b^-)/(2S+1) \tag{6.3.33}$$

is the collision length associated with this amplitude. This is the result we were looking for.[7]

3.1.8 Incident and scattered flux: differential cross-section

The incident and scattered fluxes are determined by the experiment. They are related to the previously defined collision amplitude.

Let us consider the stationary collision state

$$|\Psi\rangle)\} = |\Phi\rangle)\} + |\zeta\rangle)\} \tag{6.3.34}$$

where $|\Phi\rangle$ is the initial state (incident wave) and $|\zeta\rangle$ the scattering state (outgoing wave). These states depend on initial conditions: wave vector $\vec{k}_0$, spins, and orbits.

We associate a neutron flux with each of these states. The neutron flux operator at point $\vec{r}$ is

$$\frac{\hbar}{2m_n}[\delta(\vec{r}-\vec{r}_{op.})\vec{k}_{op}+\vec{k}_{op.}\delta(\vec{r}-\vec{r}_{op.})]$$

where $\vec{r}_{op}$ and $\vec{k}_{op}$ are operators (acting on the neutron states). For a state $|\mathcal{O}\rangle)\}$, the neutron flux $J(\vec{r})$ is by definition the mean value of the flux operator:

$$(2\pi)^{-3}\vec{J}(\vec{r}) = -\frac{i\hbar}{2m_n}[\{(\langle\mathcal{O}|\vec{r}\rangle\,\vec{\nabla}_r\langle\vec{r}|\mathcal{O}\rangle)\} - c.c.]. \tag{6.3.35}$$

Let us calculate the incident flux: the corresponding state is

$$\langle\vec{r}|\Phi\rangle)\} = (2\pi)^{-3/2}\,e^{i\vec{k}_0\cdot\vec{r}}|\alpha)|i\}$$

where $|\alpha)$ is the initial orbital state of the nucleus, $|i\}$ the initial spin state of the neutron–nucleus system. Let us replace $\langle\vec{r}|\mathcal{O}\rangle)\}$ by $\langle\vec{r}|\Phi\rangle)\rangle$ in (6.3.35). We obtain

$$J(\vec{r}) = \frac{\hbar k_0}{m_n} = J_0. \tag{6.3.36}$$

Let us calculate the scattered flux, using the factorizations corresponding to the static approximation. The projection of the scattering state in real space is [see (6.3.26)]

$$\langle\vec{r}|\zeta\rangle)\} = \sum_{\gamma,j}|\gamma)|j\}\,A(\vec{k}_0,\hat{k};\,i,j)\,\delta_{\gamma\alpha}\frac{e^{ik_0 r}}{r} \tag{6.3.37}$$

where γ denotes the final state of the nucleus and where j denotes the final spin state of the neutron–nucleus system (i denotes the initial one). Let us replace $\langle\vec{r}|\mathcal{O}\rangle)\}$ by $\langle\vec{r}|\zeta\rangle)\}$ in (6.3.35). We obtain

$$\vec{J}_I(\vec{r}) = \frac{\hbar k_0}{m_n}\frac{\hat{r}}{r^2}\sum_j A(\vec{k}_0,\hat{k};i,j)A^*(\vec{k}_0,\hat{k};i,j).$$

The scattered fluxes $J_I(\vec{r})$ are the measured quantities. However, they do not enter directly in the results because they depend not only on the incident flux J_0 but also on the distance r between the nucleus and the detector. The differential cross-section characterizes more directly the scattering and the target: it is defined by

$$\Sigma_I(\vec{q}) = \lim_{\vec{r}\to\infty}\frac{\vec{J}_I(\vec{r})\cdot\hat{r}}{J_0}r^2 \tag{6.3.38}$$

where

$$\vec{q} = k_0\hat{r} - \vec{k}_0 \qquad (\text{with } \hat{k}_0\cdot\hat{r} = \cos\theta) \qquad J_0 = \frac{\hbar k_0}{m_n},$$

and it is proportional to the number of neutrons scattered in direction $\hat{r}$, per unit solid angle. Using this definition, we obtain

$$\Sigma_I(\vec{q}) = \sum_j A(\vec{k}_0, \hat{k}; i, j) A^*(\vec{k}_0, \hat{k}; i, j) \qquad (6.3.39)$$

where I indicates a dependence with respect to the initial conditions and where the sum over j corresponds to a sum over final states.

3.1.9 Cross-section for any final spin state

In the Born approximation, the amplitude reads

$$A(\vec{k}_0, \hat{k}; i, j) = \{j|\mathfrak{b}^+ \mathfrak{p}^+ + \mathfrak{b}^- \mathfrak{p}^-|i\}.$$

By substitution in (6.3.37) and with the help of the identity

$$(\mathfrak{b}^+ \mathfrak{p}^+ + \mathfrak{b}^- \mathfrak{p}^-)^2 = (\mathfrak{b}^+)^2 \mathfrak{p}^+ + (\mathfrak{b}^-)^2 \mathfrak{p}^-$$

we find that the differential cross-section associated with an initial state I (for the radiation and the target) has a value which does not depend on $\vec{q}$

$$\Sigma_I(\vec{q}) = \Sigma_I = \{i|(\mathfrak{b}^+)^2 \mathfrak{p}^+ + (\mathfrak{b}^-)^2 \mathfrak{p}^-|i\}. \qquad (6.3.40)$$

When the spin of the scattering nucleus is zero ($S = 0$, $\mathfrak{b}^+ = \mathfrak{b}^- = \mathfrak{b}$), we get the trivial result

$$\Sigma_I = \mathfrak{b}^2.$$

However, if the target is a proton ($S = 1/2$), the situation is more complicated and in the following sections, two cases corresponding to particular but realistic initial states will be studied.

3.1.9.1 *Non-polarized incident beam and target*

When the incident beam is non-polarized, if the orientation of the nucleus spin is at random, each state has the same weight. Under such conditions, the average differential cross-section is

$$\Sigma = \frac{1}{2(2S+1)} \sum_I (\Sigma_I) = \frac{1}{2(2S+1)} \mathrm{Tr}\{(\mathfrak{b}^+)^2 \mathfrak{p}^+ + (\mathfrak{b}^-)^2 \mathfrak{p}^-\}. \qquad (6.3.41)$$

Let us calculate this expression: the number of states with total spin $(S + 1/2)$ is $2S + 2$. Thus,

$$\mathrm{Tr}\, \mathfrak{p}^+ = 2S + 2.$$

The number of states with total spin $(S - 1/2)$ is $2S$. Thus,

$$\mathrm{Tr}\, \mathfrak{p}^- = 2S$$

and therefore

$$\Sigma = \left[\frac{S+1}{2S+1}(b^+)^2 + \frac{S}{2S+1}(b^-)^2\right].$$

Setting

$$B^+ = \frac{(S+1)b^+ + Sb^-}{2S+1}$$

$$B^- = \frac{2(b^+ - b^-)}{2S+1} \qquad (6.3.42)$$

we obtain

$$\Sigma = (B^+)^2 + \left(\frac{B^-}{2}\right)^2 S(S+1) \qquad (6.3.43)$$

where, as will be shown later (Section 4), B^+ is an interesting length whereas B^- is a parasitical length.

3.1.9.2 *Polarized incident beam and polarized target*

When the scattering target contains nuclei with non-zero spin, the best experimental conditions are realized when the incident neutrons and the target nuclei are polarized simultaneously. Two important situations have to be considered: either the spins of the neutrons and of the protons point in the same direction or they point in the opposite direction.

The first situation corresponds to a state with total spin $(S + 1/2)$. It is not especially interesting because the corresponding collision length b^+ is not very different from the collision length of spinless nuclei like carbon also present in the target.

The second situation corresponds to a mixture of states of spin $(S + 1/2)$ and of spin $(S - 1/2)$. In this case, the collision length b^- which is strongly negative plays an important and useful role, by creating contrast (see Chapter 7). It is the latter situation which will now be studied.

For this situation the initial polarization state can be written

$$|i\} = |\uparrow\}_n |\downarrow\}_N.$$

The differential cross-section is given by (6.3.38)

$$\Sigma_I = \{i|(b^+)^2 p^+ + (b^-)^2 p^- |i\} = \Sigma$$

where p^+ and p^- are defined by (6.3.25). To calculate this cross-section, we have only to use the relation

$${}_n\{\uparrow|_N\{\downarrow|(\vec{s}\cdot\vec{S})|\downarrow\}_N|\uparrow\}_n = {}_n\{\uparrow|_N\{\downarrow|s_z\cdot S_z|\downarrow\}_N|\uparrow\}_n = -S/2.$$

Thus, one finds

$$\Sigma = \frac{(b^+)^2 + 2S(b^-)^2}{2S+1}. \qquad (6.3.44)$$

3.1.10 Cross-section for final state with spin-flip

Let us examine the case where the incident beam is polarized so that all the neutron spins point in the same direction (OX for instance). Let us consider collisions with spin-flip, the initial spin state

$$|i\} = |i'\}_{\mathrm{N}}|\uparrow\}_{\mathrm{n}} \quad \text{(N, for nuclei; n, for neutron)}$$

being transformed into states of the form

$$|j\} = |j'\}_{\mathrm{N}}|\downarrow\}_{\mathrm{n}}.$$

Let us calculate the corresponding scattered flux. In such conditions, the projection of the scattered state in real space can be written as follows [see (6.3.21)]

$$(2\pi)^{3/2}\langle\vec{r}|\zeta_{\mathrm{I}}\rangle)\} = |\downarrow\}_{\mathrm{n}}|\alpha)\sum_{j'}|j'\}_{\mathrm{N}}A(\vec{k}_0,\hat{k};\, i', j';\, \uparrow, \downarrow)\frac{e^{ik_0 r}}{r}$$

where α denotes a nucleus orbital state and where $A(\vec{k}_0, \hat{k};\, i', j';\, \uparrow, \downarrow)$ is the amplitude. The differential cross-section associated with this state is [see (6.3.39)]

$$\Sigma_{\mathrm{I}}(\uparrow\downarrow) = \sum_{j'} A(\vec{k}_0, \hat{k};\, i', j';\, \uparrow, \downarrow)\, A^*(\vec{k}_0, \hat{k};\, i', j';\, \uparrow, \downarrow). \tag{6.3.45}$$

The collision amplitude is given by (6.3.31), and by substitution, we obtain

$$\Sigma_{\mathrm{I}}(\uparrow\downarrow) = {}_{\mathrm{N}}\left\{i'\left|\left(\frac{B^-}{2}\right)^2 (S_x^2 + S_y^2)\right|i'\right\}_{\mathrm{N}}, \tag{6.3.46}$$

and we recall that $B^- = 2(\mathfrak{b}^+ - \mathfrak{b}^-)/(2S+1)$. Thus, when $\mathfrak{b}^+ = \mathfrak{b}^- = \mathfrak{b}$, $B^- = 0$; there is no collision-induced spin-flip.

Let us calculate the average cross-section over all the initial spin states of the nucleus. The number of these states is $2S + 1$ and each one has the same weight. We thus get

$$\Sigma(\uparrow\downarrow) = \frac{1}{2S+1}\sum_{\mathrm{I}}(\Sigma_{\mathrm{I}}(\uparrow\downarrow)) = \frac{1}{6}(B^-)^2 S(S+1). \tag{6.3.47}$$

This result must be considered as auxiliary; it will enable us to determine the parasitical quantity in (6.3.39).

3.2 Light scattering

A light beam propagating in matter interacts with the electrons of the medium. In insulators, the electrons are bound and the incident radiation induces a local polarization. Therefore, a scattering centre is a small polarizable element with the size of a monomer; this element can be assimilated to a dipole in forced oscillation regime. The radiation produced by this dipole is the scattered radiation. This is Rayleigh scattering.

3.2.1 Rayleigh scattering by a polarizable element

Let us set the polarizable element at the origin of the reference system and let ${}^0\vec{E}(z, t)$ be the incident electric field, propagating in OZ direction, with pulsation ω_0.

The polarizable element behaves like a dipole with electrical moment $D(\mathrm{t})$

$$\vec{D}(\mathrm{t}) = \vec{D}\mathrm{e}^{-i\omega_0 \mathrm{t}} \qquad (6.3.48)$$

In principle, this dipole is a functional of ${}^0\vec{E}(z, \mathrm{t})$ but the retardation effects will be neglected. Thus, the components of these vectors obey the linear relation

$$D_I^- = \sum_J \alpha_{IJ}\, {}^0\vec{E}_J^-$$

$$D_I^+ = \sum_J \alpha_{IJ}^*\, {}^0E_J^+ (6.3.49)$$

where the α_{IJ} are the elements of the polarizability tensor. Actually, we shall be interested only in the case of scattering without photon adsorbtion, i.e. in the case where

$$\alpha_{IJ} = \alpha_{IJ}^*.$$

In the electrostatic system of units, the α_{IJ} have the dimension of a volume.*

The electric fields associated with a radiation scattered at point $\vec{r}$ in direction $\hat{r}$ are identified with the electric field $\vec{E}(\vec{r}, \mathrm{t})$ produced by the dipole oscillations. We have [see (6.1.2)]

$$\vec{E}(\vec{r}, \mathrm{t}) = \vec{E}^-(\vec{r}, \mathrm{t}) + \vec{E}^+(\vec{r}, \mathrm{t})$$

where

$$\vec{E}^{\pm}(\vec{r}, \mathrm{t}) = \vec{E}^+(\vec{r})\mathrm{e}^{\pm i(\omega_0 \mathrm{t} - \vec{k}_0 \cdot \vec{r})}. \qquad (6.3.50)$$

Vectors $\vec{E}^{\pm}$ associated with the scattered field are determined by the relations (see Appendix D)

$$\vec{E}^{\pm}(\vec{r}) = \frac{1}{r} k_0^2 [\vec{D}^{\pm} - \hat{r}(\hat{r} \cdot \vec{D}^{\pm})]. \qquad (6.3.51)$$

Equations (6.3.51) and (6.3.49) show that the scattered field is related to the incident field. Let us express this relation in a proper manner.

Vectors ${}^0\vec{E}^{\pm}$ and $\vec{E}^{\pm}$ are defined in the same coordinate system. However, it will be convenient to introduce *two* orthogonal systems of coordinates associated respectively with the incident and scattered beams (see Fig. 6.1).

1. the orthogonal system OX, OY, OZ: OZ is pointing in the direction of propagation of the incident radiation and OX is perpendicular to the direction of propagation $\hat{r}$ of the scattered radiation (therefore $\hat{r}_X = 0$, $\hat{r}_Y = \sin\theta$, $\hat{r}_Z = \cos\theta$, where θ is the scattering angle);

* Indeed, in this system, $\vec{E}$ and $\vec{D}$ have the dimensions $E = \text{charge}/L$ and $D = \text{charge} \times L$.

2. the orthogonal system $Ox(\hat{r})$, $Oy(\hat{r})$, $Oz(\hat{r})$: $Oz(\hat{r})$ is directed along $\hat{r}$, and $Ox(\hat{r})$ coincides with OX (consequently, the axis $OY(\hat{r})$ is in the scattering plane).

The incident field has thus two components ${}^0E_X$, ${}^0E_Y$ and the scattered field two components E_x, E_y. The relation between these fields can be written in the form

$$E_i^-(\vec{r}, t) = \frac{e^{ik_0 r}}{r} \sum_J T_{iJ}(\vec{k}_0, \hat{r})\, {}^0E_J^-(\vec{0}, t) \tag{6.3.52}$$

with $i = x, y$ and $J = X, Y$; moreover, $E_i^+(\vec{r}, t)$ is given by the complex conjugate equation. To determine the $T_{iJ}(\vec{k}_0, \vec{r})$, we can use relations (6.3.51) which lead to the following equalities ($\hat{x}$ being the unit vector on OX)

$$E_x^-(\vec{r}) = \hat{X}\cdot\vec{E}^-(\vec{r}) = \frac{1}{r} k_0^2 D_X^-$$

$$E_y^-(\vec{r}) = -\hat{X}\cdot[\hat{r} \wedge \vec{E}^-(\vec{r})] = -\frac{1}{r} k_0^2 (\cos\theta\, D_Y^- - \sin\theta\, D_Z^-).$$

The following table is a consequence of them:

$$\begin{Vmatrix} T_{xX} & T_{yX} \\ T_{yX} & T_{yY} \end{Vmatrix} = k_0^2 \begin{Vmatrix} \alpha_{XX} & \alpha_{XY} \\ \alpha_{XY}\cos\theta - \alpha_{ZX}\sin\theta & \alpha_{YY}\cos\theta - \alpha_{ZY}\sin\theta \end{Vmatrix} \tag{6.3.53}$$

It will be convenient to express in a more compact way the table $T(\vec{k}_0, \hat{r})$ whose elements are $T_{iI}(\vec{k}_0, \hat{r})$ $(i = x, y; I = x, y)$. Let us introduce the matrices ${}^0\boldsymbol{\tau}$ and $\boldsymbol{\tau}$. The matrix ${}^0\tau$ expresses the fact that a vector of the plane OX, OY is also a vector of the space OX, OY, OZ. It reads

$${}^0\tau = \begin{vmatrix} 1 & 0 \\ 0 & 1 \\ 0 & 0 \end{vmatrix}. \tag{6.3.54}$$

The matrix τ expresses the fact that a vector of the plane $Ox(\hat{r})$, $Oy(\hat{r})$ is also a vector of the space OX, OY, OZ. It reads

$$\tau = \begin{vmatrix} 1 & 0 \\ 0 & \cos\theta \\ 0 & -\sin\theta \end{vmatrix}. \tag{6.3.55}$$

Taking into account the definition (6.3.53) of the elements T_{iJ}, we find

$$T(\vec{k}_0, \hat{r}) = k_0^2\, \tilde{\boldsymbol{\tau}}\, \boldsymbol{\alpha}\, {}^0\boldsymbol{\tau} \tag{6.3.56}$$

which is the compact form we were looking for.

3.2.2 Rayleigh scattering cross-section

Let us compare the P_μ ($\mu = 1, 2, 3, 4$) respectively associated with the incident and scattered radiations [see (6.1.11) and (6.3.52)]. From this comparison, we shall deduce the Rayleigh scattering cross-sections. For the incident radiation, we have

$$^0P_\mu = \frac{\epsilon}{\hbar\omega_0} {}^0E^+ \sigma_\mu {}^0E^- \tag{6.3.57}$$

where the $^0E^\pm$ represents vectors in the OX, OY reference system (see preceding section).

For the scattering of a radiation by a polarizable element, we have at point $\vec{r}$

$$P_\mu(\vec{r}) = \frac{\epsilon}{\hbar\omega_0} E^+(\vec{r})\sigma_\mu E^-(\vec{r}). \tag{6.3.58}$$

The modulation $e^{i(\omega t - \vec{k}_0 \cdot \vec{r})}$ that appears in (6.3.50) does not contribute to $P_\mu(\vec{r})$ and the $E^\pm(\vec{r})$ represent the vectors associated with the electric field, in the system $Ox(\vec{r})$, $Oy(\vec{r})$

The parameters $P_\mu(\vec{r})$ associated with the scattered radiation may fluctuate in time; for instance, as a result of a random rotation motion of the polarizable element. In general, the experimentalist observes the time average of the parameters P_μ. Then, he has to extract from this average, the contribution of the orientation fluctuations, which he considers either as a nuisance or as a study-deserving effect. We shall perform the calculation by using the formula [see (6.1.13) and (6.3.58)]

$$P_\mu(\vec{r}) = \frac{\epsilon}{\hbar\omega_0} \langle\langle E^+(\vec{r})\sigma_\mu E^-(\vec{r})\rangle\rangle. \tag{6.3.59}$$

However, let us first study instantaneous scattering. We shall examine the average (6.3.59) in the next section. Let us evaluate the parameters P_μ in terms of the parameters $^0P_\mu$. To do this, we use a relation deduced from (6.3.52)

$$E_i^- = \frac{1}{r}\sum_J \mathrm{T}_{iJ}(\vec{k}_0, \hat{r})\, {}^0E_J^-$$

(where $i = x, y$, and $J = X, Y$) which can be written in matrix form

$$E^- = \frac{1}{r}\mathrm{T}(\vec{k}_0, \hat{r})\, {}^0E^-.$$

Let us bring this expression in (6.3.58) and let us adopt the simplified notation $\mathrm{T} \equiv \mathrm{T}(\vec{k}_0, \hat{r})$. We obtain

$$P_\mu(\vec{r}) = \frac{\epsilon}{\hbar\omega_0}\frac{1}{r^2}\, {}^0E^+ \mathrm{T}^\dagger \sigma_\mu \mathrm{T}\, {}^0E^- \tag{6.3.60}$$

where $\mathrm{T}^\dagger$ is the adjoint of T.

The matrices $\boldsymbol{\sigma}_\mu$ constitute a complete basis for two-by-two matrices. On the other hand,

$$\mathrm{Tr}(\boldsymbol{\sigma}_\mu \boldsymbol{\sigma}_\nu) = 2\delta_{\mu\nu} \qquad \mu, \nu = 1, \ldots, 4.$$

Thus,

$$\mathrm{T}^{\dagger}\boldsymbol{\sigma}_{\mu}\mathrm{T} = \frac{1}{2}\sum_{\nu}[\mathrm{Tr}\{\mathrm{T}^{\dagger}\boldsymbol{\sigma}_{\mu}\mathrm{T}\boldsymbol{\sigma}_{\nu}\}]\boldsymbol{\sigma}_{\nu}.$$

Let us use this expansion in (6.3.60) and let us introduce the parameters ${}^{0}P_{\mu}$ defined by (6.3.57). We get

$$P_{\mu}(\vec{r}) = \frac{1}{2r^2}\sum_{\nu}[\mathrm{Tr}\{\mathrm{T}^{\dagger}\boldsymbol{\sigma}_{\mu}\mathrm{T}\boldsymbol{\sigma}_{\nu}\}]\,{}^{0}P_{\nu}. \tag{6.3.61}$$

This expression is the result we are looking for, and the quantities

$$\Sigma_{\mu\nu}(k_0, \theta) = \tfrac{1}{2}\mathrm{Tr}\{\mathrm{T}^{\dagger}\boldsymbol{\sigma}_{\mu}\mathrm{T}\boldsymbol{\sigma}_{\nu}\} \tag{6.3.62}$$

are the instantaneous differential cross-sections.

We then write

$$P_{\mu}(\vec{r}) = \frac{1}{r^2}\sum_{\nu}(\Sigma_{\mu\nu}(k_0, \theta)\,{}^{0}P_{\nu}). \tag{6.3.63}$$

3.2.3 Discussion

Let us express more explicitly the cross-sections $\Sigma_{\mu\nu}(k_\nu, \theta)$ [see (6.3.62)] in terms of the polarizability tensor $\boldsymbol{\alpha}$, of the wave number k_0, and of the scattering direction $\hat{r}$.

The polarizability tensor of a monomer is essentially anisotropic and we shall determine the characteristics of the corresponding $\Sigma_{\mu\nu}(k_0, \theta)$. In particular, we shall establish a difference between instantaneous cross-section and average cross-section, the averaging being made over the orientations of the polarizable element. But before this, we shall prove two general properties of the cross-sections $\Sigma_{\mu\nu}(k_0, \theta)$ and we shall discuss the case of an isotropic polarizable element.

3.2.3.1 *General properties of the Rayleigh scattering cross-sections*

1st property

At a given time, a totally polarized incident wave can only produce a totally polarized scattered wave. In fact, from eqn (6.3.61), we deduce the relation

$$\begin{aligned}\sum_{\mu} P_{\mu}\boldsymbol{\sigma}_{\mu} &= \frac{1}{2r^2}\sum_{\mu}\left[\mathrm{Tr}\left\{\mathrm{T}^{\dagger}\boldsymbol{\sigma}_{\mu}\mathrm{T}\left(\sum_{\nu}{}^{0}P_{\nu}\boldsymbol{\sigma}_{\nu}\right)\right\}\right]\boldsymbol{\sigma}_{\mu}\\ &= \frac{1}{2r^2}\sum_{\mu}\left[\mathrm{Tr}\left(\mathrm{T}\left(\sum_{\nu}{}^{0}P_{\nu}\boldsymbol{\sigma}_{\nu}\right)\mathrm{T}^{\dagger}\boldsymbol{\sigma}_{\mu}\right)\right]\boldsymbol{\sigma}_{\mu}\\ &= \frac{1}{r^2}\mathrm{T}\left(\sum_{\nu}{}^{0}P_{\nu}\boldsymbol{\sigma}_{\nu}\right)\mathrm{T}^{\dagger},\end{aligned} \tag{6.3.64}$$

and this relation leads to the result we want.

On the other hand, it is easy to see that

$$\text{dét}\left(\sum_{\nu=1}^{4} P_\nu \boldsymbol{\sigma}_\nu\right) = P_4^2 - P_1^2 - P_2^2 - P_3^2.$$

Let us calculate the determinant of both sides of equation (6.3.64) and let use the above result. We find

$$P_4^2 - P_1^2 - P_2^2 - P_3^2 = \frac{1}{r^4}|\det \mathrm{T}|^2[{}^0P_4^2 - {}^0P_1^2 - {}^0P_2^2 - {}^0P_3^2].$$

If the incident beam is totally polarized $\varpi_0 = 0$), the right-hand side vanishes. Consequently, the left-hand side also vanishes and the scattered beam is also totally polarized which is the predicted result.

2nd property

When the elements T_{iJ} are real, scattering cannot create ellipticity; if $P_2 = 0$ for the incident beam, then $P_2 = 0$ for the scattered beam. To prove this property, it is sufficient to establish that $\Sigma_{\nu,2} \equiv \Sigma_{2,\nu} = 0$, $\nu \neq 2$. Then let us examine the expressions

$$\Sigma_{\nu,2}(k_0, \theta) = \tfrac{1}{2}\mathrm{Tr}[\mathrm{T}^\dagger \sigma_\nu \mathrm{T}\boldsymbol{\sigma}_2]$$

when the elements of T are real. In this case, $\mathrm{T}^\dagger = \tilde{\mathrm{T}}$ where $\tilde{\mathrm{T}}$ is transposed T. Thus, we have

$$\begin{aligned}\mathrm{Tr}[\mathrm{T}^\dagger \boldsymbol{\sigma}_\nu \mathrm{T}\boldsymbol{\sigma}_2] &= \mathrm{Tr}[\tilde{\mathrm{T}}\boldsymbol{\sigma}_\nu \mathrm{T}\boldsymbol{\sigma}_2] \\ &= \mathrm{Tr}[\tilde{\boldsymbol{\sigma}}_2 \tilde{\mathrm{T}}\tilde{\boldsymbol{\sigma}}_\nu \mathrm{T}]. \qquad (6.3.65)\end{aligned}$$

But $\tilde{\boldsymbol{\sigma}}_\nu = \boldsymbol{\sigma}_\nu$, $\nu \neq 2$ and $\tilde{\boldsymbol{\sigma}}_2 = -\boldsymbol{\sigma}_2$; consequently,

$$\begin{aligned}\mathrm{Tr}[\tilde{\boldsymbol{\sigma}}_2 \tilde{\mathrm{T}}\tilde{\boldsymbol{\sigma}}_\nu \mathrm{T}] &= -\mathrm{Tr}[\boldsymbol{\sigma}_2 \tilde{\mathrm{T}}\boldsymbol{\sigma}_\nu \mathrm{T}] \\ &= -\mathrm{Tr}[\tilde{\mathrm{T}}\boldsymbol{\sigma}_\nu \mathrm{T}\boldsymbol{\sigma}_2] = -\mathrm{Tr}[\mathrm{T}^\dagger \boldsymbol{\sigma}_\nu \mathrm{T}\boldsymbol{\sigma}_2], \quad \nu \neq 2. \qquad (6.3.66)\end{aligned}$$

Equations (6.3.65) and (6.3.66) can be satisfied only if $\mathrm{Tr}[\mathrm{T}^\dagger \boldsymbol{\sigma}_\nu \mathrm{T}\boldsymbol{\sigma}_2] = 0$ and this condition establishes the property.

3.2.3.2 *Isotropic element*

When the polarizable element is isotropic, the elements α_{IJ} of the polarizability tensor can be written in the form

$$\alpha_{IJ} = \alpha\delta_{IJ}.$$

Let us calculate the cross-sections $\Sigma_{\mu\nu}(\vec{k}_0, \hat{r})$ in terms of α, with the help of eqns (6.3.62). The matrix (6.3.53) reads

$$\mathrm{T} = k_0^2\alpha\begin{Vmatrix}1 & 0 \\ 0 & \cos\theta\end{Vmatrix}. \qquad (6.3.67)$$

Let us set

$$\mathrm{b} = k_0^2 \boldsymbol{\alpha} \tag{6.3.68}$$

where b has the dimension of a length. This quantity is analogous to the collision length (6.3.21), which characterizes the neutron–nucleus interaction.

Let us introduce (6.3.67) in (6.3.62). The $\Sigma_{\mu\nu}$ thus obtained constitute the following table

$$\Sigma(k_0, \theta) = \mathrm{b}^2 \left\| \begin{array}{cccc} \cos\theta & 0 & 0 & 0 \\ 0 & \cos\theta & 0 & 0 \\ 0 & 0 & \dfrac{1+\cos^2\theta}{2} & \dfrac{\sin^2\theta}{2} \\ 0 & 0 & \dfrac{\sin^2\theta}{2} & \dfrac{1+\cos^2\theta}{2} \\ (1) & (2) & (3) & (4) \end{array} \right\| \begin{array}{c} (1) \\ (2) \\ (3) \\ (4) \\ \end{array} \tag{6.3.69}$$

and this is the result we were looking for.

Scattering experiments are often made with *linearly* polarized light. In this case, it is convenient to present the table $\Sigma(k_0, \theta)$ in a different way. Let us replace P_3 and P_4 by new variables[8]

$$P_x = \frac{1}{\sqrt{2}}(P_4 + P_3) \qquad P_y = \frac{1}{\sqrt{2}}(P_4 - P_3),$$

while keeping P_1 and P_2 as they are; we find

$$P_x = \frac{\sqrt{2}\varepsilon}{\hbar\omega}|E_x|^2$$

$$P_y = \frac{\sqrt{2}\varepsilon}{\hbar\omega}|E_y|^2 \tag{6.3.70}$$

and the cross-section table can be written in the form

$$\Sigma(k_0, \theta) = \mathrm{b}^2 \left\| \begin{array}{cccc} \cos\theta & 0 & 0 & 0 \\ 0 & \cos\theta & 0 & 0 \\ 0 & 0 & 1 & 0 \\ 0 & 0 & 0 & \cos^2\theta \\ (1) & (2) & (x) & (y) \end{array} \right\| \begin{array}{c} (1) \\ (2) \\ (x) \\ (y) \\ \end{array} \tag{6.3.71}$$

The parameter $P_x(\vec{r})$, obtained with the help of eqn (6.3.63) and of the preceding table, does not depend on the scattering angle. Thus, when the incident beam is polarized along OX, the photon flux that is scattered in the plane passing through the origin perpendicularly to OX, is isotropic; then it depends only on ^0P_x and on b.

3.2.3.3 *Anisotropic element*

The polarizable elements (monomers) associated with the polymer chains are in general anisotropic. Let us calculate the cross-sections $\Sigma_{I\mu\nu}(k_0, \theta)$ of such an element of which the orientation is defined with respect to the reference system of the laboratory. Here, the index I indicates a dependence with respect to the orientation. In the reference system OX, OY, OZ, the elements α_{JL} of the polarizability tensor, correspond to a given orientation I. Now let us introduce the elements T_{iJ} [see (6.3.53)] in the expression (6.3.62) of the cross-sections $\Sigma_{I,\mu\nu}(k_0, \theta)$. For the non-vanishing elements, we obtain

$$
\begin{aligned}
\Sigma_{I,11}(k_0, \theta) &= k_0^4[(\alpha_{XX}\alpha_{YY} + \alpha_{XY}^2)\cos\theta - (\alpha_{XX}\alpha_{YZ} + \alpha_{XY}\alpha_{ZX})\sin\theta] \\
\Sigma_{I,13}(k_0, \theta) &= k_0^4[\alpha_{XY}(\alpha_{XX} - \alpha_{YY})\cos\theta - (\alpha_{XX}\alpha_{ZX} - \alpha_{XY}\alpha_{YZ})\sin\theta] \\
\Sigma_{I,14}(k_0, \theta) &= k_0^4[\alpha_{XY}(\alpha_{XX} + \alpha_{YY})\cos\theta - (\alpha_{XX}\alpha_{ZX} + \alpha_{XY}\alpha_{YZ})\sin\theta] \\
\Sigma_{I,22}(k_0, \theta) &= k_0^4[(\alpha_{XX}\alpha_{YY} - \alpha_{XY}^2)\cos\theta - (\alpha_{XX}\alpha_{YZ} - \alpha_{XY}\alpha_{ZX})\sin\theta] \\
\Sigma_{I,33}(k_0, \theta) &= \frac{k_0^4}{2}[\alpha_{XX}^2 - \alpha_{XY}^2 - (\alpha_{XY}\cos\theta - \alpha_{ZX}\sin\theta)^2 + (\alpha_{YY}\cos\theta - \alpha_{YZ}\sin\theta)^2] \\
\Sigma_{I,34}(k_0, \theta) &= \frac{k_0^4}{2}[\alpha_{XX}^2 + \alpha_{XY}^2 - (\alpha_{XY}\cos\theta - \alpha_{ZX}\sin\theta)^2 - (\alpha_{YY}\cos\theta - \alpha_{YZ}\sin\theta)^2] \\
\Sigma_{I,41}(k_0, \theta) &= k_0^4[\alpha_{XX}\alpha_{XY} - (\alpha_{XY}\cos\theta - \alpha_{ZX}\sin\theta)(\alpha_{YY}\cos\theta - \alpha_{YZ}\sin\theta)] \\
\Sigma_{I,43}(k_0, \theta) &= \frac{k_0^4}{2}[\alpha_{XX}^2 - \alpha_{XY}^2 + (\alpha_{XY}\cos\theta - \alpha_{ZX}\sin\theta)^2 - (\alpha_{YY}\cos\theta - \alpha_{YZ}\sin\theta)^2] \\
\Sigma_{I,44}(k_0, \theta) &= \frac{k_0^4}{2}[\alpha_{XX}^2 + \alpha_{XY}^2 + (\alpha_{XY}\cos\theta - \alpha_{ZX}\sin\theta)^2 + (\alpha_{YY}\cos\theta - \alpha_{YZ}\sin\theta)^2].
\end{aligned}
\tag{6.3.72}
$$

The cross-sections table $\Sigma_{I,\mu\gamma}\theta(k_0, \theta)$ is now more complex than in the isotropic case. However, the general properties established in the preceding section, in particular concerning polarizations, remain valid.

3.2.3.4 *Averaging cross-sections over orientations*

In a polymer solution, a monomer which is an anisotropic polarizable element, has time-dependent random orientations with respect to the direction of the incident light beam. The scattered photon flux in a direction $\hat{r}$ is measured over a time interval which is large compared to the lifetime of an orientation of the polarizable element. Thus, the measured quantity is a *statistical* average over all the orientations of the scattering system.

Accordingly, let us calculate the mean cross-section of an anisotropic polarizable element by averaging over the orientations. We write

$$\Sigma_{\mu\nu}(k_0, \theta) = \langle\!\langle \Sigma_{\mathrm{I},\mu\nu}(k_0, \theta) \rangle\!\rangle$$
$$= \tfrac{1}{2} \operatorname{Tr}[\langle\!\langle \mathrm{T}^\dagger \sigma_\mu \mathrm{T} \sigma_\nu \rangle\!\rangle] \qquad (6.3.73)$$

where the brackets indicate this averaging.

Let us consider the expansion of the trace

$$\sum_{\substack{ij\\IJ}} \langle\!\langle \mathrm{T}^\dagger_{Ii} \sigma^{ij}_\mu \mathrm{T}_{jI} \sigma^{JI}_\nu \rangle\!\rangle = \sum_{\substack{ij\\IJ}} \langle\!\langle \mathrm{T}^\dagger_{Ii} \mathrm{T}_{jJ} \rangle\!\rangle \sigma^{ij}_\mu \sigma^{IJ}_\nu .$$

The calculation of the average over orientations amounts to determining the quantities [see (6.3.56)]

$$k_0^{-4} \langle\!\langle \mathrm{T}^\dagger_{Ii} \mathrm{T}_{jI} \rangle\!\rangle = \langle\!\langle (\tilde{\tau} \alpha\, {}^0\tau)^\dagger_{Ii} (\tilde{\tau} \alpha\, {}^0\tau)_{jJ} \rangle\!\rangle$$

$$= \sum_{\substack{KK'\\LL'}} \langle\!\langle \tilde{\tau}_{IK} \alpha^\dagger_{KK'}\, {}^0\tau_{K'i}\, \tilde{\tau}_{jL,}\, \alpha_{L'L}\, {}^0\tau_{LJ} \rangle\!\rangle . \qquad (6.3.74)$$

Inside the brackets, the only term that depends on the orientation is the product

$$\alpha^\dagger_{KK'} \alpha_{L'L} .$$

We shall restrict ourselves to the case where the polarizability tensor is real and symmetrical: $\alpha^\dagger = \tilde{\alpha} = \alpha$. It is possible then to find two numbers A and B such that[9]

$$\langle\!\langle \alpha_{KK'} \alpha_{LL'} \rangle\!\rangle = A \delta_{KK'} \delta_{LL'} + B(\delta_{KL} \delta_{K'L'} + \delta_{KL'} \delta_{LK'}) \qquad (6.3.75)$$
$$(K = 1,2,3 \qquad L = 1,2,3).$$

We neglected the case where the tensor α would depend on the wave vector and could be non-symmetrical (rotatory power). Thus, there is an orthogonal system of axes for which the matrix representation of the polarizability is diagonal.

$$\begin{Vmatrix} \alpha_1 & 0 & 0 \\ 0 & \alpha_2 & 0 \\ 0 & 0 & \alpha_3 \end{Vmatrix}$$

where $\alpha_1, \alpha_2, \alpha_3$ are constants. Let us fix the values of A and B by means of the invariants

$$\operatorname{Tr}[\alpha^2] = 3(A + 4B)$$
$$(\operatorname{Tr}[\alpha])^2 = 3(3A + 2B). \qquad (6.3.76)$$

Consequently,

$$10B = \operatorname{Tr}[\alpha^2] - \tfrac{1}{3}(\operatorname{Tr}[\alpha])^2$$
$$15A = 2(\operatorname{Tr}[\alpha])^2 - \operatorname{Tr}[\alpha^2] \qquad (6.3.77)$$

(In the isotropic case, $\alpha_1 = \alpha_2 = \alpha_3 = \alpha$, $\mathscr{A} = \alpha^2$ and $\mathscr{B} = 0$.)

Let us express the average cross-sections in terms of the parameters A and B. By introducing (6.3.75) in (6.3.74), we obtain

$$\begin{aligned} k_0^{-4}\langle\langle \mathrm{T}_{Ii}^{+}\,\mathrm{T}_{jJ}\rangle\rangle = {} & A({}^0\tilde{\tau}\,\tilde{\tau})_{Ii}(\tilde{\tau}\,{}^0\tau)_{jJ} \\ & + B({}^0\tilde{\tau}\,{}^0\tau)_{IJ}(\tilde{\tau}\,\tau)_{jJ} \\ & + B({}^0\tilde{\tau}\,\tau)_{Ij}(\tilde{\tau}\,{}^0\tau)_{iJ}. \end{aligned}$$

Then the cross-sections are given by [see (6.3.73)]

$$\begin{aligned} \Sigma_{\mu\nu}(k_0, \theta) = {} & \frac{k_0^4}{2}(A\,\mathrm{Tr}\{{}^0\tilde{\tau}\,\tau\,\sigma_\mu\,\tilde{\tau}\,{}^0\tau\,\sigma_\nu\} \\ & + B\,\mathrm{Tr}\{{}^0\tilde{\tau}\,\tau\tilde{\sigma}_\mu\tilde{\tau}\,{}^0\tau\,\sigma_\nu\} \\ & + B\,\mathrm{Tr}\{\tilde{\tau}\,\tau\,\sigma_\mu\}\,\mathrm{Tr}\{{}^0\tilde{\tau}\,{}^0\tau\,\sigma_\nu\}). \end{aligned} \tag{6.3.78}$$

We note the explicit expressions of the products

$$\begin{aligned} {}^0\tilde{\tau}\,\tau = \tilde{\tau}\,{}^0\tau &= \begin{Vmatrix} 1 & 0 \\ 0 & \cos\theta \end{Vmatrix} \\ {}^0\tilde{\tau}\,{}^0\tau = \tilde{\tau}\,\tau &= \begin{Vmatrix} 1 & 0 \\ 0 & 1 \end{Vmatrix}. \end{aligned} \tag{6.3.79}$$

Thus, the first term on the right-hand side of (6.3.78) has the same structure as the expression that was obtained with (6.3.66) for an isotropic element. The other terms can be calculated by using the definitions (6.1.8) and the expressions (6.3.79). The cross-section matrix appears in the form

$$\Sigma(k_0, \theta) =$$

$$k_0^4 \begin{Vmatrix} (A+B)\cos\theta & 0 & 0 & 0 \\ 0 & (A-B)\cos\theta & 0 & 0 \\ 0 & 0 & (A+B)\dfrac{(1+\cos^2\theta)}{2} & (A+B)\dfrac{\sin^2\theta}{2} \\ 0 & 0 & (A+B)\dfrac{\sin^2\theta}{2} & (A+B)\dfrac{(1+\cos^2\theta)}{2} \end{Vmatrix} \begin{matrix} (1) \\ (2) \\ (3) \\ (4) \end{matrix}$$

$$\begin{matrix} (1) & \qquad (2) & \qquad (3) & \qquad (4) \end{matrix} \tag{6.3.80}$$

The scattering represented by this matrix is depolarizing. In fact, let us consider a linearly polarized incident radiation. For an incident polarization direction along OY (see Fig. 6.1), the parameters ${}^0P_\mu$ ($\mu = 1, \ldots, 4$) have the following characteristics (see Table 6.1): ${}^0P_1 = {}^0P_2 = 0$; ${}^0P_3 = -{}^0P_4$. Let us calculate the parameters ${}^0P(\vec{r})$ of the scattered radiation, by using eqn (6.3.63) and matrix

(6.3.80). The result gives the polarization ratio

$$\varpi = |P_3|/P_4 = \frac{(A+B)\cos^2\theta}{(A+B)\cos^2\theta + 2B} < 1. \qquad (6.3.81)$$

The scattered radiation is thus depolarized if $B \neq 0$. It is totally depolarized if $\theta = \pi/2$.

For an incident polarization direction along axis OX, we have

$$^0P_1 = {}^0P_2 = 0 \qquad {}^0P_3 = {}^0P_4.$$

In this case, the polarization factor of the scattered beam is

$$\varpi = |P_2|/P_3 = \frac{A+B}{A+3B} < 1,$$

which shows that here also the scattered radiation is depolarized.

We may again introduce the transformation

$$P_x = \frac{1}{\sqrt{2}}(P_4 + P_3) \qquad P_y = \frac{1}{\sqrt{2}}(P_4 - P_3).$$

In the abstract space defined by P_1, P_2, P_x, P_y, the cross-sections are

$$\Sigma(k_0, \theta) = k_0^4 \left\| \begin{matrix} (A+B)\cos\theta & 0 & 0 & 0 \\ 0 & (A-B)\cos\theta & 0 & 0 \\ 0 & 0 & A+2B & B \\ 0 & 0 & B & (A+B)\cos^2\theta + B \end{matrix} \right\| \begin{matrix} (1) \\ (2) \\ (x) \\ (y) \end{matrix}$$

$$\begin{matrix} (1) & (2) & (x) & (y) \end{matrix} \qquad (6.3.82)$$

The parameter $P_x(\vec{r})$ obtained with the help of eqn (6.3.63) and of the above table does not depend on the scattering angle θ. Thus, if a beam is linearly polarized along OX, it is scattered into an isotropic beam in the plane passing by the origin and orthogonal to OX.

3.3 Summary and application to scattering experiments.

We have studied the scattering of a light-ray by a polarizable element. It has been assumed that the incident light was propagating along the axis OZ of the reference system, and that the polarizable element was located at the origin O. The scattered radiation has been characterized at each point of position vector $\vec{r}$, by the Stokes parameters [see (6.3.63)]

$$P_\mu(\vec{r}) = \frac{\epsilon}{\hbar\omega_0} \langle\langle E^+(\vec{r})\sigma_\mu E^-(\vec{r})\rangle\rangle \qquad (\mu = 1, \ldots, 4) \qquad (6.3.83)$$

where $E^{\pm}(\vec{r})$ are vectors representing the scattered electric field (the components of these vectors are defined in (6.3.52)). In particular, $P_4(\vec{r})$, which is proportional to the photon flux, is directly measurable.

We established relations between the parameters $P_\mu(\vec{r})$ of the scattered radiation and the parameters ${}^0P_\mu$ of the incident radiation [see (6.3.53)].

We shall keep in mind the following result: when incident light is linearly polarized along OX, the scattered photon flux in the plane (OY, OZ) has the following properties:

1. If the polarizable element is isotropic, this flux is isotropic in the plane and the radiation is linearly polarized along OX. In this case

$$P_4(\vec{r}) = \frac{1}{r^2}\Sigma_{44}(k_0, \theta)\,{}^0P_4 = \frac{1}{r^2}\mathfrak{b}^2\,{}^0P_4$$

where

$$\mathfrak{b} = k_0^2\boldsymbol{\alpha} \tag{6.3.84}$$

$\boldsymbol{\alpha}$ being the polarizability and $\Sigma_{44}(k, \theta)$ one of the elements of the cross-section table.

2. If the polarizable element is anisotropic, the mean flux is isotropic and it is obtained by averaging over all the orientations of the polarizable element; in this case, the scattered radiation is partially depolarized.* It is convenient to express P_3 and P_4 in terms of the contributions P_x and P_y

$$P_x = \frac{\sqrt{2\varepsilon}}{\hbar\omega}|E_x|^2 = \frac{1}{\sqrt{2}}(P_3 + P_4)$$

$$P_y = \frac{\sqrt{2\varepsilon}}{\hbar\omega}|E_y|^2 = \frac{1}{\sqrt{2}}(-P_3 + P_4)$$

Then, the relation between the incident and the scattered flux is

$$P_x(\vec{r}) = \frac{1}{r^2}k_0^4(A + 2B)\,{}^0P_x$$

$$P_y(\vec{r}) = \frac{1}{r^2}k_0^4 B\,{}^0P_x, \tag{6.3.85}$$

and here A and B are constants which are given, in terms of the polarizability tensor $\boldsymbol{\alpha}$ by

$$A = \tfrac{2}{15}[(\mathrm{Tr}[\boldsymbol{\alpha}])^2 - \tfrac{1}{2}\mathrm{Tr}[\boldsymbol{\alpha}^2]]$$

$$B = \tfrac{1}{10}[\mathrm{Tr}[\boldsymbol{\alpha}^2] - \tfrac{1}{3}(\mathrm{Tr}[\boldsymbol{\alpha}])^2] \tag{6.3.86}$$

In the following, we shall study radiation scattering only in the plane which is orthogonal to the direction of the (linear) incident polarization. A scattering

* And the polarization ratio is $\varpi = |P_3|/P_4$.

experiment will be interpreted with the help of eqns (6.3.84), (6.3.85), and (6.3.86) relating the scattered flux to the incident one.

4. SCATTERING BY A MACROSCOPIC SAMPLE

Let us consider a sample intended for a scattering experiment. It is made up of a large number of scattering centres. We can associate a cross-section with each isolated centre. We saw

(1) that the scattering of slow neutrons is isotropic;

(2) that the scattering of a linearly polarized light encountering a polarizable element, is on the average isotropic in the plane $\mathfrak{P}$ perpendicular to the polarization direction (the averaging is made over all orientations of the element).

In general, the cross-section of an assembly of scattering centres is not a multiple of the cross-section of one centre, even if all the centres are identical. Thus, neutron scattering is no longer isotropic. In the same way, the scattering of a light beam in planes parallel to $\mathfrak{P}$ is also no longer isotropic. There is interference from the scattered rays, and as a consequence, the scattered flux varies with the scattering angle θ. A study of this variation can be used to determine spatial correlations in the system.

It is observed that the interference effects are strongly dependent on the scatterer concentration and on the collision amplitudes. There are two extreme cases:

(1) simple scattering when the incident article collides only with one centre;

(2) refraction, when the incident particle interacts with many centers (multiple scattering)

Radiation scattering by an assembly of centres is also characterized by another effect related to time fluctuations concerning either the sample or the incident radiation. In fact, in the course of time, the total spin state of the neutron–nuclei system may change, and the same remark applies to the orientations of the anisotropic polarizable elements. Thus, the cross-section of the assembly of scattering centres is averaged over a period of time. Two contributions appear. The first one, which is called coherent, reveals interference effects between scattered rays. The second one, which is called incoherent, is the sum of the cross-section of the various centres (considered as isolated).

4.1 Simple neutron scattering by a set of nuclei with spin zero

In this section, we shall study the simple scattering of neutron radiation by a sample containing a large number N of elementary centres j with zero spin. With this as our aim we shall naïvely assume that the scattered wave is only the sum of the wavelets resulting from the scattering of the incident plane wave by the centres. Actually, in many cases, this assumption is quite realistic and its validity will be discussed later (Section 4.2). Moreover, we shall use the static approximation and Born's approximation. Thus, generalizing (6.3.21) and (6.3.23), we obtain the following expression of the stationary collision wave

$$\langle \vec{r}|\psi\rangle = (2\pi)^{-3/2}\, \mathrm{e}^{i\vec{k}_0\cdot\vec{r}} + \langle \vec{r}|\zeta\rangle$$

where

$$(2\pi)^{3/2}\langle \vec{r}|\zeta\rangle = \sum_{j=1}^{N} \mathrm{b}\, \mathrm{e}^{i\vec{k}_0\cdot\vec{r}_j} \frac{\mathrm{e}^{ik_0|\vec{r}-\vec{r}_j|}}{|\vec{r}-\vec{r}_j|}. \tag{6.4.1}$$

To calculate the scattered flux, far from the target, we shall evaluate $\langle \vec{r}|\zeta\rangle$ for $k_0 r \gg 1$. Let use the approximation

$$|\vec{r}-\vec{r}_j| - r = \frac{(\vec{r}-\vec{r}_j)^2 - r^2}{|\vec{r}-\vec{r}_j| + r} \simeq -\frac{\vec{r}\cdot\vec{r}_j}{r} = -\hat{r}\cdot\vec{r}_j.$$

This gives

$$(2\pi)^{3/2}\langle \vec{r}|\zeta\rangle \simeq \frac{\mathrm{e}^{ik_0 r}}{r} \sum_j \mathrm{b}_j \mathrm{e}^{i(\vec{k}_0 - k_0\hat{r})\cdot\vec{r}_j}.$$

The flux associated with the outgoing wave is [see (6.3.34)]

$$\vec{J}_1(\vec{r}) = -\frac{i\hbar}{2m_\mathrm{n}}[\langle\zeta|\vec{r}\rangle\vec{\nabla}_r\langle\vec{r}|\zeta\rangle - \langle\vec{r}|\zeta\rangle\vec{\nabla}_r\langle\zeta|\vec{r}|\rangle]$$

The gradient of $\zeta(\vec{r})$ is, to order $1/r$

$$(2\pi)^{3/2}\vec{\nabla}_r\langle\vec{r}|\zeta\rangle = ik_0\frac{\hat{r}}{r}\langle\vec{r}|\zeta\rangle.$$

Let us introduce the transfer vector

$$\vec{q} = k_0\hat{r} - \vec{k}_0.$$

Then the scattered flux reads

$$\vec{J}_1(\vec{r}) = \frac{\hbar k_0}{2m_\mathrm{n}}\frac{\hat{r}}{r}\sum_{j,l} \mathrm{b}_j \mathrm{b}_l \mathrm{e}^{i\vec{q}\cdot(\vec{r}_j - \vec{r}_l)} \tag{6.4.2}$$

and we recall that the incident flux is

$$J_0 = \frac{\hbar k_0}{2m_\mathrm{n}}.$$

The cross-section

$$\Sigma(\vec{q}) = \lim_{r\to\infty} r^2 \frac{\hat{r}\cdot\vec{J}(k_0\hat{r})}{J_0}$$

is given by the expression*†

$$\Sigma(\vec{q}) = \sum_{j,l} \mathrm{b}_j \mathrm{b}_l e^{i\vec{q}\cdot(\vec{r}_j - \vec{r}_l)} = \left| \sum_{j=1}^{N} \mathrm{b}_j e^{i\vec{q}\cdot\vec{r}_j} \right|^2. \tag{6.4.3}$$

We assumed that the atoms have fixed position vectors $\vec{r}_j (j = 1, \ldots, N)$. In reality, in a macroscopic sample, these points have a time-dependent distribution (and at a given time, this distribution depends on the sample). In this case, the cross-section $\Sigma(\vec{q})$ is an average over all the configurations of the system

$$\Sigma(\vec{q}) = \sum_{j,l} \mathrm{b}_j \mathrm{b}_l \langle\langle e^{i\vec{q}\cdot(\vec{r}_j - \vec{r}_l)} \rangle\rangle \tag{6.4.4}$$

which can also be written in the form

$$\Sigma(\vec{q}) = \sum_j \mathrm{b}_j^2 + \int \mathrm{d}^3 r \, e^{-i\vec{q}\cdot\vec{r}} \sum_{j,l\pm} \mathrm{b}_j \mathrm{b}_l \langle\langle \delta(\vec{r} - \vec{r}_j + r_l) \rangle\rangle. \tag{6.4.5}$$

Let us consider the case where the b_j $(j = 1, \ldots, N)$ are identical, and let us define the pair correlation function

$$g(\vec{r}) = \frac{V}{N^2} \sum_{j,l\neq} \langle\langle \delta(\vec{r} - \vec{r}_j + \vec{r}_l) \rangle\rangle \tag{6.4.6}$$

* Incidentally, let us remark that the crystallographers use a different notation when they study the cross-section anisotropy. In their language, relation (6.4.3) becomes

$$\Sigma(\vec{s}) = \left| \sum_{j=1}^{N} \mathrm{b}_j e^{i2\pi\vec{s}\cdot rj} \right|^2$$

where $\vec{s} = \frac{1}{2\pi}(k_0 \hat{r} - \vec{k}_0)$. The scattering angle is $\theta' = \theta/2$. Then, the modulus of the transfer wave vector is

$$J = \frac{2}{s} \sin\theta'.$$

This definition is associated with Bragg's relation $\lambda = 2\mathfrak{d} \sin\theta'$ where $\mathfrak{d}$ is the distance between crystalline planes. In what follows, we shall always adopt the convention (6.4.3), though certain experimentalists choose to express their results as the crystallographers do.

† A close examination of formula (6.4.3) shows that the sign of the lengths b_j is experimentally important. The relation (6.4.3) has been applied in various connections. First, it enables the experimentalist to deduce the sign of a length b_j (associated with a given chemical species) from the observation of diffraction lines. For this purpose, one uses a crystalline compound, the atoms of which have known relative positions $\vec{r}_j - \vec{r}_l$. For instance, a study of the diffraction lines of LiF irradiated with neutrons leads to the conclusion that the collision length of Li is negative. The second application proceeds from a reverse approach. The lengths b_j are known, but in a compound (for instance, a macromolecular one), the differences $\vec{r}_j - \vec{r}_l$ of the position vectors of the atoms are unknown. These relative positions are determined all the better as the numbers b_j for the various species constituting the material are contrasted. We shall come back to this point in the following chapter.

where V is the volume of the sample, in such a way that

$$\lim_{r \to \infty} g(\vec{r}) = 1,$$

which expresses the absence of long-distance correlations.

We may write

$$\Sigma(\vec{q}) = N b^2 + b^2 \frac{N^2}{V} \int d^3r \, e^{i\vec{q}\cdot\vec{r}} g(\vec{r}). \tag{6.4.7}$$

The second term is a function of the transfer vector $\vec{q}$. This function is not smooth: in particular

$$\lim_{q \to 0} \Sigma(\vec{q}) \neq \Sigma(0).$$

This singularity is related to the asymptotic behaviour of $g(\vec{r})$. Let us separate the corresponding singularity $\delta(\vec{q})$ from the rest of the contribution

$$\Sigma(\vec{q}) = N b^2 + (2\pi)^3 \frac{N^2}{V} b^2 \delta(\vec{q}) + \frac{N^2}{V} b^2 \int d^3r \, e^{i\vec{q}\cdot\vec{r}} [g(\vec{r}) - 1]. \tag{6.4.8}$$

The third term is now regular with respect to $\vec{q}$ everywhere in the interval $(0, +\infty)$. For a perfect gas, $g(\vec{r}) \equiv 1$, and, in this case, $\Sigma(\vec{q})$ reduces to the first two terms. In general, the third term does not vanish. Then, the correlation function $g(\vec{r})$ is the interesting quantity which is to be experimentally determined.

4.2 Simplicity or multiplicity of neutron scattering by an assembly of nuclei with zero spin

The relation between the pair correlation function of diffracting centres and the differential cross-section is *simple*, when the stationary collision wave obeys the 'naïve' equation (6.4.1). However, this equation is only a limiting case, because a wave which is scattered by an elementary centre can be scattered a second time, then a third time, and so on, before leaving the sample. This section will be devoted to a discussion on *multiple* scattering, a phenomenon which in general is to be avoided when scattering is used to study the structure of matter. In particular, we shall see that for a radiation impinging on a sample, strong multiple scattering implies refraction.

4.2.1 Phase shift of a neutron radiation and criterion of scattering simplicity

Here, the neutron scattering experiments by polymers will be interpreted only with eqn (6.4.1). Consequently, we must determine the conditions that have to be fulfilled, in order to ensure the validity of this equation. The right-hand side in (6.4.1) is correct only if, at each point in the sample, the modulus of the incident wave is much larger than the modulus of the outgoing wave; it is only in that case that multiple diffraction effects can be neglected. Let us express this condition, by comparing two waves at the centre of the sample of positive vector $\vec{r} = 0$. Let us set $\mathrm{b} = \mathrm{b}_j$ and

$$\delta = \mathrm{b} \sum_{i=1}^{\mathrm{N}} \mathrm{e}^{i\vec{k}_0 \cdot \vec{r}_i} \frac{\mathrm{e}^{ik_0 r_i}}{r_i}. \tag{6.4.9}$$

For $N \gg 1$ and $k_0^3\,(V/N) \gg 1$, we may go to the continuous limit:

$$\delta = \frac{\mathrm{b}N}{V} \int_V \mathrm{d}^3 r \, \mathrm{e}^{i\vec{k}_0 \cdot \vec{r}} \frac{\mathrm{e}^{ik_0 r}}{r}.$$

Let us assume that the sample is a sphere of radius R, centred at the origin, we obtain

$$\delta = \frac{3\mathrm{b}N}{k_0 R^3} \int_0^R \mathrm{d}r \sin k_0 r \, \mathrm{e}^{ik_0 r} = \frac{3\mathrm{b}N}{k_0 R^2} f(k_0 R)$$

where

$$f(x) = i\left[1 - \frac{\sin x}{x}\mathrm{e}^{ix}\right].$$

The modulus of $f(x)$ is bounded, and we see that number 2 is a trivial bound.

Thus, eqn (6.4.1) is valid only if

$$\delta = \frac{3\mathrm{b}N}{k_0 R^2} \ll 1. \tag{6.4.10}$$

Moreover, an explicit calculation of the scattering amplitude $A\,(\vec{k}_0, \hat{k})$ by a sphere of radius R, containing N centres would show that the amplitude is a function of δ, which can be considered as an 'average' phase shift. It is convenient to write the phase shift in the form

$$\delta = \frac{3\mathrm{b}N}{k_0 R^2} = \frac{4\pi \mathrm{b} C R}{k_0}. \tag{6.4.11}$$

For a fixed concentration of scattering centres, the phase shift is proportional to the volume of the sample.

One could try to extend this formula to a disordered material (for instance, a binary mixture). In this case, it is necessary to be careful because there exist several characteristic lengths. The length R which appears in (6.4.11) may be the size of the sample, but also the correlation length of the system. In all cases, the condition $\delta \ll 1$ is required. Thus, it cannot be satisfied in the vicinity of the demixtion point, when the correlation length and the size of the sample are simultaneously very large.

4.2.2 Multiple scattering

Multiple scattering is always a troublesome phenomenon for the interpretation of experimental results. However, in an important and rather simple case, the experimentalist is able to evaluate the contribution of multiple scattering and to make the correction needed to analyse the results in terms of pair correlation.

These very peculiar conditions are as follows. First, every collision produces a strong forward scattering (this is, for instance, the case with neutron scattering by polymers); then, the length of the trajectory of an incident neutron in the sample is nearly equal to the thickness L of this sample.

Before impinging on the sample, at $z = 0$, the flux of incident neutrons propagates without angular divergence in the direction OZ. For $z > 0$, the flux has a weak angular divergence which increases in proportion to the number of collisions with the centres. Let θ be the angular divergence. We shall consider its 'components' on the axes X and Y (See Fig. 6.2)

$$\theta^2 = \theta_x^2 + \theta_y^2 \quad \text{(by assimilating the angle to its tangent)}$$

$$d\theta_x d\theta_y = \text{element of solid angle.}$$

For $z > 0$, it is proper to introduce the angular flux $I(\theta_x, \theta_y, z)$ which at point z and in the direction determined by θ_x and θ_y is the neutron flux per unit time, area, and solid angle. In the isotropic case, we set

$$I(\theta_x, \theta_y; z) = I(\theta, z). \tag{6.4.12}$$

In particular, we have

$$I(\theta_x, \theta_y; 0) = J_0 \delta(\theta_x)\delta(\theta_y). \tag{6.4.13}$$

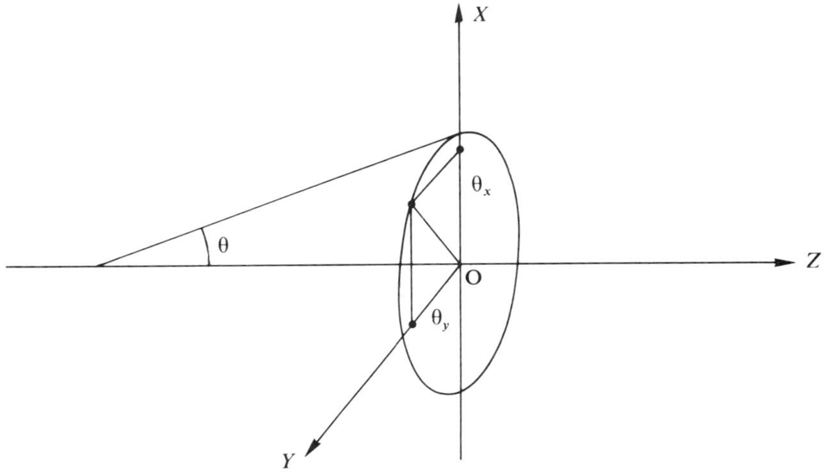

Fig. 6.2. 'Components' of the angle θ on the axes OX and OY.

With each scattering centre, we associate an angular cross-section $\Sigma_1(\theta_x, \theta_y)$. For a concentration N/V of scatterers, the angular collision rate is

$$\mathscr{K}(\theta_x, \theta_y) = \frac{N}{V}\Sigma_1(\theta_x, \theta_y),$$

a function which is appreciable only for $|\theta_x| \ll 1$, $|\theta_y| \ll 1$. It should also be noted that the quantity

$$\int \mathrm{d}\theta_x \int \mathrm{d}\theta_y \, \mathscr{K}(\theta_x, \theta_y) = \Xi \tag{6.4.14}$$

is called turbidity. The angular flux obeys the (approximate) master equation

$$\frac{\partial}{\partial z} I(\theta_x, \theta_y; z) = \int \mathrm{d}\theta'_x \, \mathrm{d}\theta'_y \, \mathscr{K}(\theta'_x, \theta'_y)[I(\theta_x - \theta'_x, \theta_y - \theta'_y; z) - I(\theta_x, \theta_y; z)].$$

Let us look for the solution of this equation of $z = L$. For this purpose, let us first perform a Fourier transformation in the space of the angular variables. We get

$$\frac{\partial}{\partial z}\tilde{I}(k_x, k_y; z) = [\mathscr{K}(k_x, k_y) - \mathscr{K}(0, 0)]\,\tilde{I}(k_x, k_y; z)$$

and

$$\tilde{I}(k_x, k_y{:}\ 0) = J_0.$$

The solution is

$$\tilde{I}(k_x, k_y; z) = J_0 \, \mathrm{e}^{z[\mathscr{K}(k_x, k_y) - \mathscr{K}(0, 0)]}. \tag{6.4.15}$$

Let us note that

$$\mathscr{K}(0, 0) = \Xi.$$

Moreover, the quantity

$$\lim_{k \to \infty} I(k_x, k_y; z) = J_0 \mathrm{e}^{-z\mathscr{K}(0, 0)} = J(z)$$

is the flux of neutrons at point z. Let us observe that turbidity is just the inverse of the mean free path of the neutron. The angular flux is the inverse Fourier transform of (6.4.21). We obtain

$$I(\theta_x, \theta_y; z) = J_0 \, \mathrm{e}^{-z\Xi}\left\{\delta(\theta_x)\delta(\theta_y) + \frac{1}{2\pi}\int \mathrm{d}k_x \mathrm{d}k_y \, \mathrm{e}^{-i(k_x\theta_x + k_y\theta_y)}\,[\mathrm{e}^{z\tilde{\mathscr{K}}(k_x, k_y)} - 1]\right\}. \tag{6.4.16}$$

This expression is written in a form similar to the one of the cross-section (6.4.16). The second term is a regular function denoted by

$$I_0(\theta_x, \theta_y; z).$$

It represents the scattered radiation. The first term contains the singularity: it represents the damping of the direct beam. From the function $I_0(\theta_x, \theta_y)$, we obtain the contributions of sample and multiple scattering. Let us expand the exponential of argument $\tilde{\mathscr{K}}(k_x, k_y)$. We get:

$$I_0(\theta_x,\theta_y;z) = \frac{J_0}{2\pi}\,\mathrm{e}^{-\Xi z}\int \mathrm{d}k_x\mathrm{d}k_y\,\mathrm{e}^{-i(k_x\theta_x + k_y\theta_y)}\sum_{n=1}^{\infty}\frac{[z\mathscr{K}(k_x,k_y)]^n}{n!}$$

The terms $n = 1, 2, \ldots$ and so on, represent single collisions, double collisions, etc. Let us attribute a Gaussian distribution to the angular collision rate

$$\tilde{\mathscr{K}}(k_x,k_y) = \Xi\,\mathrm{e}^{-k^2/2\zeta} \tag{6.4.17}$$

where $k^2 = k_x^2 + k_y^2$ and where ζ is a constant characterizing the angular broadening. Taking (6.4.2) into account, we obtain

$$I_0(\theta;z) = J_0\,\mathrm{e}^{-\Xi z}\,\zeta\sum_{n=1}^{\infty}\frac{(\Xi z)^n}{n n!}\,\mathrm{e}^{-\theta^2/2n\zeta}. \tag{6.4.18}$$

The half-height width $\bar{\theta}$ of the outgoing angular flux is determined by the equalities

$$I_0(\bar{\theta};L) = \tfrac{1}{2}I_0(0;L). \tag{6.4.19}$$

When ΞL is small, it is reasonable to keep only the first two terms of the expansion

$$I_0(\theta;L) = J_0\,\mathrm{e}^{-\Xi L}\,\zeta\left[\Xi L\,\mathrm{e}^{-\theta^2/2\zeta} + \frac{(\Xi L)^2}{4}\,\mathrm{e}^{-\theta^2/4\zeta}\right] \tag{6.4.20}$$

hence

$$\mathrm{e}^{-\bar{\theta}^2/2\zeta} + \frac{\Xi L}{4}\,\mathrm{e}^{-\bar{\theta}^2/4\zeta} = \frac{1}{2}\left(1 + \frac{\Xi L}{4}\right).$$

The half-height broadening $\bar{\theta}_0$ produced by simple scattering is given by

$$\mathrm{e}^{-\bar{\theta}_0^2/2\zeta} = \tfrac{1}{2}$$

(by dropping the second term).
An elementary calculation then shows that (for $\Xi L \ll 1$)

$$\frac{\bar{\theta} - \bar{\theta}_0}{\bar{\theta}_0} = \frac{\sqrt{2} - 1}{8\ln 2}\,\Xi L \simeq 0.075\,\Xi L,$$

and it must be noted that Ξ and L are measurable. For $\Xi L = 0.22$, which corresponds to the transmission factor $\mathscr{T} = \exp(-\Xi L) = 0.8$, the parasitical angular divergence is $(\bar{\theta} - \bar{\theta}_0)/\bar{\theta}_0 = 0.02$.

4.2.3 Refraction: an extreme case

When multiple scattering is intense, the wave impinging on a centre of position vector $\vec{r}'$ is already strongly perturbed. This wave approximately obeys the equation

$$\psi(\vec{r}) = \Phi(\vec{r}) - \mathrm{b}\int_V \mathrm{d}^3r\,\psi(\vec{r})\,\frac{\mathrm{e}^{ik_0|\vec{r} - \vec{r}'|}}{|\vec{r} - \vec{r}'|}\,C(\vec{r}') \tag{6.4.21}$$

where $\Phi(\vec{r}) \equiv <\vec{r}|\Phi>$ is the incident wave, $\Psi(\vec{r}) \equiv <\vec{r}|\Psi>$ is the stationary collision wave, and $C(\vec{r})$ represents the number of scatterers per unit volume at point $\vec{r}$ (the density). This function is obtained by coarse-graining of the real density

$$\sum_{i=1}^{N} \delta(\vec{r} - \vec{r}_i)$$

which cannot be directly introduced in (6.4.21) because this would produce divergences. In this limit, the scattering phenomenon appears as a refraction. To show this, we shall replace (6.4.21) with an equivalent but more compact equation. The operator $(\Delta + k_0^2)$ is applied to both sides of eqn (6.4.21). As, by definition,

$$(\Delta + k_0^2)\Phi(\vec{r}) = 0,$$

we get

$$(\Delta + k_0^2)\Psi(\vec{r}) = 4\pi \mathfrak{b}\,\Psi(\vec{r})C(\vec{r}).$$

Let us now define $n(\vec{r}) > 0$ by setting

$$n^2(\vec{r}) = 1 - \frac{4\pi}{k_0^2}\mathfrak{b}\,C(\vec{r}). \tag{6.4.22}$$

The above equation can be rewritten in the form

$$[\Delta + n^2(\vec{r})k_0^2]\psi(\vec{r}) = 0,$$

which is the propagation equation of a wave in a medium with a variable index $n(\vec{r})$. The coarse-graining operation eliminates the contribution of the fluctuating part of the real density. It is justified when the wavelength k_0^{-1} is large as compared to the distance between scatterers. In particular, for a homogeneous sample

$$n^2(\vec{r}) = n^2 \equiv 1 - \frac{4\pi \mathfrak{b} N}{k_0^2 V}. \tag{6.4.23}$$

In general, the length $\mathfrak{b}$* is positive, and in this case a neutron beam impinging on a material surface is totally reflected if the incidence angle is smaller than a limiting angle θ_c defined by the condition

$$\cos^2\theta_c = n^2,$$

i.e. if n is close to unity

$$\theta_c = \frac{2}{k_0}\left(\frac{\pi \mathfrak{b} N}{V}\right)^{1/2}. \tag{6.4.24}$$

The critical angles for interfaces are also small (typically = 10^{-1} angular degree/nm). The refraction phenomenon is used for various purposes: transportation of neutrons far from their source (high-efficiency wave guide), polar-

* However, $\mathfrak{b}$ is negative for hydrogen, lithium, and a few other elements.

ization of a neutron beam by total refraction on a magnetized metallic plate, study of surfaces.

4.3 Simple neutron scattering by an assembly of nuclei with non-zero spin: coherent and incoherent scattering

The parameters characterizing the interaction of a radiation with a scattering centre are able to fluctuate. When we deal with spins of neutrons and nuclei, we have quantum fluctuations; when we deal with anisotropic polarizable elements, we have time-dependent orientation fluctuations. For a macroscopic target, containing a large number of scatterers, we also have spatial fluctuations. The scattered fluxes are measured over time intervals that are very long with respect to the lifetime of a fluctuation. Moreover, they result from interactions with a large number of scatterers. Therefore, the measured quantities are statistical averages.

The average is taken over the initial states of the radiation-target system. Consequently, the cross-section appears as the sum of two well-defined contributions, called coherent and incoherent.

4.3.1 A general formula

In this section, we give a general expression of the differential cross-section for simple neutron scattering by a sample consisting of a number N of nuclei j with non-zero spin S_j.

The total initial spin state for a neutron and the nuclei is denoted by $|i\}$, and the final spin state by $|f\}$. By generalizing (6.3.21), it is possible to express the stationary collision wave in the form

$$(2\pi)^{3/2}\langle\vec{r}|\psi\rangle = |\alpha)|i\}\,\mathrm{e}^{i\vec{k}_0\cdot\vec{r}} + \sum_j\sum_f |\alpha)|f\}\,A_j(\vec{k}_0,\hat{k};i,f)\,\mathrm{e}^{i\vec{k}_0\cdot\vec{r}_j}\,\frac{\mathrm{e}^{ik_0|\vec{r}-\vec{r}_j|}}{|\vec{r}-\vec{r}_j|}$$

where $A_j(\vec{k}_0, \hat{k}; i, f)$ is the collision amplitude for nucleus j. Of course, this amplitude is non-zero only if $|f\}$ is identical to $|i\}$ or if $|f\}$ differs from $|i\}$ by a simultaneous spin-flip of the spins of nucleus j and of the neutron. More precisely, we have [see (6.3.27)]

$$A_j(\vec{k}_0,\hat{k};i,f) = -\{f|(\mathrm{b}_j^+\,\mathfrak{p}_j^+ + \mathrm{b}_j^-\,\mathfrak{p}_j^-)|i\} \qquad (6.4.25)$$

where

$$\mathfrak{p}_j^+ = (S_j + 2\vec{s}\cdot\vec{S}_j)/2S_j + 1$$

$$\mathfrak{p}_j^- = (S_j - 2\vec{s}\cdot\vec{S}_j)/2S_j + 1. \qquad (6.4.26)$$

By setting

$$B_j^+ = \frac{(S_j+1)\mathrm{b}_j^+ + S_j\mathrm{b}_j^-}{2S_j+1}$$

$$B_j^- = \frac{2(\mathrm{b}_j^+ - \mathrm{b}_j^-)}{2S_j+1} \qquad (6.4.27)$$

We may also write

$$A_j(\vec{k}_0,\hat{k};i,f) = -\{f|(B_j^+ + B_j^- \vec{s}\cdot\vec{S}_j)|i\}. \qquad (6.4.28)$$

Let us now come back to the scattered flux. For $k_0 r \gg 1$, we shall use the following approximation (see Section 4.1).

$$e^{i\vec{k}_0\cdot\vec{r}_j}\frac{e^{k_0|\vec{r}-\vec{r}_j|}}{|\vec{r}-\vec{r}_j|} \simeq e^{i(\vec{k}_0 - k_0\hat{r})\cdot\vec{r}_j} = e^{-i\vec{q}\cdot\vec{r}_j}$$

which is used to write the scattered flux in the form

$$\vec{J}_I(\vec{r}) = \frac{\hbar k_0}{2m_n}\frac{\hat{r}}{r}\sum_{j,l} e^{i\vec{q}\cdot(\vec{r}_j-\vec{r}_l)}\sum_f A_j(\vec{k}_0,\hat{k};i,f)\,A_l^*(\vec{k}_0,\hat{k};i,f).$$

The differential cross-section $\Sigma_I(\vec{q})$ associated with the initial state $|i\}$ is again defined by the equality

$$\Sigma_I(\vec{q}) = \sum_{j,l} e^{i\vec{q}\cdot(\vec{r}_j-\vec{r}_l)}\sum_f A_j(\vec{k}_0,\hat{k};i,f)\,A_l^*(\vec{k}_0,\hat{k};i,f). \qquad (6.4.29)$$

Thus, by taking (6.4.25) and (6.4.28) into account we obtain the fundamental formula

$$\begin{aligned}\Sigma_I(\vec{q}) &= \sum_{j,l} e^{i\vec{q}\cdot(\vec{r}_j-\vec{r}_l)}\{i|(\mathfrak{b}_j^+\mathfrak{p}_j^+ + \mathfrak{b}_j^-\mathfrak{p}_j^-)(\mathfrak{b}_l^+\mathfrak{p}_l^+ + \mathfrak{b}_l\mathfrak{p}_l)|i\}\\ &= \sum_{j,l} e^{i\vec{q}\cdot(\vec{r}_j-\vec{r}_l)}\{i|(B_j^+ + B_j^-(\vec{s}\cdot\vec{S}_j))(B_l^+ + B_l^-(\vec{s}\cdot\vec{S}_l))|i\} \qquad (6.4.30)\end{aligned}$$

4.3.2 Non-polarized beam and target

The differential cross-section $\Sigma_I(\vec{q})$ calculated above depends on the initial spin state $|i\}$ of the neutron–nuclei system. Let us average over these initial states $|i\}$. We allow, that the spin states of the nuclei j ($j = 1, \ldots, N$) are distributed at random and that these states and the initial state of the incident neutron are totally uncorrelated. In such a condition, the distribution of spin states for the incident neutron has no effect on the cross-section: it will be assumed here that the orientation of these spins is at random (non-polarized incident beam). The total number of spin states is

$$2\prod_{j=1}^{N}(2S_j + 1).$$

Thus, the average $\Sigma(\vec{q})$ of $\Sigma_I(\vec{q})$ reads

$$\Sigma(\vec{q}) = \sum_{j,l} e^{i\vec{q}\cdot(\vec{r}_j-\vec{r}_l)}\frac{1}{2\prod_{m=1}^{N}(2S_m+1)} \times \mathrm{Tr}\,\{(B_j^+ + B_j^+\vec{s}\cdot\vec{S}_j)(B_l^+ + B_l^-\vec{s}\cdot\vec{S}_l)\}.$$

$$(6.4.31)$$

The diagonal terms $j = l$ can easily be calculated with the help of the result obtained in Section 3, for the scattering by one nucleus. In this simple case, the cross-section was

$$\Sigma(\vec{q}) = (B^+)^2 + (B^-)^2 S(S+1).$$

Here, we obtain in the same way

$$\Sigma(\vec{q}) = \sum_{j=1}^{N} [(B_j^+)^2 + (B_j^-)^2 S_j(S_j+1)] +$$

$$+ \sum_{j,l\neq} e^{(i\vec{q}\cdot(\vec{r}_j - \vec{r}_l)} \frac{1}{2\prod_{m=1}^{N}(2S_m+1)} \mathrm{Tr}\{(B_j^+ + B_j^- \vec{s}\cdot\vec{S}_j)(B_l^+ + B_l^- \vec{s}\cdot\vec{S}_l)\}. \tag{6.4.32}$$

Let us now evaluate the last term. The neutron–nuclei spin states being uncorrelated, we have

$$\mathrm{Tr}\{\vec{s}\cdot\vec{S}_j\} = 0,$$

and also for $j \neq l$

$$\mathrm{Tr}\{(\vec{s}\cdot\vec{S}_j)(\vec{s}\cdot\vec{S}_j)\} = 0.$$

This shows that the operators $\vec{s}\cdot\vec{S}_j$ contained in the right-hand side of (6.4.32) do not contribute to the cross-section. Consequently, this cross-section is

$$\Sigma(\vec{q}) = \sum_{j=1}^{N} [(B_j^+)^2 + (B_j^-)^2 S_j(S_j+1)] + \sum_{j,l\neq} B_j^+ B_l^- e^{i\vec{q}\cdot(\vec{r}_j - \vec{r}_l)}. \tag{6.4.33}$$

Then, $\Sigma(\vec{q})$ can be written as the sum of a part which is called coherent because it contains all the $\vec{q}$-dependent terms, and of an incoherent part which does not depend on q

$$\Sigma(\vec{q}) = \Sigma^{\mathrm{coh}}(\vec{q}) + \Sigma^{\mathrm{inc}}(\vec{q}) \tag{6.4.34}$$

where

$$\Sigma^{\mathrm{coh}}(\vec{q}) = \sum_{j,l} \mathfrak{b}_j^{\mathrm{coh}} \mathfrak{b}_l^{\mathrm{coh}} e^{i\vec{q}\cdot(\vec{r}_j - \vec{r}_l)} \tag{6.4.35}$$

$$\Sigma^{\mathrm{inc}}(\vec{q}) = \sum_{j=1}^{N} (B_j^-)^2 S_j(S_j+1), \tag{6.4.36}$$

and obviously *in the present case*, we have

$$\mathfrak{b}_j^{\mathrm{coh}} = B_j^+. \tag{6.4.37}$$

For protons (and this type of scattering), this quantity equals

$$\mathfrak{b}_j^{\mathrm{coh}} = (-0.37409 \pm 0.00011) \times 10^{-5}\ \mathrm{nm}$$

(see Section 3.1.7).

Of course, this is the coherent part that will interest us for the study of the polymer configurations. Let us, however, note that for protons $B_j^- = 5.825 \times 10^{-5}$ nm, which means that B_j^- has a rather large value as compared to $\mathfrak{b}_j^{\mathrm{coh}}$. This fact causes a huge random noise which appears in the cross-sections for protons: we shall come back to these considerations in Section 4.3.4.

4.3.3 Polarized incident beam and polarized target

Let us calculate the differential cross-section of a target consisting of N polarized protons j, in the case where the incident beam itself is polarized. It will be assumed that the nuclei and the protons have opposite polarization directions.

The initial polarization state will be written in the symbolic form

$$\{i| = {}_{\mathrm{N}}\{\downarrow|_{\mathrm{n}}\{\uparrow|$$

where the index N refers to the proton nuclei and the index n to the neutrons.

The expression of the cross-section associated with this state is [see (6.4.28) and (6.4.29)]

$$\Sigma_{\mathrm{I}}(\vec{q}) = \sum_{j,l} \mathrm{e}^{i\vec{q}\cdot(\vec{r}_j - \vec{r}_l)} \{i|(B_j^+ + (\vec{s}\cdot\vec{S}_j)B_j^-)(B_l^+ + (\vec{s}\cdot\vec{S}_l)B_l^-)|i\}.$$

The terms $j = l$ in this expression have already been determined in Section 3.1.9.2, and the result is given by (6.3.44). Let us determine the terms of type $j \neq l$. We have

$$_{\mathrm{N}}\{\downarrow|_{\mathrm{n}}\{\uparrow|\vec{s}\cdot\vec{S}_j|\uparrow\}_{\mathrm{n}}|\downarrow\}_{\mathrm{N}} = -S_j/2$$

$$_{\mathrm{N}}\{\downarrow|_{\mathrm{n}}\{\uparrow|(\vec{s}\cdot\vec{S}_j)(\vec{s}\cdot\vec{S}_l)|\uparrow\}_{\mathrm{n}}|\downarrow\}_{\mathrm{N}} = S_jS_l/4, \qquad j \neq l$$

Taking the result (6.3.44) into account, we get

$$\Sigma(\vec{q}) = \Sigma_{\mathrm{I}}(\vec{q}) = \sum_j \frac{(\mathrm{b}_j^+)^2 + (\mathrm{b}_j^-)^2}{2}$$

$$+ \sum_{j,l\neq} (\mathrm{e}^{i\vec{q}\cdot(\vec{r}_j - \vec{r}_l)}\left(B_j^+ - \frac{S_j}{2}B_j^-\right)\left(B_l^+ - \frac{S_l}{2}B_l^-\right).$$

The coherent part of the cross-section is the term which depends on the transfer $\vec{q}$.

Let us set

$$\mathrm{b}_j^{\mathrm{coh}} = \left(B_j^+ - \frac{S_j}{2}B_j^-\right). \tag{6.4.38}$$

We have

$$\Sigma^{\mathrm{coh}}(\vec{q}) = \sum_{j,l} \mathrm{b}_j^{\mathrm{coh}}\,\mathrm{b}_l^{\mathrm{coh}}\,\mathrm{e}^{i\vec{q}\cdot(\vec{r}_j - \vec{r}_l)}.$$

Starting from (6.4.38) and using relations (6.4.27), we may, *in this case*, write the coherent collision length in the form

$$\mathrm{b}_j^{\mathrm{coh}} = \frac{\mathrm{b}_j^+ + 2S_j\mathrm{b}_j^-}{2}. \tag{6.4.39}$$

For protons ($S = 1/2$), this length is -1.82×10^{-5} nm (see Section 3.7.7.1) and we see that its modulus is about five times larger than the coherent collision length of protons in a non-polarized target. For the coherent cross-section, a factor 25 is won and this is why this type of experiment is valuable. The nuclear polarization of a sample has been attained, for the first time, by Abragam and his team.[10]

4.3.4 Incident polarized beam and analysis of the polarization of the scattered beam. Cross-section with spin-flip

Let us now consider the case where the spins of the incident neutrons point in one direction, and where, from the scattered flux, one extracts the neutrons with spins pointing in the opposite direction. In reality, such a selection can be made by total reflection on magnetized metallic plates.

Let $|i\} = |\uparrow\}_n |i'\}_N$ be the initial total spin state, where $|i'\}_N$ represents the initial spin state of the nuclei; on the other hand, let $|f\} = |\downarrow\}_n |f'\}_N$ be the final total spin state, where $|f'\}_N$ represents the final spin state of the nuclei. The corresponding collision amplitude is given by a formula very similar to (6.4.28) [see also (6.3.32)]

$$A_j(\vec{k}_0, \hat{k}; i, f) = -{}_N\{f'|\tfrac{1}{2}B_j^-(S_{xj} + iS_{yj})|i'\}_N.$$

Let us bring this expression in (6.4.2). We obtain the cross-section

$$\Sigma_I(\vec{q}; \uparrow\downarrow) = \sum_{j,l} e^{i\vec{q}\cdot(\vec{r}_j - \vec{r}_l)}{}_N\{i'|B_j^-(S_{xj} - iS_{yj})B_l^-(S_{xl} + iS_{yl})|i'\}_N. \tag{6.4.40}$$

This cross-section depends on the initial spin state of the nuclei. The number of these states is

$$\prod_{j=1}^{N}(2S_j + 1).$$

Let us calculate $\Sigma(\vec{q}; \uparrow\downarrow)$, the average of $\Sigma_I(\vec{q}; \uparrow\downarrow)$ over all these states

$$\Sigma(\vec{q}; \uparrow\downarrow) = \frac{1}{\prod_{j=1}^{N}(2S_j + 1)}\sum_I \Sigma_I(\vec{q}; \uparrow\downarrow)$$

$$= \frac{1}{\prod_{m=1}^{N}(2S_m + 1)}\sum_{j,l} e^{\vec{q}\cdot(\vec{r}_j - \vec{r}_l)}\mathrm{Tr}\{B_j^-(S_{xj} - iS_{yj})B_l^-(S_{xl} + iS_{yl})\}$$

The orientations of the nuclear spins are uncorrelated. Thus for $j \neq l$

$$\mathrm{Tr}\{S_{\alpha j}S_{\beta l}\} = 0 \qquad \alpha = x, y \quad \text{and} \quad \beta = x, y.$$

Therefore, the only non-vanishing terms in the expression giving the average flux are the diagonal terms $j = l$. A diagonal term is given by (6.3.47). Thus, their sum does not depend on the transfer wave vector. It is an incoherent cross-section, denoted by $\Sigma_{\uparrow\downarrow}^{\text{inc}}$

$$\Sigma_{\uparrow\downarrow}^{\text{inc}} = \Sigma(\vec{q}; \uparrow\downarrow) = \sum_{j=1}^{N}\frac{1}{6}(B_j^-)^2 S_j(S_j + 1). \tag{6.4.41}$$

We can use the result of such an experiment jointly with the result of another experiment performed with the same sample but without polarizing the neutrons. In fact, the contribution (6.4.41) is proportional to the incoherent cross-section given by (6.3.6). Thus, using two measurements it is possible by subtraction to obtain the coherent cross-section which is defined by (6.4.35) and which is the quantity of interest.

4.3.5 Numerical values of collision lengths (neutrons)

Values of $\mathfrak{b}^{coh}$, $\mathfrak{b}^{coh\,2}$ and Σ^{inc} are listed below (see Table 6.2), for nuclei usually found in polymers. The unity of area is the barn (1 barn $= 10^{-24}$ cm^2). We note that the incoherent scattering cross-section is relatively very large. This is the main origin of the experimental 'random noise'.

Let us also recall (see Section 4.4.1.2) that the coherent collision length associated with a proton belonging to a target made up of polarized protons is

$$\mathfrak{b}^{coh} = -1.82 \times 10^{-5} \text{ nm}$$

when the incident beam is polarized in a direction which is the opposite of the polarization direction of the target.

Table 6.2

Atoms	Spin S of the nucleus	$\mathfrak{b}^{coh} \times 10^5$ [nm]	$\Sigma^{coh} = (\mathfrak{b}^{coh})^2$ [barn]	Σ^{inc} [barn]
H	1/2	-0.37409 ± 0.00011	0.1398	5.65
D	1	0.6672 ± 0.0007	0.4451	0.24
C	0	0.66535 ± 0.00014	0.4427	0
N	1	0.936 ± 0.002	0.876	0.1
O	0	0.5805 ± 0.0005	0.337	0
S	0	0.285 ± 0.001	0.0812	0

4.4 Simple scattering of light by an assembly of polarizable elements

A propagating electric field ${}^0E(\vec{r}, t)$ acts on an assembly of polarizable elements, bound to an orthogonal reference system OX, OY, OZ. Let us calculate the value of the scattered field at a point of position vector $\vec{r}$. To express the components of this field, we shall use the orthogonal system O$x(\hat{r})$, O$y(\hat{r})$, O$z(\hat{r})$ where O$z(\hat{r})$ has the same direction as $\hat{r}$ and where O$x(\hat{r})$ coincides with Ox (see Fig. 6.1). We shall study the scattering in the plane OY, OZ. For a target consisting of only one polarizable element, we obtained [see (6.3.32)]

$$E_i^-(\vec{r}, t) = \frac{e^{ik_0 r}}{r} \sum_J T_{iJ}(\vec{k}_0, \hat{r})\, {}^0E_J^-(0, t). \tag{6.4.42}$$

For a set of N elements, with position vectors $r_j (j = 1, \ldots, N)$, and, in the simple collision approximation, the scattered field is given by the sum

$$E_i^-(\vec{r}, t) = \sum_{j=1}^{N} \frac{e^{ik_0|\vec{r} - \vec{r}_j|} e^{i\vec{k}_0 \cdot \vec{r}_j}}{|\vec{r} - \vec{r}_j|} \sum_J {}^jT_{iJ}\left(\vec{k}_0, \frac{\vec{r} - \vec{r}_j}{|\vec{r} - \vec{r}_j|}\right) {}^oE_j^-(0, t)$$

where ${}^jT_{iJ}(\vec{k}_0, \vec{r})$ is the matrix associated with the polarizable element j. We note that here the modulation $e^{ik_0 r}$ in (6.4.42) is essential. Let us introduce the elastic transfer wave vector $\vec{q} = k_0 \hat{r} - \vec{k}_0$ and let us use the following approximation which is valid for long distances

$$\frac{e^{ik_0|\vec{r} - \vec{r}_j|} e^{i\vec{k}_0 \cdot \vec{r}_j}}{|\vec{r} - \vec{r}_j|} \simeq \frac{e^{ik_0 r}}{r} e^{-i\vec{q} \cdot \vec{r}_j} \tag{6.4.43}$$

Thus, it is possible to write eqn (6.4.42) in the form

$$E_i^-(\vec{r}, t) \simeq \frac{e^{ik_0 r}}{r} \sum_J \sum_j e^{-i\vec{q} \cdot \vec{r}_j}\, {}^jT_{iJ}(\vec{k}_0, \hat{r})\, {}^oE_j^-(0, t).$$

Let us calculate the scattering cross-sections. We get [see (6.3.61)]

$$\Sigma_{I,\mu\nu}(k_0, \theta) = \frac{1}{2} \sum_{j,l}^{N} e^{-i\vec{q} \cdot (\vec{r}_j - \vec{r}_l)} \operatorname{Tr}\{{}^jT^\dagger \boldsymbol{\sigma}_\mu T \boldsymbol{\sigma}_\nu\} \tag{6.4.44}$$

(where ${}^jT$ is a function of k_0 and of θ, and where $q = 2k_0 \sin\theta/2$).

Here, the index I symbolizes the orientation of the polarizable elements j (in their initial state and μ and ν are indices associated with the Stokes parameters (μ, $\nu = 1, 2, 3, 4$). The dependence with respect to I concerns only systems of anisotropic polarizable elements oriented at random, but for such systems it is essential. In this case, an average is taken over all the orientations I, and each average cross-section is the sum of a coherent and of an incoherent contribution.

4.4.1 Scattering by an assembly of isotropic polarizable elements

Let us calculate the cross-sections $\Sigma_{\mu\nu}(k_0, \theta)$ in the very special case where all the polarizable elements are identical and isotropic. As, in this case, the matrices ${}^jT$ do not depend on the index j, eqn (6.4.44) reads

$$\Sigma_{\mu\nu}(k_0, \theta) = \frac{1}{2} \sum_{j,l}^{N} e^{-i\vec{q} \cdot (\vec{r}_j - \vec{r}_l)} \operatorname{Tr}\{T^\dagger \boldsymbol{\sigma}_\mu T \boldsymbol{\sigma}_\nu\}. \tag{6.4.45}$$

Let us consider the case where the incident radiation is linearly polarized in direction OX. Then, the scattering cross-section in the plane OY, OZ [see

(6.3.71)] is

$$\Sigma_{xx}(k_0, \theta) = \mathfrak{b}^2 \sum_{j,l} e^{-i\vec{q}\cdot(\vec{r}_j - \vec{r}_l)}$$

where $\mathfrak{b} = k_0^2 \boldsymbol{\alpha}$, $\boldsymbol{\alpha}$ being the polarizability, and omitting the dependence of $\mathfrak{b}$ with respect to k_0^2, we may write

$$\Sigma_{xx}(k_0, \theta) = \Sigma(\vec{q}). \tag{6.4.46}$$

4.4.2 Scattering by an assembly of identical anisotropic polarizable elements

When the polarizable elements are anisotropic, in each site j, the polarizable element has its own orientation.

In the expression giving the cross-section associated with an initial state I

$$\Sigma_{\mathrm{I},\mu\nu}(k_0, \theta) = \frac{1}{2} \sum_{j,l} e^{-i\vec{q}\cdot(\vec{r}_j - \vec{r}_l)} \operatorname{Tr}\{{}^j\mathrm{T}^\dagger \boldsymbol{\sigma}_\mu {}^l\mathrm{T} \boldsymbol{\sigma}_\nu\}$$

the elements of the table ${}^j\mathrm{T}(\vec{k}_0, \vec{r})$ depend on the orientations of the polarizable element. As the orientation varies from site to site, the exponentials cannot be factorized anymore.

Let us calculate $\Sigma_{\mu\nu}(k_0, \theta)$, the average of $\Sigma_{\mathrm{I},\mu\nu}(k_0, \theta)$ over all orientations. The diagonal ($j = l$) and non-diagonal ($j \neq l$) terms will be treated separately

$$\Sigma_{\mu\nu}(k_0, \theta) = \frac{1}{2} \sum_{j=l}^{N} \langle\!\langle \operatorname{Tr}\{{}^j\mathrm{T}^\dagger \boldsymbol{\sigma}_\mu {}^l\mathrm{T} \boldsymbol{\sigma}_\nu\} \rangle\!\rangle \tag{6.4.47}$$

$$+ \frac{1}{2} \sum_{j,l \neq} \langle\!\langle e^{-i\vec{q}\cdot(\vec{r}_j - \vec{r}_l)} \operatorname{Tr}\{{}^j\mathrm{T}^\dagger \boldsymbol{\sigma}_\mu {}^l\mathrm{T} \boldsymbol{\sigma}_\nu\} \rangle\!\rangle.$$

The calculation of the first term is given by eqn (6.3.78). Let us examine the term $j, l \neq$. We admit that the averaging over the orientations concerns only the factor $\operatorname{Tr}\{{}^j\mathrm{T}^\dagger \boldsymbol{\sigma}_\mu {}^l\mathrm{T} \boldsymbol{\sigma}_\nu\}$ and we shall show that the average is independent of the indices j and l; then we shall be able to factorize the exponential as in the isotropic case.

$$\langle\!\langle \operatorname{Tr}\{{}^j\mathrm{T}^\dagger \boldsymbol{\sigma}_\mu {}^l\mathrm{T} \boldsymbol{\sigma}_\nu\} \rangle\!\rangle = \sum_{ij,IJ} \langle\!\langle {}^j\mathrm{T}^\dagger_{Ii} {}^l\mathrm{T}_{jJ} \rangle\!\rangle \boldsymbol{\sigma}^{ij}_\mu \boldsymbol{\sigma}^{JI}_\nu.$$

Let us denote by $\boldsymbol{\alpha}$, the polarizability tensor associated with the element j, and by $\boldsymbol{\beta}$ the polarizability tensor associated with the element l. By definition [see (6.3.74)]

$$k_0^{-4} \langle\!\langle {}^j\mathrm{T}^\dagger_{Ii} {}^l\mathrm{T}_{jJ} \rangle\!\rangle = \sum_{KK',LL'} \langle\!\langle {}^0\tilde{\tau}_{IK} \boldsymbol{\alpha}_{KK'} \boldsymbol{\tau}_{K'i} \tilde{\boldsymbol{\tau}}_{jL'} \boldsymbol{\beta}_{L'L} {}^0\boldsymbol{\tau}_{LJ} \rangle\!\rangle. \tag{6.4.48}$$

Here the only term depending on orientation is the product

$$\boldsymbol{\alpha}_{KK'} \boldsymbol{\beta}_{L'L}.$$

Let us admit that the orientations of the elements $j, l \neq$ are not correlated: in this case

$$\langle\!\langle \alpha_{KK'} \beta_{L'L} \rangle\!\rangle = \langle\!\langle \alpha_{KK'} \rangle\!\rangle \langle\!\langle \beta_{L'L} \rangle\!\rangle .$$

Moreover $\langle\!\langle \alpha \rangle\!\rangle$ must be invariant by rotation. Thus

$$\langle\!\langle \alpha_{KK'} \rangle\!\rangle = \frac{1}{3} \mathrm{Tr}\{\alpha\} \delta_{KK'} = \frac{\alpha_1 + \alpha_2 + \alpha_3}{3} \delta_{KK'} .$$

The average polarization is isotropic. Therefore, as the polarizable elements (j, l) are identical [see (6.3.76)]

$$\langle\!\langle \alpha_{KK'} \beta_{L'L} \rangle\!\rangle \left(\frac{\alpha_1 + \alpha_2 + \alpha_3}{3} \right)^2 \delta_{KK'} \delta_{LL'} = \frac{3A + 2B}{3} \delta_{KK'} \delta_{LL'} .$$

By bringing this result in (6.4.48), we get

$$k_0^{-4} \langle\!\langle {}^j\mathrm{T}_{Ij}\, {}^l\mathrm{T}_{jJ} \rangle\!\rangle = \frac{3A + 2B}{3} ({}^0\tilde{\tau}\ \tau)_{Ii} (\tilde{\tau}\ {}^0\tau)_{jJ} .$$

We thus obtain

$$k_0^{-4} \langle\!\langle \mathrm{Tr}\{{}^j\mathrm{T} \sigma_\mu {}^l\mathrm{T} \sigma_\nu\} \rangle\!\rangle = \frac{3A + 2B}{3} \mathrm{Tr}\{{}^0\tilde{\tau}\ \tau\ \sigma_\mu\ \tilde{\tau}\ {}^0\tau\ \sigma_\nu\} .$$

This term no longer depends on the indices j, l; we may now factorize the exponential in the contribution of the terms $j \neq l$ to the cross-section (6.4.47). This contribution is

$$\Sigma^{\mathrm{II}}_{\mu\nu} = k_0^4 \left(\frac{A}{2} + \frac{B}{3} \right) \mathrm{Tr}\{{}^0\tilde{\tau}\ \tau \sigma_\mu\ \tilde{\tau}\ {}^0\tau\ \sigma_\nu\} \sum_{j,l} \mathrm{e}^{-i\vec{q} \cdot (\vec{r}_j - \vec{r}_l)} \tag{6.4.49}$$

The cross-sections $\Sigma_{\mu\nu}(k_0, \theta)$ are given by

$$\Sigma_{\mu\nu}(k_0, \theta) = \Sigma^{\mathrm{I}}_{\mu\nu} + \Sigma^{\mathrm{II}}_{\mu\nu} \tag{6.4.50}$$

where $\Sigma^{\mathrm{I}}_{\mu\nu}$ is just the $\Sigma_{\mu\nu}(k_0, \theta)$ given by eqn (6.3.78).

We can also express the same result in the more convenient form

$$\Sigma_{\mu\nu}(k_0, \theta) = \Sigma^{\mathrm{inc}}_{\mu\nu} + \Sigma^{\mathrm{coh}}_{\mu\nu} . \tag{6.4.51}$$

The coherent cross-section consists of the terms $j \neq l$ and also of contributions of the same form which appear in the terms $j = l$.

$$\Sigma^{\mathrm{coh}}_{\mu\nu} = k_0^4 \left(\frac{A}{2} + \frac{B}{3} \right) \mathrm{Tr}\{{}^0\tilde{\tau}\ \tau\ \sigma_\mu\ \tilde{\tau}\ {}^0\tau\ \sigma_\nu\} \sum_{j,l} \mathrm{e}^{-i\vec{q} \cdot (\vec{r}_j - \vec{r}_l)} \tag{6.4.52}$$

The incoherent cross-section contains only non-interfering terms j (hence its name). Taking (6.3.78), (6.4.51), and (6.4.52), into account, we get

$$\Sigma^{\mathrm{inc}}_{\mu\nu} = k_0^4 B\big(-\tfrac{1}{3} \mathrm{Tr}\{{}^0\tilde{\tau}\, \tau\ \sigma_\mu\ \tilde{\tau}\ {}^0\tau\ \sigma_\nu\} + \tfrac{1}{2} \mathrm{Tr}\{{}^0\tilde{\tau}\ \tau\ \tilde{\sigma}_\mu\ \tilde{\tau}\ {}^0\tau\ \sigma_\nu\}$$
$$+ \tfrac{1}{2} \mathrm{Tr}\{\tilde{\tau}\ \tau\ \sigma_\mu\} \mathrm{Tr}\{{}^0\tilde{\tau}\ {}^0\tau\ \sigma_\mu\}\big).$$

Let us explicitly calculate the traces associated with these cross-sections. For this purpose, we use the definitions (6.1.8) and the matrices (6.1.79). The result is

$$\Sigma^{coh} = \left(A + \frac{2B}{3}\right)k_0^4 \begin{Vmatrix} \cos\theta & 0 & 0 & 0 \\ 0 & \cos\theta & 0 & 0 \\ 0 & 0 & \dfrac{1+\cos^2\theta}{2} & \dfrac{\sin^2\theta}{2} \\ 0 & 0 & \dfrac{\sin^2\theta}{2} & \dfrac{1+\cos^2\theta}{2} \end{Vmatrix} \sum_{j,l}^{N} e^{-i\vec{q}\cdot(\vec{r}_j - \vec{r}_l)}$$

$$\begin{array}{cccc} (1) & (2) & (3) & (4) \end{array}$$

(1)
(2)
(3)
(4)

(6.4.5.)

and

$$\Sigma^{inc} = \frac{2B}{3}k_0^4 \begin{Vmatrix} \cos\theta & 0 & 0 & 0 \\ 0 & -5\cos\theta & 0 & 0 \\ 0 & 0 & \dfrac{1+\cos^2\theta}{2} & \dfrac{\sin^2\theta}{2} \\ 0 & 0 & \dfrac{\sin^2\theta}{2} & \dfrac{1+\cos^2\theta}{2} + 6 \end{Vmatrix} \times N$$

$$\begin{array}{cccc} (1) & (2) & (3) & (4) \end{array}$$

(1)
(2)
(3)
(4)

(6.4.54)

Let us recall that the cross-sections enable us to express the Stokes parameters $P_\mu(\vec{r})$ ($\mu = 1, \ldots, 4$) of the scattered flux in terms of the parameters $^0P_\mu$ of the incident flux. The radiation is [see (6.3.63)]

$$P_\mu(\vec{r}) = \frac{1}{r^2}\sum_\nu (\Sigma_{\mu\nu}(k_0, \theta)\, ^0P_\nu).$$

The polarization factor corresponds to the P_μ [see (6.1.15)]

$$\bar{\omega} = (P_1^2 + P_2^2 + P_3^2)^{1/2}/P_4.$$

The scattering associated with a matrix $\Sigma(k_0, \theta)$ is non-depolarizing if for any incident flux with $^0\varpi = 1$, we obtain a scattered flux with $\varpi = 1$. It is easy to verify that $\Sigma_{\mu\nu}^{coh}$ is non-depolarizing whereas $\Sigma_{\mu\nu}^{inc}$ is polarizing.

Let us consider the scattering of a light beam that is linearly polarized in direction OX. It will be convenient to calculate the scattered flux in the representation defined by the parameters

$$P_x = \frac{1}{\sqrt{2}}(P_4 + P_3), \qquad P_y = \frac{1}{\sqrt{2}}(P_4 - P_3).$$

In this representation, the matrix (6.4.53) becomes identical to the matrix

(6.3.71). The matrix associated with the incoherent scattering (6.4.54) becomes

$$\Sigma^{\text{inc}}(k_0, \theta) = \frac{2B}{3} k_0^4 \left\| \begin{matrix} \cos\theta & 0 & 0 & 0 \\ 0 & -5\cos\theta & 0 & 0 \\ 0 & 0 & 4 & 3 \\ 0 & 0 & 3 & \cos^2\theta + 3 \\ (1) & (2) & (x) & (y) \end{matrix} \right\| \begin{matrix} (1) \\ (2) \\ (x) \\ (y) \\ \end{matrix}$$

In particular, forgetting to denote explicitly the dependence of the cross-sections with respect to k_0, we may write

$$\Sigma_{xx}(k_0, \theta) = \Sigma(q) = \Sigma^{\text{coh}}(q) + \Sigma^{\text{inc}} \tag{6.4.55}$$

where

$$\Sigma^{\text{inc}} = \frac{8B}{3} k_0^4 N$$

and

$$\Sigma^{\text{coh}}(q) = \left(A + \frac{2B}{3}\right) k_0^4 \sum_{j,l}^{N} e^{i\vec{q}\cdot(\vec{r}_j - \vec{r}_l)}. \tag{6.4.56}$$

In order to get a result similar to eqn (6.4.35) which applies to neutron scattering, we set

$$\mathfrak{b}^{\text{coh}} = \left(A + \frac{2B}{3}\right)^{1/2} k_0^2 \tag{6.4.57}$$

which gives

$$\Sigma^{\text{coh}}(q) = (\mathfrak{b}^{\text{coh}})^2 \sum_{j,l} e^{i\vec{q}\cdot(\vec{r}_j - \vec{r}_l)}. \tag{6.4.58}$$

Such formulae are convenient because they enable us to compare the results of neutron scattering and of light scattering. However, let us recall that the above cross-section corresponds to very peculiar polarizations for the incident and for the scattered beams, whereas the notation $\Sigma(q)$ would suggest that there are no polarization effects. This restriction is not a nuisance. The experimental situation is as follows: the incident radiation is linearly polarized, and the scattering is observed in the perpendicular plane. But, as we saw, the coherent contribution is non-depolarized and accordingly the coherent scattered light has a polarization parallel to that of the incident light. On the other hand, the partially depolarized incoherent contribution is very small for large polymers: actually, the 'useful' values of q are small, and this corresponds to large values of $\Sigma^{\text{coh}}(q)$ as can be seen in formula (6.4.58). The incoherent contribution can be neglected, and it is not necessary to set a polarizer on the scattered beam to separate the coherent from the incoherent contributions.

4.4.3 Numerical values of polarizabilities (light)

The refraction index n is the experimentally measurable quantity which determines the interaction between light and a homogeneous condensed material. From the value of this index, one determines the polarizability $\boldsymbol{\alpha}$ of the polarizable element. In the case of isotropic elements, with concentration N/V in a homogeneous material, the relation is given by the Lorentz–Lorenz[1] formula

$$\boldsymbol{\alpha} = \frac{3}{4\pi}\frac{V}{N}\frac{n^2-1}{n^2+2}.$$

Setting $\mathfrak{b} = k_0^2\boldsymbol{\alpha}$ [see (6.3.68)], we get

$$\mathfrak{b} = \frac{3}{4\pi}\frac{Vk_0^2}{N}\frac{n^2-1}{n^2+2}. \tag{6.4.59}$$

This formula also applies to the case of anisotropic elements, if we set $\mathfrak{b} = \mathfrak{b}^{\text{coh}}$ [see (6.4.57)].

Table 6.3 gives the values[1] of these quantities for a few materials, and also the values of the coherent differential cross-section $\Sigma = (\boldsymbol{\alpha}k_0^2)^2$ for $k_0 = 1.06 \times 10^{-2}\ \text{nm}^{-1}$ (double D ray of sodium).

For comparison, let us remember that the neutron coherent and incoherent total cross-sections section for a water molecule are respectively 2.8×10^{-2} barn and 12.7 barn, for neutrons with wave number $k_0 = 10\ \text{nm}^{-1}$. However, this difference in cross-section values is compensated by the fact that, in practice, the photon fluxes are much more intense than the neutron fluxes.

Table 6.3

Molecule	Density	n	$\boldsymbol{\alpha}[\text{nm}^3]$	Σ^{coh}[barn]
H_2O	1	1.334	1.47×10^{-3}	0.272×10^{-3}
CS_2	1.264	1.628	8.34×10^{-3}	8.8×10^{-3}
C_3H_6O (acetone)	0.791	1.3589	6.4×10^{-3}	5.2×10^{-3}

(1 barn = 10^{-24} cm^2)

5. EXPERIMENTAL STUDY OF A SCATTERED RADIATION (NEUTRONS, LIGHT)

From a scattering experiment, one deduces the value of the coherent scattering cross-section. This is the result which is sought. First, the incident wave number k_0 is determined for neutrons, by Bragg effect or by time of flight; for light, by spectrometry. The incident flux J_0 is then measured by counting particles. A scattering angle θ is chosen and one measures the scattered flux $J(\vec{r})$ which falls on a detector set at a given distance from the sample. The quantity

$$\frac{r^2 J(\vec{r})}{J_0} \tag{6.5.1}$$

is by definition [see (6.3.38)] a differential cross-section. However, it is not possible to identify this result directly with the coherent differential cross-section which we want, i.e.

$$\Sigma^{\text{coh}}(\vec{q}) = \sum_{j,l} \mathrm{b}_j^{\text{coh}} \mathrm{b}_l^{\text{coh}} \langle\!\langle e^{i\vec{q}\cdot(\vec{r}_j - \vec{r}_l)} \rangle\!\rangle, \tag{6.5.2}$$

and which corresponds to simple scattering. In order to obtain the value of this cross-section, corrections have, to be made for the following reasons:

(1) the beam is neither perfectly collimated nor perfectly monoenergetic;

(2) the sample and the detector have both a finite size and the distance r between them is large but not infinite;

(3) multiple scattering occurs in the sample;

(4) when the neutrons are scattered by organic samples, the incoherent contribution is always large.

Let us now examine how to go about determining the coherent differential cross-section corresponding to simple scattering, from measurements of the scattered fluxes.

5.1 Angular resolution: collimation uncertainties and distribution of the incident wave vector

Let us consider an experimental setting for the measurement of the flux scattered by a sample in order to determine $\Sigma^{\text{coh}}(\vec{q})$ (see Fig. 6.3). The defects (1) and (2)

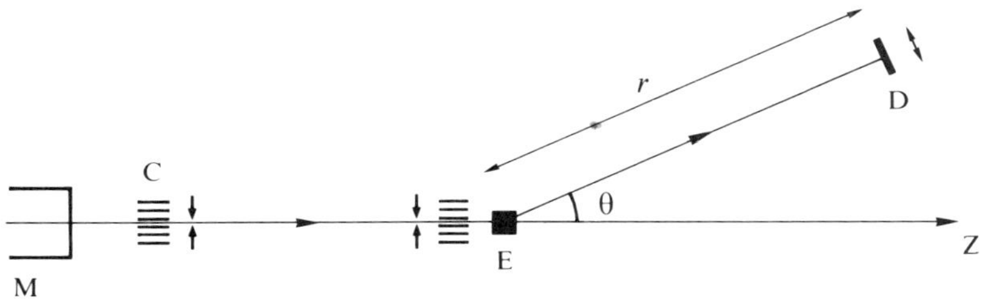

Fig. 6.3. Sketch of a neutron diffractometer. M – monochromator; C – collimator; E – sample; d – detector, with area. $\odot_{\mathrm{d}}$. Distance between the slits of the collimator $\sim$ 0.2 cm; the distance between the collimators is of the order of 1 m; θ, scattering angle; r, distance between the sample and the detector – this distance varies from 2 m to 20 m, depending on the required resolution.

mentioned above produce two noticeable effects: an uncertainty as regards the transfer wave vector, and another one as regards the cross-section.

5.1.1 Uncertainty on the wave vector transfer

In ideal conditions, the (elastic) wave vector transfer $\vec{q}$ is related to the incident wave vector $\vec{k}_0$ and to the scattering angle θ by the equality [see (6.3.18)]

$$q = 2k_0 \sin\theta/2. \tag{6.5.3}$$

when the incident wave vector and the collimation orientation are not perfectly defined, k_0 and θ are mean values. The amplitudes of the fluctuations can be characterized by the mean square deviations $(\delta k_0)^2$ and $(\delta\theta)^2$. The mean square deviation $(\delta q)^2$ of the wave vector transfer, can be deduced directly from (6.5.3)

$$\frac{(\delta q)^2}{q^2} = \frac{(\delta k_0)^2}{k_0^2} + (\operatorname{cotg}\theta/2)^2 \frac{(\delta\theta)^2}{4}. \tag{6.5.4}$$

For a given experiment $\delta k_0/k_0$ is determined by the quality of the incident beam and $\delta\theta$ by a compromise between luminosity and resolution of the diffractometer. The result (6.5.4) indicates that δq increases when the diffraction angle θ decreases: there exists a $q_{\min}$ for which $\delta q = q_{\min}$. For values of q smaller than $q_{\min}$, the resolution becomes rather poor. Thus, the reciprocal space interval which can be used by the experimentalist is limited by instrumental widths and in particular by $\delta\theta$. For a neutron diffractometer, we have, typically,

$$q_{\min} \simeq 10^{-2}\,\text{nm}^{-1}$$

because the aperture $\delta\theta$ must be fairly large. For a photon diffractometer

$$q_{\min} \simeq 10^{-4}\,\text{nm}^{-1}.$$

The reciprocal space domain also has an upper bound. Actually, for $\theta = \pi$, we have [see (6.5.3)]

$$q_{\max} \simeq 2k_0.$$

Figure 6.4 shows the two reciprocal space intervals that can be explored respectively by photons ($k_0 = 1.5 \times 10^{-2}\,\text{nm}^{-1}$) and by neutrons ($k_0 = 5\,\text{nm}^{-1}$).

5.1.2 Uncertainty on the cross-section

The collimation defects and the wavelength dispersion are such that the measured cross-section must be considered only as a relative quantity.

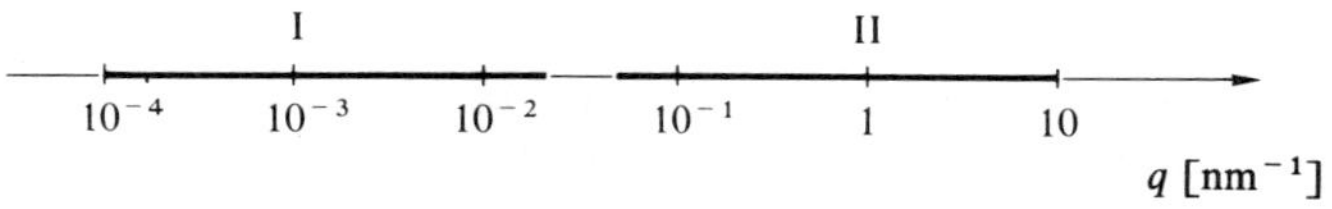

Fig. 6.4. Accessible domains in reciprocal space (I) by light scattering $k_0 = 1.5 \times 10^{-2}\,\text{nm}^{-1}$; (II) by neutron scattering $k_0 = 5\,\text{nm}^{-1}$.

Let us calculate the number $I(k_0, \theta)$ of particular impinging during a time interval t on the detector whose surface of area $\odot_d$ is at a distance r from the sample, in a direction defined by an angle θ. The incident flux distribution is $J_0(\vec{k})$ [see (6.1.19)] and the mean wave vector is $\vec{k}_0$.

The probability that an incident particle with wave number k will be scattered by the sample with a deviation angle θ' is

$$\Sigma(k,\theta')/\odot_s$$

where $\Sigma(k, \theta)$ is the cross-section and $\odot_s$ the area of the sample.

The collimation is characterized by the function $\mathscr{P}(k, \theta, \theta')$ which is the density probability that an incident particle with wave number k, deviated in the sample by an angle θ', encounters the detector whose direction is given by θ. In such conditions, the number of scattered particles that encounter the sample in a time interval t is

$$I(k_0,\theta) = \frac{1}{r^2}\,\mathrm{t}\,\odot_d \int dk\, J_0(k) \int d\theta'\, \Sigma(k,\theta')\, J(k,\theta',\theta). \qquad (6.5.5)$$

When the particles are neutrons, the widths of the distribution $J_0(k)$ and of the density $J(k,\theta',\theta)$ are rather large. In this case, to obtain $\Sigma(k_0, \theta)$ by measuring $I(k_0, \theta)$, it is not absolutely necessary to know $J_0(k)$ and $J(k,\theta',\theta)$. In principle, one could measure the effects of these widths directly. For this, it would be sufficient to scatter the beam by a perfect monocrystal with known properties. The cross-section of this crystal is different from zero, only if

$$\theta = \arc\sin\frac{1}{2k\mathfrak{d}} = \theta_B \qquad \text{(Bragg's angle)}$$

where $\mathfrak{d}$ is an inter-reticular distance. Then, let $I(k_0,\theta) = K(k_0,\theta)$ be the result of the experiment for various angles θ in the vicinity of θ_B. A ray of width $\delta\theta$, centred on θ_B is observed. This width results exclusively from collimation defects and from dispersion of the incident wave vector. Then, assuming that (6.5.5) has the form of a convolution integral, we get

$$I(q) = K(q)\Sigma(\vec{q}),$$

i.e. the result we want.

In practice, until now, it has been impossible to determine $K(q)$ in this way for values of the wave vector k_0 adapted to scattering experiments on polymers: it would be necessary to use a perfect crystal with a period of the order of 1 nm.

The experimental result is interpreted, to a first approximation, with the help of the formula

$$I(q) = K_A\,\Sigma(\vec{q}) \qquad (6.5.6)$$

where K_A is the instrument constant

$$K_A = \frac{\mathrm{t}}{r^2}\odot_d J_0 \mathscr{R} \qquad (6.5.7)$$

where q is the modulus of the mean transfer vector and J_0 the incident flux; $\mathscr{R}$ is a constant related to the angular resolution.

When the constant K_A is known with precision the measurement of the cross-section is said to be absolute. Otherwise, and this is the most common case, it is said to be relative.

5.2 Effect of the finite thickness of the sample

The number $I(q)$ of scattered particles impinging on the detector [see (6.5.6)]

$$I(q) = K_A \Sigma(\vec{q}) \tag{6.5.8}$$

is proportional to the cross-section $\Sigma(\vec{q})$; in general, this cross-section increases with the size of the sample and in particular with its thickness L. Thus, one could increase L to have better counting. However, when L is larger than the mean free path of the incident particle in the sample, we have multiple scattering. The effects of this phenomenon are a nuisance. In order to avoid them, L is reduced until the contribution of simple collisions is dominant. When each molecule contained in the sample produces strong forward scattering, this situation occurs when

$$\Xi L \leqslant 2$$

where Ξ is given in terms of the cross-sections by (6.4.19); Ξ is the turbidity. In this case, we must also introduce a correction for the constant K_A. In fact, it is necessary to take into account the attenuation of the incident flux J_0 in the incident direction. This attenuation is given by the transmission factor

$$\mathscr{T} = e^{-\Xi L}$$

Let us then replace J_0 by $J_0 \mathscr{T}$. We get

$$K_A = \frac{1}{r^2} t \odot_d J_0 \mathscr{T} \mathscr{R} \tag{6.5.9}$$

As we shall see, the transmission factor is an important parameter and the physicist who wants to determine the coherent cross-section experimentally must take it into account.

5.3 Separation of coherent and incoherent neutron contributions

We have shown that the cross-section $\Sigma(\vec{q})$ of a macroscopic sample which scatters neutrons is the sum of coherent and incoherent contributions

$$\Sigma(\vec{q}) = \Sigma^{\text{inc}} + \Sigma^{\text{coh}}(\vec{q}). \tag{6.5.10}$$

Let us see how the coherent contribution $\Sigma^{\text{coh}}(\vec{q})$ can be determined. We note that, in general. Σ^{inc} is a known quantity. It is deduced from the composition of the sample and from tables of cross-sections.

The cross-section $\Sigma(\vec{q})$ is proportional to the number of particles $I(q)$ falling on the detector

$$\Sigma(\vec{q}) = K_{\mathrm{A}}^{-1} I(q)$$

K_{A} being the instrument constant [see (6.5.7)]. We thus obtain

$$\Sigma^{\mathrm{coh}}(\vec{q}) = K_{\mathrm{A}}^{-1} I(q) - \Sigma^{\mathrm{inc}}. \tag{6.5.11}$$

Unfortunately, K_{A} is not known with great precision. When the coherent and incoherent contributions are of the same order, the term on the left has a value which is uncertain; then, it is necessary to obtain $\Sigma^{\mathrm{coh}}(\vec{q})$ in a different way. For this purpose *two successive experiments* are performed in such a way that the weights of the coherent and incoherent contributions are different in each experiment. The constants K_{A} associated with these experiments remain unknown, but their ratio is considered as known. The first experiment is always the same; we described it earlier. The second experiment can be performed in various ways. We shall describe three of them; the first two are the most frequently used, the third one seems more efficient but is not so common.

We shall use the indices 1 and 2 to denote what concerns the first and the second experiment.

$$I(q) = K_{\mathrm{A}i}(\Sigma_i^{\mathrm{coh}}(\vec{q}) + \Sigma_i^{\mathrm{inc}}) \qquad i = 1, 2. \tag{6.5.12}$$

5.3.1 Method of chemical substitution

This method applies only to very dilute systems. A second sample is used (of the same size as the first one) in which the polymer molecules are replaced by equivalent sets of non-chemically-bound monomers. The system being dilute, the correlations between monomers can be neglected and, in this case, the coherent cross-section reads [see (6.4.33)]

$$\Sigma_2^{\mathrm{coh}}(\vec{q}) = \sum_j \sum_{i=1}^{N_j} (\mathrm{b}_{ij}^{\mathrm{coh}})^2$$

where b_{ij} is the coherent collision length of the monomer i of the chain j whose number of monomers is N_j. On the other hand, we have

$$\Sigma_2^{\mathrm{inc}} = \Sigma_1^{\mathrm{inc}},$$

and for the apparatus constants

$$K_{\mathrm{A}2}/K_{\mathrm{A}1} = \mathscr{T}_2/\mathscr{T}_1$$

where $\mathscr{T}$ is the transmission factor.

Then, from the two experiments, one deduces

$$\Sigma_1^{\mathrm{coh}}(\vec{q}) = K_{\mathrm{A}1}^{-1}\left[I_1(q) - \frac{\mathscr{T}_2}{\mathscr{T}_1} \frac{I_2(q)}{1 + \sum_j \sum_i (\mathrm{b}_{ij}^{\mathrm{coh}})^2} \right] \tag{6.5.13}$$

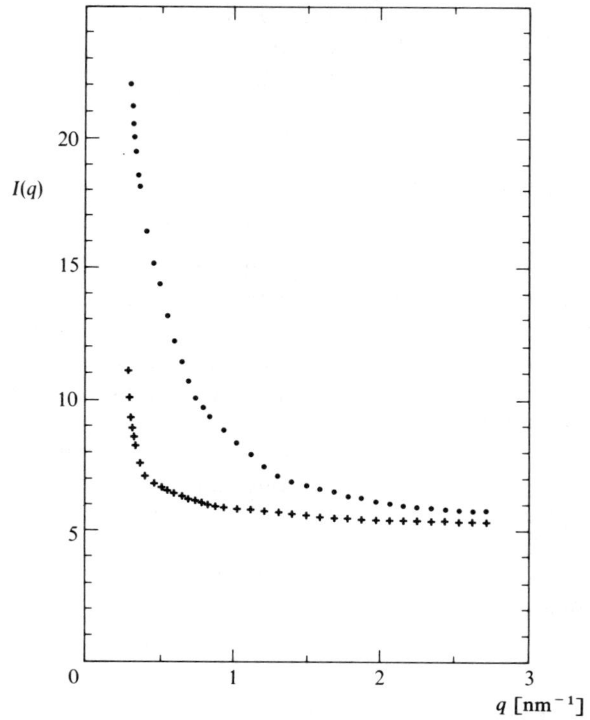

Fig. 6.5. Determination of the coherent signal in a neutron scattering experiment. (From Cotton, Decker, Benoit, Farnoux, Higgins, Jannink, Ober, Picot, and des Cloizeaux.[11]) An illustration of the chemical substitution method and of the asymptotic limit method. The sample (sample 1) is a mixture of deuterated and non-deuterated polystyrene. The measured intensity $I_1(q)$ is represented by points (●).

Method of chemical substitution. Sample 2 consists exclusively of non-deuterated polystyrene. It is assumed that the contribution of the protons to the scattering signal is completely incoherent. The intensity $I_2(q)$ is represented by crosses (+).

Method of the asymptotic limit. The figure shows the asymptotic behaviour of $I_1(q)$ for $q \to \infty$. Let us note that the random noise is large.

which determines $\Sigma_1^{\mathrm{coh}}(\vec{q})$ except for a factor [which is not true for eqn (6.5.11), though Σ^{inc} is known] (see Fig. 6.5).

5.3.2 Method of the asymptotic limit

It is admitted that asymptotically the coherent contribution reaches a constant limit when the transfer wave number q becomes much larger than the inverse of the 'correlation length' of the system. This correlation length is, in dilute solutions, the size of the chains; in semi-dilute solutions, the screening length. In the second experiment, one uses a value $q = Q$ such that $Q\xi \gg 1$. In this limit

$$\Sigma_2^{\text{coh}}(\vec{Q}) = \sum_j (\mathrm{b}_j^{\text{coh}})^2$$

$$\Sigma_2^{\text{inc}} = \Sigma_1^{\text{inc}}$$

and

$$I_2 = \Sigma_1^{\text{inc}} + \sum_j (\mathrm{b}_j^{\text{coh}})^2.$$

Combining the results of the two experiments, we get

$$\Sigma_1^{\text{coh}}(\vec{q}) = K_{\text{A}1}^{-1}\left[I_1(q) - \frac{I_2(q)}{1 + \sum_j (\mathrm{b}_j^{\text{coh}})^2}\right] \tag{6.5.14}$$

if we admit that the transmission factor $\mathscr{T}$ does not depend on the wave vector transfer (see Fig. 6.5).

5.3.3 Method of the polarized beams

Two successive scattering experiments are made with the same sample and beams with different polarizations. In the sample, the nuclei have random orientations.

In the first experiment, the incident neutrons are totally depolarized (see Section 4.4.1.1). In this case

$$K_{\text{A}1}^{-1} I_1(q) = \Sigma_1^{\text{inc}} + \Sigma_1^{\text{coh}}(\vec{q}). \tag{6.5.15}$$

Here [see (6.4.34)],

$$\Sigma_1^{\text{inc}} = \sum_j \left(\frac{B_j^-}{2}\right)^2 S_j(S_j + 1)$$

where S_j is the spin of the nucleus j and where B_j^- is defined by (6.4.27). Moreover,

$$\Sigma_1^{\text{coh}}(q) = \sum_{j,l} \mathrm{b}_j^{\text{coh}}\, \mathrm{b}_l^{\text{coh}}\, \mathrm{e}^{i\vec{q}\cdot(\vec{r}_j - \vec{r}_l)}.$$

In the second experiment, the incident beam is totally polarized: the spin of each neutron points in the same direction, denoted by $\uparrow$ ($s_z = 1/2$). Moreover, the scattered flux is filtered through an analyzer before falling on the detector: only the neutrons, with spins pointing in the direction denoted by $\downarrow$ ($s_z = -1/2$), opposite to the direction $\uparrow$, are counted. In this case, as was shown in a preceding section [see (6.4.38) and (6.4.34)], we have

$$\Sigma_2^{\text{inc}} = \Sigma_{\uparrow\downarrow}^{\text{inc}} = \tfrac{2}{3}\,\Sigma_1^{\text{inc}}$$

$$\Sigma_2^{\text{coh}}(\vec{q}) = \Sigma_{\uparrow\downarrow}^{\text{coh}} = 0$$

and

$$K_{\text{A}2}^{-1} I_2 = \tfrac{2}{3}\,\Sigma_1^{\text{inc}}.$$

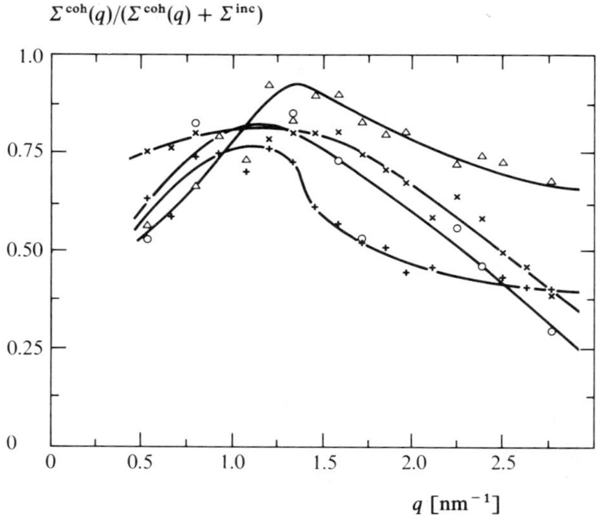

Fig. 6.6. Values of the ratio $\Sigma^{coh}(q)/(\Sigma^{coh}(q) + \Sigma^{inc})$ obtained by scattering a polarized neutron beam and by analysing the polarization of the scattered neutrons. (From Nallet, Jannink, Hayter, Oberthur, and Picot.[12]) Sample of polystyrene sulphonate, ionized in water. The counter-ion is tetramethyl ammonium (TMA).
+ deuterated PSS, non-deuterated TMA
× non-deuterated PSS, non-deuterated TMA
△ non-deuterated PSS, deuterated TMA
○ deuterated PSS, deuterated TMA

Combining the results of the two experiments, we get

$$\Sigma_1^{coh}(\vec{q}) = K_{A1}^{-1}\left[I_1(q) - \frac{3}{2}\frac{\mathscr{T}_1}{\mathscr{T}_2} I_2\right] \tag{6.5.16}$$

(see Fig. 6.6).

5.4 Conversion of the measured intensity into a scattering cross-section

The experimental set up of a scattering experiment includes collimation devices for angles and wave lengths. These collimations depend on a choice made by the experimentalist who must try to reconcile contradictory requirements of resolution and luminosity.

Thus to each set up corresponds a constant K_A whose approximate expression is

$$K_A = \frac{1}{r^2} \mathrm{t} \odot_{\mathrm{d}} J_0 \mathscr{T} \mathscr{R}$$

where r is the distance between sample and detector, t the duration of measurement, J_0 the flux of incident radiation, $\mathscr{T}$ the transmission factor and $\mathscr{R}$ the resolution constant. The constant K_A establishes a relation between the scattered intensity in a given direction $\hat{r}$, and the scattering cross-section of the sample

$$I(\hat{r}) = K_A \Sigma(\hat{r}).$$

Thus, to convert the measured quantity $I(q)$ into a cross-section, one performs the operation

$$\Sigma(\hat{r}) = I(\hat{r})/K_A.$$

In general, it is impossible to calculate K_A directly by starting from the elements of an experimental set-up. In practice, this constant is determined by calibration. A 'reference' sample with a known cross-section is placed in the experimental apparatus. A measurement of the scattered intensity I_t gives the value of the constant

$$K_A = I_t/\Sigma_t.$$

This value enables the experimentalist to convert a scattering intensity into a cross-section, and therefore it is used to exploit the experiment itself.

When dealing with neutrons, the reference sample is a piece of vanadium, a metal whose cross-section is purely incoherent and therefore independent of the direction $\hat{r}$ of scattering.

6. SUMMARY

The characteristic quantity associated with the scattering of a radiation is the differential cross-section Σ of the scattering sample. This is the number of particles scattered in a direction $\hat{r}$, per incident particle and per solid angle.

By measuring the scattered intensity I in a direction $\hat{r}$, one determines the cross-section of the sample for a given transfer wave $\vec{q}$, whose modulus is $q = 2k_0 \sin \theta/2$, where θ is the scattering angle and $\vec{k}_0$ the incident wave vector. The correspondence is given by the equality

$$I(\hat{r}) = K_A \Sigma(\vec{q})$$

where K_A is the instrument constant [see (6.5.7)].

We studied the collision of a neutron with nuclei, and subsequently the interaction of a light radiation with polarizable elements. A relation between the cross-section $\Sigma(\vec{q})$ and position correlations of the scatterers was deduced from this study. The formula which expresses this relation is the same for neutron and photon radiations. The cross-section consists of two contributions, which are called coherent and incoherent

$$\Sigma(\vec{q}) = \Sigma^{\mathrm{coh}}(\vec{q}) + \Sigma^{\mathrm{inc}} \tag{6.6.1}$$

where

$$\Sigma^{\mathrm{coh}}(\vec{q}) = \sum_{jl} \mathfrak{b}_j^{\mathrm{coh}}\, \mathfrak{b}_l^{\mathrm{coh}}\, \mathrm{e}^{i\vec{q}\cdot(\vec{r}_j - r_l)} \tag{6.6.2}$$

and where Σ^{inc} represents a random noise.

The correlations between sites $\vec{r}_j$ and $\vec{r}_l$ come into play in (6.6.2). The collision lengths $\mathfrak{b}_j^{\mathrm{coh}}$ are determined independently; for instance, by measuring a refractive index n (photon or neutron refraction):

photon radiation: see (6.4.59).

neutron radiation: see (6.4.22).

Formula (6.6.2) is used under restrictive conditions. First, it assumes that the incident particle collides once, at most, with a scatterer. The other conditions depend on the nature of the incident radiation:

neutron radiation: factorization approximation (6.3.17), static approximation (6.3.19) and Born approximation (6.3.21).

photon radiation: approximation of dipolar re-emission (Appendix D); moreover, with this radiation, the scattering experiment must be effected in the following way: the incident light is linearly polarized in a direction OX perpendicular to the direction of propagation OZ and to the scattering plane (plane parallel to the incident direction and to the direction $\hat{r}$ of scattering).

The reciprocal space interval (q_{min}, q_{max}) that can be explored is indicated in Fig. 6.4. It depends on the type of radiation used. In the case of neutrons and of a sample containing nuclei with non-zero spin (for instance, protons), the coherent collision length associated with these nuclei depends on the state of polarization of the neutron–nuclei system. The incoherent random noise (Σ^{inc}) can then be large.

REFERENCES

1. Born, M. and Wolf, E. (1983). *Principles of optics.* Pergamon Press, Oxford.
2. Jauch, J.M. and Rohrlich, F. (1959). *The theory of photons and electrons.* Addison-Wesley, Reading, Mass.
3. Marshall, W. and Lovesey, S.W. (1971). *Theory of thermal neutron scattering*, Chapter 10. Clarendon Press, Oxford.
4. Messiah, A. (1959). *Mécanique quanique*, Vol. 1. Dunod, Paris.
5. Lippmann, B.A. and Schwinger, J. (1950). *Phys. Rev.* **79**, 469.
6. Bacon, G.E. (1975). *Neutron diffraction.* Clarendon Press, Oxford.
7. Moon, R.M., Riste, T., and Koehler, W.C. (1969). *Phys. Rev.* **181**, 920.
8. Chandrasekhar, S. (1960). *Radiative transfer.* Dover Publications, New York.

9. Berne, B.J. and Pecora, R. (1976). *Dynamic light scattering*. John Wiley, New York.
10. Abragam, A. and Goldman, M. (1982). *Nuclear magnetism: order and disorder*. Clarendon Press, Oxford.
11. Cotton, J.P., Decker, D., Benoit, H., Farnoux, B., Higgins, J., Jannink, G., Ober, R., Picot, C., and des Cloizeaux, J. (1974). *Macromolecules* **7**, 863.
12. Nallet, F., Jannink, G., Hayter, J.B., Oberthur, R. and Picot, C. (1983). *J. Physique* **44**, 87.

7

STUDY OF THE STRUCTURE OF A SOLUTION BY SMALL-ANGLE SCATTERING

GENESIS

Small-angle scattering is defined by the fact that the intensity of a radiation scattered by a target varies strongly over a few angular degrees in the vicinity of the incident direction. According to A. Guinier, this forward scattering was observed for the first time in 1934 by C.V. Raman, in an experiment of X-ray scattering by carbon black. It has since been observed that colloidal substances and polymers in solution also scatter X-rays in the forward direction (this is called central scattering); moreover, it appeared that the width of the central peak is inversely proportional to the size of the scattering objects.

In the years following 1945, the phenomenon of forward scattering has been observed for light. However, the decrease of the signal with the angle is not as spectacular as in the case of X-rays. Actually, the scattering is only weakly anisotropic: this is the famous angular dissymmetry of Debye (1948). A measurement of this angular dissymmetry gives the size of the object in the same manner as genuine small-angle scattering. This is why these light scattering experiments were found to be very useful and are now commonly called 'small-angle scattering experiments'. Their application to the study of molecules has greatly developed since 1954 (H. Benoit), and at the present time (1988), the characteristic properties of macromoecules in solution are determined in this way. Moreover, other effects have been found. For instance, Debye noted, in 1959, that a binary mixture of small molecules in the vicinity of the demixtion point scatters light in a dissymetric way, like a polymer solution far from the demixtion point.

During the 1955s, genuine small-angle scattering experiments were carried out with neutrons. The first of these at Brookhaven (USA), consisted in sending a beam on a colloidal suspension and in measuring the angular broadening produced by multiple scattering. It is in this way that R. Weiss succeeded in measuring the still unknown collision length of bismuth.

Small-angle neutron scattering with simple collisions developed during the 1965s when incident beams with large wavelengths ($\lambda \geqslant 0.5$ nm) became available. At first, this technique was applied to the study of critical phenomena and of defects in alloys. From 1969 onwards, it was used at Saclay for the determination of the configurations of polymers in various environments. In particular, J.P. Cotton noted, in 1970, that neutron scattering is a very powerful method for the investigation of correlations in isotopically labelled solutions. The advantage of this labelling is that it does not perturb the systems (unless there is phase separation!). The labelling is effected by deuteration of the protonated molecules. The contrast thus obtained can be very large, and in such conditions,very precise observations can be made with neutron scattering, even if the incident flux is rather low. The same remark was made by R. Kirste in Mainz and by G.D. Wignall in London. Consequently, in the following years, a large number of small-angle neutron scattering experiments were performed in various laboratories with labelled molecular systems.

What made the use of neutrons even more appealing was the possibility of carrying out a project which Benoit and Wippler thought up in 1960 but which remained utopian for a long time. The idea was to use techniques of selective contrast and of variation of contrast to study the internal configurations of macromolecules. For that, it was necessary to break free from the contrast conditions resulting from the electronic structure of the system. In practice, the project was effected by isotopic labelling of a fraction of the molecules, which could belong either to the solvent or to the solute, or to both of them.

1. CONDITIONS OF OBSERVATION

In this chapter, we shall study the simple coherent scattering of a radiation by a system, in the static approximation.

1.1 Long-distance correlations in a sample

Small-angle scattering can be observed when the size of the scattering objects is large as compared to the radiation wavelength. Thus, a crystal or a simple liquid do not give any contribution to the cross-section in the vicinity of the zero angle. On the contrary, a polymer solution produces a strong forward scattering.

In other terms, the differential cross-section varies strongly in the vicinity of the zero angle if but only if long-range correlations exist between the positions of the scatterers in the system. For instance, the cross-section $\Sigma(\vec{q})$ of an atomic gas or of small molecules is a quantity which does not depend on the wave vector transfer

$$\vec{q} = \vec{k}_{\mathrm{f}} - \vec{k}_0$$

where $\vec{k}_{\mathrm{f}}$ and $\vec{k}_0$ are respectively the scattered and the incident wave vectors. Conversely, a material in which there are long-range correlations (for instance, a binary mixture in the vicinity of the demixtion point, or a dilute solution of macromolecules) has a typical anisotropic cross-section. However, the existence of long-range correlations is not sufficient to produce small-angle scattering. Let us consider, for instance, a homogeneous solution of macromolecules. In spite of the existence of long-range correlations, the cross-section of the liquid is practically zero in the vicinity of $q = 0$. To give rise to forward scattering in this system, it is necessary to introduce a heterogeneity; for instance, by labelling a fraction of the macromolecules. Contrast is necessary for a homogeneous condensed system to scatter through small angles.

1.2 Contrasts of amplitude, of phase, of polarization; stimulated contrast

Contrast particularizes the materials interacting with a radiation. The observation of scattering effects in molecular mixtures depends crucially on this parameter.[1]

For light scattering, contrast results from a difference in polarizability of the molecules. For neutron scattering, it results from a difference in nuclear interaction (different collision lengths). In a general way, the contrast can be defined from the refraction indices corresponding to the radiation and to the materials constituting the mixture.

The contrast techniques relating to the observation of biological structures under a microscope have been developing over a long time. In this case, the contrast can be obtained by selective attenuation of the intensity of the light beam as it passes through the sample. For instance, the attenuation can be obtained by metallic impregnation of one or several constituents of the compound. Such a contrast is called amplitude contrast. In 1934, Zernike[2] improved it by shifting the phase of one part of the incident light beam. Then came phase contrast, reinforcing amplitude contrast, and making possible the observation of transparent biological structures.

When polarization of light is taken into account, other forms of contrast appear. For instance, let us consider spherolites in solid polyethylene. These structures are clusters of crystallites which can be seen under an ordinary microscope. However, they appear[3] much more clearly if a polarizing microscope is used, the sample being observed between crossed polarizers. Here, the contrast is related to the optical anisotropy of the material.

Looking for the best contrast possible is a major task for the experimentalist who prepares a small-angle scattering experiment. For X-rays, an increase of contrast in a mixture is obtained by replacing light atoms by heavy atoms in certain molecules. The drawback to this technique is that it strongly perturbs the initial state, and a change in the interactions may produce a precipitation of the labelled molecules. For neutrons, an increase of contrast in an organic material is obtained by replacing (isotopically) protons by deuterons in molecules. Now, the 'chemical' perturbation of the initial system is very weak. The deuteration technique enables the experimentalist to master the contrasts and to adopt them to the study of a mixture by small-angle neutron scattering. Let us also note the existence of a daring method[4, 5] (studied in 1986) which consists in polarizing the protons of the target. In this case, for polarized neutrons, the contrast between the solvent and the organic solute can be even larger than with the method of isotopic labelling (see Chapter 6, Section 3.1.8.3).

Finally, let us mention the method of contrast stimulation which has been developed for Rayleigh light-scattering.[6] It consists in modifying periodically in time, the polarizability of certain types of atoms in the target by exciting electrons belonging to these atoms. Important observations have been made concerning the dynamics of polymers, by using this technique.

2. STRUCTURE FUNCTION

Let us consider a small-angle scattering experiment. The radiation falls on molecules in a container at fixed pressure. Let us study the scattering in a volume V which is small with respect to the volume of the container but not microscopic (see Fig. 7.1). An elastic transfer wave vector $\vec{q}$ corresponds to each scattering angle θ with $q = 2k_0 \sin \theta/2$, ($\vec{k}_0$ being the incident wave vector) and the scattering is characterized by the differential cross-section

$$\Sigma(\vec{q}) = \left\langle\left\langle \sum_{jl} \mathrm{b}_j \mathrm{b}_l \mathrm{e}^{i\vec{q}\,.\,(\vec{r}_j - \vec{r}_l)} \right\rangle\right\rangle. \tag{7.2.1}$$

where the brackets $\langle\langle\ \rangle\rangle$ indicate that we take the average over all the configuration of the molecules that are present in volume V, where b_j is the collision length and $\vec{r}_j$ the position of the scattering centre j).

Here, we assume that the scattering is entirely coherent: the incoherent contribution has been subtracted (see Chapter 6, Section 5.3). The cross-section is an extensive quantity, but we can introduce the cross-section per unit volume*

$$\Xi(q) = \frac{1}{V} \Sigma(\vec{q}). \tag{7.2.2}$$

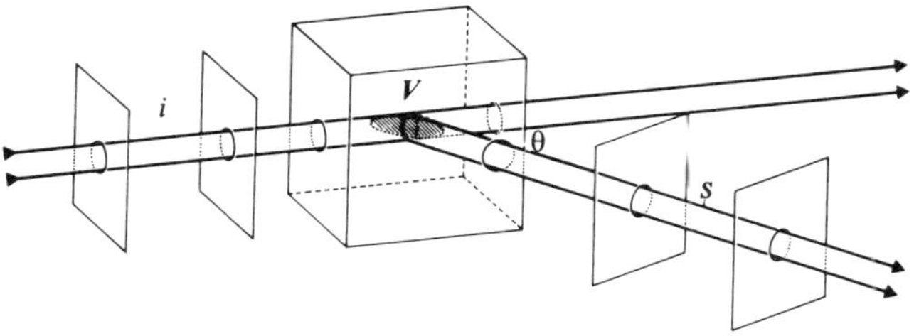

Fig. 7.1. Representation of the scattering volume V for which the cross-section is calculated. This is the grey part of the volume of the sample. It is defined by the intersection of the incident (i) and scattered (s) beams. These beams are limited by collimators which are indicated schematically on the figure. The angle θ between the directions of the beams is the scattering angle.

* The cross-section can be used to calculate the turbidity Ξ defined by eqn (6.4.14).

The intensive quantity thus defined depends on the $\mathfrak{b}_j$ and on the concentrations of the scatterers. It depends also on the internal geometrical structure of the sample. This is this structure which we wish to characterize, and this desire will lead us later to define the structure function.

2.1 Indexing the scatterers

The sample contains a large number of scattering centres belonging to various species. Let us rewrite (7.2.1) with the following notation. We use the index $\boldsymbol{a}(\boldsymbol{a} = 1, \ldots, N)$ to mark a molecule, the index $i\,(i = 1, \ldots N_{\boldsymbol{a}})$ to mark a monomer on a molecule $\boldsymbol{a}$ made up of $N_{\boldsymbol{a}}$ links ($N_{\boldsymbol{a}}$ monomers),* and the index $l(l = 1, \ldots, p_i)$ to mark a scattering centre on a monomer i. The cross-section per unit volume can thus be written in the form

$$\Xi(q) = \frac{1}{V}\left\langle\!\!\left\langle \sum_{\boldsymbol{a}=1}^{N} \sum_{\boldsymbol{b}=1}^{N} \sum_{i=1}^{N_{\boldsymbol{a}}} \sum_{j=1}^{N_{\boldsymbol{b}}} \sum_{l=1}^{p_i} \sum_{m=1}^{p_j} \mathfrak{b}_{\boldsymbol{a}il}\, \mathfrak{b}_{\boldsymbol{b}jm}\, \mathrm{e}^{i\vec{q}\cdot(\vec{r}_{\boldsymbol{a}il} - \vec{r}_{\boldsymbol{b}jm})} \right\rangle\!\!\right\rangle .$$

In Chapter 6, we defined the centres which scatter neutrons or photons as being (respectively) nuclei or electrons. In this chapter, the set of all the centres belonging to a monomer will be considered as a unique scattering centre. This more economical definition is justified by the fact that, in the interval $(0, q_{\max})$ where we study the behaviour of scattering cross-sections, the length q^{-1} is always large as compared to atomic distances. Let $\vec{r}_{\boldsymbol{a}i}$ be the position vector of the centre of mass of monomer i. Then, suppose

$$\vec{q}\cdot\vec{r}_{\boldsymbol{a}il} \simeq \vec{q}\cdot\vec{r}_{\boldsymbol{a}i} .$$

The collision length of monomer i (of molecule $\boldsymbol{a}$) is defined by

$$\mathfrak{b}_{\boldsymbol{a}i} = \sum_{l=1}^{p_i} \mathfrak{b}_{\boldsymbol{a}il} . \tag{7.2.3}$$

The expression of the cross-section per unit volume becomes (for $q \leqslant q_{\max}$):

$$\Xi(q) = \frac{1}{V}\left\langle\!\!\left\langle \sum_{\boldsymbol{a}=1}^{N} \sum_{\boldsymbol{b}=1}^{N} \sum_{i=0}^{N_{\boldsymbol{a}}} \sum_{j=0}^{N_{\boldsymbol{b}}} \mathfrak{b}_{\boldsymbol{a}i}\, \mathfrak{b}_{\boldsymbol{b}j}\, \mathrm{e}^{i\vec{q}\cdot(\vec{r}_{\boldsymbol{a}i} - \vec{r}_{\boldsymbol{b}j})} \right\rangle\!\!\right\rangle .$$

This notation is very general but it does not take into account the fact that a large number of molecules may be of the same kind. For instance, the molecules of solvent are practically identical, and the same is true for sets of molecules constituting the solute. By taking this fact into account, it will be possible to group certain terms of the preceding equation.

First, let $\mathscr{E}$ be the set of all the molecules that are able to scatter the radiation and let $\mathscr{A}, \mathscr{B}, \ldots$ be subsets of $\mathscr{E}$: for instance, $\mathscr{A} = 0$ is the set of the molecules of solvent. To indicate that a molecule belongs to $\mathscr{A}$, we shall simply write (in a

* Here, for convenience, we do not make any distinction between links and monomers.

symbolic way)

$$\boldsymbol{a} \in \mathscr{A}.$$

The number of molecules $\boldsymbol{a}$ belonging to $\mathscr{A}$ is

$$N_{\mathscr{A}} = \sum_{\boldsymbol{a} \in \mathscr{A}} 1,$$

and the average number of these molecules in volume V is

$$\langle\langle N_{\mathscr{A}} \rangle\rangle = \left\langle\left\langle \sum_{\boldsymbol{a} \in \mathscr{A}} 1 \right\rangle\right\rangle \tag{7.2.4}$$

where the bracketing $\langle\langle\ \rangle\rangle$ indicates that we take the average over the number of molecules in V. This volume V can be seen as a part of a larger volume and, in what follows the molecules belonging to V will be considered as constituting a grand canonical ensemble. Thus, we have $\langle\langle (N_{\mathscr{A}})^2 \rangle\rangle \neq \langle\langle N_{\mathscr{A}} \rangle\rangle^2$. On the other hand, the number $N_{\boldsymbol{a}}$ of monomers of a molecules $\boldsymbol{a}$ belonging to species $\mathscr{A}$ is by definition independent of $\boldsymbol{a}$ and can be represented by the symbol $N_{\mathscr{A}}$.

Let us now consider the monomers of a chain $\boldsymbol{a}$ and let $\alpha, \beta, \ldots,$ be the types to which they belong (for instance, deuterated or non-deuterated). To indicate that a monomer belongs to α, we simply write (again in a symbolic way)

$$i \in \alpha.$$

In general, the monomers belonging to the same type α have the same collision length

$$\mathfrak{b}_{ai} \equiv \mathfrak{b}_{\alpha}, \qquad i \in \alpha.$$

The cross-section per unit volume thus reads

$$\Xi(q) = \frac{1}{V} \sum_{\mathscr{A}} \sum_{\mathscr{B}} \sum_{\alpha} \sum_{\beta} \mathfrak{b}_{\alpha} \mathfrak{b}_{\beta} \left\langle\left\langle \sum_{\boldsymbol{a} \in \mathscr{A}} \sum_{\boldsymbol{b} \in \mathscr{B}} \sum_{i \in \alpha} \sum_{j \in \beta} \mathrm{e}^{i\vec{q} \cdot (\vec{r}_{ai} - \vec{r}_{bj})} \right\rangle\right\rangle \tag{7.2.5}$$

where $\sum_{\mathscr{A}}$, $\sum_{\alpha}$ mean respectively that sums are performed over the elements of the subsets $\mathscr{A}$ and α.

This grouping has the advantage of generating new factors which are independent of the collision lengths. Let us set

$$\bullet H_{\mathscr{A}\alpha\mathscr{B}\beta}(\vec{q}) = \frac{1}{V C_{\mathscr{A}\alpha} C_{\mathscr{B}\beta}} \left\langle\left\langle \sum_{\boldsymbol{a} \in \mathscr{A}} \sum_{\boldsymbol{b} \in \mathscr{B}} \sum_{i \in \alpha} \sum_{j \in \beta} \mathrm{e}^{i\vec{q} \cdot (\vec{r}_{ai} - \vec{r}_{bj})} \right\rangle\right\rangle$$

where

$$C_{\mathscr{A}\alpha} = \frac{1}{V} \left\langle\left\langle \sum_{\boldsymbol{a} \in \mathscr{A}} \sum_{i \in \alpha} 1 \right\rangle\right\rangle \tag{7.2.6}$$

is the mean concentration of monomers of type α belonging to chains of the subset $\mathscr{A}$. Finally, the cross-section per unit volume is given by

$$\Xi(q) = \sum_{\mathscr{A}} \sum_{\mathscr{B}} \sum_{\alpha} \sum_{\beta} \mathrm{b}_{\alpha\beta}\, C_{\mathscr{A}\alpha} C_{\mathscr{B}\beta}\, {}^{\bullet}\boldsymbol{H}_{\mathscr{A}\alpha,\mathscr{B}\beta}(\vec{q}). \tag{7.2.7}$$

using this formula, Leibler and Benoit[7] derived useful expressions for the scattered intensities of 'solutions' of copolymers in polymer melts.

2.2 Structure and concentration fluctuations

The function ${}^{\bullet}\boldsymbol{H}_{\mathscr{A}\alpha\mathscr{B}\beta}$ accounts for position correlations between monomers. Let us express these correlations by introducing local concentrations (in monomers)

$$C_{\mathscr{A}\alpha}(\vec{r}) = \sum_{a \in \mathscr{A}} \sum_{i \in \alpha} \delta(\vec{r} - \vec{r}_{ai}) \tag{7.2.8}$$

Consequently,

$$\left\langle\!\!\left\langle \sum_{a \in \mathscr{A}} \sum_{b \in \mathscr{B}} \sum_{i \in \alpha} \sum_{j \in \beta} \delta(\vec{r} - \vec{r}_{ai} + \vec{r}_{bj}) \right\rangle\!\!\right\rangle = \int_V \mathrm{d}^3 r' \langle\!\langle C_{\mathscr{A}\alpha}(\vec{r}\,') C_{\mathscr{B}\beta}(\vec{r} - \vec{r}\,') \rangle\!\rangle.$$

Taking into account the translation invariance (with cyclic conditions), we can write the right-hand side in the form

$$V \langle\!\langle C_{\mathscr{A}\alpha}(\vec{0}) C_{\mathscr{B}\beta}(\vec{r}) \rangle\!\rangle.$$

Thus (see 7.2.6)

$${}^{\bullet}\boldsymbol{H}_{A\alpha\mathscr{B}\beta}(\vec{q}) = \frac{1}{C_{\mathscr{A}\alpha} C_{\mathscr{B}\beta}} \int_V \mathrm{d}^3 r\, \mathrm{e}^{i\vec{q}\cdot\vec{r}} \langle\!\langle C_{\mathscr{A}\alpha}(\vec{0}) C_{\mathscr{B}\beta}(\vec{r}) \rangle\!\rangle. \tag{7.2.9}$$

The functions ${}^{\bullet}\boldsymbol{H}_{\mathscr{A}\alpha\mathscr{B}\beta}(\vec{q})$ are related to correlations of monomer concentrations.

2.2.1 Partial structure functions

Polymers in solution constitute a disordered system and the asymptotic behaviour

$$\lim_{r \to \infty} \langle\!\langle C_{\mathscr{A}\alpha}(\vec{0}) C_{\mathscr{B}\beta}(\vec{r}) \rangle\!\rangle = C_{\mathscr{A}\alpha} C_{\mathscr{B}\beta} \tag{7.2.10}$$

expresses the absence of correlations at large distances. This entails that ${}^{\bullet}\boldsymbol{H}_{\mathscr{A}\alpha\mathscr{B}\beta}(\vec{q})$ is singular for $q = 0$.

Let us regularize this function by subtracting the asymptotic contribution. Let us consider eqn (7.2.9) and let us go to the limit $V \to \infty$. Taking (7.2.10) into account, we write

$${}^{\bullet}\boldsymbol{H}_{\mathscr{A}\alpha\mathscr{B}\beta}(\vec{q}) = \boldsymbol{H}_{\mathscr{A}\alpha\mathscr{B}\beta}(q) + (2\pi)^3 \delta(\vec{q})$$

where

$$\boldsymbol{H}_{A\alpha\mathscr{B}\beta}(\vec{q}) = \int \mathrm{d}^3 r\, \mathrm{e}^{i\vec{q}\cdot\vec{r}} \left(\frac{\langle\!\langle C_{\mathscr{A}\alpha}(\vec{0}) C_{\mathscr{B}\beta}(\vec{r}) \rangle\!\rangle}{C_{\mathscr{A}\alpha} C_{\mathscr{B}\beta}} - 1 \right) \tag{7.2.11}$$

is a non-singular function, called *structure function*. Thus, for $q \neq 0$, the cross-section reads

$$\Sigma(q) = \sum_{\mathscr{A}} \sum_{\mathscr{B}} \sum_{\alpha} \sum_{\beta} \mathrm{b}_\alpha \mathrm{b}_\beta C_{\mathscr{A}\alpha} C_{\mathscr{B}\beta} \boldsymbol{H}_{\mathscr{A}\alpha\mathscr{B}\beta}(\vec{q}). \tag{7.2.12}$$

2.2.2 Partial and global structure functions

The structure function

$$\boldsymbol{H}_{\mathscr{A}\alpha\mathscr{B}\beta}(\vec{q}) = \int \mathrm{d}^3 r \, \mathrm{e}^{i\vec{q}\cdot\vec{r}} \left(\frac{\langle\!\langle C_{\mathscr{A}\alpha}(\vec{0}) C_{\mathscr{B}\beta}(\vec{r}) \rangle\!\rangle}{C_{\mathscr{A}\alpha} C_{\mathscr{B}\beta}} - 1 \right)$$

accounts for correlations between monomers of types α and β, belonging to subsets of molecules $\mathscr{A}$ and $\mathscr{B}$. This is a *partial* structure function.

Any two monomers belonging to the set of all monomers are correlated. A global structure function corresponds to this more general set. Let us define the mean monomer concentration by

$$\hat{C} = \sum_{\mathscr{A}} \sum_{\alpha} C_{\mathscr{A}\alpha}$$

and the local monomer concentration by

$$\hat{C}(\vec{r}) = \sum_{\mathscr{A}} \sum_{\alpha} C_{\mathscr{A}\alpha}(\vec{r}).$$

The global structure function is

$$\hat{\boldsymbol{H}}(\vec{q}) = \sum_{\mathscr{A}} \sum_{\mathscr{B}} \sum_{\alpha} \sum_{\beta} \frac{C_{\mathscr{A}\alpha} C_{\mathscr{B}\beta}}{\hat{C}^2} \boldsymbol{H}_{\mathscr{A}\alpha\mathscr{B}\beta}(\vec{q}). \tag{7.2.13}$$

2.2.3 Cross-section (per unit volume) of a set of homogeneous but different polymers

Let assume that $\mathrm{b}_\alpha = \mathrm{b}_{\mathscr{A}}$ for every monomer of the molecular species $\mathscr{A}$. This assumption is generally appropriate for solutions of polymers whose chemical structure is of the form

$$\mathrm{Y}_1 \!-\!\!(\mathrm{X})\!\!-_N \mathrm{Y}_2$$

where N is the number of links, where X represents the 'current' monomer, and where Y_1, and Y_2 are the terminal monomers. In this case, we neglect the end-effects, considering that Y_1 and Y_2 are identical to X.

In such conditions, the expression (7.2.13) of the cross-section simplifies. Let us introduce the mean concentration

$$C_{\mathscr{A}} = \frac{1}{V} \left\langle\!\!\left\langle \sum_{a \in \mathscr{A}} \sum_{j=1}^{N_{\mathscr{A}}} 1 \right\rangle\!\!\right\rangle$$

and the local concentration

$$C_{\mathscr{A}}(\vec{r}) = \sum_{a \in \mathscr{A}} \sum_{j=1}^{N_{\mathscr{A}}} \delta(\vec{r} - \vec{r}_{aj}) \tag{7.2.14}$$

of monomers belonging to species $\mathscr{A}$. We have

$$\Xi(q) = \sum_{\mathscr{A}} \sum_{\mathscr{B}} \mathfrak{b}_{\mathscr{A}} \mathfrak{b}_{\mathscr{B}} C_{\mathscr{A}} C_{\mathscr{B}} \boldsymbol{H}_{\mathscr{A}\mathscr{B}}(\vec{q}) \tag{7.2.15}$$

where

$$\boldsymbol{H}_{\mathscr{A}\mathscr{B}}(\vec{q}) = \int \mathrm{d}^3 r \; \mathrm{e}^{i\vec{q}\cdot\vec{r}} \left(\frac{\langle\langle C_{\mathscr{A}}(\vec{0}) C_{\mathscr{B}}(\vec{r}) \rangle\rangle}{C_{\mathscr{A}} C_{\mathscr{B}}} - 1 \right) \tag{7.2.16}$$

is the partial structure function. For $q \neq 0$, we may also write

$$\boldsymbol{H}_{\mathscr{A}\mathscr{B}}(\vec{q}) = \frac{1}{V C_{\mathscr{A}} C_{\mathscr{B}}} \sum_{a \in \mathscr{A}} \sum_{b \in \mathscr{B}} \sum_{i=1}^{N_{\mathscr{A}}} \sum_{j=1}^{N_{\mathscr{B}}} \mathrm{e}^{i\vec{q}\cdot(\vec{r}_{ai} - \vec{r}_{bj})}. \tag{7.2.17}$$

It is useful to separate, in the cross-section, the contribution of the global structure function $\hat{H}(q)$ [see (7.2.13)] from the other contributions. Let $\hat{C} = \sum_{\mathscr{A}} C_{\mathscr{A}}$ be the total concentration and $\hat{\mathfrak{b}} = \sum_{\mathscr{A}} \mathfrak{b}_{\mathscr{A}} C_{\mathscr{A}} / \hat{C}$ the mean collision length. By substituting $\mathfrak{b}_{\mathscr{A}} = \hat{\mathfrak{b}} + (\mathfrak{b}_{\mathscr{A}} - \hat{\mathfrak{b}})$ in (7.2.15), we get (for $q \neq 0$)

$$\Xi(q) = \hat{b}^2 \hat{C}^2 \boldsymbol{H}(\vec{q}) + \sum_{\mathscr{A}} \sum_{\mathscr{B}} (\mathfrak{b}_{\mathscr{B}} - \hat{\mathfrak{b}})(\mathfrak{b}_{\mathscr{B}} + \hat{\mathfrak{b}}) C_{\mathscr{A}} C_{\mathscr{B}} \boldsymbol{H}_{\mathscr{A}\mathscr{B}}(\vec{q}), \tag{7.2.18}$$

which is the result we were looking for.

2.2.4 Cross-section (per unit volume) of a monodisperse set of heterogeneous polymers.

The set is made here of N identical molecules. The molecules contain various types α, β . . . of monomers. The collision length of a monomer of type α is $\mathfrak{b}_\alpha$. The concentration of monomers of type α is C_α. The cross-section reads

$$\Xi(q) = \sum_{\alpha} \sum_{\beta} \mathfrak{b}_\alpha \mathfrak{b}_\beta C_\alpha C_\beta \boldsymbol{H}_{\alpha\beta}(\vec{q}) \tag{7.2.19}$$

where

$$\boldsymbol{H}_{\alpha\beta}(\vec{q}) = \int \mathrm{d}^3 r \; \mathrm{e}^{i\vec{q}\cdot\vec{r}} \left(\frac{\langle\langle C_\alpha(\vec{0}) C_\beta(\vec{r}) \rangle\rangle}{C_\alpha C_\beta} - 1 \right)$$

or, for $q \neq 0$

$$\boldsymbol{H}_{\alpha\beta}(\vec{q}) = \frac{1}{V C_\alpha C_\beta} \left\langle\!\!\left\langle \sum_{a=1}^{N} \sum_{b=1}^{N} \sum_{i \in \alpha} \sum_{j \in \beta} \mathrm{e}^{i\vec{q}\cdot(\vec{r}_{ai} - \vec{r}_{bj})} \right\rangle\!\!\right\rangle.$$

2.2.5 Intramolecular and intermolecular structure functions

A pair of monomers may belong either to the same chain or to two different chains. Thus, two types of correlations correspond to these two kinds of pairs.

Then, let us write the structure function as the sum of two contributions. In particular, we shall consider here the case of homogeneous but different molecules: We can write (7.2.17) in the form

$$H_{\mathscr{A}\mathscr{B}}(\vec{q}) = \delta_{\mathscr{A}\mathscr{B}} H^{\mathrm{I}}_{\mathscr{A}}(\vec{q}) + H^{\mathrm{II}}_{\mathscr{A}\mathscr{B}}(\vec{q}) \tag{7.2.20}$$

where

$$H^{\mathrm{I}}_{\mathscr{A}}(\vec{q}) = \frac{1}{VC^2_{\mathscr{A}}} \left\langle\!\!\left\langle \sum_{a \in \mathscr{A}} \sum_{i=1}^{N_{\mathscr{A}}} \sum_{j=1}^{N_{\mathscr{A}}} \mathrm{e}^{i\vec{q}\cdot(\vec{r}_{ai} - \vec{r}_{aj})} \right\rangle\!\!\right\rangle \tag{7.2.21}$$

is the intramolecular structure function of the chains of $N_{\mathscr{A}}$ moments, and where for $q = 0$

$$H^{\mathrm{II}}_{\mathscr{A}\mathscr{B}}(\vec{q}) = \frac{1}{VC_{\mathscr{A}}C_{\mathscr{B}}} \left\langle\!\!\left\langle \sum_{a \in \mathscr{A}} \sum_{b \in \mathscr{B}} \sum_{i=1}^{N_{\mathscr{A}}} \sum_{j=1}^{N_{\mathscr{B}}} \mathrm{e}^{i\vec{q}\cdot(\vec{r}_{aj} - \vec{r}_{bj})}(1 - \delta_{\mathscr{A}\mathscr{B}}\delta_{ab}) \right\rangle\!\!\right\rangle \tag{7.2.22}$$

is the intermolecular structure function of two chains.

2.2.5.1 *Correlations functions in real space*

It is convenient to express $H^{\mathrm{I}}_{\mathscr{A}}(\vec{q})$ and $H^{\mathrm{II}}_{\mathscr{A}}(\vec{q})$ in terms of correlation functions in ordinary space. First, let us introduce the following function related to the form of a molecule of species α, in ordinary space

$$\tilde{H}_{\mathscr{A}}(\vec{r}) = \frac{1}{N^2_{\mathscr{A}}\langle\!\langle N_{\mathscr{A}} \rangle\!\rangle} \left\langle\!\!\left\langle \sum_{a \in \mathscr{A}} \sum_{i=1}^{N_{\mathscr{A}}} \sum_{j=1}^{N_{\mathscr{A}}} \delta(\vec{r} - \vec{r}_{ai} + \vec{r}_{aj}) \right\rangle\!\!\right\rangle. \tag{7.2.23}$$

Its Fourier transform is called the *form function* of the molecule

$$\begin{aligned} H_{\mathscr{A}}(\vec{q}) &= \int \mathrm{d}^3 r \, \mathrm{e}^{i\vec{q}\cdot\vec{r}} \tilde{H}_{\mathscr{A}}(\vec{r}) \\ &= \frac{1}{N^2_{\mathscr{A}}\langle\!\langle N_{\mathscr{A}} \rangle\!\rangle} \left\langle\!\!\left\langle \sum_{a \in \mathscr{A}} \sum_{i=1}^{N_{\mathscr{A}}} \sum_{j=1}^{N_{\mathscr{A}}} \mathrm{e}^{i\vec{q}\cdot(\vec{r}_{ai} - \vec{r}_{bj})} \right\rangle\!\!\right\rangle \end{aligned} \tag{7.2.24}$$

and we have [see (7.2.2)]

$$H^{\mathrm{I}}_{A}(\vec{q}) = \frac{N_{\mathscr{A}}}{C_{\mathscr{A}}} H_{\mathscr{A}}(\vec{q}). \tag{7.2.25}$$

Let us now introduce the *pair correlation function* of monomers belonging to two different chains

$$g_{\mathscr{A}\mathscr{B}}(r) = \frac{1}{VC_{\mathscr{A}}C_{\mathscr{B}}} \sum_{a \in \mathscr{A}} \sum_{b \in \mathscr{B}} \sum_{i=1}^{N_{\mathscr{A}}} \sum_{j=1}^{N_{\mathscr{B}}} \delta(\vec{r} - \vec{r}_{aj} + \vec{r}_{bj})(1 - \delta_{\mathscr{A}\mathscr{B}}\delta_{ab}). \tag{7.2.26}$$

This function has the simple asymptotic property ($V \to \infty$)

$$\lim_{r \to \infty} g_{\mathscr{A}\mathscr{B}}(r) = 1,$$

and, for the intermolecular structure function, we have

$$H^{\mathrm{II}}_{\mathscr{A}\mathscr{B}}(\vec{q}) = \int \mathrm{d}^3 r\, \mathrm{e}^{i\vec{q}\cdot\vec{r}}(g_{\mathscr{A}\mathscr{B}}(r) - 1), \tag{7.2.27}$$

which is the relation we were looking for.

2.2.5.2 *Decomposition of the cross-section per unit volume*

In order to interpret the preceding results in detail, we shall consider the case where the polymers are homogeneous but different. The cross-section per unit volume is given by eqn (7.2.18)

$$\Xi(q) = \hat{\mathfrak{b}}^2 \hat{C}^2 \hat{H}(\vec{q}) + \sum_{\mathscr{A}} \sum_{\mathscr{B}} (\mathfrak{b}_{\mathscr{A}} - \hat{\mathfrak{b}})(\mathfrak{b}_{\mathscr{B}} + \hat{\mathfrak{b}}) C_{\mathscr{A}} C_{\mathscr{B}} H_{\mathscr{A}\mathscr{B}}(\vec{q}) \tag{7.2.28}$$

where the circumflex accent refers to the whole set of monomers in the mixture (see Section 2.2.2). Moreover, the structure functions can be written as the sum of two terms [see (7.2.20) and (7.2.25)]

$$H_{\mathscr{A}\mathscr{B}}(\vec{q}) = \delta_{\mathscr{A}\mathscr{B}} \frac{N_{\mathscr{A}}}{C_{\mathscr{A}}} H_{\mathscr{A}}(\vec{q}) + H^{\mathrm{II}}_{\mathscr{A}\mathscr{B}}(\vec{q})$$

where $N_{\mathscr{A}}$ is the number of links, $H_{\mathscr{A}}(\vec{q})$ the form function of a molecule of species $\mathscr{A}$. By introducing this expression in the second right-hand side term of (7.2.28), we obtain

$$\begin{aligned} \Xi(q) = \hat{\mathfrak{b}}^2 \hat{C}^2 H(\vec{q}) &+ \sum_{\mathscr{A}} (\mathfrak{b}^2_{\mathscr{A}} - \hat{\mathfrak{b}}^2) C_{\mathscr{A}} N_{\mathscr{A}} H_{\mathscr{A}}(\vec{q}) \\ &+ \sum_{\mathscr{A}} \sum_{\mathscr{B}} (\mathfrak{b}_{\mathscr{A}} - \hat{\mathfrak{b}})(\mathfrak{b}_{\mathscr{B}} + \hat{\mathfrak{b}}) C_{\mathscr{A}} C_{\mathscr{B}} H^{\mathrm{II}}_{\mathscr{A}\mathscr{B}}(\vec{q}). \end{aligned} \tag{7.2.29}$$

When the size of the molecules is small with respect to the wavelength $2\pi/k_0$ of the incident radiation, the form function $H_{\mathscr{A}}(\vec{q})$ is constant and equal to unity. For a binary mixture, the second term is called Laue's term. The third term mirrors the local order of monomers.

2.3 Contrasts

The partial structure function $H_{\mathscr{A}\mathscr{B}}(\vec{q})$ can be determined by radiation scattering only if the subsets $\mathscr{A}$ and $\mathscr{B}$ are sufficiently contrasted with respect to the rest of the sample. Accordingly, we shall see that the concept of contrast length[8,9] can be introduced in a very natural manner.

2.3.1 Contrast shown by a binary system of homogeneous isotopic molecules

Let us consider a monodisperse binary mixture of two species $\mathscr{A} = 1, 2$ of molecules with the same number N of links and the same form but with different collision lengths $\mathfrak{b}_{\mathscr{A}}$. For instance, the mixture may consist of organic molecules and of their deuterated isotopes. The molecules can be polymers.

The molecules of species $\mathscr{A} = 1, 2$ are distinguishable* only through their collision length $\mathfrak{b}_{\mathscr{A}}$. In this case

$$\boldsymbol{H}^{\mathrm{II}}_{\mathscr{A}\mathscr{B}}(\vec{q}) = \hat{\boldsymbol{H}}^{\mathrm{II}}(\vec{q}) \qquad \text{and} \qquad H_{\mathscr{A}}(\vec{q}) \equiv H(\vec{q}).$$

Thus, the expression of the cross-section (7.2.29) is

$$\begin{aligned} \Xi(q) = \hat{\mathfrak{b}}^2 \hat{C}^2 \hat{\boldsymbol{H}}(\vec{q}) &+ \sum_{\mathscr{A}=1}^{2} (\mathfrak{b}_{\mathscr{A}}^2 - \hat{\mathfrak{b}}^2) C_{\mathscr{A}} N H(\vec{q}) + \\ &+ \sum_{\mathscr{A}=1}^{2} \sum_{\mathscr{B}=1}^{2} (\mathfrak{b}_{\mathscr{A}} - \hat{\mathfrak{b}})(\mathfrak{b}_{\mathscr{B}} + \hat{\mathfrak{b}}) C_{\mathscr{A}} C_{\mathscr{B}} \hat{\boldsymbol{H}}^{\mathrm{II}}(\vec{q}) \end{aligned}$$

where $\hat{\boldsymbol{H}}(\vec{q})$ is the global structure function and

$$\hat{\mathfrak{b}} = \mathfrak{b}_1 \frac{C_1}{\hat{C}} + \mathfrak{b}_2 \frac{C_2}{\hat{C}}.$$

By definition

$$\sum_{\mathscr{A}=1}^{2} (\mathfrak{b}_{\mathscr{A}} - \hat{\mathfrak{b}}) C_{\mathscr{A}} = 0.$$

Moreover,

$$\sum_{\mathscr{A}=1}^{2} (\mathfrak{b}_{\mathscr{A}}^2 - \hat{\mathfrak{b}}^2) C_{\mathscr{A}} = (\mathfrak{b}_1 - \mathfrak{b}_2)^2 \frac{C_1 C_2}{C}.$$

Thus, the cross-section per unit volume is

$$\Xi(q) = (\mathfrak{b}_1 C_1 + \mathfrak{b}_2 C_2)^2 \hat{\boldsymbol{H}}(\vec{q}) + (\mathfrak{b}_1 - \mathfrak{b}_2)^2 \frac{C_1 C_2}{C} N H(\vec{q}). \qquad (7.2.30)$$

The difference $(\mathfrak{b}_1 - \mathfrak{b}_2)$ is by definition the 'contrast length'. When the molecules are monomers $H(\vec{q}) = 1$: then the second term is equal to Laue's contribution.

2.3.2 Contrast shown by a homogeneous mixture of chains in solution

First let us consider a pure ideal liquid, without density fluctuations

$$\langle\!\langle C(\vec{0}) C(\vec{r}) \rangle\!\rangle = C^2.$$

Then, the structure function $\boldsymbol{H}(\vec{q})$ is identically zero. In this case, the cross-section of the isotopic mixture (of molecules with the same molecular volume) reduces in expression (7.2.30) only to the contrast term.

Let us generalize this result to the case of a homogeneous mixture of several species $\mathscr{A}$ ($\mathscr{A} = 0, 1, \ldots, m \geqslant 1$). Species $\mathscr{A} = 0$, for instance, is the solvent; species $\mathscr{A} = 1, \ldots, m$ are the solutes. The partial volumes of the monomers of

* This assumption is appropriate for experiments performed with polystyrene samples. The case where the isotopic species are not compatible is discussed, chapter 16, Section 7.

species $\mathscr{A}$ are given by

$$\frac{v_{\mathscr{A}}}{N_{\mathscr{A}}} = \frac{1}{\mathbf{N}_{\mathscr{A}}} \frac{\partial \boldsymbol{V}}{\partial \langle\langle \mathbf{N}_{\mathscr{A}} \rangle\rangle} \qquad \mathscr{A} = 0, \ldots, m$$

(where $N_{\mathscr{A}}$, $\mathbf{N}_{\mathscr{A}}$ are respectively the number of links and of polymers) and they are not necessarily equal to one another.

Let us introduce the volume fraction

$$\varphi_{\mathscr{A}}(\vec{r}) = C_{\mathscr{A}}(\vec{r}) \frac{v_{\mathscr{A}}}{N_{\mathscr{A}}}$$

occupied, locally, by monomers of species $\mathscr{A}$. We assume that the mixture is 'homogeneous', which means that, at each point $\vec{r}$

$$\sum_{\mathscr{A}=0}^{m} \varphi_{\mathscr{A}}(\vec{r}) = 1. \tag{7.2.31}$$

It is then convenient to write the cross-section

$$\Xi(q) = \sum_{\mathscr{A}} \sum_{\mathscr{B}} \mathfrak{b}_{\mathscr{A}} \mathfrak{b}_{\mathscr{B}} C_{\mathscr{A}} C_{\mathscr{B}} \boldsymbol{H}_{\mathscr{A}\mathscr{B}}(\vec{q})$$

in the form

$$\Xi(q) = \sum_{\mathscr{A}=0}^{m} \sum_{\mathscr{B}=0}^{m} \dot{\mathfrak{b}}_{\mathscr{A}} \dot{\mathfrak{b}}_{\mathscr{B}} \varphi_{\mathscr{A}} \varphi_{\mathscr{B}} \boldsymbol{H}_{\mathscr{A}\mathscr{B}}(\vec{q}) \tag{7.2.32}$$

where

$$\dot{\mathfrak{b}}_{\mathscr{A}} = \mathfrak{b}_{\mathscr{A}}(N_{\mathscr{A}}/v_{\mathscr{A}})$$

$$\varphi_{\mathscr{A}} = C_{\mathscr{A}}(v_{\mathscr{A}}/N_{\mathscr{A}}).$$

on the other hand, we have

$$\boldsymbol{H}_{\mathscr{A}\mathscr{B}}(\vec{q}) = \int \mathrm{d}^3 r \, \mathrm{e}^{i\vec{q}\cdot\vec{r}} \left(\frac{\langle\langle \varphi_{\mathscr{A}}(\vec{0}) \varphi_{\mathscr{B}}(\vec{r}) \rangle\rangle}{\varphi_{\mathscr{A}} \varphi_{\mathscr{B}}} - 1 \right). \tag{7.2.33}$$

The constraint (7.2.31) enables us to express one of the volume fractions, for instance $\varphi_0(\vec{r})$ in terms of the others

$$\varphi_0(\vec{r}) = 1 - \sum_{\mathscr{A}=1}^{m} \varphi_{\mathscr{A}}(\vec{r}). \tag{7.2.34}$$

If we substitute this expression in (7.2.33) and (7.2.32), we obtain

$$\Xi(q) = \sum_{\mathscr{A}=1}^{m} \sum_{\mathscr{B}=1}^{m} \left(\dot{\mathfrak{b}}_{\mathscr{A}} - \dot{\mathfrak{b}}_0 \right) \left(\dot{\mathfrak{b}}_{\mathscr{B}} - \dot{\mathfrak{b}}_0 \right) \varphi_{\mathscr{A}} \varphi_{\mathscr{B}} \boldsymbol{H}_{\mathscr{A}\mathscr{B}}(\vec{q}). \tag{7.2.35}$$

we contrasted the solutes $\mathscr{A} = 1, \ldots, m$ with respect to the solvent. Let us introduce the contrast length $\mathfrak{b}^{\looparrowright}_{\mathscr{A}}$ of the monomer with respect to the solvent

$$\mathfrak{b}^{\looparrowright}_{\mathscr{A}} = \mathfrak{b}_{\mathscr{A}} - \mathfrak{b}_0 \frac{v_{\mathscr{A}}}{v_0} \frac{N_0}{N_{\mathscr{A}}}. \tag{7.2.36}$$

The cross-section per unit volume then reads

$$\Xi(q) = \sum_{\mathscr{A}=1}^{m} \sum_{\mathscr{B}=1}^{m} \mathfrak{b}_{\mathscr{A}}^{\circ\to} \mathfrak{b}_{\mathscr{B}}^{\circ\to} C_{\mathscr{A}} C_{\mathscr{B}} \boldsymbol{H}_{\mathscr{A}\mathscr{B}}(\vec{q}). \tag{7.2.37}$$

2.3.3 Variation of contrast

Certain solutions consist of two solvents $\mathscr{A} = 0, 1$ and of several solutes ($m \geqslant 2$). Then, it is appropriate to contrast a solute $\mathscr{B}$ with respect to the solvent mixture. We shall consider the case where the two solvents are isotopes of each other and we shall assume that each species has the same molecular volume v_s.

Let us set $C_s = C_0 + C_1$ and $x = C_0/C_s$. Then, we can write the contribution of the solvents to the cross-section in the form [see (7.2.30)]

$$\sum_{\mathscr{A}=0}^{1} \sum_{\mathscr{B}=0}^{1} \mathfrak{b}_{\mathscr{A}} \mathfrak{b}_{\mathscr{B}} C_{\mathscr{A}} C_{\mathscr{B}} \boldsymbol{H}_{\mathscr{A}\mathscr{B}}(\vec{q}) = \mathfrak{b}^2(x) C_s^2 \boldsymbol{H}_s(\vec{q}) + (\mathfrak{b}_0 - \mathfrak{b}_1)^2 x(1-x) C_s$$

where $\boldsymbol{H}_s(\vec{q})$ is the *global* structure function of the solvents, and where

$$\mathfrak{b}(x) = x\mathfrak{b}_0 + (1-x)\mathfrak{b}_1. \tag{7.2.38}$$

The contrast length of a solute $\mathscr{A}$ with respect to the mixture of solvents is

$$\mathfrak{b}_{\mathscr{A}}^{\circ\to}(x) = \mathfrak{b}_{\mathscr{A}} - \frac{\mathfrak{b}(x)}{v_s} v_{\mathscr{A}} \frac{N_s}{N_{\mathscr{A}}}. \tag{7.2.39}$$

A detailed calculation, similar to the one solvent calculation gives the result

$$\Xi(q) = \sum_{\mathscr{A}=2}^{m} \sum_{\mathscr{B}=2}^{m} \mathfrak{b}_{\mathscr{A}}^{\circ\to}(x) \mathfrak{b}_{\mathscr{A}}^{\circ\to}(x) C_{\mathscr{A}} C_{\mathscr{B}} \boldsymbol{H}_{\mathscr{A}\mathscr{B}}(\vec{q}) + (\mathfrak{b}_0 - \mathfrak{b}_1)^2 x(1-x) C_s. \tag{7.2.40}$$

By changing the composition of the solvent, we modify x and therefore also the contrast length $\mathfrak{b}_{\mathscr{A}}^{\circ\to}(x)$. This is the method of variation of contrast.[8,9]

2.3.4 Characteristic values of contrast lengths

The possibility of controlling the contrast length $\mathfrak{b}_{\mathscr{A}}^{\circ\to}$ enables the experimentalist to deduce the partial structure functions of a mixture from small-angle scattering experiments.

Let us give characteristic values of the $\mathfrak{b}_{\mathscr{A}}^{\circ\to}$ and let us discuss the range of variation of these lengths.

2.3.4.1 *Contrast lengths for neutron radiation*

A value of the contrast of species $\mathscr{A}$ with respect to $\mathscr{B}$ is given by

$$N_{\mathscr{A}} \frac{\mathfrak{b}_{\mathscr{A}}^{\circ\to}}{v_{\mathscr{A}}} = \mathfrak{b}_{\mathscr{A}} \frac{N_{\mathscr{A}}}{v_{\mathscr{A}}} - \mathfrak{b}_{\mathscr{B}} \frac{N_{\mathscr{B}}}{v_{\mathscr{B}}} = \dot{\mathfrak{b}}_{\mathscr{A}} - \dot{\mathfrak{b}}_{\mathscr{B}} = \dot{\mathfrak{b}}_{\mathscr{A}}^{\circ\to} \tag{7.2.41}$$

which depends on collision lengths and partial volumes. Let us examine the influence of these two contributions. For species $\mathscr{A}$, let $w_{\mathscr{A}}$ be the measured volume per mass (see Chapter 5, Section 2). The partial volume (volume per molecule) is

$$v_{\mathscr{A}} = w_{\mathscr{A}} M_{\mathscr{A}} / Ⓐ$$

where $M_{\mathscr{A}}$ is the molecular mass and Ⓐ the Avogadro number. Table 7.1 contains values of $M_{\mathscr{A}}$, $w_{\mathscr{A}}$, $\mathfrak{b}$, and $\mathfrak{b}_{\mathscr{A}}/v_{\mathscr{A}}$, measured[10] at two temperatures for *monomers* corresponding to a common species $\mathscr{A}$.

In Table 7.1, the index $\mathscr{A}$ refers either to the monomer or to the solvent molecule. Let us note that the ratios $\mathfrak{b}_{\mathscr{A}}/v_{\mathscr{A}}$ vary with temperature and that the variation is not always negligible.

Table 7.1. Constants concerning a few usual monomers; M = molecular mass, $\mathfrak{b}$ = collision length, w = volume per mass, $\dot{\mathfrak{b}}$ = collision length per unit volume, $\mathfrak{b}_{\mathscr{A}}$. is calculated with the help of formula (7.2.3) by using as coherent collision lengths for the nuclei, the values of $\mathfrak{b}_{\mathrm{H}}^{\mathrm{coh}}$, $\mathfrak{b}_{\mathrm{D}}^{\mathrm{coh}}$, $\mathfrak{b}_{\mathrm{C}}^{\mathrm{coh}}$ given by Table 6.12 in Chapter 6. (non-polarized sample). The partial volumes (per mass) $w_{\mathscr{A}}$ are extracted from ref. 9, and we have $v_{\mathscr{A}} = w_{\mathscr{A}} M_{\mathscr{A}} / Ⓐ$ (see Chapter 5, Section 2.1).

Monomers of species $\mathscr{A}$	$M_{\mathscr{A}}$ [dalton]	$\mathfrak{b}_{\mathscr{A}}$ $\times 10^2$ [cm]	$w_{\mathscr{A}}$ [$cm^3\,g^{-1}$]	$\dot{\mathfrak{b}}_{\mathscr{A}} = \mathfrak{b}_{\mathscr{A}}/v_{\mathscr{A}}$ [cm^{-2}]
Solvants				
Cyclohexanes				
C_6H_{12}	84	− 0.498	1.284 at 20 °C	-0.279×10^{10} at 20 °C
			1.315 at 40 °C	-0.272×10^{10} at 40 °C
C_6D_{12}	96.24	11.99	1.1213 at 20 °C	6.92×10^{10} at 20 °C
			1.2113 at 40 °C	6.40×10^{10} at 40 °C
Toluenes				
C_7H_8	92	1.66	1.156 at 22 °C	0.972×10^{10} at 22 °C
C_7D_8	100	10	1.0598 at 20 °C	5.878×10^{10} at 20 °C
Solutes				
Styrenes				
C_8H_8	104	2.32	0.95	1.46×10^{10}
C_8D_8	112	10.7	0.87 at 20 °C in cyclohexane	6.84×10^{10}
			0.875 at 40 °C in cyclohexane	6.80×10^{10}
			0.847 at 22 °C in toluene	7.02×10^{10}
Isoprenes				
C_5H_8	68.1	0.33	1.092 at 20 °C in cyclohexane	0.276×10^{10}
			1.106 at 22 °C in toluène	0.272×10^{10}
C_5D_8	76.1	8.67	1.33	5.33×10^{10}

The values of the $\mathfrak{b}_{\mathscr{A}}$ that appear in Table 7.1 for 20 °C are plotted in Fig. 7.2. Let us note that the specific contrast $\mathfrak{b}_{\mathscr{A}}$ of a species $\mathscr{A}$ with respect to a solvent is given by the difference $\dot{\mathfrak{b}}_{\mathscr{A}} - \dot{\mathfrak{b}}_{\text{solvent}}$.

Thus, for monomers, the contrast length is

$$\mathfrak{b}^{\circ\!\!\to}_{\mathscr{A}} = \dot{\mathfrak{b}}^{\circ\!\!\to}_{\mathscr{A}} v_{\mathscr{A}}.$$

2.3.4.2 *Contrast lengths for light*

In this case, the collision lengths $\mathfrak{b}_{\mathscr{A}}$ are directly related to the polarizabilities $\boldsymbol{\alpha}$ [see (6.3.68)] and to the wave number k

$$\mathfrak{b}_{\mathscr{A}} = k^2 \boldsymbol{\alpha}_{\mathscr{A}}. \tag{7.2.42}$$

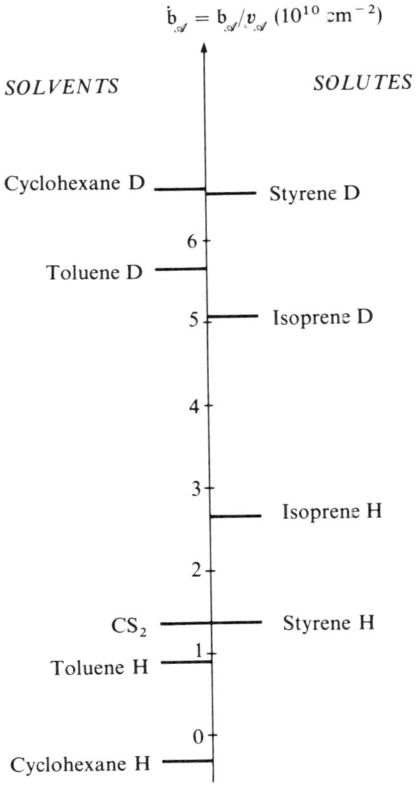

Fig. 7.2. Collision length $\dot{\mathfrak{b}}_{\mathscr{A}}$ per unit volume for a few molecular species (non-polarized sample). The contrast lengths (per unit volume) of a solute with respect to a solvent are obtained by subtraction of the quantities that appear respectively to right and to the left of the central line. (From Ionescu.[10])

The contrast length of a *monomer* of species $\mathscr{A}$ with respect to the solvent (species 0) is

$$\mathrm{b}_{\mathscr{A}}^{\hookrightarrow} = \mathrm{b}_{\mathscr{A}} - \mathrm{b}_0 v_{\mathscr{A}}/v_0 = k^2(\boldsymbol{\alpha}_{\mathscr{A}} - \boldsymbol{\alpha}_0 v_{\mathscr{A}}/v_0). \tag{7.2.43}$$

However, it will be more profitable to express $\mathrm{b}_{\mathscr{A}}^{\hookrightarrow}$ by introducing the 'index increment' of the solution

$$\frac{\partial n}{\partial \rho_{\mathscr{A}}}$$

where $\rho_{\mathscr{A}}$ is the mass per unit volume. The reason is that this increment can be determined experimentally with high accuracy, and for the experimentalist it is interesting to know the contrast length as precisely as possible, before launching a scattering experiment.

The refraction index n of an assembly of N identical particules in a volume V is given by the Lorentz–Lorenz formula

$$\frac{n^2 - 1}{n^2 + 2} = \frac{4\pi}{3} C\boldsymbol{\alpha}$$

where

$$C = N/V. \tag{7.2.44}$$

In the case where two species $\mathscr{A}$ and 0 of monomers are mixed, the polarizabilities are additive, and the refraction index of the medium reads

$$\frac{n^2 - 1}{n^2 + 2} = \frac{4\pi}{3}(C_{\mathscr{A}}\boldsymbol{\alpha}_{\mathscr{A}} + C_0\boldsymbol{\alpha}_0) \tag{7.2.45}$$

Let us make the contrast $(\boldsymbol{\alpha}_{\mathscr{A}} - \boldsymbol{\alpha}_0 v_{\mathscr{A}}/v_0)$ appear in the formula. We get

$$\frac{n^2 - 1}{n^2 + 2} = \frac{4\pi}{3}\{C_{\mathscr{A}}(\boldsymbol{\alpha}_{\mathscr{A}} - \boldsymbol{\alpha}_0 v_{\mathscr{A}}/v_0) + (C_{\mathscr{A}} v_{\mathscr{A}} + C_0 v_0)\boldsymbol{\alpha}_0/v_0\}. \tag{7.2.46}$$

Moreover, eqn (5.1.40) can be written in the form

$$C_{\mathscr{A}} v_{\mathscr{A}} + C_0 v_0 = 1.$$

Thus, we obtain

$$\frac{n^2 - 1}{n^2 + 2} = \frac{4\pi}{3}\left(\mathrm{b}_{\mathscr{A}}^{\hookrightarrow} k^{-2} C_{\mathscr{A}} + \frac{\boldsymbol{\alpha}_0}{v_0}\right). \tag{7.2.47}$$

If, at a given pressure, the partial volumes and the polarizabilities are independent of the concentration $C_{\mathscr{A}}$, which we admit, then the ratio $(n^2 - 1)/(n^2 + 2)$ is a linear function of $C_{\mathscr{A}}$. In this case, the index increment is given by

$$\frac{\partial n}{\partial C_{\mathscr{A}}} = \frac{2\pi}{9}\frac{(n^2 + 2)^2}{n}\mathrm{b}_{\mathscr{A}}^{\hookrightarrow}/k^2.$$

Thus,

$$\mathrm{b}_{\mathscr{A}}^{\hookrightarrow} = \frac{9n}{2\pi(n^2 + 2)^2} k^2 \frac{\partial n}{\partial C_{\mathscr{A}}}. \tag{7.2.48}$$

Actually, values of the increment $\partial n/\partial \rho_{\mathscr{A}}$ can be found in tables,[11] where

$$\rho_{\mathscr{A}} = C_{\mathscr{A}} \frac{M_{\mathscr{A}}}{Ⓐ}$$

(here $M_{\mathscr{A}}$ is the molecular mass and Ⓐ the Avogadro number). In what follows, it will be convenient to write (7.2.48) in the form

$$\dot{\mathfrak{b}}^{\circ\rightarrow}_{\mathscr{A}} = \frac{\mathfrak{b}^{\circ\rightarrow}_{\mathscr{A}}}{v_{\mathscr{A}}} = \frac{9n}{2\pi(n^2+2)^2} \frac{k^2\, M_{\mathscr{A}}}{Ⓐ\, v_{\mathscr{A}}} \frac{\partial n}{\partial \rho_{\mathscr{A}}}. \qquad (7.2.49)$$

Values of increments $\partial n/\partial \rho_{\mathscr{A}}$ and of contrast lengths $\dot{\mathfrak{b}}^{\circ\rightarrow}_{\mathscr{A}}$ are listed in Table 7.2 for a few characteristic solvent–solute couples and for the incident wave number $k_0 = 1.151 \times 10^{-2}\ \text{nm}^{-1}$.

Table 7.2. Index increments and contrast lengths per unit volume for light with wave number $k_0 = 1.151 \times 10^{-2}\ \text{nm}^{-1}$ (ref. 11)

SYSTEM				
Type $\mathscr{A}$ of monomer	Solvent	$n(\rho = 0)$ (solvent)	$\partial n/\partial \rho_{\mathscr{A}}$ [cm³/g]	$\dot{\mathfrak{b}}^{\circ\rightarrow}_{\mathscr{A}} = \mathfrak{b}^{\circ\rightarrow}_{\mathscr{A}}/v_{\mathscr{A}}$ [cm⁻²]
Styrene	benzene	1.498	0.1034	1.45×10^8
	cyclohexane	1.424	0.1682	2.94×10^8
	toluene	1.494	0.109	5.82×10^8
	bromobenzene	1.557	0.042	0.66×10^8
Isoprene	benzene	1.498	0.0184	0.26×10^8
	cyclohexane	1.424	0.104	3.44×10^8
	toluene	1.494	0.0308	1.41×10^8
	bromobenzene	1.557	– 0.041	-0.65×10^8

2.3.5 Advantages and disadvantages resulting from the use of neutron radiation

We note that for the same solution, the contrasts obtained with neutrons may be larger by two orders of magnitude than the contrasts obtained with photons. This property explains the fact that even a weak neutron beam can be as efficient for the observation of small-angle scattering as a light beam.

Moreover, there are situations in which using neutrons is particularly appropriate because, in this case, it is possible to modify contrasts without perturbing the system, simply by changing the isotopic composition of one or several species. Two examples will illustrate the merits of this method

1. Let us consider a mixture of one solute and one solvent. In order to separate, for *fixed concentrations*, the contribution of the form function (of a

solute molecule) from the contribution of the pair correlation function, the experimentalist modifies the isotopic composition of the solute.

2. Let us consider a mixture of two solutes and one solvent. The isotopic composition of the solvent and of the solutes are matched in such a way that only the contribution of one of the solutes shows up.

Important experimental facts determine the success of such scattering experiments. They concern, for each solute, the domain of variation of the contrast length when the isotopic concentration is changed. In fact, it may occur that an experimental aim cannot be attained because the proper contrast length cannot be obtained.

Figure 7.2 gives values of the ratio (collision length)/(partial volume) for various solvents and various solutes. The specific contrast of a solute with respect to a solvent can be read directly from the figure. In the same way, the variation domains of a specific contrast obtained by isotopic modification of the solvent or of the solute, can be obtained from this figure.

2.4 Cross-section (per unit volume) of homogeneous polymers in solution

We shall now consider polymers made of chemically identical monomers in solution in one solvent. In this case, expression (7.2.37) for the cross-section can be simplified because, then, there exists only one contrast length, independently of the fact that the system may be either monodisperse or polydisperse.

The average concentration of monomers (of solute) is given by

$$C = \sum_{\mathscr{A}=1}^{m} C_{\mathscr{A}},$$

and, in the same way, the local average concentration of monomers (of solute) is

$$C(\vec{r}) = \sum_{\mathscr{A}=1}^{m} C_{\mathscr{A}}(\vec{r}).$$

Taking this notation and eqn (7.2.16) into account, we can write

$$\sum_{\mathscr{A}=1}^{m} \sum_{\mathscr{B}=1}^{m} C_{\mathscr{A}} C_{\mathscr{B}} \boldsymbol{H}_{\mathscr{A}\mathscr{B}}(\vec{q}) = \int \mathrm{d}^3 r \, \mathrm{e}^{i\vec{q}\cdot\vec{r}} (\langle\langle C(\vec{0}) C(\vec{r}) \rangle\rangle - C^2) = C^2 \boldsymbol{H}(\vec{q}) \tag{7.2.50}$$

where $\boldsymbol{H}(\vec{q})$ is the global structure function of the *solute* [see (7.2.13)]

Now, the cross-section per unit volume can simply be expressed in terms of this structure function

$$\Xi(q) = (\mathrm{b}^{\circ\rightarrow})^2 C^2 \boldsymbol{H}(\vec{q}). \tag{7.2.51}$$

Here, the decomposition into intramolecular and intermolecular contributions gives

$$\Xi(q) = (b^{\multimap})^2 C^2 [H^{\mathrm{I}}(\vec{q}) + H^{\mathrm{II}}(\vec{q})] \tag{7.2.52}$$

where

$$C^2 H^{\mathrm{I}}(\vec{q}) = \sum_{\mathscr{A}} C_{\mathscr{A}} N_{\mathscr{A}} H_{\mathscr{A}}(\vec{q}). \tag{7.2.53}$$

In the monodisperse case, the right-hand side equals $CNH(\vec{q})$.

2.5 Relation between structure function and scattered intensity

The quantities measured during a scattering experiment are the intensities $I(q)$ of the scattered radiation, for various transfer vectors $\vec{q}$ (see Chapter 6). From these quantities, we can deduce the structure function of the solutes.

Let us express $I(q)$ in terms of the structure functions previously defined. Let V be the irradiated volume of the sample. From relations (6.5.6) and (7.2.2), we get

$$I(q) = K_{\mathrm{A}} V \Xi(q) \tag{7.2.54}$$

where K_{A} is the instrument constant (6.5.7). Now, we have only to bring the general result, (7.2.37) or one of its particular formulations, in the preceding expression, to obtain the result we are looking for.

On the other hand, it is convenient to express the amounts of monomers in terms of mass concentrations

$$\rho_{\alpha} = C_{\mathscr{A}} m_{\mathscr{A}} / Ⓐ \tag{7.2.55}$$

where $m_{\mathscr{A}}$ is the molecular mass of the monomer belonging to the molecule of species $\mathscr{A}$ and where Ⓐ is Avogadro's number.

The intensity can thus be written in the form

$$I(q) = K_{\mathrm{A}} V Ⓐ^2 \sum_{\mathscr{A} \geqslant 1} \sum_{\mathscr{B} \geqq 1} b^{\multimap}_{\mathscr{A}} b^{\multimap}_{\mathscr{B}} \frac{\rho_{\mathscr{A}} \rho_{\mathscr{B}}}{m_{\mathscr{A}} m_{\mathscr{B}}} H_{\mathscr{A}\mathscr{B}}(\vec{q}). \tag{7.2.56}$$

With the help of such a formula, the structure functions can be deduced from the measured intensities. Let now examine the most usual cases.

2.5.1 Scattering by a homogeneous solute

The solution consists of only one solvent and one solute. In this case, it is not necessary to use any index for the solute and we write

$$I(q) = K_{\mathrm{A}} V Ⓐ^2 (b^{\multimap})^2 \frac{\rho}{m} H(\vec{q}).$$

Let us set

$$K = K_{\mathrm{A}} V Ⓐ (b^{\multimap})^2 \frac{1}{m^2} \tag{7.2.57}$$

which gives

$$I(q) = K \rho^2 Ⓐ H(\vec{q}). \tag{7.2.58}$$

Then, we write the structure function as the sum of intramolecular and intermolecular contributions [see (7.2.25)]

$$\boldsymbol{H}(\vec{q}) = \frac{N}{C} H(\vec{q}) + \boldsymbol{H}^{\mathrm{II}}(\vec{q}) \tag{7.2.59}$$

where $H(\vec{q})$ is the form function of the solute molecules, N the number of monomers per molecule and $\boldsymbol{H}^{\mathrm{II}}(\vec{q})$ the intermolecular structure function.

Let $M = Nm$ be the molecular mass of the solute molecules. Now, the result (7.2.58) reads

$$I(q) = K\rho\,[M\,H(\vec{q}) + \rho\,Ⓐ\,\boldsymbol{H}^{\mathrm{II}}(\vec{q})]. \tag{7.2.60}$$

The appearance of Avogadro's number in this expression results from the definition (7.2.57) of the constant K. This can be justified as follows. In ordinary space, the correlation function $g(r) - 1$ differs practically from zero only in a range of the same order as the size R of the molecules, let us say for instance 10 nm. Therefore, the Fourier transform $\boldsymbol{H}^{\mathrm{II}}(\vec{0})$ is a volume of the order of 10^3 nm^3, but the product Ⓐ $\boldsymbol{H}^{\mathrm{II}}(\vec{0})$ is a macroscopic volume ($\sim 10^4$ cm^3) which can be more directly measured, for instance, by looking at the osmotic pressure.

2.5.2 Scattering by an isotopic mixture with two solutes

Equation (7.2.60) applies also to a mixture of two isotopic varieties D and H of the same monodisperse species.

First, let us admit that the solvent ($\mathscr{A} = 0$) is unique. Thus, the cross-section per unit volume is

$$\Xi(q) = \sum_{\mathscr{A}=1}^{2} \sum_{\mathscr{B}=1}^{2} \mathrm{b}^{\circ\!\to}_{\mathscr{A}}\, \mathrm{b}^{\circ\!\to}_{\mathscr{B}}\, C_{\mathscr{A}}\, C_{\mathscr{B}}\, \boldsymbol{H}_{\mathscr{A}\mathscr{B}}(\vec{q})$$

where subscript 1 corresponds to the deuterated variety D and subscript 2 to the non-deuterated variety H. We have here

$$\boldsymbol{H}^{\mathrm{I}}_{\mathscr{A}}(\vec{q}) = \frac{N}{C_{\mathscr{A}}} H(\vec{q}) \qquad \mathscr{A} = 1, 2$$

$$\boldsymbol{H}^{\mathrm{II}}_{\mathscr{A}\cdot\mathscr{B}}(\vec{q}) = \boldsymbol{H}^{\mathrm{II}}(\vec{q}) \qquad \mathscr{A}, \mathscr{B} = 1, 2$$

where $H(\vec{q})$ is the form function of a chain and where $\boldsymbol{H}^{\mathrm{II}}(\vec{q})$ is the intermolecular structure function of the solute. Let us write the cross-section more explicitly by replacing index 1 by D, and index 2 by H. We get

$$\Xi(q) = [(\mathrm{b}^{\circ\!\to}_{\mathrm{D}})^2 C_{\mathrm{D}} + (\mathrm{b}^{\circ\!\to}_{\mathrm{H}})^2 C_{\mathrm{H}}]\,N\,H(\vec{q}) + [\mathrm{b}^{\circ\!\to}_{\mathrm{D}}\, C_{\mathrm{D}} + \mathrm{b}^{\circ\!\to}_{\mathrm{H}}\, C_{\mathrm{H}}]^2\, \boldsymbol{H}^{\mathrm{II}}(\vec{q}). \tag{7.2.61}$$

Now, we can replace the unique solvent by a binary isotopic mixture the composition of which is given by

$$x = C_{\mathrm{sD}}/C_{\mathrm{s}}$$

where C_{sD} is the concentration of the deuterated solvent. The contrast lengths of the solute are [see (7.2.39)]

$$\mathfrak{b}_{\mathrm{D}}^{\circ\!\!\rightarrow}(x) \qquad \text{for the deuterated solute}$$

$$\mathfrak{b}_{\mathrm{H}}^{\circ\!\!\rightarrow}(x) \qquad \text{for the non-deuterated solute}$$

Now let us consider the case where there exists a composition x_0 for which $\mathfrak{b}_{\mathrm{H}}^{\circ\!\!\rightarrow}(x_0) = 0$. In agreement with (7.2.40) and (7.2.61), the cross-section per unit volume reads

$$\Xi(q) = [\mathfrak{b}_{\mathrm{D}}^{\circ\!\!\rightarrow}(x_0)]^2\,[C_{\mathrm{D}} N\,H(\vec{q}) + C_{\mathrm{D}}^2\,\boldsymbol{H}^{\mathrm{II}}(\vec{q})] + (\mathfrak{b}_0 - \mathfrak{b}_1)^2\,x_0(1 - x_0)\,C_{\mathrm{s}}$$

where $\mathfrak{b}_{\mathrm{D}}$, $\mathfrak{b}_{\mathrm{H}}$ are the collision lengths of the molecules of deuterated and non-deuterated solvent. Let us set [see (7.2.57)]

$$K(x_0) = K_{\mathrm{A}}\,V[\mathfrak{b}_{\mathrm{D}}^{\circ\!\!\rightarrow}(x_0)]^2\,\frac{Ⓐ}{m_{\mathrm{D}}}$$

where m_{D} is the molecular mass of a monomer in a deuterated chain. Let

$$\rho_{\mathrm{D}} = C_{\mathrm{D}} m_{\mathrm{D}}/Ⓐ$$

be the mass concentration of deuterated monomers and let

$$\rho_{\mathrm{s}} = C_{\mathrm{s}}\,\frac{x_0 m_0 + (1 - x_0) m_1}{Ⓐ}$$

be the mass concentration of solvent (here m_0 and m are the molecular masses of deuterated and non-deuterated solvent respectively). The scattered intensity [see (7.2.54)]

$$I(q) = K_{\mathrm{A}}\,V\,\Xi(q)$$

can be written in the form

$$\Xi(q) = K(x_0)\,\rho_{\mathrm{D}}\{M_{\mathrm{D}}\,H(\vec{q}) + \rho_{\mathrm{D}}Ⓐ\,\boldsymbol{H}^{\mathrm{II}}(\vec{q})\} + I_0 \tag{7.2.62}$$

where I_0 is a constant

$$I_0 = \frac{(\mathfrak{b}_0 - \mathfrak{b}_1)}{[\mathfrak{b}_{\mathrm{D}}^{\circ\!\!\rightarrow}(x_0)]^2}\,\frac{m_{\mathrm{D}}^2}{x_0 m_0 + (1 - x_0) m_1}\,x_0(1 - x_0)\,\rho_{\mathrm{s}}.$$

This is the result we were looking for and (7.2.62) is analogous to (7.2.60). (In the polydisperse case when the deuterated and non-deuterated chain molecules have the same distribution, the first term between braces is replaced by

$$\sum_{\mathscr{A}} \frac{\rho_{\mathscr{A}}\,M_{\mathscr{A}}\,H_{\mathscr{A}}(\vec{q})}{\rho}.$$

where $\rho_{\mathscr{A}}$ is the mass concentration of chains with $\mathscr{A}$ links.)

3. THE RULES OF SMALL-ANGLE SCATTERING

The structure function $\boldsymbol{H}(\vec{q})$ of a solute can be determined by measuring the intensity $I(q)$ scattered by the solution. However, certain techniques enable the experimentalist to make separate measurements of the intramolecular and intermolecular structure functions, namely $\boldsymbol{H}^{\mathrm{I}}(\vec{q})$ and $\boldsymbol{H}^{\mathrm{II}}(\vec{q})$. These functions have different behaviours which are schematically represented in Fig. 7.3. It is from the intramolecular structure function $\boldsymbol{H}^{\mathrm{I}}(\vec{q})$ that the scattered intensity gets its characteristic appearance (see Fig. 7.4).

There are three types of laws according to the values of q under consideration (we always assume that $q < 1/l_a$ where l_a is an interatomic distance)

1. *Scattering in the zero angle limit* ($q \to 0$)
The structure function is related to the mean square fluctuation of the number of molecules in the irradiated volume V (see Section 3.2.1). Moreover, the limit of $\boldsymbol{H}^{\mathrm{I}}(\vec{q})$ for $\vec{q} \to 0$ gives the mean molecular mass of the solute. On the other hand, the limit of $\boldsymbol{H}^{\mathrm{II}}(\vec{q})$ for $q \to 0$ and $\rho \to 0$ gives a value of the interaction between molecules of solute.

2. *Scattering for small values of* q
In the domain defined by the condition $qR < 1$ where R is the size of the molecules (Guinier domain), the dependence of $\boldsymbol{H}^{\mathrm{I}}(\vec{q})$ with respect to q gives the size of the molecules.

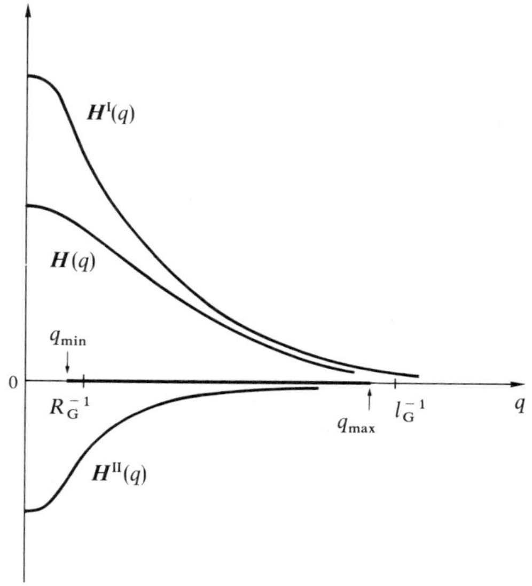

Fig. 7.3. A schematic representation of the structure functions $\boldsymbol{H}(\vec{q})$, $\boldsymbol{H}^{\mathrm{I}}(\vec{q})$, and $\boldsymbol{H}^{\mathrm{II}}(\vec{q})$. The observation window is the interval $q_{\mathrm{min}}, q_{\mathrm{max}}$. Here R_G is the radius of gyration and l_a the inter atomic distance.

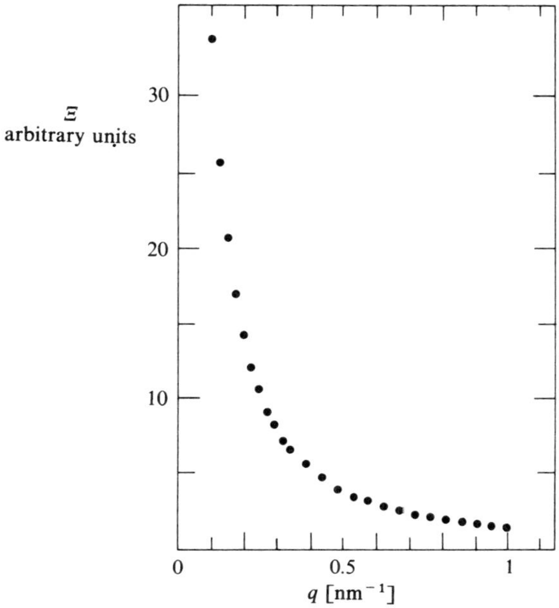

Fig. 7.4. Small-angle neutron scattering. Cross-sections of deuterated polystyrenes ($M_W = 5 \times 10^5$, $\rho = 0.005$ g/cm^3) in solution in CS_2. (From Farnoux.[12])

3. *Scattering in the asymptotic region*
In the domain

$$q \gg 1/R \quad (\text{but } q \ll 1/l_a)$$

the decrease of $\boldsymbol{H}^{\mathrm{I}}(\vec{q})$ when q increases is determined by the singularity of the intramolecular correlation function $\tilde{H}(r)$ at the origin [$\tilde{H}(r)$ appears in (7.2.24)]. On the other hand, the decrease of the modules of $H^{\mathrm{II}}(\vec{q})$ when q increases is determined by the pair correlation function $[g(r) - 1]$.

3.1 Determination of the intramolecular and intermolecular structure functions

The intensity $I(q)$ scattered by a polymer solution can always be written in the form

$$I(q) = I^{\mathrm{I}}(q) + I^{\mathrm{II}}(q) \tag{7.3.1}$$

where the first term is the sum of the contributions of each chain and where the second one is the sum of the pair contributions associated with different chains. When the solute is homogeneous, we obtain the explicit expression

$$I(q) = K\rho^2 \text{Ⓐ} [\boldsymbol{H}^{\mathrm{I}}(\vec{q}) + \boldsymbol{H}^{\mathrm{II}}(\vec{q})]$$

where ρ is the mass concentration of chains, $\boldsymbol{H}^{\mathrm{I}}(\vec{q})$ the intramolecular structure function (7.2.21), and $\boldsymbol{H}^{\mathrm{II}}(\vec{q})$ the intermolecular structure function (7.2.22). Let us examine for this case, two techniques that can be used to determine these contributions.

3.1.1 Classical technique of extrapolation a zero concentration and at zero transfer vector

The quantities $\rho \boldsymbol{H}^{\mathrm{I}}(\vec{q})$ and $\boldsymbol{H}^{\mathrm{II}}(\vec{q})$ are functions of ρ, which have finite limits when ρ goes to zero. Thus,

$$\lim_{\rho \to 0} \frac{I(q)}{K\rho} = \lim_{\rho \to 0} \frac{I^{\mathrm{I}}(q)}{K\rho} = \lim_{\rho \to 0} \rho \boldsymbol{H}^{\mathrm{I}}(\vec{q}).$$

From this relation, we extract the mean function [see (7.2.53)]

$$\rho \boldsymbol{H}^{\mathrm{I}}(\vec{q}) = \frac{\sum_{\mathscr{A}} \rho_{\mathscr{A}} M_{\mathscr{A}} H_{\mathscr{A}}(\vec{q})}{\rho}. \tag{7.3.2}$$

Let us note the normalization [see (7.2.23)]

$$H_{\mathscr{A}}(\vec{0}) = 1$$

which entails

$$\rho \boldsymbol{H}^{\mathrm{I}}(\vec{0}) = \frac{\sum \rho_{\mathscr{A}} M_{\mathscr{A}}}{\rho} = M_{\mathrm{W}}, \tag{7.3.3}$$

a quantity which is independent of ρ.

The identities (7.3.2) and (7.3.3) show that

$$\rho \text{Ⓐ} \boldsymbol{H}^{\mathrm{II}}(\vec{0}) = \lim_{q \to 0} \left[\frac{I(q)}{K\rho} - \lim_{\rho \to 0} \frac{I(q)}{K_{\rho}} \right], \tag{7.3.4}$$

and, from this relation, we deduce an experimental value of $\boldsymbol{H}^{\mathrm{II}}(\vec{0})$.

Unfortunately, this technique cannot be extended to the whole domain $(q;\rho)$ that the experimentalist can explore.

3.1.2 Technique of labelled molecules

The technique of labelled molecules[13] provides the possibility of determining the structure functions $\boldsymbol{H}^{\mathrm{I}}(\vec{q})$ and $\boldsymbol{H}^{\mathrm{II}}(\vec{q})$ of the solute for any concentration. We shall use the more explicit notation

$$\boldsymbol{H}(\vec{q}) = \boldsymbol{H}(\vec{q};\rho)$$

to indicate that the structure depends on concentration.

The method consists in measuring the apparent structure function associated with different labelled solute fractions. Let ρ_{D} be the mass concentration of the labelled chains. For a solvent that occults the non-labelled molecules, the scattered intensity is [see (7.2.62) and the comment]

$$I(q) = K(x_0)\rho_{\mathrm{D}}\left[\sum_{\mathscr{A}} \frac{\rho_{\mathscr{A}} M_{\mathscr{A}}}{\rho} \boldsymbol{H}_{\mathscr{A}}(\vec{q};\rho) + \rho_{\mathrm{D}} \text{Ⓐ}\, \boldsymbol{H}^{\mathrm{II}}(\vec{q};\rho)\right] + I_0,$$

and we note that the intensity is a function of ρ and ρ_{D}, whereas the structure functions depend only on ρ.

From this relation, we deduce

$$\sum_{\mathscr{A}} \frac{\rho_{\mathscr{A}}}{\rho} M_{\mathscr{A}} \boldsymbol{H}_{\mathscr{A}}(\vec{q};\rho) = \lim_{\rho_{\mathrm{D}} \to 0} \frac{I(q) - I_0}{K(x_0)\rho_{\mathrm{D}}} \tag{7.3.5}$$

and

$$\text{Ⓐ}\, \boldsymbol{H}^{\mathrm{II}}(\vec{q};\rho) = \lim_{\rho_{\mathrm{D}} \to 0} \frac{\partial}{\partial \rho_{\mathrm{D}}}\left[\frac{I(q) - I_0}{K(x_0)\rho_{\mathrm{D}}}\right], \tag{7.3.6}$$

which is the result we are looking for.

Figure 7.5 shows functions $\boldsymbol{H}^{\mathrm{I}}(\vec{q}, \rho)$, $\boldsymbol{H}^{\mathrm{II}}(\vec{q}, \rho)$ and $\boldsymbol{H}(\vec{q}, \rho)$ obtained from neutron scattering experiments and associated with the structure of polystyrene sulphonates (polyelectrolytes).

Incidentally, we note that for neutral polymer solutions, the form function can also be determined in a simpler way, by choosing the solvent composition so as to get zero average contrast $(C, \mathfrak{b}_1^{\circ\to} + C_2\, \mathfrak{b}_2^{\circ\to} = 0)$.

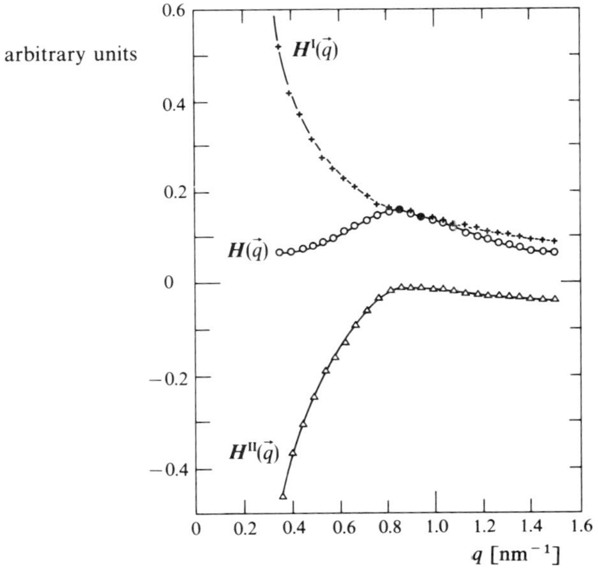

Fig. 7.5. Structure function of $\boldsymbol{H}$, $\boldsymbol{H}^{\mathrm{I}}$ and $\boldsymbol{H}^{\mathrm{II}}$ determined by small-angle scattering of neutrons (polystyrene sulphonate, $M_{\mathrm{W}} = 2 \times 10^5$, $\rho = 0.04\ \mathrm{g/cm^3}$). (From Nierlich, Boué, Lapp, and Oberthur.[14])

3.2 Scattering in the zero angle limit

When the scattering angle goes to zero, the cross-section becomes a thermodynamic quantity which mirrors the fluctuations of the number of chains and of the number of monomers per chain. Let us establish this fact by looking at the structure functions at $\vec{q} = 0$.

3.2.1 Partial structure functions at $q = 0$

The structure function of monomers of types α and β belonging respectively to chains of species $\mathscr{A}$ and $\mathscr{B}$ is

$$H_{\mathscr{A}\alpha\mathscr{B}\beta}(\vec{q}) = \int \mathrm{d}^3 r\, \mathrm{e}^{i\vec{q}\cdot\vec{r}} \left(\frac{\langle\langle C_{\mathscr{A}\alpha}(\vec{0})\, C_{\mathscr{B}\beta}(\vec{r}) \rangle\rangle}{C_{\mathscr{A}\alpha} C_{\mathscr{B}\beta}} - 1 \right) \tag{7.3.7}$$

where $C_{\mathscr{A}\alpha}$ is the mean concentration of monomers of type α belonging to chains of species $\mathscr{A}$.

Let us recall that all the chains belonging to species $\mathscr{A}$ are identical. Using this observation, we shall prove the simplified relation

$$H_{\mathscr{A}\alpha\mathscr{B}\beta}(\vec{0}) \equiv H_{\mathscr{A}\mathscr{B}}(\vec{0})$$

where

$$H_{\mathscr{A}\mathscr{B}}(\vec{q}) = \int \mathrm{d}^3 r\, \mathrm{e}^{i\vec{q}\cdot\vec{r}} \left(\frac{\langle\langle C_{\mathscr{A}}(\vec{0})\, C_{\mathscr{B}}(\vec{r}) \rangle\rangle}{C_{\mathscr{A}} C_{\mathscr{B}}} - 1 \right)$$

and $C_{\mathscr{A}}$ is the concentration of monomers belonging to the chains of species $\mathscr{A}$

In fact, let us consider the identity [see (7.2.8)]

$$\langle\langle C_{\mathscr{A}\alpha}(\vec{0})\, C_{\mathscr{B}\beta}(\vec{r}) \rangle\rangle = \frac{1}{V} \left\langle\!\!\left\langle \sum_{a\in\mathscr{A}} \sum_{b\in\mathscr{B}} \sum_{i\in\alpha} \sum_{j\in\beta} \delta(\vec{r} - \vec{r}_{ai} + \vec{r}_{bj}) \right\rangle\!\!\right\rangle$$

and let $N_{\mathscr{A}}$ be the number of molecules of species $\mathscr{A}$ in V, and $N_{\mathscr{A}\alpha}$ the number of monomers of type α in a chain of species $\mathscr{A}$.

For a grand canonical system

$$\int \mathrm{d}^3 r \langle\langle C_{\mathscr{A}\alpha}(\vec{0})\, C_{\mathscr{B}\beta}(\vec{r}) \rangle\rangle = \frac{1}{V} \langle\langle N_{\mathscr{A}} N_{\mathscr{B}} N_{\mathscr{A}\alpha} N_{\mathscr{B}\beta} \rangle\rangle.$$

As, by definition, the numbers $N_{\mathscr{A}\alpha}$, $N_{\mathscr{B}\beta}$ do not fluctuate, we can write the right-hand side in the form

$$\frac{1}{V} \langle\langle N_{\mathscr{A}} N_{\mathscr{B}} \rangle\rangle N_{\mathscr{A}\alpha} N_{\mathscr{B}\beta}.$$

Thus,

$$H_{\mathscr{A}\alpha\mathscr{B}\beta}(\vec{0}) = V \left(\frac{\langle\langle N_{\mathscr{A}} N_{\mathscr{B}} \rangle\rangle}{\langle\langle N_{\mathscr{A}} \rangle\rangle \langle\langle N_{\mathscr{B}} \rangle\rangle} \right) = H_{\mathscr{A}\mathscr{B}}(\vec{0}) \tag{7.3.8}$$

(where, in spite of appearances, $H_{\mathscr{A}\mathscr{B}}$ is a quantity which becomes independent of the volume for large V).

3.2.2 Scattered intensity in the zero angle limit

At $q = 0$, the expressions of the cross-section per unit volume (7.2.5) and of the scattered intensity (7.2.54) are divergent. It is proper to consider here the finite quantities

$$\Xi = \lim_{q \to 0} \Xi(q) \qquad \text{and} \qquad I = \lim_{q \to 0} I(q). \tag{7.3.9}$$

Starting from (7.2.12) and taking (7.3.38) into account, we can write

$$\Xi = \sum_{\mathscr{A}} \sum_{\mathscr{B}} \sum_{\alpha} \sum_{\beta} \mathrm{b}_\alpha \mathrm{b}_\beta C_{\mathscr{A}\alpha} C_{\mathscr{B}\beta} H_{\mathscr{A}\mathscr{B}}(\vec{0})$$

where the indices $\mathscr{A}\alpha$, $\mathscr{B}\beta$ indicate species of molecules and types of monomers. Let us introduce the mean collision length

$$\mathrm{b}_{\mathscr{A}} = \sum_{\alpha} \mathrm{b}_\alpha C_{\mathscr{A}\alpha}/C_{\mathscr{A}}.$$

For the cross-section per unit volume, we obtain

$$\Xi = \sum_{\mathscr{A}} \sum_{\mathscr{B}} \mathrm{b}_{\mathscr{A}} \mathrm{b}_{\mathscr{B}} C_{\mathscr{A}} C_{\mathscr{B}} H_{\mathscr{A}\mathscr{B}}(\vec{0}) \tag{7.3.10}$$

Taking (7.3.8) into account and considering that the monomer concentration is $C_{\mathscr{A}} = N_{\mathscr{A}} \langle\langle N_{\mathscr{A}} \rangle\rangle / V$, we can also write

$$\Xi = \sum_{\mathscr{A}} \sum_{\mathscr{B}} \mathrm{b}_{\mathscr{A}} \mathrm{b}_{\mathscr{B}} \frac{N_{\mathscr{A}} N_{\mathscr{B}}}{V} \{ \langle\langle \boldsymbol{N}_{\mathscr{A}} \boldsymbol{N}_{\mathscr{B}} \rangle\rangle - \langle\langle \boldsymbol{N}_{\mathscr{A}} \rangle\rangle \langle\langle \boldsymbol{N}_{\mathscr{B}} \rangle\rangle \}. \tag{7.3.11}$$

For an incompressible solution, we have

$$\Xi = \sum_{\mathscr{A} \neq 0} \sum_{\mathscr{A} \neq 0} \mathrm{b}_{\mathscr{A}} \mathrm{b}_{\mathscr{B}} \frac{N_{\mathscr{A}} N_{\mathscr{B}}}{V} \{ \langle\langle \boldsymbol{N}_{\mathscr{A}} \boldsymbol{N}_{\mathscr{B}} \rangle\rangle - \langle\langle \boldsymbol{N}_{\mathscr{A}} \rangle\rangle \langle\langle \boldsymbol{N}_{\mathscr{B}} \rangle\rangle \}. \tag{7.3.12}$$

where $\mathrm{b}_{\mathscr{A}}^{\circ\!\!\rightarrow}$ is the contrast length of the monomer of species $\mathscr{A}$ with respect to the main solvent [see (7.2.36)].

The scattered intensity $I = K_{\mathrm{A}} V \Xi$ can be expressed in the form

$$I = K_{\mathrm{A}} \sum_{\mathscr{A} \neq 0} \sum_{\mathscr{B} \neq 0} \mathrm{b}_{\mathscr{A}}^{\circ\!\!\rightarrow} \mathrm{b}_{\mathscr{B}}^{\circ\!\!\rightarrow} N_{\mathscr{A}} N_{\mathscr{B}} \{ \langle\langle \boldsymbol{N}_{\mathscr{A}} \boldsymbol{N}_{\mathscr{B}} \rangle\rangle - \langle\langle \boldsymbol{N}_{\mathscr{A}} \rangle\rangle \langle\langle \boldsymbol{N}_{\mathscr{B}} \rangle\rangle \}. \tag{7.3.13}$$

Thus, in the zero angle limit, the scattered intensity is related to the mean square fluctuations of the number of molecules (and also of the number of monomers) in the irradiated volume.

3.2.2.1 *Scattering by a homogeneous solute in the zero-angle limit*

The solute is homogeneous if all the monomers are the same.

$$\mathfrak{b}^{\circ\rightarrow}_{\mathscr{A}} \equiv \mathfrak{b}^{\circ\rightarrow}.$$

In this case, we write [see (7.2.58)]

$$I = K\rho^2 \text{Ⓐ}\, \boldsymbol{H}(\vec{0})$$

where ρ is the mass concentration, $\boldsymbol{H}(\vec{q})$ the structure function of the solute (7.2.50), and K a constant defined by (7.2.57), Ⓐ being the Avogadro number. The result (7.3.13) can be written more simply in the form

$$I = K\rho m \frac{\sum_{\mathscr{A}\neq 0}\sum_{\mathscr{B}\neq 0} N_{\mathscr{A}} N_{\mathscr{B}} \{\langle\!\langle N_{\mathscr{A}} N_{\mathscr{B}} \rangle\!\rangle - \langle\!\langle N_{\mathscr{A}} \rangle\!\rangle \langle\!\langle N_{\mathscr{B}} \rangle\!\rangle\}}{\sum_{\mathscr{A}\neq 0} N_{\mathscr{A}} \langle\!\langle N_{\mathscr{A}} \rangle\!\rangle} \tag{7.3.14}$$

where m is the molecular mass of a monomer.

The fluctuation of the number of links per chain is determined separately. In fact, we showed that [see (7.3.1) and (7.3.2)]

$$\lim_{\rho\to 0} \frac{I}{K\rho} = \lim_{\rho\to 0} \rho \boldsymbol{H}^{\mathrm{I}}(\vec{0}) = \frac{\sum_{\mathscr{A}} \rho_{\mathscr{A}} M_{\mathscr{A}}}{\rho} \tag{7.3.15}$$

where $M_{\mathscr{A}}$ is the molecular mass of species $\mathscr{A}$. The right-hand side is, by definition, the mean molecular mass M_{w} and we have

$$M_{\mathrm{W}} = m \frac{\sum_{\mathscr{A}\neq 0} N_{\mathscr{A}}^2 \langle\!\langle N_{\mathscr{A}} \rangle\!\rangle}{\sum_{\mathscr{A}\neq 0} N_{\mathscr{A}} \langle\!\langle N_{\mathscr{A}} \rangle\!\rangle}. \tag{7.3.16}$$

On the other hand, the intermolecular structure function at $q = 0$, is given by

$$\rho \text{Ⓐ}\, \boldsymbol{H}^{\mathrm{II}}(\vec{0}) = \frac{I}{K\rho} - \lim_{\rho\to 0} \frac{I}{K\rho}.$$

Let us substitute (7.3.14) and (7.3.15) in this expression. We get

$$\boldsymbol{H}^{\mathrm{II}}(\vec{0}) = V \frac{\sum_{\mathscr{A}\neq 0}\sum_{\mathscr{B}\neq 0} N_{\mathscr{A}} N_{\mathscr{B}} \{\langle\!\langle N_{\mathscr{A}} N_{\mathscr{B}} \rangle\!\rangle - \langle\!\langle N_{\mathscr{A}} \rangle\!\rangle \langle\!\langle N_{\mathscr{B}} \rangle\!\rangle - \delta_{\mathscr{A}\mathscr{B}} \langle\!\langle N_{\mathscr{A}} \rangle\!\rangle\}}{\left(\sum_{\mathscr{A}\neq 0} N_{\mathscr{A}} \langle\!\langle N_{\mathscr{A}} \rangle\!\rangle\right)^2} \tag{7.3.17}$$

and we recall that this quantity does not depend on V for large V.

Here, the fluctuations of numbers of molecules and of numbers of links contribute to the result in a non-dissociable way.

3.2.2.2 *Looking for a solvent giving zero contrast*

We noted (Section 3.1) that it is sometimes profitable to completely conceal the scattering contribution of a species $\mathscr{A}$. The result is obtained by choosing the composition of the solvent in such a way that the contrast length $\mathfrak{b}_{\mathscr{A}}^{\circ\rightarrow}$ of the monomer of species $\mathscr{A}$ vanishes.

Thus, let us consider a solvent consisting of a mixture of two isotopic varieties of the same species (for instance, deuterated and non-deuterated toluene) and let x be the number fraction of deuterated molecules in the solvent. The contrast length of the monomer of species $\mathscr{A}$ is [see (7.2.39)]

$$\begin{aligned}\mathfrak{b}_{\mathscr{A}}^{\circ\rightarrow}(x) &= b_{\mathscr{A}} - [x\mathfrak{b}_0 + (1-x)\mathfrak{b}_1]v_{\mathscr{A}}/v_s \\ &= \mathfrak{b}_{\mathscr{A}} - \mathfrak{b}_1 v_{\mathscr{A}}/v_s - x(\mathfrak{b}_0 - \mathfrak{b}_1)v_{\mathscr{A}}/v_s. \qquad (73.18)\end{aligned}$$

If there exists a composition x_0 for which $\mathfrak{b}_{\mathscr{A}}^{\circ\rightarrow}(x_0)$ vanishes, it can be calculated theoretically if the collision lengths $\mathfrak{b}_{\mathscr{A}}$ of the monomer, $\mathfrak{b}_0$ of the deuterated solvent, and $\mathfrak{b}_1$ of the non-deuterated solvent are known and if the partial volumes $v_{\mathscr{A}}$ of a monomer and v_s of a solvent molecule are also known. Nevertheless, it is necessary to verify that the signal really vanishes for this composition x_0. In fact, it happens quite often, that a poor knowledge of the partial volumes leads to an uncertainty as to the value of x_0.

The verification can be made, as follows: the solution consisting of only one solute and of a solvent of composition x, the experimentalist determines the zero angle limit of the scattering intensity

$$I = K_A \text{Ⓐ}^2 \, V\left(\frac{\mathfrak{b}_{\mathscr{A}}^{\circ\rightarrow}(x)}{m_{\mathscr{A}}}\right)^2 \rho_{\mathscr{A}}^2 \boldsymbol{H}_{\mathscr{A}\mathscr{A}}(\vec{0}).$$

This quantity is a quadratic function of x. However, $\sqrt{I}$ is a linear function of x. By plotting the values of $\sqrt{I}$ corresponding to various compositions x, one obtains a straight line which cuts the x-axis at x_0, and in this way, one obtains the desired value of x_0.

3.3 A study of the intramolecular structure function

The intramolecular structure function associated with the monomers of types α and β on chains of species $\mathscr{A}$ are

$$\boldsymbol{H}^{\mathrm{I}}_{\mathscr{A}\alpha\beta}(\vec{q}) = \frac{1}{VC_{\mathscr{A}\alpha}C_{\mathscr{A}\beta}}\left\langle\!\!\left\langle \sum_{a\in\mathscr{A}} \sum_{i\in\alpha} \sum_{j\in\beta} \mathrm{e}^{i\vec{q}\cdot(\vec{r}_{ai} - \vec{r}_{aj})} \right\rangle\!\!\right\rangle$$

where $C_{\mathscr{A}\alpha}$ is the average concentrations of monomers of type α belonging to chains of type $\mathscr{A}$. The contribution of all these functions to the scattering signal can be written in the form [see (7.2.56)]

$$I^{\mathrm{I}}(q) = K_A V \text{Ⓐ}^2 \sum_{\mathscr{A}\neq 0} \sum_{\alpha} \sum_{\beta} \frac{\mathfrak{b}_{\alpha}^{\circ\rightarrow}}{m_\alpha} \frac{\mathfrak{b}_{\beta}^{\circ\rightarrow}}{m_\beta} \rho_{\mathscr{A}\alpha}\rho_{\mathscr{A}\beta} \boldsymbol{H}^{\mathrm{I}}_{\mathscr{A}\alpha\beta}(\vec{q}) \qquad (7.3.19)$$

where K_A is the instrument constant (6.5.7), $\mathfrak{b}_\alpha^{\to}$ the contrast length for monomers of type α, m_α their molecular mass and $\rho_{\mathscr{A}\alpha}$ the mass concentration of monomers of type α belonging to molecules of species $\mathscr{A}$.

Let us describe here the methods that enable the experimentalist to deduce the functions $\boldsymbol{H}^{\mathrm{I}}_{\mathscr{A}\alpha\beta}(\vec{q})$ from the intensity $I^{\mathrm{I}}(q)$ and let us point out a few essential properties of these functions.

3.3.1 Monodisperse assembly of homogeneous chains: intramolecular structure function and form function

When the chains are homogeneous and constitute a monodisperse assembly, there is only one species $\mathscr{A}$ and we write $N_{\mathscr{A}} = N$ and $\mathfrak{b}_{\mathscr{A}} = \mathfrak{b}$. The relation between the scattered intensity and the intramolecular structure function is then [see (7.3.19)]

$$I^{\mathrm{I}}(q) = K\rho^2 \text{Ⓐ}\, \boldsymbol{H}^{\mathrm{I}}(\vec{q}) \tag{7.3.20}$$

where K is the constant (7.2.57), ρ the mass concentration of chains, and Ⓐ the A vogadro number. The intramolecular structure function $\boldsymbol{H}^{\mathrm{I}}(\vec{q})$ is directly related to the form function of the polymer chain with N links

$$\boldsymbol{H}^{\mathrm{I}}(\vec{q}) = \frac{N}{C} H(\vec{q}) \tag{7.3.21}$$

where

$$H(\vec{q}) = \frac{1}{N^2}\left\langle\!\!\left\langle \sum_{i=1}^{N}\sum_{j=1}^{N} e^{i\vec{q}\cdot(\vec{r}_i - \vec{r}_j)} \right\rangle\!\!\right\rangle \tag{7.3.22}$$

The average $\langle\!\langle\ \rangle\!\rangle$ being taken over the whole set of configurations of the system.

3.3.1.1 *Porod's sum rule*

The expression

$$\tilde{H}(\vec{r}) = \frac{1}{N^2}\left\langle\!\!\left\langle \sum_{j=1}^{N}\sum_{i=1}^{N} \delta(\vec{r} - \vec{r}_j + \vec{r}_l)\right\rangle\!\!\right\rangle$$

defined in real space has a Fourier transform which is the form function

$$H(\vec{q}) = \int \mathrm{d}^3 r\, e^{i\vec{q}\cdot\vec{r}}\tilde{H}(\vec{r})$$

and by definition $\int \mathrm{d}^3 r \tilde{H}(\vec{r}) = H(\vec{0}) = 1$

Porod's sum rule concerns the real space expression of the form function for $r \to 0$. It is related the 'volume' of the molecule. In fact, let us consider the expression

$$\tilde{H}(\vec{0}) = N^{-2}\left\langle\!\!\left\langle \sum_{i=1}^{N}\sum_{j=1}^{N} \delta(\vec{r}_i - \vec{r}_j)\right\rangle\!\!\right\rangle$$
$$= \lim_{r\to 0}\frac{1}{(2\pi)^3}\int \mathrm{d}^3 q\, e^{-i\vec{q}\cdot\vec{r}} H(\vec{q}). \tag{7.3.23}$$

This quantity is the average of the inverse 'volume' of a molecule and eqn (E.7) of Appendix E gives

$$\tilde{H}(\vec{0}) = v^{-1}. \tag{7.3.24}$$

Of course, this volume is finite in the case of 'compact' structures in space. Conversely, it may vanish for certain 'homogeneous' (i.e. self-similar) structures. This happens for structures with fractal dimensions and in particular for Brownian chains. For such structures

$$\int \mathrm{d}^3 q\, H(\vec{q})$$

does not exist and $\tilde{H}(\vec{r})$ diverges when $r \to 0$.

3.3.1.2 *Expansion of the form function in the vicinity of $q = 0$*

Let us expand the form function of a polymer molecule with respect to $\vec{q}$. By definition, $H(\vec{0}) = 1$. We also note that the contribution of the first-order terms vanishes. We get

$$H(\vec{q}) = 1 - \frac{1}{2N^2} \sum_{j=1}^{N} \sum_{l=1}^{N} \langle\!\langle [\vec{q}\cdot(\vec{r}_j - \vec{r}_l)]^2 \rangle\!\rangle.$$

As the system is, in average, isotropic

$$\langle\!\langle [\vec{q}\cdot(\vec{r}_j - \vec{r}_l)]^2 \rangle\!\rangle = \tfrac{1}{3} q^2 \langle\!\langle (\vec{r}_j - \vec{r}_l)^2 \rangle\!\rangle.$$

Let us introduce the square radius of gyration

$$R_G^2 = \frac{1}{2N^2} \sum_{j=1}^{N} \sum_{l=1}^{N} \langle\!\langle (\vec{r}_j - \vec{r}_l)^2 \rangle\!\rangle. \tag{7.3.25}$$

The preceding expansion reads

$$H(\vec{q}) = 1 - q^2 \frac{R_G^2}{3} + \ldots \tag{7.3.26}$$

which is the result we are looking for.

3.3.1.3 *Behaviour of the form function for large transfer vectors*

When the transfer wave vector $\vec{q}$ becomes infinite, the form function of a molecule

$$H(\vec{q}) = \frac{1}{N^2} \sum_{j=1}^{N} \sum_{l=1}^{N} \langle\!\langle \mathrm{e}^{i\vec{q}\cdot(\vec{r}_j - \vec{r}_l)^2} \rangle\!\rangle$$

tends to zero in a characteristic way, related to the nature of the configuration of this molecule. The profile of $H(\vec{q})$ for $\vec{q} \to \infty$ is called asymptotic behaviour. This behaviour can be determined experimentally in the case where the 'forward' scattering is well 'centred', i.e. when the incident wave number k_0 is large as compared to the inverse radius of gyration and smaller than the inverse interatomic distance l_a. There then exists a domain of transfer wave number q, in which the asymptotic behaviour of $H(\vec{q})$ can be observed: this domain is called the intermediary domain and is defined by the inequalities

$$1/R_G \ll q < \frac{1}{l_a}. \tag{7.3.27}$$

Let us examine the characteristics of this behaviour. We shall assume that the orientation of the molecule with respect to the direction of the incident beam is random. In this case, the form function $\tilde{H}(\vec{r})$, in real space, depends only on the distance r, and we write

$$H(\vec{q}) = \int d^3r\, e^{i\vec{q}\cdot\vec{r}}\, \tilde{H}(\vec{r})$$

under the form

$$H(\vec{q}) = \frac{4\pi}{q} \int_0^\infty dr\, r \sin qr\, \tilde{H}(\vec{r})$$

where $\tilde{H}(\vec{r})$ and its successive derivatives tends to zero when $r \to \infty$. The behaviour of the form function $H(\vec{q})$ in the interval (7.3.27) is determined by the analyticity of $\tilde{H}(\vec{r})$, and more precisely by its singularity at $r = 0$.

In Appendix E, we study the asymptotic behaviour of $H(\vec{q})$ for a few structures embedded in a d-dimensional space: rods, compact bodies, Brownian and Kuhnian chains. Then, two kinds of behaviour can be observed according to whether the structure has a finite molecular volume or not. Of course, at a monomer scale, the polymer has a structure and a finite molecular volume; however, in the interval (7.3.27), the polymer can be represented by a model with a zero molecular volume. Thus, it is interesting to study both cases.

Concerning these asymptotic types of behaviour, we shall bear in mind the following laws taken from Appendix E:

1. *Compact convex D-dimensional body, embedded in a D-dimensional space*
Let s be the area and v the volume of this body. We have

$$H(\vec{q}) = \frac{s}{v^2 q^{D+1}} 2^{D-1} \pi^{-1+D/2} \Gamma(D/2) \tag{7.3.28}$$

2. *Compact Convex D-dimensional body embedded in a d-dimensional space* (with $d > D$).
Let v be the 'volume' of the body. We have

$$H(\vec{q}) = \frac{1}{vq^D} (4\pi)^{D/2} \frac{\Gamma\left(\dfrac{d}{2}\right)}{\Gamma\left(\dfrac{d-D}{2}\right)}. \tag{7.3.29}$$

3. *Polymer chain in a d-dimensional space.*
Let $D = 1/\nu$ be the Hausdorff dimension of the chain and X_G the length which is related to the radius of gyration R_G by

$$R_G^2 = X_G^2 d.$$

we have

$$H(\vec{q}) = \frac{\mathrm{h}_G}{(qX_G)^D} \tag{7.3.30}$$

where h_G is a constant.

3.3.2 Homogeneous polydisperse solute: mean form function

Here the monomers have only one contrast length $\mathfrak{b}^{\circ\rightarrow}$, but the number $N_{\mathscr{A}}$ of monomers per chain varies ($\mathscr{A} = 1, \ldots$). The intensity (7.3.19) reads

$$I^1(q) = K_A V(\mathfrak{b}^{\circ\rightarrow})^2 \sum_{\mathscr{A} \neq 0} C_{\mathscr{A}}^2 \boldsymbol{H}^1_{\mathscr{A}}(\vec{q}).$$

Let us express the intensity in terms of the form functions $H_{\mathscr{A}}(\vec{q})$ of the molecules with $\mathscr{A}$ links. By definition [see (7.2.25)]

$$\boldsymbol{H}^1_{\mathscr{A}}(\vec{q}) = \frac{N_{\mathscr{A}}}{C_{\mathscr{A}}} H_{\mathscr{A}}(\vec{q})$$

which gives

$$I^1(q) = K_A V(b^{\circ\rightarrow})^2 \sum_{\mathscr{A}} N_{\mathscr{A}} C_{\mathscr{A}} H_{\mathscr{A}}(\vec{q}).$$

Let us define the mean form function

$$H_Z(\vec{q}) = \frac{\sum_{\mathscr{A}} N_{\mathscr{A}} C_{\mathscr{A}} H_{\mathscr{A}}(\vec{q})}{\sum_{\mathscr{A}} N_{\mathscr{A}} C_{\mathscr{A}}}. \tag{7.3.31}$$

This function is normalized to unity for $q = 0$.
Now, we can write

$$I^1(q) = K_A V(\mathfrak{b}^{\circ\rightarrow})^2 \left(\frac{\sum_{\mathscr{A}} N_{\mathscr{A}} C_{\mathscr{A}}}{C} \right) C H_Z(\vec{q})$$

or

$$I^1(q) = K_A V(\mathfrak{b}^{\circ\rightarrow})^2 \frac{Ⓐ}{m^2} M_W \rho H_Z(\vec{q})$$

where m is the molecular mass of a monomer and $M_W = \Sigma_{\mathscr{A}} \rho_{\mathscr{A}} M_{\mathscr{A}} / \Sigma \rho_{\mathscr{A}}$.

Setting $K = K_A V(\mathfrak{b}^{\circ\rightarrow})^2 Ⓐ/m^2$, we obtain

$$I^1(q) = K \rho M_W H_Z(\vec{q}), \tag{7.3.32}$$

which is the formulation generally used by experimentalists. Let us expand $H_Z(\vec{q})$ in the vicinity of $q = 0$; for each $\mathscr{A}$

$$H_{\mathscr{A}}(\vec{q}) = 1 - q^2 R^2_{G,\mathscr{A}}/3 + \ldots$$

where

$$R^2_{G,\mathscr{A}} = \frac{1}{2N^2_{\mathscr{A}}} \left\langle\!\!\left\langle \sum_{j=1}^{N_{\mathscr{A}}} \sum_{l=1}^{N_{\mathscr{A}}} (\vec{r}_{j\mathscr{A}} - \vec{r}_{l\mathscr{A}})^2 \right\rangle\!\!\right\rangle$$

and for $H_Z(\vec{q})$

$$H_Z(\vec{q}) = 1 - q^2 R^2_{G,Z}/3$$

where

$$R^2_{G,Z} = \frac{\sum_{\mathscr{A}} N_{\mathscr{A}} C_{\mathscr{A}} R^2_{G\mathscr{A}}}{\sum_{\mathscr{A}} N_{\mathscr{A}} C_{\mathscr{A}}} \tag{7.3.33}$$

is the mean square radius of gyration.

In brief, for small values of q, we can express the results in the form

$$\frac{K\rho}{I^{\mathrm{I}}(q)} = \frac{1}{M_{\mathrm{W}}}\left(1 + \frac{q^2}{3} R^2_{G,Z}\right)$$

and this expression does not depend on the model used to represent the molecule configurations.

Let us now consider the asymptotic behaviour of the function $H_Z(\vec{q})$. For a Brownian chain, we have [see (7.3.14) and (7.3.20)]

$$H_{\mathscr{A}}(\vec{q}) = \frac{2}{x^2_{\mathscr{A}}}(\mathrm{e}^{-x_{\mathscr{A}}} - 1 + x_{\mathscr{A}}) \tag{7.3.34}$$

where

$$x_{\mathscr{A}} = {}^0R^2_{G,\mathscr{A}} q^2 \qquad (\text{for } d = 3).$$

When $x \gg 1$, the function $H_{\mathscr{A}}(\vec{q})$ has an asymptotic behaviour given by

$$H_{\mathscr{A}}(\vec{q}) = 2\left(\frac{1}{x_{\mathscr{A}}} - \frac{1}{x^2_{\mathscr{A}}}\right) \qquad q\,{}^0R_{G,\mathscr{A}} \gg 1. \tag{7.3.35}$$

The behaviour of $H_Z(\vec{q})$ becomes asymptotic when

$$\frac{\sum_{\mathscr{A}=1}^{\infty} M_{\mathscr{A}} \rho_{\mathscr{A}} x_{\mathscr{A}}}{\sum_{\mathscr{A}=1}^{\infty} M_{\mathscr{A}} \rho_{\mathscr{A}}} \gg 1.$$

Then, the asymptotic behaviour of the polydisperse system is

$$H_Z(\vec{q}) \simeq 2\left(\frac{1}{q^2} \sum_{\mathscr{A}=1}^{\infty} \frac{M_{\mathscr{A}} \rho_{\mathscr{A}}}{{}^0R^2_{G,\mathscr{A}}} - \frac{1}{q^4} \sum_{\mathscr{A}=1}^{\infty} \frac{M_{\mathscr{A}} \rho_{\mathscr{A}}}{({}^0R^2_{G,\mathscr{A}})^2}\right) \Bigg/ \sum_{\mathscr{A}=1}^{\infty} M_{\mathscr{A}} \rho_{\mathscr{A}}. \tag{7.3.36}$$

with $q\,{}^0R_{G,Z} \gg$.

In the Brownian case, the square radius of gyration is proportional to the molecular mass. Let ${}^0\Lambda$ be the proportionality constant

$$ {}^0R^2_{G,\mathscr{A}} = {}^0\Lambda\, M_{\mathscr{A}}. \tag{7.3.37}$$

Taking this relation into account, we can write (7.3.36) as follows

$$H_Z(\vec{q}) = \frac{2}{{}^0\Lambda\, q^2}\left(\sum_{\mathscr{A}} \rho_{\mathscr{A}} - \frac{1}{{}^0\Lambda\, q^2}\sum_{\mathscr{A}} \frac{\rho_{\mathscr{A}}}{M_{\mathscr{A}}}\right)\Bigg/\sum_{\mathscr{A}} M_{\mathscr{A}}\rho_{\mathscr{A}}.$$

But, by definition,

$$\frac{\sum_{\mathscr{A}} \rho_{\mathscr{A}}}{\sum_{\mathscr{A}} \rho_{\mathscr{A}}/M_{\mathscr{A}}} \simeq \langle M \rangle = M_n$$

$$\frac{\sum_{\mathscr{A}} \rho_{\mathscr{A}} M_{\mathscr{A}}}{\sum_{\mathscr{A}} \rho_{\mathscr{A}}} \simeq \frac{\langle M^2 \rangle}{\langle M \rangle} = M_W.$$

Thus, for ${}^0R^2_{G,Z}\, q^2 \gg 1$

$$H_Z(\vec{q}) = \frac{2}{{}^0\Lambda\, q^2 M_W}\left(1 - \frac{1}{{}^0\Lambda\, q^2 M_n}\right). \tag{7.3.38}$$

Therefore, the asymptotic behaviour of the function $K\rho/I^1(q)$ is [see (7.3.32)]

$$\frac{K\rho}{I^1(q)} = \frac{{}^0\Lambda\, q^2}{2}\left(1 + \frac{1}{{}^0\Lambda\, q^2 M_n}\right) = \frac{1}{2M_n} + \frac{{}^0\Lambda\, q^2}{2} \qquad {}^0R^2_{G,z}q^2 \gg 1. \tag{7.3.39}$$

The polydispersion effects concerning the behaviour of the function $K\rho/I^1(q)$ associated with a Brownian chain, are as follows:

1. The slope of the straight line representing $K\rho/I^1(q)$ versus q^2 is ${}^0\Lambda/2$ *for any polydispersion* (for ${}^0R^2_G\, q^2 \gg 1$).
2. The extrapolation at $q = 0$ of the asymptotic behaviour of $K\rho/I^1(q)$ gives $1/2M_n$ (whereas $K\rho/I^1(0) = 1/2M_W$). Figure 7.6 shows the experimental effects of polydispersion at $q = 0$, in the Guinier domain and in the asymptotic domain, for a chain which can be considered as Brownian.

3.3.3 Heterogeneous chains: apparent form functions

Heterogeneous chains are made of monomers of various types ($\alpha = 1, \ldots$). Heterogeneity is produced either by chemical synthesis or by isotopic substitution. Then, the intramolecular contribution to the scattered intensity is

$$I^1(q) = K_A V \sum_{\mathscr{A}\neq 0}\sum_{\alpha}\sum_{\beta} \mathrm{b}_{\mathscr{A}}^{\circ\!\to} \mathrm{b}_{\mathscr{B}}^{\circ\!\to}\, C_{\mathscr{A}\alpha} C_{\mathscr{A}\beta} H^1_{\mathscr{A}\alpha\beta}(\vec{q}) \tag{7.3.40}$$

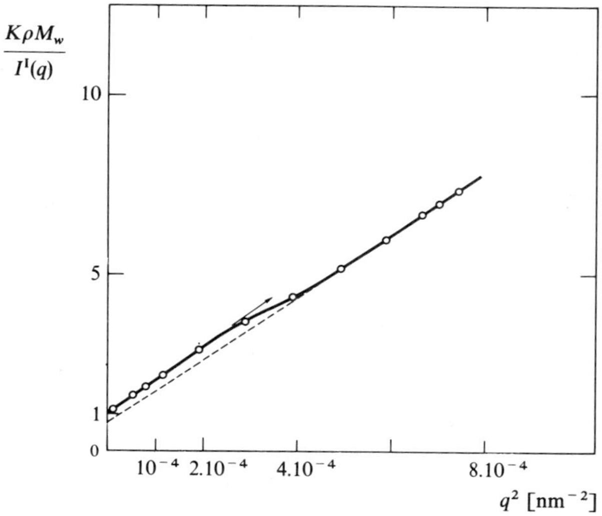

Fig. 7.6. Representation of the inverse of the mean form function corresponding to a sample of atactic polystyrene ($M_W = 7.9 \times 10^6$) in cyclohexane at 35°C (from Loucheux, Weill, and Benoit[15]): values measured by light scattering. The slope at the origin equals ${}^0R^2_{G,z}/3$ [See (7.3.33)]. The asymptotic slope equals ${}^0R^2_{G,w}/2 = M_W{}^0\Lambda/2$ [see (7.3.38)]. Its prolongation is represented by a dashed line. Here the value of this prolongation for $q = 0$, i.e. $M_W/2M_n$ is equal to 0.8. Thus, for this sample $M_W/M_n = 1.6$.

where $\mathscr{A}$ is a chain index (two chains belong to the same species if they are identical).

Let us show how a molecular form function can appear in this expression.

3.3.3.1 *Effective form function associated with a monodisperse set*

In the case under consideration, the monomers of type α are distributed in the same way on each molecule. The chains contain N links

$$\sum_\alpha N_\alpha = N$$

and the number of monomers of type α per unit volume is

$$C_\alpha = \frac{\rho_\alpha}{m_\alpha} \text{Ⓐ} .$$

In such conditions, eqn (7.3.40) simplifies. There is only one kind of chain; thus,

$$H^{\text{I}}_{\mathscr{A}\alpha\beta}(\vec{q}) \equiv H^{\text{I}}_{\alpha\beta}(\vec{q}),$$

and the intensity reads

$$I^{\mathrm{I}}(q) = K_{\mathrm{A}}\, V \sum_{\alpha} \sum_{\beta} \mathrm{b}_{\alpha}^{\circ\to}\, \mathrm{b}_{\beta}^{\circ\to}\, C_{\alpha} C_{\beta} \boldsymbol{H}^{\mathrm{I}}_{\alpha\beta}(\vec{q}). \tag{7.3.41}$$

Let us express this intensity by introducing an apparent form function $H_{\mathrm{app}}(\vec{q})$, i.e. a function which is equal to unity for $q = 0$. From definition (7.2.19), we obtain

$$\boldsymbol{H}^{\mathrm{I}}_{\alpha\beta}(\vec{q}) = \frac{1}{VC_{\alpha}C_{\beta}} \left\langle\!\!\left\langle \sum_{a=1}^{N} \sum_{j \in \alpha} \sum_{l \in \beta} \mathrm{e}^{i\vec{q}\cdot(\vec{r}_{aj} - \vec{r}_{al})} \right\rangle\!\!\right\rangle$$
$$= \frac{C}{NC_{\alpha}C_{\beta}} \left\langle\!\!\left\langle \sum_{j \in \alpha} \sum_{l \in \beta} \mathrm{e}^{i\vec{q}\cdot(\vec{r}_{1j} - \vec{r}_{1l})} \right\rangle\!\!\right\rangle$$

where

$$C = \sum_{\alpha} C_{\alpha} = \textcircled{A} \sum_{\alpha} \rho_{\alpha}/m_{\alpha}.$$

Let us introduce mean contrast lengths

$$\mathrm{b}^{\circ\to} = \sum_{\alpha} \mathrm{b}_{\alpha}^{\circ\to}\, N_{\alpha}/N.$$

We can write

$$I^{\mathrm{I}}(q) = K_{\mathrm{A}}\, V(\mathrm{b}^{\circ\to})^2 CN H_{\mathrm{app}}(\vec{q}) \tag{7.3.42}$$

where

$$H_{\mathrm{app}}(\vec{q}) = \frac{1}{N^2} \sum_{\alpha} \sum_{\beta} \frac{\mathrm{b}_{\alpha}^{\circ\to}\, \mathrm{b}_{\beta}^{\circ\to}}{(\mathrm{b}^{\circ\to})^2} \sum_{j \in \alpha} \sum_{l \in \beta} \left\langle\!\!\left\langle \mathrm{e}^{i\vec{q}\cdot(\vec{r}_j - \vec{r}_l)} \right\rangle\!\!\right\rangle$$

is the apparent form function of the molecule. By definition, this function satisfies the condition

$$H_{\mathrm{app}}(\vec{0}) = 1.$$

The intensity (7.3.42) is proportional to the mass concentration ρ and to the molecular mass M. The experimentalists are lead to write it in the form

$$I^{\mathrm{I}}(q) = K\rho M H_{\mathrm{app}}(\vec{q}) \tag{7.3.43}$$

where

$$K = K_{\mathrm{A}}\, V(\mathrm{b}^{\circ\to})^2 \frac{CN}{\rho M},$$

now

$$C = \sum_{\alpha} C_{\alpha} = \textcircled{A} \sum_{\alpha} \rho_{\alpha}/m_{\alpha}$$

and

$$M = \sum_{\alpha} m_\alpha N_\alpha = N \sum_{\alpha} m_\alpha C_\alpha / C = \frac{N\rho}{\sum_{\alpha} \rho_\alpha / m_\alpha}.$$

Thus,

$$K = K_A V(\mathfrak{b}^{\circ\rightarrow})^2 \, Ⓐ \left(\sum_{\alpha} \rho_\alpha / m_\alpha \rho \right)^2.$$

3.3.3.2 *Apparent mean square radius and variation of contrast*

Let us expand $H_{\text{app}}(\vec{q})$ in the vicinity of $\vec{q} = 0$. We get

$$H_{\text{app}}(\vec{q}) = 1 - q^2 R^2_{\text{app}}/3$$

where

$$R^2_{\text{app}} = \frac{1}{2N^2} \sum_{\alpha} \sum_{\beta} \frac{\mathfrak{b}^{\circ\rightarrow}_\alpha \mathfrak{b}^{\circ\rightarrow}_\beta}{(\mathfrak{b}^{\circ\rightarrow})^2} \sum_{j \in l\alpha} \sum_{l \in \beta} \left\langle\!\!\left\langle (\vec{r}_j - \vec{r}_l)^2 \right\rangle\!\!\right\rangle \tag{7.3.44}$$

is the apparent mean square radius. This radius is a function of the contrast lengths $\mathfrak{b}^{\circ\rightarrow}_\alpha$ ($\alpha = 1, 2, \ldots$). Consequently, R^2_{app} depends on the composition x of the solvent. It is interesting to study this dependence. For instance, let x be the fraction of deuterated molecules in the solvent. The collision length of a molecule of solvent reads [see (7.2.38)]

$$\mathfrak{b}(x) = x\mathfrak{b}_D + (1 - x)\mathfrak{b}_H$$

where $\mathfrak{b}_D$, $\mathfrak{b}_H$ are the collision lengths of the deuterated and non-deuterated molecules, respectively. Thus, the contrast length of a monomer of type α is

$$\mathfrak{b}^{\circ\rightarrow}_\alpha (x) = \mathfrak{b}_\alpha - \mathfrak{b}(x) v_0 / v_\alpha \tag{7.3.45}$$

where $\mathfrak{b}_\alpha$ is the collision length and v_α the partial volume of a monomer of type α whereas v_0 is the partial volume of a solvent molecule. (We assume that these quantities do not depend on composition.) In this case, the apparent mean square radius is a function of x of the form

$$R^2_{\text{app}}(x) = \frac{1}{2N^2} \sum_{\alpha} \sum_{\beta} \frac{\mathfrak{b}^{\circ\rightarrow}_\alpha (x) \mathfrak{b}^{\circ\rightarrow}_\beta (x)}{[b^{\circ\rightarrow}(x)]^2} \sum_{j \in \alpha} \sum_{l \in \beta} \left\langle\!\!\left\langle (\vec{r}_j - \vec{r}_l)^2 \right\rangle\!\!\right\rangle \tag{7.3.46}$$

where

$$\mathfrak{b}^{\circ\rightarrow}(x) = \sum_{\alpha} \mathfrak{b}^{\circ\rightarrow}_\alpha (x) N_\alpha / N.$$

If there exists a composition x_0 for which

$$\mathfrak{b}^{\circ\rightarrow}_\alpha (x_0) \equiv \mathfrak{b}^{\circ\rightarrow}(x_0),$$

then the apparent square radius coincide with the radius of gyration R^2_G [see (7.3.25)].

3.3.3.3 *Apparent mean square radius of Brownian block copolymers*

The linear block copolymers are made of different successive subchains $\alpha = 1,2, \ldots$ and so on. For the Brownian case, Benoit and Wippler[16] obtained an expression of ${}^0R^2_{\text{app}}(x)$ in terms of rather simple contrast parameters. Their result can be recovered as follows. First, let us write (7.3.46) in the form

$$ {}^0R^2_{\text{app}}(x) = \frac{1}{2N^2} \sum_\alpha \sum_\beta \frac{\mathrm{b}^{\circ\to}_\alpha \, \mathrm{b}^{\circ\to}_\beta}{[\mathrm{b}^{\circ\to}(x)]^2} N_\alpha N_\beta R^2_{\alpha\beta} \tag{3.47} $$

where

$$ {}^0R^2_{\alpha\beta} = \frac{1}{N_\alpha N_\beta} \sum_{j \in \alpha} \sum_{l \in \beta} \left\langle\!\!\left\langle (\vec{r}_j - \vec{r}_l)^2 \right\rangle\!\!\right\rangle $$

are the partial square radii.

Let $\vec{r}_{\mathrm{B}\alpha}$ be the position vector of the centre of gravity B_α of the monomers of subchains α. We can write

$$ {}^0R^2_{\alpha\beta} = \frac{1}{N_\alpha N_\beta} \sum_{j \in \alpha} \sum_{l \in \beta} \left\langle\!\!\left\langle <[(\vec{r}_j - \vec{r}_{B\alpha}) + (\vec{r}_{B\alpha} - \vec{r}_{B\beta}) + (\vec{r}_{B\beta} - \vec{r}_l)]^2 \right\rangle\!\!\right\rangle . \tag{7.3.48} $$

Now, let us introduce the radius of gyration of the subchain α

$$ {}^0R^2_{G\alpha} = \frac{1}{N_\alpha} \left\langle\!\!\left\langle \sum_{j \in \alpha} (\vec{r}_j - \vec{r}_{B\alpha})^2 \right\rangle\!\!\right\rangle \tag{7.3.49} $$

and the mean square distance between B_α and B_β

$$ {}^0R^2_{B\alpha B\beta} = \langle\!\langle (\vec{r}_{B\alpha} - \vec{r}_{B\beta})^2 \rangle\!\rangle . \tag{7.3.50} $$

We immediately see by expanding (7.3.48) that

$$ {}^0R^2_{\alpha\beta} = \; = {}^0R^2_{G\alpha} + {}^0R^2_{G\beta} + {}^0R^2_{B\alpha B\beta} . \tag{7.3.51} $$

On the other hand, let us introduce the contrast parameter associated with the subchains α

$$ Y_\alpha(x) = \frac{N_\alpha \mathrm{b}^{\circ\to}_\alpha (x)}{N \mathrm{b}^{\circ\to} (x)} . \tag{7.3.52} $$

Equation (7.3.47) becomes

$$ {}^0R^2_{\text{app}} = \frac{1}{2} \sum_\alpha \sum_\beta Y_\alpha(x) Y_\beta(x) ({}^0R^2_{G\alpha} + {}^0R^2_{G\beta} + {}^0R^2_{\mathrm{B}\alpha\mathrm{B}\beta}) . \tag{7.3.53} $$

As by definition

$$ \sum_\alpha Y_\alpha(x) = 1, $$

we have

$$ {}^0R^2_{\text{app}} = \sum_\alpha Y_\alpha(x) {}^0R^2_{G\alpha} + \frac{1}{2} \sum_\alpha \sum_\beta Y_\alpha(x) Y_\beta(x) {}^0R^2_{\mathrm{B}\alpha\mathrm{B}\beta} . \tag{7.3.54} $$

In particular, when the copolymer is made of two subchains ($\alpha = 1, 2$), we can set

$$Y_1(x) = Y(x). \tag{7.3.55}$$

Thus, in this case,

$$^0R^2_{\text{app}} = Y(x)\,^0R^2_{G1} + [1 - Y(x)]\,^0R^2_{G2} + Y(x)[1 - Y(x)]\,^0R^2_{B1B2} \quad . \tag{7.3.56}$$

By plotting $^0R^2_{\text{app}}(Y)$ against Y, one obtains a parabola (see Fig. 7.7).

Note that the limit $Y \to \pm\infty$ corresponds to zero average contrast. In this situation, the scattered intensity directly gives the coefficient of the quadratic term.

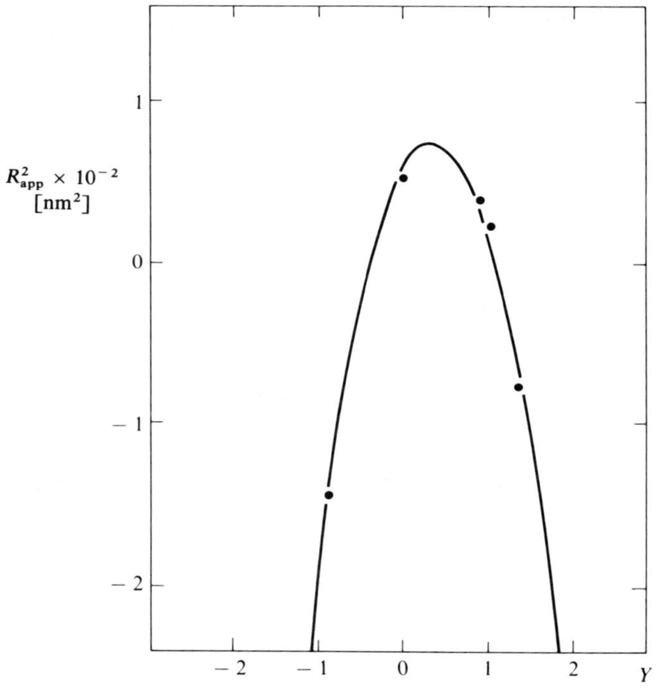

Fig. 7.7. Plot of the apparent radius of gyration against the contrast parameter $Y(x)$. Solid line: parabola (7.3.56).

Points: measured values (from Ionescu[10]) for a deuterated polystyrene–polyisoprene copolymer in solution in cyclohexane at 40° C.

Usually, the values of the collision lengths are such that the experimentalist cannot explore the interval $0 \leqslant Y \leqslant 1$ by changing x (see 7.3.52). However, this difficulty can be overcome by polarizing the sample (see Chapter 6; C. Fermon and H. Glattli, to be published).

We can now remark that the quadratic contribution of Y in (7.3.56) occurs because the centres of gravity B_1 and B_2 do not coincide. Linear, flexible copolymers belong to this class. The quadratic term disappears when the molecules are rigid and centrosymmetric.

3.4 Comments concerning the study of the structure functions of the solute

The structure function $\boldsymbol{H}(\vec{q})$ of the solute accounts for the correlations of monomer concentrations. In a natural way, it can be expressed as the sum of an intramolecular structure function $\boldsymbol{H}^{\mathrm{I}}(\vec{q})$ and of an intermolecular structure function $\boldsymbol{H}^{\mathrm{II}}(\vec{q})$. One may wonder whether one of these functions is more suitable than the others for the study of polymers in solution.

In dilute solution, the system is not very homogeneous: the easiest way to get information on the structure of the solution is to look at the intermolecular and intramolecular structure functions, $\boldsymbol{H}^{\mathrm{I}}(\vec{q})$ and $\boldsymbol{H}^{\mathrm{II}}(\vec{q})$. In particular, the study of the function

$$\boldsymbol{H}^{\mathrm{II}}(\vec{q}) = \int \mathrm{d}^3 r \, \mathrm{e}^{i\vec{q}\cdot\vec{r}}[g(r) - 1]$$

where $g(r)$ is the pair correlation function of monomers belonging to different chains [see (7.2.26)], enables us to determine the repulsive interaction between chains.

On the contrary, when the concentration is larger than the overlap concentration, the solution becomes more homogeneous and the study of the function $\boldsymbol{H}(\vec{q})$ more revealing. In fact, let us consider the definition

$$\boldsymbol{H}(\vec{q}) = \int \mathrm{d}^3 r \, \mathrm{e}^{i\vec{q}\cdot\vec{r}}\left(\frac{\langle\!\langle C(\vec{0})C(\vec{r})\rangle\!\rangle}{C^2} - 1\right)$$

where $C(\vec{r})$ and C are respectively the local and global monomer concentrations. The decrease of the modulus of the function

$$\langle\!\langle C(\vec{0})C(\vec{r})\rangle\!\rangle C^{-2} - 1$$

when r increases, results from screening effects in homogeneous structures. This property will be studied in Chapters 13 and 15.

Before the emergence of the isotopic labelling methods and of the neutron-scattering technique, it hardly seemed possible to separate the contributions $\boldsymbol{H}^{\mathrm{I}}(\vec{q})$ and $H^{\mathrm{II}}(\vec{q})$. Thus, very little has been said concerning these functions in the case of solutions with polymer overlap. However, they are directly related in the case of polymer liquids (a limiting case). Then, the structure function $\boldsymbol{H}(\vec{q})$ becomes identical to the global structure function $\hat{\boldsymbol{H}}(\vec{q})$ of the system. In agreement with the assumption that the system can be considered as homogeneous in the whole range

$$0 \leqslant q < 1/l_a$$

where l_a is the interatomic distance, we have

$$\hat{H}(\vec{q}) = H(\vec{q}) \equiv 0. \tag{7.3.57}$$

(This postulate has been used in the case of real polymer solutions to make the solute–solvent contrast more distinct.) The consequence of equality (7.3.57) is the following. In the liquid system

$$H^{\text{II}}(\vec{q}) = -H^{\text{I}}(\vec{q}) = -\frac{N}{C}H(\vec{q}). \tag{7.3.58}$$

The intermolecular structure function is proportional to the form function!

In Chapters 13 and 15 we shall see how $H^{\text{II}}(\vec{q})$ differs from $H(\vec{q})$ when the polymer liquid is diluted by a solvent. In particular, we shall find that in the zero concentration limit, $H^{\text{II}}(\vec{q})$ is proportional to the square of the form function!

An experimental verification of the fact that in the absence of solvent the structure function of the solute vanishes, i.e.

$$H(\vec{q}) = 0$$

is not easy to accomplish, because of parasitical signals.

On the other hand, when the polymer liquid is made of heterogeneous chains, a very spectacular signal $I(q)$ is obtained, and this situation is typical. Let us consider, for instance, a liquid made of monodisperse di-block copolymers, whose subchains are isotopic varieties (and therefore chemically identical). Let us assume that these subchains have the same length

$$N_\alpha = N/2 \qquad \alpha = 1, 2$$

where N is the total number of monomers per chain. Let $\mathfrak{b}_\alpha$ be the collision length of a monomer belonging to a subchain α. The intensity scattered by the liquid reads

$$I = K_A V \sum_\alpha \sum_\beta \mathfrak{b}_\alpha \mathfrak{b}_\beta C_\alpha C_\beta H_{\alpha\beta}(\vec{q})$$

where C_α is the concentration of monomers of type α. According to our assumptions,

$$C_\alpha = C/2 \qquad \alpha = 1, 2$$

where C is the total monomer concentration. By definition, for $q \neq 0$

$$VC_\alpha C_\beta H_{\alpha\beta}(\vec{q}) = \left\langle\!\!\left\langle \sum_a \sum_b \sum_{j \in \alpha} \sum_{l \in \beta} e^{i\vec{q}\cdot(\vec{r}_{aj} - \vec{r}_{bl})} \right\rangle\!\!\right\rangle$$

where a and b are chain indices, and j, l are monomer indices.

Let us introduce the mean collision length

$$\mathfrak{b} = (\mathfrak{b}_1 + \mathfrak{b}_2)/2$$

and let us write

$$I(q) =$$

$$K_A V\left\{\mathrm{b}^2 \sum_{\alpha\beta} C_\alpha C_\beta \boldsymbol{H}_{\alpha\beta}(\vec{q}) + \sum_\alpha \sum_\beta (\mathrm{b}_\alpha - \mathrm{b})(\mathrm{b}_\beta + \mathrm{b}) C_\alpha C_\beta [\boldsymbol{H}^{\mathrm{I}}_{\alpha\beta}(\vec{q}) + \boldsymbol{H}^{\mathrm{II}}_{\alpha\beta}(\vec{q})]\right\}.$$

The first term vanishes in agreement with (7.3.57)

$$\sum_{\alpha\beta} C_\alpha C_\beta \boldsymbol{H}_{\alpha\beta}(\vec{q}) = C^2 \boldsymbol{H}(\vec{q}) = 0.$$

Let us note that, in the second term, the intermolecular contributions do not depend on the indices α, β

$$\boldsymbol{H}^{\mathrm{II}}_{\alpha\beta}(\vec{q}) \equiv \boldsymbol{H}^{\mathrm{II}}(\vec{q})$$

since the subchains are identical. Thus,

$$\sum_{\alpha\beta} (\mathrm{b}_\alpha - \mathrm{b})(\mathrm{b}_\beta + \mathrm{b}) C_\alpha C_\beta \boldsymbol{H}^{\mathrm{II}}_{\alpha\beta}(q\Gamma) = 0.$$

The only non-vanishing contribution comes from the intramolecular term. We get

$$I(q) = K_A V \sum_{\alpha\beta} (\mathrm{b}_\alpha - \mathrm{b})(\mathrm{b}_\beta + \mathrm{b}) C_\alpha C_\beta \boldsymbol{H}^{\mathrm{I}}_{\alpha\beta}(\vec{q}).$$

Let $H(\vec{q})$ be the form function for the whole copolymer and $H_{1/2}(\vec{q})$ for a subchain. We obtain

$$I(q) = K_A V(\mathrm{b}_1 - \mathrm{b}_2)^2 \frac{CN}{2} [H_{1/2}(\vec{q}) - H(\vec{q})]. \qquad (7.3.59)$$

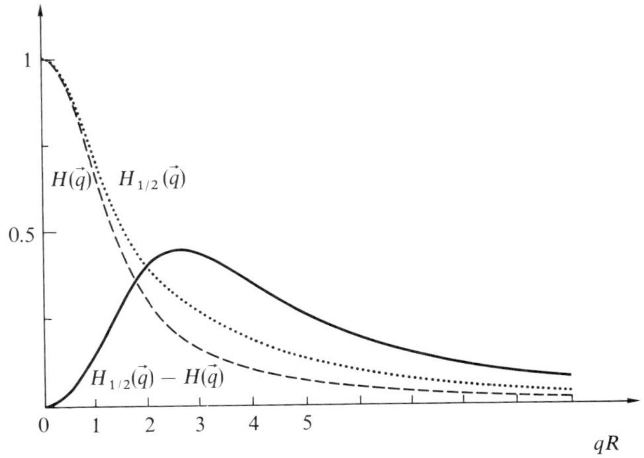

Fig. 7.8. Structure function of diblock copolymers. $I(q)$ is proportional to this function: dashed line: $H(\vec{q})$; dotted line: $H_{1/2}(\vec{q})$; solid line: $4\,[H_{1/2}(\vec{q}) - H(\vec{q})]$.

Such a function is represented in Fig. 7.8. It has a typical aspect characterizing what is called 'diffuse scattering'.

The result can be generalized for copolymers made of monodisperse unequal subchains. As can be seen in Fig. 7.8, the scattered intensity vanishes in the zero angle limit. This result is in agreement with (7.3.59). For $q \neq 0$, the scattered intensity is finite (non-zero) and this effect is related to the existence of fluctuations of intramolecular composition.

The fact that the function $I(q)$ has a maximum has been verified by neutron scattering on a sample, consisting of 'pseudo' copolymers obtained by linking two polystyrene chains, one deuterated and the other non-deuterated. This result shows clearly how compensations between intramolecular and intermolecular contributions come into play.

REFERENCES

1. Cotton, J.P. (1973). Thesis, Paris VI.
2. Zernike, F. (1935). *Z. Tech. Phys.* **16**, 454.
3. Brenschede, W. (1955). In *Die Physik des Hochpolymer* (Sci. ed. H.A. Stuart), Vol. 3, Chapter 8, Section 45, p. 500. Springer-Verlag, Berlin.
4. Abragam, A. and Goldman, M. (1982). *Nuclear magnetism: order and disorder.* Clarendon Press, Oxford.
5. Stuhrmann, H.B. (1985). *Neutron scattering in the nineties.* International Atomic Energy Agency, Vienna.
6. Léger, L., Hervet, H., and Rondelez, F. (1981). *Macromolecules* **14**, 1732.
7. Leibler, L. and Benoit, H. (1981). *Polymer* **22**, 195.
8. Stuhrmann, H.B. and Kirste, R.G. (1967). *Z. Phys. Chem.* (Frankfurt) **56**, 335.
9. Cotton, J.P. and Benoit, H. (1975). *J. Physique* **36**, 905.
10. Ionescu, L. (1976). Thesis. Université Louis Pasteur, Strasbourg.
11. Brandrup, J. and Immergut, E.H. (1975). *Polymer handbook*, Chapter VII. John Wiley, New York.
12. Farnoux, B. (1975). Thesis. Université Louis-Pasteur, Strasbourg.
13. Williams, C.E., Nierlich, M., Cotton, J.P., Jannink, G., Boué, F., Daoud, M., Farnoux, B., Picot, C., de Gennes, P.G., Rinaudo, M., Moan, M., and Wolff, C. (1979). *Polym. Sci.: Polym. Lett.* **17**, 279.
14. Nierlich, M., Boué, F., Lapp, A., and Oberthur, R. (1985). *Colloid & Polymer Sci.* **263**, 955.
15. Loucheux, C., Weill, G., and Benoit, H. (1958). *J. Chim. Phys.* **55**, 540.
16. Benoit, H. and Wippler, C. (1960). *J. Chim. Phys.* **57**, 524.

8

REPULSIVE CHAINS – OLD THEORIES

GENESIS

The strong influence of the excluded volume effect on the behaviour of a polymer in a good solvent was recognized in 1949 by P.J. Flory, who calculated the swelling of a chain and defined the Θ point ($T = T_F$). Flory's theory was accepted around 1951 and subsequently had great success; in fact, the value $\nu = 3/5$ of the three-dimensional size exponent ν, which it predicted, has been considered for a long time as being in agreement with the experiments. The method was improved later. Thus, in 1966, M. Fisher, observing that Flory's theory could be applied in d-dimensional space, pointed out that, for $d \leqslant 4$, the theory gives $\nu = \nu_F = 3/(d+2)$ and for $d \geqslant 4$, $\nu = 1/2$. Simultaneously, other approaches were developed and, in particular, the self-consistent field method introduced by S.F. Edwards in 1965. All these theories agreed with Flory's theory concerning the value of ν.

It was then generally admitted, at that time, that Flory's theory was crude but fundamentally correct. Soon, however, new difficulties emerged. In 1968, a new theory appeared, the Gaussian approximation (more general than the Brownian approximation!) which expressed in a calculable form the ideas that Flory had put forward twenty years before. But J. des Cloizeaux showed that for $4 \geqslant d \geqslant 2$ this method developed by Edwards and by himself does not lead to the Flory value $\nu_F = 3/(d+2)$ but to a larger value $\nu_s = 2/d$ previously suggested by W.H. Stockmayer and considered to be unrealistic. This strange and unexpected result was the sign of greater difficulties. In 1969, des Cloizeaux pointed out that Flory's calculation is fundamentally ill-founded, and he also showed that the self-consistent field method is quite unrealistic for long chains. Thus, these criticisms cast a doubt on the validity of Flory's result $\nu_F = 3/(d+2)$, but the physicists and chemists working in that field were still holding it to be true. As happens frequently in physics, the solution of these problems came from researches made in very different domains.

In 1970, K. Wilson discovered that to describe certain critical properties correctly it was necessary to resort to renormalization techniques borrowed from field theory. In 1972, P.G. de Gennes showed that a long polymer chain is a critical object and that the same techniques are applicable to its study. This new approach proved fruitful. Moreover, it explained the errors contained in the old theories. These errors arise partly from the fact that they take into account only one exponent, namely the size exponent ν, overlooking the fact that other exponents may exist. Actually, renormalization theory shows that there are two fundamental exponents ν and γ. All the older theories implicitly state that γ has the value $\gamma = 1$, which is not true. Thus, the fact that older theories lead to Flory's exponent, which is not very different from the exact value, can be explained only by compensations resulting from multiple errors. However, as the value $\nu = \nu_F = 3/(d+2)$ is a fairly good approximation of reality, it seems that there should exist a quite realistic approximation leading to this value. Actually, until now (1990), it has not been possible to find any *reasonable* approximation of this kind, and this seems to be a paradox if one thinks of the numerous articles which for twenty years, have been devoted to Flory's theory.

1. THE ISOLATED REPULSIVE CHAIN

1.1 The principle of free-energy minimization

Most older theories rely more or less explicitly on the principle of free-energy minimization. This principle is well-known but, in order to apply it to polymer theory, we shall reformulate it precisely.

Let us consider a system having the property that any configuration is defined by a set of vectors $\vec{u}_1, \ldots, \vec{u}_N$ and let us assume that a potential energy $\mathscr{U}(\vec{u}_1, \ldots, \vec{u}_N)$ can be associated with each configuration.

Thus, in the case of a chain with N links, the vector $\vec{u}_j$ may define the direction and the length of the link of order j (with $j = 1, \ldots, N$). In this case, the potential is frequently represented as the sum of two terms

$$\mathscr{U}(\vec{u}_1, \ldots, \vec{u}_N) = {}^{c}\vec{\mathscr{U}}_1, \ldots, \vec{u}_N) + {}^{I}\mathscr{U}(\vec{u}_1, \ldots, \vec{u}_N)$$

where ${}^{c}\mathscr{U}$ is the chain potential which connects each point of the chain to its neighbours, and ${}^{I}\mathscr{U}$ an interaction potential between all these points.

In general, a probability law $P(\vec{u}_1, \ldots, \vec{u}_N)$ is associated with the various configurations

$$\int d^d u_1 \ldots d^d u_N \, P(\vec{u}_1, \ldots, \vec{u}_N) = 1. \tag{8.1.1}$$

A mean energy $\mathbb{E}_N$ and an entropy $\mathbb{S}$ characterizing the disorder of the system correspond to this probability distribution

$$\mathbb{E}_N = \int d^d u_1 \ldots d^d u_N P(\vec{u}_1, \ldots, \vec{u}_N)\, \mathscr{U}(\vec{u}_1, \ldots, \vec{u}_N)$$

$$\mathbb{S}_N = -\int d^d u_1 \ldots d^d u_N P(\vec{u}_1, \ldots, \vec{u}_N) \ln P(\vec{u}_1, \ldots, \vec{u}_N) \tag{8.1.2}$$

The free energy is given by

$$\mathbb{F}_N = \mathbb{E}_N - \boldsymbol{\beta}^{-1} \mathbb{S}_N \tag{8.1.3}$$

In the following, we shall set $\boldsymbol{\beta} = 1$: in other terms we shall express energies and free energies in thermal units, the unit being $\beta^{-1} = k_B T$ where T is the absolute temperature and k_B Boltzmann's constant.*

The quantity $\mathbb{E}_N$ is a functional of $P(\vec{u}_1, \ldots, \vec{u}_N)$. At equilibrium, the function $P(\vec{u}_1, \ldots, \vec{u}_N)$ must minimize the free energy. In fact, the minimization of $\mathbb{F}_N$ corresponds to a maximization of the entropy $\mathbb{S}_N$ at fixed energy, and β can be considered as a Lagrange multiplier. Simultaneously, the normalization condition (8.1.1) must be obeyed, and this leads to the introduction of a second Lagrange multiplier λ.

The minimization of

$$\mathbb{F}_N - \lambda \int \mathrm{d}^d u_1 \ldots \mathrm{d}^d u_N P(\vec{u}_1, \ldots, \vec{u}_N)$$

with respect to $P(\vec{u}_1, \ldots, \vec{u}_N)$ thus gives

$$\frac{\partial \mathbb{F}_N}{\partial P(\vec{u}_1, \ldots, \vec{u}_N)} - \lambda = 0 \tag{8.1.4}$$

which leads to Boltzmann's law

$$P(\vec{u}_1, \ldots, \vec{u}_N) = \frac{1}{Z} \exp\left[-\mathscr{U}_N(\vec{u}_1, \ldots, \vec{u}_N)\right]. \tag{8.1.5}$$

(Here $Z = \mathrm{e}^{1-\lambda}$.)

The constant Z is determined by the normalization condition (8.1.1); this is the partition function, and by bringing (8.1.5) in (8.1.2), we find that at equilibrium [see (8.1.3)]

$$\mathbb{F}_N = -\ln Z, \tag{8.1.6}$$

a well-known relation.

In principle, Boltzmann's equation enables us to calculate all the properties of the system at equilibrium and in particular the mean values of all functions $\mathcal{O}(\vec{u}_1, \ldots, \vec{u}_N)$ of the configurations of the system. However, in practice, it is impossible to perform the calculation and approximations have to be made. The most classical one consist in choosing a priori a simpler trial probability $P_T(\vec{u}_1, \ldots, \vec{u}_N; \alpha_1, \ldots, \alpha_p)$ depending on parameters $\alpha_1, \ldots, \alpha_p$. This trial probability is introduced in eqns (8.1.2) and (8.1.3), and in this way, a free energy $\mathbb{F}_N(\alpha_1, \ldots, \alpha_N)$ is obtained. The parameters $\alpha_1, \ldots, \alpha_p$ are now determined by minimization of $\mathbb{F}_N(\alpha_1, \ldots, \alpha_p)$

$$\frac{\partial \mathbb{F}_N(\alpha_1, \ldots, \alpha_p)}{\partial \alpha_i} = 0. \tag{8.1.7}$$

The corresponding values of these parameters are brought back in

* Incidentally, we note that β is a more fundamental quantity than T and that the constant k_B does not possess any universal character but depends essentially on specific properties of water in solid, liquid, and gas states.

$P(\vec{u}_1, u_2, \ldots, \vec{u}_N; \alpha_1, \ldots, \alpha_N)$. In this manner, a trial probability is determined and for this trial probability it can be relatively easy to perform the calculation of simple mean values.

Still, the efficiency of this method is not warranted. The assumptions concerning the form of $P_T(\vec{u}_1, \ldots, \vec{u}_N)$ are made a priori and they have to be realistic enough. Otherwise, it is not possible to obtain correct results (even qualitatively). Unfortunately, the required conditions of simplicity and realism are difficult to reconcile in practice, and, as will be seen, this has been the source of many difficulties in polymer theory.

1.2 Flory's theory

Flory's theory has been formulated in various nearly equivalent manners[1,2,3,4] and there does not exist any classical approach. Consequently, we shall begin with a rough description of the basic principles, and afterwards we shall expound this theory more precisely by using the trial probability technique in a rather novel way.

Fundamentally, Flory admits that the excluded volume constraints produce a global swelling $\mathfrak{X}$. The free energy is estimated in terms of $\mathfrak{X}$. The minimization of $\mathbb{F}(\mathfrak{X})$ with respect to $\mathfrak{X}$ determines the swelling and leads to Flory's exponent $\nu = \nu_F = 3/(d+2)$ (d = dimension of the space in which the chain is embedded).

In order to reach this result, let us try to evaluate the free energy stored by a Brownian chain when this chain is stretched. The initial probability distribution associated with a Brownian chain at equilibrium can be written as follows

$$^{0}P(\vec{u}_1, \ldots, \vec{u}_N) = \frac{1}{[2\pi l^2]^{Nd/2}} \exp\left[-\frac{1}{2l^2} \sum_{j=1}^{N} u_j^2 \right]. \tag{8.1.8}$$

Thus, at equilibrium

$$\langle\!\langle \vec{u}_j \rangle\!\rangle = 0 \qquad \langle\!\langle u_j^2 \rangle\!\rangle = dl^2 \qquad \langle\!\langle (\vec{r}_N - \vec{r}_0)^2 \rangle\!\rangle = Ndl^2 \tag{8.1.9}$$

$$\left(\vec{r}_N - \vec{r}_0 \equiv \sum_j \vec{u}_j \right)$$

and, of course, the chain potential is

$$^{c}\mathscr{U} = \frac{1}{2l^2} \sum_{j=1}^{N} u_j^2. \tag{8.1.10}$$

Let us imagine that we pull at both extremities of the chain with a force $\vec{\mathscr{F}}$. In order to describe the state of the stretched chain, we must add a term

$$^{e}\mathscr{U} = -\vec{\mathscr{F}} \cdot (\vec{r}_N - \vec{r}_0) = -\vec{\mathscr{F}} \cdot \left(\sum_{j=1}^{N} \vec{u}_j \right). \tag{8.1.11}$$

to the chain potential.

The probability law for the stretched chain can be written in a form which does not differ much from (8.1.8)

$$P(\vec{\mathscr{F}}; \vec{u}_1, \ldots, \vec{u}_N) = \frac{1}{[2\pi l^2]^{Nd/2}} \exp\left[-\frac{1}{2l^2} \sum_{j=1}^{N} (\vec{u}_j - l^2 \vec{\mathscr{F}})^2 \right] \quad (8.1.12)$$

and in this case, we have

$$\langle\langle \vec{u}_j \rangle\rangle = l^2 \vec{\mathscr{F}} \qquad \langle\langle \vec{u}_j \cdot \vec{u}_l \rangle\rangle = l^2 (d\delta_{jl} + l^2 \mathscr{F}^2)$$

and

$$\langle\langle (\vec{r}_N - \vec{r}_0)^2 \rangle\rangle = l^2 (N^2 l^2 \mathscr{F}^2 + Nd). \quad (8.1.13)$$

In the case under consideration, the first term is dominant and therefore the relation between the size of the polymer and the force is

$$R \simeq N l^2 \mathscr{F}.$$

Consequently, the corresponding variation of free energy is*

$$\delta \mathbb{F} = \delta^c \mathbb{E} = \langle\langle \delta^c \mathscr{U} \rangle\rangle = N l^2 \mathscr{F}^2 \propto R^2/N. \quad (8.1.14)$$

Let us now consider the interaction energy ${}^1\mathscr{U}$. If we assume that the N interacting points have a uniform density in a ball of radius proportional to R, we can write

$$\delta^1 \mathbb{E} \propto N^2/R^d. \quad (8.1.15)$$

The total variation of free energy is given by

$$\delta \mathbb{F} = \delta^c \mathbb{E} + \delta^1 \mathbb{E}$$

and must be minimized with respect to R. This implies that $\delta^0 \mathbb{F}$ and $\delta^1 \mathbb{F}$ are of the same order of magnitude and the relation

$$\frac{R^2}{N} \propto \frac{N^2}{R^d} \quad (8.1.16)$$

leads immediately to the conclusion

$$R \propto N^{\frac{3}{d+2}}, \quad (8.1.17)$$

which is Flory's result for large values of N.

The same result can be recovered in a more precise manner by using the trial probability method. In fact, let us assume that the statistical state of the chain can be represented as a superposition of chains stretched by random forces which, for reasons of simplicity, will be assumed to be Gaussian.

The probability law associated with the force $\vec{\mathscr{F}}$ is of the form

$$p(\vec{\mathscr{F}}) = \left(\frac{l^2}{2\pi G^2}\right)^{d/2} \exp\left[-\mathscr{F}^2 l^2 / 2G^2 \right] \quad (8.1.18)$$

* If the chain links were not Gaussian but had a definite length, the variation of free energy would be a difference in entropy but the result would be the same.

where G is a parameter related to the swelling and we have

$$\langle\!\langle(\vec{\mathscr{F}})^2\rangle\!\rangle = dG^2 l^{-2}$$

Thus, a trial probability can be defined by setting.

$$P_{\mathrm{T}}(\vec{u}_1, \ldots, \vec{u}_N) = \int \mathrm{d}^d\mathscr{F}\; p(\vec{\mathscr{F}})P(\vec{\mathscr{F}}; \vec{u}_1, \ldots, \vec{u}_N)$$

$$= {}^0P(\vec{u}_1, \ldots, \vec{u}_N)\left(\frac{l^2}{2\pi G^2}\right)^{d/2} \int \mathrm{d}^d\mathscr{F} \exp\left[-\frac{1}{2}l^2\mathscr{F}^2\left(N + \frac{1}{G^2}\right) + \sum_j \vec{\mathscr{F}}\cdot\vec{u}_j\right]$$

which gives

$$P_{\mathrm{T}}(\vec{u}_1, \ldots, \vec{u}_N) = {}^0P(\vec{u}_1, \ldots, \vec{u}_N)\frac{1}{(1 + NG^2)^{d/2}} \exp\left[\frac{1}{2l^2}\frac{G^2}{1 + NG^2}(\vec{r}_N - \vec{r}_0)^2\right]$$

where

$$\vec{r}_N - \vec{r}_0 = \sum_1^N \vec{u}_j \tag{8.1.19}$$

Let us now calculate the mean values of the scalar products $\vec{u}_j\cdot\vec{u}_l$

$$\langle\!\langle\vec{u}_j\cdot\vec{u}_l\rangle\!\rangle = l^2(d\delta_{jl} + l^2\langle\!\langle\mathscr{F}^2\rangle\!\rangle) = dl^2[\delta_{jl} + G^2]. \tag{8.1.20}$$

As a consequence, we find

$$\left\langle\!\!\left\langle\sum_1^N u_l^2\right\rangle\!\!\right\rangle = Ndl^2(1 + G^2)$$

$$\langle\!\langle(\vec{r}_N - \vec{r}_0)^2\rangle\!\rangle = Ndl^2(1 + NG^2). \tag{8.1.21}$$

The chain swelling $\mathfrak{X}$ is defined by*

$$\langle\!\langle(\vec{r}_N - \vec{r}_0)^2\rangle\!\rangle = \mathfrak{X}Ndl^2 \tag{8.1.22}$$

which gives

$$\mathfrak{X} = 1 + NG^2. \tag{8.1.23}$$

Let us now calculate the mean value of the chain potential ${}^{\mathrm{c}}\mathscr{U}$ and the entropy $\mathbb{S}$ associated with $P_{\mathrm{T}}(\vec{u}_1, \ldots, \vec{u}_N)$ by using equalities (8.1.2)

$${}^{\mathrm{c}}\mathbb{E}(\mathfrak{X}) = \langle\!\langle{}^{\mathrm{c}}\mathscr{U}\rangle\!\rangle = \frac{1}{2l^2}\left\langle\!\!\left\langle\sum_j u_j^2\right\rangle\!\!\right\rangle = \frac{Nd}{2}(1 + G^2)$$

$$\begin{aligned} \mathbb{S}(\mathfrak{X}) &= \langle\!\langle -\ln P_{\mathrm{T}}(\vec{u}_1, \ldots, \vec{u}_N)\rangle\!\rangle = \frac{Nd}{2}\ln(2\pi l^2) + \frac{d}{2}\ln(1 + NG^2) \\ &\quad + \frac{1}{2l^2}\left\langle\!\!\left\langle\sum_j u_j^2\right\rangle\!\!\right\rangle - \frac{1}{2l^2}\frac{G^2}{1 + NG^2}\langle\!\langle(\vec{r}_N - \vec{r}_0)^2\rangle\!\rangle \\ &= \frac{Nd}{2}\ln(2\pi l^2) + \frac{d}{2}\ln(1 + NG^2) + \frac{Nd}{2} \end{aligned} \tag{8.1.24}$$

* In this book, the quantity denoted by $\mathfrak{X}$ (or $\mathfrak{X}_0$) defines the swelling; the quantity denoted α by Flory is equal to $\mathfrak{X}^{1/2}$.

By taking (8.1.21) into account, we find*

$$
\begin{aligned}
{}^{c}\mathbb{E}(\mathfrak{X}) - {}^{c}\mathbb{E}(1) &= \frac{Nd}{2} G^2 = \frac{d}{2}(\mathfrak{X} - 1) \\
\mathbb{S}(\mathfrak{X}) - \mathbb{S}(1) &= \frac{d}{2} \ln \mathfrak{X}.
\end{aligned}
\qquad (8.1.25)
$$

We now evaluate the mean interaction energy ${}^{\mathrm{I}}\mathbb{E}(\mathfrak{X})$ resulting from excluded volume constraints: we have

$$
{}^{\mathrm{I}}\mathcal{U}(\vec{u}_1, \ldots, \vec{u}_N) = \frac{1}{2} \sum_{i,j} V(\vec{r}_i - \vec{r}_j).
$$

It is convenient to introduce the excluded volume parameter

$$
v_{\mathrm{e}} = \int \mathrm{d}^d r [1 - \mathrm{e}^{-V(\vec{r})}]. \qquad (8.1.26)
$$

Let $C(\vec{r})$ be the mean number of links per unit volume at a given point. We set

$$
{}^{\mathrm{I}}\mathbb{E}(\mathfrak{X}) = \langle\!\langle {}^{\mathrm{I}}\mathcal{U} \rangle\!\rangle \simeq \frac{1}{2} \int \mathrm{d}^d r\, v_{\mathrm{e}}\, C^2(\vec{r}). \qquad (8.1.27)
$$

We can consider that $C(\vec{r})$ represents the link density around the centre of gravity of the chain (of position vector $\vec{r}_{\mathrm{G}}$)

$$
N^{-1} \int \mathrm{d}^d r\, C(\vec{r}) = 1
$$

$$
N^{-1} \int \mathrm{d}^d r\, \vec{r}\, C(\vec{r}) = \vec{r}_{\mathrm{G}}
$$

$$
N^{-1} \int \mathrm{d}^d r (\vec{r} - \vec{r}_{\mathrm{G}})^2 C(\vec{r}) = \mathfrak{X}\, {}^{0}R_{\mathrm{G}}^2 = \mathfrak{X} \frac{{}^{0}R^2}{6} = d\mathfrak{X} \frac{Nl^2}{6}. \qquad (8.1.28)
$$

We shall represent $C(\vec{r})$ by a Gaussian function, and this is a reasonable assumption

$$
C(\vec{r}) = N \left(\frac{3}{N\pi\mathfrak{X}l^2} \right)^{d/2} \exp[-3(\vec{r} - \vec{r}_{\mathrm{G}})^2 / N\mathfrak{X}l^2]. \qquad (8.1.29)
$$

Let us bring back this expression in (8.1.27). We obtain

$$
\begin{aligned}
{}^{\mathrm{I}}\mathbb{E}(\mathfrak{X}) &= \frac{v_e N^2}{2} \left(\frac{3}{N\pi\mathfrak{X}l^2} \right)^d \int \mathrm{d}^d r \exp[-6(\vec{r} - \vec{r}_{\mathrm{G}})^2 / N\mathfrak{X}l^2] \\
&= \frac{1}{2} 3^{d/2} (2\pi)^{-d/2} v_{\mathrm{e}} N^{2-d/2} \mathfrak{X}^{-d/2} l^{-d}.
\end{aligned}
\qquad (8.1.30)
$$

* We note that for $\mathfrak{X} > 1$, $\mathbb{S}(\mathfrak{X}) - \mathbb{S}(1)$ is positive and not negative as in the case where the length of the links remains constant; what is important here is that ${}^{c}\mathbb{E}(\mathfrak{X}) - \mathbb{S}(\mathfrak{X}) - {}^{c}\mathbb{E}(1) + \mathbb{S}(1)$ is positive when $\mathfrak{X} \neq 1$.

Thus, the interaction energy is given by

$$^{\mathrm{I}}\mathbb{E}(\mathfrak{X}) = \tfrac{1}{2}(3)^{d/2} z\, \mathfrak{X}^{-d/2} \tag{8.1.31}$$

where z is a dimensionless parameter which plays an essential role in polymer theory

$$z = (2\pi)^{-d/2} v_{\mathrm{e}} l^{-d} N^{2-d/2}. \tag{8.1.32}$$

(Note that a simpler and more fundamental definition of Z is given by eqn. (10.1.71).)

Finally, we obtain an expression of the free energy $\mathbb{F}(\mathfrak{X})$ by grouping the results (8.1.25) and (8.1.3)

$$\begin{aligned}\mathbb{F}(\mathfrak{X}) - \mathbb{F}_0 &= {}^{\mathrm{c}}\mathbb{E}(\mathfrak{X}) - {}^{\mathrm{c}}\mathbb{E}(1) - [\mathbb{S}(\mathfrak{X}) - \mathbb{S}(1)] + {}^{\mathrm{I}}\mathbb{E}(\mathfrak{X}) \\ &= \frac{d}{2}(\mathfrak{X} - 1 - \ln \mathfrak{X}) + \frac{1}{2}\, 3^{d/2} z \mathfrak{X}^{-d/2}\end{aligned} \tag{8.1.33}$$

Minimizing the free energy leads to the equation

$$\mathfrak{X}^{1+d/2} - \mathfrak{X}^{d/2} = \tfrac{1}{2}\, 3^{d/2} z. \tag{8.1.34}$$

For $d = 3$, this equation coincides with Flory's equation. When N goes to infinity, the first term on the left becomes dominant with respect to the second one, and in this limit

$$\mathfrak{X} = [\tfrac{1}{2}\, 3^{d/2} z]^{\frac{2}{d+2}} = [\tfrac{1}{2}\, 3^{d/2} (2\pi)^{-d/2} v_{\mathrm{e}} l^{-d}]^{\frac{2}{d+2}} N^{\frac{4-d}{d+2}}. \tag{8.1.35}$$

When N is very large, we thus obtain for the mean square end-to-end distance

$$\langle\!\langle (\vec{r}_N - \vec{r}_0)^2 \rangle\!\rangle = \mathfrak{X} N l^2 d = [\tfrac{1}{2}\, 3^{d/2} (2\pi)^{-d/2} v_{\mathrm{e}} l^{-d}]^{\frac{2}{d+2}} N^{\frac{6}{d+2}} l^2 d, \tag{8.1.36}$$

a relation which gives Flory's value to the exponent ν

$$\nu = \nu_{\mathrm{F}} \equiv \frac{3}{d+2}. \tag{8.1.37}$$

The excluded volume parameter depends on temperature and, in the vicinity of the Flory point, can be approximated by a linear function of $1/T$

$$v_{\mathrm{e}} = v_{\mathrm{e}\square}(T_{\mathrm{F}}/T - 1). \tag{8.1.38}$$

Thus, Flory's theory is able to predict how the size of an isolated polymer varies with the number of chain links and with temperature.

The value of ν thus obtained seems rather realistic. This explains why this theory was a great success in spite of its approximate character.

However, this theory has very serious defects.[4,5] In fact, let us consider the quantities $^{\mathrm{I}}\mathbb{E}(\mathfrak{X})$ and let us set

$$\mathfrak{X} = AN^{2\nu - 1} \qquad B = \tfrac{1}{2}\, 3^{d/2} (2\pi)^{-d/2}$$

In agreement with the definitions of $\mathfrak{X}$ *and* ν. Equations (8.1.25) and (8.1.31) give

$$
\begin{aligned}
{}^{c}\mathbb{E}(\mathfrak{X}) - {}^{c}\mathbb{E}(1) &= \tfrac{1}{2}(\mathfrak{X} - 1) = \tfrac{1}{2}[AN^{2\nu-1} - 1] \\
{}^{I}\mathbb{E}(\mathfrak{X}) &= B(v_e l^{-d})\mathfrak{X}^{-d/2}N^{2-d/2} = Bv_e l^{-d}A^{-d/2}N^{2-\nu d} \\
\mathbb{S}(\mathfrak{X}) - \mathbb{S}(1) &= \frac{d}{2}\ln \mathfrak{X} = \frac{d}{2}\ln A + \frac{d}{2}(2\nu - 1)\ln N.
\end{aligned}
\tag{8.1.39}
$$

We observe that all these terms are very small with respect to N, for any value of ν belonging to the interval $\frac{1}{2} < \nu < 1$ and, in particular, for $d = 3$ and $\nu = 3/5$. This is quite abnormal. As the excluded volume introduces interactions between monomers, it is clear that the difference $\mathbb{F}(\mathfrak{X}) - \mathbb{F}_0$ has to be proportional to N and in general, the same remark should apply to ${}^{c}\mathbb{E}(\mathfrak{X}) - {}^{c}\mathbb{E}(1)$, ${}^{I}\mathbb{E}(\mathfrak{X})$ and $\mathbb{S}(\mathfrak{X}) - \mathbb{S}(1)$. However, we note that the calculations of ${}^{c}\mathbb{E}(\mathfrak{X})$ and of $\mathbb{S}(\mathfrak{X})$ are exact. Consequently, the basic error comes from the calculation of ${}^{I}\mathbb{E}$. First, it is certain that the probability distribution $P_T(\vec{u}_1, \ldots, \vec{u}_N)$ does not represent in a realistic manner a chain swollen in all directions but only a stretched chain; still this defect is not the worst one.

Actually, eqn (8.1.27) is fundamentally wrong [as well as (8.1.16)] and the error is the following. When calculating ${}^{I}\mathbb{E}(\mathfrak{X})$, we ignored the crucial fact that the monomers constituting the chain are linked to each other. In fact, let us consider on the chain two points, separated by n links, and the probability distribution associated with the vector $\vec{r}$ joining these points. In first approximation,* we can write the scaling law

$$P_n(\vec{r}) = \frac{1}{n^{\nu d}} f(\vec{r}/n^{\nu}). \tag{8.1.40}$$

The probability of contact of these points is proportional to $P(\vec{0})$, and if $f(\vec{0})$ does not vanish, an assumption which will be naïvely made here,† we may write

$$P_n(\vec{0}) = \frac{f(\vec{0})}{n^{\nu d}}. \tag{8.1.41}$$

In this case, the mean interaction energy ${}^{I}\mathbb{E}$ is given by

$$
\begin{aligned}
{}^{I}\mathbb{E} &= \frac{1}{2}\sum_{ij} v_e P_{|i-j|}(\vec{0}) \\
&= v_e f(\vec{0}) \sum_{n=1}^{N} \frac{N - n}{n^{\nu d}}.
\end{aligned}
\tag{8.1.42}
$$

Now, for $d = 3$, $\nu = 3/5$, and for values of ν close to 3/5 we have $1 < \nu d < 2$.

* But only to a first approximation: see Chapter 4 and the discussion at the end of Section 2.2.

† Actually, $f(\vec{0}) = 0$, as various studies have shown (see Chapter 4, Sections 2.1.4 and 2.2.2) but this fact does not destroy the validity of the argument.

Thus, the sum $\sum_{n=1}^{N} n^{-\nu d}$ converges when $N \to \infty$ but the sum $\sum_{n=1}^{N} n^{1-\nu d}$ diverges. Consequently, the dominant terms in ${}^{\mathrm{I}}\mathbb{E}$ can be evaluated as follows

$$\begin{aligned} {}^{\mathrm{I}}\mathbb{E} &= v_{\mathrm{e}} f(\vec{0}) \left[N \sum_{1}^{\infty} \frac{1}{n^{\nu d}} - N \int_{N}^{\infty} \frac{\mathrm{d}n}{n^{\nu d}} - \int_{0}^{N} \mathrm{d}n \frac{n}{n^{\nu d}} \right] \\ &= v_{\mathrm{e}} f(\vec{0}) \left[N \sum_{1}^{\infty} \frac{1}{n^{\nu d}} - \frac{1}{(2-\nu d)(1-\nu d)} N^{2-\nu d} \right]. \end{aligned} \qquad (8.1.43)$$

As we expected, in this expression there is a term proportional to N, which is the dominant term. The term proportional to $N^{1-\nu d}$ appears here only as a correction term. The dependence of the latter term with respect to N is the same as the dependence of ${}^{\mathrm{I}}\mathbb{E}(\mathfrak{X})$ deduced from (8.1.31) with the help of (8.1.36) but here the term appears as a *negative* correction.

Flory's exponent $\nu = \nu_{\mathrm{F}}$ is obtained by writing that the terms ${}^{\mathrm{c}}\mathbb{E}(\mathfrak{X}) - {}^{\mathrm{c}}\mathbb{E}(1)$ and ${}^{\mathrm{I}}\mathbb{E}(\mathfrak{X})$ given by (8.1.39) have the same power in N. We now see that such an approach is totally unrealistic since the dominant terms of ${}^{\mathrm{I}}\mathbb{E}$ are actually proportional to N.

In brief, a realistic calculation of $\mathbb{F}$ should lead to the appearance of terms proportional to N in the expressions giving the energy and the entropy: moreover, it is by minimizing the sum of these terms that the swelling should be determined.*

Indeed, we shall see that this is exactly the method which is used when one tries to calculate the exponent ν in the framework of the Gaussian approximation. Actually, it turns out that ν is determined by contributions proportional to N, resulting from interactions of monomers that are far apart on the chain.

Thus, Flory's approach appears to be fundamentally incorrect; however, it is useful as an approximation; the agreement with experiments does not seem accidental but the exact nature of the approximation remains mysterious, in spite of many attempts made to explain it.

1.3 The self-consistent field method

The idea behind the replacement of the excluded volume interactions by a self-consistent potential was put forward by Edwards[6] and subsequently developed by various other physicists.[5, 7] Concerning the critical exponent ν, this method gives the same result as Flory's method ($\nu = 3/(d + 2)$).

It can be assumed that the chain potential is the sum of two terms, a chain term and an interaction term

$$\mathscr{U} = {}^{\mathrm{c}}\mathscr{U} + {}^{\mathrm{I}}\mathscr{U} \qquad (8.1.44)$$

* Nevertheless, none of the modern theories of renormalization calculates the exponent ν in this way (see Chapter 12)!

where we may choose (setting $\vec{u}_j = \vec{r}_j - \vec{r}_{j-1}$)

$$
\begin{aligned}
{}^{c}\mathscr{U}(\vec{u}_1, \ldots, \vec{u}_N) &= \sum_j V_c(\vec{u}_j) \\
{}^{I}\mathscr{U}(\vec{u}_1, \ldots, \vec{u}_N) &= \frac{1}{2}\sum_{ij} V(\vec{r}_i - \vec{r}_j).
\end{aligned} \tag{8.1.45}
$$

The self-consistent field method consists in replacing the true probability law

$$P(\vec{u}_1, \ldots, \vec{u}_N) = \frac{1}{Z}\exp[-\mathscr{U}(\vec{u}_1, \ldots, \vec{u}_N)] \tag{8.1.46}$$

by a trial probability

$$P_T = \frac{1}{Z_T}\exp\left[-{}^{c}\mathscr{U}(\vec{u}_1, \ldots, \vec{u}_N) - \sum_{j=1}^{N} \bar{V}(\vec{r}_j)\right] \tag{8.1.47}$$

where $\bar{V}(\vec{r})$ is a self-consistent potential which can be determined by minimization of the free energy.

In his initial article, Edwards studied the case of a semi-infinite chain starting from a point O (of position vector $\vec{r}_O = 0$) and, consequently he admitted that the potential is spherically symmetric around O. He described the method in a manner which differs somewhat from the approach presented here. Actually, we are indebted to Reiss[8] for a clear formulation of the method in terms of trial probability. In reality, the calculation of the exponent ν made by Reiss does not lead to the value $\nu = 3/5$) found by Edwards and Flory but to a larger value ($\nu = 2/3$); however, as was shown by Yamakawa,[9] the lack of agreement arises from the fact that certain terms were unduly omitted in Reiss's calculation and, finally, the correct calculation made by Yamakawa gives for ν the result found by Edwards.

The self-consistent field method can be summarized in the following simple manner which shows clearly the principle and the deficiencies of this approach.

We consider a chain with N independent links in a potential $\bar{V}(\vec{r})$ which acts on its points. We assume that one end of the chain coincides with the origin of coordinates and therefore we have

$$\vec{r}_j = \sum_{i=1}^{j} \vec{u}_i. \tag{8.1.48}$$

Let us examine how the chain progresses at a space point. We assume that all the vectors $\vec{u}_l$ with $l \neq j$ are fixed and we consider the conditional probability $p(\vec{u}_j)$ associated with $\vec{u}_j$. Let ${}^{0}p(\vec{u})$ be the a priori probability corresponding to each chain link (${}^{0}p(\vec{u}) \propto \exp(-V_c(\vec{u}))$; see (8.1.45). For $i \geqslant j$, we may set

$$\vec{r}_i = \vec{r}_i' + \vec{u}_j \tag{8.1.49}$$

where $\vec{r}_i'$ is a constant vector, independent of $\vec{u}_j$.

The conditional probability $p(\vec{u}_j)$ can thus be written in the form

$$p(\vec{u}_j) \propto {}^0p(\vec{u}_j)\exp\left[-\sum_{i=j}^{N} \bar{V}(\vec{r}_i' + \vec{u}_j)\right]. \tag{8.1.50}$$

When the number N of links is large, the size of the polymer is also large and $\bar{V}(\vec{r})$ can be considered as a slowly varying function. In this case, we may write

$$p(\vec{u}_j) \propto {}^0p(\vec{u}_j)\exp\left[-\sum_{i=j}^{N} \vec{u}_j \cdot \vec{\nabla}\,\bar{V}(\vec{r}_i')\right]$$

or more simply

$$p(\vec{u}_j) \simeq {}^0p(\vec{u}_j)\left[1-\sum_{i=j}^{N} \vec{u}_j \cdot \vec{\nabla}\bar{V}(\vec{r}_i')\right]. \tag{8.1.51}$$

In the absence of interaction, we have

$$\langle\!\langle \vec{u}_j \rangle\!\rangle_0 = 0$$

$$\langle\!\langle u_j^2 \rangle\!\rangle = \int \mathrm{d}^d u\, u^2\, {}^0p(\vec{u}) = dl^2, \tag{8.1.52}$$

and in the presence of the potential

$$\langle\!\langle \vec{u}_j \rangle\!\rangle = -\,l^2 \sum_{l=j}^{N} \vec{\nabla}\,\bar{V}(\vec{r}_l') \simeq -\,l^2 \sum_{l=j}^{N} \vec{\nabla}\,\bar{V}(\vec{r}_l).$$

$$\langle\!\langle u_j^2 \rangle\!\rangle \simeq dl^2 \tag{8.1.53}$$

Thus, in the field, the link vectors $\vec{u}_j$ can be expressed in the form

$$\vec{u}_j = -\,l^2 \sum_{i=j}^{N} \vec{\nabla}\,\bar{V}(\vec{r}_i) + \vec{\varepsilon}_j \tag{8.1.54}$$

where the $\vec{\varepsilon}_j$ are independent random vectors

$$\langle\!\langle \vec{\varepsilon}_j \cdot \vec{\varepsilon}_l \rangle\!\rangle = l^2 d\, \delta_{jl}. \tag{8.1.55}$$

The above equation can be transformed into the following equality

$$\vec{u}_{j+1} - \vec{u}_j = [\vec{r}_{j+1} + \vec{r}_{j-1} - 2\vec{r}_j] = l^2\, \vec{\nabla}\,\bar{V}(\vec{r}_j) + \vec{\varepsilon}_{j+1} - \vec{\varepsilon}_j. \tag{8.1.56}$$

Now, let us take the continuous limit, which is valid for a long chain. We set $\vec{r}(n) = \vec{r}_n$, $\vec{\varepsilon}(n) = \varepsilon_n$ considering n as a continuous variable. The continuous form of eqns (8.1.56) and (8.1.55) is

$$\frac{\mathrm{d}^2\vec{r}(n)}{\mathrm{d}n^2} = l^2\, \vec{\nabla}(\vec{r}) + \frac{\partial}{\partial n}\vec{\varepsilon}(n) \tag{8.1.57}$$

$$\langle\!\langle \vec{\varepsilon}(n) \cdot \vec{\varepsilon}(n') \rangle\!\rangle = l^2 d\, \delta(n-n'). \tag{8.1.58}$$

In particular, the last equation is easily obtained by observing that in the absence of interaction, we must have

$$\langle\langle [\vec{r}(N) - \vec{r}(0)]^2 \rangle\rangle = \left\langle\left\langle \left[\int_0^N \mathrm{d}n \, \vec{\varepsilon}(n) \right]^2 \right\rangle\right\rangle = Nl^2 d \tag{8.1.59}$$

In eqn (8.1.57) the first term in the right-hand side defines a trajectory and the second one a fluctuation around this trajectory.

In order to simplify the problem, we shall consider the case studied by Edwards, where the chain is semi-infinite and starts from O; moreover, it will be assumed that the potential $\bar{V}(\vec{r})$ is of the form

$$\bar{V}(\vec{r}) = \frac{\bar{V}}{r^{\iota}} \tag{8.1.60}$$

where ι is an exponent which we must determine, to a first approximation, the fluctuation terms which appear in (8.1.57) can be neglected and, in this way, we get the equation (with $r = |\vec{r}|$)

$$\frac{\mathrm{d}^2 r}{\mathrm{d}n^2} = -\frac{\iota \bar{V} l^2}{r^{\iota+1}} \tag{8.1.61}$$

which can be solved as follows

$$\left(\frac{\mathrm{d}r}{\mathrm{d}n}\right)^2 = \frac{2\bar{V}}{r^{\iota}} l^2$$

$$[r(n)]^{(\iota/2)+1} = (2\bar{V})^{1/2}(1 + \iota/2) n l. \tag{8.1.62}$$

Thus, the size exponent ν characterizing the stretching of the chain is related to ι by

$$\nu = \frac{2}{\iota + 2}. \tag{8.1.63}$$

We can now determine $\bar{V}(\vec{r})$ in a self-consistent way. Let v_{e} be the excluded volume and $C(\vec{r})$ the mean number of links per unit volume. The (mean field) approximation consists in setting

$$\bar{V}(\vec{r}) = v_{\mathrm{e}} C(\vec{r}) \tag{8.1.64}$$

On the other hand, since $\vec{r}(n)$ progresses steadily with n, neglecting fluctuations, we can write [see (8.1.62)]

$$C(\vec{r}) = \frac{1}{\Omega(d)\, r^{d-1}} \left(\frac{\mathrm{d}r}{\mathrm{d}n}\right)^{-1} = \frac{1}{\Omega(d)\, l\, r^{d-1}} \left(\frac{2\bar{V}}{r^{\iota}}\right)^{-1/2} \tag{8.1.65}$$

where $\Omega(d)$ is the surface of a ball in a d-dimensional space (see Appendix F). From the above equations, we deduce

$$\bar{V}(\vec{r}) = \frac{v_{\mathrm{e}}}{\Omega(d) l} (2\bar{V})^{-1/2} \frac{1}{r^{d-1-\iota/2}}, \tag{8.1.66}$$

and to determine ι and $V(r)$, we have only to identify this expression with eqn (8.1.60). In particular, this identification gives

$$\iota = \tfrac{2}{3}(d-1),$$

and, by bringing back this value of ι in (8.1.63), we find

$$\nu = \frac{3}{d+2} \tag{8.1.67}$$

which is (for $d = 3$) Edwards's result. However, the theory is not realistic when the number of links is large. Indeed, in this case, the chain follows the lines of force of the potential. The distance covered by the chain along a line of force increases as N^{ν} whereas the side-fluctuations resulting from the random terms which appear in (8.1.54) and (8.1.57) are of the order of $N^{1/2}$. Thus, asymptotically, the chain takes the form of a needle (see Fig. 8.1) and in the present case, since the potential has a spherical symmetry, the chain is practically rectilinear.

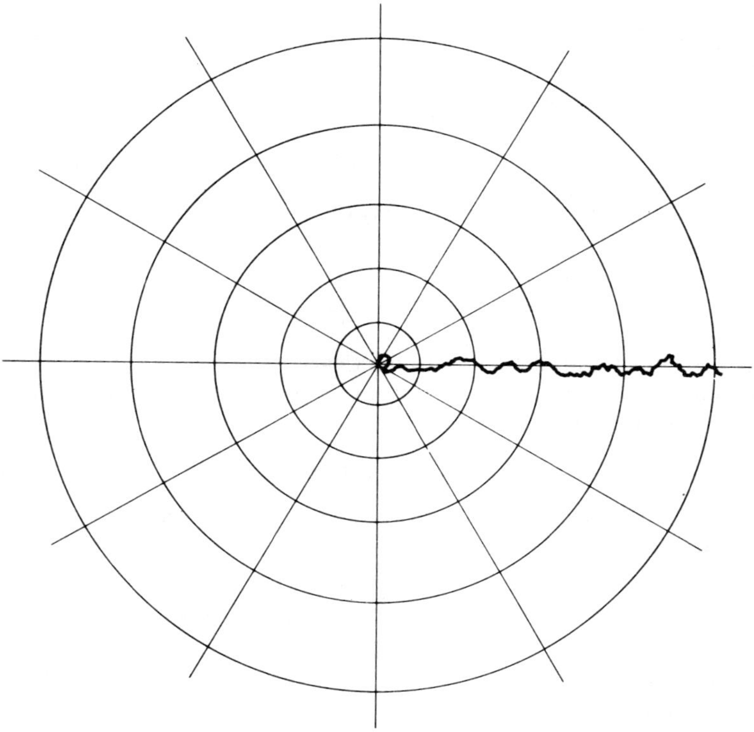

Fig. 8.1. Chain progressing in a potential along the lines of force. Here, the potential has a spherical symmetry and the lines of force are straight lines.

Consequently, the approximation which has been made leads to the following unacceptable situation:

1. The interaction energy is underestimated, because the mean density is obtained by assuming that the chain has an isotropic distribution around the origin, an assumption which in principle looks reasonable but which here is very seriously at variance with the result.
2. On the other hand, the interaction energy is also overestimated because internal correlations are neglected, which means that eqn (8.1.64) is not right. However, in practice both errors compensate partially, and the result is a value of ν which is close to the exact value.
3. There is an unacceptable dissymmetry between the two chain ends. Indeed, for a rectilinear chain, if one end of the chain chosen as the origin of the chain coincides with the origin of space, we should have $r(n/2) = \frac{1}{2}r(n)$ an equation which for realistic values of ι and ν contradicts (8.1.67)

Of course, this dissymmetry can be remedied by replacing the semi-infinite chain by a finite chain and by introducing a self-consistent field without spherical symmetry; however, the remedy would be only partial. In fact, in all cases, when N is large, in any self-consistent potential $\vec{V}(\vec{r})$, the chain has a tendency to follow the lines of force. Thus, in the asymptotic limit, the chain always has the shape of a needle, which may be curved. This means that the action of an external potential cannot produce a swelling of the chain in all directions. For this, it is necessary to introduce true correlations between chain links.

In brief, the self-consistent field method appears here as completely unrealistic.

1.4 The Gaussian approximation

1.4.1 Presentation of the method

The Gaussian approximation was developed around 1968 by S.F. Edwards (unpublished) and des Cloizeaux.[5] Contrary to the preceding approaches, this method takes explicitly into account correlations between chain links. Thus, it expresses mathematically the very ideas of Flory and in a manner which initially seemed quite realistic.

From a conceptual point of view, this theory seems much better than those that had previously been propounded. However, the method gives a value of the size exponent ν which is not Flory's value $\nu_F = 3/(d+2)$ but Stockmayer's value $\nu = 2/d$, which is too large to be realistic.

This unexpected result seemed very surprising when the method was developed but we now believe that we understand its non-trivial origin. This strange behaviour seems related to the fact that, like all theories prior to 1972, the Gaussian approximation neglects subtle correlations, the importance of which was recognized only after 1971, and then only gradually. We shall come back to this question later on.

The Gaussian approximation will be described in the simple and usual case where the potential is the sum of two terms

$$\mathscr{U} = {}^{\mathrm{c}}\mathscr{U} + {}^{\mathrm{I}}\mathscr{U}$$

$$ {}^{\mathrm{c}}\mathscr{U} = \frac{1}{2l^2} \sum_{j=1}^{N} u_j^2 \tag{8.1.68}$$

(where $\vec{u}_j = \vec{r}_j - \vec{r}_{j-1}$).

$$ {}^{\mathrm{I}}\mathscr{U} = \sum_{0 \leqslant i < j \leqslant N} V(\vec{r}_i - \vec{r}_j)$$

This Gaussian approximation consists in choosing a trial probability

$$P_{\mathrm{T}}(\vec{u}_1, \ldots, \vec{u}_N) = \frac{1}{Z} \exp[-\mathscr{Q}\{\vec{u}_j\}] \tag{8.1.69}$$

where $\mathscr{Q}\{\vec{u}_j\}$ is an arbitrary quadratic function of the vectors $\vec{u}_j$. The quadratic form $\mathscr{Q}\{\vec{u}_j\}$ is determined by minimization of the corresponding free-energy and the properties of the chain can be deduced from it.

Incidentally, we observe that the trial probability which we used previously to derive Flory's result is a restricted form of the general Gaussian probability which is considered here.

In order to study the Gaussian approximation, we have to diagonalize the quadratic form $\mathscr{Q}\{u_j\}$, and in general this is not an easy task. However, this can be done immediately if the chain is a closed ring, i.e. if $\vec{r}_0$ coincides with $\vec{r}_N$. Mathematically, the operation consists in replacing the probability distribution $P(\vec{u}_1, \ldots, \vec{u}_N)$ with the distribution $P(\vec{u}_1, \ldots, \vec{u}_N)\delta(\vec{u}_1 + \ldots + \vec{u}_N)$. Then, the position of a point on the chain is labelled by an integer j defined modulo N. Incidentally, we note that the ring thus defined is not a true ring but a 'phantom' ring which is not submitted to any topological constraint and which may form knots.

The dimensions of the ring will be characterized by the mean values

$$\langle\!\langle (\vec{r}_{j+n} - \vec{r}_j)^2 \rangle\!\rangle = dl^2 A(n). \tag{8.1.70}$$

In the framework of the Gaussian approximation, the quantities $A(n)$ determine the properties of the chain completely. Indeed, all mean values of products of components of the $\vec{u}_j$ can be expressed in terms of the $A(n)$ by means of Wick's theorem (see Appendix A). In particular, we have

$$\langle\!\langle \vec{u}_{j+n} \cdot \vec{u}_j \rangle\!\rangle = dl^2 B(n) \tag{8.1.71}$$

and we observe that the $A(n)$ and $B(n)$ obey the relations

$$A(n) = 2 \sum_{p=1}^{n} (n-p) B(p) + nB(0) \tag{8.1.72}$$

$$B(n) = \tfrac{1}{2}[A(n+1) + A(n-1) - 2A(n)]. \quad (8.1.73)$$

The quantities $A(n)$ and $B(n)$ can also be considered as parameters defining the quadratic form $\mathcal{Q}\{\vec{u}_j\}$

To diagonalize $\mathcal{Q}\{\vec{u}_j\}$, we can introduce new variables $\vec{\rho}_q$ (with $q = 0, 1, \ldots, N-1$)

$$\vec{\rho}_q = N^{-1/2} \sum_{j=1}^{N} \exp[i2\pi jq/N]\vec{u}_j$$

$$\vec{u}_j = N^{-1/2} \sum_{q=0}^{N-1} \exp[-i2\pi jq/N]\vec{\rho}_q \quad (8.1.74)$$

and we observe that the preceding transformation is orthogonal. Thus,

$$\mathrm{d}^d\rho_0 \ldots \mathrm{d}^d\rho_{N-1} = \mathrm{d}^d u_1 \ldots \mathrm{d}^d u_N. \quad (8.1.75)$$

In particular, we have

$$\vec{\rho}_0 = N^{-1/2}(\vec{u}_1 + \ldots + \vec{u}_N).$$

Expressing that the chain is a ring amounts to setting $\vec{\rho}_0 = 0$ and therefore the variable $\vec{\rho}_0$ is eliminated from the calculation. Each variable $\vec{\rho}_q$ corresponds to an irreducible representation of the cyclic transformations of the chain ($j \to j+1$, for instance). Thus, for reasons of cyclic invariance, the new variables $\vec{\rho}_q$ diagonalize the quadratic form $\mathcal{Q} \equiv \mathcal{Q}\{u_j\}$ and $\mathcal{Q}$ can be written in the form

$$\mathcal{Q} = \frac{1}{2l^2} \sum_{q=1}^{N-1} \mathfrak{H}_N(2\pi q/N)\vec{\rho}_q \cdot \vec{\rho}_{-q}. \quad (8.1.76)$$

and we introduced here the variable

$$k = 2\pi q/N \quad (8.1.77)$$

in order to go more easily to the continuous limit. Starting from the trial probability $P_{\mathrm{T}} = Z^{-1}\exp(-\mathcal{Q})$, we directly obtain the mean value

$$\langle\!\langle \vec{\rho}_q \cdot \vec{\rho}_{-q} \rangle\!\rangle = dl^2\, \mathfrak{H}_N^{-1}(2\pi q/N) \quad (8.1.78)$$

and from it, with the help of (8.1.74) (taking symmetries into account), we deduce

$$B(n) = \frac{1}{N} \sum_{q=1}^{N-1} \cos(2\pi qn/N)\mathfrak{H}_N^{-1}(2\pi q/N). \quad (8.1.79)$$

It will be especially convenient to study the limit where the number N of links becomes infinite.

Moreover, in this limit, the fact that the chain is closed becomes irrelevant and it is easy to see that the infinite chain and the infinite 'phantom' ring must be characterized by the same values of $A(n)$ and $B(n)$. In this case, we have

$$\lim_{N\to\infty} \mathfrak{H}_N(2\pi q/N) = \mathfrak{H}(k) \quad (8.1.80)$$

and

$$B(n) = \frac{1}{2\pi}\int_{-\pi}^{+\pi} \mathrm{d}k \cos nk [\mathfrak{H}(k)]^{-1}, \tag{8.1.81}$$

hence we deduce with the help of (8.1.72) and (8.1.73)

$$A(n) = \frac{1}{2\pi}\int_{-\pi}^{+\pi} \mathrm{d}k \frac{1-\cos nk}{1-\cos k}[\mathfrak{H}(k)]^{-1}. \tag{8.1.82}$$

When n is large, $A(n)$ is expected to have the following behaviour

$$A(n) \simeq An^{2\nu}, \tag{8.1.83}$$

which leads to

$$B(n) \simeq \frac{A\nu(2\nu-1)}{n^{2-2\nu}}. \tag{8.1.84}$$

Conversely, if $\nu > 1/2$, (8.1.84) entails (8.1.83) (the condition $\nu > 1/2$ implies that the sum $\sum_{p=1}^{N} B(p)$ which appears in (8.1.72) diverges when $N \to \infty$).

It is easy to see that $A(n)$ has the expected behaviour if, for small values of k,

$$\mathfrak{H}(k) \simeq \mathfrak{H}|k|^{2\nu-1}. \tag{8.1.85}$$

Thus, for large values of n ($n \gg 1$)

$$\begin{aligned} A(n) &\simeq \frac{2}{\pi}\mathfrak{H}^{-1}\int_0^{\infty} \mathrm{d}k \frac{1-\cos nk}{k^{1+2\nu}} \\ &= \frac{2}{\pi}\mathfrak{H}^{-1} n^{2\nu}\int_0^{\infty} \mathrm{d}x \frac{1-\cos x}{x^{1+2\nu}} = \frac{\mathfrak{H}^{-1}}{(2\nu)!\sin \pi\nu} n^{2\nu}, \end{aligned} \tag{8.1.86}$$

from which we deduce

$$\mathfrak{H}^{-1} = A(2\nu)!\sin \pi(1-\nu) \tag{8.1.87}$$

by identification with (8.1.83).

1.4.2 Calculation and minimization of the free energy: critical exponent

We shall now calculate the free energy associated with the trial probability.

Let us first evaluate the mean energy of the system [see (8.1.68) and (8.1.70)]. The chain energy is immediately obtained

$${}^{c}E/N = \langle\langle {}^{c}\mathcal{U} \rangle\rangle/N = \frac{d}{2}A(1). \tag{8.1.88}$$

To evaluate the interaction energy, the distribution law for end-to-end distances must be used

$$P_n(\vec{r}) = \frac{1}{(2\pi l^2 A(n))^{d/2}} \exp[-r^2/2l^2 A(\mathring{n})]$$

$$^{\mathrm{I}}\mathbb{E}/N = \langle\langle ^{\mathrm{I}}\mathcal{U} \rangle\rangle/N = \sum_{n=1}^{N} \int \mathrm{d}^d r\, P_n(\vec{r})\, V(\vec{r}). \tag{8.1.89}$$

When N is large, the contributions corresponding to large values of n play a crucial role. But, for such values of n, the law $P_n(\vec{r})$ is very flat; defining the excluded volume parameter z_1 by the dimensionless expression*

$$z_1 = \frac{1}{(2\pi l^2)^{d/2}} \int \mathrm{d}^d r\, V(\vec{r}); \tag{8.1.90}$$

we may thus write

$$^{\mathrm{I}}\mathbb{E}/N = \sum_{n=1}^{N} \frac{z_1}{[A(n)]^{d/2}} \tag{8.1.91}$$

without making an important error. Therefore, in the limit $N \to \infty$, we have

$$^{\mathrm{I}}\mathbb{E}/N = 2 \sum_{1}^{\infty} \frac{z_1}{[A(n)]^{d/2}}. \tag{8.1.92}$$

Let us now evaluate the entropy

$$\mathbb{S} = -\int \mathrm{d}^d u_1 \ldots \mathrm{d}^d u_{N-1}\, P\{\vec{\rho}_q\} \ln P\{\vec{\rho}_q\} \tag{8.1.93}$$

where

$$P\{\vec{\rho}_q\} = \frac{1}{Z} \exp[-\mathcal{Q}\{\vec{\rho}_q\}]$$

and

$$\mathcal{Q}\{\vec{\rho}_q\} = \frac{1}{2l^2} \sum_{q=1}^{N-1} \mathfrak{H}(2\pi q/N)\, \vec{\rho}_q \cdot \vec{\rho}_{-q}$$

$$Z = (2\pi l^2)^{(N-1)d/2} \prod_{q=1}^{N-1} [\mathfrak{H}(2\pi q/N)]^{-d/2}. \tag{8.1.94}$$

On the other hand, from (8.1.78 we get

$$\langle\langle \mathcal{Q}\{\vec{\rho}_q\} \rangle\rangle = (N-1)d/2. \tag{8.1.95}$$

Thus, for $N \gg 1$

$$\mathbb{S}/N = -\frac{d}{2N} \sum_{q=1}^{N-1} \ln[\mathfrak{H}(2\pi q/N)] + \frac{d}{2}[1 + \ln(2\pi l^2)] \tag{8.1.96}$$

* In the same way, we might set $z_N = N^{2-d/2} z_1$; then $z_N \simeq z$ where z is the classical interaction parameter [see (10.1.7)].

or, by taking the limit $N \to \infty$,

$$\mathbb{S}/N = -\frac{d}{4\pi}\int_0^{2\pi} \mathrm{d}k \ln[\mathfrak{H}(k)] + \frac{d}{2}[1 + \ln(2\pi l^2)].$$

The total free energy is given by the sum

$$\mathbb{F} = {}^{c}\mathbb{E} + {}^{I}\mathbb{E} - \mathbb{S}$$

which can be written in the explicit form

$$\mathbb{F}/N = \frac{d}{4\pi}\int_0^{2\pi} \mathrm{d}k \ln(\mathfrak{H}(k)) + 2\sum_1^{\infty}\frac{z_1}{[A(n)]^{d/2}} + \frac{d}{2}[A(1) + 1 + \ln(2\pi l^2)]. \tag{8.1.97}$$

Let us minimize the free energy by writing the condition $\partial\mathbb{F}/\partial\mathfrak{H}(k) = 0$. Since according to (8.1.82)

$$\frac{\partial A(n)}{\partial[\mathfrak{H}(k)]^{-1}} = \frac{1}{2\pi}\left(\frac{1 - \cos nk}{1 - \cos k}\right) \tag{8.1.98}$$

we have to solve the coupled equations

$$\mathfrak{H}(k) = 1 - 2z_1\sum_1^{\infty}\left(\frac{1 - \cos nk}{1 - \cos k}\right)\frac{1}{[A(n)]^{1+d/2}}, \tag{8.1.99}$$

$$A(n) = \frac{1}{2\pi}\int_{-\pi}^{+\pi} \mathrm{d}k\left(\frac{1 - \cos nk}{1 - \cos k}\right)\frac{1}{\mathfrak{H}(k)}. \tag{8.1.100}$$

The quantity $\mathfrak{H}(k)$ is positive, (but smaller than unity) for any value of k. Thus, in all cases, the sum $\sum_1^{\infty}\frac{n^2}{[A(n)]^{1+d/2}}$ has to be convergent. (to see this, assume that k is small). Now, for large values of n, $A(n) \propto A\,n^{2\nu}$. Thus, the fact that the sum converges is equivalent to the condition

$$\nu > \frac{3}{d+2}$$

which also reads

$$\nu - \frac{1}{2} > \frac{4-d}{2(d+2)}.$$

Thus, for $d > 4$ and $\nu = 1/2$, the above condition is satisfied; in this case, the chain is asymptotically Gaussian, and it is possible to find a solution $\mathfrak{H}(k)$ for which $\mathfrak{H}(0)$ is finite (non-zero).

On the contrary, for $d < 4$ the value of ν must be larger than 1/2 and even larger than Flory's value $\nu_F = 3/(d+2)$, which is strange. In this case, eqn (8.1.85) shows that $\mathfrak{H}(k)$ must vanish when k goes to zero and this fact [(according to (8.1.99)] implies the equality

$$2z_1\sum_1^{\infty}\frac{n^2}{[A(n)]^{1+d/2}} = 1. \tag{8.1.101}$$

Taking this identity into account, we can rewrite (8.1.99) in the form

$$\mathfrak{H}(k) = 2z_1 \sum_1^{\infty} \left[n^2 - \left(\frac{1 - \cos nk}{1 - \cos k} \right) \right] \frac{1}{[A(n)]^{1+d/2}}. \tag{8.1.102}$$

To study the singularity of $\mathfrak{H}(k) \propto k^{2\nu-1}$, it is convenient to differentiate the preceding equation with respect to k. For small values of k, we have

$$k^2 \frac{\partial}{\partial k} \mathfrak{H}(k) \simeq -4z_1 \sum_1^{\infty} \sin nk \frac{n}{[A(n)]^{1+d/2}}. \tag{8.1.103}$$

For small k, this expression is proportional to $k^{2\nu}$. In order to determine the singularity, we differentiate twice

$$\frac{\partial^2}{\partial k^2}\left[k^2 \frac{\partial}{\partial k} \mathfrak{H}(k) \right] \simeq 4z_1 \sum_1^{\infty} \sin nk \frac{n^3}{[A(n)]^{1+d/2}}. \tag{8.1.104}$$

This expression becomes infinite when $k \to 0$ and the sum $\sum_1^{\infty} \frac{n^3}{[A(n)]^{1+d/2}}$ diverges. Thus small values of k correspond to large values of n. To determine ν, we have only to replace $A(n)$ and $\mathfrak{H}(k)$ by their asymptotic expressions, namely (8.1.83) and (8.1.85). Moreover, the sum can be replaced by an integral, and we obtain

$$\begin{aligned} 2\nu(2\nu - 1)^2 \mathfrak{H} k^{2\nu-2} &= 4z_1 A^{-1-d/2} \int_0^{\infty} \mathrm{d}n \frac{\sin nk}{n^{\nu(d+2)-3}} \\ &= 4z_1 A^{-1-d/2} k^{\nu(d+2)-4} \int_0^{\infty} \mathrm{d}x \frac{\sin x}{x^{\nu(d+2)-3}}. \end{aligned} \tag{8.1.105}$$

From this equality, we deduce immediately

$$2\nu - 2 = \nu(d + 2) - 4 \tag{8.1.106}$$

so again we find

$$\nu = 2/d,$$

and obviously this result is valid only in the range $4 > d > 1/2$.

Cases $d = 4$ and $d = 1$ are limiting cases for this model. For $d = 4$, one finds $\nu = 1/2$ and for $d = 4$, $\nu = 1/2$, but in both cases the functions $A(n)$ and $\mathfrak{H}(k)$ cannot be represented by pure power laws in the asymptotic domain and logarithmic corrections occur quite naturally.[3]

The above equation also gives a relation between $\mathfrak{H}$ and A, namely

$$\mathfrak{H}^{-1} = \frac{1}{2\pi z_1}(2\nu - 1)(2\nu)! \sin\left[\pi\left(\nu - \frac{1}{2}\right)\right] A^{1+1/\nu} \tag{8.1.107}$$

(with $\nu = 2/d$).

From this equation and from (8.1.87), we get an expression of A which determines the asymptotic properties of the chain

$$A = (2\pi z_1)^\nu \left[\frac{\operatorname{tg} \nu\pi}{1 - 2\nu}\right]^\nu \tag{8.1.108}$$

(with $\nu = 2/d$).

Thus, in dimension three, we find

$$A = 3(2\pi z_1)^{2/3}.$$

1.4.3 Continuous limit and scaling laws

We can now consider the continuous limit. In this case, the length l which appears in (8.1.70) becomes very small. Consequently, one must look at large values of x, i.e. small values of k.

From a physical point of view, it is interesting to consider this limit when the interaction is small. Thus, when the number of links increases, there is a continuous transition from Brownian behaviour to excluded volume behaviour. In this limit, the coupled equations (8.1.99) and (8.1.100) read

$$\mathfrak{H}(k) = 1 - 4z_1 \int_0^\infty \mathrm{d}n \left(\frac{1 - \cos nk}{k^2}\right) \frac{1}{[A(n)]^{1+d/2}}$$

$$A(n) = \frac{1}{\pi} \int_{-\infty}^{+\infty} \mathrm{d}k \left(\frac{1 - \cos nk}{k^2}\right) \frac{1}{\mathfrak{H}(k)}, \tag{8.1.109}$$

and we see at once that the solutions of these equations can be written in scaling form

$$\mathfrak{H}(k) = \mathscr{H}(kz_1^{-2/\varepsilon})$$

$$A(n) = n\mathscr{A}(nz_1^{2/\varepsilon}) \tag{8.1.110}$$

where $\varepsilon = 4 - d$.
The functions $\mathscr{H}(x)$ and $\mathscr{A}(y)$ are given by the equations

$$\mathscr{H}(x) = 1 - \frac{4}{x^2} \int_0^\infty \mathrm{d}y \frac{(1 - \cos xy)}{[y\mathscr{A}(y)]^{1+d/2}}$$

$$\mathscr{A}(y) = \frac{2}{\pi y} \int_0^\infty \mathrm{d}x \frac{(1 - \cos xy)}{x^2 \mathscr{H}(x)} \tag{8.1.111}$$

and by the sum rule [see (8.1.101)]

$$2 \int_0^\infty \mathrm{d}y\, y^{1-d/2} [\mathscr{A}(y)]^{-d/2} = 1.$$

When n is small ($nz_1^{2/\varepsilon} \ll 1$) which corresponds to large values of k ($kz_1^{-2/\varepsilon} \gg 1$), the chain looks quasi-Browmian. In fact, from the above equations, we derive

$$\mathscr{H}(\infty) = 1 \quad \text{and} \quad \mathscr{A}(0) = 1. \tag{8.1.112}$$

On the contrary, when n is large ($nz_1^{2/\varepsilon} \gg 1$) and k small ($kz_1^{-2/\varepsilon} \ll 1$), we are in the excluded volume asymptotic domain and in accordance with eqns (8.1.111), (8.1.110), and (8.1.109), we find [see also (8.1.83) and (8.1.85)]

$$\mathscr{H}(x) = \mathscr{H}x^{-1+4/d} \qquad x \ll 1 \tag{8.1.113}$$

$$\mathscr{A}(y) = \left[2\pi\frac{\operatorname{tg}(\pi d/2)}{1-4/d}\right]^{d/2} y^{-1+4/d} \qquad y \gg 1.$$

1.4.4 Large-size rings: scaling law

In a very analogous manner, remaining in the asymptotic domain, we can study the dimensions of a large but finite ring. The calculation of $\langle\langle(\vec{r}_{j+n} - \vec{r}_j)^2\rangle\rangle$ leads directly to a scaling law of the form

$$\langle\langle(\vec{r}_{j+n} - \vec{r}_j)^2\rangle\rangle = dl^2(N^2 z_1)^\nu A_0(n/N) \tag{8.1.114}$$

where $A_0(x)$ is an even and periodical function

$$A_0(x) = A_0(-x) = A_0(x+1)$$

which, for small values of x, is of the form

$$A_0(x) = A_0|x|^{2\nu}$$

as could be expected.

1.4.5 Large-size rings: free energy correction terms

The free energy of a finite ring with N links ($N \gg 1$) can be written as the sum

$$\mathbb{F}_N = N\mathbb{F} + \delta\mathbb{F}_N, \tag{8.1.115}$$

and one may wonder how $\delta\mathbb{F}_N$ varies with respect to N.

The additional free energy $\delta\mathbb{F}_N$ is the sum of two terms

$$\delta\mathbb{F}_N = \delta\mathbb{E}_N - \delta\mathbb{S}_N.$$

According to general arguments which have previously been expounded (see Section 1.2), $\delta\mathbb{E}_N$ and $\delta\mathbb{S}_N$ should behave as follows

$$\delta\mathbb{E}_N \propto N^{2-\nu d}$$

$$\delta\mathbb{S}_N \propto N^{2\nu-1}. \tag{8.1.116}$$

Let us examine these terms in the framework of the Gaussian approximation. As this theory gives $\nu d = 2$, the term $\delta\mathbb{E}_N$ must be proportional to a constant or to $\ln N$.

On the contrary, the second term $\delta\mathbb{S}_N$ seems dominant. However a detailed calculation[3] of the coefficient of $N^{2\nu-1}$ in $\mathbb{S}_N$ shows that, owing to a sum rule, this coefficient vanishes. Thus Flory's terms (8.1.116) disappear, and one finds that $\delta\mathbb{F}_N$ is proportional to $\ln N$.

This remarkable result may look surprising. Actually, we shall see later that renormalization theory leads to exactly the same result, and this shows that when trying to evaluate the interaction energy of a chain or contact probabilities one has to be very prudent, and that intuitive arguments cannot be trusted without discrimination.

1.4.6 Criticism of the Gaussian approximation

In spite of appealing features, the Gaussian approximation has a very serious defect: it predicts a size exponent $\nu = 2/d$ which is too big, in contradiction to the results of computer simulations and experiments for real polymers.

This phenomenon can be analysed as follow. When a chain swells, the interaction between links is reduced, but simultaneously the entropy diminishes, and there is a balance between these two effects. But, in the framework of the Gaussian approximation, the entropy is calculated exactly; as a consequence, the fact that the calculated swelling is too large implies that the interaction energy is overestimated.

Now, we observe that for a Gaussian chain, the probability distribution of the end-to-end vector is

$$P_n(\vec{r}) = \langle\!\langle \delta(\vec{r} - \vec{r}_{j+n} + \vec{r}_j) \rangle\!\rangle$$
$$= \frac{1}{[2\pi d^{-1} \langle\!\langle (x_{j+n} - x_j)^2 \rangle\!\rangle]^{d/2}} \exp\left[- \frac{r^2}{2\langle\!\langle (x_{j+n} - x_j)^2 \rangle\!\rangle} \right]. \quad (8.1.117)$$

For such a model, the probability of contact of the points M_j and M_{j+n} belonging to the chain is proportional to

$$P_n(\vec{0}) \propto \frac{1}{[\langle\!\langle (x_{j+n} - x_j)^2 \rangle\!\rangle]^{d/2}} \propto \frac{1}{n^{\nu d}}. \quad (8.1.118)$$

In the same way, and for the same reasons, the Gaussian approximation predicts that the closure probability p_N of a finite chain is proportional to $N^{-\nu d}$.

Computer simulations (see Chapter 4) show that this assumption is wrong. Let Z_N be the number of configurations of an open chain and U_N the number of configurations of a closed chain. Using the notation of Chapter 4 (Table 4.1), we can write

$$p_N = \frac{U_N}{Z_N} \propto \frac{N^{\alpha-2}}{N^{\gamma-1}}.$$

On the other hand, we saw in this chapter that the relation $\alpha = 2 - \nu d$ is well verified, in agreement with the most recent theories which indicate that this result must be exact. We thus obtain the result

$$p_N \propto N^{-\nu d - (\gamma - 1)} \quad (8.1.119)$$

which is inconsistent with the Gaussian approximation.

The error arises from the fact that in reality the distribution probability of the end-to-end vector is not Gaussian but is given by a law of the form

$$P_N(\vec{r}) = \frac{1}{N^{\nu d}} f(\vec{r}/N^{\nu})$$

with the property that $f(\vec{0}) = 0$ and $f(x) \simeq f x^{(\gamma-1)/\nu}$ $(x \ll 1)$ (see Chapters 3 and 10).

In brief, the Gaussian approximation does not take into acccunt the existence of the exponent γ (and of similar exponents: see Chapter 12) and this is why it is unrealistic.

In reality, all the older theories have the same defect and, consequently, if some of them lead to more exact values of ν, this is only the result of an accumulation of errors which cancel partially.

As will be seen later on, the renormalization theory provides a solution to these difficulties but unfortunately at the cost of complications which, until now, explicitly prevented the construction of a reasonably realistic model of a chain in a good solvent.

2. REPULSIVE CHAINS AT FINITE CONCENTRATIONS

As we saw in the preceding sections, the theory of isolated chains in a good solvent faced many difficulties and it is clear that, in principle, it is even more difficult to work out a theory of polymer solutions at finite concentrations.

However, the problem was tackled a long time ago, and as early as 1942 Flory and Huggins independently presented a fairly simple theory for polymer solutions. This theory explained the large increase of osmotic pressure that is observed when the concentration goes up. Moreover, it predicted correctly that, at large concentrations, the properties of the solution become independent of the molecular mass of the polymers and then depend only on the mass concentration of solute.

The problem was reconsidered in 1966 by Edwards, who introduced a field theory formalism in order to treat this problem, and who used self-consistent methods to calculate the screening effect.

However, we must note that none of these theories predicted correct scaling laws for long polymers. It was only in 1975 that the application of renormalized Lagrangian theories to polymer solutions, led to a correct description of these systems and provided a means for accurately calculating some of their properties.

2.1 The Flory–Huggins theory

The starting point of the Flory–Huggins theory (1942)[10,11] is a model in which the polymer solution is represented by a set of non-intersecting chains drawn on a lattice. The entropy is calculated in an approximate manner and the thermodynamic properties of the system are deduced from it. Admitting the existence of a lattice is not essential, but choosing this representation is rather convenient.

2.1.1 Approximate calculation of the entropy of mixing

The monomers are in the sites of a periodic lattice of coordination number $\mathfrak{f}$ consisting of M sites.

First, we consider the simple case where N particles occupy the M lattice sites at random without overlap (see Fig. 8.2). The number of configurations of the particles is $M!/N!\,(M-N)!$ and the entropy of the system can be written in the form

$$\mathbb{S} = \ln(M!) - \ln(N!) - \ln((M-N)!). \tag{8.2.1}$$

We can also consider the system as consisting of a mixture of $N_0 = M - N$ particles of species 0 (solvent) and of N particles of solute. Then, the entropy is the entropy of mixing which can be written in the form.

$$\mathbb{S} = \ln((N_0 + N)!) - \ln(N_0!) - \ln(N!) \tag{8.2.2}$$

The problem becomes more complicated if polymers are laid on the lattice (see Fig. 8.3). The number of polymers will be N and it will be assumed that each polymer is made of N monomers. Each monomer occupies a lattice site and two successive monomers are adjacent on the lattice.

Let $Z(N, M, \mathfrak{f})$ be the total number of configurations. In order to evaluate this quantity, we can set the polymers one by one on the lattice. When $(j-1)$ polymers are already on the lattice, there are $\mathcal{N}_j$ ways of adding another one. Moreover, as the same configuration can be obtained by setting the polymers on

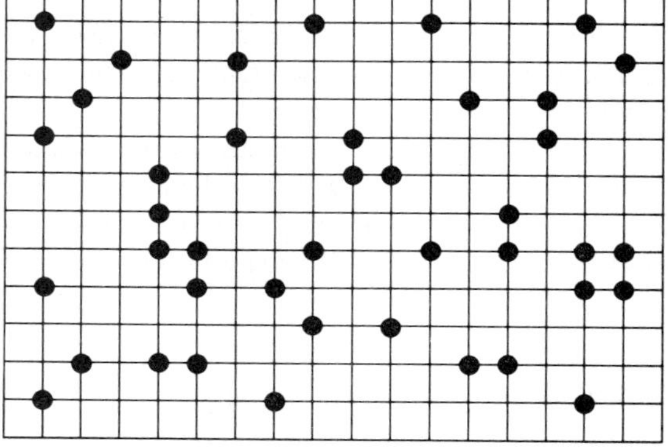

Fig. 8.2. Monomers on a lattice.

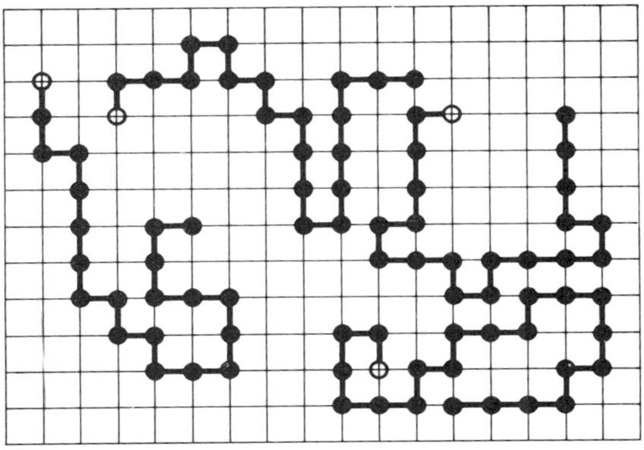

Fig. 8.3. Polymers on a lattice.

the lattice in an arbitrary order

$$Z(N, M, \mathfrak{f}) = \frac{1}{N!} \prod_{j=1}^{N} \mathcal{N}_j. \tag{8.2.3}$$

Let us set the monomers of the j-th polymer one by one. When there are just $(j-1)$ polymers on the lattice, $(M-(j-1)N)$ free sites remain and the first monomer can be set on any of these sites. The second monomer can be set in $(\mathfrak{f}-1)$ possible ways provided that the accessible sites are free, and the average probability that a site is free can be considered as equal to $\dfrac{M-(j-1)N-1}{M}$. Similarly, the third monomer can be set in $(\mathfrak{f}-1)$ different ways provided that the accessible sites are free and the average probability that a site is free can be considered as equal to $\dfrac{M-(j-1)N-2}{M}$. We go on in the same way up to the N-th monomer and we see that $\mathcal{N}_j$ can be approximately written in the form

$$\begin{aligned} \mathcal{N}_j &= (\mathfrak{f}-1)^{N-1} M^{-N+1} \prod_{q=1}^{N} [M-(j-1)N-q+1] \\ &= (\mathfrak{f}-1)^{N-1} M^{-N+1} \frac{[M-(j-1)N]!}{[M-jN]!}. \end{aligned}$$

By bringing this expression of $\mathcal{N}_j$ in $Z(N, M, \mathfrak{f})$, we find

$$Z(N, M, \mathfrak{f}) = [(\mathfrak{f}-1)^{N-1} M^{-N+1}]^N \frac{M!}{N!(M-NN)!}. \tag{8.2.4}$$

The entropy of the system is thus given by

$$\mathbb{S} = \ln Z(N, M, \mathfrak{f})$$

and we have

$$\boldsymbol{M} = N_0 + NN \tag{8.2.5}$$

where, as above, N_0 represents the number of sites that are occupied by solvent molecules.

The numbers N_0, N, and $\boldsymbol{M}$ are assumed to be large. For calculating the dominant terms in $\mathbb{S}$, we use Moivre's approximation $M! \simeq (\boldsymbol{M}/\mathrm{e})^{\boldsymbol{M}}$ which amounts to neglecting logarithmic terms in the entropy. Thus, disregarding minor terms,* we obtain

$$Z(N, \boldsymbol{M}, \mathfrak{f}) \simeq \left(\frac{\mathfrak{f} - 1}{\mathrm{e}}\right)^{N(N-1)} \left(\frac{\boldsymbol{M}}{N}\right)^{N_0} \left(\frac{\boldsymbol{M}}{N_0}\right)^{N_0},$$

and the entropy

$$\mathbb{S} = -N \ln\left(\frac{N}{N_0 + NN}\right) - N_0 \ln\left(\frac{N_0}{N_0 + NN}\right) + N(N-1)\left[\ln(\mathfrak{f} - 1) - 1\right]. \tag{8.2.6}$$

The expression thus obtained is especially simple, but we see immediately that the method allows us to do only very approximate calculations, because the excluded volume constraints are introduced only in a very global way which does not take properly into account correlations between links, especially on the same chain.

2.1.2 Free energy and osmotic pressure

Let J_{00}, J_{11}, and J_{01} be the energies that can be associated respectively to solvent–solvent, solute–solute, and solute–solvent contacts. The change of energy which takes place when the solute is dissolved in the solvent can be characterized by the quantity

$$\chi = \boldsymbol{\beta}(\mathfrak{f} - 2)\left[J_{01} - \tfrac{1}{2} J_{11} - \tfrac{1}{2} J_{00}\right]. \tag{8.2.7}$$

For $\chi > 0$, we have a repulsive solute–solvent interaction which amounts to an attraction between chains, and this repulsion counterbalances the excluded volume effects.

Thus, the energy associated with the solution is approximately (to a constant and in thermal units $1/\boldsymbol{\beta}$)

$$\mathbb{E} = \chi \frac{N_0 NN}{\boldsymbol{M}} = \chi \frac{N_0 NN}{N_0 + NN}. \tag{8.2.8}$$

Indeed, each point of a polymer chain (there are NN of them) may have $(\mathfrak{f} - 2)$ contacts with the solvent.

* For the number of configurations of a chain covering the whole lattice, Flory's calculation gives the approximate value $Z(\boldsymbol{M}, \mathrm{t}) = ((\mathrm{t} - 1)/\mathrm{e})^{\boldsymbol{M}}$, but, as will be shown in Chapter 11, Section 6.2, a more exact calculation by Orland, Itzykson, and De Dominicis gives $Z(\boldsymbol{M}, \mathfrak{f}) = (\mathfrak{f}/\mathrm{e})^{\boldsymbol{M}}$.

From (8.2.6) and (8.2.8), we immediately derive the free energy which will be expressed in thermal units $1/\beta$

$$\mathbb{F} = \mathbb{E} - \mathbb{S} = \frac{\chi N_0 NN}{N_0 + NN} + N \ln\left(\frac{N}{N_0 + NN}\right) + N_0 \ln\left(\frac{N_0}{N_0 + NN}\right) - \\ - N(N-1)[\ln(\mathfrak{f} - 1) - 1]. \tag{8.2.9}$$

To simplify the calculation it will be convenient to consider the system as a gas of polymers in a volume V. Let us set

$$V = (N_0 + NN)v_r \tag{8.2.10}$$

where v_r is the volume per site (for the lattice), and let us eliminate N_0 from $\mathbb{F}$. We obtain

$$\mathbb{F} = -\chi N^2 N^2 v_r/V + N \ln(Nv_r/V) \\ + (V/v_r)(1 - NNv_r/V)\ln(1 - NNv_r/V) + N\mu_0 \tag{8.2.11}$$

where μ_0 is a constant.

From the free energy, we deduce the chemical potential μ and the osmotic pressure Π which can be conveniently expressed in terms of the number of polymers per unit volume

$$C = \frac{N}{V} \tag{8.2.12}$$

$$\mu = \frac{\partial \mathbb{E}(V, N)}{\partial N} = -2\chi NCv_r + \ln(Cv_r) - N\ln(1 - NCv_r) - N + 1 + \mu_0 \tag{8.2.13}$$

$$\Pi\beta = -\frac{\partial \mathbb{F}(V, N)}{\partial V} = C\left[1 - N - \frac{1}{Cv_r}\ln(1 - NCv_r) - \chi N^2 Cv_r\right]. \tag{8.2.14}$$

In these equations, Π is expressed in mechanical units and $\mathbb{F}$ in thermal units $1/\beta$, μ_0 is a constant. Let us now examine the result.

For very dilute polymer solutions ($N^2 Cv_r \ll 1$), we obtain Van't Hoff's law as could be expected

$$\Pi\beta = C. \tag{8.2.15}$$

On the contrary, for large concentrations, the osmotic pressure increases very quickly with C and becomes infinite when the polymer fills the whole lattice $NCv_r = 1$.

For low concentrations, we have the virial expansion

$$\Pi\beta = C\left[1 - \chi N^2 Cv_r + N\sum_{p=1}^{\infty}\frac{(NCv_r)^p}{p+1}\right]. \tag{8.2.16}$$

For large values of N and small values of C, the osmotic pressure, as expected, depends only on the monomer concentration, i.e. on the product $C = NC$.

$$\boldsymbol{\Pi\beta} = \frac{1}{v_r}\ln\left[\frac{1}{1 - Cv_r}\right] - C - \chi C^2 v_r. \tag{8.2.17}$$

Such behaviour is reasonable when the chains have a large overlap. However, this formula ignores the modification of critical exponents resulting from excluded volume effects and therefore it cannot lead to proper scaling laws.

Flory's temperature corresponds to the vanishing of the second virial coefficient (the coefficient of $\boldsymbol{C}^2$ in the expansion of $\boldsymbol{\Pi\beta}$)

$$A_2 = N^2 v_r (\tfrac{1}{2} - \chi). \tag{8.2.18}$$

As the parameter χ is temperature-dependent, we can write

$$\frac{1}{2} - \chi = A\left(1 - \frac{T_F}{T}\right) \tag{8.2.19}$$

which defines the Flory temperature T_F.

Below T_F, we are in the poor solvent region and demixtion may occur. This region will be studied in Chapter 14 and, in particular, it will be shown that the Flory–Huggins theory allows us to calculate demixtion curves that are in qualitative agreement with the experiment.

2.2 Edwards's theory

In Edwards's theory (1966),[12] the chains are represented by continuous curves. The interaction between chains is transformed so as to be replaced by an external random potential whose effect is mathematically equivalent. The chain behaviour in this external potential is studied in an approximate manner, and subsequently, the result is averaged over the set of potentials.

2.2.1 Formulation of the problem and introduction of auxiliary random potentials

We consider a set of $\mathcal{N}$ Gaussian chains. Each chain is labelled with an index j, with $j = 1, \ldots, \mathcal{N}$. It will be assumed that all chains have the same number N of links and that the vector position of a point belonging to a chain j is denoted by the function $r_j(n)$ with $0 < n < N$.

If this set of chains is Brownian without interaction, it is possible to associate with it (see Chapter 2, Section 1.2) a weight ${}^0\mathscr{W}$ given by

$${}^0\mathscr{W} = \exp\left\{-\sum_{j=1}^{\mathcal{N}} \frac{1}{2l^2}\int_0^N \mathrm{d}n\left[\frac{\partial r_j(n)}{\partial n}\right]^2\right\}. \tag{8.2.20}$$

The length l, which is arbitrary, is introduced as usual for reasons of homogeneity, and N can be considered as the number of links.*

* This corresponds to Edwards's presentation; actually l could be eliminated by introducing 'Brownian areas' (see, for instance, Chapters 2 and 3).

The weight associated with the chains in interaction is given by Boltzmann's law, and, expressing the energies in thermal units $1/\beta$, we can express it in the form

$$\mathscr{W} = {}^{0}\mathscr{W} \exp(-{}^{1}\mathscr{U}) \tag{8.2.21}$$

where ${}^{1}\mathscr{U}$ is the interaction energy given by

$${}^{1}\mathscr{U} = \frac{1}{2}\sum_{j'j''}\int_0^N \mathrm{d}n' \int_0^N \mathrm{d}n''\, V(\vec{r}_{j'}(n') - \vec{r}_{j''}(n'')) \tag{8.2.22}$$

Thus, the free energy of the system can be defined by the equality

$$\mathrm{e}^{-(\mathbb{F} - {}^{0}\mathbb{F})} = \langle\!\langle \exp(-{}^{1}\mathscr{U})\rangle\!\rangle_0 \tag{8.2.23}$$

The index zero indicates that the average is taken by choosing the Gaussian weight ${}^{0}\mathscr{W}$ for the configurations. The quantity ${}^{0}\mathbb{E}$ is the free energy of the non-interacting system. Then let ${}^{0}Z$ be the partition function of an isolated chain. We have

$$\mathrm{e}^{-{}^{0}\mathbb{F}} = \frac{({}^{0}Z)^{\mathcal{N}}}{\mathcal{N}!} \simeq ({}^{0}Z\mathrm{e}/\mathcal{N})^{\mathcal{N}} \tag{8.2.24}$$

(there is a factor $1/\mathcal{N}!$ because the chains are undistinguishable). Consequently, as ${}^{0}Z$ is obviously proportional to $\mathcal{V}$, we may write

$${}^{0}\mathbb{F} = \mathcal{N}[\ln(\mathcal{N}l^d/\mathcal{V})] + \mathcal{N}\mu_0 \tag{8.2.25}$$

where μ_0 is a constant (μ_0 depends only on N; l is introduced in order to get a pure number in the logarithm).

We can now consider the quantities $V(\vec{r} - \vec{r}\,')$ as the matrix elements of an operator V in the space defined by the vectors $\vec{r}$. The inverse operator V^{-1} will be defined by matrix elements $V^{-1}(\vec{r} - \vec{r}\,')$ determined by the relations

$$\int \mathrm{d}^d r''\, V(\vec{r} - \vec{r}\,'')\, V^{-1}(\vec{r}\,'' - \vec{r}\,') = \delta(\vec{r} - \vec{r}\,'). \tag{8.2.26}$$

We note that the relation between V and V^{-1} can be expressed in a very simple way by introducing Fourier transforms. Indeed, setting

$$\tilde{V}(\vec{k}) = \int \mathrm{d}^d r\, \mathrm{e}^{-i\vec{k}\cdot\vec{r}}\, V(\vec{r})$$

$$\tilde{V}^{-1}(\vec{k}) = \int \mathrm{d}^d r\, \mathrm{e}^{-i\vec{k}\cdot\vec{r}}\, V^{-1}(\vec{r}) \tag{8.2.27}$$

we see immediately that (8.2.25) gives

$$\tilde{V}(\vec{k})\,\tilde{V}^{-1}(\vec{k}) = 1. \tag{8.2.28}$$

This relation shows that $V(\vec{r})$ and $V^{-1}(\vec{r})$ cannot both be genuine functions and that the inverse transformations corresponding to the transformations (8.2.27) can converge only in the sense of distributions. We recall indeed that the Fourier transform $\tilde{f}(\vec{k})$ of a function $f(\vec{r})$ of bounded variation goes to zero when k goes to infinity, a condition which cannot be simultaneously fulfilled by $V(\vec{k})$ and $V^{-1}(\vec{k})$. Thus, if for example $V(\vec{r})$ is of the form

$$V(\vec{r}) = \frac{A}{4\pi}\frac{e^{-mr}}{r} \qquad d = 3$$

we have

$$V(\vec{k}) = \frac{A}{k^2 + m^2}$$

$$V^{-1}(\vec{k}) = \frac{k^2 + m^2}{A},$$

and by making the inverse Fourier transformation, we find

$$V^{-1}(\vec{r}) = \frac{1}{A}[-\Delta + m^2]\delta(\vec{r}).$$

The operator V^{-1} corresponding to V being now well defined, we shall associate a Gaussian random potential $\chi(\vec{r})$ with the chains. To introduce this potential, we start from the trivial identity

$$\mathscr{Z}^{-1}\int \mathrm{d}\{\chi\}\exp\left\{-\frac{1}{2}\int \mathrm{d}^d r\int \mathrm{d}^d r'\left[\chi(\vec{r}) - i\sum_{\mathrm{M}} V(\vec{r} - \vec{r}_{\mathrm{M}})\right]V^{-1}(\vec{r} - \vec{r}\,')\right.$$
$$\left.\times\left[\chi(\vec{r}\,') - i\sum_{\mathrm{M}'} V(\vec{r}\,' - \vec{r}_{\mathrm{M}'})\right]\right\} = 1 \qquad (8.2.29)$$

where

$$\mathscr{Z} = \int \mathrm{d}\{\chi\}\exp\left\{-\frac{1}{2}\int \mathrm{d}^d r\int \mathrm{d}^d r'\,\chi(\vec{r})V^{-1}(\vec{r} - \vec{r}\,')\chi(\vec{r}\,')\right\}$$

and where the set of points denoted by M and M' is arbitrary. Here $\mathrm{d}\{\chi\}$ is a volume element of the set of all functions $\chi(\vec{r})$, and the above integral is a functional integral. This formula leads to the identity

$$\exp\left[-\frac{1}{2}\sum_{\mathrm{M,M}'} V(\vec{r}_{\mathrm{M}} - \vec{r}_{\mathrm{M}'})\right] = \mathscr{Z}^{-1}\int \mathrm{d}\{\chi\}\exp\left\{-i\sum_{\mathrm{M}}\chi(\vec{r}_{\mathrm{M}})\right.$$
$$\left.-\frac{1}{2}\int \mathrm{d}^d r\int \mathrm{d}^d r'\,\chi(\vec{r})V^{-1}(\vec{r} - \vec{r}\,')\chi(\vec{r}\,')\right\} \qquad (8.2.30)$$

Thus, by associating the weight

$$\mathscr{W}_\chi = \exp\left\{-\frac{1}{2}\int \mathrm{d}^d r\int \mathrm{d}^d r'\,\chi\{\vec{r}\}V^{-1}(\vec{r} - \vec{r}\,')\chi(\vec{r}\,')\right\} \qquad (8.2.31)$$

with the random field, we can write

$$\exp\left[-\sum_{\mathrm{M,M}'} V(\vec{r}_{\mathrm{M}} - \vec{r}_{\mathrm{M}'})\right] = \left\langle\!\!\left\langle \exp\left\{-i\sum_{\mathrm{M}}\chi(\vec{r}_{\mathrm{M}})\right\}\right\rangle\!\!\right\rangle_\chi \qquad (8.2.32)$$

the index χ indicating that the averages are calculated by using the weight $\mathscr{W}_\chi$. This identity allows us to write $\exp(-{}^{\mathrm{I}}\mathscr{U})$ in the form

$$\exp(-{}^{\mathrm{I}}\mathscr{U}) = \left\langle\!\!\left\langle \exp\left\{ -i \int_0^N \mathrm{d}n \sum_{\mathrm{j}} \chi(\vec{r}_{\mathrm{j}}(n)) \right\} \right\rangle\!\!\right\rangle_\chi . \tag{8.2.33}$$

Let us bring this expression in (8.2.33) and let us change the order in which the averages are taken. Now, it looks as if we had N independent chains in a random potential $\chi(\vec{r})$. Let $\mathbb{F}\{\chi\}$ be the free energy of one chain. Thus, the free energy of the set of chains in the potential is $N\mathbb{F}$ for obvious reasons, and we have

$$\mathrm{e}^{-(\mathbb{F} - {}^0\mathbb{F})} = \langle\!\langle \mathrm{e}^{-N\mathbb{F}\{\chi\}} \rangle\!\rangle_\chi$$

$$\mathrm{e}^{-\mathbb{F}\{\chi\}} = \left\langle\!\!\left\langle \exp\left\{ -i \int_0^N \mathrm{d}n\, \chi(\vec{r}(n)) \right\} \right\rangle\!\!\right\rangle_0 . \tag{8.2.34}$$

2.2.2 Free energy of a chain in a random complex potential

We assume that the system is contained in a box having the shape of a torus and volume V (a parallelepiped with periodical boundary conditions). Let us set

$$\tilde{\chi}_{\vec{k}} = \frac{1}{V} \int \mathrm{d}^d r\, e^{-i\vec{k}\cdot\vec{r}} \chi(\vec{r}) \tag{8.2.35}$$

($\vec{k}$ discrete)

$$\chi(\vec{r}) = \sum_{\vec{k}} \mathrm{e}^{i\vec{k}\cdot\vec{r}} \chi_{\vec{k}}$$

The weight $\mathscr{W}_\chi$ can be written in the form

$$\mathscr{W}_\chi = \exp\left\{ -\frac{V}{2} \sum_{\vec{k}} [\chi_{\vec{k}} \chi_{-\vec{k}} / \tilde{V}(\vec{k})] \right\} \tag{8.2.36}$$

from which we get immediately

$$\langle\!\langle \chi_{\vec{k}} \chi_{\vec{k}'} \rangle\!\rangle_\chi = \frac{1}{V} \tilde{V}(\vec{k}) \delta(\vec{k} + \vec{k}'). \tag{8.2.37}$$

All transformations made so far are exact but the problem becomes more complicated with the calculation of $\mathbb{F}\{\chi\}$. The approximation used by Edwards consists in expanding $\mathbb{F}\{\chi\}$ to second-order in χ. Accordingly, starting from (8.2.34), we write

$$\mathrm{e}^{-\mathbb{F}\{\chi\}} = 1 - i \int_0^N \mathrm{d}n \langle\!\langle \chi(\vec{r}(n)) \rangle\!\rangle_0 - \frac{1}{2} \int_0^N \mathrm{d}n \int_0^N \mathrm{d}n' \langle\!\langle \chi(\vec{r}(n)) \chi(\vec{r}(n')) \rangle\!\rangle_0 . \tag{8.2.38}$$

In the second term on the right-hand side, we have

$$\langle\!\langle \chi(\vec{r}(n)) \rangle\!\rangle_0 = \chi_0 .$$

In the third term in the right-hand side, we have

$$\langle\langle \chi(\vec{r}(n))\chi(\vec{r}(n'))\rangle\rangle_0 = \sum_{\vec{k}} \chi_{\vec{k}}\chi_{-\vec{k}}\langle\langle \exp\{i\vec{k}\cdot[\vec{r}(n) - \vec{r}(n')]\}\rangle\rangle_0. \quad (8.2.39)$$

Moreover, $\vec{r}(n)$ is the position vector of a point on a Brownian chain; consequently, as was shown in Chapter 2 [see (2.1.5) and (2.1.8)], we have

$$\langle\langle \exp\{i\vec{k}\cdot[\vec{r}(n) - \vec{r}(n')]\}\rangle\rangle_0 = \mathrm{e}^{-k^2l^2|n-n'|/2}. \quad (8.2.40)$$

Grouping all these results, we obtain

$$\begin{aligned}\mathrm{e}^{-\mathbb{F}\{\chi\}} &= 1 - iN\chi_0 - \frac{1}{2}\sum_{\vec{k}} \chi_{\vec{k}}\chi_{-\vec{k}} \int_0^N \mathrm{d}n \int_0^N \mathrm{d}n'\mathrm{e}^{-k^2l^2|n-n'|/2} \\ &= 1 - iN\chi_0 - \frac{N^2}{2}\sum_{\vec{k}} {}^0H(\vec{k})\chi_{\vec{k}}\chi_{-\vec{k}} \end{aligned} \quad (8.2.41)$$

where

$${}^0H(\vec{k}) = \frac{2}{N^2}\int_0^N \mathrm{d}n[N-n]\mathrm{e}^{-k^2l^2n/2} \quad (8.2.42)$$

[${}^0H(\vec{k})$ is the form function of a Brownian chain, and we have ${}^0H(\vec{0}) = 1$].

From these results, we deduce the expression of $\mathbb{F}\{\chi\}$ up to second-order

$$\mathbb{F}\{\chi\} = iN\chi_0 - \frac{N^2}{2}\chi_0^2 + \frac{N^2}{2}\sum_{\vec{k}} {}^0H(\vec{k})\chi_{\vec{k}}\chi_{-\vec{k}}. \quad (8.2.43)$$

Let us bring this value into the first equation (8.2.34). We obtain

$$\mathrm{e}^{-(\mathbb{F} - {}^0\mathbb{F})} = \left\langle\!\!\left\langle \exp\left[-iNN\chi_0 + \frac{NN^2}{2}\chi_0^2 - \frac{NN^2}{2}\sum_{\vec{k}} {}^0H(\vec{k})\chi_{\vec{k}}\chi_{-\vec{k}}\right]\right\rangle\!\!\right\rangle_\chi;$$

i.e. when (8.2.36) is taken into account

$$\mathrm{e}^{-(\mathbb{F} - {}^0\mathbb{F})} = \frac{\displaystyle\int \mathrm{d}\{\chi\}\exp\left\{-iNN\chi_0 + \frac{NN^2}{2}\chi_0^2 - \frac{1}{2}\sum_{\vec{k}}\left[\frac{V}{\tilde{V}(\vec{k})} + NN^2\;{}^0H(\vec{k})\right]\chi_{\vec{k}}\chi_{-\vec{k}}\right\}}{\displaystyle\int \mathrm{d}\{\chi\}\exp\left\{-\frac{1}{2}\sum_{\vec{k}}\frac{V}{\tilde{V}(\vec{k})}\chi_{\vec{k}}\chi_{-\vec{k}}\right\}}. \quad (8.2.44)$$

The terms containing χ_0 in the numerator can be written as follows

$$-iNN\chi_0 - \frac{1}{2}\frac{V}{\tilde{V}(\vec{0})}\chi_0^2 = -\frac{V}{2\tilde{V}(\vec{0})}\left(\chi_0 + iN\frac{N\tilde{V}(\vec{0})}{V}\right)^2 - \frac{N^2N^2\tilde{V}(\vec{0})}{2V} \quad (8.2.45)$$

Thus, $\mathrm{e}^{-(\mathbb{F} - {}^0\mathbb{F})}$ is given by

$$\mathrm{e}^{-(\mathbb{F} - {}^0\mathbb{F})} = \exp\left[-\frac{N^2N^2\tilde{V}(\vec{0})}{2V}\right]\prod_{\vec{k}}\left[1 + \frac{NN^2}{V}\tilde{V}(\vec{k}){}^0H(\vec{k})\right]^{-1/2}$$

and therefore

$$\mathbb{F} - {}^0\mathbb{F} = \frac{\mathcal{N}^2 N^2 \tilde{V}(\vec{0})}{2V} + \frac{1}{2}\sum_{\vec{k}} \ln\left[1 + \frac{\mathcal{N} N^2 \tilde{V}(\vec{k})\, {}^0H(\vec{k})}{V}\right]. \qquad (8.2.46)$$

When the volume of the box is large, the sum can be replaced by an integral and the well-known correspondence is

$$\frac{1}{V}\sum_{\vec{k}} \ldots = \frac{1}{(2\pi)^d}\int \mathrm{d}^d k \ldots \qquad (8.2.47)$$

(Actually, in three-dimensional space, we can consider a rectangular box with sides of lengths a, b, c and of volume $V = abc$). Then for periodical boundary conditions, the quantification conditions on the three axes give $k_x a = 2\Pi p_1$, $k_y b = 2\Pi p_2$, $k_z c = 2\Pi p_3$, where p_1, p_2, p_3 are integers; whence the above equality).

$$\mathbb{F} - {}^0\mathbb{F} = \frac{\mathcal{N}^2 N^2}{2}\frac{\tilde{V}(\vec{0})}{V} + \frac{V}{2(2\pi)^d}\int \mathrm{d}^d k \ln\left[1 + \frac{\mathcal{N} N^2}{V}\tilde{V}(\vec{k})\, {}^0H(\vec{k})\right]. \qquad (8.2.48)$$

On the other hand, the function ${}^0H(\vec{k})$ can be expressed in the form [see (8.2.42)]

$${}^0H(\vec{k}) = {}^0\mathrm{h}\left(\frac{k^2 l^2 N}{2}\right) \qquad (8.2.49)$$

where ${}^0\mathrm{h}(x)$ is the Debye function

$$\begin{aligned} {}^0\mathrm{h}(x) &= 2\int_0^1 \mathrm{d}t(1-t)\mathrm{e}^{-x^2t/2} \\ &= \frac{2}{x^2}[\mathrm{e}^{-x} - 1 + x]. \end{aligned} \qquad (8.2.50)$$

Starting from (8.2.25) and (8.2.48), we obtain the free energy

$$\begin{aligned} \mathbb{F} = \mathcal{N}\left[\ln\left(\frac{\mathcal{N} l^d}{V}\right) + \mu_0\right] + \frac{\mathcal{N}^2 N^2 \tilde{V}(\vec{0})}{2V} + \\ + \frac{V}{2(2\pi)^d}\int \mathrm{d}^d k \ln\left[1 + \frac{\mathcal{N} N}{V}\tilde{V}(\vec{k})\, {}^0\mathrm{h}(k^2 l^2 N/2)\right]. \end{aligned} \qquad (8.2.51)$$

When the chain is long and the interaction well localized, $\tilde{V}(\vec{k})$ varies much more slowly than ${}^0\mathrm{h}/(k^2 l N/2)$, and in the preceding expressions we may try to replace $\tilde{V}(\vec{k})$ by the covolume v_e

$$\tilde{V}(\vec{0}) = \int \mathrm{d}^d r V(\vec{r}) \simeq v_e. \qquad (8.2.52)$$

However, when this replacement is made directly in (8.2.51), the expression giving $\mathbb{F}$ diverges (for $d = 3$) because ${}^0\mathrm{h}(k^2 l^2 N/2)$ decreases as $1/k^2$ when $\vec{k} \to \infty$. Still, the divergence coming from chain self-intersections can be absorbed by renormalization of the constant μ_0. Let us set

$$\mu_1 - \mu_0 = \frac{N^2}{2(2\pi)^d} \int \mathrm{d}^d k \tilde{V}(\vec{k})\, {}^0\mathrm{h}(k^2 l^2 N/2) \tag{8.2.53}$$

where μ_1 is a constant independent of N and V. We can rewrite (8.2.51) in the form

$$\mathbb{F} = N\left[\ln\left(\frac{Nl^d}{V}\right) + \mu_1\right] + \frac{N^2 N^2 \tilde{V}(\vec{0})}{2V} + \frac{V}{2(2\pi)^d}\int \mathrm{d}^d k$$
$$\times \left\{\ln\left[1 + \frac{N}{V} N^2 \tilde{V}(\vec{k})\,{}^0\mathrm{h}(k^2 l^2 N/2) - \frac{N}{V} N^2 \tilde{V}(\vec{k})\,{}^0\mathrm{h}(k^2 l^2 N/2)\right]\right\}. \tag{8.2.54}$$

Replacing $\tilde{V}(\vec{k})$ by the covolume v_e is now legitimate, and this gives

$$\mathbb{F} = N\left[\ln\left(\frac{Nl^d}{V}\right) + \mu_1\right] + \frac{N^2 N^2 v_e}{2V} + \frac{\Omega(d)\, V}{2(2\pi l)^d} N^{-d/2} \int_0^\infty \mathrm{d}y\, y^{d-1}$$
$$\left\{\ln\left[1 + \frac{N N^2 v_e\,{}^0\mathrm{h}(y^2/2)}{V} - \frac{N}{V} N^2 v_e\,{}^0\mathrm{h}(y^2/2)\right]\right\} \tag{8.2.55}$$

($\Omega(d)$ = 'area' of a ball of unit radius in a d-dimensional space).

From the free energy, we easily deduce the chemical potential μ and the osmotic pressure Π since

$$\mu = \frac{\partial \mathbb{F}(V, N)}{\partial N} \qquad \Pi\beta = -\frac{\partial \mathbb{F}(V, N)}{\partial V}.$$

These quantities can conveniently be expressed in terms of $C = N/V$ and N. We obtain

$$\mu - \mu_1 = 1 + \ln(Cl^d)$$
$$+ CN^2 v_e \left[1 - \frac{\Omega(d)}{2(2\pi l)^d} v_e N^{2-d/2} \int_0^\infty \mathrm{d}y \frac{y^{d-1}\,{}^0\mathrm{h}(y^2/2)}{1 + CN^2 v_e \mathrm{h}(y^2/2)}\right]$$
$$\Pi\beta = C + \frac{1}{2} C^2 N^2 v_e - \frac{\Omega(d)}{2(2\pi l)^d} N^{-d/2} \int_0^\infty \mathrm{d}y\, y^{d-1}$$
$$\times \left\{\ln\left[1 + CN^2 v_e\,{}^0\mathrm{h}(y^2/2) - \frac{CN^2 v_e\,{}^0\mathrm{h}(y^2/2)}{1 + CN^2 v_e\,{}^0\mathrm{h}(y^2/2)}\right]\right\} \tag{8.2.56}$$

where $\Omega(d) = 2\Pi^{d/2}/\Gamma(d/2)$ (see Appendix F).

A simpler result can be obtained if we replace ${}^0\mathrm{h}(x)$ by its asymptotic value $2/x$, as was done by Edwards, and, for $d = 3$, this simplification, which is valid for $CN^2 v_e \gg 1$ (semi-dilute regime) gives

$$\mathbb{F}/V = C[\ln(Cl^3) + \mu_1] + \frac{1}{2} C^2 N^2 v_e - \frac{2}{3\pi l^3}(CNv_e)^{3/2}$$

$$\Pi\beta = C + \frac{1}{2} C^2 N^2 v_e - \frac{1}{3\pi l^3}(CNv_e)^{3/2}. \tag{8.2.57}$$

For relatively small values of C (very dilute solutions), the first term is dominant and, of course, one gets the perfect gas law $\Pi\beta = C$. For relatively large values of C (semi-dilute solutions), the osmotic pressure now depends only on $C = NC$.

$$\Pi\beta = \frac{1}{2} C^2 v_e - \frac{1}{3\pi l^3}(Cv_e)^{3/2} \tag{8.2.58}$$

Between these limits, the approximation does not lead to any true scaling law. Moreover, formula (8.2.57) cannot be expanded with respect to C so as to give a

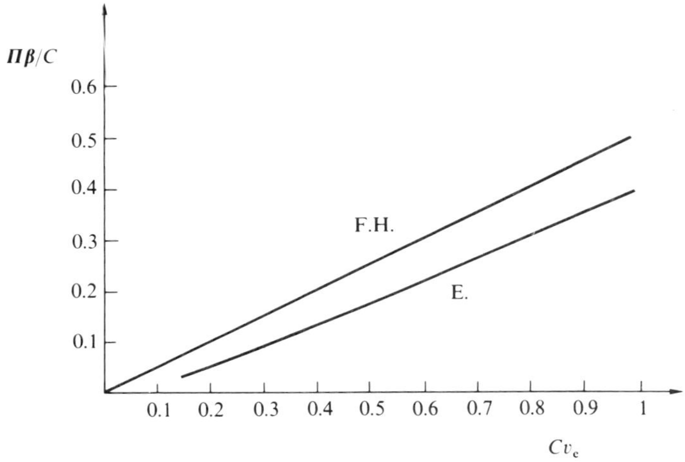

Fig. 8.4. Osmotic pressure in the semi-dilute regime: a comparison between the values calculated by Flory and Huggins

$$\Pi\beta = \tfrac{1}{2} C^2 v_e$$

and by Edwards

$$\Pi\beta = \tfrac{1}{2} C^2 v_e - \frac{1}{3\pi l^3}(Cv_e)^{3/2}$$

C = monomer concentration
v_e = excluded volume
l = length of the statistical element.

The curve corresponds to the case where $l^3 = v_e$.

regular series, contrary to our expectations. Thus, the result cannot be a good approximation, and Edwards's method, in spite of its merit, has to be used cautiously.

In the semi-dilute domain, Edwards's result can be compared with the result obtained by Flory and Huggins, which, in the semi-dilute case [$Cv_r \ll 1$ in eqn (8.2.17)], is

$$\Pi\beta = (\tfrac{1}{2} - \chi)\, C^2 v_r. \tag{8.2.59}$$

To compare the results, we have to identify the dominant terms

$$v_e = (1 - 2\chi) v_r. \tag{8.2.60}$$

We then observe that the osmotic pressure in good solvent, calculated by Edwards, is smaller than the osmotic pressure calculated by Flory and Huggins (see Fig. 8.4).

This property of Edwards's approximation is also a result of the more exact theory established by des Cloizeaux, but only in the domain where the theory applies (see Chapter 13). It can be interpreted by observing that the excluded volume interactions are screened in the semi-dilute regime. This effect thus diminishes the probability of contact between segments.

REFERENCES

1. Flory, P.J.(1971). *Principles of polymer chemistry*, Chapter IV, p. 596. Cornell University Press;
   Yamakawa, H. (1971). *Modern theory of polymer solutions*, Chapter 3, p. 69. Harper & Row.
2. Fisher, M.E. (1968). Discussion following the communication made by J. des Cloizeaux at the Kyoto Conference on Statistical Mechanics, 1968; see reference 5 below.
3. des Cloizeaux, J. (1970). *J. Physique* **31**, 715. The criticism contained in the article concerning the self-consistent field method is essentially correct; however, eqn (VIII.32) in the article does not really describe a random chain but, in reality, a random walk, as obtained by direct use of the Monte Carlo method. To describe a random chain, one has to replace (III.32) in the article by an equation of the form

$$\frac{d^2\vec{r}(L)}{dL^2} = S^{-1}\vec{\nabla}V(\vec{r}) + \frac{d}{dL}\vec{\varepsilon}(L)$$

   (P.G. de Gennes, private communication).
4. Ullman, R. (1978). *Macromolecules* **11**, 1292.
5. des Cloizeaux, J. (1969). Kyoto Conference on Statistical Mechanics. *J. Phys. Soc. Japan* Suppl. 26, 42.
6. Edwards, S.F. (1965). *Proc. Phys. Soc. London* **85**, 613.
7. Freed, K.F. (1971). *J. Chem. Phys.* **55**, 3910.
   Freed, K.F. and Gillis, H.P. (1971). *Chem. Phys. Lett.* **8**, 384.
   Gillis, H.P. and Freed, K.F. (1975). *J. Chem. Phys.* **63**, 852.
8. Reiss, H. (1967). *J. Chem. Phys.* **47**, 186.

9. Yamakawa, H. (1968). *J. Chem. Phys.* **48**, 3845.
10. Huggins, M.L. (1942). *J. Phys. Chem.* **46**, 151.
11. Flory, P.J. (1942). *J. Chem. Phys.* **10**, 51; see also reference 1.
12. Edwards, S. F. (1966). *Proc. Phys. Soc. London* **88**, 1265. (Coefficients in certain formulas need to be corrected.)

9

THE GRAND CANONICAL FORMALISM

1. GENERALITIES

Whenever the Hamiltonian of a system obeys a global conservation law, like the conservation of energy or of the number of particles, one might be tempted to describe the statistics of the system by giving a well-defined value to the conserved quantity. However, such an approach would be clumsy because, in this case, any local fluctuation of the conserved quantity must be compensated elsewhere by fluctuations with the opposite sign, and thus the conservation law would entail correlations devoid of interest and difficult to take into account.

In order to break free from these correlations, larger ensembles are used in which a chemical potential or fugacity is associated with each conserved quantity so as to fix only averages of these conserved quantities. Thus, when the energy is conserved in the system, canonical ensembles are used; when, moreover, the system consists of particles whose number is also conserved, we use grand canonical ensembles.

This is the reason why the grand canonical formalism is used to study polymer solutions, and this fact has important practical consequences. Actually, it is in the grand canonical framework that the simplest perturbation expansions can be obtained, and we note that these expansions are not only useful by themselves; they also serve as a basis for all the extensions that are derived from prescriptions of the renormalization theories.

Thus, to study polymer solutions, the grand canonical formalism plays a role which, from a theoretical point of view, is quite essential.

2. THE MODELS

The grand canonical formalism can be developed for various polymer models.

The physical nature of polymers leads us, quite naturally, to represent a polymer by a chain with N links in continuous space. Then two important parameters characterize a set of polymers: the number $\mathbf{C}$ of polymers per unit volume and the number C of monomers per unit volume. This is the representation that has been introduced in Chapters 5 and 8, and we shall also use it in the following. However, one must be aware of the fact that this model is neither the only possible one, nor even the best one.

For the mathematician or the physicist who uses a computer to simulate and count polymer configurations, the best model is that of chains with discrete links on a lattice. In this case, all the interesting quantities are finite and are pure numbers. In particular, the number φ of polymers per site and the number φ of monomers per site are important parameters.

Conversely, for the physicist who wants to make analytical calculations, the best model is that of continuous chains in continuous space. This type of model is studied in detail in Chapter 10 and in the following chapters.

3. CANONICAL AND GRAND CANONICAL ENSEMBLES

3.1 Canonical partition functions of a set of molecules

Let us consider a gas or a compressible liquid made of $(m + 1)$ species of molecules occupying a volume V. A configuration $\mathbf{\Omega}$ of the system is given by the set of positions of the atoms in the system. With each configuration is associated a weight $W(\Omega)$ given by a law *à la* Boltzmann. The corresponding partition function is obtained by summation over all configurations

$$Q = \sum_{\mathbf{\Omega}} W(\mathbf{\Omega}). \tag{9.3.1}$$

We can also assume that these configurations are subject to a constraint. For instance, the constraint may be that a set of points $\mathscr{P}$ belonging to the molecules have definite space positions and then $\mathscr{R}$ represents the set of these positions. We thus define the restricted partition function $Q(\mathscr{R}, \mathscr{P})$ by

$$Q(\mathscr{R}, \mathscr{P}) = \sum_{\mathbf{\Omega} \in (\mathscr{R}, \mathscr{P})} W(\mathbf{\Omega}) \tag{9.3.2}$$

which shows that the sum is made over the only configurations that obey the constraint $\mathscr{R}$. Thus, for instance, the probability $p(\mathscr{R}, \mathscr{P})$ that the constraint be realized is given by

$$p(\mathscr{R}, \mathscr{P}) = \frac{Q(\mathscr{R}, \mathscr{P})}{Q}. \tag{9.3.3}$$

In a more general way, the weight $W(\mathbf{\Omega})$ can be used to calculate mean values. By definition, we have

$$\langle\!\langle \mathscr{O} \rangle\!\rangle = \frac{\sum_{\mathbf{\Omega}} W(\mathbf{\Omega}) \mathscr{O}(\mathbf{\Omega})}{\sum_{\mathbf{\Omega}} W(\mathbf{\Omega})} = \frac{1}{Q} \sum_{\mathbf{\Omega}} W(\mathbf{\Omega}) \mathscr{O}(\mathbf{\Omega}). \tag{9.3.4}$$

Let us observe that the probabilities $p(\mathscr{R}, \mathscr{P})$ can be considered as mean values. Actually, representing by $\mathscr{R}(\mathscr{P}, \mathbf{\Omega})$ the positions of the points $\mathscr{P}$ in the configuration $\mathbf{\Omega}$, we can write (9.3.2) in the symbolic form

$$Q(\mathscr{R}, \mathscr{P}) = \sum_{\mathbf{\Omega}} W(\mathbf{\Omega}) \, \delta[\mathscr{R}, \mathscr{R}(\mathscr{P}, \mathbf{\Omega})] \tag{9.3.5}$$

and accordingly

$$p(\mathscr{R}, \mathscr{P}) = \langle\!\langle \delta(\mathscr{R}, \mathscr{R}(\mathscr{P}, \mathbf{\Omega})) \rangle\!\rangle. \tag{9.3.6}$$

When giving the preceding definitions, we implicitly assumed that the number of configurations of the system is finite. This is not so in the case which we are considering since the positions of the atoms can vary continuously. However, it is still possible to define finite partition functions by replacing discrete sums with integrals over the positions of the atoms constituting the molecules. In particular, for the time being, let us consider only configurations for which the numbers $N_{\mathscr{A}}$ of molecules of each species are fixed, and let Q and $Q(\mathscr{R}, \mathscr{P})$ be the corresponding partition functions. Replacing sum by integrals over the positions of the atoms amounts to dividing Q and $Q(\mathscr{R}, \mathscr{P})$ by a factor independent of $\mathscr{R}$ but infinite. This multiplicative renormalization thus keeps $p(\mathscr{R}, \mathscr{P})$ invariant, as it must.

More precisely $\hat{Q}_G(V; N, \ldots, N_m)$ represents the partition function associated with the configurations for which the volume V contains $N_{\mathscr{A}}$ particles of species $\mathscr{A}$ ($\mathscr{A} = 0, \ldots, m$). The circumflex means that we deal with all the constituents of the solution, i.e. the solvent as well as the solutes; we shall drop it later after eliminating the parameters concerning the solvent. The index G (for global) has been added here for reasons of convenience which will appear more clearly later.*

Similarly, the function $\hat{Q}_G(V, \mathscr{R}, \mathscr{P}; N_0, \ldots, N_m)$ represents the 'restricted' partition functions associated with the configurations defined by V, by the $N_{\mathscr{A}}$ and by restrictive conditions $(\mathscr{R}, \mathscr{P})$.

A configuration is given by the positions $\vec{r}_j$ of the atoms or monomers constituting the solution, and the associated weight can be written in the form

$$W\{\vec{r}\} = \exp\left[-\sum_{jl} V_{j,l}(\vec{r}_j - \vec{r}_l)\right] \tag{9.3.7}$$

where $V_{j,l}(\vec{r})$ is a quasi-potential which depends on temperature and also on the nature of the atoms and monomers but only globally. In fact, the problem is not to calculate the properties of polymers completely by starting from their chemical structure, but to find a reasonably realistic model so as to calculate the properties of long polymers and, in particular, universal properties which do not depend on the chemical structure.

3.2 A grand canonical ensemble of molecules: fugacities and chemical potentials

In a grand canonical ensemble, the numbers $N_{\mathscr{A}}$ of molecules of species $\mathscr{A}$ are fixed only on average, and statistical mechanics shows that the weight

$$\hat{f}_0^{N_0} \ldots \hat{f}_m^{N_m} \hat{Q}_G(V; N_0, \ldots, N_m)$$

must be attributed to the set of configurations for which the volume V contains,

* The index G reminds us that the partition function to which it refers is a non-local quantity (in perturbation theory it is called non-connected)

for each species $\mathscr{A}$, a well-defined number $N_{\mathscr{A}}$ of particles. Moreover, by definition, we set

$$\hat{Q}_{\mathrm{G}}(V;0,\ldots,0)=1. \tag{9.3.8}$$

The coefficients $\hat{f}_0,\ldots,\hat{f}_m$ are parameters which are called 'fugacities' and which are related to the mean values $\langle\langle N_{\mathscr{A}}\rangle\rangle$ calculated for the solute–solvent ensemble. Incidentally, renormalized fugacities $f_{\mathscr{A}}$ will be defined later on.

Accordingly, the grand partition function $\hat{Q}_{\mathrm{G}}(V,\{\hat{f}\})$ is defined as the sum

$$\hat{Q}_{\mathrm{G}}(V;\{\hat{f}\})=\sum_{N_0=0}^{\infty}\cdots\sum_{N_m=0}^{\infty}\hat{f}_0^{N_0}\cdots\hat{f}_m^{N_m}\,\hat{Q}_{\mathrm{G}}(V;N_0,\ldots,N_m). \tag{9.3.9}$$

We then see immediately that the mean values $\langle\langle N_{\mathscr{A}}\rangle\rangle$ of the numbers of particles of species $\mathscr{A}$ contained in V are given by

$$\langle\langle N_{\mathscr{A}}\rangle\rangle=\frac{1}{\hat{Q}(V;\{\hat{f}\})}\sum_{N_0=0}^{\infty}\cdots\sum_{N_m=0}^{\infty}N_{\mathscr{A}}\hat{f}_0^{N_0}\cdots\hat{f}_m^{N_m}\,Q_{\mathrm{G}}(V;N_0,\ldots,N_m). \tag{9.3.10}$$

Now, let us set

$$V\hat{Q}(V;\{\hat{f}\})=\ln\hat{Q}_{\mathrm{G}}(V;\{\hat{f}\}). \tag{9.3.11}$$

We obtain

$$C_{\mathscr{A}}=\frac{\langle\langle N_{\mathscr{A}}\rangle\rangle}{V}=\hat{f}_{\mathscr{A}}\frac{\partial}{\partial\hat{f}_{\mathscr{A}}}\hat{Q}(V;\{\hat{f}\}). \tag{9.3.12}$$

Thus, relation (9.3.11) shows that $V\hat{Q}(V;\{\hat{f}\})$ is the generating function of the cumulants $V\hat{Q}(V;N_0,\ldots,N_m)$ of the set of functions $\hat{Q}_{\mathrm{G}}(V;N_0,\ldots,N_m)$.

These new partition functions are especially interesting. In fact, it is easy to show, either by using physical arguments or by studying the structure of diagram expansions, that the function $\hat{Q}(V;N_0,\ldots,N_m)$ has a finite limit when $V\to\infty$

$$\lim_{V\to\infty}\hat{Q}(V;N_0,\ldots,N_m)=\hat{Q}(N_0,\ldots,N_m), \tag{9.3.13}$$

$$\lim_{V\to\infty}\hat{Q}(V;\{\hat{f}\})=\hat{Q}\{\hat{f}]. \tag{9.3.14}$$

Therefore, in the limit of large volumes

$$C_{\mathscr{A}}=\hat{f}_{\mathscr{A}}\frac{\partial}{\partial\hat{f}_{\mathscr{A}}}\hat{Q}\{\hat{f}\}. \tag{9.3.15}$$

We must observe that the fluctuations of the number $N_{\mathscr{A}}$ of particles $\mathscr{A}$ around the mean value $\langle\langle N_{\mathscr{A}}\rangle\rangle$ are relatively small. Indeed,

$$\langle\langle N_{\mathscr{A}}^2\rangle\rangle=\frac{1}{\hat{Q}_{\mathrm{G}}(V;\{\hat{f}\})}\left(\hat{f}_{\mathscr{A}}\frac{\partial}{\partial\hat{f}_{\mathscr{A}}}\right)^2\hat{Q}_{\mathrm{G}}(V;\{\hat{f}\}). \tag{9.3.16}$$

Using the equality

$$\hat{Q}_G(V;\{\hat{f}\}) = \exp\,[V\hat{Q}(V;\{\hat{f}\})]$$

and taking relation (9.3.14) into account, we find

$$\langle\langle (N_{\mathscr{A}} - \langle\langle N_{\mathscr{A}}\rangle\rangle)^2 \rangle\rangle = \langle\langle N_{\mathscr{A}}^2\rangle\rangle - \langle\langle N_{\mathscr{A}}\rangle\rangle^2 = V\left(\hat{f}_{\mathscr{A}}\frac{\partial}{\partial \hat{f}_{\mathscr{A}}}\right)^2 \hat{Q}(V;\{\hat{f}\}),$$

and in the limit of large volumes

$$\langle\langle (N_{\mathscr{A}} - \langle\langle N_{\mathscr{A}}\rangle\rangle)^2 \rangle\rangle = V\left(\hat{f}_{\mathscr{A}}\frac{\partial}{\partial \hat{f}_{\mathscr{A}}}\right)^2 \hat{Q}\{\hat{f}\}. \tag{9.3.17}$$

The fluctuations around $\langle\langle N_{\mathscr{A}}\rangle\rangle$ are thus of the order of $\langle\langle N_{\mathscr{A}}\rangle\rangle^{1/2}$ and consequently are relatively small. This result is an essential property of the grand canonical systems. It shows that the canonical and grand canonical representations are nearly equivalent in the limit of large systems.

Let us study this point more precisely. The connection between $\hat{Q}_G(V; N_0, \ldots, N_m)$ and the free energy is given by

$$\mathscr{F}(V;\{N\}]) = -\ln \hat{Q}_G(V; N_0, \ldots, N_m)$$

and, for large volumes, we may write

$$\mathscr{F}(V;\{N\}) = V\Phi\left(\frac{N_0}{V}, \ldots, \frac{N_m}{V}\right). \tag{9.3.18}$$

Consequently, in this limit, the grand canonical partition function takes the form,

$$\hat{Q}_G(V;\{\hat{f}\} =$$

$$\sum_{N_0=0}\cdots\sum_{N_m=0}\exp\left[-V\Phi\left(\frac{N_0}{V},\ldots,\frac{N_m}{V}\right) + N_0\ln\hat{f}_0 + \ldots + N_m\ln\hat{f}_m\right]$$

or

$$\hat{Q}_G(V;\{\hat{f}]) \simeq$$

$$V^m\int_0^\infty dx_0,\ldots\int_0^\infty dx_m \exp\{-V[\Phi(x_0,\ldots,x_m + x_0\ln\hat{f}_0 + \ldots + x_m\ln\hat{f}_m]\}.$$

This expression can be evaluated by using the steepest descent method. Let ${}^c x_{\mathscr{A}}$ be the value of $x_{\mathscr{A}}$ at the saddle point; we obtain

$$\ln\hat{f}_{\mathscr{A}} = \frac{\partial}{\partial x_{\mathscr{A}}}\Phi({}^c x_0, \ldots, {}^c x_m). \tag{9.3.19}$$

In first approximation, we have

$$\ln \hat{Q}_G(V;\{\hat{f}\}]) = V[-\Phi({}^c x_0,\ldots,{}^c x_m) + {}^c x_0\ln\hat{f}_0 + \ldots + {}^c x_m\ln\hat{f}_m]$$

and, comparing this expression with equalities (9.3.11) and (9.3.14), we find

$$\hat{Q}\{\hat{f}\} = -\Phi({}^cx_0, \ldots, {}^cx_m) + {}^cx_0 \ln \hat{f}_0 + \ldots + {}^cx_m \ln \hat{f}_m.$$

Moreover, as $x_{\mathscr{A}} = N_{\mathscr{A}}/V$, we have, in the limit of large volumes

$$ {}^cx_{\mathscr{A}} = C_{\mathscr{A}}. \tag{9.3.20}$$

This calculation also allows us to find a relation between the chemical potentials and the fugacities. By definition

$$\beta \mu_{\mathscr{A}}(V;\{N\}) = \frac{\partial}{\partial N_{\mathscr{A}}} \mathscr{F}(V;\{N\}).$$

For a canonical system, eqn (9.3.18) gives

$$\beta \mu_{\mathscr{A}} = \frac{\partial}{\partial x_{\mathscr{A}}} \Phi(x_0, \ldots, x_m)$$

where

$$x_{\mathscr{A}} = N_{\mathscr{A}}/V$$

(then $N_{\mathscr{A}}$ is fixed).
For a grand canonical ensemble, we have

$$\beta \mu_{\mathscr{A}} = \frac{\partial}{\partial x_{\mathscr{A}}} \Phi({}^cx_0, \ldots, {}^cx_m). \tag{9.3.21}$$

comparing (9.3.19) with (9.3.20), we finally obtain

$$\beta \mu_{\mathscr{A}} = \ln \hat{f}_{\mathscr{A}} = \frac{\partial}{\partial C_{\mathscr{A}}} \Phi(C_0, \ldots, C_m).$$

Thus, in the limit of large volumes, the chemical potentials are directly related to the fugacities.

Let us now consider the correlation functions, i.e. the quantities $p(\mathscr{R}, \mathscr{P})$ defined by relations of the form (9.3.3). We can define a 'restricted' grand canonical function by setting

$$\hat{Q}_G(V, \mathscr{R}, \mathscr{P};\{\hat{f}\}) = \sum_{N_0} \cdots \sum_{N_m} \hat{f}_0^{N_0} \ldots \hat{f}_m^{N_m} \hat{Q}_G(V, \mathscr{R}, \mathscr{P}; N_0, \ldots, N_m). \tag{9.3.22}$$

In the framework of the grand canonical formalism, we define the correlations corresponding to $\mathscr{R}$, $\mathscr{P}$ by means of the probability (or the probability distribution)

$$\hat{p}(V, \mathscr{R}, \mathscr{P};\{\hat{f}\}) = \frac{\hat{Q}_G(V, \mathscr{R}, \mathscr{P};\{\hat{f}\})}{\hat{Q}_G(V;\{\hat{f}\})} \tag{9.3.23}$$

which in general has a finite limit when $V \to \infty$

$$\hat{p}(\mathscr{R}, \mathscr{P};\{\hat{f}\}) = \lim_{V \to \infty} \hat{p}(V, \mathscr{R}, \mathscr{P};\{\hat{f}\}).$$

On the other hand, the concentrations $C_{\mathscr{A}}$ can be expressed in terms of the $\hat{f}_{\mathscr{A}}$ by means of (9.3.15).

In the grand canonical framework, the dependence of the correlation functions with respect to the concentrations is thus given in parametric form.

3.3 Pressure and osmotic pressure in the grand canonical formalism

Let us now calculate the pressure in the system. For a canonical system, we have

$$P = -\frac{\partial}{\partial V}\mathscr{F}(V; N_0, \ldots, N_m).$$

where

$$\beta\mathscr{F}(V; N_0, \ldots, N_m) = -\ln \hat{Q}_G(V; N_0, \ldots, N_m).$$

which can be written in the form

$$P\beta = \frac{1}{\hat{Q}_G(V; N_0, \ldots, N_m)}\frac{\partial}{\partial V}\hat{Q}_G(V; N_0, \ldots, N_m).$$

In an analogous way, for a grand canonical system, we can define the pressure by setting

$$P\beta = \frac{1}{\hat{Q}_G(V; \{\hat{f}\})}\frac{\partial}{\partial V}\hat{Q}_G(V; \{\hat{f}\})$$

which is valid in the limit of large volumes since the numbers of particles do not fluctuate much around their mean values.

This equation can be written in the form

$$P\beta = \frac{\partial}{\partial V}[V\hat{Q}(V; \{\hat{f}\})],$$

and, in the limit of large volumes, we have

$$P\beta = \hat{Q}\{\hat{f}\}. \tag{9.3.24}$$

Relations (9.3.15) and (9.3.24) define, in parametric form, the variation of P in terms of the numbers $C_{\mathscr{A}}$ of molecules per unit volume.

Let us now examine how this formalism can be used to calculate osmotic pressures. Let us consider a solvent made of several chemical species denoted by indices $\mathscr{A} = 0, \ldots, l$, and a solution made both of the same chemical species and of a few other ones (at least one) denoted by indices $\mathscr{A} = l + 1, \ldots, m$. Let us assume that the solvent and the solution are in osmotic equilibrium on both sides of a membrane which is permeable for the species $\mathscr{A} = 0, \ldots, l$ but not for the species $\mathscr{A} = l + 1, \ldots, m$. Cell I contains the solution and cell II contains the solvent. Let P^{I} be the pressure in the solution and P^{II} the pressure in the solvent. As was shown in Chapter 5, the osmotic pressure of the solution with respect to the solvent is

$$\boldsymbol{\Pi} = \boldsymbol{P}^{\mathrm{I}} - \boldsymbol{P}^{\mathrm{II}}. \tag{9.3.25}$$

Thus, to evaluate $\boldsymbol{\Pi}$, we have only to calculate the pressures of homogeneous mixtures. The osmotic equilibrium condition can be expressed by writing that the chemical potentials, and therefore the fugacities of species $\mathscr{A} = 0, \ldots, l$, are equal on both sides of the membrane. On the solution side, we have

$$\{\hat{f}^{\mathrm{I}}\} = (\hat{f}_0, \ldots, \hat{f}_l, \hat{f}_{l+1}, \ldots, \hat{f}_m),$$

and, on the solvent side, we have

$$\{\hat{f}^{\mathrm{II}}\} = (\hat{f}_0, \ldots, \hat{f}_l, 0, \ldots, 0),$$

since on the solvent side

$$\hat{f}_{l+1} = \ldots = \hat{f}_m = 0.$$

Now, eqn (9.3.24) gives

$$\boldsymbol{P}^{\mathrm{I}}\boldsymbol{\beta} = \hat{Q}\{\hat{f}^{\mathrm{I}}\} \tag{9.3.26}$$

$$\boldsymbol{P}^{\mathrm{II}}\boldsymbol{\beta} = \hat{Q}\{\hat{f}^{\mathrm{II}}\} \tag{9.3.27}$$

(the volumes of the cells are assumed to be large). Thus, the osmotic pressure is

$$\boldsymbol{\Pi\beta} = \hat{Q}\{f^{\mathrm{I}}\} - \hat{Q}\{f^{\mathrm{II}}\} \tag{9.3.28}$$

The fugacities are implicitly determined by the concentrations of the species in each cell

$$\boldsymbol{C}^{\mathrm{I}}_{\mathscr{A}} = \hat{f}_{\mathscr{A}}\frac{\partial}{\partial \hat{f}_{\mathscr{A}}}\hat{Q}\{\hat{f}^{\mathrm{I}}\} \qquad \text{for } \mathscr{A} = 0, \ldots, l, \ldots, m \tag{9.3.29}$$

$$\boldsymbol{C}^{\mathrm{II}}_{\mathscr{A}} = \hat{f}_{\mathscr{A}}\frac{\partial}{\partial \hat{f}_{\mathscr{A}}}\hat{Q}\{\hat{f}^{\mathrm{II}}\} \qquad \text{for } \mathscr{A} = 0, \ldots, l. \tag{9.3.30}$$

If we know the functions $\hat{Q}\{\hat{f}^{\mathrm{I}}\}$ and $\hat{Q}\{\hat{f}^{\mathrm{I}}\}$ and the basic experimental data, we can determine the osmotic equilibrium.

First, we can fix the composition of the solvent $\{\boldsymbol{C}^{\mathrm{II}}\}$ and, for instance, we can imagine that cell II is connected to a large tank (dialysis equilibrium, see Chapter 5, Section 1.8). Equations (9.3.3.) determine the fugacities $\hat{f}^{\mathrm{II}}$ and eqn (9.3.27) gives the pressure in the solvent.

Let us now fix the numbers $\boldsymbol{N}_{\mathscr{A}}$ ($\mathscr{A} = l+1, \ldots, m$) of molecules of solute in cell I (all of them are in cell I). We can also fix the volume V^{I} of the solution contained in this cell since, in principle, we can always apply the pressure $\boldsymbol{P}^{\mathrm{I}}$ required to ensure that $\boldsymbol{V}^{\mathrm{I}}$ has the described value. This amounts to fixing the quantities

$$\boldsymbol{C}^{\mathrm{I}}_{\mathscr{A}} = \frac{\langle\!\langle \boldsymbol{N}^{\mathrm{I}}_{\mathscr{A}} \rangle\!\rangle}{\boldsymbol{V}} \qquad \mathscr{A} = l+1, \ldots m.$$

The equations (9.3.29) for $\mathscr{A} = l+1, \ldots, m$) and the fugacities $\hat{f}^{\mathrm{II}}$, give the values of all the fugacities $\hat{f}^{\mathrm{I}}$. Everything is now determined. We deduce $\boldsymbol{P}^{\mathrm{I}}$ from

(9.3.22) and Π from (9.3.28). Moreover, the equations (9.3.29) for $\mathscr{A} = 0, \ldots, l$ give us the concentrations in cell I, for the species constituting the solvent. We observe that a priori for $\mathscr{A} = 0, \ldots, l$ the concentrations $C^{\mathrm{I}}_{\mathscr{A}}$ and $C^{\mathrm{II}}_{\mathscr{A}}$ are not necessarily proportional: this is the preferential adsorption phenomenon (see Chapter 5, Section 1.8).

3.4 Structure function in the grand canonical formalism

Let us consider the simple case where the solute consists of homogeneous (but different) polymers (see Chapter 7, Section 2.2.3). In this case, the monomer concentration is given by [See (9.2.14)]

$$C_{\mathscr{A}}(\vec{r}) = \sum_{a \in \mathscr{A}} \sum_{j=1}^{N_a} \delta(\vec{r} - \vec{r}_{aj}). \tag{9.3.31}$$

The partial structure function $\boldsymbol{H}_{\mathscr{A}\mathscr{B}}(\vec{q})$ concerning species $\mathscr{A}$ and $\mathscr{B}$ is defined by [see (9.2.16)]

$$\boldsymbol{H}_{\mathscr{A}\mathscr{B}}(\vec{q}) = \int \mathrm{d}^3 r \, \mathrm{e}^{i\vec{q}\cdot\vec{r}} \left[\frac{\langle\!\langle C_{\mathscr{A}}(\vec{0}) \, C_{\mathscr{B}}(\vec{r}) \rangle\!\rangle}{C_{\mathscr{A}} \, C_{\mathscr{B}}} - 1 \right]$$

where

$$C_{\mathscr{A}} = \langle\!\langle C_{\mathscr{A}}(\vec{r}) \rangle\!\rangle = \frac{N_{\mathscr{A}} \langle\!\langle N_{\mathscr{A}} \rangle\!\rangle}{V}.$$

In particular, in Chapter 7 [see (9.3.8)] we proved the equality

$$\boldsymbol{H}_{\mathscr{A}\mathscr{B}}(\vec{0}) = V\left(\frac{\langle\!\langle N_{\mathscr{A}} N_{\mathscr{B}} \rangle\!\rangle}{\langle\!\langle N_{\mathscr{A}} \rangle\!\rangle \langle\!\langle N_{\mathscr{B}} \rangle\!\rangle} - 1 \right]. \tag{9.3.32}$$

For reasons expounded at the beginning of this chapter, it is convenient to express these mean values in the grand canonical formalism. Thus, for instance, the quantity $\langle\!\langle C_{\mathscr{A}}(\vec{r}) \, C_{\mathscr{B}}(\vec{r}) \rangle\!\rangle$ is just one of these functions $\hat{p}\,(\mathscr{R}, \mathscr{P}, \{\hat{f}\})$ defined by (9.3.23) in a very general way.

We observe that in the limit of large volumes, $\boldsymbol{H}_{\mathscr{A}\cdot\mathscr{B}}$ can be expressed in terms of $\hat{Q}_{\mathrm{G}}\{\hat{f}\}$. Indeed, using the definition (9.3.9) of $\hat{Q}_{\mathrm{G}}$, we find

$$\langle\!\langle N_{\mathscr{A}} \rangle\!\rangle = \frac{1}{\hat{Q}_{\mathrm{G}}(V; \{\hat{f}\})} \hat{f}_{\mathscr{A}} \frac{\partial}{\partial \hat{f}_{\mathscr{A}}} \hat{Q}_{\mathrm{G}}(V, \{\hat{f}\})$$

$$\langle\!\langle N_{\mathscr{A}} N_{\mathscr{B}} \rangle\!\rangle = \frac{1}{\hat{Q}_{\mathrm{G}}(V; \{\hat{f}\})} \left(\hat{f}_{\mathscr{A}} \frac{\partial}{\partial \hat{f}_{\mathscr{A}}} \right) \left(\hat{f}_{\mathscr{B}} \frac{\partial}{\partial \hat{f}_{\mathscr{B}}} \right) \hat{Q}_{\mathrm{G}}(V, \{\hat{f}]).$$

Using now (9.3.11), let us go to the limit $V \to \infty$; we obtain

$$\begin{aligned} \langle\!\langle (N_{\mathscr{A}} - \langle\!\langle N_{\mathscr{A}} \rangle\!\rangle)(N_{\mathscr{B}} - \langle\!\langle N_{\mathscr{B}} \rangle\!\rangle) \rangle\!\rangle &= \langle\!\langle N_{\mathscr{A}} N_{\mathscr{B}} \rangle\!\rangle - \langle\!\langle N_{\mathscr{A}} \rangle\!\rangle \langle\!\langle N_{\mathscr{B}} \rangle\!\rangle \\ &= V \left(\hat{f}_{\mathscr{A}} \frac{\partial}{\partial \hat{f}_{\mathscr{A}}} \right) \left(\hat{f}_{\mathscr{B}} \frac{\partial}{\partial \hat{f}_{\mathscr{B}}} \right) \hat{Q}\{\hat{f}\}. \end{aligned} \tag{9.3.33}$$

Consequently,

$$\boldsymbol{H}_{\mathscr{A}\mathscr{B}}(\vec{0}) = \frac{\left(\hat{f}_{\mathscr{A}} \frac{\partial}{\partial \hat{f}_{\mathscr{A}}}\right)\left(\hat{f}_{\mathscr{B}} \frac{\partial}{\partial \hat{f}_{\mathscr{B}}}\right) \hat{Q}\{\hat{f}\}}{\left[\hat{f}_{\mathscr{A}} \frac{\partial}{\partial \hat{f}_{\mathscr{A}}} \hat{Q}\{\hat{f}\}\right]\left[\hat{f}_{\mathscr{B}} \frac{\partial}{\partial \hat{f}_{\mathscr{B}}} \hat{Q}\{\hat{f}\}\right]}. \tag{9.3.34}$$

Taking (9.3.15) into account, i.e.

$$\boldsymbol{C}_{\mathscr{A}} = \hat{f}_{\mathscr{A}} \frac{\partial}{\partial \hat{f}_{\mathscr{A}}} \hat{Q}\{\hat{f}\},$$

we can also write

$$\boldsymbol{H}_{\mathscr{A}\mathscr{B}}(\vec{0}) = \frac{1}{\boldsymbol{C}_{\mathscr{A}} \boldsymbol{C}_{\mathscr{B}}} \hat{f}_{\mathscr{A}} \frac{\partial}{\partial \hat{f}_{\mathscr{A}}} \boldsymbol{C}_{\mathscr{B}}. \tag{9.3.35}$$

3.5 Pure solvent: structure function and compressibility

For pure solvent, the situation is especially simple; in this case, we have only one concentration $\boldsymbol{C}_0$, one fugacity $\hat{f}$, and one structure function $\boldsymbol{H}_{00}(\vec{q})$ (zero is the solvent index).

Then, there exists a relation between $\boldsymbol{H}_{00}(\vec{0})$ and the compressibility. In fact, (9.3.15), (9.3.24), and (9.3.34) take the simple form

$$\boldsymbol{C}_0 = \hat{f} \frac{\partial}{\partial \hat{f}} \hat{Q}(\hat{f})$$

$$\boldsymbol{P}\boldsymbol{\beta} = \hat{Q}(\hat{f})$$

$$\boldsymbol{H}_{00}(\vec{0}) = \frac{\left(\hat{f} \frac{\partial}{\partial \hat{f}}\right)^2 \hat{Q}(\hat{f})}{\left[\hat{f} \frac{\partial}{\partial \hat{f}} \hat{Q}(\hat{f})\right]^2}. \tag{9.3.36}$$

where $\hat{Q}(\hat{f})$ is the grand partition function. From these equations, we deduce

$$\mathrm{d}\boldsymbol{C}_0 = \left[\left(\hat{f} \frac{\partial}{\partial \hat{f}}\right)^2 \hat{Q}\right] \frac{\mathrm{d}\hat{f}}{\hat{f}},$$

$$\boldsymbol{\beta}\,\mathrm{d}\boldsymbol{P} = \left[\hat{f} \frac{\partial}{\partial \hat{f}} \hat{Q}\right] \frac{\mathrm{d}\hat{f}}{\hat{f}}. \tag{9.3.37}$$

By combining eqns (9.3.36) and (9.3.37) we obtain

$$\boldsymbol{H}_{00}(\vec{0}) = \frac{1}{\boldsymbol{\beta}\boldsymbol{C}_0} \frac{\partial \boldsymbol{C}_0}{\partial \boldsymbol{P}} \tag{9.3.38}$$

where $\boldsymbol{C}_0^{-1}\, \partial \boldsymbol{C}_0 / \partial \boldsymbol{P}$ is the compressibility.

When a radiation is scattered by density fluctuations, the scattered intensity is directly proportional to the structure function (see Chapter 7). Thus in the zero angle limit, the scattered intensity is proportional to the compressibility.

This relation will be generalized in the following sections and we shall find another important relation between the structure function and the osmotic pressure.

3.6 Solution composed of one solvent and of several solutes: elimination of the solvent fugacity in the incompressible case

When the solution is incompressible (or only slightly compressible), the situation simplifies and the fugacity of the solvent can be eliminated.

For any configuration of an incompressible solution, we have

$$\sum_{\mathscr{A}=0}^{m} N_{\mathscr{A}} v_{\mathscr{A}} = V,$$

and this relation can be written in the form

$$N_0 = \frac{V}{v_0} - \sum_{\mathscr{A}=0}^{m} N_{\mathscr{A}} \frac{v_{\mathscr{A}}}{v_0} \tag{9.3.39}$$

where $v_{\mathscr{A}}$ is the partial volume of a molecule belonging to species $\mathscr{A}$. Thus, the number N_0 if solvent particles is fixed as soon as the other numbers are determined.

The canonical partition function corresponding to possible configurations, can be written as follows

$$Q_G(V; N_1, \ldots, N_m) \equiv \hat{Q}_G\left(V; \frac{V}{v_0} - \sum_{\mathscr{A}=1}^{m} \frac{v_{\mathscr{A}}}{v_0} N_{\mathscr{A}},\ N_1, \ldots, N_m\right), \tag{9.3.40}$$

and, in particular, we have

$$Q_G(V; 0, \ldots, 0) \equiv \hat{Q}_G\left(V; \frac{V}{v_0}, 0, \ldots, 0\right) = 1. \tag{9.3.41}$$

Moreover, the grand partition function defined by (9.3.9) can be written in the form

$$\hat{Q}_G(V; \{\hat{f}\}) = (\hat{f}_0)^{V/v_0} \sum_{N_1=0}^{\infty} \ldots \sum_{N_m=0}^{\infty} \left(\frac{\hat{f}_1}{(\hat{f}_0)^{v_1/v_0}}\right)^{N_1} \ldots \left(\frac{\hat{f}_m}{(\hat{f}_0)^{v_M/v_0}}\right)^{N_m} \times$$

$$\times \hat{Q}_G\left(V; \frac{V}{v_0} - \sum_{\mathscr{A}=1}^{m} \frac{v_{\mathscr{A}} N_{\mathscr{A}}}{v_0}, N_1, \ldots, N_m\right). \tag{9.3.42}$$

Thus, we can associate the grand partition function

$$Q_G(V; \{f\}) = \sum_{N_1=0}^{\infty} \ldots \sum_{N_m=0}^{\infty} f_1^{N_1} \ldots f_1^{N_m} Q_G(V; N_1, \ldots, N_m) \tag{9.3.43}$$

with the new partition functions $Q_G(V; N_1, \ldots, N_m)$, and this formula, very similar to (9.3.9), differs from it only because the contribution of the solvent disappeared.

Let us now compare (9.3.42) and (9.3.43) by taking (9.3.40) into account. We see that, by setting

$$f_{\mathscr{A}} = \frac{\hat{f}_{\mathscr{A}}}{(\hat{f}_0)^{v_{\mathscr{A}}/v_0}} \qquad \mathscr{A} \neq 0 \tag{9.3.44}$$

we obtain the correspondence

$$\hat{Q}_G(V; \{\hat{f}\}) = (\hat{f}_0)^{V/v_0} Q_G(V; \{\hat{f}\}). \tag{9.3.45}$$

Of course, we set

$$VQ(V; \{f\}) = \ln Q_G(V; \{f\}), \tag{9.3.46}$$

a relation which is analogous to (9.3.11) and, for $V \to \infty$

$$Q\{f\} = \lim_{V \to \infty} Q(V; \{f\}). \tag{9.3.47}$$

Thus, we obtain

$$Q(V; \{f\}) = \hat{Q}(V; \{\hat{f}\}) - \frac{1}{v_0} \ln \hat{f}_0.$$

According to (9.3.44), for $\mathscr{A} \neq 0$, we also have

$$\hat{f}_{\mathscr{A}} \frac{\partial}{\partial \hat{f}_{\mathscr{A}}} = f_{\mathscr{A}} \frac{\partial}{\partial f_{\mathscr{A}}} \tag{9.3.48}$$

Taking (9.3.41) and (9.3.42) into account, we can rewrite (9.3.12) and (9.3.33) in the form

$$\boldsymbol{C}_{\mathscr{A}} = \frac{\langle\langle N_{\mathscr{A}} \rangle\rangle}{V} = f_{\mathscr{A}} \frac{\partial}{\partial f_{\mathscr{A}}} Q(V; \{f\}) \qquad \mathscr{A} \neq 0 \tag{9.3.49}$$

$$\left\langle\!\!\left\langle (N_{\mathscr{A}} - \langle\langle N_{\mathscr{A}} \rangle\rangle)(N_{\mathscr{B}} - \langle\langle N_{\mathscr{B}} \rangle\rangle) \right\rangle\!\!\right\rangle = V\left(f_{\mathscr{A}} \frac{\partial}{\partial f_{\mathscr{A}}}\right)\left(f_{\mathscr{B}} \frac{\partial}{\partial f_{\mathscr{B}}}\right) Q(V; \{f\})$$
$$\mathscr{A} \neq 0 \quad \mathscr{B} \neq 0. \tag{9.3.50}$$

The same correspondence can be established when a condition $\mathscr{R}, \mathscr{P}$ is enforced. Thus, we can define restricted partition functions $Q_G(V, \mathscr{R}, \mathscr{P}; N_1, \ldots, N_m)$ with which we associate the grand partition function

$$Q_G(V, \mathscr{R}, \mathscr{P}; \{f\}) = \sum_{N_q = 0}^{\infty} \cdots \sum_{N_m = 0}^{\infty} f_1^{N_1} \ldots f_m^{N_m} Q_G(V, \mathscr{R}, \mathscr{P}; N_1, \ldots, N_m). \tag{9.3.51}$$

Thus, requiring again the validity of relation (9.3.44) which expresses the $f_{\mathscr{A}}$ in terms of the $\hat{f}_{\mathscr{A}}$, we obtain a correspondence analogous to (9.3.45)

$$\hat{Q}_G(V, \mathscr{R}, \mathscr{P}; \{f\}) = \hat{f}_0^{\,V/v_0} Q_G(V, \mathscr{R}, \mathscr{P}; \{f\}). \tag{9.3.52}$$

Defining the correlation $p(\mathscr{R},\mathscr{P};\{f\})$ by

$$p(\mathscr{R},\mathscr{P};\{f\}) = \frac{Q_G(V,\mathscr{R},\mathscr{P};\{f\})}{Q_G(V;\{f\})}, \tag{9.3.53}$$

we see, by comparing with (9.3.19), that we have

$$p(\mathscr{R},\mathscr{P};\{f\}) = \hat{p}(\mathscr{R},\mathscr{P};\{f\}). \tag{9.3.54}$$

For instance, let us consider $\boldsymbol{H}_{\mathscr{A}\mathscr{B}}(\vec{0})$ which is given by (9.3.35), and let us assume $\mathscr{A} \neq 0$ and $\mathscr{B} \neq 0$. In (9.3.35), $\boldsymbol{C}_{\mathscr{B}}$ is considered as a function of the $\hat{f}_{\mathscr{A}}$ ($\mathscr{A} = 0, \ldots, m$) but, if the solution is incompressible, $\boldsymbol{C}_{\mathscr{B}}$ can be re-expressed in terms of the $f_{\mathscr{A}}$ ($\mathscr{A} = 1, \ldots, m$); thus taking (9.3.48) into account, we can rewrite (9.3.35) in the form

$$\boldsymbol{H}_{\mathscr{A}\mathscr{B}}(\vec{0}) = \frac{1}{\boldsymbol{C}_{\mathscr{A}}\boldsymbol{C}_{\mathscr{B}}} f_{\mathscr{A}} \frac{\partial}{\partial f_{\mathscr{A}}} \boldsymbol{C}_{\mathscr{B}}, \tag{9.3.55}$$

or, by using (9.3.49),

$$\boldsymbol{H}_{\mathscr{A}\mathscr{B}}(\vec{0}) = \frac{\left(f_{\mathscr{A}} \dfrac{\partial}{\partial f_{\mathscr{A}}}\right)\left(f_{\mathscr{B}} \dfrac{\partial}{\partial f_{\mathscr{B}}}\right) Q\{f\}}{\left[f_{\mathscr{A}} \dfrac{\partial}{\partial f_{\mathscr{A}}} Q\{f\}\right]\left[f_{\mathscr{B}} \dfrac{\partial}{\partial f_{\mathscr{B}}} Q\{f\}\right]}. \tag{9.3.56}$$

We thus obtain an expression which formally looks like (9.3.34) but which depends only on m variables instead of $(m + 1)$.

In brief, if the solution is incompressible, the parameters concerning the solvent can be eliminated very simply. However, no completely incompressible system exists, and the mathematical transformations presented above may look rather crude. More refinements[1] could be introduced and the elimination could be done in a more rigorous way but the results would not be very different. Moreover, it is always possible to take the pressure variations into account empirically by allowing variations of the 'quasi-potentials' that appear in the theoretical expressions (see Chapter 10).

3.7 Osmotic pressure of an incompressible solution composed of one solvent and of several solutes

Let us consider a solution in equilibrium with pure solvent, through a semi-permeable membrane. The pressure in cell I, which contains the solution is given by (9.3.24), and, for reasons of convenience, $\hat{Q}\{f\}$ will be written in a form in which the fugacity of the solvent is separated from the fugacities of the solutes. Thus,

$$\boldsymbol{P}^{\mathrm{I}}\boldsymbol{\beta} = \hat{Q}\{\hat{f}\} = \hat{Q}(\hat{f}_0, \{\hat{f}_{\mathscr{A}}\}_{\mathscr{A} \neq 0}). \tag{9.3.57}$$

As was shown in Section 5, the fugacity $\hat{f}_0$ must be the same on both sides of the membrane. Thus, the pressure $\boldsymbol{P}^{\mathrm{II}}$ in cell II, which contains pure solvent, is given by

$$\boldsymbol{P}^{\mathrm{II}}\boldsymbol{\beta} = \hat{Q}(\hat{f}_0, \{\hat{f}_{\mathscr{A}} = 0\}_{\mathscr{A} \neq 0}). \tag{9.3.58}$$

Let us now assume that solution and solvent are both incompressible. In this case, we can use (9.3.47) (for $V \to \infty$)

$$Q\{f\} = Q\{\hat{f}\} - \frac{1}{v_0}\ln \hat{f}_0. \tag{9.3.59}$$

We observe that, if for every $\mathscr{A} = 1, \ldots, \mathrm{m}$, $\hat{f}_{\mathscr{A}} = 0$, then we also have $f_{\mathscr{A}} = 0$, according to (9.3.44) and, in this case, $Q\{f\} = 0$ because $Q_{\mathrm{G}}(V, \{f\}) = 1$ in agreement with the expansion (9.3.43) and the condition (9.3.41). Thus, bringing (9.3.59)) in (9.3.57) and (9.3.58), we obtain

$$\boldsymbol{P}^{\mathrm{I}}\boldsymbol{\beta} = Q\{f\} + \frac{1}{v_0}\ln \hat{f}_0,$$

$$\boldsymbol{P}^{\mathrm{II}}\boldsymbol{\beta} = \frac{1}{v_0}\ln \hat{f}_0.$$

The osmotic pressure is given by the difference between these pressures [see (9.3.28)]. In this way, we find

$$\boldsymbol{\Pi}\boldsymbol{\beta} = Q\{f\}. \tag{9.3.60}$$

Moreover, since according to (9.3.44) (for $V \to \infty$), we have

$$\boldsymbol{C}_{\mathscr{A}} = f_{\mathscr{A}}\frac{\partial}{\partial f_{\mathscr{A}}}Q\{f\} \tag{9.3.61}$$

we see that if we know $Q\{f\}$, we can express the osmotic pressure as a function of the concentrations in parametric form, and the parameters are the fugacities $f_{\mathscr{A}}$ where $\mathscr{A} = 1, \ldots, m$.

3.8 Structure function and osmotic compressibility of an incompressible solution composed of one solvent and one solute

Mathematically, the present situation is very similar to the situation considered in Section 3.5 when we were dealing with a pure compressible solvent. The physical situation is different, since here we consider an incompressible solution with only one solute. However, as before, there is only one concentration $\boldsymbol{C}$, one fugacity f and one structure function $\boldsymbol{H}_{11}(\vec{q})$ (the index 1 is the solute index).

In this case, there is a relation between $\boldsymbol{H}_{11}(\vec{0})$ and the osmotic pressure. Indeed, (9.3.61), (9.3.60) and (9.3.56) read

$$C_1 = f\frac{\partial}{\partial f}Q(f)$$

$$\Pi\beta = Q(f) \tag{9.3.62}$$

$$\boldsymbol{H}_{11}(\vec{0}) = \frac{\left(f\frac{\partial}{\partial f}\right)^2 Q(f)}{\left[f\frac{\partial}{\partial f}Q(f)\right]^2}. \tag{9.3.63}$$

From these equations, we deduce

$$\mathrm{d}C_1 = \left[\left(f\frac{\partial}{\partial f}\right)^2 Q\right]\frac{\mathrm{d}f}{f}$$

$$\beta\,\mathrm{d}\Pi = \left[f\frac{\partial}{\partial f}Q\right]\frac{\mathrm{d}f}{f}. \tag{9.3.63}$$

By combining eqns (9.3.62) and (9.3.63) we obtain

$$\boldsymbol{H}_{11}(\vec{0}) = \frac{1}{\beta C_1}\frac{\partial C_1}{\partial \Pi} \tag{9.3.64}$$

where $(C_1)^{-1}\partial C_1/\partial\Pi$ is the osmotic compressibility. This relation is very similar to (9.3.38) but it applies in a different situation. It is very important because it enables the experimentalist to measure the osmotic compressibility by scattering, in the zero angle limit. In fact, for $\vec{q}\to 0$, relation (9.2.37) gives

$$\Xi(0) = (\mathfrak{b}^{\circ\!\to})^2 C_1^2 \boldsymbol{H}_{11}(\vec{0})$$

where $\Xi(\vec{q})$ is the cross-section per unit volume of solution, $\mathfrak{b}^{\circ\!\to}$ the contrast length and C_1 the number of scatterers (or monomers) per unit volume. Equation, (9.3.64) applies only in the case where the solute contains only one kind of molecules. It is valid for a monodisperse polymer sample not for a polydisperse one. However, in the semi-dilute limit, the polydispersion effects disappear because of screening effects. Thus, as in the monodisperse case, (9.3.64) can also be written in the form

$$\boldsymbol{H}_{11}(\vec{0}) = \frac{1}{\beta C_1}\frac{\partial C_1}{\partial \Pi} \tag{9.3.65}$$

which remains meaningful in the semi-dilute limit. It is to be expected that (9.3.65) is valid in this limit, even in the polydisperse case. On the contrary, for dilute solutions, polydispersion plays a role which will be studied in Section 4 of the present chapter

3.9 Structure function and osmotic compressibility of an incompressible solution made of two solvents and one solute

Frequently, when samples are prepared for scattering experiments, a second solvent is added to the main solvent, in order to increase the solvent–solute contrast, and this operation is also used to study effects related to the quality of the solvent.

Thus, let us consider a solution made of two solvents (characterized by indices 0 and 1) and one solute characterized by index 2). It will be assumed that the solution is incompressible: consequently, the parameters associated with the main solvent can be eliminated (index 0).

According to (7.2.37), for a given radiation, the scattering cross-section per unit volume can be written in the following form

$$\Xi(q) = (\mathfrak{b}_1^{\circ\rightarrow} C_1)^2 \boldsymbol{H}_{11}(\vec{q}) + 2\mathfrak{b}_1^{\circ\rightarrow} \mathfrak{b}_2^{\circ\rightarrow} C_1 C_2 \boldsymbol{H}_{12}(\vec{q}) + \mathfrak{b}_2^{\circ\rightarrow} C_2^2 \boldsymbol{H}_{22}(\vec{q}). \qquad (9.3.66)$$

Now there are three different structure functions which can be studied by scattering experiments. In what follows, we shall be interested in the zero angle limit. More precisely, it will be shown that there is a relation between $\boldsymbol{H}_{22}(\vec{0})$ and the osmotic compressibility, which will now be defined.

First, we note that, in principle, it is possible to measure the osmotic pressure of the solution put in cell I with respect to a mixture of two solvents put in cell II. It is assumed that both solvents can freely cross the membrane between the cells. Thus, the second solvent must have the same fugacity on both sides of the membrane; consequently, we set

$$f_1^{\mathrm{I}} = f_1^{\mathrm{II}} = f_1$$
$$f_2^{\mathrm{I}} = f_2 \qquad f_2^{\mathrm{II}} = 0.$$

Moreover, the elimination of the fugacity $\hat{f}_0$ in (9.3.28) leads to the equality

$$\boldsymbol{\Pi\beta} = Q(f_1, f_2) - Q(f_1, 0)$$

which is a simple generalization of (9.3.60). On the other hand, according to (9.3.61), we have

$$\boldsymbol{C}_2 = f_2 \frac{\partial}{\partial f_2} Q(f_1, f_2). \qquad (9.3.67)$$

Keeping f_1 fixed and letting f_2 vary, we can write

$$\mathrm{d}\boldsymbol{C}_2 = \left[\left(f_2 \frac{\partial}{\partial f_2}\right)^2 Q(f_1, f_2)\right] \frac{\mathrm{d}f_2}{f_2}$$

$$\boldsymbol{\beta}\,\mathrm{d}\boldsymbol{\Pi} = \left[f_2 \frac{\partial}{\partial f_2} Q(f_1, f_2)\right] \frac{\mathrm{d}f_2}{f_2}.$$

Considering $\boldsymbol{C}_2$ as a function of $\boldsymbol{\Pi}$ and f_1, we find

$$\frac{\partial \boldsymbol{C}_2(\boldsymbol{\Pi}, f_1)}{\boldsymbol{\beta}\,\partial \boldsymbol{\Pi}} = \frac{\left(f_2 \dfrac{\partial}{\partial f_2}\right)^2 Q(f_1, f_2)}{\left[f_2 \dfrac{\partial}{\partial f_2} Q(f_1, f_2)\right]}. \tag{9.3.68}$$

Now, we observe that (for $\mathscr{A} = \mathscr{B} = 2$), (9.3.56) reads

$$\boldsymbol{H}_{22}(\vec{0}) = \frac{\left[\left(f_2 \dfrac{\partial}{\partial f_2}\right)^2 Q(f_1, f_2)\right]}{\left[f_2 \dfrac{\partial}{\partial f_2} Q(f_1, f_2)\right]^2}. \tag{9.3.69}$$

From eqns (9.3.68), (9.3.67), and (9.3.69), we get

$$\boldsymbol{H}_{22}(\vec{0}) = \frac{1}{\boldsymbol{\beta} \boldsymbol{C}_2} \frac{\partial \boldsymbol{C}_2(\boldsymbol{\Pi}, f_1)}{\partial \boldsymbol{\Pi}}. \tag{9.3.70}$$

Let us now come back to $\Xi(0)$ and let us consider the special case where the solvent molecules are monomers ($C_1 = \boldsymbol{C}_1$) and where each polymer molecule contains N monomers ($C_2 = N\boldsymbol{C}_2$). Taking (9.3.70) and (9.3.55) into account, we can write $\Xi(0)$ in the form

$$\Xi(0) = (\mathfrak{b}_1^{\circ\!\to})^2 f_1 \frac{\partial \boldsymbol{C}_1}{\partial f_1} + 2\mathfrak{b}_1^{\circ\!\to} \mathfrak{b}_2^{\circ\!\to} N f_1 \frac{\partial}{\partial f_1} \boldsymbol{C}_2 + (\mathfrak{b}_2^{\circ\!\to} N)^2 \frac{\boldsymbol{C}_2}{\boldsymbol{\beta}} \frac{\partial \boldsymbol{C}_2(\boldsymbol{\Pi}, f_1)}{\partial \boldsymbol{\Pi}}.$$

Unfortunately, the result is rather complicated. However, when the contrast length $\mathfrak{b}^{\circ\!\to}$, vanishes, the cross-section is directly proportional to the inverse of the derivative of the osmotic pressure with respect to the solute concentration.

There is also another interesting case, namely, when the two solvents 0 and 1 are isotopic varieties of the same chemical species (for instance, H_2O and D_2O). In fact, such mixtures are used to study general properties of polymer solutions by small-angle scattering. Then, one may admit that the partial volumes of the molecules belonging to the two varieties are the same. In this case, the expression of the cross-section in the zero angle limit simplifies. Equation (7.2.40) gives

$$\Xi(0) = (\mathfrak{b}_0 - \mathfrak{b}_1)^2 x(1 - x) C_s + (\mathfrak{b}_2^{\circ\!\to} C_2)^2 \boldsymbol{H}_{22}(\vec{0})$$

where $C_s = \boldsymbol{C}_s = \boldsymbol{C}_0 + \boldsymbol{C}_1$ and $x = \boldsymbol{C}_1/(\boldsymbol{C}_0 + \boldsymbol{C}_1)]$.

Thus, using (9.3.70), one obtain

$$\Xi(0) = (\mathfrak{b}_0 - \mathfrak{b}_1)^2 x(1 - x) \boldsymbol{C}_s + (\mathfrak{b}_2^{\circ\!\to} N)^2 \frac{\boldsymbol{C}_2}{\boldsymbol{\beta}} \frac{\partial \boldsymbol{C}_2(\boldsymbol{\Pi})}{\partial \boldsymbol{\Pi}}. \tag{9.3.71}$$

More explicit results can be found in an article by des Cloizeaux and Jannink (1980)[1].

4. HOMOGENEOUS POLYMERS: EXPANSIONS IN POWERS OF THE CONCENTRATIONS

4.1 Representation of the partition functions and symmetry factors

A set $\mathscr{E}$ of molecules belonging to m types can be defined by giving for every $\mathscr{A}$ ($\mathscr{A} = 1, \ldots, m$), the number $N_{\mathscr{A}}$ of molecules of type $\mathscr{A}$ belonging to the set. This is what we have done until now and consequently we represented the partition function of such a set by using the symbol $Q_G(V; N_1, \ldots, N_m)$ where V is the volume of space occupied by the molecules.

It is also possible to define a set $\mathscr{E}$ of $\boldsymbol{N}$ molecules in a different way, by listing the molecules contained in $\mathscr{E}$, and this other representation is more adapted to the study of sets of homogeneous polymers.

In fact, a homogeneous polymer can be characterized by its number N of monomers (or links): polymers having different numbers of monomers belong to different species, and therefore it is possible to identify $\mathscr{A}$ with $N(N_{\mathscr{A}} \to N_N)$. Thus $\mathscr{E}$ can be defined by the set $(N_1, \ldots, N_{\boldsymbol{N}})$ of the numbers of monomers contained in the chains constituting $\mathscr{E}$; obviously, the corresponding N_N are thus given by

$$N_N = \sum_{j=1}^{\boldsymbol{N}} \delta_{N, N_j}$$

$$\boldsymbol{N} = \sum_{N=1}^{\infty} N_N . \tag{9.4.1}$$

In the same spirit, we can represent the partition function associated with a set $\mathscr{E}$ of $\boldsymbol{N}$ polymers by the symbol Z_G $(V; N_1, \ldots, N_{\boldsymbol{N}})$. Everywhere in the following pages of this book Z_G $(V; N_1, \ldots, N_{\boldsymbol{N}})$ are always calculated as if the polymers were distinguishable (actually they may or may not be so), and as if they had a distinguishable origin (dissymmetrical polymers).

Actually, the synthetic polymers which are used in experiments are generally dissymmetrical, but in both cases, the properties of solutions of long polymers are the same. On the other hand, the theoretical implications of the presence or the absence of a symmetry are trivial.

When certain polymers in $\mathscr{E}$ have the same number of monomers, the true partition function is

$$Q_G(V; N_1, N_2, \ldots) = \frac{Z_G(V; N_1, \ldots, N_p)}{\$(N_1, \ldots, N_p)} \tag{9.4.2}$$

where $\$(N_1, \ldots, N_p)$ is a symmetry number. When the chains of $\mathscr{E}$ are different and dissymmetrical $\$(N_1, \ldots, N_p) = 2^p$.

In a general way, when the chains of $\mathscr{E}$ are dissymmetrical, which will be assumed in the future, $\mathcal{S}(N, \ldots, N_p)$ is given by

$$\mathcal{S}(N_1, \ldots, N_p) = \prod_{N=1}^{\infty} N_N! \tag{9.4.3}$$

These symmetry factors are important: neglecting them would lead to Gibbs's paradox, and their role appears clearly in the following simple example.

Let us consider a set $\mathscr{E}$ of $\boldsymbol{N}$ identical chains made of N independent monomers. The partition function of one chain is of the form

$$Z_{\mathrm{G}}(N) = \boldsymbol{V}Z^N$$

where Z is a constant. Thus, the total partition function is

$$Q_{\mathrm{G}}(\boldsymbol{V}; \boldsymbol{N}) = \frac{(\boldsymbol{V}Z^N)^{\boldsymbol{N}}}{\boldsymbol{N}!}.$$

The free energy is given by

$$\beta\mathscr{F} = -\ln Q_{\mathrm{G}}(\boldsymbol{V}; \boldsymbol{N}).$$

Now, let us consider the limit $\boldsymbol{V} \to \infty$, $\boldsymbol{N} \to \infty$. Recalling Moivre's formula $\boldsymbol{N}! \sim (\boldsymbol{N}/\mathrm{e})^{\boldsymbol{N}}$ and definition $\boldsymbol{C} = \boldsymbol{N}/\boldsymbol{V}$, we find

$$\beta\mathscr{F}/\boldsymbol{V} = \boldsymbol{C}(\ln \boldsymbol{C} - N \ln Z - 1).$$

We see that the symmetry factor ensures the proportionality of $\mathscr{F}$ to $\boldsymbol{V}$, as we want, for large values of $\boldsymbol{V}$.

We also note that the grand canonical formalism deals with symmetry factors in a very elegant way. In fact, with the help of (9.4.2) and of (9.4.3), it is possible to rewrite (9.3.43) in the form

$$Q_{\mathrm{G}}(\boldsymbol{V}; \{f\}) = 1 + \sum_{\boldsymbol{N}=1}^{\infty} \frac{1}{\boldsymbol{N}!} \sum_{N_1, \ldots, N_{\boldsymbol{N}}} f_{N_1} \ldots f_{N_{\boldsymbol{N}}} Z_{\mathrm{G}}(\boldsymbol{V}; N_1, \ldots, N_{\boldsymbol{N}}), \tag{9.4.4}$$

and, in the same way, the function

$$Q(\boldsymbol{V}; \{f\}) = \boldsymbol{V}^{-1} \ln Q_{\mathrm{G}}(\boldsymbol{V}; \{f\})$$

takes the form

$$Q(\boldsymbol{V}; \{f\}) = \sum_{\boldsymbol{N}=1}^{\infty} \frac{1}{\boldsymbol{N}!} \sum_{N_1, \ldots, N_{\boldsymbol{N}}} f_{N_1} \ldots f_{N_{\boldsymbol{N}}} Z(\boldsymbol{V}; N_1, \ldots, N_{\boldsymbol{N}}) \tag{9.4.5}$$

where the $Z(\boldsymbol{V}; N_1, \ldots, N_{\boldsymbol{N}})$ are the cumulants of the $Z_{\mathrm{G}}(\boldsymbol{V}; N_1, \ldots, N_{\boldsymbol{N}})$ and where the factor $1/\boldsymbol{N}!$ plays an essential and complex role. It avoids over-counting when the N_j are different, but it plays the role of a symmetry factor when certain monomer numbers are equal.

Incidentally, we observe that owing to the symmetry factor the true partition functions $Z_{\mathrm{G}}(\boldsymbol{V}, N_1, \ldots, N_{\boldsymbol{N}})/\mathcal{S}(N_1, \ldots, N_{\boldsymbol{N}})$ vary discontinuously when two numbers of monomers become equal. On the contrary, the dependence of the quantities $Z_{\mathrm{G}}(\boldsymbol{V}; N_1, \ldots, N_{\boldsymbol{N}})$ and $Z(\boldsymbol{V}; N_1, \ldots, N_{\boldsymbol{N}})$ with respect to the monomer number is smooth; this is why these quantities are especially interesting.

The restricted partition functions can be represented in the same way. The true restricted partition function can be written in the form

$$\mathcal{Q}_G(V; \mathcal{R}, \mathcal{P}; N_1, N_2, \ldots) = \frac{Z_G(V; \mathcal{R}, \mathcal{P}; N_1, \ldots, N_N)}{\$(\mathcal{P}; N_1, \ldots, N_N)}, \tag{9.4.6}$$

an equality which generalizes (9.4.2). Here, $\$(\mathcal{P}; N_1, \ldots, N_N)$ is also a symmetry number. This number depends on the points marked on the chains* and symbolized by $\mathcal{P}$. On the contrary, it does not depend on the positions $\mathcal{R}$ assigned in space to these marked points, because these positions are a priori considered as distinguishable.

The corresponding restricted grand partition function [see (9.3.51)] is of the form

$$\mathcal{Q}_G(V; \mathcal{R}, \mathcal{P}; \{f\}) = \sum_{N=1}^{\infty} \frac{1}{\$_N(\mathcal{P})} \sum_{N_1, \ldots, N_N} f_{N_1} \ldots f_{N_N} Z_G(V; \mathcal{R}, \mathcal{P}; N_1, \ldots, N_N) \tag{9.4.7}$$

where $\$_N(\mathcal{P})$ is a symmetry number which depends on the marked points and which is calculated as if all chains were identical.

Moreover, as we associated $\mathcal{Q}(V; \{f\})$ with $\mathcal{Q}_G(V; \{f\})$, we can associate a function $\mathcal{Q}(V, \mathcal{R}, \mathcal{P}; \{f\})$ with $\mathcal{Q}_G(V, \mathcal{R}, \mathcal{P}; \{f\})$. This function is

$$\mathcal{Q}(V; \mathcal{R}, \mathcal{P}; \{f\}) = \mathcal{Q}_G(V, \mathcal{R}, \mathcal{P}; \{f\})/\mathcal{Q}_G(V; \{f\}), \tag{9.4.8}$$

and it can be expanded in the form

$$\mathcal{Q}(V, \mathcal{R}, \mathcal{P}; \{f\}) = \sum_{N=1}^{\infty} \frac{1}{\$_N(\mathcal{P})} \sum_{N_1} \cdots \sum_{N_N} f_{N_1} \ldots f_{N_N} Z(V, \mathcal{R}, \mathcal{P}; N_1, \ldots, N_N) \tag{9.4.9}$$

where the $Z(V, \mathcal{R}, \mathcal{P}; N_1, \ldots, N_N)$ have simple representations in terms of perturbation diagrams (see Chapter 10). Moreover, when $V \to \infty$, the correlation functions $Z(V, \mathcal{R}, \mathcal{P}; N_1, \ldots, N_N)$ have finite limits, namely $Z(\mathcal{R}, \mathcal{P}; N_1, \ldots, N_N)$, and $\mathcal{Q}(V; \mathcal{R}, \mathcal{P}; \{f\})$ also has a finite limit, namely $\mathcal{Q}(\mathcal{R}, \mathcal{P}; \{f\})$.

4.2 Expansion of the osmotic pressure in powers of the concentration

The grand canonical formalism provides expansions of the osmotic pressure and of the concentrations in powers of the fugacities but, in principle, it is possible to eliminate these fugacities so as to obtain an expansion of the osmotic pressure in powers of the concentrations; moreover, in practice, the calculations are nearly always performed in the limiting case $V \to \infty$.

* For instance, let us consider three polymers with N links. A function $Z_G(V; N, N, N)$ is associated with them and the corresponding symmetry number is $\$(N, N, N) = 1$. Now, let us mark a point on one polymer. Then, the $Z_G(V, \bullet; N, N, N) = Z_G(V; N, N, N)$ is associated with this situation and the corresponding symmetry number is $\$(\bullet; N, N, N) = 2!$.

Wher the polymers in solution are monodisperse, i.e. when all the chains have the same number of links, the problem simplifies, because the polymers are then identical (or can be considered as such). In this case, the connected part of the partitition function of $\boldsymbol{N}$ chains (each one with N links), is to be denoted by

$$\boldsymbol{Q}(\boldsymbol{N}) = Z(\boldsymbol{N}\times N)/\boldsymbol{N}!$$

[thus, for instance, $Z(1\times N) \equiv Z(N)$ and $Z(2\times \mathrm{N}) \equiv Z(N, N)$].

The osmotic pressure $\boldsymbol{\Pi}$ is given in terms of the polymer concentration $\boldsymbol{C}$ by the formulae [see (9.3.60) and (9.3.61)]

$$\boldsymbol{\Pi\beta} = Q(f) \tag{9.4.10}$$

$$\boldsymbol{C} = f\frac{\partial}{\partial f}Q(f) \tag{9.4.11}$$

where

$$Q(f) = \sum_{\boldsymbol{N}=1}^{\infty} \frac{f^{\boldsymbol{N}}}{\boldsymbol{N}!} Z(\boldsymbol{N}\times N). \tag{9.4.12}$$

The expansion of $\boldsymbol{C}$ in powers of f given by (9.4.11) can be inverted, so as to give an expansion of f in powers of $\boldsymbol{C}$; then, by bringing the latter in (9.4.10), one obtains an expansion of $\boldsymbol{\Pi\beta}$ in powers of $\boldsymbol{C}$; it is sometimes called the virial expansion.*

* Historically, this name comes from the fact that the pressure $\boldsymbol{P}$ of a gas of particles in a tank can be calculated by applying the virial theorem. This theorem is established by writing that the motion of a particle of mass m_j and position r_j is a consequence of the force $\mathscr{F}$; which is exerted on it

$$m_j\frac{\mathrm{d}^2 r_j}{\mathrm{d}r^2} = \mathscr{F}_j \qquad j = 1, \ldots, \boldsymbol{N}.$$

This entails that

$$\sum_{j=1}^{\boldsymbol{N}} \vec{\mathscr{F}}_j\cdot\vec{r}_j = \frac{\mathrm{d}}{\mathrm{d}t}\left[\sum_{j=1}^{\boldsymbol{N}} m_j\vec{r}_j\cdot\frac{\mathrm{d}\vec{r}_j}{\mathrm{d}t}\right] - \sum_{j=1}^{\boldsymbol{N}} m_j\left(\frac{\mathrm{d}\vec{r}_j}{\mathrm{d}t}\right)^2.$$

By time averaging of the preceding expression and by using the fact that the mean value of the kinetic energy has the value $3/2\boldsymbol{\beta}$, we find that

$$\left\langle\!\!\left\langle \sum_{j=1}^{\boldsymbol{N}} \vec{\mathscr{F}}_j\cdot\vec{r}_j \right\rangle\!\!\right\rangle = -3\boldsymbol{N}/\boldsymbol{\beta}.$$

There are two kinds of forces: first, the molecular forces which can be represented by ${}^{\mathrm{I}}\vec{\mathscr{F}}_j$, and secondly the forces exerted on the molecules by the walls of the tank; these external forces are equal and opposite to the pressure forces.

By separating the two kinds of forces, we find

$$-3\boldsymbol{PV} + \left\langle\!\!\left\langle \sum_{j=1}^{\boldsymbol{N}} \vec{\mathscr{F}}_j\cdot\vec{r}_j \right\rangle\!\!\right\rangle = -3\boldsymbol{N}/\boldsymbol{\beta}$$

and by setting $\boldsymbol{C} = \boldsymbol{N}/\boldsymbol{V}$

$$\boldsymbol{P\beta} = \boldsymbol{C} + \frac{\boldsymbol{\beta}}{3\boldsymbol{V}}\left\langle\!\!\left\langle \sum_{j=1}^{\boldsymbol{N}} {}^{\mathrm{I}}\vec{\mathscr{F}}_j\cdot\vec{r}_j \right\rangle\!\!\right\rangle.$$

By definition, the sum $\sum_j {}^{\mathrm{I}}\vec{\mathscr{F}}_j\cdot\vec{r}_j$ is the virial associated with the system. For a system composed of simple (point-like, molecules, the expansion of the virial in powers of the concentration can be expressed directly in a simple manner. This is not the same when one deals with polymers.

Using the notation

$$Y(N \times N) = \frac{Z(N \times N)}{[Z(N)]^N} \tag{9.4.13}$$

we can write the expansion of f in the form

$$f = \frac{C}{Z(N)}\{1 - CY(2\times N) - C^2[\tfrac{1}{2}Y(3\times N) - 2Y^2(2\times N)] -$$

$$- C^3[\tfrac{1}{6}Y(4\times N) - \tfrac{5}{2}Y(3\times N)Y(2\times N) + 5Y^3(2\times N)] + \ldots\}, \tag{9.4.14}$$

and, in the same way, the first terms of the expansion of $\Pi\beta$ are

$$\Pi\beta = C\{1 - \frac{C}{2}Y(2\times N) - C^2[\tfrac{1}{3}Y(3\times N) - Y^2(2\times N)] -$$

$$- C^3[\tfrac{1}{8}Y(4\times N) - \tfrac{3}{2}Y(3\times N)Y(2\times N) + \tfrac{5}{2}Y^3(2\times N)] + \ldots\}. \tag{9.4.15}$$

Incidentally, let us note that these formulae which apply in the limit $V \to \infty$ can be extended to the finite case by replacing* everywhere $Y(N\times N)$ by $Y(V; N\times N)$.

* An expansion of $\Pi\beta$ in powers of C could have been obtained in the canonical framework by starting from the formulae

$$\Pi\beta = \frac{\partial}{\partial V}\ln Q_G(V; N) \qquad C = N/V$$

and by re-expressing $Q_G(V; N)$ in terms of the $Z(V; N\times N)$ in the form of a polynomial in V. For this purpose, one starts from the equation

$$Q_G(V; f) = \exp[VQ(V; f)].$$

$Q_G(V; f)$ and $Q(V; f)$ are replaced by the expansion

$$Q_G(V; f) = 1 + \sum_{N=1}^{\infty} \frac{f^N}{N!} Q_G(V; N)$$

$$Q(V; f) = \sum_{N=1}^{\infty} \frac{f^N}{N!} Q(V; N)$$

and the terms in f_N are identified with one another. The first terms are

$$Q_G(V; N) = V^N \frac{[Z(V; N)]^N}{N!} + V^{N-1}\frac{Z(V; N, N)[Z(V; N)]^{N-2}}{2(N-2)!},$$

and by expanding with respect to $1/V$, we obtain

$$\ln Q_G(V; N) = N\ln[VZ(V; N)] - \ln N! + \frac{N(N-1)}{2V} Y(V; N, N)$$

from which a value of the pressure associated with the canonical system can be deduced

$$\Pi\beta = C\left[1 - \frac{1}{2}\left(C - \frac{1}{V}\right)Y(V; N, N)\right].$$

We observe that this formula coincides with the canonical result

$$\Pi\beta = C[1 - \tfrac{1}{2}CY(V; N, N) + \ldots]$$

only in the limit $V \to \infty$. The same would occur at all orders.

We can now come back to the polydisperse case

$$\Pi\beta = Q\{f\}$$

$$C_N = f_N \frac{\partial}{\partial f_N} Q\{f\}$$

$$Q\{f\} = \sum_{N=1}^{\infty} \frac{1}{\mathbf{N}!} \sum_{N_1, \ldots, N_{\mathbf{N}}} f_{N_1} \ldots f_{N_{\mathbf{N}}} Z(N_1, \ldots, N_{\mathbf{N}}) \tag{9.4.16}$$

which gives

$$C_N = f_N \sum_{\mathbf{N}=0}^{\infty} \frac{1}{\mathbf{N}!} \sum_{N_1, \ldots, N_{\mathbf{N}}} f_{N_1} \ldots f_{N_{\mathbf{N}}} Z(N, N_1, \ldots, N_{\mathbf{N}}).$$

The f_N can be expressed in the form of expansions in powers of the concentrations and by bringing these expansions in $Q\{f\}$, one finds an expansion of Π in powers of the concentrations.

Using the notation

$$Y(N_1, \ldots, N_{\mathbf{N}}) = \frac{Z(N_1, \ldots, N_{\mathbf{N}})}{Z(N_1) \ldots Z(N_{\mathbf{N}})} \tag{9.4.17}$$

we find for the fugacities

$$\begin{aligned} f_N = \frac{C_N}{Z_N} \{1 &- \sum_A C_A Y(N, A) - \sum_{AB} C_A C_B [\tfrac{1}{2} Y(N, A, B) - 2Y(N, A) Y(N, B)] \\ &- \sum_{ABC} C_A C_B C_C (\tfrac{1}{6} Y(N, A, B, C) - Y(N, A, B) Y(N, C) - Y(N, A, B) Y(A, C) \\ &- \tfrac{1}{2} Y(N, A) Y(A, B, C) + Y(N, A) Y(N, B) Y(N, C) + 2Y(N, A) Y(N, B) Y(A, C) \\ &+ Y(N, A) Y(A, B) Y(A, C) + Y(N, A) Y(A, B) Y(B, C)] + \ldots\} \end{aligned} \tag{9.4.18}$$

and for the osmotic pressure

$$\begin{aligned} \Pi\beta = C &- \tfrac{1}{2} \sum_{AB} C_A C_B Y(A, B) - \sum_{ABC} C_A C_B C_C [\tfrac{1}{3} Y(A, B, C) - Y(A, B) Y(B, C)] \\ &- \sum_{ABCD} C_A C_B C_C C_D \Big[\tfrac{1}{8} Y(A, B, C, D) - \tfrac{3}{2} Y(A, B, C) Y(A, D) \\ &+ \tfrac{5}{2} Y(A, B) Y(B, C) Y(C, D)\Big] + \ldots \end{aligned} \tag{9.4.19}$$

4.3 Expansion of the structure function in the zero angle limit in powers of concentrations

For a monodisperse system, the expansion of $H(\vec{0})$ in powers of the concentration is immediately obtained by using (9.3.64)

$$H(\vec{0}) = \left(\beta C \frac{\partial \Pi}{\partial C}\right)^{-1}.$$

Now (9.4.15) gives

$$\beta C \frac{\partial \Pi}{\partial C} = C\{1 - CY(2\times N) - C^2[Y(3\times N) - 3Y^2(2\times N)] -$$

$$- C^3[\tfrac{1}{2}Y(4\times N) - 6Y(3\times N)Y(2\times N) + 10Y^3(2\times N)] + \dots\}.$$

Therefore, we get

$$H(\vec{0}) = \frac{1}{C}\{1 + CY(2\times N) + C^2[Y(3\times N) - 2Y^2(2\times N)] +$$

$$+ C^3[\tfrac{1}{2}Y(4\times N) - 4Y(3\times N)Y(2\times N) + 5Y^3(2\times N)] + \dots\}. \quad (9.4.20)$$

The question is more complicated when the system is polydisperse. In this case, the structure function $H(\vec{0})$ is defined by

$$C^2 H(\vec{0}) = \sum_{A,B} C_A C_B H_{AB}(\vec{0}) \quad (9.4.21)$$

where $C_A = AC_A$ and $C = \sum_A C_A$.

Moreover, the terms $H_{AB}(\vec{0})$ are given by (9.3.55)

$$C_A C_B H_{AB} = f_A \frac{\partial}{\partial f_A} C_B.$$

Thus, using (9.4.16), we find

$$C_A C_B H_{AB}(\vec{0}) = \delta_{AB} C_A$$

$$+ f_A f_B \sum_{N=0}^{\infty} \frac{1}{N!} \sum_{N_1, \dots, N_N} f_{N_1}, \dots f_{N_N} Z(A, B, N_1, \dots, N_N).$$

Now, we can re-express $H_{AB}(\vec{0})$ in terms of the concentrations by using (9.4.18). Let us calculate only the first terms; we find

$$C_A C_B H_{AB}(\vec{0}) = \delta_{AB} C_A + C_A C_B \Big\{ Y(A, B)$$

$$+ \sum_N C_N[Y(A, B, N) - Y(A, B)Y(A, N) - Y(A, B)Y(B, N)] + \dots \Big\}.$$

Let us bring back this result in (9.4.21); we obtain

$$C^2 H(\vec{0}) = \sum_A A^2 C_A + \sum_{AB} AB\, C_A C_B$$

$$\Big\{ Y(A, B) + \sum_A C_N[Y(A, B, N) - 2Y(A, B)\, Y(A, N)] \Big\} + \dots \quad (9.4.22)$$

This formula is in agreement with (9.4.20) in the monodisperse case, but we see that in the polydisperse case $H(\vec{0})$ does not depend only on $C = \sum_A AC_A$ but also on polydispersion in a rather complex manner. In particular, to first-order in $\boldsymbol{C}$, we have

$$\boldsymbol{H}(\vec{0}) \simeq \frac{\langle A^2 \rangle}{\boldsymbol{C}\langle A \rangle^2}. \tag{9.4.23}$$

5. AN APPROXIMATE GRAND CANONICAL ENSEMBLE: THE EQUILIBRIUM ENSEMBLE

The introduction of many fugacities is a source of complications which one would like to avoid. Thus, in order to apply field theory and renormalization principles to polymer solutions, des Cloizeaux used a simpler ensemble[2] admitting that for long polymers, this ensemble has the same scaling properties as the exact grand canonical ensemble.

The approximation consists in giving a simple form to the fugacity f_N which appears in the definition of the grand canonical partition function. Thus, f_N is represented a priori by the expression

$$f_N = f ¢^N$$

where f and $¢$ are two factors related to the mean number of polymers and to the mean number of monomers in the solution.

In this case, we write as before ($V \to \infty$)

$$Q_G(V; f, ¢) \simeq \exp[VQ(f; ¢)]$$

where

$$Q(f; ¢) = \sum_{N=1}^{\infty} \frac{f^N}{N!} \sum_{N_1 \ldots N_N} ¢^{N_1} \ldots ¢^{N_N} Z(N_1, \ldots, N_N). \tag{9.5.1}$$

Of course, the osmotic pressure is given by the equation

$$\boldsymbol{\Pi\beta} = Q(f; ¢). \tag{9.5.2}$$

The factors f and $¢$ are determined by fixing the mean polymer concentration $\boldsymbol{C}$ and the mean monomer concentration C. We have

$$\boldsymbol{C} = f \frac{\partial}{\partial f} Q(f, ¢)$$

$$C = ¢ \frac{\partial}{\partial ¢} Q(f, ¢). \tag{9.5.3}$$

It is still possible to expand f, $¢$, and $\boldsymbol{\Pi}$ in powers of $\boldsymbol{C}$. We thus obtain

$$f = \frac{\boldsymbol{C}}{\sum ¢^N Z(N)} \left[1 - \boldsymbol{C} \frac{\sum_{BC} ¢^{B+C} Z(B, C)}{\left[\sum_A ¢^A Z(A) \right]^2} + \ldots \right]$$

$$\frac{C}{\mathbf{C}} = \frac{\sum_A A ¢^A Z(A)}{\sum_A ¢^A Z(A)}$$

$$+ \mathbf{C}\left[\frac{\sum_{BC} B¢^{B+C} Z(B, C)}{\left[\sum_A ¢^A Z(A)\right]^2} - \frac{\sum_A A¢^A Z(A) \sum_{BC} ¢^{B+C} Z(B, C)}{\left(\sum_A ¢^A Z(A)\right)^2}\right] + \ldots.$$

In principle, the second equation enables us to determine $¢$ as a function of $\mathbf{C}$ and C. The first equation gives f. The probability $p_N = \mathbf{C}_N/\mathbf{C}$ can be calculated by starting from the general formula (9.3.61)

$$p_N = \frac{\mathbf{C}_N}{\mathbf{C}} =$$

$$\frac{¢^N Z(N)}{\sum_A ¢^A Z(A)}\left\{1 + \mathbf{C}\left[\frac{\sum_A ¢^A Z(N, A)}{Z(N)\sum_A ¢^A Z(A)} - \frac{\sum_{AB} ¢^{A+B} Z(A, B)}{\left(\sum_A ¢^A Z(A)\right)^2}\right] + \ldots\right\}.$$

On the other hand, up to second-order, the osmotic pressure takes the form

$$\boldsymbol{\Pi\beta} = \mathbf{C} - \frac{\mathbf{C}^2}{2}\frac{\sum_{AB} ¢^{A+B} Z(A, B)}{\left[\sum_A ¢^A Z(A)\right]^2} + \ldots \qquad (9.5.4)$$

which to this order coincides with the exact formula (9.4.19) provided that we replace C_A and C_B in the latter formula by the values calculated in terms of f and $¢$.

However, the approximation is not very realistic because the polydispersion, given by the values of p_N, vary with $\mathbf{C}$ and C in an uncontrolled manner. It has mainly a historical value, since it is with the help of this method that in 1975, for the first time, correct scaling laws have been written for polymer solutions.[2]

REFERENCES

1. des Cloizeaux, J. and Jannink, G. (1980). *Physica* **102**A, 120
2. des Cloizeaux, J. (1975). *J. Physique* **36**, 281.

10

STANDARD CONTINUOUS MODEL AND PERTURBATION CALCULATIONS

GENESIS

The interactions existing between segments of the same polymer or different polymers modify the properties of the system in a manner which can be treated by perturbation.

The perturbation expansions that are used to study polymer solutions are generalizations of the Ursell–Mayer–Yvon expansions which apply to gas theory. This technique has been developed from 1972 onwards, especially in the United States and in Japan. The expansions start from a model. The most simple and common one is the famous continuous model which will be described in this chapter in great detail.

During the following decades, a rather large number of results have been obtained by various physicists and, among them, let us mention M. Fixman in the United States and H. Yamakawa in Japan. However, these calculations can be used only to study short chains or chains with weak interactions, and we know now that these perturbation series are divergent.

Nevertheless, concerning long chains, great progress has been made from 1972 onwards, since at that time P.G. de Gennes was able to establish a connection between polymer theory and field theory. Then it was realized that the renormalization principles and techniques could be applied to polymers. Thus, perturbation calculations acquired a new meaning and an increased importance. New results were obtained, for this purpose, which had to be used in a very different way.

Perturbation theory can thus be considered as a basis for the theory of polymers in dilute or semi-dilute solutions.

1. DESCRIPTION OF THE STANDARD CONTINUOUS MODEL

The lattice models or the models with discrete links in continuous space do not enable the theoretician to pursue theoretical calculations very far. At best, they can be used if the theoretician is content with the tree approximation which will be studied later on. On the contrary, the continuous models are more adapted to analytical calculations. As a counterpart, divergences appear, but, as will be shown, they can be remedied by regularizations, and eliminated by renormalizations; thus these operations do not constitute a serious obstacle and do not prevent the continuous model from being very useful.

Thus, we shall study the properties of a set of N Brownian chains, interacting in d-dimensional space.

The weight associated with a configuration of a Brownian chain reads

$$ {}^{0}\mathscr{W} = \exp\left[-\frac{1}{2}\int_0^S \mathrm{d}s \left(\frac{\mathrm{d}\vec{r}}{\mathrm{d}s}\right)^2 \right] \tag{10.1.1} $$

where $\vec{r}(s)$ (with $0 \leqslant s \leqslant S$) determines the position of a point on the curve. It is immediately seen that s and S have the dimensions of an area. Moreover, it is easy to show that the mean square end-to-end distance is

$$ {}^{0}R^2 \equiv \langle\langle[\vec{r}(S) - \vec{r}(0)]^2\rangle\rangle_0 = Sd. \tag{10.1.2} $$

A Brownian chain is characterized by its 'area' S in agreement with the fact that its Hausdorff dimension is two (see Chapter 2). The area is proportional to the number of links. The area of a Brownian chain made of a succession of Brownian segments is the sum of the areas of these segments

Let us now consider N interacting Brownian chains with areas $S_1, \ldots, S_N$. The position of a point on the chain of order $\boldsymbol{a}$ is defined by a function $\vec{r}_{\boldsymbol{a}}(s)$ with $0 \leqslant s \leqslant S_{\boldsymbol{a}}$. The weight associated with a configuration can be written as follows

$$ \mathscr{W}_N\{\vec{r}\} = {}^{0}\mathscr{W}_N\{\vec{r}\}\exp[-{}^{1}\mathfrak{H}_N\{\vec{r}\}] $$

with

$$ {}^{0}\mathscr{W}_N\{\vec{r}\} = \exp\left\{ -\frac{1}{2}\sum_{\boldsymbol{a}=1}^{N}\int_0^{S_{\boldsymbol{a}}} \mathrm{d}s\left[\frac{\mathrm{d}r_{\boldsymbol{a}}(s)}{\mathrm{d}s}\right]^2\right\} $$

$$ {}^{1}\mathfrak{H}_N\{\vec{r}\} = \frac{b}{2}\sum_{\boldsymbol{a}=1}^{N}\sum_{\boldsymbol{b}=1}^{N}\int_0^{S_{\boldsymbol{a}}}\mathrm{d}s'\int_0^{S_{\boldsymbol{b}}}\mathrm{d}s''\,\delta(\vec{r}_{\boldsymbol{a}}(s') - \vec{r}_{\boldsymbol{b}}(S'')) $$

$$ + \frac{c}{6}\sum_{\boldsymbol{a}=1}^{N}\sum_{\boldsymbol{b}=1}^{N}\sum_{\boldsymbol{c}=1}^{N}\int_0^{S_{\boldsymbol{a}}}\mathrm{d}s'\int_0^{S_{\boldsymbol{b}}}\mathrm{d}s''\int_0^{S_{\boldsymbol{c}}}\mathrm{d}s'''\,\delta[\vec{r}_{\boldsymbol{a}}(s') - \vec{r}_{\boldsymbol{b}}(s'')]\,\delta[\vec{r}_{\boldsymbol{a}}(s') - \vec{r}_{\boldsymbol{c}}(s''')] \tag{10.1.3} $$

where b and c define the two-body and three-body interactions respectively. Let us emphasize that we deal only with a model and that eqn (10.1.3) cannot be identified with Boltzmann's equation. Thus, the interactions b and c are to be interpreted as empirical data depending on temperature; their physical meaning will be discussed in more details in Chapter 15. On the contrary, the areas $S_{\boldsymbol{a}}$ can be considered as practically independent of temperature.

In addition, we set

$$ \mathscr{C}(\vec{r}) = \sum_{\boldsymbol{a}=1}^{N}\int_0^{S_{\boldsymbol{a}}}\mathrm{d}s\,\delta[\vec{r} - \vec{r}_{\boldsymbol{a}}(S)] \tag{10.1.4} $$

when $\mathscr{C}(\vec{r})$ is an 'area' per unit volume (monomer concentration). Thus, using this notation and the identity

$$\delta(\vec{r}_1 - \vec{r}_2) = \int \mathrm{d}^d r\, \delta(\vec{r} - \vec{r}_1)\, \delta(\vec{r} - \vec{r}_2)$$

we can write the simple formula

$$^1\mathfrak{H}_N\{\vec{r}\} = \int \mathrm{d}^d r \left\{\frac{b}{2}[\mathscr{C}(\vec{r})]^2 + \frac{c}{6}[\mathscr{C}(\vec{r})]^3\right\} \tag{10.1.5}$$

(the integration is made over the volume in which the chains are contained).

In particular, it is possible to assume that the system is monodisperse $S_1 = S_2 = \ldots = S_N$, and in this case the number of parameters is greatly reduced.

As can easily be verified with the help of (10.1.3), 10.1.4), and (10.1.5), the dimensional formulae of these parameters are

$$S \sim L^2 \quad \mathscr{C}(\vec{r}) \sim L^{2-d} \quad b \sim L^{d-4} \quad c \sim L^{2d-6}. \tag{10.1.6}$$

Thus, in the monodisperse case, the interaction is characterized by two dimensionless parameters which we define by setting

$$z = b\, S^{2-d/2}(2\pi)^{-d/2} \tag{10.1.7}$$

$$y = c\, S^{3-d}(2\pi)^{-d}. \tag{10.1.8}$$

When the problem is only the study of the behaviour of polymers in good solvents, we may content ourselves with the model with two-body repulsive interaction ($z > 0$, $y = 0$). Then, by looking at (9.1.7), we immediately see that dimension $d = 4$ plays a special role. In this case, there is only one dimensionless parameter. In particular, the size of an isolated chain is given by the mean square end-to-end distance which can be written as follows

$$R^2 = \langle\!\langle[\vec{r}(S) - \vec{r}(0)]^2\rangle\!\rangle = {}^{\mathrm{e}}S\, d = S\, \mathfrak{X}_0(z) d \tag{10.1.9}$$

where $\mathfrak{X}_0(z)$ is (at fixed d) a well-defined function. This is the swelling, which will be interpreted as a renormalization factor in Chapter 12, and we claim that, for $z \to \infty$, we have

$$\mathfrak{X}_0(z) \propto z^{2(2\nu-1)/(4-d)}.$$

On the contrary, when the problem is the study of the behaviour of polymers in poor solvents, we must keep two parameters. z and y ($|z| \ll 1$, $y > 0$). In this case, dimension $d = 3$ plays a special role.

Let us now consider polymer solutions. The number of chains per unit volume is denoted by $\boldsymbol{C}$ and, in the polydisperse case, $\boldsymbol{C}(S)$ represents the number of chains with area S, per unit volume.

$$\boldsymbol{C} = \int_0^\infty \mathrm{d}S\, \boldsymbol{C}(S). \tag{10.1.10}$$

On the other hand, in the monodisperse case, $\mathbf{C}$ is sufficient to characterize the solution and a new dimensionless parameter $\mathbf{C}S^{d/2}$ appears.

Before finishing this description of the standard continuous model, we must note that the partition functions which will be calculated by using the weight (10.1.3) may diverge. In order to avoid such divergences a 'cut-off area' s_0 will be introduced, and it will be assumed that in the integrals that appear in (10.1.3), we always have $|s' - s''| \geqslant s_0$, $|s' - s'''| \geqslant s_0$, $|s'' - s'''| \geqslant s_0$. Thus, this regularization requires a new parameter, but it will be possible to eliminate it afterwards by renormalization (ultraviolet renormalization).

2. COMPARISON BETWEEN VARIOUS POLYMER MODELS

Exact correspondences do not exist between the various polymer models, except in the limit of very long chains because, then, a universal behaviour appears. Nevertheless, it is possible to compare the discrete lattice model (*DD*), the discrete link model in continuous space (*DC*) and the purely continuous model (*CC*). In the absence of interaction, the (*D*, *D*) and (*DC*) chains are chains with independent links. Then let l^2 be the mean square projection (on an axis) of the length of a link. The correspondence with (*CC*) can be established by setting

$$S = N l^2 \tag{10.2.1}$$

where N is the number of links of the discrete chain and S the area of the continuous chain.

The polymer concentrations are measured as follows. With the (*DC*) and (*CC*) models, we associate the number $\mathbf{C}$ of polymers per unit volume, and with the (*DD*) model, the number $\boldsymbol{\varphi}$ of polymers per site. Thus, we have

$$\boldsymbol{\varphi} = \mathbf{C} v_{\mathrm{r}}$$

where v_{r} is the volume per site for the lattice.

The monomer concentrations are defined in three different ways. With the (*D*, *D*) model we associate, the number φ of monomers per site, with the (*DC*) model, the number C of monomers per unit volume and with the (*CC*) model, the area $\mathscr{C}$ per unit volume. Thus for a monodisperse set

$$\varphi = \boldsymbol{\varphi} N \qquad C = \mathbf{C} N \qquad \mathscr{C} = \mathbf{C} S. \tag{10.2.2}$$

Let us now compare the interactions, when we have only two-body repulsive interactions. For each model, let us consider a set of chains. A weight

$$\exp\left[-\sum_{m,n} V_{A(m),\,B(n)}\right]$$

can be attributed to each possible configuration of (*DD*). Here A and B represent sites occupied by different monomers m and n (we may have $A = B$) and the sum $\sum_{m,n}$ is performed over all couples of monomers. Moreover $V_{A,B}$ depends only on the relative positions of the sites. If an Ursell–Mayer–Yvon expansion[1,2] is to be made, one writes

$$\exp\left[-\sum_{m,n} V_{A(m),B(n)}\right] = \prod_{m,n}[1-(1-\exp[-V_{A(m),B(n)}])]$$

and this leads to define the fraction φ_e of excluded sites

$$\varphi_e = \sum_B [1 - \exp(-V_{A,B})].$$

Similarly, for the (DC) model, when the chains interact, the weight of each configuration is multiplied by a factor of the form

$$\exp\left[-\sum_{m,n} V(\vec{r}_m - \vec{r}_n)\right] = \prod_{m,n}[1-(1-\exp[-V(\vec{r}_m - \vec{r}_n)])]$$

where summations and products are performed over all couples (m,n) of monomers. Thus, one defines the excluded volume v_e which keeps a meaning when $V(\vec{r}) = \infty$ in a finite domain (covolume)

$$v_e = \int d^3r\,[1 - e^{-V(r)}].$$

Thus, we establish the following correspondences, which are compatible with dimension equations

$$v_e = \varphi_e v_r$$
$$v_e = b\,l^4. \qquad (10.2.3)$$

All these correspondences are approximate and we observe that to each model correspond its own parameters. Thus, we note that the excluded volume only has a meaning for the discrete link model in continuous space. This model is not particularly useful in polymer theory. In fact, the best models are the lattice model, which is simple and well-adapted to numerical simulations, and the purely continuous model, which is very useful for analytic calculations. Thus, the famous excluded volume is not as interesting in polymer theory as many people think (in 1988).

3. ANOTHER FORMULATION OF THE CONTINUOUS MODEL: EDWARDS'S TRANSFORMATION

The standard continuous model ($b \neq 0.\ c = 0$) can be transformed into an equivalent model which is valuable for certain applications. This transformation was introduced in polymer theory by Edwards[3] around 1965 (See Chapter 7, Section 2.2.1) and has been used by him or by others in various articles.

As was seen in Section 1, when a repulsive interaction is present, the weight associated with a configuration can be written in the form

$$\mathscr{W}_N\{\vec{r}\} = {}^0\mathscr{W}_N\{\vec{r}\} \exp[-{}^1\mathfrak{H}_N\{\vec{r}\}]$$

where according to (10.1.5)

$$^1\mathfrak{H}_N\{\vec{r}\} = \frac{b}{2}\int \mathrm{d}^d r\, [\mathscr{C}(\vec{r})]^2. \qquad (10.3.1)$$

Now, an auxiliary random field $\chi(\vec{r})$ can be associated with the function $\mathscr{C}(\vec{r})$; the probability distribution of this field is defined by the weight

$$W\{\chi\} = \exp\left[-\frac{1}{2b}\int \mathrm{d}^d r\, \chi^2\{\vec{r}\}\right]$$

which can be used to calculate mean values. Thus, we easily see that

$$\exp\left[-\frac{b}{2}\int \mathrm{d}^d r\,[\mathscr{C}(\vec{r})]^2\right] = \left\langle\!\!\left\langle \exp\left[-i\int \mathrm{d}^d r\, \mathscr{C}(\vec{r})\,\chi(\vec{r})\right]\right\rangle\!\!\right\rangle_\chi .$$

By introducing this field $\chi(\vec{r})$, we can replace the weight $\mathscr{W}_N(\vec{r})$ by the weight $W_N\{\vec{r}, \chi\}$ defined by

$$W_N\{\vec{r},\chi\} = {}^0\mathscr{W}_N\{\vec{r}\}\exp\left[-i\int \mathrm{d}^d r\, \mathscr{C}(\vec{r})\,\chi(\vec{r}) - \frac{1}{2b}\int \mathrm{d}^d r\, \chi^2(\vec{r})\right]. \qquad (10.3.2)$$

We note that this formalism is convenient to calculate quantities like $\langle\!\langle \mathscr{C}(\vec{r})\, \mathscr{C}(\vec{r}\,')\rangle\!\rangle$

In a more explicit manner, we can also write

$$W_N\{\vec{r},\chi\} = \exp\left\{-\frac{1}{2}\sum_{a=1}^{N}\int_0^{S_a} \mathrm{d}s\left[\left(\frac{\mathrm{d}\,r_a(s)}{\mathrm{d}s}\right)^2 - i\,\chi(\vec{r}_a(s))\right] - \frac{1}{2b}\int \mathrm{d}^d r\, \chi^2(\vec{r})\right\}. \qquad (10.3.3)$$

Thus, one has to study independent chains in a random field and this approach corresponds to the spirit of Edwards's method (see Chapter 8).

In spite of its value, this method will not be used much in what follows. We shall prefer more expressive diagrammatic representations. In particular, we shall encounter again the correlation $\langle\!\langle \mathscr{C}(\vec{r})\, \mathscr{C}(\vec{r}\,')\rangle\!\rangle$ in Chapter 13, when dealing with the structure function.

4. DIAGRAM EXPANSIONS: BASIC RULES

4.1 The continuous limit

We start with a discrete link model and we try to evaluate a correlation function or, in other words, the mean value of an observable. The grand canonical formalism gives us [see (9.1.4) and (9.4.6)]

$$Q_G(V; \{f\}) = 1 + \sum_{N=1}^{\infty} \frac{1}{N!} \sum_{N_1, \ldots, N_N} f_{N_1} \ldots f_{N_N} Z_G(V; N_1, \ldots, N_N),$$

$$Q_G(V, \mathscr{R}, \mathscr{P}; \{f\}) = \sum_{N=1}^{\infty} \frac{1}{\mathcal{S}_N(\mathscr{P})} \sum_{N_1, \ldots, N_N} f_{N_1} \ldots f_{N_N} Z_G(V, \mathscr{R}, \mathscr{P}; N_1, \ldots, N_N), \tag{10.4.1}$$

and we also defined more physical quantities

$$Q(V; \{f\}) = V^{-1} \ln Q_G(V; \{f\}$$

$$Q(V, \mathscr{R}, \mathscr{P}; \{f\}) = \frac{Q_G(V, \mathscr{R}, \mathscr{P}; \{f\})}{Q_G(V; \{f\})}, \tag{10.4.2}$$

the latter function being the generating function of the restricted portion functions $Z(V, \mathscr{R}, \mathscr{P}; N_1, \ldots, N_N)$.

We approach the continuous limit by increasing the number of links, but in such a way as to ensure that the size of non-interacting chains remains constant. However, as the number of degrees of freedom increases, the value of the partition function becomes infinite even in the absence of interaction. For periodical boundary conditions and a large volume, we have

$$^0Z(V; N) \equiv {}^0Z(N) = \mu^N$$

in agreement with our general definitions. In order to go to the continuous limit, we shall introduce a new variable by setting

$$S = N l^2$$

where $l^2 d$ is the mean square length of a link of the discrete chain. Moreover, we shall renormalize the partition functions by setting

$$^+\mathscr{Z}_G(V; S_1, \ldots, S_N) = \frac{Z_G(V; N_1, \ldots, N_N)}{^0Z(N_1) \ldots {}^0Z(N_N)}$$

$$^+\mathscr{Z}_G(V, \mathscr{R}, \mathscr{P}; S_1, \ldots, S_N) = \frac{Z_G(V, \mathscr{R}, \mathscr{P}; N_1, \ldots, N_N)}{^0Z(N_1) \ldots {}^0Z(N_N)}. \tag{10.4.3}$$

Thus, on condition that regularizations exist (they are recalled to mind by the symbol $+$), the new partition functions have continuous limits. In the same way, the fugacities will be renormalized by setting

$$f(S) = f_N \, {}^0Z(N) \, l^2, \tag{10.4.4}$$

and the sum will be replaced by integrals. Thus, we write

$$\sum_{N=1}^{\infty} \cdots = \int_0^{\infty} \frac{dS}{l^2} \cdots \,. \tag{10.4.5}$$

The grand partition functions remain unchanged, but they must be re-expressed in terms of the new variables. We set

$$^{+}\mathscr{Q}_{\mathrm{G}}(V;\{f(S)\}) \equiv Q_{\mathrm{G}}(V;\{f_N\}) \tag{10.4.6}$$

and the same change of notation will be used for all restricted and non-restricted partition functions.

In particular, the cumulants $^{+}\mathscr{Z}(V; S_1, \ldots, S_N)$ of the $^{+}\mathscr{Z}_{\mathrm{G}}(V; S_1, \ldots, S_N)$ are given by

$$^{+}\mathscr{Z}(V; S_1, \ldots, S_N) = \frac{Z(V; N_1, \ldots, N_N)}{{}^{0}Z(N_1)\ldots{}^{0}Z(N_N)}. \tag{10.4.7}$$

By definition, for $V \to \infty$, the limits of these quantities are $^{+}\mathscr{Z}(S_1, \ldots, S_N)$ and in the case where all the S_j are equal to S, the symbol $^{+}\mathscr{Z}(N \times S)$ will also be used (the + sign is a regularization sign and can be omitted in the absence of interaction)

$$\begin{aligned} {}^{0}\mathscr{Z}(S) &= 1 \\ {}^{0}\mathscr{Z}(S_1, \ldots, S_N) &= 0 \qquad N > 1 \end{aligned} \tag{10.4.8}$$

(in this case ${}^{0}\mathscr{Z}_{\mathrm{G}}(S_1, \ldots, S_N) = V^N\,{}^{0}\mathscr{Z}(S_1)\ldots{}^{0}\mathscr{Z}(S_N) = V^N$)

Moreover, in the continuous case, we shall introduce the following connected partition functions

$$^{+}\mathscr{Z}(V, \mathscr{R}, \mathscr{P}; S_1, \ldots, S_N) = \frac{Z(V, \mathscr{R}, \mathscr{P}; N_1, \ldots, N_N)}{Z(N_1)\ldots Z(N_N)} \tag{10.4.9}$$

whose limits for $V \to 0$ are $^{+}\mathscr{Z}(\mathscr{R}, \mathscr{P}; S_1, \ldots, S_N)$. Now, eqns (10.4.1) and (10.4.2) take the form

$$^{+}\mathscr{Q}_{\mathrm{G}}(V;\{f\}) = 1 + \sum_{N=1}^{\infty} \frac{1}{N!} \int \mathrm{d}S_1 \ldots \mathrm{d}S_N f(S_1)\ldots f(S_N)\,^{+}\mathscr{Z}(V; S_1, \ldots, S_N)$$

$$\begin{aligned} ^{+}\mathscr{Q}_{\mathrm{G}}(V, \mathscr{R}, \mathscr{P}; \{f\}) = \sum_{N=1}^{\infty} \frac{1}{\mathcal{S}_N(\mathscr{P})} \times \\ \int \mathrm{d}S_1 \ldots \mathrm{d}S_N f(S_1)\ldots f(S_N)\,^{+}\mathscr{Z}(V, \mathscr{R}, \mathscr{P}; S_1, \ldots, S_N) \end{aligned}$$

$$^{+}\mathscr{Q}(V;\{f\}) = V^{-1}\ln {}^{+}\mathscr{Q}_{\mathrm{G}}(V;\{f\})$$

$$^{+}\mathscr{Q}(V, \mathscr{R}, \mathscr{P}; \{f\}) = \frac{^{+}\mathscr{Q}_{\mathrm{G}}(V, \mathscr{R}, \mathscr{P}; \{f\})}{^{+}\mathscr{Q}_{\mathrm{G}}(V;\{f\})}. \tag{10.4.10}$$

4.2 Isolated chain: diagram expansion

4.2.1 Partition function and restricted partition functions of a continuous chain

The partition function $^{+}\mathscr{Z}(S)$ has been defined as the ratio (see (10.4.7)); N and V are assumed to be very large)

$$^{+}\mathscr{Z}(S) = \frac{^{+}Z(N)}{^{0}Z(N)} = \frac{\int \mathrm{d}\{\vec{r}\}\, \mathscr{W}_1\{\vec{r}\}\, \delta(\vec{r}(0))}{\int \mathrm{d}\{\vec{r}\}\, {}^{0}\mathscr{W}_1\{\vec{r}\}\, \delta(\vec{r}(0))} \tag{10.4.11}$$

where $\mathscr{W}_1\{\vec{r}\}$ and ${}^0\mathscr{W}_1\{\vec{r}\}$ are weights for the chain, with or without interaction. The sums are performed over all the configurations.

Then, equality (10.4.11) can be rewritten in the form

$$ {}^+\mathscr{Z}(S) = \frac{\langle\!\langle \delta(\vec{r}(0)) \exp[-{}^{\mathrm{I}}\mathfrak{H}_1\{\vec{r}\}] \rangle\!\rangle_0}{\langle\!\langle \delta(\vec{r}(0)) \rangle\!\rangle_0} \tag{10.4.12} $$

and here the mean values are calculated with the weight

$$ {}^0\mathscr{W}\{\vec{r}\} = \exp\left\{ - \int_0^S \mathrm{d}s \left[\frac{\mathrm{d}\vec{r}(s)}{\mathrm{d}s} \right]^2 \right\}. $$

The Hamiltonian ${}^I\mathfrak{H}_1$ is given by (10.1.3) for $N = 1$ and depends on the two-body interaction b and on the three-body interaction c.

The restricted partition functions ${}^+\mathscr{Z}(\mathscr{R}, \mathscr{P}; S)$ can be defined in a similar way but, first, let us introduce more precise notation. On a chain, we define E correlations points given by their coordinates along the chain, i.e. the areas $s_1, \ldots, s_E$ and we attribute definite values $\vec{r}_1, \ldots, \vec{r}_E$ to the position vectors of these points. The corresponding restricted correlation functions will be represented by the symbol

$$ {}^+\mathscr{Z}(\mathscr{R}, \mathscr{P}; S) \equiv {}^+\mathscr{Z}(\vec{r}_1, \ldots, \vec{r}_E; s_1, \ldots, s_E; S). $$

Now, eqn (10.4.12) can be generalized in a very simple manner and the preceding restricted correlation function can be rewritten in the explicit form

$$ {}^+\mathscr{Z}(\vec{r}_1, \ldots, \vec{r}_E; s_1, \ldots, s_E; s) = \frac{\langle\!\langle \delta[\vec{r}_1 - \vec{r}(S_1)] \ldots \delta(\vec{r}_E - \vec{r}(S_E)] \exp[-{}^I\mathfrak{H}_1\{\vec{r}\}] \rangle\!\rangle}{\langle\!\langle \delta[\vec{r}(0)] \rangle\!\rangle}. \tag{10.4.13} $$

Equations (10.4.13) serve as a starting point for perturbation calculations concerning the isolated continuous chain. Incidentally, let us note that since (10.4.12) is only a special case of (10.4.13), we have

$$ {}^+\mathscr{Z}(S) \equiv {}^+\mathscr{Z}(\vec{r}_1; s_1; S). $$

4.2.2 Chains with two-body interaction: real space expansions

For simplicity, let us first consider the simple case $b > 0$, $c = 0$.

Let us expand ${}^+\mathscr{Z}(S)$ in powers of b. For this purpose, we choose a curve $\vec{r}(s)$ and express $\exp[-{}^I\mathfrak{H}(\vec{r})]$ in the form

$$ \exp[-{}^I\mathfrak{H}_1\{\vec{r}\}] = \sum_{p=0}^{\infty} \frac{1}{p!} [-{}^I\mathfrak{H}_1\{\vec{r}\}]^p. $$

The term of order p reads

$$ -b^p \int_{\mathfrak{Z}} \mathrm{d}s_1' \, \mathrm{d}s_1'' \ldots \mathrm{d}s_p' \, \mathrm{d}s_p'' \, \delta(\vec{r}(s_1') - \vec{r}(s_1'')) \ldots \delta(\vec{r}(s_p') - \vec{r}(s_p'')) $$

the integration domain $\mathfrak{J}$ being defined by the ordering relations

$$0 < s'_1 < s''_1 < S \ \ldots \ 0 < s'_p < s''_p < S$$

and

$$s'_1 < s'_2 < \ \ldots \ < s'_p.$$

The interesting result is that these conditions lead to the elimination of the factor $1/2^p p!$. Now, we can give well-defined values to $s'_1, s''_1, \ldots, s'_p, s''_p$ and, for a while, we can also fix the position vectors $\vec{r}(s'_1), \vec{r}(s''_1) \ldots \vec{r}(s'_p), \vec{r}(s''_p)$. These interaction points cut the polymer line into $(2p + 1)$ successive segments with areas $s_1, \ldots s_{2p+1}$ respectively.

The position of the origin of segment j is defined by $\vec{r}_j$ and the position of its end point by $\vec{r}_{j+1}$. Thus the terms of order p lead to diagrams of order p and each diagram corresponds to a peculiar setting of the interaction points on the polymer line (see Fig. 10.1).

Now, one has to sum over the configurations of all the intermediate points, of position vector $\vec{r}(s)$ where $s \neq s'_1, s \neq s''_1, \ldots, s \neq s'_p, s \neq s''_p$. In other words, a factor ${}^0\mathscr{Z}(\vec{r}_{j+1} - \vec{r}_j; s_j)$ must be associated with each segment j, and we have

$$ {}^0\mathscr{Z}(\vec{r}; S) = \langle\!\langle \delta(\vec{r}(S) - \vec{r}(0) - \vec{r}) \rangle\!\rangle_0. \qquad (10.4.14)$$

Actually, this function is the distribution probability of the end-to-end vector for a Brownian chain and the result can be proved again by proceeding as follows

$$ {}^0\mathscr{Z}(\vec{r}; S) = \frac{1}{(2\pi)^d} \int \mathrm{d}^d k \, {}^0\tilde{\mathscr{Z}}(\vec{k}; S)$$

$$ {}^0\tilde{\mathscr{Z}}(\vec{k}; S) = \langle\!\langle \exp[i\vec{k} \cdot [\vec{r}(S) - \vec{r}(0)]] \rangle\!\rangle_0. \qquad (10.4.15)$$

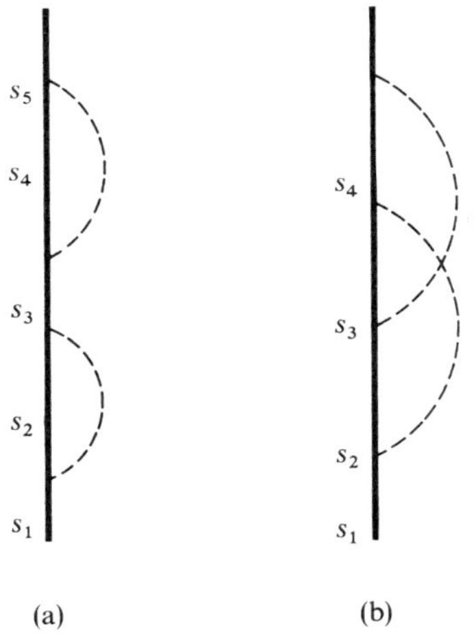

Fig. 10.1. Second-order diagrams, in the continuous case, with two-body interactions. The solid line represents the polymer; the dashed lines indicate the interactions.

Let us calculate ${}^{0}\tilde{\mathscr{Z}}(\vec{k};S)$ which is given by the functional integral

$$ {}^{0}\tilde{\mathscr{Z}}(\vec{k};S) = \frac{\displaystyle\int \mathrm{d}\{\vec{r}\}\exp\left(i\vec{k}\cdot[\vec{r}(S)-\vec{r}(0)] - \frac{1}{2}\int_0^S \mathrm{d}s\left[\frac{\mathrm{d}\vec{r}(s)}{\mathrm{d}s}\right]^2\right)}{\displaystyle\int \mathrm{d}\{\vec{r}\}\exp\left(-\frac{1}{2}\int_0^S \mathrm{d}s\left[\frac{\mathrm{d}\vec{r}(s)}{\mathrm{d}s}\right]^2\right)}. $$

In the integral, let us write $[\vec{r}(S)-\vec{r}(0)]$ in the form

$$ \int_0^S \mathrm{d}s\left[\frac{\mathrm{d}\vec{r}(s)}{\mathrm{d}s}\right]. $$

In this way, we obtain the identity

$$ {}^{0}\tilde{\mathscr{Z}}(\vec{k};S) = \frac{\displaystyle\int \mathrm{d}\{\vec{r}\}\exp\left(-\frac{1}{2}\int_0^S \mathrm{d}s\left[\left(\frac{\mathrm{d}\vec{r}(s)}{\mathrm{d}s} - i\vec{k}\right)^2 + k^2\right]\right)}{\displaystyle\int \mathrm{d}\{\vec{r}\}\exp\left(-\frac{1}{2}\int_0^S \mathrm{d}s\left[\frac{\mathrm{d}\vec{r}(s)}{\mathrm{d}s}\right]^2\right)}. $$

A simple generalization of the identity

$$ \int_{-\infty}^{+\infty} \mathrm{d}x\, \mathrm{e}^{-A(x-ik)^2} = \int_{-\infty}^{+\infty} \mathrm{d}x\, \mathrm{e}^{-Ax^2} $$

shows that the Gaussian integrals cancel out. Thus, we find

$$ {}^{0}\tilde{\mathscr{Z}}(\vec{k};S) = \mathrm{e}^{-k^2S/2} $$

and therefore

$$ {}^{0}\mathscr{Z}(\vec{r};S) = (2\pi S)^{-d/2}\mathrm{e}^{-r^2/2S} \tag{10.4.16} $$

$$ {}^{0}\mathscr{Z}({}^{\mathrm{I}}\vec{r}_j - {}^{\mathrm{I}}\vec{r}_{j-1};s_j) = (2\pi s_j)^{-d/2}\exp[-({}^{\mathrm{I}}\vec{r}_j - {}^{\mathrm{I}}\vec{r}_{j-1})^2/2s_j]. \tag{10.4.17} $$

we can now summarize the rules of construction and calculation of diagrams

1. Each diagram consists of a polymer line and of interaction lines* which join points of the polymer two by two. These points are the interaction points. On a diagram of order p, the interaction points separate $2p+1$ segments. The positions of the successive points on the chain are defined by the vectors ${}^{\mathrm{I}}\vec{r}_1, \ldots, {}^{\mathrm{I}}\vec{r}_{2p}$. It is assumed that the origin of the chain coincides with the origin of coordinates. The areas of the successive segments of the chain are $s_1, s_2, \ldots, s_{2p+1}$ with $s_1 + s_2 + \ldots + s_{2p+1} = S$ where S is the area of the chain.

2. With each internal segment j of area s_j, joining points of position vectors ${}^{\mathrm{I}}\vec{r}_{j-1}$ and ${}^{\mathrm{I}}\vec{r}_j$, we associate a factor $(2\pi)^{-d/2}\exp[-({}^{\mathrm{I}}\vec{r}_j - {}^{\mathrm{I}}\vec{r}_{j-1})^2/2s_j]$. With the extreme segments in contact either with the origin or with the end-point of the chain, we associate a factor unity.

* Let us note that in the framework of Edwards's formalism (Section 3), these interaction lines correspond to the 'propagator' of the auxiliary field.

3. With each line joining two interaction points of position vectors ${}^{\mathrm{I}}\vec{r}_j$ and ${}^{\mathrm{I}}\vec{r}_l$, we associate a factor $-b\,\delta({}^{\mathrm{I}}\vec{r}_j - {}^{\mathrm{I}}\vec{r}_l)$.
4. The contribution of a diagram is obtained by summing up the product of all factors. A summation is made over the positions of the interaction points by performing the integration $\int d^{\mathrm{d}}({}^{\mathrm{I}}\vec{r}_2) \ldots \int d^{\mathrm{d}}({}^{\mathrm{I}}\vec{r}_{2p})$. Then, a summation is made over the areas by performing the integration
$$\int_{s_0}^{S} \mathrm{d}s_1 \ldots \int_{s_0}^{S} \mathrm{d}s_{2p+1}\, \delta(S - s_1 - \ldots s_{2p+1})$$
where s_0 is the cut-off area ($s_0 < S$).
5. The partition function ${}^{+}\mathscr{Z}(s)$ is the sum of the contributions of all the diagrams.

The restricted correlation functions are calculated in the same manner [see (10.4.12) and (10.4.13)]. For instance, if we want to calculate
$${}^{+}\mathscr{Z}(\vec{r}_1, \ldots, \vec{r}_E; s_1, \ldots, s_E; S)$$
we consider a chain of area S defined by a vector function $\vec{r}(s)$ and we impose the conditions $\vec{r}(s_1) = \vec{r}_1, \ldots, \vec{r}(s_E) = \vec{r}_E$. The calculation of a diagram of order p can be made by fixing, on the chain, the E correlation points and the $2p$ interaction points. The correlation and interaction points cut the chain into segments. The number of internal segments is $2p + E - 1$ and the number of external segments is 0, 1, or 2. The contributions of the diagrams are calculated as before by summing over all possible positions of the interaction points in space and along the chain.

The sum of the contributions of all the diagrams give ${}^{+}\mathscr{Z}(\vec{r}_1, \ldots, \vec{r}_E; s_1, \ldots, s_E; S)$.

4.2.3 Chain with two-body and three-body interactions: real space expansions

The case where the chain is submitted to two-body and three-body interactions can be treated in a similar way. In this case, (10.4.12) and (10.4.13) remain valid and ${}^{\mathrm{I}}\mathfrak{H}_1\{\vec{r}]$ is given by (10.1.3) with $c > 0$. The expansion is now made with respect to b and c and can be expressed in the form of diagrams.

A two-body interaction is represented on the diagram by two interaction points connected by one interaction line; in the same way, a three-body interaction will be represented by three interaction points connected by two interaction lines (see Fig. 10.2). With each three-body interaction applied to the interaction points of position vectors ${}^{\mathrm{I}}\vec{r}_i$, ${}^{\mathrm{I}}\vec{r}_j$, and ${}^{\mathrm{I}}\vec{r}_k$, we associate a factor $-c\,\delta({}^{\mathrm{I}}\vec{r}_i - {}^{\mathrm{I}}\vec{r}_j)\,\delta({}^{\mathrm{I}}\vec{r}_i - {}^{\mathrm{I}}\vec{r}_k)$.

The calculation is performed just as in the case where only two-body interactions are present.

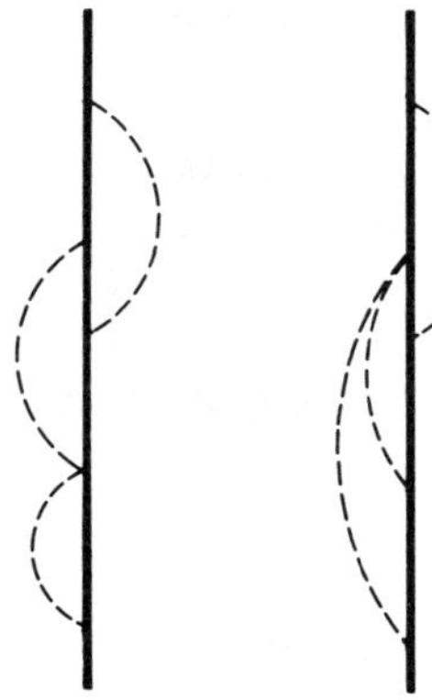

(a) (b)

Fig. 10.2. (a) A second-order diagram in the continuous case with two-body and three-body interactions. (b) The *same* diagram represented in a slightly different way.

4.2.4 Reciprocal space expansions for an isolated chain

Diagrams can also correspond to Fourier transforms, and such a representation is often convenient. We can use it either to calculate ${}^{+}\mathscr{Z}(s)$ or to calculate the Fourier transform of a restricted partition function ${}^{+}\mathscr{Z}(\vec{r}_1, \ldots, \vec{r}_E; s_1, \ldots, s_E; S)$, defined by

$$\tilde{\mathscr{Z}}(\vec{k}_1, \ldots, \vec{k}_E; s_1, \ldots, s_E; S) =$$

$$= \int \mathrm{d}^d r_1 \ldots \int \mathrm{d}^d r_E \, \mathrm{e}^{i(\vec{k}_1 \cdot \vec{r}_1 + \ldots + \vec{k}_E \cdot \vec{r}_E)} \mathscr{Z}(\vec{r}_1, \ldots, \vec{r}_E; s_1, \ldots, s_E; S) \qquad (10.4.18)$$

(the index $+$ is omitted because the relation remains valid after elimination of the short-distance cut-off. We note that $\mathscr{Z}(\vec{r}_1, \ldots, \vec{r}_E; s_1, \ldots s_E; S)$ depends only on the differences between the vectors $\vec{r}_1, \ldots, \vec{r}_E$. Consequently, when the system occupies a large rectangular box with periodical boundary conditions (a torus), a fact which implies that the vectors $\vec{k}$ have discrete values, we can write

$$\tilde{\mathscr{Z}}(\vec{k}_1, \ldots, \vec{k}_E; s_1, \ldots, s_E; S) = V \delta_{\vec{k}_1 + \ldots + \vec{k}_E} \bar{\mathscr{Z}}(\vec{k}_1, \ldots, \vec{k}_E; s_1, \ldots, s_E; S). \qquad (10.4.19)$$

When the volume is infinite, this relation takes the form

$$\tilde{\mathscr{Z}}(\vec{k}_1, \ldots, \vec{k}_E; s_1, \ldots, s_E; S) = (2\pi)^d \delta(\vec{k}_1 + \ldots + \vec{k}_E) \bar{\mathscr{Z}}(\vec{k}_1, \ldots, k_E; s_1, \ldots, s_E; S). \qquad (10.4.20)$$

In all cases, $\bar{\mathscr{Z}}(\vec{k}_1, \ldots, \vec{k}_E; s_1, \ldots, s_E; S)$ is defined only for $\vec{k}_1 + \ldots + \vec{k}_E = 0$.

Moreover, we observe that $\mathscr{Z}(S)$ can be calculated as a function $\bar{\mathscr{Z}}(\ldots; S)$ since according to (10.4.18) and (10.4.19), we have

$$\mathscr{Z}(S) = \bar{\mathscr{Z}}(\vec{0}, \vec{0}; 0, S; S). \qquad (10.4.21)$$

To determine the diagrammatic representation of ${}^{+}\mathscr{Z}(S)$ or of ${}^{+}\mathscr{Z}(k_1, \ldots, \vec{k}_E; \mathscr{Z}_1, \ldots, \mathscr{Z}_E; S)$ in reciprocal space, we must introduce the Fourier transform of the interaction terms

$$
\begin{aligned}
-b\,\delta(\vec{r}_i - \vec{r}_j) &= -\frac{b}{(2\pi)^d}\int \mathrm{d}^d q \exp[i\vec{q}\cdot(\vec{r}_i - \vec{r}_j)] \\
-c\,\delta(\vec{r}_i - \vec{r}_j)\,\delta(\vec{r}_i - \vec{r}_l) &= -\frac{c}{(2\pi)^{2d}}\int \mathrm{d}^d q' \int \mathrm{d}^d q'' \exp\{i[\vec{q}' + \vec{q}'')\cdot\vec{r}_i - \vec{q}'\cdot\vec{r}_j - \vec{q}''\cdot\vec{r}_l]\}
\end{aligned}
\tag{10.4.22}
$$

In the same way, we must express the factors associated with the polymer segments in the form of Fourier transforms. Thus, the factor associated with a segment of area s_j, joining two successive interaction points with position vectors ${}^{\mathrm{I}}\vec{r}_{j-1}$ and ${}^{\mathrm{I}}\vec{r}_j$, has the form

$$
\begin{aligned}
&\frac{1}{(2\pi s_i)^{d/2}} \exp[-({}^{\mathrm{I}}\vec{r}_j - {}^{\mathrm{I}}\vec{r}_{j-1})^2/2\,s_j] \\
&\qquad = \frac{1}{(2\pi)^d}\int \mathrm{d}^d q \exp[i\vec{q}\cdot({}^{\mathrm{I}}\vec{r}_j - {}^{\mathrm{I}}\vec{r}_{j-1})]\exp[-q^2 s_j/2].
\end{aligned}
\tag{10.4.23}
$$

This formula can be interpreted by saying that the vector $\vec{q}$ is transferred from interaction point j–1 to interaction point j (or $-\vec{q}$ from interaction point j to interaction point j) and eqn (10.4.22) can be interpreted in the same manner.

Thus, with each polymer segment, we can associate an orientation and a wave vector $\vec{q}$. In the same way, with each interaction line, we can associate an orientation and a wave vector. Each wave vector can be considered to be transferred in the direction of the arrow which defines the corresponding orientation. We note that a two-body interaction transfers one wave vector $\vec{q}$, whereas a three-body interaction transfers two wave vectors $\vec{q}'$ and $\vec{q}''$. We also remark that the orientation of each line is arbitrary, but if the vector $\vec{q}$ is associated with an orientation, the vector $-\vec{q}$ must be associated with the opposite orientation. On a polymer, it is generally convenient to choose the direction in which s increases to define the orientations. Finally, we note that a factor $(2\pi)^{-d}$ corresponds to the introduction of each wave vector.

However, these wave vectors are not independent. In fact, let us consider (for instance) a two-body interaction point. Two polymer segments and one interaction line converge at that point (see Fig. 10.3). Let $\vec{q}''$ be the sum of the wave vectors, pointing towards the interaction point and $\vec{q}'$ the sum of the wave vectors pointing in the opposite direction. The interaction point gives the contribution $\int \mathrm{d}^d r \exp[-i(\vec{q}'' - \vec{q}')\cdot\vec{r}] = (2\pi)^d\,\delta(\vec{q}'' - \vec{q}')$. The same kind of argument applies to any correlation point where an external wave vector $\vec{k}$ is injected and in particular at the free ends of the polymer where a zero wave vector is injected (see Fig. 10.4).

Thus, each relation between wave vectors leads to a factor $(2\pi)^d$. In particular, the sum of the external wave vectors must vanish and the factor

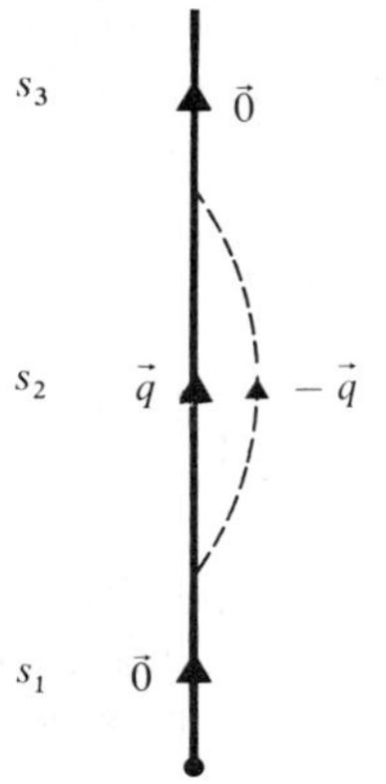

Fig. 10.3. Diagram concerning $\mathscr{Z}(s)$ in reciprocal space.

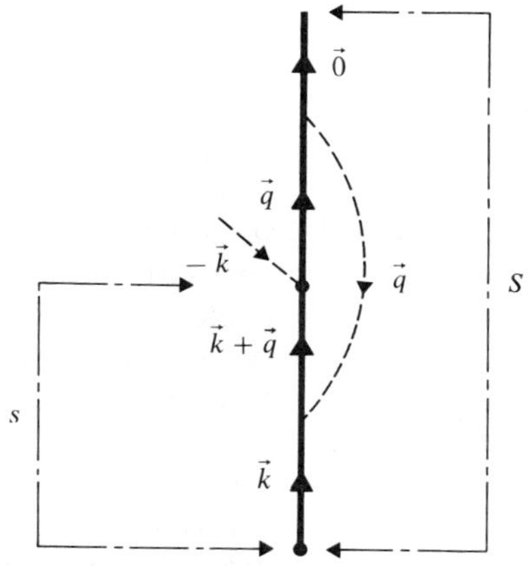

Fig. 10.4. Diagram concerning $\bar{\mathscr{Z}}(\vec{k}, -\vec{k}; 0, s; S)$ in reciprocal space.

$(2\pi)^d\,\delta(\vec{k}_1 + \ldots + \vec{k}_E)$ must be found in the definition (10.4.20) of the $\bar{\mathscr{Z}}(\ldots)$. Finally, there remains a factor $(2\pi)^{\div d \circlearrowleft}$ where $\circlearrowleft$ is the number of loops in the diagram, i.e. the number of independent wave vectors. In other terms, a factor $(2\pi)^{-d}$ is associated with each independent vector appearing in the diagram.

In brief the expansion of $\bar{\mathscr{Z}}(k_1, \ldots, \vec{k}_E; s_1, \ldots, s_E; S)$ or that of $\mathscr{Z}(S)$, can be represented by diagrams and the calculation of the various terms can be performed as follows:

1. Each diagram consists of one polymer line representing a chain of 'area' S, and of interaction lines connecting the interaction points of the chain.

2. On a chain, there are E correlation points whose coordinates along the chain are $s_1, \ldots, s_E$ $(0 \leqslant s_1 \leqslant s_2 \ldots \leqslant s_E \leqslant S)$.

3. Correlation points and interaction points cut the chain into segments and an 'area' is associated with each segment. The sum of the areas of all the segments is the area S of the polymer chain.

4. Each segment and each interaction line is oriented and bears a wave vector.

5. Wave vectors $\vec{k}_1, \ldots, \vec{k}_E$ with $\vec{k}_1 + \ldots + \vec{k}_E = 0$ are injected at the E correlation points.

6. When one extremity of a segment coincides with an extremity of the chain, and is not a correlation point, the segment bears a zero wave vector.

7. At each correlation point and at each interaction point, the sum of the wave vectors vanish (flow conservation).

8. With each segment of area s, bearing a wave vector $\vec{q}$, we associate the factor $\exp(-q^2 s/2)$.

9. With each two-body interaction transferring a vector $\vec{q}$, we associate a factor $-b$. With each three-body interaction transferring two vectors $\vec{q}'$ and $\vec{q}''$, we associate a factor $-c$.

10. The contribution associated with a diagram is (generally) obtained by multiplying all the factors defined above. Then, the expression is integrated over all the independent wave vectors $\vec{q}$ (integrals of the form $(2\pi)^{-d/2} \int d^d q \ldots$) and over all the independent areas (integral of the form $\int ds$)

The summation over the internal wave vectors can be performed without difficulty. The product of the factors associated with the polymer segments is the exponential of a homogeneous quadratic form of the external vectors $\vec{k}_1, \ldots, \vec{k}_E$ and of the internal vectors $\vec{q}_1, \ldots, \vec{q}_{\circlearrowleft}$ where $\circlearrowleft$ is the number of loops in the diagram.

The independent areas s have then to be integrated, but it must be noted that, when the preceding rules are applied, the integrals do not always converge for small areas. In order to cure these divergences a cut-off s_0 is introduced, and s_0 acts as a lower bound for the integration variables. We denote the presence of a cut-off by using an index $+$; we thus write ${}^{+}\mathscr{Z}(S)$ for the partition function, and, in a similar way ${}^{+}\mathscr{Z}(\ldots, S)$ for the restricted partition functions and ${}^{+}\bar{\mathscr{Z}}(\ldots, S)$ for their Fourier transforms.

However, as will be shown later, these divergences can be suppressed by renormalization,* and then the cut-off s_0 can be eliminated.

4.2.5 Summations for an isolated chain: Fixman's method

As we saw, a restricted partition function can be calculated either in real space or in reciprocal space.

* This renormalization is equivalent to an analytic continuation with respect to dimension d.

If we work in real space and if we want to calculate a diagram (see Figs. 10.1 and 10.2), we start by fixing the correlation points and the interaction points both in space and on the polymer line. Subsequently, the interaction points are displaced in every possible way.

It is possible to proceed more directly by observing that the interaction energy does not really depend on the relative positions of all the interaction points, but only on the relative positions of the points which belong to the same doublet or the same triplet of interaction points. This observation is the basis of Fixman's method.[4]

However, in general, it is more convenient to work in reciprocal space by calculating the Fourier transforms of the restricted partition functions. In this case, the contribution of a diagram depends on a certain number of internal vectors. Applying Fixman's method is equivalent to summing the contributions over these wave vectors. The number $\circlearrowleft$ of independent wave vectors is called the number of loops. In particular, for a diagram with p two-body interactions and p' three-body interactions, we have

$$\circlearrowleft = p + 2p'.$$

Now, considering a diagram $\mathfrak{D}$, let us denote the segments of polymer line by a subscript j, and the internal independent vectors by a subscript α. The area of segment j is denoted by s_j. An independent wave vector is represented by $\vec{q}_\alpha$ (with $\alpha = 1, \ldots, \circlearrowleft$) and such a vector will always be borne by an interaction line, when all the wave vectors $\vec{q}_\alpha$ vanish, a flow of wave vectors remains in the diagram, resulting from the fact that external wave vectors are injected at the correlation points; in this case, matters are so arranged that each interaction line bears a zero wave vector. Then let $\vec{k}_j$ be the wave vector borne by segment j when all $\vec{q}_\alpha$ vanish.

When the $\vec{q}_\alpha$ do not vanish, the segment j bears the wave vector

$$\vec{k}_j + \sum_{\alpha=1}^{\circlearrowleft} \ominus_{j\alpha} \vec{q}_\alpha$$

where $\ominus_{j\alpha} = 1$ if $\vec{q}_\alpha$ is transferred through j in the positive direction;
$\ominus_{j\alpha} = -1$ if $\vec{q}_\alpha$ is transferred through j in the negative direction;
$\ominus_{j\alpha} = 0$ if $\vec{q}_\alpha$ is not transferred through j.

Before summation over the areas of the polymer segments, the contribution $\mathscr{D}\{s\}$ of a polymer diagram can be written in the form

$$\mathscr{D}\{s\} = (-b)^p(-c)^{p'}\mathscr{I}\{s\} \tag{10.4.24}$$

$$\mathscr{I}\{s\} = \int \prod_{\alpha=1}^{\circlearrowleft} \left[\frac{d^d q_\alpha}{(2\pi)^d}\right] \exp\left[-\frac{1}{2}\sum_j s_j \left(\vec{k}_j + \sum_\alpha \ominus_{j\alpha}\vec{q}_\alpha\right)^2\right]$$

where $\circlearrowleft$ is the number of independent vectors (number of loops). Let us set

$$\mathscr{S}_{\alpha\beta} = \sum_j s_j \ominus_{j\alpha} \ominus_{j\beta}$$

$$\vec{k}_\alpha = \sum_j s_j \ominus_{j\alpha} \vec{k}_j. \tag{10.4.25}$$

By summing over repeated indices, we have

$$\mathscr{I}\{s\} = \int \prod_{\alpha=1}^{\circlearrowright} \left[\frac{\mathrm{d}^d q_\alpha}{(2\pi)^d}\right] \exp\left\{-\frac{1}{2}\left[\sum_{\alpha\beta} \vec{q}_\alpha \mathscr{S}_{\alpha\beta} \vec{q}_\beta + 2\sum_\alpha \vec{k}_\alpha \cdot \vec{q}_\beta + \sum_j s_j k_j^2\right]\right\}.$$

where $\vec{q}_\alpha \mathscr{S}_{\alpha\beta} \vec{q}_\beta$ means $\mathscr{S}_{\alpha\beta} \vec{q}_\alpha \cdot \vec{q}_\beta$

A matrix $\mathscr{S}$ corresponds to the coefficients $\mathscr{S}_{\alpha\beta}$, a column vector $\vec{q}$ correspond to the $\vec{q}_\alpha$ and a column vector $\vec{k}$ to the $\vec{k}_\alpha$. Using this notation, we can write the preceding equation in the form

$$\mathscr{I}\{s\} = (2\pi)^{-d\circlearrowright} \int \prod_{\alpha=1}^{\circlearrowright} \mathrm{d}^d q_\alpha \exp\{-\tfrac{1}{2}[(\vec{q} + \vec{k}\mathscr{S}^{-1})\mathscr{S}(\vec{q} + \mathscr{S}^{-1}\vec{k}) - \vec{k}\mathscr{S}^{-1}\vec{k} + s_j k_j^2]\}$$

$$= (2\pi)^{-d\circlearrowright} \int \prod_{\alpha=1}^{\circlearrowright} \mathrm{d}^d q_\alpha \exp\{-\tfrac{1}{2}[\vec{q}\mathscr{S}\vec{q} - \vec{k}\mathscr{S}^{-1}\vec{k} + s_j k_j^2]\}.$$

Let us integrate; we get

$$\mathscr{I}\{s\} = (2\pi)^{-d\circlearrowright/2} |\det \mathscr{S}|^{-d/2} \exp\left[-\tfrac{1}{2}\sum_j s_i k_j^2 + \tfrac{1}{2}\vec{k}\mathscr{S}^{-1}\vec{k}\right].$$

Finally, we obtain

$$\mathscr{D}\{s\} = (-b)^p(-c)^{p'}(2\pi)^{-d\circlearrowright/2} |\det \mathscr{S}|^{-d/2}$$
$$\exp\left[-\frac{1}{2}\sum_j s_j k_j^2 + \frac{1}{2}\sum_{\alpha\beta} \vec{k}_\alpha \mathscr{S}^{-1}_{\alpha\beta} \vec{k}_\beta\right]. \tag{10.4.26}$$

Let us recall [see (10.4.25)] that $\mathscr{S}_{\alpha\beta}$ is the sum of the s_j subtended by α and β and that $\vec{k}_\alpha$ is the sum of the $s_j\vec{k}_j$ subtended by α.

The internal wave vectors are now eliminated and eqn (10.4.26) summarizes Fixman's method.

4.2.6 Elimination of short-range divergences for a chain with two-body interactions

Let us consider the calculation of a diagram. After summing over all the space positions of the interaction points, we have to calculate the integrals concerning the areas of the polymer segments that appear on the diagram. However, an integral of the form $\int_0^S \mathrm{d}s \ldots$ is not always convergent for $s = 0$. A cut-off s_0 has to be introduced, and the integral has to be written in the form $\int_{s_0}^S \mathrm{d}s \ldots$. However, physically interesting quantities are often independent of the cut-off in the limit $s_0 = 0$. Consequently, the quantities which need regularization must be separated from the others.

To analyse the nature of the divergences, it is useful to introduce the P-reducibility concept, i.e. the reducibility of the diagrams *with respect to polymer lines*. A diagram is said to be m-P-reducible if, one has to cut at least m polymer segments to separate the diagram into two non-trivial disconnected parts (this means that each part must contain interaction points). A diagram which is not m-P-irreducible, is m-P-reducible (see Fig. 10.5). In what follows, we shall be especially interested in 1-P-irreducible diagrams. Actually, the importance of this class of diagrams, comes from the fact that the divergences of the 1-P-reducible diagrams can immediately be deduced from those of their 1-P-irreducible subdiagrams.

The origin of the divergences associated with diagrams contributing to $\mathscr{Z}(\vec{k}, -\vec{k}; S)$ appears in a simple way by first considering a 2-P-irreducible diagram, i.e. essentially a 1-P-irreducible diagram that does not contain any 1-P-irreducible insertion (see Fig. 10.6). Let s_1 be the coordinate of the first interaction point on the chain, and s_2 the coordinate of the last one. To calculate the contribution of the diagram, we fix s_1 and s_2 and, in the interval (s_1, s_2), we let the interaction points vary in space, by displacing them along the polymer line without crossing. It is, easy to see that the corresponding integrals converge when $s_0 \to 0$. Then, let p be the order of the diagram (as the diagram contains only one polymer line, p also represents the number of loops, i.e. the number of internal independent wave vectors). The contribution $\mathscr{D}_p$ of the diagram is given by an integral of the form

$$\mathscr{D}_p(k) = \int_{0<s'<s''<S} \mathrm{d}s'\,\mathrm{d}s''\, f[s''-s', k^2(s''-s')].$$

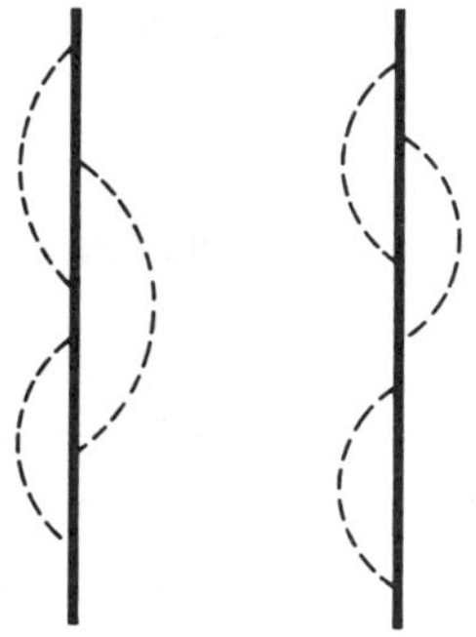

(a) (b)

Fig. 10.5. (a) A 1-P-irreducible diagram. (b) A 1-P-reducible diagram.

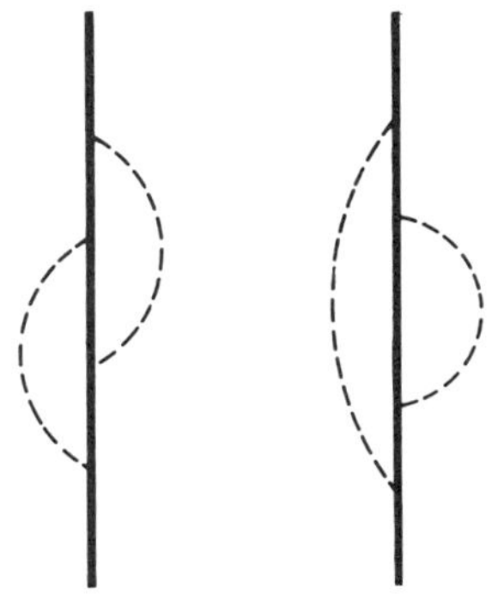

(a) (b)

Fig. 10.6. (a) A 2-P-irreducible diagram. (b) A 1-P-irreducible but 2-P-reducible diagram.

Like $\bar{\mathscr{Z}}(\vec{k}, -\vec{k}; s)$ (or $\mathscr{Z}(s)$), $\mathscr{D}_p(k)$ is a pure number; consequently, in the absence of divergences, one should find

$$\mathscr{D}_p(0) \propto (bS^{2-d/2})^p.$$

Therefore, for reasons of homogeneity, the preceding integral must read

$$\mathscr{D}_p(k) = \int_{0<s'<s''<S} ds'\, ds''\, (s'' - s')^{-2+p(2-d/2)} f[k^2(s'' - s')]$$

where $f(x)$ is a function which is regular at the point $x = 0$, as is shown by (10.4.26). The exponent $[-2 + p(2 - d/2)]$ is called the superficial degree of divergence.

More precisely, let us study the case where $2 < d < 4$. We observe that, for $p = 1$, the integral diverges when $s' \to s''$. In this case, we must write

$$^{+}\mathscr{D}_p(k) = b^p \int_{s_{01}}^{S} ds\, (S - s)\, s^{-2+p(2-d/2)} f(k^2 s) \qquad (10.4.27)$$

and the sign + indicates that we had to introduce a cut-off area s_{01}, which depends on the diagram but which is proportional to the cut-off s_0 characterizing the model.

We observe that the terms which diverge when $s_{01} \to 0$ are proportional to $f(0)$. Let us separate these contributions

$$^{+}\mathscr{D}_p(k) = b^p \int_{s_{01}}^{S} ds\, (S - s)\, s^{-2+p(2-d/2)} [f(k^2 s) - f(0)] +$$

$$+ b^p f(0) \int_{s_{01}}^{S} ds\, (S - s)\, s^{-2+p(2-d/2)}.$$

Dropping the terms which vanish when s_{01} tends to zero, we can rewrite the preceding expression in the form

$$
\begin{aligned}
{}^{+}\mathscr{D}_p(k) = b^p \int_0^S \mathrm{d}s\,(S-s)\,s^{-2+p(2-d)}\,[f(k^2 s) - f(0)] \\
- b^p f(0) \frac{S^{p(2-d/2)}}{p(2-d/2)\,[1-p(2-d/2)]} + b^p f(0) \frac{S\,(s_{01})^{-1+p(2-d/2)}}{1-p(2-d/2)}
\end{aligned}
\tag{10.4.28}
$$

(assuming that $[1 - p(2-d/2)] \neq 0$.

The first and the second term are regular terms whose dependence in S and k^2 is in accordance with the scaling predictions. The last term is an additional term proportional to S. Let us set

$$
\begin{aligned}
z_0 &= b\,(s_0)^{2-d/2}\,(2\pi)^{-d/2} \\
z_{01} &= b\,(s_{01})^{2-d/2}\,(2\pi)^{-d/2}
\end{aligned}
\tag{10.4.29}
$$

in agreement with (10.1.7). Thus, we obtain

$$
\begin{aligned}
{}^{+}\mathscr{D}_p(k) = {}^{-}\mathscr{D}_p(k) - f(0) \frac{z^p}{p(2-d/2)[1-p(2-d/2)]} \\
+ f(0)\left(\frac{S}{s_{01}}\right) \frac{(z_{01})^p}{[1-p(2-d/2)]}
\end{aligned}
\tag{10.4.30}
$$

where $\mathscr{D}_p(k)$ represents the integral which appears in the right-hand side of (10.4.28) and where s_{01} is proportional to s_0 and z_{01} to z_0.

Of course, it happens that d is of the form $d = 4 - 2/p$ where p is an integer; this is so for $d = 2$, $p = 1$ and for $d = 3$, $p = 2$. Then the term ${}^{+}\mathscr{D}(k)$ reads

$$
\begin{aligned}
{}^{+}\mathscr{D}_p(k) &= {}^{-}\mathscr{D}_p(k) + b^p f(0) \int_{s_0'}^{S} \mathrm{d}s\,(S-s)\,s^{-1} \\
&\simeq {}^{-}\mathscr{D}(k) + b^p f(0)\,S\,[\ln(S/s_{01}) - 1].
\end{aligned}
$$

In this case, (10.4.30) must be replaced by the equation

$$
{}^{+}\mathscr{D}_p(k) = {}^{-}\mathscr{D}_p(k) + f(0)\,z^p\left[\frac{\ln z}{2-d/2} - 1\right] - f(0)\frac{S}{s_{01}}\frac{(z_{01})^p \ln z_{01}}{2-d/2}. \tag{10.4.31}
$$

In all cases, the last term (a function of z_{01}) can be subtracted from ${}^{+}\mathscr{D}_p(k)$ and can be considered as a point insertion in the diagram. Then, the remaining (renormalized) contribution can be written in the symbolic form

$$
\mathscr{D}_p(k) = b^p \int_{\odot}^{S} \mathrm{d}s\,(S-s)s^{-2+p(2-d/2)} f(k^2 s) \tag{10.4.32}
$$

which reminds us that the divergent terms are to be forgotten. In other words, this is the Hadamard finite part. Thus, using the preceding notation, we can write for any $n \neq 1$

$$\int_{\odot}^{S} \mathrm{d}s\, s^n = \frac{S^{n+1}}{n+1}$$

and for $n = 1$

$$\int_{\odot}^{S} \mathrm{d}s\, s^{-1} = \frac{1}{2-d/2} \ln(b\, S^{2-d/2}(2\pi)^{-d/2}) = \frac{\ln z}{2-d/2}.$$

Moreover, we observe that for $d < 2$, ${}^{+}\mathscr{D}_p(k) \simeq \mathscr{D}_p(k)$ where ${}^{+}\mathscr{D}_p(k)$ can now be defined by the convergent integral

$$\mathscr{D}_p(k) = b^p \int_0^S \mathrm{d}S\,(S-s)\, s^{-2+p(2-d/2)} f(k^2 s).$$

Thus, for $4 > d > 2$, the value of $\mathscr{D}_p(k)$ given by (10.4.2) is just the analytic continuation with respect to d, of the integral giving $\mathscr{D}_p(k)$ for $d < 2$. As was seen above, this analytic continuation can be obtained with the help of subtractions, and the process is quite general (see Appendices G and H); in field theory, it is called 'dimensional regularization'.

Of course, the cut-off dependent terms which are subtracted cannot be eliminated without further examination; we have to show that these terms can be taken into account by recalibration of certain physical quantities. This is the renormalization process which will be applied here to short range divergences. It will be shown that, in the present case, the divergent terms can be taken into account by performing a multiplicative renormalization of ${}^{+}\mathscr{Z}(\vec{k}, -\vec{k}; S)$.

First, we observe that, in any diagram (contributing to the partition function) all the divergent parts can be subtracted, step by step, by starting with the most internal sub-diagrams. Thus, with each 1-P-irreducible part, one can associate a contribution calculated by subtracting the divergent parts of the 1-P-irreducible sub-diagrams. The contribution of the diagram contains a term proportional to the area S, which comes from the divergence corresponding to the diagram itself and a regular term $\mathscr{D}(k)$ which is independent of s_{01} (and therefore of the cut-off area s_0).

By considering eqns (10.4.30) and (10.4.31) we see that we can represent by $(S/s_0)\,\mathrm{C}(z_0)$ the sum of the terms proportional to S, appearing in the contributions of the 1-P-irreducible diagrams after 'renormalization' of the sub-diagrams. The divergent parts can be represented by point insertions (see Fig. 10.7). The total weight of an insertion is equal to $s_0^{-1}\,\mathrm{C}(z_0)$. Now, let us consider a diagram with a regular contribution $\mathscr{D}(k)$. By making of point insertions in this diagram we obtain the contribution.

$$\mathscr{D}(k) \int_{0<s_1<\,\ldots\,<s_q<S} \mathrm{d}s_1,\ldots \mathrm{d}s_q\,[s_0^{-1}\,\mathrm{C}(z_0)]^q = \mathscr{D}(k)\frac{[\mathrm{C}(z_0)(S/s_0)]^q}{q!}$$

Fig. 10.7. Point insertions in a diagram.

The contribution $\mathscr{D}(k)$ of the diagram with all possible insertions thus reads

$$^{+}\mathscr{D}(k) = \mathscr{D}(k) \sum_{q=0}^{\infty} \frac{[C(z_0)(S/s_0)]^q}{q!} = \mathscr{D}(k) \exp[C(z_0)].$$

The number C (z_0) is a quantity without any universal feature. In the model, it accounts only for the microstructure of the chain. Moreover, $^{+}\mathscr{D}$ is only the contribution obtained for the diagram by regularization (by means of the cut-off s_0).

Finally, the preceding discussion shows that $^{+}\mathscr{Z}(\vec{k}, -\vec{k}; S)$ can be written in the form*

$$^{+}\mathscr{Z}(\vec{k}, -\vec{k}; S) = e^{C(z_0)S/s_0}\, \bar{\mathscr{Z}}(\vec{k}, -\vec{k}; S) \tag{10.4.33}$$

where $\bar{\mathscr{Z}}(\vec{k}, -\vec{k}; S)$ is a quantity which depends only on z and k^2S. The same relation applies to $\mathscr{Z}(S)$ since $\mathscr{Z}(S) \equiv \bar{\mathscr{Z}}(\vec{0}, \vec{0}; S)$.

In the same way, renormalized restricted partition functions $\mathscr{Z}(\ldots; S)$ and $\bar{\mathscr{Z}}(\ldots; S)$ correspond to $^{+}\mathscr{Z}(\ldots; S)$ and $^{+}\mathscr{Z}(\ldots; S)$

$$^{+}\mathscr{Z}(\ldots; S) = e^{C(z_0)S/s_0}\, \mathscr{Z}(\ldots; S)$$

$$^{+}\bar{\mathscr{Z}}(\ldots; S) = e^{C(z_0)S/s_0}\, \bar{\mathscr{Z}}(\ldots; S) \tag{10.4.34}$$

4.2.7 Elimination of the short-range divergences for a chain with two-body and three-body interactions

When the chains are submitted to three-body interactions (or two-body and three-body interactons), the contribution of the diagrams concerning

* Actually, the relation would be more exact if a factor $A(z_0)$ had been introduced but $A(z_0) \to 1$ when $s_0 \to 0$.

$\overline{\mathscr{Z}}$ $(\vec{k}, -\vec{k}; S)$ present new short-range divergences and a cut-off has to be introduced to regularize these contributions. However, the finite part of the contributions of the diagrams can be obtained by dimensional regularization and in this respect the situation is the same as in the case where there are only two-body interactions (see Section 4.3.5 and Appendices H and G). Moreover, it appears that the terms which depend on s_0 can be absorbed by various renormalizations. However, the situation is much more complex than in the case where only two-body interactions are present, and a simple renormalization of the partition function is not sufficient to eliminate the terms depending on s_0. In particular, combinations of three-body interactions may generate cut-off dependent two-body interactions. Thus, in this case a renormalization of the two-body interaction becomes necessary.

In practice, the study of the properties of polymers in good solvent does not require the introduction of three-body forces. But, in the case of poor solvents, we have to consider such interactions. This will be done in Chapter 14 and, then, we shall re-examine the problems concerning the appearance of short-range divergences generated by three-body interactions.

4.3 Set of chains: diagram expansions

4.3.1 Partition function and restricted partition functions of a set of continuous chains

Let us consider, in a volume V, a set $\mathscr{E}$ of N chains with areas $S_1, \ldots, S_N$, the configuration of each chain a being determined by the function $\vec{r}_a(s)$ (with $0 \leqslant s \leqslant S_a$).

The partition function ${}^{+}\mathscr{Z}_{\mathrm{G}}(V; S_1, \ldots, S_N)$ has been defined as the ratio [see (10.4.3); the N_j are assumed to be very large]

$$
\begin{aligned}
{}^{+}\mathscr{Z}_{\mathrm{G}}(V; S_1, \ldots, S_N) &= \frac{{}^{+}Z_{\mathrm{G}}(V; N_1, \ldots, N_N)}{{}^{0}Z(N_1) \ldots {}^{0}Z(N_N)} \\
&= \frac{\displaystyle\int \mathrm{d}\{\vec{r}\}\, \mathscr{W}_N\{\vec{s}\}}{\displaystyle\int \mathrm{d}\{\vec{r}\}\, {}^{0}\mathscr{W}_N\{\vec{r}\}\, \delta[\vec{r}_1(0)] \ldots \delta[\vec{r}_N(0)]}
\end{aligned}
$$

where $\mathscr{W}_N\{\vec{r}\}$ and ${}^{0}\mathscr{W}_N\{\vec{r}\}$ are given by (10.1.3). This equality can be rewritten in the form

$$
{}^{+}\mathscr{Z}_{\mathrm{G}}(V; S_1, \ldots, S_N) = \frac{\langle\!\langle \exp[-{}^{\mathrm{I}}\mathfrak{H}_N\{\vec{r}\}] \rangle\!\rangle_0}{\langle\!\langle \delta(\vec{r}_1(0)) \ldots \delta(\vec{r}_N(0)) \rangle\!\rangle_0} \tag{10.4.35}
$$

the mean values being calculated with the weight ${}^{0}\mathscr{W}\{\vec{r}\}$. Incidentally, we note that in the absence of interaction, we have

$$
{}^{0}\mathscr{Z}_{\mathrm{G}}(V; S_1, \ldots, S_N) = V^N.
$$

The partition functions ${}^{+}\mathscr{Z}_{\mathrm{G}}(\mathscr{R}, \mathscr{P}; S_1, \ldots, S_{\boldsymbol{N}})$ can be defined in a similar manner, but let us first specify the notation. On the chains, we choose E correlation points $\vec{r}_{a_1}(s_1), \ldots, \vec{r}_{a_E}(s_E)$ and we want them to occupy positions defined by the vectors $\vec{r}_1, \ldots, \vec{r}_E$. The corresponding restricted partition function will be represented by the function

$$ {}^{+}\mathscr{Z}_{\mathrm{G}}(\mathscr{R},\mathscr{P};S_1,\ldots,S_{\boldsymbol{N}}) \equiv {}^{+}\mathscr{Z}_{\mathrm{G}}(\vec{r}_1,\ldots,\vec{r}_E;\boldsymbol{a}_1,\ldots,a_E;\, s_1,\ldots,s_E;\, S_1,\ldots,S_{\boldsymbol{N}}). $$

The points $\vec{r}_1, \ldots, \vec{r}_E$ may be called anchoring points. Relation (9.4.35) generalizes immediately, and we can write

$$ {}^{+}\mathscr{Z}_{\mathrm{G}}(\vec{r}_1,\ldots,\vec{r}_E;\, \boldsymbol{a}_1,\ldots,\boldsymbol{a}_E;\, s_1,\ldots,s_E;\, S_1,\ldots,S_{\boldsymbol{N}}) = \frac{\langle\!\langle \delta(\vec{r}_1 - \vec{r}_{a_1}(s_1)) \ldots \delta(\vec{r}_E - \vec{r}_{a_E}(s_E)) \exp(-{}^{\mathrm{I}}\mathfrak{H}_{\boldsymbol{N}}\{\vec{r}\})\rangle\!\rangle_0}{\langle\!\langle \delta(\vec{r}_1(0) \ldots \delta(\vec{r}_{\boldsymbol{N}}(0))\rangle\!\rangle_0}. \quad (10.4.36) $$

4.3.2 Set of chains: expansions and connexity of the diagrams

For a set of chains, expansions in powers of the interactions are performed in the same manner as for an isolated chain. The various terms of the expansion of ${}^{+}\mathscr{Z}_{\mathrm{G}}(S_1, \ldots, S_{\boldsymbol{N}})$ or of a function ${}^{+}\mathscr{Z}_{\mathrm{G}}(\mathscr{R}, \mathscr{P}; S_1, \ldots, S_{\boldsymbol{N}})$ are represented by diagrams on which N polymer lines appear (see Figs. 10.8 and 10.9). Now, interaction lines may connect two different polymer lines. Otherwise, N-chain diagrams are very similar to one-chain diagrams. However, let us note that a diagram can now be composed of several disconnected parts. The contributions associated with each of the connected parts factorize. If a connected part is bound to one or several anchoring points, its contribution is finite. On the contrary, if a connected part is not bound to any anchoring point, it can be displaced entirely, and consequently it gives a contribution proportional to V.

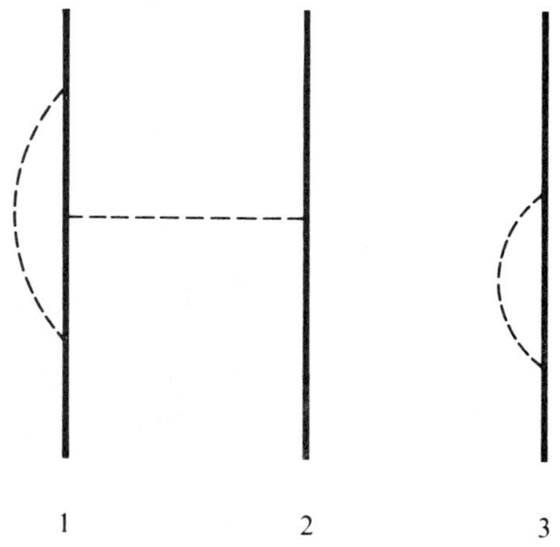

Fig. 10.8. A diagram contributing to ${}^{+}\mathscr{Z}_{\mathrm{G}}(V; S_1, S_2, S_3)$.

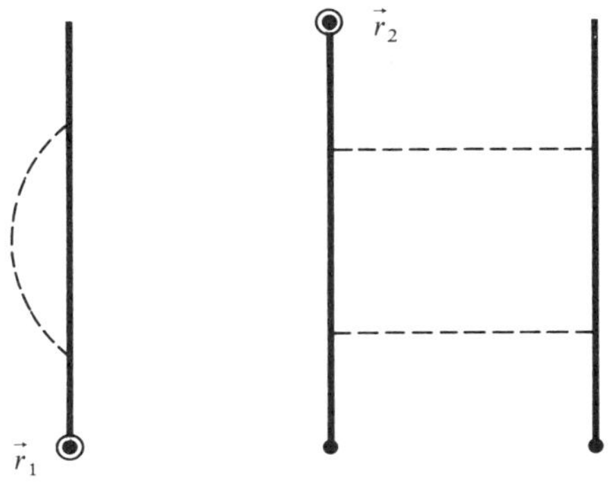

Fig. 10.9. A diagram contributing to ${}^{+}\mathscr{Z}_{\mathrm{G}}(\vec{r}_1, \vec{r}_2; 1, 2; 0, S_2; S_1, S_2, S_3)$.

We define

$$^{+}\mathscr{Z}_{\mathrm{c}}(V; S_1, \ldots, S_N) = V^{-1}[{}^{+}\mathscr{Z}_{\mathrm{G}}(V; S_1, \ldots, S_N)]_{\text{connected}}, \tag{10.4.37}$$

and, in the next section it will be shown that the ${}^{+}\mathscr{Z}_{\mathrm{c}}(V; S_1, \ldots, S_N)$ coincide with the cumulants ${}^{+}\mathscr{Z}(V; S_1, \ldots, S_N)$ of the ${}^{+}\mathscr{Z}_{\mathrm{G}}(V; S_1, \ldots, S_N)$.

In the same way, the contribution of the diagrams bound to anchoring points can be represented by

$$^{+}\mathscr{Z}_{\mathrm{a}}(V, \mathscr{R}, \mathscr{P}; S_1, \ldots, S_N) = [{}^{+}\mathscr{Z}_{\mathrm{G}}(V, \mathscr{R}, \mathscr{P}; S_1, \ldots, S_N)]_{\text{anchored}} \tag{10.4.38}$$

and we shall see that they coincide with the functions ${}^{+}\mathscr{Z}(V, \mathscr{R}, \mathscr{P}; S_1, \ldots, S_N)$ which appear in the expansion of the function $\mathscr{Q}(V, \mathscr{R}, \mathscr{P}; \{f\})$ defined by

$$\mathscr{Q}(V, \mathscr{R}, \mathscr{P}; \{f\}) = \frac{\mathscr{Q}_{\mathrm{G}}(V, \mathscr{R}, \mathscr{P}; \{f\})}{\mathscr{Q}_{\mathrm{G}}(V, \{f\})}.$$

This expansion reads

$$\mathscr{Q}(V, \mathscr{R}, \mathscr{P}; \{f\}) = \sum_{N=1}^{\infty} \frac{1}{\mathscr{S}_N(\mathscr{P})} \int_0^{\infty} \mathrm{d}S_1 \ldots \int_0^{\infty} \mathrm{d}S_N \times f(S_1) \ldots f(S_N) \mathscr{Z}(V, \mathscr{R}, \mathscr{P}; S_1, \ldots, S_N). \tag{10.4.39}$$

This enables us to define the contributions of the connected diagrams bound to anchoring points

$$^{+}\mathscr{Z}_{\mathrm{ac}}(V, \mathscr{R}, \mathscr{P}; S_1, \ldots, S_N) = [{}^{+}\mathscr{Z}_{\mathrm{G}}(V, \mathscr{R}, \mathscr{P}; S_1, \ldots, S_N)]_{\text{anchored connected}} \tag{10.4.40}$$

Connected diagrams are the only important diagrams from a physical point of view. The contributions of connected diagrams with N polymers can be calculated exactly in the same manner as the contributions of the diagrams with one polymer (see Sections 4.1.2 and 4.1.3).

4.3.3 General diagrams and connected diagrams

The contributions of the general diagrams can be expressed in terms of those of the connected diagrams, and precise relations between these contributions will be given here in order to justify the results announced in the preceding section. For this purpose, it is convenient to use a very general simplified notation. Indeed, the results are not only true for the continuous model but also for the discrete-link model in continuous space (Ursell–Mayer–Yvon expansions) and for lattice models.

A set of chains in a volume V is characterized by the set of 'areas' of the non-perturbed chains $\mathscr{E} \equiv (S_1, \ldots, S_N)$; moreover, two sets corresponding to a different ordering of the same chains are considered to be identical.

The partition function associated with the set is represented by

$$\frac{\mathscr{Z}_G(\mathscr{E})}{\$(\mathscr{E})}$$

where $\$(\mathscr{E})$ is the symmetry number of the set (the sign + which reminds us of the existence of a cut-off is omitted here).

In the same way, the restricted partition function obtained by fixing the space positions $\mathscr{R}$ of a set of points $\mathscr{P}$ marked on the chains is given by

$$\frac{\mathscr{Z}_G(\mathscr{R}, \mathscr{P}, \mathscr{E})}{\$(\mathscr{P}, \mathscr{E})}.$$

First, let us consider the diagram expansion of the partition function. Let $V\mathscr{Z}_c(\mathscr{E})$ be the contribution of the connected diagrams to $\mathscr{Z}_G(\mathscr{E})$. Let us try to express the $\mathscr{Z}_G(\mathscr{E})$ in terms of the $\mathscr{Z}_c(\mathscr{E})$. Every general diagram made of m connected parts corresponds to a partition $\mathfrak{P}(\mathscr{E})$ of the set $\mathscr{E}$, i.e. to a partition of the areas $S_1, \ldots, S_N$. Two partitions $\mathfrak{P}(\mathscr{E})$ are equivalent if each part of one of them corresponds to a part of the other one with the same content. Thus, the partitions $\mathfrak{P}(\mathscr{E})$ can be grouped into equivalence classes $\mathfrak{P}^*(\mathscr{E})$. The number of equivalent partitions belonging to class $\mathfrak{P}^*(\mathscr{E})$ will be denoted by $\$(\mathscr{E}, \mathfrak{P}^*)$ and $(\mathscr{E}, \mathfrak{P}^*)$ can be considered as the symmetry number of the partition. For every class $\mathfrak{P}^*(\mathscr{E})$, we can write

$$\mathscr{E} = \sum_j m_j \mathscr{E}_j \qquad \text{with} \quad \sum_j m_j = m.$$

The function $\mathscr{Z}_G(\mathscr{E})$ can be expressed as a sum over all partitions or more simply over all inequivalent partitions, i.e. over all partition classes

$$\mathscr{Z}_G(\mathscr{E}) = \sum_{\mathfrak{P}^*(\mathscr{E})} \$(\mathscr{E}, \mathfrak{P}^*) \prod_i [V\mathscr{Z}_c(\mathscr{E}_i)]^{m_i}. \tag{10.4.41}$$

Moreover, for every $\mathfrak{P}^*(\mathscr{E})$, the symmetry number $\mathcal{S}(\mathscr{E})$ can be written in the form

$$\mathcal{S}(\mathscr{E}) = \mathcal{S}(\mathscr{E}, \mathfrak{P}^*) \prod_j m_j! \, [\mathcal{S}(\mathscr{E}_j)]^{m_j} \tag{10.4.42}$$

and, taking (10.4.27) into account, we find

$$\frac{\mathscr{Z}_G(\mathscr{E})}{\mathcal{S}(\mathscr{E})} = \sum_{\mathfrak{P}^*(\mathscr{E})} \prod_j \frac{1}{m_j!} \left[\frac{V \mathscr{Z}_c(\mathscr{E}_j)}{\mathcal{S}(\mathscr{E}_j)} \right]^{m_j}. \tag{10.4.43}$$

This relation can be written in a more compact form by using generating functions; that is to say, grand partition functions. With each set $\mathscr{E}$, let us associate the quantity

$$f(\mathscr{E}) = f(S_1) \ldots f(S_N).$$

Then, we write

$$\mathscr{Q}_G\{f\} = 1 + \sum_{\mathscr{E}} f(\mathscr{E}) \frac{\mathscr{Z}_G(\mathscr{E})}{\mathcal{S}(\mathscr{E})}$$

$$\mathscr{Q}_c\{f\} = \sum_{\mathscr{E}} f(\mathscr{E}) \frac{\mathscr{Z}_c(\mathscr{E})}{\mathcal{S}(\mathscr{E})}.$$

Now, we can express (10.4.43) in the form

$$\mathscr{Q}_G(V; \{f\}) = \exp\, [V \mathscr{Q}_c(V; \{f\})] \tag{10.4.44}$$

which shows that the $\mathscr{Z}_c(\mathscr{E})$ are indeed the cumulants of the $\mathscr{Z}_G(\mathscr{E})$.

Let us now consider restricted partition functions. Their expansions are generally made of several connected parts of which some are free and some are anchored at the positions chosen for the correlation points.

Each general diagram corresponds to a partition $\mathfrak{P}(\mathscr{P}, \mathscr{E})$ of the set $\mathscr{E}$ fitted out with marked points $\mathscr{P}$ (a set denoted by $(\mathscr{P}, \mathscr{E})$).

The equivalence classes $\mathfrak{P}^*(\mathscr{P}, \mathscr{E})$ are associated with the partitions $\mathfrak{P}(\mathscr{P}, \mathscr{E})$ and the corresponding symmetry number is $\mathcal{S}(\mathscr{P}, \mathscr{E}, \mathfrak{P}^*)$. We can formally write

$$(\mathscr{P}, \mathscr{E}) = \sum_i n_i(\mathscr{P}_i, \mathscr{E}_i) + \sum_j m_j \mathscr{E}_j.$$

Here the subsets $\mathscr{E}_j$ consist of chains on which there are no correlation points. $\mathscr{Z}_G(\mathscr{R}, \mathscr{P}, \mathscr{E})$ can be expressed as a sum over all partitions of $(\mathscr{P}, \mathscr{E})$

$$\mathscr{Z}_G(\mathscr{R}, \mathscr{P}, \mathscr{E}) = \sum_{\mathfrak{P}^*(\mathscr{P}, \mathscr{E})} \mathcal{S}(\mathscr{P}, \mathscr{E}, \mathfrak{P}^*) \prod_i [\mathscr{Z}_{ac}(\mathscr{R}_i, \mathscr{P}_i, \mathscr{E}_i)]^{n_i} \prod_j [V \mathscr{Z}_c(\mathscr{E}_j)]^{m_j}. \tag{10.4.45}$$

Moreover, the symmetry number $\mathcal{S}(\mathscr{P}, \mathscr{E})$ can be expressed in the form

$$\mathcal{S}(\mathscr{P}, \mathscr{E}) = \mathcal{S}(\mathscr{P}, \mathscr{E}, \mathfrak{P}^*) \prod_i n_i! [\mathcal{S}(\mathscr{P}_i, \mathscr{E}_i)]^{n_i} \prod_j m_j! [\mathcal{S}(\mathscr{E}_j)]^{m_j}. \tag{10.4.46}$$

By combining (10.4.45) and (10.4.46), we obtain

$$\frac{\mathscr{Z}_{\mathrm{G}}(\mathscr{R}, \mathscr{P}, \mathscr{E})}{\$(\mathscr{P}, \mathscr{E})} = \sum_{\mathfrak{P}^*(\mathscr{P},\mathscr{E})} \prod_i \frac{1}{n_i!}\left[\frac{\mathscr{Z}_{\mathrm{ac}}(\mathscr{R}_i, \mathscr{P}_i, \mathscr{E}_i)}{\$(\mathscr{P}_i, \mathscr{E}_i)}\right]^{n_i} \prod_j \frac{1}{m_j!}\left[V\frac{\mathscr{Z}_{\mathrm{c}}(\mathscr{E}_j)}{\$(\mathscr{E}_j)}\right]^{m_j}. \tag{10.4.47}$$

Consequently, we introduce the restricted grand partition function

$$\mathscr{Q}_{\mathrm{G}}(\mathscr{R}, \mathscr{P}; \{f\}) = \sum_{\mathscr{E}} f(\mathscr{E})\frac{\mathscr{Z}_{\mathrm{G}}(\mathscr{R}, \mathscr{P}, \mathscr{E})}{\$(\mathscr{P}, \mathscr{E})}.$$

Then, setting

$$(\mathscr{P}, \mathscr{E}') = \sum_i n_i(\mathscr{P}_i, \mathscr{E}_i)$$

we obtain

$$\mathscr{Q}(\mathscr{R}, \mathscr{P}; \{f\}) \equiv \frac{\mathscr{Q}_{\mathrm{G}}(\mathscr{R}, \mathscr{P}; \{f\})}{\mathscr{Q}_{\mathrm{G}}\{f\}}$$

$$= \sum_{\mathfrak{P}^*(\mathscr{P},\mathscr{E}')} \prod_i \frac{1}{n_i!}\left[\frac{f(\mathscr{E}_i)\mathscr{Z}_{\mathrm{ac}}(\mathscr{R}_i, \mathscr{P}_i, \mathscr{E}_i)}{\$(\mathscr{P}_i, \mathscr{E}_i)}\right]^{n_i} \tag{10.4.48}$$

(the index ac means anchored connected), an equality which justifies the statements made in the preceding section, concerning the restricted partition functions.

4.3.4 Rules for the calculation of N-chain diagrams in real space

As a result of the preceding discussion, it appears that, from a physical point of view, the connected partition functions or the connected restricted partition functions are the only interesting quantities. They have a limit when $V \to \infty$, and in what follows we shall study only this limit, representing a partition function by the symbol ${}^{+}\mathscr{Z}(S_1, \ldots, S_N)$ and a restricted partition function by*

$${}^{+}\mathscr{Z}(\vec{r}_1, \ldots, \vec{r}_E; \boldsymbol{a}_1, \ldots, \boldsymbol{a}_E; s_1, \ldots, s_E; S_1, \ldots, S_N)$$

In practice, the rules for the calculation of a connected diagram are similar to those that apply to an isolated chain (see Sections 4.1.2, 4.1.3, and 4.1.4). The only difference comes from the fact that, in general, several polymer lines connected by interaction lines appear in the diagrams. Thus, on each polymer line, there is always at least one interaction point. Correlation points may also be present. Interaction points and correlation points cut each polymer line into polymer segments. The diagrams can be drawn either in real space or in reciprocal space, but in practice the calculations are more often made in reciprocal space. Later on, we shall give precise rules for the calculation of diagrams in reciprocal space. Here, we shall only point out that, for the calculation of diagrams in real space, the factors that have to be taken into

* In Chapter 11, we shall especially study the functions ${}^{+}\mathscr{Z}(\vec{r}_1, \ldots, \vec{r}_{2N}; S_1, \ldots, S_N) \equiv {}^{+}\mathscr{Z}(\vec{r}_1, \ldots, \vec{r}_{2N}; 0, S_1, \ldots, 0, S_N; S_1, \ldots, S_N)$.

account are:

(1) a factor b for each two-body interaction;

(2) a factor c for each three-body interaction;

(3) a factor $(2\pi)^{-d/2} \exp[-(\vec{r}_1 - \vec{r}_2)^2/2s^2]$ for each polymer segment with area s, joining points with position vectors $\vec{r}_1$ and $\vec{r}_2$ (interaction points or correlation points).

4.3.5 Elimination of short-range divergences for *N* chains with two-body interactions: elementary scaling laws

Let us now consider partition functions in reciprocal space. The Fourier transform of a (connected) restricted partition function can be expressed as follows

$$\begin{aligned}
&\tilde{\mathscr{Z}}(\vec{k}_1, \ldots, \vec{k}_E; \boldsymbol{a}_1, \ldots, \boldsymbol{a}_E; s_1, \ldots, s_E; S_1, \ldots, S_N) \\
&= \int \mathrm{d}^d r_1 \ldots \int \mathrm{d}^d r_E \exp[i(\vec{k}_1 \cdot \vec{r}_1 + \ldots + \vec{k}_E \cdot \vec{r}_E)] \\
&\times \mathscr{Z}(\vec{r}_1, \ldots, \vec{r}_E; \boldsymbol{a}_1, \ldots, \boldsymbol{a}_E; s_1, \ldots, s_E; S_1, \ldots, S_N) \\
&= (2\pi)^d \delta(\vec{k}_1 + \ldots + \vec{k}_E) \bar{\mathscr{Z}}(\vec{k}_1, \ldots, \vec{k}_E; \boldsymbol{a}_1, \ldots, \boldsymbol{a}_E; s_1, \ldots, s_E; S_1, \ldots, S_N)
\end{aligned} \tag{10.4.49}$$

(the index + has been omitted because the relation remains valid for renormalized functions).

We want to calculate the Fourier transforms ${}^{+}\mathscr{Z}(\ldots; S_1, \ldots, S_N)$ with the help of diagrams (see Fig. 10.10). In particular, the rules allow us to calculate the

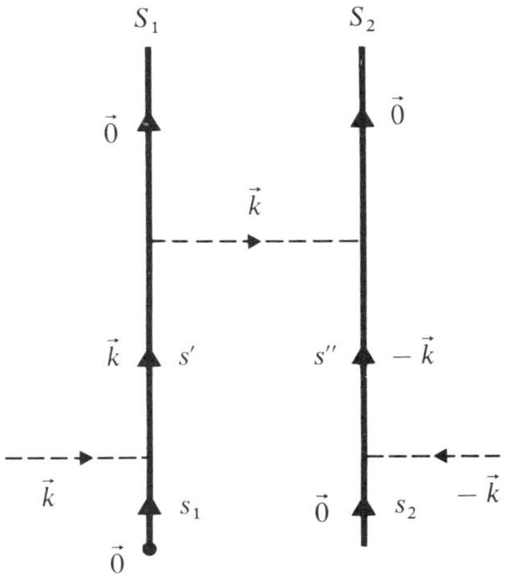

Fig. 10.10. A diagram associated with $\bar{\mathscr{Z}}(\vec{k}, -\vec{k}; 1, 2; s_1, s_2; S_1, S_2)$.

connected partition function ${}^{+}\mathscr{Z}(S_1, \ldots, S_N)$, which belong to the same class as the preceding functions, since

$$ {}^{+}\mathscr{Z}(S_1, \ldots, S_N) \equiv $$
$$ {}^{+}\bar{\mathscr{Z}}(\vec{0}, \ldots, \vec{0}; 1, 1, 2, 2, \ldots, N, N; 0, S_1, 0, S_2, \ldots, 0, S_N; S_1, \ldots, S_N). $$

The elimination of the short-distance divergences are performed for a set of chains in the same way as for an isolated chain. Here, as in Section 4.1.5, we shall consider the case where there are only two-body interactions ($b \neq 0$, $c = 0$). In this case, we have only to set

$$ {}^{+}\mathscr{Z}(\ldots; S_1, \ldots, S_N) = \exp\,[\mathrm{C}(z_0)(S_1 + \ldots + S_N)/s_0]\,\mathscr{Z}(\ldots; S_1, \ldots, S_N) \tag{10.4.50} $$

where $\mathrm{C}(z_0)$ was defined in Section 4.2.6 as a sum of short-range contributions. In the same way, for Fourier transforms, we set

$$ \bar{\mathscr{Z}}(\ldots; S_1, \ldots, S_N) = \exp\,[\mathrm{C}(z_0)(S_1 + \ldots + S_N)/s_0]\,\bar{\mathscr{Z}}(\ldots; S_1, \ldots, S_N). \tag{10.4.51} $$

The renormalized quantities $\mathscr{Z}(\ldots)$ or $\bar{\mathscr{Z}}(\ldots)$ do depend only on the polymer areas and on the interaction. But we know that they have definite dimensions. Thus, (10.4.36) gives

$$ {}^{+}\mathscr{Z}_{\mathrm{G}}(\vec{r}_1, \ldots, \vec{r}_E; \boldsymbol{a}_1, \ldots, \boldsymbol{a}_E; s_1, \ldots, s_E; S_1, \ldots, S_N) \sim L^{d(N-E)}. $$

The renormalizations and the fact that we deal with connected functions do not change the dimensions, but the situation is different for Fourier transforms. Thus, (10.4.49) leads to the equality

$$ {}^{+}\bar{\mathscr{Z}}(\vec{k}_1, \ldots, \vec{k}_E; \boldsymbol{a}_1, \ldots, \boldsymbol{a}_E; s_1, \ldots, s_E; S_1, \ldots, S_N) \sim L^{d(N-1)}. \tag{10.4.52} $$

Thus, as $S \sim L^2$, the (connected) partition function of a monodisperse set can be written in the form ($b \neq 0$, $c \neq 0$)

$$ \mathscr{Z}(N \times S) = (2\pi S)^{(N-1)d/2}\,\mathfrak{z}_N(z). \tag{10.4.53} $$

In the same way, the end-to-end correlation function of an isolated chain can be defined as follows

$$ \bar{\mathscr{Z}}(\vec{k}, -\vec{k}; 0, S; S) = \mathfrak{z}(z, k^2 S). \tag{10.4.54} $$

4.3.6 Rules for the calculation of N-chain diagrams in reciprocal space

The rules which must be applied to calculate diagrams for a set of chains, in reciprocal space, are very similar to the ones given in Section 4.2.4 for an isolated chain, and they can be deduced in the same manner. The difference is that now we have several chains connected with one another by interaction lines whose ends on the chains are the interaction points. To calculate restricted partition functions, one has to mark correlation points on the chains. At these points, external wave vectors $\vec{k}_i$ are injected and their total sum is zero.

Interaction points and correlation points cut the chains into segments. The latter, like the interaction lines, are oriented and bear a wave vector; the flux of these wave vectors is conserved at the interaction points and at the correlation points.

Each chain segment has an area and the sum of the areas of the segments of a chain j is equal to S_j; the S_j are quantities given a priori.

The factors associated with the diagrams are the same here as they are for an isolated chain. The rules can be summarized as follows.

1. With each polymer segment having an area s and bearing a wave vector $\vec{q}$, we associate a factor $\exp(-sq^2/2)$.
2. With each two-body interaction, we associate a factor $-b$, and with each three-body interaction, a factor $-c$.
3. With the whole diagram made of N polymer lines, we associate a global factor $(2\pi)^{(N-1)d}$
4. The product of all the factors is integrated over all the independent wave vectors; each integral is written in the form $\frac{1}{(2\pi)^d}\int d^dq \ldots$.
5. Finally, the preceding result is integrated over all the independent segment areas; each integral is written in the form $\int ds \ldots$, with boundary conditions that are determined by the constraints.

The function ${}^{+}\mathscr{Z}(\ldots, S_1, \ldots, S_N)$ is obtained by introducing a lower bound s_0 for the length of each segment, in order to avoid divergences. Conversely, $\bar{\mathscr{Z}}(\ldots; S_1, \ldots, S_N)$ is the finite part of the integrals.

4.3.7 Application of Fixman's method to the case of several chains

Fixman's method (see Section 4.1.5) can be applied without difficulty when several chains interact. However, the situation is then slightly more complicated because the vectors which in ordinary space define the relative positions of the correlation and interaction points, can now join points belonging to different chains.

Here, as in the case of an isolated chain, it is convenient to work in reciprocal space and to calculate the Fourier transforms of the restricted partition functions. The method can be generalized quite simply.

The number $\circlearrowleft$ of loops (i.e. of internal independent wave vectors) is now given by

$$\circlearrowleft = p + 2p' - N + 1 \tag{10.4.55}$$

where p is the number of two-body interactions and p' the number of three-body interactions, in the (connected) diagram.

First, among the interaction lines joining the N polymers, we choose $N - 1$ of them in such a way that the set of these interaction lines and of the polymer lines

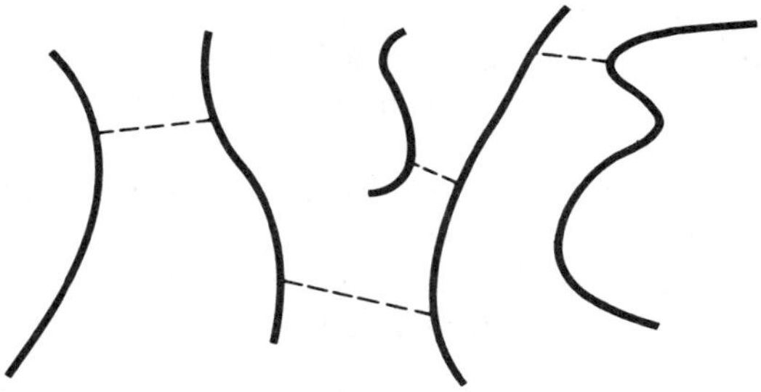

Fig. 10.11. A simple-tree diagram.

constitute a simple-tree diagram (see Fig. 10.11). These interaction lines will be called structure lines; in general, the choice of such lines is somewhat arbitrary. We can now generalize the notation which we used for isolated chains (see Section 4.1.4).

As before, the area of a segment j of polymer chain is denoted by s_j, but now the segments belong to various polymer lines. The wave vectors $\vec{q}_\alpha$ corresponding to the internal independent wave vectors are always borne by certain interaction lines. In fact, when the $\vec{q}_\alpha$ vanish, we can always manage in such a way that the interaction lines which are not structure lines, bear a zero wave vector. Then, when the $\vec{q}_\alpha$ vanish the segment j bears a wave vector which we denote by $\vec{k}_j$. When the $\vec{q}_\alpha$ do not vanish, the segment j bears the wave vector

$$\vec{k}_j + \sum_{\alpha=1}^{\circlearrowleft} \ominus_{j\alpha} \vec{q}_\alpha$$

where $\ominus_{j\alpha} = 1$ if $\vec{q}_\alpha$ is transferred along segment j in the positive direction;
$\ominus_{j\alpha} = -1$ if $\vec{q}_\alpha$ is transferred along segment j in the negative direction;
$\ominus_{j\alpha} = 0$ if $\vec{q}_\alpha$ is not transferred along segment j.

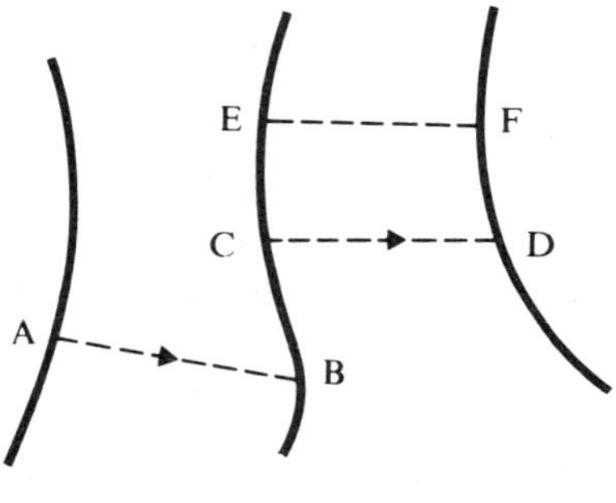

Fig. 10.12. In this diagram, the structure lines are indicated by arrows. We see that AB has to be a structure line but that either CD or EF can be chosen as structure line. Here, we chose CD.

The contribution of a connected diagram $\mathfrak{D}$ containing N polymer lines, can be written, *before summation* over the areas of the polymer segments, in the form

$$\mathscr{D}\{s\} = (-b)^{P}(-c)^{P'}\mathscr{I}\{s\}$$

$$\mathscr{I}\{s\} = \int\left[\prod_{\alpha=1}^{\mathfrak{O}} \frac{d^d q_\alpha}{(2\pi)^d}\right]\exp\left[-\frac{1}{2}\sum_j s_j\left(\vec{k}_j + \sum_\alpha \ominus_{j\alpha}\vec{q}_\alpha\right)^2\right].$$

Thus the formal expression of $\mathscr{I}\{s\}$ remains unchanged [see (10.4.24)].

The quantities $\mathscr{I}_{\alpha\beta}$ and $\vec{k}_\alpha$ are still defined by the equalities

$$\mathscr{S}_{\alpha\beta} = \sum_j s_j \ominus_{j\alpha}\ominus_{j\beta},$$

$$\vec{k}_\alpha = \sum_j s_j \ominus_{j\alpha}\vec{k}_j. \tag{10.4.56}$$

Arguing as in Section 4.1.5, we find

$$\mathscr{D}\{s\} = (-b)^{P}(-c)^{P'}(2\pi)^{-d\mathfrak{O}/2}|\det\mathscr{S}|^{-d/2}\exp\left[-\frac{1}{2}\sum_j s_j k_j^2 + \frac{1}{2}\vec{k}\,\mathscr{S}^{-1}\,\vec{k}\right] \tag{10.4.57}$$

where $\mathscr{S}$ the 'Fixman matrix' whose elements are given by (10.4.56). Formally, this formula coincides with (10.4.26). Here, as elsewhere, $\mathscr{S}_{\alpha\beta}$ is the sum, in magnitude and sign, of the s_j that are simultaneously subtended by the interaction lines α and β (of course α and β are not structure lines); in the same way, $\vec{k}_\alpha$ is the sum, in magnitude and sign, of the products $s_j\vec{k}_j$ subtended by the interaction line α.

5. GRAND CANONICAL ENSEMBLES: OSMOTIC PRESSURE AND STRUCTURE FUNCTIONS OF CONTINUOUS CHAINS

Everywhere in this book, renormalizations of partition functions play a basic role: in general, this prevents divergences from appearing. We thus renormalized in order to eliminate the solvent, in order to be able to reach the continuous limit, and finally in order to eliminate short-range divergences; we shall renormalize again in Chapter 12 when dealing with the properties of long chains. All these *multiplicative* renormalizations are of the same kind. The grand partition functions remain formally invariant in the course of the renormalization process, because the renormalization factors of the partition functions are absorbed by simultaneous renormalizations of the fugacities. The following formulae are just trivial transcriptions of formulae previously established in a different context.

In the monodisperse case, the (connected) grand partition function reads [see (9.4.12)]

$$\mathcal{Q}(f) = \sum_{N=1}^{\infty} \frac{f^N}{N!} \mathcal{Z}(N \times S). \tag{10.5.1}$$

The osmotic pressure $\boldsymbol{\Pi}$ and the number of polymers per unit volume are given by [see (9.4.10) and (9.4.11)]

$$\begin{aligned} \boldsymbol{\Pi\beta} &= \mathcal{Q}(f) \\ \boldsymbol{C} &= f \frac{\partial}{\partial f} \mathcal{Q}(f). \end{aligned} \tag{10.5.2}$$

The total area per unit volume is

$$\mathscr{C} = \boldsymbol{C}S \tag{10.5.3}$$

($\boldsymbol{C}$ is the number of polymers per unit volume).

In the polydisperse case, the (connected) grand partition function reads

$$\mathcal{Q}\{f\} = \sum_{N=1}^{\infty} \frac{1}{N!} \int_0^{\infty} \mathrm{d}S_1 \ldots \int_0^{\infty} \mathrm{d}S_{\boldsymbol{N}}\, f(S_1) \ldots f(S_{\boldsymbol{N}}) \mathcal{Z}(S_1, \ldots, S_{\boldsymbol{N}}). \tag{10.5.4}$$

The osmotic pressure $\boldsymbol{\Pi}$ and the number $\boldsymbol{C}(S)$ of polymers per unit volume and per unit area are given by

$$\begin{aligned} \boldsymbol{\Pi\beta} &= \mathcal{Q}\{f\} \\ \boldsymbol{C}(S) &= f(S) \frac{\partial \mathcal{Q}\{f\}}{\partial f(S)}. \end{aligned} \tag{10.5.5}$$

The total number $\boldsymbol{C}$ of polymers per unit volume is

$$\boldsymbol{C} = \int_0^{\infty} \mathrm{d}S\, \boldsymbol{C}(S)$$

and the total 'area' per unit volume is

$$\mathscr{C} = \int_0^{\infty} \mathrm{d}S\, S\, \boldsymbol{C}(S)$$

(this quantity plays the role of monomer concentration).

Consequently, the first terms of the virial expansion are obtained by eliminating the fugacities in terms of which the osmotic pressure and the concentrations are expressed. In the monodisperse case, we set

$$\mathcal{Y}(\boldsymbol{N} \times S) = \frac{\mathcal{Z}(\boldsymbol{N} \times S)}{[\mathcal{Z}(S)]^{\boldsymbol{N}}},$$

and we obtain

$$\boldsymbol{\Pi\beta} = \boldsymbol{C} - \tfrac{1}{2} \boldsymbol{C}^2 \mathcal{Y}(2 \times S) - \boldsymbol{C}^3 \left[\tfrac{1}{3} \mathcal{Y}(3 \times S) - [\mathcal{Y}(2 \times S)]^2\right] + \ldots. \tag{10.5.6}$$

In the polydisperse case, we set

$$\mathscr{Y}(S_1, \ldots, S_N) = \frac{\mathscr{Z}(S_1, \ldots, S_N)}{\mathscr{Z}(S_1) \ldots \mathscr{Z}(S_N)},$$

and in this case

$$\Pi\beta = C - \frac{1}{2}\int_0^\infty \mathrm{d}A \int_0^\infty \mathrm{d}B\, C(A)C(B)\mathscr{Y}(A, B)$$

$$- \int_0^\infty \mathrm{d}A \int_0^\infty \mathrm{d}B \int_0^\infty \mathrm{d}C\, C(A)C(B)C(C)\,[\tfrac{1}{2}\mathscr{Y}(A, B, C) - \mathscr{Y}(A, B)\mathscr{Y}(B, C)]. \qquad (10.5.7)$$

These formulae are just transpositions of eqns (9.4.17) and (9.4.19) to the continuous case.

Moreover, we are going to show that the osmotic pressure of a solution of monodisperse polymers obeys an elementary scaling law. Let us start from (10.4.53)

$$\mathscr{Z}(N \times S) = (2\pi S)^{(N-1)d/2}\, \mathfrak{Z}_N(z). \qquad (10.5.8)$$

By bringing this expression in (10.5.1), we find that the function $\mathscr{Q}(f)$ is of the form

$$\mathscr{Q}(f) = f\, \mathfrak{Z}(f S^{d/2}, z). \qquad (10.5.9)$$

Thus, from eqns (10.5.5), we derive the law

$$\Pi\beta = C F_0(C S^{d/2}, z). \qquad (10.5.10)$$

In the same way, the structure function at $\vec{q} = 0$ and for a monodisperse system is given by

$$H(\vec{0}) = \frac{\left(f\dfrac{\partial}{\partial f}\right)^2 \mathscr{Q}(f)}{\left[f\dfrac{\partial}{\partial f}\mathscr{Q}(f)\right]^2} = \frac{1}{C^2} f \frac{\partial C}{\partial f}$$

$$C = f\frac{\partial}{\partial f}\mathscr{Q}(f). \qquad (10.5.11)$$

This leads to the expansion

$$H(\vec{0}) = \frac{1}{C}\{1 + C\mathscr{Y}(2\times S) + C^2[\mathscr{Y}(3\times S) - 2\mathscr{Y}^2(2\times S)]$$

$$+ C^3\,[\tfrac{1}{2}\mathscr{Y}(4\times S) - 4\mathscr{Y}(3\times S)\mathscr{Y}(2\times S) + 5\mathscr{Y}^3(2\times S)] + \ldots\} \qquad (10.5.12)$$

which is a transcription of (9.4.20). It is also easy to see that $CH(\vec{0})$ is a function depending only on the dimensionless numbers z and $CS^{d/2}$.

6. A STUDY OF THE TREE STRUCTURE IN POLYMER DIAGRAMS

6.1 Generalities

In general, the systems that have a tree structure, i.e. without cycle, can easily be studied, and in particular, all mean field approximations are obtained by replacing the initial system by a system with such a structure.

Actually, in perturbation theory, two approaches are possible. Either we select a class of diagrams having a rather simple tree structure, we drop the other ones and, in this way, we obtain an interesting approximation; or we consider the most general diagrams and, recognizing that they have an overall tree structure, we reduce them by a summation which in reality amounts to a Legendre transformation.

For example, let us consider simple molecules in solution. Using the grand canonical formalism, we can find a parametric representation of the dependence of the osmotic pressure with respect to the concentration

$$\boldsymbol{\Pi\beta} = Q(f)$$

$$\boldsymbol{C} = f\frac{\partial}{\partial f}Q(f)$$

where

$$Q\,(f) = \sum_{N} f^{N}\frac{Z_N}{N!}. \tag{10.6.1}$$

In this case, the Z_N are represented by cactus-shaped diagrams (see Fig. 10.13) and, by summation, it is possible to eliminate the fugacity and to re-express $\boldsymbol{\Pi\beta}$, in diagrammatic form, as a direct expansion in powers of the concentration.[1,2]

Unfortunately, the method does not apply to polymers because the structure of these molecules is not simple. The fact that diagrams have a tree structure does not directly imply that summations are possible: the contribution of a tree

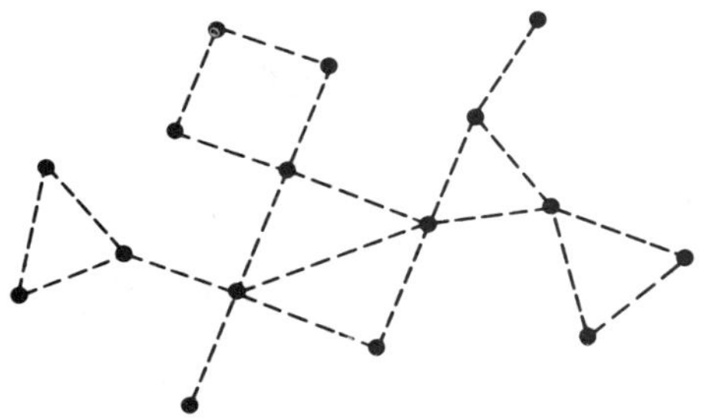

Fig. 10.13. Cactus diagram representing a contribution to the partition function of point-like molecules. This diagram is made of eight irreductible parts.

diagram has also to be factorizable. In other words, a summation is possible only if the contribution of a tree diagram is the product of the contributions of the branches of the tree.

Thus for example, field theory diagrams are structurally very analogous to polymer diagrams (see Chapter 11 for more details) but the contributions of polymer diagrams cannot be factorized as easily. In field theory, to pass from Green's functions to vertex functions, we thus apply the P-reducibility concept which can be defined as follows. A connected diagram is P-reducible if, by cutting one of the lines corresponding to a field propagator (polymer line), the diagram separates into two pieces.

In field theory, the contribution of a P-reducible diagram can be factorized, but not in polymer theory. However, in polymer theory, there is another way of considering diagrams as trees; namely, by applying the I-reducibility concept which can be defined as follows. A diagram is I-reducible if, by cutting an interaction line, the diagram separates into two pieces. In polymer theory, the contribution of an I-reducible diagram can be factorized and this property enables us to perform summations which will be studied in the following sections.

First, we shall describe the simple-tree approximation by using a very direct method. Later, we shall analyse the structure of the I-reducible diagrams in a general way by using a slightly different method, more powerful but also more difficult to work out.

6.2 The simple-tree approximation ($b \neq 0$, $c \neq 0$)

To calculate the osmotic pressure or the correlation functions of polymers in solution, the most straightforward approximation consists is summing up the diagrams having a simple-tree structure with respect to the interactions (see

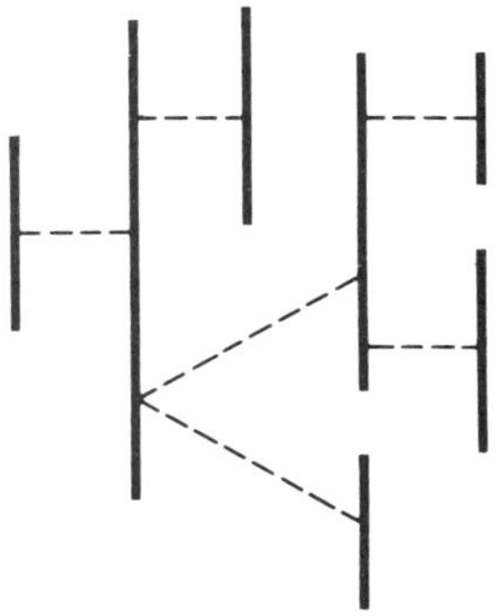

Fig. 10.14. A simple-tree diagram (contributing to $\mathscr{Z}(7 \times S)$) with two-body and three-body interactions (in simple-tree diagrams, there are no loops).

Fig. 10.14). This will be done here in the case where the polymers are identical dissymmetrical chains submitted to a two-body interaction of intensity b and to a three-body interaction of intensity c.

We shall deal here only with the calculation of the osmotic pressure Π, but, in Chapter 13, the same approximation will be used to calculate the structure function $H(\vec{q})$. Consequently, in this section, we have to evaluate $\mathscr{Z}(N \times S)$, since according to (10.5.4) and (10.5.5) we have

$$\mathscr{Q}(f) = \sum_{N=1}^{\infty} \frac{\mathscr{Z}(N \times S)}{N!} f^N$$

$$C = f \frac{\partial}{\partial f} \mathscr{Q}(f)$$

$$\Pi\beta = \mathscr{Q}(f) \tag{10.6.2}$$

Then let us consider the simple-tree diagrams made up of N polymer lines. Applying the rules given in Sections 4.2.4 and 4.2.5, we see immediately that the weight of a diagram containing p two-body interactions and p' three-body interactions equals

$$(-bS^2)^p(-cS^3)^{p'}[{}^0\mathscr{Q}(S)]^{p+2p'}$$

where $N = p + 2p' + 1$ is the number of polymer lines on the diagram (see Fig. 10.14). Thus, $\mathscr{Z}(N \times S)$ is of the form

$$\mathscr{Z}(N \times S) = \sum_{\substack{p,p' \\ (p+2p'=N-1)}} n_{p,p'}(-bS^2)^p(-cS^3)^{p'}[{}^0\mathscr{Z}(S)]^N$$

where $n_{pp'}$ is obtained by counting the number of settings; the chains are assumed to be distinguishable. However, the counting simplifies when the chains are assumed to be indistinguishable. In this case, a symmetry number is associated with each diagram; it is defined as the number of identical settings that can be obtained by permuting polymers on the diagram without changing the topology (see Fig. 10.15). Thus, a symmetry factor (the inverse of the symmetry number) is introduced for each diagram. Moreover, it is easy to see that $\frac{1}{N!} \sum_{\substack{p,p' \\ (p+2p'=N-1)}} n_{p,p'}$ is the sum of the symmetry factors of the trees built with N polymer lines. In fact, let us consider a diagram composed of N indistinguishable polymers and let $\$$ be its symmetry number: by labelling the polymers of the initial diagram, one obtains $N!/\$$ different labelled diagrams.

In what follows, we shall also use the equality ${}^0\mathscr{Z}(S) = 1$. In fact, when the partition function of one chain is not unity, the situation can always be reduced to this case by renormalization of the fugacity f; in fact, a product $f\,{}^0\mathscr{Z}(S)$ is associated with each polymer line, and renormalizing f does not destroy the validity of eqns (10.6.1).

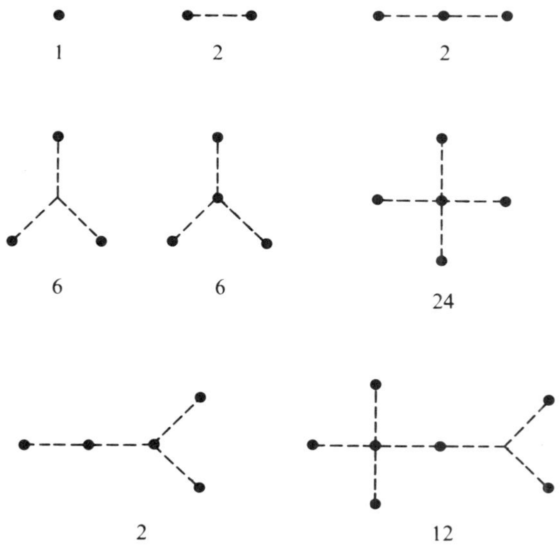

Fig. 10.15. Rootless trees and symmetry numbers. Each point symbolizes a polymer. On one of the trees (on the left) there is a three-body interaction which is represented in a symmetric form.

Actually, instead of evaluating $\mathscr{Z}(N\times S)$, one prefers calculating $\mathscr{Q}(f)$ in a self-consistent way. This possibility is a consequence of the structure of the diagrams: in general, the sum of the factorizable contributions of tree diagrams is obtained with the help of a Legendre transformation (see Appendix I).

Here the transformation enables us to express $\boldsymbol{\Pi}$ directly in terms of $\boldsymbol{C}$ and the operation can be described in an elementary way. First, we observe that the diagrams contributing to $\boldsymbol{C} = f\dfrac{\partial}{\partial f}\mathscr{Q}(f)$ are same as the ones contributing to $\mathscr{Q}(f)$ but are counted with different weights. More precisely, the operator $f\dfrac{\partial}{\partial f}$ has the property of marking successively each polymer chain on the diagram. The marked chain is the 'root' of the diagram (see Fig. 10.16). The function $\mathscr{Q}(f)$ is the sum of the contributions of the rootless diagrams whereas $\boldsymbol{C} = f\dfrac{\partial}{\partial f}\mathscr{Q}(f)$ is the sum of the contributions of the diagrams with a root.

Now, let us consider the trees whose root is connected to the other polymers by p two-body interactions and by p' three-body interactions. The $(p + 2p')$ polymers which are directly connected to the root can now be considered as the roots of sub-diagrams having the same kind of structure as the initial diagram. Then let $\mathscr{D}_{p,p'}$ be the contribution of the trees whose root is connected to the

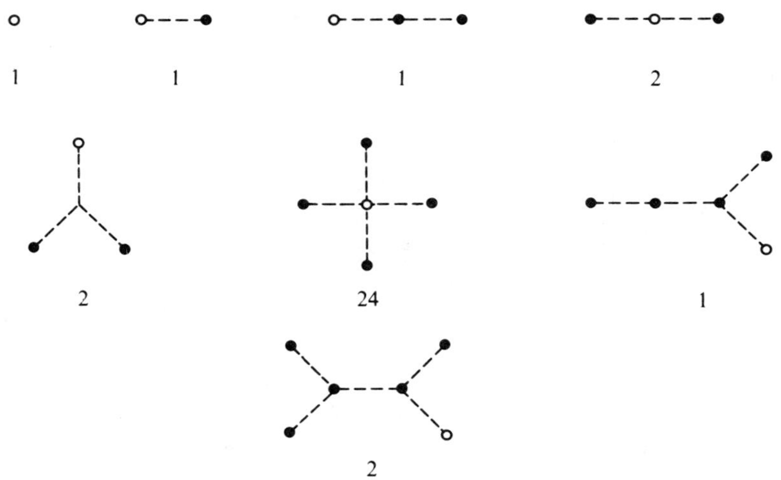

Fig. 10.16. Trees with a root and symmetry numbers. The root is represented by a small circle. On one of the trees a three-body interaction appears which is represented in a symmetric form.

neighbouring polymers by p two-body interactions and p' three-body interactions; we have

$$\mathscr{D}_{p,p'} = f\frac{(-bS^2C)^p}{p!}\frac{(-cS^3C^2/2)^{p'}}{p'!}$$

and

$$C = \sum_{p=0}^{\infty}\sum_{p'=0}^{\infty}\mathscr{D}_{p,p'}.$$

Therefore,

$$f = C\exp\left(bS^2C + \frac{c}{2}S^3C^2\right). \tag{10.6.3}$$

But, on the other hand, considering (10.6.2), we can write

$$\frac{\partial\mathscr{Q}}{\partial C} = \frac{\partial\mathscr{Q}}{\partial f}\frac{\partial f}{\partial C} = \frac{C}{f}\frac{\partial f}{\partial C}. \tag{10.6.4}$$

Let us bring in (10.6.4), the value of f given by (10.6.3)

$$\frac{\partial\mathscr{Q}}{\partial C} = 1 + bS^2C + cS^3C^2.$$

Moreover, if $f = 0$, $C = 0$, and $\mathscr{Q} = 0$. Therefore, by integrating the preceding equation, one obtains

$$\mathscr{Q} = C + \frac{b}{2}S^2C^2 + \frac{c}{3}S^3C^3 \tag{10.6.5}$$

which, according to (10.6.2), gives

$$\Pi\beta = C\left[1 + \frac{b}{2}S^2C + \frac{c}{3}S^3C^2\right]. \qquad (10.6.6)$$

We can also obtain the value of the structure function $H(\vec{q})$ for $\vec{q} = 0$ by using (9.3.64)

$$H(\vec{0}) = \left[C\frac{\partial(\Pi\beta)}{\partial C}\right]^{-1} = \frac{1}{C[1 + bS^2C + cS^3C^2]}. \qquad (10.6.7)$$

6.3 Diagrams and tree-structure with respect to the interaction ($b \neq 0$, $c \neq 0$)

The grand canonical formalism, presented in Section 5 allows us to express the osmotic pressure and the concentrations in parametric form as functions of the fugacities $f(S)$. Moreover, perturbation theory, expounded in Section 4, enables us to construct connected diagrams representing the expansions of these physical quantities in powers of interactions and fugacities.

It turns out that these connected diagrams can be expressed in terms of more elementary diagrams and thus we are induced to sum up the diagrams that have a tree structure with respect to the interaction. The problem consists in decomposing $\mathscr{Z}(S_1, \ldots, S_N)$ into simpler elements since, according to (10.5.4) and (10.5.5), we have

$$\Pi\beta = \mathscr{Q}\{f\}$$

$$C(S) = f(S)\frac{\partial\mathscr{Q}\{f\}}{\partial f(S)}$$

$$\mathscr{Q}\{f\} = \sum_{N=1}^{\infty}\frac{1}{N!}\int_0^{\infty} dS_1 \ldots \int_0^{\infty} dS_N\, f(S_1)\ldots f(S_N)\mathscr{Z}(S_1, \ldots, S_N). \qquad (10.6.8)$$

6.3.1 Reduction of tree diagrams and summation of their contributions ($b \neq 0$, $c = 0$)

A diagram contributing to $\mathscr{Z}(S_1, \ldots, S_N)$ is said to be reducible with respect to the interaction lines, if the diagram can be separated into two disconnected pieces by cutting a particular interaction line (see Fig. 10.17). In this case the interaction line bears a zero wave vector and its end points can be freely displaced along the polymer lines to which they belong; thus, the contribution of the total diagram is obtained by factorization of the contributions of the two sub-diagrams and of the interaction line.

The diagrams that cannot be separated into two disconnected parts by cutting an interaction line are called I-irreducible.

The sum of the contributions of the I-reducible and I-irreducible diagrams $\mathfrak{D}$ containing N polymer lines give $\mathscr{Z}(S_1, \ldots, S_N)$. In a similar way, the sum of the

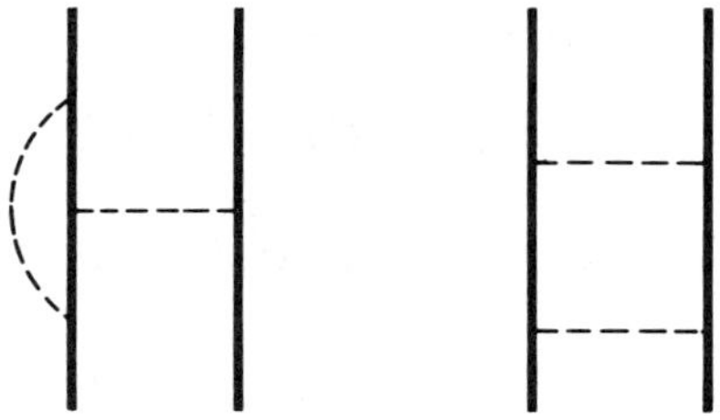

(a) (b)

Fig. 10.17. Reducible (a) and irreducible (b) diagrams with respect to the interactions.

contributions of the I-irreducible diagrams $\mathfrak{D}_{\mathrm{I}}$ containing N polymer lines gives $\mathscr{Z}_{\mathrm{I}}(S_1, \ldots, S_N)$

$$\mathscr{Z}(S_1, \ldots, S_N) = \sum_{\mathfrak{D}} \mathscr{D}(S_1, \ldots, S_N)$$

$$\mathscr{Z}_{\mathrm{I}}(S_1, \ldots, S_N) = \sum_{\mathfrak{D}_{\mathrm{I}}} \mathscr{D}_{\mathrm{I}}(S_1, \ldots, S_N).$$

A I-reducible diagram is made of p I-irreducible parts and of interaction lines connecting these irreducible parts. Each I-irreducible part is considered as an m-leg vertex ($m \geqslant 0$, see Fig. 10.18), when m interaction lines connect this irreducible part to other irreducible parts. The corresponding tree has no external legs (see Fig. 10.15); consequently, its p vertices are connected by $(p - 1)$ interaction lines*. Each vertex of the skeleton tree can be labelled by an index j ($j = 1, \ldots, p$). Here, the symbol m_j represents the number of legs of

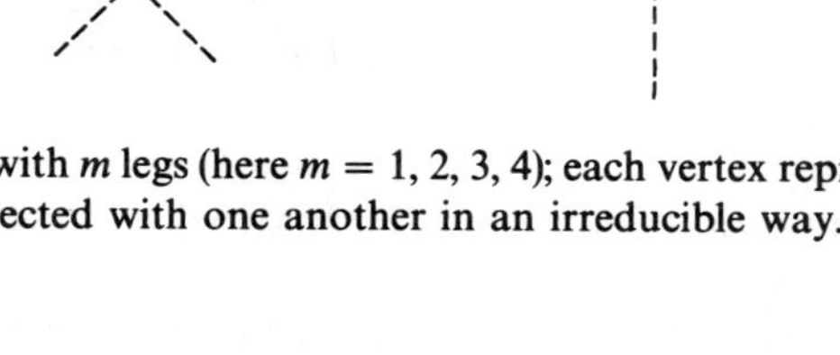

Fig. 10.18. Vertices with m legs (here m = 1, 2, 3, 4); each vertex represent a polymer or a set of polymers connected with one another in an irreducible way.

* For a chemist, the legs would be valences, the interaction lines would be bonds!

vertex j (also called sub-diagram j). Moreover, N_j is the number of polymers appearing in sub-diagram j). Thus, we have

$$\sum_{j=1}^{p} m_j = 2(p-1)$$

$$\sum_{j=1}^{p} N_j = N. \tag{10.6.9}$$

The areas of the polymer lines belonging to the sub-diagram j will be denoted by $S_{j,1}, \ldots, S_{j,N_j}$ and their union constitutes the set $S_1, \ldots, S_N$ of the areas of the whole diagram.

In what follows, we shall study just the case where there are only two-body interactions but a generalization is possible.

Let $\mathscr{D}(S_1, \ldots, S_N)$ be the contribution (to $\mathscr{Z}(S_1, \ldots, S_N)$) of the whole diagram and $\mathscr{D}_{\mathrm{I},j}$ $(S_{j,1}, \ldots, S_{N_j})$ that of the irreducible sub-diagram j. Interaction lines connect sub-diagram j to other sub-diagrams, and originate from m_j interaction points belonging to sub-diagram j; in the polymer of area $S_{j,q}$, the number of such interaction points is $m_{j,q}$. Thus,

$$\sum_{q=1}^{N_j} m_{j,q} = m_j.$$

By applying the rules prescribed in Sections 4.3.3 and 4.2.5, one finds

$$\mathscr{D}(S_1, \ldots, S_N) = (-b)^{p-1} \prod_{j=1}^{p} \mathscr{D}_{\mathrm{I},j}(S_{j,1}, \ldots, S_{j,N_j}) \prod_{q=1}^{N_j} (S_{j,q})^{m_{j,q}}. \tag{10.6.10}$$

The factors on the right-hand side come from the displacement of the interaction points along the polymer lines (see Fig. 10.19). The grand partition function can be rewritten in the form

$$\mathscr{Q}\{f\} = \sum_{N=1}^{\infty} \frac{1}{N!} \int_0^{\infty} \mathrm{d}S_1 \ldots \int_0^{\infty} \mathrm{d}S_N f(S_1) \ldots f(S_N) \sum_{\mathfrak{D}} \mathscr{D}(S_1, \ldots, S_N). \tag{10.6.11}$$

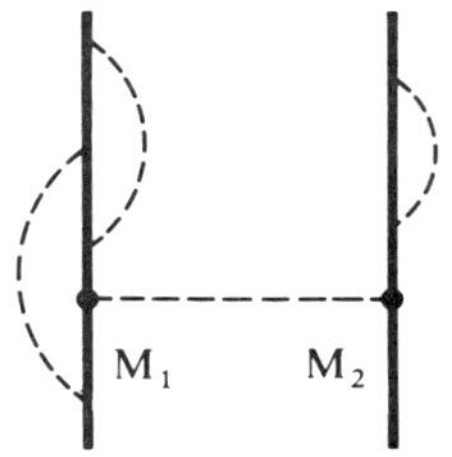

Fig. 10.19. In this diagram contributing to $\mathscr{Z}(S_1, S_2)$, the interaction line gives a contribution $-b$ independently of the positions of points M_1, and M_2 on the polymers. These points can be displaced along each polymer line. Thus, the contributions of the sub-diagrams factorize and an extra factor $(-bS_1S_2)$ is produced.

We try to express $\mathscr{Q}\{f\}$ directly in terms of the quantities $\mathscr{Z}_{\mathrm{I}}(S_1, \ldots S_N)$, in the form of a sum over the trees $\mathfrak{T}$. This operation must be done by summing each diagram once and only once.

Thus, symmetry factors enter automatically and the following observations indicate their origin.

1. Every change in numbering that modifies the distribution of polymers among the sub-diagrams is sufficient to produce new diagrams (see Fig. 10.20) and it is easy to see that the factor which is introduced in this way equals $N!/\prod_{j=1}^{p} N_j!$ (see Fig. 10.20).
2. The vertices of a tree are not a priori supposed to be numbered, and consequently each tree $\mathfrak{T}$ in invariant with respect to transformations belonging to the group $\mathrm{G}(\mathfrak{T})$ of automorphisms of the tree. By definition, the number of elements of $\mathrm{G}(\mathfrak{T})$ is the symmetry number $\mathcal{S}(\mathfrak{T})$. A diagram can be built by attributing a given content to the vertices of the corresponding tree $\mathfrak{T}$. However, any permutation of vertices followed by a permutation of contents give a diagram which coincide with the initial diagram. Thus, a factor $1/\mathcal{S}(\mathfrak{T})$ must be introduced to avoid over-counting.
3. In a sub-diagram j, a permutation of the end points of the lines connecting this diagram to other sub-diagrams produces different diagrams whose contributions are the same (see Fig. 10.15). The factor to be taken into account is the number of these diagrams, namely $m_j!/\prod_q m_{j,q}!$ (see Fig. 10.21).

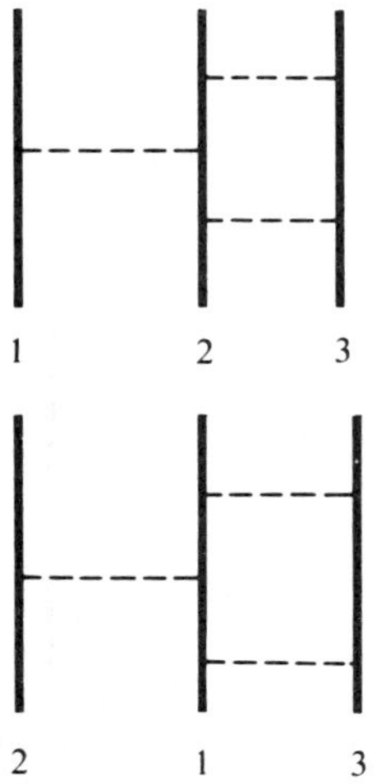

Fig. 10.20. A simple change in the numbering of the polymer chains can be sufficient to produce new diagrams (here $N = 3$, $p = 2$).

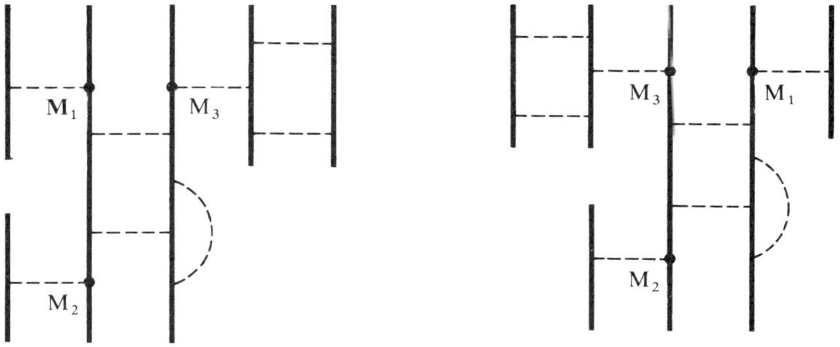

Fig. 10.21. In a sub-diagram, a permutation of the end points M_1, M_2, M_3 of the interaction lines connecting this sub-diagram to other sub-diagrams leads to different diagrams having the same contribution. (Here, $N = 7, p = 4$, and for the central diagram to which M_1, M_2, M_3 belong, we have $m = 3$.)

Now we observe that we can set any I-irreducible diagram at each vertex of $\mathfrak{T}$. Thus a vertex function $v_m\{f\}$ is associated with each vertex of a tree; the following expression defines $v_m\{f\}$

$$v_m\{f\} = \sum_{N=1}^{\infty} \frac{m!}{N!} \sum_{m_1,\ldots m_N} \delta(m_1 + \ldots + m_N - m)$$

$$\times \int_0^{\infty} \mathrm{d}S_1 \ldots \int_0^{\infty} \mathrm{d}S_N \frac{(S_1)^{m_1}}{m_1!} \ldots \frac{(S_N)^{m_N}}{m_N!} \times f(S_1) \ldots f(S_N) \mathscr{L}_{\mathrm{I}}(S_1, \ldots, S_N) \tag{10.6.12}$$

By taking into account (10.6.10), (10.6.11), (10.6.12), and symmetry factors, we can thus express $\mathscr{Q}\{f\}$ in terms of products of vertex functions

$$\mathscr{Q}\{f\} = \sum_{\mathfrak{T}} \frac{1}{S(\mathfrak{T})} (-b)^{p(\mathfrak{T})-1} \prod_{j=1}^{p(\mathfrak{T})} v_{m(\mathfrak{T},j)}\{f\} \tag{10.6.13}$$

where $p(\mathfrak{T})$ is the number of vertices of the tree $\mathfrak{T}$ and $m(\mathfrak{T}, j)$ the number of legs of the vertex j on the tree. The contributions are calculated by attributing a factor $-b$ to each interaction line and a factor $v_m\{f\}$ to each vertex with m legs.

We observe that the generating function of the vertex functions has a very simple form

$$V(y, \{f\}) = \sum_{m=0}^{\infty} v_m\{f\} y^m/m!$$

$$= \sum_{N=1}^{\infty} \frac{1}{N!} \int_0^{\infty} \mathrm{d}S_1 \ldots \int_0^{\infty} \mathrm{d}S_N e^{y(S_1 + \ldots + S_N)} f(S_1) \ldots f(S_N) \mathscr{L}_{\mathrm{I}}(S_1, \ldots, S_N). \tag{10.6.14}$$

It obeys the equation

$$\frac{\partial}{\partial y} V(y, \{f\}) = \int_0^\infty \mathrm{d}S\, S f(S) \frac{\partial}{\partial f(S)} V(y, \{f\}) \tag{10.6.15}$$

and this formula shows that the vertex functions are related to one another by the useful equality

$$v_{m+1}\{f\} = \int_0^\infty \mathrm{d}S\, S f(S) \frac{\partial}{\partial f(S)} v_m\{f\}. \tag{10.6.16}$$

Moreover, $V(y, \{f\})$ can be expressed, in a simple manner in terms of the new variables

$$\mathfrak{f}(S) = f(S) \mathrm{e}^{Sy}. \tag{10.6.17}$$

In fact,

$$V(y, \{f\}) = v_0\{\mathfrak{f}\}, \tag{10.6.18}$$

and we also have

$$\begin{aligned} \frac{\partial}{\partial y} V(y, \{f\}) &= \int_0^\infty \mathrm{d}S\, S f(S) \frac{\partial}{\partial f(S)} V(y, \{f\} \\ &= \int_0^\infty \mathrm{d}S\, S \mathfrak{f}(S) \frac{\partial}{\partial \mathfrak{f}(S)} v_0\{\mathfrak{f}\} = v_1\{\mathfrak{f}\}). \end{aligned} \tag{10.6.19}$$

6.3.2 Summation of tree diagrams by Legendre transformation

Let us now calculate $\mathcal{Q}\{f\}$. Equation (10.6.13) shows that this function can be written in the form

$$\mathcal{Q}\{f\} = v_0\{f\} + A(v_1\{f\}) \tag{10.6.20}$$

where $v_0\{f\}$ is the first term of the expansion (10.6.13), whereas $A(x)$ is an implicit function of b and of the vertex functions $v_m\{f\}$ (with $m > 1$). The function $A(x)$ is a generating function associated with trees $\mathfrak{T}$ (containing at least two one-leg vertices). It is obtained by replacing all the one-leg vertices by the variable x.

Thus, the contribution corresponding to each tree $\mathfrak{T}(p(\mathfrak{T}) \geqslant 2)$ is the product of all the factors resulting from the remaining vertices ($m > 1$), from the interaction lines (powers of $-b$), from the external legs (powers of x) and from the symmetry of the diagram (see Appendix I).

The structure of $A(x)$ is rather complex, but, as we show in Appendix I, the Legendre transform $B(y)$ of $A(x)$ has a much simpler structure than $A(x)$. Thus, a Legendre transformation allows us to express $\mathcal{Q}\{f\}$ directly in terms of the $v_m\{f\}$.

This Legendre transformation can be defined by the coupled equations

$$\begin{aligned} A(x) + B(y) - xy &= 0 \\ y &= A'(x) \\ x &= B'(y) \end{aligned} \tag{10.6.21}$$

and the function $B(y)$ is given by (I.16)

$$B(y) = -\frac{1}{2b}y^2 - \sum_{m=2}^{\infty} \frac{v_m}{m!} y^m. \tag{10.6.22}$$

Now, let us set $x = v_1\{f\}$. Equations (10.6.21) and (10.6.22) give

$$A(v_1\{f\}) = yB'(y) - B(y) = -\frac{1}{2b}y^2 - \sum_{m=2}^{\infty} v_m\{f\}\frac{(m-1)}{m!}y^m.$$

$$v_1\{f\} = -\frac{y}{b} - \sum_{m=2}^{\infty} v_m\{f\}\frac{y^{m-1}}{(m-1)!} \tag{10.6.23}$$

and, by combining these two equations, we get

$$A(v_1\{f\}) = \sum_{m=1}^{\infty} v_m\{f\}\left(1 - \frac{m}{2}\right)\frac{y^m}{m!}.$$

Bringing this results in (10.6.20) and recalling (10.6.23), we obtain the coupled equations

$$\mathcal{Q}\{f\} = \sum_{m=0}^{\infty} v_m\{f\}\left(1 - \frac{m}{2}\right)\frac{y^m}{m!}$$

$$y = -b\sum_{m=1}^{\infty} v_m\{f\}\frac{y^{m-1}}{(m-1)!}. \tag{10.6.24}$$

In agreement with (10.6.14), we can also write

$$\mathcal{Q}\{f\} = \left(1 - \frac{1}{2}y\frac{\partial}{\partial y}\right)V(y, \{f\})$$

$$y = -b\frac{\partial}{\partial y}V(y, \{f\}).$$

Thus (10.6.18) and (10.6.19) give

$$\mathcal{Q}\{f\} = v_0\{\mathfrak{f}\} - \frac{y}{2}v_1\{\mathfrak{f}\}$$

$$y = -bv_1\{\mathfrak{f}\}.$$

Taking definition (10.6.17) into account, we can eliminate y, and we obtain the coupled equations

$$\Pi\beta = \mathcal{Q}\{f\} = v_0\{\mathfrak{f}\} + \frac{b}{2}[v_1\{\mathfrak{f}\}]^2, \tag{10.6.25}$$

$$f(S) = \mathfrak{f}(S)\exp(bSv_1\{\mathfrak{f}\}). \tag{10.6.26}$$

Now, we have to express the polymer concentrations in terms of the new parameters $\mathfrak{f}(S)$. For this purpose, let us express $f(S)\dfrac{\partial}{\partial f(S)}$ in terms of the new variables.

Let us start from the identity

$$\frac{\partial}{\partial \mathfrak{f}(S)} = \int_0^\infty \mathrm{d}S' \frac{\partial f(S')}{\partial \mathfrak{f}(S)} \frac{\partial}{\partial f(S')}$$

and let us use (10.6.26) to get the value of $\dfrac{\partial f(S')}{\partial \mathfrak{f}(S)}$

$$\mathfrak{f}(S)\frac{\partial f(S')}{\partial \mathfrak{f}(S)} = \delta(S'-S)f(S) + b\,\mathfrak{f}(S)\frac{\partial v_1\{\mathfrak{f}\}}{\partial \mathfrak{f}(S)} S' f(S').$$

$$\mathfrak{f}(S)\frac{\partial}{\partial \mathfrak{f}(S)} = f(S)\frac{\partial}{\partial f(S)} + b\mathfrak{f}(S)\frac{\partial v_1\{\mathfrak{f}\}}{\partial \mathfrak{f}(S)} \int_0^\infty \mathrm{d}S' S' f(S') \frac{\partial}{\partial f(S')}.$$

Finally, from this equality and from (10.6.16) we deduce

$$\int_0^\infty \mathrm{d}S\, S f(S) \frac{\partial}{\partial f(S)} = \frac{1}{1+bv_2\{\mathfrak{f}\}} \int_0^\infty \mathrm{d}S\, S\mathfrak{f}(S) \frac{\partial}{\partial \mathfrak{f}(S)}$$

$$f(S)\frac{\partial}{\partial f(S)} = \mathfrak{f}(S)\frac{\partial}{\partial \mathfrak{f}(S)} - \frac{b}{1+bv_2\{\mathfrak{f}\}} \mathfrak{f}(S)\frac{\partial}{\partial \mathfrak{f}(S)} v_1\{\mathfrak{f}\} \left(\int_0^\infty \mathrm{d}S' \mathfrak{f}(S') \frac{\partial}{\partial \mathfrak{f}(S')}\right).$$

With the help of these identities and by combining eqns (10.6.8) and (10.6.25), we obtain

$$\boldsymbol{\Pi\beta} = v_0\{\mathfrak{f}\} + \frac{b}{2}[v_1\{\mathfrak{f}\}]^2$$

$$\boldsymbol{C}(S) = \mathfrak{f}(S)\frac{\partial}{\partial \mathfrak{f}(S)} v_0\{\mathfrak{f}\}. \tag{10.6.27}$$

The area concentration is also given by a simple expression [see (10.6.16)]

$$\mathscr{C} = \int_0^\infty \mathrm{d}S\, S\, \boldsymbol{C}(S) = v_1\{\mathfrak{f}\}. \tag{10.6.28}$$

Using this equality we can write

$$\boldsymbol{\Pi\beta} = v_0\{\mathfrak{f}\} + b\mathscr{C}^2/2$$

$$\boldsymbol{C}(S) = \mathfrak{f}(S)\frac{\partial}{\partial \mathfrak{f}(S)} v_0\{\mathfrak{f}\}. \tag{10.6.29}$$

These results apply immediately to the monodisperse case, when each polymer has the same area. Then, there is only one fugacity f and the vertex functions $v_m(f)$ are defined by the following equations which are special cases of (10.6.12) and (10.6.16).

$$v_m(f) = \sum_{N=1}^{\infty} \frac{(NS)^m}{N!} f^N \mathscr{Z}_{\mathrm{I}}(N \times S)$$

$$v_{m+1}(f) = Sf \frac{\partial}{\partial f} v_m(f). \tag{10.6.30}$$

Moreover, eqns (10.6.29) gives

$$\Pi\beta = \mathcal{Q}_{\rm I}(\mathfrak{f}\} + b\mathscr{C}^2/2$$

$$C = \mathfrak{f} \frac{\partial}{\partial \mathfrak{f}} \mathcal{Q}_{\rm I}(\mathfrak{f})$$

$$\mathscr{C} = CS$$

where

$$\mathcal{Q}_{\rm I}(\mathfrak{f}) = \sum_{N=1} \frac{\mathfrak{f}^N}{N!} \mathscr{Z}_{\rm I}(N\times S) \tag{10.6.31}$$

and the initial fugacity f equals

$$f = \mathfrak{f} \exp(bS^2 C).$$

In particular, the simple-tree approximation described in Section 6.2 can be recovered by retaining from among the irreducible diagrams only those which contain one polymer line and no more. This is equivalent to the approximation in which one keeps only the term $Z_{\rm I}(1 \times S)$ and this result is obtained by setting

$$v_0(\mathfrak{f}) = \mathfrak{f}\mathscr{Z}_{\rm I}(1\times S)$$

$$v_1(\mathfrak{f}) = Sv_0(\mathfrak{f}).$$

Then eqn (10.6.25) gives

$$\Pi\beta = \mathscr{C}/S + b\mathscr{C}^2/2 \tag{10.6.32}$$

which coincide with (10.6.6) for $c = 0$.

For monodisperse solutions, we can find elementary scaling laws, as we did in Section 5. As $\mathscr{Z}_{\rm I}(N\times S)$ has the same dimension as $\mathscr{Z}(N\times S)$, we may set

$$\mathscr{Z}_{\rm I}(N \times S) = (2\pi S)^{(N-1)d/2} \mathfrak{z}_{{\rm I},N}(z), \tag{10.6.33}$$

and this formula is analogous to (10.5.8). Taking in account eqn (10.6.31) and the definition (10.1.7) of z, we can write eqns (10.6.29) in the simple form

$$\Pi\beta(2\pi S)^{d/2} = \mathfrak{z}_{\rm I}(t, z) + \tfrac{1}{2}C^2(2\pi S)^d z$$

$$C(2\pi S)^{d/2} = t\frac{\partial}{\partial t}\mathfrak{z}_{\rm I}(t, z)$$

$$\mathfrak{z}_{\rm I}(t, z) = \sum_{N=1}^{\infty} \frac{t^N}{N!} \mathfrak{z}_{{\rm I},N}(z)$$

by setting

$$t = (2\pi S)^{-d/2}\mathfrak{f}. \tag{10.6.34}$$

7. EFFECTIVE CALCULATIONS: ISOLATED CHAIN WITH TWO-BODY INTERACTION

In the present section, we review only results of perturbation calculations performed directly by starting from polymer diagrams.

Calculations of polymer swelling have been made more recently (1984)[5] with the help of computers, by using the correspondence which exists between polymer theory and field theory. They will be described in Chapter 12, Section 3.2.5.

7.1 Partition function of a chain and mean square end-to-end distance ($b \neq 0$, $c = 0$)

Let us consider an isolated chain in an infinite volume and the restricted partition function

$$\bar{\mathscr{Z}}(\vec{k}, -\vec{k}; 0, S; S) = \langle\!\langle \exp[i\vec{k}\cdot(\vec{r}(S) - \vec{r}(0))] \exp[-{}^{\prime}\mathfrak{H}_1\{\vec{r}\})] \rangle\!\rangle_0. \tag{10.7.1}$$

This function gives interesting results concerning the chain and, in particular, the partition function $\mathscr{Z}(S)$ and the mean square end-to-end distance

$$\mathscr{Z}(S) = \bar{\mathscr{Z}}(0, 0; 0, S; S)$$

$$R^2 = \langle\!\langle [\vec{r}(S) - \vec{r}(0)]^2 \rangle\!\rangle = -\frac{2d}{\mathscr{Z}(S)} \frac{\mathrm{d}}{\mathrm{d}k^2} \bar{\mathscr{Z}}(\vec{k}, -\vec{k}; 0, S; S)\Big|_{k^2=0}. \tag{10.7.2}$$

We shall expand $\bar{\mathscr{Z}}(\vec{k}, -\vec{k}; 0, S; S)$ in powers of the interaction by setting

$$\bar{\mathscr{Z}}(\vec{k}, -\vec{k}; 0, S; S) = {}^0\mathscr{Z}(k, S) + {}^1\mathscr{Z}(k, S) + {}^2\mathscr{Z}(k, S). \tag{10.7.3}$$

We can now calculate the first two terms explicitly. The first term corresponds to the trivial diagram (see Fig. 10.22)

$${}^0\mathscr{Z}(k, S) = \exp[-k^2 S/2] = 1 - k^2 S/2 + \ldots.$$

The second term is given by the one-loop diagram (see Fig. 10.22). Applying the rules prescribed above, we find

$$\begin{aligned}
{}^1\mathscr{Z}(k, S) &= -b \int_0^S \mathrm{d}s\,(S - s) \exp[-(S - s)k^2/2](2\pi)^{-d} \int \mathrm{d}^d q \exp[-sq^2/2] \\
&= -b(2\pi)^{-d/2} \int_0^S \mathrm{d}s\,(S - s) s^{-d/2} \exp[-(S - s)k^2/2]
\end{aligned}$$

$${}^1\mathscr{Z}(k, S) = -z \int_0^1 \mathrm{d}x\,(1 - x) x^{-d/2} \exp[-(1 - x)k^2 S/2].$$

We observe that the integral diverges; this simply means, as we have seen, that ${}^1\mathscr{Z}(k, S)$ is given by the principal value of the integral (dimensional regulariz-

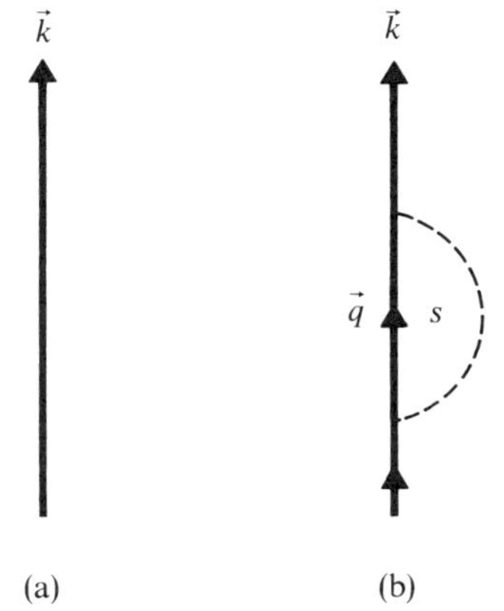

Fig. 10.22. Diagrams corresponding to ${}^0\mathscr{Z}(k, S)$ and to ${}^1\mathscr{Z}(k, S)$ respectively.

ation). We can write formally

$$\begin{aligned}{}^1\mathscr{Z}(k, S) = & -z\exp[-k^2S/2]\int_0^1 \mathrm{d}x\,(1-x)x^{-d/2} \\ & -z\int_0^1 \mathrm{d}x\,(1-x)x^{-d/2}\{\exp[-(1-x)k^2S/2] - \exp[-k^2S/2]\}.\end{aligned}$$

We thus obtain

$$\begin{aligned}{}^1\mathscr{Z}(k, S) = & -z\exp[-k^2S/2]\frac{1}{(1-d/2)(2-d/2)} \\ & -z\int_0^1 \mathrm{d}x(1-x)x^{-d/2}\{\exp[-(1-x)k^2S/2] - \exp[-k^2S/2]\}.\end{aligned} \tag{10.7.4}$$

Let us expand with respect to k^2

$${}^1\mathscr{Z}(k, S) = -z\left[\frac{1}{(1-d/2)(2-d/2)} - \frac{k^2S}{2}\int_0^1 \mathrm{d}x(1-x)x^{-d/2} + \dots\right].$$

Finally, we find

$${}^1\mathscr{Z}(k, S) = -z\frac{1}{(1-d/2)(2-d/2)}\left[1 - \frac{k^2S}{2}\left(\frac{2}{3-d/2}\right) + \dots\right]. \tag{10.7.5}$$

From the preceding formulae (10.7.2), (10.7.3), and (10.7.5), and to first-order with respect to z, we deduce

$$\mathscr{Z}(S) = 1 - \frac{z}{(1-d/2)(2-d/2)} + \dots$$

$$R^2/Sd = 1 + \frac{z}{(2 - d/2)(3 - d/2)} + \ldots . \tag{10.7.6}$$

Thus, for $d = 3$, we have

$$\mathscr{X}(S) = 1 + 4z + \ldots$$

$$R^2 = \langle\!\langle [\vec{r}(S) - \vec{r}(0)]^2 \rangle\!\rangle = 3S[1 + \tfrac{4}{3}z + \ldots], \tag{10.7.7}$$

a very simple and classical result.

The second-order contributions to ${}^2\mathscr{X}(k, S)$ are represented by three diagrams (see Fig. 10.23)

$${}^2\mathscr{X}(k, S) = {}^{2a}\mathscr{X}(k, S) + {}^{2b}\mathscr{X}(k, S) + {}^{2c}\mathscr{X}(k, S).$$

These quantities can be easily calculated by using Fixman's method. The areas s_1, s_2, and s_3 appearing in Fig. 10.3 can be expressed in terms of dimensionless variables

$$s_1 = St_1 \qquad s_2 = St_2 \qquad s_3 = St_3,$$

and, in this way, one finds (in the sense of principal parts)

$${}^{2a}\mathscr{X}(k, S) = z^2 \int_{t_1 + t_2 < 1} \mathrm{d}t_1 \mathrm{d}t_2 \, \tfrac{1}{2}(1 - t_1 - t_2)^2 (t_1 t_2)^{-d/2}$$
$$\times \exp\left[-\frac{k^2 S}{2}(1 - t_1 - t_2) \right]$$

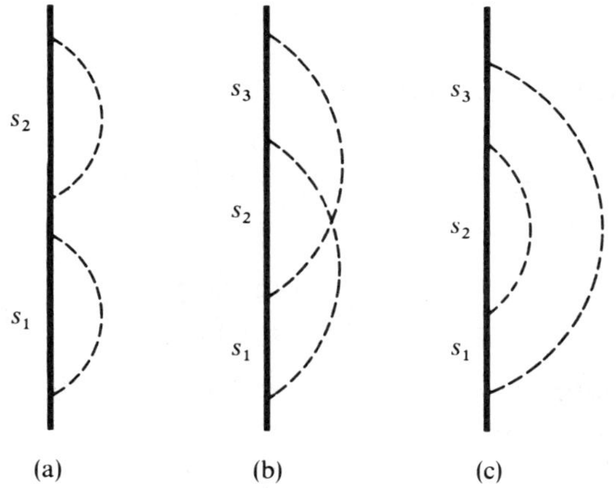

Fig. 10.23. Diagram corresponding to ${}^2\mathscr{X}(k, S)$.

$$^{2b}\mathscr{Z}(k, S) =$$

$$z^2 \int_{t_1+t_2+t_3<1} dt_1 dt_2 dt_3 (1 - t_1 - t_2 - t_3)(t_1 t_2 + t_2 t_3 + t_1 t_3)^{-d/2}$$

$$\times \exp\left[-\frac{k^2 S}{2}\left(1 - \frac{(t_1 + t_2)(t_2 + t_3)(t_1 + t_3)}{t_1 t_2 + t_2 t_3 + t_1 t_3}\right)\right]$$

$$^{2c}\mathscr{Z}(k, S) = z^2 \int_{t_1+t_2+t_3<1} dt_1 dt_2 dt_3 (1 - t_1 - t_2 - t_3)(t_1 + t_3)^{-d/2}(t_2)^{-d/2}$$

$$\times \exp\left[-\frac{k^2 S}{2}(1 - t_1 - t_2 - t_3)\right]. \tag{10.7.8}$$

For example, let us show how the expression which gives $^{2b}\mathscr{Z}(k, S)$ can be obtained. We have

$$^{2b}\mathscr{Z}(k, S) = \int_0^S ds_1 \ldots \int_0^S ds_5 \, \delta(S - s_1 - s_2 - s_3 - s_4 - s_5)\, {}^{b}\mathscr{D}\{s\}$$

where ${}^{b}\mathscr{D}\{s\}$ is given by (10.4.26)

$${}^{b}\mathscr{D}\{s\} = b^2(2\pi)^{-d}|\text{dét}\,\mathscr{S}|^{-d/2}\exp\left[-\frac{1}{2}k^2 S + \frac{1}{2}\sum_{\alpha\beta} k_\alpha \mathscr{S}^{-1}_{\alpha\beta} k_\beta \right].$$

The orientations which we chose and the indices α and β are indicated in Fig. 10.24.

From definitions (10.4.25) we deduce

$$\mathscr{S} = \begin{Vmatrix} s_1 + s_2 & s_2 \\ s_2 & s_2 + s_3 \end{Vmatrix} \begin{matrix} (\alpha) \\ (\beta) \end{matrix}$$
$$\qquad (\alpha) \qquad\qquad (\beta)$$

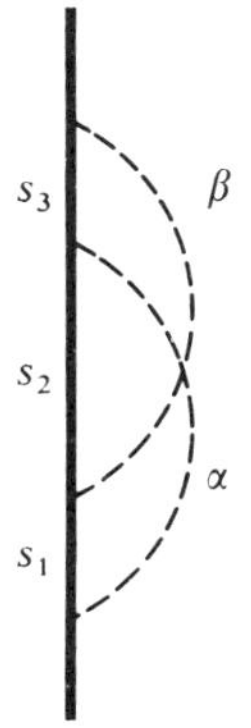

Fig. 10.24. Diagram corresponding to $^{2b}\mathscr{Z}(k, S)$: notation

$$k_\alpha = (s_1 + s_2)k \qquad k_\beta = (s_2 + s_3)k$$

$$|\det \mathscr{S}| = s_1 s_2 + s_2 s_3 + s_1 s_3$$

$$\sum_{\alpha\beta} k_\alpha \mathscr{S}^{-1}_{\alpha\beta} k_\beta = k^2 \frac{(s_1 + s_2)(s_2 + s_3)(s_1 + s_3)}{s_1 s_2 + s_2 s_3 + s_1 s_3}.$$

Moreover, we have

$$\int_0^S ds_4 \int_0^S ds_5\, \delta(S - s_1 - s_2 - s_3 - s_4 - s_5) = S - s_1 - s_2 - s_3.$$

By combining these elements and by setting

$$s_1 = St_1 \qquad s_2 = St_2 \qquad s_3 = St_3$$

and

$$z = b\, S^{2-d/2}(2\pi)^{-d/2},$$

we obtain the expression of ${}^{2b}\mathscr{Z}(k, S)$ given above.

For $d = 3$, various theoreticians calculated the first terms of the expansion

$$R^2 \equiv 3^e S = 3S(1 + \tfrac{4}{3}z + a_2 z^2 + a_3 z^3 + 4 \ldots) \tag{10.7.9}$$

The second order term was calculated by Fixman (1955); this term equals

$$a_2 = -\frac{16}{3} + \frac{28}{27}\pi = -2.075\,385. \tag{10.7.10}$$

The third-order term was calculated by Yamakawa and Tanaka (1967)[6]; they found

$$a_3 = \frac{64}{3} - \pi\left(\frac{73\,679}{8\,100} - \frac{13\,202}{2025}\ln 2 + \frac{1\,616}{405}\ln\frac{3}{2}\right) + \frac{512}{45} I_A$$

where

$$I_A = \int_0^{\pi/2} d\theta \frac{\theta^2}{1 + 3\sin^2\theta} = 0.403\,25, \tag{10.7.11}$$

which gives

$$a_3 = 6.459.$$

More recently, Barrett[7] found a slightly different value (more exact)

$$a_3 = 6.296\,88.$$

Moreover, recent calculations made on a computer by using the Green's function formalism allowed theoreticians to reach the sixth-order[5] (see Chapter 12, Section 3.2.5).

The mean values of other functions of $[\vec{r}(S) - \vec{r}(0)]$ can also be expanded in powers of z and to obtain them to first-order in z, we can start from (10.7.3) and (10.7.4).

Thus, for $d = 3$, we find[8]

$$\langle\!\langle [\vec{r}(S) - \vec{r}(0)]^{2p} \rangle\!\rangle = \langle\!\langle [\vec{r}(S) - \vec{r}(0)]^{2p} \rangle\!\rangle_0 \left[1 + 2z\left(\frac{\pi^{1/2}(p+1)!}{(p+1/2)!} - 2 \right) + \ldots \right.$$

where

$$\langle\!\langle [\vec{r}(S) - \vec{r}(0)]^{2p} \rangle\!\rangle_0 = \frac{2^{p+1} S^p}{\pi^{1/2}} (p + 1/2)! \qquad (10.7.12)$$

a result which is due to Yamakawa and Kurata[9] and which is not difficult to recover (the formulae are valid for $p > -3/2$).

However, it is clear that such expansions are useful only for weak interactions (poor solvents) and short chains. In fact, in three dimensions, the expansion has a meaning only if z is small, a condition which does not apply to long chains. In the latter case, we are interested in the behaviour of chains for large values of z. It is not easy to deduce this behaviour from an expansion in powers of z.

A serious study of the asymptotic properties of long chains is possible, but only by resorting to special techniques, as will be shown later on, when renormalization theory will be expounded.

7.2 Correlations between internal points of a chain ($b \neq 0$, $c = 0$)

The correlations between two points M_1 and M_2 with coordinates s_1 and s_2 along the chain are given by the partition function

$$\mathscr{Z}(\vec{k}, -\vec{k}; s_1, s_2; S) = \langle\!\langle \exp[i\vec{k}\cdot[\vec{r}(s_2) - \vec{r}(s_1)]] \exp[-{}^1\mathfrak{H}_1\{\vec{r}\}] \rangle\!\rangle \qquad (10.7.13)$$

which will be used to calculate

$$\langle\!\langle [\vec{r}(s_2) - \vec{r}(s_1)]^2 \rangle\!\rangle.$$

At order zero, we have (see Fig. 10.25)

$${}^0\mathscr{Z}(\vec{k}, -\vec{k}; s_1, s_2; S) = \exp[-k^2(s_2 - s_1)/2].$$

At order one, the contributions are represented by six diagrams (see Fig. 10.25)

$${}^1\mathscr{Z}_{12}(\vec{k}, -\vec{k}; s_1, s_2; S) = {}^{1a}\mathscr{Z}_{12}(k, S) + {}^{1b}\mathscr{Z}_{12}(k, S) + \ldots + {}^{1f}\mathscr{Z}_{12}(k, S). \qquad (10.7.14)$$

The first two terms are immediately obtained

$${}^{1a}\mathscr{Z}_{12}(k, S) = {}^1\mathscr{Z}(s_1) \exp[-k^2(s_2 - s_1)/2]$$

$${}^{1b}\mathscr{Z}_{12}(k, S) = {}^1\mathscr{Z}(S - s_2) \exp[-k^2(s_2 - s_1)/2]. \qquad (10.7.15)$$

${}^1\mathscr{Z}(S)$ is the term of order one in the expansion of $\mathscr{Z}(S)$ and consequently, according to (10.7.6) and to the definition of z (10.1.7), we have

$${}^1\mathscr{Z}(s) = -\frac{b(2\pi)^{-d/2} s^{2-d/2}}{(1 - d/2)(2 - d/2)}.$$

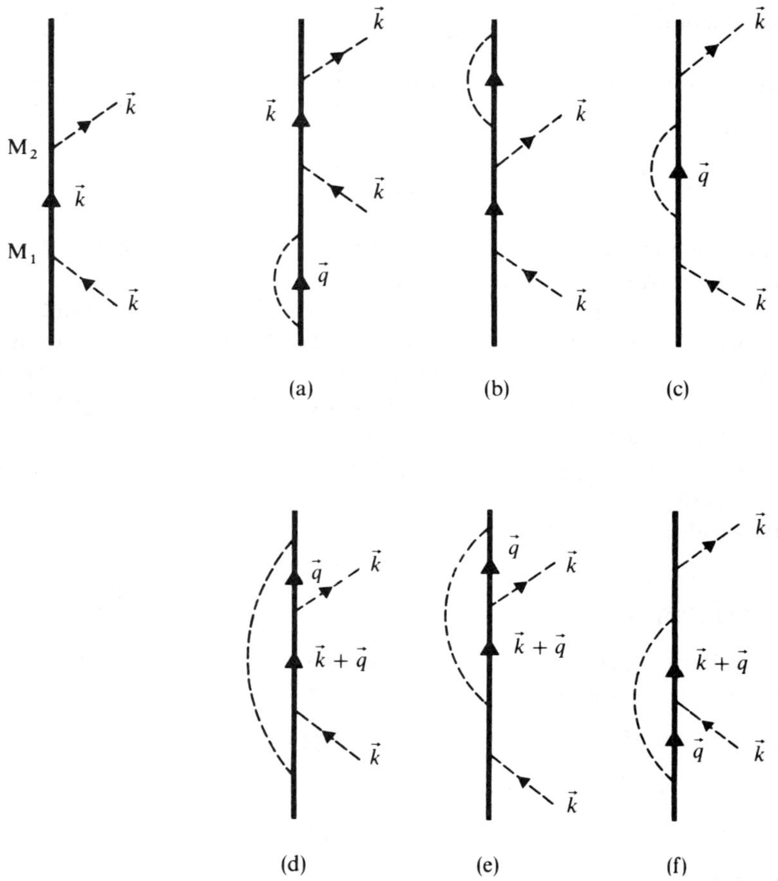

Fig. 10.25. Diagrams associated with $\mathscr{Z}(\vec{k}, -\vec{k}; s_1, s_2; S)$ to order zero and to order one (a,b,c,d,e, f). The coordinates of M_1 and M_2 along the polymer are s_1 and s_2 respectively.

Moreover, we find

$$
\begin{aligned}
{}^{1c}\mathscr{Z}_{12}(k, S) = & -b \int_{s_1 < s' < s'' < s_2} \mathrm{d}s'\,\mathrm{d}s'' \int \frac{\mathrm{d}^d q}{(2\pi)^d} \\
& \times \exp[-(s_2 - s'' + s' - s_1)k^2/2] \exp[-(s'' - s')q^2/2] \\
= & -b(2\pi)^{-d/2} \exp[-(s_2 - s_1)k^2/2] \\
& \times \int_{s_1 < s' < s'' < s_2} \mathrm{d}s'\,\mathrm{d}s''(s'' - s')^{-d/2} \exp[(s'' - s')k^2/2]
\end{aligned}
$$

$$
\begin{aligned}
{}^{1d}\mathscr{Z}_{12}(k,S) &= -b\int_{0<s'<s_1<s_2<s''<S} ds'ds''\int\frac{d^dq}{(2\pi)^d} \\
&\quad\times\exp[-(s_2-s_1)(\vec{k}+\vec{q})^2/2]\exp[-(s''-s_2+s_1-s')q^2/2] \\
&= -b(2\pi)^{-d/2}\exp[-(s_2-s_1)k^2/2] \\
&\quad\times\int_{0<s'<s_1<s_2<s''<S} ds'ds''(s''-s')^{-d/2}\exp\left[\frac{(s_2-s_1)^2}{s''-s'}\frac{k^2}{2}\right] \\
{}^{1e}\mathscr{Z}_{12}(k,S) &= -b\int_{s_1<s'<s_2<s''<S} ds'ds''\int\frac{d^dq}{(2\pi)^d}\exp[-(s_2-s')(\vec{k}+\vec{q})^2/2] \\
&\quad\times\exp[-(s'-s_1)k^2/2]\exp[-(s''-s_2)q^2/2] \\
&= -b(2\pi)^{-d/2}\exp[-(s_2-s_1)k^2/2] \\
&\quad\times\int_{s_1<s'<s_2<s''<S} ds'ds''(s''-s')^{-d/2}\exp\left[\frac{(s_2-s')^2}{s''-s'}\frac{k^2}{2}\right] \\
{}^{1f}\mathscr{Z}_{12}(k,S) &= -b\int_{0<s'<s_1<s''<s_2} ds'ds''\int\frac{d^dq}{(2\pi)^d}\exp[-(s_2-s'')k^2/2] \\
&\quad\times\exp[-(s''-s_1)(\vec{k}+\vec{q})^2/2]\exp[-(s_1-s')q^2/2] \\
&= -b(2\pi)^{-d/2}\exp[-(s_2-s_1)k^2/2] \\
&\quad\times\int_{0<s'<s_1<s''<s_2} ds'ds''(s''-s')^{d/2}\exp\left[\frac{(s''-s_1)^2}{s''-s'}\frac{k^2}{2}\right]
\end{aligned}
\tag{10.7.16}
$$

Obviously, we have

$$\bar{\mathscr{Z}}(\vec{0},\vec{0};s_1,s_2;S) = \mathscr{Z}(S)$$

and

$$\langle\langle[\vec{r}(s_2)-\vec{r}(s_1)]^2\rangle\rangle = -2d\left[\frac{d}{dk^2}\bar{\mathscr{Z}}(\vec{k},-\vec{k};s_2,s_1;S)\right]_{k=0}. \tag{10.7.17}$$

To evaluate $\langle\langle[\vec{r}(s_2)-\vec{r}(s_1)]^2\rangle\rangle$, we have only to expand the partition function to first order in k^2. We note that the factor $\exp[-k^2(s_2-s_1)/2]$ appears in each term. This leads us to write the expansion in the form

$$
\begin{aligned}
\bar{\mathscr{Z}}(\vec{k},-\vec{k};s_1,s_2;S) = \mathscr{Z}(S)&\left[1-\frac{k^2}{2}(s_2-s_1)\right] \\
&-\frac{k^2}{2}b(2\pi)^{-d/2}\mathscr{Z}^{\cdot}(s_1,s_2;S)+\ldots.
\end{aligned}
\tag{10.7.18}
$$

The first term is obtained by putting $k=0$ in all the integrals of (10.7.16), the second term is the remaining.

From relations (10.7.16), we deduce immediately

$$\mathscr{Z}^{\cdot}(s_1, s_2; S) = \int_{s_1 < s' < s'' < s_2} ds'\, ds''(s'' - s')^{1-d/2}$$
$$+ \int_{0 < s' < s_1 < s_2 < s'' < S} ds'\, ds''(s'' - s')^{-1-d/2}(s_2 - s_1)^2$$
$$+ \int_{s_1 < s' < s_2 < s'' < s} ds'\, ds''(s'' - s')^{-1-d/2}(s_2 - s')^2$$
$$+ \int_{0 < s' < s_1 < s'' < s_2} ds'\, ds''(s'' - s')^{-1-d/2}(s'' - s_1)^2. \quad (10.7.19)$$

Calculating the integrals gives

$$\mathscr{Z}^{\cdot}(s_1, s_2; S) = \frac{4}{d(d-2)}\left\{\frac{8}{(4-d)(6-d)}(s_2 - s_1)^{3-d/2} + (s_2 - s_1)^2\, S^{1-d/2}\right.$$
$$- \frac{4(s_2 - s_1)}{4-d}[(S - s_1)^{2-d/2} + (s_2)^{2-d/2}]$$
$$\left.- \frac{8}{(4-d)(6-d)}[(S - s_2)^{3-d/2} - (S - s_1)^{3-d/2} + (s_1)^{3-d/2} - (s_2)^{3-d/2}]\right\} \quad (10.7.20)$$

and moreover, from (10.7.17) and (10.7.18), we get

$$\langle\!\langle [\vec{r}(s_2) - \vec{r}(s_1)]^2 \rangle\!\rangle = d(s_2 - s_1) + b(2\pi)^{-d/2}\frac{\mathscr{Z}^{\cdot}(s_1, s_2; S)}{\mathscr{Z}(S)}. \quad (10.7.21)$$

Let us now set

$$s_1 = S\lambda_1 \qquad S - s_2 = S\lambda_2.$$

The preceding equations lead to the result

$$\langle\!\langle [\vec{r}(s_2) - \vec{r}(s_1)]^2 \rangle\!\rangle / Sd =$$
$$= (1 - \lambda_1 - \lambda_2) + \frac{4z}{d(d-2)}\left\{\frac{8}{(4-d)(6-d)}(1 - \lambda_1 - \lambda_2)^{3-d/2}\right.$$
$$+ (1 - \lambda_1 - \lambda_2)^2 - \frac{4(1 - \lambda_1 - \lambda_2)}{4-d}[(1 - \lambda_1)^{2-d/2} + (1 - \lambda_2)^{2-d/2}]$$
$$\left.- \frac{8}{(4-d)(6-d)}[\lambda_1^{3-d/2} + \lambda_2^{3-d/2} - (1 - \lambda_1)^{3-d/2} - (1 - \lambda_2)^{3-d/2}]\right\}.$$

In particular, for $d = 3$, one finds the following expression, obtained for the first time in 1958 by Yamakawa and Kurata[9]

$$\langle\langle[\vec{r}(s_2) - \vec{r}(s_1)]^2\rangle\rangle/3S = 1 - \lambda_1 - \lambda_2 + \frac{4z}{3}\{\tfrac{8}{3}(1 - \lambda_1 - \lambda_2)^{3/2}$$
$$+ (1 - \lambda_1 - \lambda_2)^2 - 4(1 - \lambda_1 - \lambda_2)[(1 - \lambda_1)^{1/2} + (1 - \lambda_2)^{1/2}]$$
$$- \tfrac{8}{3}(\lambda_1^{3/2} + \lambda_2^{3/2} - (1 - \lambda_1)^{3/2} - (1 - \lambda_2)^{3/2})\} + \ldots . \quad (10.7.23)$$

It is worthwhile to examine how the swelling of a segment varies according to its place on the chain; for this purpose, we use the preceding equation. Let us set

$$z_{12} = b(2\pi)^{-d/2}(s_2 - s_1)^{2-d/2} = z(1 - \lambda_1 - \lambda_2)^{2-d/2}.$$

If the segment constitute the whole chain ($s_1 = 0, s_2 = S$), one recovers the result [see (10.7.7)]

$$\langle\langle[\vec{r}(s_2) - \vec{r}(s_1)]^2\rangle\rangle = 3(s_2 - s_1)\left[1 + \frac{4}{3}z_{12} + \ldots\right] \quad (10.7.24)$$

and, as will be seen in Section 7.3, one finds that the radius of gyration of the segment is given by

$$R_G^2(s_2, s_1) = \frac{1}{2}(s_2 - s_1)\left[1 + \frac{134}{105}z_{12} + \ldots\right]. \quad (10.7.25)$$

If the segment of area $(s_2 - s_1)$ is at one end of a very long chain of area S ($s_1 = 0$ with $S \gg s_2$), one finds

$$\langle\langle[\vec{r}(s_2) - \vec{r}(s_1)]^2\rangle\rangle = 3(s_2 - s_1)\left[1 + \frac{16}{9}z_{12} + \ldots\right] \quad (10.7.26)$$

and for the radius of gyration

$$R_{G,1}^2(s_1, s_2) = \frac{1}{2}(s_2 - s_1)\left[1 + \frac{32}{21}z_{12} + \ldots\right] \quad (10.7.27)$$

(we note that $16/9 > 4/3$ and that $32/21 > 134/105$). On the contrary, when the segment is really in the interior of a long chain of area S ($S/s_1 \gg 1$, $s_1 \gg s_2 - s_1$, $S - s_2 \gg s_2 - s_1$), we find

$$\langle\langle[\vec{r}(s_2) - \vec{r}(s_1)]^2\rangle\rangle = 3(s_2 - s_1)\left[1 + \frac{32}{9}z_{12} + \ldots\right] \quad (10.7.28)$$

and for the radius of gyration

$$R_{G,2}^2(s_1, s_2) = \frac{1}{2}(s_2 - s_1)\left[1 + \frac{256}{105}z_{12} + \ldots\right]. \quad (10.7.29)$$

From these expressions, we can deduce the following observations* concerning

* These observations remain valid in the asymptotic limit $z \gg 1$ (see Chapter 13, Section 1.7).

the swelling, i.e. the ratio $\dfrac{\langle\langle[\vec{r}(s_2) - \vec{r}(s_1)]^2\rangle\rangle}{3(s_2 - s_1)}$, in various cases

1. First, it is clear that a short segment swells *relatively* less than a long segment, as can be verified by looking at the preceding expressions. Thus, every chain segment swells proportionally less than the whole chain.

2. However, if one examines the swelling *at constant 'area'*, the results obtained from the same formulae look rather different. Thus, a segment of area s located at the extremity of a chain of area S ($S > s$) swells more than a chain of area s. In the same way, a segment of area s located in the middle of a chain swells more than a segment of area s located at one extremity of the same chain. These observations can easily be interpreted by taking into account the fact that two chain segments always have a tendency to repel one another, owing to excluded volume effects. Briefly speaking, in contradiction to what was believed by certain physico-chemists and in contradiction to statements contained in a standard textbook[10], it turns out that the excluded volume effects are stronger in the central part of the polymer than near its end points.

The two effects described above are obviously complementary but different. They have to be well-understood, otherwise the excluded volume effects cannot be interpreted correctly.

Moreover, we verify that the square of the radius of gyration of a chain (or of a segment) does not increase as fast with z than the mean square end-to-end distance, and this fact also appears to be quite normal.

7.3 Radius of gyration of an isolated chain ($b \neq 0$, $c = 0$)

The size of a polymer is defined experimentally by its radius of gyration, which can be determined by *static* scattering; it is therefore an important quantity.

The radius of gyration of a chain of area S can be calculated directly by starting from its definition

$$R_G^2 = \frac{1}{2N^2} \sum_{i=1}^{N} \sum_{j=1}^{N} \langle\langle(\vec{r}_i - \vec{r}_j)^2\rangle\rangle$$

or in the continuous limit

$$R_G^2 = \frac{1}{2S^2} \int_0^S ds_1 \int_0^S ds_2 \langle\langle[\vec{r}(s_2) - \vec{r}(s_1)]^2\rangle\rangle. \tag{10.7.30}$$

By bringing, in this expression, the explicit values of $\langle\langle[\vec{r}(s_2) - \vec{r}(s_1)]^2\rangle\rangle$ calculated to first-order with respect to the interaction, we obtain, after elementary integrations,

$$R_G^2 = \frac{Sd}{6}\left[1 + 2z\frac{d^2 - 26d + 136}{(4-d)(6-d)(8-d)(10-d)}\right]. \qquad (10.7.31)$$

Thus, for $d = 3$, we find

$$R_G^2/{}^0R_G^2 = 1 + \frac{134}{105}z + \ldots \qquad (10.7.32)$$

a result obtained as early as 1953 by Zimm, Stockmayer, and Fixman.[11] The second-order calculation in z was made later on, for $d = 3$, by Yamakawa, Aoki, and Tanaka (1966)[12] who found

$$R_G^2/{}^0R_G^2 = 1 + \frac{134}{105}z - \left(\frac{536}{105} - \frac{1247}{1296}\pi\right)z^2 + \ldots \qquad (10.7.33)$$

$$\simeq 1 + 1.276z - 2.082z^2 + \ldots$$

This result can be compared to the result previously obtained for the mean square end-to-end distance, which is [see (10.7.9) and (10.7.10)]

$$R^2/{}^0R^2 = 1 + \frac{4}{3}z - \left(\frac{16}{3} - \frac{28}{27}\pi\right)z^2 + \ldots. \qquad (10.7.34)$$

As ${}^0R_G^2/{}^0R^2 = 1/6$, we find the ratio

$$\aleph = 6R_G^2/R^2 = 1 - \frac{2}{35}z + \left(\frac{32}{105} - \frac{97}{1296}\pi\right)z^2 + \ldots$$

which can be written in the form of a Padé approximant as follows

$$\aleph = \frac{1 + \left(\frac{554}{105} - \frac{3395}{2592}\pi\right)z}{1 + \left(\frac{16}{3} - \frac{3395}{2592}\pi\right)z}.$$

This ratio decreases when z increases, as can be observed here for small values of z; this effect results from the fact that long chains swell more than short chains, when subjected to excluded volume interactions.

7.4 Hydrodynamic radius of an isolated chain ($b \neq 0$, $c = 0$)

As was shown by Kirkwood,[13] if dynamical measurements are carried out on polymers in dilute solutions, a new characteristic length appears, the hydrodynamic radius R_H. By definition

$$\frac{1}{R_H} = \frac{1}{2N^2}\sum_{i=1}^{N}\sum_{j=1}^{N}\left\langle\!\!\left\langle \frac{1}{|\vec{r}_i - \vec{r}_j|} \right\rangle\!\!\right\rangle,$$

or, for the continuous model,

$$\frac{1}{R_{\mathrm{H}}} = \frac{1}{2S^2}\int_0^S \mathrm{d}s' \int_0^S \mathrm{d}s'' \left\langle\!\!\left\langle \frac{1}{|\vec{r}(s') - \vec{r}(s'')|} \right\rangle\!\!\right\rangle. \tag{10.7.35}$$

Let us calculate R_{H} for a Brownian chain. First, we observe that for such a chain

$$\left\langle\!\!\left\langle \frac{1}{|\vec{r}(S) - \vec{r}(0)|} \right\rangle\!\!\right\rangle = \frac{1}{(2\pi S)^{d/2}}\int \mathrm{d}^d r \frac{1}{r}\mathrm{e}^{-r^2/2S} = \frac{\Gamma\left(\frac{d}{2}-\frac{1}{2}\right)}{\Gamma\left(\frac{d}{2}\right)}(2S)^{-1/2}.$$

Thus, the hydrodynamic radius of a Brownian chain is given by

$$\frac{1}{{}^0R_{\mathrm{H}}} = \frac{\Gamma\left(\frac{d}{2}-\frac{1}{2}\right)}{\Gamma\left(\frac{d}{2}\right)S^2}\int_0^S \mathrm{d}s(S-s)(2s)^{-1/2}$$

$${}^0R_{\mathrm{H}} = \frac{3}{4}(2S)^{1/2}\frac{\Gamma\left(\frac{d}{2}\right)}{\Gamma\left(\frac{d}{2}-\frac{1}{2}\right)}. \tag{10.7.36}$$

Thus for $d = 3$ $\quad {}^0R_{\mathrm{H}} = \frac{3}{8}(2\pi S)^{1/2}$

for $d = 4$ $\quad {}^0R_{\mathrm{H}} = \frac{3}{2}\left(\frac{2S}{\pi}\right)^{1/2}$

The hydrodynamic radius can be expanded in powers of z. The first-order term was found, for $d = 3$, by Stockmayer and Albrecht (1958)[14]

$$\begin{aligned} R_{\mathrm{H}}/{}^0R_{\mathrm{H}} &= 1 + \left[4 - \pi\left(\frac{27}{16} - \frac{3}{2}\ln\frac{3}{2}\right)\right]z \\ &\simeq 1 + 0.609z + \ldots \\ &\qquad (d = 3). \end{aligned} \tag{1}$$

The question was studied again by Barrett (1983)[15] who calculated the value of $\langle\!\langle |\vec{r}(s_2) - \vec{r}(s_1)|^{-1} \rangle\!\rangle$ to first-order in z. For $0 < s_1 < s_2 < S$, let us set

$$\mu_1 = \frac{s_1}{s_2 - s_1} \qquad \mu_2 = \frac{S - s_2}{s_2 - s_1}$$

$$z_{12} = b(s_2 - s_1)^{1/2}(2\pi)^{-3/2}.$$

We can thus write Barrett's result[15] in a somewhat simplified form

$$\frac{\langle\langle|\vec{r}(s_2)-\vec{r}(s_1)|^{-1}\rangle\rangle}{\langle\langle|\vec{r}(s_2)-\vec{r}(s_1)|^{-1}\rangle\rangle_0} =$$

$$1 - 2z_{12}\Big\{2(1+\mu_1+\mu_2)^{1/2} - (\mu_1+\mu_2)^{1/2}$$

$$+ (1+\mu_1+\mu_2) \quad \text{Arctg}\,[(\mu_1+\mu_2)^{1/2}] - \frac{\pi}{2} + U((\mu_1)^{1/2}) + U((\mu_2)^{1/2})\Big\} \tag{10.7.38}$$

where

$$U(x) = -x - (1+x^2)\,\text{Arctg}\,x + 2\int_x^{2x}\frac{\mathrm{d}y}{y}\,\text{Arctg}\,y. \tag{10.7.39}$$

8. EFFECTIVE CALCULATIONS: SET OF CHAINS WITH TWO-BODY INTERACTIONS

8.1 Interaction of two chains in solution; second virial coefficient ($b \neq 0$, $c = 0$)

As was shown above [see (10.5.7)], the expansion of the osmotic pressure up to second-order reads

$$\boldsymbol{\Pi\beta} = \boldsymbol{C} - \frac{1}{2}\int_0^\infty \mathrm{d}A \int_0^\infty \mathrm{d}B\, \boldsymbol{C}(A)\boldsymbol{C}(B)\,\mathscr{Y}(A,B) + \ \ldots \tag{10.8.1}$$

with

$$\mathscr{Y}(A,B) = \frac{\mathscr{Z}(A,B)}{\mathscr{L}(A)\mathscr{L}(B)}.$$

By using the perturbation techniques described previously, we can expand $\mathscr{Y}(A, B)$ in powers of the interaction parameter b

$$\mathscr{Y}(A,B) = {}^1\mathscr{Y}(A,B) + {}^2\mathscr{Y}(A,B) + {}^3\mathscr{Y}(A,B) + \ \ldots$$

where ${}^p\mathscr{Y}(A, B)$ is proportional to b^p.

First, we observe that, among the diagrams associated with $\mathscr{Z}(A, B)$, those in which only one interaction line joins A and B, give a contribution which is just equal to $-b\mathscr{Z}(A)\mathscr{Z}(B)$; this appears on Fig. 10.26, when proper account is taken of the rules given in Section 3.2.2. Now, let $\mathscr{Z}_{\text{I}}(A, B)$ be the sum of the contributions of the diagrams in which at least two interaction lines join A and B. We have

$$\mathscr{Z}(A,B) = -bAB\,\mathscr{Z}(A)\mathscr{Z}(B) + \mathscr{Z}_{\text{I}}(A,B)$$

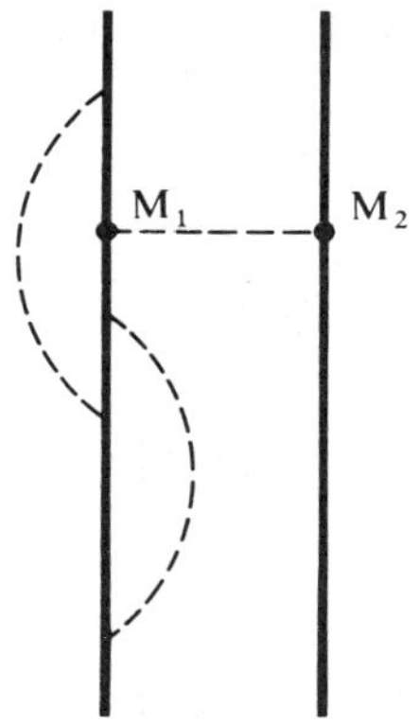

Fig. 10.26. In this diagram contributing to $\mathscr{Z}(A, B)$, the interaction line joining the two polymers bears a zero wave vector. Thus the contribution of the diagram is independent of the position of points M_1 and M_2 on the polymer lines. Thus, there is factorization and a factor $-bAB$ is produced.

and consequently,

$$\mathscr{Y}(A, B) = -bAB + \frac{\mathscr{Z}_{\rm I}(A, B)}{\mathscr{Z}(A)\mathscr{Z}(B)}. \tag{10.8.2}$$

The expansion of $\mathscr{Z}(A, B)$ is a consequence of the following expansions

$$\mathscr{Z}_{\rm I}(A, B) = {}^2\mathscr{Z}_{\rm I}(A, B) + {}^3\mathscr{Z}_{\rm I}(A, B) + \ldots$$
$$\mathscr{Z}(A) = 1 + {}^1\mathscr{Z}(A) + \ldots$$

from which we deduce

$${}^1\mathscr{Y}(A, B) = -bAB$$
$${}^2\mathscr{Y}(A, B) = {}^2\mathscr{Z}_{\rm I}(A, B)$$
$${}^3\mathscr{Y}(A, B) = {}^3\mathscr{Z}_{\rm I}(A, B) - {}^2\mathscr{Z}_{\rm I}(A, B)[{}^1\mathscr{Z}(A) + {}^1\mathscr{Z}(B)].$$

Moreover, we find (see Fig. 10.27)

$${}^2\mathscr{Z}_{\rm I}(A, B) = 2b^2(2\pi)^{1/2}\int_0^A {\rm d}a \int_0^B {\rm d}b\,(A-a)(B-b)(a+b)^{-d/2},$$

or more explicitly

$${}^2\mathscr{Z}_{\rm I}(A, B) = \frac{32(2\pi)^{-d/2}b^2}{(d-2)(4-d)(6-d)(8-d)}$$
$$\times\,[A^{4-d/2} + B^{4-d/2} - (A+B)^{4-d/2} + (4-d/2)AB(A^{2-d/2} + B^{2-d/2})]. \tag{10.8.3}$$

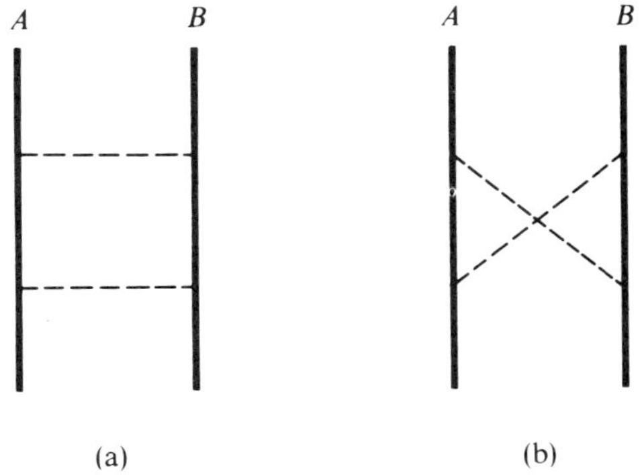

Fig. 10.27. The diagrams (a) and (b) contributing to $^2\mathscr{X}(A, B)$ are equivalent.

To second-order, we have according to (10.8.2)

$$\mathscr{Y}(A, B) = -bAB + {}^2\mathscr{X}_{\mathrm{I}}(A, B) \tag{10.8.4}$$

Now, let us set

$$\langle A^p \rangle = \frac{1}{C}\int_0^\infty \mathrm{d}A\, C(A) A^p$$

$$\langle (A+B)^q \rangle = \frac{1}{C^2}\int_0^\infty \mathrm{d}A \int_0^\infty \mathrm{d}B\, C(A) C(B)(A+B)^q.$$

For $d = 3$, combining (10.8.1), (10.8.4), and (10.8.3), we find the expansion of the osmotitc pressure up to second-order in C

$$\Pi\beta = C + \frac{1}{2}C^2 b\left\{\langle A \rangle^2 - \frac{32(2\pi)^{-d/2} b}{(d-2)(4-d)(6-d)(8-d)}\right.$$
$$\left.\times[2\langle A^{4-d/2}\rangle - \langle (A+B)^{4-d/2}\rangle + (8-d)\langle A^{3-d/2}\rangle\langle B\rangle]\right\}. \tag{10.8.5}$$

In the monodisperse case, $A = B = S$ and one obtains

$$\Pi\beta = C + \frac{1}{2}C^2 bS^2\left[1 - z\frac{32(10 - d - 2^{4-d/2})}{(d-2)(4-d)(6-d)(8-d)}\right] + \ldots \tag{10.8.6}$$

Consequently, for $d = 3$, we have [see (10.5.8)]

$$^2\mathscr{X}_{\mathrm{I}}(S, S) = (2\pi S)^{3/2}\, {}^2\mathfrak{Z}_{\mathrm{I}}(z) = (2\pi S)^{3/2} z^2 \frac{32}{15}(7 - 2^{5/2})$$

$$\Pi\beta = C + \frac{1}{2}C^2(2\pi S)^{3/2} z\left[1 - \frac{32}{15}(7 - 2^{5/2})z + \ldots\right] \tag{10.8.7}$$

$$\text{(where } \frac{32}{15}(7 - 2^{5/2}) = 2.865\text{).}$$

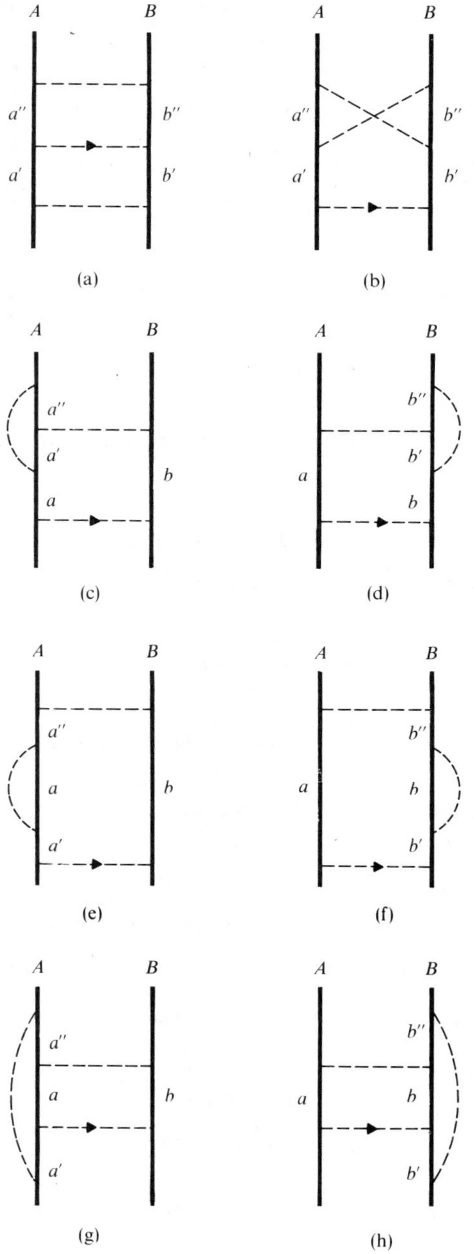

Fig. 10.28. Diagrams giving ${}^{3}\mathscr{L}(A, B)$. The 'areas' between the interaction points are denoted by a, a', a'' and b, b', b''. The interaction line chosen as 'structure line' appears on the diagram with an arrow. The indices of the 'Fixman matrices' are associated with the interaction lines that do not have an arrow. The chain segments subtended by interaction lines are obtained by following the polymer lines and the structure line exclusively.

The calculation of the third-order terms with respect to the interaction can be made in a similar way. The contribution ${}^{3}\mathscr{Z}_{\mathrm{I}}(A, B)$ corresponds to the diagrams of Fig. 10.28 and to the equivalent crossed diagrams. The corresponding matrices obtained by applying Fixman's method are

$$^{3a}\mathscr{S} = \begin{vmatrix} a' + b' & 0 \\ 0 & a'' + b'' \end{vmatrix}$$

$$^{3b}\mathscr{S} = \begin{vmatrix} a' + a'' + b' & a' + b' \\ a' + b' & a' + b' + b'' \end{vmatrix}$$

$$^{3c}\mathscr{S} = \begin{vmatrix} a' + a'' & a' \\ a' & a + a' + b \end{vmatrix} \qquad ^{3d}\mathscr{S} = \begin{vmatrix} b' + b'' & b' \\ b' & b + b' + a \end{vmatrix}$$

$$^{3e}\mathscr{S} = \begin{vmatrix} a & a \\ a & a + a' + a'' + b \end{vmatrix} \qquad ^{3f}\mathscr{S} = \begin{vmatrix} b & b \\ b & b + b' + b'' + a \end{vmatrix}$$

$$^{3g}\mathscr{S} = \begin{vmatrix} a + a' + a'' & a \\ a & a + b \end{vmatrix} \qquad ^{3h}\mathscr{S} = \begin{vmatrix} b + b' + b'' & b \\ b & b + a \end{vmatrix}.$$

For $d = 3$ and a monodisperse set ($A = S$, $B = S$), the complete calculation has been performed by Yamakawa and Kurata,[16] and has been corrected firstly by Kurata *et al.*,[17] secondly by Tagami and Casassa.[18]

The result is expected to be an expression of the form

$$\boldsymbol{\Pi\beta} = \boldsymbol{C} + \tfrac{1}{2}\boldsymbol{C}^2(2\pi S)^{d/2} z j(z) + \ldots$$

with

$$j(z) = 1 - \frac{\mathfrak{z}_{\mathrm{I},2}(z)}{z[\mathfrak{z}_{\mathrm{I},1}(z)]^2} \tag{10.8.8}$$

where $\mathfrak{z}_{\mathrm{I},N}(z)$ correspond to $\mathscr{Z}_{\mathrm{I}}(N \times S)$ (see eqn. (10.5.8)). Thus, for $d = 3$, these authors obtained the expansion

$$j(z) = 1 - 2.865z + 14.278z^2 + \ldots. \tag{10.8.9}$$

Let us note that the series representing $j(z)$ may not converge even for small z. Actually, there are reasons to believe that it diverges for all z. This can be understood by observing that a continuous theory with a purely attractive interaction ($z < 0$) does not exist 1 (see Chapter 15). However, it seems likely that the Borel transform* of the series exists for small z and that $j(z)$ can be defined for all z by analytic continuation of the sum of the transformed series (see Chapter 12).

The area S appears explicitly in the expression (10.8.8) of $\boldsymbol{\Pi\beta}$. An alternative formula can be obtained by introducing the swollen area ${}^{e}S \equiv X^2$. Then, we

* We give in Chapter 12, Section 3.2.2, a definition of the Borel transform.

write

$$\Pi\beta = C + \tfrac{1}{2}C^2(2\pi\,{}^{e}S)^{d/2}g(z)$$

where

$$g(z) = ({}^{e}S/S)^{-d/2}\,zj(z). \tag{10.8.10}$$

Now, for $d = 3$, according to (10.7.9) and (10.7.10), we have

$${}^{e}S/S = 1 + 1.333z - 2.075\,385z^2 + \ldots .$$

Thus, by using the expansion (10.8.9) of $j(z)$, we find for $d = 3$

$$g(z) = z - 4.865z^2 + 26.454z^3 + \ldots . \tag{10.8.11}$$

Moreover, for all d, the value of $g(z)$ up to second-order in z can be obtained by comparing (10.8.6) and (10.8.10) which gives

$$({}^{e}S/S)^{d/2}\,g(z) = z - z^2\frac{32(10 - d - 2^{4-d/2})}{(d-2)(4-d)(6-d)(8-d)}. \tag{10.8.12}$$

As, according to (10.7.6) (since ${}^{e}S/S = R^2/Sd$)

$${}^{e}S/S = 1 + \frac{2d}{(4-d)(6-d)}z + \ldots$$

we finally find

$$g(z) = z - z^2\frac{2}{(4-d)(6-d)}\left[d + \frac{16(10 - d - 2^{4-d/2})}{(d-2)(8-d)}\right] + \ldots . \tag{10.8.13}$$

8.2 Interaction of three chains in a solvent, third virial coefficient ($b \neq 0$, $c = 0$)

The expansion of the osmotic pressure with respect to concentration can be written up to third-order in the form [see (10.5.7)]

$$\begin{aligned}\Pi\beta = C - \frac{1}{2}\int_0^\infty \mathrm{d}A \int_0^\infty \mathrm{d}B\, C(A)C(B)\mathscr{Y}(A, B)\\ - \frac{1}{3}\int_0^\infty \mathrm{d}A \int_0^\infty \mathrm{d}B \int_0^\infty \mathrm{d}C\, C(A)C(B)C(C)[\mathscr{Y}(A,B,C) - 3\mathscr{Y}(A,B)\mathscr{Y}(C)] + \ldots\end{aligned} \tag{10.8.14}$$

where

$$\mathscr{Y}(A, B) = \frac{\mathscr{Z}(A, B)}{\mathscr{Z}(A)\mathscr{Z}(B)}$$

$$\mathscr{Y}(A, B, C) = \frac{\mathscr{Z}(A, B, C)}{\mathscr{Z}(A)\mathscr{Z}(B)\mathscr{Z}(C)}.$$

Few articles have been devoted to the calculation of the last term; only a few approximate expressions have been obtained as mentioned by Yamakawa.[19] However, let us note that the expansion of this term in powers of b can be evaluated up to third-order without great difficulty. In fact, the calculation of a large number of third-order diagrams can be reduced to the calculation of first- or second-order diagrams. Actually, the second- and third-order terms can be written as follows.

$$\begin{aligned}\mathscr{Y}(A, B, C) = {} & b^2 ABC(A + B + C) \\ & - b\left[A(B + C)\frac{\mathscr{Z}_{\mathrm{I}}(B,C)}{\mathscr{Z}(B)\mathscr{Z}(C)} + B(A + C)\frac{\mathscr{Z}_{\mathrm{I}}(A, C)}{\mathscr{Z}(A)\mathscr{Z}(C)} \right. \\ & \left. + C(A + B)\frac{\mathscr{Z}_{\mathrm{I}}(A,B)}{\mathscr{Z}(A)\mathscr{Z}(B)} \right] + \mathscr{Z}_{\mathrm{I}}(A, B, C),\end{aligned}$$

Assuming, as we always do, that the chains A, B, C are distinguishable and dissymmetrical, we see that $\mathscr{Z}_{\mathrm{I}}(A,B,C)$ equals eight times the contribution of the diagram represented in Fig. 10.29.

Taking (10.8.2) into account, we can also write

$$\begin{aligned}& \mathscr{Y}(A, B, C) - \mathscr{Y}(A, B)\mathscr{Y}(A, C) - \mathscr{Y}(A, B)\mathscr{Y}(B, C) \\ & \quad - \mathscr{Y}(A, C)\mathscr{Y}(B, C) \simeq {}^3\mathscr{Z}_{\mathrm{I}}(A, B, C). \qquad (10.8.15)\end{aligned}$$

Thus, to third-order in b, eqn (10.8.14) can be rewritten in the form

$$\begin{aligned}\boldsymbol{\Pi\beta} = \boldsymbol{C} & - \frac{1}{2}\int_0^\infty \mathrm{d}A \int_0^\infty \mathrm{d}B\, \boldsymbol{C}(A)\boldsymbol{C}(B)\mathscr{Y}(A, B) \\ & - \frac{1}{3}\int_0^\infty \mathrm{d}A \int_0^\infty \mathrm{d}B \int_0^\infty \mathrm{d}C\, \boldsymbol{C}(A)\boldsymbol{C}(B)\boldsymbol{C}(C)\, {}^3\mathscr{Z}_{\mathrm{I}}(A, B, C).\end{aligned}$$

The matrix ${}^3\mathscr{S}_{\mathrm{I}}$ corresponding to the diagram of Fig. 10.29 is a number

$${}^3\mathscr{S}_{\mathrm{I}} = A + B + C.$$

After integrating over the wave vector which goes around the loop, we get

$$- b^3(2\pi)^{-d/2}(A + B + C)^{-d/2}$$

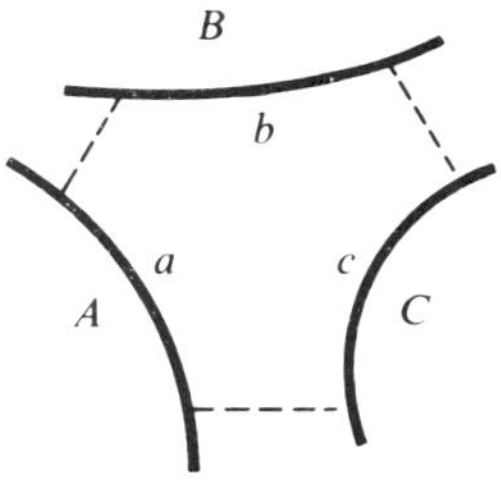

Fig. 10.29. Diagram giving ${}^3\mathscr{Z}_{\mathrm{I}}(A, B, C)$.

according to (10.4.56). Consequently, we find

$$\frac{1}{8}\,{}^{3}\mathscr{Z}_{\mathrm{I}}(A,B,C) =$$

$$-\,b^{3}(2\pi)^{-d/2}\int_{0}^{A}\mathrm{d}a\int_{0}^{B}\mathrm{d}b\int_{0}^{C}\mathrm{d}c(A-a)(B-b)(C-c)(a+b+c)^{-d/2},$$

or more precisely,

$$\frac{1}{8}\,{}^{3}\mathscr{Z}_{\mathrm{I}}(A,B,C) = -\frac{16b^{3}(2\pi)^{-d/2}}{(d-2)(4-d)(6-d)(8-d)}\times$$

$$\times\left\{\frac{4}{(10-d)(12-d)}[-(A+B+C)^{6-d/2}+(B+C)^{6-d/2}\right.$$

$$+(A+C)^{6-d/2}+(A+B)^{6-d/2}-A^{6-d/2}-B^{6-d/2}-C^{6-d/2}]$$

$$+\frac{2}{10-d}[A(B+C)^{5-d/2}+B(A+C)^{5-d/2}+C(A+B)^{5-d/2}$$

$$-(B+C)A^{5-d/2}-(A+C)B^{5-d/2}-(A+B)C^{5-d/2}]$$

$$\left.-[BCA^{4-d/2}+ACB^{4-d/2}+ABC^{4-d/2}]\right\}. \tag{10.8.16}$$

Thus, by grouping the results given by eqns (10.8.2), (10.8.14), (10.8.15), and (10.8.16), we find (for d = 3)

$$\boldsymbol{\Pi\beta} = \boldsymbol{C}$$

$$+\frac{1}{2}\boldsymbol{C}^{2}b\left\{\langle AB\rangle-\frac{32}{15}b(2\pi)^{-3/2}[2\langle A^{5/2}\rangle+5\langle BA^{3/2}\rangle-\langle (A+B)^{5/2}\rangle]\right\}+$$

$$+\frac{128}{45}\boldsymbol{C}^{3}b^{3}(2\pi)^{-3/2}\left\{\frac{4}{63}[-\langle (A+B+C)^{9/2}\rangle+3\langle (A+B)^{9/2}\rangle-3\langle A^{9/2}\rangle]\right.$$

$$\left.+\frac{6}{7}[\langle (A+B)^{7/2}C\rangle-\langle (A+B)C^{7/2}\rangle]-3\langle A^{5/2}BC\rangle\right\}. \tag{10.8.17}$$

For a monodisperse system $A = B = C = S$, the third-order term of the irreducible partition function of three chains reads

$$\mathscr{Z}_{\mathrm{I}}(S,S,S) = (2\pi S)^{3}\mathfrak{z}_{1,z}(z)$$

$$= (2\pi S^{3})z^{3}\frac{128}{315}[-108\times 3^{1/2}+208\times 2^{1/2}-103], \tag{10.8.18}$$

and the osmotic pressure is [see (10.8.17)]

$$\boldsymbol{\Pi\beta} = \boldsymbol{C}+\frac{1}{2}\boldsymbol{C}^{2}(2\pi S)^{3/2}z\left[1-\frac{32}{15}(7-2^{5/2})z\right]$$

$$+\,\boldsymbol{C}^{3}(2\pi S)^{3}z^{3}\,\frac{128}{945}[-108\times 3^{1/2}+208\times 2^{1/2}-103]+\ \ldots. \tag{10.8.19}$$

Taking into account (10.8.8) and (10.8.9), we can also write

$$\mathfrak{z}_{1,3} = -z^3(1.6640)$$

$$\begin{aligned}\Pi\beta = C + \tfrac{1}{2}C^2(2\pi S)^{3/2}z(1 - 2.865z + 14.278z^2 + \ldots) \\ + \tfrac{1}{3}C^3(2\pi S)^3 z^3(1.6640).\end{aligned} \qquad (10.8.20)$$

8.3 Parametric representation of the virial expansion ($b \neq 0$, $c = 0$)

The results obtained in the preceding sections can be synthesized by starting from the general formulae (10.6.34) obtained by summation of trees and by bringing in them the expressions of $\mathfrak{z}_{1,2}(z)$ and $\mathfrak{z}_{1,3}(z)$ to first-orders. Actually, according to (10.6.33) and (10.7.7)

$$\mathfrak{z}_{1,1}(z) = 1 + 4z + \ldots$$

according to (10.8.7) and (10.8.8)

$$\mathfrak{z}_{1,2}(z) = 2.865z^2 + 8.642z^3,$$

and according to (10.8.20)

$$\mathfrak{z}_{1,3}(z) = -z^3(1.6640).$$

By bringing these values in (10.6.34), we obtain

$$\begin{aligned}\Pi\beta/C = 1 + \tfrac{1}{2}C(2\pi S)^{3/2}z - [C(2\pi S)^{3/2}]^{-1}t^2[1.4325z + 4.341z^2 - 0.5547\,tz] \\ C(2\pi S)^{3/2} \simeq t[1 + 4z + (2.865z + 8.642z^2)t - 0.8320z^3t^2].\end{aligned} \qquad (10.8.21)$$

8.4 Osmotic pressure of dilute and semi-dilute solutions in moderately good solvent: one-loop approximation ($b \neq 0$, $c = 0$)

The one-loop diagrams are diagrams with only one independent wave vector (see Fig. 10.30). The one-loop approximation consists in calculating $\mathscr{Z}_1(N \times S)$ by keeping only zero-loop and one-loop diagrams. As will be seen, it is valid only for moderately long chains or not very good solvents. This approximation was introduced and used, in an implicit way, by S.F. Edwards[20] in 1966; it was explicitly used in a special case by M.A. Moore[21] in 1977 and was developed by J. des Cloizeaux[22] in 1980.

The calculation of one-loop diagrams is performed by applying the rules given in Section 3.4.2 and can be made as easily in the polydisperse case as in the monodisperse one. In the latter case

$$^1\mathscr{Z}_1(N\times S) = \frac{1}{(2\pi)^d}(-b)^N 2^{N-1}(N-1)! \int d^dk \left[{}^{+,0}H(k)\frac{S^2}{2}\right]^N \qquad (10.8.22)$$

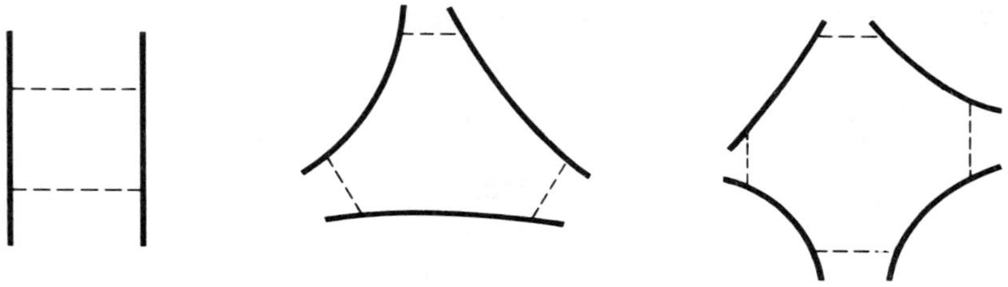

Fig. 10.30. One-loop diagrams.

where $2^{N-1}(N-1)!$ is a counting factor. The form function ${}^{+,0}H(k)$ is given by

$$
\begin{aligned}
{}^{+,0}H(k)\frac{S^2}{2} &= \int_{s_0}^{S} \mathrm{d}s\,(S-s)\exp[-sk^2/2] \\
&= \frac{(S-s_0)\exp[-s_0k^2/2]}{k^2/2} - \frac{\exp[-s_0k^2/2]-\exp[-Sk^2/2]}{(k^2/2)^2}
\end{aligned}
$$

where s_0 is the cut-off area.

For $N>1$, we can calculate $\mathscr{Z}_{\mathrm{I}}(N\times S)$ by putting $s_0=0$ and by replacing ${}^{+,0}H$ by ${}^{0}H(k)$ in (10.8.22) (${}^{0}H(k)$ is the Brownian form-function)

$$
{}^{0}H(k)\frac{S^2}{2} = \frac{S}{k^2/2} - \frac{1-\exp[-Sk^2/2]}{(k^2/2)^2}.
$$

The same cannot be done for $N=1$ because $\int \mathrm{d}^dk\ {}^{+,0}H(k)$ diverges (see also Chapter 7; Section 3.3.1.1); the terms of $\int \mathrm{d}^dk\ {}^{+,0}H(k)$ that become infinite in the limit $s_0\to 0$ are

$$
(S-s_0)\int \mathrm{d}^dk\frac{\exp[-s_0k^2/2]}{k^2/2} \propto \frac{S-s_0}{s_0^{-1+d+2}}.
$$

Thus, for calculating $\mathscr{Z}_{\mathrm{I}}(S)$, we have to subtract this term as indicated in Section 4.1.5; this can be done by replacing ${}^{+,0}H(k)$ by $[{}^{0}H(k)-4/k^2S]$. Taking this remark into account, we see that

$$
{}^{1}\mathscr{Z}_{\mathrm{I}}(N\times S) =
$$

$$
\frac{1}{(2\pi)^d}(-b)^N2^{N-1}(N-1)!\int \mathrm{d}^dk\left[\frac{2S}{k^2}(1-\delta_{N,1})-\frac{4}{k^4}+\frac{4}{k^4}\exp[-Sk^2/2]\right]^N.
$$

In agreement with (10.6.33), we set

$$
\begin{aligned}
{}^{0}\mathscr{Z}_{\mathrm{I}}(N\times S) &= (2\pi S)^{(N-1)d/2}\quad {}^{0}\mathfrak{z}_{\mathrm{I},N}(z) \\
{}^{1}\mathscr{Z}_{\mathrm{I}}(N\times S) &= (2\pi S)^{(N-1)d/2}\quad {}^{1}\mathfrak{z}_{\mathrm{I},N}(z)
\end{aligned}
$$

where

$$
\begin{aligned}
{}^{0}\mathfrak{z}_{\mathrm{I},N}(z) &= \delta_{N,1} \\
{}^{1}\mathfrak{z}_{\mathrm{I},N}(z) &= \frac{(-2z)^N}{\Gamma(d/2)} \int_0^{\infty} \mathrm{d}x\, x^{d-1} \left[\left(\frac{1}{x^2} - \frac{1}{x^4} + \frac{\mathrm{e}^{-x^2}}{x^4} \right)^N - \frac{\delta_{N,1}}{x^2} \right].
\end{aligned} \tag{10.8.24}
$$

Consequently, in agreement with the notation used in (10.6.34), we can write

$$
\begin{aligned}
\mathscr{Q}(f) &= (2\pi S)^{-d/2}\, \mathfrak{z}_{\mathrm{I}}(t, z) \\
t &= (2\pi S)^{-1/2}\mathfrak{f} \\
\mathscr{L}_{\mathrm{I}}(t,z) &\simeq {}^{0}\mathfrak{z}_{\mathrm{I}}(t,z) + {}^{1}\mathfrak{z}_{\mathrm{I}}(t, z) \\
&= \sum_{N=1}^{\infty} \frac{t^N}{N!} [{}^{0}\mathfrak{z}_{\mathrm{I},N}(z) + {}^{1}\mathfrak{z}_{\mathrm{I},N}(z)].
\end{aligned} \tag{10.8.25}
$$

Thus, the one-loop approximation gives

$$
\begin{aligned}
\mathfrak{z}_{\mathrm{I}}(t,z) &= t + {}^{1}\mathfrak{z}_{\mathrm{I}}(t\; z)] \\
{}^{1}\mathfrak{z}_{\mathrm{I}}(\tau) &= \frac{1}{\Gamma(d/2)} \int_0^{\infty} \mathrm{d}x\, x^{d-1} \left\{ \frac{2\tau}{x^2} - \ln\left[1 + 2\tau\left(\frac{1}{x^2} - \frac{1}{x^4} + \frac{\mathrm{e}^{-x^2}}{x^4} \right) \right] \right\}.
\end{aligned} \tag{10.8.26}
$$

Moreover, the osmotic pressure and the concentration are given in terms of $\mathfrak{z}_{\mathrm{I}}(t,z)$ by

$$
\begin{aligned}
\Pi\beta(2\pi S)^{d/2} &= \mathfrak{z}_{\mathrm{I}}(t,z) + \tfrac{1}{2} C^2 (2\pi S)^d z \\
C(2\pi S)^{d/2} &= t\frac{\partial}{\partial t}\mathfrak{z}_{\mathrm{I}}(t, z)
\end{aligned}
$$

which can be written in the form (remember the definition (10.1.7) of z)

$$
\begin{aligned}
\Pi\beta(2\pi S)^{d/2} &= \tau z^{-1} + {}^{1}\mathfrak{z}_{\mathrm{I}}(\tau) + \tfrac{1}{2} C^2 (2\pi S)^d z \\
C(2\pi S)^{d/2} &= \tau z^{-1} + \tau\frac{\partial}{\partial \tau}\, {}^{1}\mathfrak{z}_{\mathrm{I}}(\tau)
\end{aligned} \tag{10.8.27}
$$

with

$$
z(2\pi S)^{d/2} = bS^2
$$

In particular, for $d = 3$

$$
\begin{aligned}
{}^{1}\mathfrak{z}_{\mathrm{I}}(\tau) &= 2\pi^{-1/2} \int_0^{\infty} \mathrm{d}x \left\{ 2\tau - x^2 \ln\left[1 + 2\tau\left(\frac{1}{x^2} - \frac{1}{x^4} + \frac{\mathrm{e}^{-x^2}}{x^4} \right) \right] \right\} \\
&= 2\pi^{-1/2}\tau^{3/2} \int_0^{\infty} \mathrm{d}y \left\{ 2 - y^2 \ln\left[1 + \frac{2}{y^2} - \frac{2}{\tau y^2} + \frac{\mathrm{e}^{-\tau y^2}}{\tau y^2} \right] \right\}.
\end{aligned} \tag{10.8.28}
$$

For small τ, we have

$$
{}^{1}\mathfrak{z}_{\mathrm{I}}(\tau) = 4\tau + \frac{16}{15}(7 - 4 \times 2^{1/2})\tau^2 + \ldots.
$$

For large τ, we have

$$^{1}\mathfrak{z}_{\mathrm{I}}(\tau) = \tfrac{4}{3}(2\pi)^{1/2}\tau^{3/2}.$$

We verify that the one-loop approximation is valid only for small values of z (poor solvent).

For rather small concentrations ($\tau \ll 1$, $\boldsymbol{C}bS^2z = \boldsymbol{C}(2\pi S)^{3/2} \ll 1$), we have

$$\boldsymbol{\Pi\beta} = \boldsymbol{C}\left\{\frac{1}{2} + \frac{1}{2}\boldsymbol{C}bS^2 + \left[1 + \frac{128(7 - 4 \times 2^{1/2})}{15(1+4z)^2}z\boldsymbol{C}bS^2\right)^{1/2}\right]^{-1}\right\}$$

and the first two terms of the expansion of this formula with respect to b or z would obviously coincide with those obtained, at the same order, by eliminating τ between the two equations (10.8.27).

For rather large concentrations ($\tau \gg 1$, $\boldsymbol{C}bS^2 \gg 1$), we find

$$\boldsymbol{\Pi\beta}bS^2 - \tfrac{1}{2}(\boldsymbol{C}bS^2)^2 = \tau + \tfrac{4}{3}(2\pi)^{1/2}z\tau^{3/2}$$

$$\boldsymbol{C}bS^2 = \tau + 2(2\pi)^{1/2}z\tau^{3/2} \qquad (10.8.29)$$

and if z is small enough, there remains two limiting cases.

First, there is an intermediary domain ($z \ll \boldsymbol{C}bS^2$, $z \ll 1$, $z\tau^{1/2}$ small) for which we have

$$\boldsymbol{\Pi\beta} = \boldsymbol{C}[1 + \tfrac{1}{2}\boldsymbol{C}bS^2 - \tfrac{2}{3}(2\pi)^{1/2}z(\boldsymbol{C}bS^2)^{1/2}]. \qquad (10.8.30)$$

In this case, taking into account the definition (10.1.7) of z, we exactly reproduce the result obtained by Edwards[20] and Moore[21] [see (8.2.57)]. As can easily be shown, this equation can easily be extended to the polydisperse case[22], in the following way. Introducing the total 'area' of polymer per unit volume (proportional to the monomer concentration)

$$\mathscr{C} = \int_0^\infty \mathrm{d}S\,\mathscr{C}(S)S \quad \left(\boldsymbol{C} = \int \mathrm{d}S\,\boldsymbol{C}(S)\right) \qquad (10.8.31)$$

we can rewrite (10.8.30) in the form

$$\boldsymbol{\Pi\beta} = \boldsymbol{C} + \frac{1}{2}\mathscr{C}^2b - \frac{1}{3\pi}(\mathscr{C}b)^{3/2} + \ \dots \qquad (10.8.32)$$

which remains meaningful for a polydisperse solution.

On the contrary, for rather large concentrations ($\boldsymbol{C}bS^2z \gg 1$, $z\tau^{1/2}$ large), eqns (10.8.29) give

$$\boldsymbol{\Pi\beta} = \boldsymbol{C}\left[\tfrac{2}{3} + \tfrac{1}{2}\boldsymbol{C}bS^2 - \frac{\pi^{-1/3}}{6}z^{-2/3}(\boldsymbol{C}bS^2)^{-1/3}\right]. \qquad (10.8.33)$$

Thus, in the limit where the chains are widely overlapping, a situation characterizing semi-dilute solutions, we have

$$\boldsymbol{\Pi\beta} \simeq \tfrac{1}{2}\mathscr{C}^2b \qquad (10.8.34)$$

(Edwards's law).* Now, the osmotic pressure depends only on the monomer concentration.

The crossover domain corresponds to concentrations of the order of C^+ where C^+ is given by the following condition deduced from (10.8.29)

$$C^+ bS^2 = 2\tau = 4(2\pi)^{1/2} z \tau^{3/2};$$

that is to say,

$$C^+ = \frac{2\pi^2}{b^3 S^3}. \qquad (10.8.35)$$

9. SUMMARY

The partition functions of a chain or of a set of chains are expanded in powers of the effective two-body interaction (or eventually in powers of the effective two-body and three-body interactions). The elementary tool is the diagram made of a set of polymer lines (on which correlation points are fixed) and of interaction lines joining interaction points on the polymer lines. Each term in the expansion corresponds to a given number of interaction lines. The calculation of such a term is performed by summing up all the contributions of the diagrams associated with this number of lines.

However, applying the method presents difficulties related to the fact that each molecule may bear a large number of interaction points.

A first approximation used to calculate connected diagrams with N polymer lines consists in introducing only $N-1$ interaction lines. This is the simple-tree approximation. The sum of the contributions is obtained by a Legendre transformation.

The grand canonical formalism is used to calculate the osmotic pressure of chains as a function of the chain concentrations. Then, the physical quantities are given in parametric form as functions of the fugacities. Moreover, it is convenient to express these expansions in reciprocal space. A number $\circlearrowleft$ of independent internal wave vectors (number of loops) corresponds to each diagram. The contribution of a diagram is obtained by summations over all these wave vectors and by summations over all the positions of the interaction points on the chains.

The following physical quantities are calculated: mean square end-to-end distance, correlations between internal points of a chain, radius of gyration, second and third virial coefficient, and so on. The parameters for this model are the area S of the equivalent Brownian chain, the two-body interaction b with dimension L^{d-4} (eventually the three-body interaction c with dimension L^{2d-6}), and, for a monodisperse sample, the chain concentration C. The swelling of

* We shall see in Chapter 13, how this law must be modified (for a semi-dilute assembly of long chains in good solvent).

an isolated chain is expressed by the dimensionless parameter $\mathfrak{X}_0 = R^2/{}^0R^2$ (and eventually $\mathfrak{X}_G = R_G^2/{}^0R_G^2$)

In the present chapter, we ignored the three-body interactions, but we shall come back to them in chapter 14, where we deal with poor solvents. The quantities calculated here by perturbation are expanded to first-orders in powers of z. For a chain ensemble, the other dimensionless parameter is $C\,S^{d/2}$. An expression of the osmotic pressure corresponding to one-loop approximation and valid for small values of z is given above.

REFERENCES

1. Hill, T.L. (1956). *Statistical Mechanics.* McGraw-Hill, New York.
2. Fisher, I.Z. (1964). *Statistical theory of liquids.* University of Chicago Press.
3. Edwards, S.F. (1965). *Proc. Phys. Soc.* **85**, 613; (1966) **88**, 265.
4. Fixman, M. (1955). *J. Chem. Phys.* **23**, 1656.
5. Muthukumar, M. and Nickel, B.G. (1984). *J. Chem. Phys.* **80**, 5839.
6. Yamakawa, H. and Tanaka, G. (1967). *J. Chem. Phys.* **47**, 3991.
7. Barrett, A.J. (1975). Thesis. University of London. Lax, M., Barrett, A.J., and Domb, C. (1978). *J. Phys. A: Math. Gen.* **11**, 361.
8. Yamakawa, H. (1971). *Modern theory of polymer solutions*, Chapter III, p. 91. Harper & Row.
9. Yamakawa, H. and Kurata, M. (1958). *J. Phys. Soc. Japan.* **13**, 78. Kurata, M., Yamakawa, H., and Teramoto, E. (1958). *J. Chem. Phys.* **28**, 785.
10. Yamakawa, H. (1971). *Modern theory of polymer solutions*, Chapter III, p. 92. Harper & Row.
11. Zimm, B.H., Stockmayer, W.H., and Fixman, M. (1953). *J. Chem. Phys.* **21**, 1716.
12. Yamakawa, H., Aoki, A., and Tanaka, G. (1966). *J. Chem. Phys.* **45**, 1938.
13. Kirkwood, J.G. (1953). *J. Polymer. Sci.* **12**, 1; see also des Cloizeaux, J. (1978). *J. Physique Lett.* **39**, L151.
14. Stockmayer, W.H. and Albrecht, A.C. (1958). *J. Polymer Sci.* **32**, 215.
15. Barrett, J. (1983). *J. Phys.* A: *Math. Gen.* **16**, 2321.
16. Yamakawa, H. (1958). *J. Phys. Soc. Japan* **13**, 87. Kurata, M. and Yamakawa, H. (1958). *J. Chem. Phys.* **29**, 311.
17. Kurata, M., Fukatsu, M., Sotobayashi, H., and Yamakawa, H. (1964). *J. Chem. Phys.* **41**, 139.
18. Casassa, E.F. (1959). *J. Chem. Phys.* **31**, 800.
19. Yamakawa, H. (1971). *Modern theory of polymer solutions*, Chapter IV, Sect. 22, p. 170.
20. Edwards, S.F. (1966). *Proc. Phys. Soc.* **88**, 265; see also Section 2.2 in Chapter 8.
21. Moore, M.A. (1977). *J. Physique* **38**, 265.
22. des Cloizeaux, J. (1980). *J. Physique* **41**, 749 and 761.

11

RELATIONS BETWEEN CHAIN THEORY AND FIELD THEORY: LAPLACE–DE GENNES TRANSFORMATION

GENESIS

Temperley observed, as early as 1956, that the perturbation diagrams associated with magnetic systems looked like sets of interacting chains. For years, the existence of this formal analogy has been well known but the correspondence remained too vague to be useful.

The situation changed in 1972 when P.G. de Gennes established a precise correspondence between a polymer and a magnetic system in zero magnetic field. In a general way, the properties of magnetic systems depend on the number n of components of the magnetization vector and this dependence is particularly crucial at the critical point. Thus, it is currently admitted that the critical exponents are universal and depend only on n and on the space dimension d. In principle, n and d are positive integers but the structure of the diagrams associated with the Green's functions enables one to extend the definition of these functions to any values of n and d. Thus, de Gennes observed that the properties of a magnetic system with n components correspond precisely, in the limit $n = 0$, to those of an isolated chain with excluded volume. Later on, this correspondence has been re-established by numerous physicists in various ways, either with the help of diagrams or by using analytical means.

The de Gennes's discovery allowed us to extend to polymers all the results that had been obtained a few years before, by applying renormalization theory to the study of critical phenomena.

In particular, these results provided a test of the validity of Flory's formula $\nu = \nu_F \equiv 3/(d + 2)$. At that time, this formula was generally considered to be exact. However, the criticisms formulated by J. des Cloizeaux in 1970 could induce physicists not only to reject the proofs of this formula but also to doubt its complete validity. Actually, renormalization theory enabled us to calculate in a seemingly exact manner, the first-orders of the expansion of ν in terms of $\varepsilon = 4 - d$: it gives $\nu = \frac{1}{2}\left(1 + \frac{\varepsilon}{8} + \ldots\right)$ whereas $\nu_F = \frac{1}{2}\left(1 + \frac{\varepsilon}{6} + \ldots\right)$. Thus, it was established that ν_F is only an approximate value of ν, and that the true exponent is smaller, a result which was later confirmed.

The next step was taken by des Cloizeaux who showed, in 1975, that, in the same way as an isolated polymer corresponds to a zero-component magnetic system in zero field, a polymer solution corresponds approximately to a magnetic system in an external field. Thus, it became clear that the Lagrangian theory could provide a lot of information concerning the properties of dilute and semi-dilute solutions in good solvent.

This broader correspondence led to the discovery of new scaling laws; in particular, for the osmotic pressure. Then, Daoud, de Gennes, and Jannink who wanted to know how, in the semi-dilute regime, the radius of gyration and the screening length decrease with concentration, proposed a coherent set of simple laws to describe the situation. Experiments were systematically performed in order to test these laws and they produced favourable results.

Moreover, these results remain valid when the two-body repulsive interaction between polymer segments diminishes. In particular, de Gennes pointed out in 1975 that for polymer solutions, Flory's Θ point ($T = T_F$) is a tri-critical point and must be treated as such. Thus, the Lagrangian theory can be used not only to describe the behaviour of solutions in the repulsive domain where excluded volume is dominant, but also in the attractive domain.

1. GENERALITIES

In 1972, de Gennes[1] discovered a transformation relating the partition functions of a polymer system to the Green's functions of a zero-component field theory. This correspondence was initially established for an isolated polymer but it can be extended to a set of polymers without great difficulty.

In the present book, we deal essentially with linear polymers. However, it must be noted that it is also possible to establish correspondences between branched polymer theories and appropriate field theories. This question has been studied especially by Lubensky and Isaacson.[2]

Concerning linear polymers, the correspondence can be formulated in various ways, and two kinds of methods can be used to establish it: the diagrammatic approach and the direct analytic approach. The transformation applies to various types of chains and fields. In this chapter, we describe this transformation by choosing significant examples. However, before looking at details, we shall examine from a general point of view the analogy that exist between the behaviour of a long polymer in solution and the properties of a magnetic system (in zero magnetic field) near the Curie point; this comparison will thus enable us to establish a correspondence between the critical exponents γ and ν associated with these two systems.

2. ANALOGY BETWEEN AN ISOLATED POLYMER AND A FERROMAGNETIC SYSTEM IN ZERO EXTERNAL FIELD: A COMPARISON BETWEEN CRITICAL EXPONENTS

In order to establish the polymer-ferromagnet analogy in a qualitative way, we shall show that a 'Green's function' can be associated with the two-point 'restricted' partition function of a polymer and that formally this Green's function looks very much like the correlation function of a magnetic system in zero external field.

We consider the partition functions corresponding to the configurations of a repulsive chain on a lattice, assuming that the chain has N links and a fixed origin O. In particular, let ${}^0Z_N$ be the total number of non-interacting chains and let Z_N be the total number of self-avoiding chains. Moreover, let $Z_N(\vec{r})$ be the number of self-avoiding chains ending at the point of position vector $\vec{r}$. The Fourier transform of this function is

$$\tilde{Z}_N(\vec{k}) = \sum_{\vec{r}} e^{i\vec{k}\cdot\vec{r}} Z_N(\vec{r}) \tag{11.2.1}$$

where we sum over all lattice sites and where $\tilde{\mathscr{Z}}_N(k)$ has the same periods as the reciprocal lattice.

The Laplace transform of $\tilde{\mathscr{Z}}_N(k)$ is a function

$$\tilde{G}(\vec{k}, A) = \sum_{N=1}^{\infty} e^{-AN} \tilde{Z}_N(\vec{k}) \tag{11.2.2}$$

which, as we shall see, can be considered as a Green's function. In particular, for a non-interacting chain, we have

$${}^0\tilde{Z}_N(\vec{k}) = [\tilde{Z}_1(\vec{k})]^N$$

$${}^0\tilde{G}(\vec{k}, A) = \frac{1}{1 - e^{-A}\tilde{Z}_1(\vec{k})}. \tag{11.2.3}$$

Let us consider long chains. For small values of k

$$\tilde{Z}_1(\vec{k}) = \mathfrak{k}(1 - Bk^2)$$

where $\mathfrak{k}$ is the coordination number of the lattice and let us set

$$e^{A_c} = \mathfrak{k}. \tag{11.2.4}$$

Long chains correspond to small values of $A - A_c$ and the Brownian limit is obtained for small values of $\vec{k}$. In this case, we have

$${}^0\tilde{G}(\vec{k}, A) \propto \frac{1}{A - A_c + Bk^2}. \tag{11.2.5}$$

This 'Green's function' looks like the Green's function of a free particle.

For a self-avoiding chain, $\tilde{Z}_N(\vec{k})$ is much more complicated but for large values of N and small values of $\vec{k}$ (i.è. large values of $\vec{r}$), we expect, for this quantity, a scaling law of the form.

$$\tilde{Z}_N(\vec{k}) \simeq N^{\gamma-1} \exp(N A_c)\, \tilde{\mathfrak{f}}(k N^{\nu}) \tag{11.2.6}$$

[see Chapter 2, eqn. 2.3.7)]
where $\langle\langle r^2 \rangle\rangle$ is given by

$$\langle\langle r^2 \rangle\rangle / d = N^{2\nu} \tilde{\mathfrak{f}}''(0)/\tilde{\mathfrak{f}}(0),$$

a relation obtained from (11.2.1) by expanding in k^2.

For small values of $(A - A_c)$, the sum over N in (11.2.2) can be replaced by an integral and we can write

$$\tilde{G}(\vec{k}, A) \simeq \int_0^\infty dN \exp[-(A - A_c)N] N^{\gamma-1} \tilde{f}(k N^\nu)$$

$$= \frac{1}{(A - A_c)^\gamma} \int_0^\infty du\, e^{-u} u^{\gamma-1} \tilde{f}\left(\frac{k u^\nu}{(A - A_c)^\nu}\right)$$

By setting

$$\int_0^\infty du\, e^{-u} u^{\gamma-1} f(x u^\nu) = 1/\phi(x)$$

we obtain

$$\tilde{G}(\vec{k}, A) = \frac{1}{(A - A_c)^\gamma \phi(k(A - A_c)^{-\nu})}. \tag{11.2.7}$$

Thus, we see how the exponents γ and ν defined by (11.2.6) appear in (11.2.7).

Let us now consider a ferromagnetic system made of spins $\vec{\sigma}(\vec{r})$ attached to the lattice sites. We shall admit that the spin–spin interaction is given by a Hamiltonian $\mathscr{H}\{\vec{\sigma}\}$ which will be assumed to be translationally invariant. At equilibrium, the weight of a configuration of the system is given by the Boltzmann's factor

$$W = \exp(-\mathscr{H}\{\vec{\sigma}\})$$

where $\mathscr{H}\{\vec{\sigma}\}$ can be considered as a linear function of $\boldsymbol{\beta} = 1/k_B T$.

The correlation function is the mean value

$$\langle\langle \vec{\sigma}(\vec{o}) \cdot \vec{\sigma}(\vec{r}) \rangle\rangle$$

which can also be expressed in the form of Fourier transform. We can thus introduce a Green's function $\boldsymbol{G}(\vec{k})$ by setting

$$\tilde{G}(\vec{k}, A) = \sum_{\vec{r}} e^{i\vec{k}\cdot\vec{r}} \langle\langle \vec{\sigma}(\vec{o}) \cdot \vec{\sigma}(\vec{r}) \rangle\rangle$$

where the domain of k can be restricted to the Brillouin zone associated with the lattice.

Let us now consider the system at a temperature only slightly larger than the Curie temperature T_c and let us introduce the variable

$$t = (T - T_c)/T = 1 - \boldsymbol{\beta}/\boldsymbol{\beta}_c. \tag{11.2.8}$$

For $t > 0$, $\langle\langle \vec{\sigma}(\vec{o}) \cdot \vec{\sigma}(\vec{r}) \rangle\rangle$ decreases exponentially like $\exp(-r/\xi(t))$ where $\xi(t)$ is the correlation length but $\xi(t) \to \infty$ when $t \to 0$.

The susceptibility χ of the system[3] is given by the sum

$$\chi = \sum_{r} \langle\langle \vec{\sigma}(\vec{o}) \cdot \vec{\sigma}(\vec{r}) \rangle\rangle = \tilde{G}(\vec{o}, A), \tag{11.2.9}$$

and we can postulate that for small values of t

$$\chi = \tilde{G}(\vec{o}, A) \propto 1/t^{\gamma}. \tag{11.2.10}$$

Moreover, for small t, we have by definition

$$\xi \propto 1/t^{\nu}. \tag{11.2.11}$$

For small values of t, correlations extend to large distances. In order to study them, one looks at small values of k (i.e. $k \simeq 1/\xi$). These conditions define the critical domain and it is expected that in this domain $\tilde{G}(\vec{k}, A)$ is given by a scaling law of the form

$$\tilde{G}(\vec{k}, A) \simeq \frac{1}{t^{\gamma}\psi(kt^{-\nu})} \tag{11.2.12}$$

It is clear that eqn (11.2.7) which describes the behaviour of a long polymer is very similar to eqn (11.2.12) which describes the behaviour of a magnetic system in the absence of external magnetic field but in the vicinity of the Curie point.

In particular, we remark that the exponents γ and ν are exactly homologous for these systems, and this is the fact we wanted to establish here.

3. FIELDS AND SETS OF CHAINS: DIAGRAMMATIC METHODS

As an example, we shall apply diagrammatic methods to the study of discrete models on a lattice. The polymer system will consist of $\mathbf{N}$ chains with fixed lengths, drawn on a lattice. We assume that each chain, defined by its order $\boldsymbol{j}$, has a fixed origin of position vector r_{2j-1} and a fixed end point of position vector $\vec{r}_j$. By definition, the weight associated with a configuration $\mathbf{\Omega}$ equals $\exp[A_0 N - bp(\mathbf{\Omega})]$ where N is the total number of links ($N = N_1 + \ldots + N_{\mathbf{N}}$) and $p(\mathbf{\Omega})$ is the number of couples of different points belonging to the same chain or to two chains, and located on the same site. The partition function of the system is defined by

$$Z_{\mathrm{G}}(\vec{r}_1, \ldots, \vec{r}_{2\mathbf{N}}; N_1, \ldots, N_{\mathbf{N}})$$

(where the index G means global or, in perturbation theory, non-connected).

The ferromagnetic system consists of spins attached to the sites of the lattice, interacting with one another. With each site are associated $\mathbf{N}'$ classical spins, i.e. $\mathbf{N}'$ vectors $\vec{\sigma}_j(\vec{r})$ with $\boldsymbol{j} = 1, \ldots, \mathbf{N}'$. Each spin $\vec{\sigma}_j(\vec{r})$ has $\mathfrak{n}$ components $\sigma_{j,\alpha}(\vec{r})$ in an abstract space (with $\alpha = 1, \ldots, \mathfrak{n}$). The spins are random variables and we can associate with any function $\mathcal{O}\{\sigma\}$ of these spins, the mean value $\langle\!\langle \mathcal{O}\{\sigma\} \rangle\!\rangle$.

These mean values are calculated by attributing a weight

$$W = {}^{I}W \exp[-{}^{1}\mathscr{H}\{\sigma\}] \tag{11.3.1}$$

to each spin configuration; moreover ${}^{1}\mathscr{H}\{\sigma\}$ and ${}^{I}W$ are defined as follows.

The term ${}^1\mathscr{H}\{\sigma\}$ is given by

$$ {}^1\mathscr{H}\{\sigma\} = -\frac{1}{2}\sum_{\vec{r},\vec{r}',j,\alpha} V_{\vec{r},\vec{r}'} \exp(A - A_0)\, \sigma_{j,\alpha}(\vec{r})\, \sigma_{j,\alpha}(\vec{r}') \tag{11.3.2} $$

where $V_{\vec{r},\vec{r}'} = 1$ if the points of position vectors $\vec{r}$ and $\vec{r}'$ are nearest neighbours on the lattice and $V_{\vec{r},\vec{r}'} = 0$ in the other cases (for instance if $\vec{r} = \vec{r}'$).

The weight ${}^1W$ factorizes into weights associated with each site. Thus, in the absence of ${}^1\mathscr{H}\{\sigma\}$, the spins on each site are independent from one another. Actually, we write

$$ {}^1W = \prod_{\vec{r}} w(\{\sigma\}(\vec{r})) \tag{11.3.3.} $$

where the symbol $\{\sigma\}(\vec{r})$ represents the set of all the vectors $\vec{\sigma}_j(\vec{r})$ that are attached to the site of position vector $\vec{r}$.

The various models differ only by the form of $w\{\sigma\}$ which will be discussed in each case, later on. Here, we shall be interested in correlations involving $2N$ sites, where N is an arbitrary integer. However, the following Green's function which determines such correlations, is defined only for $N \leqslant N'$

$$ G_{\mathrm{G}}(\vec{r}_1, \ldots, \vec{r}_{2N}; \{A_j\}) = \langle\!\langle [\sigma_{1,1}(\vec{r}_1)\sigma_{1,1}(\vec{r}_2)] \ldots [\sigma_{N,1}(\vec{r}_{2N-1})\sigma_{N,1}(\vec{r}_{2N})] \rangle\!\rangle \tag{11.3.4} $$

(the index G for global always has the same meaning).

This function is defined for integer values of n but it will be shown that the definition can be extended, by analytic continuation, to all values of n. In particular, it is possible to go to the limit $n = 0$, and it turns out that, in this case, the Green's function is a function of $A_1, \ldots, A_N$ only and not of the other A_j. For $n \to 0$, we write

$$ G_{\mathrm{G}}(\vec{r}_1, \ldots, \vec{r}_{2N}; \{A_j\}) = G_{\mathrm{G}}(\vec{r}_1, \ldots, \vec{r}_1, A_1 \ldots, A_N). \tag{11.3.5} $$

Finally, we shall establish the correspondence existing between a system of interacting chains on a lattice and a zero-component field theory, by proving the equality

$$ G_{\mathrm{G}}(\vec{r}_1, \ldots, \vec{r}_{2N}; A_1, \ldots, A_N) = $$
$$ = \sum_{N_1} \ldots \sum_{N_N} \exp[-(A_1 N_1 + \ldots + A_N N_N)]\, Z_{\mathrm{G}}(\vec{r}_1, \ldots, \vec{r}_{2N}; N_1, \ldots, N_N) \tag{11.3.6} $$

(Laplace–de Gennes transform). This correspondence between Green's functions and partition functions remains valid for their cumulants (also called connected parts). Thus, eqn (11.3.6) applies also to connected Green's functions and connected partition functions (the fact that they are connected is denoted by the absence of the index G).

3.1 Continuous spin model (de Gennes)

In this model, the weight $w\{\sigma\}$ at one site is given by

$$w\{\sigma\} = \exp\left\{ -\frac{1}{2}\sum_{j=1}^{N'}\sum_{\alpha=1}^{n}(\sigma_{j,\alpha})^2 - \frac{b}{8}\left[\sum_{j=1}^{N'}\sum_{\alpha=1}^{n}(\sigma_{j,\alpha})^2\right]^2\right\} \tag{11.3.7}$$

Then ${}^{I}W$ reads

$${}^{I}W = \exp(-{}^{0}\mathscr{H}\{\sigma\} - {}^{I}\mathscr{H}\{\sigma\})$$

where

$${}^{0}\mathscr{H}\{\sigma\} = \frac{1}{2}\sum_{\vec{r}}\sum_{j=1}^{N'}\sum_{\alpha=1}^{n}[\sigma_{j,\alpha}(\vec{r})]^2$$

$${}^{I}\mathscr{H}\{\sigma\} = \frac{b}{8}\sum_{\vec{r}}\left[\sum_{j=1}^{N'}\sum_{\alpha=1}^{n}(\sigma_{j,\alpha}(\vec{r}))^2\right]^2. \tag{11.3.8}$$

In the absence of interaction between spins belonging to the same site ($b = 0$), the weight ${}^{I}W$ reduces to the weight ${}^{00}W$

$${}^{00}W = \exp(-{}^{0}\mathscr{H}\{\sigma\})$$

and the weight W given by (11.3.1) becomes

$${}^{0}W = {}^{00}W\exp(-{}^{1}\mathscr{H}\{\sigma\}).$$

It can immediately be verified that, when there is no nearest-neighbour coupling and no interaction,

$$\langle\!\langle \sigma_{j,\alpha}(\vec{r})\,\sigma_{j',\alpha'}(\vec{r}\,')\rangle\!\rangle_{00} = \delta_{jj'}\,\delta_{\alpha,\alpha'}\,\delta_{\vec{r},\vec{r}\,'}.$$

Let us now consider the Green's function corresponding to the case without interaction

$${}^{0}G(\vec{r}_1,\vec{r}_2;\{A_j\}) = \langle\!\langle \sigma_{1,1}(\vec{r}_1)\,\sigma_{1,1}(\vec{r}_2)\rangle\!\rangle_0 \tag{11.3.9}$$

Expanding ${}^{0}W$ in powers of ${}^{1}\mathscr{H}$ and applying Wick's theorem in the averages (see Appendix A), we can represent ${}^{0}G(\vec{r}_1,\vec{r}_2;\,A)$ by diagrams. The diagrams correspond to random walks on the lattice. It can be verified that this function depends only on A_1; the result is the same as in the case where there is only one field

$${}^{0}G(\vec{r}_1,\vec{r}_2;\{A_j\}) = {}^{0}G(\vec{r}_1,\vec{r}_2;A_1) \tag{11.3.10}$$

and we can write

$${}^{0}G(\vec{r}_1,\vec{r}_2;A) = \sum_{N=1}^{\infty} e^{-AN}\,{}^{0}Z(\vec{r}_1,\vec{r}_2;N) \tag{11.3.11}$$

where

$${}^{0}Z(\vec{r}_1,\vec{r}_2;1) = \exp(A_0)\,V_{\vec{r}_1,\vec{r}_2}.$$

Incidentally, the Fourier transform of this result is very simple. Indeed, let us set

$$\tilde{V}(\vec{k}) = \sum_{\vec{r}} e^{i\vec{k}\cdot\vec{r}}\, V_{\vec{o},\vec{r}}$$

$$\bar{Z}(\vec{k}, -\vec{k}; N) = \sum_{\vec{r}} e^{i\vec{k}\cdot\vec{r}}\, Z(\vec{o},\vec{r}; N).$$

It is easy to verify that

$$^0\tilde{\boldsymbol{G}}(\vec{k}, -\vec{k}; A) = \sum_{\vec{r}} e^{i\vec{k}.\vec{r}}\; {}^0\boldsymbol{G}(\vec{o},\vec{r}; A) =$$

$$= \sum_{N=0}^{\infty} e^{-AN}\,[{}^0Z(\vec{k}, -\vec{k}; 1)]^N = [1 - \exp[-(A - A_0)]\,\tilde{V}(\vec{k})]^{-1} \tag{11.3.12}$$

Now, we want to generalize (11.3.11) by extending it to the case with interaction. Let us first consider the case where $\mathfrak{n}$ is an integer. By expanding W simultaneously in powers of $^1\mathscr{H}$ and $^I\mathscr{H}$ [see (11.3.2) and (11.3.8)] and by using Wick's theorem (see Appendix A), we can expand $\boldsymbol{G}(\vec{r}_1, \ldots, \vec{r}_{2\boldsymbol{N}}; \{A_j\})$ and the expansion can be represented by diagrams very similar to the diagrams studied in Chapter 10. In these diagrams, $\boldsymbol{N}$ solid lines join the points of position vectors $\vec{r}_1$ and $\vec{r}_2$, $\vec{r}_3$ and $\vec{r}_4, \ldots, \vec{r}_{2\boldsymbol{N}-1}$ and $\vec{r}_{2\boldsymbol{N}}$, respectively, but there are also closed solid lines: this is a difference between these diagrams and polymer diagrams, (see Fig. 11.1). The contacts are represented, as usual, by interaction lines. Each open solid line $\boldsymbol{j}$ joining r_{2j-1} to r_{2j} corresponds to the *first* component of the vectorial fields $\vec{\sigma}_j(\vec{r})$. Thus, the weight associated with line $\boldsymbol{j}$ depends only on A_j. Moreover, each closed solid line is labelled with two indices $\boldsymbol{j}'$ and α'; the weight associated with such a line depends on $A_{j'}$, but not on α'. To calculate the contribution of a diagram, one has to sum over the indices $\boldsymbol{j}'$ and α' associated with each closed line. This remark shows that the contribution of any diagram containing $\boldsymbol{N}'$ closed solid lines is the product of $\mathfrak{n}^{N'}$ (sum over all α' and $\boldsymbol{j}'$) by a function of the A_j, which does not depend on $\mathfrak{n}$ (sum over the $\boldsymbol{j}$). Thus, the value of a diagram can be defined for any real (or complex) value of $\mathfrak{n}$. Moreover, in the limit $\mathfrak{n} \to 0$, the only diagrams that contribute are the ones which *do not contain any closed solid line* (see Fig. 11.1); the contribution of these diagrams depend only on $A_1, \ldots, A_{\boldsymbol{N}}$ and b.

We also note that the line joining $\vec{r}_{2j-1}$ and $\vec{r}_{2j}$ contains an arbitrary number of links and that a weight $\exp(-A_j)$ is associated with each link. Formula (11.3.6) is a direct consequence of these observations, and, for a given lattice, it establishes an exact correspondence between an interacting-chain theory and a zero-component field theory.

3.2 Models with spins of definite lengths (Sarma, Hilhorst).

The ferromagnetic models proposed by Sarma[4] and Hilhorst[5] correspond to self-avoiding polymers drawn on a lattice.

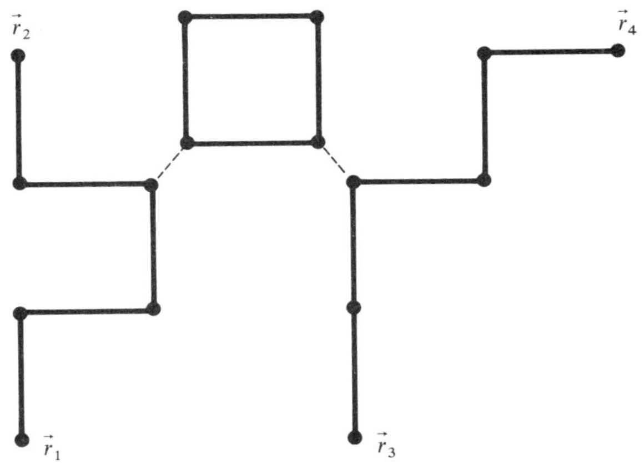

Fig. 11.1. Diagram containing a closed solid line. In the limit $n \to 0$, the contribution of this diagram vanishes.

In both models, the weight ${}^1W$ has the following fundamental property. For a given $\vec{r}$ (i.e. for a given site), the mean value of any product of components $\sigma_{j,\alpha}(\vec{r})$ vanish in the limit $n \to 0$, except the quantities

$$\langle\langle [\sigma_{j,\alpha}(\vec{r})]^2 \rangle\rangle_I = 1. \qquad (11.3.13)$$

We shall see later on how these conditions can be realized in practice.

The Green's function $G(\vec{r}_1, \ldots, \vec{r}_{2N}; \{A_j\})$, defined by (11.3.4) as a mean value calculated with the weight W,

$$G\{\vec{r}\} \equiv \langle\langle \mathcal{O}\{\vec{r}\} \rangle\rangle$$

can also be expressed in the form

$$G\{\vec{r}\} = \frac{\langle\langle \mathcal{O}\{\vec{r}\} \exp[-{}^1\mathcal{H}\{\sigma\}] \rangle\rangle_I}{\langle\langle \exp[-{}^1\mathcal{H}\{\sigma\}] \rangle\rangle_I}$$

where index I means that the average is made with the weight ${}^1W$.

In the latter expression, the numerator and the denominator can be expanded in powers of ${}^1\mathcal{H}\{\sigma\}$ and diagrammatic representations of their expansions are obtained. Taking (11.3.13) into account, we see that only one piece of chain may end at the points of position vectors $\vec{r}_1, \ldots, \vec{r}_{2N}$ and only two (or zero) pieces of chain may end at the other points. Thus, the diagrams contributing to the numerator consist of N chains joining $(\vec{r}_1, \vec{r}_2) \ldots (\vec{r}_{2N-1}, \vec{r}_{2N}]$ respectively and of closed chains, without intersections; the denominator consists only of closed chains, without intersections. Now, we observe that the closed chains do not contribute, because each closed chain introduces a factor n which tends to zero. Consequently,

$$\langle\langle \exp[-{}^1\mathcal{H}\{\sigma\}] \rangle\rangle_I = 1.$$

Thus, the expansion of $G(\vec{r}_1, \ldots, \vec{r}_{2N}; \{A_j\})$ can be represented by diagrams consisting only of non-intersecting lines drawn on the lattice from $\vec{r}_1$ to $\vec{r}_2$, from $\vec{r}_3$ to $\vec{r}_4$ and so on. Each line j joining the points of position vectors $\vec{r}_{2j-1}$ and $\vec{r}_{2j}$ may have any number of links and a factor $V_{\vec{r},\vec{r}'} \exp[A_\alpha - A_0]$ is associated with each link. These remarks directly establish the validity of eqns (11.3.5) and (11.3.6) in the case under consideration.

Now we have only to define $w\{\sigma\}$ for each model, and to prove the validity of the fundamental property mentioned above, concerning the mean values of products of components $\sigma_{j\,\alpha}(\vec{r})$ [where $\vec{r}$ is fixed; see (11.3.13)].

Sarma's model deals with'continuous spins of fixed length; it is defined by the probability law

$$w\{\sigma\} = \delta\left(\sum_{j=1}^{N'} \sum_{\alpha=1}^{\mathfrak{n}} (\sigma_{j,\alpha})^2 - \mathfrak{n}N'\right) \tag{11.3.14}$$

$$\langle\!\langle (\sigma_{j,\alpha})^2 \rangle\!\rangle = 1 \tag{11.3.15}$$

in agreement with eqn (11.3.13).

The variables $\sigma_{j,\alpha}$ can be considered as the components of a vector $\vec{\sigma}$ in a space with $\mathfrak{n}N'$ dimensions. In the same space, let us consider a vector $\vec{k}$ with components $k_{j,\alpha}$. Taking (11.3.15) into account, we can calculate

$$\varphi(k) = \langle\!\langle e^{i\vec{k}.\vec{\sigma}} \rangle\!\rangle = 1 - k^2/2 + \ldots \tag{11.3.16}$$

The Laplacian of $\varphi(k)$ in the $\mathfrak{n}N'$-dimensional space is given by

$$\Delta\varphi(k) = \langle\!\langle -(\vec{\sigma})^2 e^{i\vec{k}.\vec{\sigma}} \rangle\!\rangle = -\mathfrak{n}N'\varphi(k),$$

and this equality can be written in the more explicit form

$$\left[\frac{\partial^2}{\partial k^2} + \frac{\mathfrak{n}N' - 1}{k}\frac{\partial}{\partial k} + \mathfrak{n}N'\right]\varphi(k) = 0.$$

In the limit $\mathfrak{n} = 0$, $\varphi(k)$ is determined by the equation

$$\left(\frac{\partial^2}{\partial k^2} - \frac{1}{k}\frac{\partial}{\partial k}\right)\varphi(k) = 0 \tag{11.3.17}$$

and by the conditions (11.3.16). The corresponding solution is simply

$$\varphi(k) = 1 - k^2/2. \tag{11.3.18}$$

Then, by expanding the formal expression of $\varphi(k)$ in powers of the $k_{j,\alpha}$, it can easily be shown that all products of components vanish except the quantities

$$\langle\!\langle (\sigma_{j,\alpha})^2 \rangle\!\rangle = 1.$$

Hilhorst's model deals with discrete spins and each spin has $2\mathfrak{n}N'$ equally probable states. A state is defined by a couple of values j_0, α_0 (where $j_0 = 1, \ldots, N'$ and $\alpha_0 = 1, \ldots, \mathfrak{n}$) and a number $\varepsilon = \pm 1$. By definition, for this state

$$\sigma_{j,\alpha} = \varepsilon \delta_{jj_0} \delta_{\alpha\alpha_0} (\mathfrak{n} N')^{1/2}. \qquad (11.3.19)$$

From this equality, we immediately derive

$$\langle\!\langle (\sigma_{j,\alpha})^2 \rangle\!\rangle = 1 \quad \langle\!\langle \sigma_{j,\alpha} \rangle\!\rangle = 0. \qquad (11.3.20)$$

Moreover, we observe that the mean value of the product of $2p$ components $\sigma_{j,\alpha}$ is always proportional to $(\mathfrak{n}N')^{-1+p}$. Thus, in the limit $\mathfrak{n} \to 0$, all such mean values vanish except the quantities $\langle\!\langle (\sigma_{j,\alpha})^2 \rangle\!\rangle$

Thus, the property that we used above is now proved both for Sarma's and Hilhorst's models.

4. FIELDS AND SETS OF CHAINS: ANALYTICAL METHOD

The correspondence existing between polymer theory and zero-component field theory can also be established analytically as was shown by Emery[6] and also by Kosmas and Freed.[7] A more general approach has been found by Duplantier.[8] We shall follow it here but with a broader point of view. We shall choose a continuous model,* ignoring all possible divergences. More precisely, in what follows we define partition functions $\mathscr{Z}_G(\vec{r}_1, \ldots, \vec{r}_{2N}; S_1, \ldots, S_N)$ analogous to the $\mathscr{Z}_G(\vec{r}_1, \ldots, \vec{r}_{2N}; N_1, \ldots, N_N)$ introduced at the beginning of Section 3. With them, we associate the Laplace transforms

$$\mathscr{G}_G(\vec{r}_1, \ldots, \vec{r}_{2N}; a_1, \ldots, a_N) = \int_0^\infty dS_1 \ldots \int_0^\infty dS_N e^{-a_1 S_1 \ldots - a_N S_N} \mathscr{Z}_G(\vec{r}_1, \ldots, \vec{r}_{2N}; S_1, \ldots, S_N), \qquad (11.4.1)$$

and we show that these quantities $\mathscr{G}_G(\vec{r}_1, \ldots, \vec{r}_{2N}; S_1, \ldots, S_N)$ can be interpreted as the Green's functions of a field theory.

Moreover, it is clear that such a relation also implies that the connected Green's functions are the Laplace transforms of connected partition functions (without index G). We call it the Laplace–de Gennes transformation.

4.1 Continuous chain model: a fundamental identity

We consider N chains of areas $S_1, \ldots, S_N$ in a d-dimensional continuous space. Each chain j (where $j = 1, \ldots, N$) is characterized by a function $r_j(s_j)$ where

* One could choose a discrete model as well.

$0 \leqslant s_j \leqslant S_j$. A weight

$$\mathscr{W} = \exp(-{}^{0}\mathfrak{H}_N\{\vec{r}\} - {}^{1}\mathfrak{H}_N\{\vec{r}\})$$

$${}^{0}\mathfrak{H}_N\{\vec{r}\} = \frac{1}{2}\sum_{j=1}^{N}\int_0^{S_j} \mathrm{d}s \left[\frac{\mathrm{d}\vec{r}_j(s)}{\mathrm{d}s}\right]^2 \tag{11.4.2}$$

is associated with this set.

The interaction energy can be expressed in terms of the area concentration

$$\mathscr{C}(\vec{r}) = \sum_{j=1}^{N} \mathscr{C}_j(\vec{r})$$

$$\mathscr{C}_j(\vec{r}) = \int_0^{S_j} \mathrm{d}s\, \delta[\vec{r} - \vec{r}_j(s)] \tag{11.4.3}$$

and of potentials $V(\vec{r}_1, \vec{r}_2)$ (two-body interaction), $V(\vec{r}_1, \vec{r}_2, \vec{r}_3)$ (three-body interaction) and so on; in general, these potentials depend only on the differences $(\vec{r}_i - \vec{r}_j)$

$${}^{1}\mathfrak{H}_N\{\vec{r}\} \equiv {}^{1}\mathfrak{H}\{\mathscr{C}\} = \sum_{p=1}^{\infty} \int \mathrm{d}^d r_1 \ldots \mathrm{d}^d r_p\, V_p(\vec{r}_1, \ldots \vec{r}_p)\mathscr{C}(\vec{r}_1) \ldots \mathscr{C}(\vec{r}_p) \tag{11.4.4}$$

The partition functions $\mathscr{Z}(\vec{r}_1, \ldots, r_{2N}; S_1, \ldots, S_N)$ are given by

$$\mathscr{Z}_G(\vec{r}_1, \ldots, \vec{r}_{2N}; S_1, \ldots, S_N) =$$

$$V^N \left\langle\!\!\left\langle \prod_{j=1}^{N} \delta(\vec{r}_{2j-1} - \vec{r}_j(0))\delta(\vec{r}_{2j} - \vec{r}_j(S_j))\exp[{}^{1}\mathfrak{H}_N\{\vec{r}\}] \right\rangle\!\!\right\rangle. \tag{11.4.5}$$

(Note that the V^N factor is a consequence of the equation $\langle\!\langle \delta(\vec{r} - \vec{r}_j(0)\rangle\!\rangle_0 = 1/V$.)

Let us now introduce an inhomogeneous source $J(\vec{r})$ coupled to the concentration $\mathscr{C}(\vec{r})$. (This reminds us of Edwards's transformation provided we put $J = -i\chi$: see Chapter 10, Section 3.) Setting

$$J\cdot\mathscr{C} = \int \mathrm{d}^d r\, J(\vec{r})\mathscr{C}(\vec{r})$$

we define the partition function of a chain j in the field of this source by

$$\mathscr{Z}(\vec{r}, \vec{r}\,'; S_j; J) = V\langle\!\langle \delta[\vec{r} - \vec{r}_j(0)]\,\delta[\vec{r}\,' - \vec{r}_j(S_j)]\exp(J\cdot\mathscr{C}_j)\rangle\!\rangle_0 \tag{11.4.6}$$

Now, we remark that any functional $\mathscr{O}\{\mathscr{C}\}$ of $\mathscr{C}(\vec{r})$ can be written in the (symbolic)* form

$$\mathscr{O}\{\mathscr{C}\} = \mathscr{O}\left\{\frac{\partial}{\partial J}\right\}\exp(J\cdot\mathscr{C})\bigg|_{J=0}. \tag{11.4.7}$$

* $\left\{\frac{\partial}{\partial J}\right\}$ represents the set of the functional derivatives $\frac{\partial}{\partial J(\vec{r})}$; condition $J = 0$ means that $J(\vec{r}) = 0$ for any $\vec{r}$.

In particular, we may have $\mathcal{O}\{\mathscr{C}\} = \exp[-{}^{\mathrm{I}}\mathfrak{H}\{\mathscr{C}\}]$ and, in this way, we deduce the fundamental identity

$$\mathscr{Z}_{\mathrm{G}}(\vec{r}_1, \ldots, \vec{r}_{2N}; S_1, \ldots, S_N) = \exp\left(-{}^{\mathrm{I}}\mathfrak{H}\left\{\frac{\partial}{\partial J}\right\}\right) \prod_{j=1}^{N} \mathscr{Z}(\vec{r}_{2j-1}, \vec{r}_{2j}; S_j; J)|_{J=0} \tag{11.4.8}$$

from (11.4.4) and (11.4.7).

4.2 Functions associated with the partition functions

Let us introduce the Laplace transforms of the partition functions

$$\mathscr{G}_{\mathrm{G}}(\vec{r}_1, \ldots, \vec{r}_{2N}; a_1, \ldots, a_N) =$$
$$\int_0^\infty \mathrm{d}S_1 \ldots \int_0^\infty \mathrm{d}S_N \exp[-a_1 S_1 \ldots - a_N S_N] \mathscr{Z}_{\mathrm{G}}(\vec{r}_1, \ldots, \vec{r}_{2N}; S_1, \ldots, S_N).$$

For one chain, we have

$${}^0\mathscr{G}(\vec{r}, \vec{r}\,'; a) = \int_0^\infty \mathrm{d}S\, \mathrm{e}^{-aS}\, {}^0\mathscr{Z}(\vec{r}, \vec{r}\,'; S)$$

which corresponds to the non-interacting case, and

$$\mathscr{G}(\vec{r}, \vec{r}\,'; a, J) = \int_0^\infty \mathrm{d}S\, \mathrm{e}^{-aS} \mathscr{Z}(\vec{r}, \vec{r}\,'; S; J) \tag{11.4.9}$$

which corresponds to the case where the chain interacts only with a fixed potential $J(\vec{r})$.

By Laplace transformation, (11.4.8) can be written in the form

$$\mathscr{G}_{\mathrm{G}}(\vec{r}_1, \ldots, \vec{r}_{2N}; a_1, \ldots, a_N) = \exp\left(-{}^{\mathrm{I}}\mathfrak{H}\left\{\frac{\partial}{\partial J}\right\}\right) \prod_{j=1}^{N} \mathscr{G}(\vec{r}_{2j-1}, \vec{r}_{2j}; a_j; J)|_{J=0} \tag{11.4.10}$$

where, as we shall see, the function $\mathscr{G}(\vec{r}_1, \vec{r}\,'; a; J)$ has a simple expression in terms of the functions $\mathscr{G}_0(\vec{r}, \vec{r}\,'; a)$ and $J(\vec{r})$

According to (11.4.5) and (11.4.2), we have

$$\mathscr{Z}(\vec{r}, \vec{r}\,'; S; J) = V\left\langle\!\!\left\langle \delta[\vec{r} - \vec{r}(0)]\,\delta[\vec{r}\,' - \vec{r}(S)] \exp\left[\int_0^S \mathrm{d}s\, J[\vec{r}(s)]\right]\right\rangle\!\!\right\rangle_0$$

where $\delta[\vec{r}(S)]$ and $J[\vec{r}(S)]$ have the same meaning as $\delta(\vec{r}(S))$ and $J(\vec{r}(S))$

$$= \frac{\displaystyle\int \mathrm{d}\{r\}\,\delta[\vec{r} - \vec{r}(0)]\,\delta[\vec{r}\,' - \vec{r}(S)] \exp\left[\int_0^S \mathrm{d}s\, J[\vec{r}(s)]\right] \exp[-{}^0\mathfrak{H}_1\{\vec{r}\}]}{\displaystyle\int \mathrm{d}\{r\}\,\delta[\vec{r} - \vec{r}(0)] \exp[-{}^0\mathfrak{H}_1\{\vec{r}\}]}. \tag{11.4.11}$$

The identity

$$\exp \int_0^S \mathrm{d}s\, J[\vec{r}(s)] = 1 + \int_0^S \mathrm{d}s\, J[\vec{r}(s)] \exp \int_0^s \mathrm{d}s'\, J[\vec{r}(s')]$$
$$= 1 + \int \mathrm{d}^d r'' \int_0^S \mathrm{d}s\, J(\vec{r}'')\,\delta[\vec{r}'' - \vec{r}(s)] \exp \int_0^s \mathrm{d}s'\, J[\vec{r}(s')]$$

can be directly verified for $S = 0$ and can be proved for $S \neq 0$ by differentiation with respect to S. Let us use it to transform (11.4.11). In this way, we obtain

$$\mathscr{Z}(\vec{r}, \vec{r}'; S; J) = {}^0\mathscr{Z}(\vec{r}, \vec{r}'; S; J) + \int \mathrm{d}^d r''\, J(r'')$$
$$\times \int_0^S \mathrm{d}s\, V \left\langle\!\!\left\langle \delta[\vec{r} - \vec{r}(0)]\,\delta[\vec{r}'' - \vec{r}(s)]\,\delta[\vec{r}' - \vec{r}(S)] \exp \int_0^s \mathrm{d}s'\, J[\vec{r}(s')] \right\rangle\!\!\right\rangle. \tag{11.4.12}$$

The integrand can also be written in the form

$$V \left\langle\!\!\left\langle \delta[\vec{r} - \vec{r}(0)]\,\delta[\vec{r}'' - \vec{r}(s)]\,\delta[\vec{r}' - \vec{r}(S)] \exp \int_0^s \mathrm{d}s'\, J[\vec{r}(s')] \right\rangle\!\!\right\rangle$$
$$= \frac{\int \mathrm{d}\{r\}\,\delta[\vec{r} - \vec{r}(0)]\,\delta[\vec{r}'' - \vec{r}(s)]\,\delta[\vec{r}' - \vec{r}(S)] \exp\left[\int_0^s \mathrm{d}s'\, J[\vec{r}(s')]\right] \exp[-{}^0\mathfrak{H}_1\{\vec{r}\}]}{\int \mathrm{d}\{r\}\,\delta[\vec{r} - \vec{r}(0)] \exp[-{}^0\mathfrak{H}_1\{\vec{r}\}]} \tag{11.4.13}$$

This quantity is equal to the product $\mathscr{Z}(\vec{r}, \vec{r}''; J)\ {}^0\mathscr{Z}(\vec{r}'', \vec{r}'; S - s)$. Indeed, by definition, we have

$$\mathscr{Z}(\vec{r}, \vec{r}''; s; J)\,{}^0\mathscr{Z}(\vec{r}'', \vec{r}'; S - s) =$$
$$\frac{\int \mathrm{d}\{r_1\} \int \mathrm{d}\{r_2\}\, A(\vec{r}, \vec{r}''; \{\vec{r}_1\}; J)\, B(\vec{r}'', \vec{r}'; \{\vec{r}_2\})}{\int \mathrm{d}\{r_1\} \int \mathrm{d}\{r_2\}\,\delta[\vec{r} - \vec{r}_1(0)]\,\delta[\vec{r}'' - \vec{r}_2(0)] \exp[-{}^0\mathfrak{H}_1\{\vec{r}_1\} - {}^0\mathfrak{H}_1\{\vec{r}_2\}]} \tag{11.4.14}$$

where

$$A(\vec{r}, \vec{r}'', \{\vec{r}_1\}; J) = \delta[\vec{r} - \vec{r}_1(0)]\,\delta[\vec{r}'' - \vec{r}_1(s)]$$
$$\times \exp\left[\int_0^s \mathrm{d}s\, J[\vec{r}_1(s)\right] \exp[-{}^0\mathfrak{H}_1\{\vec{r}_1\}]$$
$$B(\vec{r}'', \vec{r}'; \{\vec{r}_2\}) = \delta[\vec{r}'' - \vec{r}_2(0)]\,\delta[\vec{r}' - \vec{r}_2(S - s)] \exp[-{}^0\mathfrak{H}_1\{\vec{r}_2\}].$$

Let us now identify the following vectors

$$\vec{r}_1(s') = \vec{r}(s) \qquad 0 \leqslant s' < s$$

$$\vec{r}_2(s') = \vec{r}(s+s') + \vec{a} \qquad 0 \leqslant s' \leqslant S - s$$

where $\vec{a}$ is an arbitrary vector. This leads to

$$\mathrm{d}\{r_1\}\,\mathrm{d}\{r_2\} = \mathrm{d}\{r\}\,\mathrm{d}^d a.$$

In the right-hand side of (11.4.14), we can now replace the variables $\vec{r}_1(s')$, $\vec{r}_2(s')$ by the variables $\vec{r}(s)$, $\vec{a}$. In this way, we get an expression which coincides with the right-hand side of (11.4.12).

Thus, (11.4.12) can be expressed in the form

$$\mathscr{Z}(\vec{r},\vec{r}\,';S;J) = {}^0\mathscr{Z}(\vec{r},\vec{r}\,';S) + \int \mathrm{d}^d r'' \int_0^S \mathrm{d}s\, \mathscr{Z}(\vec{r},\vec{r}\,'';s;J)\,J(\vec{r}\,'')\,{}^0\mathscr{Z}(\vec{r}\,'',\vec{r}\,';S-s). \qquad (11.4.15)$$

The Laplace transform of this equation reads

$$\mathscr{G}(\vec{r},\vec{r}\,';a;J) = {}^0\mathscr{G}(\vec{r},\vec{r}\,';a) + \int \mathrm{d}^d r''\,\mathscr{G}(\vec{r},\vec{r}\,'';a;J)\,J(\vec{r}\,'')\,{}^0\mathscr{G}(\vec{r}\,'',\vec{r}\,';a)$$

and in operator form

$$\mathscr{G}(a;J) = {}^0\mathscr{G}(a) + \mathscr{G}(a;J)\,\boldsymbol{J}\,{}^0\mathscr{G}(a),$$

or

$$[\mathscr{G}(a;J)]^{-1} = [{}^0\mathscr{G}(a)]^{-1} - \boldsymbol{J} \qquad (11.4.16)$$

where the matrix elements of operator $\boldsymbol{J}$ are

$$\boldsymbol{J}(\vec{r},\vec{r}\,') = J(\vec{r})\,\delta(\vec{r}-\vec{r}\,'). \qquad (11.4.17)$$

These equalities define the function $\mathscr{G}(\vec{r},\vec{r}\,';a;J)$ which appear in (11.4.10). Now, it remains to be shown that the Laplace transforms $\mathscr{G}(\vec{r}_1,\ldots,\vec{r}_{2N};a_1,\ldots,a_N)$ of the partition functions are Green's functions associated with a set of fields.

4.3 Correspondence with a set of fields

Let us consider N free fields $\phi_j(j = 1,\ldots,N)$, each of them with $\mathfrak{n}$ components φ_{jl} $(l = 1,\ldots,\mathfrak{n})$ taking values $\varphi_{jl}(\vec{r})$ at each point $\vec{r}$. We can write these functions in the form $\varphi_{jl}(\vec{r}) = (\vec{r}\,|\varphi_{jl})$, by considering them as the components of a vector $|\varphi_{jl})$ which is real.

Let us associate the following weights with the configurations of the system

$${}^0\mathscr{W} = \exp[-{}^0\mathscr{H}\{\varphi\}]$$

$${}^0\mathscr{H}\{\varphi\} = \frac{1}{2}\sum_{j,1}(\varphi_{jl}|[{}^0\mathscr{G}(a_j)]^{-1}|\varphi_{jl}) \qquad (11.4.18)$$

and

$$^{J}\mathscr{W} = {}^{0}\mathscr{W}\exp[-{}^{J}\mathscr{H}\{\varphi\}]$$

$$^{J}\mathscr{H}\{\varphi\} = -\frac{1}{2}\sum_{j,l}(\varphi_{jl}|J|\varphi_{jl}). \tag{11.4.19}$$

Thus, by using (11.4.16), are can write

$$^{J}\mathscr{W} = \exp\left(-\frac{1}{2}\sum_{j,l}(\varphi_{jl}|[\mathscr{G}(a, J)]^{-1}/\varphi_{jl})\right).$$

From this Gaussian distribution (see Appendix A), we deduce the mean value

$$\mathscr{G}(\vec{r}, \vec{r}'; a_j; J) = \langle\!\langle \varphi_{jl}(\vec{r})\varphi_{jl}(\vec{r}')\rangle\!\rangle_J. \tag{11.4.20}$$

By bringing this result in (11.4.10) and by remarking that the fields are independent of one another, we obtain

$$\mathscr{G}_{\mathrm{G}}(\vec{r}_1, \ldots, \vec{r}_{2N}; a_1, \ldots, a_j) =$$

$$\exp\left(-{}^{\mathrm{I}}\mathfrak{H}\left\{\frac{\partial}{\partial J}\right\}\right)\langle\!\langle \varphi_{11}(\vec{r}_1)\varphi_{11}(\vec{r}_2)\ldots\varphi_{N1}(\vec{r}_{2N-1})\varphi_{N1}(\vec{r}_{2N})\rangle\!\rangle_J\Big|_{J=0}. \tag{11.4.21}$$

Moreover,

$$\langle\!\langle \varphi_{11}(\vec{r}_1)\varphi_{11}(\vec{r}_2)\ldots\varphi_{N1}(\vec{r}_{2N-1})\varphi_{N1}(\vec{r}_{2N})\rangle\!\rangle =$$

$$= \langle\!\langle \varphi_{11}(\vec{r}_1)\varphi_{11}(\vec{r}_2)\ldots\varphi_{N1}(\vec{r}_{2N-1})\varphi_{N1}(\vec{r}_{2N})\,\mathrm{e}^{-{}^{J}\mathscr{H}\{\varphi\}}\rangle\!\rangle_0/\langle\!\langle \mathrm{e}^{-{}^{J}\mathscr{H}\{\varphi\}}\rangle\!\rangle_0. \tag{11.4.22}$$

The quantity $\langle\!\langle\exp(-{}^{J}\mathscr{H}\{\varphi\}\rangle\!\rangle$ is formally given by

$$\langle\!\langle\exp(-{}^{J}\mathscr{H}\{\varphi\})\rangle\!\rangle_0 = \frac{\int \mathrm{d}\{\varphi\}\exp[-{}^{0}\mathscr{H}\{\varphi\} - {}^{J}\mathscr{H}\{\varphi\}]}{\int \mathrm{d}\{\varphi\}\exp[-{}^{0}\mathscr{H}\{\varphi\}]}$$

where ${}^{0}\mathscr{H}\{\varphi\}$ and ${}^{J}\mathscr{H}\{\varphi\}$ are quadratic forms of the fields. Consequently, this quantity is the ratio of two determinants (see Appendix A, Section 4).

$$\langle\!\langle\exp(-{}^{J}\mathscr{H}\{\varphi\})\rangle\!\rangle_0 = \prod_{j=1}^{N}\left[\frac{\det(\mathscr{G}_0^{-1}(a_j))}{\det(\mathscr{G}_0^{-1}(a_j) - \boldsymbol{J})}\right]^{\mathfrak{n}/2}$$

$$= \prod_{j=1}^{N}[\det(1 - \boldsymbol{J}\mathscr{G}_0(a_j))]^{-\mathfrak{n}/2} \tag{11.4.23}$$

The theory is defined for integer-values of $\mathfrak{n}$; however, it can be extended to all values of $\mathfrak{n}$ by analytic continuation. Thus, in the limit $\mathfrak{n} \to 0$, we find

$$\langle\!\langle\exp(-{}^{J}\mathscr{H}\{\varphi\})\rangle\!\rangle_0 = 1. \tag{11.4.24}$$

Consequently, in the limit $n \to 0$, (11.4.21), (11.4.22) and (11.4.23) give

$$\mathscr{G}_{\mathrm{G}}(\vec{r}_1, \ldots, \vec{r}_{2N}; a_1, \ldots, a_N) = \exp\left(- {}^{\mathrm{I}}\mathfrak{H}\left\{\frac{\partial}{\partial J}\right\}\right)$$
$$\times \langle\langle \varphi_{11}(\vec{r}_1)\varphi_{11}(\vec{r}_2) \ldots \varphi_{N1}(\vec{r}_{2N1})\varphi_{N1}(\vec{r}_{2N})\exp(-{}^{J}\mathscr{H}\{\varphi\})\rangle\rangle_0|_{J=0}. \tag{11.4.25}$$

Let us set

$$\exp(-{}^{\mathrm{I}}\mathfrak{H}\{\tfrac{\partial}{\partial J}\})\exp(-{}^{J}\mathscr{H}\{\varphi\})|_{J=0} = \exp(-{}^{\mathrm{I}}\mathscr{H}\{\varphi\}). \tag{11.4.26}$$

From (11.4.24) we deduce the relation ($n \to 0$)

$$\langle\langle \exp(-{}^{\mathrm{I}}\mathscr{H}\{\varphi\})\rangle\rangle_0 = 1. \tag{11.4.27}$$

Moreover, we find

$$\mathscr{G}_{\mathrm{G}}(\vec{r}_1, \ldots, \vec{r}_{2N}; a_1, \ldots, a_N) =$$
$$\langle\langle \varphi_{11}(\vec{r}_1)\varphi_{11}(\vec{r}_2) \ldots \varphi_{N}(\vec{r}_{2N-1})\varphi_{N1}(\vec{r}_{2N})\exp(-{}^{\mathrm{I}}\mathscr{H}\{\varphi\})\rangle\rangle_0. \tag{11.4.28}$$

Now, let us introduce the weight

$$\mathscr{W} = \exp[-{}^{0}\mathscr{H}\{\varphi\} - {}^{\mathrm{I}}\mathscr{H}\{\varphi\}]. \tag{11.4.29}$$

Taking (11.4.27) into account, we see that the Laplace transform of the partition function of a set of chains is equal to the average of a product of fields, calculated with the weight $\mathscr{W}$

$$\mathscr{G}_{\mathrm{G}}(\vec{r}_1, \ldots, \vec{r}_{2N}; a_1, \ldots, a_N) =$$
$$\langle\langle \varphi_{11}(\vec{r}_1)\varphi_{11}(\vec{r}_2) \ldots \varphi_{N1}(\vec{r}_{2N-1})\varphi_{N1}(\vec{r}_{2N})\rangle\rangle. \tag{11.4.30}$$

Thus, this function is a Green's function of a field theory, and the above equality establishes a correspondence between polymer theory and field theory in the limit $n \to 0$. To define this correspondence more precisely, we have only to calculate ${}^{\mathrm{I}}\mathscr{H}\{\varphi\}$ which is determined by (11.4.26). We remark that ${}^{J}\mathscr{H}\{\varphi\}$, which is defined by (11.4.19) can be written in the form [see (11.4.19)) and (11.4.17)]

$${}^{J}\mathscr{H}\{\varphi\} = J\cdot K \equiv \int \mathrm{d}^d r\ J(\vec{r})\cdot K(\vec{r})$$

where

$$K(\vec{r}) = \frac{1}{2}\sum_{j,l}[\varphi_{jl}(\vec{r})]^2. \tag{11.4.31}$$

${}^{\mathrm{I}}\mathscr{H}\{\varphi\}$ is given by (11.4.26) and consequently

$$\mathrm{e}^{-{}^{\mathrm{I}}\mathscr{H}\{\varphi\}} = \exp\left(-{}^{\mathrm{I}}\mathfrak{H}\left\{\frac{\partial}{\partial J}\right\}\right)\mathrm{e}^{J\cdot K}|_{J=0} = \mathrm{e}^{-{}^{\mathrm{I}}\mathfrak{H}\{K\}}.$$

In other words, taking definition (11.4.4) into account, we find

$$^{\mathrm{I}}\mathscr{H}\{\varphi\} = {}^{\mathrm{I}}\mathfrak{H}\{K\} = \sum_{p=2}^{\infty} \int \mathrm{d}^d r_1 \ldots \mathrm{d}^d r_p\, V(\vec{r}_1, \ldots, \vec{r}_p)\, K(\vec{r}_1) \ldots K(\vec{r}_p) \quad (11.4.32)$$

where $K(\vec{r})$ is given by (11.4.31).

The set of eqns (11.4.1), (11.4.4), and (11.4.32) defines the correspondence we were looking for, in the limit $\mathfrak{n} \to 0$.

4.4 Remark: limits for $\mathfrak{n} = 0$ and $\mathfrak{n} = -2$

In the preceding sections, we showed that the Landau–Ginzburg field theory corresponds to a polymer theory in the limit $\mathfrak{n} \to 0$. Incidentally, we observe that the limit $\mathfrak{n} \to -2$ is also very interesting. As was shown by Fisher[9] (and before him, by Balian and Toulouse[10] in a special case), for $N = 1$ and $\mathfrak{n} = -2$, the Green's functions of the system are the same as those of a non-interacting system. Thus, for $\mathfrak{n} = -2$, the exponents ν and γ [see (11.2.7)] take the classical values $\nu = 1/2$ and $\gamma = 1$. This result can be obtained by introducing an auxiliary source $J(\vec{r})$ as was done in Section 4.3. In particular, we observe that, in the general case, the partition function $\mathscr{Z}$ of the set of interacting fields is given by

$$\frac{\mathscr{Z}}{{}^{0}\mathscr{Z}} = \exp\left(-{}^{\mathrm{I}}\mathfrak{H}\left\{\frac{\partial}{\partial J}\right\}\right) \langle\!\langle \mathrm{e}^{-{}^{J}\mathscr{H}\{\varphi\}} \rangle\!\rangle_0 \big|_{J=0}.$$

But, for $N = 1$ and $\mathfrak{n} = -2$, we have

$$\langle\!\langle \mathrm{e}^{-{}^{J}\mathscr{H}\{\varphi\}} \rangle\!\rangle_0 = \det[\mathbf{1} - \boldsymbol{J}\boldsymbol{\mathscr{G}}_0(a_1)]$$

according to (11.4.23) and, taking (11.4.17) into account, we see immediately that the determinant is a linear function of each $J(\vec{r})$ (considered separately) and therefore a multilinear function of the set of the values taken by $J(\vec{r})$.

Thus, if ${}^{\mathrm{I}}\mathfrak{H}\{\mathscr{H}\}$ is of the form

$${}^{\mathrm{I}}\mathfrak{H}\{\mathscr{C}\} = \sum_{p=2}^{\infty} V_p \int \mathrm{d}^d r\, [\mathscr{C}(\vec{r})]^p$$

[a special case of (11.4.4)]

it is clear that

$$\frac{\mathscr{Z}}{{}^{0}\mathscr{Z}} = 1 \quad (\mathfrak{n} = -2), \qquad (11.4.33)$$

and we see that, in this particular case, the interactions do not play any role.

4.5 Fields associated with the standard continuous model

We established the correspondence between polymers and fields in the most general case. In practice, we shall work as most physicists do, within the framework of the standard continuous model defined in Chapter 10, Section 4. In the absence of interaction, we have [see (10.7.3)]

$$^{0}\overline{\mathscr{Z}}(\vec{k}, -\vec{k}; S) = \exp(-k^2 S/2).$$

The corresponding Green's function is given by (11.4.9)

$$^{0}\overline{\mathscr{G}}(\vec{k}, -\vec{k}; S) = \int_0^\infty \mathrm{d}a^{-aS} \exp(-k^2 S/2) = \frac{1}{a + k^2/2}. \qquad (11.4.34)$$

Thus, in the case without interaction, we set

$$^{0}\mathscr{W} = \exp[-{}^{0}\mathscr{H}\{\varphi\}]$$

$$^{0}\mathscr{H}\{\varphi\} = \frac{1}{2}\int \mathrm{d}^d r \sum_{j=1}^{N} \sum_{l=1}^{n} \left\{\frac{1}{2}[\vec{\nabla}\varphi_{jl}(\vec{r})]^2 + a_j[\varphi_{jl}(\vec{r})]^2\right\} \qquad (11.4.35)$$

in agreement with (10.4.18). In the same way, in the case with interaction, we set

$$\mathscr{W} = \exp[-{}^{0}\mathscr{H}\{\varphi\} - {}^{1}\mathscr{H}\{\varphi\}]$$

$$^{1}\mathscr{H}\{\varphi\} = \frac{b}{8}\,\mathrm{d}^d r \left\{\sum_{j=1}^{N} \sum_{l=1}^{n} [\varphi_{jl}(\vec{r})]^2\right\}^2 \qquad (11.4.36)$$

in agreement with (11.4.26) and (11.4.27). In particular, the Landau–Ginzburg model, which will be studied in detail later, corresponds to the case $N = 1$.

In all cases, in order to describe polymers, one has to take the limit $n \to 0$.

5. THE ONE-FIELD LANDAU–GINZBURG MODEL: APPLICATION IN POLYMER THEORY

5.1 Generalities

In the preceding sections, we associated N fields with N polymers; each field had n components, and we showed that a correspondence exists between partition functions and Green's functions in the limit $n \to 0$. Of course, to calculate critical exponents only one-field is sufficient because, in this case, only isolated polymers can be considered. However, it is also possible to find a correspondence between a one-field model and a polymer ensemble. This correspondence played a decisive role in its time, because it provided the means by which renormalization theory could be applied to polymer solutions, and it led to the discovery of new scaling laws. The correspondence can be established by using a lattice model, but here we shall follow the historical approach. Thus, we shall deal with a continuous model, more useful for practical applications, without caring too much about the problems concerning short-distance divergences.

Let us start from relation (11.4.1) defining the Laplace transforms of the partition functions, and let us consider the case where the a_j are equal for all j. In this case, we write

$$\mathscr{G}(\vec{r}_1, \ldots, \vec{r}_{2N}; a) =$$

$$\int_0^\infty \mathrm{d}S_1 \ldots \int_0^\infty \mathrm{d}S_N \exp[-a(S_1 + \ldots + S_N)]\, \mathscr{Z}(\vec{r}_1, \ldots, \vec{r}_{2N}; S_1, \ldots, S_N). \tag{11.5.1}$$

The fact that all a_j are equal introduces a new symmetry. Previously, we associated a field $\varphi_j(\vec{r})$ with each polymer j. Now the components $\varphi_{j,m}(\vec{r})$ of the N fields can be considered as the components of one field in a space with $n_0 = Nn$ dimensions. In particular, we may set

$$l = j + N(m-1)$$

$$j = 1, \ldots, N \quad m = 1, \ldots, n \quad l = 1, \ldots, n_0$$

$$\varphi_l(\vec{r}) = \varphi_{j,m}(\vec{r}).$$

For this one-field model 'ordered' Green's function are defined by the equality

$$\mathscr{G}(\vec{r}_1, \ldots, \vec{r}_{2N}; a) = \langle\langle [\varphi_1(\vec{r}_1)\varphi_1(\vec{r}_2)] \ldots [\varphi_N(\vec{r}_{2N-1})\varphi_N(\vec{r}_{2N})] \rangle\rangle. \tag{11.5.2}$$

Let us emphasize that these functions remain well-defined in the limit $n_0 = 0$ in spite of the fact that, a priori, the preceding mean values are defined only for $n_0 \geqslant N$ (see Fig. 11.2). It is by taking the limit $n^0 \to 0$ that we can describe, in an approximate but rather simple manner, the properties of polymer solutions.

The one-field model corresponding to the standard continuous representation of polymers is the Landau–Ginzburg model which we shall now review.

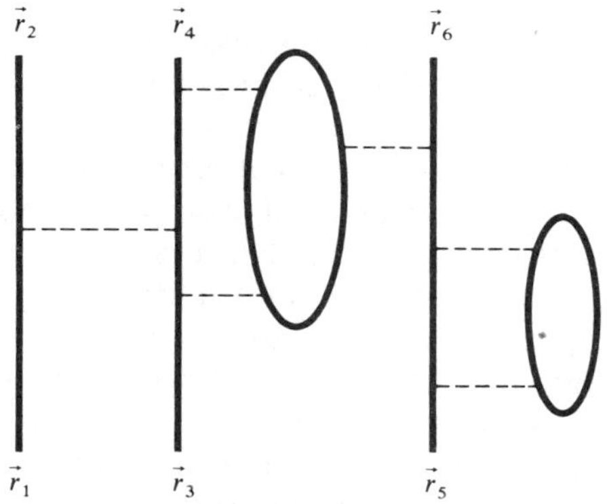

Fig. 11.2. Diagram contribution to $\mathscr{G}(r_1, \ldots, r_6; a)$ and containing two closed solid lines. As a consequence the contribution of the diagram is proportional to n^2.

5.2 Description of the Landau–Ginzburg model

5.2.1 Green's functions and diagrams

With each space point $\vec{r}$, let us associate the $\mathfrak{n}$ components $\varphi_j(\vec{r})$ of a random field $\vec{\varphi}$. The weight corresponding to a configuration of the field is [see (11.4.32) and (11.4.33)]

$$\mathscr{W}\{\varphi\} = \exp\left[-\int \mathrm{d}^d r\, \mathscr{L}(\vec{r})\right]$$

$$\mathscr{L}(\vec{r}) = \frac{1}{2}\sum_{j=1}^{\mathfrak{n}}\left[a\varphi_j^2(\vec{r}) + \frac{1}{2}\sum_{i=1}^{d}(\partial_i\varphi_j(\vec{r}))^2\right] + \frac{b}{8}\left[\sum_{j=1}^{\mathfrak{n}}\varphi_j^2(\vec{r})\right]^2 \qquad (11.5.3)$$

A short-distance cut-off is also introduced in order to ensure convergence. In polymer theory, we are interested in the limit $\mathfrak{n} \to 0$, but here we assume that $\mathfrak{n}$ is a positive integer.

The problem is to calculate the Green's functions

$$\mathscr{G}_{\mathrm{G}}(\vec{r}_1, \ldots, \vec{r}_p; j_1, \ldots, j_p; a) = \langle\!\langle \varphi_{j_1}(\vec{r}) \ldots \varphi_{j_p}(\vec{r}_p)\rangle\!\rangle. \qquad (11.5.4)$$

The index G indicates that here we are dealing with general Green's functions (i.e. connected or non-connected).

The generating function of the Green's functions is by definition a functional of an auxiliary field with $\mathfrak{n}$ components $B_j(\vec{r})$

$$\mathbf{G}_{\mathrm{G}}\{B\} = \left\langle\!\!\left\langle \exp\left[\int \mathrm{d}^d r \sum_j B_j(\vec{r})\,\varphi_j(\vec{r})\right]\right\rangle\!\!\right\rangle. \qquad (11.5.5)$$

The Green's functions can be expanded in powers of B and represented by diagrams. Connected diagrams correspond to connected Green's functions $\mathscr{G}(\vec{r}_1, \ldots, \vec{r}_p; j, \ldots, j_p; a)$ which are the cumulants* of the general Green's functions $\mathscr{G}_{\mathrm{G}}(\vec{r}, \ldots, \vec{r}_p; j_1, \ldots, j_p; a)$. The generating function of the connected Green's functions

$$\mathbf{G}\{B\} =$$

$$\sum_{p=1}^{\infty}\frac{1}{p!}\sum_{j_1\ldots j_p}\int \mathrm{d}^d r_1 \ldots \int \mathrm{d}^d r_p\, B_{j_1}(\vec{r}_1)\ldots B_{j_p}(\vec{r}_p)\mathscr{G}(\vec{r}_1, \ldots, \vec{r}_p; j_1, \ldots j_p; a) \qquad (11.5.6)$$

is given by†

$$\mathbf{G}\{B\} = \ln \mathbf{G}_{\mathrm{G}}\{B\}. \qquad (11.5.7)$$

All these quantities can easily be expressed in Fourier transforms. We set

$$\tilde{\varphi}_j(\vec{k}) = \int \mathrm{d}^d r\, \mathrm{e}^{\mathrm{i}\vec{k}\cdot\vec{r}}\,\varphi_j(\vec{r})$$

* This concept was already introduced in Chapter 10 for partition functions (see Chapter 10, Section 4.3.2).

† The vector with components B_j can be considered as a magnetic field.

which leads us to define the transforms $\bar{\mathscr{G}}(\vec{k}_1, \ldots, \vec{k}_p; j_1, \ldots, j_p; a)$ of the connected Green's functions. By definition, for large volumes

$$\langle\!\langle \tilde{\varphi}_{j_1}(\vec{k}_1) \ldots \tilde{\varphi}_{j_p}(\vec{k}_p) \rangle\!\rangle_c \sim V\bar{\mathscr{G}}\left(\vec{k}_1, \ldots, \vec{k}_{p-1}, -\sum_1^{p-1} \vec{k}_j; j_1, \ldots, j_p; a\right),$$

and in the infinite limit

$$\langle\!\langle \tilde{\varphi}_{j_1}(\vec{k}_1) \ldots \tilde{\varphi}_{j_p}(\vec{k}_p) \rangle\!\rangle_c = (2\pi)^d \delta(\vec{k}_1 + \ldots + \vec{k}_p) \bar{\mathscr{G}}(\vec{k}_1, \ldots, \vec{k}_p; j_1, \ldots, j_p; a) \quad (11.5.8)$$

[these definitions are similar to those that were given for the partition functions of polymers (see 10.4.49)].

Let us note that as a consequence of the invariance of the Landau–Ginzburg Lagrangian for the transformation $\vec{\varphi}(\vec{r}) \to -\varphi(\vec{r})$ for all $\vec{r}$, the non-vanishing Green's functions correspond to even values of p. The unperturbed one-body Green's function is

$$^0\mathscr{G}(\vec{k}) \equiv {}^0\bar{\mathscr{G}}(\vec{k}, -\vec{k}; j, j; a) = \frac{1}{a + k^2/2} \quad (11.5.9)$$

The diagrams representing the functions $\mathscr{G}(\vec{r}_1, \ldots, \vec{r}_p; j_1, \ldots, j_p; a)$ are nearly the same as the polymer diagrams but differ from them in two respects: (1) they may contain closed solid lines; (2) each open solid line and each closed solid line bears an index j. The indices of the open lines are determined by the indices $j_1, \ldots j_p$ of the corresponding Green's function $\mathscr{G}(\vec{r}_1, \ldots, \vec{r}_p; j_1, \ldots, j_p; a)$. Thus, let us consider a diagram contributing to $\mathscr{G}(\vec{r}_1, \ldots, \vec{r}_p; j_1, \ldots, j_p; a)$; if an open solid line joins the points of position vectors $\vec{r}_a$ and $\vec{r}_b$ on this diagram ($1 \leqslant a < b \leqslant p$), we must have $j_a = j_b$ and, in this case, the line under consideration bears an index $j = j_a = j_b$. In the same way, each closed solid line bears an index but this index can take any value from 1 to n. Let us note however, that the contributions of the diagrams do not explicitly depend on the indices [see for instance (11.5.9)].

The Green's functions $\mathscr{G}(\vec{r}_1, \ldots, \vec{r}_p; j_1, \ldots, j_p; a)$ are defined for integer values of n. In order to extend the concept of Green's function to non-integer values of n, it is convenient to introduce 'ordered' Green's functions[11] $\mathscr{G}(\vec{r}_1, \ldots \vec{r}_{2N}; a)$. For $N \leqslant n$, these functions can be defined by the equality

$$\mathscr{G}(\vec{r}_1, \ldots, \vec{r}_{2N}; a) = \mathscr{G}(\vec{r}_1, \ldots, \vec{r}_{2N}; 1,1,2,2, \ldots, N, N; a). \quad (11.5.10)$$

The diagrams corresponding to the 'ordered' Green's functions look very much like polymer diagrams (see Chapter 10, Section 4); the only difference comes from the fact that they may include closed solid lines (see Fig. 11.2).

The contributions of the diagrams giving the Fourier transforms $\bar{\mathscr{G}}(\vec{k}_1, \ldots k_{2N}; a)$ of the ordered Green's functions (here $\vec{k}_1 + \ldots \vec{k}_{2N} = 0$), are calculated according to the following rules

1. Each solid-line segment and each interaction line bears a wave vector. The wave vectors $\vec{k}_1$ and $\vec{k}_2$ are injected at the end points of the first open solid

line, $\vec{k}_3$ and $\vec{k}_4$ at the end points of the second one, and so on. At each interaction point, the flow of wave vectors is conserved.

2. A factor $1/(a + k^2/2)$ is associated with each solid line segment bearing a wave vector $\vec{k}$.
3. A factor $-b$ is associated with each interaction line.
4. A factor $\mathfrak{n}$ is associated with each closed solid line.
5. The product of all factors is performed.
6. The wave vectors appearing on the diagram are expressed in terms of independent vectors; the number of these independent vectors is called the number of loops. One integrates with respect to all these vectors.

It should be noted that in order to ensure convergence, one has to introduce a cut-off k_0 at large wave vectors.

More generally, the ordered Green's functions are defined by diagrams (see Fig. 11.2) in which the open solid lines always connect $\vec{r}_1$ to $\vec{r}_2$, $\vec{r}_3$ to $\vec{r}_4, \ldots \vec{r}_{2N-1}$ to $\vec{r}_{2N}$. Thus, for $N \leqslant \mathfrak{n}$ the calculation does not depend on the indices but only on $\mathfrak{n}$. It is therefore trivial to generalize the ordered Green's function to the case $N \geqslant \mathfrak{n}$, by giving a diagrammatic definition of these functions.

Simpler than the ordinary Green's functions, the ordered Green's functions are also more general, since the usual Green's functions can be deduced from them: in fact, in all cases, we can write

$$\mathscr{G}(\vec{r}_1, \ldots, \vec{r}_{2N}; j_1, \ldots, j_{2N}; a) =$$

$$\frac{1}{2^N N!} \sum_{\wp} \delta_{j_{\wp 1}, j_{\wp 2}} \ldots \delta_{j_{\wp(2N-1)}, j_{\wp(2N)}} \mathscr{G}(\vec{r}_{\wp 1}, \ldots, \vec{r}_{\wp(2N)}; a) \qquad (11.5.11)$$

where $\wp$ is a permutation of the numbers $1, \ldots, 2N$ and $\wp_j$ the transform of j in this permutation. Thus, the dependence of the Green's functions with respect to the indices can be eliminated, since the functions $\mathscr{G}(\vec{r}_1, \ldots, \vec{r}_{2N}; a)$ depend only on their number $\mathfrak{n}$. The same is true for $\mathbf{G}\{B\}$, which can now be written in the form [see (11.5.6)] and (11.5.10)]

$$\mathbf{G}\{B\} =$$

$$\sum_p \frac{1}{2^p p!} \int \mathrm{d}^d r_1 \ldots \int \mathrm{d}^d r_p [\vec{B}(\vec{r}_1)\cdot\vec{B}(\vec{r}_2)] \ldots [\vec{B}(\vec{r}_{2p-1})\cdot\vec{B}(\vec{r}_{2p})] \mathscr{G}(\vec{r}_1, \ldots, \vec{r}_{2p}; a). \qquad (11.5.12)$$

5.2.2 Vertex functions

Perturbation and renormalization calculations can be performed without taking care of any possible reducibility of the diagrams; still it is convenient and usual to take the P-reducibility* into account. A diagram is P-reducible if by cutting a solid line (between two interaction points), it is possible to separate the diagram into two disconnected parts; in the opposite case, it is P-irreducible. The P-*irreducible diagrams deprived of their external legs* define vertices and the corresponding contributions are vertex functions.

The generating function of the functions $\boldsymbol{\Gamma}(\vec{r}_1, \ldots, \vec{r}_p; j_1, \ldots, j_p; a)$ is denoted by $\boldsymbol{\Gamma}\{M\}$

$$\boldsymbol{\Gamma}\{M\} =$$

$$\sum_{p=1}^{\infty} \frac{1}{p!} \sum_{j_1, \ldots, j_p} \mathrm{d}^d r_1 \ldots \mathrm{d}^d r_p M_{j1}(\vec{r}_1) \ldots M_{jp}(\vec{r}_p) \boldsymbol{\Gamma}(\vec{r}_1, \ldots, \vec{r}_p; j_1, \ldots, j_p; a). \tag{11.5.13}$$

In order to define the vertex functions more precisely we write that $\mathbf{G}\{B\}$ and $\boldsymbol{\Gamma}\{M\}$ are Laplace transforms of each other (see Appendix I)

$$\mathbf{G}\{B\} + \boldsymbol{\Gamma}\{M\} - \sum_{j=1}^{n} \int \mathrm{d}^d r\, B_j(\vec{r}) M_j(\vec{r}) = 0$$

$$M_j(\vec{r}) = \frac{\partial \mathbf{G}\{B\}}{\partial B_j(\vec{r})}$$

$$B_j(\vec{r}) = \frac{\partial \boldsymbol{\Gamma}\{M\}}{\partial M_j(\vec{r})}. \tag{11.5.14}$$

This transformation shows that the reducible diagrams contributing to Green's functions have a tree structure and that the vertices of these trees are P-irreducible diagrams which define the vertex functions (see Appendix I).

Of course, we can introduce not only vertex functions $\boldsymbol{\Gamma}(\vec{r}_1, \ldots, \vec{r}_{2N}; j_1, \ldots, j_{2N}; a)$ but also 'ordered' vertex functions $\boldsymbol{\Gamma}(\vec{r}_1, \ldots, r_{2N}; a)$ in terms of which the vertex functions can be directly expressed [the formula is analogous to (11.5.10)]. In particular, from eqn (11.5.4) we deduce (see also Appendix I)

$$\bar{\boldsymbol{\Gamma}}(\vec{k}, -\vec{k}; a) = [\mathscr{G}(\vec{k}, -\vec{k}; a)]^{-1} = [{}^0\mathscr{G}(\vec{k}, -\vec{k}; a)]^{-1} - \boldsymbol{\Sigma}(\vec{k}, -\vec{k}; a)$$

$$\bar{\boldsymbol{\Gamma}}(\vec{k}_1, \ldots, \vec{k}_4; a) = -\mathscr{G}(\vec{k}_1, \ldots, \vec{k}_4; a) \prod_{i=1}^{4} [\mathscr{G}(\vec{k}_i, -\vec{k}_i; a)]^{-1} \tag{11.5.15}$$

where $\boldsymbol{\Sigma}$ is the 'self-energy', and the calculation of these functions can be performed directly.

5.2.3 Dimensionalities

Equations (11.5.3) show that the dimensionality of the Lagrangian $\mathscr{L}(\vec{r})$ is given by

$$\mathscr{L}(\vec{r}) \sim L^{-d} \qquad (L = \text{length}).$$

* The P-reducibility is defined with respect to polymer lines, the I-reducibility with respect to interaction lines.

From it, we can deduce the dimensionalities of $\varphi(\vec{r})$ and of the constants a and b

$$\begin{aligned} \varphi_j(\vec{r}) &\sim L^{1-d/2} \\ a &\sim L^{-2} \\ b &\sim L^{d-4} \end{aligned} \tag{11.5.16}$$

The fact that $b \sim L^{d-4}$ indicates that the value $d = 4$ is 'marginal'. Indeed, all expansions are expressed in terms of a dimensionless constant. Here, this constant may be $b(a - a_c)^{-2+d/2}$ where a_c is the critical value of a. We see that for $d < 4$, this parameter becomes infinite if $a \to a_c$. This is the origin of 'infra-red divergences' related to the existence of non-classical exponents.

It is also easy to find out the dimensionalities of Green's functions and vertex functions

$$\begin{aligned} \bar{\mathscr{G}}(\vec{k}_1, \ldots, \vec{k}_{2N}; a) &\sim L^{2+(N-1)(d+2)} \\ \bar{\Gamma}(\vec{k}_1, \ldots, \vec{k}_{2N}; a) &\sim L^{-2+(N-1)(d-2)} \end{aligned} \tag{11.5.17}$$

(the difference comes from the contribution of the external legs).

5.2.4 Model with external magnetic field

The Landau–Ginzburg model defined by (11.5.3) pictures a magnetic system in the absence of any external interaction. It can be completed by introducing a coupling between the internal field $\vec{\varphi}$ and an external magnetic field B_0 directed along the first component of the field $\vec{\varphi}$. This amounts to introducing the weight

$${}^B\mathscr{W}\{\varphi\} = \exp\left[-\int \mathrm{d}^d r \; {}^B\mathscr{L}(\vec{r})\right] \tag{11.5.18}$$

with

$${}^B\mathscr{L}(\vec{r}) = \mathscr{L}(\vec{r}) + B_0 \int \mathrm{d}^d r \; \varphi_1(\vec{r}) \tag{11.5.19}$$

where $\mathscr{L}(\vec{r})$ is given by (11.5.3).

'General' Green's functions and their generating function ${}^B\mathbf{G}_\mathrm{G}\{B\}$ can be associated with this new weight ${}^B\mathscr{W}\{\varphi\}$. Let us set

$${}^0B_j(\vec{r}) = B_0 \delta_{j1}.$$

Then, definition (11.5.5) shows that

$${}^B\mathbf{G}_\mathrm{G}\{B\} = \mathbf{G}_\mathrm{G}\{B + {}^0B\}. \tag{11.5.20}$$

As always, the connected Green's functions are the cumulants of the Green's functions, and consequently their generating function ${}^B\mathbf{G}\{B\}$ is given by

$${}^B\mathbf{G}\{B\} = \ln \mathbf{G}_\mathrm{G}\{B + {}^0B\} = \mathbf{G}\{B + {}^0B\}. \tag{11.5.21}$$

In particular, we have

$${}^B\mathbf{G}\{o\} = \ln \mathbf{G}_\mathrm{G}\{{}^0B\} = \mathbf{G}\{{}^0B\} = V\mathbf{G}(B_0) \tag{11.5.22}$$

where $\mathbf{G}(B_0)$ remains finite when $V \to \infty$. The connected Green's functions associated with ${}^{B}\mathscr{W}\{\varphi\}$ can thus be very easily deduced from the connected Green's functions associated with $\mathscr{W}\{\varphi\}$.

In particular, there exists two two-leg Green's functions, the longitudinal Green's function (in the direction of the magnetic field)

$$\mathscr{G}_{\mathrm{L}}(\vec{x}, \vec{y}; a) = \langle\!\langle \varphi_1(\vec{x})\varphi_1(\vec{y}) \rangle\!\rangle_{B,\,\mathrm{conn.}} \tag{11.5.23}$$

and the transverse Green's function

$$\mathscr{G}_{\mathrm{T}}(\vec{x}, \vec{y}; a) = \langle\!\langle \varphi_2(\vec{x})\varphi_2(\vec{y}) \rangle\!\rangle_{B,\,\mathrm{conn.}} \tag{11.5.24}$$

(see Fig. 11.3).

Incidentally, we note that these functions are important in polymer theory: in fact, the field-polymer correspondence and Fig. 11.3 show that the transverse Green's function is related only to intrachain correlations, and the longitudinal Green's function both to interchain *and* intrachain correlations.

Let us now consider the vertex functions, and their generating function which we shall determine. First, we can introduce the Legendre transform $\Gamma(M_0)$ of the function ${}^{B}\mathbf{G}(B_0)$ defined by

$$\begin{aligned} \mathbf{G}(B_0) + \Gamma(M_0) - M_0 B_0 &= 0 \\ M_0 &= \frac{\partial}{\partial B_0}\mathbf{G}(B_0) \\ B_0 &= \frac{\partial}{\partial M_0}\Gamma(M_0). \end{aligned} \tag{11.5.25}$$

This transformation is a special case of (11.5.14). It corresponds to the choice

$$\begin{aligned} {}^{0}B_j(\vec{r}) &= B_0 \delta_{j1} \\ {}^{0}M_j(\vec{r}) &= M_0 \delta_{j1}. \end{aligned}$$

Moreover, a simple change of variables enables us to write (11.5.14) in the form

$$\mathbf{G}\{B + {}^{0}B\} + \Gamma\{M + {}^{0}M\} - \sum_j \int \mathrm{d}^d r\, [B_j(\vec{r}) + \delta_{ji} B_0]\,[M_j(\vec{r}) + \delta_{ji} M_0] = 0$$

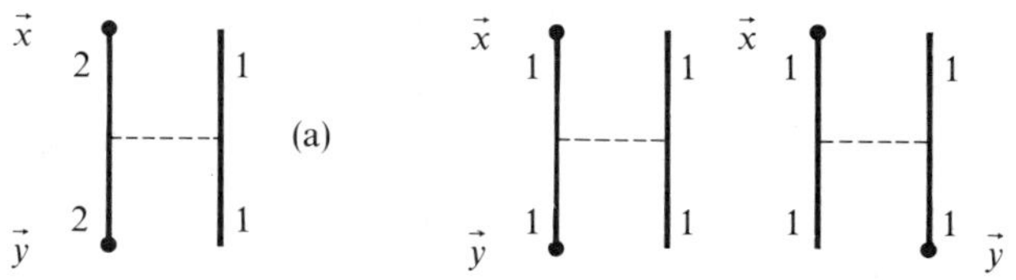

Fig. 11.3. (a) Diagram contributing to $\mathscr{G}_{\mathrm{T}}(\vec{x},\vec{y};\ a)$; (b) Diagrams contributing to $\mathscr{G}_{\mathrm{L}}(\vec{x}, \vec{y}; a)$. The numbers on the solid lines refer to the components of the internal field.

$$M_j(\vec{r}) + \delta_{j1} M_0 = \frac{\partial \mathbf{G}\{B + {}^0B\}}{\partial B_j(\vec{r})}$$

$$B_j(\vec{r}) + \delta_{j1} B_0 = \frac{\partial \mathbf{\Gamma}\{M + {}^0M\}}{\partial M_j(\vec{r})}. \tag{11.5.26}$$

Now, let us set

$$\mathbf{G}^{\perp}\{B\} = \mathbf{G}\{B + {}^0B\} - V\mathbf{G}(B_0) - \frac{\partial \mathbf{G}(B_0)}{\partial B_0} \int \mathrm{d}^d r\, B_1(\vec{r})$$

$$\mathbf{\Gamma}^{\perp}\{M\} = \mathbf{\Gamma}\{M + {}^0M\} - V\mathbf{\Gamma}(M_0) - \frac{\partial \mathbf{\Gamma}(M_0)}{\partial M_0} \int \mathrm{d}^d r\, M_1(\vec{r}). \tag{11.5.27}$$

By combining (11.5.26) and (11.5.25) we see that $\mathbf{G}^{\perp}\{B\}$ and $\mathbf{\Gamma}^{\perp}\{M\}$ obey the relations

$$\mathbf{G}^{\perp}\{B\} + \mathbf{\Gamma}^{\perp}\{M\} - \sum_j \int \mathrm{d}^d r B_j(\vec{r}) M_j(\vec{r}) = 0$$

$$M_j(\vec{r}) = \frac{\partial \mathbf{G}^{\perp}\{B\}}{\partial B_j(\vec{r})}$$

$$B_j(\vec{r}) = \frac{\partial \mathbf{\Gamma}^{\perp}\{M\}}{\partial M_j(\vec{r})}. \tag{11.5.28}$$

$\mathbf{\Gamma}^{\perp}\{M\}$ is the Laplace transform of $\mathbf{G}^{\perp}\{B\}$; this fact shows that $\mathbf{G}^{\perp}\{B\}$ must be considered as the genuine generating function of the Green's functions and not ${}^B\mathbf{G}\{B\} = \mathbf{G}\{B + B_0\}$. In fact, the diagrams contributing to ${}^B\mathbf{G}^{\perp}$ have a tree structure. The weights associated with the vertices of the tree are the vertex functions; therefore it is $\mathbf{\Gamma}^{\perp}\{M\}$ which must be considered as the true generating function of the vertex functions. However, we note that the difference between $\mathbf{\Gamma}^{\perp}\{M\}$ and $\mathbf{\Gamma}\{M + {}^0M\}$ contains only terms which are linear with respect to $M_1(\vec{r})$ [see (11.5.27)].

5.3 Equilibrium ensembles and Landau–Ginzburg model

5.3.1 Osmotic pressure and generating function of the Green's functions

The one-field model gives a description of the approximate grand canonical ensembles introduced in Chapter 11, Section 5. These ensembles, called 'equilibrium ensembles', depend on only two fugacities, which determine respectively the average number of polymers and the average number of monomers (constituting the polymers). In the continuous case, a connected partition function $\mathcal{Q}(f, a)$ can be associated with this ensemble; it is defined by

$$\mathcal{Q}(f, a) = \sum_{N=1}^{\infty} \frac{f^N}{N!} \int_0^{\infty} \mathrm{d}S_1 \ldots \mathrm{d}S_N \exp\left[-a(S_1 + \ldots + S_N)\right] \mathscr{Z}(S_1, \ldots, S_N). \tag{11.5.29}$$

Here f and e^{-a} are the two fugacities determining the ensemble. This equation is analogous to (11.5.1) (since $S \propto N$ and $e^{-a} \propto ¢$). When $\mathcal{Q}(f, a)$ is known, the osmotic pressure $\boldsymbol{\Pi}$ can be expressed, in parametric form, as a function of the average number $\boldsymbol{C}$ of polymers per unit volume and of the average total area $\mathscr{C}$ per unit volume ($\mathscr{C}$ can be considered as proportional to the monomer concentration).

$$\begin{aligned} \boldsymbol{\Pi}\boldsymbol{\beta} &= \mathcal{Q}(f, a) \\ \boldsymbol{C} &= f\frac{\partial}{\partial f}\mathcal{Q}(f, a) \\ \mathscr{C} &= -\frac{\partial}{\partial a}\mathcal{Q}(f, a) \end{aligned} \tag{11.5.30}$$

The function $\mathcal{Q}(f, a)$ has a simple interpretation in field theory. In fact, by Fourier transformation, (11.5.1) becomes

$$\bar{\mathscr{G}}(\vec{k}_1, \ldots, \vec{k}_{2N}; a) = \int_0^\infty \mathrm{d}S_1 \ldots \mathrm{d}S_N \exp\left[-a(S_1 + \ldots + S_N)\right] \tilde{\mathscr{Z}}(\vec{k}_1, \ldots, \vec{k}_{2N}; S_1, \ldots, S_N),$$

and we know that

$$\tilde{\mathscr{Z}}(\vec{0}, \ldots, \vec{0}; S_1, \ldots, S_N) = \mathscr{Z}(S_1, \ldots, S_N).$$

Thus, we may set

$$\bar{\mathscr{G}}(N\times\vec{0}; a) = \bar{\mathscr{G}}(\vec{k}_1, \ldots, \vec{k}_N; a)|_{\vec{k}_1 = \ldots = \vec{k}_N = 0}, \tag{11.5.31}$$

and in an analogous way

$$\bar{\mathscr{G}}_1(N\times\vec{0}; N\times 1; a) = \bar{\mathscr{G}}(\vec{k}_1, \ldots, \vec{k}_N; 1, \ldots, 1; a)|_{\vec{k}_1 = \ldots = \vec{k}_N = 0}. \tag{11.5.32}$$

Taking the first definition into account, we can write

$$\bar{\mathscr{G}}(2N\times\vec{0}; a) = \int \mathrm{d}S_1 \ldots \mathrm{d}S_N \exp\left[-a(S_1 + \ldots + S_N)\right] \mathscr{Z}(S_1, \ldots, S_N). \tag{11.5.33}$$

By bringing this expression in eqn (11.5.29), which defines $\mathcal{Q}(f, a)$, we find

$$\mathcal{Q}(f, a) = \sum_{N=1}^{\infty} \frac{f^N}{N!} \bar{\mathscr{G}}(2N\times\vec{0}; a) \tag{11.5.34}$$

Moreover, as we can directly see on diagrams, the true Green's function is related to the ordered Green's function by the equality

$$\mathscr{G}(2N\times\vec{0}; 2N\times 1; a) = \frac{(2N)!}{N!2^N} \bar{\mathscr{G}}(2N\times\vec{0}; a), \tag{11.5.35}$$

and by bringing this expression in (11.5.4) we find

$$\mathscr{Q}(f, a) = \sum_{N=1}^{\infty} \frac{(2f)^N}{(2N)!} \bar{\mathscr{G}}(2N\times\vec{0};\, 2N\times 1;\, a) \tag{11.5.36}$$

or

$$\mathscr{Q}(f, a) = \sum_{p=1}^{\infty} \frac{(2f)^{p/2}}{p!} \bar{\mathscr{G}}(p\times\vec{0};\, p\times 1;\, a),$$

the terms of uneven order being zero in the latter equation.

Finally, taking the definition of $\bar{\mathscr{G}}(p\times\vec{0};\, p\times 1;\, a)$ into account see (11.5.32)] and (11.5.4), we find

$$\mathscr{Q}(f, a) = \frac{1}{V} \left\langle\!\!\left\langle \exp\left[(2f)^{1/2} \int \mathrm{d}^d r\, \varphi_1(\vec{r}) \right] \right\rangle\!\!\right\rangle_{\text{conn.}} \tag{11.5.37}$$

the mean value being taken with the weight (11.5.3). Moreover, we know that the generating function of the general Green's functions, and the generating function of the connected Green's functions, is given by

$$\mathbf{G}_{\mathrm{G}}\{B\} = \left\langle\!\!\left\langle \exp\left[\sum_j \int \mathrm{d}^d r\, B_j(\vec{r})\varphi_j(\vec{r}) \right] \right\rangle\!\!\right\rangle$$

$$\mathbf{G}\{B\} = \ln \mathbf{G}_{\mathrm{G}}\{B\} = \left\langle\!\!\left\langle \exp\left[\sum_j \int \mathrm{d}^d r\, Bj(\vec{r})\varphi_j(\vec{r}) \right] \right\rangle\!\!\right\rangle \tag{11.5.38}$$

In particular, for $B_j(\vec{r}) = B\delta_{1j}$, we can write

$$\mathbf{G}\{B\} = V\,\mathbf{G}(B).$$

Comparing (11.5.37) and (11.5.38), we find

$$\mathscr{Q}(f, a) = \mathbf{G}((2f)^{1/2})$$

where $\mathbf{G}(B)$ can be defined by

$$\mathbf{G}(B) = \sum_{N=1}^{\infty} \frac{B^{2N}}{(2N)!} \bar{\mathscr{G}}(2N\times\vec{0};\, 2N\times 1;\, a)$$

$$= \sum_{N=1}^{\infty} \frac{B^{2N}}{N!\,2^N} \bar{\mathscr{G}}(2N\times\vec{0};\, a) \tag{11.5.39}$$

The osmotic pressure is thus given by the generating function of the Green's functions $\mathbf{G}(B)$

$$\Pi\beta = \mathbf{G}(B)$$

$$C = \frac{B}{2}\frac{\partial}{\partial B}\mathbf{G}(B)$$

$$\mathscr{C} = -\frac{\partial}{\partial a}\mathbf{G}(B). \tag{11.5.40}$$

5.3.2 Osmotic pressure and generating function of the vertex functions

As was previously shown (Section 5.2.2), the Green's functions can be expressed in terms of simpler functions called vertex functions. The generating function $\Gamma\{M\}$ of these vertex functions is a functional of the 'magnetization' $M_j(\vec{r})$, and is related to $\mathbf{G}\{B\}$ by a Legendre transformation (11.5.14).

We saw above that the osmotic pressure can be expressed in terms of the function $\mathbf{G}(B)$ associated with $\mathbf{G}\{B\}$. In the same way for

$$M_j(\vec{r}) = M\delta_{1j}$$

we set

$$\Gamma\{M\} = V\,\Gamma(M). \tag{11.5.41}$$

The functions $\mathbf{G}(B)$ and $\Gamma(M)$ are connected to each other by a Legendre transformation (11.5.25)

$$\mathbf{G}(B) + \Gamma(M) - MB = 0$$

$$M = \frac{\partial}{\partial B}\mathbf{G}(B)$$

$$B = \frac{\partial}{\partial M}\Gamma(M). \tag{11.5.42}$$

Consequently, by combining (11.5.40) and (11.5.42), we can express the osmotic pressure and the concentrations, in terms of the parameter M. Thus, we obtain (des Cloizeaux 1975).[11]

$$\Pi\beta = M\frac{\partial\Gamma(M)}{\partial M} - \Gamma(M)$$

$$C = \frac{1}{2}M\frac{\partial\Gamma(M)}{\partial M}$$

$$\mathscr{C} = \frac{\partial\Gamma(M)}{\partial a}. \tag{11.5.43}$$

5.3.3 Calculation of the osmotic pressure for an equilibrium ensemble in the zero-loop approximation

The osmotic pressure is obtained by calculating $\Gamma(M)$. To lowest order in b, the function $\mathbf{G}(B)$ associated with the Lagrangian (11.5.3) is given by

$$\mathbf{G}(B) = \frac{B^2}{2a} - \frac{b}{8a^4}B^4 + \ldots$$

according to (11.5.42). By Legendre transformation (11.5.42), we thus find

$$\Gamma(M) = \frac{a}{2}M^2 + \frac{b}{8}M^4 + \ldots .$$

As $\Gamma(M)$ is the generating function associated with the P-irreducible diagrams, we see that to zero-loop order, we have

$$\Gamma(M) = {}^0\Gamma(M) = \frac{a}{2} M^2 + \frac{b}{8} M^4. \tag{11.5.44}$$

We note that this result can also be (very directly) deduced from (11.5.15).

Let us bring the preceding expression in (11.5.43); we find

$$\begin{aligned} \Pi\beta &= \frac{a}{2} M^2 + \frac{3b}{8} M^4 \\ C &= \frac{a}{2} M^2 + \frac{b}{4} M^4 \\ \mathscr{C} &= \frac{M^2}{2} \end{aligned} \tag{11.5.45}$$

and by eliminating M

$$\Pi\beta = C + \frac{b}{2}\mathscr{C}^2. \tag{11.5.46}$$

In practice, this formula coincides with (10.6.32), but (10.6.32) corresponds to a monodisperse ensemble. Thus, to that order, polydispersion effects are not felt.

5.3.4 One-loop calculation of the osmotic pressure for the equilibrium ensemble

The function $\Gamma(M)$ can be written in a general form analogous to (11.5.33)

$$\Gamma(M) = \sum_1^\infty \frac{M^{2N}}{N!\,2^N} \bar{\Gamma}(2N \times \vec{0};\, a). \tag{11.5.47}$$

Let ${}^1\Gamma(M)$ be the contribution of the one-loop diagrams. Each term ${}^1\bar{\Gamma}(2N \times \vec{0};\, a)$ is represented by a series of diagrams whose contribution is calculated according to the rules formulated in this chapter, Sections 5.2.1 and 5.2.2. For each N, there exists a bare diagram (of lowest order) and dressed diagrams. The bare diagrams are represented (for the first values of N), in Fig. 11.4. Dressed diagrams are represented in Fig. 11.5. To sum up the contributions of these diagrams, we have only to start from the diagram of lowest order and, for each polymer segment, to replace the propagator $1/(a + q^2/2)$ by the propagator $1/(a + bM^2/2 + q^2/2)$. Thus, we find

$${}^1\bar{\Gamma}(2N \times \vec{0};\, a) = -(N-1)!\; 2^{N-1} \frac{(-b)^N}{(2\pi)^d} \int \mathrm{d}^d q \left[\frac{1}{a + \frac{b}{2} M^2 + \frac{q^2}{2}} \right]^N. \tag{11.5.48}$$

The factor $(N-1)!2^{N-1}$ is obtained by labelling each polymer line and by

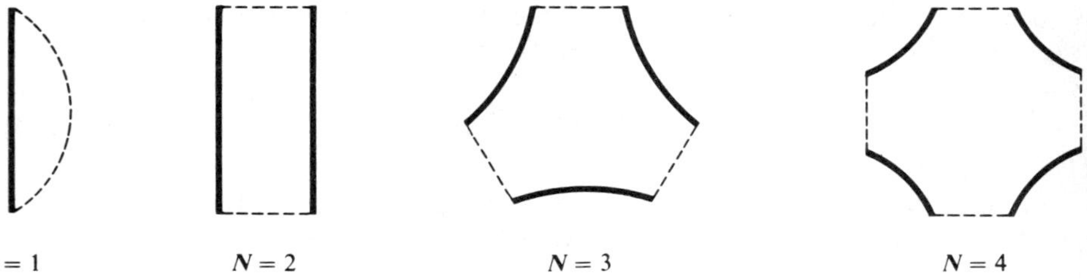

Fig. 11.4. Bare one-loop diagrams contributing to ${}^{1}\bar{\Gamma}(2N\times\vec{0};\,a)$ for various values of N. The contributions are calculated by labelling the polymer lines with the numbers $1,2,\ldots,N$ and by marking the origin and the end point of each polymer line.

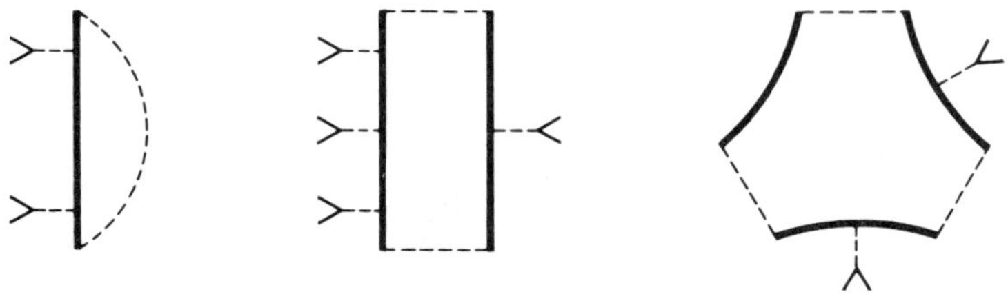

Fig. 11.5. Dressed one-loop diagrams contributing to $\bar{\Gamma}(2N\times\vec{0};\,a)$.

marking its origin and its end point. Incidentally, we note that the same calculation could be made by using non-labelled, non-marked diagrams. In this case, the symmetry factor $1/N!2^N$ would not appear in (11.5.47); however, in (11.5.48) the factor $(N-1)2^{N-1}$ of the labelled diagrams would be replaced by the symmetry factor $1/2N$ of the non-labelled diagrams. The same ${}^{1}\Gamma(M)$ would be obtained.

Let us now observe that (11.5.48) is valid only for $N > 1$. For $d = 3$ and $N = 1$, the integral diverges when $q \to \infty$. This 'ultraviolet' divergence can be suppressed by introducing a cut-off for $|\vec{q}| > K_0$. This is a 'regularization', but, in this case, ${}^{1}\Gamma(2\times\vec{0};\,a)$ depends on K_0. This dependence can be absorbed by renormalization. In fact, let us consider the Lagrangian $\mathscr{L}(\vec{r})$ defined by (11.5.3). Instead of taking

$$\mathscr{L}_0(\vec{r}) = \frac{1}{2}\sum_{j=1}^{n}\left[a\varphi_j^2(\vec{r}) + \tfrac{1}{2}\sum_{i=1}^{d}(\partial_i\varphi_j(\vec{r}))^2\right]$$

as the unperturbed Lagrangian, we can use

$$\mathscr{L}'_0(\vec{r}) = \mathscr{L}_0(\vec{r}) - \frac{1}{2}a_c\sum_{j=1}^{n}\varphi_j^2(\vec{r}),$$

and transfer $\frac{1}{2}a_c\sum_j \varphi^2(\vec{r})$ to the perturbation in the form of a counter-term. To one-loop order, we may set

$$a_c = \int_{|q|<K_0} d^d q \, \frac{1}{(q^2/2)}.$$

Consequently,

$$^1\bar{\Gamma}(2 \times \vec{0}; a) = \frac{1}{(2\pi)^d} \int_{|q|<K_0} d^d q \left[\frac{1}{a - a_c + \frac{b}{2} M^2 + \frac{q^2}{2}} - \frac{1}{\frac{q^2}{2}} \right].$$

It is now possible to take the limit $K_0 \to \infty$ and to one-loop order, we obtain the result

$$^1\bar{\Gamma}(2N \times \vec{0}; a) = - (N-1)!\, 2^{N-1} \frac{(-b)^N}{(2\pi)^d} \int d^d q \left\{ \left[\frac{1}{a - a_c + \frac{b}{2} M^2 + \frac{q^2}{2}} \right]^N - \frac{\delta_{1,N}}{\frac{q^2}{2}} \right\}. \quad (11.5.49)$$

In the same way $^0\Gamma(M)$ is modified since the unperturbed Lagrangian is modified: now it must read

$$^0\Gamma(M) = \frac{(a - a_c)}{2} M^2 + \frac{b}{8} M^4, \quad (11.5.50)$$

and, moreover, (11.5.47) and (11.5.49) give

$$^1\Gamma(M) = \frac{1}{2(2\pi)^d} \int d^d q \left[\ln \left(1 + \frac{bM^2}{a - a_c + \frac{b}{2} M^2 + \frac{q^2}{2}} \right) - \frac{bM^2}{\frac{q^2}{2}} \right].$$

This integral can be calculated without difficulty, and we obtain

$$^1\Gamma(M) = -I_d \left[\left(a - a_c + \frac{3b}{2} M^2 \right)^{d/2} - \left(a - a_c + \frac{b}{2} M^2 \right)^{d/2} \right] \quad (11.5.51)$$

where

$$I_d = \frac{2\bar{\Gamma}(2 - d/2)}{(2\pi)^{d/2} d(d-2)} \left(I_3 = \frac{2^{-1/2}}{3\pi} \simeq 0.0750 \right).$$

Thus, to one-loop order, we have

$$\Gamma(M) = {}^0\Gamma(M) + {}^1\Gamma(M) \quad (11.5.52)$$

where $^0\Gamma(M)$ and $^1\Gamma(M)$ are given by (11.5.50) and (11.5.51) (a result published by Moore in 1977).[12]

Finally, by bringing (11.5.52) in eqns (11.5.43) which give the osmotic pressure in terms of the concentrations, and by setting

$$a - a_c = t$$

$$M^2/2 = m$$

we find, for $d = 3$, the following representation:

$$\Pi\beta = tm + \frac{3}{2}bm^2 - \frac{2^{-1/2}}{\pi}bm[3(t + 3bm)^{1/2} - (t + bm)^{1/2}]$$
$$+ \frac{2^{-1/2}}{3\pi}[(t + 3bm)^{3/2} - (t + bm)^{3/2}]$$
$$C = tm + bm^2 - \frac{2^{-3/2}}{\pi}bm[3(t + 3bm)^{1/2} - (t + bm)^{1/2}]$$
$$\mathscr{C} = m - \frac{2^{-3/2}}{\pi}[(t + 3bm)^{1/2} - (t + bm)^{1/2}]. \qquad (11.5.53)$$

These formulae have the merit of expressing Π, C, and $\mathscr{C}$ in simple algebraic form as functions of the parameters t, b, and m. However, they have two defects. They describe ensembles whose polydispersion varies in an uncontrolled manner, whereas, to the same order of perturbation, the calculation can be made for a monodisperse ensemble, as was shown in Chapter 10, Section 8.4 [see (10.8.27) and (10.8.28)].

Moreover, the preceding formulae are valid only for short chains or weak interactions. In order to explore the behaviour of long chains, we have to use renormalization theories, as will be explained in Chapter 12.

6. HAMILTONIAN CIRCUITS AND FIELD MODELS

Hamiltonian circuits are a mathematical idealization of polymer melts. General considerations concerning these circuits are given here. Exact results, in two dimensions will be found at the end of the next chapter.

6.1 Field representation of non-intersecting circuits covering a lattice

The correspondence established in the present chapter between chain theory and field theory can be generalized (Orland, Itzykson, and De Dominicis 1985)[13] in order to count configurations of self-avoiding chains drawn on a lattice so as to cover it completely. The lattice under consideration consists of M sites and of bonds between neighbouring sites. In the following, we consider only self-avoiding circuits (closed chains) covering the lattice, but generalizations are possible.[14] The symbol Z_N will denote the number of configurations of N self-avoiding circuits covering the lattice. Here, we are especially interested in Z_1 which is the number of Hamiltonian circuits on the lattice (see Chapter 3, Section 2.1). With each lattice site A, we associate a vector $\vec{\varphi}_A$ with $\mathfrak{n}$ components $\varphi_{A\alpha}$ ($\alpha = 1, \ldots, \mathfrak{n}$) and we postulate that the components of all these vectors constitute a set of Gaussian variables (see Appendix A); we set

$$\langle\!\langle \varphi_{A\alpha}\varphi_{B\beta} \rangle\!\rangle = \delta_{\alpha\beta} H_{AB} \tag{11.6.1}$$

where H_{AB} are matrix elements of an operator H. Equations (11.6.1) define the parameters of the Gaussian probability law whose formal expression is

$$^{0}W\{\varphi\} = (2\pi)^{-M}\,[\det H]^{-n/2} \exp\left[-\frac{1}{2}\sum_{AB} \vec{\varphi}_A \cdot \vec{\varphi}_B\,(H^{-1})_{AB}\right]. \tag{11.6.2}$$

In what follows it will be assumed that H_{AB} equals one if A and B are nearest neighbours and that otherwise H_{AB} equals zero. Of course, this postulate is questionable, because, in this case, the matrix H_{AB} is not positive definite, in contradiction to the assumptions we always make concerning Gaussian variables (see Appendix A). However, we shall show later on that this difficulty is not as serious as it looks, and for the time being, we shall ignore it. Now, let us introduce the quantity

$$Q(n) = \sum_{N=1}^{\infty} n^N Z_N. \tag{11.6.3}$$

A straightforward application of Wick's theorem (see Appendix A) shows that

$$Q(n) = \left\langle\!\!\left\langle \prod_A \frac{(\vec{\varphi}_A)^2}{2} \right\rangle\!\!\right\rangle \tag{11.6.4}$$

provided that a weight 1/2 is attributed to the circuits containing only two sites). This equality defines a correspondence between Hamiltonian circuits and field theory. Moreover, we postulate that (11.6.3) defines $Q(n)$ for any n (not only for integers); in other words, we admit that $Q(n)$ verifies the conditions required by Carlson's theorem.[15] Accordingly, the number Z_1 of Hamiltonian circuits drawn on the lattice is given by

$$Z_1 = \lim_{n \to 0} n^{-1} Q(n). \tag{11.6.5}$$

Let us now come back to the definition of the Gaussian ensemble. The real symmetric operator H has eigenvectors $\mathcal{O}_m$ (with real components $\mathcal{O}_{Am}$) corresponding to the real eigenvalues E_m

$$H\mathcal{O}_m = E_m \mathcal{O}_m \tag{11.6.6}$$

and it is easy to verify that

$$\sum_A \mathcal{O}_{Am}\mathcal{O}_{Am'} = \delta_{mm'}. \tag{11.6.7}$$

Thus, new variables $\vec{\varphi}_m$ can be associated with the variables $\vec{\varphi}_A$ by means of the orthogonal transformation which diagonalizes H

$$\vec{\varphi}_A = \sum_m \mathcal{O}_{Am}\vec{\varphi}_m.$$

Consequently, we may formally write

$$\sum_{AB} \vec{\varphi}_A \cdot \vec{\varphi}_B \, (H^{-1})_{AB} = \sum_m \vec{\varphi}_m \cdot \vec{\varphi}_m \frac{1}{E_m}. \tag{11.6.8}$$

Let us transform this quadratic form in order to make it positive. To do that, we have only to introduce variables $\vec{\psi}_m$ defined as follows

$$\vec{\psi}_m = \varepsilon_m \vec{\varphi}_m$$

where

$$\varepsilon_m = 1 \quad \text{if} \quad E_m \geqslant 0$$

$$\varepsilon_m = i \quad \text{if} \quad E_m < 0$$

Consequently, the quadratic expression in (11.6.2) can be written in the form

$$\sum_{AB} \vec{\varphi}_A \cdot \vec{\varphi}_B \, (H^{-1})_{AB} = \sum_m \vec{\psi}_m \cdot \vec{\psi}_m \frac{1}{|E_m|}. \tag{11.6.9}$$

We can now consider the $\vec{\psi}_m$ as the real basic Gaussian variables. We note that the fact that E_m may vanish does not present any difficulty: in this case, we may just put $\vec{\psi}_m = 0$. Then, for the set of variables $\vec{\psi}_m$, the probability law can be written in the form

$$P\{\psi\} = \prod_{\substack{m'' \\ (E_m = 0)}} \delta(\vec{\psi}_{m''}) \prod_{\substack{m' \\ (E_{m'} \neq 0)}} \frac{1}{(2\pi |E_{m'}|)^{n/2}} \exp\left[-\frac{1}{2} \vec{\psi}_{m'} \vec{\psi}_{m'} \frac{1}{|E_{m'}|} \right]. \tag{11.6.10}$$

The vectors $\vec{\varphi}_A$ are given in terms of the $\vec{\psi}_m$ by

$$\vec{\varphi}_A = \sum_m \mathcal{O}_{Am} \varepsilon_m \vec{\psi}_m$$

where the $\vec{\psi}_m$ are real vectors. On the contrary, the $\vec{\varphi}_A$ are, in general, complex; but this is not troublesome because formula (11.6.1) remain true. The mathematical correspondence characterized by (11.6.3) and (11.6.4) is now perfectly defined.

6.2 An approximate value of the number of Hamiltonian circuits covering a homogeneous lattice

Let us try to calculate the entropy per site

$$\mathbb{S}/M = \frac{1}{M} \ln Z_1$$

in the case where the lattice is homogeneous, and let $\mathfrak{k}$ be the coordination number of this lattice. In this case, all the sites are equivalent; the eigenvector $\mathcal{O}_0$ of H corresponding to the eigenvalue with the largest modulus is given by

$$\mathcal{O}_{A0} = M^{-1/2},$$

and the equation

$$\sum_{\mathrm{B}} H_{\mathrm{AB}} \mathcal{O}_{\mathrm{BO}} = E_0 \mathcal{O}_{\mathrm{AO}}$$

leads immediately to the result $E_0 = \mathfrak{k}$. It is reasonable to assume that this zero mode is dominant. Consequently, the probability $P\{\psi\}$ will be replaced by the approximate probability

$$P_{\mathrm{a}}\{\psi\} = \prod_{m \neq 0} \delta(\vec{\psi}_m) \frac{1}{(2\pi E_0)^{\mathfrak{n}/2}} \exp\left[-\frac{\psi_0^2}{2E_0} \right] \tag{11.6.11}$$

Now let us set $\vec{\psi}_0 = \vec{\varphi}$; we thus obtain

$$\vec{\varphi}_{\mathrm{A}} = \frac{\vec{\varphi}}{M^{1/2}}$$

$$Q(\mathfrak{n}) = \left\langle\!\!\left\langle \prod_{\mathrm{A}} \frac{(\vec{\varphi}_{\mathrm{A}})^2}{2} \right\rangle\!\!\right\rangle \simeq \frac{\displaystyle\int \mathrm{d}^{\mathfrak{n}}\varphi \left(\frac{\varphi^2}{2M}\right)^M \exp(-\varphi^2/2\mathfrak{k})}{\displaystyle\int \mathrm{d}^{\mathfrak{n}}\varphi \exp(-\varphi^2/2\mathfrak{k})}$$

$$= \frac{\Omega(\mathfrak{n})}{(2\pi\mathfrak{k})^{\mathfrak{n}/2}} \int_0^\infty \mathrm{d}\varphi\, \varphi^{\mathfrak{n}-1} \left(\frac{\varphi^2}{2M}\right)^M \exp(-\varphi^2/2\mathfrak{k})$$

where $\Omega(\mathfrak{n})$ is the 'area' of a ball in an $\mathfrak{n}$-dimensional space (see Appendix F)

$$\Omega(\mathfrak{n}) = \frac{2\pi^{\mathfrak{n}/2}}{\Gamma(\mathfrak{n}/2)}.$$

Finally, we obtain

$$Q(\mathfrak{n}) = \frac{\Gamma(M + \mathfrak{n}/2)}{\Gamma(\mathfrak{n}/2)} \left(\frac{\mathfrak{k}}{M}\right)^M. \tag{11.6.12}$$

Now, Z_1 is given by (11.6.5); thus, by gaining to the limit $\mathfrak{n} \to 0$, we find

$$Z_1 = \frac{(M-1)!}{2} \left(\frac{\mathfrak{k}}{M}\right)^M,$$

and for $M \to \infty$, we find

$$\mathbb{S}/M = \frac{\ln Z_1}{M} = \mathfrak{k}/\mathrm{e}.$$

The approximation is very good as can be seen by comparison with exact results or with precise numerical results obtained with the strip method.[16] Let us give two examples

For a square lattice ($\mathfrak{k} = 4$), numerical computations (ref. 16 and B. Derrida, unpublished) lead to the result $\mathbb{S}/M = 1.472$ (or 1.473) and the approximation gives $\mathbb{S}/M = 4/e = 0.4715$.

For a hexagonal lattice ($\mathfrak{k} = 3$), numerical calculations (B. Derrida, unpublished) lead to the result $\mathbb{S}/\boldsymbol{M} = 1.1408$ and the approximation gives $\mathbb{S}/\boldsymbol{M} = 3/e = 1.1036$.

These results are very encouraging* and the method could lead to interesting developments.

REFERENCES

1. de Gennes, P.G. (1972). *Phys. Lett.* **38A**, 339.
2. Lubensky, T.C. and Isaacson, J. (1981). *J. Physique* **42**, 175.
   Obukhov, S.P. (1981). *J. Physique* **42**, 1591.
3. See, for instance, Stanley, H.E. (1971). *Introduction to phase transitions and critical phenomena.* Clarendon Press, Oxford.
4. Daoud, M., Cotton, J.P., Farnoux, B., Jannink, G., Sarma, G., Benoit, H., Duplessix, R., Picot, C., and de Gennes, P.G. (1975). *Macromolecules* **8**, 804.
5. Hilhorst, H.J. (1976). *Phys. Lett.* **56A**, 153.
6. Emery, V.J. (1975). *Phys. Rev.* B **11**, 239.
7. Kosmas, M.K. and Freed, K.F. (1978). *J. Chem. Phys.* **68**, 4878.
8. Duplantier, B. (1980). *C.R. hebd. Séanc. Acad. Sci., Paris* **290B**, 199.
9. Fisher, M.E. (1973). *Phys. Rev. Lett.* **30**, 679.
10. Balian, R. and Toulouse, G. (1973). *Phys. Rev. Lett.* **30**, 544.
11. des Cloizeaux, J. (1975). *J. Physique* **36**, 281.
12. Moore, M.A. (1977). *J. Physique* **38**, 265.
13. Orland, H., Itzykson, C., and De Dominicis, C. (1985). *J. Physique Lett.* **46**, L-353.
14. Bawendi, M.G., Freed, K.F., and Mohanty, U. (1986). *J. Chem. Phys.* **84**, 7036.
15. Titchmarsh, E.C. (1939). *The theory of functions*, 2nd edn, Sect. 5.8, p. 185. Oxford University Press.
16. Schmalz, T.G., Hite, G.E., and Klein, D.J. (1984). *J. Phys.* A: *Math. Gen.* **17**, 445.

* Remember that the Flory approximation produced results which were much less exact (see Chapter 8, Section 2.1.1).

12

RENORMALIZATION AND CRITICALITY

Undoubtedly philosophers are in the right when they tell us that nothing is great or little otherwise than by comparison.

Jonathan Swift
Gulliver's Travels

GENESIS

The concept of renormalization covers several different processes. Its origins are old and, for instance, dielectric constants and effective masses can be considered as renormalized quantities.

In Quantum Electrodynamics, renormalization was developed around 1950, especially by Schwinger, Feynman, and Dyson, in order to eliminate non-physical ultraviolet divergences.

In Statistical Mechanics, renormalization was introduced in order to describe critical systems. Physicists began to study critical phenomena seriously around 1965. Scaling laws were written and simple relations between critical exponents were found.

The real breakthrough occurred in 1972 when K. Wilson showed how renormalization theory can be applied to magnetic systems. In particular, he introduced the fixed point concept. Moreover, he showed that perturbation series can be defined for non-integer values of the space dimension d, and by using renormalization principles he could calculate exact expansions of the critical exponents in powers of $\varepsilon = 4 - d$. At first, Wilson thought that by combining renormalization principles and computer calculations, it would be possible to solve strong coupling problems, admitting that, in this case, perturbation calculations would be useless.

However, very interesting results were obtained by applying analytic renormalization methods to perturbation series. These techniques initiated by Wilson himself, were developed by many physicists, and, among them, Brézin, Le Guillou, and Zinn-Justin played an important role.

The same approach was used for polymers after 1972, owing to the analogy existing between polymer theory and the zero-component field theory. In the following years, the renormalization techniques became more adapted to the specific problems of polymer theory. Thus, renormalization techniques could be applied directly to polymers. In this way, difficulties related to the fact that polydispersion cannot be controlled with a one-field theory were solved, and with these new techniques the monodispersion and polydispersion effects arising from the chain length distribution could be accounted for precisely.

Moreover, in 1970, Polyakov discovered that the correlation functions of a critical magnetic system have invariance properties for transformations belonging to the special conformal group. Then, in 1984, Belavin, Polyakov, and Zamolodchikov showed that these conformal transformations are really important in two dimensions. Since that time, research in this domain led to very interesting results. Thus 'exact values' of the main exponents associated with two-dimensional polymer solutions, were found in 1986 by Saleur and Duplantier.

In this domain, new progress and a better understanding of the phenomena can still be expected (1989).

1. INTRODUCTION

A system with a complex microstructure is a critical system if its large-scale physical properties depend only on very few macroscopic parameters, and if their dependence with respect to these parameters presents universal features. In the critical domain, a critical system depends (normally) on only one length, and at the critical point it becomes completely scale invariant, i.e. invariant for space dilatations.*

This means that a critical system can be described in various equivalent ways by choosing different scales. This fact leads to the concept of renormalization: in principle, renormalizing amounts to re-expressing physical parameters by using a different scale.

Consequently, to describe a critical system, one has to use renormalization techniques[1,2] which come into play quite naturally; however, the renormalization concept itself can be considered from quite different points of view. Roughly, we can distinguish two kinds of approach: iterative renormalizations and analytic renormalizations.

In this chapter, we describe both types successively by showing how these techniques can be applied to the determination of properties of polymers in solution. However, one must be aware of the fact that this separation into two kinds of renormalization is somewhat arbitrary; there are connections between the various points of view and bridges between the various methods, and all of them follow from the seminal work of Wilson.[3,4]

* Actually, at the critical point the system is also invariant for transformations belonging to the conformal group; dilatation is only one element of this group. However, this more powerful invariance has been taken into account only very recently and only in two dimensions (see Section 4.2).

2. ITERATIVE RENORMALIZATIONS

2.1 General description of renormalization techniques in real space

Iterative renormalizations were introduced around 1970 by Wilson and were widely used by him; they were inspired by an earlier article by Kadanoff.[5] To perform such renormalizations it is really necessary to use a computer.

The spirit of the method is as follows. A large system depending on a parameter λ is to be studied. This system contains a very large set $\{{}^0\sigma\}$ of degrees of freedom ${}^0\sigma$; these degrees of freedom are random variables defined by the weight W_0. In the vicinity of the value $\lambda = \lambda_c$, collective effects appear in the system and correlations extend to long distances; the order of magnitude of these distances, which are much larger than the interatomic distances, is defined by the correlation length ξ. When $\lambda \to \lambda_c$, ξ becomes infinite. To perform a renormalization transformation, one groups the degrees of freedom belonging to $\{{}^0\sigma\}$ into bunches; in this way, one produces a new set $\{{}^1\sigma\}$ of degrees of freedom possessing more collective characteristics; then the process is repeated and one obtains a series $\{{}^1\sigma\}, \{{}^2\sigma\} \ldots \{{}^N\sigma\}$ of sets, the iteration corresponding to a progressive reduction of the number of degrees of freedom. Practically, one transforms a set $\{{}^N\sigma\}$ into a set $\{{}^{N+1}\sigma\}$. Now, we must show how the weight W_{N+1} corresponding to $\{{}^{N+1}\sigma\}$ is related to the weight W_N corresponding to $\{{}^N\sigma\}$. We are thus induced to define the probability $P(\{\sigma'\}, \{\sigma\})$ that the degrees of freedom ${}^{N+1}\sigma$ take the values σ' when the degrees of freedom ${}^N\sigma$ have the values σ. By definition, we have

$$\sum_{\sigma'} P(\{\sigma'\}, \{\sigma\}) = 1. \tag{12.2.1}$$

The initial set is described by a model Hamiltonian $\mathscr{H}_0\{{}^0\sigma\}$ which will be assumed to be classical. Then, the statistical state is defined by the weight $W_0 = \exp[-\mathscr{H}_0\{{}^0\sigma\}]$. For reasons of simplicity, the factor $\boldsymbol{\beta} = 1/k_B T$ is considered as included in the Hamiltonian. Actually, this model Hamiltonian can be rather different from the true Hamiltonian and it may depend on temperature in a complex manner through λ. The weight W_0 is associated with the configurations of the systems; in other words, it is a function of the values taken by the degrees of freedom.

The set can also be defined through a new Hamiltonian $\mathscr{H}_1\{{}^1\sigma\}$, which is a function of the new degrees of freedom constructed from the initial degrees of freedom ${}^0\sigma$. Thus, we form a series of Hamiltonians $\mathscr{H}_0\{{}^0\sigma\}, \mathscr{H}_1\{{}^1\sigma\}, \ldots, \mathscr{H}_N\{{}^N\sigma\}$ resulting from iterations defined by setting (for each N):

$$\exp[-\mathscr{H}_{N+1}\{{}^{N+1}\sigma\}] = \sum_{\{{}^N\sigma\}} P(\{{}^{N+1}\sigma\}, \{{}^N\sigma\})\exp[-\mathscr{H}_N\{{}^N\sigma\}]. \tag{12.2.2}$$

This definition can easily be justified: in particular we remark that owing to (12.2.1), the partition function

$$Z = \sum_{\{^N\sigma\}} \exp[-\mathscr{H}_N\{^N\sigma\}] \tag{12.2.3}$$

does not depend on N.

Of course, each transformation from N to N + 1, is implemented by proper scale changes and by a re-numbering of the degrees of freedom. In this way, for an infinite system, it is possible to establish a one-to-one correspondence between the degrees of freedom of $\{^N\sigma\}$ and the degrees of freedom of $\{^{N+1}\sigma\}$. Thus, the Hamiltonians $\mathscr{H}_N\{\sigma\}$ and $\mathscr{H}_{N+1}\{\sigma\}$ (associated with the same set $\{\sigma\}$) can be compared, and the renormalization transformation can be expressed formally as follows

$$\mathscr{H}_{N+1}\{\sigma\} = \mathfrak{R}\mathscr{H}_N\{\sigma\}.$$

Now, let us assume that we are at the critical point $\lambda = \lambda_c$. By definition, a critical system is invariant for dilatations. Thus, we expect that there exists a limit Hamiltonian[5]

$$\lim_{N\to\infty} \mathscr{H}_N\{\sigma\} = \mathscr{H}^*\{\sigma\}. \tag{12.2.4}$$

The existence of such a Hamiltonian was discussed (in 1979) by Griffiths.[6] The successive transformations define 'the renormalization group' of the system. The Hamiltonian $\mathscr{H}^*\{\sigma\}$ is the 'fixed point' of these transformations. This Hamiltonian $\mathscr{H}^*\{\sigma\}$, and the transformations which leave it invariant, defines the critical properties of the system.

For instance, let us examine how the critical exponents come into play. Consider a value of the parameter λ, very close to λ_c: in order to eliminate the microscopic degrees of freedom, we perform a sufficient number of transformations. In this way, we obtain a Hamiltonian $\mathscr{H}(\{\sigma\}, \mu)$ where μ is a constant (a function of λ and N) defining the distance from the critical point (for which $\mu = \mu_c$).

$$\mathscr{H}^*\{\sigma\} = \mathscr{H}(\{\sigma\}, \mu_c) \tag{12.2.5}$$

A correlation length ξ corresponds to the Hamiltonian $\mathscr{H}(\{\sigma\}, \mu)$.

Now, let us perform a renormalization transformation

$$\mathfrak{R}\mathscr{H}(\{\sigma\}, \mu) = \mathscr{H}(\{\sigma\}, \mu')$$

$$\mathfrak{R}\xi = \xi'.$$

We have

$$\mu' - \mu_c = A(\mu - \mu_c)$$

$$\xi' = B\xi. \tag{12.2.6}$$

The initial transformation (12.2.2) which changes $\mathscr{H}_N\{^N\sigma\}$ into $\mathscr{H}_{N+1}\{^{N+1}\sigma\}$ is associated with a reduction of the (quasi-infinite) number of degrees of freedom.

Thus, the renormalization process which establishes a correspondence between the old degrees of freedom and the new ones goes along with a space contraction. In other words, in (12.2.6), we have $\xi' > \xi$ and therefore $B > 1$. The transformation $\mathfrak{R}$ drives us away from the critical point. Consequently, in (12.2.6), we have $A < 1$.

In the critical domain, the function ξ varies as a power of $(\mu - \mu_c)$

$$\xi = \xi_0 (\mu - \mu_c)^{-\nu}.$$

The preceding equations (12.2.6) define ν; in fact, we have

$$B = A^{-\nu} \qquad \nu = -\frac{\ln B}{\ln A}. \tag{12.2.7}$$

This kind of approach was used as early as 1973 by Niemeijer and Van Leeuwen[7] to study the behaviour of an Ising system associated with a two-dimensional lattice in the vicinity of the critical point.

Similar techniques have been applied to the study of long polymers and they will be described briefly in the following sections.

2.2 Iterative renormalization of polymers on a lattice

The iterative renormalization method on lattices was extended to the case of polymers by H. Hilhorst in 1976.[8] The technique used by Hilhorst relies on the polymer–magnetic system correspondence for $n \to 0$, as it is described in Chapter 11, Section 3.2. Hilhorst introduced spins located on the sites of a cyclic triangular lattice: a spin $\vec{\sigma}_M$ with components σ_M^i corresponds to each lattice site M. The components take the values $-n^{1/2}$, 0 or $n^{1/2}$ and have to fulfil the condition that only one of these components is different from zero. The Hamiltonian of the system is given by the sum

$$\mathscr{H} = K \sum_{(A,B)} \vec{\sigma}_A \vec{\sigma}_B$$

over all pairs of nearest-neighbour sites (A, B), and Hilhorst was particularly interested in the partition function

$$Z = \sum_{\{\sigma\}} \exp(-\mathscr{H}).$$

Renormalization consists in grouping the lattice sites three by three so as to form a new triangular lattice whose mesh is $\sqrt{3}$ times larger (see Fig. 12.1). Thus, let us consider three sites A, B, C bearing spins $\vec{\sigma}_A$, $\vec{\sigma}_B$, $\vec{\sigma}_C$; we can associate with them a site bearing a spin $\vec{\sigma}_{ABC}$ which is a random vector with the same properties as the $\vec{\sigma}_M$. The probability law which Hilhorst attributes to $\vec{\sigma}_{ABC}$

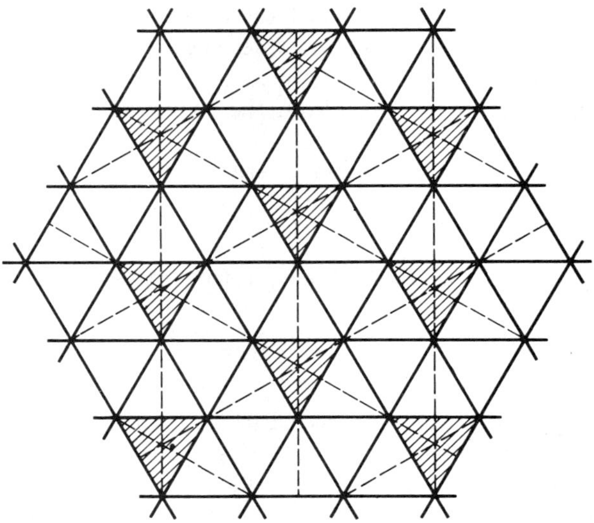

Fig. 12.1. Grouping the sites of a triangular lattice 3 by 3 so as to form a new lattice.

depends on $\vec{\sigma}_A$, $\vec{\sigma}_B$, $\vec{\sigma}_C$ in the following way

$$P(\vec{\sigma}_{ABC};\vec{\sigma}_A, \vec{\sigma}_B, \vec{\sigma}_C) = 1 - q_2 + \frac{q_2}{3n} \sum_j (\sigma^j_{ABC})^2 [(\sigma^j_A)^2 + (\sigma^j_B)^2 + (\sigma^j_C)^2]$$

$$+ q_1 \sum_j \sigma^j_{ABC}[\sigma^j_A + \sigma^j_B + \sigma^j_C]$$

where q_1 and q_2 are constants which must be chosen as well as possible.

Thus, the value of the exponent ν found in the limit $n \to 0$ by Hilhorst is $\nu = 0.740$. However, this approach is not completely satisfactory. As happens for many of these iterative methods, the difficulty comes from the fact that the choice of new variables at each iteration is somewhat arbitrary. A rule must be adopted (as was done by Hilhorst) but no criterion exists for determining this rule in a unique way. Thus, the method lacks efficiency and reliability.

It is also possible to ignore the polymer–magnetic system correspondence and to apply the renormalization process directly to the polymer–lattice system,[9] but the results thus obtained have not been very stimulating.

2.3 Decimation along the polymer chain

It is also possible to consider a polymer chain as being made of discrete links embedded in a d-dimensional space, and one can try to perform direct renormalization on polymers. This decimation method consists in representing a polymer of N links as a succession of sequences made of n links, and in considering each straight segment joining the extremities of a sequence as a link of a new chain. Thus, this new 'renormalized' chain is made of N/n links. We have now to evaluate the effective interaction that exists between these new links. The process is then repeated. By using proper units, we can represent the effective interaction by a pure number u which after successive iterations tends to the limiting value u^*. Thus, one reaches the critical domain.

This renormalization operation thus transforms a chain with N links of length l into a chain with N/n links of length l'.

Let R^2 be the mean square end-to-end distance of the chain. In the critical domain, we can write

$$R = AN^{\nu}l = A(N/n)^{\nu}l'.$$

Consequently, the exponent ν is given by the ratio

$$\nu = \frac{\ln(l'/l)}{\ln n}. \tag{12.2.8}$$

This computation program, which looks straightforward, was described by de Gennes in 1977,[10,11] but it was not easy to put it into practice. However, in 1978, Gabay and Garel[12] succeeded in obtaining, with this method, the first term in the expansion of ν as a function of $\varepsilon = 4 - d$ (d = space dimension) namely $\nu = 1/2(1 + \varepsilon/8 + \ldots)$. However, this result was already known, and at the present time (1988), no algorithm exists which could be used to proceed much further with the decimation method.[13]

2.4 'Phenomenological renormalization' of a polymer chain; the strip method

'Phenomenological renormalization' is a method which is adapted to the study of critical phenomena in two and three dimensions. It consists in performing calculations on lattices that are infinite in one direction and finite in the others, by using the scaling properties of the system as well as possible. It was invented by Nightingale[14] in 1976 and applied by him to the study of the Ising model. Since that time, it has been used successfully to study other critical systems. In 1981, Derrida[15] applied it to the case of self-avoiding chains on a two-dimensional lattice. The value of ν obtained with this methods equals

$$\nu = 0.7503 \pm 0.0002 \tag{12.2.9}$$

and therefore the precision seems excellent.*

The principle of the calculation is as follows. Let us consider a two-dimensional lattice, infinite in one dimension and of width n in the other one (this lattice is a strip: see Fig. 12.2). Let us associate random variables with the lattice

* The 'exact' result (obtained by Nienhuis in 1982) is $\nu = 3/4$; see Section 4 of the present chapter.

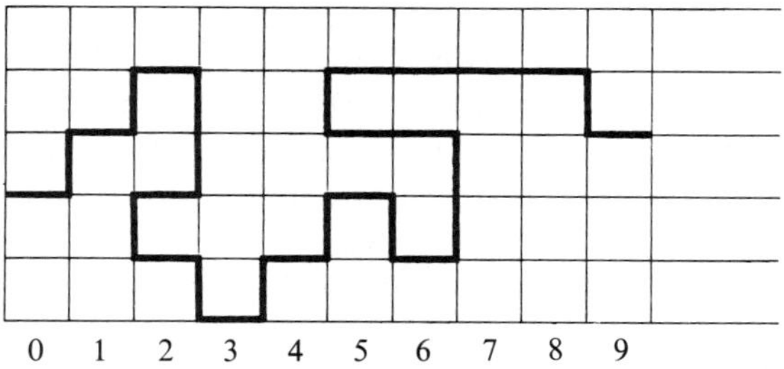

Fig. 12.2. A semi-infinite lattice (here $n = 6$) and a polymer drawn on this lattice from column 1 to column 9. The possible configurations of such polymers define a random system.

sites. Thus, we define 'a system', and it will be assumed that this system depends on a parameter x. Let us also assume that for $n = \infty$ and $x = x_c$ this system is critical. We can admit that for values of x close to x_c and n finite, any macroscopic average value $A_n(x)$ is given by a scaling law of the form

$$A_n(x) = A_\infty(x) f(n/\xi_\infty(x))$$

where $\xi_\infty(x)$ is the correlation length in the system for $n = \infty$ (thus $\xi_\infty(x_c) = \infty$).

Let us calculate the same quantity for other values m and y of n and x. We have

$$\frac{A_n(x)}{A_m(y)} = \frac{A_\infty(x) f(n/\xi_\infty(x))}{A_\infty(y) f(m/\xi_\infty(y))}.$$

A correspondence can be established between the lattices of widths n and m in such a way that

$$\frac{\xi_\infty(x)}{\xi_\infty(y)} = \frac{n}{m} \tag{12.2.10}$$

which implies

$$\frac{A_n(x)}{A_m(y)} = \frac{A_\infty(x)}{A_\infty(y)}. \tag{12.2.11}$$

Moreover, by definition, we have

$$\frac{\xi_\infty(x)}{\xi_\infty(y)} = \left(\frac{x - x_c}{y - y_c}\right)^{-\nu}. \tag{12.2.12}$$

In particular, the quantity $A_n(n)$ could be the correlation length $\xi_n(x)$ of the system. In this case, (12.2.11) gives

$$\frac{\xi_n(x)}{\xi_m(y)} = \frac{\xi_\infty(x)}{\xi_\infty(y)}.$$

Let us combine this equality with eqns (12.2.10) and (12.2.12); we obtain

$$\frac{\xi_n(x)}{\xi_m(y)} = \frac{n}{m} \tag{12.2.13}$$

$$\frac{\xi_n(x)}{\xi_m(y)} = \left(\frac{x - x_c}{y - y_c}\right)^{-\nu} \tag{12.2.14}$$

and from these equations we also derive

$$\frac{\mathrm{d}\xi_n(x)/\mathrm{d}x}{\mathrm{d}\xi_m(y)/\mathrm{d}y} = \left(\frac{x - x_c}{y - y_c}\right)^{-1-\nu} = \left(\frac{n}{m}\right)^{1+1/\nu}. \tag{12.2.15}$$

Let us now observe that the system corresponding to a finite value of n is not critical for $x = x_c$. This entails that $\xi_n(x)$ and $\mathrm{d}\xi_n(x)/\mathrm{d}x$ are regular at $x = x_c$. Now x and y may tend to x_c and, in this way, we obtain

$$\frac{\xi_n(x_c)}{\xi_m(x_c)} = \frac{n}{m}$$

$$\frac{\xi'_n(x_c)}{\xi'_m(x_c)} = \left(\frac{n}{m}\right)^{1+1/\nu}. \tag{12.2.16}$$

These relations, which are valid for $n \gg 1$ and $m \gg 1$, show how the phenomenological renormalization method can be used to calculate ν and x_c from the functions $\xi_n(x)$. One has only to set $m = n - 1$ and to solve the preceding equations for successive values of n. In this way, approximate values $x_c(n)$ and $\nu(n)$ are found, and they are extrapolated so as to obtain $x_c = x(\infty)$ and $\nu = \nu(\infty)$.

Let us now examine how the method applies when the critical system consists of a polymer drawn on the lattice (see Fig. 12.2). Of course, we must define $\xi_n(x)$ for this system and we must show how this function can be calculated.

Each column of the lattice consists of n horizontal segments belonging to the lattice and we can label the position of the column by a positive integer R. Let $Z_n(N, R)$ be the number of self-avoiding polymers of N links (and with different configurations) that can be drawn in such a way that the first chain segment belongs to column zero, that the last chain segment belongs to column R and that the end point of the chain is on the border between columns R and $R + 1$ (and not between columns $R - 1$ and R).

We can set

$$\boldsymbol{G}_n(x,R) = \sum_{N=0}^{\infty} x^N Z_n(N,R),$$

and this definition of the Green's function $\boldsymbol{G}_n(x,R)$ is very similar to the definition given in Chapter 11. For large values of R

$$\boldsymbol{G}_n(x,R) \propto \exp(-R/\xi_n(x)) \tag{12.2.17}$$

which defines $\xi_n(x)$.

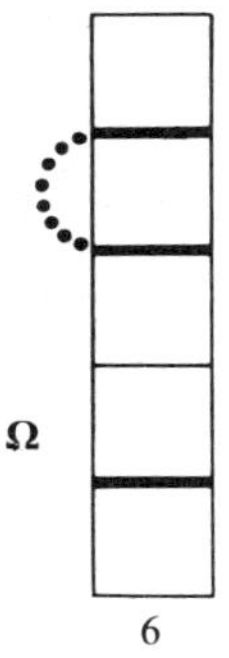

Fig. 12.3. Configuration $\boldsymbol{\Omega}$ corresponding to column 6 of Fig. 12.2.

Let us now see how $G_n(x,R)$ can be evaluated. We consider column r where $0 < r < R$. With a given chain and a given column r, we can associate the configuration $\boldsymbol{\Omega}$ of the chain segments contained in column r. $\boldsymbol{\Omega}$ is defined as follows (see Fig. 12.3):

(a) We mentally erase all portions of chain that belong to columns of rank larger than r.

(b) We mark in column r, the segment $\mathcal{O}$ that is directly connected to the origin.

(c) We mark in column r, the pairs of segments that are connected by portions of chains belonging to the interval $(0, r)$.

For any configuration $\boldsymbol{\Omega}$, we can define the quantity

$$H_n(r,\boldsymbol{\Omega}) = \sum_{N=0}^{\infty} x^N Z_n(N,r;\boldsymbol{\Omega})$$

where $Z_n(N,r;\boldsymbol{\Omega})$ is the number of configurations of chain fragments belonging to the columns $0, \ldots, r$, made on the whole of N links and compatible with $\boldsymbol{\Omega}$.
$H_n(r+1,\boldsymbol{\Omega})$ is related to $H_n(r,\Omega')$ by the transfer matrix $\mathcal{T}_n$

$$H_n(r+1,\boldsymbol{\Omega}) = \sum_{\boldsymbol{\Omega}'} \mathcal{T}_n(\boldsymbol{\Omega},\boldsymbol{\Omega}')H_n(r,\boldsymbol{\Omega}'). \qquad (12.2.18)$$

If $\boldsymbol{\Omega}$ and $\boldsymbol{\Omega}'$ are incompatible $\mathcal{T}_n(\boldsymbol{\Omega},\boldsymbol{\Omega}') = 0$; if $\boldsymbol{\Omega}$ and $\boldsymbol{\Omega}'$ are compatible $\mathcal{T}_n(\boldsymbol{\Omega},\boldsymbol{\Omega}') = x^{t(\boldsymbol{\Omega},\boldsymbol{\Omega}')}$ where $t(\boldsymbol{\Omega},\boldsymbol{\Omega}')$ is the number of links that have to added on the right of column r, to configuration $\boldsymbol{\Omega}'$, to obtain configuration $\boldsymbol{\Omega}$ in column $r+1$.

Thus, the finite matrix $\mathcal{T}_n$ is a function of x and n. It is clear that when R is large

$$G_n(x,R) \propto [\lambda_n(x)]^R$$

where $\lambda_n(x)$ is the largest eigenvalue of $\mathcal{T}_n$. From the calculation of $\mathcal{T}_n(\boldsymbol{\Omega},\boldsymbol{\Omega}')$, we extract $\xi_n(x)$

$$\xi_n(x) = -\frac{1}{\ln \lambda_n(x)}.$$

This makes it possible to calculate the critical value x_c and the exponent ν as was explained above, and this leads to the result mentioned at the beginning of this section.

More recently, strip methods have also been used by H. Saleur and D. Derrida (unpublished) to calculate γ/ν. They found results compatible with the value $\gamma/\nu = 43/24$ ($= 1.79167$) predicted by Nienhuis (see Section 4).

3. ANALYTIC RENORMALIZATIONS

3.1 General principles common to all analytical renormalization techniques

Analytic renormalization techniques have been borrowed from field theory. In field theory, perturbation calculations exhibit short-range divergences, called ultraviolet divergences. These divergences can be hidden by the introduction of a short distance cut-off: this is a *regularization.* Unfortunately, in this way, one introduces a new parameter devoid of physical interest. This cut-off can be eliminated if the theory is renormalizable. For this, it is necessary that divergences appear only in a very special series of diagrams; then the divergences can be eliminated after a finite number of *renormalizations*. These renormalizations are either subtractions or multiplications; thus 'renormalization factors' come into play, and these quantities are strongly dependent on the cut-off.

Because of these renormalizations, the physical quantities are no longer expressed in terms of the 'bare' parameters of the system but in terms of new 'physical' parameters which are observables.

When the cut-off tends to zero, the renormalization factors, which in field theory are not observables, become infinite; on the contrary, the renormalized system remains finite, and this property enables us to find relations between the various observable physical quantities.

In each case, renormalizability has to be proved and the proof is generally made in the framework of perturbation theory. For instance, the electromagnetic theory and the Landau–Ginzburg theory are renormalizable,[16,17] in this sense, for $d = 4$. In a similar way, it is not difficult to show that for $d < 4$, the Landau–Ginzburg theory is renormalizable with respect to short-range divergences.

In another connection, we explicitly showed in Chapter 10 that, in polymer theory, for $d < 4$, it is possible to construct a theory independent of any short-range cut-off. In fact, we started from a regularized theory, introducing a cut-off 'area' s_o (Chapter 10, Section 1), and we explained how a simple renormalization of the partition function (Chapter 10, Section 4.1.5) can be used to eliminate s_0.

Now, we are interested in long-range divergences. In the Landau–Ginzburg model, these divergences come from the fact that the correlation length becomes infinite at the critical point; in polymer theory, divergences occur because the size of polymers becomes infinite when the number of links becomes infinite. The study of the long-range divergences, called infra-red divergences, can be made, by using renormalization techniques again. This possibility can be understood in different ways.

First, we remark that the dimension $d = 4$ is marginal for the Landau–Ginzburg model and for polymers. This means that when d increases up to the value $d = 4$, logarithmic ultraviolet divergences appear. In this case, we encounter integrals of the form $I = \int_1^A \frac{\mathrm{d}x}{x}$ which diverge at large distances when the cut-off parameter A becomes infinite. By setting $x = Ay$, we can also write this integral in the form $I = \int_{1/A}^1 \frac{\mathrm{d}y}{y}$. Now, the integral divergences at short distances, when $1/A$ tends to zero.

For $d = 4$, the ultraviolet divergences and the infra-red divergences are thus related to each other, and it is not surprising that the same techniques can be applied in both cases.

For $d < 4$, the situation is different. In a sense, the infra-red divergences for $d < 4$, are analogous to the ultraviolet divergences for $d > 4$. But, it is often said that in respect of the ultraviolet divergences, the Landau–Ginzburg model is not renormalizable for $d > 4$; if we adopted this point of view entirely, we would be inclined to believe, that, in the same way, in respect of infra-red divergences, the Landau–Ginzburg model is not renormalizable for $d < 4$. Actually, perturbation calculations show that the order of the infra-red divergences increases with the order of the diagrams. The same phenomenon occurs in polymer theory; in this case, the contribution of a diagram of order p is proportional to z^p and z becomes infinite when the size of the polymers become infinite. Thus, we are required to sum up series of more and more divergent terms.

However, the situation is not as serious as it may look. For physical reasons, we expect that scaling laws exist in the critical domain (small $(T - T_c)$ or large N). In other words, it can be expected that, in this limit, the various physical quantities depend only on one fundamental parameter, which is the characteristic length of the system, and eventually on a few finite dimensionless parameters. For the Landau–Ginzburg model, the fundamental length is the correlation length, and for a polymer-system, the fundamental length is the size of an isolated swollen polymer.

The principle requiring that all calculated quantities can be expressed in terms of this fundamental length does not hold in general. In order to ensure that this principle applies, we must absorb the divergences of the theory by renormalizing the calculated values. Actually, it is necessary to verify that if the fundamental length of the system is chosen as a scale, all macroscopic observable quantities can be expressed in terms of renormalized quantities.

The method works because there are only a few renormalization factors and subtraction coefficients, which absorb all the divergences and, in particular, suppress all dependence with respect to the cut-off.

In this way, the theory acquires universal features, since the effects of the microstructure are absorbed in the renormalization process. Let us note, however, that certain observables, such as the transition temperature T_c of a magnetic system, have no universal character. Their values depend on microscopic details and, consequently, it is not easy to calculate them.

In brief, the analytic renormalization techniques apply to continuous models depending on very few 'bare' parameters, and they aim at expressing measurable quantities directly in terms of 'macroscopic' and 'observable' fundamental parameters. Thus, in spite of a certain mathematical complexity, this approach appears as essentially realistic.

3.2 Landau–Ginzburg model: renormalization and critical exponents

In Chapter 11, we showed that polymer models correspond to field models.

To determine the critical exponents, it is sufficient to study the properties of an isolated polymer. For this purpose, we can content ourselves with a one field theory: this is the Landau–Ginzburg model described in Chapter 11. Incidentally, we may recall that the Landau–Ginzburg model with external magnetic field corresponds to the 'equilibrium ensembles' defined in Chapters 9 and 11; consequently, it can be used to calculate approximately the properties of polymer solutions.[18]

3.2.1 Existence of a renormalized Landau–Ginzburg theory

We assume that, when a is close to a_c ($(a - a_c)b^{2/(d-4)} \ll 1$) the system, which then belongs to the critical domain, can be described by a limiting theory. In fact, the existence of this theory can be proved by perturbation in the vicinity of $d = 4$.[17] The physical quantities, in this limit, are the renormalized Green's functions $\mathscr{G}_R(\vec{r}_1, \ldots, \vec{r}_p)$ which by definition are proportional to the Green's functions $\mathscr{G}(\vec{r}_1, \ldots, \vec{r}_p) \equiv \mathscr{G}(\vec{r}_1, \ldots, \vec{r}_p; a)$. We have

$$\mathscr{G}_R(\vec{r}_1, \ldots, \vec{r}_{2p}) = (\mathfrak{Z}_1)^{-2p} \mathscr{G}(\vec{r}_1, \ldots, \vec{r}_{2p}) \tag{12.3.1}$$

where $\mathfrak{Z}_1$, a coefficient which will be defined below, is a function of the interactions;* it is introduced to absorb the divergences of the Green's functions which occur when $a \to a_c$ (the dependence on a is implicit).

The coefficient $\mathfrak{Z}_1$ is by definition the 'renormalization factor of the field'; in fact, we may set $\varphi(\vec{r}) = \mathfrak{Z}_1 \varphi_R(\vec{r})$ in agreement with (12.3.1).

* In the literature, other symbols are commonly used, $(\mathfrak{Z}_1)^2$ being denoted by Z or Z_3.

We can associate renormalized vertex functions with the renormalized Green's functions, by proceeding as was done in Chapter 11, Section 5.2.2. It is easy to verify that

$$\Gamma_R(\vec{r}_1, \ldots, \vec{r}_{2p}) = (\mathfrak{Z}_1)^{2p}\, \Gamma(\vec{r}_1, \ldots, \vec{r}_{2p}). \tag{12.3.2}$$

In particular, we have

$$\overline{\Gamma}_R(\vec{k}, -\vec{k}) = (\mathfrak{Z}_1)^{2p}\, \overline{\Gamma}(\vec{k}, -\vec{k}). \tag{12.3.3}$$

We can now fix $\mathfrak{Z}_1$ by imposing the renormalization condition

$$2\frac{\partial}{\partial k^2}\overline{\Gamma}_R(\vec{k}, -\vec{k}) = 1 \tag{12.3.4}$$

which gives

$$(\mathfrak{Z}_1)^{-2} = 2\frac{\partial}{\partial k^2}\overline{\Gamma}(\vec{k}, -\vec{k})\Big|_{k=0}. \tag{12.3.5}$$

Thus, the expansion of $\overline{\Gamma}_R(\vec{k}, -\vec{k})$ with respect to k reads

$$\overline{\Gamma}_R(\vec{k}, -\vec{k}) = a_R + k^2/2 + \ldots \tag{12.3.6}$$

where a_R is a renormalized variable. Now let us set

$$\overline{\Gamma}(\vec{0},\vec{0}) = a_B. \tag{12.3.7}$$

We see that

$$a_B = (\mathfrak{Z}_1)^{-2}\, a_R. \tag{12.3.8}$$

(B for 'bare')

In order to define a theory that remains finite when $a \to a_c$, we have of course to express the renormalized vertex functions in terms of quantities having a physical meaning, and we shall study this point below more precisely.

3.2.2 Three-dimensional renormalization and calculation of critical exponents

Analytic renormalization methods were applied to the Landau–Ginzburg method (around 1970) but, originally for a space dimension close to 4, dimension 4 being marginal. However, it appeared later that analytical methods could also be used directly in dimension 3, as G. Parisi suggested in 1973 (Cargèse Summer School).

Calculations of the critical exponents ν and γ (or more precisely γ and $\eta = 2 - \gamma/\nu$) have been performed step by step. Around 1975, Nickel succeeded in obtaining,[19] the coefficients of perturbation series up to order six or seven for several vertex functions.

The diagrams were counted and their contributions were calculated with the help of a computer. Results were obtained for $n = 0, 1, 2, 3, 4$. Incidentally we note that the study of the dependence in n of the terms of the series is not a serious problem: in fact, the contribution of any diagram is just proportional to $\mathfrak{n}^f$ where f is the number of closed solid lines contained in the diagram.*

The results were later exploited by Baker, Nickel, Green, and Meiron[20] who deduced from them values of the critical exponents γ, ν and also of the exponent ω related to corrections to scaling laws. These results were reconsidered and re-analysed in a more refined manner by Le Guillou and Zinn-Justin in 1977.[21] Actually, a theory due to Lipatov[22] leads to predictions concerning the form of the perturbation series at large orders. Le Guillou and Zinn-Justin, making explicit use of the knowledge they had concerning these series, were thus able to improve the precision of the results substantially.

For $\mathfrak{n} = 0$, which corresponds to the case of polymers (see Chapter 11), the values obtained by Le Guillou and Zinn-Justin[21] for the critical exponents, are (for $d = 3$)

$$\nu = 0.5880 \pm 0.0010$$

$$\gamma = 1.1615 \pm 0.0011$$

$$\omega = 0.790 \pm 0.015$$

$$\Delta_1 \equiv \omega\nu = 0.465 \pm 0.009 \qquad (\mathfrak{n} = 0). \tag{12.3.9}$$

The case $\mathfrak{n} = 1$ is also useful in polymer theory. As will be shown in Chapter 14, the point at the top of the demixtion curve (coexistence curve) of a polymer solution is indeed a critical point belonging to the class: $\mathfrak{n} = 1$, $d = 3$. The corresponding exponents calculated by Le Guillou and Zinn-Justin are

$$\nu = 0.6300 \pm 0.0008$$

$$\gamma = 1.2402 \pm 0.0009$$

$$\omega = 0.782 \pm 0.010 \qquad (\mathfrak{n} = 1). \tag{12.3.10}$$

On the other hand, by analysing the series in a slightly different way, Baker, Nickel, and Meiron found,[23] for $\mathfrak{n} = 0$

$$\nu = 0.588 \pm 0.001$$

$$\gamma = 1.161 \pm 0.003.$$

Let us now consider these methods in more detail, starting with the series calculated by Nickel.[24]

First, we consider the diagrams contributing to $\bar{\Gamma}(\vec{k}, -\vec{k})$ (see Fig. 12.4). A diagram containing $\circlearrowleft$ loops (i.e. by definition $\circlearrowleft$ independent internal wave vectors) is made of $\circlearrowleft$ interaction lines and $(2\circlearrowleft - 1)$ segments. The contribution is calculated by integrating the product of $(2\circlearrowleft - 1)$ denominators of the form $1/(a + q^2/2)$; the number of integrations to be performed in reciprocal space over independent wave vectors equals $\circlearrowleft$. Therefore, the superficial degree of divergence is $\delta = d\circlearrowleft - 2(2\circlearrowleft - 1) = 2 - (4 - d)\circlearrowleft$.

* Not to be confused with the number of loops denoted by $\circlearrowleft$!

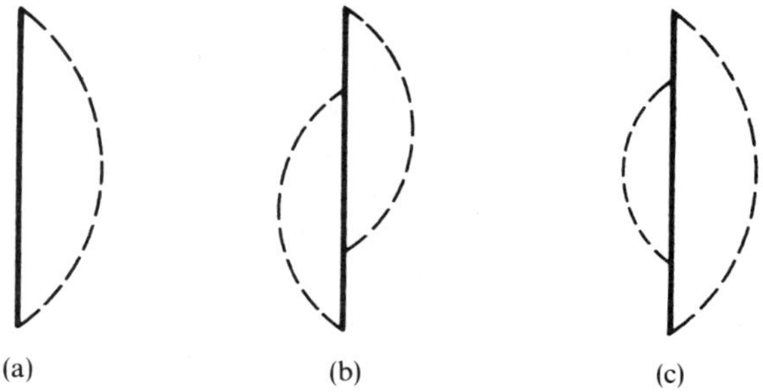

Fig. 12.4. A few diagrams contributing to $\bar{\Gamma}(\vec{k}, -\vec{k})$. (a) 1 loop and 1 segment; (b) and (c) 2 loops and 3 segments.

A diagram is primitively divergent for large wave vectors, if its degree of divergence δ is positive or zero. Thus one-loop and two-loop diagrams are primitively divergent (for instance, for ⟲ $= 1$, one finds $\int d^3k(a + k^2/2)^{-1} = \infty$). The diagrams whose number of loops is larger than two are primitively convergent but may still diverge for the following reason. A diagram with several loops can be 2-P-reducible, which means that by cutting two properly chosen polymer lines, one can separate this diagram into two disconnected parts. The part of the primitive diagram included between the two sections is a sub-diagram which has the same structure as one of the diagrams contributing to $\bar{\Gamma}(\vec{k}, -\vec{k})$. Such a sub-diagram can be primitively divergent. Thus, a primitively convergent diagram may diverge if it contains sub-diagrams which are primitively divergent.

In order to avoid such divergences a cut-off K_0 must be introduced at large wave vectors and $\bar{\Gamma}(\vec{k}, -\vec{k})$ must depend on this cut-off. But we are interested in values of a close to the value $a = a_c(K_0)$ for which $\bar{\Gamma}(\vec{0},\vec{0}) = 0$. We could try to eliminate this dependence by re-expressing $\bar{\Gamma}(\vec{k}, -\vec{k})$ as a function of $(a - a_c)$, an operation which is equivalent to a renormalization of a. However, it is *much more advantageous* to proceed otherwise.

The convenient method consists in setting $\bar{\Gamma}(\vec{0},\vec{0}) = a_B$ (B for 'bare') and in expressing all quantities in terms of a_B instead of expressing them in terms of a.

With diagrams, this can be done in a very simple way. The recipe consists in rewriting the unperturbed part

$$^0\mathscr{L}(\vec{r}) = \frac{1}{2}\sum_{j=1}^{n}\left[a\,\varphi_j^2 + \frac{1}{2}\sum_{i=1}^{d}(\partial_i\varphi_j)^2\right]$$

of $\mathscr{L}(\vec{r})$ [see (11.5.3)] in the form

$$^0\mathscr{L}(\vec{r}) = \frac{1}{2}\sum_{j=1}^{n}\left[a_B\,\varphi_j^2 + \frac{1}{2}\sum_{i=1}^{d}(\partial_i\varphi_j)^2\right] + \frac{1}{2}\sum_{j=1}^{n}(a - a_B)\varphi_j^2 \qquad (12.3.11)$$

and in considering the first term of the second formula as the non-perturbed term and the second one as a counter-term leading to point insertions in the diagrams (see Fig. 12.5). Of course, the elements constituting these counter-terms can be calculated order by order. It is not difficult to convince oneself that the contributions of the diagrams associated with $\overline{\Gamma}(\vec{k}, -\vec{k})$ can be obtained as follows. The simplest situation corresponds to the case where the diagram is 2-P-irreducible (i.e. without sub-diagrams similar to a diagram of $\overline{\Gamma}(\vec{k}, -\vec{k})$; see Fig. 12.6). Then, let $\mathscr{D}(\vec{k}, \vec{q}_1, \ldots, \vec{q}_{\circlearrowleft})$ be the weight of the diagram calculated before integration over the $\circlearrowleft$ internal wave vectors. The weight is calculated by using the quantity $1/(a_{\mathrm{B}} + q^2/2)$ as propagator [and not $1/(a + q^2/2)$].

The total contribution of the 2-P-irreducible diagram is given by

$$\mathscr{D}(\vec{k}) = \int \mathrm{d}^3 q_1 \ldots \int \mathrm{d}^3 q_{\circlearrowleft} \, [\mathscr{D}(\vec{k}, \vec{q}_1, \ldots, \vec{q}_{\circlearrowleft}) - \mathscr{D}(\vec{0}, \vec{q}_1, \ldots, \vec{q}_{\circlearrowleft})]. \qquad (12.3.12)$$

The second term corresponds to a counter-term, and now the integral is always

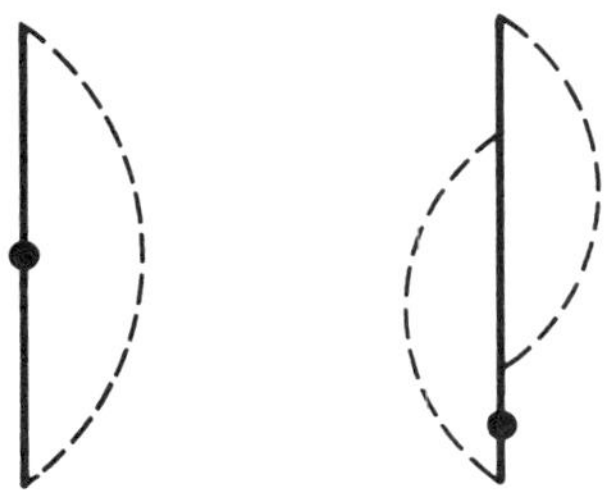

(a) (b)

Fig. 12.5. (a) and (b). Point insertion on a diagram.

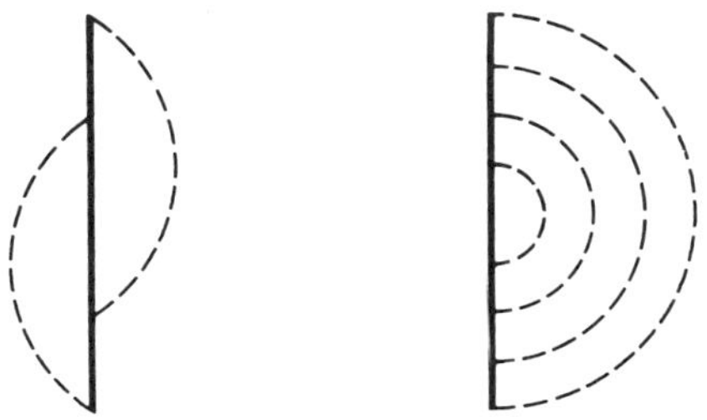

(a) (b)

Fig. 12.6. (a) 2-P-irreducible diagram, (b) 2-P-reducible diagram, containing a cascade of insertions (like Russian dolls!).

convergent (i.e. even when $K_0 \to 0$, K_0 being the length of the cut-off wave vector).

Thus, to first-order (◯ = 1), we find

$$\Gamma(\vec{k}, -\vec{k}) \simeq a_B + k^2/2 - \frac{b}{(2\pi)^3}\int d^3q\left[\frac{1}{a_B + (\vec{k}+\vec{q})^2/2} - \frac{1}{a_B + q^2/2}\right].$$

Let us add and subtract $\frac{2}{(\vec{k}+\vec{q})^2} - \frac{2}{q^2}$ in the integrand. We observe that

$$\int d^3q\left[\frac{1}{2a_B + (\vec{k}+\vec{q})^2} - \frac{1}{(\vec{k}, +\vec{q})^2}\right] = \int d^3q\left[\frac{1}{2a_B + q^2} - \frac{1}{q^2}\right].$$

Consequently, we have

$$\bar{\Gamma}(\vec{k}, -\vec{k}) \simeq a_B + k^2/2 - \frac{b}{2\pi^3}\int d^dq\left[\frac{1}{(\vec{k}+\vec{q})^2} - \frac{1}{q^2}\right].$$

$$\simeq a_B + \frac{k^2}{2}$$

We may do the same for higher orders.

The characteristic length ξ of the system† in the vicinity of the critical point can be defined by

$$\xi^2 = 1/a_R = (\mathfrak{Z}_1)^{-2}/a_B \qquad (12.3.13)$$

[see (12.3.6), (12.3.8), and (12.3.5)].

To determine ξ, one calculates $2\frac{\partial}{\partial k^2}\bar{\Gamma}(\vec{k}, -\vec{k})$ for $k = 0$. The expansion is of the form

$$\xi^2 a_B = 2\frac{\partial}{\partial k^2}\bar{\Gamma}(\vec{k}, -\vec{k})\bigg|_{k=0} = \sum_0^\infty A_p(b\, a_B^{-1/2})^p \qquad (12.3.14)$$

since $b\, a_B^{-1/2}$ is dimensionless for $d = 3$; actually, a_B has the same dimension as a and therefore for $d = 3$, $b \sim L^{-1}$, and $a_B \sim L^{-2}$ [see eqn (12.3.14)].

The parameter a_B, which is the exact value of the bare vertex $\bar{\Gamma}(\vec{0}, \vec{0})$ must now be related to the initial parameter a. For this purpose, we may calculate

$$\frac{da_B}{da} = \frac{d}{da}\bar{\Gamma}(\vec{0}, \vec{0}).$$

In order to calculate this quantity, we have only to let a vary slightly in $\mathscr{L}(\vec{r})$; this adds a new term which can either be treated by perturbation or incorporated in the propagator; in both cases, this term appears in diagrams as a point insertion (see Fig. 12.5) since

$$\frac{1}{a_B - \delta a + k^2/2} = \frac{1}{a_B + k^2/2} + \frac{1}{a_B + k^2/2}\delta a\frac{1}{a_B + k^2/2} + \ldots\ .$$

The factor which must be associated with this insertion equals unity. The

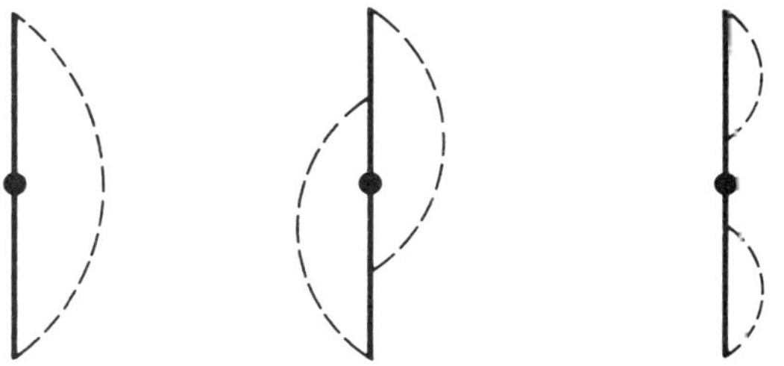

Fig. 12.7. Diagrams contributing to $\mathrm{d}a_{\mathrm{B}}/\mathrm{d}a$.

diagrams which enable us to calculate $\mathrm{d}a_{\mathrm{B}}/\mathrm{d}a$ are simply obtained by adding one point to the diagrams contributing to $\bar{\Gamma}(\vec{0},\vec{0})$ (see Fig. 12.7). The only diagrams to be taken into account are the 2-P-irreducible diagrams (ignoring the point). A diagram with $\circlearrowleft$ loops is made of $\circlearrowleft$ interaction lines and of $2\circlearrowleft$ segments (see Fig. 12.7). Therefore the superficial degree of divergence of such diagrams is

$$\delta = 3\circlearrowleft - 2(2\circlearrowleft) = -\circlearrowleft.$$

The diagrams are always superficially convergent, and since they do not contain divergent insertions, they are always convergent. We thus obtain a series

$$\frac{\mathrm{d}a_{\mathrm{B}}}{\mathrm{d}a} = \frac{\mathrm{d}}{\mathrm{d}a}\bar{\Gamma}(\vec{0},\vec{0}) = \sum_{0}^{\infty} B_p (b\, a_{\mathrm{B}}^{-1/2})^p. \tag{12.3.15}$$

In this way, to first-order, we get

$$\frac{\mathrm{d}a_{\mathrm{B}}}{\mathrm{d}a} = 1 - \frac{b}{(2\pi)^3}\int \frac{\mathrm{d}^3 q}{(a_{\mathrm{B}} + q^2/2)^2} = 1 - \frac{2^{1/2}}{4\pi}(b\, a_{\mathrm{B}})^{-1/2}.$$

If we know both series with coefficients A_p and B_p, we can in principle determine the exponents γ and ν which define the asymptotic behaviour of these series. In fact, if a_{c} is the critical value of a (a value which depends on the cut-off K_0), we have

$$a_{\mathrm{B}} \propto (a - a_{\mathrm{c}})^{\gamma} \qquad \xi \propto (a - a_{\mathrm{c}})^{-\nu}. \tag{12.3.16}$$

To become convinced of the validity of these relations, one has only to look at Chapter 11 and to remark that, in the vicinity of the critical point, we may write $a - a_{\mathrm{c}} \propto t$ where t is the parameter $(T - T_{\mathrm{c}})/T$ appearing in eqns (11.2.10) and (11.2.11) [see also (11.2.8)].

Consequently, when $b\, a_{\mathrm{B}}^{-1/2} \to \infty$, the functions corresponding to the series (12.3.13) and (12.3.14) must have the following behaviour

$$\frac{\mathrm{d}a}{\mathrm{d}a_{\mathrm{B}}} \simeq B_{\infty}(b\, a_{\mathrm{B}}^{-1/2})^{2(1-1/\gamma)}$$

$$\xi^2 a_{\mathrm{B}} \simeq A_{\infty}(b\, a_{\mathrm{B}}^{-1/2})^{2(-1+2\nu/\gamma)}. \tag{12.3.17}$$

Actually, the physicists who aimed at calculating these critical exponents did not study these series directly, but they re-expressed them in terms of a new parameter u, incorrectly called the 'renormalized coupling constant'. The advantage comes from the fact that this parameter has a physical meaning and that it has a finite limit u^* when $b\,a_B^{-1/2}$ becomes infinite.

For this purpose, they make a perturbation calculation of a third quantity, the value $\bar{\Gamma}(\vec{0},\vec{0},\vec{0},\vec{0})$ of the four-leg vertex function for vanishing external wave vectors. We must note that this operation increases the computation work only moderately because all the diagrams contributing to $\mathrm{d}a_B/\mathrm{d}a$ contribute also to a; the only difference is that different weights are attributed to them (see Fig. 12.8).

In this way, one obtains a series

$$\bar{\Gamma}(\vec{0},\vec{0},\vec{0},\vec{0}) = b\sum_0^\infty C_p(b\,a_B^{-1/2})^p. \tag{12.3.18}$$

To first-order, we have

$$\bar{\Gamma}(\vec{0},\vec{0},\vec{0},\vec{0}) = b - \frac{4b^2}{(2\pi)^3}\int \frac{\mathrm{d}^3 q}{(a_B + q^2/2)^2} = b\left(1 - \frac{2^{1/2}}{\pi}(b\,a_B)^{-1/2}\right).$$

For $d = 3$, the calculation made by Nickel gives the three series of coefficients A_p, B_p and C_p [see (12.3.14), (12.3.15), and (12.3.16)].

For $n = 0$ (polymer case), Nickel[24] finds the results recorded in the following tables. The factor $(-4\pi\,2^{1/2})$ comes from differences in notation. We note that the coefficients A_p were calculated by Nickel in two slightly different ways, which should be equivalent. The differences observed in the results constitute a test of the precision of the calculations.

p	$A'_p = (-4\pi\,2^{1/2})^p A_p$
0	1
1	0
2	0.296 296 296 400
3	0.302 888 582 080 × 10
4	0.320 045 910 704 × 10^2
5	0.371 839 423 405 × 10^3
6	0.474 095 980 564 × 10^4

determined from two-leg diagrams

p	$B'_p = (-4\pi\, 2^{1/2})^p B_p$	
0	1	1
1	2	2
2	12	12
3	$0.101\,116\,626\,927 \times 10^3$	$0.101\,116\,626\,927 \times 10^3$
4	$0.104\,084\,697\,578 \times 10^4$	$0.104\,084\,697\,574 \times 10^4$
5	$0.123\,526\,676\,139 \times 10^5$	$0.123\,526\,676\,988 \times 10^5$
6	$0.163\,894\,411\,479 \times 10^6$	$0.163\,894\,413\,062 \times 10^6$
	determined from four-leg diagrams	determined from two-leg diagrams

p	$C'_p = (-4\pi\, 2^{1/2})^p C_p$
0	1
1	8
2	$0.786\,666\,666\,696 \times 10^2$
3	$0.890\,777\,336\,883 \times 10^3$
4	$0.112\,621\,348\,984 \times 10^5$
5	$0.156\,110\,497\,289 \times 10^6$
6	$0.234\,457\,142\,821 \times 10^7$

determined from four-leg diagrams

Let us now examine how values of critical exponents can be extracted from the following series

$$\xi^2 a_{\mathrm{B}} = 2\frac{\partial}{\partial k^2}\bar{\Gamma}(\vec{k}, -\vec{k})\bigg|_{k=0} = \sum_{p=0}^{\infty} A_p (b\, a_{\mathrm{B}}^{-1/2})^p, \tag{12.3.19}$$

$$\frac{\mathrm{d}a_{\mathrm{B}}}{\mathrm{d}a} = \frac{\mathrm{d}}{\mathrm{d}a}\bar{\Gamma}(\vec{0},\vec{0}) = \sum_{p=0}^{\infty} B_p (b\, a_{\mathrm{B}}^{-1/2})^p, \tag{12.3.20}$$

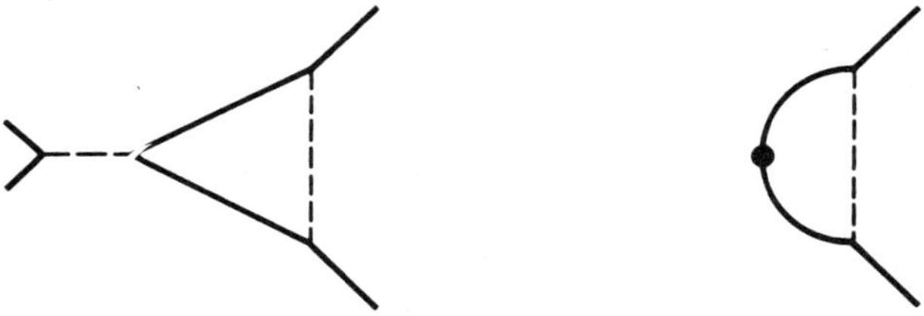

(a) (b)

Fig. 12.8. (a) Diagram contributing to $\overline{\Gamma}(\vec{0},\vec{0},\vec{0},\vec{0})$. (b) Diagram contributing to da_B/da. The contributions of these diagrams are clearly equal.

$$\overline{\Gamma}(\vec{0},\vec{0},\vec{0},\vec{0}) = b \sum_{p=0}^{\infty} C_p(b\, a_B^{-1/2})^p. \tag{12.3.21}$$

We can first extract the renormalization coefficient $\mathfrak{Z}_1$ from eqns (12.3.5) and (12.3.6)

$$(\mathfrak{Z}_1)^{-2} = \sum_{p=0}^{\infty} A_p(b\, a_B^{-1/2})^p$$

and this result enables us to derive the expansion of the renormalized vertex $\overline{\Gamma}_R(\vec{0},\vec{0},\vec{0},\vec{0})$

$$\overline{\Gamma}_R(\vec{0},\vec{0},\vec{0},\vec{0}) = (\mathfrak{Z}_1)^2 \overline{\Gamma}(\vec{0},\vec{0},\vec{0},\vec{0}) = b \sum_{p=0}^{\infty} D_p(b\, a_B^{-1/2})^p. \tag{12.3.22}$$

We can now introduce the dimensionless variable u by setting

$$\overline{\Gamma}_R(\vec{0},\vec{0},\vec{0},\vec{0}) = \xi^{-1} u \tag{12.3.23}$$

where u is a pure number which is incorrectly called 'renormalized coupling constant' [see the dimension equations (11.5.17)]. By comparing the two preceding equations and by using eqn (12.3.19), one obtains

$$u = b\,\xi . \sum_{p=0}^{\infty} D_p(b\, a_B^{-1/2})^p$$

where $b\,\xi$ can be deduced from (11.3.14)

$$b\,\xi = b\, a_B^{-1/2} \left[\sum_{p=0}^{\infty} A_p(b\, a_B^{-1/2})^p \right]^{1/2}.$$

Thus, u can be expressed in the form of a series

$$u = \sum_{p=1}^{\infty} E_p(b\, a_B^{-1/2})^p \qquad \text{where } E_1 = 1. \tag{12.3.24}$$

This quantity u plays an essential role in the theory. In fact, eqn (12.3.23) expresses $\bar{\Gamma}_R(\vec{0},\vec{0},\vec{0},\vec{0})$ as a function of the characteristic length ξ which defines the size of the system in the critical domain. Therefore, eqn (12.3.23) is a scaling law and u a genuine physical quantity. Since the renormalized model exists even in the critical domain, as a consequence of the fact that all the divergences are eliminated by renormalization, the value of u always remains finite. Thus, whereas $b\,a_B^{-1/2} \to \infty$ when one reaches the critical point, the quantity u has a finite limit u^*. Moreover, as the Landau–Ginzburg model becomes classical above four dimensions (the mean field theory applies in this case), we expect that $u^* = 0$ for $d > 4$. Thus the variable u is a very good expansion parameter.

We can now invert eqn (12.3.24) and, in this way, we obtain

$$b\,a_B^{-1/2} = \sum_{p=1}^{\infty} F_p u^p. \tag{12.3.25}$$

All series can be re-expressed in terms of u. Now let us see how u^* can be determined. Let us set

$$w(u) = -\,b\frac{\partial u}{\partial b} \qquad \left(= -\frac{\partial u}{\partial \ln b}\right).$$

When $b\,a_B^{-1/2} \to \infty$, $u \to u^*$ and therefore u^* is a solution of the equation

$$w(u^*) = 0. \tag{12.3.26}$$

The effective exponents $\gamma(u)$ and $\nu(u)$ can be determined from eqns (12.3.19) and (12.3.20). In agreement with (12.3.17), we may set

$$2[1 - 1/\gamma(u)] = b\frac{\partial}{\partial b}\ln(\mathrm{d}a/\mathrm{d}a_B)$$

$$2[-1 + 2\nu(u)/\gamma(u)] = b\frac{\partial}{\partial b}\ln(\xi^2 a_B)$$

and express the right-hand side in terms of u with the help of (12.3.25). When $b\,a_B^{-1/2} \to \infty$ and $u \to u^*$, the limits of the effective exponents $\gamma(u)$ and $\nu(u)$ are the critical exponents γ and ν

$$\gamma(u^*) = \gamma$$
$$\nu(u^*) = \nu. \tag{12.3.27}$$

Moreover, the value of the exponent ω which characterizes the critical approach is given by

$$\omega = w'(u^*) \tag{12.3.28}$$

Thus, the functions $w(u)$, $\gamma(u)$ and $\nu(u)$ are given in terms of u by three series whose first terms are known. Unfortunately, these series are not convergent (even for very small values of u). This phenomenon is very general, the number of diagrams of order p increasing very strongly with p. Here, the contribution of order p increases, to a first approximation, like $p!$ and a similar behaviour appears even in very simple cases.*

More precisely, Brezin, Le Guillou, and Zinn-Justin showed[25] that the coefficients $\mathcal{O}_p$ of the expansion of any physical quantity associated with the Landau–Ginzburg model

$$\mathcal{O}(u) = \sum_{p=0}^{\infty} \mathcal{O}_p u^p,$$

have an asymptotic behaviour of the form

$$\mathcal{O}_p \simeq (-)^p p! \, p^{\alpha_0} K^p \tag{12.3.29}$$

where α_0 is an exponent and K a constant which they could calculate. In particular, for $n = 0$, they found $\alpha_0 = 4$. The method found by Lipatov relies on a generalization of the steepest descent method which is applied to functional integrals.

The series giving $\mathcal{O}(u)$ (which might be $\omega(u)$. $\gamma(u)$ or $v(u)$) is never convergent. However, a Borel transformation[26] of $\mathcal{O}(u)$ gives a function $B(u)$, which is defined by a more convergent series in u. This Borel transformation is given by the following set of equations

$$\mathcal{O}(u) = \int_0^{\infty} \mathrm{d}t \, t^{\beta_0} \mathrm{e}^{-t} B(tu)$$

$$B(u) = \sum_{p=0}^{\infty} B_p u^p = \sum_{p=0}^{\infty} \frac{\mathcal{O}_p u^p}{(\beta_0 + p)!} \tag{12.3.30}$$

* For instance, let us consider the integral

$$Q(b) = \int_{-\infty}^{+\infty} \mathrm{d}x \, \mathrm{e}^{-ax^2 - bx^4}$$

which can be considered as the partition function of a classical harmonic oscillator. We see immediately that

$$Q(b) = \sum_{p=0}^{\infty} Z_p(-b)^p$$

where

$$Z_p = \int_{-\infty}^{+\infty} \mathrm{d}x \frac{x^{4p}}{p!} \mathrm{e}^{-ax^2} = \frac{(2p - 1/2)!}{p!} a^{-2p-1/2}.$$

Thus when $p \to \infty$

$$Z_p \simeq \frac{\pi^{-1/2}}{2} (p-1)! \, (a/2)^{-2p-1/2}$$

The asymptotic behaviour of Z_p could also be obtained by the steepest descent method and the latter approach is more general. In fact, we can write

$$(-1)^p Z_p = \frac{1}{2i\pi} \int_c \mathrm{d}b \frac{1}{b^{p+1}} \int_{-\infty}^{+\infty} \mathrm{d}x \, \mathrm{e}^{-ax^2 - bx^4}$$

where C is a contour circling the origin in the complex plane of u.

In these formulas, β_0 is an arbitrary exponent (but as will be seen later on, it is advantageous to give it the value $\beta_0 = \alpha_0 + 1$).

If the asymptotic behaviour of $\mathcal{O}_p$ is of the form (12.3.29), we have for $p \gg 1$

$$B_p \simeq (-)^p K^p p^{\alpha_0 - \beta_0}. \tag{12.3.31}$$

Thus, the series representing $B(u)$ has a finite radius of convergence ($u = 1/K$) but, for physical reasons, we believe that the functions $B(u)$ studied here remain well-defined for any real positive value of u. Thus, an analytic continuation of the series corresponding to $B(u)$ can be performed along the real axis. From the asymptotic form of B_p, one deduces that $B(u)$ has a singularity at $u = -1/K$ and that this singularity is the closest one to the origin. In the vicinity of this point, $B(u)$ behaves like $(1 + Ku)^{\beta_0 - \alpha_0 + 1}$ if $\beta_0 \neq \alpha_0 + 1$ or like $\ln(1 + Ku)$ if $\beta_0 = \alpha_0 + 1$. Thus, the physicist who uses a Borel transformation finds it advantageous to choose for β_0 the value $\beta_0 = \alpha_0 + 1$ (or a close value) so as to get only a weak singularity at $u = -1/K$.

Le Guillou and Zinn-Justin[21] assumed that $B(u)$ is analytic in the complex plane of u with a cut from $u = -1/K$ to $-\infty$. By conformal transformation, they mapped this cut plane onto the interior of a circle by setting $u = x(1 + Kx/4)^{-2}$. This mapping enabled them to represent $B(u)$ for all positive values of u by a uniformly convergent series in x. Some time later, Baker, Nickel, and Meiron[23] used a slightly different approach representing $B(u)$ by a Padé approximant[27] form

$$B(u) = \frac{P(u)}{Q(u)}$$

where $P(u)$ and $Q(u)$ are polynomials of the same degree or nearly the same.

These techniques give means of calculating $B(u)$ for all real positive values of u. Then, $\mathcal{O}(u)$ is deduced from $B(u)$ by integration. In this way, it is possible to calculate $\omega(u)$, $\gamma(u)$, and $\nu(u)$ from series expansions. Equations (12.3.26) and (12.3.27) show that, subsequently, it is not difficult to determine u^* and the critical exponents γ, ν, and ω.

3.2.3 Expansions with respect to $\varepsilon = 4 - d$

For Landau–Ginzburg theory as for polymer theory, the space dimension $d = 4$ plays a pivotal role. Indeed, for $d > 4$, the corresponding systems have a classical behaviour, and accordingly for $d \geqslant 4$, the critical exponents ν, γ, and ω have the classical values $\nu = 1/2$, $\gamma = 1$, and $\omega = 0$. For $d < 4$, the behaviour changes and with it, the values of the critical exponents which become functions of d. These functions can be expanded, in the vicinity of $d = 4$, in powers of $\varepsilon = 4 - d$.

Various methods were used to obtain such expansions in the framework of Landau–Ginzburg theory, but all of them are founded on common principles: the existence of critical systems described by renormalized theories and the necessity of transforming the expansions with respect to the interaction, into more convergent series. The new expansion parameter is chosen in such a way that it reaches a limit (a 'fixed point') in the asymptotic critical domain. This limit, which vanishes for $d = 4$, can now be expanded with respect to $\varepsilon = 4 - d$. Thus, expansions in powers of the interaction and of ε can be transformed into expansions with respect to ε alone.

Interpreting the 'fixed point' is not obvious, if one adopts the point of view and the methods originally developed by K. Wilson. On the contrary, the interpretation is transparent, if one uses the most recent renormalization methods. Then the expansion parameter is a number defining the intensity of a physical quantity: this quantity itself is expressed by using the characteristic length of the system as the length unit. Therefore, if, as we believe, there exists a limiting critical system, the expansion parameter which is a physical observable must have a finite limit.

In the following, we shall not explain the technical points more precisely. In fact, in the preceding section we already described a renormalization technique in detail, and at the end of this chapter we shall show, also in detail, how we can obtain ε-expansions by direct renormalization of polymers. Consequently, only results will be given here, and the reader will be referred to the relevant original articles.

Wilson was the first to obtain an exact expansion[28] of the critical exponents to first- and second-order in ε. From his results, (and also by using the relation $\eta = 2 - \gamma/\nu$), one gets, for $n = 0$,

$$\nu = \frac{1}{2}\left(1 + \frac{\varepsilon}{8} + \frac{15}{256}\varepsilon^2 + \dots\right)$$

$$\gamma = \left(1 + \frac{\varepsilon}{8} + \frac{13\varepsilon^2}{256} + \dots\right) \qquad (12.3.32)$$

We observe that even for the first-order in ε, there is a discrepancy between this expansion and Flory's formula

$$\nu_{\mathrm{F}} = \frac{3}{d+2} \simeq \frac{1}{2}\left(1 + \frac{\varepsilon}{6} + \frac{\varepsilon^2}{36} + \dots\right).$$

This fact shows clearly that Flory's formula for ν cannot be exact and that it predicts a value which is too large.

Now, let us put $\varepsilon = 1$ in expressions (12.3.32). We find $\nu = 0.592$ and $\gamma = 1.176$. These values are not very different from the more precise values $\nu = 0.588$ and $\gamma = 1.161$ found directly, for $d = 3$, by Le Guillou and Zinn-Justin [see formula (12.3.9)].

Let us put $\varepsilon = 2$, in these expressions. We find $\nu = 0.742$ and $\gamma = 1.453$. These values are similar to those which the British school found by exact counting of polymer configurations on lattices, namely $\nu = 3/4 = 0.75$ and $\gamma = 4/3 = 1.333$ (see Chapter 4).

Finally, for $\varepsilon = 3$ ($d = 1$), one finds $\nu = 0.951$ and $\gamma = 1.832$. Of course, we cannot expect these values to be close to reality; however, they are not very far from it, since the exact value of ν is $\nu = 1$, and since the limit of γ when $d \to 1$ is probably $\gamma = 2$ (but in reality* for $d = 1$, $\gamma = 1$).

The critical exponents γ, ν, and ω were calculated to third-order in ε independently by Brézin, Le Guillou, and Zinn-Justin, and by Nickel. The outcome was published jointly by these physicists in 1973.[29] To obtain the result, Nickel used Wilson's original method; conversely, Brézin, Le Guillou, and Zinn-Justin used another approach;[30] they eliminated the cut-off by renormalization, and in order to calculate simpler integrals, they worked directly at the critical point; the critical theory and the renormalization constants were then defined by introducing adequate renormalization conditions.[2]

Unfortunately, the critical exponents are represented by ε-expansions which obviously do not converge, and for $\varepsilon = 1$, the terms of order ε^3 are not small with respect to the terms of order ε^2. Thus, the expansion cannot be directly used.

The calculation was extended to fourth-order by Vladimirov, Kazakov, and Tarasov in 1979.[31] This was already an achievement but new progress was made subsequently in this direction since Chetyrkin, Larin, and Tkachov[32,33] succeeded in going to fifth-order in ε. For $n = 0$, the polymer case, the result is the following

$$\frac{1}{\nu} = 2 - \frac{\varepsilon}{4} - \varepsilon^2\left(\frac{11}{128}\right) - \frac{\varepsilon^3}{256}\left(\frac{83}{8} - 33\,\zeta(3)\right)$$
$$- \frac{\varepsilon^4}{1024}\left(\frac{1345}{128} - \frac{609}{8}\zeta(3) - 99\,\zeta(4) + 465\,\zeta(5)\right) + \varepsilon^5(0.913\,484\ldots)$$

* In the critical domain, we expect the Green's function $\mathscr{G}(\vec{k}, -\vec{k})$ to be of the form

$$\mathscr{G}(\vec{k}, -\vec{k}) = \frac{1}{(a - a_c)^{\gamma}\,\Phi(k(a - a_c)^{-\nu})}$$

where $\Phi(x)$ obeys the conditions

(1) $\Phi(0)$ is finite;

(2) $\Phi(x) \propto x^{\gamma/\nu}$ for $x \gg 1$.

Thus at the critical point $\mathscr{G}(\vec{k}, -\vec{k}) \propto 1/k^{\gamma/\nu}$.

This is not what occurs for $d = 1$. In fact, an elementary calculation gives

$$\mathscr{G}(\vec{k}, -\vec{k}) = \frac{A}{(a - a_c)\left[1 + A_0 \dfrac{k^2}{(a - a_c)^2}\right]}$$

and we observe that in this case $\mathscr{G}(\vec{k}, -\vec{k})$ has no limit when $a \to a_c$. Thus, the case $d = 1$ looks pathological, but for d slightly larger than one, we may hope for normal behaviour. Thus, we might have

$$\lim_{d \to 1} \gamma(d) \neq \gamma(1).$$

$$\eta = 2 - \gamma/\nu = \frac{\varepsilon^2}{64}\left(1 + \varepsilon\left(\frac{17}{16}\right) + \frac{\varepsilon^2}{32}\left(\frac{721}{32} - 33\,\zeta(3)\right)\right)$$

$$+ \frac{\varepsilon^3}{128}\left(\frac{11\,923}{128} - \frac{669}{4}\zeta(3) - \frac{153}{2}\zeta(4) + 465\,\zeta(5)\right) + \ldots$$

$$\omega = \varepsilon - \varepsilon^2\left(\frac{21}{32}\right) + \frac{\varepsilon^3}{512}(299 + 528\,\zeta(3))$$

$$- \frac{\varepsilon^4}{4096}\left(\frac{20\,309}{8} + 2\,904\,\zeta(3) - 3\,168\,\zeta(4) + 22\,320\,\zeta(5)\right) \tag{12.3.33}$$

where $\zeta(x)$ is Riemann's function

$$\zeta(x) = \sum_{n=1}^{\infty} \frac{1}{n^x}$$

$$(\zeta(3) = 1.2020 \quad \zeta(4) = 1.0823 \quad \zeta(5) = 1.0369).$$

Incidentally, it seems that in $1/\nu$ the coefficient of ε^5 can be expressed as the sum of rational numbers and of terms of the form $\zeta(x)$ where x is an integer. Thus, the only non-integer numbers appearing in the expansion may be of the form $\zeta(x)$ with x = integer, and this very remarkable result seems also to be true for $\mathfrak{n} \neq 0$ (magnetic systems and other ones). This might suggest that the critical exponents are given by exact simple expressions that are still unknown.

The series corresponding to $1/\nu$ can be written in the explicit form

$$1/\nu = 2 - (0.25)\,\varepsilon - (0.0860)\,\varepsilon^2 + (0.1144)\,\varepsilon^3 + (0.2871)\,\varepsilon^4 + (0.9135)\,\varepsilon^5 + \ldots.$$

We see that the convergence of this series is very poor and the same is true for any $\mathfrak{n}$. Still, by using summation methods *à la* Borel (see Section 3.2.2), it is possible to extract precise values of the exponents from these divergent series. In particular, Le Guillou and Zinn-Justin[34] found the following values

for $d = 3$ ($\mathfrak{n} = 0$)

$$\nu = 0.5885 \pm 0.0025$$
$$\gamma = 1.160 \pm 0.004$$
$$\omega = 0.82 \pm 0.04 \tag{12.3.34}$$

for $d = 2$ ($\mathfrak{n} = 0$)

$$\nu = 0.76 \pm 0.03$$
$$\gamma = 1.39 \pm 0.04$$
$$\omega = 1.7 \pm 0.2 \tag{12.3.35}$$

We observe that, for $d = 3$, the values given by (12.3.9) and by (12.3.34) are very close to one another. For $d = 2$, the values given by (12.3.35) are consistent with the 'exact' ones (see Section 4.1).

3.2.4 Partition functions of a continuous chain: anomalous behaviour for space dimension $d = 4 - 2/p$ (p integer) and in particular for $d = 3$

3.2.4.1 *Origin of the anomaly*

In Chapter 9, Section 3.3.5, we noted that for $d = 4 - 2/p$ where p is an integer, the elimination of short-range divergences has the effect that not only powers of z but also powers of $\ln z$ could appear in the expansions of partition functions. However, we also observed that all physical quantities that can be expressed as ratios of partition functions could be expanded in powers of z, without logarithmic terms.

It will be shown in the present section that this phenomenon is quite general; and we shall follow a method due to one of the authors (1982)[35] and later re-introduced independently by Muthukumar and Nickel (1984).[36] For this purpose, one uses the correspondence between field-theory and polymer theory, described in Chapter 11. The result is that the logarithmic terms which appear in the partition functions can be summed up in the form of an anomalous factor. This factor is the same for all partition functions and therefore cancels out when the calculated quantity is the ratio of two different partition functions.

Let us now describe the method which leads to this result.

3.2.4.2 *Structure of the perturbation series*

We are especially interested in the partition function ${}^{+}\mathscr{Z}(S) = {}^{+}\bar{\mathscr{Z}}(\vec{0}, \vec{0}; S)$ of an isolated polymer chain; this function is calculated in the framework of the standard continuous model and the index $+$ indicates that a short-range cut-off is introduced for regularization (see Chapter 10, Section 4.1). As was shown in Chapter 11, ${}^{+}\bar{\mathscr{Z}}(\vec{k}, -\vec{k}; S)$ is the inverse Laplace–de Gennes transform of $\bar{\mathscr{G}}(\vec{k}, -\vec{k}; a)$

$$\bar{\mathscr{G}}(\vec{k}, -\vec{k}; a) = \int_0^\infty \mathrm{d}s\, \mathrm{e}^{-as}\, {}^{+}\bar{\mathscr{Z}}(\vec{k}, -\vec{k}; S) \tag{12.3.36}$$

which is the Green's function of a Landau–Ginzburg model (in the limit $n \to 0$). The Landau–Ginzburg Lagrangian is given by (11.5.3), and reads

$$\mathscr{L}(\vec{r}) = {}^{0}\mathscr{L}(\vec{r}) + {}^{1}\mathscr{L}(\vec{r})$$

where

$${}^{0}\mathscr{L}(\vec{r}) = \frac{1}{2}\sum_{j=1}^{n}\left\{a\,\varphi_j^2(\vec{r}) + \frac{1}{2}\sum_{i=1}^{d}\,[\partial_i\varphi_j(\vec{r})]^2\right\}$$

is 'usually' considered as the unperturbed Lagrangian. The corresponding unperturbed Green's function reads

$$\bar{\mathscr{G}}_0(\vec{k}, -\vec{k}; a) = \frac{1}{a + k^2/2}.$$

This function diverges at $\vec{k} = 0$ when $a \to 0$. However, the true critical value a_c of

a does not vanish; in fact, a_c depends simultaneously on the interaction parameter b and on the short-range cut-off. This is why expanding $\mathscr{G}(k, -\vec{k}: a)$ in powers of $b\,a^{-2+d/2}$ would not be proper. It would be better to expand the Green's function in powers of the parameter $b(a - a_c)^{-2+d/2}$ and to do that, we could subtract the term $(a - a_c) \sum_{j=1}^{n} \varphi_j^2(\vec{r})$ from ${}^0\mathscr{L}(\vec{r})$ and add this term to ${}^1\mathscr{L}(\vec{r})$ (it would act as a counter-term). However, the best solution is to arrange things in such a way that the unperturbed Green's function takes the form

$$\bar{\mathscr{G}}_1(\vec{k}, -\vec{k}; a) = \frac{1}{a_B + k^2/2}$$

where $\dfrac{1}{a_B} = \bar{\mathscr{G}}(\vec{0}, \vec{0}; a)$ is the exact value of the (non-renormalized) Green's function at $\vec{k} = 0$. To obtain this result, we have only to choose the following expression

$${}^1\mathscr{L}(\vec{r}) = {}^0\mathscr{L}(\vec{r}) + \frac{1}{2}(a_B - a) \sum_{j=1}^{n} \varphi_j^2(\vec{r})$$

as the unperturbed Lagrangian and to introduce in ${}^1\mathscr{L}(\vec{r})$ the corresponding counter-terms.

The effect of these counter-terms in the contributions of the diagram is the elimination all the 'self energy' terms at $k = 0$; indeed, no such term should appear, since the value a_B is exact; therefore the process consists only in performing the necessary subtractions, order by order. Thus, when this prescription is followed, the contributions of the diagrams are obtained in a simple and convergent manner.

Actually, this approach is the classical method used by B.G. Nickel in three dimensions and described at the beginning of this chapter, Section 3.3.2. As we saw, it is thus possible to express a in terms of a_B by using an expansion of the following form [see (12.3.15)]

$$\frac{da_B}{da} = 1 + \sum_{q=1}^{\infty} B_q [b(a_B)^{-2+d/2}]^q \tag{12.3.37}$$

where B_q depends on the space dimension. The expression does not depend on the cut-off, which we let go to zero. On the other hand, by putting $\vec{k} = 0$ in (12.3.36), we find

$$\frac{1}{a_B} = \int_0^\infty dS\, e^{-aS}\, {}^+\mathscr{Z}(S). \tag{12.3.38}$$

Consequently, the partition function ${}^+\mathscr{Z}(S)$ we are looking for is the inverse Fourier transform of $1/a_B$

$${}^+\mathscr{Z}(S) = \frac{1}{2\pi i} \int_{c-i\infty}^{c+i\infty} \frac{da}{a_B} e^{aS}. \tag{12.3.39}$$

Thus, from the expansion (12.3.37), we can derive a formal expansion of ${}^{+}\mathscr{Z}(S)$. In fact, eqn (12.3.37) can be rewritten in the form

$$\frac{\mathrm{d}a}{\mathrm{d}a_{\mathrm{B}}} = 1 + \sum_{q=1}^{\infty} C_q[b(a_{\mathrm{B}})^{-2+d/2}]^q \tag{12.3.40}$$

which by integration gives

$$a - a_{\mathrm{c}} = a_{\mathrm{B}}\left\{1 + \sum_{q=1}^{\infty} \frac{C_q}{1 + q(-2 + d/2)}[b(a_{\mathrm{B}})^{-2+d/2}]^q\right\}, \tag{12.3.41}$$

provided that $d \neq 4 - 2p$ where p is an integer; in particular, this condition excludes the physically important cases $d = 3$ ($p = 2$) and $d = 2$ ($p = 1$).

In the favourable case, where no integer p exists such that $d = 4 - 2/p$, we can invert (12.3.41), and in this way we obtain an expansion of the form

$$\frac{1}{a_{\mathrm{B}}} = \frac{1}{a - a_{\mathrm{c}}}\left\{1 + \sum_{q=1}^{\infty} D_q[b(a - a_{\mathrm{c}})^{-2+d/2}]^q\right\} \tag{12.3.42}$$

which can be brought in (12.3.39). Thus, with the help of the coupled equations ($\alpha > -1$)

$$\int_0^{\infty} \mathrm{d}S \exp[-S(a - a_{\mathrm{c}})]S^{\alpha} = \alpha!\,(a - a_{\mathrm{c}})^{-\alpha-1}$$

$$\frac{1}{2\pi i}\int_{c-i\infty}^{c+i\infty} \mathrm{d}a\, \mathrm{e}^{aS}(a - a_{\mathrm{c}})^{-\alpha-1} = \frac{1}{\alpha!}S^{\alpha}\exp[a_{\mathrm{c}}S] \tag{12.3.43}$$

we find the following formal expansion of ${}^{+}\mathscr{Z}(S)$

$${}^{+}\mathscr{Z}(S) = \exp[a_{\mathrm{c}}S]\left[1 + \sum_{q=1}^{\infty} \frac{D_q}{[q(2 - d/2)]!}(b\,S^{2-d/2})^q\right]$$

and consequently, since $z = b\,S^{2-d/2}(2\pi)^{-d/2}$ according to (10.1.7), we thus find the series expansion of the renormalized partition function $\mathscr{Z}(S)$

$$\mathscr{Z}(S) = 1 + \sum_{q=1}^{\infty} \frac{D_q(2\pi)^{dq/2}}{[q(2 - d/2)]!}z^q. \tag{12.3.44}$$

3.2.4.3 *A study of the anomalies*

The preceding reasoning does not apply if there exists an integer p such that $d = 4 - 2/p$, because in this case (eqn 12.3.41) becomes meaningless. Let us then examine the situation in which

$$d = 4 - \frac{2}{p}(1 - \eta)$$

where η is very small. In this case, (12.3.41) reads

$$a - a_{\mathrm{c}} = \frac{C_p b^p a_{\mathrm{B}}^{\eta}}{\eta} + a_{\mathrm{B}}\left\{1 + \sum_{\substack{q=1\\ q\neq p}}^{\infty} \frac{C_q}{1 + q(-2 + d/2)}[b(a_{\mathrm{B}})^{-2+d/2}]^q\right\}.$$

In order to eliminate the singularity at $\eta = 0$ (formally), we introduce

$$a_0 = a_c + \frac{C_p b^{p(1+\eta)}}{\eta}.$$

Going to the limit $\eta \to 0$, we find*

$$ {}^{\cdot}a \equiv a - a_0 - C_p b^p \ln(a_B b^{-p}) = a_B \left\{ 1 + \sum_{\substack{q=1\\ q\neq p}}^{\infty} \frac{C_q}{1 - q/p} [b(a_B)^{-1/p}]^q \right\}. \tag{12.3.45}$$

The effect produced by the logarithmic term can be analysed by using a_B as integrating variable, instead of a, in formula (12.3.39). Consequently, from (12.3.39) and with an integration by parts, we deduce

$$ {}^{+}\mathscr{L}(S) = \frac{1}{2\pi i} \int_{c-i\infty}^{c+i\infty} \mathrm{d}a \frac{\mathrm{e}^{aS}}{S} \left(\frac{\mathrm{d}a_B}{\mathrm{d}a} \right) \frac{1}{(a_B)^2} = \frac{1}{2\pi i} \int_{c_N - i\infty}^{c_N + i\infty} \mathrm{d}a_B \frac{\mathrm{e}^{aS}}{S(a_B)^2}. \tag{12.3.46}$$

Let us re-express a in terms of ${}^{\cdot}a$ by using (12.3.45)

$$ {}^{+}\mathscr{L}(S) = \frac{\mathrm{e}^{Sa_0}}{2\pi i} \int_{c_N - i\infty}^{c_N + i\infty} \frac{\mathrm{d}a_B}{S a_B^2} (a_B b^{-p})^{C_p b^p S} \mathrm{e}^{{}^{\cdot}aS}. \tag{12.3.47}$$

Moreover, by inverting the following expansion

$$ {}^{\cdot}a = a_B \left\{ 1 + \sum_{\substack{q=1\\ q\neq p}}^{\infty} \frac{C_q}{1 - q/p} [b(a_B)^{-1/p}]^q \right\}$$

we obtain an expansion analogous to (12.3.42)

$$\frac{1}{a_B} = \frac{1}{{}^{\cdot}a} \left[1 + \sum_{\substack{q=1\\ q\neq p}}^{\infty} {}^{\cdot}D_q (b\, {}^{\cdot}a^{-1/p})^q \right]$$

and by differentiation

$$\frac{\mathrm{d}a_B}{a_B^2} = \frac{\mathrm{d}{}^{\cdot}a}{{}^{\cdot}a^2} \left[1 - \sum_{\substack{q=1\\ q\neq p}}^{\infty} (1 - q/p)\, {}^{\cdot}D_q (b\, {}^{\cdot}a^{-1/p})^q \right].$$

Finally, using these expansions, we can rewrite (12.3.47) in the form

$$ {}^{+}\mathscr{L}(S) = \frac{\mathrm{e}^{Sa_0}}{2\pi i S} \int_{c' - i\infty}^{c' + i\infty} \frac{\mathrm{d}{}^{\cdot}a}{{}^{\cdot}a^2} ({}^{\cdot}a\, b^{-p})^{C_p b^p S} \mathrm{e}^{{}^{\cdot}aS} \times \left[1 + \sum_{q=1}^{\infty} F_q (b\, {}^{\cdot}a^{-1/p})^q \right]$$

where the F_q are coefficients which can be calculated from the D_q and therefore

* As there is a cut-off, a_0 remains finite even in the limit $\eta \to 0$.

from the C_q. In this way, using (12.3.43), we find

$$^+\mathscr{Z}(S) = e^{Sa_0}(b^pS)^{-C_pb^pS}\left[\frac{1}{(1-C_pb^pS)!} + \sum_{q=1}^{\infty}\frac{F_q}{(1-C_pb^pS+q/p)!}(b\,S^{1/p})^q\right]$$

where, according to (10.1.7) and to the equality $4 - d = 2/p$, we have

$$b^p\,S = (2\pi)^{2p-1}\,z^p. \tag{12.3.48}$$

The anomalous factor in front of the expansion can be eliminated by renormalization. For this purpose, we set

$$^+\mathscr{Z}(S) = \frac{e^{Sa_0}(b^pS)^{-C_pb^pS}}{(1-C_pb^pS)!}\mathscr{Z}_p(S) \tag{12.3.49}$$

where $\mathscr{Z}_p(S)$ is the renormalized function

$$\mathscr{Z}_p(S) = 1 + \sum_{q=1}^{\infty} F_q\frac{[1 - C_p(2\pi)^{2p-1}z^p]!}{[1 + q/p - C_p(2\pi)^{2p-1}z^p]!}[(2\pi)^{2-1/p}z]^q. \tag{12.3.50}$$

It is easy to see that $\mathscr{Z}_p(S)$ has a regular expansion in powers of z. Thus, the anomaly at dimension $d = 4 - 2/p$ simply results in an additional renormalization factor which is singular with respect to z and of the form

$$z^{\mu z^p} \qquad (\text{where } \mu = -\,p(2\pi)^{2p-1}C_p)$$

as can be seen by looking at (12.3.49) and (12.3.48).

This result does not apply to $^+\mathscr{Z}(S)$ alone but also to all partition functions, restricted or not, concerning one or several polymers. In all cases, the renormalization factor has the same form, and it is not difficult to understand why. This is a consequence of the fact that in field theory, all the Green's functions have regular series expansions in a_B. The anomaly comes only from the fact that, to get the partition functions, one has to re-express a_B as a function of a.

This explains why anomalies do not appear when one calculates physical quantities given by ratios of the form

$$\frac{^+\mathscr{Z}(\mathscr{R},\mathscr{P};S_1,\ldots,S_N)}{^+\mathscr{Z}(S_1,\ldots,S_N)} \quad \text{or} \quad \frac{^+\mathscr{Z}(S_1,\ldots,S_N)}{^+\mathscr{Z}(S_1)\ldots{}^+\mathscr{Z}(S_N)}.$$

The renormalization factors cancel out, and for the physical quantities under consideration, perfectly regular expansions in powers of $z_1,\ldots,z_N$ are obtained. They are given by

$$\frac{\mathscr{Z}(\mathscr{R},\mathscr{P};S_1,\ldots,S_N)}{\mathscr{Z}(S_1,\ldots,S_N)} \quad \text{or} \quad \frac{\mathscr{Z}(S_1,\ldots,S_N)}{\mathscr{Z}(S_1)\ldots\mathscr{Z}(S_N)}.$$

3.2.5 Landau–Ginzburg model

In Section 3.2.4.1, we showed that in two and three dimensions, singularities appear in the partition functions but that they are eliminated when physical quantities are calculated. This fact has been exploited by Muthukumar and Nickel[36] who used results obtained in field theory, in order to calculate the expansion of the swelling in powers of z, directly in two and three dimensions. Starting from series expansions obtained by Nickel[24] for a zero-component Landau–Ginzburg model, they obtained the following results

For $d = 3$,

$$R^2/3S = 1 + \tfrac{4}{3}z - 2.075\,385\,396\,z^2 + 6.296\,879\,676\,z^3 - 25.057\,250\,72\,z^4 + 116.134\,785\,z^5 - 594.716\,63\,z^6 + \dots \quad (12.3.51)$$

For $d = 2$,

$$R^2/2S = 1 + \frac{z}{2} - 0.121\,545\,25\,z^2 + 0.026\,631\,36\,z^3 - 0.132\,236\,03\,z^4 + \dots \quad (12.3.52)$$

These series are obviously divergent, but by applying adequate renormalization techniques, it is possible to deduce from them precise evaluations of the swelling which is a well-defined function of z. We shall come back to this topic in Chapter 13, Section 1.4.1.

3.3 Direct renormalization of polymers

3.3.1 Basis and extent of the method

The principles and the techniques of renormalization theory were first developed in the framework of field theory. Later, de Gennes showed that these techniques could be applied to polymers but only as a consequence of his discovery of a correspondence between polymers and fields.

However, *the principles and the techniques of renormalization theory are not directly related to the existence of fields.* They apply whenever one deals with a critical system, i.e. whenever one has to describe large-scale phenomena which depend only globally on the chemical microstructure. Thus, because an ensemble of long polymers in a solution constitutes a critical system, renormalization principles and renormalization techniques must be directly applicable to their study. Actually, this idea appeared quite naturally. It led to the decimation method which has been described previously and which lacks efficiency. However, the same idea can be applied in a much better way. This direct renormalization method (des Cloizeaux 1980)[37,38] consists in adapting to polymers methods which had been successful in field theory.[39] In other words, the aim is to bypass the Laplace–de Gennes transformation (see Chapter 11). *This method applies to semi-dilute solutions as well as to dilute solutions.*

In field theory, one renormalizes vertex functions or, in an equivalent way, Green's functions. The vertex functions of field theory have no simple counterparts in polymer theory because the p-reducible diagrams which are factorizable in field theory have not this property in polymer theory (see Chapter 10, Section 6.1). This difference is related to the fact that an area is associated with each polymer segment and that the sum of these areas has to be equal to the area of the polymer. However, Green's functions are directly related to partition functions through the Laplace–de Gennes transformation. This remark suggests the idea that, in order to apply renormalization principles directly to polymers, one has only to renormalize the partition functions: this is exactly what we are going to do.

The advantages to be expected from such a method are important. In fact, we shall see that this approach is very flexible. It enables us to treat, without special difficulties, monodisperse systems as well as systems with a given polydispersion, linear polymers and branched polymers, homopolymers and copolymers. Thus, a large investigation domain is covered by this method.

Of course, approximations have to be made. Perturbation calculation provides a rather convenient framework, but other approximations are possible, and the renormalization techniques which we are going to describe have a very wide scope.

3.3.2 Principles of direct renormalization

3.3.2.1 *Basic partition functions*

We consider that for a set of polymers the connected restricted partition functions $\mathscr{Z}(\ldots)$ constitute the starting data. These functions were defined in Chapter 10, Section 4. When the polymers are monodisperse, these $\mathscr{Z}(\ldots)$ depend only on the area S, on the dimensionless parameter z, and on the external constraints corresponding to the $\mathscr{Z}(\ldots)$ under consideration. The area S [see (10.1.2)] defines the size of an isolated non-interacting polymer; the parameter z, with $0 \leqslant z < \infty$, defines the interaction [see (10.1.7)], and its value indicates precisely whether the behaviour of the system is nearly critical or not. Perturbation calculation gives expressions of the functions $\mathscr{Z}(\ldots)$ for small values of z. The critical behaviour is obtained for large values of z, and we want to study here this critical domain by using renormalization principles. In essence, we postulate the existence of a limiting theory, describing sets of very long polymers.

The most important quantities, from a theoretical point of view, are:

(a) The connected partition functions $\mathscr{Z}(S)$ and $\mathscr{Z}(S, S)$ concerning respectively one and two polymers.

(b) The Fourier transform $\overline{\mathscr{Z}}(\vec{k}, -\vec{k}; S)$ of the partition function $\mathscr{Z}(\vec{r}_1, \vec{r}_2; S)$, which can be defined by

$$\overline{\mathscr{Z}}(\vec{k}, -\vec{k}; S) = \mathscr{Z}(S)\langle\langle \exp[i\vec{k}\cdot\vec{r}(S) - \vec{r}(0)]\rangle\rangle.$$

In particular, this function leads to the mean square end-to-end distance of an isolated chain, since we have

$$R^2 = \langle\!\langle [\vec{r}(S) - \vec{r}(0)]^2 \rangle\!\rangle = -\frac{2d}{\mathscr{Z}(S)}\left[\frac{\partial}{\partial k^2}\overline{\mathscr{Z}}(\vec{k}, -\vec{k}; S)\right]_{k=0}. \quad (12.3.53)$$

On the other hand, the second virial coefficient is given by $\mathscr{Z}(S, S)$ and $\mathscr{Z}(S)$ since according to (10.5.6) we have

$$\boldsymbol{\Pi\beta} = \boldsymbol{C} - \frac{1}{2}\boldsymbol{C}^2 \frac{\mathscr{Z}(S, S)}{[\mathscr{Z}(S)]^2} + \dots . \quad (12.3.54)$$

3.3.2.2 *Introduction of a new scale*

The area S characterizes the size of non-interacting Brownian chains corresponding to $z = 0$. When z becomes large, the size of the chains becomes also very large and in order to represent the physical situation in the domain $z \gg 1$, it is necessary to use another scale.*

For an isolated self-interacting polymer, we can define an area ${}^{e}S$ as follows

$$R^2 \equiv \langle\!\langle [\vec{r}(S) - \vec{r}(0)]^2 \rangle\!\rangle = {}^{e}S d, \quad (12.3.55)$$

and sometimes we shall also use the notation

$${}^{e}S = X^2. \quad (12.3.56)$$

The area ${}^{e}S$ (or the length X) defines the new scale.

Of course, we have

$${}^{e}S/S = X^2/S = \mathfrak{X}_0(z) \quad (12.3.57)$$

where $\mathfrak{X}_0(z)$ is the well-known swelling factor.

Let us now examine how $\mathfrak{X}_0(z)$ increases when $z \to \infty$. We may consider a long self-avoiding chain drawn on a lattice; there exists an approximate correspondence with our model

$$S \propto N$$

$${}^{e}S \propto N^{2\nu}$$

and consequently,

$${}^{e}S/S \propto N^{2\nu - 1} \propto S^{2\nu - 1}.$$

Moreover, according to (10.1.7), when the interaction b is fixed, we have

$$z \propto S^{\varepsilon/2} \qquad \text{with } \varepsilon = 4 - d.$$

Thus, the asymptotic behaviour of $\mathfrak{X}_0(z)$ must be of the form

$$\mathfrak{X}_0(z) \propto z^{2(2\nu - 1)/\varepsilon} \qquad (z \to \infty). \quad (12.3.58)$$

When S remains fixed and when z becomes infinite, ${}^{e}S$ tends to infinity. This is

* Obviously, from a theoretical point of view, this change of scale is imperative, because for $z \gg 1$ the swelling becomes enormous; nevertheless, in practice, for a good solvent, the swelling can be 5 or 6 but not much larger

why ${}^{e}S$ *which is a physical quantity* must be chosen to define the new scale. Consequently, $\mathfrak{X}_0(z)$ must be considered as a renormalization factor.

3.3.2.3 *Third renormalization; renormalization factors*

The partition functions $\mathscr{Z}(\ldots)$ which are functions of S and z can be re-expressed in terms of ${}^{e}S$ and z, and if ${}^{e}S$ is kept fixed, they must converge to *a finite limit* when $z \to \infty$ (which implies $S \to 0$).

This limiting behaviour characterizes 'Kuhnian' chains which are continuous limits of chains with excluded volume, just as 'Brownian' chains are continuous limits of chains with independent links. Kuhnian chains ($z \to \infty$), like Brownian chains ($z = 0$), are critical objects which depend only on one length X. Non-zero finite values of z correspond to a crossover domain between two different critical types of behaviour.

These principles come up against a difficulty. All functions $\mathscr{Z}(\ldots)$ become infinite when $z \to \infty$, and this is easy to see.

For example, let us consider $\mathscr{Z}(S)$. This quantity is dimensionless (see (10.4.52)] and, consequently, we may set

$$\mathscr{Z}(S) = [\mathfrak{X}_1(z)]^2. \qquad (12.3.59)$$

On the other hand, for a self-avoiding chain with N links, drawn on a lattice we have

$$Z(N) \propto N^{\gamma-1}\mu^N.$$

In the framework of our model, this relation amounts to the assertion that $\mathscr{Z}(S)$ is proportional to $S^{\gamma-1}$, the factor μ^N being absorbed by the first and the second renormalization. In fact, as we saw previously (Chapter 10, Section 4.1), going to the continuous limit implies a first renormalization, even in the absence of interactions, and a second renormalization, related to the interactions. These renormalizations aim at eliminating short-range divergences which are called 'ultraviolet' divergences from analogy with electromagnetism. As $z \propto S^{\varepsilon/2}$, in accordance with the preceding remarks, we expect $\mathfrak{X}_1(z)$ to have an asymptotic behaviour of the form

$$\mathfrak{X}_1(z) \propto z^{(\gamma-1)/\varepsilon} \qquad (z \gg 1). \qquad (12.3.60)$$

We must now describe the system in such a way that it remains well-defined in the limit $z \to \infty$. Therefore, it is necessary to renormalize $\mathscr{Z}(S)$. Let $\mathscr{Z}_{\mathrm{R}}({}^{e}S)$ be the corresponding renormalized partition function, which like $\mathscr{Z}(S)$ must be dimensionless. By definition, in the limit $z \to \infty$, $\mathscr{Z}_{\mathrm{R}}({}^{e}S)$ must depend only on the physical quantity ${}^{e}S = X^2$, which defines the length scale in the system. Consequently, $\mathscr{Z}_{\mathrm{R}}({}^{e}S)$ must be a constant and we may set

$$\mathscr{Z}_{\mathrm{R}}({}^{e}S) = 1. \qquad (12.3.61)$$

In agreement with (12.3.59), we also have

$$\mathscr{Z}(S) = [\mathfrak{X}_1(z)]^2 \mathscr{Z}_{\mathrm{R}}({}^{e}S). \qquad (12.3.62)$$

Previously, we defined the renormalization factor $\mathfrak{X}_0(z)$; we now see that $\mathfrak{X}_1(z)$ can be considered as another renormalization factor defined by the renormalization condition (12.3.61) in connection with (12.3.62). These renormalization factors (and we shall see that there are others!) are associated with the third renormalization.

This third renormalization differs from the two preceding ones, the aim of which was to eliminate (ultraviolet) short-range divergences. On the contrary, the third renormalization aims at absorbing long-range divergences of the infra-red type, related to the swelling of the chains. These infra-red divergences are especially severe for $d < 4$, whereas the ultraviolet divergences are especially severe (i.e. non-renormalizable) for $d > 4$. The space dimension $d = 4$ is marginal both for infra-red and ultraviolet divergences, a fact which has sometimes produced some confusion.

These facts imply that the renormalization techniques which have been so useful in field theory (ultraviolet renormalization) should also be applicable here. The direct renormalization method[38] which we expound here can be partly justified by invoking the following considerations. In field theory, for a Landau–Ginzburg system with N fields characterized by parameters $a_1, \ldots, a_N$ (square masses; see Chapter 11), the ultraviolet divergences can be eliminated by renormalization, and it is possible to find renormalization processes *which do not depend* on the a_j. Thus, the latter commute with Laplace–de Gennes transformation and consequently they can be used *directly* to renormalize the partitions functions of polymers. For instance, Schäfer and Witten[40] noted that in field-theory the zero-mass renormalization satisfies this criterion. More recently, a different and perhaps more elegant approach was found by Benhamou[41] who observed that the dimensional renormalization, introduced by 't Hooft and Veltman[42] is valid for non-zero masses and still commutes with the Laplace–de Gennes transformation. These mathematical justifications will not be developed in the present book, and we shall content ourselves here with a more pragmatic approach.[38]

The fundamental idea is that, in the limit $z \to \infty$, everything should depend only on ${}^{e}S$. However, this hope is only partially realized. To obtain such a behaviour for all quantities, we must renormalize all partition functions, as we renormalized $\mathscr{Z}(S)$ by defining $\mathscr{Z}_{\mathrm{R}}({}^{e}S)$. We assert that to perform this operation, we need only a few renormalization factors. These renormalization factors are always pure numbers that are functions of z. Let us now compare polymer theory and field theory (Landau–Ginzburg model: see Chapter 11); relying on similarity arguments, we claim that, in polymer theory, the renormalization factors are related to the presence of chain ends (vertex insertions in field theory). More precisely, we shall consider a set of polymer configurations in which M pieces of polymers with areas $S_1, \ldots, S_M$ are tied together at the same point (see Fig. 12.9) and, in agreement with (10.1.7) we can introduce the parameters

$$z_j = b\, S_j^{2-d/2}\, (2\pi)^{-d/2}.$$

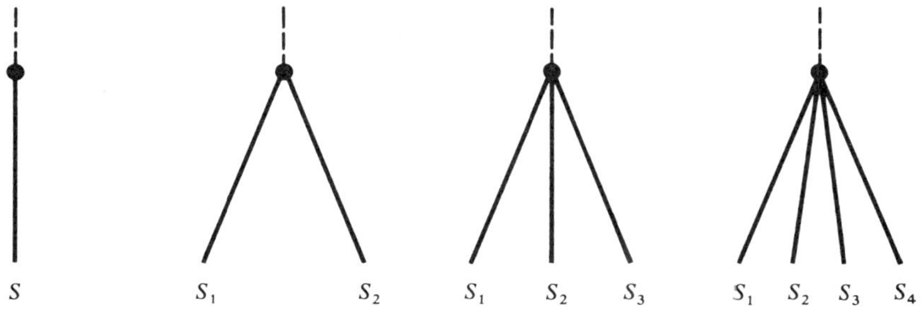

Fig. 12.9. Vertices or insertions requiring the introduction of renormalization factors. A wave vector is injected by the (dashed) interaction line at each vertex. If this wave vector vanishes, the interaction line may be omitted.

To renormalize the partition functions associated with such configurations, it is necessary to introduce a renormalization factor $\mathfrak{X}(z_1, \ldots, z_M)$ for the vertex. Such renormalization factors must be used

(1) to renormalize the partition functions of star polymers and comb polymers, which can be treated in the same way as linear polymers;

(2) when one tries to evaluate probabilities of contact between polymers.

In a sense, these vertex renormalization factors play about the same role (in a different connection) as the dielectric constants in classical physics.

When all the areas $S_1, \ldots, S_M$ of a vertex are of the same order of magnitude $S_1 = m_1 S, \ldots, S_M = m_M S$ where $m_1, \ldots, m_M$ are finite numbers, we can introduce simpler renormalization factors $\mathfrak{X}_M(z)$.

A partition function $\mathcal{Z}(\mathbf{\Omega})$ coresponding to a set $\mathbf{\Omega}$ of configurations can be renormalized by writing[37,38]

$$\mathcal{Z}(\mathbf{\Omega}) = \prod_i \mathfrak{X}_{M_i} \mathcal{Z}_{\mathrm{R}}(\mathbf{\Omega}) \tag{12.3.63}$$

where the index i is used to label the vertices of the system.*

For example, we renormalize the connected partition function $\mathcal{Z}(S_1, \ldots, S_M)$ by setting

$$\mathcal{Z}(S_1, \ldots, S_N) = \mathfrak{X}_1^2(z_1) \ldots \mathfrak{X}_1^2(z_N) \mathcal{Z}_{\mathrm{R}}({}^{\mathrm{e}}S_1, \ldots, {}^{\mathrm{e}}S_N) \tag{12.3.64}$$

(for $N = 1$, this equation coincides with (12.3.62)). Renormalization factors can be precisely defined, in all cases, by introducing adequate renormalization conditions which the renormalized partition functions must satisfy.

One of these, the renormalization factor $\mathfrak{X}_2(z)$ is trivial. To see this we shall consider a diagram (connected or non-connected). At one point of this diagram, on a polymer line, we attach an external interaction line bearing a zero wave

* In particular, this fundamental formula can be used to count the number of configurations of a network with fixed topology.[43]

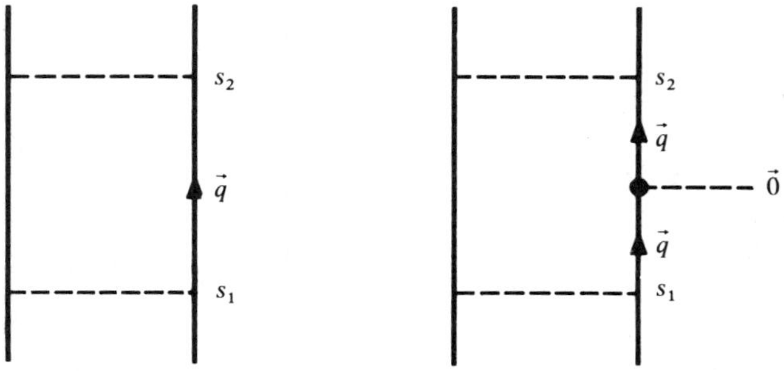

Fig. 12.10. The contributions of these two diagrams are identical; the insertion does not play any role.

vector (see Fig. 12.10). Let $\vec{q}$ be the wave vector borne by the polymer segment on which we grafted the external line. Let s_1 and s_2 be the coordinates of the end points of the segment along the polymer and s_i the coordinate of the insertion point with $s_1 < s_i < s_2$. The contribution of the diagram including the insertion is

$$\exp[-q^2(s_2 - s_i)/2] \exp[-q^2(s_i - s_1)/2] = \exp[-q^2(s_2 - s_1)/2].$$

Therefore, it is strictly equal to the contribution of the corresponding diagram without insertion. The insertion does not play any role, and consequenlty we may set

$$\mathfrak{X}_2(z_1, z_2) = 1 \qquad \mathfrak{X}_2(z) = 1. \tag{12.3.65}$$

This result may look surprising to certain physicists; indeed, in field theory, the renormalization of the two-leg diagram is not at all trivial since the exponent ν is associated with it.

In polymer theory, the situation is different; it is necessary to introduce another renormalization factor $\mathfrak{X}_0(z)$ which renormalizes the scale of the system and ν is the critical exponent associated with this renormolization factor, as was emphasized in the preceding Section 3.3.2.2.

3.3.2.4 *Effective critical exponents and critical exponents*

'Effective' critical exponents can be associated with renormalization factors, and we define them as follows

$$\sigma_M(z) = \frac{\varepsilon}{2} \frac{\partial \ln \mathfrak{X}_M(z)}{\partial \ln z}. \tag{12.3.66}$$

Taking into account the definition (10.1.7) of z, we can also write

$$\sigma_M(z) = S \frac{\partial}{\partial S} \ln \mathfrak{X}_M(z) \bigg|_{b = \text{const.}} \tag{12.3.67}$$

and, as S must be considered as proportional to the number of links, the meaning of these exponents $\sigma_M(z)$ is obvious.

The critical exponents σ_M are given by the limits

$$\sigma_M = \lim_{z\to\infty} \sigma_M(z). \tag{12.3.68}$$

The most important critical exponents are σ_0 and σ_1. Actually, eqns (12.3.58), (12.3.60) and (12.3.65) show that

$$\begin{aligned}\sigma_0 &= 2\nu - 1\\ \sigma_1 &= (\gamma - 1)/2\\ \sigma_2 &= 0\end{aligned} \tag{12.3.69}$$

The behaviour of $\sigma_M(z)$ for $z \gg 1$ will be more precisely studied in the following section.

3.3.2.5 *Renormalization strategy; fixed point and behaviour in the vicinity of the fixed point*

To study the most general properties of solutions of linear polymers, only two renormalization factors, $\mathfrak{X}_0(z)$ and $\mathfrak{X}_1(z)$, are needed. The renormalization condition which determines $\mathfrak{X}_1(z)$ is

$$\mathscr{Z}_{\mathrm{R}}({}^{\mathrm{e}}S) = 1 \qquad \mathfrak{X}_1^2(z) = \mathscr{Z}(S). \tag{12.3.70}$$

The renormalization condition which determines $\mathfrak{X}_0(z)$ is

$$\mathfrak{X}_0{}^0(z) = {}^{\mathrm{e}}S/S = -\frac{2}{\mathscr{Z}(S)}\left[\frac{\partial}{\partial(k^2S)}\overline{\mathscr{Z}}(\vec{k}, -\vec{k}; S)\right]\Bigg|_{k=0}. \tag{12.3.71}$$

Thus, if we know the partition functions $\mathscr{Z}(S)$ and $\mathscr{Z}(\vec{k}, -\vec{k}; S)$, we can determine the renormalization factors $\mathfrak{X}_0(z)$ and $\mathfrak{X}_1(z)$, and, consequently, the critical exponents $\sigma_0(z)$ and $\sigma_1(z)$. Unfortunately, in general, these quantities are given by z-expansions and we are interested in large values of z, the critical exponents σ_0 and σ_1 being defined for $z \to \infty$. This means that z is not a good expansion parameter. On the contrary, any number representing a physical quantity that vanishes when $z = 0$ and has a finite limit when $z \to \infty$, can be considered as a good expansion parameter.

In field theory, one uses a parameter related to the behaviour of the renormalized four-leg vertex and incorrectly called the 'renormalized interaction'. In a very similar way, in polymer theory, the second virial coefficient can be used to define the interaction between two polymers. We proceed as follows.

The dimension of $\mathscr{Z}_{\mathrm{R}}({}^{\mathrm{e}}S, {}^{\mathrm{e}}S)$ is given by (10.4.52)

$$\mathscr{Z}_{\mathrm{R}}({}^{\mathrm{e}}S, {}^{\mathrm{e}}S) \sim L^d \qquad (L = \text{length})$$

since the successive renormalizations that are performed to transform ${}^{+}\mathscr{Z}(S, S)$

into $\mathscr{Z}_{\mathbf{R}}({}^{\mathrm{e}}S, {}^{\mathrm{e}}S)$ do not change the dimension of the quantities to which they are applied; in fact, all the renormalization factors that we define are pure numbers.

Consequently, $\mathscr{Z}_{\mathbf{R}}({}^{\mathrm{e}}S, {}^{\mathrm{e}}S)$ can be written in the form

$$\mathscr{Z}_{\mathbf{R}}({}^{\mathrm{e}}S, {}^{\mathrm{e}}S) = -(2\pi)^{d/2}({}^{\mathrm{e}}S)^{d/2} g(z) \qquad (\text{with } {}^{\mathrm{e}}S = X^2) \tag{12.3.72}$$

where $g(z)$ is a pure number. This number is directly related to the second virial coefficient, since according to (12.3.54), we have

$$\begin{aligned}\boldsymbol{\Pi\beta} &= \boldsymbol{C} - \frac{1}{2}\boldsymbol{C}^2 \frac{\mathscr{Z}(S, S)}{[\mathscr{Z}(S)]^2} + \ldots \\ &= \boldsymbol{C} - \frac{1}{2}\boldsymbol{C}^2 \frac{\mathscr{Z}_{\mathbf{R}}({}^{\mathrm{e}}S, {}^{\mathrm{e}}S)}{[\mathscr{Z}_{\mathbf{R}}({}^{\mathrm{e}}S)]^2} + \ldots\end{aligned}$$

and therefore according to (12.3.70) and (12.3.72)

$$\boldsymbol{\Pi\beta} = \boldsymbol{C}[1 + \tfrac{1}{2}(2\pi)^{d/2} g(z) \boldsymbol{C} X^d + \ldots]. \tag{12.3.73}$$

This expansion of $\boldsymbol{\Pi\beta}$ has the form of a scaling law.

When $z \to \infty$, the renormalized theory must remain finite and consequently

$$\lim_{z \to \infty} g(z) = g^*. \tag{12.3.74}$$

Moreover, for $d \geqslant 4$, long chains are expected to behave like Brownian chains in spite of the interactions between chain links; thus $g^* = 0$ for $d \geqslant 4$. Therefore, g is a good expansion parameter which remains finite when $z \to \infty$ and which is also rather small for $d = 3$.

The function $g(z)$ can be obtained by calculating the partition functions $\mathscr{Z}(S)$, $\mathscr{Z}(S, S)$ and $\mathscr{Z}(\vec{k}, -\vec{k}; S)$. This function is given by a series expansion in powers of z and the series can be inverted; in this way, we define a function $z(g)$ such that $z(g^*) = \infty$. All the physical quantities, such as, for instance, the effective exponents $\sigma_M(z)$, can be re-expressed as functions of g. We have

$$\sigma_M[g] = \sigma_M[z(g)] \tag{12.3.75}$$

and the critical exponents σ_M are given by

$$\sigma_M = \sigma_M[g^*]. \tag{12.3.76}$$

Now g^* has to be evaluated. Let us define the function

$$w(z) = 2S \frac{\partial}{\partial S} g \bigg|_{b=\mathrm{cte}} = \varepsilon z \frac{\partial g}{\partial z} \tag{12.3.77}$$

where $w(z)$ is the effective exponent associated with g. When $z \to \infty$, $g \to g^*$ and therefore $w(z) \to 0$. This condition means that we reach the critical limit, and this is precisely what we want.

Now, in $w(z)$, z can be re-expressed in terms of g

$$w[g] = w(z(g)) \tag{12.3.78}$$

and consequently g^* is given by the implicit equation

$$w[g^*] = 0. \tag{12.3.79}$$

Critical exponents can thus be calculated, and a direct application of this method is given in the following section.

Let us now examine the behaviour of $g(z)$ and of the critical exponents in the vicinity of the fixed point. It is admitted that all the functions of g which we consider are regular in the vicinity of g^* (note that similar assumptions are made in field theory). In particular, we define the exponent $\sigma_a[g]$ by

$$\sigma_a[g] = -\frac{\mathrm{d}w[g]}{\mathrm{d}g} \tag{12.3.80}$$

(the index 'a' means asymptotic) and the limiting exponent σ_a by

$$\sigma_a = \sigma_a[g^*]. \tag{12.3.81}$$

This exponent determines the critical approach; it is related to the exponents ω and Δ_1 which are currently used in statistical field theory. The relation between these exponents is as follows

$$\sigma_a = 2\omega\nu = 2\Delta_1. \tag{12.3.82}$$

Thus, for g close to g^*, we have

$$w[g] \sim -(g - g^*)\sigma_a \tag{12.3.83}$$

and eqn (12.3.77) reads

$$\frac{\mathrm{d}z}{z} \simeq \frac{\varepsilon}{\sigma_a}\frac{\mathrm{d}g}{g^* - g}$$

$$z \simeq \frac{z_\vdash}{(g^* - g)^{\varepsilon/\sigma_a}} \tag{12.3.84}$$

where $z_\vdash$ is an integration constant. Thus,

$$g = g^* - (z/z_\vdash)^{-\sigma_a/\varepsilon} \tag{12.3.85}$$

In the same way, we have

$$\sigma_M(z) \simeq \sigma_M + (g - g^*)\left.\frac{\mathrm{d}\sigma_M}{\mathrm{d}g}\right|_{g = g^*}$$

$$\simeq \sigma_M - (z/z_\vdash)^{-\sigma_a/\varepsilon}\left.\frac{\mathrm{d}\sigma_M}{\mathrm{d}g}\right|_{g = g^*}. \tag{12.3.86}$$

More generally, the relation between z and g is given by

$$\frac{\mathrm{d}z}{z} = \varepsilon\frac{\mathrm{d}g}{w[g]} \tag{12.3.87}$$

and this equation can be integrated, in order to calculate the constant $z_\vdash$ which appears in (12.3.84). We note that for small values of z, we have $g = z + \ldots$ In fact, for z small

$$\mathscr{L}_R({}^eS, {}^eS) = -(2\pi)^{d/2}({}^eS)^{d/2} g \simeq -(2\pi)^{d/2} S^{d/2} g = gbS^2/z$$

$$\mathscr{L}_R({}^eS, {}^eS) = \frac{\mathscr{L}(S, S)}{[\mathscr{L}(S)]^2} \simeq bS^2$$

[see (12.3.73), (10.1.7), and Chapter 10, Section 4.3].

By bringing the equation $g = z + \ldots$ in (12.3.77), we find $w(z) = \varepsilon z + \ldots$ and, consequently, for g small $w[g] = \varepsilon g + \ldots$.

Thus eqn (12.3.87) can be integrated in the form

$$\ln\left(\frac{z}{g}\right) = \int_0^g dg' \left(\frac{\varepsilon}{w[g']} - \frac{1}{g'}\right). \tag{12.3.88}$$

We can also write

$$\ln\left(\frac{z}{g}\right) = -\frac{\varepsilon}{\sigma_a}\ln\left(1 - \frac{g}{g^*}\right) + \int_0^g dg' \left(\frac{\varepsilon}{w[g']} - \frac{1}{g'} - \frac{\varepsilon}{(g^* - g')\sigma_a}\right)$$

which gives

$$z = g\left(1 - \frac{g}{g^*}\right)^{-\varepsilon/\sigma_a} \exp\left\{\int_0^g dg' \left(\frac{\varepsilon}{w[g']} - \frac{1}{g'} - \frac{\varepsilon}{(g^* - g')\sigma_a}\right)\right\}. \tag{12.3.89}$$

Now the integrand is regular at $g' = 0$ and at $g' = g^*$. By comparing (12.3.84) and (12.3.89), we find

$$z_\vdash = (g^*)^{1 + \varepsilon/\sigma_a} \exp\left\{\int_0^{g^*} dg \left(\frac{\varepsilon}{w[g]} - \frac{1}{g} - \frac{\varepsilon}{(g^* - g)\sigma_a}\right)\right\}. \tag{12.3.90}$$

Thus, formula (12.3.89) gives us the dependence of z with respect to g near the fixed point and therefore defines the critical approach.

Moreover, we remark that it is possible to express the renormalization factors directly as functions of z

$$\mathfrak{X}_M(z) = \mathfrak{X}_M[g].$$

By combining eqns (12.3.66) and (12.3.77), we obtain

$$\frac{\partial}{\partial g}\ln \mathfrak{X}_M[g] = 2\frac{\sigma_M[g]}{w[g]} \tag{12.3.91}$$

and we note that the equation no longer contains ε. Its integration gives

$$\ln \mathfrak{X}_M[g] = 2\int_0^g dg' \frac{\sigma_M[g']}{w[g']}. \tag{12.3.92}$$

The integrand has a pole at $g' = g^*$ and therefore it diverges when $g \to g^*$. We can separate the contribution of this pole. When g is close to g^*, we have

$$\frac{\sigma_M[g]}{w[g]} \simeq \frac{\sigma_M}{(g^* - g)\sigma_a}.$$

Consequently, eqn (12.3.92) can also be written in the form

$$\mathfrak{X}_M[g] = (1 - g/g^*)^{-2\sigma_M/\sigma_a} \exp\left\{2\int_0^g dg' \left(\frac{\sigma_M[g']}{w[g']} - \frac{\sigma_M}{(g^* - g')\sigma_a}\right)\right\}. \quad (12.3.93)$$

In the exponential, the integrand is now regular for $g = g^*$. Thus, the replacement of $\sigma_M[g]$ and $w[g]$ by simple approximations can lead to reasonable results.

We shall end this section by remarking that it is not at all necessary to choose g as the 'physical parameter'. For instance, we could choose the exponent $\sigma = \sigma_0(z)$ as the physical parameter and such a choice would be also reasonable, since $\sigma_0(0) = 0$ and since $\sigma_0(\equiv \sigma_0(\infty))$ vanishes for $d \geqslant 4$. In this case, we would define a function

$$w_0(z) = \varepsilon \frac{d\sigma_0(z)}{d \ln z}.$$

Then, by inverting the equality $\sigma = \sigma_0(z)$, we would get $z(\sigma)$ [with $z(0) = 0$] and re-expressing $w_0(z)$ in terms of σ, we would write

$$w_0(z) = w_0[\sigma].$$

The limiting value of σ when z becomes infinite is the critical exponent σ_0; thus σ_0 must be identified with the fixed point σ^* of σ

$$w_0[\sigma^*] = 0 \qquad \sigma_0 = \sigma^*.$$

In this way, the exponent σ_0 would be obtained. Actually, in Chapter 13, Section 1.4.1, we shall take advantage of this method to calculate the swelling of a polymer and the exponent σ_0 directly at $d = 3$.

In the same way, the other physical parameters can be re-expressed in terms of σ; for instance, we could write

$$g(z) = g[\sigma] \qquad \text{and} \qquad g^* = g[\sigma^*].$$

However, it is often more convenient to choose g as the 'physical expansion parameter'.

3.3.3 Direct renormalization calculation of the second virial coefficient and of critical exponents to second-order in ε

The renormalization principles expounded above, can be used to study the behaviour of polymer solutions in the vicinity of dimension $d = 4$. In particular, we shall show how the first terms of the series expansions of critical exponents in powers of $\varepsilon = 4 - d$ can be obtained. This approach is successful because renormalized theories exist in four dimensions, a fact which has been proved in field theory and has been the subject of many articles.[16,17] As an exact correspondence exists between polymer theory and field theory, it turns out that these renormalizability properties are also valid for polymers as was pointed out in Section 3.3.2.3.

By using methods described in Chapter 10, one calculates the first terms of the expansions of the partition functions $\mathscr{Z}(\vec{k}, -\vec{k}; S)$ and $\mathscr{Z}(S, S)$ in powers of z and ε. The diagrams corresponding to $\overline{\mathscr{Z}}(\vec{k}, -\vec{k}; S)$ are indicated in Fig. 12.11.

As $\overline{\mathscr{Z}}(\vec{k}, -\vec{k}; S)$ is used essentially to determine the mean square end-to-end distance of an isolated polymer, we content ourselves with the linear terms in k^2. The result to second-order in z and ε is

$$\overline{\mathscr{Z}}(\vec{k}, -\vec{k}; S) = 1 + z\frac{1}{\varepsilon}(2 + \varepsilon) - z^2\frac{1}{\varepsilon^2}(6 + 7\varepsilon) - \\ - \frac{k^2 S}{2}\left[1 + z\frac{1}{\varepsilon}(4 + 0\varepsilon) - z^2\frac{1}{\varepsilon^2}\left(8 + \frac{3}{2}\varepsilon\right)\right]. \tag{12.3.94}$$

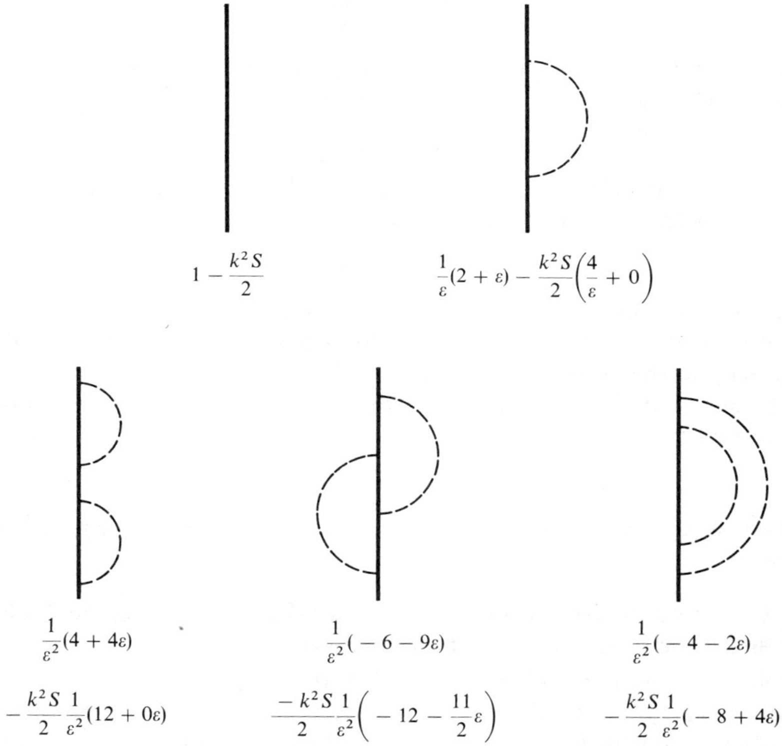

Fig. 12.11. The diagrams of order one and two contributing to $\overline{\mathscr{Z}}(\vec{k}, -\vec{k}; S)$. We give only linear contribution in $k^2 S$ to first-orders in ε.

The diagrams contributing to $\mathscr{Z}(S, S)$ are either I-reducible (by cutting an interaction line: see Fig. 12.12) or I-irreducible. The contributions of I-reducible diagrams can be factorized and are therefore easy to calculate. The simplest I-irreducible diagrams are shown in Fig. 12.13.

By summing the contributions of the I-reducible and I-irreducible diagrams, to first-orders in z and ε, we find

$$\mathscr{Z}(S, S) = -bS^2\left[1 + z(1 + 4\ln 2) + z^2\frac{1}{\varepsilon}(2 - 32\ln 2)\right]$$

$$= (2\pi)^{d/2} S^{d/2}\left[z + z^2(1 + 4\ln 2) + z^3\frac{1}{\varepsilon}(2 - 32\ln 2)\right]. \qquad (12.3.95)$$

We note the absence of terms proportional to z^2/ε and z^3/ε^2, as a consequence of cancellations between diagrams.

Equations (12.3.71) and (12.3.94) give

$${}^{e}S/S = \frac{1 + z\frac{1}{\varepsilon}(4 + 0\varepsilon) - z^2\frac{1}{\varepsilon^2}\left(8 + \frac{3}{2}\varepsilon\right)}{1 + z\frac{1}{\varepsilon}(2 + \varepsilon) - z^2\frac{1}{\varepsilon^2}(6 + 7\varepsilon)},$$

and consequently

$$\mathfrak{X}_0(z) = {}^{e}S/S = 1 + z\frac{1}{\varepsilon}(2 - \varepsilon) + z^2\frac{1}{\varepsilon^2}\left(-6 + \frac{11}{2}\varepsilon\right). \qquad (12.3.96)$$

Moreover, eqns (12.3.59) and (12.3.94) give

$$\mathfrak{X}_1(z) = [\mathscr{Z}(S)]^{1/2} \equiv [\overline{\mathscr{Z}}(\vec{0}, \vec{0}; S)]^{1/2}$$

$$= 1 + z\frac{1}{\varepsilon}\left(1 + \frac{\varepsilon}{2}\right) + z^2\frac{1}{\varepsilon^2}\left(-\frac{7}{2} - 4\varepsilon\right). \qquad (12.3.97)$$

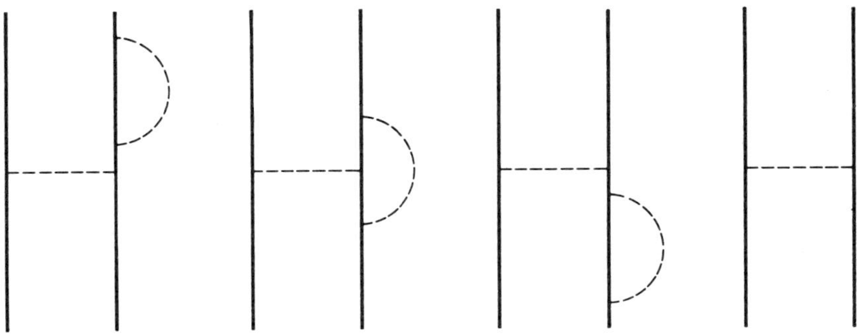

Fig. 12.12. I-reducible diagrams contributing to $\mathscr{Z}(S, S)$. The contributions of these diagrams are factorized and can be obtained from the contribution of diagrams corresponding to $\mathscr{Z}(S)$, represented in Fig. 12.11.

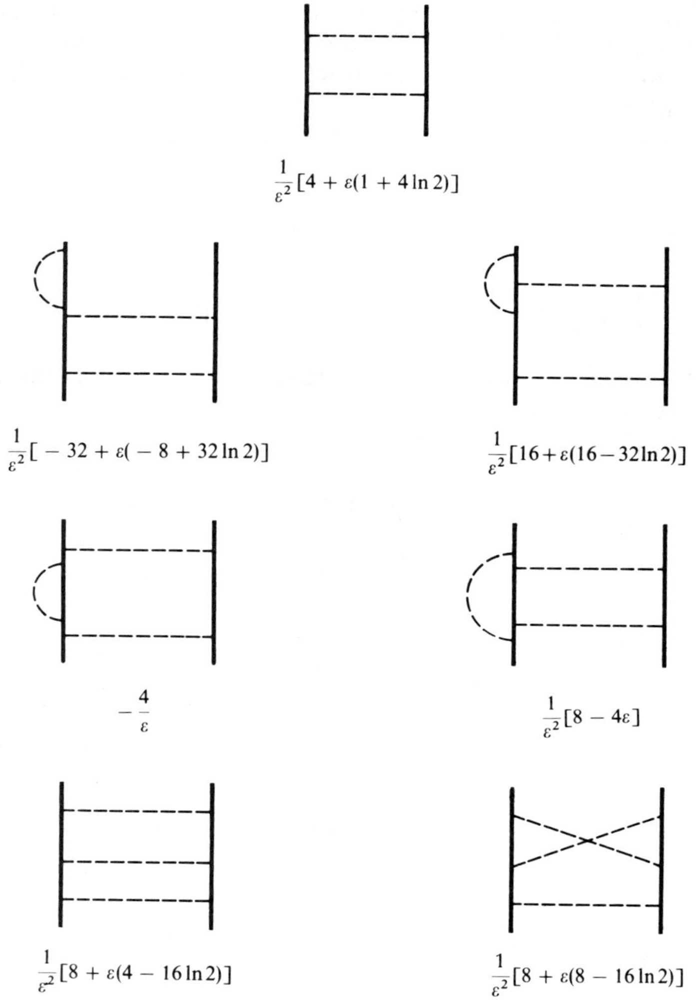

Fig. 12.13. I-irreducible diagrams of order two and three corresponding to $\mathscr{X}(S, S)$ and their contributions to first-orders in ε.

Thus, the effective exponents $\sigma_0(z)$ and $\sigma_1(z)$ defined by (12.3.66) have the following expansions

$$\sigma_0(z) = z\left(1 - \frac{\varepsilon}{2}\right) + z^2\frac{1}{\varepsilon}\left(-8 + \frac{15}{2}\varepsilon\right) + \dots$$

$$\sigma_1(z) = z\left(\frac{1}{2} + \frac{\varepsilon}{4}\right) + z^2\frac{1}{\varepsilon}\left(-4 - \frac{9}{2}\varepsilon\right) + \dots \quad (12.3.98)$$

Let us now express g in terms of z; we start from the coupled equations [see (12.3.72) and (12.3.64)]:

$$(2\pi)^{-d/2}\mathscr{Z}_{\mathbf{R}}({}^{\mathrm{e}}S, {}^{\mathrm{e}}S) = -({}^{\mathrm{e}}S)^{d/2} g$$
$$\mathscr{Z}_{\mathbf{R}}({}^{\mathrm{e}}S, {}^{\mathrm{e}}S) = [\mathfrak{X}_1(z)]^{-4}\mathscr{Z}(S, S),$$

and, with the help of (12.3.95), this equation leads to

$$\begin{aligned} g &= [\mathfrak{X}_1(z)]^{-4}({}^{\mathrm{e}}S)^{-d/2}[-(2\pi)^{-d/2}\mathscr{Z}(S, S)] \\ &= [\mathfrak{X}_1(z)]^{-4}[\mathfrak{X}_0(z)]^{-2+\varepsilon/2}\left[z + z^2(1 + 4\ln 2) + z^3\frac{1}{\varepsilon}(2 - 32\ln 2)\right]. \end{aligned}$$

Finally, by using (12.3.96) and (12.3.97), we find

$$g = z + z^2\frac{1}{\varepsilon}[-8 + \varepsilon(2 + 4\ln 2)] + z^3\frac{1}{\varepsilon^2}[64 + \varepsilon(-15 - 64\ln 2)]. \qquad (12.3.99)$$

From this expression, we deduce $w(z)$ [defined by (12.3.77)]

$$\begin{aligned} w(z) = \varepsilon z\frac{\mathrm{d}g}{\mathrm{d}z} &= \varepsilon z + z^2[-16 + \varepsilon(4 + 8\ln 2)] \\ &\quad + z^3\frac{1}{\varepsilon}[192 + \varepsilon(-45 - 192\ln 2)]. \end{aligned}$$

Inverting the expansion (12.3.99), we find

$$z = g + \frac{g^2}{\varepsilon}[8 + \varepsilon(-2 - 4\ln 2)] + \frac{g^3}{\varepsilon^2}[64 + \varepsilon(-49 - 64\ln 2)]. \qquad (12.3.100)$$

Let us now re-express $w(z)$ in terms of g. We obtain

$$w[g] = \varepsilon g + g^2[-8 + \varepsilon(2 + 4\ln 2)] + g^3(34) + \ldots. \qquad (12.3.101)$$

We note that all singularities in $1/\varepsilon$ disappear. The value g^* associated with the fixed point is given by the equation $w[g^*] = 0$. Thus, from (12.3.101), we deduce the first two terms of the expansion of g^* with respect to ε:

$$g^* = \frac{\varepsilon}{8} + \frac{\varepsilon^2}{16}\left(\frac{25}{16} + \ln 2\right) + \ldots. \qquad (12.3.102)$$

This result is interesting because g^* is a physical quantity which defines the second virial coefficient of a polymer solution in good solvent and for very long chains. In other terms, g^* defines the second virial coefficient of a solution of Kuhnian chains. For $d = 3$ ($\varepsilon = 1$), the preceding formula gives $g^* = 0.266$, a result which, apparently, is not very very precise, because the second term in (12.3.102) is not small with respect to the first one. This question is discussed, in more detail, in Chapter 13.

Let us now come back to the effective exponents. Re-expressing $\sigma_0(z)$ and $\sigma_1(z)$ [given by (12.3.98)] in terms of g, we find

$$\sigma_0[g] = g\left(1 - \frac{\varepsilon}{2}\right) + g^2\left(\frac{3}{2} - 4\ln 2\right) + \ldots$$

$$\sigma_1[g] = g\left(\frac{1}{2} + \frac{\varepsilon}{4}\right) + g^2\left(-\frac{7}{2} - 2\ln 2\right) + \ldots \qquad (12.3.103)$$

with the help of (12.3.100). Again, we remark that the singularities in $1/\varepsilon$ disappear as they must. The expansion of $\sigma_a[g]$ can be deduced from definition (12.3.80) and from expansion (12.3.101)

$$\sigma_a[g] = -\varepsilon + g[16 + \varepsilon(-4 - 8\ln 2)] - g^2(102) + \ldots . \qquad (12.3.104)$$

Finally, replacing g by g^* in these expressions, we find the critical exponents σ_0, σ_1, and σ_a. Consequently, eqns (12.3.102), (12.3.103), and (12.3.104), give

$$\sigma_0 = \frac{\varepsilon}{8} + \frac{15}{256}\varepsilon^2 + \ldots$$

$$\sigma_1 = \frac{\varepsilon}{16} + \frac{13}{512}\varepsilon^2 + \ldots$$

$$\sigma_a = \varepsilon - \frac{17}{32}\varepsilon^2 + \ldots . \qquad (12.3.105)$$

The usual critical exponents ν, γ, and ω are related to σ_0, σ_1, and σ_a by (12.3.69) and (12.3.82); using these relations, we find

$$\nu = \frac{\sigma_0 + 1}{2} = \frac{1}{2}\left(1 + \frac{\varepsilon}{8} + \frac{15}{256}\varepsilon^2 + \ldots\right)$$

$$\gamma = 2\sigma_1 + 1 = 1 + \frac{\varepsilon}{8} + \frac{13}{256}\varepsilon^2 + \ldots$$

$$\Delta_1 = \omega\nu = \frac{\sigma_a}{2} = \frac{\varepsilon}{2} - \frac{17}{64}\varepsilon^2 + \ldots \qquad (12.3.106)$$

and, in this way, we recover results which have initially been derived from field theory, in the limit of zero-component fields [see (12.3.33)].

More recently (1989), B. Duplantier[44] calculated the expansion of σ_M up to second order for any M ($M \geqslant 0$) and he found

$$\sigma_M = \left(\frac{\varepsilon}{8}\right)\frac{M}{2}(2 - M) + \left(\frac{\varepsilon}{8}\right)^2\frac{M}{8}(M - 2)(8M - 21).$$

3.3.4 Direct renormalization in four dimensions

3.3.4.1 *Marginality of dimension four*

As has been pointed out in this chapter and in Chapter 10, the space dimension $d = 4$ is a limiting dimension. This can easily be seen, for instance, by looking at the definition (10.1.7) of the parameter z. For $d = 4$, b is a pure number; we have

$$z = (2\pi)^{-2} b$$

and now z does not depend on the area S of the unperturbed Brownian chains. For $d = 4$, the critical exponents take their classical values: $\nu = 1/2$, $\gamma = 1$, and the corresponding exponents σ_0, σ_1 vanish.

This can be seen by putting $\varepsilon = 0$ in the expansions (12.3.106). In the same way, g^* vanishes for $d = 4$. Chains, for $d = 4$, have a quasi-Brownian behaviour.

There still remain logarithmic factors which appear when the critical domain is reached. It is interesting to study the origin of these logarithmic terms and we shall verify that, strictly speaking, they cannot be considered as mere scaling law corrections since they are dominant when $N \to \infty$.

Of course, studying the properties of 'good' polymer solutions in a four-dimensional space may look rather academic, but studying the properties of true polymers in solution at Flory's temperature T_F, is a task of great importance, and both problems are very analogous. Indeed, as de Gennes pointed out as early as 1975,[45] T_F corresponds to a tricritical point, and we note that a tricritical system is marginal for $d = 3$, just as a critical system is marginal for $d = 4$.

The fact that logarithmic terms appear in dimension four has been recognized, in field theory, a fairly long time ago. Such terms were already calculated in 1969 for various models by Larkin and Khmel'nitskii[46] who used the 'parquet' approximation. Analogous results were recovered in 1973 by Wegner and Riedel[47] in the framework of a renormalization process. Since that time, other authors have taken an interest in this question. Concerning polymers, this matter was studied more recently (1981) by Tanaka.[48]

3.3.4.2 *Logarithmic corrections in four dimensions*

In order to understand the situation in four dimensions, we place ourselves just below four dimensions and we let $\varepsilon = 4 - d$ go to zero. Looking at the results of Section 3.3.3, we observe

1. That the series expansions in z of the partition functions become infinite when $\varepsilon \to 0$ and this fact was also conspicuous in Chapter 10 [see, for instance, (10.7.7)].

2. That the expressions of all physical quantities in terms of g have a finite limit when $\varepsilon \to 0$.

Thus, for instance, the renormalization factors $\mathfrak{X}_0(z)$ and $\mathfrak{X}_1(z)$ given by (12.3.96) and (12.3.97) become infinite when $\varepsilon \to 0$. On the other hand, when

$\varepsilon \to 0$, the corresponding effective exponents $\sigma_0[g]$ and $\sigma_1[g]$ remain finite and according to (12.3.101), (12.3.103), and (12.3.104), we have

$$w[g] = -8g^2 + 34g^3 + \ldots$$

$$\sigma_0[g] = g + g^2\left[\frac{3}{2} - 4\ln 2\right] + \ldots$$

$$\sigma_1[g] = \frac{g}{2} + g^2\left[-\frac{7}{2} - 2\ln 2\right] + \ldots$$

$$\sigma_a[g] = 16g - 102g^2 + \ldots. \tag{12.3.107}$$

As the renormalization factors are related to the effective exponents by equations of the form (12.3.91), we see that the problem in four dimensions is to express g in terms of the initial parameters.

Now, we observe that, in four dimensions, according to (10.1.7)

$$z = \frac{b}{4\pi^2}. \tag{12.3.108}$$

In this limiting case, z remains constant when $S \to \infty$. The fact that new ultraviolet divergences appear when $\varepsilon \to 0$, implies that again the cut-off area s_0 must be introduced. A new dimensionless parameter $S(s_0)$ comes into play and now the critical behaviour corresponds to the limit $S/s_0 \to \infty$.

Then, for $\varepsilon \neq 0$, keeping b fixed, we let S vary. Taking the definition of z into account [see (10.1.7)], we can rewrite eqn (12.3.87) in the form

$$\frac{\mathrm{d}S}{S} = 2\frac{\mathrm{d}g}{w[g]} \qquad (b = \text{const.}) \tag{12.3.109}$$

which remains valid when $\varepsilon \to 0$. In this limit, according to (12.3.107), we have

$$\frac{1}{w[g]} = -\frac{1}{8g^2} - \frac{17}{32g}. \tag{12.3.110}$$

Because of divergences, it is impossible to integrate (12.3.109) by starting from the values $S = 0$ and $g = 0$. We admit that for $S = s_0$, then $g = g_0$, where g_0 is a function of b. Replacing $w[g]$ in (12.3.109) by the value given by (12.3.110) we can now integrate (12.3.109), and we obtain

$$\ln(S/s_0) = \frac{1}{4g} - \frac{1}{4g_0} - \frac{17}{16}\ln(g/g_0),$$

or to a first approximation

$$g \simeq \frac{1}{4\ln(S/s_0)}. \tag{12.3.111}$$

The asymptotic behaviour of a renormalization factor is obtained in the same way by integrating (12.3.91) in the form

$$\ln(\mathfrak{X}_M[g]/\mathfrak{X}_M[g_0]) = 2\int_{g_0}^{g} dg' \frac{\sigma_M[g']}{w[g']}. \qquad (12.3.112)$$

Thus, for instance, taking (12.3.103) into account, we find (for $M = 0$)

$$\begin{aligned}\ln(\mathfrak{X}_0[g]/\mathfrak{X}_0[g_0]) &= -\frac{1}{4}\int_{g_0}^{g}\frac{dg'}{g'}\left[1 + g'\left(\frac{23}{4} - 4\ln 2\right) + \dots\right] \\ &= -\frac{1}{4}\ln(g/g_0) - \frac{(g-g_0)}{4}\left(\frac{23}{4} - 4\ln 2\right) + \dots\end{aligned}$$

or to a first approximation

$$\frac{\mathfrak{X}_0[g]}{\mathfrak{X}_0[g_0]} \simeq (g/g_0)^{-1/4} \simeq [4g_0 \ln(S/s_0)]^{1/4}. \qquad (12.3.113)$$

Therefore, using the definition (12.3.57) of the swelling $\mathfrak{X}_0$, we obtain

$$\mathfrak{X}_0 = {}^{e}S/S \propto [\ln(S/s_0)]^{1/4} \propto [\ln N]^{1/4}. \qquad (12.3.114)$$

Other quantities can be examined and similar logarithmic corrections can be found for them.

In practice, it seems difficult to verify the existence of such corrections. Nevertheless, by computer simulation Havlin and Ben-Avraham (1981),[49] and also Aragão de Carvalho and Caracciolo (1983),[50] have been able to observe logarithmic effects; for that purpose, the first-named authors used Monte Carlo methods to construct self-avoiding walks on a four-dimensional lattice.

4. EXACT EXPONENTS IN TWO DIMENSIONS

Around 1984, it became clear that critical phenomena in two dimensions can be described exactly and that two different approaches are possible.

As early as 1970, Polyakov[51] showed that at the critical point, a critical system is invariant for transformations belonging to the conformal group. Thus, as this group is rather rich in two dimensions, these invariance properties could later[52] be used to make precise studies of critical phenomena and, in particular, to find exact values of critical exponents.

On the other hand, it appeared that the renormalization properties of a Coulomb gas in two dimensions can be considered as (relatively) trivial[53] and this property can also be used to derive the critical properties of a series of equivalent (or nearly equivalent) models.

A whole book would be necessary to describe the more recent developments of these methods. Here, we shall give only the main results for polymers, indicating only briefly how these results have been obtained.

4.1 Early conjectures

The exact values of the exponents ν and ω corresponding to polymers interacting in a two-dimensional space can be considered as known. They have been found by means of field theory. The formulae given by Cardy and Hamber[54] and confirmed by Nienhuis (1982)[52] led to the following results

$$\nu = 3/4$$
$$\omega = 2. \tag{12.4.1}$$

Moreover, according to a conjecture by Nienhuis, the exponent η is

$$\eta = 5/24;$$

and consequently the value of γ must be

$$\gamma = (2 - \eta)\nu = 43/32 \qquad (= 1.34\,375). \tag{12.4.2}$$

The values of ν and (to a lesser degree) of γ given above are in good agreement with those which the British school obtained by counting chains on a lattice (see Chapter 4). The value 3/4 is also in very good agreement with the value obtained by Derrida by means of the phenomenological renormalization method (see Section 2.2.4); we also note that strangely enough, it coincides with Flory's formula $\nu_F = 3/(d+2)$ for $d = 2$.

The results found by Cardy and Hamber, and by Nienhuis, apply more generally to models of the Landau–Ginzburg type with n-component fields. In the general case, the exponents γ, ν, and ω are given in terms of n (with $-2 < n < 2$) in the parametric form

$$\begin{aligned} n &= -2\cos(2\pi\tau) \\ 1/\nu &= 4 - 2/\tau \\ \omega &= -6 + 6/\tau \qquad 1/2 < \tau \leqslant 1 \\ 2 - \eta/2 &= 1 + 3\tau/4 + 1/4\tau \quad \text{(conjecture)} \end{aligned} \tag{12.4.3}$$

and, of course, to find the exponents concerning polymers we have only to set $n = 0$, $\tau = 3/4$.

In order to justify the conjecture made by Cardy and Hamber, Nienhuis used a cascade of models (including an unsolved six-vertex model)* and equivalences that are more or less exact; finally, he came down to a two-dimensional Coulomb gas. This gas is made of positively and negatively charged particles in interaction, the interaction potential being proportional to $\ln r$ where r is the distance between two charges. Then, Nienhuis could apply to this system approximate renormalization techniques which enabled him to predict the 'critical' properties of the system.

* Six-vertex models have been solved by E. Lieb and followers[55,56] but the six-vertex model considered by Nienhuis is defined on a kagome lattice.

These subtle and complicated arguments are not rigorous and will not be reproduced here. One reason is that the same results can be found by using a very promising method based on properties of conformal invariance. We shall come back to this question in Section 4.2. In what follows we only show how Nienhuis began to tackle the problem.

The starting model is the O(n) model defined as follows. We associate random vectors $\vec{S}_{\mathrm{M}}$ with the sites M of a hexagonal lattice (honeycomb lattice, see Fig. 12.14). Each vector $\vec{S}_{\mathrm{M}}$ has n components S^j_{M}, which have the following stochastic properties

$$\langle\!\langle S^j_{\mathrm{M}} \rangle\!\rangle = 0$$

$$\langle\!\langle S^j_{\mathrm{M}} S^l_{\mathrm{N}} \rangle\!\rangle = \delta_{\mathrm{MN}}\delta_{jl}. \tag{12.4.4}$$

A partition function associated with these fields can be defined by

$$Z(x) = \left\langle\!\!\left\langle \prod_{(\mathrm{A},\,\mathrm{B})} (1 + x\vec{S}_{\mathrm{A}} \cdot \vec{S}_{\mathrm{B}}) \right\rangle\!\!\right\rangle \tag{12.4.5}$$

the product being made over all pairs A, B of nearest neighbours on the lattice. The model depends on n and on another parameter x. For $n = 1$, the model reduces to the Ising model. For any n, we can postulate that in the vicinity of the critical value $x_c(n)$, this model belongs to the same universality class as the Landau–Ginzburg model with n-component field.

The corresponding free energy is defined by the equality

$$\mathbb{F}(x) = -\ln Z(x)$$

and, if we know $\mathbb{F}$, we can deduce the exponents ν and ω from it. Indeed, the classical theories of magnetism[2,57] indicate that in the vicinity of the critical point, the singular part $\mathbb{F}_s(x)$ of $\mathbb{F}(x)$ is of the form

$$\mathbb{F}_s(x) \simeq |x - x_c|^{2\nu} (f_1 + |x - x_c|^{\omega\nu} f_2 + \ldots).$$

The exponents ν and ω which appear in this formula are those which Nienhuis calculates exactly by transforming the initial O(n) model.

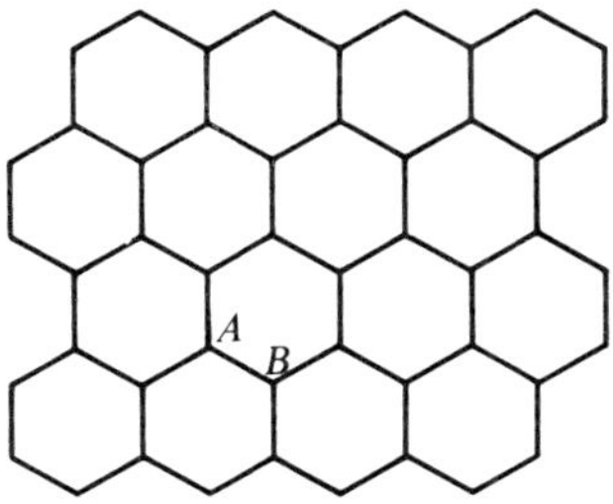

Fig. 12.14. Hexagonal lattice (honeycomb) serving as a basis for the O(n) model considered by Nienhuis.[53] A and B are nearest neighbours.

The first transformation and the simplest one establishes a correspondence between the O(n) model and a 'solid-on-solid' model. In particular, this transformation shows us why in the parametric representation (12.4.3), the number n appears in the form of a cosine. For this reason, the end of this section will be devoted to a description of this transformation.

Let us first remark that Z can be expressed as a sum of contributions corresponding to graphs. Their structure is easily deduced from the x-expansions of Z, obtained from (12.4.5) and (12.4.4). Each graph $\mathfrak{G}$ is made up of a certain number $\mathcal{N}(\mathfrak{G})$ of closed lines (cycles) drawn on the lattice.

Let $N(\mathfrak{G})$ be the total number of links constituting the lines. The graph $\mathfrak{G}$ are constructed in all possible ways. In principle, an index $j\,(j = 1, \ldots, \mathfrak{n})$ must be attributed to each cycle, but the contribution of a graph does not depend on the value of these indices. Therefore, we can sum over each j independently. A factor n will thus be associated with each cycle of $\mathfrak{G}$. Consequently, we have

$$Z(x) = \sum_{\mathfrak{G}} x^{N(\mathfrak{G})}\mathfrak{n}^{\mathcal{N}(\mathfrak{G})}. \tag{12.4.6}$$

Let us now pass on to the 'solid-on-solid' model. With each graph $\mathfrak{G}$, we associate graphs $\mathfrak{G}'$ obtained by orienting the cycles which constitute $\mathfrak{G}$, in all possible ways. Therefore, a graph $\mathfrak{G}$ generates $2^{\mathcal{N}(\mathfrak{G})}$ graphs $\mathfrak{G}'$ (see Fig. 12.15). In this way, we get the solid-on-solid model, and the domains bounded by the oriented cycles of $\mathfrak{G}'$ can be considered as terraces of heights h where h is an integer. Two terraces separated by a cycle have heights which differ by one unit. The walker who goes along a cycle sees the higher terrace on the right, the lower one on the left. The partition function Z' corresponding to the solid-on-solid models is the sum of the contributions associated with the graphs $\mathfrak{G}'$ and calculated as follows. A factor is attributed to each site M belonging to a cycle; if the oriented line turns to the right at M, the factor is $x\,\mathrm{e}^{i\alpha}$; if it turns to the left, the factor is $x\,\mathrm{e}^{-i\alpha}$. The contribution associated with a graph $\mathfrak{G}'$ is the product of the contributions associated with all the sites constituting the graph. Thus a factor $x^N\mathrm{e}^{i6\alpha}$ or $x^N\mathrm{e}^{-i6\alpha}$ is attributed to each N-link graph $\mathfrak{G}'$. Therefore, the value of the partition function corresponding to the model is

$$Z'(x) = \sum_{\mathfrak{G}} x^{N(\mathfrak{G})}\,(\mathrm{e}^{i6\alpha} + \mathrm{e}^{-i6\alpha})^{\mathcal{N}(\mathfrak{G})}. \tag{12.4.7}$$

See Fig. 12.16.

Let us now compare (12.4.6) and (12.4.7). We see that provided we set

$$\mathfrak{n} = 2\cos 6\alpha \tag{12.4.8}$$

we can write

$$Z(x) = Z'(x).$$

This equality establishes a correspondence between the O(n) model and the solid-on-solid model. Similar results could be obtained with a square lattice. It seems likely that all these solid-on-solid models belong to the same universality class and have similar critical properties. They may also be more fundamental than the O(n) models, and this is also why they look especially interesting.

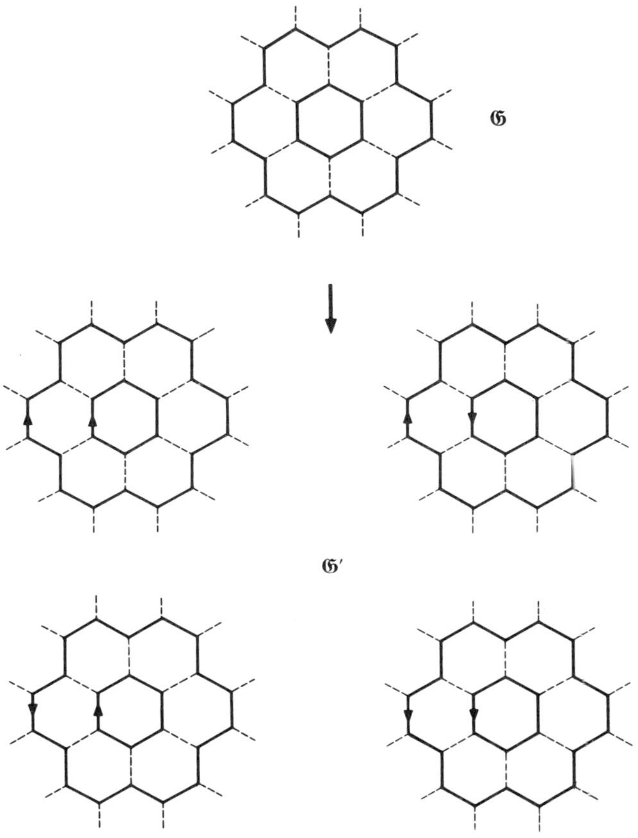

Fig. 12.15. A set a graphs $\mathfrak{G}'$ generated by a graph $\mathfrak{G}$.

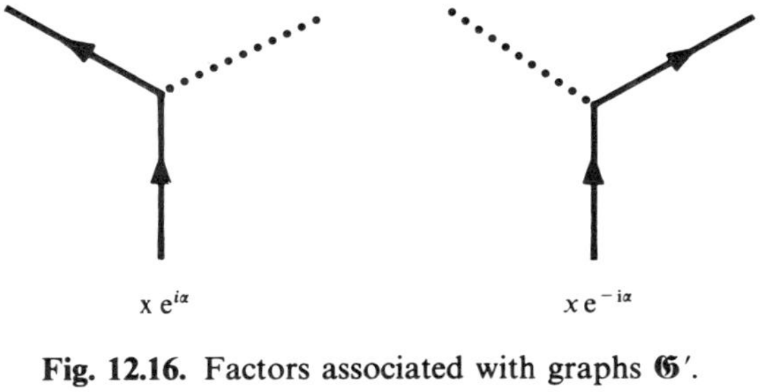

Fig. 12.16. Factors associated with graphs $\mathfrak{G}'$.

4.2 Critical exponents and conformal transformations

4.2.1 Conformal transformations and field theory

As we saw previously, the correlation functions of a magnetic system remain invariant for translations and rotations. When the system becomes critical a new invariance appears for dilatations. Thus, the simplest correlation function associated with a field component $\varphi_p(\vec{r})$ reads

$$\langle\langle \varphi_p(\vec{r}_1)\varphi_p(\vec{r}_2) \rangle\rangle = \frac{\mathscr{A}}{|\vec{r}_1 - \vec{r}_2|^{2D}} \tag{12.4.9}$$

where $-D$ can be considered as the dimensionality of the field. The invariance for dilatations is observed by making the simultaneous transformations

$$\begin{aligned} \vec{r} &\to \vec{r}' = \lambda\vec{r} \\ \varphi_p &\to \varphi_p' = \lambda^{-D}\varphi_p. \end{aligned} \tag{12.4.10}$$

This invariance is trivial, but as was shown by Polyakov in 1970,[51] a critical system is also invariant for transformations belonging to the conformal group.

The conformal group consists of a set of point transformations preserving angles. In other words, if a conformal transformation transforms a point M into M′, the neighbourhood of M′ can be deduced from the neighbourhood of M by local rotation and dilatation.

The special conformal group is a subgroup of the conformal group and it is generated by the following transformations ($\vec{r} \to \vec{r}'$)

- rotation and translation (the matrix $\mathscr{G}$ is orthogonal; the vector $\vec{a}$ is a constant)

$$r_j' = \sum_l \mathscr{G}_{jl} r_l + a_j$$

- dilatation

$$r_j' = \lambda r_j$$

- product of two inversions with different centres (the vector $\vec{b}$ is a constant)

$$\frac{\vec{r}'}{r'^2} = \frac{\vec{r}}{r^2} + \vec{b}. \tag{12.4.11}$$

Now, let us perform the conformal transformation

$$\begin{aligned} &\vec{r} \to \vec{r}' \\ &\varphi_p(\vec{r}) \to \varphi_p'(\vec{r}') = \left| \frac{\mathrm{d}\{\vec{r}'\}}{\mathrm{d}\{\vec{r}\}} \right|^{-D/d} \varphi_p(\vec{r}') \end{aligned} \tag{12.4.12}$$

where $\left| \frac{\mathrm{d}\{r'\}}{\mathrm{d}\{\vec{r}\}} \right|$ is the Jacobian of the transformation. Conformal invariance appears in the fact that (12.4.9) entails the equality

$$\langle\!\langle \varphi'_j(\vec{r}_1)\varphi'_j(\vec{r}_2)\rangle\!\rangle = \frac{A}{|\vec{r}'_1 - \vec{r}'_2|^{2D}}$$

as can easily be verified by using the infinitesimal conformal invariance

$$r'_j = r_j + b_j r^2 - 2r_j \sum_l b_l r_l$$

[this transformation is deduced from (12.4.11) for $|\vec{b}| \ll r$].

In two dimensions, the conformal graph is a continuous graph whose transformations in the complex plane, associated with the variable $z = x + iy$, can be written in the form

$$z' = f(z)$$

[$f(z)$ is analytic in a certain domain].

The homographic transformations

$$z' = \frac{az + b}{cz + d}$$

constitute a subgroup which contains the special conformal group. This subgroup is important because the homographic transformations are the only conformal transformations that are one-to-one applications of the entire space on itself.

The conformal invariance properties of critical systems are certainly important for any space dimension, but even more so in two dimensions because the two-dimensional conformal group is very rich. This essential fact was clearly recognized and exploited by Belavin, Polyakov, and Zamolodchikov in 1984.[52] Their theory is not simple, but very interesting; it is presently developing (1990) and consequently certain points remain obscure. Here, we shall give only an idea of this method, in spite of the fact that it has already led to fundamental results. In particular, as we shall see, it provided the means for the calculation the exact values of the exponents σ_M defined in Section 3.3 of the present chapter [see (12.3.68)].

A study of the properties of the impulsion-energy tensor is the starting point of their method. The impulsion-energy tensor can be defined as follows. The weight W of a configuration of a magnetic system has the form

$$W = \mathrm{e}^{-\mathscr{H}\{\varphi\}}$$

where $\mathscr{H}\{\varphi\}$ is the integral of a Lagrangian

$$\mathscr{H}\{\varphi\} = \int \mathrm{d}^d r\, \mathscr{L}(\vec{r}).$$

The Lagrangian $\mathscr{L}(\vec{r})$ is a function of the fields $\varphi_j(\vec{r})$ and of their partial derivatives $\partial_l \varphi_j(\vec{r})$. When an infinitesimal transformation

$$r_l \to r_l + \delta r_l$$
$$\varphi_j \to \varphi_j + \delta\varphi_j$$

is performed, the Hamiltonian is modified and its variation $\delta\mathscr{H}$ can be written in the form

$$\delta\mathscr{H} = \int \mathrm{d}^d\vec{r} \sum_{jl} T_{jl}(\vec{r})\,\partial_j(\delta r^l) \qquad (12.4.13)$$

which defines the impulsion-energy tensor $T_{jl}(\vec{r})$.

If $T_{jl}(\vec{r}) = T_{lj}(\vec{r})$, then $\delta\mathscr{H} = 0$ for rotations; if $\sum_j T_{jj}(\vec{r}) = 0$, then $\delta\mathscr{H} = 0$ for dilatations as can be seen by considering (12.4.13). Moreover, if both conditions hold simultaneously, then $\delta\mathscr{H} = 0$ for conformal transformations.

Theoreticians have begun to exploit this invariance.[52] They are interested in the correlation functions which are the cumulants of the mean values of products of observables $\mathscr{O}_p(\vec{r}_p)$. Let us, for instance, consider the cumulant

$$G(\vec{r}_1, \ldots, \vec{r}_n) = \langle\!\langle \mathscr{O}_1(\vec{r}_1) \ldots \mathscr{O}_n(\vec{r}_n) \rangle\!\rangle_{\mathrm{c}}$$

which is a function of the distances $|\vec{r}_j - \vec{r}_k|$ (here c means cumulant or connected). Let us make a change of coordinates

$$\vec{r}_j = \vec{r}'_j + \delta\vec{r}_j$$
$$\mathscr{O}_p = \mathscr{O}'_p + \delta\mathscr{O}_p.$$

This operation does not change the value of $G(\vec{r}_1, \ldots, \vec{r}_n)$ but expresses this quantity in a different manner. In the new expression, we may now suppress the 'primes' and this is equivalent to the transformation

$$\vec{r}_j \to \vec{r}_j + \delta\vec{r}_j$$
$$\mathscr{O}_p \to \mathscr{O}_p + \delta\mathscr{O}_p.$$

This operation leads to the equation

$$\sum_{q=1}^{p} \langle\!\langle \mathscr{O}_1(\vec{r}_1) \ldots \delta\mathscr{O}_q(\vec{r}_q) \ldots \mathscr{O}_p(\vec{r}_p) \rangle\!\rangle_{\mathrm{c}}$$
$$- \int \mathrm{d}^d r \sum_{jl} \langle\!\langle \mathscr{O}(\vec{r}_1) \ldots \mathscr{O}(\vec{r}_n)\, T_{jl}(\vec{r}) \rangle\!\rangle_{\mathrm{c}}\, \partial_j(\delta r^l) = 0 \qquad (12.4.14)$$

(r^l is a component of $\vec{r}$).

We say that $G(\vec{r}, \ldots, \vec{r}_n)$ is invariant for a conformal transformation if each term vanishes separately for every infinitesimal conformal transformation. Moreover, we see that the properties of $G(\vec{r}, \ldots, \vec{r}_n)$ are strongly related to those of the tensor $T_{jl}(\vec{r})$ which is of great importance.

One of the most elementary properties is the equality

$$\sum_j \partial_j T_{jl}(\vec{r}) = 0 \tag{12.4.15}$$

which can be deduced from the fact that the first term of (12.4.14) depends only linearly on the $\delta\vec{r}_p$ (with $p = 1, \ldots, n$).

In two dimensions, the study of $T_{jl}(\vec{r})$ can be reduced to the study of $T(\vec{r})$ defined by

$$T(\vec{r}) = T_{xx}(\vec{r}) - i\,T_{xy}(\vec{r})$$

(where $T_{xx}(\vec{r}) + T_{yy}(\vec{r}) = 0$ and $T_{xy}(\vec{r}) - T_{yx}(\vec{r}) = 0$).
Let us set $z = x + iy \qquad \bar{z} = x - iy$.

Using (12.4.15), we easily verify that $\dfrac{\partial}{\partial \bar{z}} T(\vec{r}) = 0$. Consequently,

$$T(\vec{r}) = T(z),$$

and now we have only to study $T(z)$.

It is not difficult to convince ourselves that in the initial state

$$\langle\!\langle T(z) \rangle\!\rangle = 0.$$

On the other hand, for reasons of dimensionality, we may set

$$\langle\!\langle T(z_1) T(z_2) \rangle\!\rangle_c = \frac{C}{2(z_1 - z_2)^4} \tag{12.4.16}$$

and the pure number C, called central charge, plays an essential role in the theory because it defines the universality class of the system.*

On the other hand, looking at (12.4.12), we could expect that for a conformal transformation $z \to z'$

$$T(z) \to T'(z') = \left(\frac{dz'}{dz}\right)^2 T(z').$$

Actually, this is true only for homographic transformations for which there is complete conformal invariance. Thus, for example, the equality $\langle\!\langle T(z) \rangle\!\rangle = 0$ remains true only for homographic transformations.

In order to develop theoretical aspects and also in order to make numerical calculations, the physicists who studied $T(z)$ found it convenient to perform the

* The coefficient C does not depend on the length scale, whereas the coefficients appearing in the correlation functions do depend on it. This crucial point can be understood by examining the effect of a simple change of scale on the left-hand side of (12.4.16). (Be careful: a change of scale is not a dilatation!) Changing the scale amounts to changing the Hamiltonian and consequently to modifying $T(z)$ as a consequence of (12.4.13). However, the mean values must now be calculated with a different weight. These two effects cancel out. This can be verified in the trivial case where $\mathscr{L} = \lambda \sum_{j,l} (\partial_j \varphi_l)^2$.

transformation

$$w = a \ln z \qquad (\text{with } w = w' + iw'' \quad \text{and} \quad z = x + iy)$$

which maps the complex plane into a tubular ribbon, and $T(z)$ into $T[w]$. Then, any system on this tube can be treated by using the transfer matrix formalism (the strip method: see Chapter 12, Section 2.4). This operator acts over states which represent all the local configurations of the system in a given cross-section of the tube. Now, an operator $\hat{T}[w]$ can be associated with $T[w]$. Let $|0\rangle$ be the ground state of the transfer operator $\mathscr{T}$. Any average of products of observables in the initial space corresponds to the mean value of an ordered product of operators in the ground state $|0\rangle$. The advantage of the new representation comes from the fact that at the present time, physicists know better how to manipulate operators than products of random observables.

In the same way, in the complex plane, an operator $\hat{T}(z)$ can be associated with $T(z)$. This operator can be expanded in the form of Laurent series

$$\hat{T}(z) = \sum_{n=-\infty}^{+\infty} z^{-n-2} L_n \tag{12.4.17}$$

where the L_n are operators. The variation of $T(z)$ for infinitesimal transformations being known, we can deduce from them algebraic relations between the L_n. The commutators are

$$[L_p, L_q] = (p - q)L_{p+q} + \frac{C}{12}p(p^2 - 1)\delta_{p+q}. \tag{12.4.18}$$

These relations define an algebra which was introduced by Virasoro[59] as early as 1970, but only in particle physics and in the framework of the 'duality theory'.

We are thus brought to consider states Φ such that

$$L_0 \Phi = h\Phi \qquad L_p \Phi = 0 \qquad p > 0$$

It can also be shown that with each h it is possible to associate an observable $\mathcal{O}(\vec{r})$ for which

$$\langle\!\langle \mathcal{O}(\vec{r}_1)\mathcal{O}(\vec{r}_2) \rangle\!\rangle \propto \frac{1}{|\vec{r}_1 - \vec{r}_2|^{4h}}. \tag{12.4.19}$$

Starting from a state Φ, we can generate a series of states of the form

$$L_{-p_1} \ldots L_{-p_n} \Phi.$$

The number of these states is finite, only if h takes special values and Kac showed[60] that these eigenvalues $h_{m,n}$ depend on two integers m and n.

Moreover, it has been shown[61] that for n = 0 (polymers), C takes the value C = 0. The corresponding value of $h_{m,n}$ is given by

$$h_{m,n} = \frac{(3m - 2n)^2 - 1}{24}. \tag{12.4.20}$$

To study polymer problems, it is convenient to introduce the composite operator

$$\mathcal{O}_M(\vec{r}) = \prod_{j=1}^{N} \varphi_1(\vec{r} + \vec{\eta}_j)$$

where $\varphi_1(\vec{r})$ is the first component of the field and where the vectors $\vec{\eta}_j$ are considered as small. However, we assume that they do not vanish because otherwise divergences would occur and $\mathcal{O}_M(\vec{r})$ would not be well defined. At the critical point and for $|\vec{r}| > |\vec{\eta}_j|$ (for all j), we have

$$\langle\!\langle \mathcal{O}_M(\vec{0})\mathcal{O}_M(\vec{r}) \rangle\!\rangle \propto \frac{1}{r^{4h_M}} \tag{12.4.21}$$

in agreement with (12.4.19) (see Fig. 12.17). Following conjectures made by Dotsenko and Fateev,[62] Saleur showed,[63] by using the strip method (see Fig. 12.18), that the h_M can be identified with certain $h_{m,n}$. So, his result can be written in the form

$$h_M = h_{M/2,0} = \frac{9M^2 - 4}{96}, \tag{12.4.22}$$

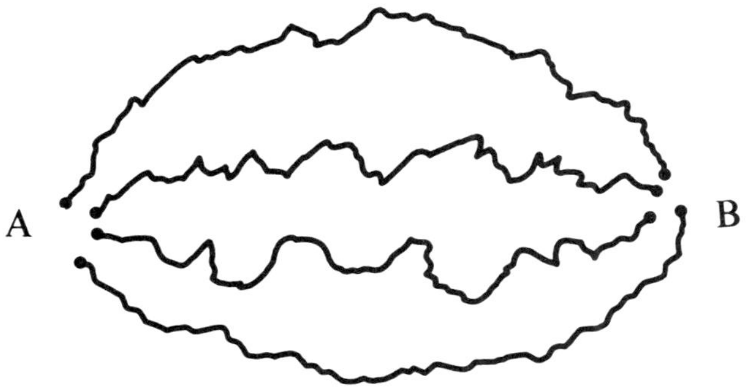

Fig. 12.17. In two-dimensional space, we see **M** polymers whose origins are very close to one another and whose end points have the same property.

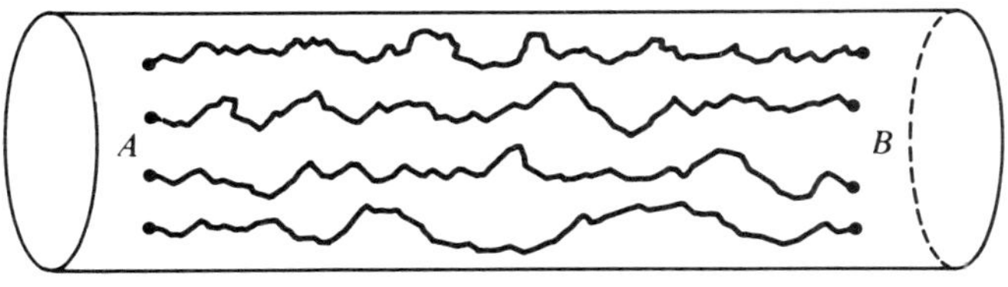

Fig. 12.18. The conformal transformation $w = a \ln z$ of a planar space gives a tubular space and the present figure is the transform of Fig. 12.17.

but other forms are possible. However, the fact that this formula m may take only half-integer values ($m = M/2$) has not been explained yet (1990).

4.2.2 Derivation of the exponents σ_M and ν (polymer theory) from the exponents h_M

In polymer theory, the exponents σ_M characterize the asymptotic behaviour of the renormalization factors $\mathfrak{x}_M(z)$. As was shown by Duplantier,[43] we can derive the values of these exponents σ_M from the values of the exponents h_M found by H. Saleur.[62]

First, let us observe that in the vicinity of the critical point, relation (13.4.19) reads

$$\langle\!\langle \mathcal{O}_M(\vec{0})\mathcal{O}_M(\vec{r})\rangle\!\rangle = \frac{1}{r^{4h_M}} f_M(r/\xi)$$

where ξ is the correlation length

$$\xi \propto (a - a_c)^{-\nu}.$$

Let us now set

$$G_M(a) = \int d^2r \langle\!\langle \mathcal{O}_M(\vec{0})\mathcal{O}_M(\vec{r})\rangle\!\rangle.$$

We see immediately that

$$G_M(a) \propto \xi^{2-4h_M} \propto (a - a_c)^{\nu(4h_M - 2)} \tag{12.4.23}$$

Moreover, $G_M(a)$ is the Laplace–de Gennes transform of the partition function $\mathscr{Z}_{M,M}(S_1, \ldots, S_M)$ corresponding to configurations of M polymers with the same origin and the same end point.

$$G_M(a) =$$

$$\int_0^\infty dS_1 \ldots \int_0^\infty dS_M \exp[-(a - a_c)(S_1 + \ldots + S_M)]\, \mathscr{Z}_{MM}(S_1, \ldots, S_M).$$

We can also write

$$G_M(a) = \int_0^\infty dS \int_0^\infty dS_1 \ldots \int_0^\infty dS_M \exp[-(a - a_c)S]$$
$$\times\, \mathscr{Z}_{M,M}(S_1, \ldots, S_M)\,\delta(S - S_1 \ldots - S_M).$$

Then, setting $S_j = x_j S$, we obtain

$$G_M(a) = \int_0^\infty dS \exp[-(a - a_c)S]\, S^{M-1} \times$$

$$\times \int_0^\infty dx_1 \ldots \int_0^\infty dx_M\, \mathscr{Z}_{M,M}(Sx_1, \ldots, Sx_M)\,\delta(x_1 + \ldots + x_M - 1). \tag{12.4.24}$$

As the areas $S_1, \ldots, S_M$ have (by definition) the same order of magnitude, the partition function obey the scaling law

$$\mathscr{Z}_{M,M}(Sx_1, \ldots, Sx_M) = S^{\zeta_M} f(x_1, \ldots, x_M) \tag{12.4.25}$$

where $f(x_1, \ldots, x_M)$ also depends on the interaction b. By bringing (12.4.25) into (12.4.24), we find

$$G_M(a) \propto (a - a_c)^{-M-\zeta_M}$$

Thus, comparing this expression with (12.4.23) we obtain the first result

$$\zeta_M = -M - 2\nu(2h_M - 1). \tag{12.4.26}$$

Let us now examine the properties of the partition function $\mathscr{Z}_{M,M}(S, \ldots, S)$ of $\boldsymbol{M}$ identical polymers. In the long-chain limit, we can write

$$\mathscr{Z}_{M,M}(S, \ldots, S) \propto S^{\zeta_M}. \tag{12.4.27}$$

Moreover, the dimensional formula of $\mathscr{Z}_{M,M}(S, \ldots, S)$ is easily obtained by remarking that $\mathscr{Z}(S)$ is a pure number and that the same property holds true for the partition function $\mathscr{Z}_M(S, \ldots, S)$ associated with $\boldsymbol{M}$ polymers having the same origin but end points at random. Thus, by tying together the end points of these polymers, we find

$$\mathscr{Z}_{M,M}(S, \ldots, S) \sim L^{-(M-1)d}$$

where L is a length and $(\boldsymbol{M} - 1)$ the number of loops of the 'water-melon' diagram represented in Fig. 12.19.

However, in the Kuhnian limit, the renormalized partition function $\mathscr{Z}_{M,M,\mathrm{R}}({}^{\mathrm{e}}S, \ldots, {}^{\mathrm{e}}S)$ depends only on the size X of the isolated polymer. Thus

$$\mathscr{Z}_{M,M,\mathrm{R}}({}^{\mathrm{e}}S, \ldots, {}^{\mathrm{e}}S) \propto X^{-(M-1)d},$$

or if the interaction is kept constant

$$\mathscr{Z}_{M,M,\mathrm{R}}({}^{\mathrm{e}}S, \ldots, {}^{\mathrm{e}}S) \propto S^{-(M-1)\nu d}. \tag{12.4.28}$$

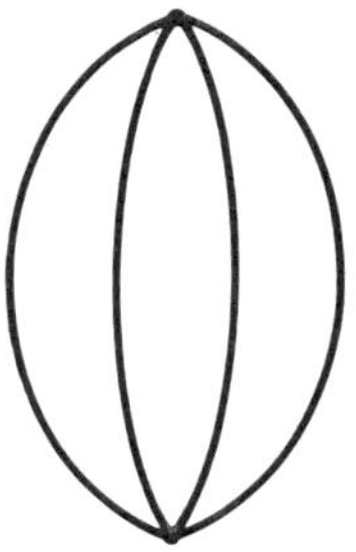

Fig. 12.19. 'Water-melon' diagram (of order zero) for **M** polymers. This water-melon diagram corresponds to the physical situation represented in Fig. 12.17.

Moreover, $\mathscr{Z}_{M,M}(S, \ldots, S)$ is related to $\mathscr{Z}_{M,M,\mathrm{R}}({}^{\mathrm{e}}S, \ldots, {}^{\mathrm{e}}S)$ by the renormalization equation [see (12.3.63) and Fig. 12.19]

$$\mathscr{Z}_{M,M,}(S, \ldots, S) = (\mathfrak{X}_M(z))^2 \mathscr{Z}_{M,M,\mathrm{R}}({}^{\mathrm{e}}S, \ldots, {}^{\mathrm{e}}S) \tag{12.4.29}$$

where, according to (12.3.67) and (12.3.68), we have

$$\mathfrak{X}_M(z) \propto S^{\sigma_M}. \tag{12.4.30}$$

By combining (12.4.28), (12.4.29), and (12.4.30), we obtain

$$\mathscr{Z}_{M,M,}(S, \ldots, S) \propto S^{2\sigma_M - (M -)\nu d}.$$

Let us set $d = 2$ in this formula and let us compare with (12.4.27); taking (12.4.26) into account, we get

$$\sigma_M = \nu(M - 2h_M) - M/2. \tag{12.4.31}$$

Now, we can use the results derived from conformal theory and we can replace h_M by the value given by Saleur [see (12.4.22)]. In this way, we obtain

$$\sigma_M = \nu\left(\frac{4 + 48M - 9M^2}{48}\right) - M/2. \tag{12.4.32}$$

We also know that $\sigma_2 = 0$ [see (12.3.69)]. Setting $M = 2$ in the preceding formula, we obtain the result

$$\nu = 3/4,$$

a value previously obtained by Nienhuis (see Section 4.1) and verified by Derrida with precision (see Section 2.4). Bringing this value of ν in (11.4.32), we finally find

$$\sigma_M = \frac{(2 - M)(9M + 2)}{64} \qquad (M = 1, 2, \ldots). \tag{12.4.33}$$

Thus, it is possible to obtain the 'exact' values of the critical exponents σ_M in two dimensions, and, in polymer theory, these exponents are, with ν, the most important critical exponents. However, the method used here is neither very direct nor illuminating.

4.2.3 Conformal invariance of a Brownian chain

We may wonder whether there is a direct relation between the structure of polymers and this conformal invariance which proved so useful for the study of the properties of correlation functions in field theory. Thus, in order to cast some light on this question, we shall show here directly how the property of conformal invariance applies to a two-dimensional Brownian chain.

The weight is

$$W = \exp\left[-\frac{1}{2}\int_0^S \mathrm{d}s\left(\frac{\mathrm{d}\vec{r}(s)}{\mathrm{d}s}\right)^2\right].$$

Let us set $z = x + iy \quad \bar{z} = x - iy$. We see that

$$W = \exp\left[-\frac{1}{2}\int_0^S \mathrm{d}s \left(\frac{\mathrm{d}z}{\mathrm{d}s}\right)\left(\frac{\mathrm{d}\bar{z}}{\mathrm{d}s}\right)\right]. \tag{12.4.34}$$

Now let us make a conformal transformation by setting

$$z_{\mathrm{T}} = f(z).$$

we get

$$\left(\frac{\mathrm{d}z}{\mathrm{d}s}\right)\left(\frac{\mathrm{d}\bar{z}}{\mathrm{d}s}\right) = \left(\frac{\mathrm{d}z_{\mathrm{T}}}{\mathrm{d}s}\right)\left(\frac{\mathrm{d}\bar{z}_{\mathrm{T}}}{\mathrm{d}s}\right)|f'(z)|^{-2}.$$

The term $|f'(z)|^2$ is a local dilatation factor. Just as an infinite chain is statistically invariant for dilatations, a local dilatation amounts to a local increase of matter, i.e. of Brownian area. Consequently, we have to set

$$\mathrm{d}s_{\mathrm{T}} = \mathrm{d}s\,|f'(z(s))|^2$$

$$S_{\mathrm{T}} = \int_0^S dt\,|f'(z(t))|^2.$$

Now, the weight can be written in the form

$$W = \exp\left(-\int_0^{S_{\mathrm{T}}} \mathrm{d}s_{\mathrm{T}} \left(\frac{\mathrm{d}z_{\mathrm{T}}}{\mathrm{d}s_{\mathrm{T}}}\right)\left(\frac{\mathrm{d}z_{\mathrm{T}}}{\mathrm{d}s_{\mathrm{T}}}\right)\right),$$

an equality very similar to (12.4.34). We see that W keeps the same form and this fact expresses (statistically) the property of conformal invariance. However, this invariance is not complete, since the final area S_{T} differs from the initial area S. It is complete only if S is infinite.

Now, we feel that Kuhnian chains must have very similar invariance properties for conformal transformations.

4.3 Exact critical exponents for a polymer melt

It is well-known that in a polymer melt as in a semi-dilute solution, the size exponent has the trivial value $\nu = 1/2$. However, for a melt, it is also possible to define exponents σ_M as was done for dilute solutions (see Section 3.3): as before, these exponents are directly related to the partition functions associated with a network [see eqns (12.3.64) and (12.3.65)]. The values of σ_M in the present case are not obvious. However, the exact values in two-dimensions

$$\sigma_M = \frac{4 - M^2}{32} \qquad (M \geqslant 1) \tag{12.4.35}$$

were found by Duplantier and Saleur in 1987.[64] These exponents σ_M are obtained by using the same kind of method as in the dilute case. The method has been expounded in Section 4.2. For the melt, an exponent h_M can be defined by an equation analogous to (12.4.2) and its value is[63]

$$h_M = \frac{M^2 - 4}{32}.$$

Eqn. (12.4.32) remains valid.
Thus, by using eqn (12.4.31) with $\nu = 1/2$, we find (12.4.35).

REFERENCES

1. Toulouse, G. and Pfeuty, P. (1975). *Introduction au groupe de renormalisation et à ses applications. Presse Universitaines de Grenoble.* Ma Shang-Keng, (1976). *Modern theory of critical phenomena.* Benjamin, New York.
2. Brezin, E., Le Guillou, J.C., and Zinn–Justin, J. (1976). Field theoretical approach to critical phenomena. In *Phase transitions and critical phenomena* (eds. C. Domb and M.S. Green), Vol. 6. p. 125. Academic Press.
   Amit, D.J. (1978). *Field theory, the renormalization group and critical phenomena.* McGraw-Hill. Also 2nd edn (1984), World Scientific, Singapore.
3. Wilson, K.G. (1971). *Phys. Rev.* **B 4**, 3174, 3184.
4. Wilson, K.G. and Kogut, J. (1974). *Phys. Rep.* **12-C**(2), 76.
5. Kadanoff, L.P. (1966). *Physics* **2**, 263.
6. Griffiths, R.B. and Pierce, P.A. (1979). *J. Stat. Phys.* **20**, 499.
7. Niemeijer, Th. and van Leeuwen, J.M.J. (1973). *Phys. Rev. Lett.* **31**, 1411; (1974) *Physica* **71**, 17.
8. Hilhorst, H.J. (1976). *Phys. Lett.* **56A**, 153; (1977). *Phys. Rev.* **B 16**, 1253.
9. Shapiro, B. (1978). *J. Phys. C: Solid State Phys.* **11**, 2829.
10. de Gennes, P.G. (1977). *Riv. Nuovo Cimento* **7**, 363.
11. de Gennes, P.G. (1979). *Scaling concepts in polymer physics*, p. 290. Cornell University Press, Ithaca, NY.
12. Gabay, M. and Garel, T. (1978). *J. Physique Lett.* **39**. L-123.
13. Oono, Y. (1979). *J. Phys. Soc. Japan* **47**, 683.
14. Nightingale, M.P. (1975). *Physica* **83A**, 561.
15. Derrida, B. (1981). *J. Phys. A: Math. Gen.* **14**, L5.
16. For electromagnetism see:
    Bogoliubov, N.N. and Shirkov, D.V. (1959). *Introduction to the theory of quantized field.* Interscience, New York.
    Itzykson, C. and Zuber, J.B. (1980). *Quantum field theory.* McGraw-Hill, New York.
17. For more recent developments, and especially for Landau–Ginzburg theory, see:
    Zimmerman, W. (1970). Local operator product and renormalization in quantum field theory. *Lectures on elementary particles and quantum field theory*, Brandeis Summer Institute in Theoretical Physics (eds. S. Deser and M. Grisaru), p. 399. MIT Press, Cambridge, Mass.
    Callan, C.G. (1975). Introduction to renormalization theory. In *Méthodes en théorie des champs*, Les Houches (eds. R. Balian and J. Zinn-Justin). North-Holland, Amsterdam.
18. des Cloizeaux, J. (1975). *J. Physique* **36**, 281.
19. Concerning this topic, see:
    Nickel, B.G. (1978). *J. Math. Phys.* **19**, 542.

20. Baker, G.A., Nickel, B.G., Green, M.S., and Meiron, D.I. (1976). *Phys, Rev. Lett.* **36**, 1351.
21. Le Guillou, J.C. and Zinn-Justin, J. (1977). *Phys. Rev. Lett.* **39**, 95: (1980) *Phys. Rev.* B **21**, 3976.
22. Lipatov, L.N. (1977). *Pis'ma Zh. Eksp. teor. Fiz.* **25**, 116: *JETP Lett.* **25**, 104. Lipatov, L.N. (1977). *Zh. Eksp. Teor. Fiz.* **72**, 411; *Sov. Phys. JETP* **45**, 216.
23. Baker, G.A., Nickel, B.G., and Meiron, D.I. (1978). *Phys. Rev.* B **17**, 1365.
24. Nickel, B.G., Meiron, D.I., and Baker, G.A. (1977). *Compilation of* 2pt *and* 4pt *graphs for continuous spin models.* University of Guelph Report, Canada.
25. Brézin, E., Le Guillou, J.C., and Zinn-Justin, J. (1977). *Phys. Rev.* **D15**, 1544, 1558.
26. Borel, E. (1886). *J. Math.* 5éme Serie 2, 103; (1899). *Ann. Ec. Norm. Sup.* 3éme serie **16**, 9. Leroy, E. (1900). *Ann. Fac. Univ. Toulouse* 2, 317.
27. Baker, G.A. and Gammel, J.L. (1970). *The Padé approximant in theoretical physics.* Academic Press.
28. Wilson, K.G. (1972). *Phys. Rev. Lett.* **28**, 548.
29. Brézin, E., Le Guillou, J.C., Zinn-Justin, J., and Nickel, B.G. (1973). *Phys. Lett.* **44A**, 227.
30. Brézin, E., Le Guillou, J.C., and Zinn-Justin, J. (1974). *Phys. Rev.* D **9**, 1121.
31. Vladimirov, A.A., Kazakov, D.I., and Tarasov, O.V. (1979) *Zh. Eksp. Teor. Fiz.* **77**, 1035; *Sov. Phys. JETP* **50**, 521.
32. Chetyrkin, K.G., Kataev, A.L., and Tkachov, F.V. (1981). *Phys. Lett.* **99B**, 147; (1981) **101B**, 457 (E); also unpublished work. Chetyrkin, K.G., Gorishny, S.G., Larin, S.A., and Tkachov, F.V. (1983). *Phys. Lett.* **132B**, 351 (be wary of errors!). Kazakov, D.I. (1983). *Phys. Lett.* **133B**, 406.
33. Gorishny, S.G., Larin, S.A., and Tkachov, F.V. (1984). *Phys. Lett.* **101A**, 120.
34. Le Guillou, J. C. and Zinn-Justin, J. (1985). *J. Physique Lett.* **46**, L137.
35. des Cloizeaux, J. (1982). *J. Physique* **43**, 1743.
36. Muthukumar, M. and Nickel, B.G. (1984). *J. Chem Phys.* **80**, 5839.
37. des Cloizeaux, J. (1980). *J. Physique Lett.* **41**, L151.
38. des Cloizeaux, J. (1981). *J. Physique* **42**, 635.
39. Levy, M., Le Guillou, J.C., and Zinn-Justin, J. (eds.) (1982). *Phase transitions.* Course of the Summer School of Cargèse, 1980. Plenum Press, New York.
40. Schäfer, L. and Witten, T.A. (1980). *J. Physique* **41**, 459.
41. Benhamou, M. and Mahoux, G. (1986). *J. Physique* **47**, 559. See also: Duplantier, B. (1986). *J. Physique* **47**, 569.
42. 't Hooft, G. and Veltman, M. (1972). *Nucl. Phys.* B **44**, 189. 't Hooft, G. (1973). *Nucl. Phys.* B **61**, 455. See also: Collins, J.C. (1974). B **80**, 341.
43. Duplantier, B. (1986). *Phys. Rev. Lett.* **57**, 941; E (1986) 2332.
44. Duplantier, B. (1989). *J. Stat. Phys.* **54**, 581.
45. de Gennes, P.G. (1975). *J. Physique Lett.* **36**, L-55.
46. Larkin, A.I. and Khmel'nitskii, D.E. (1969). *Zh. Eksp. Teor. Fiz.* **56**, 2087: (1969) *Sov. Phys. JETP* 29, 1123.
47. Wegner, F.J. and Riedel, E.K. (1973). *Phys. Rev.* B **7**, 248.
48. Tanaka, G. (1982). *Macromolecules* **15**, 1525.
49. Havlin, S. and Ben-Avraham, D. (1982). *J. Phys.* A **15**, L317, L321.
50. Argão de Carvalho, C. and Caracciolo, S. (1983). *J. Physique* **44**, 323.
51. Polyakov, A.M. (1970). *Zh. Eksp. Teor. Fiz. Lett.* **12**, 538; (1970) *Sov. Phys. JETP Lett.* **12**, 381.

52. Belavin, A.A., Polyakov, A.M. and Zamolodchikov, A.B. (1984). *J. Stat. Phys.* **34**, 763; (1984) *Nucl. Phys.* B **241**, 333.
53. Nienhuis, B. (1982). *Phys. Rev. Lett.* **49**, 1062.
Nienhuis, B. (1984). *J. Stat. Phys.* **34**, 731.
54. Cardy, J.L. and Hamber, H.W. (1980). *Phys. Rev. Lett.* **45**, 499.
55. Lieb, E.H. and Wu, F.Y. (1972). Two dimensional ferroelectric models. In *Phase transitions and critical phenomena* (eds. C. Domb and M.S. Green), Vol. 1, p. 331. Academic Press.
56. Gaudin, M. (1953). *La fonction d'onde de Bethe*, Chapter 7. Masson, Paris.
57. Stanley, E. (1971). *Introduction to phase transitions and critical phenomena.* Clarendon Press, Oxford.
58. Cardy, J.L. (1987). Conformal variance. In *Phase transitions and critical phenomena* (eds. C. Domb and M.S. Green), Vol. 2. Academic Press.
Itzykson, C. (1986). DEA course, Marseille.
59. Virasoro, M.A. (1970). *Phys. Rev.* D 1, 2933.
60. Kac, V.G. (1979). Group theoretical methods in physics. In *Lecture Notes in Physics*, Vol. 94 (eds. W. Beiglbock, A. Bohm, and E. Takasuji).
61. Blöte, H.W., Cardy, J.L., and Nightingale, M.P. (1986). *Phys. Rev. Lett.* **56**, 742. (The value of C is zero when the number of field components goes to zero, for a box with any shape.)
62. Dotsenko, VI.S. and Fateev, V.A. (1984). *Nucl. Phys.* B **240**, 312.
63. Saleur, H. (1986). *J. Phys.* A
64. Duplantier, B. and Saleur, H. (1987). *Nucl. Phys.* B **290**, 291.

13

POLYMERS IN SOLUTION IN GOOD SOLVENTS: THEORETICAL RESULTS

GENESIS

Many theoretical results concerning the behaviour of polymers in good solvents have been found in the framework of the standard continuous model by using perturbation and renormalization techniques. It seems that S.F. Edwards was the first to realize (before 1965) that it is fruitful to use continuous curves to model polymer chains, and he was also the first to understand (around 1965) that interacting polymers in solution screen one another when they overlap. However, this theoretical progress acquired its full meaning only when it was fully recognized (between 1975 and 1985) that a set of long polymers can be considered as a critical system and therefore may have universal properties. Of course, the existence of critical exponents had been accepted much earlier, but the existence of universal functions and universal ratios was really recognized only after 1975. At the same time appeared the concept of semi-dilute solution. Such a solution has a low concentration with a strong chain overlap; the screening effects then play an essential role in agreement with Edwards's ideas.

Thus, all these developments led us to a rather satisfactory understanding of the structure of 'good' polymer solutions, either dilute or semi-dilute.

1. ISOLATED POLYMER IN GOOD SOLVENT

In this chapter, polymers are represented by means of the standard continuous model, described in Chapter 10. To calculate their properties in good solvent, we shall use the renormalization techniques described in Chapter 12. The principles and methods which serve to determine the properties of isolated polymers or of a small number of polymers (dilute solutions) also apply quite naturally to the study of polymer ensembles (solution with overlap) which we consider in the second part of the present chapter. In this way, we obtain all kinds of results* which can be compared to experimental data.

* The reader may also look at the review article by Oono[1] (1985) which deals both with statics and dynamics.

1.1 Effective interaction and second virial coefficient

The effective interaction of the various parts of an isolated polymer with one another can be 'measured' by the second virial coefficient which defines the repulsion of two polymers in very dilute solutions. This is a physically important quantity which we also encounter when studying the structure functions of dilute polymer solutions.

We showed in the preceding chapter, that from a physical point of view, it is reasonable to write the osmotic pressure by using as a length scale the size X of an isolated chain, defined by

$$R^2 \equiv \langle\!\langle [\vec{r}(S) - \vec{r}(0)]^2 \rangle\!\rangle = {}^{c}Sd = X^2 d. \tag{13.1.1}$$

This leads us to the expansion [see (13.3.73)]

$$\boldsymbol{\Pi\beta} = \boldsymbol{C}[1 + \tfrac{1}{2}(2\pi)^{d/2} g(z)(\boldsymbol{C}X^d) + \tfrac{1}{3}(2\pi)^d h(z)(\boldsymbol{C}X^d)^2 + \ldots] \tag{13.1.2}$$

where $g(z)$ represents the effective interaction. Here, we must calculate $g(z)$, first in the limit $z \to \infty$ (this limit being g^*) and then for all z. We are able to do so only approximately, and consequently various methods will be examined. The expansion of $g(z)$ up to third-order in powers of z is [see (10.8.11)]

$$g(z) = z - 4.865\, z^2 + 26.454\, z^3. \tag{13.1.3}$$

We can also write the first two terms in the form of a [1/1] Padé approximant

$$g(z) = \frac{z}{1 + 4.865\, z} \tag{13.1.4}$$

which, in the limit $z \to \infty$, gives an approximate value of g^*

$$g^* = 0.205. \tag{13.1.5}$$

A better method consists of calculating g^* as a root of an equation in g, by proceeding as indicated in Chapter 11, Section 3.3. We define $w(z)$ by the equality

$$w(z) = z\frac{\partial g(z)}{\partial z} \qquad \text{where} \quad w(z) \to 0 \quad \text{when} \quad z \to \infty. \tag{13.1.6}$$

Now, we obtain $z(g)$ by inverting expansion (13.1.3)

$$z(g) = g + 4.865\, g^2 + 20.882\, g^3 \tag{13.1.7}$$

and we can re-express $w(z)$ in terms of g by setting: $w[g] = w(z)$. Equation (12.1.16) can be rewritten in the form

$$w[g] = \frac{z(g)}{\partial z(g)/\partial g}. \tag{13.1.8}$$

From the expansion (13.1.7) of $z(g)$, we deduce the expansion of $w[g]$

$$w[g] = g(1 - 4.865g + 5.572g^2 + \ldots). \tag{13.1.9}$$

The number g^* is given by the equation

$$w[g^*] = 0.$$

Contenting ourselves with the first three terms of the expansion, we find

$$g^* = 0.331, \tag{13.1.10}$$

a value which is not quite the same as (13.1.5).

To calculate $g(z)$ and g^* in three dimensions, we can proceed in a different way, by using ε-expansions ($\varepsilon = 4 - d$) and renormalization theory, as was explained in Chapter 12, Section 3.3. The expansion of g^* in terms of ε was calculated up to second-order [see (12.3.102)]

$$g^* = \frac{\varepsilon}{8} + \frac{\varepsilon^2}{16}\left(\frac{25}{16} + \ln 2\right) + \ldots. \tag{13.1.11}$$

Thus, for $\varepsilon = 1$, the first two terms give

$$g^* = 0.266. \tag{13.1.12}$$

A better result can be obtained by taking into account the fact that, for $\varepsilon = 3$ ($d = 1$), the value of g^* is known. Indeed, when z goes to infinity, a polymer solution in one dimension behaves like a gas of hard rods of length X. The osmotic pressure for such a system is given by (here ${}^{e}S = X^2$)

$$\boldsymbol{\Pi\beta} = \frac{C}{1 - \boldsymbol{C}X} = \boldsymbol{C}(1 + \boldsymbol{C}X + \ldots) \tag{13.1.13}$$

which must be compared to

$$\boldsymbol{\Pi\beta} = \boldsymbol{C}[1 + \tfrac{1}{2}g^*(2\pi)^{1/2}\boldsymbol{C}X + \ldots].$$

Hence, for $d = 1$, we get

$$g^* = (2/\pi)^{1/2}. \tag{13.1.14}$$

Thus, we can interpolate g^* between $d = 4$ and $d = 1$, by representing g in the form of a polynomial of degree 3 in ε as follows: the first two terms of the polynomial coincide with (13.1.11) and the value of this polynomial for ε gives the exact result (13.1.14). Thus, we get

$$g^* = \frac{\varepsilon}{8} + \frac{\varepsilon^2}{16}\left(\frac{25}{16} + \ln 2\right) + \frac{\varepsilon^2}{27}\left[(2/\pi)^{1/2} - \frac{321}{256} - \frac{9}{16}\ln 2\right] \tag{13.1.15}$$

and for $\varepsilon = 1$, this expression gives the value

$$g^* = 0.233 \tag{13.1.16}$$

which seems to be the most reasonable one.[2]

We can also try to determine the curve representing the function $g(z)$, i.e. how $g(z)$ approaches g^*. For that, we use eqn (12.3.89)

$$z = g\left(1 - \frac{g}{g^*}\right)^{-\varepsilon/\sigma_a} \exp\left\{\int_0^g \mathrm{d}g'\left[\frac{\varepsilon}{w[g']} - \frac{1}{g'} - \frac{\varepsilon}{(g^* - g')\sigma_a}\right]\right\} \tag{13.1.17}$$

where by definition

$$w[g^*] = 0 \qquad \sigma_a = -\left.\frac{\partial w[g]}{\partial g}\right|_{g = g^*}. \tag{13.1.18}$$

The function $w[g]$ appearing in these equations is given by expansion (12.3.101)

$$w[g] = \varepsilon g + g^2[-8 + \varepsilon(2 + 4\ln 2)] + g^3(34) + \ldots . \tag{13.1.19}$$

It is convenient to write $w[g]$ in the form

$$w[g] = \frac{\varepsilon g}{1 + Ag}[1 - g/g^*] \tag{13.1.20}$$

where according to (13.1.11) and (13.1.8)

$$A \simeq \frac{17}{4}. \tag{13.1.21}$$

Moreover, (13.1.8) and (13.1.20) give

$$\sigma_a = \frac{\varepsilon}{1 + Ag^*}. \tag{13.1.22}$$

We find, approximately,

$$\frac{\varepsilon}{w[g]} - \frac{1}{g} - \frac{\varepsilon}{(g^* - g)\sigma_a} = 0$$

and consequently (13.1.17)

$$z = g\left(1 - \frac{g}{g^*}\right)^{-\varepsilon/\sigma_a}. \tag{13.1.23}$$

As according to (13.1.21) and (13.1.11), we have

$$Ag^* \simeq \frac{17}{32}\varepsilon$$

we obtain [see also (12.3.105)]

$$\sigma_a \simeq \frac{\varepsilon}{1 + \frac{17}{32}\varepsilon} \tag{13.1.24}$$

a formula which, for $\varepsilon = 1$, gives[2]

$$\sigma_a = 0.653. \tag{13.1.25}$$

However, a more precise value of σ_a can be deduced from the equality $\sigma_a = 2\omega\gamma$, and from the values directly obtained in three dimensions for ω and ν by Le Guillou and Zinn-Justin (12.3.9)

$$\sigma_a = 0.929 \pm 0.020. \tag{13.1.26}$$

Thus, using the results (13.1.23), (13.1.16), and (13.1.26), we can write the relation between z and g in the approximate form[2]

$$z = 0.233 \frac{g}{g^*}\left(1 - \frac{g}{g^*}\right)^{-1.076} \tag{13.1.27}$$

which is represented in Fig. 13.1.

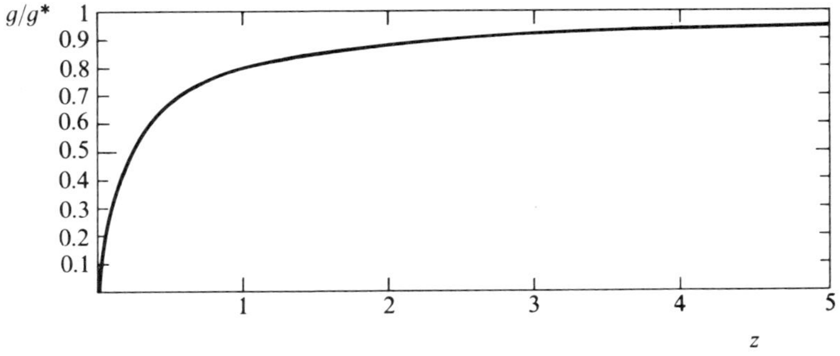

Fig. 13.1. The second virial coefficient of a polymer solution is directly proportional to the ratio $g(z)/g^*$ whose dependence with respect to the parameter z is represented above. The value of g^* corresponding to this curve is $g^* = 0.233$.

1.2 Third virial coefficient

The number $h(z)$ which appears in (13.1.2) can serve to define the third virial coefficient. However, from a theoretical point of view $h(z)$ is not as important as $g(z)$. Moreover, the third virial coefficient is not easily measurable, and this is a pity because, in principle, the ratio $h(z)/g^2(z)$ can be measured independently of the swelling.

For $d = 3$, $h(z)$ is obtained to lowest order in z, by comparing (13.1.2) and (10.8.19). In this way, we find

$$h(z) = z^3 \frac{128}{315}(-108 \times 3^{1/2} + 208 \times 2^{1/2} - 103) = z^3(1.6640).$$

The number h can be re-expressed in terms of g

$$h(z) \equiv h[g]$$

and (13.1.7) gives

$$h[g] = g^3(1.6640).$$

Thus, going to the limit $z \to \infty$, $g \to g^*$ ($= 0.233$), we obtain

$$\frac{h(z \to \infty)}{g^2(z \to \infty)} = \frac{h[g^*]}{g^{*2}} \simeq 0.388. \qquad (13.1.28)$$

1.3 Entropy of an isolated chain

Let us consider on a lattice the self-avoiding chains with N links which start from the origin of the lattice. The entropy $\mathbb{S}_N$ of these chains is defined in a very simple way. The energy of such chains is obviously zero. Consequently, the entropy is directly deduced from the partition function Z_N

$$\mathbb{S}_N = \ln Z_N.$$

Thus, the computer calculations made for chains on lattices, (described in Chapter 4) lead to the following asymptotic result

$$\mathbb{S}_N = N \ln \mu + (\gamma - 1) \ln N + \text{const.} \qquad (13.1.29)$$

which is valid for long chains ($N \gg 1$).

Entropy can also be calculated by starting from the continuous model and, in this case, we expect a similar result. However, we must note that in the continuous case, the entropy of the system is really infinite and that a finite entropy can be obtained only after performing a subtractive renormalization. Actually, this question has been tackled several times, either indirectly with the help of a zero component field theory,[3,4] or by using the direct renormalization method.[2] We shall now describe in detail the latter approach.

It has been shown in Chapter 10, Section 1, that the weight $\mathscr{W}\{\vec{r}\}$ associated with a chain defined by the function $\vec{r}(s)$ can be written in the form

$$\mathscr{W}\{\vec{r}\} = \exp[-\mathfrak{H}(S, b, s_0)] \qquad (13.1.30)$$

where

$$\mathfrak{H}(S, b, s_0) = \frac{1}{2}\int_0^S \mathrm{d}s \left(\frac{\mathrm{d}r}{\mathrm{d}s}\right)^2 + b \int_{\substack{0 < s' < s'' < S \\ s'' - s' > s_0}} \mathrm{d}s'\, \mathrm{d}s''\, \delta[\vec{r}(s'') - \vec{r}(s')]. \qquad (13.1.31)$$

In this formula, s_0 is a cut-off, and we saw in Chapter 9 that the introduction of this cut-off enables us to define a partition function $^+\mathscr{Z}(s)$ and the corresponding free energy

$$^+\mathbb{F} = -\ln {}^+\mathscr{Z}(S).$$

To evaluate the energy $^+\mathbb{F}$ (and consequently the entropy $^+\mathbb{S}$), we shall use the following trick.

Let us replace $\mathfrak{H}(S, b, s_0)$ by $\boldsymbol{\beta}\mathfrak{H}(S, b, s_0)$

$$\mathscr{W}_{\boldsymbol{\beta}}[\vec{r}\} = \mathrm{e}^{-\ \mathfrak{H}(S,b,s_0)}.$$

We can associate with this weight a partition function $^{+}\mathscr{Z}(S, \boldsymbol{\beta})$ with $\mathscr{Z}(S, 1) \equiv {}^{+}\mathscr{Z}(S)$. We see immediately that

$$^{+}\mathbb{E} = -\frac{\partial}{\partial \boldsymbol{\beta}}[\ln {}^{+}\mathscr{Z}(S, \boldsymbol{\beta})]_{\beta = 1}.$$

Making the transformations $s \to \boldsymbol{\beta} S$, $s' \to \boldsymbol{\beta} S'$ and $s'' \to \boldsymbol{\beta} s''$ in $\mathfrak{H}(S, b, s_0)$ given by (13.1.31), we see immediately that

$$\boldsymbol{\beta}\,\mathfrak{H}(S, b, s_0) = \mathfrak{H}(S/\boldsymbol{\beta}, b\boldsymbol{\beta}^3, s_0/\boldsymbol{\beta}).$$

In the same way, $^{+}\mathscr{Z}(S, \boldsymbol{\beta})$ is obtained from $^{+}\mathscr{Z}(S)$ by replacing S by $S/\boldsymbol{\beta}$, s_0 by $s_0/\boldsymbol{\beta}$ and b by $b\boldsymbol{\beta}^3$. Consequently, the energy $^{+}\mathbb{E}$ is given by

$$^{+}\mathbb{E} = \left(S\frac{\partial}{\partial S} - 3b\frac{\partial}{\partial b} + s_0\frac{\partial}{\partial s_0}\right)\ln {}^{+}\mathscr{Z}(S).$$

Finally, the entropy $^{+}\mathbb{S}$ is defined by the equality

$$^{+}\mathbb{S} = {}^{+}\mathbb{E} - {}^{+}\mathbb{F}$$

and we note that this entropy is partially renormalized since our model is continuous). In order to eliminate the cut-off,* we have to perform a multiplicative renormalization of $^{+}\mathscr{Z}(S)$. This transformation leads simultaneously to a subtractive renormalization of $^{+}\mathbb{F}$, $^{+}\mathbb{E}$, and $^{+}\mathbb{S}$. We thus obtain $\mathbb{F}$, $\mathbb{S}$, and $\mathbb{E}$, by subtracting from $^{+}\mathbb{F}$, $^{+}\mathbb{S}$, and $^{+}\mathbb{E}$, terms which depend on s_0 and are proportional to S.

$$\begin{aligned}\mathbb{F} &= -\ln \mathscr{Z}(S)\\ \mathbb{E} &= \left(S\frac{\partial}{\partial S} - 3b\frac{\partial}{\partial b}\right)\ln \mathscr{Z}(S)\\ \mathbb{S} &= \mathbb{E} - \mathbb{F}.\end{aligned} \tag{13.1.32}$$

Let us now remark that, according to (12.3.70) and (10.1.7), we have

$$\mathscr{Z}(S) = [\mathfrak{X}_1(z)]^2$$
$$z = bS^{\varepsilon/2}(2\pi)^{-d/2}$$

and, consequently, we can write

$$\mathbb{F} = -2\ln \mathfrak{X}_1(z)$$

$$\mathbb{E} = (\varepsilon - 6)z\frac{\mathrm{d}}{\mathrm{d}z}\ln \mathfrak{X}_1(z).$$

* Minor difficulties still occur when $d = 4 - 2/p$ where p is an integer, and especially when $d = 3$ (see Chapter 10, Section 4.2.6).

Moreover, according to (12.3.66),

$$\sigma_1(z) = \frac{\varepsilon}{2} z \frac{\mathrm{d}}{\mathrm{d}z} \ln \mathfrak{X}_1(z),$$

and consequently

$$\mathbb{F}(z) = -\frac{4}{\varepsilon}\int_0^z \mathrm{d}z' \frac{\sigma_1(z')}{z'}$$

$$\mathbb{E}(z) = 2\left(1 - \frac{6}{\varepsilon}\right)\sigma_1(z)$$

$$\mathbb{S}(z) = -2\left(\frac{6}{\varepsilon} - 1\right)\sigma_1(z) + \frac{4}{\varepsilon}\int_0^z \mathrm{d}z' \frac{\sigma_1(z')}{z'} \qquad (13.1.33)$$

when $z \to \infty$, $\sigma_1(z) \to \sigma_1 \equiv (\gamma - 1)/2$ [see (12.3.69)] and, in this limit, we can write

$$\mathbb{S} = \frac{4\sigma_1}{\varepsilon} \ln z + \mathbb{S}_0 \qquad (13.1.34)$$

where $\mathbb{S}_0$ is the constant

$$\mathbb{S}_0 = -2\left(\frac{6}{\varepsilon} - 1\right)\sigma_1 - \frac{4}{\varepsilon}\int_0^\infty \mathrm{d}z' \ln z' \frac{\mathrm{d}\sigma_1(z')}{\mathrm{d}z'},$$

which can be evaluated approximately without any particular difficulty. (For instance, the integral can be transformed by choosing the osmotic parameter g as the variable, as was done in Chapter 12.) Taking into account the definitions of z (10.1.7) and of σ_1 (12.3.69), we can also write (13.1.34) in the form

$$\mathbb{S} = (\gamma - 1)\left[\ln(Sb^{2/\varepsilon}) - \frac{d}{\varepsilon}\ln(2\pi)\right] + \mathbb{S}_0. \qquad (13.1.35)$$

The result is in agreement with (13.1.29), since we can still consider that $\mathbb{S}_0$ is proportional to N.

Unfortunately, it is experimentally difficult to deduce $(\gamma - 1)$ from measurements of the entropy of the system. We shall see later (Section 1.5.5) that there exists another experimental method by which to measure γ. It has not been put in practice yet (1988) but it seems more attractive.

1.4 Swelling

In a good solvent, a polymer swells; this is an experimental fact. To evaluate the swelling qualitatively, we must model the polymer in the form of a chain the elements of which repel one another, and compare its average size with the size an equivalent non-interacting chain.

In order to define the swelling, the theoretician finds it convenient to use R^2, i.e. the mean square end-to-end distance.* In the absence of interaction, this quantity will be denoted by ${}^0R^2$. Thus, the swelling of a chain can be represented by the ratio $R^2/{}^0R^2$.

On the other hand, for the experimentalist the most directly measurable length is the mean square radius of gyration R_G^2 which in the absence of interaction is denoted by ${}^0R_G^2$. Thus, it is also possible to define the swelling as the ratio $R_G^2/{}^0R_G^2$.

These definitions are not equivalent. They depend, in particular, on the model we choose. In what follows, these quantities will be studied in the framework of the standard continuous model. Incidentally, we may note[5] than the ratio $R_G^2/{}^0R_G^2$ reaches its asymptotic behaviour more slowly than $R^2/{}^0R^2$ (see in Chapter 10, the discussion in Section 3.6.2).

However, in all cases, when the chain length becomes infinite, the ratio R_G^2/R^2 has a limit which is a universal constant. This ratio can easily be calculated for simple models. Thus, let us represent by $t = X/N$ the fraction of length (along the chain) of a chain element; we find

for a Brownian chain (a critical chain with $\nu = 1/2$)

$${}^0R_G^2/{}^0R^2 = \int_0^1 \mathrm{d}t\,(1-t)t = \frac{1}{6};$$

for a rigid rod (a critical chain with $\nu = 1$)

$$R_G^2/R^2 = \int_0^1 \mathrm{d}t\,(1-t)t^2 = \frac{1}{12}.$$

Of course, for a Kuhnian chain R_G^2/R^2 takes an intermediate value, which depends only on d. This constant plays an essential role when we study the properties long polymers; in fact, it allows us to establish a correspondence between experimental results and theoretical results. We shall come back later on to this important question.

1.4.1 Mean square end-to-end distance

Until recently (1984), for $d = 3$, the expansion of the swelling $R^2/{}^0R^2 = {}^eS/S = \mathfrak{X}_0(z)$ in powers z, was known only up to third-order in z [see (10.7.9),

* Let us observe that if we stretch a chain by applying (weak) forces $\vec{\mathscr{F}}$ and $-\vec{\mathscr{F}}$ at the end points, the stretching is directly related to R. In fact,

$$\langle\!\langle \vec{r}_N - \vec{r}_0 \rangle\!\rangle_{\mathscr{F}} = \frac{\langle\!\langle (\vec{r}_N - \vec{r}_0)\exp(\vec{\mathscr{F}}.(\vec{r}_N - \vec{r}_0)) \rangle\!\rangle}{\langle\!\langle \exp(\vec{\mathscr{F}}.(\vec{r}_N - \vec{r}_0)) \rangle\!\rangle} \simeq \frac{\vec{\mathscr{F}}}{d} \langle\!\langle (\vec{r}_N - \vec{r}_0)^2 \rangle\!\rangle$$

(forces are expressed in thermal units).

(10.7.10) and (10.7.11)]

$$R^2/{}^0R^2 \equiv \mathfrak{X}_0(z) = 1 + \frac{4}{3}z - 2.0754\, z^2 + 6.3\, z^3 + \dots .$$

Now it happens that the convergence of the series is poor. (Actually, it does not converge at all!) We represented the first approximations of $\mathfrak{X}_0(z)$ on Fig. 13.3; we see that for $z > 0.1$ the approximation becomes really bad. In 1984, Muthukumar and Nickel[6] succeeded in obtaining expansion of $\mathfrak{X}_0(z)$ up to sixth-order but the series in itself is practically useless.

To obtain results that are valid for large values of z, one must take as a guide the principles which lead to renormalization theory. In this way; it is possible to construct an expression of $\mathfrak{X}_0(z)$ which is valid for all values of z. The results are given in parametric form. Various dimensionless parameters can be used, but the basic parameter thus chosen is always a physical number which a finite limit when $z \to \infty$.

Two approaches will be presented here. The first method (des Cloizeaux 1981)[2] is used to calculate the swelling for all z from expansions in $\varepsilon = 4 - d$ and

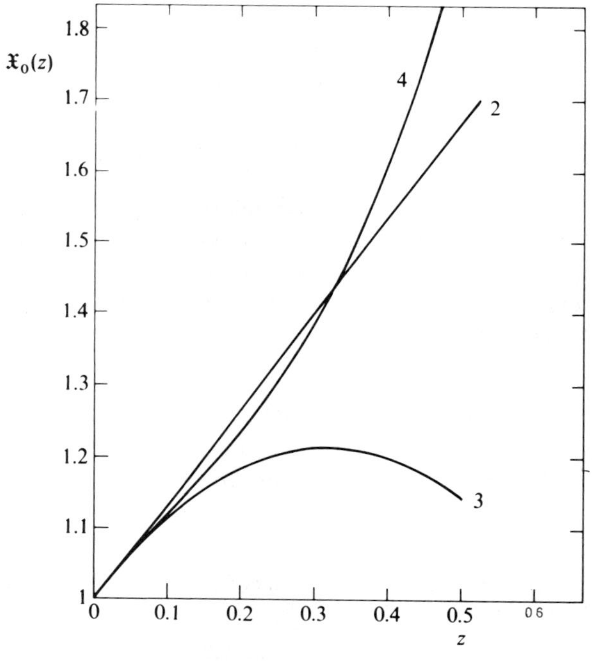

Fig. 13.2. Representation of the swelling $\mathfrak{X}_0(z)$ of a chain in three-dimensional space, by a series in z. The straight line 2 corresponds to the first two terms of the series, curve 3 to the first three terms, curve 4 to the first four terms.

z (up to second-order). The second method (des Cloizeaux, Conte, and Jannink 1985)[7] is used to calculate the swelling for all z from the series obtained directly at $d = 3$ by Muthukumar and Nickel.

First method

The starting point of the first method is the following: the *basic physical parameter* we choose is the osmotic parameter g. Consequently, to begin with, we try to estimate $\mathfrak{X}_0[g]$ and $z[g]$.

The swelling $\mathfrak{X}_0[g]$ is given by (12.3.93)

$$\mathfrak{X}_0[g] = \left(1 - \frac{g}{g^*}\right)^{-2\sigma_0/\sigma_a} \exp\left\{2\int_0^g dg' \left[\frac{\sigma_0[g']}{w[g']} - \frac{\sigma_0}{(g^* - g')\sigma_a}\right]\right\} \tag{13.1.36}$$

and in a similar way, $z[g]$ is given by (12.3.89)

$$z = g\left(1 - \frac{g}{g^*}\right)^{-\varepsilon/\sigma_a} \exp\left\{\int_0^g dg' \left[\frac{\varepsilon}{w[g']} - \frac{1}{g'} - \frac{\varepsilon}{(g^* - g')\sigma_a}\right]\right\}. \tag{13.1.37}$$

These expressions may look complicated, but we must emphasize the fact that the functions which appear in the integrals are regular and vary only slowly with g.

We content ourselves with second-order expansions in g and we write [see (13.1.20) and (13.1.22)]

$$w[g] = \frac{\varepsilon g}{1 + Ag}\left(1 - \frac{g}{g^*}\right)$$

$$\sigma_a = \frac{\varepsilon}{1 + Ag^*}$$

$$\sigma_0[g] = a_0 g + b_0 g^2$$

where A is given by (13.1.21) and a_0, b_0 by (12.3.103).

By definition, the quantity $\left(\frac{\sigma_0[g]}{w[g]} - \frac{\sigma_0}{(g^* - g)\sigma_2}\right)$ has no pole for $g = g^*$, and, according to the preceding approximations, it is a linear function of g. Thus, to obtain its expression, we can calculate it for $g = 0$ and for $g \to \infty$. In this way, we find

$$\frac{\sigma_0[g]}{w[g]} - \frac{\sigma_0}{(g^* - g)\sigma_a} \simeq \left(\frac{a_0}{\varepsilon} - \frac{\sigma_0}{g^*\sigma_a}\right) + \left(\frac{1}{\varepsilon} - \frac{1}{\sigma_a}\right) b_0 g.$$

Let us bring this expression in (13.1.36); we find:

$$\mathfrak{X}_0[g] = \left(1 - \frac{g}{g^*}\right)^{-2\sigma_0/\sigma_a} \exp\left\{2\left(\frac{a_0}{\varepsilon} - \frac{\sigma_0}{g^*\sigma_a}\right)g + \left(\frac{1}{\varepsilon} - \frac{1}{\sigma_a}\right)b_0 g^2\right\}. \tag{13.1.38}$$

For $d = 3$ (i.e. for $\varepsilon = 1$), eqn (12.3.103) gives:

$$a_0 = 0.5 \qquad b_0 = -1.272,$$

and in the same way (12.3.105) gives

$$\sigma_0 = 47/256 = 0.183.$$

Finally, we choose for g^* and σ_a, the values (with $d = 3$) corresponding to (13.1.12) and (13.1.25), namely*

$$g^* = 0.266 \qquad \sigma_a = 0.653.$$

By bringing these values into (13.1.38), we find

$$\mathfrak{X}_0[g] = \left(1 - \frac{g}{0.266}\right)^{-0.562} \exp[-(1.113)g + (0.676)g^2]. \tag{13.1.39}$$

The calculation of $z[g]$ is similar and has been performed previously; eqn (13.1.23) gives

$$z = g\left(1 - \frac{g}{0.266}\right)^{-1.531}. \tag{13.1.40}$$

Thus (13.1.39) and (13.1.40) together constitute a parametric representation of $\mathfrak{X}_0(z)$.

In particular, we can examine the behaviour of $\mathfrak{X}_0(z)$ when $z \to \infty$. From eqns (13.1.38) and (13.1.23), we deduce (with $g = g^*$)

$$\mathfrak{X}_0(z)/z^{2\sigma_0/\varepsilon} = (g^*)^{-2\sigma_0/\varepsilon} \exp\left[\frac{2a_0 g^* + b_0(g^*)^2}{\varepsilon} - \frac{2\sigma_0 + b_0(g^*)^2}{\sigma_a}\right].$$

Replacing $a_0, b_0, \sigma_0, \sigma_a$ and g^* by the values of these quantities for $d = 3$, we obtain

$$\mathfrak{X}_0 \simeq (1.27)z^{0.366}, \tag{13.1.41}$$

but for the more exact value $\nu = 0.588$ the exponent of z would be

$$2\sigma_0 = 2(2\nu - 1) = 0.352. \tag{13.1.42}$$

The swelling $\mathfrak{X}_0(z)$ obtained in various ways is plotted against $\ln(1 + z)$ in Fig. 13.3. The function $\mathfrak{X}_0(z)$ given by (13.1.39) and (13.1.40) is represented by curve R; we also drew the result corresponding to Flory's formula[5] (curve F) which is derived in Chapter 8 [see (8.1.34)] and which, for $d = 3$, is

$$\mathfrak{X}_0^{5/2} - \mathfrak{X}_0^{3/2} = \tfrac{1}{2}3^{1/2}z. \tag{13.1.43}$$

We observe that these two results are represented by rather similar curves. The exact result lies in between. This fact could have been predicted in 1981 by examining how $\mathfrak{X}_0(z)$ behaves for $0 < z \ll 1$ and $z \gg 1$; now we can check it because more recent studies[6] which we are going to describe, allow us to draw the real curve very precisely.

* These values correspond to a second-order expansion with respect to ε, and thus are consistent with the approximation which is used here.

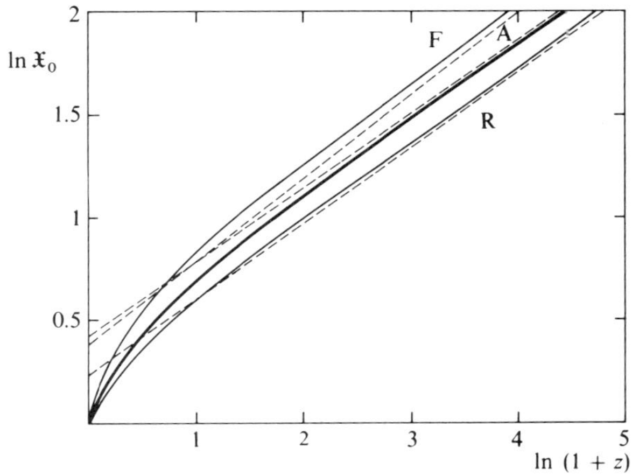

Fig. 13.3. Curves representing various approximations of the swelling $\mathfrak{X}_0(z)$ of a three-dimensional chain are represented here.

A – result obtained by renormalization (re-arrangement) of the series giving $\mathfrak{X}_0(z)$ in powers of z (for $d = 3$ and upto order six); this result is the most precise result we have;

R – result of a renormalization calculation obtained by starting from second-order expansions in $\varepsilon = 4 - d$;

F – Flory's approximation.

For $z \ll 1$	for $z \gg 1$
$[\mathfrak{X}_0]_A \simeq 1 + \frac{4}{3}z + \dots$ (exact)	$[\ln \mathfrak{X}_0]_A = 0.3538 \ln z + 0.4278$
$[\mathfrak{X}_0]_R \simeq 1 + z + \dots$	$[\ln \mathfrak{X}_0]_R = 0.367 \ln z + 0.238$
$[\mathfrak{X}_0]_F \simeq 1 + \dfrac{3^{3/2}}{2} z + \dots \simeq 1 + 2.60\, z + \dots$	$[\ln \mathfrak{X}_0]_F = 0.4 \ln z + 0.506$

The asymptots are represented by dashed lines; note their positions.

Second method

As was pointed out previously [see Chapter 12, eqns (12.3.5) and (12.5.2)] for $d = 3$ and $d = 2$, Muthukumar and Nickel calculated expansions of the form

$$\mathfrak{X}_0(z) = 1 + a_1 z + a_2 z^2 + a_3 z^3 + \dots . \tag{13.1.44}$$

For $d = 3$, the calculated coefficients $a_0, \dots, a_6$ appear in Table 13.1:

This gives a great deal of information on the function $\mathfrak{X}_0(z)$. Obviously, the series is divergent, since the ratios of the moduli of successive coefficients increases with n. However, we may hope to extract interesting results from it concerning the value of $\mathfrak{X}_0(z)$ for any z. General principles have to be applied in order to rearrange the series.[7] For this, we need an adequate *physical parameter*.

Table 13.1

n	a_n	a_n/a_{n-1}
1	+ 4/3	1.333
2	− 2.075 385 396	1.556
3	+ 6.296 879 676	3.034
4	− 25.057 250 72	3.979
5	+ 116.134 785	4.634
6	− 594.716 63	5.121

The parameter g chosen previously cannot be used here because we do not know the expansion of $g(z)$ for $d = 3$ at high orders. We must look for a more accessible *physical parameter*.

For any value of $d = 4 - \varepsilon$, the exponent $\sigma_0(z)$ is defined by the equality

$$\sigma_0(z) = \frac{\varepsilon}{2} z \frac{\partial}{\partial z} \ln \mathfrak{X}_0(z) \tag{13.1.45}$$

which can be written as follows

$$\sigma_0(z) = S \frac{\partial}{\partial S} \mathfrak{X}_0(z) \bigg|_{b = \text{const.}}$$

and we know that (see Chapter 12)

$$\sigma(0) = 0 \quad \sigma(z \to \infty) = \sigma_0^* = 2\nu - 1.$$

Thus, $\sigma_0(z)$ is a slowly varying function of z, and we may choose this quantity $\sigma_0 = \sigma_0(z)$ as the basic physical parameter.

The expansion of $\sigma_0(z)$ is immediately deduced from the expansion (13.1.44) of $\mathfrak{X}_0(z)$

$$\frac{2}{\varepsilon}\sigma_0(z) = a_1 z + 2a_2 z^2 + 3a_3 z^3 + \dots . \tag{13.1.46}$$

This series can be inverted, and we find

$$z[\sigma_0] = \frac{1}{a_1}\left(\frac{2\sigma_0}{\varepsilon}\right) - 2\frac{a_2}{(a_1)^3}\left(\frac{2\sigma_0}{\varepsilon}\right)^2 + \dots . \tag{13.1.47}$$

In the same way, we can re-express the swelling by setting

$$\mathfrak{X}_0[\sigma_0] = \mathfrak{X}_0(z[\sigma_0]).$$

Let us now introduce the quantity:

$$w_0(z) = \varepsilon z \frac{\partial}{\partial z} \sigma_0(z) \tag{13.1.48}$$

which can be considered as a function of σ_0 by setting

$$w[\sigma_0] = w_0(z).$$

We observe that, from (13.1.48), (13.1.46), and (13.1.47), we can get an expansion of $w_0[\sigma_0]$ of the form

$$w_0[\sigma_0] = b_1 \sigma_0 + b_2 \sigma_0^2 + \ldots . \tag{13.1.49}$$

(where $b_1 = \varepsilon$).

Let us also note that $w_0(z)$, which is the derivative of $\sigma_0(z)$ with respect to $\ln z$ goes to zero when $z \to \infty$. Consequently, as $\sigma_0 \to \sigma_0^*$ when $z \to \infty$, we have

$$w_0[\sigma_0^*] = 0.$$

The series (13.1.49) serves to calculate σ_0^* and then the exponent ν. Moreover, to calculate the swelling, we must use (13.1.49) to obtain an approximate expression of $w_0[\sigma_0]$ in the domain $0 < \sigma_0 < \sigma_0^*$. Then, $w_0[\sigma_0]$ can be used to calculate $\mathfrak{X}_0[\sigma_0]$ and $z[\sigma_0]$. Indeed, (13.1.48) and (13.1.49) can be rewritten in the form

$$\frac{1}{z}\frac{\partial z}{\partial \sigma_0} = \frac{\varepsilon}{w[\sigma_0]}$$

$$\frac{\partial}{\partial \sigma_0} \ln \mathfrak{X}_0 = \frac{2\sigma_0}{w[\sigma_0]}. \tag{13.1.50}$$

Let us integrate these equations with the boundary conditions

$$\mathfrak{X}_0(0) = 1$$

$$z \simeq \frac{2\sigma_0}{\varepsilon a_1} \qquad 0 < \sigma_0 \ll 1.$$

We obtain

$$z = \frac{2\sigma_0}{\varepsilon a_1} \exp\left\{\int_0^{\sigma_0} d\sigma' \left(\frac{\varepsilon}{w_0[\sigma']} - \frac{1}{\sigma'}\right)\right\}$$

$$\mathfrak{X}_0 = \exp\left\{2\int_0^{\sigma_0} d\sigma' \frac{\sigma'}{w_0[\sigma']}\right\}. \tag{13.1.51}$$

This method was applied by des Cloizeaux, Conte, and Jannink[7] for $\varepsilon = 1$. The coefficients $b_1, \ldots, b_6$ which appear in (13.1.49) were calculated from the values $a_1, \ldots, a_6$ of Table 13.1. The result is

$$b_1 = 1 \qquad b_3 = 10.8060 \qquad b_5 = 112.918$$

$$b_2 = -6.66962 \qquad b_4 = -39.2626 \qquad b_6 = -556.248. \tag{13.1.52}$$

The function $w_0[\sigma_0]$ could be estimated by keeping the first known terms of the series, but this approximation does not seem to be very good (see Table 13.1). For this reason, we choose to replace the unknown terms by a geometric series, an approximation which is suggested by the fact that the ratio of the moduli of the successive coefficients increases only slowly:

$$w_0[\sigma_0] = b_1\sigma_0 + \ldots + b_4\sigma_0^4 + b_5\frac{\sigma_0^5}{1 - b_6\sigma_0/b_5}. \tag{13.1.53}$$

In other words, $w_0[\sigma_0]$ can be represented by its Padé approximant of type [5/1]. Thus, for $1/w_0[\sigma_0]$ and $\sigma_0/w_0[\sigma_0]$, we obtain the approximate representation

$$\frac{1}{w_0[\sigma_0]} = \frac{1}{\sigma_0} + \frac{A^*}{\sigma_0 - \sigma_0^*} + \frac{A^{\cdot}}{\sigma_0 - \sigma^{\cdot}} + \frac{A' + iA''}{\sigma_0 - \sigma' - i\sigma''} + \frac{A' - iA''}{\sigma_0 - \sigma' + i\sigma''}$$

$$\frac{\sigma_0}{w_0[\sigma_0]} = \frac{B^*}{\sigma_0 - \sigma_0^*} + \frac{B^{\cdot}}{\sigma_0 - \sigma^{\cdot}} + \frac{B' + iB''}{\sigma_0 - \sigma' - i\sigma''} + \frac{B' - iB''}{\sigma_0 - \sigma' + i\sigma''}.$$

By using these expressions, we find

$$z = \frac{3}{2}\sigma_0\left[1 - \frac{\sigma_0}{\sigma_0^*}\right]^{A^*}\left[1 - \frac{\sigma_0}{\sigma^{\cdot}}\right]^{A^{\cdot}}\left[1 - \frac{2\sigma_0\sigma' - \sigma_0^2}{\sigma'^2 + \sigma''^2}\right]^{A'} \times$$
$$\times \exp\left[-2A''\,\mathrm{Arctg}\left(\frac{\sigma_0\sigma''}{\sigma'^2 + \sigma''^2 - \sigma_0^2}\right)\right]$$

$$(\mathfrak{X}_0)^{1/2} = \left[1 - \frac{\sigma_0}{\sigma^*}\right]^{B^*}\left[1 - \frac{\sigma_0}{\sigma^{\cdot}}\right]^{B^{\cdot}}\left[\mathbf{1} > \frac{2\sigma_0\sigma' - \sigma_0^2}{\sigma'^2 + \sigma''^2}\right]^{B'} \times$$
$$\times \exp\left[-2B''\,\mathrm{Arctg}\left(\frac{\sigma_0\sigma''}{\sigma'^2 + \sigma''^2 - \sigma_0^2}\right)\right] \tag{13.1.54}$$

For large z, we get

$$\ln \mathfrak{X}_0 = \frac{2B^*}{A^*}\ln z + \mathsf{A} + O(z^{1/A^*})$$

(where A is a constant).

The exponents ν and $\Delta_1 = \omega\nu$ are given by

$$\nu = \frac{1 + \sigma_0^*}{2} = \frac{1}{2} + \frac{B^*}{2A^*}$$

$$\Delta_1 = -\frac{1}{2A^*}$$

in agreement with (12.3.69), (12.3.80), and (12.3.82). The numerical values of the constants are

$$\begin{array}{lll}
\sigma_0^* = 0.176\,903 & A^* = -1.056\,145 & B^* = -0.186\,836 \\
\sigma^{\cdot} = -0.218\,430 & A^{\cdot} = 0.026\,275 & B^{\cdot} = -0.005\,739 \\
\sigma' = 0.107\,488 & A' = 0.014\,935 & B' = 0.096\,287 \\
\sigma'' = 0.556\,660 & A'' = -0.170\,090 & B'' = -0.009\,969.
\end{array} \tag{13.1.55}$$

Table 13.2

	p	2	3	4	5	6
Padé	σ_0^*		0.1980	0.1823	0.1797	0.1769
$[(p-1)/1]$	ν		0.5990	0.5912	0.5899	0.5885
Padé	σ_0^*	0.1499	0.2567	0.1677	0.1902	0.1701
$[p/0]$	ν	0.5750	0.6283	0.5839	0.5951	0.5851

These results corresponding to the case $d = 3$ seem reasonable. In particular, for large z, we find

$$\ln \mathfrak{X}_0 = 0.3538 \ln z + 0.4278 + O(z^{-0.9468}).$$

For ν and Δ_1, we obtain the values

$$\nu = 0.588 \qquad \Delta_1 = 0.473 \tag{13.1.56}$$

in full agreement with those found by Le Guillou and Zinn-Justin [see Chapter 12, eqn (12.3.9)].

The convergence of the process can be tested by calculating the successive approximate values of σ_0^* obtained by keeping only p terms in the series (13.1.44). The calculation was made by R. Conte[7] who represented $w[\sigma_0]$ by

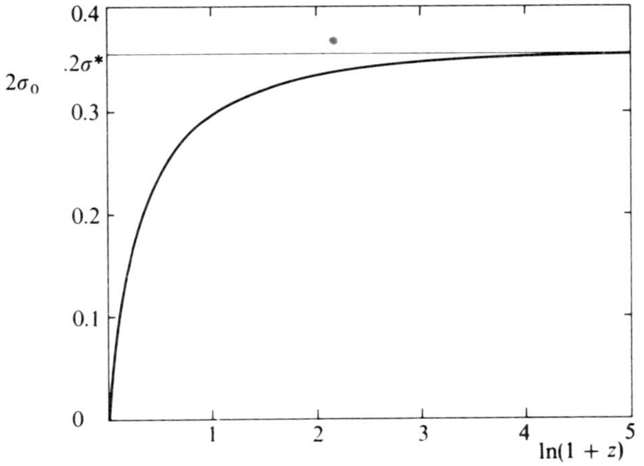

Fig. 13.4. Representation of the effective exponent $\sigma_0(z)$ of a three-dimensional chain. The result was obtained by renormalization (re-arrangement) of the series expansion of $\mathfrak{X}_0(z)$ in powers of z (for $d = 3$ up to sixth-order).

a Pade′ approximant either of type $[(p-1)/1]$ as in (13.1.51) or of type $[p/0]$, i.e. the truncated series itself.

In Fig. 13.3 we drew the curve A representing the function $\mathfrak{X}_0(z)$ obtained in this way; we believe that this curve gives a very faithful representation of the swelling, the values at the origin and at infinity being quite exact. Moreover, in Fig. 13.4, we drew the curve representing the function $\sigma_0(z)$.

The same method could be applied in the case $d = 2$, but the series given by Muthukumar and Nickel[6] does not contain as many terms for $d = 2$ as for $d = 3$ ([see (12.3.52)] and the values of the calculated coefficients look rather erratic, which leads to some, as yet, unsolved difficulties (1990).

1.4.2 Radius of gyration

As was indicated in Chapter 10, the expansion of $R_G^2/{}^0R_G^2$ in powers of z, is known for $d = 3$ up to third-order [see (10.7.33)]

$$R_G^2/{}^0R_G^2 \sim 1 + \frac{134}{105}z - \left(\frac{536}{105} - \frac{1247}{1296}\pi\right)z^2. \qquad (13.1.57)$$

As we saw previously, we can deduce from it a second-order expansion of the ratio $\aleph(z) = 6R_G^2/R^2$ in powers of z. This ratio can be written in the form of a Padé approximant.

$$\aleph(z) = 6\,R_G^2/R^2 = \frac{1 + \left(\dfrac{554}{105} - \dfrac{3395}{2592}\pi\right)z}{1 + \left(\dfrac{16}{3} - \dfrac{3395}{2592}\pi\right)z}$$

$$= \frac{1 + 1.1613\,z}{1 + 1.2185\,z}. \qquad (13.1.58)$$

This formula is valid only for small values of z. However, it can be extrapolated by letting z go to infinity. In this way, we find the rather uncertain value

$$\aleph_{d=3} = 1 - 0.047. \qquad (13.1.59)$$

This ratio can be obtained in another way. We can consider the polymer as a chain in a d-dimensional space and expand the ratio in powers of $\varepsilon = 4 - d$. To second-order in ε, we have:

$$\aleph = 1 - \frac{\varepsilon}{96} - \frac{\varepsilon^2}{384}\left(\frac{1}{24} + 10\pi\,3^{-1/2}\ln(4/3) + 30J\right). \qquad (13.1.60)$$

$$J = -\int_0^1 \mathrm{d}t\,\frac{\ln(1-t^2)}{1+3t^2} \simeq 0.216\,718$$

In other words:

$$\aleph = 1 - 0.010\,417\,\varepsilon - 0.030\,628\,\varepsilon^2 + \dots. \qquad (13.1.61)$$

The first term was obtained in 1978 by Witten and Schäfer[8] who used a zero component field theory to derive this result; it was re-obtained in 1981 by using the direct renormalization method and, as an example, we shall give this calculation at the end of this section. The term in ε^2 was calculated in 1983 by Benhamou and Mahoux[9] who used the direct renormalization method.

A direct application of (13.1.61) gives

$$\aleph_{d=3} = 1 - 0.0410$$
$$\aleph_{d=2} = 1 - 0.1433$$
$$\aleph_{d=1} = 1 - 0.3069. \qquad (13.1.62)$$

Let us note that for $d = 1$, we know the exact solution (which corresponds to rigid rods), namely:

$$\aleph_{d=1} = 1/2.$$

Thus, we can try to get better results than (12.1.62) by representing $\aleph$ as a third-order polynomial in ε, compatible with (13.1.61) and with the exact result $\aleph_{d=1} = 0.5$. In this way, we obtain the approximate formula

$$\aleph = 1 - 0.010\,417\,\varepsilon - 0.030\,628\,\varepsilon^2 - 0.007\,152\,\varepsilon^2$$

which seems reasonable, since, in it, the coefficient of ε^3 is rather small. From this equality, we deduce the values

$$\aleph_{d=3} = 1 - 0.048\,196 = 0.952$$
$$\aleph_{d=2} = 1 - 0.200\,560 = 0.799. \qquad (13.1.63)$$

By comparing (13.1.62) and (13.1.63) we see that for $d = 3$, the discrepancy is rather small; moreover, by comparing (13.1.59) and (13.1.63), we also see that there is a good agreement, a fact which leads us to trust the results. Thus, we know $\aleph$ with a good precision and this is important when we compare the theoretical predictions (expressed in terms R^2) quantitatively with the experimental results (expressed in terms of R_G^2).

Now, let us come back to the calculation of $\aleph$ to first-order in ε [see (13.1.60)]. We saw in Chapter 10 that to first order in z, $R_G^2/{}^0R_G^2$ can be calculated for any value of d. Thus (10.7.31) gives

$$R_G^2 = \frac{Sd}{6}\left[1 + 2z\frac{d^2 - 26d + 136}{(4-d)(6-d)(8-d)(10-d)} + \ldots\right].$$

In the same way, (10.7.6) gives

$$R^2 = Sd\left[1 + \frac{4z}{(4-d)(6-d)} + \ldots\right].$$

These equalities entail that, to first-order in z, we have

$$\aleph(z) = 6\,R_G^2/R^2 = 1 - \frac{2z}{(8-d)(10-d)} + \ldots . \qquad (13.1.64)$$

In other words, for $\varepsilon \ll 1$

$$\aleph(z) = 1 - \frac{z}{12} + \dots .$$

We can re-express $\aleph(z)$ in terms of the osmotic parameter g. To first-order in g, we have $z \simeq g$ [see (10.8.13)] and therefore

$$\aleph[g] = 1 - \frac{g}{12} + \dots .$$

When $z \to \infty$, $g \to g^* = \varepsilon/8$ [see (12.3.102)] and consequently

$$\aleph = 1 - \frac{g^*}{12} + \dots = 1 - \frac{\varepsilon}{96} + \dots .$$

Thus, to first-order, we find the predicted result.

1.4.3 Hydrodynamic radius

It seems that only the first-order of the expansion of $R_H/{}^0R_H$ is known [see (10.3.37)]

$$R_H/{}^0R_H = 1 + \left[4 - \pi\left(\frac{27}{16} - \frac{3}{2}\ln\frac{3}{2}\right)\right]z + \dots = 1 + (0.6093)z + \dots . \quad (13.1.65)$$

This expansion could be re-expressed in terms of the osmotic parameter g (defining the second virial coefficient) and by taking the limit $g \to g^*$, one could try to evaluate the limit of the ratio $R_H/{}^0R_H$ when $z \to \infty$ (Kuhnian limit). However, we must be aware of the fact that, in practice, for real polymers, the domain where the Kuhnian limit is valid is never reached for the hydrodynamic radius, whereas it can be reached for the radius of gyration. This fact was pointed out by Weill and des Cloizeaux[10] in 1979 and was explained by them as follows. Let i and j be the indices ($0 < i < j < N$) labelling two points on an isolated chain and let r_{ij} be the distance between these points. The definitions of R_G and R_H [see (10.7.30) and (10.7.35)] show that R_G^2 is obtained by averaging r_{ij}^2, and $1/R_H$ by averaging $1/r_{ij}$, simultaneously over the configurations and over the values of i and j. Thus, when $N \to \infty$, short sequences ($|j - i|\, |N \ll 1$) have more influence on the behaviour of R_H than on the behaviour of R_G. On the other hand, as was observed in Chapter 10, the relative swelling of a piece of chain increases with the number of links. We thus understand why R_H reaches the asymptotic regime much more slowly than R_G, and an elementary calculation[10] shows this effect in a striking manner. It demonstrates that, experimentally, it is very difficult to measure the exponent ν by using dynamical methods because, in general, they give us only effective exponents that are smaller than the asymptotic value. This conclusion is in good agreement with experiments.[11]

1.5 Limit of the probability distribution of the end-to-end vector

1.5.1 Definition of $P(\vec{r})$

The form of the probability distribution $P(\vec{r})$ of the vector $\vec{r}$ joining the ends of a long polymer in a good solvent has been much studied[12,13] between 1970 and 1974. For a long polymer, the function $P(\vec{r})$ appears in scaling form

$$P(\vec{r}) = \frac{1}{X^d}\,\mathrm{f}(r/X) \tag{13.1.66}$$

where X defines the mean size of a chain. Let R^2 be the mean square end-to-end distance. We set [see (12.3.55) and (12.3.56)]

$$R^2 = {}^{\mathrm{c}}S d = X^2 d. \tag{13.1.67}$$

The normalization of the probability and the preceding equation lead to the conditions

$$\int \mathrm{d}^d x\, \mathrm{f}(x) = 1,$$

$$\int \mathrm{d}^d x\, x^2\, \mathrm{f}(x) = d. \tag{13.1.68}$$

These conditions define $\mathrm{f}(x)$ unambiguously and $\mathrm{f}(x)$ can be considered as a universal function. It must depend only on the dimension d of the space in which the polymer is embedded. Of course, we are especially interested in the case where $d = 3$. Two approaches are possible: either we try to determine $\mathrm{f}(x)$ for any value of x, or we try to determine $\mathrm{f}(x)$ with great precision for small x and for large x, interpolating $\mathrm{f}(x)$ for intermediate values of x so as to fulfil the conditions (13.1.68). We believe that the second method is the most efficient one and we shall use it here.

The function $P(\vec{r})$ can be obtained from the partition function

$$\bar{\mathscr{Z}}(\vec{k}, -\vec{k}; S) \equiv \bar{\mathscr{Z}}(\vec{k}, -\vec{k}; 0, S; S)$$

and $\mathrm{f}(x)$ is deduced from it by going to the limit $z \to \infty$. In fact, let $\tilde{\mathrm{f}}(u)$ be the Fourier transform of $\mathrm{f}(x)$

$$\tilde{\mathrm{f}}(u) = \int \mathrm{d}^d x\, \mathrm{e}^{i\vec{u}\cdot\vec{x}}\, \mathrm{f}(x). \tag{13.1.69}$$

Conditions (13.1.68) give

$$\tilde{\mathrm{f}}(0) = 1$$

$$\frac{1}{u}\frac{\mathrm{d}}{\mathrm{d}u}\tilde{\mathrm{f}}(u)\Big|_{u=0} = -1 \tag{13.1.70}$$

and the Fourier transform of (13.1.66) reads

$$\tilde{P}(\vec{k}) = \tilde{\mathrm{f}}(Xk). \tag{13.1.71}$$

The function $\tilde{\mathrm{f}}(u)$ is directly related to the renormalized partition function

$$\bar{\mathscr{Z}}_{\mathrm{R}}(\vec{k}, \; -\vec{k}; {}^{\mathrm{e}}S).$$

Actually, this partition function fulfils the condition [see (12.3.53) and (12.3.61)]

$$\frac{1}{k}\frac{\partial}{\partial k}\bar{\mathscr{Z}}_{\mathrm{R}}(\vec{k}, -\vec{k}; {}^{\mathrm{e}}S)|_{k=0} = -{}^{\mathrm{e}}S = -X^2. \tag{13.1.72}$$

Thus, $\tilde{P}(\vec{k})$ is given by

$$\tilde{P}(\vec{k}) = \overline{\mathscr{Z}}_{\mathrm{R}}(\vec{k}, -\vec{k}; {}^{\mathrm{e}}S). \tag{13.1.73}$$

Consequently, taking into account eqns (13.1.70) and (13.1.72), we can write

$$\tilde{\mathrm{f}}(u) = \overline{\mathscr{Z}}_{\mathrm{R}}(\vec{u}X^{-1}, -\vec{u}X^{-1}; X^2)$$

and we verify that conditions (13.1.70) and (13.1.72) correspond to each other.

1.5.2 Function $P(\vec{r})$ and Green's function

In practice, to study the properties of $P(\vec{r})$ and especially of $\mathrm{f}(x)$, it is convenient to use the correspondence existing between polymer theory and field theory (see Chapter 11). The Green's function $\mathscr{G}(\vec{k}, -k; a)$ of the Landau–Ginzburg model (zero component field) is connected to the partition function of an isolated chain by the Laplace transform introduced by de Gennes.

$$\mathscr{G}(\vec{0}, \vec{r}; a) = \int_0^\infty \mathrm{d}S\, \mathrm{e}^{-aS}\, \mathscr{Z}(\vec{0}, \vec{r}; S). \tag{13.1.74}$$

It is then convenient to describe the situation in the context of the renormalized theory (see Chapter 12, Section 3.2.1).

$$\mathscr{G}_{\mathrm{R}}(\vec{0}, \vec{r}; a_{\mathrm{R}}) = (\mathfrak{Z}_1)^{-2}\mathscr{G}(\vec{0}, \vec{r}; a). \tag{13.1.75}$$

The behaviour of $\mathscr{G}_{\mathrm{R}}(\vec{0}, \vec{r}; a_{\mathrm{R}})$ for large values of $\vec{r}$ is related to the singularities of its Fourier transform $\bar{\mathscr{G}}_{\mathrm{R}}(\vec{k}, -\vec{k}; a_{\mathrm{R}})$

$$\mathscr{G}_{\mathrm{R}}(\vec{0}, \vec{r}; a_{\mathrm{R}}) = \frac{1}{(2\pi)^d}\int \mathrm{d}^d k\, \mathrm{e}^{-i\vec{k}\cdot\vec{r}}\, \bar{\mathscr{G}}_{\mathrm{R}}(\vec{k}, -\vec{k}; a_{\mathrm{R}}). \tag{13.1.76}$$

Moreover, in the critical domain ($0 < a - a_{\mathrm{c}} \ll b^{2/\varepsilon}$), $\bar{\mathscr{G}}_{\mathrm{R}}(\vec{k}, -\vec{k}; a_{\mathrm{R}})$ is given by a scaling law of the form

$$\bar{\mathscr{G}}_{\mathrm{R}}(\vec{k}, -\vec{k}; a_{\mathrm{R}}) = \frac{1}{\bar{\Gamma}_{\mathrm{R}}(\vec{k}, -\vec{k}; a_{\mathrm{R}})} = \frac{1}{a_{\mathrm{R}}\Phi(k a_{\mathrm{R}}^{-1/2})} \tag{13.1.77}$$

where $a_{\mathrm{R}}^{-1/2}$ is a length proportional to the correlation length ξ which determines the exponential decrease of the Green's function $\mathscr{G}_{\mathrm{R}}(\vec{0}, \vec{r}; a_{\mathrm{R}})$ when $r \to \infty$ (see Chapter 12, Section 3).

In agreement with (12.3.6), the renormalization condition reads

$$\Phi(x) = 1 + \frac{x^2}{2} + \ldots \tag{13.1.78}$$

and this condition leads to an exact definition of $\Phi(x)$.

Now, let us come back to the bare Green's function. We know that

$$a_{\mathrm{R}}^{-1/2} \simeq A(a - a_{\mathrm{c}})^{-\nu} \tag{13.1.79}$$

$$\bar{\mathscr{G}}(\vec{0}, \vec{0}; a) = B(a - a_{\mathrm{c}})^{-\gamma} \tag{13.1.80}$$

[see, for instance, (11.2.7) where for reasons of homogeneity, we have

$A = \mathsf{A}b^{(2\nu - 1)/(4 - d)}$

$B = \mathsf{B}b^{(2\gamma - 1)/(4 - d)}$ (b = interaction, A = const., B = const.).].

The renormalization equation (13.1.75) can be written in the form

$$\frac{\bar{\mathscr{G}}(\vec{k}, -\vec{k}; a)}{\bar{\mathscr{G}}_{\mathrm{R}}(\vec{k}, -\vec{k}; a_{\mathrm{R}})} = \frac{\bar{\mathscr{G}}(\vec{0}, -\vec{0}; a)}{\bar{\mathscr{G}}_{\mathrm{R}}(\vec{0}, -\vec{0}; a_{\mathrm{R}})} = a_{\mathrm{R}} B(a - a_{\mathrm{c}})^{-\gamma}, \tag{13.1.81}$$

or, with the help of (13.1.19)

$$\bar{\mathscr{G}}(\vec{k}, -\vec{k}; a) = BA^{-2}(a - a_{\mathrm{c}})^{-\gamma + 2\nu}\, \bar{\mathscr{G}}_{\mathrm{R}}(\vec{k}, -\vec{k}; a_{\mathrm{R}}). \tag{13.1.82}$$

Thus, eqns (13.1.77), (13.1.80), and 13.1.82) give

$$\bar{\mathscr{G}}(\vec{k}, -\vec{k}; a) = \frac{B(a - a_{\mathrm{c}})^{-\gamma}}{\Phi(Ak(a - a_{\mathrm{c}})^{-\nu})}.$$

The partition function $\bar{\mathscr{Z}}(\vec{k}, -\vec{k}; S)$ is obtained by inversion of the Laplace–de Gennes transformation:

$$\bar{\mathscr{Z}}(\vec{k}, -\vec{k}; S) = \frac{1}{2i\pi} \int_{c - i\infty}^{c + i\infty} \mathrm{d}a\, \exp(aS)\, \bar{\mathscr{G}}(\vec{k}, -\vec{k}; a), \tag{13.1.83}$$

the integration path running in the complex plane on the right of the singularities of $\bar{\mathscr{G}}(\vec{k}, -\vec{k}; a)$. Thus, in the present case

$$\bar{\mathscr{Z}}(\vec{k}, -\vec{k}; S) = \frac{B}{2i\pi} \int_{-i\infty}^{i\infty} \mathrm{d}a\, \exp(aS)\, \frac{(a - a_{\mathrm{c}})^{-\gamma}}{\Phi(Ak(a - a_{\mathrm{c}})^{-\nu})}. \tag{13.1.84}$$

In particular, for small values of k, we must find

$$\bar{\mathscr{Z}}(\vec{k}, -\vec{k}; S) = \mathscr{Z}(S)\left(1 - \frac{k^2 X^2}{2} + \ldots\right) \tag{13.1.85}$$

where $X^2 = {}^{\mathrm{e}}S$ defines the mean square end-to-end distance. Using condition

(13.1.78) and expanding $\bar{\mathscr{G}}(\vec{k}, -\vec{k}; a)$ in (13.1.83), we obtain

$$\bar{\mathscr{Z}}(\vec{k}, -\vec{k}; S) = \frac{B}{2i\pi}\int_{-i\infty}^{+i\infty} da\ \exp(aS)\left[(a-a_c)^{-\gamma} - \frac{A^2}{2}k^2(a-a_c)^{-\gamma-2\nu} + \dots\right]$$

which gives

$$\bar{\mathscr{Z}}(\vec{k}, -\vec{k}; S) = B\exp(a_c S)S^{\gamma-1}\left[\frac{1}{(\gamma-1)!} - \frac{A^2k^2S^{2\nu}}{2(\gamma+2\nu-1)!} + \dots\right].$$

Thus, by identifying this expression with (13.1.85), we find

$$X = \left[\frac{(\gamma-1)!}{(\gamma-2\nu-1)!}\right]^{1/2} AS^{\nu}. \tag{13.1.86}$$

1.5.3 Behaviour of $P(\vec{r})$ for large values of r

In this section, we show that the function $f(x)$ which appears in the expression (10.1.66) of $P(\vec{r})$ has the following asymptotic behaviour for large values of x

$$f(x) \propto x^{\kappa}\exp(-Dx^{\delta})$$

Here δ and κ are critical exponents which depend only on the dimension d of the space in which the polymer chain is embedded. We already indicated, in Chapter 3, Section 3.3, how, in 1966, Fisher[14] found the relation

$$\delta = \frac{1}{1-\nu}.$$

In this section, we study this question again by starting from different premises.

First, let us consider field theory. If the Lagrangian theory defined in Chapter 11 is a genuine field theory, as we believe, the Green's function has a pole for $k^2 = -m^2$ (where m can be considered as the mass of the particle which the field describes) and also a cut for more negative values of k^2. Thus, for large values of r, we must attribute a dominant role to this pole. At the pole, $\Phi(x)$ vanishes, and in the vicinity of this point we may set

$$\Phi(x) = (x^2 + x_0^2)\Phi. \tag{13.1.87}$$

In order to study the behaviour of $P(\vec{r})$ for large values of r, we can represent $\bar{\mathscr{G}}_R(\vec{k}, -\vec{k}; a_R)$ by the following expression derived from (13.1.77) and (13.1.87)

$$\bar{\mathscr{G}}_R(\vec{k}, -\vec{k}; a_R) = \frac{\Phi^{-1}}{k^2 + a_R x_0^2} = \Phi^{-1}\frac{\xi^2}{k^2\xi^2 + 1}$$

where we have

$$\xi = 1/x_0 a_R^{1/2} \tag{13.1.88}$$

(consequently $m = 1/\xi$).
Calculating now

$$\bar{\mathscr{G}}_R(\vec{0}, \vec{r}; a) = \frac{1}{(2\pi)^d}\int d^d k\, e^{-i\vec{k}\cdot\vec{r}}\bar{\mathscr{G}}_R(\vec{k}, -\vec{k}; a).$$

Using the identity

$$\frac{1}{k^2\xi^2+1} = \int_0^{\infty} d\tau\, e^{-\tau(k^2\xi^2+1)}$$

we see that

$$\bar{\mathscr{G}}_R(\vec{0}, -\vec{r}; a) = \Phi^{-1}(4\pi)^{-d/2}\,\xi^{2-d}\int_0^{\infty} d\tau\, \tau^{-d/2}\exp(-\tau - r^2/4\tau\xi^2).$$

As we are interested only in large values of r/ξ, we can evaluate the integral* directly by using the steepest descent method. At the saddle point, $\tau = r/2\xi$. We thus obtain

$$\bar{\mathscr{G}}_R(\vec{0}, \vec{r}; a_R) = \frac{\xi^{(3-d)/2}}{2\Phi}(2\pi r)^{(1-d)/2}\, e^{-r/\xi} \tag{13.1.89}$$

and we note that ξ is indeed the correlation length.

Let us now return to the bare Green's function, which we need to invert the Laplace–de Gennes transformation. First, we remark that for large values of ξ, according to (13.1.79) and (13.1.88), we have

$$\xi = x_0^{-1} A(a - a_c)^{-\nu}.$$

Moreover, we can also write (13.1.82) in the form

$$\mathscr{G}(\vec{0}, \vec{r}; a) = B\,A^{-2}(a - a_c)^{-\gamma+2\nu}\,\bar{\mathscr{G}}_R(\vec{0}, -\vec{r}; a_R).$$

From these equations, we deduce the following result which is valid for large values of r

$$\bar{\mathscr{G}}_R(\vec{0}, \vec{r}; a) \propto (a - a_c)^{-\gamma+\nu(1+d)/2}\, r^{(1-d)/2}\exp[-r x_0(a - a_c)^{\nu}/A]. \tag{13.1.90}$$

We did not give the value of the proportionality constant explicitly for reasons of simplicity. Thus, in what follows the asymptotic behaviour will be obtained only up to a proportionality factor. However, it would be possible to calculate this constant.

Let us calculate $\mathscr{Z}(\vec{0}, \vec{r}; S)$ up to a numerical factor by performing the inverse Laplace transformation

$$\mathscr{Z}(\vec{0}, \vec{r}; S) = \frac{1}{2\pi i}\int_{-i\infty}^{+i\infty} da\, \exp(aS)\,\mathscr{G}(\vec{0}, \vec{r}; a)$$

$$\propto r^{(1-d)/2}\, I\left(-\gamma + \frac{\nu}{2} + \frac{\nu d}{2}, S, \frac{r x_0}{A}\right) \tag{13.1.91}$$

where by definition

$$I(n, S, v) = \frac{1}{2\pi i}\int_{-i\infty}^{+i\infty} da\, e^{aS}(a - a_c)^n \exp[-v(a - a_c)^{\nu}],$$

* See Bessel functions of imaginary argument.

which can be written in the simpler form

$$I(n, S, v) = \frac{S^{-n-1}}{2\pi i} \exp(a_c S) \int_{-i\infty}^{+i\infty} d\alpha\, \alpha^n \exp[\alpha - vS^{-\nu}\alpha^\nu]. \qquad (13.1.92)$$

The integral is to be calculated over a path in the complex plane, cut along the negative half-axis (see Fig. 13.5). We must study $I(n, S, v)$ for large values of $vS^{-\nu}$, and for this purpose we shall use the steepest descent method. The relevant saddle point corresponds to the minimum of the function

$$\varphi(\alpha) = \alpha - vS^{-\nu}\alpha^\nu.$$

The value α_c of α at the saddle point is given by

$$\varphi'(\alpha_c) = 0$$

$$\alpha_c = (\nu v S^{-\nu})^{1/(1-\nu)},$$

and, at the saddle point, the constant is given by

$$\varphi''(\alpha_c) = (1 - \nu)(\nu v S^{-\nu})^{-1/(1-\nu)}.$$

Thus, we find

$$I(n, S, v) = \frac{S^{-n-1}\, e^{a_c S}}{[2\pi(1-\nu)]^{1/2}} (\nu v S^{-\nu})^{(n+1/2)/(1-\nu)}$$
$$\times \exp\left[-\left(\frac{1}{\nu} - 1\right)(\nu v S^{-\nu})^{1/(1-\nu)}\right]. \qquad (13.1.93)$$

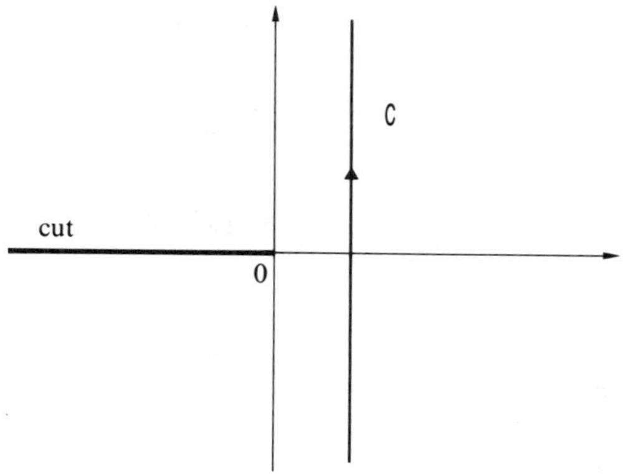

Fig. 13.5. Calculation of $\mathscr{Z}(\vec{0}, \vec{r}; S)$ by inverse Laplace transformation of $\mathscr{G}(\vec{0}, \vec{r}; a)$ for large values of a. In the complex plane of the variable a, we drew the path C which runs through the saddle point.

Finally, by putting $n = -\gamma + \frac{\nu}{2} + \frac{\nu d}{2}$ and $v = r x_0/A$ in (13.1.93) and using (13.1.91), we obtain the result

$$\mathscr{Z}(\vec{0},\vec{r};S) \propto$$

$$e^{a_c S} S^{\gamma-1-\nu d} (rS^{-\nu})^{\frac{1-\gamma+\nu d-d/2}{1-\nu}} \exp\left\{-\left(\frac{1}{\nu}-1\right)\left(\frac{\nu x_0 r S^{-\nu}}{A}\right)^{1/(1-\nu)}\right\}. \quad (13.1.94)$$

To calculate $P(\vec{r})$, we must also know

$$\mathscr{Z}(S) \equiv \bar{\mathscr{Z}}(\vec{0},\vec{0};S) = \int \mathrm{d}^d r\, \mathscr{Z}(\vec{0},\vec{r};S).$$

By using (13.1.83), we immediately find

$$\bar{\mathscr{Z}}(\vec{0},\vec{0};S) = \frac{1}{2\pi i}\int_{-i\infty}^{+i\infty} \mathrm{d}S\, e^{aS}\, \mathscr{G}(\vec{0},\vec{0};a) = \frac{B}{2i\pi}\int_{-i\infty}^{+i\infty} \mathrm{d}S\, e^{aS}(a-a_c)^{-\gamma},$$

i.e.

$$\mathscr{Z}(S) = \mathscr{Z}(\vec{0},\vec{0};S) = \frac{B}{(\gamma-1)!} S^{\gamma-1}\exp(a_c S). \quad (13.1.95)$$

Finally, from (13.1.94) and (13.1.95), we deduce

$$P(\vec{r}) = \frac{\mathscr{Z}(\vec{0},\vec{r};S)}{\mathscr{Z}(S)}$$

$$\propto S^{-\nu d}(rS^{-\nu})^{\frac{1-\gamma+\nu d-d/2}{1-\nu}} \exp\left\{-\left(\frac{1}{\nu}-1\right)\left(\frac{\nu x_0 r S^{-\nu}}{A}\right)^{1/(1-\nu)}\right\} \quad (13.1.96)$$

Now, we can express $P(\vec{r})$ in terms of X by using (13.1.86). We see that for large values of r, $P(\vec{r})$ can be written in the form

$$P(\vec{r}) \equiv X^{-d} \mathrm{f}(r/X) \simeq \mathrm{f}_\infty X^d (r/X)^{\frac{1-\gamma+\nu d-d/2}{1-\nu}} \exp(-\mathrm{D}(r/X)^{1/(1-\nu)}) \quad (13.1.97)$$

where D is deduced from (13.1.86) and (13.1.96)

$$\mathrm{D} = \left(\frac{1}{\nu}-1\right)\left[\frac{\nu^2 x_0^2(\gamma-1)!}{(\gamma+2\nu-1)!}\right]^{1/[2(1-\nu)]} \quad (13.1.98)$$

Moreover, f_∞ is a constant which could be expressed in terms of A, B, Φ, and x_0. Thus, in principle, the two constants D and f_∞ could be calculated (for instance, by ε-expansion).

Incidentally, let us remark that for $d = 4$ we have

$$\gamma = 1 \qquad \nu = 1/2 \qquad x_0^2 = 2$$

$$\mathrm{f}(x) = (2\pi)^{-d/2} e^{-x^2/2}.$$

This is compatible with (13.1.97) and (13.1.98) and, for $d = 4$, we recover the trivial results

$$\frac{1}{1-\nu} = 2$$

$$\mathrm{D} = \tfrac{1}{2}$$

$$\frac{1-\gamma+\nu d - d/2}{1-\nu} = 0.$$

We observe that in (13.1.97), we have the exponential factor $\exp[-\mathrm{D}(r/X)^{1/(1-\nu)}]$, the existence of which was demonstrated by Fisher[14] in 1966 (see Chapter 3, Section 3.3, and Chapter 10, Section 2.2.2). Moreover, we must note that the existence of the prefactor $(r/X)^{\frac{1-\gamma+\nu d-d/2}{1-\nu}}$ was predicted by McKenzie and Moore[12] as early as 1971, i.e. one year before de Gennes established an exact correspondence between partition functions and Green's functions (for zero-component fields). Nevertheless, the proof given by McKenzie and Moore[12] is very similar to the preceding demonstration, because these authors were guided by intuitions which further developments confirmed.[13]

The characteristic exponents can be expanded in ε; up to second-order, we find [see (12.3.33) and (12.3.106)]

$$\delta \equiv \frac{1}{1-\nu} = 2 + \frac{\varepsilon}{4} + \frac{19}{128}\varepsilon^2 + \ldots$$

$$\kappa \equiv \frac{1-\gamma+\nu d - d/2}{1-\nu} = \frac{\varepsilon}{4} + \frac{5}{128}\varepsilon^2 + \ldots . \tag{13.1.99}$$

We can also try to evaluate the exponents δ and κ directly at $d = 3$ and $d = 2$. For $d = 3$, we can replace γ and ν by the values calculated directly by Le Guillou and Zinn-Justin [see (12.3.9)]. In this manner we obtain

$$\delta = 2.427 \pm 0.006$$

$$\kappa = 0.249 \pm 0.011. \tag{13.1.100}$$

For $d = 2$, we can use the 'exact' values $\nu = 3/4$ and $\gamma = 43/32$ given in Chapter 2, Sections 4.1 and 4.2 [we note that the value $\nu = 0.75$ seems to be compatible with the computer experiments made by Derrida (1981); see (12.2.9)]. In this way, we obtain

$$\delta = 4$$

$$\kappa = 5/8.$$

Finally, we observe that all these values of δ are compatible with less accurate results obtained, in 1965 by Domb, Gillis, and Wilmers,[15] and in 1966 by Fisher,[14] namely $\delta = 2.5$ for $d = 3$, and $\delta = 4.0$ for $d = 2$.

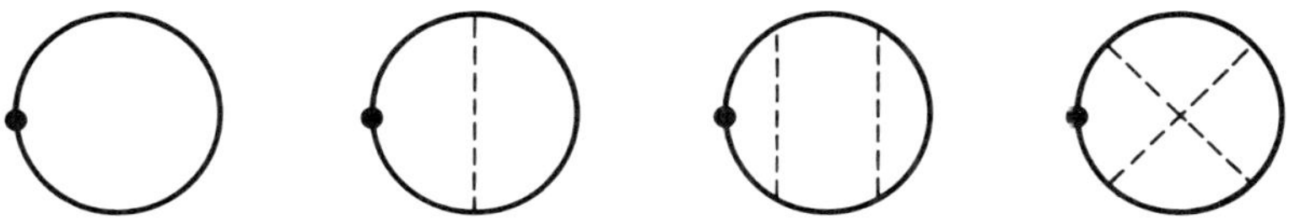

Fig. 13.6. Ring diagrams contributing to $\overline{\mathscr{Z}}(\vec{0},\vec{0};S)$.

1.5.4 Behaviour of $P(\vec{r})$ for small values of r

First, let us try to evaluate $P(\vec{0}) = \dfrac{\overline{\mathscr{Z}}(\vec{0},\vec{0};S)}{\mathscr{Z}(S)}$. In the absence of interaction, we have

$$^{0}\mathscr{Z}(S) = 1$$

$$^{0}\mathscr{Z}(\vec{0},\vec{0};S) = \frac{1}{(2\pi)^d}\int \mathrm{d}^d k\ \mathrm{e}^{-k^2S/2} = (2\pi S)^{-d/2} \qquad (13.1.101)$$

and therefore

$$^{0}P(\vec{0}) = (2\pi S)^{-d/2}.$$

Let us now assume that the chain interacts with itself. $\overline{\mathscr{Z}}(\vec{0},\vec{0};S)$ is the sum of the contributions of the ring diagrams (see Fig. 13.6). As these diagrams have no free ends, their renormalization is trivial;* the only renormalization factor is the swelling $\mathfrak{X}_0(z)$. In fact, these ring diagrams are characterized by the insertion of one two-leg vertex (represented by a point in Fig. 13.6) but, as was shown in Chapter 12, Section 3.3.3.4, the renormalization factor of two-leg vertices can be chosen equal to unity.

Consequently, we write

$$\mathscr{Z}(\vec{0},\vec{0};S) = \mathscr{Z}_{\mathrm{R}}(\vec{0},\vec{0};{}^{\mathrm{e}}S) \qquad (13.1.102)$$

As $\mathscr{Z}_{\mathrm{R}}(\vec{0},\vec{0};{}^{\mathrm{e}}S)$ has obviously the same dimensionality as $\mathscr{Z}(S)$ and depends only on $X = ({}^{\mathrm{e}}S)^{1/2}$, we have (for $z \to \infty$)

$$\mathscr{Z}(\vec{0},\vec{0};S) = \mathscr{Z}_{\mathrm{R}}(\vec{0},\vec{0};{}^{\mathrm{e}}S) \simeq \mathscr{Z}_0 X^{-d} \qquad (13.1.103)$$

where $\mathscr{Z}_0$ is a universal constant which depends only on the space dimension d.

We shall ponder a while over this result because for a fixed interaction (b = const.), it tells us how the partition function of a polymer whose end points coincide, depends on S. Indeed, (12.3.59) and (12.3.58) show that for $z \gg 1$

$$X^2 \propto S z^{2(2\nu-1)/\varepsilon}. \qquad (13.1.104)$$

Moreover, as definition (10.1.7) says that

$$z \propto b S^{\varepsilon/2} \qquad (13.1.105)$$

* Let us note that this conclusion is not completely obvious. In 1973, Fisher[16] discussed possible violations of the 'hyperscaling'.

we deduce the asymptotic expression

$$\mathscr{Z}(\vec{0},\vec{0};S) \propto X^{-d} \propto S^{-\nu d} b^{-(2\nu-1)d/(4-d)} \qquad (13.1.106)$$

from these equations. Now, returning to Table 13.1 of Chapter 4, we see that the number U_N of closed self-avoiding chains drawn on a lattice from a given origin has been written in the form

$$U_N \mu^{-N} \propto N^{\alpha-2} \qquad (13.1.107)$$

which defines the exponent α. We also recall that the area S in the continuous model can be considered as proportional to the number N of links in the discrete model. Moreover, we must admit that, for large values of N and S, the quantities $\mathscr{Z}(\vec{0},\vec{0};S)$ and $U_N\mu^{-N}$ are proportional. Thus the equalities (13.1.106) and (13.1.107) can be compared. By writing that, in these expressions, the exponents of S and N are equal, we obtain the relation*

$$\alpha = 2 - \nu d$$

Let us now come back to the calculation of $P(\vec{0})$ which is equal to the ratio $\mathscr{Z}(\vec{0},\vec{0};S)/\mathscr{Z}(S)$ where $\mathscr{Z}(\vec{0},\vec{0};S)$ is given by (13.1.106). Let us recall that

$$\mathscr{Z}(S) = [\mathfrak{X}_1(z)]^2$$

where $\mathfrak{X}_1(z)$ is the renormalization factor of the one-leg vertex [see (13.3.62)]. Thus, by combining the preceding equality and (13.1.103), we obtain the result

$$P(\vec{0}) = \frac{\mathscr{Z}(\vec{0},\vec{0};S)}{\mathscr{Z}(S)} = \mathscr{Z}_0\,[\mathfrak{X}_1(z)]^{-2} X^{-d}. \qquad (13.1.108)$$

Moreover, for $z \gg 1$, we have according to (12.3.60)

$$\mathfrak{X}_1(z) \propto z^{(r-1)/\varepsilon}.$$

Consequently, for $z \gg 1$, we have

$$P(\vec{0}) \propto z^{-2(\gamma-1)/\varepsilon} X^{-d}. \qquad (13.1.109)$$

Moreover, we expect $P(\vec{r})$ to obey the scaling law

$$P(\vec{r}) = X^{-d} \mathrm{f}(r/X). \qquad (13.1.110)$$

Let us now compare (13.1.109) and (13.1.110): consider the limits $z \to \infty$ in (13.1.109) and $r \to 0$ in (13.1.110). As $(\gamma - 1) > 0$, the equation

$$\mathrm{f}(0) = 0 \qquad (13.1.111)$$

must be right. This interesting result is a subtle consequence of renormalization

* This relation appears in the same form in the critical theory of liquid–gas transitions and in ferromagnetic transitions. It is very clear that the definitions of α, γ, and ν used here have been chosen *ad hoc*.

principles. It agrees with the computer experiments made by Renardy (1971)[17] and McKenzie (1973).[18]

Let us now examine how $P(\vec{0})$ goes to zero when, the interaction remaining fixed, the length, or more precisely the area S of the chain, becomes infinite. For $z \gg 1$, (12.3.56) and (12.3.57) give

$$X^2 = S\mathfrak{X}_0(z) \propto S z^{2(2\nu - 1)/\varepsilon},$$

and consequently (13.1.109) reads as follows

$$P(\vec{0}) \propto S^{-d/2} z^{2(1-\gamma-\nu d + d/2)/\varepsilon}. \tag{13.1.112}$$

As, definition (10.1.7) says that

$$z \propto b S^{\varepsilon/2}$$

we can re-express $P(\vec{0})$ in terms of S and b. We obtain

$$P(\vec{0}) \propto S^{1-\gamma-\nu d} b^{2(1-\gamma-\nu d + d/2)/\varepsilon}. \tag{13.1.113}$$

In the same way, considering a self-avoiding chain of N links on a lattice, we can define its closure probability p_N. Admitting the universality principle, we can now deduce the asymptotic behaviour of p_N for large N, from (13.1.113); we obtain

$$p_N \propto N^{1-\gamma-\nu d}. \tag{13.1.114}$$

Now let us examine how the universal function $\mathrm{f}(x)$ varies for small values of x. As $\mathrm{f}(0) = 0$, we can admit that, for $x \ll 1$, $\mathrm{f}(x)$ is proportional to a power of x

$$\mathrm{f}(x) \propto x^{\theta} \tag{13.1.115}$$

where θ is an exponent which has to be determined (let us note that in Section 1.6 analogous exponents θ_1 and θ_2 will be defined and that in this section we write $\theta \equiv \theta_0$).

Thus for $r/X \ll 1$, we write

$$P(\vec{r}) \propto X^{-d-\theta} r^{\theta}. \tag{13.1.116}$$

On the other hand, (13.1.112) and (13.1.113) show that for large values of z

$$P(\vec{0}) \propto z^{-2(\gamma-1)/\varepsilon} X^{-d} \tag{13.1.117}$$

where

$$z \propto b S^{\varepsilon/2}.$$

For large values of z and small values of r/X, there is a crossover between the limits defined by (13.1.116) and (13.1.111). Remarking that $b^{-1/\varepsilon}$ has always the dimension of a length, we can represent $P(\vec{r})$ in the crossover region by an expression of the form

$$P(\vec{r}) = X^{-d-\theta} \psi(r b^{1/\varepsilon}) r^{\theta}. \tag{13.1.118}$$

For $x \to 0$, we must have $\psi(x) \propto x^{-\theta}$ in order to eliminate the dependence with respect to r and to find (12.1.117); for $X \to \infty$, we must have $\psi(x) \to$ constant in order to obtain a dependence of the form r^θ as in (13.1.116). Thus (13.1.118) leads to the result

$$P(\vec{0}) \propto X^{-d}(X\,b^{1/\varepsilon})^{-\theta}. \tag{13.1.119}$$

Moreover, for a finite self-avoiding chain drawn on a lattice, this result can be written in the form

$$p_{0,N} \propto N^{-\nu(\theta+d)} \tag{13.1.120}$$

where p_{0N} is the closure probability. Let us now compare (13.1.117) and (13.1.119): these expressions must be identical. In other words the value of θ must be such that

$$(X^\varepsilon b)^\theta \propto z^{2(\gamma-1)}.$$

Moreover, we know that [see (13.1.97)]

$$X^\varepsilon \propto S^{\varepsilon/2}\,[\mathfrak{X}_0(z)]^{\varepsilon/2} \propto S^{\varepsilon/2}\,z^{2\nu-1} \propto b^{-1}\,z^{2\nu}.$$

Thus, by comparing the preceding equations, we obtain

$$\theta = (\gamma - 1)/\nu \tag{13.1.121}$$

[compare also (13.1.114) and (13.1.120)].

This result is obtained here by direct renormalization. It is also possible to proceed indirectly by means of field theory; it is in this way that relation (13.1.121) was first established[13] and, at that time, direct renormalization did not exist. The essence of the proof is the same in both cases. Actually, in field theory, the Green's functions $\mathscr{G}(\vec{x}, \vec{y}; a)$ and $\mathscr{G}(\vec{x}, \vec{x}; a)$ are defined as follows

$$\mathscr{G}(\vec{x}, \vec{y}; a) = \langle\!\langle \varphi_1(\vec{x})\varphi_1(\vec{y}) \rangle\!\rangle,$$
$$\mathscr{G}(\vec{x}, \vec{x}; a) = \langle\!\langle [\varphi_1(x)]^2 \rangle\!\rangle,$$

and the change in the behaviour of these functions which is observed when $\vec{y} \to \vec{x}$ can be explained by remarking that the field φ and the field φ^2 renormalize in a different manner. As usual in field theory, this difference can be expressed, by introducing *the anomalous dimensions* d_{a1} and d_{a2} of the fields φ and φ^2; these exponents which correspond to the direct renormalization exponents σ_1 and σ_2 (see Chapter 12) are given by the equalities

$$2d_{a1} = d - \gamma/\nu$$
$$d_{a2} = d - 1/\nu$$

and vanish for $d = 4$. Then, we can observe that θ is directly related to these exponents by the equality

$$\theta = d_{a2} - 2d_{a1}.$$

However, let us note that when one tries to use this approach through field theory, a technical difficulty appears. When $\vec{r} \to 0$ $\mathscr{G}(\vec{0}, \vec{r}; a) \to \infty$ and therefore $\mathscr{G}(\vec{0}, \vec{0}; a)$ is infinite, even in the absence of interaction. Actually, we can formally write

$$^0\mathscr{G}(\vec{0},\vec{r};a) = \int_0^\infty \mathrm{d}S\, \mathrm{e}^{-aS}\; ^0\mathscr{Z}(\vec{0},\vec{r};S)$$

where according to (13.1.101)

$$^0\mathscr{Z}(\vec{0},\vec{0};S) = (2\pi S)^{-d/2},$$

and the integral $\int_0^\infty \mathrm{d}S\, \mathrm{e}^{-aS} S^{-d/2}$ is divergent for $d = 3$. This divergence has no deep meaning and can be avoided by working with the function $\frac{\partial}{\partial a}\mathscr{G}(\vec{0},\vec{r};a)$ which is always well-defined. Nevertheless, this divergence explains why, in field theory, in order to define $\mathscr{G}(\vec{x},\vec{x};a)$, it is necessary to introduce a supplementary subtractive renormalization.

Let us now come back to our results. We showed that for small values of r/X

$$P(\vec{r}) = \mathrm{f}_1 X^{-d}(r/X)^\theta \tag{13.1.122}$$

where

$$\theta = (\gamma - 1)/\nu.$$

This conclusion has not yet been verified for $d = 3$, but this could be done as will be explained in the following section. On the contrary, in two dimensions, Monte Carlo computer experiments[17] on lattices enable us to check this point as was indicated in Chapter 4, Section 3.2. The exponent θ can be expanded with respect to ε, and we find up to second-order

$$\theta = \frac{\varepsilon}{4} + \frac{9}{128}\varepsilon^2 + \ldots \qquad (\varepsilon = 4 - d). \tag{13.1.123}$$

Moreover, if we replace γ and ν by the values directly obtained for $d = 3$ [see (12.3.9)], we find

$$\theta = (\gamma - 1)/\nu = 0.275 \pm 0.002. \tag{13.1.124}$$

It is also possible to calculate the universal coefficient f_1. Calculations made with field theory by Oono, Ohta, and Freed,[19] and also the direct calculation given in Appendix K, lead to the expansion

$$\mathrm{f}_1 = \frac{1}{4\pi^2}\left[1 + \frac{\varepsilon}{8}(-2 + 4\ln(2\pi) + \mathfrak{C}) + \ldots\right] \tag{13.1.125}$$

where $\mathfrak{C}$ is Euler's constant $\mathfrak{C}$($\mathfrak{C} = 0.577$) (see Section 1.5.6 below).

1.5.5 Possible measurement of θ

At the present time, the exponent γ has not been measured for a polymer solution and a direct measurement would be very difficult (see Section 1.2). However, we may hope to obtain γ indirectly by measuring θ. For this purpose, one would use a dilute solution of polymers with marked end points. These marked end points should be elements able to selectively scatter a well-chosen monochromatic radiation falling on the solution. The scattered intensity $I(q)$ corresponding to a wave vector transfer $\vec{q}$ would be measured and we would get*

$$I(q) = A + B\tilde{P}(\vec{q})$$

where

$$\tilde{P}(\vec{q}) \equiv \langle\langle e^{i\vec{q}\cdot[\vec{r}(S) - \vec{r}(0)]} \rangle\rangle$$

is the Fourier transform of the probability distribution $P(\vec{r})$ of the end-to-end vector $\vec{r}$. We expect $(I(q) - A)$ to decrease like a power of q when $q \to \infty$

$$I(q) - A \propto q^{-\tau} \tag{13.1.126}$$

and by measuring τ, one would obtain a value of θ. Indeed, as we shall now show, the singularity of $P(\vec{r})$ at the origin, determines the manner in which $\tilde{P}(\vec{q})$ decreases at infinity.

Using (13.1.66), we can write

$$\tilde{P}(\vec{q}) = \int d^d r\, e^{i\vec{q}\cdot\vec{r}}\, P(\vec{r}) = \int d^d x\, e^{iX\vec{q}\cdot\vec{x}}\, f(x).$$

This formula can also take the form

$$\tilde{P}(\vec{q}) = -\frac{1}{X^2 q^2}\int d^d x\;\; f(x)\,\Delta_x\, e^{iX\vec{q}\cdot\vec{x}}$$

and, integrating by parts we find

$$\tilde{P}(\vec{q}) = -\frac{1}{X^2 q^2}\int d^d x\, e^{iX\vec{q}\cdot\vec{x}}\left[f''(x) + \frac{d-1}{x}f'(x)\right].$$

The function $\left[f''(x) + \frac{d-1}{x}f'(x)\right]$ is singular at the origin and gives the dominant contribution for large q. Consequently, in the preceding formula, we may put

$$f(x) \simeq f_1 x^{\theta}. \tag{13.1.127}$$

We obtain, for $qX \gg 1$

$$\tilde{P}(\vec{q}) = -f_1\frac{\theta(\theta + d - 2)}{X^2 q^2}\int d^d x\,\frac{e^{iX\vec{q}\cdot\vec{x}}}{x^{2-\theta}}. \tag{13.1.128}$$

For $d = 3$, the integral converges and a simple change of variable $(\vec{x} \to \vec{x}'/qX)$

* More precisely, the intensity scattered by chains whose end monomers (0 and N) are marked, has the form (see Chapter 7, Section 3.3.3)

$$\lim_{C\to 0} = \frac{I(q)}{K_A C V} = (b_1^{\circ\!\to})^2\, C(H^1_{00}(\vec{q}) + H^1_{NN}(\vec{q}) + 2H^1_{0N}(\vec{q}))$$

where $b_1^{\circ\!\to}$ is the contrast length of the marked monomers and where it is admitted that the contrast length of the non-marked monomers is zero. The structure function $H^1_{0N}(\vec{q})$ is equal to $\tilde{P}(\vec{q})/C$.

gives

$$\tilde{P}(\vec{q}) \propto -\frac{1}{(qX)^{d+\theta}}.$$

Thus, as $I(q)$ is proportional to $\tilde{P}(\vec{q})$, the value of τ in (13.1.126) is

$$\tau = d + \theta.$$

Incidentally, we note that the integral which appears in (13.1.28) can be easily calculated (see Appendix M). One finds

$$\tilde{P}(\vec{q}) \simeq f_1\, \pi^{d/2} \frac{\Gamma\left(\frac{\theta}{2} + \frac{d}{2}\right)}{\Gamma\left(-\frac{\theta}{2}\right)} \left(\frac{2}{qX}\right)^{d+\theta}. \tag{13.1.129}$$

Conversely, such a dependence for large values of qX, implies that $P(\vec{r})$ has a singularity of type (13.1.122).

1.5.6 Curve representing $P(\vec{r})$

In the preceding sections, we showed that the probability distribution $P(\vec{r})$ of the vector $\vec{r}$ joining the end points of a polymer in a good solvent is given by the scaling law

$$P(\vec{r}) = X^{-d} f(r/X)$$

$$\int d^d x\, f(x) = 1$$

$$\int d^d x\, x^2 f(x) = d \tag{13.1.130}$$

and we studied characteristic properties of $f(x)$. Now, we can try to represent $f(x)$ for any value of x and $d = 3$.

Unfortunately, $P(\vec{r})$ has never been measured in three dimensions for a real polymer. There exist only results of calculations concerning rather short chains drawn on lattices ($N = 18$ for a cubic lattice). We must content ourselves with rather coarse estimations.

For $d = 3$ according to (12.3.9), we have

$$\theta = \frac{\gamma - 1}{\nu} = 0.275 \pm 0.002$$

$$\frac{1 - \gamma + \nu d - d/2}{1 - \nu} = 0.249 \pm 0.011$$

and we observe that the values of these exponents which characterize $f(x)$ are not very different (at order ε, both are equal to $\varepsilon/4$).

Moreover, as $\dfrac{1}{1-\nu} \simeq 2.427 \pm 0.006$, we can represent f(x) by a simple approximate expression of the form

$$f(x) = f_1 x^{\theta} \exp(-Dx^{1(1-\nu)}). \tag{13.1.131}$$

As the coefficients f_1 and D are not very well-known for $d = 3$, we choose to determine them by requiring that f(x) obeys the definition conditions (13.1.130) which give

$$D = \left(\frac{(4 - 5\nu + \theta - \nu\theta)!}{3(2 - 3\nu + \theta - \nu\theta)!} \right)^{1/2(1-\nu)}$$

$$f_1 = \frac{1}{4\pi(1-\nu)} \frac{[\frac{1}{3}(4 - 5\nu + \theta + \nu\theta)!]^{(3+\theta)/2}}{[(2 - \nu + \theta - \nu\theta)!]^{(5+\theta)/2}}. \tag{13.1.132}$$

Taking for γ and ν the values (12.3.9) calculated directly for $d = 3$ ($n = 0$), we obtain the final result

$$f(x) = (0.0460)\, x^{0.275} \exp(-0.318)\, x^{2.427}. \tag{13.1.133}$$

This function is represented in Fig. 13.7.

Moreover, a calculation of $P(\vec{r})$ to first order in ε was performed for any r, by Oono *et al.*[19] who used the formalism of field theory. This calculation can be made directly as we show in Appendix K. The Fourier transform $\tilde{P}(\vec{k})$ of $P(\vec{r})$ to

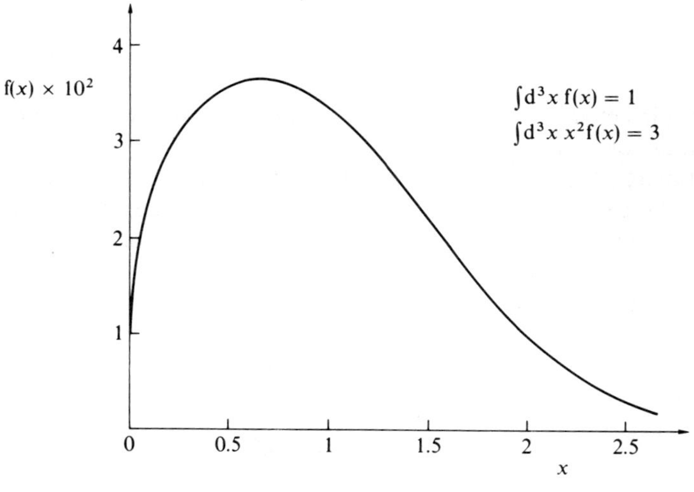

Fig. 13.7. Probability distribution $P(r)$ of the vector $\vec{r}$ joining the end points of a long polymer ($d = 3$) in a good solvent: $P(\vec{r}) = X^{-3} f(r/X)$. Representation of f(x).

order ε is given by (see eqn K.1)

$$\tilde{P}(\vec{k}) = e^{-k^2X^2/2}\left[1 - g\int_0^1 dx(1-x)x^{-2}\left(e^{xk^2X^2/2} - 1 - \frac{xk^2X^2}{2}\right)\right] \quad (13.1.134)$$

where $R^2 = dX^2$ is the mean square end-to-end distance, and g the osmotic parameter which for $z \to \infty$ has the limit $g^* = \frac{\varepsilon}{8} + \ldots$.

A Fourier transformation in d-dimensional space gives the corresponding value of $P(\vec{r})$ [see eqn (K.2)]

$$P(\vec{r})(2\pi)^{d/2}X^d = e^{-y}\{1 + g(\ln y + \mathfrak{E} - 2) - gy(\ln y + \mathfrak{E} - 2)\}$$

$$y = r^2/2X^2$$

where $\mathfrak{E}$ is the Euler constant $\mathfrak{E} = 0.5772$. Thus, exponentiating the logarithms, we obtain

$$P(\vec{r})(2\pi)^{d/2}X^d = [1 + g(\mathfrak{E} - 2)]y^g \exp\{-[1 + g(\mathfrak{E} - 2)]y^{1+g}\} \quad (13.1.135)$$

This result has the same form as (13.1.131). The calculation of $P(\vec{r})$ to order ε does not give more information than the approximate calculation which led to formula (13.1.133).

Finally, we note that the preceding theoretical results seem to be in good agreement[19] with the computer calculations of Domb *et al.*[15]

1.6 Correlations between points of a chain and probabilities of contact

Correlations between elements of an isolated polymer in a good solvent are given by scaling laws and the fundamental length which we choose to write these laws is always $X = (R^2/d)^{1/2}$ where R^2 is the mean square end-to-end distance of the polymer; on the other hand, a polymer is characterized by its Brownian area S or its number of links N.

In particular, we can define the probability distribution of the vector joining the end points of a polymer segment of area S or of N links. More precisely, three special cases denoted by the exponent ι (with $\iota = 1,2,3$) can be considered (see Fig. 13.8):

case $\iota = 0$: the segment coincide with the polymer itself; this case has been studied in Section 1.4 (but in that section, the index 0 was omitted),

case $\iota = 1$: the segment constitute one extremity of an infinite polymer,

case $\iota = 2$: the segment belongs to the central part of an infinite polymer.

The corresponding distribution probabilities $P_\iota(\vec{r})$ obey scaling laws of the form

$$P_\iota(\vec{r}) = X^{-d}f_\iota(r/X) \quad (13.1.136)$$

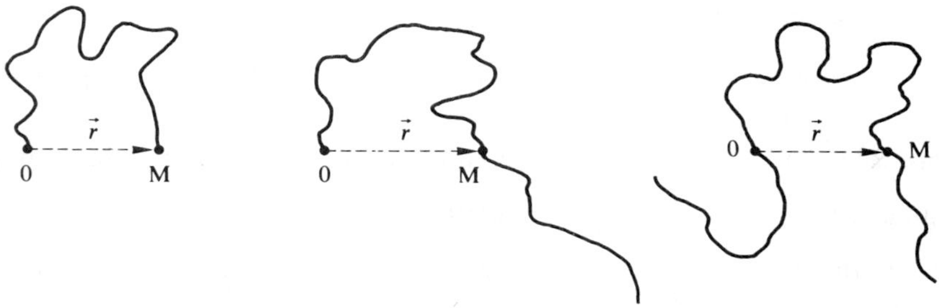

Fig. 13.8. Correlations between two points of a polymer chain, in cases $\imath = 0$, $\imath = 1$ and $\imath = 2$. In all cases, the chain segment OM has the same area S or the same number of links N.

with

$$\int d^3x\, \mathrm{f}_\imath(x) = 1.$$

Obviously, the various functions $\mathrm{f}_\imath(x)$ are universal but have different properties. In this section, we study characteristic features of these functions (critical exponents) and we shall deal later (Section 1.7) with their effective calculation.

It is to be expected that for large x, the functions $\mathrm{f}_1(x)$ and $\mathrm{f}_2(x)$ decrease like $\mathrm{f}(x)$, as follows (see Section 1.5.3)

$$\mathrm{f}_\imath(x) \sim \exp[-\mathrm{D}_\imath x^{\delta_\imath}]$$

and we feel that the exponents δ_1 and δ_2 must be larger or equal to $\delta = 1/(1-\nu)$. Indeed, for $x \gg 1$, we must have

$$\mathrm{f}_2(x) < \mathrm{f}_1(x) < \mathrm{f}(x)$$

owing to the additional repulsions produced by the semi-infinite parts of chains attached to the end points of the segment under consideration (whose end-to-end distance is xX).

For small values of x, we must have

$$\mathrm{f}_\imath(x) \propto x^{\theta_\imath}$$

and, of course, the exponents θ_1 and θ_2 differ from the exponent $\theta_0 \equiv \theta$ introduced in Section 1.4. This comes from the fact that contacts involving two strands, three strands and four strands are renormalized by introducing different renormalization factors. Thus, for the continuous model, these factors are $\mathfrak{X}_2(z) = 1$, $\mathfrak{X}_3(z)$, and $\mathfrak{X}_4(z)$ defined in Chapter 12, Section 3.3.2.3.

In fact, let us consider the contact probability $P_\imath(\vec{0})$ for large but finite values of z. In each case, this quantity is expressed as the ratio of two partition functions, and these functions can be re-expressed in terms of renormalized partition functions which (written in scaling form) have a finite limit when $z \to \infty$. The renormalization principles expounded in Chapter 12 and summarized by (12.3.63) show that we can write

$$P_\iota(\vec{0}) \propto \frac{\mathfrak{X}_{\iota+2}(z)\, X^{-d}}{[\mathfrak{X}_2(z)]^\iota\, (\mathfrak{X}_1(z))^{2-\iota}} \propto z^{(2/\varepsilon)[\sigma_{\iota+2} - \iota\sigma_2 + (\iota - 2)\sigma_1]}\, X^{-d} \tag{13.1.137}$$

in agreement with (12.3.66) and (12.3.68).

To deduce the exponents θ_ι from the exponents σ_ι, we proceed as a in Section 1.5.4, thus, in the crossover region which corresponds to small values of r and large values of z, we represent $P_\iota(\vec{r})$ in the following form which is a generalization of (13.1.118)

$$P_\iota(\vec{r}) = X^{-d-\theta_\iota}\, \psi_\iota(rb^{1/\varepsilon})\, r^{\theta_\iota}$$

where $\psi_\iota(x)$ has a finite limit when $x \to \infty$. On the contrary, when $x \to 0$, the dependence in r must disappear, and in this case

$$\psi_\iota(x) \propto x^{-\theta_\iota}.$$

This formula leads to the result

$$P_\iota(\vec{0}) \propto X^{-d-\theta_\iota}\, b^{-\theta_\iota/\varepsilon} \tag{13.1.138}$$

which is similar to (13.1.119).

Of course, this result established in the framework of the standard continuous model must remain true for any model, according to the universality principle. In particular, for a chain drawn on a lattice, the result (13.1.138) can easily be transcribed; thus, we see that in this case, the probabilitity $p_{\iota,N}$ of forming a loop of N links, either at the end of the chain ($\iota = 1$) or in the middle of the chain ($\iota = 2$), is given by

$$p_{\iota,N} \propto N^{-\nu(\theta_\iota + d)}, \tag{13.1.139}$$

a relation which is analogous to (13.1.120).

Now, in order to reduce eqn (13.1.138) to eqn (13.1.139), we must express z in terms of x and b. The definition (10.1.7) of z gives

$$z \propto b\, S^{\varepsilon/2}$$

and, moreover, we have [see (12.3.56), (12.3.57), (12.3.67), and (12.3.68)]

$$X^2 \equiv {}^{\mathrm{e}}S = S\mathfrak{X}_0(z) \propto S\, z^{2\sigma_0/\varepsilon}.$$

We thus get

$$z \propto (bX^\varepsilon)^{1/(1+\sigma_0)}.$$

Let us bring this result in (13.1.137) and let us identify the result with (13.1.138). Moreover, let us recall that $\sigma_2 = 0$, as was shown in Chapter 12, eqn (12.3.69). We obtain

$$\theta_\iota = \frac{2[(2-\iota)\sigma_1 - \sigma_{\iota+2}]}{1 + \sigma_0}. \tag{13.1.140}$$

The values of the exponents σ_l have now to be evaluated and this can be done by calculating the corresponding renormalization factors. The exponent $\theta_0 \equiv \theta = (\gamma - 1)/\nu$ is well-known (see Section 1.1.4). Moreover, expansions of θ_1 and θ_2 were calculated by des Cloizeaux[20] in the framework of field theory. The result for these three exponents is as follows:

$$\theta_0 = \frac{\varepsilon}{4} + \frac{9}{128}\varepsilon^2 + \ldots$$

$$\theta_1 = \frac{\varepsilon}{2} - \frac{3}{64}\varepsilon^2 + \ldots$$

$$\theta_2 = \varepsilon - \frac{15}{32}\varepsilon^2 + \ldots . \tag{13.1.141}$$

For $d = 3$, putting $\varepsilon = 1$ in the preceding expansions, we obtain the following estimates

$$\theta_0 = 0.32$$

$$\theta_1 = 0.45$$

$$\theta_2 = 0.53.$$

However, we know that for $d = 3$, θ_0 is close to 0.275. Thus, the preceding estimates are not really good but this should not surprise us. We know that the perturbation series which we calculate, are divergent (see Chapter 12) and there is no reason to believe that critical exponents are analytic functions of ε at $\varepsilon = 0$. Nevertheless, one may think that their Borel transforms (see Chapter 12) are analytic in the vicinity of $\varepsilon = 0$ and, for $\varepsilon > 0$, it seems reasonable to represent this transform in the form of a Padé approximant.

Consequently, it is reasonable to represent an exponent θ whose expansion reads

$$\theta = A\varepsilon - B\varepsilon^2 + \ldots$$

by the expression

$$\theta = \int_0^\infty \mathrm{d}t\, \mathrm{e}^{-t} \frac{A\varepsilon t}{1 + B\varepsilon t/2A} = \frac{2A^2}{B}\left[1 - \frac{2A}{B\varepsilon}\mathrm{e}^{2A/B\varepsilon}\, E_\mathrm{i}\left(-\frac{2A}{B\varepsilon}\right)\right]$$

$$\left(E_\mathrm{i}(x) = \int_{-\infty}^{x} \mathrm{d}t\, \frac{\mathrm{e}^t}{t} \quad \text{exponential integral}\right),$$

which leads to the somewhat uncertain values

$$d = 3 \qquad \theta_1 = 0.46 \qquad \theta_2 = 0.71$$

$$d = 2 \qquad \theta_1 = 0.85 \qquad \theta_2 = 1.14. \tag{13.1.142}$$

As was noted in Section 1.4.5, it is, in principle, possible to measure θ by making experiments with real polymers and the exponents θ_1 and θ_2 could also be measured by applying similar techniques. However, measuring these exponents would not be easy and it would be even more difficult to measure correlation functions.

Actually, it is easier to obtain 'experimental' values by studying the behaviour of self-avoiding chains on a lattice. Thus, in 1972, Trueman and Whittington[21] calculated $p_{1,N}$ for a square lattice and they found $p_{1,N} \propto p_{1,N} \propto N^{-2/3}$, i.e. $\alpha_1 = 2.15 \pm 0.25$ for the square lattice. For $d = 3$, they found $\alpha_1 = 2.15 \pm 0.15$ for the face-centred cubic lattice and $\alpha_1 = 2.10 \pm 0.15$ for the simple cubic lattice. Finally, in 1975, Whittington, Trueman, and Wilker[22] resumed such calculations and, using a Monte Carlo method, they found $\alpha_1 = 2.18$ for the face-centred cubic lattice. All these results are thus compatible with one another and with the universality principle. Moreover, using an exact counting method, Redner,[23] in 1980, evaluated the distributions $P_0(\vec{r})$ and $P_1(\vec{r})$; from them, be deduced values of the exponents θ_ι and δ_ι, and presented a synthesis of the results.

More recently, as we saw in Chapter 12, Section 4, exact values of the exponents σ_M have been found for $d = 2$

$$\sigma_0 = 2\nu - 1 = 1/2$$

$$\sigma_M = \frac{(2 - M)(9M + 2)}{64}$$

Thus, using (13.1.140), we obtain for $d = 2$ the exact values of the θ_ι

$$\theta_\iota = \frac{1}{48}(9\iota^2 + 9\iota + 22) \tag{13.1.143}$$

and, in particular, this formula gives $\theta \equiv \theta_0 = \dfrac{11}{24}\left(= \dfrac{\gamma - 1}{\nu}\right)$.

A list of results is given in the following tables.

***Exponents* δ_ι**

For $d = 3$

$\delta_1 = 2.6 \pm 0.06$ (Redner 1980)[23]

to be compared with the theoretical value

$$\delta_0 \equiv \delta = \frac{1}{1 - \nu} = 2.427 \pm 0.006$$

[see (12.3.9)].

For $d = 2$

$\delta_1 = 4.5 \pm 0.4$ (Redner 1980)[23]

$\delta_2 = 4.6 \pm 0.6$

to be compared with the 'exact' value

$$\delta_0 \equiv \delta = \frac{1}{1-\nu} = 4$$

[see (12.2.9) and Section 4 of Chapter 12].

Exponents θ_ι

For $d = 3$

$$\theta_1 = \begin{cases} 0.67 \pm 0.17 & \text{(Guttmann and Sykes 1973)}^{24} \\ 0.70 \pm 0.12 & \text{(Whittington, Trueman, and Wilker 1973)}^{22} \end{cases}$$

to be compared with the rather approximate theoretical result

$\theta_1 = 0.46$ [see (13.1.143)]

$\theta_2 = 0.67 \pm 0.34$ (Redner 1980)[23]

to be compared with the rather approximate theoretical result

$\theta_2 = 0.71$ [Borel–Padé approximant to order ε^2; see (13.1.143)].

For $d = 2$

$$\theta_1 = \begin{cases} 0.84 \pm 0.10 & \text{(Trueman and Whittington 1972)}^{21} \\ 0.84 \pm 0.13 & \text{(Guttmann and Sykes 1973)}^{24} \end{cases}$$

to be compared with the 'exact' result

$\theta_1 = 5/6 = 0.833\,333$ [see (13.1.143)]

$\theta_2 = 1.93 \pm 0.27$ (Redner 1980)[23]

to be compared with the exact result

$\theta_2 = 19/12 = 1.583\,333$ [(see 13.1.143)].

Until now, the exponents θ_ι have been calculated (in powers of ε) only from field theory; however, it is also possible to use the direct renormalization method. Thus, as an example, we shall use this method to explicitly calculate the exponents θ_ι to first-order in ε. For this purpose we shall use formula (13.1.140) which connects the θ_ι to the σ_ι. Thus, we have only to calculate $\mathfrak{X}_\iota(z)$ to first-order in z.

Now, let $\mathscr{Z}_\iota(S)$ be the partition function for a star made of ι polymers of area S, these polymers constituting the branches of the star. We can renormalize by setting

$$\mathscr{Z}_{\iota,\mathbf{R}}(^{\mathrm{e}}S) = 1,$$

and this condition defines the renormalization factors $\mathfrak{X}_\iota(z)$. Indeed, eqn (12.3.64) gives

$$\mathscr{Z}_\iota(S) = \mathfrak{X}_\iota(z)\,[\mathfrak{X}_1(z)]^\iota,$$

and from it we deduce

$$\mathfrak{X}_\iota(z) = \mathscr{Z}_\iota(S)\,[\mathscr{Z}(S)]^{-\iota/2}. \tag{13.1.144}$$

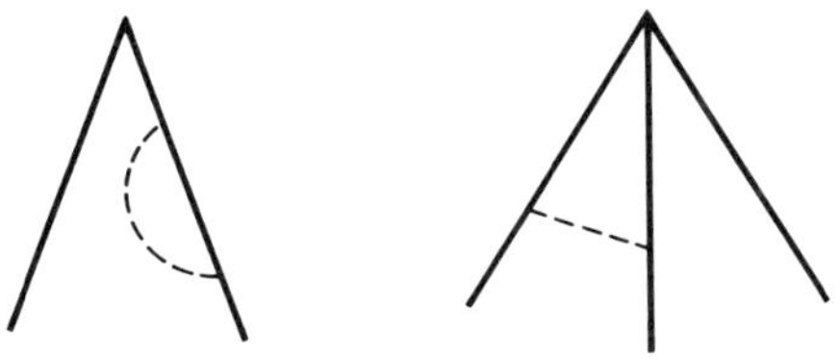

(a) (b)

Fig. 13.9. Calculation of $\mathfrak{X}_3(z)$: first-order diagrams.

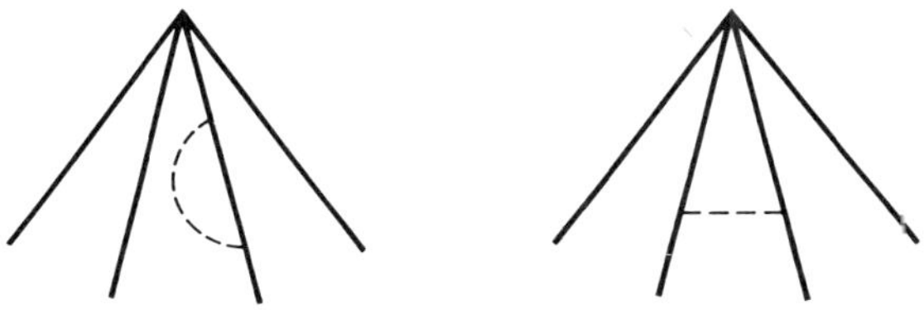

(a) (b)

Fig. 13.10. Calculation of $\mathfrak{X}_4(z)$: first-order diagrams.

The diagrams concerning $\mathscr{L}_3(S)$ and $\mathscr{L}_4(S)$ and contributing to first-order in z are represented in Figs 13.9 and 13.10. We see immediately that, at this order, we have

$$\mathscr{L}_\iota(S) = 1 + {}^1\mathscr{L}_\iota(S) + \dots$$

$${}^1\mathscr{L}_\iota(S) = \iota a(z) + \frac{\iota(\iota - 1)}{2} b(z). \qquad (13.1.145)$$

Moreover we also see that

$$\mathscr{L}_1(S) = \mathscr{L}(S) = [\mathfrak{X}_1(z)]^2$$

$$\mathscr{L}_2(S) = \mathscr{L}(2S)$$

To first-order in ε, and for small ε, we have according to (12.3.97),

$$\mathscr{L}_1(S) = 1 + \frac{2}{\varepsilon} z + \dots$$

$$\mathscr{L}_2(S) = 1 + \frac{2}{\varepsilon} z + \dots$$

which can be compared with (13.1.145), and gives

$$a(z) = -\, b(z) \simeq \frac{2z}{\varepsilon}.$$

Consequently, in the general case, we have

$$\mathscr{X}_\iota(S) = 1 + \iota\,(3 - \iota)\frac{z}{\varepsilon}.$$

Let us bring this result in (13.1.144); we obtain

$$\mathfrak{X}_\iota(z) = 1 + \iota\,(2 - \iota)\frac{z}{\varepsilon}.$$

The effective exponent $\sigma_\iota(z)$ can be deduced from this equality by applying the definition (12.3.66)

$$\sigma_\iota[g] = \iota\,(2 - \iota)\,z/2.$$

This exponent can be re-expressed in terms of the osmotic parameter g [see (12.3.100)]. We find

$$\sigma_\iota[g] = \iota\,(2 - \iota)g/2.$$

Finally, in the limit $z \to \infty$, where

$$g \to g^* \simeq \frac{\varepsilon}{8} \cdots$$

according to (12.3.102), we obtain

$$\sigma_\iota = \sigma_\iota[g^*] = \frac{\iota(2 - \iota)}{16}\varepsilon \qquad (\iota > 0). \tag{13.1.146}$$

Incidentally, we note that $\mathfrak{X}_\iota(z)$ is the characteristic renormalization factor of star polymers with ι branches where as σ_ι is the corresponding exponent.

Now let us come back to the exponents θ_ι which are given by eqn (13.1.140). Combining (13.1.146) with (13.1.140) and observing that σ_0 is of order ε, we find to first-order in ε

$$\theta_\iota = \frac{\varepsilon}{8}\,(\iota^2 + \iota + 2) \qquad \iota = 0, 1, 2,$$

i.e.

$$\theta_0 \simeq \varepsilon/4 \qquad \theta_1 \simeq \varepsilon/2 \qquad \theta_2 = \varepsilon,$$

in agreement with (13.1.141).

1.7 Form function of an isolated chain

1.7.1 General properties

The form function $H(\vec{q})$ of a chain made of N links is defined by the equality

$$H(\vec{q}) = \frac{1}{(N + 1)^2}\left\langle\!\!\left\langle \sum_{j=0}^{N}\sum_{l=0}^{N} \exp[i\vec{q}\cdot(\vec{r}_j - \vec{r}_l)] \right\rangle\!\!\right\rangle \tag{13.1.147}$$

where $\vec{r}_j$ is the position vector of the point M_j of the chain. In the case where a continuous model is used, the definition of the form function becomes

$$H(\vec{q}) = \frac{1}{S^2}\left\langle\!\!\left\langle \int_0^S \mathrm{d}s' \int_0^S \mathrm{d}s'' \exp\{i\vec{q}\cdot[\vec{r}(s') - \vec{r}(s'')]\} \right\rangle\!\!\right\rangle. \qquad (13.1.148)$$

This function is especially important because it is the quantity that is measured when a scattering experiment is performed.

For small values of $q(q \ll S^{-1/2})$, we have

$$H(\vec{q}) = 1 - \frac{q^2}{d} R_G^2 + \ldots \qquad (13.1.149)$$

where R_G^2 is the mean square radius of gyration

$$R_G^2 = \frac{1}{2(N+1)^2}\left\langle\!\!\left\langle \sum_{j,l} (\vec{r}_j - \vec{r}_l)^2 \right\rangle\!\!\right\rangle.$$

When the chain is Brownian, the form function is easily calculated; actually, in this case,

$$\left\langle\!\!\left\langle e^{i\vec{q}\cdot[\vec{r}(s') - \vec{r}(s'')]} \right\rangle\!\!\right\rangle = \exp\left[-\frac{q^2}{2d}\langle\!\langle[\vec{r}(s') - \vec{r}(s'')]^2\rangle\!\rangle\right]$$
$$= \exp[-q^2|s' - s''|/2].$$

Consequently, by bringing this expression in (13.1.148), we find

$$^0H(\vec{q}) = \frac{2}{S^2}\int_0^S \mathrm{d}s(S - s)e^{-q^2 s/2}, \qquad (13.1.150)$$

i.e.

$$^0H(\vec{q}) = {}^0\mathrm{h}\left(\frac{q^2 S}{2}\right)$$

where

$$^0\mathrm{h}(x) = \frac{2}{x^2}[e^{-x} - 1 + x]. \qquad (13.1.151)$$

This is Debye's formula[25] (see Fig. 13.11).

When interactions are present, the chain is swollen and $H(\vec{q})$ can be written in the form

$$H(\vec{q}) = h\left(\frac{q^2 X^2}{2}, z\right) \qquad (13.1.152)$$

where $X^2 = R^2/d = {}^{e}S$ (see Chapter 12) and z the dimensionless parameter characterizing the interaction. We can also use the osmotic parameter g, related to the second virial coefficient, and write

$$H(\vec{q}) = h\left[\frac{q^2 X^2}{2}, g\right]. \qquad (13.1.153)$$

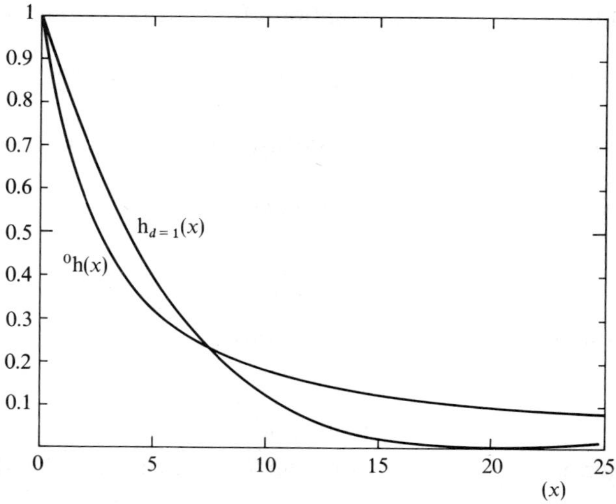

Fig. 13.11. Form functions. Functions ${}^0\mathrm{h}(x) \equiv \mathrm{h}_{d=4}(x)$ (Brownian case) and $\mathrm{h}_{d=1}(x)$ (Kuhnian chain for $d = 1$, i.e. rigid rod). The function $\mathrm{h}_{d=3}(x)$ corresponding to the Kuhnian case for $d = 3$ would be represented by an intermediate curve.

Let us now consider the long-chain limit. We set

$$h(x, \infty) = h[x, g^*] \equiv \mathrm{h}(x).$$

In this limit, we obtain the scaling law

$$H(\vec{q}) = \mathrm{h}(q^2X^2/2) \qquad (X^2 = R^2/d)$$

where $\mathrm{h}(x)$ is a universal function characterizing Kuhnian chains in d dimensions. In particular, for $d = 1$, an exact expression of $\mathrm{h}(x)$ can be calculated. In this case, we have

$$H_{d=1}(\vec{q}) = 2\int_0^1 \mathrm{d}\alpha\,(1-\alpha)\cos(\alpha qX) = 2\frac{1-\cos qX}{q^2X^2}$$

$$\mathrm{h}_{d=1}(x) = \frac{1}{x}[1-\cos(2x)^{1/2}]. \qquad (13.1.154)$$

The function $\mathrm{h}_{d=1}(x)$ is represented in Fig. 13.11.

Let us now come back to the general problem and particularly to the case $d = 3$ which corresponds to polymers in solution. The limit where $q \gg X^{-1}$ is especially interesting. For a Brownian chain, (13.1.151) gives

$${}^0H(\vec{q}) = 4/q^2S.$$

More generally, in this infinite chain limit, (13.1.148) leads to the expression

$$H(\vec{q}) \simeq \frac{1}{S}\left\langle\!\!\left\langle \int_{-\infty}^{+\infty} \mathrm{d}s \exp\{i\vec{q}\cdot[\vec{r}(s) - \vec{r}(0)]\} \right\rangle\!\!\right\rangle. \qquad (13.1.155)$$

Thus, when $S \to \infty$ at constant $q\,(q \neq 0)$, $S \times H(\vec{q})$ has a finite limit, and in a similar way, for a discrete model, $N \times H(\vec{q})$ remains finite when $N \to \infty$. Let us now examine what happens in the limit $z \to \infty$. Then, $H(\vec{q})$ is given by the scaling low $H(\vec{q}) = \mathrm{h}(q^2x^2/2)$ and, in the limit $x \gg q^{-1}$, for a fixed interaction (b = constant), $H(\vec{q})$ must be proportional to S^{-1} and therefore to $X^{-1/\nu}$. Consequently, in this case we have

$$H(\vec{q}) \simeq \frac{\mathrm{h}_\infty}{(q^2X^2/2)^{1/2\nu}} \qquad \mathrm{h}(x) \simeq \mathrm{h}_\infty\, x^{-1/2\nu}, \qquad (13.1.156)$$

a result which is compatible with the asymptotic behaviour of $\mathrm{h}(x)$ for $d = 1$ ($\nu = 1$) and also the asymptotic behaviour of ${}^0\mathrm{h}(x)$ ($\nu = 1/2$). Consequently, a measurement of the q dependence of the form function, for large values of q, enables us to obtain an experimental value of the exponent ν.

We also note that $\mathrm{h}_\infty = 2 + O(\varepsilon)$ (since $\mathrm{h}_\infty = 2$ for a Brownian chain).

1.7.2 Form function of an isolated chain: exact results

The form function $H(\vec{q})$ associated with the continuous model is a function $h(x, z)$ [see (13.1.152)] and this function can be expanded in powers of z

$$h(x, z) = {}^0\mathrm{h}(x) + z\,{}^1\mathrm{h}(x) + \ldots$$

where ${}^0\mathrm{h}(x)$ is given by (13.1.151).

In principle, ${}^1\mathrm{h}(x)$ can be calculated directly by starting from (13.1.148) and (13.1.153). It is also possible to get the result through field theory.

Using renormalization theory, Witten and Schäfer[26] calculated the expansion of the form factor, in the limit $z \to \infty$, to first order in $\varepsilon = 4 - d$ and to second-order in x. Using the notation introduced above, we can express their result as follows

$$\mathrm{h}(x) = 1 - \frac{x}{3}\left(1 - \frac{\varepsilon}{96}\right) + \frac{x^2}{12}\left(1 - \frac{19\varepsilon}{240}\right). \qquad (13.1.157)$$

On the other hand, it is not difficult to calculate h_∞ to first-order in ε, by starting from eqn (13.1.153). In this way, one obtains (des Cloizeaux and Duplantier)[27] for $x \gg 1$

$$\mathrm{h}(x) = \frac{2}{x}\left[1 + \frac{\varepsilon}{8}(\mathfrak{E} - 5 + \ln x) + \ldots\right] \qquad (13.1.158)$$

where $\mathfrak{E}$ is Euler's constant ($\mathfrak{E} = 0.577$). This result can be rewritten in the form

$$\mathrm{h}(x) = \frac{2}{x^{1-\varepsilon/8}}\left[1 - \frac{\varepsilon}{8}(5 - \mathfrak{E})\right] \qquad (13.1.159)$$

It is compatible with the formula

$$\nu = \frac{1}{2}\left(1 + \frac{\varepsilon}{8} + \ldots\right),$$

and from it, we deduce

$$\mathrm{h}_\infty = 2\left[1 - \frac{\varepsilon}{8}(5 - \varepsilon) + \ldots\right] \simeq 2(1 - 0.5528\,\varepsilon + \ldots), \qquad (13.1.160)$$

i.e. for $d = 3$, $\mathrm{h}_\infty = 0.89$ a value which is very different from the Brownian value. The preceding result had been obtained earlier by Witten but in a slightly different form[28] and through field theory.

Incidentally, we would like to know the value of h_∞ for $d = 1$. Now, according to (13.1.154) we have

$$\mathrm{h}_{d=1}(x) = \frac{1}{x}[1 - \cos(2x)^{1/2}],$$

and according to (13.1.156), for $x \gg 1$, we must have

$$\mathrm{h}(x) = \mathrm{h}_\infty\, x^{-1/2}.$$

Thus, we find the result

$$\mathrm{h}_\infty = 0 \qquad \text{for } d = 1.$$

This observation enables us to obtain a better result than (13.1.160). We can thus represent h_∞ by a polynomial in ε, requiring this polynomial to be compatible with (13.1.142) and with the fact that h_∞ vanishes for $d = 1$. In this way, we obtain

$$\mathrm{h}_\infty = 2(1 - 0.5528\varepsilon + 0.0731\varepsilon^2) \qquad (13.1.161)$$

a formula which, for $d = 3$ ($\varepsilon = 1$), gives

$$\mathrm{h}_\infty = 1.04. \qquad (13.1.162)$$

Moreover, the calculation of $\mathrm{h}(x)$ was first performed for any value of x by Ohta, Oono, and Freed (1981)[29] and also more recently by Duplantier.[30] In the latter calculation, the parameters g and d appear explicitly; therefore in spite of the fact, that, strictly speaking, the formula is only valid to first-order in ε, we may hope that it constitutes a reasonable approximation even in three dimensions. This result is

$$\mathrm{h}(x) = {}^0\mathrm{h}(x) + 2gx^{-(4-d/2)}[I_1(x) + 2I_2(x) + I_3(x)]$$

$${}^0\mathrm{h}(x) = \frac{2}{x^2}(x - 1 + \mathrm{e}^{-x})$$

$$I_1(x) = \int_0^x \mathrm{d}t\; t^{2-d/2}(x - t)\,\mathrm{e}^{-t} \int_0^1 \mathrm{d}u\; u^{-d/2}(1 - u)(1 - \mathrm{e}^{tu})$$

$$I_2(x) = \int_0^x \mathrm{d}t\; t^{1-d/2}(1 + t - x - \mathrm{e}^{t-x}) \int_0^1 \mathrm{d}u\; e^{-ut}(\mathrm{e}^{tu^2} - 1)$$

$$I_3(x) = \int_0^x \mathrm{d}t\; t^{2-d/2}(t - x) \int_0^I \mathrm{d}u\;(1 - u)\,\mathrm{e}^{-ut}\,(\mathrm{e}^{tu^2} - 1). \qquad (13.1.163)$$

1.7.3 Form function of an isolated chain: semi-phenomenological approaches, 'thermic sequences'

An approximate expression of $\mathrm{h}(x)$ for a chain with a critical exponent ν can be obtained in a very simple way as was done by Peterlin (1955)[31], and also Ptitsyn (1957)[32] and Benoit (1957).[33] The basic assumptions are that

$$\langle\!\langle [\vec{r}(s') - \vec{r}(s'')]^2 \rangle\!\rangle = (|s' - s''|/S)^{2\nu} R^2, \qquad (13.1.164)$$

and that the following relation

$$\langle\!\langle [\exp\{i\vec{q}[\vec{r}(s') - \vec{r}(s'')]\} \rangle\!\rangle = \exp\left\{ -\frac{q^2}{2d} \langle\!\langle [\vec{r}(s') - \vec{r}(s'')]^2 \rangle\!\rangle \right\} \qquad (13.1.165)$$

which is valid for Gaussian chains remains nearly true for chains with excluded volume. These assumptions give

$$H(\vec{q}) = \frac{2}{S^2}\int_0^S ds\,(S - s)\,\exp\left[-\frac{q^2X^2}{2}\,(s/S)^{2\nu}\right] \qquad (13.1.166)$$

and taking (13.1.152) into account, we can write

$$\mathrm{h}(x) = 2\int_0^1 d\alpha\,(1 - \alpha)\,\exp[-x\alpha^{2\nu}]. \qquad (13.1.167)$$

In particular, for $x \gg 1$, we find

$$\mathrm{h}(x) \simeq 2\int_0^\infty d\alpha\,\exp[-x\alpha^{2\nu}] = \frac{2}{x^{1/2\nu}}\,(1/2\nu)!$$

and

$$\mathrm{h}_\infty = 2\,(1/2\nu)! \qquad (13.1.168)$$

Unfortunately, the result (13.1.167) is not as good as could be expected, for reasons already explained in connection with the calculation of the radius of gyration (Chapter 4, Section 2.1.2). In fact, eqns (13.1.164) and (13.1.165) represent reality only very crudely (see Section 1.5.6 and Chapter 10, Section 7.2). Actually, the equality (13.1.166) can be tested by comparison with exact results.

Thus, the value of h_∞, which can be deduced from it, is given by (13.1.167) and for the exact value $\nu = \frac{1}{2}\left(\frac{1}{2} + \frac{\varepsilon}{8} + \ldots\right)$, this equality leads to the result

$$\mathrm{h}_\infty = 2[1 - \varepsilon(1 - \mathfrak{C}) + \ldots],$$

which is very different from the exact result (13.1.160).

Moreover, starting from (13.1.167) and expanding in powers of x, we find

$$\mathrm{h}(x) = 1 - \frac{x}{(2\nu + 1)(\nu + 1)} + \frac{x^2}{(4\nu + 1)(4\nu + 2)} + \dots .$$

Now, writing $\nu = \frac{1}{2}\left(1 + \frac{\varepsilon}{8}\right)$, we can expand $\mathrm{h}(x)$ to first-order in ε and we obtain

$$\mathrm{h}(x) = 1 - \frac{x}{3}\left(1 - \frac{5}{48}\varepsilon\right) + \frac{x^2}{12}\left(1 - \frac{7}{48}\varepsilon\right) + \dots .$$

Then, we note that this formula is very different from the exact result (13.1.157). Thus, equality (13.1.166) cannot be considered as a good approximation.

Anyway, it is difficult to measure $\mathrm{h}(x)$. Such a measurement would require very long chains. In fact, as was previously noted (in particular in Chapter 10), a portion of chain swells less than the whole chain, and the smaller the portion the lesser the swelling. Thus, it is difficult to reach the asymptotic domain for $H(\vec{q})$. In general, it is necessary to take into account the existence of a crossover domain between a Brownian behaviour and a Kuhnian behaviour. Such is the aim of the method of sequences* developed in France between 1975 and 1980 especially by Daoûd, de Gennes, and Jannink. Concerning the form function, the basic postulates are the following. There exists an area S_t defined by the conditions

$$\text{for } S \leqslant S_t \quad \langle\!\langle [\vec{r}(s') - \vec{r}(s' + s)]^2 \rangle\!\rangle = sd$$

$$\text{for } S \geqslant S_t \quad \langle\!\langle [\vec{r}(s') - r(s' + s)]^2 \rangle\!\rangle = S_t\,(s/S_t)^{2\nu} d$$

(index t for 'thermic'). (13.1.169)

The area S_t is a function of the interaction but does not depend on S. Thus, in agreement with (10.1.77), we may write

$$S_t = ((2\pi)^{d/2}\, b^{-1} z_t)^{2/\varepsilon} = A_t\, b^{-2/\varepsilon} \tag{13.1.170}$$

where z_t and A_t are constants which can be estimated. In fact, from (13.1.147) we deduce

$$X^2 = R^2/d = S\,(S/S_t)^{2\nu - 1} = S\,(z/z_t)^{2(2\nu - 1)/\varepsilon} \tag{13.1.171}$$

and for $d = 3$ with $\nu = 0.588$, we obtain

$$R^2 = 3S(z/z_t)^{0.35}.$$

This formula can be compared with (13.1.41), (13.1.42) which reads

$$\mathfrak{X}_0(z) = R^2/3S = 1.27\, z^{0.35}$$

* A sequence is also commonly called a blob, but we believe that the word blob does not give a very good picture of reality.

and consequently for $d = 3$, we obtain

$$z_t = 0.51 \qquad A_t = 4.1. \tag{13.1.172}$$

Thus, for a polymer and a given solvent, S_t is a function of temperature, and the area S_t defines the curvilinear dimension (number of links) of a *thermic sequence.* Moreover, the size of this sequence is the *thermic length*

$$\xi_t = (S_t)^{1/2} \simeq 2.0 \ b^{-1/\varepsilon}. \tag{13.1.173}$$

To calculate $H(\vec{q})$, Farnoux, Boué, Cotton, Daoud, Jannink, Nierlich, and de Gennes[34] like Peterlin[31] use the approximation (13.1.165). Consequently, $H(\vec{q})$ is represented by the expression

$$H(\vec{q}) = \frac{2}{S^2}\int_0^{S_t} ds\,(S-s)\,e^{-sq^2/2} + \frac{2}{S^2}\int_{S_t}^{S} ds\,(S-s)\exp[-\tfrac{1}{2}S_t q^2 (s/S_t)^{2\nu}].$$

Setting $\tau = S_t/S$ and using the equality (13.1.171) which can be written in the form $S = X^2\tau^{2\nu-1}$, we find

$$H(\vec{q}) = H\left(\frac{q^2X^2}{2}, \tau\right)$$

where

$$H(x, \tau) = 2\int_0^{\tau} d\alpha(1-\alpha)\exp(-x\alpha\tau^{2\nu-1}) + 2\int_{\tau}^{1} d\alpha(1-\alpha)\exp(-x\alpha^{2\nu})$$

Consequently, $H(x, \tau)$ can be expressed with the help of the incomplete gamma function

$$\Gamma(x, y) = \int_0^{y} dt\, t^{x-1}\, e^{-t}.$$

Then, a simple calculation gives

$$\begin{aligned} H(x, \tau) = {} & \frac{2\tau}{(x\tau^{2\nu})^2}\{x\tau^{2\nu} - \tau + [\tau - (1-\tau)x\tau^{2\nu}]\exp(-x\tau^{2\nu}) \\ & + \frac{1}{\nu}x^{-1/2\nu}\left[\Gamma\left(\frac{1}{2\nu}, x\right) - \Gamma\left(\frac{1}{2\nu}, x\tau^{2\nu}\right)\right] \\ & - \frac{1}{\nu}x^{-1/\nu}\left[\Gamma\left(\frac{1}{\nu}, x\right) - \Gamma\left(\frac{1}{2\nu}, x\tau^{2\nu}\right)\right] \end{aligned} \tag{13.1.174}$$

and this is the result of Farnoux *et al.*[34] expressed with different notation. For $\tau = 1$, we recover the Brownian case; for $\tau = 0$, we get the Peterlin approximation which has been discussed at the beginning of this section. Therefore, the result obtained by Farnoux *et al.*[34] is expected to be far from good for large swelling. By employing cleverly all the information contained in the present book it would, of course, be possible to find much better approximations. However, we shall avoid here complicated refinements.

1.8 Limit of the probability distribution of vectors joining internal points of a chain; moments of the distribution

The probability distribution $P_{12}(\vec{r})$ of the vector joining two points of the chain is defined by

$$P_{12}(\vec{r}) = \langle\langle\, \delta[\vec{r} - \vec{r}(s_2) + \vec{r}(s_1)] \,\rangle\rangle$$

and we may assume that $0 < s_1 < s_2 < S$. In order to define the position of points on the chain, we find it convenient to use dimensionless parameters

$$\lambda_1 = s_1/S \qquad \lambda_2 = (S - s_2)/S.$$

The distribution $P_{12}(\vec{r})$ was calculated in the asymptotic limit $z \gg 1$ to order ε by Oono and Ohta (1981)[35]. Their result can be expressed in terms of the reduced variable

$$\bar{x} = 2(r^2/R^2)\left(1 - \frac{3\varepsilon}{8}\right)$$

in the form

$$P_{12}(\vec{r}) \propto \bar{x}^{\varepsilon/8} \exp\left\{ - (\bar{x})^{1+\varepsilon/8} - \frac{\varepsilon}{8}(\mathfrak{E} - 1)\bar{x} - \frac{\varepsilon}{8}[f_1(\bar{x}, \lambda_1) + f_1(\bar{x}, \lambda_2) + f_2(\bar{x}, \lambda_1, \lambda_2)] \right\} \tag{13.1.175}$$

where $\mathfrak{E}$ is Euler's constant, $\mathfrak{E} = 0.577$, and where

$$f_1(x, \lambda) = 1 - \mathfrak{E} - \ln x - \ln\lambda + x^{-1}\int_0^1 \frac{dt}{t^2}\left[\exp\left(-\frac{xt^2}{\lambda + t(1-t)}\right) - 1\right]$$

$$f_2(x, \lambda_1, \lambda_2) = \varphi\left(\frac{x}{\lambda_1 + \lambda_2}\right) - \varphi\left(\frac{x}{\lambda_1}\right) - \varphi\left(\frac{x}{\lambda_2}\right)$$

$$\varphi(y) = \int_y^\infty dt\, \frac{e^{-t}}{t^2}. \tag{13.1.176}$$

This result is in good agreement with the predictions made in Section 1.6 concerning the exponents θ and δ.

The calculation of the moments of the distribution was performed by Duplantier (1965).[36] For the mean square distance, his result is expressed as a function of the parameters

$$t = \frac{s_2 - s_1}{S} = 1 - \lambda_1 - \lambda_2$$

$$y = \frac{s_1}{S - s_2 + s_1} = \frac{\lambda_1}{\lambda_1 + \lambda_2} \tag{13.1.177}$$

and reads

$$\langle\langle [[\vec{r}(s_2) - \vec{r}(s_1)]^2 \rangle\rangle = d\, X^2 t^{1+\varepsilon/8}[1 + g^* A(y, t)] \tag{13.1.178}$$

where g is the osmotic parameter studied in Section 1.1 and $A(y, t)$ an additional swelling coefficient

$$A(y, t) = \left(\frac{1-t}{t}\right)\left\{y \ln\left(1 + \frac{t}{y(1-t)}\right) + (1-y)\ln\left(1 + \frac{t}{(1-y)(1-t)}\right)\right\} - \left(\frac{1-t}{2}\right). \quad (13.1.179)$$

The function $A(y, t)$ is represented in Fig. 13.12 for various values of y and t.

Then, we observe the following facts. The swelling of a very small sequence is uniform along the chain (the end parts being excluded)

$$A(y, 0) = 3/2 \qquad y \neq 0 \qquad y \neq 1$$

and it is larger than the swelling at the end of the chain

$$A(0, t \to 0) = 1/2.$$

Moreover the swelling of such small sequences is relatively larger than the swelling of the entire chain since we have

$$A(1/2, 1) = 0.$$

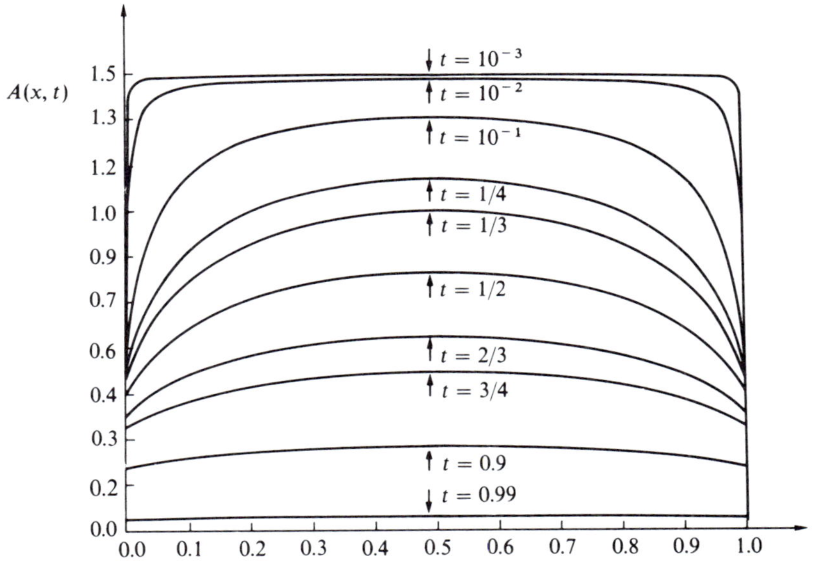

Fig. 13.12. A representation of the additional swelling $A(y, t)$ in terms of y for various values of t. Here y defines the position of the sequence on the chain and t its relative length.

However, we observe that; if the sequence constitutes a finite fraction of the chain, its swelling depends on its position on the chain: it is larger at the centre of the chain than on the sides. This result confirms the remarks made in Chapter 10, Section 7.2.

Moreover, Duplantier calculated the mean values of other powers of the distance between points of a chain. Three cases characterized by the index ι have been considered.

$\iota = 0$ the sequence coincides with the chain,

$\iota = 1$ the sequence is located at one extremity of a semi-infinite chain,

$\iota = 2$ the sequence is located in the middle of an infinite chain.

The result at order ε reads

$$\langle\langle [\vec{r}(s_2) - \vec{r}(s_1)]^{2p} \rangle\rangle = X^{2p}(t^{1+\varepsilon/8})^p \, 2^p \frac{\Gamma(p + d/2)}{\Gamma(d/2)} (1 + g^* A_\iota(p)) \qquad (13.1.180)$$

with

$$A_\iota(p) = 2p - (p+1)[\psi(p) + \mathcal{E}] + \frac{\iota(\iota+1)}{2}\left[\psi(p) + \mathcal{E} + \frac{1}{p+1} - 1\right] \qquad (13.1.181)$$

where

$$\psi(x) = \frac{\partial}{\partial x}\ln(x!) = -\mathcal{E} + \sum_{m=1}^{\infty}\left(\frac{1}{m} - \frac{1}{m+x}\right).$$

Thus, $A_\iota(p)$ can be considered as an analytic function of p with poles for $p = -2, -3$ and so on. Of course, for $p = 0$ and $p = 1$, we recover the results

$$A_\iota(0) = 0$$

$$A_1(1) = A(0, t \to 0) = 1/2$$

$$A_2(1) = A(y, 0) = 3/2.$$

2. PROPERTIES OF 'GOOD POLYMER SOLUTIONS'

2.1 Dilute and semi-dilute solutions. A comprehensive description of concentration effects: swollen sequences

When physicists tried to apply the scaling law concept and the principles of renormalization theory to the study of polymer solutions, they encountered serious technical problems. However, previous investigations of Flory and especially of Edwards,[37] had already shown that the overlap of polymers in a solution produces a new phenomenon which, with Edwards, we may call the screening effect.

This screening can roughly be described as follows. In a solution, in which polymers strongly overlap, the repulsions acting between elements of the same chain try to induce a swelling of the chain, but they are counter-balanced by repulsions produced by other chains. In other words, if the concentration is large enough, the chains screen one another. It was necessary to examine the implications of this fact in the framework of the new theories. However, a simultaneous calculation of swelling and concentration effects is not easy to make, and originally one tried to explain these effects by taking into account the screening phenomenon in an intuitive and qualitative manner. We shall first describe this approach developed mainly by Daoud[38,39] and de Gennes,[39,40,41] and this will give us the opportunity to introduce the most basic concepts. Then, later, we shall discuss difficulties that are encountered when one deals with real polymers.

In the dilute domain, the polymers remain far apart and they repel one another weakly; their size is then about the same as the size of an isolated polymer. The concentration effects really appear when the concentration C of the polymers becomes large enough so that they begin to overlap. By letting the size of the chain grow, we finally obtain the semi-dilute regime in which the chains strongly overlap. The transition occurs in the vicinity of a concentration C^*. Various rather equivalent definitions of C^* are possible. The theoretician likes to refer to the mean square end-to-end distance of the isolated polymer. This leads to the definition

$$C^* = X^{-d} = (R^2/d)^{-d/2}, \tag{13.2.1}$$

i.e. for $d = 3$

$$C^* = 3^{3/2}\, R^{-3} \simeq 5.196\, R^{-3}.$$

On the contrary, the experimentalist* likes to refer to the radius of gyration R_G. Thus, we also define

$$C_G^* = (6R_G^2/d)^{-d/2},$$

i.e, for $d = 3$,

$$C_G^* = 2^{-3/2}\, R_G^{-3} \simeq 0.353\, R_G^{-3}.$$

Thus the values of C_G^* and C^* are very close to each other

$$C_G^* = \aleph^{-d/2}\, C^* \tag{13.2.2}$$

where $\aleph^{-d/2}$ varies between 1($d = 4$ and Brownian chain) and 1.4($d = 1$ and rigid rod).

* Certain experimentalists prefer the definition

$$C^* = 1/(4\pi\, R_G^3/3)$$

but this definition is less natural than it looks and, in practice, the value thus defined seems too small; a Brownian chain can hardly be compared to a sphere!

The polymer concentration C^* is related to the number N of monomers per chain. We can also tackle the problem by introducing the monomer concentration $\mathbf{C} = \mathbf{C}N$, and in order to define the transition between the dilute and the semi-dilute domain at N constant, we can introduce C^* ($C^* = \mathbf{C}^*N$).

However, this definition cannot be applied to the continuous model and it must be understood that a 'reasonable' definition of the monomer concentration depends essentially on the model under consideration.

For instance, let us consider an ensemble of Brownian chains. In this case the *monomer concentration* is actually the area concentration

$$\mathscr{C} = \mathbf{C}S \tag{13.2.3}$$

(since S is proportional to N) and the transition between the dilute, and the semi-dilute domain occurs for a concentration

$$\mathscr{C}^* \propto S^{1-d/2}.$$

We see that the dimensional formula of $\mathscr{C}$ is $\mathscr{C} \sim L^{2-d}$ and this remark induces us to define the *Brownian overlap length.* ξ_0 which for a semi-dilute solution is something like the distance between Brownian chains. Thus, we set

$$\xi_0 = \mathscr{C}^{-1/(d-2)}. \tag{13.2.4}$$

Incidentally, we note that in two dimensions $\mathscr{C}$ is a pure number; in this case, there is no characteristic length for the semi-dilute solution.

Let us now consider the ideal case where the isolated chain is a Kuhnian chain (i.e. a chain which is uniformly swollen). We can study the properties of a monodisperse ensemble of Kuhnian chains. The only available parameters are the chain concentration and the length X characterizing the size of an isolated chain ($R^2 = X^2 d$). In this case, *the monomer concentration* can be represented by the quantity

$$\mathscr{C}_{\mathrm{k}} = \mathbf{C}X^{1/\nu} \tag{13.2.5}$$

and accordingly we set

$$\mathscr{C}^*_{\mathrm{k}} = \mathbf{C}^*X^{1/\nu}. \tag{13.2.6}$$

For the experimentalist, it is convenient to use a fourth quantity, the mass concentration ρ

$$\rho = \mathbf{C}M/Ⓐ$$

where M is the molecular mass of the polymers (which are supposed to be monodisperse). In this case, ρ^* corresponds to C^*.

In a good solvent, long isolated polymers can be represented by Kuhnian chains characterized by only one length (R_{G} or $X = (R^2/d)^{1/2}$).

In this ideal case, the properties of polymer solutions are characterized by only one dimensionless parameter, namely

$$\mathbf{C}X^d = \mathbf{C}/\mathbf{C}^* = \rho/\rho^*$$

or

$$\mathbf{C}/\mathbf{C}^*_{\mathrm{G}} = \rho/\rho^*_{\mathrm{G}},$$

which is proportional to it, in the Kuhnian limit. In particular, for $d = 3$, according to (13.1.63) we have

$$C^*_G/C^* = (\aleph)^{-3/2} = (0.952)^{-3/2} = 1.076. \qquad (13.2.7)$$

In this chapter we consider only solutions whose concentration is small ($\rho/\rho_0 \ll 1$ where ρ_0 is the mass of solvent per unit volume). However, the fact that this condition is fulfilled does not prevent the chains from overlapping (even strongly) if they are long enough. In fact, at constant ρ, ρ^* can be made arbitrarily small by increasing the molecular mass M of the polymers (of course, there are experimental limitations). Thus, for large M

$$R \propto M^{\nu} \qquad \text{(isolated polymer).}$$

Consequently, according to (3.2.1)

$$C^* \propto M^{-\nu d}$$

and according to (13.2.6)

$$\rho^* \propto M^{1-\nu d} \qquad (\text{where } \nu d > 1) \qquad (13.2.8)$$

Thus, for constant ρ, $\rho^* \to 0$ when $M \to \infty$. Nevertheless, we must realize that it is difficult to make and to manipulate polymers of high degree of polymerization. Actually, if a solution of very long polymers undergoes constraints, even weak ones, the polymer molecules break spontaneously.*

In brief, in the ideal case ($\rho/\rho_0 \ll 1$, $z \gg 1$), we can distinguish three regimes:

for $C \ll C^*$, the dilute regime

for $C \gg C^*$, the semi-dilute regime

for $C \sim C^*$, a crossover domain.

In the semi-dilute domain, the chains overlap strongly and screen one another. This fact has been known for a long time[37] but it has been exploited only recently (1977).[38] We thus note that for $C \gg C^*$, the *static* properties of the solution must no longer depend on the size of the chains but only on the monomer concentration. In the ideal case, the only parameter characterizing the properties of the solution is the Kuhnian concentration $\mathscr{C}_k$ which plays the role of monomer concentration. Now, we note that eqn (13.2.5) leads to the dimensional equation

$$\mathscr{C}_k \sim L^{-d+1/\nu}$$

This entails that in the problem, there is only one length (the *Kuhnian overlap length*)

$$\xi_k = \mathscr{C}_k^{-1/(d-1/\nu)}. \qquad (13.2.9)$$

* Special precautions must be taken when manipulating polymers having a high degree of polymerization; for instance, edges of pipettes must be rounded.

Later, we shall give a more precise definition of the screening length ξ_e associated with the screening effect described above. However, we can already remark that, in the present case, ξ_e must be proportional to ξ_k and in the following chapters, we write

$$\xi_e = \Gamma \xi_k \tag{13.2.10}$$

where Γ (alif) is a universal constant. Owing to this screening effect, at a scale larger than ξ_k, the chain recovers a Brownian behaviour. Thus, we may consider that, in the semi-dilute regime, a chain is made of *swollen sequences* whose size is defined by ξ_k and which play the role of independent links.

Now, let R^2 be the mean square end-to-end distance of such a chain. We can write

$$R^2 = A N_\xi \xi_k^2 \tag{13.2.11}$$

where A is a constant. The number of links is proportional, to $X^{1/\nu}$ for a chain, to $\xi_k^{1/\nu}$ for a sequence. Consequently, we have

$$N_\xi = (X/\xi_k)^{1/\nu} \tag{13.2.12}$$

and finally, we obtain

$$R^2 = A\, \xi_k^{2-1/\nu} X^{1/\nu} \tag{13.2.13}$$

where A can be calculated.

Now, let us come back to the variables C and X. From eqns (13.2.5) and (13.2.9), we deduce

$$\xi_k = (CX^{1/\nu})^{-1/(d-1/\nu)}, \tag{13.2.14}$$

$$R^2 = A(CX^{1/\nu})^{-(2-1/\nu)/(d-1/\nu)}\, X^{1/\nu}. \tag{13.2.15}$$

In particular, the first formula shows that for a given solution, the length ξ_k decreases when X increases, i.e. when the temperature increases.

Moreover, if we consider the polymer chain as made of N links, we can write [see (10.2.1)]

$$X = N^{\nu} l_e .$$

We thus recover formulae introduced by Daoud[38]

$$\xi_k = l_e(CN)^{-1/(d-1/\nu)} \propto (CN)^{-0.77} \qquad \text{for } d = 3$$

$$R^2 = A l_e^2\, N(CN)^{-(2-1/\nu)/(d-1/\nu)} \propto N(CN)^{-0.23} \qquad \text{for } d = 3. \tag{13.2.16}$$

Finally, let us note that formulae (13.2.16) may look more concrete than the initial formulae (13.2.9), (13.2.12), and (13.2.13) but it would be a mistake to believe it. In fact, the quantities N and l_e are neither intrinsic nor universal. The same remark does not apply to C and X which, in principle, are directly measurable quantities. Thus, for very long chains in good solvent, $\mathscr{C}_k$ defines the monomer density in a manner which may look abstract but which is non-ambiguous and realistic.

Of course, the question becomes more complicated when one deals with partially swollen chains and, more precisely, when one studies real polymers. In this case, the behaviour of a sequence of links belonging to a long chain can depend strongly on the number of links constituting the sequence.

To characterize this dependence mathematically, it is convenient to introduce the effective exponent $\nu(n)$. Let $R^2(n)$ be the mean square end-to-end distance of a portion of chain. We set

$$\nu(n) = \frac{n}{R(n)} \frac{\mathrm{d}R(n)}{\mathrm{d}n}.$$

For a real and rather long chain, it is possible to distinguish up to four kinds of behaviour according to the value of n

(a) $n < n_1$: the number n_1 defines a domain in which $X(n)$ is of the order of the persistence length; then $\nu(n) = 1$.

(b) $n_1 < n < n_2$: the persistence effects disappeared and n_2 defines a domain in which the excluded volume effects remain very small; then $\nu(n) = 1/2$.

(c) $n_2 < n < n_3$: the excluded volume effects are large without screening because n_3 defines a domain in which $X(n) < \xi_e$; then $\nu \simeq 0.59$;

(d) $n_3 < n < N$: in this domain, the excluded volume effects are hidden by the screening effect; then $\nu = 1/2$.

Let us add that the properties of polymer solutions also depend on the polydispersion of the molecular masses in the polymer sample, at least as long as the semi-dilute regime is not reached. Of course, we cannot think of studying each case in detail. A proper understanding of the physics of polymer solutions requires only that theoreticians and experimentalists compare their results in limiting cases, which should be defined as precisely as possible. It is from this point of view that we intend to study the detailed properties of polymer solutions, in the remainder of this chapter.

2.2 A discussion concerning the nature of universal functions in polymer theory

As was shown in Chapter 10, a mathematical correspondence exists between polymer models and models that are used to study critical magnetic systems (and others), as, for instance, the Ising and the Heisenberg models. Thus, one might think that for all these critical phenomena, the universal functions in the scaling laws appear in a very similar way. However, the analogy is not complete, and in what follows it will be shown that, in reality, the universal functions appear in a more pleasant form in polymer theory than in magnetic theories.

In fact, it is always possible to compare the properties of polymers in moderately dilute or semi-dilute solutions with those which the same polymers would have in a very dilute solution. In other words, the size of an isolated N-link polymer in a solvent can always be used as a scale length to describe the properties of moderately dilute or semi-dilute solutions of N-link polymers interacting in the same solvent. In this way, we can always manage to write scaling laws with universal functions in which the variables also have a universal character (see, for instance, Section 2.8 concerning the osmotic pressure of a solution). This cannot be done as easily in the theory of magnetism; in this case, the variables appearing in the universal functions do not present a universal character. This is, for instance, what happens for the reduced temperature $t = (T - T_c)/T$ which is a pure number without universal meaning since T_c itself depends on the chemical microstructure. The origin of this difference can be understood as follows. Consider a solution of N-link polymers and let C be the link concentration. In the plane $(1/N, C)$, we can associate a point A with the polymer solution. Now consider a ferromagnetic system of magnetization M at temperature T; in the plane (T, M), we can associate a point B with the corresponding magnetic system. On these diagrams, the points A and B correspond to each other (see Fig. 13.13) and the same correspondence exists for the critical domains. The essential difference comes from the fact that the *coexistence curve* $M = f(T)$ in the plane (T, M) corresponds to the *straight line* $1/N = 0$ in the plane $(1/N, C)$. Thus, in the plane $(1/N, C)$, all polymer solutions corresponding to the same value of N are represented by points belonging to a straight line parallel to the *coexistence straight line*. With any point A, we can thus associate its projection A_0 on the axis $C = 0$, and this association has a clear physical meaning. The same simplicity does not occur in the theory of magnetism, because the natural variables M and T are such that the coexistence curve has a non-trivial shape.

This fact also explains why the concept of broken symmetry which seems crucial in the theory of magnetism (and also in field theory) does not appear at all in polymer theory.

Of course, these differences have no intrinsic character. The theoretical framework is fundamentally the same but the physical context is different. The difference comes only from the fact that in the two models, the physical quantities correspond to peculiar mathematical entities and that, in polymer

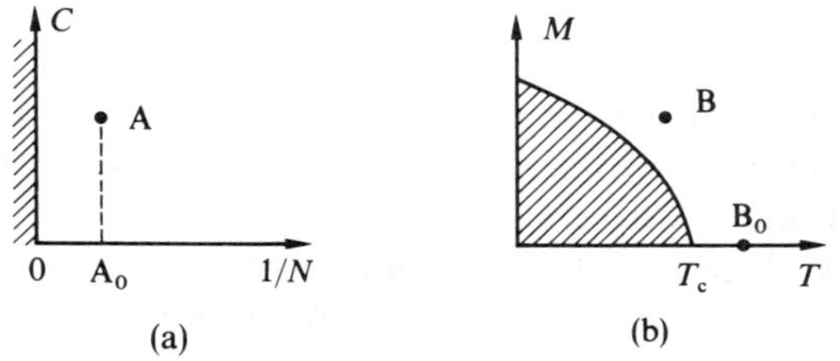

Fig. 13.13. Diagrams for (a) a polymer solution; (b) a ferromagnetic system. The point A and B correspond to each other.

theory, the correspondence is simpler. This is why, in polymer theory, the universal functions appear in a slightly more universal form that in the theory of magnetism. Unfortunately, this point has not always been fully appreciated. Thus, physicists who borrowed renormalization techniques from the theory of magnetism (field theory) and applied them to polymers, did not always exploit their results as much as they could. Fortunately, it is often possible, to cure the defect, by re-interpreting their results.

2.3 Calculations of correlation functions for polymer solutions: a description of the proper formalism

To calculate correlations between polymers of a solution, it is especially convenient to use the *grand ensemble* formalism which was introduced in Chapters 9 and 10.

It can be defined in the framework of the standard continuous model, as follows. Let us consider N polymers with fixed areas $S_1, \ldots, S_N$ and let $\mathbf{\Omega}_N$ be a configuration of these polymers. We attribute the weight $\mathscr{W}_N(S_1, \ldots, S_N; \mathbf{\Omega}_N)$ to this configuration and the weight must be identified with the functional $\mathscr{W}_N(\vec{r})$ defined by (10.1.3). In a grand ensemble, the number N of chains is not fixed and therefore any configuration $\mathbf{\Omega}_N$ (for any N) is acceptable. Therefore, the weight attributed to $\mathbf{\Omega}_N$ in a grand ensemble is

$$f(S_1)\ldots f(S_N)\, \mathscr{W}_N(S_1, \ldots, S_N; \mathbf{\Omega}_N) \qquad (13.2.17)$$

where $f(S)$ is the fugacity of a polymer of area S.

In what follows, we shall be especially interested in monodisperse ensembles; in this case, all chains have the same area and the weight associated with a configuration $\mathbf{\Omega}_N$ can be written in the simplified form

$$f^N\, \mathscr{W}_N(N \times S; \mathbf{\Omega}_N). \qquad (13.2.18)$$

Now, there is only one fugacity and, as was shown in Chapter 10, the grand partition function reads

$$\mathscr{Q}_G(f; S) = 1 + \sum_{N=1}^{\infty} \frac{f^N}{N!} \mathscr{Z}_G(N \times S),$$

or, if we introduce connected functions,

$$\mathscr{Q}_G(f; S) = \exp[V\mathscr{Q}(f; S)]$$

$$\mathscr{Q}(f; S) = \sum_{N=1}^{\infty} \frac{f^N}{N!} \mathscr{Z}(N \times S).$$

Here, the factor $1/N!$ avoids an overcounting of the configurations. In fact, when we calculate $\mathscr{Z}(N \times S)$, we find it convenient to assume that the N polymers are distinguishable and this is why the factor must be introduced.

Moreover, the average number of polymers per unit volume is given by

$$C = f\frac{\partial}{\partial f}\mathscr{Q}(f;\, S). \tag{13.2.19}$$

Let us now see how correlations can be calculated. In a general way, we consider a function $\mathscr{O}(\mathbf{\Omega})$ of the configuration of the system. In practice, this function imposes that certain points belonging to the polymer chain are fixed at definite space points which we call anchoring points. For instance, the function $\mathscr{O}(\boldsymbol{L})$ could be the density of area at the point of position vector $\vec{r}$, namely

$$\mathscr{C}(\vec{r}) = \sum_j \int \mathrm{d}s\; \delta[\vec{r} - \vec{r}_j(s)]$$

where j is a chain index and s the coordinate along the chain. Here $\vec{r}$ defines the anchoring point. In general, we try to calculate the mean value $\langle\!\langle\, \mathscr{O}(\mathbf{\Omega})\rangle\!\rangle$. For instance, we could calculate the mean area concentration $\mathscr{C} = \langle\!\langle\, \mathscr{C}(\vec{r})\rangle\!\rangle$.

Thus, restricted partition functions $\mathscr{Z}_{\mathrm{G}}(\mathscr{O};\, N \times S)$ are associated with $\mathscr{O}(\mathbf{\Omega})$ and we have

$$\langle\!\langle\, \mathscr{O}(\mathbf{\Omega})\rangle\!\rangle = \frac{\sum_{N=1}^{\infty} (f^N/\$_N)\; \mathscr{Z}_{\mathrm{G}}(\mathscr{O};\, N \times S)}{1 + \sum_{N=1}^{\infty} (f^N/N!)\; \mathscr{Z}_{\mathrm{G}}(N \times S)} \tag{13.2.20}$$

where $\$_N$ is the adequate symmetry number. The diagrams contributing to $\mathscr{Z}_{\mathrm{G}}(\mathscr{O};\, N \times S)$ are generally non-connected. Thus, it is proper to introduce *anchored* diagrams which, by definition, are made of one or several connected parts, each one being attached to at least one anchoring point (see Fig. 13.14). We associate the partition function $\mathscr{Z}_{\mathrm{a}}(\mathscr{O};\, N \times S)$ with these anchored diagrams and it is not difficult to show that

$$\langle\!\langle\, \mathscr{O}(\mathbf{\Omega})\rangle\!\rangle = \sum_{N=1}^{\infty} \frac{f^N}{\$_N}\, \mathscr{Z}_{\mathrm{a}}(\mathscr{O};\, N \times S).$$

In general, this quantity can be written in the form

$$\langle\!\langle\, \mathscr{O}(\mathbf{\Omega})\rangle\!\rangle = F(\vec{r}_1, \ldots, \vec{r}_p).$$

where $\vec{r}_1, \ldots, \vec{r}_p$ are vectors defining the p anchoring points. If these points dc not belong to the same polymers, we can introduce the cumulants $F_{\mathrm{c}}(\vec{r}_1, \ldots \vec{r}_p$ of the functions $F(\vec{v}_1, \ldots \vec{v}_p)$ and, in this way, we define the correlation function $\langle\!\langle\, \mathscr{O}(\mathbf{\Omega})\rangle\!\rangle_{\mathrm{c}} = F_{\mathrm{c}}(\vec{r}_1, \ldots \vec{r}_p)$

Thus, for example, the cumulant $\langle\!\langle\, \mathscr{C}(\vec{x})\, \vec{\mathscr{C}}(\vec{y}) - \mathscr{C}^2\rangle\!\rangle$ corresponds to the mea value $\langle\!\langle \mathscr{C}(\vec{x})\; \mathscr{C}(\vec{y})\rangle\!\rangle$ and this cumulant is just the density–density correlatio function studied in Chapter 7.

We see immediately that the correlation function are given by partitio functions $\mathscr{Z}(\mathscr{O};\, N \times S)$ associated with *totally connected* diagrams (se

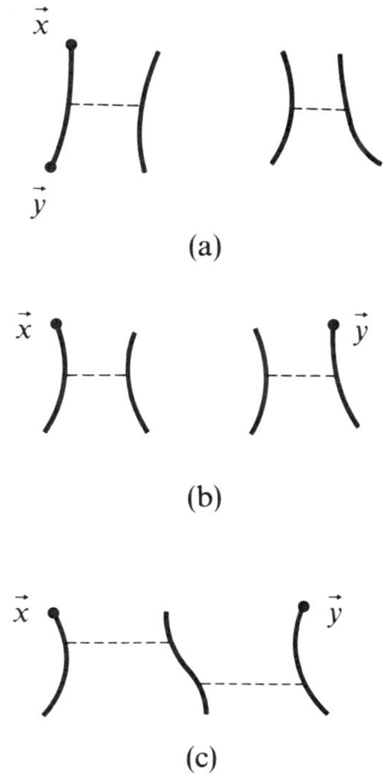

Fig. 13.14. Diagrams corresponding to density–density correlations at points of position vectors $\vec{x}$ and $\vec{y}$; (a) non-connected; (b) partially connected diagram; (c) totally connected diagram. The points of position vectors $\vec{x}$ and $\vec{y}$ are the anchoring points.

Fig. 13.12). Thus, we get the result

$$\langle\!\langle \mathcal{O}(\mathbf{\Omega}) \rangle\!\rangle_c = \sum_{N=1} \frac{f^N}{\$_N} \mathscr{Z}(\mathcal{O};\, N \times S) \tag{13.2.21}$$

where $\$_N$ is the proper symmetry number.

Of course, the Fourier transforms of the correlation functions are calculated by applying the same principles. Moreover, the generalization to the polydisperse case is trivial. Finally, we recall that $\mathscr{Z}(\mathcal{O};\, N \times S)$ can be easily expressed as a sum of contributions associated with anchored diagrams.

2.4 Polydispersion corrections

For reasons of simplicity, theoreticians like to assume that, in polymer solutions, all the chains have the same number of links. It is also quite clear that experimentalists try to use polymer samples that are as monodisperse as possible. Nevertheless, it is true that at the present time (1988), the best samples are far from being monodisperse, as was discussed in Chapter 1. This fact, leads to uncertainties in the experimental results and to small discrepancies. To take polydispersion into account, it is convenient and probably realistic to use the Schulz–Zimm probability distribution (see Chapter 1) in the framework of the standard continuous model. In other words, the probability distribution of the area A of a polymer is represented by the law

$$P(A) = \frac{1}{(-1+1/p)!}\,\frac{1}{pS}\left(\frac{A}{pS}\right)^{-1+1/p} \exp[-A/pS] \qquad (13.2.22)$$

which gives

$$\langle A \rangle = S$$

$$\langle A^n \rangle = \frac{(n-1+1/p)!}{(-1+1/p)!}\, p^n S^n. \qquad (13.2.23)$$

The distribution is thus characterized by the area S and the polydispersity p

$$\frac{\langle A^2 \rangle}{\langle A \rangle^2} = 1 + p \qquad (13.2.24)$$

where, in general, p is a small number.

Here $S^{1/2}$ defines the size of the *average polymer* in the absence of interaction. In the same way, in the presence of interaction, the size ${}^{e}S^{1/2} = X$ of the average polymer in an infinitely dilute solution can be considered as a proper reference length.

Thus, each time we want to take polydispersion into account, we use the Schulz–Zimm law and we choose the size of the isolated average polymer as the reference length. Consequently, all formulae are expressed in the polydisperse case in the same way as in the non-disperse case, the only difference being that now a new dimensionless parameter p defining the polydispersion of the system has to be introduced.

2.5 Osmotic pressure

2.5.1 Generalities

In Chapter 5, we defined the osmotic pressure of a polymer solution, we indicated how it can be measured, and we described various effects concerning the compressibility and the preferential adsorption. When the polymers are very long and when the volume fraction occupied by the polymers in the solution is small, the complex reality can be represented by a simple model which is the standard continuous model, studied in Chapter 10 in the context of perturbation theory. This model is especially useful because it allows us to perform effective calculations. In particular, it can be used in the limit of long polymers to determine universal quantities because, then, the general properties of long polymers become independent of the chemical microstructure. Calculations are always made in the framework of the grand canonical formalism which is well-adapted to the problem. Moreover, the theory of critical phenomena enables us to get a clear view, first of the swelling phenomena, and secondly of the overlap effects in polymer solutions. Now, we have to bring together all these elements in order to obtain theoretical expressions for the osmotic pressure and to compare these expressions to experimental results.

The osmotic pressure $\boldsymbol{\Pi}$ of a very dilute polymer solution obeys Van 't Hoff's law

$$\boldsymbol{\Pi\beta} = \boldsymbol{C}$$

where $\boldsymbol{C}$ is the number of polymers per unit volume, and this law is valid for good and poor solvents. When $\boldsymbol{C}$ increases the situation becomes much more complex. Traditionally, the experimentalist represents the osmotic pressure of a solution as a series in powers of ρ (ρ = mass concentration)

$$\boldsymbol{\Pi}/\boldsymbol{RT} = \sum_{p=1}^{\infty} A_p \rho^p. \tag{13.2.25}$$

This representation is not the best one, since the coefficients A_p have dimensions and depend on many parameters which actually have little to do with the physical problem. (For instance, the A_p depend on the isotopic composition of the atoms constituting the polymers!)

A description more adequate to theoretical calculations can be given in the context of the standard continuous model, First, let us consider the case where all polymers have the same area (monodisperse ensemble). Then, there are only two dimensionless parameters, namely z which defines the strength of the interaction, and $\boldsymbol{C}S^{d/2}$ which defines the rate of overlap. We can thus write

$$\boldsymbol{\Pi\beta}/\boldsymbol{C} = F_0(\boldsymbol{C}S^{d/2}, z) \tag{13.2.26}$$

and an approximate parametric representation of $F_0(x, z)$ for $d = 3$ has been given in Chapter 10, formula (10.8.27).

Nevertheless, from a physical point of view, it is better to use other parameters. First, instead of choosing z to define the interaction, we can use the osmotic parameter g which is related to it. In this connection, let us recall that the function $g = g(z)$ was studied in detail at the beginning of this chapter, Section 1.1. Secondly, instead of choosing the size $S^{1/2}$ of the non-interacting polymer as the length scale, we can use the size $x = (R^2/d)^{1/2}$ of the isolated polymer in solution [the variable x is defined by (13.1.1)]. Thus, the osmotic pressure can be expressed in the form

$$\boldsymbol{\Pi\beta} = \boldsymbol{C}F(\boldsymbol{C}X^d, g) = \boldsymbol{C}\left[1 + \sum_{p=1}^{\infty} {}^{p}F(g)\,(\boldsymbol{C}X^d)^p\right] \tag{13.2.27}$$

and we recall that by definition (see (13.1.2) and (12.3.72)]

$${}^{1}F(g) = \tfrac{1}{2}\,(2\pi)^{d/2}\, g. \tag{13.2.28}$$

We understand why this representation is especially useful: in fact, the variables

C, X, and g are observables and therefore are a priori measurable. Thus (in principle), any expression of $F(x, y)$ can be tested experimentally.

In practice, the experimentalist measures the mean square radius of gyration R_G^2, and not the mean square end-to-end distance R^2. On the contrary, for the theoretician R^2 is a much more elementary quantity than R_G^2. Fortunately, the ratio $\aleph(g) = 6R^2/R_G^2$ is given by theoretical considerations with enough precision (see Section 1.4.2).

The coefficients ${}^pF(g)$ can be expanded in powers of g and, for small g, ${}^pF(g) \propto g^p$. This fact can easily be shown by remarking that

(1) each time a polymer line is added to a connected diagram, a supplementary factor $\boldsymbol{C}$ and (at least) a supplementary factor b must be introduced;

(2) for small $\boldsymbol{C}$, $f \propto \boldsymbol{C}$;

(3) for small g, $b \propto z \propto g$.

Formula (12.2.27) describes the behavious of $\boldsymbol{\Pi\beta}$ in the whole concentration domain which extends from dilute to semi-dilute solutions, but, in this domain, it is not universal, and therefore it gives only an approximate image of reality. On the contrary, when $z \to \infty$ and $g \to g^*$ (large swelling but X is kept fixed), the Kuhnian limit is reached

$$F(x, g^*) = \mathrm{F}(x).$$

In other words, the coefficients ${}^pF(g)$ are not universal but their limits ${}^p\mathrm{F} \equiv {}^pF(g^*)$ are universal.

In the Kuhnian limit, $\boldsymbol{\Pi}$ is given by the universal scaling law (des Cloizeaux 1975 and 1980).[42,43]

$$\boldsymbol{\Pi\beta} = \boldsymbol{C}\,\mathrm{F}(\boldsymbol{C}X^d) = \boldsymbol{C}\left[1 + \sum_{p=1}^{\infty} {}^p\mathrm{F}\,(\boldsymbol{C}X^d)^p\right] \tag{13.2.29}$$

Now, the coefficients ${}^p\mathrm{F}$ depend only on the dimensionality d of the space in which the chains are embedded. They can be expanded in powers of $\varepsilon = 4 - d$. As $g^* \simeq \varepsilon/8$ for $0 < \varepsilon \ll 1$, we find that, for small ε, ${}^p\mathrm{F} \propto \varepsilon^p$.

The semi-dilute regime corresponds to the domain $\boldsymbol{C}X^d \gg 1$. In this domain $\boldsymbol{\Pi\beta}$ is expected to depend only on the monomer concentration. As was pointed out in Section 2.1, the number of monomers of a chain is proportional to $X^{1/\nu}$, and consequently, in this Kuhnian limit, the monomer concentration must be defined as the Kuhnian concentration.

$$\mathscr{C}_{\mathrm{k}} = \boldsymbol{C}X^{1/\nu} \tag{13.2.30}$$

Thus when $\boldsymbol{C}X^d \to \infty$, $\boldsymbol{\Pi\beta}$ should become a function of $\mathscr{C}_{\mathrm{k}}$, which implies the following consequences (des Cloizeaux 1975 and 1980)[42,43]

$$\mathrm{F}(x) \simeq \mathrm{F}_\infty\, x^{1/(\nu d - 1)} \qquad x \gg 1$$

$$\boldsymbol{\Pi\beta} = \mathrm{F}_\infty\, \mathscr{C}_{\mathrm{k}}^{d/(d-1/\nu)}. \tag{13.2.31}$$

Thus, we find the law which gives the osmotic pressure as a function of concentration for semi-dilute solutions.

For $d = 3$, we have $\nu = 0.588$ [see (12.3.9)] and, consequently,

$$\frac{d}{d - 1/\nu} = 2.31. \tag{13.2.32}$$

The order of magnitude of F_∞ can be found as follows. We remark that to first-order in ε [see (12.3.32) and (12.3.106)]

$$\frac{d}{d - 1/\nu} = 2\left(1 + \frac{\varepsilon}{8} + \ldots\right) \tag{13.2.33}$$

In other words, in the limit $\varepsilon \to 0$, the coefficient F_∞ must coincide with the second coefficient of the expansion of $\mathrm{F}(x)$, namely

$${}^2\mathrm{F} = {}^2F(g^*) = \tfrac{1}{2}\,(2\pi)^{d/2}\, g^*. \tag{13.2.34}$$

Now, we recall that for $\varepsilon \ll 1$, $g^* \simeq \varepsilon/8$ [see (12.3.102)] and, consequently, at order ε, we have

$$\mathrm{F}_\infty \simeq \frac{\pi^2}{4}\,\varepsilon + \ldots . \tag{13.2.35}$$

Let us note that we can also calculate, without difficulty, the first term (in ε^p) of the expansion of ${}^p\mathrm{F}$ with respect to ε. All these elements enable us to compare theoretical predictions and experimental results in a satisfactory manner (des Cloizeaux and Noda 1982)[44]. It is also possible to calculate $\mathrm{F}(x)$ for all values of x. L. Schäfer took a special interest in this problem and, of course, used the grand canonical formalism which gives the dependence of $\boldsymbol{\Pi}$ with respect to $\boldsymbol{C}$ in parametric form. In a first article (1981)[45] written in collaboration with Knoll and Witten, Schäfer derived expansions to order ε^2; then, later (1982),[46] he summarized the complicated results he had obtained, with the help of a semi-phenomenological formula, and he showed that they were in good agreement with experimental results. His formula which takes polydispersity into account can be written in the following way (here $d = 3$)

$$\mathrm{F}(x) = 1 + B\frac{t}{4}\left[\frac{40 + 27t + t^2}{22 + t}\right]^{.30}$$

where

$$B = 1 + 0.06\left(2 - \frac{M_\mathrm{w}}{M_\mathrm{n}}\right)(0.68)^t$$

$$t \propto x. \tag{13.2.36}$$

This formula does not appear in the universal form adopted in the present book (see Section 2.2). However, it is possible to fix the proportionality coefficient between t and x so as to insure an exact determination of the second virial coefficient: this requirement leads to the equality

$$\frac{t}{4}\left(\frac{40}{22}\right)^{0.30} = \frac{1}{2}(2\pi)^{3/2} g^*$$

where $g^* = 0.233$ [see (12.1.16)] and therefore

$$t = 6.13x. \tag{13.2.37}$$

Schäfer's formula is deduced from various one-loop and two-loop expansions: the methods used are not very clearly explained and the calculations are complicated. They will not be expounded here. We shall content ourselves with more elementary approaches which give also good results.

2.5.2 Renormalization and asymptotic form of $\mathscr{Q}(f)$

Let us consider monodisperse polymers in solution and let us study this system in the framework of the standard continuous model. As was shown in Chapter 10 [see (10.5.2)], the osmotic pressure is given in terms of the number $\boldsymbol{C}$ of polymers per unit volume by the parametric representation

$$\boldsymbol{\Pi\beta} = \mathscr{Q}(f)$$

$$\boldsymbol{C} = f\,\frac{\partial\,\mathscr{Q}(f)}{\partial f}$$

where

$$\mathscr{Q}(f) = \sum_{N=1}^{\infty} \frac{1}{N!} f^N \mathscr{Z}(N \times \mathrm{S}) \tag{13.2.38}$$

and where the fugacity f is a variable of dimension L^{-d}.

In order to determine the behaviour of the solution in the Kuhnian limit, we must re-express $\boldsymbol{\Pi\beta}$ and $\boldsymbol{C}$ in terms of renormalized quantities. Consequently, we write

$$\mathscr{Z}(N \times S) = [\mathfrak{X}_1(z)]^{2N} \mathscr{Z}_{\mathrm{R}}(N \times S)$$

$$f = [\mathfrak{X}_1(z)]^{-2} f_{\mathrm{R}}$$

$$\mathscr{Q}(f) = \mathscr{Q}_{\mathrm{R}}(f_{\mathrm{R}}) = \sum_{N=1}^{\infty} \frac{1}{N!} f_{\mathrm{R}}^N \mathscr{Z}_{\mathrm{R}}(N \times S) \tag{13.2.39}$$

and this leads to the equalities

$$\boldsymbol{\Pi\beta} = \mathscr{Q}_{\mathrm{R}}(f_{\mathrm{R}})$$

$$\boldsymbol{C} = f_{\mathrm{R}} \frac{\partial}{\partial f_{\mathrm{R}}} \mathscr{Q}_{\mathrm{R}}(f_{\mathrm{R}}).$$

Moreover, the dimensional formulae (10.4.52) and (10.4.53) show that in the limit $z \gg 1$, we must have

$$\mathscr{Z}_{\mathrm{R}}(N \times {}^{\mathrm{e}}S) = X^{(N-1)d}\,\zeta_N \tag{13.2.40}$$

where ζ_N is a pure number. Let us set

$$\mathrm{Q}(y) = \sum_1^{\infty} \frac{y^N}{N!} \zeta_N \tag{13.2.41}$$

which implies

$$\mathscr{Q}_{\mathrm{R}}(f_{\mathrm{R}}) = X^{-d}\, \mathrm{Q}(f_{\mathrm{R}} X^d). \tag{13.2.42}$$

From these equations, we deduce

$$\Pi\beta = X^{-d}\, \mathrm{Q}(f_{\mathrm{R}} X^d)$$

$$C = f_{\mathrm{R}}\, \mathrm{Q}'(f_{\mathrm{R}} X^d).$$

By setting

$$f_{\mathrm{R}} X^d = y$$

we obtain the scaling law (13.2.29) in the form

$$\Pi\beta = X^{-d}\, \mathrm{Q}(y)$$

$$C = X^{-d} y\, \mathrm{Q}'(y) \tag{13.2.43}$$

and we see immediately that $\mathrm{F}(x)$ is related to $Q(y)$. In fact, setting

$$x = CX^d = y\, \mathrm{Q}'(y)$$

we find

$$\mathrm{F}(y\, \mathrm{Q}'(y)] = \mathrm{Q}(y)/y\, \mathrm{Q}'(y). \tag{13.2.44}$$

A simple approximation, valid for small values of CX^d can be obtained by setting

$$\mathscr{Q}_{\mathrm{R}}(f_{\mathrm{R}}) \simeq f_{\mathrm{R}}\, \mathscr{Z}_{\mathrm{R}}({}^{\mathrm{e}}S) + \frac{f_{\mathrm{R}}^2}{2}\, \mathscr{Z}_{\mathrm{R}}({}^{\mathrm{e}}S, {}^{\mathrm{e}}S),$$

or, according to (12.3.61)

$$\mathscr{Z}_{\mathrm{R}}({}^{\mathrm{e}}S) = 1.$$

Moreover, according to (12.3.61),

$$\mathscr{Z}_{\mathrm{R}}({}^{\mathrm{e}}S, {}^{\mathrm{e}}S) = -\,(2\pi)^{d/2}\, X^d g\,.$$

From the preceding expression, we get

$$\mathscr{Q}_{\mathrm{R}}(f_{\mathrm{R}}) = f_{\mathrm{R}} - \tfrac{1}{2}(2\pi)^{d/2}\, X^d g\, f_{\mathrm{R}}^2$$

and by comparing with (13.2.42)

$$\mathrm{Q}(y) = y - \tfrac{1}{2}\,(2\pi)^{d/2}\, g y^2\,.$$

Bringing this expression in (13.2.43) and eliminating y, we find

$$\Pi\beta = C + \frac{2g(2\pi)^{d/2} C^2\, X^d}{[1 + (1 - 4g(2\pi)^{d/2}\, CX^d)^{1/2}]^2}\,.$$

Let us now come back to the general case and let us consider the limit

$$x = CX^d \gg 1\,.$$

We assert that $y \to \infty$ when $x \to \infty$ and that for $y \gg 1$

$$\mathcal{Q}(y) \simeq \mathcal{Q}_\infty (\ln y)^{\nu d}. \tag{13.2.45}$$

This property will be established later for various cases, but first let us show that it entails the asymptotic law (13.2.31) corresponding to the semi-dilute limit. In fact,

$$x = y\,\mathcal{Q}'(y) \simeq \nu d\,\mathcal{Q}_\infty (\ln y)^{\nu d - 1}.$$

From this equation, we get

$$\ln y \simeq \left[\frac{x}{\nu d\,\mathcal{Q}_\infty}\right]^{1/(\nu d - 1)}$$

and by bringing this expression in (13.2.45) and (13.2.44), we find

$$\mathrm{F}(x) \simeq \left(\frac{x}{\mathcal{Q}_\infty (\nu d)^{\nu d}}\right)^{1/(\nu d - 1)}$$

which can be compared with (13.2.31) and gives

$$\mathrm{F}_\infty = [\mathcal{Q}_\infty (\nu d)^{\nu d}]^{-1/(\nu d - 1)}. \tag{13.2.46}$$

In particular, for $0 < 4 - d \ll 1$, we have $\nu d \simeq 2$ and consequently

$$\mathrm{F}_\infty \simeq \frac{1}{4\mathcal{Q}_\infty}. \tag{13.2.47}$$

The validity of (13.2.45) can be directly verified for $d = 1$ and $d = 4$, and also for small values of $\varepsilon = 4 - d$, as will now be shown.

2.5.3 Direct calculation of the asymptotic form of $\mathcal{Q}(f)$ in the case $d = 1$

In one dimension, in the limit $z \to \infty$, polymers behave like rigid rods of length X. Let L be the length of the box. The partition function is

$$\mathscr{Z}_{\mathrm{G}}(N \times S) = (L - NX)^N.$$

Consequently, the grand partition function is

$$\mathcal{Q}_{\mathrm{G}}(f) = \sum_{N=1}^{N_{\max}} \frac{f^N (L - NX)^N}{N!}.$$

As $L/X \gg 1$, we may set

$$N! = (2\pi)^{1/2}\, N^{N+1/2}\, \mathrm{e}^{-N}$$

$$NX/L = t,$$

and we may replace the sum by an integral

$$\mathcal{Q}_G(f) = (2\pi)^{-1/2}\left(\frac{\boldsymbol{L}}{X}\right)^{1/2} \int_0^\infty \mathrm{d}t \, \exp[\varphi(t)]$$

where

$$\varphi(t) = \frac{\boldsymbol{L}}{X}\, t[\ln(1 - t) - \ln t + \ln(Xf) + 1] - \frac{1}{2}\ln t$$

(the last term in $\varphi(t)$ is negligible).
In the thermodynamic limit, we have

$$\mathcal{Q}(f) \simeq \frac{1}{\boldsymbol{L}} \max \varphi(t) = \frac{1}{\boldsymbol{L}}\, \varphi(\theta),$$

and, for $\boldsymbol{L}/X \gg 1$,

$$\frac{X}{\boldsymbol{L}}\, \varphi'(\theta) = \ln(1 - \theta) - \ln\theta + \ln(Xf) + 1 - \frac{1}{1-\theta} = 0.$$

We thus obtain $\mathcal{Q}(f)$ in parametric form

$$\mathcal{Q}(f) = \frac{\theta}{X(1-\theta)}$$

$$Xf = \frac{\theta}{1-\theta} \exp[\theta/(1-\theta)]$$

or, if we eliminate θ, in the form

$$f = \mathcal{Q} \exp[X\mathcal{Q}]. \tag{13.2.48}$$

As

$$\boldsymbol{C} = f\frac{\partial}{\partial f}\mathcal{Q}$$

we have

$$f = \boldsymbol{C}(1 + X\mathcal{Q}) \exp[X\mathcal{Q}].$$

By eliminating f, we obtain

$$\boldsymbol{C}(1 + X\mathcal{Q}) = \mathcal{Q},$$

and therefore

$$\boldsymbol{\Pi\beta} = \mathcal{Q} = \frac{\boldsymbol{C}}{1 - \boldsymbol{C}X}. \tag{13.2.49}$$

Thus, we find a result which is a priori evident; moreover, we see that, in the limit of large concentrations, $1 - \boldsymbol{C}X \to 0$, then $\mathcal{Q} \to \infty$; therefore, according to (13.2.48), $f \to \infty$ and we may write

$$X\,\mathcal{Q}(f) = \ln(Xf) - \ln(X\mathcal{Q}).$$

Consequently, in the limit $Xf \gg 1$, we find

$$X\mathcal{Q}(f) \simeq \ln(Xf)$$

in full agreement with (13.2.42) and (13.2.45).

2.5.4 Direct calculation of the asymptotic form of $\mathcal{Q}(f)$ in the case $d = 4$

In four dimensions, the chains are asymptotically Brownian as was shown in Chapter 12, Section 3.3.4. In this case, to calculate $\mathcal{Q}(t)$, we have only to use the simple-tree approximation described in Chapter 10, Section 6.2.

The grand partition function is defined (for $c = 0$) by the coupled equations (10.6.3) and (10.6.5)

$$f = C\exp(bS^2C)$$
$$\mathcal{Q} = C + \frac{b}{2}S^2C^2. \tag{13.2.50}$$

Now, let us set

$$x = bS^2f$$
$$y = \exp(bS^2C).$$

From (13.2.50), we deduce

$$x = y\ln y$$
$$\mathcal{Q}bS^2 = \ln y + \tfrac{1}{2}(\ln y)^2.$$

The semi-dilute case corresponds to the limits $x \to \infty$, $y \to \infty$ for which

$$\ln y \simeq \ln(x)$$
$$\mathcal{Q}bS^2 \simeq \frac{1}{2}[\ln(x)]^2,$$

and therefore

$$\mathcal{Q}(f) \simeq \frac{1}{2bS^2}[\ln(bS^2f)]^2. \tag{13.2.51}$$

We verify that this formula is compatible with (13.2.42) and (13.2.45) since in four dimensions $\nu d = 2$.

2.5.5 One-loop approximation of the osmotic pressure in the Kuhnian limit

We showed that, in the Kuhnian limit, the osmotic pressure Π given by $\mathcal{Q}(f)$ can be expressed in the form of a scaling law, provided f and $\mathcal{Q}(f)$ are properly renormalized (see Section 2.5.2).

Moreover, we showed that, in practice, it is often convenient to sum up all trees and to express Π not as a function of $\mathcal{Q}(f)$ but as a function of $\mathcal{Q}_{\text{I}}(f)$ (see

Chapter 10, Section 6.3). Equations (10.6.31) give

$$\boldsymbol{\Pi\beta} = \mathscr{Q}(f) = \mathscr{Q}_{\mathrm{I}}(\mathfrak{f}) + \frac{\boldsymbol{C}}{2}(b\boldsymbol{C}S^2)$$

$$\boldsymbol{C} = f\frac{\partial}{\partial f}\mathscr{Q}(f) = \mathfrak{f}\frac{\partial}{\partial \mathfrak{f}}\mathscr{Q}_{\mathrm{I}}(\mathfrak{f})$$

$$f = \mathfrak{f}\exp(b\boldsymbol{C}S^2) \tag{13.2.52}$$

Unfortunately, $\mathscr{Q}_{\mathrm{I}}(\mathfrak{f})$ cannot be renormalized in a simple manner. This can be seen by observing that the preceding equations contain the term $b\boldsymbol{C}S^2$. Now, according to (10.1.7) and (10.3.57), we have

$$bS^2/X^d = z(2\pi S)^{d/2}\, X^{-d} = z[\mathfrak{X}_0\,(z)]^{-d/2}\,(2\pi)^{d/2}, \tag{13.2.53}$$

and for $z \gg 1$ (see Chapter 12)

$$bS^2/X^d \propto z^{2(2-\nu d)/(4-d)}.$$

Then, we observe that $\boldsymbol{C}X^d$ remains finite and retains a meaning in the Kuhnian limit, whereas, on the contrary, $b\boldsymbol{C}S^2$ becomes infinite and loses any meaning in this limit. Accordingly, $\mathfrak{f}$ and $\mathscr{Q}_{\mathrm{I}}(\mathfrak{f})$ renormalize in a manner which is highly non-trivial; consequently, we must abandon this approach contenting ourselves with the summation of fewer diagrams.*

In fact, if we want to renormalize the diagrams to one-loop order, so as to calculate $\mathscr{X}(\boldsymbol{N} \times S)$, we must retain from the contributing diagrams only those which contain one-loop at most; this is the step we shall take. By applying renormalization techniques to the result, we shall derive an expression of $\boldsymbol{\Pi}$ valid for small values of g, i.e. actually for small values of $\varepsilon = 4 - d$ (nevertheless the value 3 for d can be considered close to 4).

At one loop order, $\mathscr{Q}(f)$ can be written as the sum of two terms

$$\mathscr{Q}(f) = {}^0\mathscr{Q}(f) + {}^1\mathscr{Q}(f). \tag{13.2.54}$$

${}^0\mathscr{Q}(f)$ is the sum of all simple-tree terms (see Chapter 10, Section 2.5.4) and ${}^1\mathscr{Q}(f)$ is the sum of all one-loop terms (the simplest diagrams) dressed with simple-tree sub-diagrams (see Fig. 13.15).

To take into account the fact that simple-tree sub-diagrams may be attached to a polymer, we have only to replace the initial fugacity f by a new fugacity $\mathfrak{f}$ which is related to f in the following self-consistent way:

$$\mathfrak{f} = f\sum_{n=0}^{\infty}\frac{(-bS\mathfrak{f})^n}{n!} = f\exp(-bS^2\mathfrak{f}), \tag{13.2.55}$$

which can be written in the form

$$f = \mathfrak{f}\exp[(2\pi S)^{d/2}\,z\,\mathfrak{f}]. \tag{13.2.56}$$

* The following text is founded on unpublished works (1985) by B. Duplantier and J. des Cloizeaux.

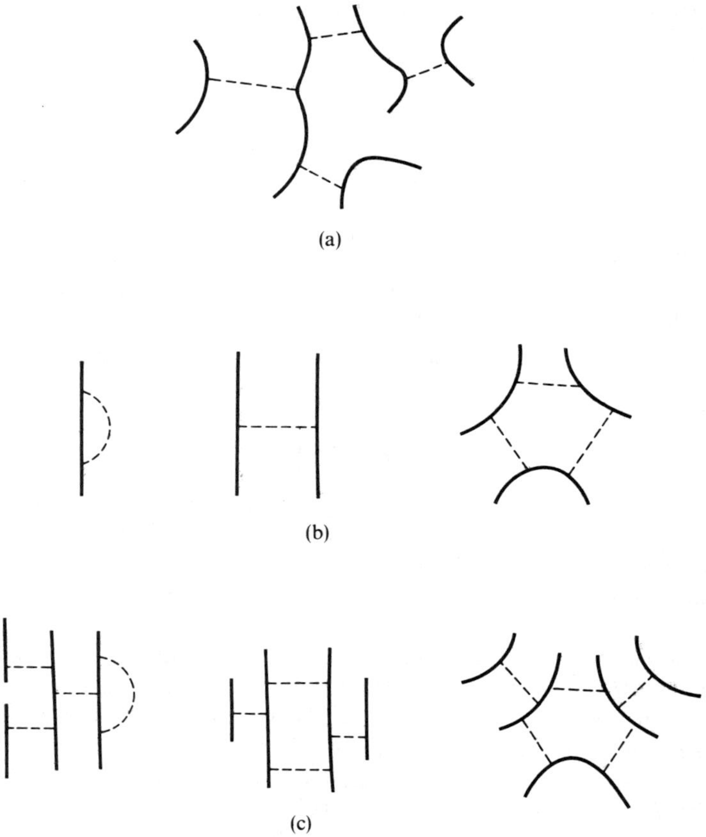

Fig. 13.15. (a) Simple-tree diagram.
(b) Elementary one-loop diagrams.
(c) Elementary one-loop diagrams dressed with simple-tree sub-diagrams.

Moreover, it is easy to see that

$$\mathfrak{f} = f \frac{\partial\, {}^0\mathcal{Q}}{\partial f}. \tag{13.2.57}$$

To determine ${}^0\mathcal{Q}(f)$, it is convenient to calculate $\partial\, {}^0\mathcal{Q}/\partial \mathfrak{f}$ by using (13.2.55)

$$\frac{\partial\, {}^0\mathcal{Q}}{\partial \mathfrak{f}} = \frac{\partial\, {}^0\mathcal{Q}}{\partial f}\,\frac{\partial f}{\partial \mathfrak{f}} = \frac{\partial\, {}^0\mathcal{Q}}{\partial f}\,[\exp[-bS^2\mathfrak{f}] + bS^2\mathfrak{f}\,\exp[-bS^2\mathfrak{f}]]$$

$$= f\frac{\partial\, {}^0\mathcal{Q}}{\partial f}\left[\frac{1}{\mathfrak{f}} + bS^2\right].$$

Then, taking (13.2.57) into account, we find

$$\frac{\partial\,{}^0\mathcal{Q}}{\partial \mathfrak{f}} = 1 + bS^2\mathfrak{f}.$$

If $\mathfrak{f} = 0$, ${}^0\mathcal{Q} = 0$ because in this case $f = 0$. Consequently,

$${}^0\mathcal{Q} = \mathfrak{f} + \tfrac{1}{2}bS^2\,\mathfrak{f}^2 = \mathfrak{f} + \tfrac{1}{2}(2\pi S)^{d/2}\; z\,\mathfrak{f}^2. \tag{13.2.58}$$

Thus (13.2.56) and (13.2.50) determine ${}^0\mathcal{Q}$ implicitly in terms of z (or bS^2) and $\mathbf{C}$.

On the other hand, the second term ${}^1\mathcal{Q}(f)$ is obtained from ${}^1\mathcal{Q}_{\mathrm{I}}(f)$, which is the generating function of the I-irreducible one-loop diagrams. We have only to replace f by $\mathfrak{f}$ to dress these diagrams, and accordingly we may write here

$${}^1\mathcal{Q}(f) = {}^1\mathcal{Q}_{\mathrm{I}}(\mathfrak{f}).$$

The function ${}^1\mathcal{Q}_{\mathrm{I}}(\mathfrak{f})$ has an expansion

$${}^1\mathcal{Q}_{\mathrm{I}}(\mathfrak{f}) = \sum_{N=1}^{\infty} \frac{\mathfrak{f}^N}{N!}\; {}^1\mathcal{Z}_{\mathrm{I}}(N \times S)$$

where ${}^1\mathcal{Z}_{\mathrm{I}}\,(N \times S)$ is given by (10.8.23).

Thus, by using the same notation as in (10.8.24), (10.8.25), (10.8.26), we find

$${}^1\mathcal{Q}_{\mathrm{I}}(\mathfrak{f}) = (2\pi S)^{-d/2}\;{}^1\mathfrak{z}_{\mathrm{I}}((2\pi S)^{d/2} z\,\mathfrak{f}),$$

or, according to (10.8.26) (and by setting $y = x^2$,

$${}^1\mathfrak{z}_{\mathrm{I}}(\tau) = \frac{1}{2\Gamma(d/2)} \int_0^{\infty} \mathrm{d}y\; y^{-1+d/2} \left\{ \frac{2\tau}{y} - \ln[1 + \tau\;{}^0\mathrm{h}(y)] \right\}$$

$${}^0\mathrm{h}(y) = 2\left(\frac{1}{y} - \frac{1}{y^2} + \frac{\mathrm{e}^{-y}}{y^2}\right). \tag{13.2.59}$$

Finally, by grouping the preceding results, we obtain $\mathcal{Q}(f)$ in the following parametric form

$$\mathcal{Q}(f) = \mathfrak{f} + \tfrac{1}{2}(2\pi S)^{d/2}\; z\,\mathfrak{f}^2 + (2\pi S)^{-d/2}\;{}^1\mathfrak{z}_{\mathrm{I}}\,[(2\pi S)^{d/2}\, z\,\mathfrak{f}]$$

$$f = \mathfrak{f}\;\exp[(2\pi S)^{d/2}\; z\mathfrak{f}]. \tag{13.2.60}$$

When the chain becomes very long, this expression cannot be used directly; we must re-express S in terms of ${}^{\mathrm{e}}S = X^2$, and $\mathcal{Q}(f)$ in terms of the renormalized fugacity $f_{\mathbf{R}}$ [see (13.2.39)]

$$f_{\mathbf{R}} = [\mathfrak{X}_1(z)]^2\, f. \tag{13.2.61}$$

Now, we observe that ${}^1\mathfrak{z}_{\mathrm{I}}(\tau)$ diverges when $\varepsilon \to 0$ (for $y \to \infty$ in the integral) but the only divergent terms are those proportional to τ and τ^2. These divergences are fictitious and they must disappear by renormalization. The divergent terms are given by a simple calculation

$$ {}^{1}\mathfrak{z}_1'(0) = \frac{1}{\Gamma(d/2)} \int_0^\infty \mathrm{d}y\, y^{-1+d/2} \left[\frac{1}{y} - {}^{0}\mathrm{h}(y)\right] = \frac{4}{(d-2)(4-d)} \simeq \frac{2}{\varepsilon} + 1 + \dots. $$

$$ \frac{1}{2}\,{}^{1}\mathfrak{z}_1''(0) = \frac{1}{\Gamma(d/2)} \int_0^\infty \mathrm{d}y\, y^{-1+d/2}\, {}^{0}\mathrm{h}^2(y) = $$

$$ \frac{16(10 - d - 2^{4-d/2})}{(d-2)(4-d)(6-d)(8-d)} \simeq \frac{2}{\varepsilon} + \left(\frac{1}{2} - 2\ln 2\right) + \dots. \quad (13.2.62) $$

Then, let us define the renormalized fugacity $\mathfrak{f}_{\mathbf{R}}$. According to (13.2.53) for small values of z, we have

$$ bS^2 \simeq (2\pi)^{d/2}\, gX^d. $$

It is therefore reasonable to define $\mathfrak{f}_{\mathbf{R}}$ by setting

$$ f_{\mathbf{R}} = \mathfrak{f}_{\mathbf{R}} \exp[(2\pi)^{d/2}\, gX^d\, \mathfrak{f}_{\mathbf{R}}]. \quad (13.2.63) $$

To obtain the renormalized form of (13.2.56), we must first re-express $\mathfrak{f}$ in terms of $\mathfrak{f}_{\mathbf{R}}$ (to one-loop order). By using (13.2.60), (13.2.61) and (13.2.63), we find

$$ \frac{\mathfrak{f}}{\mathfrak{f}_{\mathbf{R}}} \exp[(2\pi)^{d/2}\,(zS^{d/2}\,\mathfrak{f} - gX^d\,\mathfrak{f}_{\mathbf{R}})] = [\mathfrak{X}_1(z)]^{-2}. \quad (13.2.64) $$

Three renormalization factors play a role in the renormalization process, and clearly they are (z/g), $\mathfrak{X}_1(z)$, and $\mathfrak{X}_0(z) = X^2/S$ [see definition (12.3.57)]. Let us set

$$ \mathfrak{f}/\mathfrak{f}_{\mathbf{R}} = 1 + \eta $$

and let us expand (13.2.64) in powers of g (or z). To simplify the expression, we use the notation

$$ \tau_{\mathbf{R}} = (2\pi)^{d/2}\, gX^d\, \mathfrak{f}_{\mathbf{R}} \quad (13.2.65) $$

where $\tau_{\mathbf{R}}$ is a dimensionless number which retains a meaning in the Kuhnian limit. We find

$$ \eta(1 + \tau_{\mathbf{R}}) + \tau_{\mathbf{R}}\left(\frac{z}{g}[\mathfrak{X}_0(z)]^{-d/2} - 1\right) = [\mathfrak{X}_1(z)]^{-2} - 1. \quad (13.2.66) $$

In a similar way, applying the renormalization process to $\mathcal{Q}(f)$, we must set

$$ \mathcal{Q}(f) = \mathcal{Q}_{\mathbf{R}}(f_{\mathbf{R}}). \quad (13.2.67) $$

we observe that to this order, the products of renormalization corrections by zero-loop order terms (like $(2\pi S)^{d/2}\, zf$) serve to renormalize the one-loop term. On the contrary, the products of renormalization corrections by one-loop terms would have to be taken into consideration only if the expansion was pursued to the order of two loops. Considering these remarks and eqns (13.2.67), (13.2.60), and (13.2.65), we deduce from them

$$\mathcal{Q}_R(f_R) = \mathfrak{f}_R(1+\eta) + \tfrac{1}{2}\tau_R \mathfrak{f}_R(1+2\eta) + \tfrac{1}{2}\tau_R \mathfrak{f}_R(zg^{-1}[\mathfrak{X}_0(z)]^{-d/2} - 1) + g\mathfrak{f}_R\tau_R^{-1}\,{}^1\mathfrak{z}_I(\tau_R).$$

Let us bring into this equality, the value of η given by (13.2.66); we obtain

$$\mathcal{Q}_R(f_R) = \mathfrak{f}_R[\mathfrak{X}_1(z)]^{-2} + \tfrac{1}{2}\tau_R \mathfrak{f}_R(2 - zg^{-1}[\mathfrak{X}_0(z)]^{-d/2}) + g\mathfrak{f}_R\tau_R^{-1}\,{}^1\mathfrak{z}_I(\tau_R). \tag{13.2.68}$$

The renormalization factors zg^{-1}, $\mathfrak{X}_0(z)$ and $\mathfrak{X}_1(z)$ are expanded to first-order in g. Of course, to eliminate the singularities in $1/\varepsilon$, we could content outselves with the terms in g/ε, and such an approximation is sufficient when calculating critical exponents. The situation is different here. In the renormalization terms, we must keep not only the terms proportional to g/ε but also the terms proportional to g. Equations (12.3.100), (12.3.96), and (12.3.97) give

$$zg^{-1} \simeq 1 + \frac{g}{\varepsilon}[8 + \varepsilon(-2 - 4\ln 2)]$$

$$\mathfrak{X}_0(z) \simeq 1 + \frac{g}{\varepsilon}(2 - \varepsilon)$$

$$\mathfrak{X}_1(z) \simeq 1 + \frac{g}{\varepsilon}\left(1 + \frac{\varepsilon}{2}\right)$$

and, consequently,

$$zg^{-1}[\mathfrak{X}_0(z)]^{-d/2} \simeq 1 + \frac{g}{\varepsilon}[4 + \varepsilon(1 - 4\ln 2)].$$

Now, let us replace the renormalization factors present in (13.2.68) by their expansions with respect to g and ε. We obtain

$$\mathcal{Q}_R(f_R) = \mathfrak{f}_R + \frac{1}{2}\tau_R\mathfrak{f}_R + g\mathfrak{f}_R\tau_R^{-1}\,{}^1\mathfrak{z}_I(\tau_R) - g\mathfrak{f}_R\left(\frac{2}{\varepsilon} + 1\right) - \frac{1}{2}\tau_R\mathfrak{f}_R\left(\frac{4}{\varepsilon} + 1 - 4\ln 2\right). \tag{13.2.69}$$

Now, in ${}^1\mathfrak{z}_I(z)$, let us separate the terms proportional to τ_R and τ_R^2, which diverge when $\varepsilon \to 0$; for this purpose, we set

$${}^1\mathfrak{z}_I(\tau_R) = \tau_R\,{}^1\mathfrak{z}_I'(0) + \tfrac{1}{2}\tau_R^2\,{}^1\mathfrak{z}_I''(0) + \mathfrak{z}^\vdash(\tau_R)$$

where

$$\mathfrak{z}^\vdash(\tau) = \frac{1}{2\Gamma(d/2)}\int_0^\infty \mathrm{d}y\, y^{-1+d/2}\left\{\tau\,{}^0\mathrm{h}(y) - \frac{\tau^2}{2}[{}^0\mathrm{h}(y)]^2 - \ln[1 + \tau\,{}^0\mathrm{h}(y)]\right\}$$

$${}^0\mathrm{h}(y) = 2\left(\frac{1}{y} - \frac{1}{y^2} + \frac{\mathrm{e}^{-y}}{y^2}\right) \tag{13.2.70}$$

Let us keep in ${}^1\mathfrak{z}'(0)$ and ${}^1\mathfrak{z}''(0)$ the terms proportional to $1/\varepsilon$ and the constant terms (when $\varepsilon \to 0$). Taking (13.2.62) into account, we obtain

$$ {}^1\mathfrak{z}_1(\tau) = \left(\frac{2}{\varepsilon} + 1\right)\tau + \left(\frac{2}{\varepsilon} + \tfrac{1}{2} - 2\ln 2\right)\tau^2 + \mathfrak{z}^{\vdash}(\tau). \qquad (13.2.71) $$

Let us bring this expression in (13.2.69). We get the simple result

$$ \mathcal{Q}_{\mathrm{R}}(f_{\mathrm{R}}) = \mathfrak{f}_{\mathrm{R}} + \frac{1}{2}\tau_{\mathrm{R}}\mathfrak{f}_{\mathrm{R}} + g\mathfrak{f}_{\mathrm{R}}\tau_{\mathrm{R}}^{-1}\,\mathfrak{z}^{\vdash}(\tau_{\mathrm{R}}), $$

and the other terms eliminate. We now observe that this expression has a finite limit when $\varepsilon \to 0$.

Thus, considering (13.2.65) and (13.2.63), we see that $\mathcal{Q}_{\mathrm{R}}(f_{\mathrm{R}})$ is given in implicit form by

$$ (2\pi)^{d/2}\, gX^d\, \mathcal{Q}_{\mathrm{R}}(f_{\mathrm{R}}) = \tau_{\mathrm{R}} + \tfrac{1}{2}\tau_{\mathrm{R}}^2 + g\,\mathfrak{z}^{\vdash}(\tau_{\mathrm{R}}) $$
$$ (2\pi)^{d/2}\, gX^d f_{\mathrm{R}} = \tau_{\mathrm{R}}\exp(\tau_{\mathrm{R}}) \qquad (13.2.72) $$

It is worthwhile to note that for all g and for all d, these equalities are compatible with the expansion

$$ \mathcal{Q}_{\mathrm{R}}(f_{\mathrm{R}}) = f_{\mathrm{R}} - \tfrac{1}{2}(2\pi)^{d/2}\, g f_{\mathrm{R}}^2 + \ldots. $$

which, by definition, is exact. (Actually, we wanted to ensure this compatibility, and this is why we kept not only the terms proportional to g/ε but also the terms proportional to g.) Equations (13.2.72) can thus be used for $d < 4$ and for any g. This is the *one-loop approximation* of $\mathcal{Q}_{\mathrm{R}}(f_{\mathrm{R}})$.

The corresponding function $\mathsf{Q}(y)$ is given by the coupled equations (13.2.42)

$$ (2\pi)^{d/2}\, g\mathsf{Q}(y) = \tau + \tfrac{1}{2}\tau^2 + g\,\mathfrak{z}^{\vdash}(\tau) $$
$$ (2\pi)^{d/2}\, gy = \tau\, \mathrm{e}^{\tau}. \qquad (13.2.73) $$

The scaling law which defines the dependence of the osmotic pressure with respect to the concentration is given by the function $\mathrm{F}(x)$. The relation (13.2.43) between $\mathrm{F}(x)$ and $\mathsf{Q}(y)$ can be expressed in the form

$$ x\,\mathrm{F}(x) = \mathsf{Q}(y) \qquad (13.2.74) $$

$$ x = y\,\mathsf{Q}'(y). \qquad (13.2.75) $$

It is possible to express x and $\mathrm{F}(x)$ directly in terms of τ by eliminating y. Actually, (13.2.75) can be written in the form

$$ x = \frac{\mathrm{d}}{\mathrm{d}\tau}\mathsf{Q}(y) \Big/ \left(\frac{1}{y}\frac{\mathrm{d}y}{\mathrm{d}\tau}\right) $$

where [see (13.2.73)]

$$ \frac{1}{y}\frac{\mathrm{d}y}{\mathrm{d}\tau} = 1 + \frac{1}{\tau}. $$

Using the two preceding equalities and (13.2.73), we can calculate x in terms of τ;

moreover, eqns (13.2.73) and (13.2.74) enable us to calculate $x\mathrm{F}(x)$ in terms of τ. We obtain the coupled equations

$$(2\pi)^{d/2}\, gx\, \mathrm{F}(x) = \tau + \tfrac{1}{2}\tau^2 + g\mathfrak{z}^{\vdash}(\tau)$$

$$(2\pi)^{d/2}\, gx = \tau + \frac{g\tau}{1+\tau}\frac{\mathrm{d}}{\mathrm{d}\tau}\mathfrak{z}^{\vdash}(\tau) \qquad (13.2.76)$$

where $\mathfrak{z}^{\vdash}(\tau)$ is given by (13.2.70).

To reach the Kuhnian limit, we must put $g = g^*$. For $d = 3$, we can choose the value $g^* = 0.233$ [see (13.1.16)]. Thus, by direct renormalization we recover results obtained (partially and in a different way) by Knoll, Schäfer, and Witten (1981)[45] and by Ohta and Nakanishi.[47]

The curve representing the values of $\mathrm{F}(x)$ given for $d = 3$ by (13.2.76), is drawn in Fig. 13.16. We may think that the approximation is not bad; however, it will be shown below that when $x \to \infty$ ($\tau \to \infty$), formulae (13.2.76) do not give the

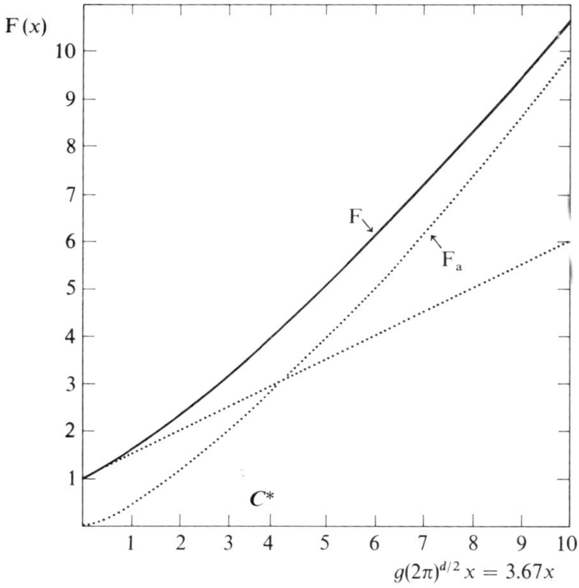

Fig. 13.16. Representation of the curve. $\Pi\beta/C = \mathrm{F}(CX^d)$ to one-loop order for $d = 3$ ($g^* = 0.233$). The curve corresponding to $\mathrm{F}(x)$ is the solid line. On the same graph, we see a dotted line representing the asymptotic function

$$\mathrm{F_a}(x) = \mathrm{F}_\infty\,[g(2\pi)^{d/2}x]^{1/(\nu d-1)} = 0.4855\,(3.67x)^{1.31}$$

[with $\nu = 0.588$ and $g = 0.233$; according to B. Duplantier (private communication); see (13.2.81)]. The slope at the origin is also indicated.

correct asymptotic behaviour for F(x) (even approximately); still, the result seems reasonable for $\tau < 15$.

We shall now study the asymptotic limit $x \to \infty$ ($y \to \infty$). Let us come back to ${}^1\mathfrak{z}_{\mathrm{I}}(\tau)$ given by (13.2.59). To find an expression of ${}^1\mathfrak{z}_{\mathrm{I}}(\tau)$ for large values of τ, it is convenient to write ${}^1\mathfrak{z}_{\mathrm{I}}(\tau)$ in the form

$$ {}^1\mathfrak{z}_{\mathrm{I}}(\tau) = \frac{\tau^{d/2}}{2\Gamma(d/2)} \int_0^\infty \mathrm{d}t\, t^{-1+d/2} \left[\frac{2}{t} - \ln\left(1 + \frac{2}{t} - \frac{2}{\tau t^2} + 2\frac{\mathrm{e}^{-\tau t}}{\tau t^2}\right)\right]. $$

Thus, for $\tau \gg 1$, we have

$$ {}^1\mathfrak{z}_{\mathrm{I}}(\tau) \simeq \frac{\tau^{d/2}}{2\Gamma(d/2)} \int_0^\infty \mathrm{d}t\, t^{-1+d/2} \left[\frac{2}{t} - \ln\left(1 + \frac{2}{t}\right)\right] \simeq 2^{1+d/2} \frac{(1-d/2)!}{d(d-2)} \tau^{d/2}. $$

It would not be very meaningful to use this formula directly, but we know that one-loop expansions are associated with expansions of order ε. Thus, for small values of ε, we can write

$$ {}^1\mathfrak{z}_{\mathrm{I}}(\tau) \simeq \frac{2\tau^2}{\varepsilon} - \tau^2 \ln \tau, \tag{13.2.77} $$

and it is clear that we must keep the term of order $\tau^2 \ln \tau$ since τ may be large. Thus, taking (13.2.71) and (13.2.77) into account, we see that, to order ε, we can write

$$ \mathfrak{z}^{\vdash}(\tau) \simeq -\tau^2 \ln \tau, $$

Let us bring back this value in (13.2.73); we find

$$ (2\pi)^{d/2}\, g\mathrm{Q}(y) \simeq \tau + \frac{\tau^2}{2} - g\tau^2 \ln \tau, \tag{13.2.78} $$

which can be exponentialized as follows

$$ (2\pi)^{d/2}\, g\mathrm{Q}(y) \simeq \tfrac{1}{2}\tau^{2-2g}. \tag{13.2.79} $$

Moreover, for $\tau \gg 1$, (13.2.73) gives

$$ \tau \simeq \ln y $$

and, for small values of ε, we have $g^* = \varepsilon/8$ [according to (13.3.102)]. We find

$$ \mathrm{Q}(y) \simeq \frac{1}{\pi^2 \varepsilon} (\ln y)^{2-\varepsilon/4}. \tag{13.2.80} $$

Thus, as we also have, to order ε,

$$ \nu d = 2 - \varepsilon/4 $$

according to (12.3.106), formula (12.2.80) is in agreement with the general formula (13.2.45).

As was seen in Section 2.5.2, the behaviour of F(x) when $x \to \infty$ is related to the behaviour of Q(y) when $y \to \infty$. In particular, we note that for small ε, by comparing (13.2.80) and (13.2.45), one finds

$$\mathrm{Q}_\infty = \frac{1}{\pi^2 \varepsilon}.$$

By bringing this result in (13.2.47), we obtain

$$\mathrm{F}_\infty = \frac{\pi^2 \varepsilon}{4}$$

which, to first-order with respect to ε, can also be expressed as follows

$$\mathrm{F}_\infty = \tfrac{1}{2}(2\pi)^{d/2}\, g^*. \tag{13.2.81}$$

Thus, we find that for small ε, F_∞ coincides with the second virial coefficient in agreement with (13.2.35). A better result was recently found by B. Duplantier, who showed that, for $x \gg 1$, $\mathrm{F}(x)$ can be written in the form

$$\mathrm{F}(x) = \bar{\mathrm{F}}_\infty\, [(2\pi)^{d/2}\, g^* x]^{1/(\nu d - 1)}$$

$$\bar{\mathrm{F}}_\infty = \frac{1}{2} + \frac{\varepsilon}{8}(\mathfrak{E} - \ln 2) \qquad (\mathfrak{E} = 0.577 = \text{Euler's constant} \tag{13.2.82}$$

i.e.

$$\bar{\mathrm{F}}_\infty = 0.4855.$$

Finally, we note that eqns (13.2.76) which are aymptotically correct for $g \ln \tau \leqslant 1$ only, could be 'corrected' so as to give (13.2.79); thus, for $\tau \to \infty$, and $g \ll 1$, $\mathrm{F}(x)$ would be proportional to a power of x with an exponent correct to order ε.

2.5.6 Semi-dilute limit and field theory

We showed that, in the semi-dilute limit we must have

$$\Pi\beta = \mathrm{F}_\infty\, \mathscr{C}_{\mathrm{k}}^{d/(d-1/\nu)} \tag{13.2.83}$$

where $\mathscr{C}_{\mathrm{k}}$ is proportional to the number C of monomers per unit volume. This formula was obtained by a simple and direct argument. However, the result was first formulated in the framework of field theory by studying the properties of *equilibrium ensembles* which are polydisperse and were defined in Chapter 11, Section 5.3. This initial approach had a heuristic value but lacks firm foundations. More recently, Knoll *et al.*[45] endeavoured to present this question in a more rigorous manner, but their line of argument holds only if one admits, more or less implicitly, the existence of a finite field theory, in the limit corresponding to the semi-dilute case. Now, we note that the naïve approach, which we described in Section 2.5.1, postulates the existence of a finite theory at $\mathscr{C}_{\mathrm{k}}$ fixed when the size $X \to \infty$. Thus, both approaches rely on very similar postulates.

Actually, a true justification of (13.2.83) could be obtained only through an elaborate study of the structure of the diagrams and of their contributions, in the semi-dilute limit. Such a study has not been made yet. In what follows, we shall only explain how, historically, field theory enabled us to discover, for the osmotic pressure, the scaling laws which we have written above.

In Chapter 11, we showed that for an *equilibrium ensemble*, the osmotic pressure $\boldsymbol{\Pi}$, the polymer concentration $\boldsymbol{C}$ and the area concentration $\mathscr{C}$ are connected by the following relations [see (11.5.43)]

$$\boldsymbol{\Pi}\boldsymbol{\beta} = M\frac{\partial \boldsymbol{\Gamma}(M)}{\partial M} - \boldsymbol{\Gamma}(M)$$

$$\boldsymbol{C} = \frac{1}{2}M\frac{\partial \boldsymbol{\Gamma}(M)}{\partial M}$$

$$\mathscr{C} = \frac{\partial \boldsymbol{\Gamma}(M)}{\partial a}. \tag{13.2.84}$$

In this formula, $\boldsymbol{\Gamma}(M)$ is the generating function of the vertex functions (in the limit $n = 0$). It depends on parameters a and M or more precisely on $(a - a_c)$ and M because the short distance cut-off can be absorbed in the constant a_c. Moreover, we note that the magnetic field B is related to $\boldsymbol{\Gamma}(M)$ by the equality [see (11.5.14)]

$$B = \frac{\partial \boldsymbol{\Gamma}(M)}{\partial M}.$$

In the limit $a \to a_c$, the vertex functions diverge, but the divergences can be absorbed by renormalization. Thus, in agreement with (12.3.3), we set

$$\bar{\Gamma}_R(\vec{k}, -\vec{k}) = (\mathfrak{Z}_1)^2\, \bar{\Gamma}(\vec{k}, -\vec{k}),$$

and in a similar way for an N-leg vertex function

$$\bar{\Gamma}_R^{(N)}(\ldots) = (\mathfrak{Z}_1)^N\, \bar{\Gamma}^{(N)}(\ldots),$$

where $\mathfrak{Z}_1$ is a dimensionless renormalization factor. Now, let $\boldsymbol{\Gamma}_R(M)$ be the generating function of the renormalized vertex functions at zero wave vectors; we have

$$\boldsymbol{\Gamma}(M) = \boldsymbol{\Gamma}^R(\mathfrak{Z}_1^{-1}M)$$

Let us now consider these renormalized vertex functions. In the critical domain, they depend only on the correlation length ξ. The nature of this dependence can be found by examining the dimension equations of various quantities. The *normal* dimensions of $\varphi(\vec{r})$ and of its Fourier transform $\tilde{\varphi}(\vec{k})$ are given by

$$\varphi(\vec{r}) \sim L^{1-d/2}$$

$$\tilde{\varphi}(\vec{k}) \sim L^{1+d/2}$$

and this comes from the fact that the integral $\int d^d r\, \mathscr{L}(\vec{r})$ must be a pure number. Consequently, for the Green's functions, we have

$$\bar{\mathscr{G}}^{(N)}(\ldots) \sim L^{-d+N(1+d/2)},$$

and for the vertex functions

$$\bar{\Gamma}^{(N)}(\ldots) \sim L^{-d+N(-1+d/2)}.$$

Thus, the vertex functions at zero wave vectors are of the form

$$\bar{\Gamma}_{\mathrm{R}}^{(N)}(\vec{0}, \ldots, \vec{0}) \propto \xi^{-d+N(-1+d/2)}. \tag{13.2.85}$$

This entails the equality

$$\Gamma(M) = \xi^{-d}\, \Upsilon(M \mathfrak{Z}_1^{-1} \xi^{-1+d/2}). \tag{13.2.86}$$

Then, equations (13.2.84) give

$$\boldsymbol{\Pi\beta} = \xi^{-d}[x\Upsilon'(x) - \Upsilon(x)]$$

$$\boldsymbol{C} = \frac{\xi^{-d}}{2} x\, \Upsilon'(x)$$

$$\mathscr{C} = \xi^{-d}\left(x\Upsilon'(x)\left(-\frac{1}{\mathfrak{Z}_1}\frac{\mathrm{d}\mathfrak{Z}_1}{\mathrm{d}a} + \left(\frac{d}{2} - 1\right)\frac{1}{\xi}\frac{\mathrm{d}\xi}{\mathrm{d}a}\right) - d\Upsilon(x)\,\frac{1}{\xi}\frac{\mathrm{d}\xi}{\mathrm{d}a}\right). \tag{13.2.87}$$

Moreover, when $a \to a_{\mathrm{c}}$, according to (12.3.16) and (12.3.13), we have

$$a_{\mathrm{N}} \propto (a - a_{\mathrm{c}})^{\gamma}$$

$$\xi \propto (a - a_{\mathrm{c}})^{-\nu}$$

$$\mathfrak{Z}_1^{-1} = \xi(a_{\mathrm{N}})^{1/2} \propto (a - a_{\mathrm{c}})^{(\gamma/2)-\nu}. \tag{13.2.88}$$

Consequently,

$$-\frac{1}{\mathfrak{Z}_1}\frac{\mathrm{d}\mathfrak{Z}_1}{\mathrm{d}a} + \left(\frac{d}{2} - 1\right)\frac{1}{\xi}\frac{\mathrm{d}\xi}{\mathrm{d}a} \simeq \left(\frac{\gamma - \nu d}{2}\right)\frac{1}{a - a_{\mathrm{c}}}.$$

It is convenient to re-express everything in terms of ξ. We set

$$\xi = l[l^2(a - a_{\mathrm{c}})]^{-\nu} \tag{13.2.89}$$

where l is a length which could be expressed in terms of the interaction b (and eventually of the cut-off). In this way, we obtain

$$\mathscr{C} = l^{2-1/\nu}\, \xi^{-d+1/\nu}\left[\frac{(\gamma - \nu d)}{2} x\, \Upsilon'(x) + \nu d\,\Upsilon(x)\right] \tag{13.2.90}$$

and for the average area S

$$S = \mathscr{C}/\boldsymbol{C} = l^2(\xi/l)^{1/\nu}[\gamma - \nu d + 2\nu d\,\Upsilon(x)/x\,\Upsilon'(x)]. \tag{13.2.91}$$

Thus, for the osmotic pressure we obtain a scaling law rather similar to (13.2.29). This law is given in parametric form by

$$\boldsymbol{\Pi\beta}/\boldsymbol{C} = [x\,\Upsilon'(x) - \Upsilon(x)]/x\,\Upsilon'(x)$$

$$\boldsymbol{C}S^{\nu d} = l^{(2\nu-1)d}\,\frac{[(\gamma - \nu d)x\,\Upsilon'(x) + 2\nu d\,\Upsilon(x)]^{\nu d}}{2[x\Upsilon'(x)]^{\nu d-1}}. \tag{13.2.92}$$

Let us now consider the semi-dilute limit. This limit corresponds, in the magnetic representation, to the vicinity of the coexistence curve. As according to (11.5.14)

$$B = \frac{\partial \Gamma(M)}{\partial M}$$

the coexistence curve is given by the condition that, for a fixed magnetization, the magnetic field vanishes.

$$\left.\frac{\partial \Gamma(M)}{\partial M}\right|_{M = M_0} = 0. \qquad (13.2.93)$$

In a similar way, we admit the existence of a value x_0 (which can be obtained by perturbation) for which

$$Y'(x_0) = 0 \qquad (\text{for } n = 0). \qquad (13.2.94)$$

The polymer concentration vanishes ($\boldsymbol{C} \to 0$ but $\mathscr{C}$ is fixed). Moreover, we obtain

$$\boldsymbol{\Pi\beta} = -\xi^{-d}\,\Upsilon(x_0)$$
$$\mathscr{C} = l^{2-1/\nu}\,\xi^{-d+1/\nu}\,\nu d\,\Upsilon(x_0). \qquad (13.2.95)$$

Thus eliminating ξ between these two equations, we find

$$\boldsymbol{\Pi\beta} \propto l^{-d}(\mathscr{C}\,l^{d-2})^{d/(d-1/\nu)} \qquad (13.2.96)$$

in good agreement with (13.2.83). Let us note, however, that in (13.2.83), $\boldsymbol{\Pi\beta}$ is expressed in terms of $\mathscr{C}_k$, and not of $\mathscr{C}$, and that , in the Kuhnian limit, $\mathscr{C}_k$ is finite and represents the monomer density whereas $\mathscr{C}$ becomes infinite and meaningless.

2.6 Structure function and screening length

2.6.1 Definitions

The structure function of a homogeneous solution is related to the density–density fluctuations of monomers in the solution. However, the definition of the concentration is model-dependent. On a lattice, the monomer concentration is the number φ of monomer per site. For the standard continuous model, it is expressed as an area per unit volume, which is denoted by $\mathscr{C}$. For a Kuhnian chain, the quantity $\mathscr{C}_k = \boldsymbol{C}X^{1/\nu}$ represents the monomer concentration. However, the definition of the structure function should not really depend on the model under consideration, and therefore we shall define this quantity in an intrinsic manner.

In the following, the structure function is defined in the framework of the standard continuous model but generalizing to other models is trivial. For the standard continuous model, the area concentration $\mathscr{C}(\vec{r})$ plays the role of monomer concentration and we have

$$\mathscr{C}(\vec{r}) = \sum_a \int_0^{S_a} \mathrm{d}s\, \delta[\vec{r} - \vec{r}_a(s)]. \qquad (13.2.97)$$

If the point of position vector $\vec{r}$ belongs to a large volume, the average is a constant

$$\mathscr{C} = \boldsymbol{C}S = \langle\!\langle \mathscr{C}(\vec{r}\,') \rangle\!\rangle .$$

The average $\langle\!\langle \mathscr{C}(\vec{r}\,')\mathscr{C}(\vec{r}\,'') \rangle\!\rangle$ is a function of the difference $(\vec{r}\,'' - \vec{r}\,')$.

The structure function $\boldsymbol{H}(\vec{q})$ can then be expressed in the form

$$\boldsymbol{H}(\vec{q}) = \frac{1}{\mathscr{C}^2} \int \mathrm{d}^d r\, \mathrm{e}^{i\vec{q}\cdot\vec{r}} [\langle\!\langle \mathscr{C}(\vec{r})\mathscr{C}(\vec{0}) \rangle\!\rangle - \mathscr{C}^2]$$

in agreement with the definition given in Chapter 7. This structure function depends only on the ratio $\mathscr{C}(\vec{r})/\mathscr{C}$, a quantity which in practice, does not depend on the model we choose. The averages are taken over a *grand ensemble* of chains. Thus $\boldsymbol{H}(\vec{q})$ is the Fourier transform of a function of $\vec{r}$ which goes to zero when $r \to \infty$. The preceding definition is valid for a polydisperse ensemble but, in the following, we study the monodisperse case more particularly.

The structure function $\boldsymbol{H}(\vec{q})$ plays a very important role because it is directly measurable by radiation scattering, as was indicated in Chapter 7. We shall now examine its analytical properties, and this study will enable us to give a rigorous definition of the screening length ξ_e.

First, let us consider the elementary situation where the *grand ensemble* is made of independent Brownian chains ($b = 0$ in the standard continuous model). Then, it is conspicuous that the only correlations that exist and contribute to $\boldsymbol{H}(\vec{q})$ are those corresponding to points on the same chain. In this case, for a monodisperse ensemble, we have

$$\boldsymbol{H}(\vec{q}) = (\boldsymbol{C}^{-1})\, {}^0H(\vec{q}) = (\boldsymbol{C}^{-1})\, {}^0\mathrm{h}(q^2S/2) \qquad (13.2.98)$$

where ${}^0H(\vec{q})$ and ${}^0\mathrm{h}(x)$ are given by (13.1.151)

$${}^0H(\vec{q}) = \frac{8}{q^4S^2}(e^{-q^2S/2} - 1 + q^2S/2).$$

The correlations extend over distances which are of the order of $S^{1/2}$; they do not depend on concentration. In particular, let us consider the limit $S \to \infty$

$${}^0H_\infty(\vec{q}) \sim \frac{4}{q^2S}$$

$${}^0\boldsymbol{H}_\infty(\vec{q}) = \frac{4}{\boldsymbol{C}q^2S}. \qquad (13.2.99)$$

This function is singular at the origin. This implies that the density–density correlations decrease when $r \to \infty$, more slowly than exponentially. For instance, for $d = 3$, we have

$$\langle\!\langle \mathscr{C}(\vec{r})\mathscr{C}(\vec{0}) - \mathscr{C}^2 \rangle\!\rangle = \frac{4\mathscr{C}}{(2\pi)^3} \int \mathrm{d}^3 q \frac{\mathrm{e}^{-i\vec{q}\cdot\vec{r}}}{q^2} = \frac{\mathscr{C}}{\pi r}. \tag{13.2.100}$$

In this case, the correlation length is infinite.

The situation is completely different when the chains repel one another. In this case, the correlations are screened and the function $\langle\!\langle \mathscr{C}(\vec{q})\mathscr{C}(\vec{0}) - \mathscr{C}^2 \rangle\!\rangle$ decrease exponentially with r. More precisely, it is possible to find a non-empty set of lengths L (large enough) such that

$$\lim_{r\to\infty} \mathrm{e}^{r/L} \langle\!\langle \mathscr{C}(\vec{r})\mathscr{C}(\vec{0}) - \mathscr{C}^2 \rangle\!\rangle = 0, \tag{13.2.101}$$

and the screening length ξ_e is defined by the condition

$$\xi_\mathrm{e} = \min L. \tag{13.2.102}$$

This screening implies that $\boldsymbol{H}(\vec{q})$ has special analyticity properties, and the converse is true. These properties are as follows: The quantity $\boldsymbol{H}(\vec{q})$ is an analytic function of the complex vector $\vec{q} = \vec{q}' + i\vec{q}''$ in a d-dimensional tube, defined by the inequality $|\vec{q}''| < 1/\xi_\mathrm{e}$. The nearest singularity to real reciprocal space ($\vec{q}'' = 0$) appears on the 'surface' of the tube. Thus, the screening length is directly related to the analytical properties of $\boldsymbol{H}(\vec{q})$. It is clear that, in the standard continuous model for fixed values of b and $\mathscr{C}$, ξ_e is a function of the chain length; however, if $S \to \infty$, which corresponds to the semi-dilute limit, the length ξ_e must have a finite limit.

Let us finally remark that the screening length ξ_e can be defined even in cross-over domains. In good solvents (Kuhnian chains), the length ξ_e and the length $\xi_\mathrm{k} = \mathscr{C}_\mathrm{k}^{-\nu/(\nu d-1)}$ which measures the "distance" between Kuhnian chains, are proportional. However, in absence of interaction (Brownian chains), ξ_e is infinite whereas the length $\xi_0 = \mathscr{C}^{-1/(d-2)}$ which measures the "distance" between Brownian chains, remains finite. In the intermediate cross-over range, ξ_e is finite but naturally larger than ξ_0.

In the preceding, we assumed implicitly that we were dealing with a monodisperse system, but if the chains form a (moderately) polydisperse ensemble the same considerations apply. Then, the structure function $\boldsymbol{H}(\vec{q})$ is defined on the average. Thus, for Brownian chains, (13.2.98) is replaced by

$${}^0\boldsymbol{H}(\vec{q}) = \langle A^2\, {}^0\mathrm{h}(q^2 A/2)\rangle / \langle A^2 \rangle \boldsymbol{C}$$

where ${}^0\mathrm{h}(x)$ is given by (13.1.151).

In particular for a Schulz–Zimm distribution, we find

$${}^0\boldsymbol{H}(\vec{q}) = {}^0\mathrm{h}\left(\frac{q^2 S}{2}, p\right) \Big/ \boldsymbol{C} \qquad (S = \langle A \rangle)$$

$${}^0\mathrm{h}(x, p) = \frac{2}{(1+p)x^2}[(1+px)^{-1/p} - 1 + x]. \tag{13.2.103}$$

Thus, for $x = \dfrac{q^2 S}{2} \gg 1$, we obtain

$$ {}^{0}\boldsymbol{H}(\vec{q}) \simeq \frac{4}{C q^2 S} \tag{13.2.104} $$

which coincide with (13.2.99).

2.6.2 Diagram representation of the structure function

The structure function $\boldsymbol{H}(\vec{q})$ is defined for the standard continuous model by the equality

$$ \boldsymbol{H}(\vec{q}) = \frac{1}{\mathscr{C}^2} \int \mathrm{d}^d r\, \mathrm{e}^{i\vec{q}\cdot\vec{r}} [\langle\!\langle \mathscr{C}(\vec{r})\mathscr{C}(\vec{0}) - \mathscr{C}^2 \rangle\!\rangle] \tag{13.2.105} $$

and to evaluate the averages, it is convenient to use the grand-ensemble formalism described in Section 2.3 (and also in Chapter 7). The quantity $\mathscr{C}^2 \boldsymbol{H}(\vec{q})$ can be expanded in powers of the interaction, and this expansion is represented by connected diagrams (see Fig. 13.17) of the same type as those introduced in Chapter 9. More precisely, $\boldsymbol{H}(\vec{q})$ is given in terms of the polymer concentrations $C(S)$ in parametric form, the parameters being the fugacities $f(s)$ already introduced in Chapter 9.

Thus in agreement with (10.6.8), we write

$$ \begin{aligned} C(S) &= f(S)\frac{\partial}{\partial f(S)} \mathscr{Q}\{f\} \equiv \\ &\equiv f(S)\frac{\partial}{\partial f(S)} \sum_{N=1}^{\infty} \frac{1}{N!} \int \mathrm{d}S_1 \ldots \mathrm{d}S_N\, f(S_1) \ldots f(S_N) \mathscr{Z}(S_1, \ldots, S_N) \\ \mathscr{C}^2 \boldsymbol{H}(\vec{q}) &= \mathscr{Q}(\textstyle\int; \vec{q}, -\vec{q}; \{f\}) \\ &\equiv \sum_{N=1} \frac{1}{N!} \int \mathrm{d}S_1 \ldots \mathrm{d}S_N\, f(S_1) \ldots f(S_N) \overline{\mathscr{Z}}(\textstyle\int; \vec{q}, -\vec{q}; S_1, \ldots, S_N). \end{aligned} \tag{13.2.106} $$

Here $\mathscr{Z}(S_1, \ldots, S_N)$ is the sum of the contributions of the connected diagrams

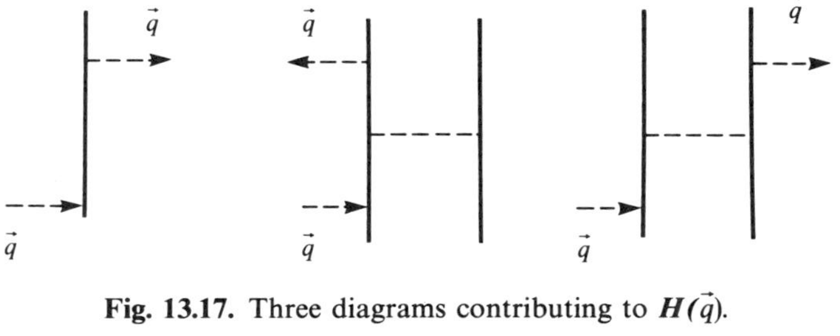

Fig. 13.17. Three diagrams contributing to $\boldsymbol{H}(\vec{q})$.

associated with N polymers of areas $S_1, \ldots, S_N$. On the other hand, the symbol $\overline{\mathscr{Z}}(\int;\vec{q}, -\vec{q}; S_1, \ldots, S_N)$ represents the sum of the connected diagrams (for N polymers of areas $S_1, \ldots, S_N$) through which a wave vector $\vec{q}$ flows from one point belonging to a polymer, to another point belonging either to the same polymer or to another one. The anchoring points at which the vectors $\vec{q}$ and $-\vec{q}$ are injected in the diagram are at random. The contributions of the diagrams are obtained by summing over all positions of these points on the polymers. For instance, referring to the general notation introduced in Chapter 10, we can write

$$\overline{\mathscr{Z}}(\int;\vec{q}, -\vec{q};S) = \int_0^S \mathrm{d}s' \int_0^S \mathrm{d}s''\, \overline{\mathscr{Z}}(\vec{q}, -\vec{q}; s'; s''; S)$$

and also

$$\overline{\mathscr{Z}}(\int;\vec{0},\vec{0};S_1, \ldots, S_N) = \left(\sum_{i=1}^{N} S_i\right)^2 \mathscr{Z}(S_1, \ldots, S_N). \tag{13.2.107}$$

Moreover, $\mathscr{Q}(\int;\vec{q}, -\vec{q}; \{f\})$ is nothing but the grand partition function associated with the $\overline{\mathscr{Z}}(\int;\vec{q}, -\vec{q}; S_1, \ldots, S_N)$.

These formulae become slightly simpler in the monodisperse case. Then, we have

$$\begin{aligned} C &= f\frac{\partial}{\partial f}\mathscr{Q}(f) = f\frac{\partial}{\partial f}\sum_{N=1}^{\infty}\frac{f^N}{N!}\mathscr{Z}(N \times S), \\ \mathscr{C}^2 H(\vec{q}) &= \mathscr{Q}(\int;\vec{q}, -\vec{q}) = \sum_{N=1}^{\infty}\frac{f^N}{N!}\overline{\mathscr{Z}}(\int;\vec{q}, -\vec{q};N \times S), \end{aligned} \tag{13.2.108}$$

and in this case $\mathscr{C} = CS$. Moreover, (13.2.107) reads

$$\overline{\mathscr{Z}}(\int;\vec{0},\vec{0};N \times S) = N^2S^2\, \mathscr{Z}(N \times S).$$

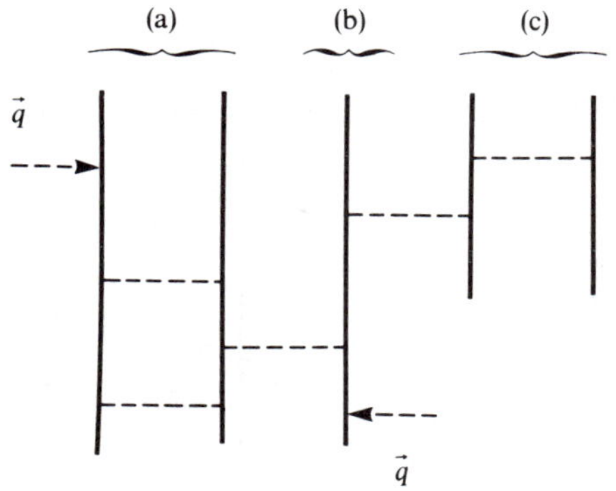

Fig. 13.18. Diagram contributing to $\overline{\mathscr{Z}}(\int;\vec{q}, -\vec{q};4 \times S)$. The main axis is made of two I-irreducible sub-diagrams (a and b) and the sub-diagram (c) is a simple-tree subdiagram.

By bringing these quantities in the expression of $\boldsymbol{H}(\vec{0})$, we find

$$\boldsymbol{H}(\vec{0}) = \boldsymbol{C}^{-2}\left(f\frac{\partial}{\partial f}\right)^2 \mathscr{Z}(f) \tag{13.2.109}$$

and since

$$\boldsymbol{\Pi}\boldsymbol{\beta} = \mathscr{Z}(f) \qquad \boldsymbol{C} = f\frac{\partial}{\partial f}\mathscr{Z}(f)$$

we recover the well-known relation (see Chapter 9, Section 3.8)

$$\boldsymbol{H}(\vec{0}) = \frac{1}{\boldsymbol{C}}\frac{\partial \boldsymbol{C}}{\partial(\boldsymbol{\Pi}\boldsymbol{\beta})}. \tag{13.2.110}$$

However, let us recall that unfortunately this relation cannot be generalized in the polydisperse case. Then, there is no simple relation between $\boldsymbol{H}(\vec{0})$ and the osmotic compressibility.

2.6.3 Calculation of the structure function and of the screening length in the context of the simple-tree approximation

In order to show the nature of the screening effect, we begin with a calculation of $\boldsymbol{H}(\vec{q})$ performed by using the simple-tree approximation (also studied in Chapter 10, Section 6.2). The shape of the diagrams contributing to $\boldsymbol{C}$ and $\boldsymbol{H}(\vec{q})$ is shown on Figs 13.19 and 13.20. We see immediately that the sum of the diagrams contributing to $\mathscr{C} = \boldsymbol{C}S$ can be calculated in a self-consistent way, which gives

$$\mathscr{C} = Sf\sum_{p=0}^{\infty}\frac{(b\mathscr{C}S)^p}{p!} = Sf\exp(b\mathscr{C}S). \tag{13.2.111}$$

In the same way, we see that the diagrams contributing to the approximate value ${}^0\boldsymbol{H}(\vec{q})$ of $\boldsymbol{H}(\vec{q})$ consist of a series of polymers supporting the wave vector $\vec{q}$ (axis of the diagram), and of side branches (see Fig. 13.20). We easily see that the factor corresponding to all the side branches attached to a polymer belonging to

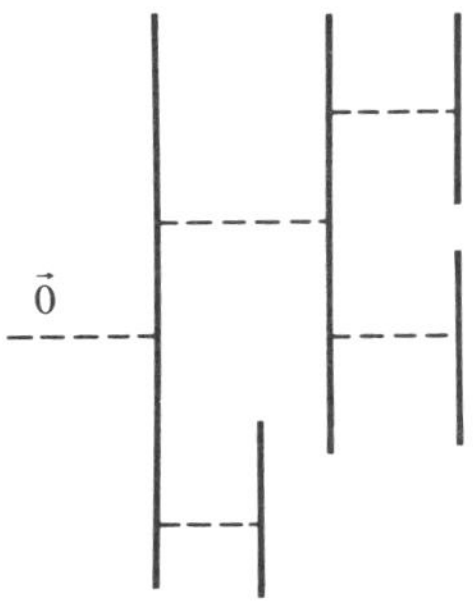

Fig. 13.19. Simple-tree diagram contributing to $\boldsymbol{C}$.

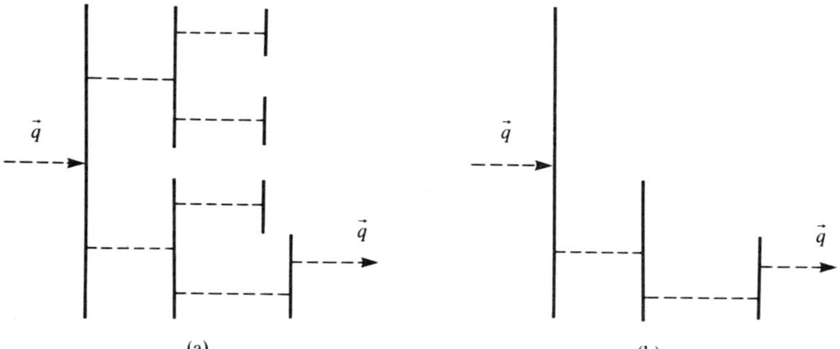

Fig. 13.20. (a) Simple-tree diagram contributing to $H(\vec{q})$. (b) Main axis of the diagram (the side branch which appears on (a) has been removed).

the axis is equal to

$$\sum_{p=0}^{\infty} \frac{(b\mathscr{C}S)^p}{p!} = \exp(b\mathscr{C}S).$$

Moreover, with each polymer we must associate the factor $f\,{}^0H(\vec{q})$ [see (13.1.151)] and with each interaction line, the factor—bS^2. Finally (in agreement with (13.2.111)), we find

$$(\mathscr{C}^2)\,{}^0\boldsymbol{H}(\vec{q}) = (fS^2)\,{}^0H(\vec{q})\exp(b\mathscr{C}S)\sum_{p=0}^{\infty}[-(bS^2f)\,{}^0H(\vec{q})\exp(b\mathscr{C}S)]^p$$
$$= \frac{fS^2\exp(b\mathscr{C}S)\,{}^0H(\vec{q})}{1 + bS^2f\,{}^0H(\vec{q})\exp(b\mathscr{C}S)}$$

and by taking (13.2.111) into account

$$\mathscr{C}^2\,{}^0\boldsymbol{H}(\vec{q}) = \frac{\mathscr{C}S\,{}^0H(\vec{q})}{1 + b\mathscr{C}S\,{}^0H(\vec{q})}. \tag{13.2.112}$$

This expression can be written as the sum of the intramolecular and intermolecular functions [see (7.2.52)]

$${}^0\boldsymbol{H}(\vec{q}) = {}^0\boldsymbol{H}^{\mathrm{I}}(\vec{q}) + {}^0\boldsymbol{H}^{\mathrm{II}}(\vec{q}).$$

The intramolecular structure function is obtain by letting $b\mathscr{C}S$ become very small in (13.2.112); we thus find

$${}^0\boldsymbol{H}^{\mathrm{I}}(\vec{q}) = \frac{S}{\mathscr{C}}\,{}^0H(\vec{q}), \tag{13.2.113}$$

a formula which, in the context of the simple-tree approximation, is valid for any concentration. Consequently,

$$ {}^0\boldsymbol{H}^{\mathrm{II}}(\vec{q}) = -\frac{bS^2[{}^0H(\vec{q})]^2}{1 + b\mathscr{C}S\,{}^0H(\vec{q})}. \tag{13.2.114}$$

The result (13.2.112) can also be written in Zimm's form [see (13.1.151)]

$$\frac{1}{{}^0\boldsymbol{H}(\vec{q})} = \frac{\boldsymbol{C}}{{}^0H(\vec{q})} + b\mathscr{C}^2 = \frac{\boldsymbol{C}}{{}^0\mathrm{h}(q^2S/2)} + b\boldsymbol{C}^2S^2 \tag{13.2.115}$$

where

$${}^0\mathrm{h}(x) = \frac{2}{x^2}(\mathrm{e}^{-x} - 1 + x).$$

We note that for $b\mathscr{C}S \gg 1$ ($\mathscr{C}$ remaining finite; semi-dilute domain)

$$ {}^0\boldsymbol{H}^{\mathrm{II}}(\vec{q}) \simeq -\frac{S}{\mathscr{C}}\,{}^0H(\vec{q}) = -\,{}^0\boldsymbol{H}^{\mathrm{I}}(\vec{q}). \tag{13.2.116}$$

In this case, the contributions ${}^0\boldsymbol{H}^{\mathrm{I}}(\vec{q})$ and ${}^0\boldsymbol{H}^{\mathrm{II}}(\vec{q}\ )$ become nearly equal with opposite signs, and very large; their difference remains finite

$${}^0\boldsymbol{H}(\vec{q}) \simeq 1/b\mathscr{C}^2.$$

The fact that ${}^0\boldsymbol{H}(\vec{q})$ does not increase like ${}^0\boldsymbol{H}^{\mathrm{I}}(\vec{q})$ is related to the screening effect. As we shall see, this effect simply amounts to a tendency of the system towards homogeneity and we know that for a really homogeneous system $\boldsymbol{H}(\vec{q}) = 0$.

Now, for real positive values of q^2, ${}^0H(\vec{q})$ remains always positive as can be easily seen from (13.1.150) and ${}^0\boldsymbol{H}(\vec{q})$ is bounded without singularity for real $\vec{q}$. In particular, in the 'extreme' semi-dilute case when q^2S is very large, we find

$$ {}^0\boldsymbol{H}(\vec{q}) = \frac{4}{\mathscr{C}(q^2 + 4\mathscr{C}b)}. \tag{13.2.117}$$

Thus contrary to ${}^0H(\vec{q})$, ${}^0\boldsymbol{H}(\vec{q})$ is an analytic function of the vector $\vec{q} = \vec{q}\,' + i\vec{q}\,''$ in a tube of the complex space, defined by the equality $|\vec{q}\,''| < q''_{\mathrm{e}}$. Thus, the screening length, defined by (13.2.102) is given by $\xi_{\mathrm{e}} = 1/q''_{\mathrm{e}}$, as a consequence of general theorems concerning Fourier transforms. In the limit $S \to \infty$ ('extreme' semi-dilute limit) ${}^0\boldsymbol{H}(\vec{q})$ is given by (13.2.117) and we immediately see that, in this case, the screening length is

$$\xi_{\mathrm{e}} = \tfrac{1}{2}(\mathscr{C}b)^{-1/2}. \tag{13.2.118}$$

Thus, we recover with different notation a result obtained by S.F. Edwards in 1966[37] (still the results differ by a numerical factor because Edwards does not define the screening length exactly as we do).

Our definition (13.2.102) of the screening length can be extended to the 'normal' semi-dilute 'case' (S finite, $\mathscr{C}S^{-1+d/2} \gg 1$). However, one must be aware of the fact that, in this case, there is no complete screening. In other words, as long as S remains finite, the real screening length is proportional to $S^{1/2}$ as will be shown now.

In the complex $\vec{q}$ vector space, the location of the singularity nearest to real space is given by a vector $\vec{q}_e = \vec{q}'_e + i\vec{q}''_e$. Let us set $q^2S/2 = y = y' + iy''$. The number y is a solution of the equation

$$^0\mathrm{h}(y) = \frac{2}{y^2}(\mathrm{e}^{-y} - 1 + y) = -\frac{1}{\mathscr{C}bS} \tag{13.2.119}$$

and since $^0\mathrm{h}(x)$ is always positive when x is real, y is *always complex*. Moreover, $\vec{q}'_e$ and $\vec{q}''_e$ obey the equations

$$S[\vec{q}'_e)^2 - (\vec{q}''_e)^2] = 2y'$$

$$S\vec{q}'_e \cdot \vec{q}''_e = y''$$

and are the solutions for which $\vec{q}''_e$ is minimal. In this case, $\vec{q}'_e$ and $\vec{q}''_e$ must be collinear and, accordingly, we find

$$(\vec{q}''_e)^2 = S^{-1}[(y'^2 + y''^2)^{1/2} - y'] \tag{13.2.120}$$

from which we get the value of the *real* screening length namely $\xi_{er} = 1/\vec{q}''_e$.

Let us come back to eqn (13.2.119); we observe that we always have $y' < 0$. In fact, we easily verify that otherwise the equation

$$\mathrm{Re}\left\{\mathrm{e}^{-y} - 1 + y + \frac{1}{2\mathscr{C}bS}y^2\right\} = 0$$

cannot be satisfied. This means that if y' becomes very negative, the term e^{-y} becomes very large and cannot be neglected. Let us note, however, that this is exactly what is done implicitly when one goes to the limit $q^2S/2 \gg 1$ which leads to (13.2.117); in fact, in this limit, eqn(13.2.119) reduces to

$$\frac{2}{y} = -\frac{1}{\mathscr{C}bS}.$$

When $q^2S/2$ remains finite, the situation is different. Actually, it is easy to see that, when $\mathscr{C}bS \to \infty$, the solution of (13.2.119) which defines the useful singularity, has a finite limit which is a solution of the equation

$$\mathrm{e}^{-y} - 1 + y = 0 \tag{13.2.121}$$

which can be written in the form

$$\mathrm{e}^{-y'}\cos y'' - 1 + y' = 0$$

$$\mathrm{e}^{-y'}\sin y'' - y'' = 0.$$

Thus, we find that (13.2.120) reads

$$(\vec{q}''_e)^2 S = (\mathrm{e}^{-2y'} - 1 + 2y')^{1/2} - y'$$

where y' ($y' < 0$) is, in absolute value, the smaller solution of the equation

$$\cos([e^{-2y'} - (1 - y')^2]^{1/2}) = e^{y'}(1 - y').$$

A numerical calculation gives

$$y' = -1.774 \qquad y'' = 5.20$$
$$(\vec{q}''_e)^2 S = 7.27$$
$$\xi_{er} = 0.371\, S^{1/2}. \tag{13.2.122}$$

We see that this length ξ_{er} which tells us how the density–density correlations decrease with r is actually proportional to the size of the chains. In a semi-dilute solution, it is larger than the length ξ_e defined by (13.2.102). Of course, ξ_e is really what we usually call the screening length but the very existence of ξ_{er} shows that, for finite chains, there is no complete screening.

Let us now come back to the extreme semi-dilute case for which ${}^0\boldsymbol{H}(\vec{q})$ is given by (13.2.117). In this expression, we can re-express b, first in terms of z and then in terms of g, since, to first-order in z, we have $z \simeq g$ [see (13.3.100)]. In this way, we obtain

$${}^0\boldsymbol{H}(\vec{q}) = \frac{4}{\mathscr{C}(q^2 + 4(2\pi)^{d/2} g\, \mathscr{C}\, S^{d/2-2})}, \tag{13.2.123}$$

a formula which, contrary to (13.2.115), should remain valid for rather large values of z. Actually, when $z \to \infty$, $g \to g^*$ and, for $d = 3$, we have $g^* \simeq 0.233$ [see (13.1.16)]. This leads to the result (for $d = 3$)

$${}^0\boldsymbol{H}(\vec{q}) \simeq \frac{4}{\mathscr{C}(q^2 + 14.7\,\mathscr{C}\, S^{-1/2})}. \tag{13.2.124}$$

Unfortunately, formula (13.2.123) is not very exact because, for $z \to \infty$, it gives

$$\xi_e = \tfrac{1}{2}(2\pi)^{-d/4}(g^*\mathscr{C}\, S^{-\varepsilon/2})^{-1/2}. \tag{13.2.125}$$

Now, we note that the general discussion in Section 2.2, led to the conclusion that in this limit, and, in agreement with (13.2.10), and (13.2.5), we must have

$$\xi_e = \Gamma \xi_K = \Gamma \mathscr{C}_K^{-\nu/(\nu d - 1)}. \tag{13.2.126}$$

Thus, we see that the approximation $\xi_e \propto \mathscr{C}^{-1/2}$ is valid only in the limit $d = 4$, $\nu = 1/2$. This is not surprising since the simple-tree approximation does not account for the swelling of the chains. Exponent corrections appear only when *loops* appear in the diagrams. Thus, comparing (13.2.125) and (13.2.126), we are inclined to believe that a more exact calculation in terms of ε would give

$$\xi_e = \tfrac{1}{2}(2\pi)^{-d/4}(g^* \boldsymbol{C}X^{1/\nu})^{-\nu/(\nu d - 1)}. \tag{13.2.127}$$

In other words, for small values of ε, we must have

$$\Gamma = \frac{1}{4\pi}(g^*)^{-1/2} \tag{13.2.128}$$

($\Gamma \simeq 0.165$ for $d = 3$).

The results obtained for a monodisperse system generalize easily to the polydisperse case. Then, S represents the average area of a polymer

$$S = \langle A \rangle$$

$$\boldsymbol{C} = \int \mathrm{d}A\, \boldsymbol{C}(A)$$

$$\mathscr{C} = \int \mathrm{d}A\, A\, \boldsymbol{C}(A). \tag{13.2.129}$$

Consequently, we can still write

$$\mathscr{C} = \boldsymbol{C}S.$$

We see that formula (13.2.114) can easily be extended to the polydisperse case; then it reads

$$\frac{1}{\boldsymbol{H}(\vec{q})} = \frac{\boldsymbol{C}S^2}{\langle\!\langle A^2\, {}^0\mathrm{h}(q^2A/2)\rangle\!\rangle} + b\mathscr{C}^2.$$

Now, we can assume that

$$\boldsymbol{C}(A) = \boldsymbol{C}\, P(A)$$

where $P(A)$ is a Schulz–Zimm distribution (13.2.22) and, in this case,

$$\frac{1}{\boldsymbol{H}(\vec{q})} = \frac{\boldsymbol{C}}{{}^0\mathrm{h}(q^2S/2, p)} + b\mathscr{C}^2 \tag{13.2.130}$$

where ${}^0\mathrm{h}(x, p)$ is given by (13.2.103).

In particular, in the limit $q^2S \gg 1$, we recover the result (13.2.117)

$$\frac{1}{{}^0\boldsymbol{H}(\vec{q})} = \frac{\mathscr{C}q^2}{4} + b\mathscr{C}^2 \tag{13.2.131}$$

2.6.4 A general expression of the structure function in terms of the contributions of I-irreducible diagrams

As was previously shown, $\boldsymbol{H}(\vec{q})$ is given in terms of $\boldsymbol{C}$ by the following parametric representation

$$\mathscr{C}^2\boldsymbol{H}(\vec{q}) = \mathscr{Q}(\textstyle\int;\vec{q}, -\vec{q}; f) = \displaystyle\sum_{N=1}^{\infty} \frac{f^N}{N!}\,\overline{\mathscr{Z}}(\textstyle\int;\vec{q}, -\vec{q}; N\times S)$$

$$\boldsymbol{C} = f\frac{\partial}{\partial f}\mathscr{Q}(f) = f\frac{\partial}{\partial f}\left[\sum_{N=1}^{\infty}\frac{f^N}{N!}\mathscr{Z}(N\times S)\right].$$

The diagrams corresponding to $\mathscr{Z}(N\times S)$ and to $\mathscr{Z}(\int:\vec{q}, -\vec{q}; f)$ are connected, but sometimes they can be separated into two parts by cutting an interaction line. In this case, it is said that they are I-reducible. In particular, such is the case for the diagram represented in Fig. 13.18. As we saw in Chapter 10, connected diagrams are trees, made of I-irreducible parts bound together by interaction lines (without formation of cycles).

Thus, it is natural to introduce the generating functions of the I-irreducible diagrams, namely

$$\mathscr{Q}_{\mathrm{I}}(\textstyle\int;\vec{q},-\vec{q};\mathfrak{f})=\sum_{N}\frac{\mathfrak{f}^{N}}{N!}\mathscr{Z}_{\mathrm{I}}(\textstyle\int;\vec{q},-\vec{q};N\times S)$$

$$\mathscr{Q}_{\mathrm{I}}(\mathfrak{f})=\sum_{N}\frac{\mathfrak{f}^{N}}{N!}\mathscr{Z}_{\mathrm{I}}(N\times S) \qquad (13.2.132)$$

[compare to (10.6.31)].

The summation of the trees made of irreducible parts can be made in a very simple way and we can proceed as in Section 2.3 when we described the simple-tree approximation. For more details, the reader may refer to Chapter 10, Section 2.5.5. We see that an I-irreducible diagram of order N can be *dressed* with side branches, in all possible ways, and that this dressing amounts to replacing the factor f by a factor

$$\mathfrak{f}=f\mathrm{e}^{-CbS^2}$$

[see (10.6.31)].

However, when dressing the diagrams, one must be careful in order to avoid overcounting; thus we have

$$\mathscr{Q}(f)\neq\mathscr{Q}_{\mathrm{I}}(\mathfrak{f})$$

the relation between $\mathscr{Q}(f)$ and $\mathscr{Q}_{\mathrm{I}}(\mathfrak{f})$ being given by [see (10.6.31)]

$$C=f\frac{\partial}{\partial f}\mathscr{Q}(f)=\mathfrak{f}\frac{\partial}{\partial\mathfrak{f}}\mathscr{Q}_{\mathrm{I}}(\mathfrak{f}).$$

(The operations $f\frac{\partial}{\partial f}$ and $\mathfrak{f}\frac{\partial}{\partial\mathfrak{f}}$ serve to mark a polymer on the diagrams.)

Let us now consider the diagrams which contribute to $\mathscr{Q}(\int;\vec{q},-\vec{q};f)$; they consist of a series of I-irreducible sub-diagrams supporting the wave vector $\vec{q}$, and of side branches. These side branches *dress* the I-irreducible sub-diagram constituting the diagram axis. Thus, $\mathscr{Q}(\int;\vec{q},-\vec{q};f)$ can be written in the form

$$\mathscr{Q}(\textstyle\int;\vec{q},-\vec{q};f)=\sum_{p=1}^{\infty}(-b)^{p-1}[\mathscr{Q}_{\mathrm{I}}(\textstyle\int;\vec{q},-\vec{q};\mathfrak{f})]^{p}.$$

In other words, $\boldsymbol{H}(\vec{q})$ is given in terms of $\boldsymbol{C}$ by the following parametric expressions

$$\mathscr{C}^2\boldsymbol{H}(\vec{q})=\frac{\mathscr{Q}_{\mathrm{I}}(\int;\vec{q},-\vec{q};\mathfrak{f})}{1+b\,\mathscr{Q}_{\mathrm{I}}(\int;\vec{q},-\vec{q};\mathfrak{f})}$$

$$\boldsymbol{C}=\mathfrak{f}\frac{\partial}{\partial\mathfrak{f}}\mathscr{Q}_{\mathrm{I}}(\mathfrak{f}). \qquad (13.2.133)$$

Of course, the simplest approximation consists in keeping only the contribution of the trivial one-polymer diagram without interaction. Then, we recover the simple-tree approximation, since in this case (zero loop approximation), we have

$$^{0}\mathcal{Q}_{\mathrm{I}}(\mathfrak{f}) = \mathfrak{f}$$

$$^{0}\mathcal{Q}_{\mathrm{I}}(\textstyle\int;\vec{q}, -\vec{q}; \mathfrak{f}) = \mathfrak{f}S^{2}\,{}^{0}H(\vec{q}). \tag{13.2.134}$$

Therefore, in this approximation

$$\mathfrak{f} = \boldsymbol{C} = \mathscr{C}/S$$

and we recover (13.2.112).

The calculation of one-loop diagrams corresponding to the next approximation can be performed without great difficulty. The one-loop diagrams contributing to $\mathcal{Q}_{\mathrm{I}}(\mathfrak{f})$ are represented in Fig. 13.21 and the sum of their contributions was calculated in Chapter 10, Section 8.4. Each one of the preceding diagrams generates several diagrams contributing to $\mathcal{Q}_{\mathrm{I}}(\int;\vec{q}, -\vec{q}; \mathfrak{f})$; some of them are represented on Fig. 13.22.

Of course, as we take a special interest in correlations, it is proper to consider the case where $q^2 S \gg 1$ which leads to simplifications (in this approximation

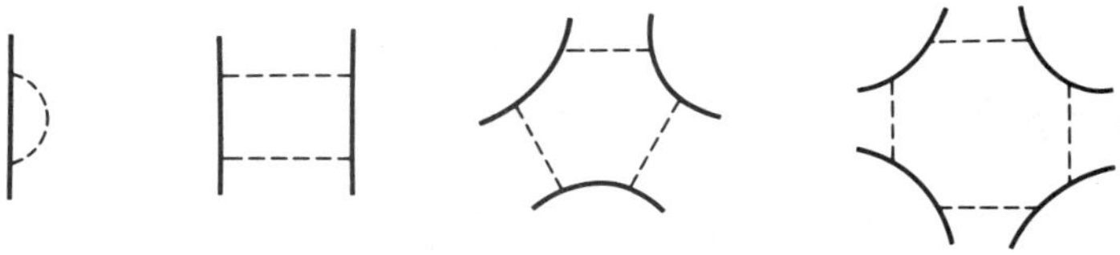

Fig. 13.21. The four first one-loop diagrams contributing to $\mathcal{Q}_{\mathrm{I}}(\mathfrak{f})$.

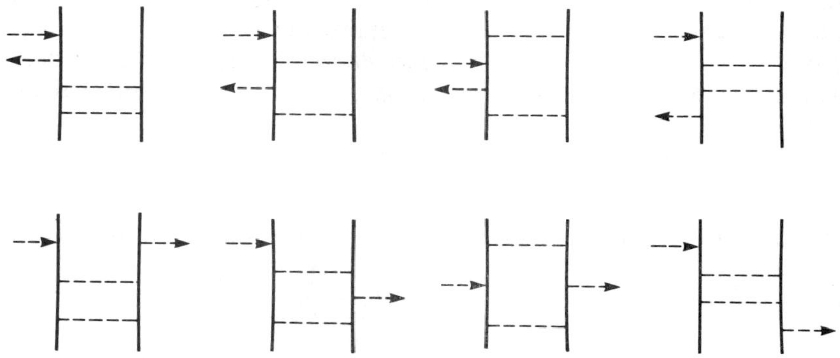

Fig. 13.22. A few one-loop diagrams contributing to $\mathcal{Q}_{\mathrm{I}}(\int;\vec{q}, -\vec{q};\mathfrak{f})$. The arrows indicate the flux of wave vector $\vec{q}$.

${}^0\boldsymbol{H}(\vec{q}) = 4/Sq^2$). It does not seem that full calculations have been performed yet (1988), in spite of attempts in this direction.[42]

In the general case (for any q^2S), any perturbation calculation must lead to a result of the form

$$y = \mathfrak{f}\, bS^2$$

$$CbS^2 = A(y,z) \qquad (\text{with } A(y,0) = y)$$

$$\boldsymbol{H}(\vec{q}) = \frac{B(y,z,q^2S)}{C[1 + CbS^2B(y,z,q^2S)]} \tag{13.2.135}$$

[see the remark concerning $H(\vec{0})$ in Chapter 10, Section 5] where the dependence of $\boldsymbol{H}(\vec{q})$ with respect to $\boldsymbol{C}bS^2$ is expressed in parametric form. (Incidentally, we note that $\boldsymbol{C}bS^2 = \boldsymbol{C}S^{d/2}(2\pi)^{d/2}z$). Here y, z, and q^2S are pure numbers; moreover, z could be re-expressed in terms of the osmotic parameter g.

2.6.5 Asymptotic form of the structure function of long chains in good solvent

Our definition of the structure function $\boldsymbol{H}(\vec{q})$ being quite simple (see 13.2.105), for a good solvent (Kuhnian limit), we can directly write the following scaling law

$$\boldsymbol{H}(\vec{q}) = \boldsymbol{C}^{-1}\,\mathrm{J}(\boldsymbol{C}X^d, qX).$$

In particular, we can find a relation between $\mathrm{J}(x,0)$ and the function $\mathrm{F}(x)$ which appears (13.2.29) and which gives the osmotic pressure, under the same conditions. We have only to apply (13.2.110) and to take (13.2.29) into account. Thus we find

$$\mathrm{J}(x,0) = \frac{1}{\mathrm{F}(x) + x\mathrm{F}'(x)}. \tag{13.2.136}$$

The case where the chains can be considered as very long is especially interesting. Then, we believe that $\boldsymbol{C}X^{1/\nu d}$ is the only relevant parameter, but one could also use the equivalent parameter ξ_{k},

$$\xi_{\mathrm{k}} = (\mathscr{C}_{\mathrm{k}})^{\nu/(\nu d-1)} = (\boldsymbol{C}X^{1/\nu})^{-\nu/(\nu d-1)},$$

which is a length (and which must not be confused with the screening length).

In the limit $qX \gg 1$, we can write

$$\boldsymbol{H}(\vec{q}) = \boldsymbol{H}(\vec{0})\mathrm{J}(q\,\xi_{\mathrm{k}}). \tag{13.2.137}$$

When $q \to 0$, we are in the semi-dilute limit and, in this case, according to (13.2.30) and (13.2.31), we have

$$\boldsymbol{\Pi\beta} = \mathrm{F}_{\infty}(\boldsymbol{C}X^{1/\nu})^{\nu d/(\nu d-1)}.$$

Let us now use (13.2.110)

$$\boldsymbol{C}\frac{\partial(\boldsymbol{\Pi\beta})}{\partial \boldsymbol{C}} = \frac{\nu d}{\nu d - 1}\mathrm{F}_{\infty}(\boldsymbol{C}X^{1/\nu})^{\nu d/(\nu d-1)} = \frac{\nu d}{\nu d - 1}\mathrm{F}_{\infty}\,\xi_{\mathrm{k}}^{-d}.$$

Consequently,

$$\boldsymbol{H}(\vec{0}) = \frac{\nu d - 1}{\nu d} \frac{1}{\mathrm{F}_{\infty}} \xi_{\mathrm{k}}^{d}. \tag{13.2.138}$$

2.7 Screening of the interaction

The interaction between polymers is given by a repulsive potential of the form

$$b_0(\vec{r}) = b\delta(\vec{r}).$$

However, the interaction between two polymers in solutions looks as if it is modified by the presence of the other polymers, and this situation leads to a special kind of screening. Actually, Edwards[48] who introduced this idea says that the real interaction can be replaced by an effective interaction $b_{\mathrm{E}}(\vec{r})$ whose essential property is to be screened in the sense that

$$\int \mathrm{d}^d r \, b_{\mathrm{E}}(\vec{r}) = 0. \tag{13.2.139}$$

In the following, the Fourier transform of $b_{\mathrm{E}}(\vec{r})$ will be denoted by

$$b(\vec{q}) = \int \mathrm{d}^d r \, \mathrm{e}^{i\vec{q}\cdot\vec{r}} \, b_{\mathrm{E}}(\vec{r}). \tag{13.2.140}$$

The effective interaction represented by $b(\vec{q})$ can be considered as the sum of two terms, a direct interaction and an indirect interaction through one or several interacting polymers. The definition of the structure factor and its diagram representation (see Fig. 13.18) shows that $b(\vec{q})$ must be defined by the equality

$$b(\vec{q}) = b - b^2 \mathscr{C}^2 \boldsymbol{H}(\vec{q}).$$

For a grand ensemble of polymers, whose concentration is determined by the fugacity $\mathfrak{f}$, $\boldsymbol{H}(\vec{q})$ is given by (13.2.133). In other words, $b(\vec{q})$ is given as a function of concentration by the coupled equations

$$b(\vec{q}) = \frac{b}{1 + b\,\mathscr{Q}_{\mathrm{I}}(\int; \vec{q}, -\vec{q}; \mathfrak{f})},$$

$$\boldsymbol{C} = \mathfrak{f} \frac{\partial}{\partial \mathfrak{f}} \mathscr{Q}_{\mathrm{I}}(\mathfrak{f}). \tag{13.2.141}$$

In particular, if we use the simple-tree approximation [see (13.2.115)], we find

$$b(\vec{q}) = \frac{b}{1 + \boldsymbol{C} b S^2 \, {}^0\boldsymbol{H}(\vec{q})}. \tag{13.2.142}$$

where ${}^0\boldsymbol{H}(\vec{q})$ is given by (13.1.151).

For $q^2 S \gg 1$, we thereby obtain

$$b(\vec{q}) = \frac{bq^2}{q^2 + 4\boldsymbol{C} b S^2}. \tag{13.2.143}$$

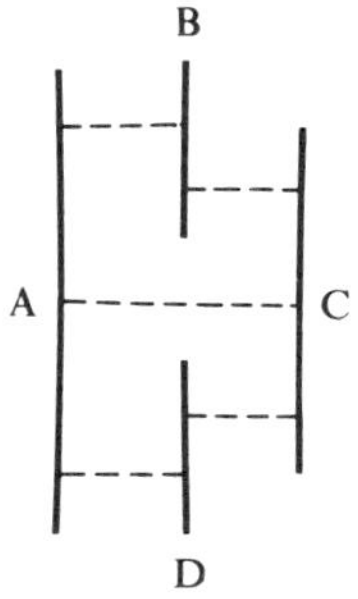

Fig. 13.23. The interactions of the polymer A with itself, induced by the polymers B, C, D as shown in this diagram, cannot be represented by a two-body screened interaction.

In this case, relation (13.2.139) is satisfied since (13.2.143) implies the equality $b(\vec{0}) = 0$. In the general case, the relation remains true provided that $\mathscr{Q}_1(\int;\vec{q}, -\vec{q};\mathfrak{f}) \to \infty$ when $\vec{q} \to 0$ (and this is likely to be true).

In brief, the function $b(\vec{q})$ defined by (13.2.141) is a screening interaction in Edwards's sense. Obviously, this screening is the reason why, on a large scale, polymers in semi-dilute solutions have a Brownian behaviour. A question still arises: is it possible to use this screened interaction in calculations? For instance, let us consider a polymer in a semi-dilute solution; can we calculate the properties of this polymer by considering it as isolated but self-interacting with a screened potential? The answer is: yes, but it is only an approximation which amounts to neglecting certain diagrams, as, for instance, the diagram represented in Fig. 13.23 (in this diagram the line A represents the polymer under consideration).

2.8 End-to-end distance of a polymer in a semi-dilute solution

In a semi-dilute solution, the chains screen one another. Then, in order to determine the end-to-end distance of a polymer, we can perform an approximate perturbation calculation, replacing the real interaction by the screened interaction. In the following, we reproduce a calculation first made by Edwards.[48] It consists in determining the mean square end-to-end distance by perturbation to first order with respect to the screened interaction $b(\vec{q})$ or more precisely its approximate value (13.2.143)

$$b(\vec{q}) = \frac{bq^2}{q^2 + 4CbS^2}.$$

By proceeding as in Chapter 10, Section 7.1, we can calculate the mean square distance R_E^2 from the equality

$$R_E^2 = \langle\langle [\vec{r}(S) - \vec{r}(\vec{0})]^2 \rangle\rangle = -\frac{2}{\mathscr{Z}_E(S)}\left[\frac{d}{dk^2}\mathscr{Z}_E(\vec{k}, -\vec{k};0,S;S)\right]_{k^2=0} \quad (13.2.144)$$

and the only difference from (10.7.2) is the introduction of an index E which indicates that we are dealing here with a polymer in solution, among others. In other words, the quantities labelled with index E are calculated by replacing the bare interaction by the screened interaction. We write [compare with (10.7.3)]

$$\overline{\mathscr{Z}}_E(\vec{k}, -\vec{k};0,S;S) = e^{-k^2S/2} + {}^1\overline{\mathscr{Z}}_E(k,S). \quad (13.2.145)$$

The first-order term ${}^1\overline{\mathscr{Z}}_E(\vec{k}, -\vec{k};0,S;S)$ corresponds to the diagram of Fig. 13.24. It is given by a formula very analogous to (10.7.4), namely

$$ {}^1\overline{\mathscr{Z}}_E(k,S) = -\int_0^S ds(S-s)\exp[-(S-s)\,k^2/2] \times (2\pi)^{-d}\int d^dq\, b(\vec{q})\exp[-s(\vec{k}-\vec{q})^2/2]$$

where $b(\vec{q})$ is given by (13.2.143).

It is convenient to represent $b(\vec{q})$ by the expression

$$b(\vec{q}) = b\left[1 - 2\mathscr{C}b\int_0^\infty d\sigma \exp[-\sigma(q^2 + 4\mathscr{C}b)/2]\right],$$

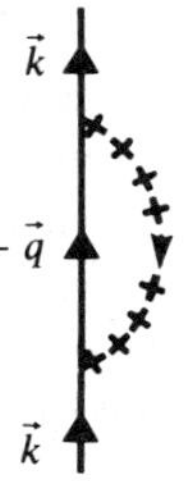

Fig. 13.24. Size of a polymer in solution. First-order diagram contributing to $\overline{\mathscr{Z}}_E(\vec{k}, -\vec{k};0,S;S)$. The interaction line is made of small crosses in order to symbolize the screened interaction $b(\vec{q})$.

and consequently $\mathscr{Z}_{\mathrm{E}}(k,S)$ takes the form

$$
\begin{aligned}
{}^1\mathscr{Z}_{\mathrm{E}}(k,S) = & -b(2\pi)^{-d/2}\int_0^S \mathrm{d}s\,(S-s)\exp[-(S-s)k^2/2] \times \\
& \times\left[s^{-d/2} - 2\mathscr{C}b\int_0^\infty \mathrm{d}\sigma(s+\sigma)^{-d/2}\exp\left[-\frac{s\sigma}{2(s+\sigma)}k^2 - 2\sigma\mathscr{C}b\right]\right] \\
= & -z\int_0^1 \mathrm{d}x(1-x)\exp[-(1-x)k^2S/2]\times \\
& \times\left[x^{-d/2} - 2\mathscr{C}bS\int_0^\infty \mathrm{d}y(x+y)^{-d/2}\exp\left[-\frac{xy}{2(x+y)}k^2S - 2y\mathscr{C}bS\right]\right].
\end{aligned}
$$

This expression diverges for small s (or x) like the expression (10.7.4) which represents ${}^1\mathscr{Z}(k,S)$; this means that ${}^1\mathscr{Z}_{\mathrm{E}}(k,S)$ is given by the principal part of the integral.

Let us expand with respect to k^2; we get

$$
\begin{aligned}
{}^1\mathscr{Z}_{\mathrm{E}}(k,S) = & \\
& -z\int_0^1 \mathrm{d}x(1-x)\left[x^{-d/2} - 2\mathscr{C}bS\int_0^\infty \mathrm{d}y(x+y)^{-d/2}\exp[-2y\mathscr{C}bS]\right] \\
& +\frac{k^2S}{2}z\int_0^1 \mathrm{d}x(1-x)\Bigg((1-x)x^{-d/2} \\
& -2\mathscr{C}bS\int_0^\infty \mathrm{d}y(x+y-x^2)(x+y)^{-1-d/2}\exp[-2y\mathscr{C}bS]\Bigg).
\end{aligned}
\tag{13.2.146}
$$

Finally, R^2 is obtained from (13.2.144), (13.2.145), and (13.2.146); thus, after integrating by parts with respect to x and to y, we find

$$
R_{\mathrm{E}}^2/Sd = 1 + z\int_0^1 \mathrm{d}x\int_0^\infty \mathrm{d}y\,x(2-3x)(x+y)^{-1-d/2}\exp[-2\mathscr{C}bSy].
\tag{13.2.147}
$$

We note that the integral is convergent for $d = 4$ and that for $\mathscr{C}bS = 0$, we recover the result (10.7.6) concerning dilute solutions.

Let us now consider the limit $\mathscr{C}bS \gg 1$; in this case, the smallest values of x and y give the largest contributions in the integral; we obtain the approximate expression

$$
R_{\mathrm{E}}^2/Sd = 1 + 2z\int_0^\infty \mathrm{d}y\exp[-2\mathscr{C}bSy]\int_0^\infty \mathrm{d}x\,x(x+y)^{-1-d/2}
$$

which can immediately be integrated and gives

$$
R_{\mathrm{E}}^2/Sd = 1 + 2(-1-d/2)!\,z(2\mathscr{C}bS)^{-2+d/2}.
\tag{13.2.148}
$$

For $d = 3$, we find

$$R_{\mathrm{E}}^{2}/3S = 1 + \tfrac{8}{3}\pi^{1/2}z(2\mathscr{C}bS)^{-1/2}$$

where

$$z = bS^{1/2}(2\pi)^{-3/2},$$

and finally

$$R_{\mathrm{E}}^{2}/3S = 1 + \frac{2}{3\pi}(b/\mathscr{C})^{1/2}. \tag{13.2.149}$$

The dependence with respect to S disappeared from the swelling, which became independent of the polymer size and proportional to the square root of the interaction; this is Edwards's result[48] (with a different factor). It is valid when the screening effect is sufficiently large to justify the assumption that all portions of chains are nearly Brownian.

2.9 Diagrams representing possible states of a polymer solution

In the previous sections, we described various regimes. It is natural to try to represent the state of a polymer solution by a point in a plane, in order to establish a map of the various domains. This has been done by Daoud and Jannink,[49] but it is possible to proceed in other ways.

For instance, we can consider the interaction as fixed and represent a solution by a point in the plane $(1/N, C)$ where N is the number of links and C the number of monomers per unit volume. In a very analogous way, in the theory of magnetism, the state of a ferromagnetic system is represented by a point in the plane (T, M) where T is the temperature and M the magnetization of the system. Thus the straight line $1/N = 0$ of the plane $(1/N, C)$ just corresponds to the spontaneous magnetization curve (coexistence curve) $M = \mathscr{M}(t)$ of the plane (T, M). In particular, the critical point $(1/N = 0,\ C = 0)$ of the plane $(1/N, C)$ corresponds to the critical point $(T = T_c, M = 0)$ of the plane (T, M).

On the contrary, with Daoud and Jannink,[49] we can consider the length of the polymers as fixed and assume that the temperature and consequently the solubility of the solution may vary. In this case, a polymer solution is represented by a point in the (C, T) plane. The good-solubility domain corresponds to temperatures $T > T_{\mathrm{F}}$ where T_{F} in the Flory temperature. By definition $T_{\mathscr{V}\mathrm{F}}(N)$ is the temperature at which the second virial coefficient vanishes and $T_{\mathrm{F}} = \lim T_{\mathscr{V}\mathrm{F}}(N \to \infty)$. Anyway, if N is large $T_{\mathscr{V}\mathrm{F}}(N)$ is close to T_{F}.

The region $T < T_{\mathrm{F}}$ corresponds to a poor solubility domain, and when T is lowered, the demixtion curve is encountered. However, in this region (even for $T \simeq T_{\mathrm{F}}$), it is impossible to faithfully represent a solution by a point in a plane,[51] because then the properties of the solution depend on too many parameters (at least three). In the following, we thus ignore the region $T < T_{\mathrm{F}}$, reserving to study it in Chapter 14.

Actually, these diagrams have a precise meaning in the context of the standard continuous model, because this model depends only on two parameters. For

$d = 3$, these parameters are

$$z = bS^{1/2}(2\pi)^{-3/2}$$
$$CS^{3/2} = \mathscr{C}S^{1/2}. \tag{13.2.150}$$

Therefore, we may consider that

$$S \propto N$$
$$\mathscr{C} \propto C$$
$$b \propto T - T_F$$

(for the model).

We then observe that the two diagrams concerning the planes $(1/N, C)$ and (C, T) are just different forms of the same diagram. Thus, for more clarity, we shall redefine the coordinates of these diagrams by using the dimensionless parameters of the standard continuous model (Figs 13.25 and 13.26). We note that the model is not defined for $z < 0$; if the interaction is negative, the chain collapses; here the curve $z = 0$ is the coexistence curve.

We can now define the various regions of the diagram by referring to the standard continuous model. Of course, between these regions, there are fuzzy *crossover* zones which correspond to the dashed lines limiting the regions in

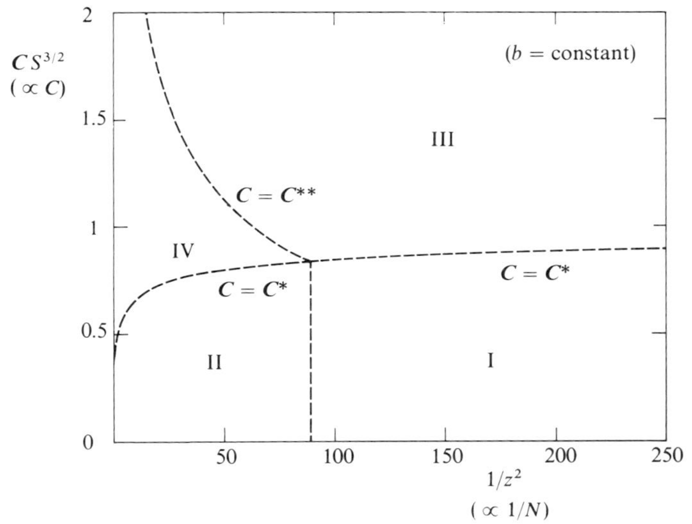

Fig. 13.25. 'Magnetic' diagram of a polymer solution. This diagram has been drawn by starting from eqns (13.2.154) and (13.2.156). It has a precise meaning in the context of the standard continuous model. For this model, the demixtion curve is at infinity. We indicated, between brackets usual (but ill-defined) variables.

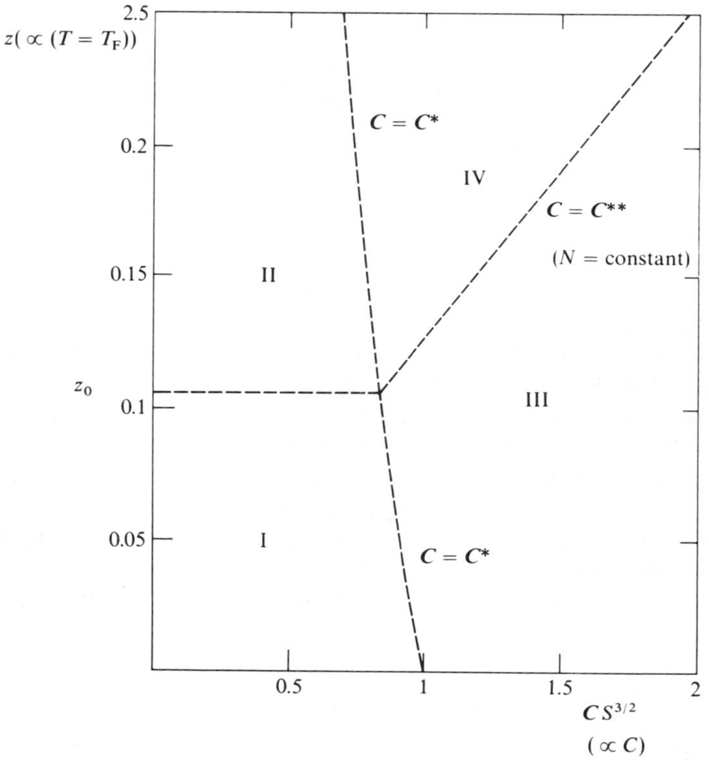

Fig. 13.26. 'Daoud–Jannink' diagram of a polymer solution ($d = 3$). This diagram has been drawn by starting from eqns (13.2.154) and (13.2.156). It is equivalent to the diagram of Fig. 13.25 and has a precise meaning for the standard continuous model. The horizontal axis ($z = 0$) is the demixtion curve for this model. Usual variables are indicated between brackets. (we note that for $d = 2$, the straight line $\boldsymbol{C} = \boldsymbol{C}^{**}$ would be vertical because in this case $\boldsymbol{C}^{**}$ does not depend on z; see the footnote on p. 644.)

Figs. 13.25 and 13.26. Nevertheless, we can write equations defining (for $z \geqslant 0$) the lines which separate, in a coarse way, the various regions.

(*a*) *Transition between dilute and semi-dilute solutions*

In the crossover zone lying between the dilute and the semi-dilute region, the interactions begin to come into play. In order to define the borderline between the dilute and the semi-dilute region, we can simply set $\boldsymbol{C}^* \, {}^{\mathrm{e}}S^{3/2} \equiv \boldsymbol{C}^* X^3 = A^*$ where A^* is a constant which must be estimated. Now, in a good solvent, we have

$$\boldsymbol{\Pi\beta} = \boldsymbol{C}[1 + \tfrac{1}{2}(2\pi)^{3/2} g^* \boldsymbol{C} X^3 + \dots]$$

where $g^* = 0.233$ [see (13.1.16)], i.e.

$$\boldsymbol{\Pi}\boldsymbol{\beta} = [1 + 1.835\,\boldsymbol{C}X^3 + \dots]. \tag{13.2.151}$$

The transition between the dilute and the semi-dilute region occurs when the second term becomes equal to the first one (or perhaps slightly larger). Thus, for a good solvent, it is reasonable from an experimental point of view, and simple from a theoretical point of view, to define $\boldsymbol{C}^*$ by the equality

$$\boldsymbol{C}^* X^3 = 1. \tag{13.2.152}$$

Moreover, for a very good solvent ($d = 3$)

$$X \propto S^{1/2} z^{2\nu - 1} = S^{1/2} z^{0.176}$$

and therefore, in this case, (13.2.152) reads

$$\boldsymbol{C}^* S^{3/2} \propto z^{-0.528}.$$

For a poor solvent ($d = 3$), according to (10.7.9), we have

$$X^2/S = {}^{\mathrm{e}}S/S = 1 + \tfrac{4}{3}z + \dots$$

and, in this case, (13.2.152) reads

$$\boldsymbol{C}^* S^{3/2} = (1 - 2z + \dots). \tag{13.2.153}$$

Thus, a good approximate representation of $\boldsymbol{C}^*$ for all values of z is obtained by writing

$$\boldsymbol{C}^* S^{3/2} = (1 + 3.8z)^{-0.53}. \tag{13.2.154}$$

This is just the function plotted in Figs 13.25 and 13.26.

(b) *Transition between poor and good solvents*

In the dilute region, the borderline between poor and good solvents can be defined by a value $z = z_0$ which will be chosen later on.

The same criterion does not apply in the semi-dilute region because, then, screening effects must be taken into account. Let us thus consider a semi-dilute solution of Brownian chains. The 'distance' ξ_0 between Brownian chains, the so-called 'overlap Brownian length', is given by (13.2.4) and, for $d = 3$,

$$\xi_0 = \mathscr{C}^{-1} \equiv (\boldsymbol{C}S)^{-1}.$$

If $\boldsymbol{C}$ and S are constant, ξ_0 is also constant. However, if the interaction increases, the thermic length ξ_t decreases and according to (13.1.174), for $d = 3$,

$$\xi_t = 2.0\, b^{-1}.$$

Thus, the transition between poor and good solvent in a semi-dilute solution corresponds to a concentration $\boldsymbol{C} = \boldsymbol{C}^{**}$ at which the screening effect begins to be effective. We admit that this happens if $\xi_0 = \xi_t$, an equality which for $d = 3$,* gives

$$\mathscr{C}^{**} = \boldsymbol{C}^{**} S = b/2. \tag{13.2.155}$$

Using the dimensionless parameters of the standard continuous model, we can also write the equation $\xi_0 = \xi_t$ in the form [see (10.1.7)]

$$\boldsymbol{C}^{**} S^{3/2} = (2\pi^3)^{1/2} z = 7.9\, z. \tag{13.2.156}$$

The borderline between dilute and semi-dilute solutions intersects the borderline between poor and good solvents at a point the ordinate of which can be chosen as the value z_0. Thus, by writing $\boldsymbol{C}^* = \boldsymbol{C}^{**}$ and by comparing (13.2.154) with (13.2.156), we find

$$7.9\, z_0 = (1 + 3.8\, z_0)^{-0.53},$$

i.e.

$$z_0 \simeq 0.106, \tag{13.2.157}$$

a value which does not seem unreasonable if one looks at Figs 13.2 and 13.3.

We can now define the various regions which appear on the maps associated with the characteristic features of a solution

Region I – Poor solvent, dilute solution

$$z < z_0 = 0.106$$

$$\boldsymbol{C} S^{3/2} < \boldsymbol{C}^* S^{3/2} = (1 + 3.8\, z)^{-0.53}.$$

Region II – Good solvent, dilute solution

$$z > z_0 = 0.106$$

$$\boldsymbol{C} S^{3/2} < \boldsymbol{C}^* S^{3/2}.$$

Region III – Semi dilute solution with moderate screening

$$\boldsymbol{C} S^{3/2} > \boldsymbol{C}^* S^{3/2}$$

$$\boldsymbol{C} S^{3/2} > \boldsymbol{C}^{**} S^{3/2} = 7.9\, z.$$

This is a region corresponding to rather poor solvents.

* Equations (13.2.4) and (13.1.174) give the general relations

$$\xi_0 = \mathscr{C}^{-1/(d-2)}$$

$$\xi_t = 2.0\, b^{-1/(4-d)}.$$

Thus, the value $\mathscr{C}^{**}$ being defined by the equality $\xi_0 = \xi_t$, we obtain

$$\mathscr{C}^{**} = \boldsymbol{C}^{**} S = (2.0)^{-(d-2)}\, b^{-(d-2)/(d-4)}.$$

In particular, for $d = 2$, we find

$$\mathscr{C}^{**} = \boldsymbol{C}^{**} S = 1.$$

Region IV – Semi-dilute solution with strong screening

$$z > z_0$$

$$C^* S^{3/2} < CS^{3/2} < C^{**} S^{3/2}.$$

This is a region corresponding to very good solvents.

The various regions are represented in Figs 13.25 and 13.26 which are completely equivalent. Of course, the numerical values given above bring in only vague information and define fuzzy crossover regions. Moreover, we note that the maps which appear in Figs 13.25 and 13.26 would probably have been more meaningful if, contrary to custom, they had been represented with logarithmic coordinates.

Finally, we observe that the discrimination between good and poor solvent, may depend on the scale at which an observation is made. In particular, neutron scattering makes it possible to explore the behaviour of short sequences in a long chain and the state, Kuhnian or Brownian, is not necessarily the same for the sequence and for the whole chain.

2.1 Behaviour of a very long polymer in a semi-dilute solution of moderately long polymers

In a semi-dilute solution, the interactions between polymers are screened but, if in a semi-dilute solution of monodisperse polymers, a very long polymer is introduced, the screening for this polymer cannot be complete. Indeed, for a scale of the same order as the size of the very long polymer, the semi-dilute solution may look practically homogeneous, but as was noticed by Flory[51] and de Gennes,[52] the very long chain is subjected to an excluded volume interaction; our aim here is to calculate this effective interaction.

Let $\mathcal{Q}(f)$ be the (connected) grand partition function associated with the solution of moderately long polymers. In what follows, we introduce functions $\mathcal{Q}(S;f)$ and $\mathcal{Q}(S,S;f)$ representing *modified* partition functions associated with one or two polymers chosen among the moderately long polymers. By definition, we have

$$\mathscr{Z}(S;f) = \frac{\partial \mathcal{Q}(f)}{\partial f},$$

$$\mathscr{Z}(S;S;f) = \frac{\partial^2 \mathcal{Q}(f)}{\partial f^2},$$

and, of course, these definitions entail the equalities

$$\mathscr{Z}(S;0) = \mathscr{Z}(S) \qquad \mathscr{Z}(S,S;0) = \mathscr{Z}(S,S).$$

We know that, for weak interactions

$$\frac{\mathscr{Z}(S,S)}{[\mathscr{Z}(S)]^2} \simeq -bS^2.$$

In an analogous way, we may set

$$b_{\text{eff}}S^2 = -\frac{\mathscr{Z}(S,S;f)}{[\mathscr{Z}(S;f)]^2} = -\frac{\partial^2 \mathscr{Q}(f)}{\partial f^2} \Bigg/ \left[\frac{\partial}{\partial f}\mathscr{Q}(f)\right]^2$$

and this equality defines the effective interaction between two polymers belonging to a solution which is not necessarily dilute. The effective interaction b_{eff} can be expressed in terms of the osmotic pressure of the moderately long chains: as we have

$$\boldsymbol{\Pi\beta} = \mathscr{Z}(f)$$

$$\boldsymbol{C} = f\frac{\partial \mathscr{Z}(f)}{\partial f}$$

it is easy to find that

$$b_{\text{eff}}S^2 = \frac{1}{\boldsymbol{C}}\left[1 - \frac{1}{\partial(\boldsymbol{\Pi\beta})/\partial\boldsymbol{C}}\right].$$

In particular, in the semi-dilute region

$$\partial(\boldsymbol{\Pi\beta})/\partial\boldsymbol{C} \gg 1,$$

and consequently

$$b_{\text{eff}} = \frac{1}{\mathscr{C}S} \qquad (\text{where } \mathscr{C} = \boldsymbol{C}S). \tag{13.2.158}$$

Now, the dimensionless parameter z_{eff} concerning the very long chains can be defined by the equality [see (10.1.7)]

$$z_{\text{eff}} = b_{\text{eff}}S_1^{2-d/2}(2\pi)^{-d/2} \tag{13.2.159}$$

where S_1 is the area of the very long chain

$$z_{\text{eff}} = \frac{1}{\mathscr{C}}\frac{S_1^{2-d/2}}{S}(2\pi)^{-d/2}. \tag{13.2.160}$$

We can assume that the formula remains true when the moderately long chains constitute a polymer melt and, in this way, we recover a result announced by de Gennes.[52]

Thus, a long chain ceases to be Brownian (like the moderately long chains) when z_{eff} becomes non-negligible. This requirement is equivalent to the inequality $z_{\text{eff}} > z_0 \simeq 0.1$ (see preceding section) and this condition can be written in the form

$$\frac{S_1/S}{(\boldsymbol{C}S^{d/2})^{2/(4-d)}} > \mathrm{A}$$

where A is a constant which can be estimated. As $CS^{d/2} \gg 1$ for a semi-dilute solution, the repulsive effect can be observed only if $S_1/S \gg 1$.

This approach gives only qualitative estimates. Thus, the definition of b_{eff} is not completely satisfactory and we could also introduce a c_{eff} corresponding to three-body forces. It is certainly possible to do better by analysing the screening effects more precisely (see Sections 2.7 and 2.9). Finally, we note that experimentally the phenomenon described here has been observed[53] as will be shown in Chapter 15.

REFERENCES

1. Oono, Y., (1985). Statistical physics of polymer solutions. *Adv. Chem. Phys.* **61**, 301.
2. des Cloizeaux, J. (1981). *J. Physique* **42**, 635.
3. Elderfield, D.J. (1981). *J. Phys.* A: *Math. Gen.* **14**, 1797.
4. des Cloizeaux, J. (1976). *J. Physique* **37**, 341.
5. Flory, P.J. (1949). *J. Chem Phys.* **17**, 303.
   Flory, P.J. and Fox, T.G. (1951). *J. Am. Chem. Soc.* **73**, 1904.
6. Muthukumar, M. and Nickel, B.G. (1984). *J. Chem. Phys.* **80**, 5839.
7. des Cloizeaux, J. Conte, R., and Jannink, G. (1985). *J. Physique Lett.* **46**, L-595.
8. Witten, T.A. and Schäfer, L. (1978). *J. Phys.* A: *Math. Gen.* **11**, 1843.
9. Benhamou, M. and Mahoux, G. (1985). *J. Physique Lett.* **46**, L-689.
10. Weill, G. and des Cloizeaux, J. (1979). *J. Physique* **40**, 99.
11. Adam, M. and Delsanti, M. (1976). *J. Physique* **37**, 1045.
12. McKenzie, D.S. and Moore, M.A. (1971). *J. Phys.* A: *Gen. Phys.* **4**, L-82.
13. des Cloizeaux, J. (1974). *Phys. Rev.* A **10**, 1665.
14. Fisher, M.E. (1966). *J. Chem. Phys.* **44**, 616.
15. Domb, C., Gillis, J., and Wilmers, G. (1965). *Proc. Phys. Soc.* **85**, 625.
16. Fisher, M.E. (1973). General scaling theory for critical points. *Collective properties of physical systems. Proc.* 24*th Nobel Symposium*, p. 16.
17. Renardy, J.F. (1971). C.E.N. Saclay, Service de Physique Theoretique. Preprint No. 24.
18. McKenzie, D.S. (1973). *J. Phys.* A: *Gen. Phys.* **6**, 338.
19. Oono, Y., Ohta, T., and Freed, K.F. (1981). *Macromolecules* **14**, 880; (1981), *J. Chem. Phys.* **74**, 6458.
20. des Cloizeaux, J. (1980). *J. Physique* **41**, 223.
21. Trueman, R.E. and Whittington, S.G. (1972). *J. Phys.* A: *Gen. Phys.* **74**, 1664.
22. Whittington, S.G., Trueman, R.E., and Wilker, J.B. (1975). *J. Phys.* A: *Math. Gen.* **8**, 56.
23. Redner, S. (1980). *J. Phys.* A: *Math. Gen.* **13**, 3525.
24. Guttmann, A.J. and Sykes, M. (1973). *J. Phys.* C: *Solid State Phys.* **6**, 945.
25. Debye, P. (1947). *J. Phys. Colloid Chem.* **51**, 18.
26. Witten, T.A. and Schäfer, L. (1981). *J. Chem. Phys.* **74**, 2582.
27. des Cloizeaux, J. and Duplantier, B. (1985). *J. Physique Lett.* **46**, L-457.
28. Witten, T.A. (1982). *J. Chem Phys.* **76**, 3300.
29. Ohta, T., Oono, Y., and Freed, K. (1981). *Macromolecules* **14**, 1588; (1982). *Phys. Rev.* A **25**, 2801.
30. Duplantier, B. (1986). *J. Physique* **47**, 1633.

31. Peterlin, A.J. (1955). *J. Chem. Phys.* **23**, 2464.
32. Ptitsyn, O.B. (1957). *Zh. Fiz. Khim.* **31**, 1091.
33. Benoit, H. (1957). *C.R. hebd. séanc. Acad. Sci.* **245**, 2244.
34. Farnoux, B., Boué, F., Cotton, J.P., Daoud, M., Jannink, G., Nierlich, M., and de Gennes, P.G. (1978). *J. Physique* **39**, 77.
35. Oono, Y. and Ohta, T. (1981). *Phys. Lett.* **85A**, 480.
36. Duplantier, B. (1985). *J. Physique Lett.* **46**, L-751.
37. Edwards, S.F. (1966). *Proc. Phys. Soc.* **88**, 265.
38. Daoud, M. (1977). Propriétés thermodynamiques des polymères flexibles. Thesis, Université de Paris VI (Feb.).
39. Daoud, M., Cotton, J.P., Farnoux, B., Jannink. G., Sarma, G., Benoit, H., Duplessix, R., Picot, C., and de Gennes, P.G. (1975). *Macromolecules* **8**, 804.
40. de Gennes, P.G. (1979). *Scaling concepts in polymer physics.* Cornell University Press.
41. Farnoux, B., Boué, F., Cotton, J.P., Daoud, M., Jannink, G., Nierlich, M., and de Gennes, P.G. (1978). *J. Physique* **39**, 77.
42. des Cloizeaux, J. (1975). *J. Physique*, **36**, 281.
43. des Cloizeaux, J. (1982). *Theory of polymers in solutions*, in *Phase transitions*, Cargèse Summer School 1980 (eds. M. Lévy, J.C. Le Guillou, and J. Zinn-Justin). Plenum Publ. Corp.
44. des Cloizeaux, J. and Noda, I. (1982). *Macromolecules* **15**, 1505.
45. Knoll, A., Schäfer, L., and Witten, T.A. (1981). *J. Physique* **42**, 767.
46. Schäfer, L. (1982). *Macromolecules* **15**, 652.
47. Ohta, T. and Nakanishi, A. (1983). *J. Phys.* A: *Math. Gen.* **16**, 4155.
48. Edwards, S.F. (1975). *J. Phys.* A: *Math. Gen.* **8**, 1670.
49. Daoud, M. and Jannink, G. (1976). *J. Physique* **37**, 973.
50. Stepanek, P. Perzynski, R., Delsanti, M., and Adam, M. (1984). *Macromolecules* **17**, 2340.
51. Flory, J.P. (1971). *Principles of polymer chemistry*, pp. 601 and 602. Cornell University Press.
52. de Gennes, P.G. (1979). *Scaling concepts in polymer physics*, Chapter II, Sect. 1.3. Cornell University Press.
53. Kirste, R.G. and Lehnen, B.R. (1976). *Makromol. Chem.* **177**, 1137.

14

PARTIALLY ATTRACTIVE CHAINS: THEORETICAL RESULTS

GENESIS

When the temperature of a polymer solution diminishes, the attractive long-range forces become more efficient than the hard-core repulsive forces. At a temperature T_F recognized by Paul Flory in 1942, attractions and repulsions compensate. In this case, for a temperature slightly smaller than T_F, the solution separates into two phases and this phenomenon is called demixtion.

In dilute solutions, demixtion is preceded by change of the polymer shape which becomes more compact: this is the coil–globule transition. Consequently, though both effects are related, two different types of studies appeared.

A systematic study of demixtion curves was undertaken as early as 1942 by Flory both from experimental and theoretical points of view. In particular, he showed that the dissymmetry of the demixtion curves is large for high molecular masses. Nevertheless, the top of the demixtion curve can be considered as an ordinary critical point. The critical opalescence associated with it was studied by P. Debye and collaborators in 1962, but correct calculations of critical exponents and of critical properties could not be made before 1972 or so, and had to wait for the renormalization methods discovered by K. Wilson.

On the other hand, the collapse of a chain in a dilute solution, when the temperature decreases, was first seen by simulation calculations on lattices (Fisher and Hiley 1961, and others) and later was observed by light scattering (Mazur and McIntyre 1976) and by neutron scattering (Cotton 1978). The question was also examined by Soviet physicists who, while studying protein denaturation, took an interest in the coil–globule transition. In particular, Ilia Lifshitz and his students studied this transition from a theoretical point of view, and developed an approximation which reduces this transformation to the liquid–gas transition.

The subject was renewed by application of renormalization theory to this problem. P.G. de Gennes pointed out in 1975 that, at the Flory temperature T_F, the polymer solution can be considered as a tricritical system. Since that time, this fact has been used by various physicists, among them, P.G. de Gennes, M.J. Stephen, and B. Duplantier (1981). However, these new approaches have to be developed; the theoretician or the experimentalist who studies the behaviour of polymer solutions in the vicinity of T_F encounters difficult problems; in fact, the question remains very complex.

1. POLYMERS IN POOR SOLVENTS: EXPERIMENTAL OBSERVATIONS. EXISTENCE OF DEMIXTION

In most cases, the polymer–solvent mixing is endothermic; to dissolve a solute in a solvent at constant temperature and pressure, one has to provide the system with heat, and since the total volume of the system does not vary much, this heat is practically equal to the internal energy. This fact shows the existence of attractive forces between the molecules of the solute in the solution, independently of the excluded volume effects. These forces express the fact that solute–solute and solvent–solvent contacts are preferred to solute–solvent contacts.

Practically, the hard core effects of quantum origin do not depend on temperature. On the contrary, the effects of the van der Waals attractive forces, which are softer and have a longer range, increase when temperature diminishes. Concurrently, the contribution of the entropy to the free energy of the system decreases.

These facts have important experimental consequences. Thus, let us consider polystyrene solutions in a poor solvent like cyclohexane at 40 °C. By lowering the temperature gradually, we can observe effects revealing the existence of attractive interactions between the polystyrene chains and also between monomers belonging to the same chain. We shall present here three significant experimental observations.

1. Let us first examine how the osmotic pressure varies with concentration ρ. We have already seen that at 40 °C, the value A_2 of the second virial coefficient

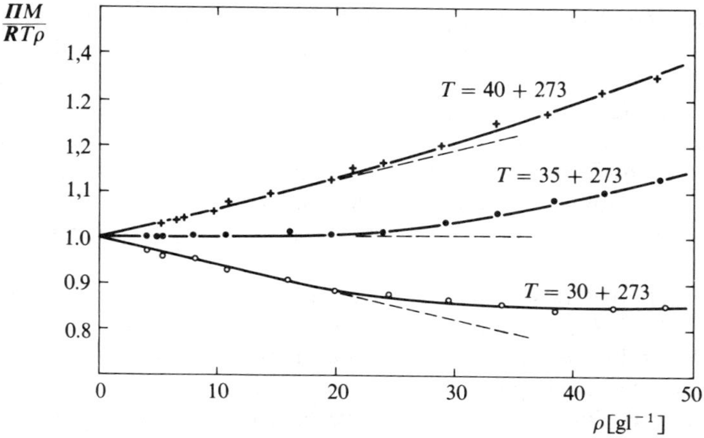

Fig. 14.1. The reduced osmotic pressure $\Pi M/RT\rho$ is plotted against the mass concentration ρ, for various temperatures (according to Strazielle;[1] polystyrene: $M_n = 130\,000$; solvent: cyclohexane; temperatures; 40 °C, 35 °C, 30 °C). The temperature T corresponding to 35 °C is close to T_F (the Flory temperature).

which is observed is weaker for cyclohexane than for benzene in the same conditions (see Fig. 5.8). We now consider the result obtained by C. Strazielle[1] at 30 °C, for cyclohexane and a molecular mass $M_n = 130\,000$ (see Fig. 14.1). We see that the curve representing $\Pi/RT\rho$ as a function of ρ has a negative slope at the origin. This shows that in the expansion of the osmotic pressure with respect to ρ, the second coefficient A_2 is negative

$$\frac{\Pi}{RT} = \frac{\rho}{M_n} + A_2\rho^2 + A_3\rho^3 + \ldots .$$

Let us come back to the definition (10.8.1), (10.8.6) of this coefficient. It is clear that a negative value for A_2 reflects the existence of attractive forces between the chains in solution.

2. When the temperature of the solution is again lowered, we can observe that the solution which was limpid and transparent becomes turbid. Then, if the temperature remains constant, a meniscus appears after some time and separates two phases in the sample. We have demixtion, and this is the second characteristic effect. Precise measurements of the demixtion temperature have been made, in particular by the Japanese school.

What we have said concerns the lowest demixtion temperature. Let us note that if we heat a good polymer solution, a phase separation may also occur, but this second demixtion domain has not been studied very much.*

3. The third observation concerns the swelling of chains in *very* dilute solutions, when T becomes smaller than the temperature for which $A_2 = 0$ (see the beginning of this section). Then the second virial coefficient becomes negative, and this phenomenon is an effect of the attractive forces existing between chains. These attractive forces produce another effect. In fact, Perzynski, Adam, and Delsanti[3] observed that the hydrodynamic swelling $\mathfrak{X}_H$ which is nearly equal to unity when $A_2 = 0$, becomes smaller than 1 when the temperature of the solution decreases (see Fig. 14.2). In other words, in this case, the chain is smaller than the equivalent Brownian chain.

2. GENERAL THEORETICAL CONSIDERATIONS

2.1 Purely attractive or repulsive chains; existence of continuous models

It would be very difficult to analyse the interaction between polymer and solvent in detail, at a given temperature. Actually, this would also be useless, if we want only to predict the properties of solutions of long polymers. In fact, the

* The demixtion obtained by heating was discussed by Zeman, Biros and Patterson (1972)[2] for polyisobutylene and polydimethylsiloxane solutions in short-chain alcanes.

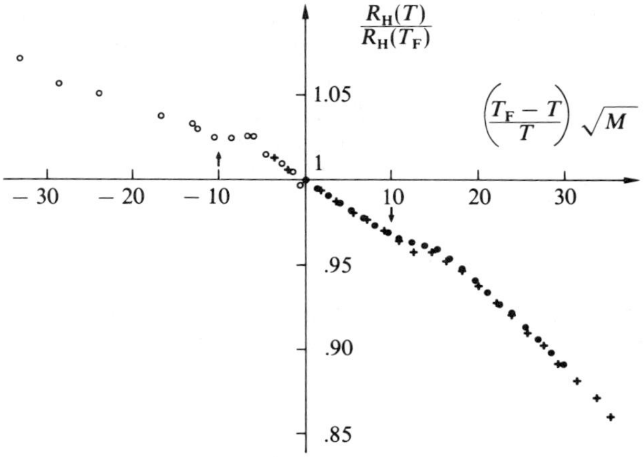

Fig. 14.2. Variation of the hydrodynamic swelling of a polystyrene chain in cyclohexane, with respect to temperature. The results come from measurements of the diffusion coefficient of a polymer, by light scattering (according to Perzynski, Adam, and Delsanti[3]) and were extrapolated to zero concentration. The figure shows how the ratio $R_H(T)/R_H(T_F)$ varies with respect to $\frac{T_F - T}{T} M^{1/2}$ for various samples (R_H: hydrodynamic radius)

Polystyrene:
- ● $M_w = 1.71 \times 10^5$ $\quad \varphi \simeq 10^{-3}$
- + $M_w = 1.26 \times 10^6$ $\quad \varphi \simeq 10^{-4}$
- ○ $M_w = 1.71 \times 10^5$ $\quad \varphi \simeq 10^{-3}$

The arrows show the limits of the Θ domain (tricritical domain).

properties of these solutions, depend only on global characteristics of the interaction and not on microscopic details. To determine them, it is sufficient to find a good model of interacting chains and to be able to predict the behaviour of the chains in definite situations.

Let us examine which model can be chosen to represent a chain the links of which are partially attractive. First, let us consider chains with N links ($N \gg 1$) on a d-dimensional square lattice, starting from an origin O. We assume that the weight associated with such a chain is

$$W = \exp\left(-\frac{u}{2} \sum_{j=1}^{N} \sum_{k=1}^{N} \delta_{\vec{r}_j, \vec{r}_k} \right) \tag{14.2.1}$$

where $\vec{r}_j$ defines the position on the lattice of the point of order j on the chain (with $j = 0, \ldots, N$) and where u is the two-body interaction. If $u = 0$, the chain consists of independent links and for $N \gg 1$ has a Brownian behaviour. If $u > 0$ the chain has about the same properties as a chain with excluded volume and for

$N \gg 1$ has a Kuhnian behaviour (chains in good solvent). What happens if $u \ll 0$?

To see this, we can try to minimize the free energy and, in this section, any energy, free energy, potential, or chemical potential will be expressed in thermal units. Consequently, the free energy reads

$$\mathbb{F}_N = \mathbb{E}_N - \mathbb{S}_N = \langle\!\langle \mathscr{U}_N \rangle\!\rangle - \mathbb{S}_N.$$

Now, we observe that the entropy $\mathbb{S}_N$ which is always a positive quantity remains always proportional to N

$$0 < \mathbb{S}_N < N \ln \mathfrak{f}$$

where $\mathfrak{f}$ is the connectivity of each site. On the contrary, we note that the energy $\mathbb{E}_N$ behaves in a very different manner, decreasing proportionally to $-N^2$ when N increases. To convince ourselves of this fact, let us consider the configuration with the lowest energy; then the chain is completely folded on itself (each link coincides with the same lattice bond). In order to simplify, we can assume that N is uneven (see Fig. 14.3); an elementary calculation shows that the interaction energy is

$$\mathscr{U}_N^0 = -|u|\left(\frac{N^2-1}{4}\right). \tag{14.2.2}$$

Thus, whereas in the purely repulsive case, $\mathbb{E}_N$ increases like N, in the purely attractive case $\mathbb{E}_N$ decreases like $-N^2$. In other words, for $u < 0$ and $N \gg 1$, in the free energy, the energy terms are always dominant with respect to the entropy terms. Moreover, the energy $\mathbb{E}_N^0$ of the ground state is much lower than the energy $\mathbb{E}_N^1$ of the first excited state (see Figs 14.3 and 14.4) and an elementary

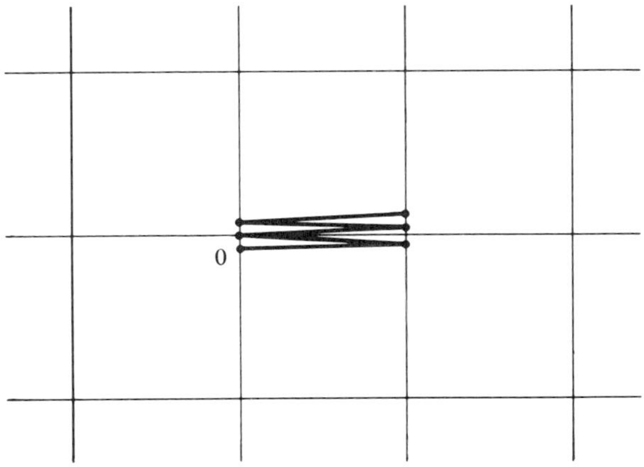

Fig. 14.3. Ground state of an attractive chain drawn on a lattice: the chain has 5 links and is folded on itself.

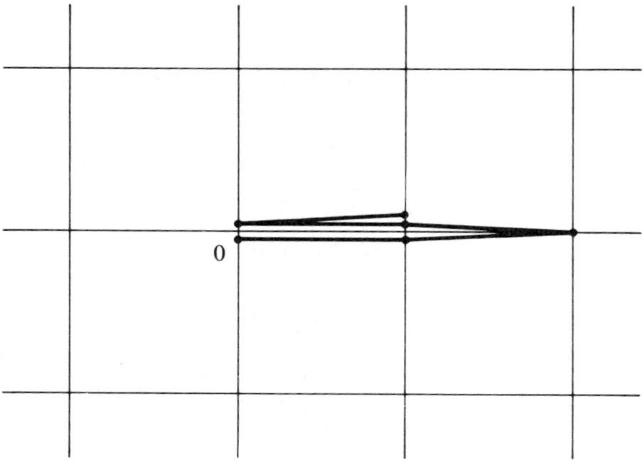

Fig. 14.4. First excited state of an attractive chain drawn on a lattice: a chain with 5 links.

calculation shows that

$$\mathbb{E}_N^1 = \mathbb{E}_N^0 + |u|\left(\frac{N-1}{2}\right). \tag{14.2.3}$$

In brief, when $N \gg 1$, the ground state has a much larger probability than any other one; on average, the chain always remains folded on itself

$$\mathbb{E}_N \simeq E_N^0 \qquad \text{and} \qquad \langle\!\langle (\vec{r}_N - \vec{r}_0)^2 \rangle\!\rangle \simeq dl^2 \tag{14.2.4}$$

where l defines the size of a lattice cell (for a d-dimensional cubic lattice $l = l_0$ where l_0 is the length of a lattice bond)

The non-realistic character of the model becomes more conspicuous in the continuous limit. Then, to any chain configuration defined by the vector $\vec{r}(s)$ $(0 \leqslant s \leqslant S)$, we attribute the weight

$$\mathscr{W} = \exp\left\{-\frac{1}{2}\int_0^S \mathrm{d}s\left(\frac{\mathrm{d}\vec{r}}{\mathrm{d}s}\right)^2 - b\int_0^S \mathrm{d}s' \int_0^S \mathrm{d}s''\,\delta(\vec{r}(s') - \vec{r}(s''))\right\} \tag{14.2.5}$$

as we did in Chapter 10, but now we assume that b is negative. As was shown in Chapter 10 (but for $b > 0$), there exists an approximate correspondence between the lattice model and the continuous model which is obtained by setting [see (10.2.1), (10.2.3)]

$$Nl^2 = S \qquad (Nl_0^2 = Sd)$$

$$bl^4 = uv_r$$

where v_r is the volume per site ($v_r = (l_0)^d$ for a cubic lattice of side length l_0); these formulae can be easily recovered by comparing (14.2.1) and (14.2.5), and by taking into account the dimensions of the quantities under consideration.

The correspondence becomes exact when $N \to \infty$ and $l \to 0$ at S fixed. Then, we see that, according to (14.2.2), $\mathbb{E}(S)/S \propto \mathbb{E}_N/Nl^2 \to -\infty$ as N does and simultaneously (14.2.4) shows that

$$\langle\!\langle [\vec{r}(S) - \vec{r}(0)]^2 \rangle\!\rangle \simeq \langle\!\langle [\vec{r}_N - \vec{r}_0)^2 \rangle\!\rangle \to 0.$$

This discussion shows that a purely attractive continuous chain reduces to a point, and this result, established here in a rather intuitive way, is confirmed by more precise calculations which will be given later. Finally, we must admit that a purely attractive model is unrealistic. Thus, the results obtained with the continuous model corresponding to the weight (14.2.5) cannot be (analytically) continued for $b < 0$.

As was previously pointed out, the expansions of physical quantities, like the swelling, in terms of z ($z = bS^{1/2}(2\pi)^{-3/2}$) are divergent for any z; they loose all meaning when z becomes negative because, in this case, the continuous chain collapses completely, reducing to a point. For this simple model, the demixtion curve is given by the equality $b = 0$. In other words, in this model, the demixtion curve is the straight line $T = T_F$ (where T_F is the Flory temperature)[4].

This result is important. It shows that a more sophisticated theory is needed to study excluded volume effects in the vicinity of the Flory temperature. Therefore, the properties of polymer solutions, above and below the Flory temperature, are not directly related, and this is an important fact which has not always been recognized.[5, 6]

We have now to find a new model giving a realistic representation of the behaviour of a polymer solution for temperatures that do not differ much from T_F. In particular, this model should serve to describe, for $N \gg 1$, the top of the demixtion curve and its vicinity.

2.2 Simultaneously attractive and repulsive chains: model with two-body and three-body interactions

It seems at first sight that the most realistic approach consists in choosing a model with two-body interaction, this interaction being the sum of a short-range repulsive part and of a long-range attractive part. At the Flory temperature T_F, the attractive and repulsive effects compensate each other (on average). For $T < T_F$, the attractive part is dominant but in all cases the hard repulsive core prevents a complete collapse of the chains. Thus, when the attraction becomes very strong the chain constitutes a roughly spherical globule whose volume is proportional to the number N of links. This is the saturation phenomenon, and in this case the size of the globule is given by

$$\langle\!\langle (\vec{r}_N - \vec{r}_0)^2 \rangle\!\rangle \simeq l_g^2 N^{2/d} \qquad (14.2.6)$$

where d is the space dimension and l_g a length defining the size of the apparent hard core.

Let us then try to represent a chain by a continuous model with an associated weight of the form

$$\mathscr{W} = \exp\left[-\frac{1}{2}\int_0^S \mathrm{d}s \left[\frac{\mathrm{d}\vec{r}(s)}{\mathrm{d}x}\right]^2 - b_0 \int_0^S \mathrm{d}s' \int_0^S \mathrm{d}s''\, \delta(\vec{r}(s') - \vec{r}(s'')) \right.$$
$$\left. - b_1 \int_0^S \mathrm{d}s' \int_0^S \mathrm{d}s''\, |\vec{r}(s') - \vec{r}(s'')|^{-\iota} \right] \tag{14.2.7}$$

where b_0 is positive (short-range repulsion) and where b_1 is positive or negative (ι is a positive exponent). In the expansions in powers of the interaction, two dimensionless constants will appear, namely

$$z_0 \propto b_0 S^{2-d/2},$$
$$z_1 \propto b_1 S^{2-\iota/2}. \tag{14.2.8}$$

The value of ι is not completely arbitrary: we require the range of the attractive force ($b_1 < 0$) to be finite, i.e. $\left(\int_{r>l} \mathrm{d}^d r\, r^{-\iota} \ll \infty\right)$; consequently, we must have $\iota < d$.

Moreover, in order to ensure that the attractive force is efficient for long chains (z_1 does not tend to zero when $S \to \infty$), we must have $2 - \iota/2 \geqslant 0$. Both conditions together give

$$4 \geqslant \iota > d. \tag{14.2.9}$$

In principle, such a model would be valid if the forces between polymers were such that $4 \geqslant \iota > 3$. In fact, this is not true; for neutral polymers, the forces with the longest range are van der Waals forces for which $\iota = 6$. We see that in this case $z_1 \to 0$ when $S \to 0$. Thus, it looks as if this attractive interaction should behave like a short range interaction and should mingle with the repulsive interaction; therefore, it seems that we come back to the purely repulsive or purely attractive model which was rejected for reasons expounded in Section 3.

Nevertheless, there is a way to solve this difficulty. We have only to claim the existence of three-body forces in addition to two-body forces. Of course, we do not necessarily believe that these three-body forces really exist; we only say that the system behaves as if three-body forces did exist.

Let us be more precise. We consider a real chain with only two-body interaction, but we assume that, in order to construct a more realistic model, we perform a finite number of renormalizations of the chain by using the decimation method. We see immediately that the two-body interactions of the initial chain generate three-body interactions (and also four-body interactions, and so on). When we discussed this kind of renormalization, we even said that this effect was a drawback of the method. Here, this effect preserves us from difficulties; it shows that two-body interactions can, in a simple way, generate three-body interactions (but the existence of a short-range cut-off has to be taken into account).

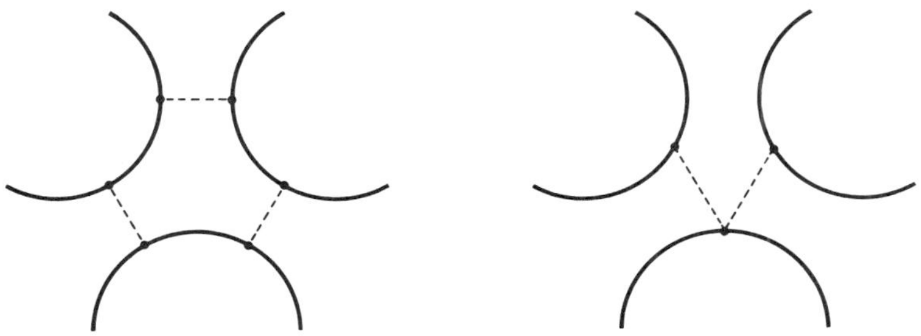

Fig. 14.5. (a) Sub-diagram made with two-body interactions. (b) Three-body interaction *equivalent* to the sub-diagram made with two-body interactions.

We can argue in a different manner and reach the same conclusion, by starting from perturbation theory. Let us consider a diagram expansion. In these diagrams, there are sub-diagrams which are constructed with two-body interaction and, in the diagrams, the sub-diagrams play the role of effective three-body interactions (or of interactions of higher degree) (see Fig. 14.5). Moreover, the contributions of these sub-diagrams are cut-off dependent.

Thus, we are led to introduce a continuous model with local two-body and three-body interactions.[7] Each configuration of a chain is represented as usual by a function $\vec{r}(s)$ (with $0 < s < S$) and the following weight $\mathscr{W}$ is associated with each configuration

$$\mathscr{W} = \exp\left\{ -\frac{1}{2}\int_0^S \mathrm{d}s \left(\frac{\mathrm{d}\vec{r}}{\mathrm{d}x}\right)^2 - \frac{b}{2}\int_0^S \mathrm{d}s' \int_0^S \mathrm{d}s'' \delta(\vec{r}(s') - \vec{r}(s'')) \right.$$
$$\left. - \frac{c}{6}\int_0^S \mathrm{d}s' \int_0^S \mathrm{d}s'' \int_0^S \mathrm{d}s''' \delta(\vec{r}(s') - \vec{r}(s''))\delta(\vec{r}(s') - \vec{r}(s''')) \right\}. \quad (14.2.10)$$

In this model, b which gives the strength of the two-body interaction is positive or negative, but c which gives the strength of the three-body interaction is always positive. The positivity of c ensures the stability of the system, i.e. the fact that $\mathscr{W}$ is meaningful.

The physical quantities corresponding to this model depend on two dimensionless parameters defined as follows

$$z = bS^{2-d/2}(2\pi)^{-d/2};$$
$$y = cS^{3-d}(2\pi)^{-d}. \quad (14.2.11)$$

We verify that for $d = 3$, y remains constant when $S \to \infty$. Thus, the three-body interaction is just marginal when $d = 3$.

The marginality of this interaction for $d = 3$ shows that the model with two-body and three-body interactions defined by (14.2.10) is a realistic model. However, in order to define the model more precisely, it is necessary to introduce a short-distance cut-off and renormalizations as we did in Chapter 10 for the model with two-body interaction. This new model is the basis of the following detailed study.

It is convenient to consider c as a constant and b as a linear function of temperature. Thus, a question arises: 'which values of z and y correspond to the Flory temperature T_F?'. The simplest and most physical definition of T_F is as follows: it is the temperature at which the second virial coefficient vanishes. This condition can be written in the form

$$g(z, y, S/S_0) = 0 \tag{14.2.12}$$

where s_0 is a cut-off area which is needed for convergence. In this way, we obtain a temperature $T_{\mathscr{V}F}(S)$ which slightly depends on the size S of the polymer, i.e. on the number of links. Thus, the true temperature T_F is defined as the limit

$$T_F = \lim_{S\to\infty} T_{\mathscr{V}F}(S). \tag{14.2.13}$$

It is also possible to define a temperature $T_{\mathscr{G}F}$ for which the swelling of an isolated chain is equal to unity. However, this temperature, which is well-defined from a theoretical point of view, is not very meaningful from an experimental point of view. In fact, it is more interesting to define the temperature $T_{\mathscr{G}\infty F}(S)$ at which the swelling of a finite chain is the same as the swelling of an infinite chain at temperature T_F. As will be shown, $T_{\mathscr{G}F}(S)$ and $T_{\mathscr{V}F}(S)$ are not very different from T_F, and we have

$$T_F = \lim_{S\to\infty} T_{\mathscr{G}F}(S) = \lim_{S\to\infty} T_{\mathscr{G}\infty F}(S). \tag{14.2.14}$$

Moreover, let $T_c(S)$ be the temperature corresponding to the critical point of demixtion (the top of the demixtion curve). This temperature is not as close to T_F as $T_{\mathscr{V}F}(S)$ is; however, we also have

$$T_F = \lim_{S\to\infty} T_c(S).$$

The properties of a polymer solution in the vicinity of temperature T_F can be studied, either directly or by using the magnetic analogy. Actually, de Gennes noted in 1975 that the properties of a polymer solution in the vicinity of T_F are similar to those of a tricritical system,[7] and this fact has been used a great deal since that time.

When $N \to \infty$, we expect to find scaling laws, and we may wonder which parameters should be used to write them. Of course, these parameters must be pure numbers. Accordingly, the experimentalists who measure demixtion curves often use the quantities

$$t = \frac{T - T_c}{T} \quad \text{and} \quad \frac{\varphi}{\varphi_c}$$

where T_c is the temperature and φ_c the volume fraction at the demixtion critical point (the top of the demixtion curve). However, T_c and φ_c depend on the microstructure of the system and, for finite values of N, they are not universal quantities. Thus, in this case, the parameters t and φ/φ_c do not enable one to write universal laws. (t is never universal!).

Let us consider the tricritical domain. When $N \to \infty$ $T_c \to T_F$ and $\varphi_c \to 0$. In this critical domain, it is proper to use the most significant parameters. For concentration, the obvious choice remains φ/φ_c. For temperature, the best choice seems to be

$$\tau = \frac{\beta_c - \beta}{\beta_c - \beta_F} = \frac{T_F}{T}\frac{T - T_c}{T_F - T_c}. \quad (14.2.15)$$

Indeed, we note that these parameters seem to acquire a universal meaning in the limit $N \to \infty$.

2.3 The state of the theory

The properties of a polymer solution become extremely complex in the vicinity of the Flory temperature T_F, when the size of the chains increases. In fact when $N \to \infty$, $T_c \to T_F$. Then, the system becomes critical in a double manner. First, a system made of very long chains is equivalent to a Landau–Ginzburg model, in which the order parameter is a zero component vector ($n = 0$). Secondly, the top of the demixtion curve, even for chains of finite length, is a critical point, and near this point the system can be described by another Landau–Ginzburg model the order parameter of which is a one-component vector ($n = 1$).

Consequently, the shape of the demixtion curve is not simple. Moreover, as we noticed, it seems that two dimensionless parameters (y and z) are needed to characterize the chains in this domain whereas only one dimensionless parameter (z) is sufficient in good solvent. In these conditions, it is not surprising that we still lack good theoretical descriptions of the properties of the system. Of course, various approaches exist but they are not very consistent.

Certain physicists tried to study only the coil–globule transition. To describe this transition, it is possible to generalize Flory's theory but, unfortunately, it is difficult to justify this approach from a theoretical point of view. Another approach developed by Lifshitz seems a priori more realistic, and it will be expounded later on. However, we can already note that both approaches represent the coil–globule transition as a transformation which may be discontinuous. It is clear that, in reality, it cannot be so. The statistics of a system with a finite number of degrees of freedom (proportional to the number of links) can vary with the parameters of the system only in a continuous way. However, without being really discontinuous, the coil–globule transition may occur in an abrupt way if the chain is long.

One might also look at the properties of solutions and try to predict the shape of the demixtion curve. Flory–Huggins theory is at the present time (1988) the only theory giving a global view of the phenomena; this is why it is especially interesting. Renormalization theories give results that are more exact but fragmentary. The critical theory thus gives the limit of the demixtion curve when the number of links becomes infinite; it also shows that, in the vicinity of T_F, long chains are not exactly Brownian but that logarithmic corrections exist. Moreover, for fixed values of N, classical results of renormalization theory enable us to describe the top of the demixtion curve in a rather precise manner and in agreement with the experimental facts.

In short, we have a good knowledge of the subject but, in practice, really good theories are still lacking.

3. ISOLATED CHAIN: FLORY'S THEORY IN THE ATTRACTIVE CASE

3.1 Purely attractive model

As we showed in Chapter 8 when we were dealing with the repulsive case, Flory's 'theory' has no sound theoretical basis. For reasons still unclear (1990), it leads to interesting results. This is why we shall examine here the predictions of this theory in the attractive case.

For $d = 3$, Flory's equation which gives the swelling of an isolated chain reads (see (8.1.34),

$$\mathfrak{X}^{5/2} - \mathfrak{X}^{3/2} = \frac{1}{2} 3^{3/2} z \tag{14.3.1}$$

and we note that the curve $z(\mathfrak{X})$ defined for $\mathfrak{X} > 0$ (see Fig. 14.6) has a minimum

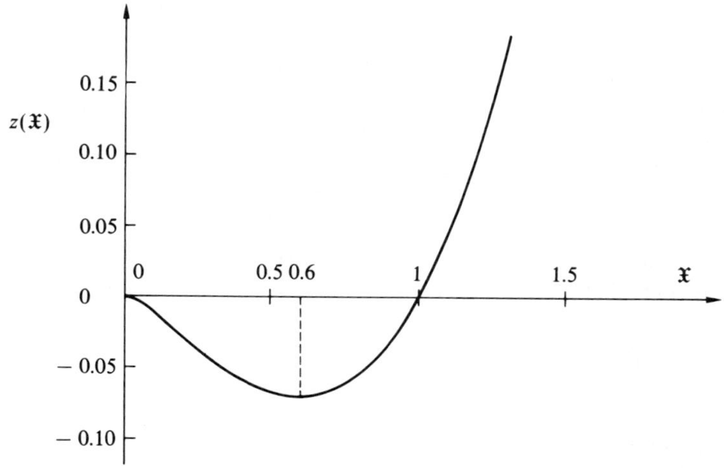

Fig. 14.6. The curve: $z = 2 \times 3^{-3/2}(\mathfrak{X}^{5/2} - \mathfrak{X}^{3/2})$.

$z_m = -4 \times 5^{-5/2} \simeq -0.0715$ for $\mathfrak{X} = 3/5$. Thus, eqn (14.3.1) has a root $\mathfrak{X}$ for $z > 0$, two roots $\mathfrak{X}'$ and $\mathfrak{X}''$ for $0 > z > -4 \times 5^{-5/2}$, and no root for $z < -4 \times 5^{-5/2}$.

Which solution must be chosen when $z < 0$? In order to answer this question, let us come back to the expression of $\mathbb{F}(\mathfrak{X})$ given by (8.1.33). For $d = 3$, we have

$$\mathbb{F}(\mathfrak{X}) - \mathbb{F}_0 = \frac{3}{2}(\mathfrak{X} - 1 - \ln \mathfrak{X}) + \frac{3^{3/2}}{2} z\mathfrak{X}^{-3/2}$$

We can see that for $z < 0$, $\mathbb{F}(\mathfrak{X})$ goes to $-\infty$ when $\mathfrak{X} \to 0$. Thus, for $z < 0$, $\mathbb{F}(\mathfrak{X})$ has no minimum and the model predicts a complete collapse of the chain ($\mathfrak{X} = 0$). This conclusion is in agreement with the more general comments made in the preceding sections.

In order to avoid this catastrophe, we must use a more sophisticated model. In fact, Flory and de Gennes[7] showed that the introduction of three-body forces was sufficient to obtain reasonable results; the corresponding model will now be studied.

3.2 A model with two-body attraction and three-body repulsion

The preceding discussions showed that it is necessary to introduce both two-body attractive forces and three-body repulsive forces. Now, as we showed in Chapter 8, Flory's results can be obtained by minimizing the free energy and we can write [see (8.1.33)].

$$\begin{aligned} \mathbb{F}(\mathfrak{X}) - \mathbb{F}(1) &= [\mathbb{E}_c(\mathfrak{X}) - \mathbb{E}_c(1)] - [\mathbb{S}(\mathfrak{X}) - \mathbb{S}(1)] + {}^{I}\mathbb{E}(\mathfrak{X}) \\ &= \frac{d}{2}(\mathfrak{X} - 1 - \ln \mathfrak{X}) + {}^{I}\mathbb{E}(\mathfrak{X}) \end{aligned} \tag{14.3.2}$$

where $\mathfrak{X}$ denotes the swelling of the chain ($\mathfrak{X}$ may be identified with $\mathfrak{X}_0$ or with $\mathfrak{X}_G$). The interaction energy ${}^{I}\mathbb{E}(\mathfrak{X})$ can be obtained by generalizing eqn (8.1.27) as follows

$${}^{I}\mathbb{E}(\mathfrak{X}) = \frac{1}{2}\int d^d r\, v_e C^2(\vec{r}) + \frac{1}{6}\int d^d r\, w_e C^3(\vec{r}). \tag{14.3.3}$$

In this formula, $C(\vec{r})$ is the average number of monomers per unit volume, v_e is the volume defining the two-body interaction, and w_e the hyper-volume defining the three-body interaction. Thus, the initial calculation given by Flory can be generalized as was shown by Flory himself and by de Gennes (1975).[7]

As in Chapter 8 and for an isolated chain, we set*

$$C(\vec{r}) = \frac{N}{\mathfrak{X}^{d/2}} P(r\mathfrak{X}^{-1/2})$$

* Let us note that $C(\vec{r})$ can be estimated in various ways which may have a strong influence on the value of the coefficients in the final formula giving the swelling of the chain.

$$P(\vec{r}) = \frac{1}{(N\pi l^2/3)^{d/2}} \mathrm{e}^{-3r^2/Nl^2}$$

we thus find

$$^{\mathrm{I}}\mathbb{E}(\mathfrak{X}) = \frac{v_\mathrm{e}}{2} N^2 \mathfrak{X}^{-d/2} \int \frac{d^d r}{(N\pi l^2/3)^d} \mathrm{e}^{-6r^2/Nl^2} + \frac{w_\mathrm{e}}{6} N^3 \mathfrak{X}^{-d} \int \frac{\mathrm{d}^d r}{(N\pi l^2/3)^{3d/2}} \mathrm{e}^{-9r^2/Nl^2}$$

and more explicitly

$$^{\mathrm{I}}\mathbb{E}(\mathfrak{X}) = \frac{1}{2} 3^{d/2} (2\pi)^{-d/2} v_\mathrm{e} l^{-d} N^{2-d/2} \mathfrak{X}^{-d/2} + \frac{1}{6} (12)^{d/2} (2\pi)^{-d} w_\mathrm{e} l^{-2d} N^{3-d} \mathfrak{X}^{-d}.$$

It is convenient to use standard notation and to set

$$z = (2\pi)^{-d/2} v_\mathrm{e} l^{-d} N^{2-d/2}$$

$$y = (2\pi)^{-d} w_\mathrm{e} l^{-2d} N^{3-d}. \tag{14.3.4}$$

Finally, we have

$$^{\mathrm{I}}\mathbb{E}(\mathfrak{X}) = \frac{1}{2} 3^{d/2} z \mathfrak{X}^{-d/2} + \frac{1}{6} (12)^{d/2} y \mathfrak{X}^{-d}. \tag{14.3.5}$$

Coming back to the result (14.3.2), we get

$$\mathbb{F}(\mathfrak{X}) - \mathbb{F}(1) = \frac{d}{2}(\mathfrak{X} - 1 - \ln \mathfrak{X}) + \frac{1}{2} 3^{d/2} z \mathfrak{X}^{-d/2} + \frac{1}{6} (12)^{d/2} y \mathfrak{X}^{-d}. \tag{14.3.6}$$

Let us minimize this expression with respect to $\mathfrak{X}$; we obtain

$$\mathfrak{X}^{1+d/2} - \mathfrak{X}^{d/2} = \frac{1}{2} 3^{d/2} z + \frac{1}{3} (12)^{d/2} y \mathfrak{X}^{-d/2}. \tag{14.3.7}$$

In this formula, y which defines the three-body interaction, can be considered as a constant, and z which defines the two-body interaction, can be considered as a variable.

Temperature $T_{\mathcal{G}\mathrm{F}}$ corresponds to the value $\mathfrak{X} = 1$ [see (14.2.14)] and at this point $z = z_{\mathcal{G}\mathrm{F}}$ where

$$z_{\mathcal{G}\mathrm{F}} = -\frac{2^{1+d}}{3} y. \tag{14.3.8}$$

Consequently, this approximation gives $T_{\mathcal{G}\mathrm{F}} < T_\mathrm{F}$.

The result appears as quite natural since the three-body terms are repulsive and finite for $T = T_\mathrm{F}$. Thus, they should contribute to the swelling in a positive way, as it appears in formula (14.3.7). However, this conclusion is not exact, as will be shown in Section 6.7. Actually, we ignored the fact that three-body interactions can create short-range (ultraviolet) divergences and that these divergences have to be renormalized. The outcome of this renormalization is that the finite contribution of the diagrams is given by an analytic continuation with respect to space dimension. As a consequence, it is not a priori possible to

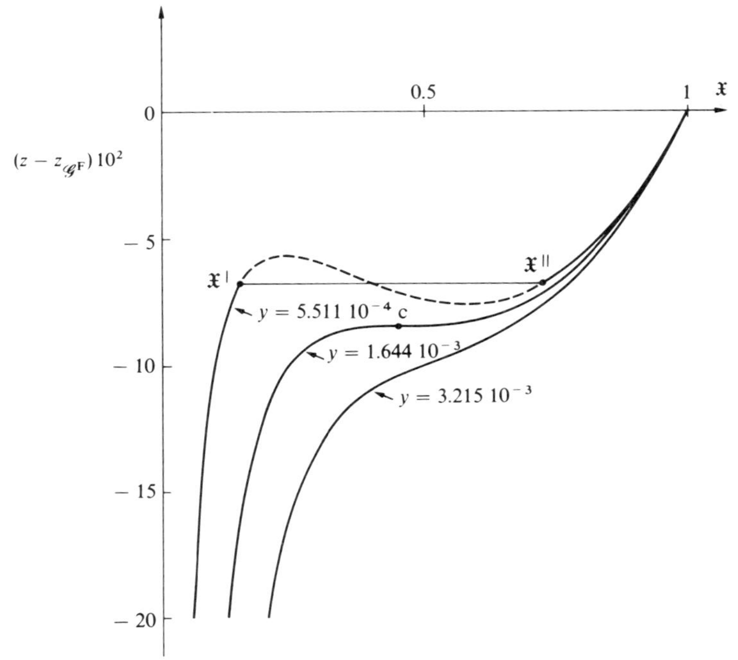

Fig. 14.7. Curve representing $z(\mathfrak{X}) - z_{\mathscr{G}F}$ for various values of y, according to Flory's theory. For $y < 0.001\,644$, the coil–globule transition occurs in a discontinuous manner. Point c is the critical point.

predict the sign of the 'renormalized' three-body interaction. We shall see later that the correct theory gives the inequality $T_{\mathscr{G}F} > T_F$ and not the contrary which is predicted by Flory's formula (14.3.7).

Nevertheless, let us examine Flory's results in greater detail. In first approximation, we can assume that $z - z_{\mathscr{G}F}$ is proportional to $T - T_{\mathscr{G}F}$. Thus, it is worthwhile to plot $z - z_{\mathscr{G}F}$ against $\mathfrak{X}$, for $d = 3$.

$$z(\mathfrak{X}) - z_{\mathscr{G}F} = 2 \times 3^{-3/2}[\mathfrak{X}^{5/2} - \mathfrak{X}^{3/2} - 8 \times 3^{1/2} y(\mathfrak{X}^{-3/2} - 1)],$$

and consequently

$$z'(\mathfrak{X}) = 3^{-3/2}\mathfrak{X}^{-5/2}(5\mathfrak{X}^4 - 3\mathfrak{X}^3 + 24 \times 3^{1/2} y).$$

Three different situations may occur

(1) $y > \dfrac{243 \times 3^{1/2}}{256\,000} = 0.001\,644.$

In this case, $z'(\mathfrak{X}) > 0$ for any $\mathfrak{X}$; when z decreases, the swelling decreases in a regular manner.

(2) $y = \dfrac{243 \times 3^{1/2}}{256\,000}$.

The value of y defines a critical curve. At the critical point $\mathfrak{X}_c = 9/20$. $z_c - z_{\mathscr{G}F} = -0.0842$, we have $z'(9/20) = 0$, $z''(9/20) = 0$.

(3) $y < \dfrac{243 \times 3^{1/2}}{256\,000}$.

In this case, the sign of $z'(\mathfrak{X})$ changes when $\mathfrak{X}$ varies. The curve representing $\mathbb{F}(\mathfrak{X})$ for a fixed value of z has two minima and a maximum: the corresponding values of $\mathfrak{X}$ are solutions of the equation

$$z(\mathfrak{X}) = z$$

which coincides with (14.3.7)

Let $\mathfrak{X}^{\text{I}}$ and $\mathfrak{X}^{\text{II}}$ be the values of $\mathfrak{X}$ corresponding to the minima (with $\mathfrak{X}^{\text{I}} < \mathfrak{X}^{\text{II}}$). Equations (14.3.6) and (14.3.7) show that, for $d = 3$,

$$\mathbb{F}'(\mathfrak{X}) = \frac{3}{4} 3^{3/2} \mathfrak{X}^{-5/2} [z(\mathfrak{X}) - z],$$

and consequently

$$\mathbb{F}(\mathfrak{X}^{\text{II}}) - \mathbb{F}(\mathfrak{X}^{\text{I}}) = \frac{3}{4} 3^{3/2} \int_{\mathfrak{X}^{\text{I}}}^{\mathfrak{X}^{\text{II}}} \mathrm{d}\mathfrak{X}\, \mathfrak{X}^{-5/2} [z(\mathfrak{X}) - z]. \qquad (14.3.9)$$

The sign of the integral enables one to determine the relative stability of these minima. The transition occurs for a value z_0 determined by the condition

$$\int_{\mathfrak{X}^{\text{I}}(z_0)}^{\mathfrak{X}^{\text{II}}(z_0)} \mathrm{d}\mathfrak{X}\, \mathfrak{X}^{-5/2} [z(\mathfrak{X}) - z_0] = 0.$$

This equality reminds us Maxwell's rule (see Appendix L). In this way, we can determine the stability domains of the curve $z(\mathfrak{X})$ (see Fig. 14.7). (In particular, for $y = 10$, we find $z_0 = 0$.)

We can also write the preceding condition in the form

$$[\mathfrak{X} - \ln \mathfrak{X} + 8 \times 3^{-1/2} y \mathfrak{X}^{-3} + 3^{1/2} z_0 \mathfrak{X}^{-3/2}]_{\mathfrak{X}^{\text{I}}(z_0)}^{\mathfrak{X}^{\text{II}}(z_0)} = 0,$$

and if we know $\mathfrak{X}(z)$, we can deduce z_0, $\mathfrak{X}^{\text{I}}$ and $\mathfrak{X}^{\text{II}}$ from this equation.

The calculation shows that for $y < 0.00164$ the coil–globule transition is discontinuous. As will be shown later, it seems that Lifshitz theory also leads to the same conclusion.

Nevertheless, we must be aware of the fact that like any system depending on a finite number of parameters, a chain with N links cannot have abrupt phase transitions. We cannot be as categorical, for finite continuous chains, but, in this case also, the coil–globule transition might always be continuous, the discontinuity appearing only when the size of the chain becomes infinite.

4. ISOLATED CHAIN: LIFSHITZ THEORY

4.1 Model and method

To describe the coil–globule transition in the vicinity of the Flory temperature, in 1968 Lifshitz[8] proposed an original method which amounts to expressing the free energy of a polymer in terms of the local monomer concentration. This method was subsequently reexamined and developed in various articles.[2] We shall describe it here but, for reasons of convenience, we shall use a slightly different formalism.

Lifshitz represents a polymer containing N monomers as a chain with N independent links subjected to interactions. Let $\mathbf{\Omega}$ be the configuration of such a polymer, ${}^{0}\mathscr{W}(\mathbf{\Omega})$ the weight associated with the free chain and $\mathscr{U}(\mathbf{\Omega})$ the interaction energy (in thermal units $1/\beta$). The partition function Z can be written in the following symbolic form

$$Z = \frac{\sum_{\mathbf{\Omega}} \mathrm{e}^{-\mathscr{U}(\mathbf{\Omega})}\, {}^{0}\mathscr{W}(\mathbf{\Omega})}{\sum_{\mathbf{\Omega}} {}^{0}\mathscr{W}(\mathbf{\Omega})} \tag{14.4.1}$$

(here the denominator acts as a single renormalization factor).

Let $C_{\mathbf{\Omega}}(\vec{r})$ be the monomer concentration corresponding to configuration $\mathbf{\Omega}$. We can a priori require this concentration to have a given value $C(\vec{r})$ at each point $\vec{r}$. In this way, we define a restricted partition function $Z\{C\}$

$$Z\{C\} = \frac{\sum_{\mathbf{\Omega}} \mathrm{e}^{-\mathscr{U}(\mathbf{\Omega})}\delta\{C_{\mathbf{\Omega}} - C\}\, {}^{0}\mathscr{W}(\mathbf{\Omega})}{\sum_{\mathbf{\Omega}} {}^{0}\mathscr{W}(\mathbf{\Omega})} \tag{14.4.2}$$

($\{C\}$ means the set of values $C(\vec{r})$)
It is convenient to write $Z\{C\}$ in the form of a product

$$Z\{C\} = Z_{E}\{C\} Z_{S}\{C\}$$

($\mathbb{E}$ for energy, $\mathbb{S}$ for entropy) where

$$Z_{E}\{C\} = \frac{\sum_{\mathbf{\Omega}} \mathrm{e}^{-\mathscr{U}(\mathbf{\Omega})}\delta\{C_{\mathbf{\Omega}} - C\}\, {}^{0}\mathscr{W}(\mathbf{\Omega})}{\sum_{\mathbf{\Omega}} \delta\{C_{\mathbf{\Omega}} - C\}\, {}^{0}\mathscr{W}(\mathbf{\Omega})}, \tag{14.4.3}$$

$$Z_{S}\{C\} = \frac{\sum_{\mathbf{\Omega}} \delta\{C_{\mathbf{\Omega}} - C\}\, {}^{0}\mathscr{W}(\mathbf{\Omega})}{\sum_{\mathbf{\Omega}} {}^{0}\mathscr{W}(\mathbf{\Omega})}. \tag{14.4.4}$$

The free energy $\mathbb{F}\{C\}$ is defined by

$$\mathbb{F}\{C\} = -\ln Z\{C\}$$

and can be expressed in the form

$$\begin{aligned}\mathbb{F}\{C\} &= \mathbb{E}\{C\} - \mathbb{S}\{C\}\\ \mathbb{E}\{C\} &= -\ln Z_E\{C\}\\ \mathbb{S}\{C\} &= \ln Z_S\{C\}.\end{aligned} \tag{14.4.5}$$

Here $\mathbb{E}\{C\}$ plays the role of an energy and $\mathbb{S}\{C\}$ of an entropy. Lifshitz's method consists in finding proper approximate expressions for $\mathbb{S}\{C\}$ and $\mathbb{E}\{C\}$, and in determining the mean concentration $C(\vec{r})$ by minimization of the free energy

$$\frac{\partial \mathbb{F}\{C\}}{\partial C(\vec{r})} = 0. \tag{14.4.6}$$

An expression of $\mathbb{S}\{C\}$ is obtained by studying how the behaviour of a Brownian chain is influenced by a potential $V(\vec{r})$ which acts on each point of this chain. Lifshitz used this technique, for a chain with N links, but here we apply it to a (practically equivalent) continuous chain. On the other hand, $\mathbb{E}\{C\}$ will be deduced from properties of solutions of *free monomers*.

Let us now show how the ideas put forward by Lifshitz can be used to calculate $\mathbb{S}\{C\}$ and $\mathbb{E}\{C\}$.

4.2 Configuration of a Brownian chain in an attractive external potential

In order to calculate the entropy $\mathbb{S}\{C\}$ of a real chain in terms of the local monomer concentration $C(\vec{r})$, we set the corresponding Brownian chain in an external potential $V(\vec{r})$, the result being that the mean link concentration at point $\vec{r}$ is $C(\vec{r})$. In this case, the entropy of the chain is a function of the potential but, on the other hand, there is a relation between $V(\vec{r})$ and $C(\vec{r})$ and consequently the entropy can be re-expressed in terms of $C(\vec{r})$. First, let us study the statistical state of the chain. The weight associated with a Brownian chain of area S is given by (see Chapter 2)

$${}^{0}\mathscr{W} = \exp\left[-\frac{1}{2}\int_0^S ds \left(\frac{d\vec{r}}{ds}\right)^2\right]. \tag{14.4.7}$$

When an attractive potential acts on this chain, the new weight is given by

$$\mathscr{W} = {}^{0}\mathscr{W} \exp\left[-\int_0^S ds\, V[\vec{r}(s)]\right]. \tag{14.4.8}$$

Let $\mathscr{Z}(\vec{x}, \vec{y}; S)$ be the (renormalized) partition function of a chain for which $\vec{x}$ is the position vector of the origin and $\vec{y}$ the position vector of the end point. We have

$$\mathscr{Z}(\vec{x}, \vec{y}; S) = \frac{\left\langle\!\!\left\langle \delta(\vec{r}(0) - \vec{x})\delta(\vec{r}(s) - \vec{y})\exp\left[-\int_0^S \mathrm{d}s\, V(\vec{r}(s))\right]\right\rangle\!\!\right\rangle_0}{\langle\!\langle \delta(\vec{r}(0) - \vec{x})\rangle\!\rangle_0} \tag{14.4.9}$$

where the mean values are taken with the weight ${}^0\mathscr{W}$. In the absence of interaction, we have

$${}^0\mathscr{Z}(\vec{x}, \vec{y}; S) = \frac{1}{(2\pi S)^{3/2}}\mathrm{e}^{-(\vec{y} - \vec{x})^2/2S}$$

and we verify that this function is defined, for $S \geqslant 0$ by the equation

$$\frac{\partial}{\partial S}{}^0\mathscr{Z}(\vec{x}, \vec{y}; S) = \frac{1}{2}\Delta_y {}^0\mathscr{Z}(\vec{x}, \vec{y}; S), \tag{14.4.10}$$

and by the condition

$${}^0\mathscr{Z}(\vec{x}, \vec{y}, 0) = \delta(\vec{y} - \vec{x}).$$

When interactions are present, we can define $\mathscr{Z}(\vec{x}, \vec{y}, S)$ as a solution of the equation

$$\mathscr{Z}(\vec{x}, \vec{y}; S) = {}^0\mathscr{Z}(\vec{x}, \vec{y}; S) - \int_0^S \mathrm{d}s \int \mathrm{d}^3z\, \mathscr{Z}(\vec{x}, \vec{z}; s)\, V(\vec{z})\, {}^0\mathscr{Z}(\vec{z}, \vec{y}; S - s) \tag{14.4.11}$$

which can be easily deduced from (14.4.9) by expanding the exponential which appears in the numerator. The solution must satisfy the condition

$$\mathscr{Z}(\vec{x}, \vec{y}; 0) = \delta(\vec{y} - \vec{x}). \tag{14.4.12}$$

Now let us calculate $\partial\mathscr{Z}(\vec{x}, \vec{y}; S)/\partial S$. For $S > 0$, equations (14.4.10) and (14.4.11) give

$$\frac{\partial}{\partial S}\mathscr{Z}(\vec{x}, \vec{y}; S) = \left[\frac{1}{2}\Delta_y - V(\vec{y})\right]\mathscr{Z}(\vec{x}, \vec{y}; S). \tag{14.4.13}$$

Finally we note that (14.4.12) and (14.4.13) can also be written as follows (for any S)

$$\frac{\partial}{\partial S}\mathscr{Z}(\vec{x}, \vec{y}; S) = \left[\frac{1}{2}\Delta_y - V(\vec{y})\right]\mathscr{Z}(\vec{x}, \vec{y}; S) + \delta(S)\delta(\vec{y} - \vec{x}) \tag{14.4.14}$$

with the condition

$$\mathscr{Z}(\vec{x}, \vec{y}; S) = 0 \qquad \text{for } S < 0.$$

Let us consider the eigenfunctions $\psi_l(\vec{r})$ which are solutions of the equations

$$\left[-\tfrac{1}{2}\Delta + V(\vec{r})\right]\psi_l(\vec{r}) = \omega_l \psi_l(\vec{r}) \tag{14.4.15}$$

where ω_l is the eigenvalue associated with $\psi_l(r)$. These functions $\psi_l(\vec{r})$ can be choosen so as to form an orthonormal basis.

We then see immediately that the solution of eqn (14.4.14) is

$$\mathscr{Z}(\vec{x}, \vec{y}; S) = \sum_l \mathrm{e}^{-S\omega_l} \psi_l(\vec{x})\psi_l(\vec{y}). \tag{14.4.16}$$

Moreover, the total partition function $\mathscr{Z}(S)$ is given by

$$\mathscr{Z}(S) = \int \mathrm{d}^3x \int \mathrm{d}^3y\, \mathscr{Z}(\vec{x}, \vec{y}; S) = \sum_l (a_l)^2 \mathrm{e}^{-S\omega_l} \tag{14.4.17}$$

where

$$a_l = \int \mathrm{d}^3x\, \psi_l(x). \tag{14.4.18}$$

When $S \to \infty$, the dominant term is given by the ground state ($l = 0$) and, moreover, it is known that $\psi_0(\vec{x})$ is the only eigenfunction remaining positive for any $\vec{x}$.

If the ground state is a bound state, we say that the chain is a globule. In this case, the free energy $\mathbb{F}(S) = -\ln \mathscr{Z}(S)$ can be approximated for $S \gg 1$ by

$$\mathbb{F}(S) \simeq S\omega_0 \tag{14.4.19}$$

which is a direct consequence of (14.4.17).

Let us now introduce the monomer concentration. More precisely, as we deal with a continuous chain, we define the area concentration by setting

$$\mathscr{C}(\vec{r}) = \int_0^S \mathrm{d}s \langle\!\langle \delta(\vec{r} - \vec{r}(s)) \rangle\!\rangle. \tag{14.4.20}$$

We verify that

$$\int \mathrm{d}^3r\, \mathscr{C}(\vec{r}) = S.$$

However, it is possible to define a genuine monomer concentration by establishing a correspondence between the continuous chain and a chain with N links. Accordingly, we set

$$S = Nl^2$$
$$C(\vec{r}) = l^{-2}\mathscr{C}(\vec{r}) \tag{14.4.21}$$

where l is a length characterizing the size of a link. Thus, we have

$$\int \mathrm{d}^3r\, C(\vec{r}) = N. \tag{14.4.22}$$

Let us come back to $\mathscr{C}(\vec{r})$ and express this function in terms of partition functions. We consider a chain of area S and a given value s, with $0 < s < S$; we assume that $\vec{r}(0) = \vec{x}$, $\vec{r}(S) = \vec{y}$ and $\vec{r}(s) = \vec{z}$. As the chain does not interact with itself, there are no correlations between the first part of area s and the second part of area $S - s$; therefore, the partition function associated with such a configuration is $\mathscr{Z}(\vec{x}, \vec{z}; s)\, \mathscr{Z}(\vec{z}, \vec{y}; S - s)$. The definition (14.4.20) of $\mathscr{C}(\vec{r})$ can thus be expressed in the form

$$\mathscr{C}(\vec{r}) = \int_0^S \mathrm{d}s \, \frac{\int \mathrm{d}^3x \int \mathrm{d}^3y \, \mathscr{Z}(\vec{x}, \vec{r}; s)\mathscr{Z}(\vec{r}, \vec{y}; S-s)}{\int \mathrm{d}^3x \int \mathrm{d}^3y \, \mathscr{Z}(\vec{x}, \vec{y}; S)} \qquad (14.4.23)$$

Taking eqns (14.4.16) and (14.4.18) into account, we also find

$$\mathscr{C}(\vec{r}) = \int_0^S \mathrm{d}s \, \frac{\sum_{lm} \exp[-s\omega_l - (S-s)\omega_m] a_l a_m \psi_l(\vec{r})\psi_m(\vec{r})}{\sum_n (a_n)^2 \mathrm{e}^{-S\omega_n}}.$$

If the ground state is a bound state and if S is large ($S \gg 1/\omega_0$), we obtain the very simple approximate formula

$$\mathscr{C}(\vec{r}) \simeq S[\psi_0(\vec{r})]^2. \qquad (14.4.24)$$

Finally, we note that the average energy of a chain in the potential V is given by

$$\mathbb{E}_V(S) = \int \mathrm{d}^3r \, \mathscr{C}(\vec{r}) V(\vec{r}) \simeq S \int \mathrm{d}^3r \, V(\vec{r})[\psi_0(\vec{r})]^2, \qquad (14.4.25)$$

and this is a first result.

4.3 Calculation of the entropy $\mathbb{S}\{C\}$

The entropy $\mathbb{S}(S)$ is given by

$$\mathbb{S}(S) = \mathbb{E}_V(S) - \mathbb{F}(S)$$

where, in first approximation, $\mathbb{E}_V(S)$ and $\mathbb{F}(S)$ are given by (14.4.25) and (14.4.19)

$$\mathbb{S} = S\left(\int \mathrm{d}^3r \, V(\vec{r})[\psi_0(\vec{r})]^2 - \omega_0\right).$$

Moreover, (14.4.15) shows that

$$\omega_0 = \int \mathrm{d}^3r \left[\frac{1}{2}(\vec{\nabla}\psi_0(\vec{r}))^2 + V(\vec{r})(\psi_0(\vec{r}))^2\right].$$

Consequently,

$$\mathbb{S}(S) = -\frac{S}{2}\int \mathrm{d}^3r [\vec{\nabla}\psi_0(\vec{r})]^2. \qquad (14.4.26)$$

Incidently, we note that this entropy which is defined only up to an additive constant is negative, and that it becomes increasingly negative as the gradient increases, and therefore as the size of the globule decreases.

Using (14.4.24), we can now express $\mathbb{S}(S)$ in terms of $\mathscr{C}(\vec{r})$. In this way, we obtain

$$\mathbb{S}(S) = -\frac{1}{8}\int \mathrm{d}^3r \frac{1}{\mathscr{C}(\vec{r})}[\vec{\nabla}\mathscr{C}(\vec{r})]^2. \qquad (14.4.27)$$

This result can be applied to a long chain made of discrete links by using the correspondence given by (14.4.21). Moreover, as $V(\vec{r})$ is arbitrary, we can consider the entropy as a functional of the monomer concentration. Thus, we obtain

$$\mathbb{S}\{C\} = -\frac{l^2}{8}\int \mathrm{d}^3 r \frac{1}{C(\vec{r})}[\vec{\nabla} C(\vec{r})]^2 \tag{14.4.28}$$

which is the result found by Lifshitz.

4.4 An estimate of the energy $\mathbb{E}\{C\}$ deduced from properties of solutions of free monomers

The interactions between monomers of a chain consist of very short-range repulsive forces (hard core) and of attractive forces with a slightly longer range (van der Waals forces). We may assume that the forces exerted between free independent monomers are the same as the forces exerted between chain monomers. Lifshitz used this idea to estimate $E\{C\}$ without performing detailed calculations of the effect of the interactions on a microscopic scale. He assumed that the mean concentration at a point does not vary much over distances of the order of the interaction range, this range being of the same order of magnitude as the size of the links. As a consequence, in order to simplify the calculation of $Z_E\{C\}$, he neglected the fact that the monomers are linked (what Lifshitz calls *linear memory*). This simplification enables one to write eqn (14.4.3) in the approximative form

$$Z_E\{C\} \simeq \frac{\sum_{\Omega} \mathrm{e}^{-\mathscr{V}(\Omega)} \delta(C_\Omega - C)}{\sum_{\Omega} \delta(C_\Omega - C)}$$

If $C(\vec{r})$ is a constant $[C(\vec{r}) = C]$, we see that

$$\mathbb{E}(C) = -\ln Z_E(C) = \mathbb{E}_{\mathrm{m}}(C) - {}^0\mathbb{E}_{\mathrm{m}}(C) \tag{14.4.29}$$

for a monomer solution of concentration C where $\mathbb{E}_{\mathrm{m}}(C)$ is the free energy of the interacting monomer and ${}^0\mathbb{E}_{\mathrm{m}}(C)$ the free energy in the absence of interaction.

For these monomers, the free energy per unit volume is

$$\boldsymbol{f}(C) = \mathbb{E}_{\mathrm{m}}(C)/V \tag{14.4.30}$$

and the corresponding chemical potential $\boldsymbol{\mu}(C)$ is given by

$$\boldsymbol{\mu}(C) = \frac{\partial \boldsymbol{f}(C)}{\partial C}. \tag{14.4.31}$$

In the following, it will be convenient to use the notation

$$\boldsymbol{f}^{\vdash}(C) = \boldsymbol{f}(C) - {}^0\boldsymbol{f}(C)$$

$$\boldsymbol{\mu}^{\vdash}(C) = \frac{\partial \boldsymbol{f}^{\vdash}}{\partial C} = \boldsymbol{\mu}^{\vdash}(C) = \boldsymbol{\mu}(C) - {}^0\boldsymbol{\mu}(C) \tag{14.4.32}$$

and, in this way, (14.4.29) can be rewritten in the form

$$\mathbb{E}(C) = V\boldsymbol{f}^{\vdash}(C). \tag{14.4.33}$$

Let us now assume that $C(\vec{r})$ is a slowly varying variable; we can see that in this case and in agreement with the preceding equation we have

$$\mathbb{E}\{C\} = -\ln Z_E\{C\} = \int d^3r\, \boldsymbol{f}^{\vdash}(C(\vec{r})). \tag{14.4.34}$$

4.5 Minimization of the free energy

The free energy of a polymer chain

$$\mathbb{F}\{C\} = -\mathbb{S}\{C\} + \mathbb{E}\{C\}$$

is given in terms of $C(\vec{r})$ by eqns (14.4.28) and (14.4.34)

$$\mathbb{F}\{C\} = \int d^3r \left[\frac{l^2}{8}\frac{1}{C(\vec{r})}[\vec{\nabla}C(\vec{r})]^2 + \boldsymbol{f}^{\vdash}(C(\vec{r}))\right]. \tag{14.4.35}$$

The first term comes from the entropy and takes implicitly into account the fact that we deal with a chain (linear memory); the second term comes from the energy and is derived from the properties of a monomer solution. With Lifshitz, we grant that the equilibrium concentration is given by the functional derivative

$$\frac{\partial}{\partial C(\vec{r})}\left[\mathbb{F}\{C\} - \lambda\int \mathrm{d}^3r' C(\vec{r}\,')\right] = 0$$

where λ is a Lagrange multiplier, introduced in order to account for condition (14.4.22). (Let us remember that by definition $\partial C(\vec{r})/\partial C(\vec{r}) = \partial(\vec{r}, \vec{r}\,')$.)

Finally, by taking eqns (14.4.35) and (14.4.32) into account, we can write the above equation in the explicit form

$$\frac{l^2}{2}\Delta[C(\vec{r})]^{1/2} - [\boldsymbol{\mu}^{\vdash}(C(\vec{r})) - \lambda]\,[C(\vec{r})]^{1/2} = 0. \tag{14.4.36}$$

4.6 A study of the chemical potential $\boldsymbol{\mu}^{\vdash}(C)$

To study the form of the solutions of (14.4.36), we must know what $\boldsymbol{\mu}^{\vdash}(C)$ looks like. In particular, we can obtain this quantity from the osmotic pressure $\boldsymbol{\Pi}$ of a monomer solution. $\boldsymbol{\Pi}$ being expressed in mechanical units and $\mathbb{F}$ in thermal units $1/\boldsymbol{\beta}$, we have

$$\boldsymbol{\Pi\beta} = -\left.\frac{\partial \mathbb{F}}{\partial \boldsymbol{V}}\right|_N = -\frac{\partial}{\partial \boldsymbol{V}}[\boldsymbol{V}\boldsymbol{f}(N/\boldsymbol{V})],$$

i.e.

$$\boldsymbol{\beta\Pi}(C) = C\boldsymbol{f}'(C) - \boldsymbol{f}(C).$$

This equation can be integrated in the form

$$\boldsymbol{f}(C) = C \int_0^C dx \left[\frac{\beta \Pi(x) - x}{x^2} \right] + C \ln(C/C_0)$$

where C_0 is a constant whose value is not really useful. In particular, in the absence of interactions

$$^{0}\boldsymbol{f}(C) = C \ln(C/C_0)$$
$$^{0}\boldsymbol{\mu}(C) = 1 + \ln(C/C_0) \tag{14.4.37}$$

which implies

$$\boldsymbol{f}^{\vdash}(C) = C \int_0^C dx \left[\frac{\beta \Pi(x) - x}{x^2} \right]. \tag{14.4.38}$$

In first approximation, we can content ourselves with the three first virial coefficients

$$\beta\Pi(C) = C - AC^2 + BC^3. \tag{14.4.39}$$

In this case (14.4.38) and (14.4.31) give

$$\boldsymbol{f}^{\vdash}(C) = -AC^2 + \frac{BC^3}{2}$$

$$\boldsymbol{\mu}^{\vdash}(C) = -2AC + \frac{3BC^2}{2}$$

and since we consider the attractive case, we can assume that *A and B* are positive.

If $A^2 - 3B < 0$, $\Pi(C)$ increases steadily with C. On the contrary, if $A^2 - 3B > 0$, demixtion occurs and $\Pi(C)$ remains constant for $C^{\text{I}} < C < C^{\text{II}}$. The concentrations C^{I} and C^{II} are given by the conditions

$$\Pi(C^{\text{II}}) = \Pi(C^{\text{I}})$$
$$\boldsymbol{\mu}(C^{\text{II}}) = \boldsymbol{\mu}(C^{\text{I}}) \tag{14.4.40}$$

and the second equality can also be transformed with the help of (14.4.32) and (14.4.37) so as to give

$$\boldsymbol{\mu}^{\vdash}(C^{\text{II}}) - \boldsymbol{\mu}^{\vdash}(C^{\text{I}}) + \ln(C^{\text{II}}/C^{\text{I}}) = 0. \tag{14.4.41}$$

The explicit form of the preceding conditions is

$$B(C^{\text{II}^2} + C^{\text{II}}C^{\text{I}} + C^{\text{I}^2}) - A(C^{\text{II}} + C^{\text{I}}) + 1 = 0$$

$$(C^{\text{II}} - C^{\text{I}})[3B(C^{\text{II}} + C^{\text{I}}) - 4A] - 2\ln(C^{\text{II}}/C^{\text{I}}) = 0.$$

Thus in the present case $\boldsymbol{\mu}(C)$ is not defined in the interval $C^{\text{I}} < C < C^{\text{II}}$. The situation is depicted in Fig. 14.8.

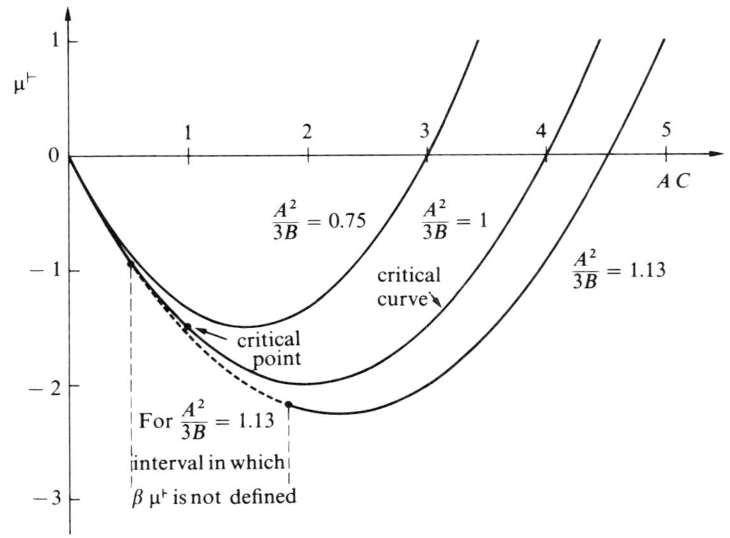

Fig. 14.8. A plot of $\mu^{\vdash}(C)$ for the simple approximation described in the text. The critical point is indicated on the critical curve. For $A^2/3B > 1$, there exists an interval $C^{\mathrm{I}} < C < C^{\mathrm{II}}$ in which $\mu(C)$ is not defined.

4.7 Study of the shape of a globule

The shape of a globule is given by the function $C(\vec{r})$, a solution of eqn (14.4.36). Let us note, with Lifshitz, that, if the term coming from the entropy were neglected, the equation would reduce to

$$[\mu^{\vdash}(C(\vec{r})) - \lambda][C(\vec{r})]^{1/2} = 0.$$

In this case, the solution is discontinuous (see Fig. 14.9). For $r > R$, $C(\vec{r}) = 0$ and for $r < R$, $C(\vec{r}) = C$.

The values of R and C are determined as follows. Equation (14.4.34) gives

$$\mathbb{E} = \frac{4\pi R^3}{3} \int_0^C \mathrm{d}x\, \mu^{\vdash}(x)$$

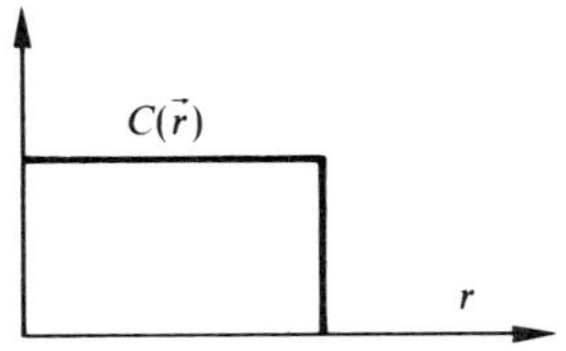

Fig. 14.9. The concentration $C(\vec{r})$ obtained when the entropy term is neglected.

[eqn (14.4.32) is taken into account] and, moreover,

$$\frac{4\pi R^3}{3} C = N. \tag{14.4.42}$$

Consequently,

$$\mathbb{E} = \frac{N}{C} \int_0^C \mathrm{d}x \, \boldsymbol{\mu}^{\vdash}(x).$$

The minimization of $\mathbb{E}$ with respect to C provides us with the condition

$$C \boldsymbol{\mu}^{\vdash}(C) = \int_0^C \mathrm{d}x \, \boldsymbol{\mu}^{\vdash}(x). \tag{14.4.43}$$

The solution of this equation is given by a process analogous to the Maxwell rule (see Fig. 14.10) and R is deduced from (14.4.42).

In reality, the entropy term cannot be neglected and actually $C(\vec{r})$ is the solution of eqn (14.4.36). Then, two cases may occur

1. $T > T_c$; then $\boldsymbol{\mu}^{\vdash}(C)$ is a continuous function defined for any C. In this case, the solution $C(\vec{r})$ is also continuous (see Fig. 14.11) and can be calculated without special difficulties.

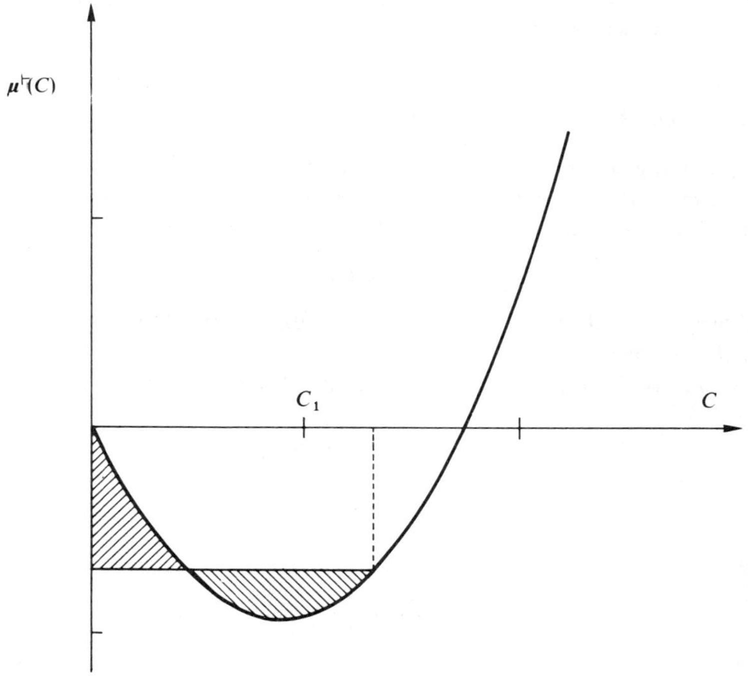

Fig. 14.10. Determination of the solution $C = C_1$ of (eqn (14.4.43). The hachured parts have the same area.

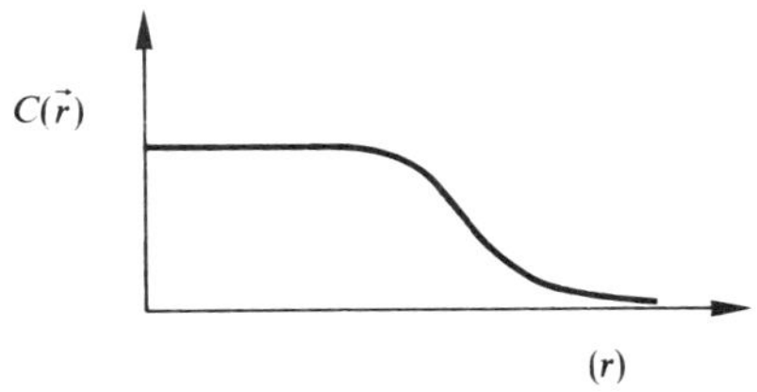

Fig. 14.11. Form of a globule when $\mu^{\vdash}(C)$ is well-defined for any C. This is so for $T > T_c$. It is also so for $T < T_c$, provided $\mu^{\vdash}(C)$ is interpolated in the interval $C^{\mathrm{I}} < C < C^{\mathrm{II}}$.

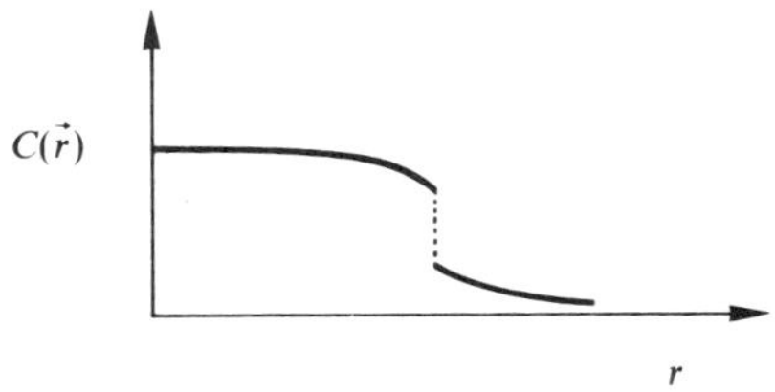

Fig. 14.12. Another possible form of a globule according to Lifshitz[8,9] for $T < T_c$.

2. $T < T_c$, then $\mu^{\vdash}(C)$ is defined only outside the interval $C^{\mathrm{I}} < C < C^{\mathrm{II}}$. In this case, Lifshitz assumes that $C(\vec{r})$ has no values in the interval and is therefore discontinuous (see Fig. 14.12). This conclusion is open to question. In fact, we can admit that $\mu^{\vdash}(C)$ remains bounded and if a solution $C(\vec{r})$ of (14.4.36) exists, it must be continuous since $\Delta[C(\vec{r})]$ must remain bounded. Now, if $C(\vec{r})$ is continuous, there exists no interval $C^{\mathrm{I}} < C < C^{\mathrm{II}}$ in which $\mu^{\vdash}(\vec{r})$ is not defined, since $\mu^{\vdash}[C(\vec{r})]$ is related to $C(\vec{r})$ by (14.4.36). This discussion shows that in fact there is no solution for (14.4.36). However, it would be possible to give a meaning to this equation for any C, by interpolating $\mu^{\vdash}(C)$. Such an operation is not difficult to perform (see Fig. 14.8, for instance) but could be done in various ways. In all cases, the solution $C(\vec{r})$ thus obtained would be continuous.

The method proposed by Lifshitz has been discussed by him and by his collaborators[9] in some detail, but it appears that until now the method has not been used to interpret precise experimental facts. This is perhaps unfortunate.

5. POLYMER SOLUTIONS: FLORY–HUGGINS THEORY IN THE ATTRACTIVE CASE

5.1 Generalities: limit $N \to \infty$

Flory–Huggins theory was studied in detail in Chapter 8, Section 2.1. We recall that in this model, $\mathcal{N}$ self-avoiding chains of $(N - 1)$ links (N monomers) are drawn on a lattice. The number of sites is defined by V/v_r where V is the total volume and v_r the volume per site. The parameter χ represent an attractive interaction between chains. The free energy is given by (8.2.11)

$$\mathbb{F} = -\chi N^2 \mathcal{N}^2 v_r/V + \mathcal{N}\ln(\mathcal{N}v_r/V) + [V/v_r - N\mathcal{N}]\ln(1 - N\mathcal{N}v_r/V) + \mathcal{N}\mu_0$$

and from it one can deduce the chemical potential μ and the osmotic pressure Π, which can be expressed in terms of the volume fraction $C = \mathcal{N}/V$ [see (8.2.13), (8.2.14)]

$$\mathbb{F}/V = -\chi N^2 C^2 v_r + C\ln(Cv_r) + [-NC + 1/v_r]\ln(1 - NCv_r) + C\mu_0 \quad (14.5.1)$$

$$\mu = -2\chi N^2 Cv_r + \ln(Cv_r) - N\ln(1 - NCv_r) + \mu_1$$

$$\Pi\beta = C\left[1 - N - \frac{1}{Cv_r}\ln(1 - NCv_r) - \chi N^2 Cv_r\right] \quad (14.5.2)$$

(here μ_0 and μ_1 are constants).

First, let us examine what happens when the number N of links becomes very large. In this case, it is convenient to express the osmotic pressure in terms of the volume fraction $\varphi = NCv_r$ (number of monomers per site)

$$\Pi\beta = C\left[1 - N\left(1 + \frac{1}{\varphi}\ln(1 - \varphi) + \chi\varphi\right)\right]$$

$$= C\left[1 + N\left(\left(\frac{1}{2} - \chi\right)\varphi + \frac{\varphi^2}{3} + \ldots\right)\right]$$

In this context, the Flory temperature T_F is defined by the condition that the second virial coefficient vanishes, a fact which occurs when $\chi = 1/2$. Consequently, in the vicinity of T_F, we have

$$\frac{1}{2} - \chi \simeq A_F\left(1 - \frac{\beta}{\beta_F}\right) = A_F\left(\frac{T - T_F}{T}\right).$$

When N goes to infinity, one part of the demixtion curve coincides with the axis $C = 0$. The limit of the demixtion curve is given by the equation $\Pi = 0$ which in this case reads

$$1 + \frac{1}{\varphi}\ln(1 - \varphi) + \chi\varphi = 0$$

(see Fig. 14.13).

In the semi-dilute limit, φ is small and the limit of the demixtion curve is given by the straight line

$$\varphi = -3\left(\frac{1}{2} - \chi\right) \simeq -3A_F\left(\frac{T - T_F}{T}\right). \quad (14.5.3)$$

It will be shown later that this law is not quite correct: in fact, renormalization theory shows that there exist additional logarithmic terms (see Section 6).

Let us now study demixtion for finite values of N.

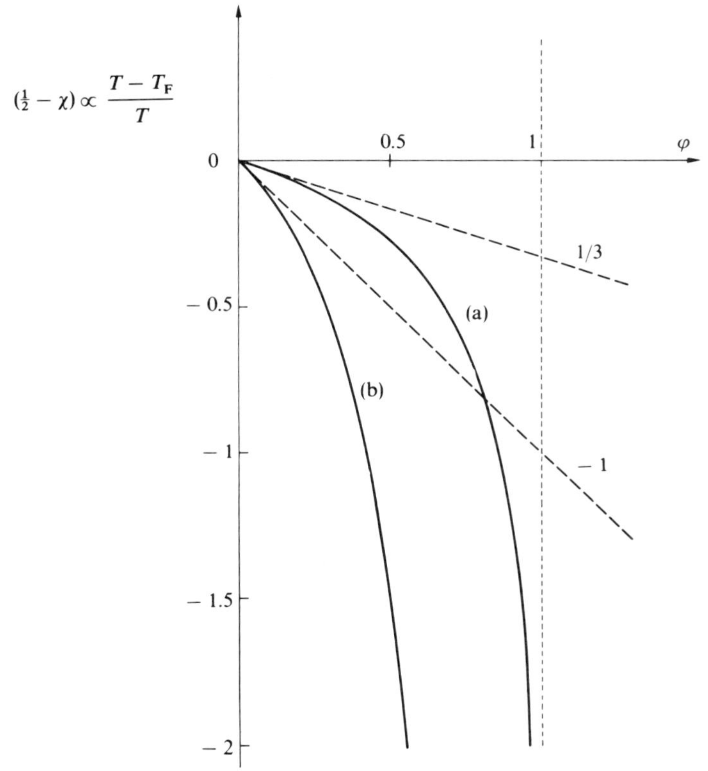

Fig. 14.13. Flory–Huggins theory.
Curve (a): limit of the demixtion curve when $N \to \infty$. The abscissa φ is the volume fraction $\varphi = NCv_r$ and the ordinate $(\frac{1}{2} - \chi)$ is proportional to $(T - T_F)/T$.
Curve (b): Curve representing the variation of the top of the demixtion curve when N varies.

5.2 A study of the demixtion curve for finite N

At a given temperature, demixtion occurs when the concavity of the curve representing the free energy $\mathbb{F}$ as a function of $\boldsymbol{C}$ is not uniformly directed upwards (see Fig. 14.14). It is also possible to argue by considering the chemical potential. Demixtion occurs at a given temperature if the chemical potential can take the same value for two different concentrations C^{I} and C^{II} (thus $\boldsymbol{\mu}^{\mathrm{I}} = \boldsymbol{\mu}^{\mathrm{II}}$; see Fig. 14.15). In this case, the points constituting the demixtion curve are given by Maxwell's rule (equal areas).

According to (14.5.2) with $\boldsymbol{\mu}_1 = \boldsymbol{\mu}_0 - \ln N + N - 1$, we have

$$\boldsymbol{\mu} - \boldsymbol{\mu}_1 = -2\chi N\varphi + \ln\varphi - N\ln(1 - \varphi)$$

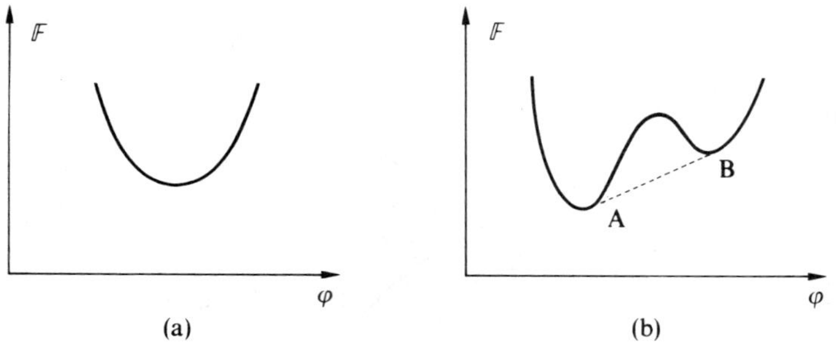

Fig. 14.14. The free energy $\mathbb{F}$ as a function of the volume fraction φ at a given temperature:
(a) when the solution is always stable;
(b) when demixtion occurs; the points A and B belong to the demixtion curve; the points with zero curvature between A and B belong to the spinodal.

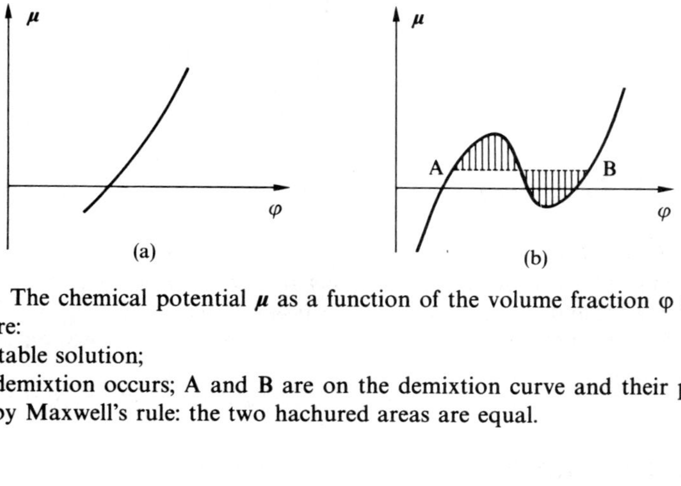

Fig. 14.15. The chemical potential μ as a function of the volume fraction φ at a given temperature:
(a) for a stable solution;
(b) when demixtion occurs; A and B are on the demixtion curve and their position is given by Maxwell's rule: the two hachured areas are equal.

and, consequently,

$$\frac{d\mu}{d\varphi} = -2\chi N + \frac{1}{\varphi} + \frac{N}{1-\varphi}.$$

The top of the demixtion curve is given by the equations

$$\frac{d\mu}{d\varphi} = 0 \qquad \frac{d^2\mu}{d\varphi^2} = 0$$

which read

$$-2\chi N + \frac{1}{\varphi} + \frac{N}{1-\varphi} = 0,$$

$$-\frac{1}{\varphi^2} + \frac{N}{(1-\varphi)^2} = 0,$$

and from them, we deduce

$$\varphi = \frac{1}{1+N^{1/2}},$$

$$\frac{1}{2} - \chi = -\frac{1}{N^{1/2}} - \frac{1}{2N}. \tag{14.5.4}$$

The top of the demixtion curve is given by the equation

$$\frac{1}{2} - \chi = -\frac{\varphi(1-\varphi/2)}{(1-\varphi)^2} \tag{14.5.5}$$

(see Fig. 14.13).

Let us now examine the shape of the demixtion curve for *small* values of φ (dilute and semi-dilute solutions). In this case, we have

$$\boldsymbol{\mu} - \boldsymbol{\mu}_1 \simeq (1-2\chi)N\varphi + \ln\varphi + N\varphi^2/2$$

$$\frac{d\boldsymbol{\mu}}{d\varphi} \simeq (1-2\chi)N + \frac{1}{\varphi} + N\varphi.$$

The law of equal areas (Maxwell's rule) which defines the demixtion curve can be expressed as follows [see Appendix L, eqns (L.1) and (L.5)]

$$\boldsymbol{\mu}(\varphi^{\mathrm{I}}) = \boldsymbol{\mu}(\varphi^{\mathrm{II}}) = \boldsymbol{\mu},$$

$$\int_{\varphi^{\mathrm{I}}}^{\varphi^{\mathrm{II}}} d\varphi\,[\boldsymbol{\mu}(\varphi) - \boldsymbol{\mu}] = 0.$$

This set of conditions can also be written in the form

$$\boldsymbol{\mu}(\varphi^{\mathrm{I}})) = \boldsymbol{\mu}(\varphi^{\mathrm{II}}) \qquad \int_{\varphi^{\mathrm{I}}}^{\varphi^{\mathrm{II}}} d\varphi\,\varphi\,\frac{\partial\boldsymbol{\mu}(\varphi)}{\partial\varphi} = 0.$$

The first equality gives

$$\ln(\varphi^{\mathrm{II}}/\varphi^{\mathrm{I}}) + 2\left(\frac{1}{2} - \chi\right)N(\varphi^{\mathrm{II}} - \varphi^{\mathrm{I}}) + \frac{N}{2}(\varphi^{\mathrm{II}^2} - \varphi^{\mathrm{I}^2}) = 0,$$

and the second one

$$1 + \left(\frac{1}{2} - \chi\right)N(\varphi^{\mathrm{II}} + \varphi^{\mathrm{I}}) + \frac{N}{3}(\varphi^{\mathrm{II}^2} + \varphi^{\mathrm{I}}\varphi^{\mathrm{II}} + \varphi^{\mathrm{I}^2}) = 0.$$

This set of equations can be solved in parametric form, the parameter being $t = \varphi^{\|}/\varphi^{|}$ with $0 < t < \infty$. The representation is the following

$$\left(\frac{1}{2} - \chi\right)N^{1/2} = -\frac{2(t^2+t+1)\ln t - 3(t^2-1)}{[6(t+1)\ln t - 12(t-1)]^{1/2}(t-1)^{3/2}}$$

$$\varphi^{|}N^{1/2} = \frac{[6(t+1)\ln t - 12(t-1)]^{1/2}}{(t-1)^{3/2}}$$

$$\varphi^{\|} = t\varphi^{|}. \tag{14.5.6}$$

In this way, we obtain a universal curve which is represented in Fig. 14.16. We observe that the top of the curve is parabolic. In fact, eqns (14.5.6) show that in the vicinity of this point

$$\left(\frac{1}{2} - \chi\right)N^{1/2} = -1 - \frac{1}{6}(\varphi N^{1/2} - 1)^2. \tag{14.5.7}$$

Let us now examine the low temperature behaviour ($t \gg 1$). Equations (14.5.6) shows that in this case

$$\varphi^{\|}N^{1/2} \simeq -3\left(\frac{1}{2} - \chi\right)N^{1/2} + \frac{1}{\left(\frac{1}{2} - \chi\right)N^{1/2}}$$

$$\varphi^{|}N^{1/2} \simeq -\frac{3}{e}\left(\frac{1}{2} - \chi\right)N^{1/2}\exp\left(-\frac{3}{2}\left(\frac{1}{2} - \chi\right)^2 N\right) \tag{14.5.8}$$

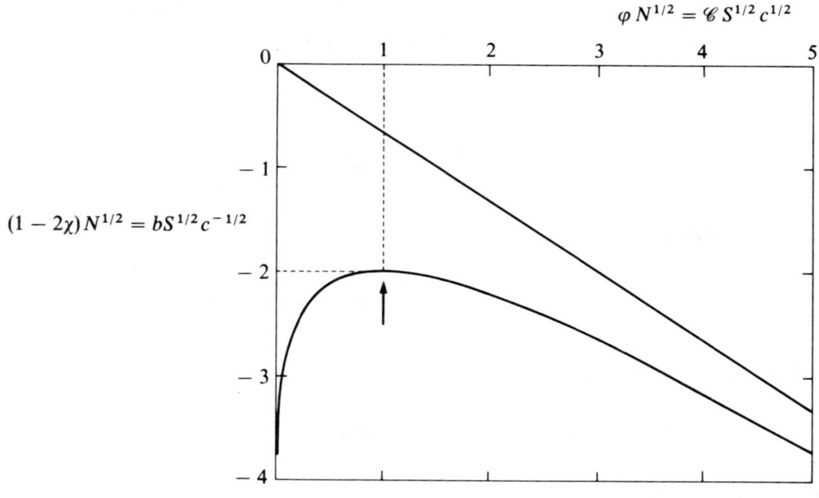

Fig. 14.16. Flory–Huggins theory. Demixtion curve in the dilute and semi-dilute domains. The critical point is indicated by an arrow. The variables used as coordinates are also given in terms of the parameters of the continuous model.

Thus, the curve has a tendency to stick to the vertical asymptote but approaches the oblique asymptote much more slowly.

In brief, it can be said that in the attractive domain, the Flory–Huggins theory describes the demixtion curve in a reasonable way. However, the renormalization theory of critical phenomena shows that the top of the demixtion curve is not parabolic but much flatter. Moreover, the tricritical theory shows that when φ increases, there is no oblique asymptote but a limiting logarithmic curve (see Section 6). Actually, experiments seem to confirm these more recent theories. Nevertheless, the Flory–Huggins theory remains a very interesting approximation whose continuous limit will now be studied.

5.3 Continuous model and the Flory–Huggins theory

The continuous model can be treated very much like the lattice model. In this case, we write

$$\mathbb{F}/V = \frac{b}{2}\mathscr{C}^2 + \frac{c}{6}\mathscr{C}^3 + C\ln(Cv_r) + C\mu_0,$$

or

$$\mathbb{F}/V = \frac{bS^2}{2} C^2 + \frac{cS^3}{6} C^3 + C\ln(Cv_r) + C\mu_0, \tag{14.5.9}$$

where $\mathscr{C} = CS = NS/V$, v_r is a constant which has the dimension of a volume and μ_0 is a dimensionless constant.

The first two terms in $\mathbb{F}$ correspond to the two-body and three-body interactions; the third term represents the entropy of the system and would be equal to the entropy if the chains were not interacting with one another (for a perfect gas we have $v_r = (2\Pi\beta\hbar^2/m)^{3/2}\mathrm{e}^{-1}$).

The expression (14.5.9) appears as the continuous limit of (14.5.1) for $CNv_r \ll 1$, if we set

$$\begin{aligned} bS^2 &= N^2(1 - 2\chi)v_r \\ cS^3 &= N^3 v_r^2 \end{aligned} \tag{14.5.10}$$

which is possible since we have $S \propto N$.

The osmotic pressure Π and the chemical potential μ can be immediately deduced from (14.5.9) where $\mathbb{F}/V$ is considered as a function of C

$$\Pi\beta = -\left.\frac{\partial \mathbb{F}}{\partial V}\right|_N = C\frac{\partial}{\partial C}(\mathbb{F}/V) - (\mathbb{F}/V)$$

$$\mu = \left.\frac{\partial \mathbb{F}}{\partial N}\right|_V = \frac{\partial}{\partial C}(\mathbb{F}/V)$$

In this way, we obtain

$$\Pi\beta = C + \frac{b}{2}S^2C^2 + \frac{c}{3}S^3C^3, \tag{14.5.11}$$

$$\mu = bS^2C + \frac{c}{2}S^3C^2 + \ln(Cv_r) + \mu_1. \qquad (14.5.12)$$

Then the demixtion curve is calculated exactly as in the general case. The result can be obtained by using the correspondence (14.5.10). Thus, using the definition $\varphi = NCv_r$, we get

$$\begin{aligned} \varphi N^{1/2} &= CS^{3/2}c^{1/2} \\ (1-2\chi)N^{1/2} &= bS^{1/2}c^{-1/2}. \end{aligned} \qquad (14.5.13)$$

By using (14.5.6), we can thus represent the demixtion curve in terms of the variables of the continuous model (see Fig. 14.16).

Unfortunately, this approximation cannot be generalized directly. We shall come back to this matter in the context of the diagram expansions of grand canonical systems and we shall note that the approximation used above coincides with the simple-tree approximation.

6. CONTINUOUS MODEL WITH ATTRACTION. APPLICATION OF RENORMALIZATION TECHNIQUES

6.1 Model, perturbation calculation and diagrams

The only difference between the model used here and the 'standard' continuous model comes from the fact that the two-body term is attractive and that we add a repulsive three-body term. The weight associated with a configuration of a set of N chains reads.

$$\mathscr{W}\{\vec{r}\} \equiv {}^0\mathscr{W}\{\vec{r}\}\exp(-{}^{\mathrm{I}}\mathfrak{H}_N\{\vec{r}\}) =$$
$$\exp\left\{-\frac{1}{2}\sum_{a=1}^{N} \mathrm{d}s\left(\frac{\mathrm{d}\vec{r}_a}{\mathrm{d}s}\right)^2 - \frac{b}{2}\sum_{a=1}^{N}\sum_{b=1}^{N}\int_0^{S_a}\mathrm{d}s'\int_0^{S_b}\mathrm{d}s''\,\delta(\vec{r}_a(s')-\vec{r}_b(s''))\right.$$
$$\left.-\frac{c}{6}\sum_{a=1}^{N}\sum_{b=1}^{N}\sum_{c=1}^{N}\int_0^{S_a}\mathrm{d}s'\int_0^{S_b}\mathrm{d}s''\int_0^{S_c}\mathrm{d}s'''\,\delta(\vec{r}_a(s')-\vec{r}_b(s''))\delta(\vec{r}_b(s')-\vec{r}_c(s'''))\right\}$$

Of course, we assume, as in Chapter 10, that there exists a short range cut-off s_0 on the chain, so as to ensure that perturbation calculations give finite results. It is expected that the properties of long chains do not really depend on s_0. Nevertheless, the situation is not as simple as in the purely repulsive case treated in Chapter 10. As we shall see later, the introduction of three-body terms leads to serious difficulties which fortunately can be overcome.

In the following, we consider only monodisperse systems and we study tricritical systems by using the direct renormalization method. It is also possible to proceed indirectly by introducing a tricritical field theory and the correspondence which exists between field theory and polymer theory. This approach introduced by de Gennes[7] was used by Stephen.[10] Afterwards, the question was re-examined and studied in more detail by Duplantier[11,12,13] who used *the so-called renormalization group method.* It relies on the fact that a finite renormalized theory must exist in the limit $s_0 \to 0$. From this fact, one then deduces that the properties of a chain can be calculated with a cut-off area s'_0 different from s_0 provided that we renormalize the interactions and the various parameters of the theory. In the asymptotic limit, the new interactions thus become well-defined functions of the scaling variable s'_0/s_0 and their evolution is given by *the equations of the renormalization group.*[12,14] Then, the various physical quantities are obtained by perturbation with respect to these renormalized interactions, in the framework of these scale transformations.

Here, we use *the direct renormalization method* which relies on a slightly different conceptual approach. Nevertheless the points of view are equivalent and the results obtained through field theory can be directly recovered, (as we shall now show) by using more recent results obtained (1985) by Duplantier.[15]

In the monodisperse case, there are two basic dimensionless parameters which define the strength of the interactions

$$z = bS^{2-d/2}(2\pi)^{-d/2}$$

$$y = cS^{3-d}(2\pi)^{-d} \tag{14.6.2}$$

where S is the area of a polymer.

We also use *physical* dimensionless parameters, g and h which are respectively related to the second and to the third virial coefficient. More precisely, we write

$$\Pi\beta = C(1 + \tfrac{1}{2}(2\pi)^{d/2}g(CX^d) + \tfrac{1}{3}(2\pi)^d h(CX^d)^2 + \ldots)$$

where $X^2 = {}^eS$ characterizes the size of the isolated swollen or deswollen polymer (thus if $y = 0$, $g = z + \ldots$ and if $z = 0$, $h = y + \ldots$)

It is assumed here that the temperature of the solution does not differ much from the Flory temperature T_F. Now, we recall that for $d = 3$, the three-body interaction is marginal, as can be seen from (14.6.2). Thus, for $|g| \ll 1$ and $d = 3$, the situation of the system reminds us of the situation of the standard continuous model for $d = 4$, and these marginal systems will be treated in very similar manners (see Chapter 12, Section 3.3.4.2). This means that in practice, for $|g| \ll 1$, the chains are nearly Brownian but that logarithmic corrections must be calculated. However, the fact that demixtion may occur when g is weakly negative introduces additional difficulties.

The diagrams are about the same as those corresponding to the standard continuous model; the difference comes from the fact that three-body interactions may occur in these diagrams (see Figs 14.17 and 14.18). The calculation rules have been given in Chapter 10, Section 3.4.2. A short-distance regularization (by means of a cut-off area s_0) is necessary; in principle, the terms depending on the cut-off should then be eliminated by renormalization, as was done in Chapter 10, but here this programme presents difficulties which will be

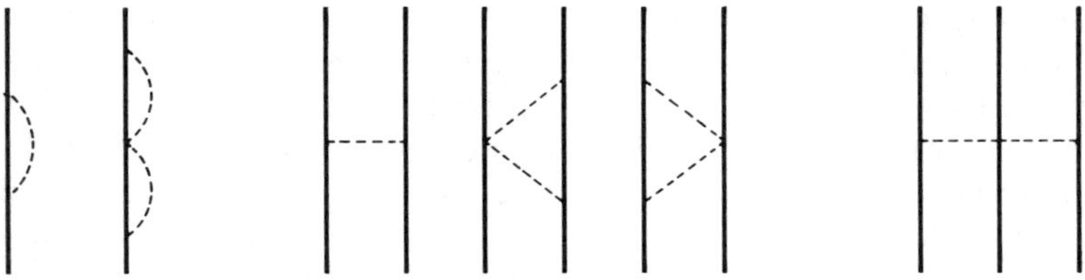

(a) (b) (c)

Fig. 14.17. Connected diagrams to first order in z and y
(a) for $\mathscr{Z}(S)$;
(b) for $\mathscr{Z}(S, S)$;
(c) for $\mathscr{Z}(S, S, S)$.

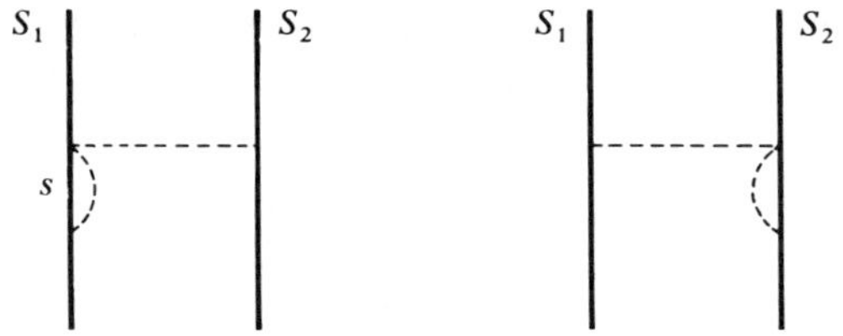

(a) (b)

Fig. 14.18. The three-body interaction appearing on these diagrams gives a divergent contribution, for $d > 2$. This short-range divergence generates a two-body interaction from a three-body interaction.

studied later. In fact, we shall circumvent these difficulties by using a dimensional regularization method. Thus, after calculating diagrams for a space dimension $d < 3$, we shall take the limit $d = 3$.

6.2 The standard continuous model with attraction and the simple-tree approximation

The calculation of the osmotic pressure of a polymer solution has been performed in the framework of the simple-tree approximation, in Chapter 10, Section 6.2. For a monodisperse system, we found [see (10.6.6)]

$$\Pi\beta = C\left(1 + \frac{b}{2}S^2C + \frac{c}{3}S^3C^2\right) \tag{14.6.3}$$

a formula identical to (14.5.11). This result thus coincides with the continuous limit of the Flory–Huggins model provided we use the equivalences (14.5.10). The demixtion curve can be directly obtained from (14.6.3) and, as before, we consider that b is a linear function of temperature and that c is a constant.

The top of the demixtion curve is a critical point for which

$$\frac{\partial \Pi}{\partial C} = 0 \qquad \frac{\partial^2 \Pi}{\partial C^2} = 0$$

which gives

$$\begin{aligned} b_c &= -\,2(c/S)^{1/2} \\ C_c &= (cS^3)^{-1/2} \\ \Pi\beta &= \tfrac{1}{3}(cS^3)^{-1/2} \end{aligned} \qquad (14.6.4)$$

Demixtion occurs for $b < b_c$. For each value of b, there are two concentrations $C^{|}$ and $C^{||}$ corresponding to the demixtion curve. They are given by the equality

$$\Pi(C^{|}) = \Pi(C^{||}) = \bar{\Pi}$$

and Maxwell's rule (see Appendix L)

$$\int_{V^{|}}^{V^{||}} \mathrm{d}V\,\Pi = \bar{\Pi}(V^{||} - V^{|}),$$

which can be written in the form

$$\int_{C^{|}}^{C^{||}} \mathrm{d}C\,\frac{\Pi(C) - \bar{\Pi}}{C^2} = 0.$$

We thus recover the curve previously calculated in Section 5.2. Accordingly, (14.5.6) and (14.5.13) lead to the parametric representations

$$\begin{aligned} \frac{b}{2}S^{1/2}c^{-1/2} &= \frac{-\,2(t^2 + t + 1)\ln t + 3(t^2 - 1)}{(t - 1)^{3/2}[6(t + 1)\ln t - 12(t - 1)]^{3/2}} \\ C^{|}S^{3/2}c^{1/2} &= \frac{[6(t + 1)\ln t - 12(t - 1)]^{1/2}}{(t - 1)^{3/2}} \\ C^{||} &= tC^{|}. \end{aligned} \qquad (14.6.6)$$

Let us recall that at the critical point $t = 1$

$$\begin{aligned} b_c &= -\,2S^{-1/2}c^{1/2} \\ C_c &= S^{-3/2}c^{-1/2}. \end{aligned} \qquad (14.6.6)$$

In the vicinity of the critical point

$$\frac{b}{2}S^{1/2}c^{-1/2} = -\left(1 + \frac{(t - 1)^2}{24} + \ldots\right)$$

and consequently

$$(t - 1)^2 = 24\left(\frac{b_c - b}{b_c}\right).$$

Finally, we obtain

$$\frac{C^{\parallel} - C^{|}}{C^{|}} = t - 1 = \left[24\left(\frac{b_c - b}{b_c}\right)\right]^{1/2} \tag{14.6.7}$$

Moreover, for large values of $CS^{1/2}$

$$b = -\frac{2c}{3}CS \tag{14.6.8}$$

(see Fig. 14.16).

6.3 Short-range divergences in diagrams and dimensional regularization

We want to calculate the swelling of a chain and the osmotic pressure of a set of chains in the vicinity of the Flory point. For this purpose, we can calculate, on one side ${}^{+}\bar{\mathscr{Z}}(\vec{k}, -\vec{k}; S)$ to first-order in k^2, on the other side the quantities ${}^{+}\mathscr{Z}(N \times S)$.

Of course, these various quantities depend on the cut-off s_0 and diverge if $s_0 \to 0$. However, these divergences cannot be eliminated by a simple renormalization of the partition functions as could be done in the purely repulsive case (see Chapter 10). The bare interaction has also to be renormalized. In particular, as will be shown later in more detail, the three-body interactions act as if they were generating cut-off dependent two-body interactions. The elimination of the cut-off dependent terms then results from a re-interpretation of the various terms of the diagram expansion; nevertheless, the process is complex.

However, a simple method, the dimensional regularization, enables us to eliminate the cut-off in a rational way. The polymer is embedded in a space of dimension d and in principle we can calculate diagrams for any (non-integer) value of d. The contribution of a diagram associated with a partition function ${}^{+}\mathscr{Z}(\ldots)$ is a function $\mathscr{D}(d, s_0)$. It is easy to show [see (10.4.57)] that if the real part of d is small enough ($\mathrm{Re}\, d < d_0$, $d_0 > 0$), then

$$\lim_{s_0 \to 0} \mathscr{D}(d, s_0) = \mathscr{D}(d).$$

where $\mathscr{D}(d)$ is a finite quantity which is an analytic function of d. Now, an analytic continuation of this function can be found for all values of d (see Appendix H) and, in general, one gets a meromorphic function which can be represented by the same symbol $\mathscr{D}(d)$. By generalizing this process to all diagrams contributing to ${}^{+}\mathscr{Z}(\ldots)$, one defines a function $\mathscr{Z}(\ldots)$ of the complex variable d, which can be calculated, at least to first-orders, by analytic continuation (with respect to d). This process is the dimensional regularization.

The importance of dimensional regularization comes from the fact that it is equivalent to a renormalization. Actually, the fact has not been rigorously proved in polymer theory, but it seems very likely to be true, and in Appendix

H we explain why. We can also verify that in the repulsive case treated in Chapter 13, this equivalence holds.

Thus, to calculate the swelling of an isolated chain and the osmotic pressure of a set of chains in the vicinity of the Flory point, a determination of the partition functions $\bar{\mathscr{Z}}(\vec{k}, -\vec{k}; S)$ and $\mathscr{Z}(N \times S)$ by dimensional renormalization is sufficient. The calculations will be performed to first orders in x and y for a space dimension $d = 3 - {}'\varepsilon$ $(0 < {}'\varepsilon \ll 1)$, and we note that the purely repulsive terms have been already calculated in Chapter 10. The partition functions will be represented by series in terms of x and y. Finally, in order to study the behaviour of long polymers, we shall treat these series by using the direct renormalization method.

6.4 Three-body interactions and counter-terms

The dimensional regularization method enables one to eliminate a large number of divergences automatically. However, these divergences have a physical meaning and this is why, in this section, and in the two following ones, their nature is briefly analysed.

In a very general way, divergences in diagrams can be eliminated with the help of counter-terms; these counter-terms can be interpreted as resulting from additional interaction terms and the latter are absorbed by renormalization of the partition functions and of the bare interactions. This is exactly what we did in Chapter 10, Section 4.2.6, for the model with purely repulsive two-body interactions, and, in this case, it is easy to see that the process amounts to dimensional regularization.

When three-body interactions are also present, the question becomes more complicated, but again we can determine the nature of the divergent terms by using the following dimension equations which can be easily deduced from (14.6.1) and (14.6.2)

$$\begin{aligned} b &\sim L^{d-4} \\ c &\sim L^{2d-6} \\ s_0 &\sim L^2. \end{aligned} \tag{14.6.9}$$

Let us first consider the renormalization of the partition functions; such a renormalization, for a polymer, can be written in the form

$$^{+}\mathscr{Z}(S) = \mathrm{e}^{AS + 2A_1}\mathscr{Z}(S)$$

where A and A_1 must diverge when $s_0 \to 0$ (A_1 concerns the end points).

The dimension of A is thus given by $A \sim L^{-2}$ and if such a term results from diagrams with p two-body interactions and q three-body interactions, we then have

$$A \propto \frac{b^p c^q}{s_0^{1 + p(-2 + d/2) + q(-3 + d)}}.$$

Of course, A must diverge when $s_0 \to 0$ which implies that

$$1 + p(-2 + d/2) + q(-3 + d) > 0.$$

In particular, for $d = 3$, this condition holds true for $p = 0$ or $p = 1$, and any value of q. In the same way, for $d = 3$, we expect the existence of a renormalization factor A, obtained from diagrams corresponding to any values of p and q.

Moreover, in a diagram, two-body interactions depending on s_0 can be generated from three-body interactions. Thus, from a purely dimensional point of view, we can write

$$b \sim \frac{c}{s_0^{-1+d/2}}. \tag{14.6.10}$$

Accordingly, we see that for $d > 2$, a three-body interaction may partially play the role of a two-body interaction and this fact will be more precisely analysed in Section 6.5.

In the same way, from a purely dimensional point of view, we can write

$$b \sim \frac{c^2}{s_0^{-4+3d/2}}. \tag{14.6.11}$$

Thus, we see that for $d > 8/3$, a three-body interaction in a divergent diagram may play the role of a two-body interaction. In particular, this is true for $d = 3$, since in this case

$$b^2 \sim \frac{c^2}{s_0^{1/2}}.$$

Similarly, we can generate two-body interactions from two three-body interactions (or from more than two of them).

Thus, a study of the dimension equations makes it possible to discover which renormalizations are necessary to construct a finite theory. Of course, it would be interesting to study more precisely how, in polymer theory, the renormalization terms are produced but this question is not simple.

In the following sections, we only show in some detail how three-body interactions generate cut-off dependent two-body interactions. However, one must be aware of the fact that these divergent terms which are eliminated by renormalization have no universal character. Their main importance is conceptual. In practice, we can ignore them by applying the dimensional regularization technique.

6.5 Generation of a two-body interaction by means of a three-body interaction in a divergent diagram

In agreement with the conclusions of the preceding section, it will be shown in this section and in the following one, how the introduction of a three-body interaction produces divergences which are to be eliminated by renormalization of the bare interactions.

Firstly, let us show how a three-body interaction may give divergent contributions which are equivalent to two-body interactions in agreement with (14.6.10). This three-body interaction acts on polymers of areas S_1 and S_2 in a manner represented in Fig. 14.18. The contribution $\mathscr{D}_a$ of the diagram (a) of this figure can be calculated easily by applying the rules given in Chapter 10; we thus find ($d < 4$, s_0 = cut-off area).

$$\mathscr{D}_a = cS_1(2\pi)^{d/2}\int_{s_0}^{S_2} ds\frac{(S_2 - s)}{s^{d/2}}$$

or, if we neglect terms which tend to zero when $s_0 \to 0$,

$$\mathscr{D}_a = cS_1S_2^{2-d/2}\frac{4(2\pi)^{d/2}}{(2-d)(4-d)} - cS_1S_2\frac{2(2\pi)^{d/2}s_0^{1-d/2}}{2-d}.$$

The sum $\mathscr{D}$ of the contributions of the diagrams (a) and (b) is thus given by

$$\mathscr{D} = c(S_1S_2^{2-d/2} + S_2S_1^{2-d/2})\frac{4(2\pi)^{d/2}}{(2-d)(4-d)} - cS_1S_2\frac{4(2\pi)^{d/2}s_0^{1-d/2}}{2-d} \tag{14.6.12}$$

The first term is the finite part of the three-body contribution; we note that for $d < 2$ this finite part represents the whole contribution, because, in this case, the second term tends to zero when $s_0 \to 0$. Then, the first term is positive. Conversely, for $2 < d < 4$, the finite part of the contribution is only the analytic continuation (with respect to d) of the expression which, for $d < 2$, represents the contribution of the three-body interaction. For $2 < d < 4$, the first term (which represents this finite part) is negative but its contribution is counterbalanced by the divergent second term and the sum remains positive. However, this second term is proportional to S_1S_2 and therefore, it appears formally as the contribution of a two-body effective interaction whose coefficient b' is given by

$$b' = \frac{c}{s_0^{-1+d/2}}\frac{4(2\pi)^{d/2}}{2-d} \tag{14.6.13}$$

in agreement with (14.6.10).

Of course, the same phenomenon would be observed if the diagrams of Fig. 14.18 were not considered as true diagrams but as sub-diagrams of a more complex diagram. In particular, the interaction joining two chains could bear a wave vector; this situation would not alter the fact that for $2 < d < 4$, the short-range divergence generates an effective two-body interaction.

This two-body interaction which depends on the cut-off can be eliminated by redefining the two-body interaction. In other words, the initial two-body interaction characterized by a coefficient b_0 must be replaced by an interaction whose coefficient is $b = b_0 + b'$. Simultaneously, the divergent terms are subtracted from the contributions of the three-body interaction. This re-definition of the two-body interaction can be called *renormalization of the bare interactions.*

The divergent dependence with respect to s_0 is transferred to b, but b is not infinite because the short-range cut-off does not physically vanish. Thus, the real aim of renormalization is not the suppression of divergences but the elimination of the dependence upon s_0, i.e. upon the chemical microstructure.

Finally, with this example, we can see that the determination of physical quantities by perturbation theory requires only a calculation of the finite contributions of the diagrams. Moreover, the above example shows us that these finite parts are functions of the space dimension and can be obtained for all values of d by analytic continuation with respect to d.

6.6 Generation of a two-body interaction by means of two three-body interactions in a divergent diagram

We have just seen, in a very simple way, how, because of short-range divergences, a three-body interaction can generate two-body interactions. Let us now give another example corresponding to (14.6.11). The diagrams (or sub-diagrams) under consideration are represented in Fig. 14.19. Let s_1, s_2, s_3, s_4 be the areas associated with the four internal polymer segments belonging to any one of the seven diagrams of Fig. 14.19. After integration over the internal wave vectors, the contribution of one of these diagrams is given by

$$\mathscr{D}\{s\} = \frac{c^2}{(2\pi)^{3/2}} \frac{1}{(s_1 s_2 s_3 + s_1 s_2 s_4 + s_1 s_3 s_4 + s_2 s_3 s_4)^{d/2}}$$

in agreement with (10.4.57)

The effective two-body interaction generated by the three-body interactions is characterized by the constant b' defined by the equality

$$\begin{aligned} b' &= -7 \int_{s_0}^{\infty} ds_1 \int_{s_0}^{\infty} ds_2 \int_0^{\infty} ds_3 \int_0^{\infty} ds_4 \; \mathscr{D}\{s\} \\ &= -\frac{c^2}{s_0^{-4+3d/2}} \frac{7}{(2\pi)^{3d/2}} I(d) \end{aligned} \tag{14.6.14}$$

where

$$I(d) = \int_1^{\infty} dx_1 \int_1^{\infty} dx_2 \int_1^{\infty} dx_3 \int_1^{\infty} dx_4 \frac{1}{(x_1 x_2 x_3 + x_1 x_2 x_4 + x_1 x_3 x_4 + x_2 x_3 x_4)^{d/2}}.$$

In particular, the quadruple integral corresponding to $I(3)$ can be expressed in the following simple form

$$I(3) = 32 \int_0^{\sqrt{2}} d\rho \left[\frac{\pi}{2} - \text{Arc}\sin\left(\frac{1}{1+\rho^2}\right)\right]\left[\frac{\pi}{4} - \Theta(\rho - 1)\,\text{Arc}\cos\left(\frac{1}{\rho}\right)\right] \tag{14.6.15}$$

which leads to $I(3)/8\pi \simeq 1.0$.

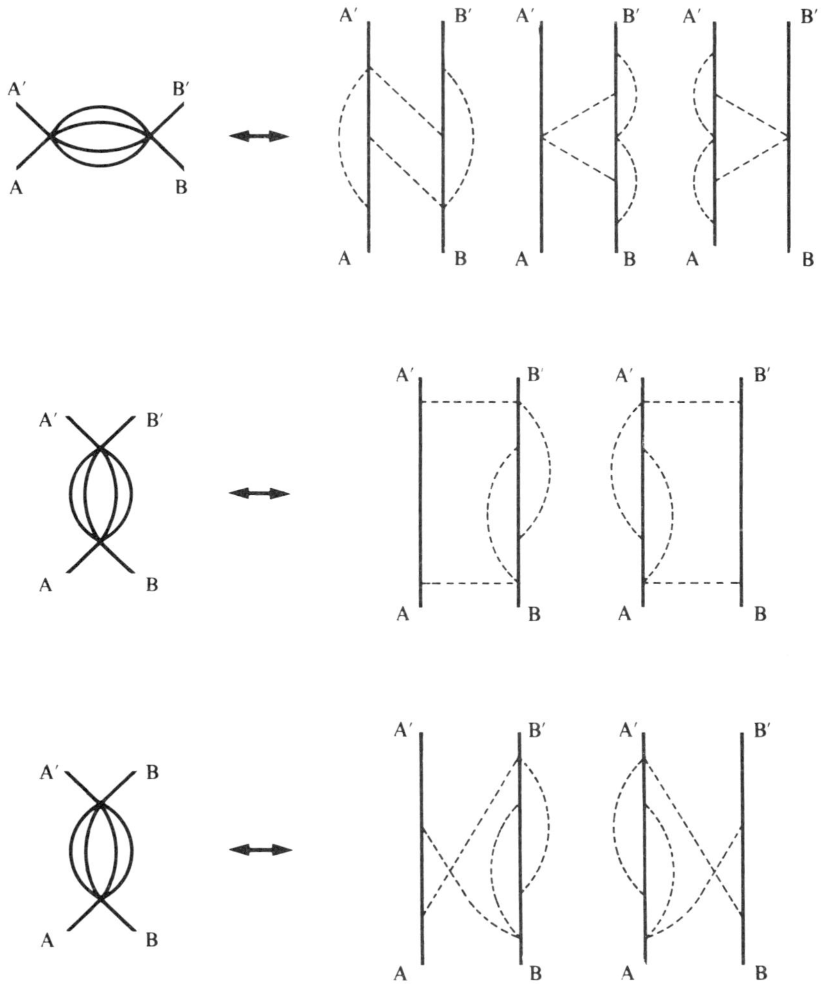

Fig. 14.19. Generation of a two-body interaction from two three-body interactions in a divergent diagram (see the discussion in Section 6.4). The figure shows seven diagrams belonging to three different types. The contributions associated with these types are the same if each external leg bears a zero wave vector and if we neglect the constraints concerning the lengths of the chains. The origins of these chains are in A and B.

6.7 Partition functions at $d = 3 - {}'\varepsilon$ for polymers with two-body and three-body interactions

In order to calculate the swelling of a chain in the vicinity of the Flory temperature, we need $\mathscr{Z}(\vec{k}, -\vec{k})$ to first order in k^2. In particular, this quantity will give us $\mathscr{Z}(S) = \mathscr{Z}(\vec{0},\vec{0};S)$. We also need $\mathscr{Z}(S, S)$ in order to obtain the second virial coefficient. Moreover, to calculate the osmotic pressure and the specific heat of the system it is also necessary to know $\mathscr{Z}(S, S, S)$. The expansions of these partition functions in powers of z and y, defined by (14.2.11), are calculated for $d = 3 - {}'\varepsilon$ and, in the following, we keep only the dominant terms. They are of the lowest degree in ${}'\varepsilon$ and generally linear with respect to z. This approximation will be sufficient here because we shall not stray very far away from the Flory point (tricritical point $z = 0$). Nevertheless, in $\mathscr{Z}(S)$ and $\mathscr{Z}(S, S)$, we keep z^2 terms which become infinite when ${}'\varepsilon = 0$ and which are needed to calculate the singular part of the specific heat at the tricritical point. The values of certain diagrams contributing to $\mathscr{Z}(\vec{k}, -\vec{k}; S)$ and $\mathscr{Z}(S, S)$ have been calculated by Duplantier[15] (those most difficult to calculate).

The calculation of the dominant contributions of the diagrams associated with $\mathscr{Z}(\vec{k}, -\vec{k}; S)$ leads to the result

$$\mathscr{Z}(\vec{k}, -\vec{k}; S) = \mathscr{Z}(S) - \frac{k^2 S}{2}\mathscr{Z}^{\bullet}(S) + \cdots$$

$$\mathscr{Z}(S) = 1 - 4\pi y + \frac{186\pi^2}{{}'\varepsilon}y^2 + 4z\left(1 - \frac{16\pi}{{}'\varepsilon}y\right) + \frac{2\pi}{{}'\varepsilon}z^2\left(1 - \frac{16\pi}{{}'\varepsilon}y^2\right) + \ldots$$

$$\mathscr{Z}^{\bullet}(S) = 1 - 8\pi y + \frac{1118\pi^2}{3{}'\varepsilon}y^2 + \frac{16}{3}z\left(1 - \frac{16\pi}{{}'\varepsilon}y\right) + \frac{2\pi}{{}'\varepsilon}z^2 + \ldots . \qquad (14.6.16)$$

On the other hand, the calculation of the dominant contributions of the diagrams associated with $\mathscr{Z}(S, S)$ and $\mathscr{Z}(S, S, S)$ leads to the results

$$\mathscr{Z}(S, S) = (2\pi S)^{d/2}\left\{8y\left(1 - \frac{44\pi}{{}'\varepsilon}y\right) - z\left(1 - \frac{16\pi y}{{}'\varepsilon}\right) + \left(\frac{0}{{}'\varepsilon}\right)z^2 + \ldots\right\} \qquad (14.6.17)$$

$$\mathscr{Z}(S, S, S) = (2\pi S)^{d}\left\{-y\left(1 - \frac{44\pi}{{}'\varepsilon}y\right) + \left(\frac{0}{{}'\varepsilon}\right)zy + \left(\frac{0}{{}'\varepsilon}\right)z^2 + \ldots\right\}. \qquad (14.6.18)$$

6.8 Expansions concerning the swelling of an isolated chain and the osmotic parameters

The swelling of an isolated chain is here a function $\mathfrak{X}_0(z, y)$ of the interaction parameters. We have

$$\mathfrak{X}_0(z, y) = {}^{e}S/S = \mathscr{Z}^{\bullet}(S)/\mathscr{Z}(S)$$

where $\mathscr{Z}(S)$ and $\mathscr{Z}^{\bullet}(S)$ are given by (14.6.16) and consequently

$$\mathfrak{X}_0(z, y) = 1 - 4\pi y + \frac{560\pi^2}{3\,{}'\varepsilon}y^2 + \ldots + \frac{4}{3}z\left(1 - \frac{16\pi}{{}'\varepsilon}y + \ldots\right) + \left(\frac{0}{{}'\varepsilon}\right)z^2 + \ldots . \qquad (14.6.19)$$

Incidentally, let us remark the anomalous sign of the second term ($-4\pi y$) and let us recall that such anomalies are related to the existence of ultraviolet divergences (in particular, see Section 6.5).

Let us now consider the osmotic pressure. It is a function of $\boldsymbol{C}$ which is given by the parametric representation

$$\Pi\beta = f\mathscr{Z}(S) + \frac{f^2}{2}\mathscr{Z}(S,S) + \frac{f^3}{6}\mathscr{Z}(S,S,S) + \ldots$$

$$C = f\mathscr{Z}(S) + f^2\mathscr{Z}(S,S) + \frac{f^3}{2}\mathscr{Z}(S,S,S) + \ldots$$

and by eliminating f, we obtain

$$\Pi\beta = C - \frac{C^2}{2}\frac{\mathscr{Z}(S,S)}{\mathscr{Z}^2(S)} - \frac{C^3}{3}\left[\frac{\mathscr{Z}(S,S,S)}{\mathscr{Z}^3(S)} - 3\frac{\mathscr{Z}^2(S,S)}{\mathscr{Z}^4(S)}\right] + \ldots . \tag{14.6.20}$$

This equality can be rewritten in the form of a scaling law (where $X^2 = {}^{\mathrm{e}}S = R^2/d$)

$$\Pi\beta = C\left[1 + \frac{g}{2}(2\pi)^{d/2}CX^d + \frac{h}{3}(2\pi)^d(CX^d)^2 + \ldots\right]. \tag{14.6.21}$$

By comparing the above equations, we obtain relations providing us with the values of the osmotic parameters g and h

$$g(2\pi)^{d/2}X^d = -\frac{\mathscr{Z}(S,S)}{\mathscr{Z}^2(S)}$$

$$h(2\pi)^d X^{2d} = -\left[\frac{\mathscr{Z}(S,S,S)}{\mathscr{Z}^3(S)} - 3\frac{\mathscr{Z}^2(S,S)}{\mathscr{Z}^4(S)}\right] \tag{14.6.22}$$

or, if we introduce the renormalization factor $\mathfrak{X}_0$

$$g(2\pi S)^{d/2} = -\frac{\mathscr{Z}(S,S)}{\mathscr{Z}^2(S)}[\mathfrak{X}_0]^{-d/2}$$

$$h(2\pi S)^{d/2} = -\left[\frac{\mathscr{Z}(S,S,S)}{\mathscr{Z}^3(S)} - 3\frac{\mathscr{Z}(S,S)}{\mathscr{Z}^4(S)}\right][\mathfrak{X}_0]^{-d} \tag{14.6.23}$$

From eqns (14.6.16), (14.6.17), (14.6.18), and (14.6.19), we deduce (by keeping the terms of second-order in $1/'\varepsilon$)

$$g = -8y\left(1 - \frac{44\pi}{'\varepsilon}y\right) + z\left(1 - \frac{16\pi}{'\varepsilon}y\right) + \left(\frac{0}{'\varepsilon}\right)z^2 \tag{14.6.24}$$

$$h = y - \frac{44\pi}{'\varepsilon}y^2 + \left(\frac{0}{'\varepsilon}\right)z^2 \tag{14.6.25}$$

The preceding equations and (14.6.19) will now serve as basic data for renormalization calculations.

6.9 Renormalization of the tricritical system $|z| \ll 1$, $y \neq 0$

6.9.1 Parametrization of the system

In this section, we consider systems at temperatures for which the *effective* two-body interaction is very small. In other words, we assume that we have $|z| \ll 1$, $y \neq 0$ and, for $d = 3 - {}'\varepsilon$, no other parameter. However, as the three-body interaction is marginal at $d = 3$, it will be necessary to calculate logarithmic effects depending on (S/s_0) and not on y, which remains finite at $d = 3$ when $S \to \infty$. In what follows, we consider the system for a space dimension $d = 3 - {}'\varepsilon$ in order to calculate various exponents, and subsequently we go to the limit ${}'\varepsilon \to 0$. Here, we use a method very similar to the method which we applied, in the purely repulsive case (good solvent), in order to calculate logarithmic corrections for $d = 4$ and $S/s_0 \gg 1$ (see Chapter 12, Section 3.3.4).

For tricritical chains, the important physical parameter is the osmotic parameter h, which is proportional to the third virial coefficient) and not g as in the purely repulsive case. In the following, we keep only linear terms in z. The values of h, g, and $\mathfrak{X}_0$ are

$$h = h(y, S/s_0)$$

$$g = {}^0g(y, S/s_0) + z\, {}^1g(y, S/s_0)$$

$$R^2/3S = \mathfrak{X}_0 = {}^0\mathfrak{X}_0(y, S/s_0) + \tfrac{4}{3} z\, {}^1\mathfrak{X}_0(y, S/s_0) \qquad (14.6.26)$$

where

$${}^1g(0, S/s_0) = 1$$

$${}^0\mathfrak{X}_0(0, S/s_0) = {}^1\mathfrak{X}_0(0, S/s_0) = 1.$$

The factors ${}^1g(y, S, s_0)$ and ${}^1\mathfrak{X}_0(y, S/s_0)$ renormalize the two-body interaction. When $S/s_0 \to \infty$, these renormalization factors must be equivalent to each other and also to the renormalization factor ${}^1\mathfrak{X}_4(y, S/s_0)$ associated with the centre of a polymer star consisting of four (very long) branches with three-body interactions (see Fig. 14.20(b)). In fact, when the space dimension is $d = 3 - {}'\varepsilon$,

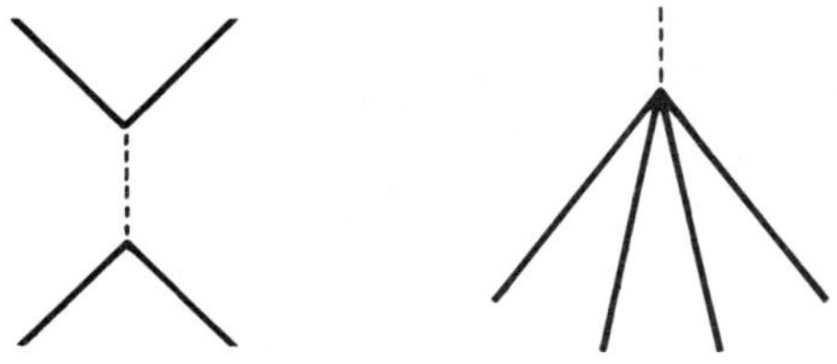

Fig. 14.20. For interacting polymers, the renormalization factor associated with a two-body interaction (see (a)) is just the renormalization factor ${}^1\mathfrak{X}_4(y)$ corresponding to a four-polymer star (see (b)).

the cut-off can be eliminated and the renormalization factors take the simple form ${}^1g(y)$, ${}^1\mathfrak{X}_0(y)$, and ${}^4\mathfrak{X}_4(y)$. Incidentally, we note that, in this case, the definition of ${}^1\mathfrak{X}_4(y)$ is very similar to those of the renormalization factors $\mathfrak{X}_N(z)$ introduced in Chapter 12, Section 3.3.3.

By comparing (14.6.26) with (14.6.27) and with (14.6.19), we obtain the perturbation expansions

$$ {}^1g(y = 1 - \frac{16\pi}{'\varepsilon} y + \dots $$

$$ {}^1\mathfrak{X}_0(y) = 1 - \frac{16\pi}{'\varepsilon} y + \dots . $$

Consequently, we may set

$$ '\mathfrak{X}_4(y) = 1 - \frac{16\pi}{'\varepsilon} y + \dots \tag{14.6.27} $$

and for large values of S/s_0, we can write (14.6.26) in the form

$$ \begin{aligned} h &= h(y, S/s_0) \\ g &= {}^0g(y, S/s_0) + z\ '\mathfrak{X}_4(y, S/s_0) \\ X^2/S = \mathfrak{X}_0 &= {}^0\mathfrak{X}_0(y, S/s_0) + \tfrac{4}{3} z\ '\mathfrak{X}_4(y, S/s_0). \end{aligned} \tag{14.6.28} $$

For $d = 3$, y is a number which remains constant when $S/s_0 \to \infty$; on the contrary, as will be shown later, the osmotic parameter h goes to zero when $S/s_0 \to \infty$. Thus, h is the good parameter and, in the following, we re-express all physical quantities in terms of h. But, first, let us examine the behaviour of h when $S/s_0 \to \infty$.

6.9.2 Tricritical behaviour of the third virial coefficient; domain of validity

In order to study how h varies when $S/s_0 \to \infty$, we consider a system for which $z = 0$, and the value of h obtained in this case will remain valid to first order in z. In fact, in agreement with (14.6.25), the expansion of h starts with terms proportional to z^2. Moreover, we assume that the space dimension $d = 3 - '\varepsilon$ is slightly smaller than three:

The exponent concerning h is a function $'w(y)$ defined by the following equation which is very analogous to (12.3.77)

$$ 'w(y) = '\varepsilon y \frac{\partial h}{\partial y}. \tag{14.6.29} $$

The expansion of h in powers of y and $'\varepsilon$ is given by (14.6.25)

$$ h = y - \frac{44\pi}{'\varepsilon} y^2 + \dots $$

which can be inverted by expanding y with respect to h

$$y = h + \frac{44\pi}{'\varepsilon} h^2 + \ldots \tag{14.6.30}$$

From this expression, we deduce the expansion of $'w(y)$

$$'w(y) = {'\varepsilon} y - 88\pi y^2 + \ldots . \tag{14.6.31}$$

Let us now set

$$'w[h] = {'w}(y) \tag{14.6.32}$$

an equality which is analogous to (12.3.78). The expansions (14.6.36) and (14.6.31) give

$$'w[h] = {'\varepsilon} h - 44\pi h^2 + \ldots . \tag{14.6.33}$$

On the other hand, as according to (14.2.15)

$$y = cS^{'\varepsilon}(2\pi)^{-d}$$

we can write

$$'w[h] = S \frac{\partial}{\partial S} h \bigg|_{c = \text{const.}} \tag{14.6.34}$$

with the help of (14.6.29) and (14.6.32). Let us combine (14.6.33) and (14.6.34); then by going to the limit $'\varepsilon \to 0$, we obtain the equation

$$S \frac{\partial}{\partial S} h = -44\pi h^2 \tag{14.6.35}$$

which is valid for $d = 3$ and small h.

Let us integrate this equation for large values of S; we obtain

$$h = \frac{1}{44\pi \ln(S/s_1)} \tag{14.6.36}$$

where s_1 is a constant area which depends on the interaction constant c and on the microstructure of the chains. In other words, for our model, we must have

$$s_1 = s_0 m_1(c)$$

where $m_1(c)$ is a function of c which could be calculated.

The result (14.6.36) coincides (with different notation) with the result obtained by Duplantier, either directly or through field theory[13,14] (equilibrium ensemble); this fact, in particular, shows that polydispersion effects do not play any role to the order under consideration.

The domain of validity of (14.6.36) can easily be obtained by observing that for very small values of y ($y = c$ for $d = 3$), we have $h \simeq y$. Compare with (14.6.36). We can see that the condition of validity of (14.6.36) can be written in the form

$$y \ln(S/s_0) \gg \frac{1}{44\pi}. \tag{14.6.37}$$

6.9.3 Swelling and partition function of an isolated chain at the tricritical point

As before, let us set $z = 0$ and $d = 3$. The swelling and the partition function of an isolated chain are functions of y and their expansions are given by (14.6.19) and (14.6.16)

$$R^2/Sd \equiv {}^0\mathfrak{X}_0(y) = 1 - 4\pi y + \frac{560\pi^2}{3'\varepsilon} y^2 + \ldots,$$

$$\mathscr{X}(S) \equiv {}^0\mathfrak{X}_1(y) = 1 - 4\pi y + \frac{186\pi^2}{'\varepsilon} y^2 + \ldots. \quad (14.6.38)$$

The functions ${}^0\mathfrak{X}_0(y)$ and ${}^0\mathfrak{X}_1(y)$ can be considered as renormalization factors and effective exponents can be associated with them: we define the exponents by the following formulae which are analogous to (12.3.66)

$${}^0\sigma_0(\mathrm{y}) = {}'\varepsilon \frac{\partial \ln {}^0\mathfrak{X}_0(y)}{\partial \ln y},$$

$${}^0\sigma_1(y) = {}'\varepsilon \frac{\partial \ln {}^0\mathfrak{X}_1(y)}{\partial \ln y}. \quad (14.6.39)$$

The expansions of these exponents with respect to y are obtained from (14.6.38)

$${}^0\sigma_0(y) = -4\pi('\varepsilon)y + \frac{1120\pi^2}{3} y^2 + \ldots,$$

$${}^0\sigma_1(y) = -4\pi('\varepsilon)y + 372\pi^2 y^2 + \ldots. \quad (14.6.40)$$

These effective exponents can be re-expressed in terms of the tricritical parameter h. Let us set

$${}^0\sigma_0(y) = {}^0\sigma_0[h] \qquad {}^0\sigma_1(y) = {}^0\sigma_1[h] \quad (14.6.41)$$

Using (14.6.30), we find

$${}^0\sigma_0[h] = -4\pi('\varepsilon)h + \frac{592\pi^2}{3} h^2 + \ldots$$

$${}^0\sigma_1[h] = -4\pi('\varepsilon)h + 196\pi^2 h^2 + \ldots \quad (14.6.42)$$

Moreover, as according to (14.2.11)

$$y = cS^{'\varepsilon}(2\pi)^{-d}$$

we can rewrite eqns (14.6.39) in the form

$${}^0\sigma_0[h] = S\frac{\partial}{\partial S} \ln {}^0\mathfrak{X}_0 \big|_{c=\mathrm{const.}}$$

$$^0\sigma_1[h] = S\frac{\partial}{\partial S}\ln {}^0\mathfrak{X}_1 \,|_{c=\text{const.}} \tag{14.6.43}$$

Let us combine (14.6.43) and (14.6.44) and go to the limit $'\varepsilon \to 0$. Thus, for $d = 3$, for $c = y =$ constant, and for small h, we obtain

$$S\frac{\partial}{\partial S}\ln {}^0\mathfrak{X}_0 = \frac{592}{3}\pi^2 h^2,$$

$$S\frac{\partial}{\partial S}\ln {}^0\mathfrak{X}_1 = 196\pi^2 h^2.$$

Taking (14.6.35) into account, we can also write

$$\frac{\partial}{\partial S}\ln {}^0\mathfrak{X}_0 = -\frac{148\pi}{33}\frac{\partial h}{\partial S}$$

$$\frac{\partial}{\partial S}\ln {}^1\mathfrak{X}_0 = -\frac{49\pi}{11}\frac{\partial h}{\partial S}$$

Integrating these equations, we obtain (for small h)

$$^0\mathfrak{X}_0 = {}^0A_0(y)\exp\left[-\frac{148\pi}{33}h\right]$$

$$^0\mathfrak{X}_1 = {}^0A_1(y)\exp\left[-\frac{49\pi}{11}h\right] \tag{14.6.44}$$

where $^0A_0(y)$ and $^0A_1(y)$ are functions which, in principle, could be calculated. In particular, we have

$$^0A_0(0) = 1 \qquad {}^0A_1(0) = 1 \tag{14.6.45}$$

In the limit $S/s_0 \gg 1$, the value of h is small and (14.6.36) leads to the result

$$R^2/3S \equiv {}^0\mathfrak{X}_0 = {}^0A_0(y)\left[1 - \frac{148\pi}{33}h\right] = {}^0A_0(y)\left(1 - \frac{37}{363\ln(S/s_0)}\right),$$

$$\mathscr{X}(S) \equiv {}^0\mathfrak{X}_1 = {}^0A_1(y)\left[1 - \frac{49\pi}{11}h\right] = {}^0A_0(y)\left(1 - \frac{49}{484\ln S/s_0)}\right). \tag{14.6.46}$$

6.9.4 Tricritical behaviour of the osmotic parameter related to the second virial coefficient

For $z = 0$ and $d = 3 - {}'\varepsilon$, the osmotic parameter g related to the second virial coefficient has the expansion [see (14.6.24)]

$$g = -8y\left(1 - \frac{44\pi}{'\varepsilon}y\right) + \ldots$$

and moreover, we observe that

$$h = y\left(1 - \frac{44\pi}{'\varepsilon}y\right) + \dots$$

In this way, we obtain

$$g = -8h + (\text{a non-infinite term in } h^2). \tag{14.6.47}$$

This is quite reasonable since h is the parameter which determines the effective strength of the three-body interaction and since this fact entails that the other virial coefficients must be given for all d by finite expansions in powers of h. Thus no renormalization is necessary to express g in terms of h and according to (14.6.36), we have

$$g \simeq -\frac{2}{11\pi \ln(S/s_1)}.$$

6.9.5 Tricritical renormalization of the two-body interaction terms that are proportional to z

As we saw in Section 6.8, in the tricritical domain, ($|z| \ll 1$, $S/s_0 \to \infty$), the two-body interaction renormalizes.

For the terms that are proportional to z, the renormalization factor is the quantity $'\mathfrak{X}_4(y)$ whose expansion to first-orders in y and $'\varepsilon$ is given by

$$'\mathfrak{X}_4(y) = 1 - \frac{16\pi}{'\varepsilon}y + \dots$$

[see (14.6.27)].

The following effective exponent is associated with it

$$'\sigma_4(y) = {'\varepsilon}\frac{\partial \ln {'\mathfrak{X}_4}}{\partial \ln y} = -16\pi y + \dots . \tag{14.6.48}$$

We set

$$'\sigma_4(y) = {'\sigma_4}[h]$$

where according to (14.6.30)

$$'\sigma_4(y) = -16\pi h + \dots .$$

Taking the definition of y into account, we can rewrite (14.6.47) in the form

$$'\sigma_4[h] = S\frac{\partial \ln {'\mathfrak{X}_4}}{\partial S}\bigg|_{c=\text{const.}},$$

i.e.

$$S\frac{\partial}{\partial S}\ln {'\mathfrak{X}_4} = -16\pi h + \dots ,$$

a formula which remains valid in the limit $'\varepsilon \to 0$, for small h.

Now, using (14.6.35), we can rewrite this equation in the form

$$\frac{\partial}{\partial S}\ln{}'\mathfrak{X}_4 = \frac{4}{11}\frac{1}{h}\frac{\partial h}{\partial S}$$

which finally gives

$$'\mathfrak{X}_4 = A_4(y)(h/y)^{4/11} \tag{14.6.49}$$

where $A_4(y)$ is a constant which, in principle, can be calculated. In particular, we have

$$A_4(0) = 1 \tag{14.6.50}$$

because for $y \ll 1$, $h \simeq y$, and $'\mathfrak{X}_4 \simeq 1$.

In the preceding expression, let us replace h by its value given by (14.6.36); we find the final formula

$$'\mathfrak{X}_4 = A_4(y)\left[\frac{1}{44\pi y \ln(S/s_1)}\right]^{4/11} \tag{14.6.51}$$

6.9.6 Osmotic pressure and swelling near the critical point

We now assume that z is small but finite and that $d = 3$. For large values of S/s_0, we can rewrite the equations (14.6.28) in explicit form by collecting the results obtained in the preceding sections [see (14.6.36), (14.6.24), (14.6.46), and (14.6.49)]

$$h = \frac{1}{44\pi \ln(S/s_1)}$$

$$g = -8h + zA_4(y)(h/y)^{4/11} \tag{14.6.52}$$

$$\Pi\beta = C\left[1 + \frac{g}{2}C(2\pi X)^{3/2} + \frac{h}{3}C^2(2\pi X)^3 + \ldots\right] \tag{14.6.53}$$

$$\mathfrak{X}_0 = X^2/S = {}^0A_0(y)\left[1 - \frac{148\pi}{33}h\right] + \frac{4}{3}zA_4(y)(h/y)^{4/11} \tag{14.6.54}$$

[where ${}^0A_0 \simeq 1$, see (14.6.45) and $A_4(0) \simeq 1$, see (14.6.50)].
These are the formulae which we shall use to study the experimental results.
The critical temperature T_F is given by the conditions

$$S/s_0 \to \infty \qquad h = 0 \qquad z = 0.$$

For a finite chain, $T_{\gamma F}(S)$ is the temperature for which $g = 0$; the corresponding value $z_{\gamma F}$ of z is given by

$$z_{\gamma F} = 8\frac{y^{4/11}}{A_4(y)}h^{7/11} = 8\frac{y^{4/11}}{A_4(y)}[44\pi \ln(S/s_1)]^{-7/11}. \tag{14.6.55}$$

For a finite chain, $T_{\mathscr{G}\mathrm{F}}(S)$ is the temperature for which $\mathfrak{X}_0 = {}^0A_0(y)$, because this swelling is also the swelling of an infinite chain at temperature T_F (see Section 2.2); the corresponding value $z_{\mathscr{G}\mathrm{F}}(S)$ of z is given by

$$z_{\mathscr{G}\infty\mathrm{F}} = \frac{37\pi}{11}\frac{{}^0A_0(y)}{A_4(y)}y^{4/11}h^{7/11} = \frac{37\pi}{11}\frac{{}^0A_0(y)}{A_4(y)}y^{4/11}\,[44\pi\ln(S/s_1)]^{-7/11}. \qquad (14.6.56)$$

Thus, we see that the tricritical theory gives

$$z_{\mathscr{V}\mathrm{F}} > 0 \qquad \text{and} \qquad z_{\mathscr{G}\infty\mathrm{F}} > 0 \qquad (14.6.57)$$

and moreover we note that the ratio of these quantities appears as independent of the ratio (S/s_0) when (S/s_0) is large (s_0 and s_1 are not very different)

$$\frac{z_{\mathscr{G}\infty\mathrm{F}}}{z_{\mathscr{V}\mathrm{F}}} = \frac{37\pi}{88}\,{}^0A_0(y) \simeq \frac{37\pi}{88} = 1.32.$$

The result (14.6.57) is not trivial and can be obtained only after elimination of the short-range divergences (for instance, by analytic continuation as we did here). The older theory of Flory, which does not recognize this necessity, gives opposite inequalities; it seems that the same effect occurs for the decimation theory presented by de Gennes in his book[6] (see, in particular, his Fig. 11.6).

6.9.7 Renormalization of the partition function of an isolated chain near the tricritical point

The dominant terms of the expansion of $\ln\mathscr{Z}(S)$ in powers of y and z are

$$\ln\mathscr{Z}(S) = -4\pi y + \frac{186\pi^2}{'\varepsilon}y^2 + 4z\left(1 - \frac{16\pi}{'\varepsilon}y\right) + \frac{2\pi}{'\varepsilon}z^2\left(1 - \frac{16\pi}{'\varepsilon}y\right) + \ldots \qquad (14.6.58)$$

in agreement with (14.6.16) and the terms proportional to z^2 have been retained because they are needed to calculate the specific heat at the tricritical point.

A priori, the renormalization of the terms porportional to z is not difficult, and these terms could be treated in the limit $S/s_0 \to \infty$, by direct application of the methods which we used to study the swelling and the virial coefficients. In particular, we see that the term proportional to z appears in the form

$$z\left(1 - \frac{16\pi}{'\varepsilon}y\right) \simeq z('\mathfrak{X}_4(y))$$

in agreement with our previsions (see Sections 6.9.1 and 6.9.5)

However, we can see that a problem arises concerning the term proportional to z^2. First, we note that even for $y = 0$, this term becomes infinite when $d \to 3$; secondly, this term appears in the form

$$z^2\left(1 - \frac{16}{'\varepsilon}y\right)$$

and not in the form

$$z^2\left(1 - \frac{32}{'\varepsilon}y\right) \simeq [z\,'\mathfrak{X}_4(y)]^2.$$

In other words, z^2 does not renormalize like z.

Duplantier, who encountered this difficulty, was able to overcome it and the solution can be presented as follows. First, it is clear that for $y = 0$, $\mathscr{X}(S)$ is singular with respect to z. This is not a new fact and we have already studied this singularity in Chapter 12, Section 3.2.4 [see in particular eqn (12.3.49)]. Moreover, the introduction of three-body forces creates new singularities and therefore there is a confluence of singularities. Consequently, the divergence of $\mathscr{X}(S)$, when $S/s_0 \to \infty$, must be treated in a global way. Thus, for $d = 3 - '\varepsilon$, we set

$$\mathscr{X}(S) = [\mathfrak{X}_1(y, z)]^2$$

and considering $\mathfrak{X}_1(y, z)$ as a renormalization factor, we can associate an effective exponent with it. However, before defining this exponent we shall make the following remark: whereas the three-body interaction is marginal in three dimension, since c becomes a dimensionless constant for $d = 3$, the two-body interaction is not marginal since b keeps a dimension for $d = 3$. We are thus convinced that it is proper to introduce a marginal two-body interaction which, by construction, is marginal in three dimensions; we define it by the equality

$$b_{\mathrm{m}} = bS^{1/2}. \tag{14.6.59}$$

Thus, b_{m} is a constant for $d = 3$ but not for $d = 3 - '\varepsilon$, and in this case, we have

$$z = b_{\mathrm{m}}S^{'\varepsilon/2}(2\pi)^{-d/2}. \tag{14.6.60}$$

Now, it seems reasonable to define the effective exponent $\sigma_1(x, y)$ by the equality

$$\sigma_1(y, z) = \left(\frac{'\varepsilon}{2}z\frac{\partial}{\partial z} + '\varepsilon y\frac{\partial}{\partial y}\right)\ln \mathfrak{X}_1(y, z) = S\frac{\partial}{\partial S}\ln \mathfrak{X}_1\bigg|_{b_{\mathrm{m}} = \text{const.},\ c = \text{const.}} \tag{14.6.61}$$

(c is the three-body interaction and $\mathfrak{X}_1(y, z) = (\mathscr{X}(S))^{1/2}$); indeed, it is only a simple generalization of the definitions (12.3.70) and (14.6.39). In this way, we obtain

$$\sigma_1(y, z) = -2\pi('\varepsilon)y + 196\pi^2y^2 + z('\varepsilon - 48\pi y) + \pi z^2\left(1 - \frac{32\pi}{'\varepsilon}y\right) + \dots .$$

This expression diverges when $'\varepsilon \to 0$ but, in order to avoid such divergences, we can re-express $\sigma_1(y, z)$ in terms of h and g, which are physical quantities. Let us set

$$\sigma_1(y, z) = \sigma_1[h, g].$$

By inverting eqns (14.6.25) and (14.6.24), we find

$$y = h + \frac{44}{'\varepsilon}\pi h^2 + \ldots$$

$$z = (g + 8h)\left(1 + \frac{16\pi}{'\varepsilon}h\right) + \ldots$$

and this leads to the result

$$\sigma_1[h, g] = -2\pi('\varepsilon)h + '\varepsilon(g + 8h) + 108\pi^2 h^2 + \pi(g + 8h)(g - 24h) + \ldots .$$

Moreover, eqn (14.6.61) can be rewritten in the form

$$S\frac{\partial}{\partial S}\ln \mathfrak{X}_1\bigg|_{b_{\mathrm{m}} = \text{const.},\, c = \text{const.}} = \sigma_1[h, g], \tag{14.6.62}$$

and now we can go to the limit $'\varepsilon \to 0$. Thus, for $d = 3$

$$\sigma_1[h, g] = 108\pi^2 h^2 + \pi(g + 8h)(g - 24h) + \ldots . \tag{14.6.63}$$

This formula is valid in the asymptotic region, and, in this domain, it will be convenient to re-express g in terms of h and z by using (14.6.52)

$$g = -8h + zA_4(y)\,y^{-4/11}h^{4/11} + \ldots$$

where z must be considered as a constant proportional to b_{m}. In this way, we obtain

$$\sigma_1[h, g] = 108\pi^2 h^2 - 32\pi z A_4(y)\,y^{-4/11}h^{15/11} + \pi[zA_4(y)]^2 y^{-8/11}h^{8/11} + \ldots .$$

Moreover, taking (14.6.35) into account, we can rewrite (14.6.62) in the form

$$\frac{\partial}{\partial h}(\ln \mathfrak{X}_1)\bigg|_{z,y} = -\frac{1}{44\pi h^2}\sigma_1[h, g]$$

and finally, we find

$$\frac{\partial}{\partial h}(\ln \mathfrak{X}_1) = -\frac{27\pi}{11} + \frac{8}{11}zA_4(y)\,y^{-4/11}h^{-7/11}$$
$$- \frac{1}{44}[zA_4(y)]^2 y^{-8/11}h^{-14/11} + \ldots$$

We can now integrate this expression. We obtain

$$\ln \mathscr{Z}(S) = 2\ln \mathfrak{X}_1(S) = -\frac{54}{11}\pi h + 4\pi z A_4(y)(h/y)^{4/11} +$$
$$+ \frac{1}{6}[zA_4(y)]^2 y^{-1}(h/y)^{-3/11} + B(z, y) \tag{14.6.64}$$

where $B(z, y)$ is of the form

$$B(z, y) = {}^0B(y) + {}^1B(y)z + {}^2B(y)z^2 \tag{14.6.65}$$

Moreover, we have

$$z = b_{\mathrm{m}}(2\pi)^{-3/2} = bS^{1/2}(2\pi)^{-3/2}$$

$$h = \frac{1}{44\pi \ln (S/s_1)}.$$

Thus, we can see that in (14.6.64) the dominant term with respect to S is the term proportional to $z^2 h^{-3/11} \propto S[\ln(S/s_0)]^{3/11}$. This is just the term which gives the dominant contribution to the specific heat.

6.9.8 Internal energy and specific heat of an isolated polymer chain

The free energy of a chain is the sum of a regular part which is proportional to S (to the number of links), and of a singular part $\mathbb{F}$

$$\mathbb{F}(S, \boldsymbol{\beta}) = \ln \mathscr{Z}(S)$$

where $\mathscr{Z}(S)$ is an implicit function of $\boldsymbol{\beta}$. The quantities S, b, and c which characterize the chain must be considered as functions of temperature ($T = 1/k_{\mathrm{B}}\boldsymbol{\beta}$). However, to a first approximation, we can assume that the area S is a fundamental feature of the chain and that it is practically independent of temperature, the bonding being ensured by quantum mechanisms which are rather unsensitive to thermal perturbations. On the contrary, the parameters y and z should appear as regular functions of temperature. In what follows, it will be assumed that y remains constant and that z is a linear function of $\boldsymbol{\beta}$ but more general assumptions could be chosen.

From the free energy of the chain, we can deduce the internal energy and the specific heat of the chain. Let $\mathbb{E}(S, \boldsymbol{\beta})$ be the singular part of the internal energy and $\mathbb{C}(S, \boldsymbol{\beta})$ the singular part of the specific heat. These quantities are given by

$$\mathbb{E} = \frac{\partial(\boldsymbol{\beta}\mathbb{F})}{\partial\boldsymbol{\beta}}$$

$$\mathbb{C}/k_{\mathrm{B}} = -\boldsymbol{\beta}^2 \frac{\partial \mathbb{E}}{\partial \boldsymbol{\beta}}.$$

The internal energy $\mathbb{E}(S, \boldsymbol{\beta})$ thus reads

$$\mathbb{E} = -\frac{\partial \ln \mathscr{Z}(S)}{\partial z}\frac{\partial z}{\partial \boldsymbol{\beta}}$$

where $\mathscr{Z}(S)$ is given by (14.6.64) and (14.6.65). Consequently,

$$\mathbb{E} =$$

$$-\frac{\mathrm{d}z}{\mathrm{d}\boldsymbol{\beta}}\left\{4\pi A_4(y)\, y^{-4/11} h^{4/11} + \tfrac{1}{3} z[A_4(y)]^2 y^{-8/11} h^{-3/11} + {}^1B(y) + 2z\, {}^2B(y)\right\} \tag{14.6.66}$$

and, the derivative $\mathrm{d}z/\mathrm{d}\boldsymbol{\beta}$ being assumed to be constant ($\mathrm{d}z/\mathrm{d}\boldsymbol{\beta} < 0$), we finally find

$$\mathbb{C}/k_{\mathrm{B}} = \boldsymbol{\beta}^2 \left(\frac{\mathrm{d}z}{\mathrm{d}\boldsymbol{\beta}}\right)^2 \left\{\frac{1}{3}[A_4(y)]^2 y^{-8/11} h^{-3/11} + 2\,{}^2B(y)\right\}.$$

The last term is a regular term and we may drop it. Thus, taking into account (14.6.36) and the equality $z = bS^{1/2}(2\pi)^{-3/2}$, we find that the singular part of $\mathbb{C}$ at the tricritical point is given by

$$\mathbb{C}/k_{\mathrm{B}} = T^2 \left(\frac{\mathrm{d}b}{\mathrm{d}T}\right)^2 \frac{(44\pi)^{3/11}}{24\pi^3} [y^{-4/11} A_4(y)]^2 S[\ln(S/s_1)]^{3/11} \qquad (14.6.67)$$

where $y = (2\pi)^3 c$ and $A_4(0) = 1$ according to (14.6.50).

We see that, at a tricritical point, the specific heat per chain length becomes infinite when $S/s_0 \to \infty$. The dependence with respect to S is in agreement with Duplantier's findings[12] (though he used somewhat different assumptions to calculate $\mathbb{C}$).

The preceding result is not surprising. By computer simulation on a lattice, Kremer, Baumgärtner, and Binder[16] showed that in the vicinity of T_{F}, the specific heat of a polymer solution has a peak which becomes more and more conspicuous when the size of the chains increases (see Chapter 4, Section 3.2). However, we note that the preceding theoretical result (14.6.67) is not sufficient to predict and to calculate the shape of this peak. A precise and detailed study of this phenomenon remains to be made (1990).

6.10 Semi-dilute solutions in the tricritical domain

6.10.1 Tricritical domain and concentrations

As we have seen [eqn (14.6.37)], the tricriticality condition in dilute solution reads

$$y \ln(S/s_0) \gg \frac{1}{44\pi}.$$

When the chains begin to overlap, a screening effect appears. The screening length ξ_{e} must be of the same order of magnitude as the Brownian overlap-length ξ_0($\xi_0 = \mathscr{C}^{-1/(d-2)}$), which means that, for $d = 3$,

$$\xi_{\mathrm{e}} \propto \mathscr{C}^{-1} \qquad (\mathscr{C} = CS).$$

The screening effect can be felt when the screening length is of the same order of magnitude as the size of the chains. Now, for dilute solutions, according to (14.6.46), we have

$$R^2 \simeq 3S\ {}^0A_0(y). \qquad (14.6.68)$$

The transition between dilute and semi-dilute solutions is thus given by an equality of the form $\xi_{\mathrm{e}} \simeq R$, i.e.

$$S \simeq \mathscr{C}^{-2}.$$

Consequently, the logarithmic term $\ln(S/s_0)$ which appears in the expressions of various physical quantities, will be transformed for semi-dilute solutions into a term $\ln(1/\mathscr{C}^2 s_0)$.

Of course, in any case, we must have $\mathscr{C}^2 s_0 \ll 1$, this condition expressing the fact that we never deal with concentrated solutions. In fact, if N is the number of links of the discrete chain corresponding to our model, we can approximately set $S = N s_0$. Thus

$$\mathscr{C} s_0^{1/12} = CS s_0^{1/2} = CN s_0^{3/2} = C s_0^{3/2}.$$

As $s_0^{3/2}$ can be considered as the volume of a link and since C is the number of links per unit volume, we can see that the product $C s_0^{3/2} = \mathscr{C} s_0^{1/2}$ is proportional and nearly equal to the volume fraction of polymer; here we assume that this quantity is always small.

Equations (14.6.52), (14.6.53) lead to the result

$$h \simeq \frac{1}{44\pi\,|\ln(\mathscr{C}^2 s_1)|} \qquad \mathscr{C}^2 s_1 < 1 \tag{14.6.69}$$

$$g + 8h = z A_4(y)(h/y)^{4/11} \tag{14.6.70}$$

(with $A_4(0) \simeq 1$).

6.10.2 Osmotic pressure of a semi-dilute solution in the tricritical domain

The osmotic pressure in the semi-dilute regime is still given by eqns (14.6.53) and, for g and h, we must use the values given by (14.6.70), (14.6.69). In order to simplify, we may drop small corrections, and we set

$$A(y) = (2\pi)^{3/2}\ {}^0\!A_0(y).$$

In this way, we obtain

$$\boldsymbol{\Pi\beta}\, A(y) S^{3/2} =$$

$$\mathscr{C} S^{1/2} A(y) + \left[-4h + \frac{1}{2} b S^{1/2} (2\pi)^{-3/2} A_4(y)(h/y)^{4/11}\right] [\mathscr{C} S^{1/2} A(y)]^2 + \frac{h}{3} [\mathscr{C} S^{1/2} A(y)]^3 \tag{14.6.71}$$

where h is given by (14.6.69).

As was shown in Section 6.2 (see also Fig. 14.16), the demixtion curve can be drawn in the plane $(bS^{1/2}, \mathscr{C} S^{1/2})$. However, for $S \to \infty$, the product $\boldsymbol{\Pi\beta}$ becomes a function of $\mathscr{C}$ only, if we disregard logarithmic corrections, the other terms becoming negligible. In this limit, (14.6.71) gives

$$\boldsymbol{\Pi\beta}/{}^0\!A_0(y) = \left[\frac{b}{2} A_4(y)(h/y)^{4/11}\right] \mathscr{C}^2 + \frac{1}{3}(2\pi)^3\ {}^0\!A_0(y) h \mathscr{C}^3. \tag{14.6.72}$$

Moreover, we note that, the pressure along the demixtion curve diminishes when the number of links increases; actually, in the plane $(b, \mathscr{C})$, one branch of the demixtion comes nearer and nearer to the vertical axis so as to coincide with it. Along this vertical axis, the osmotic pressure vanishes, and as two phases in equilibrium at a fixed temperature have always the same osmotic pressure, $\Pi\beta \to 0$ everywhere along the demixtion curve when $N \to \infty$. Then eqn (14.6.72) leads to the equality

$$b = -\frac{2}{3}(2\pi)^3 y \frac{{}^0A_0(y)}{A_4(y)}(h/y)^{7/11}\mathscr{C}.$$

Finally, giving to h the asymptotic value (14.6.69), we obtain

$$b = -\frac{2}{3}(2\pi)^3 y \frac{{}^0A_0(y)}{A_4(y)}[44\pi y\,|\ln(\mathscr{C}^2 s_0)|]^{-7/11}\mathscr{C} \tag{14.6.73}$$

where ${}^0A_0(y)$ and $A_4(y)$ can be considered as constants. Moreover, we know

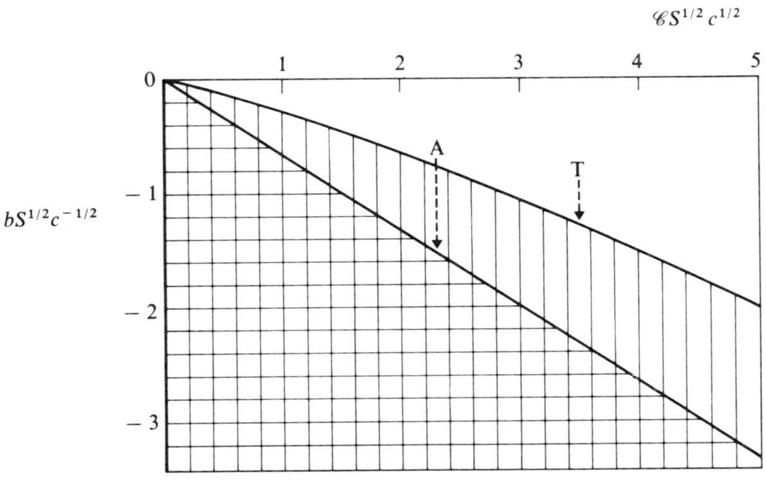

Fig. 14.21. Limiting curve of the demixtion curves
(A) for the simple-tree approximation (and the Flory–Huggins approximation), the asymptot is a straight line.
(T) for the tricritical theory; the curve (which is not universal) has always a horizontal tangent at the origin; here, we choose to represent the curve defined by the equation

$$bS^{1/2}c^{-1/2} = -\frac{2}{3}(\mathscr{C}S^{1/2}c^{1/2})\left\{\frac{1}{2}\ln\frac{2000}{\mathscr{C}^2Sc^{-1}}\right\}^{-7/11}.$$

In both cases, the zones containing the demixtion curves are hachured. (The coordinates are the same as in Fig. 14.16.)

that ${}^0A_0(0) = 1$ and $A_4(0) = 1$, but in the above formula y is not necessarily small. Again, (14.6.73) reproduces a result obtained by Duplantier (1980).[11,12]

Let us note that the simple-tree approximation (or the Flory–Huggins approximation would give [see Sections 6.2 and 5.2 and eqn (14.6.6)]

$$b = -\frac{2c}{3}\mathscr{C} = -\frac{2}{3}(2\pi)^3 y\mathscr{C}. \tag{14.6.74}$$

The difference of behaviour is depicted in Fig. 14.21.

6.10.3 Swelling in a semi-dilute solution for the tricritical domain

Of course, in a semi-dilute solution, because of screening effects, the influence of the interactions on the swelling diminishes. In other words, in the vicinity of the tricritical point, a chain is nearly Brownian with reduced logarithmic corrections. In first approximation, we can deduce an expression of the swelling from (14.6.54) by using for h the value given by (14.6.69). In this way, we obtain

$$\mathfrak{X}_0 = X^2/S = {}^0A_0(y)\left(1 - \frac{37}{363\,|\ln(\mathscr{C}^2 s_1)|}\right)$$
$$+ \frac{4}{3} z A_4(y)\,[44\pi y\,|\ln(\mathscr{C}^2 s_1)|]^{-4/11}. \tag{14.6.75}$$

Therefore, the logarithmic corrections do not produce any really spectacular effect. Moreover, in the absence of precise calculations of the constants coming with the logarithms, it is difficult to show the existence of these logarithmic corrections experimentally. It would be interesting to study more precisely how the swelling (or the deswelling) changes in the vicinity of the demixtion curve, in particular by means of a proper treatment of the one-loop approximation, but such a task has not been undertaken yet (1988).

7. PARTIALLY ATTRACTIVE POLYMERS IN TWO DIMENSIONS

7.1 Tricritical swelling exponent in two dimensions: estimations and exact values

For $d = 3$ the three-body interaction is just marginal. This is not the case for $d < 3$ (and in particular for $d = 2$) since, then, the dimensionless parameter $y = cS^{3-d}$ becomes infinite when $S \to \infty$. In this case, a system with two-body and three-body interactions is tricritical when the second virial coefficient A_2 vanishes (or $z = 0$). Then, for an isolated chain, the dependence of R^2 with respect to S (for large S) is characterized by an exponent ν_t ($\nu_t > 1/2$)

$$R^2 \propto S^{2\nu_t}.$$

Just as Flory's exponent $\nu_F = 3/(d+2)$ can be calculated from the equation

$$\frac{R^2}{S} \propto \frac{S^2}{R^d}$$

in which the two-body interaction is involved (see (8.1.16) and (8.1.17)), the tricritical exponent ν_t can be estimated from a similar equation

$$\frac{R^2}{S} \propto \frac{S^3}{R^{2d}}$$

in which the three-body interaction is involved. In this way, one obtains ($d < 3$) an approximate value ν_{tF} of ν_t

$$\nu_{tF} = \frac{2}{d+1}$$

which gives a rough estimate $\nu_{tF} = 2/3$ for $d = 2$.

The exponent ν_t has been estimated in a better way, by expanding ν_t in powers of $'\varepsilon = 3 - d$. In the framework of field theory, Stephen and McCauley[17] calculated the exponents $\eta_t = 2 - \gamma_t/\nu_t$ and γ_t to order $'\varepsilon^2$. For $n = 0$, their results are

$$\nu_t = \frac{1}{2}\left(1 + \frac{4}{363}\,'\varepsilon^2\right)$$

$$\gamma_t = 1 + \frac{5}{484}\,'\varepsilon^2$$

values which can be recovered, in polymer theory, by direct renormalization. In this way, for $'\varepsilon = 1$, one obtains $\nu_t = 0.505$.

The value of ν_t for $d = 2$ has also been determined by computer simulation on lattices[18,19] and, by using the strip method (see Chapter 12, Section 2.4), Derrida and Saleur[19] found

$$\nu_t = 0.55 \pm 0.01$$

This value is not very different from the experimental result $\nu_t = 0.56 \pm 0.01$ obtained by Vilanove and Rondelez[20] for a 'poor' two-dimensional solution (polymethyl methacrylate on a water–air interface; see Chapter 5, Section 1.11.3).

Finally, in 1987, Duplantier and Saleur found the exact values[21] of the tricritical exponents ν_t and γ_t in two dimensions. These values are

$$\nu_t = \frac{4}{7} \qquad \gamma_t = \frac{3}{7}. \tag{14.7.1}$$

These results were obtained by considering a tractable model and by applying the universality principle.

The model is as follows. A self-avoiding chain of N links is drawn on a honeycomb lattice, whose faces are marked with probability p. The edges of the marked hexagons are forbidden for the chain. The system is in equilibrium. Now consider all the configurations compatible with a given chain: their weight is $(1-p)^H$ where H is the number of hexagons sharing edges with the chain. For a stretched chain $H = N + 1$, but, in general, H is smaller: if two pieces of the chain come close to one another, one hexagon may share three or four links with the chain and H is reduced. The weight $(1-p)^H$ simulates the effect of attractive interactions. Thus, it is expected that for a particular value p_c of p, the chain belong to the universality class corresponding to the Θ point (the Flory temperature T_F).

Simple arguments now show that p_c should be equal to the site-percolation value of the dual triangular lattice, and therefore $p_c = 1/2$.

Connected sets of missing faces can be represented by their perimeters which form non-intersecting self-avoiding loops. Let us come back to the $\mathcal{O}(\mathfrak{n})$ model and to the solid and solid model described in Section 4.1. The partition function, in the absence of the chain, is given by eqn (12.4.6)

$$Z(x) = \sum_{\mathfrak{G}} x^N \mathfrak{n}^{\mathcal{N}}$$

where $\mathcal{N}$ is the number of loops and N the number of links on the graph $\mathfrak{G}$. Here, we have $x = 1$ and $\mathfrak{n} = 1$ (Ising model below T_c since $x_c = 1/\sqrt{3}$ according to Nienhuis).[22]

Now the chain can be replaced by a very large loop which consequently has the properties of a cluster perimeter (a hull) at the percolation threshold. (Incidentally, we recall that the hull of a cluster of loops is the most external loop containing all the other ones). The exponent ν_t is related to the dimensionality D_H of the hull by the equation

$$\nu_t = 1/D_H,$$

which is analogous to (2.2.17) for the same reasons.

The value $D_H = 7/4$ has been calculated by Saleur and Duplantier[23] in various ways; in particular, one method consists in establishing successive one-to-one correspondences between hull configurations, a gas of loops corresponding to the $\mathfrak{n} = 1$ Ising model and to a Coulomb gas. In this way, they obtained $\nu_t = 4/7$.

The value $\gamma_t = 8/7$ was obtained[22] by applying similar arguments to the study of the two-point correlation function associated with the model.

7.2 Multicritical swelling exponents for $3 > d > 2$

When the space dimension d decreases from three to two, more and more interactions become *relevant*. Actually, for a chain of Brownian area S, let us consider a p-body interaction energy of the form

$$b_p \int_0^S \mathrm{d}s_1 \ldots \int_0^S \mathrm{d}s_p \delta[\vec{r}(s_1) - \vec{r}(s_2)] \ldots \delta[\vec{r}(s_1) - \vec{r}(s_p)].$$

This energy being a pure number, we immediately deduce the dimensional formula of the interaction b_p from the preceding expression

$$b_p \sim L^{-2p+(p-1)d}.$$

Therefore, the dimensionless parameter corresponding to this interaction is $b_p S^{p-(p-1)d/2}$ up to a numerical factor. The interaction is relevant if this parameter goes to infinity when S becomes infinite, i.e. if

$$d < \frac{2p}{p-1}. \tag{14.7.2}$$

Thus, for $d < 3$, the three-body interaction is relevant; for $d < 8/3$, the four-body interaction is also relevant, and so on. For $d = 2$, there is an infinite number of relevant interactions.

Let us now consider a set of chains with two-body, three-body, . . . , m-body interactions. We have a p-critical system ($p \leqslant m$), if $(p - 2)$ virial coefficients vanish

$$A_2 = 0 \quad \ldots \quad A_{p-1} = 0.$$

In this case, if condition (14.7.2) is not fulfilled, the p-body interaction is not relevant; then the chains are asymptotically Brownian: in other words, the swelling exponent has the value 1/2. On the contrary, if (14.7.2) is fulfilled, the p-body interaction is relevant: then, there exists a p-critical exponent, which we denote by the symbol $\nu_p (1 > \nu_p > 1/2)$. Thus, for an isolated p-critical chain, we have

$$R^2 \propto S^{2\nu_p}$$

(in particular $\nu_2 = \nu$, $\nu_3 = \nu_t$).

Consequently, for $\dfrac{2(p+1)}{p} \leqslant d < \dfrac{2p}{p-1}$, we obtain a series of exponents*

$$\nu_2 > \nu_3 > \ldots > \nu_p.$$

For $d = 2$, this series is infinite. However, in practice, it seems difficult to show the existence of this phenomenon; in fact, for $d = 2$ and large p, the value of ν_p is very close to 1/2.

REFERENCES

1. Values given to the authors by C. Strazielle, Centre de Recherche des Macromolécules (Institut Charles Sadron), Strasbourg.
2. Zeman, L., Biros, J., Delmas, G., and Patterson, D. (1972). *J. Phys. Chem.* **76**, 1206.
3. Perzynski, R., Adam, M., and Delsanti, M. (1982). *J. Physique* **43**, 129.

* As can easily be seen, the value *à la* Flory of the exponent of order q is

$$\nu_{q,\mathrm{F}} = \frac{q+1}{(q-1)d+2}.$$

4. Flory, P. (1953). *Principles of polymer chemistry*, Chapter XII. Cornell University Press.
5. Daoud, M. and Jannink, G. (1976). *J. Physique* **37**, 973.
6. de Gennes, P.G. (1979). *Scaling concepts in polymer physics*, Chapter IV, pp. 113 and 114.
7. de Gennes, P.G. (1975). *J. Physique Lett.* **36**, L-55. For a definition of the tricritical points see: Griffiths, R.B. (1973) *Phys. Rev.* B **7**, 545.
8. Lifshitz, I.M. (1968). *J. Eksp. Teor. Fiz.* **55**, 2408; (1969) *JETP* **28**, 1280. Lifshitz, I.M. and Grosberg, A.Yu. (1973). *J. Eksp. Teor. Fiz.* **65**, 2399; (1974) *JETP* **38**, 1198.
9. See, in particular, Lifshitz, I.M., Grosberg, A.Yu., and Khokhlov, A.R. (1978). *Rev. Mod. Phys.* **50**, 683; (1978). *Uspekhi Fiz. Nauk.* **127**, 353.
10. Stephen, M.J. (1975). *Phys. Lett.* **53A**, 363 [sign error in (6)].
11. Duplantier, B. (1980). *J. Physique Lett.* **41**, L-409.
12. Duplantier, B. (1982). *J. Physique* **43**, 991.
13. Duplantier, B. (1982). Doctorat d'Etat thesis, Paris VI.
14. Brézin, E., Le Guillou, J.C., and Zinn-Justin, J. (1976). *Phase transitions and critical phenomena* (eds C. Domb and M.S. Green), Vol. 6, p. 125. Academic Press.
15. Duplantier, B. (1986). *Europhys. Lett.* **1**, 491. (This article contains results which do not appear in the present book; in particular, concerning the radius of gyration of a chain in the vicinity of the critical point.)
16. Kremer, K., Baumgärtner, A., and Binder, K. (1982). *J. Phys.* A: *Math. Gen.* **15**, 2879.
17. Stephen, M.J. and McCauley, J.L. (1973). *Phys. Lett.* **44A**, 89.
18. Baumgärtner, A. (1982). *J. Physique* **43**, 1407.
19. Derrida, B. and Saleur, H. (1985). *J. Phys.* A: *Math. Gen.* **18**, L-1075.
20. Vilanove, R. and Rondelez (1980). *Phys. Rev. Lett.* **45**, 1502.
21. Duplantier, B. and Saleur, H. (1987). *Phys. Rev. Lett.* **59**, 539.
22. Nienhuis, B. (1982). *Phys. Rev. Lett.* **49**, 1063.
23. Saleur, H. and Duplantier, B. (1987). *Phys. Rev. Lett.* **58**, 2325.

15

POLYMERS IN GOOD SOLVENTS: EXPERIMENTAL RESULTS

GENESIS

The structure and the behaviour of polymer chains in solutions have, since the 1930s, been the object of intensive experimental work. The main effort has been carried out on the isolated chain in good solvent, and the swelling caused by repulsive interaction between monomers. The effect had already been predicted by Kuhn, in an article published in 1934, describing the spatial occupancy of chains (*Raumerfüllung*). The swelling of a rubber (cross-linked chains) in a solvent can be observed directly with the naked eye. In the case of linear chains, the observation of all aspects of this effect has, however, required the use of instrumentation which has grown heavier, year by year.

Finally, precise experiments carried out on long linear polymer chains made it possible to observe the singular and universal behaviour predicted by the recent theories. There are conclusive tests, as we shall see in this chapter. The crucial experiments concerning the most significative facts in this respect, are recent; however, earlier experiments were neither systematic nor careful enough to provide evidence of these facts. Certainly, the light scattering experiments of Loucheux, Weill, and Benoit in 1958, for instance, produced results suggesting that the statistical occupancy of chains in dilute solutions is singular, but this was considered as an anomaly related only to the isolated chain. It was believed that a progressive and uniform return to the Brownian state, i.e. the normal state, should necessarily occur as a result of concentration increase.

This point of view was too narrow. In fact, until 1970, experimentalists felt no necessity to study the structure function of polymer solutions and the screening effects which it reveals. As a consequence, the physicists have been deprived of the facts that could have induced them to conceive and recognize the existence of the semi-dilute state. On the other hand, a strong interest was shown in osmotic pressure, but always in the framework of the simple-tree approximation. As a consequence, great pains were taken to let the osmotic pressure in the semi-dilute range appear to be proportional to the square of the concentration. The dangers of such a preconceived idea are now well-known.

It was in Saclay, Strasbourg, and the Collège de France (the 'Strasacol' team) that the existence of truly semi-dilute solutions was first shown, experimentally and systematically. The experiments consisted in studying the structure of solutions by neutron scattering. In 1975, Daoud, Cotton, Farnoux, Jannink, Sarma, Beniot, Duplessix Picot, and de Gennes published their results showing that, as the polymer concentration increases, the radius of gyration decreases as a power law of concentration. Later, many experimentalists made various measurements on semi-dilute solutions, using other techniques. In particular, Noda, Imai, Kitano, and Nagasawa, from Nagoya, carefully measured the osmotic pressure of such solutions and obtained a very exhaustive set of data.

All these experimental results, and in particular the osmotic pressure data, provide good reasons for believing in the correctness of the critical theory of polymer solutions. It is to be noted, however, that it was the technique of neutron scattering requiring heavy equipment which enabled the first breakthrough, and this surprising fact is easier to understand if it is recognized that the neutron scattering technique allows us in principle to determine *all* the parameters belonging to the structure of polymer solutions: intra- and intermolecular structure functions, osmotic compressibility, etc.

1. DETERMINATION OF THE BASIC QUANTITIES FROM EXPERIMENTAL RESULTS

The experimentalist measures quite easily the mass concentration ρ, an average molecular mass (i.e. M_w), an average mean square radius of gyration in the limit of zero concentration (i.e. $R^2_{G,z}$), and an average second virial coefficient $A_{2,\sim}$.

To these crude quantities correspond the basic quantities

- chain concentration

$$C = \rho \frac{Ⓐ}{M_n} \tag{15.1.1}$$

where Ⓐ is the Avogadro number;

- square radius of gyration $R^2_{G,z}$;
- effective molecular volume $A_{2,\sim}/M_n Ⓐ$.

These basic quantities are derived from the results of two fundamental experiments: the measure of the osmotic pressure Π of the solution, at membrane equilibrium with the solvent, and the measure of the intensity scattered by the solution at constant pressure.

1.1 Samples

In the experiments which will be mentioned, the polymer is generally atactic polystyrene; we shall also discuss experiments done with polymethylstyrene and polydimethylsiloxane. Polystyrene dissolves easily at room temperature. The good solvents which are used in this case, are

(a) at room temperature: benzene, toluene, and carbon disulphide

(b) at 60°C: cyclohexane and trans-decalin.

A precise definition of the good solvent state is given in Figs 13.25 and 13.26. The chains are obtained synthetically (see Chapter 1), the result being as monodisperse as possible. However, the chains in a sample used for experimental tests, do not have the same number of links. This is an important fact, and the measured quantities are averages. Let us recall the notation of Chapter 1. Let $\mathcal{O}(A)$ be a mean quantity depending upon the number A of links and let $\boldsymbol{C}_A$ be the concentration of chains with A links. We write

$$\langle A^p \mathcal{O}(A) \rangle = \sum_A \frac{C_A}{C} A^p \mathcal{O}(A), \qquad p = 0, 1, \ldots . \tag{15.1.2}$$

In particular,

$$\langle A^p \rangle = \sum_A \frac{C_A}{C} A^p$$

is the moment of order p of the distribution.

By extension, for a quantity $\mathcal{O}(A, B)$ depending upon the number of links of two polymers we define the mean value

$$\langle A^p B^q \mathcal{O}(A, B) \rangle = \sum_A \sum_A \frac{C_A}{C} \frac{C_B}{C} A^p B^q \mathcal{O}(A, B).$$

Various kinds of mean values of $\mathcal{O}$ are defined in this way. The average value $\mathcal{O}_p$ is

$$\mathcal{O}_p = \frac{\langle A^p \mathcal{O}(A) \rangle}{\langle A^p \rangle}, \qquad p = 0, 1, \ldots \tag{15.1.3}$$

and the average value $\mathcal{O}_{p,q}$ is

$$\mathcal{O}_{p,q} = \frac{\langle A^p B^q \mathcal{O}(A, B) \rangle}{\langle A^p B^q \rangle}.$$

In the usual terminology, the index $p = 0$ is written 'n' (number); the index $p = 1$ is written 'W' (weight); the index $p = 2$ is written 'Z'.

The basic quantities then are the concentration by mass ρ and the averages M_p, $R^2_{\mathrm{G},p} (p = 0, 1, \ldots)$.

The averages related to the second virial coefficient, A_2, will be defined according to the type of experiment. With the aid of this data, we shall study the structural laws of dilute solutions and solutions with overlap.

1.2 Quantities derived from osmotic pressure

The measurement of the osmotic pressure at various fixed concentrations ρ, immediately gives the basic quantity $\boldsymbol{\Pi}(\rho)$. We observed in Chapter 5 that the molecular mass and the higher virial coefficients are derived from the function $\boldsymbol{\Pi}(\rho)$, determined in the range of weak concentrations (dilute solutions). We must write the development (5.1.107) of the osmotic pressure in the form

$$\frac{\Pi}{RT} = \rho/M_n + A_{2,\Pi}\rho^2 + \ldots$$

where $M_n = \langle M \rangle = m\langle N \rangle$, N being the number of links of a chain, m the molecular mass of the monomer and $A_{2,\Pi}$ a certain average of A_2.

Note that this relation also reads

$$\boldsymbol{\Pi\beta} = \boldsymbol{C} + \boldsymbol{A}_{2,\Pi}\boldsymbol{C}^2 + \ldots$$

where

$$\boldsymbol{RT} = \text{Ⓐ}/\boldsymbol{\beta}$$

$$M_n = \rho\,\text{Ⓐ}/\boldsymbol{C} \tag{15.1.4}$$

and

$$A_{2,\Pi} = \boldsymbol{A}_{2,\Pi}\,\text{Ⓐ}/M_n^2.$$

The average $A_{2,\Pi}$ of the second virial coefficient is defined by the relation [see (9.4.19)]

$$A_{2,\Pi} = -\frac{\text{Ⓐ}}{2m^2}\frac{\langle Y(A,B)\rangle}{(\langle A \rangle)^2}. \tag{15.1.5}$$

The quantity M_n is obtained by the extrapolation

$$M_n^{-1} = \lim_{\rho \to 0} \frac{\Pi}{RT\rho}. \tag{15.1.6}$$

The highest molecular mass which can be measured by this technique is of the order of 10^5 (see Chapter 5). The coefficient $A_{2,\Pi}$ is obtained by the extrapolation

$$A_{2,\Pi} = \lim_{\rho \to 0} \frac{\partial}{\partial \rho}\left(\frac{\Pi}{RT\rho}\right). \tag{15.1.7}$$

1.3 Quantities derived from radiation scattering

The measurement of the intensity $I(q)$ scattered by a polymer solution, gives us almost directly the structure function

$$\boldsymbol{H}(\vec{q}) = \int d^3r\, e^{i\vec{q}\cdot\vec{r}}\, \frac{\langle C(\vec{r})C(\vec{0}) - 1 \rangle}{C^2}$$

where $C(\vec{r})$ is the local concentration and C the average concentration of monomers. In fact, we have established the relation [see (3.2.58)]

$$I(q) = K\rho^2\,\text{Ⓐ}\,\boldsymbol{H}(\vec{q}) \tag{15.1.8}$$

where K is the constant related to the apparatus and to the contrast between solvent and solute (see 7.2.57).

The basic quantities M, R_G^2 and A_2 characterize the structure function $H(\vec{q})$. Let us examine the manner in which these quantities are derived from experimental results. We write

$$H(\vec{q}) = H^{\mathrm{I}}(\vec{q}) + H^{\mathrm{II}}(\vec{q}) \tag{15.1.9}$$

where $H^{\mathrm{I}}(\vec{q})$ and $H^{\mathrm{II}}(\vec{q})$ are respectively intra- and intermolecular structure functions (see 7.2.20).

1.3.1 A study of the scattered intensity close to the zero angle: the Zimm representation

From measurements of the scattered intensity close to the zero angle and in the limit of zero concentration, we get certain averages of the basic quantities M, R_G^2, and A_2. Let us indeed recall the results (7.3.2) and (7.3.3)

$$\lim_{\rho\to 0} \frac{I(q)}{K\rho} = \lim_{\rho\to 0} \rho \text{Ⓐ} H^{\mathrm{I}}(\vec{q}) = M_W H_Z(\vec{q}) \tag{15.1.10}$$

where $M_W = \langle M^2\rangle/\langle M\rangle$ and where $H(\vec{q})$ is the 'Z' average of the molecular form function. Expanding

$$H_Z(\vec{q}) = 1 - q^2 R_{G,Z}^2/3 + \ldots$$

where

$$R_{G,Z}^2 = \frac{\langle A^2 R_G^2(A)\rangle}{\langle A^2\rangle}$$

we obtain

$$R_{G,Z}^2 = \lim_{\rho\to 0} \lim_{q\to 0} \frac{3}{q^2}\left[\frac{I(\to 0) - I(q)}{I(\to 0)}\right] \tag{15.1.11}$$

The experimental determination of this quantity is therefore independent of the determination of the constant K.

Consider now the intermolecular contribution. We established the relation [see (7.3.4)]

$$\rho \text{Ⓐ} H^{\mathrm{II}}(\vec{0}) = \lim_{q\to 0} \left\{\frac{I(q)}{K\rho} - \lim_{\rho\to 0} \frac{I(q)}{K\rho}\right\} \tag{15.1.12}$$

On the other hand, we can express this quantity in terms of the virial expansion (9.4.22), after changing C into ρ [see (15.1.1)]. We have

$$\rho \text{Ⓐ} H^{\mathrm{II}}(\vec{0}) = \rho \text{Ⓐ} (H(\vec{0}) - H^{\mathrm{I}}(\vec{0})) = -(2A_{2,\sim} M_W^2 \rho + \ldots)$$

$$A_{2,\sim} M_W \rho \leqslant 1 \tag{15.1.13}$$

where $A_{2,\sim}$ is a particular average of the second virial coefficient A_2 [see (9.4.22)]

$$A_{2,\sim} = -\frac{\text{Ⓐ}}{2m^2} \frac{\langle AB\, Y(A,B)\rangle}{(\langle A^2\rangle)^2} \tag{15.1.14}$$

(m is the molecular mass of the monomer; A and B are numbers of monomers). Hence, combining (15.1.13), (15.1.12), and (15.1.10), we find

$$2A_{2,\sim} = \lim_{\rho\to 0} \frac{-1}{\rho M_W^2} \lim_{q\to 0} \left\{ \frac{I(q)}{K\rho} - M_W \right\}$$

and, consequently,

$$2M_W A_{2,\sim} = \lim_{\rho\to 0} \frac{1}{\rho} \lim_{q\to 0} \left\{ 1 - \frac{I(q)}{K\rho M_W} \right\}. \quad (15.1.15)$$

The factor K appears explicitly in this relation. However, the extrapolation

$$\lim_{\rho\to 0} \lim_{q\to 0} \frac{I(q)}{\rho} = K M_W \quad (15.1.16)$$

directly gives KM_W. Hence, the measurement of $M_W A_{2,\sim}$ is independent of the measurement of the constant K. Finally, the only quantity determined by scattering and requiring the value of the constant K is the molecular mass.

In general, all the extrapolations discussed above, are made in the Zimm[1] representation of the experimental data. The latter consists in plotting on the same diagram the quantity

$$\frac{K\rho}{I(q)} = \frac{1}{\rho \textcircled{A} \boldsymbol{H}(\vec{q})} \quad (15.1.17)$$

against both the square q^2 of the transfer wave vector, and the mass concentration ρ. Indeed, it appears, experimentally and theoretically, that the inverse of the scattered intensity is a quasi-linear function of q^2 and ρ. In fact, we note first that for Brownian chains, $\rho/I(q)$ varies linearly with q^2, both for $q^2 R_G^2 \ll 1$ and for $q^2 R_G^2 \gg 1$, and we observe next that the tree approximation (see Chapter 13, Section 2.6.3) would give for $\rho/I(q)$ a linear dependence on ρ, for all values of q.

Let us therefore find the development of $K\rho/I(q)$ to first-order in q^2 and ρ. From (15.1.8), (15.1.9), and (15.1.10), we obtain

$$\frac{I(q)}{K\rho} = M_W H_Z(\vec{q}) + \rho \textcircled{A} \boldsymbol{H}^{\text{II}}(\vec{q}),$$

which can be written in the form

$$\frac{K\rho}{I(q)} = \frac{1}{M_W H_Z(\vec{q})} - \frac{\rho \textcircled{A} \boldsymbol{H}^{\text{II}}(\vec{q})}{M_W H_Z(\vec{q}) [M_W H_Z(\vec{q}) + \rho \textcircled{A} \boldsymbol{H}^{\text{II}}(\vec{q})]}, \quad (15.1.18)$$

or also

$$\frac{K\rho}{I(q)} = \frac{1}{M_W H_Z(\vec{q})} - \frac{\rho \textcircled{A} \boldsymbol{H}^{\text{II}}(\vec{0})}{M_W^2} \left(\frac{\boldsymbol{H}^{\text{II}}(\vec{q})}{\boldsymbol{H}^{\text{II}}(\vec{0}) \left(H_Z(\vec{q}) + \frac{\rho \textcircled{A}}{M_W} \boldsymbol{H}^{\text{II}}(\vec{q}) H_Z(\vec{q}) \right)} \right). \quad (15.1.19)$$

Making use of (15.1.13) and (7.2.59), we get to first-order in q^2 and ρ

$$\frac{K\rho}{I(q)} = \frac{1}{M_W} + \frac{q^2 R^2_{G,Z}}{3M_W} + 2A_{2,\sim}\,\rho \qquad (15.1.20)$$

where $R^2_{G,Z}$ is the 'Z' average of the mean square radius of gyration.

The Zimm representation is then obtained by using for the ordinate the quantities $K\rho/I(q)$, and for the abscissa the quantities $q^2 + \lambda\rho$, where λ is an arbitrary constant, of order of magnitude equal to $6M_W A_{2,\sim}/R^2_{G,Z}$. A typical result, obtained by Kirste and Lehnen,[2] is reproduced in Fig. 15.1. The extrapolation of the measured quantities to $q \to 0$ and $\rho \to 0$ is then, in principle, linear. As shown in (15.1.20), one reads directly from the figure the extrapolated ordinate $1/M_W$, and the extrapolated slopes $R^2_{G,Z}/3M_W$ ($\rho \to 0$ and then $q \to 0$). and $2A_{2,\sim}/\lambda$ ($q \to 0$ and then $\rho \to 0$). Equation (15.1.20) is only valid in the limit $\rho \to 0$. However, for a finite but small concentration ρ, it may be generalized in

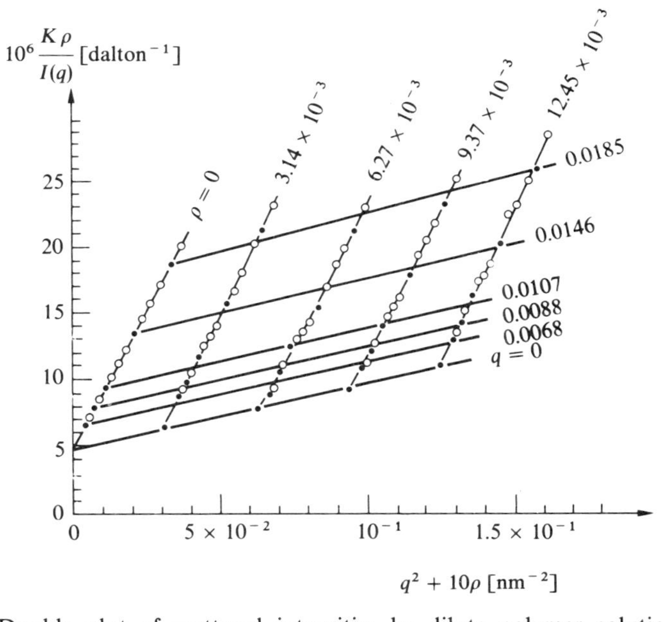

Fig. 15.1. Double plot of scattered intensities by dilute polymer solutions, against concentrations ρ[g/l] and values q [nm^{-1}] of the transfer wave vector (Zimm representation). Neutron scattering: non-deuterated polydimethylsiloxane, ($M_W = 2 \times 10^5$) in solution in liquid deuterated polydimethylsiloxane ($M_W = 3 \times 10^2$) (from Kirste and Lehnen[2]):
the lines q = constant are slightly oblique
the lines ρ = constant are nearly vertical
• measured values
o other measured values.

the form

$$\frac{K\rho}{I(q)} = \frac{1}{M_W}(1 + q^2 R^2_{G,Z}(\rho)/3) + 2A_{2,\sim}\,\rho \tag{15.1.21}$$

where $R^2_{G,Z}(\rho)$ is the radius of gyration at concentration ρ. This formula, derived for the case $q^2 R^2_{G,Z}(\rho) \ll 1$ is reminiscent of the simple-tree approximation. In fact, the radius taken from (15.1.21) is an *apparent* radius. The true radius of gyration is determined experimentally by neutron diffraction, with labelled molecules as a target (see Chapter 7, Section 3.1.2).

1.3.2 Scattered intensity in the limit of zero concentration: intra- and intermolecular structure functions

The mean form function of a chain is obtained from scattering experiments on dilute solutions, by extrapolation at zero concentration [see (15.1.10)],

$$H_Z(\vec{q}) = \lim_{\rho \to 0}\left(I(q) \Big/ \lim_{p \to 0} I(p) \right)$$

This function is by definition normalized to unity for $q = 0$. The significant reciprocal space interval for $H_Z(\vec{q})$, is $(0, 1/l_a)$ where l_a is the interatomic distance. Radiation scattering allows us to explore the intervals (see Chapter 6, Fig. 6.4, p. 234)

$$10^{-4} \text{ to } 3.8 \times 10^{-2}\,\text{nm}^{-1} \text{ (light scattering)} \tag{15.1.22}$$

$$4 \times 10^{-2} \text{ to } 7\,\text{nm}^{-1} \text{ (neutron scattering)} \tag{15.1.23}$$

Therefore, in order to determine the form function $H_Z(\vec{q})$ of a sample in the largest possible interval, one has in general to perform two experiments, using a different type of radiation in each case.

It is, however, possible to produce samples for which a unique light scattering experiment enables the experimentalist to explore the three characteristic ranges:

$qR_{G,Z} < 1$ Guinier range

$qR_{G,Z} \gg 1$ asymptotic range

$qR_{G,Z} \simeq 1$ intermediate range

For this, the inverse $1/R_{G,Z}$ of the mean radius of gyration must fall in the middle of the observation window defined by (15.1.22). The result is obtained, for instance, with chains of radius $R_{G,Z} \sim 200$ nm. In order to visualize the three correlation ranges, it is convenient to represent the scattered intensities in the form

$$\lim_{\rho \to 0} \frac{q^4 I(q)}{K\rho} = q^4 M_W H_Z(\vec{q}). \tag{15.1.24}$$

In particular, when the chains in the sample are monodisperse and Brownian

$$H_Z(\vec{q}) = {}^0H(\vec{q}).$$

Therefore, for $d = 3$ (see (7.3.34)],

$$q^4 M\, {}^0H(\vec{q}) = \frac{2M}{{}^0R_G^4}\left(e^{-q^2({}^0R_G^2)} - 1 + q^2({}^0R_G^2)\right). \tag{15.1.25}$$

Miyaki, Einaga, and Fujita[3] represented their measured scattered intensities in the form (15.1.24) and their results are reproduced in Fig. 15.2. Their sample was a dilute solution of long chains ($M_W = 8.78 \times 10^6$) of polystyrene in cyclohexane at 34.6 °C [the temperature T_F corresponds to 34 °C; see (14.2.13) for the definition of T_F]. In this situation, the polystyrene chains have a conformation close to that of equivalent Brownian chains. It is then possible to recognize the three correlation ranges in Fig. 15.2, using (15.1.25).

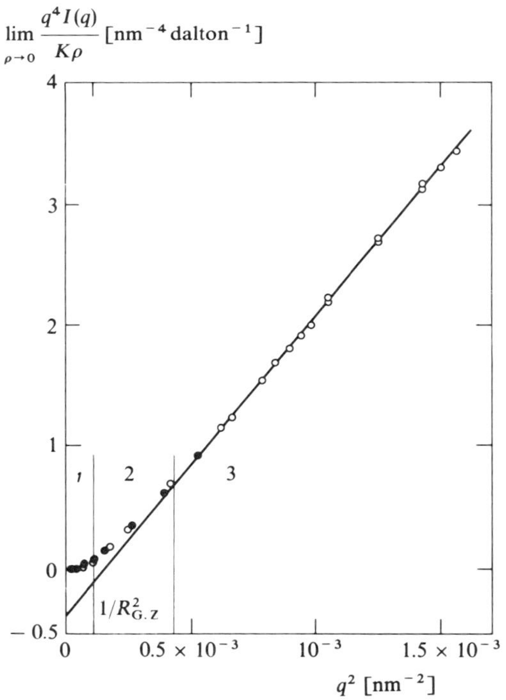

Fig. 15.2. Representation of the quantities $\lim_{\rho\to 0} \frac{q^4 I(q)}{K\rho}$, extrapolated at zero concentration from values measured for dilute solutions. Light scattering (from Miyaki *et al.*[3]). Atactic polystyrene ($M_W = 8.78 \times 10^6$) in cyclohexane at 34.5 °C ($\simeq T_F$):
1 Guinier range; *2* Intermediate range; *3* Asymptotic range.
o values measured at a wavelength 436 nm
• values measured at a wavelength 546 nm.

Finally, the light scattering experiments on long chains in solution make it possible to determine the basic quantities with the best accuracy in the largest range.

Neutron scattering is a useful technique for the study of intramolecular correlations in the following ranges:

Guinier range for smaller chains ($R_G < 20$ nm)

asymptotic range of very long flexible chains.

Let us also mention the range in which the persistence length is noticeable (see Chapter 3, Section 3). We shall not be interested in the study of local structure, but we shall have to determine its effects because they restrict the domain of exploration of the asymptotic behaviour.

The neutron scattering technique also makes it possible to study the intermolecular function $\boldsymbol{H}^{\text{II}}(\vec{q})$ in the entire observation window (15.1.23). Here, experiments are carried out on blends of deuterated and non-deuterated chains, following the method described in Chapter 7, Section 3.1.2. This has been done by Nierlich and Cotton,[4] on polystyrene samples of molecular mass $M_W = 114\,000$, in carbon disulphide (a good solvent). The resulting functions $\boldsymbol{H}^{\text{II}}(\vec{q})$ and $H_Z(q)$ are then combined to test the hypothesis derived from the

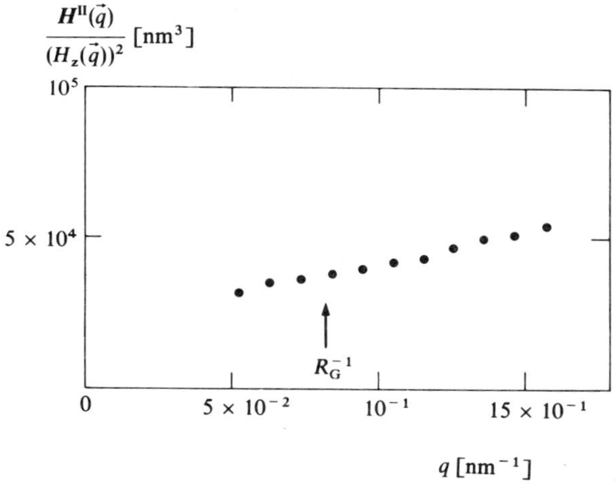

Fig. 15.3. Ratio of the intermolecular structure function, $\boldsymbol{H}^{\text{II}}(\vec{q})$ and the square form function $H_Z(\vec{q})$. (From Nierlich and Cotton.[4]) The plot of this observed ratio against transfer q contradicts the simple-tree approximation. On the original drawing the ordinates were in arbitrary units. We have set the scale by observing that $\lim_{\rho \to 0} \boldsymbol{H}^{\text{II}}(\vec{0})/(H_Z(\vec{0}))^2$ is equal to the product $2M_W^2 A_{2,\sim}/Ⓐ$ (see 15.1.20).

simple-tree approximation, which states that

$$-\lim_{\rho \to 0} Ⓐ \frac{H^{\text{II}}(\vec{q})}{(H_Z(\vec{q}))^2} = 2M_W^2 A_{2,\sim} \qquad (15.1.26)$$

where $A_{2,\sim}$ is a constant independent of q.

Figure 15.3 represents experimental values of this ratio for various values of the wave vector transfer. It appears that the left-hand side of (15.1.26) is a slightly increasing function of q, in contradiction to formula (15.1.26).

1.4 Table of basic quantities for polystyrene

There are many results giving the basic quantities for polystyrene. We have chosen to present the results obtained by five teams, using light scattering.

(A) Y. Miyaki, Y. Einaga, and H. Fujita (1978)[3]
(B) I. Noda, M. Imai, T. Kitano, and M. Nagasawa (1983)[5]
(C) M. Fukuda, M. Fukutomi, Y. Kato, and T. Hashimoto (1974)[6]
(D) T. Nose and B. Chu (1979)[7]
(E) A. Yamamoto, M. Fuji, G. Tanaka, and H. Yamakawa (1971)[8]

Teams A, B, C, and E studied the variation of swelling with solute molecular mass, for a same solvent quality.

Teams D and E studied the variation of swelling with temperature, for a given molecular mass of the solute.

The results are reported in Table 15.1. We see four types of quantities in this table

1. The chemical nature of the solvent, the temperature of the solution and the molecular mass of the solute are given in columns 3 and 4.

2. Next appear the basic quantities associated with the good solvent state: the 'Z' average $R_{G,Z}^2$ of the mean square radius of gyration in the limit of zero concentration and the $A_{2,\sim}$ average of the second virial coefficient.

3. Columns 7 and 8 contain quantities associated with the Brownian state of the samples: the 'Z' average, ${}^0R_{G,Z}^2$, of the mean square radius of gyration and the constant ${}^0\Lambda = {}^0R_{G,Z}^2/M$. These constants are measured in the limit of zero concentration in poor solvents, at a temperature $T \simeq T_F$, where T_F is defined in Chapter 14, Section 2.2; teams A, C, and E used cyclohexane, team D used trans-decalin. The reference temperatures to the Brownian state are, for cyclohexane 34.5 °C (team A) and 34.6 °C (teams C and E), for trans-decalin 20.5°C. In Chapter 16, we shall see that the experimental value of T_F is more precisely 34 °C for polystyrene in cyclohexane. The fact that T does not correspond exactly to T_F is of minor importance here.

The constant ${}^0\Lambda = {}^0R_{G,Z}^2/M$ is derived from the asymptotic behaviour of the form function (see Chapter 7, Section 3.2.2) at temperature $T \simeq T_F$

Table 15.1

Sample No.	Team	Solvent and temperature [°C]	$M_W\,10^{-6}$	$R^2_{G,z}\,10^{-2}$ [nm²]	$10^2 A_{2,\sim}$ [cm³/dalton²]	${}^0R^2_{G,z}\,10^{-2}$ [nm²]	${}^0\Lambda\,10^4$ [nm²/dalton]	M_W/M_n	M_Z/M_W	p
1	A	b 25°	56.3	2560	1.09	519	8.46		1.10	0.05
2	B	b 25°	38.5	1537	1.22	335	8.40		1.05	0.025
3	A	b 25°	31.9	1246	1.28	279	8.21		1.06	0.03
4	A	b 25°	23.4	882	1.44	210	8.51		1.05	0.025
5	B	t 25°	21	700	1.29					
6	A	b 25°	15.1	515	1.66	134	8.32		1.06	0.03
7	C	b 30°	13.4	445	1.50	118				
8	D	td 40°	11.7	306		158				
9	D	td 30°	11.7	246		158				
10	D	td 25°	11.7	207		158				
11	B	t 25°	11	346	1.49					
12	C	b 30°	9.7	301	1.60	78.5		1.03		0.03
13	A	b 25°	8.78	269	1.82	77	8.22		1.07	0.035
14	B	t 25°	8.4	240	1.68					
15	C	b 30°	7.62	230	1.73	60.3		1.02		0.02
16	C	b 30°	5.53	158	1.89	45.6		1.04		0.04
17	C	b 30°	4.59	125	2.03	36.8		1.05		0.05
18	C	b 30°	3.63	96.4	2.21	28.3		1.07		0.07
19	E	b 30°	2.87	68.7	2.25					
20	C	b 30°	2.42	62.5	2.33	19.0		1.04		0.04
21	E	b 30°	1.59	33.1	2.91					
22	C	b 30°	1.23	25.9	2.86	9.08		1.02		0.04
23	E	b 30°	0.756	14.0	3.34	6.61				
24	E	c 60.1°	0.756	8.96	0.933	6.61				
25	E	c 50.1°	0.756	8.39	0.677	6.61				
26	E	c 45.4°	0.756	7.98	0.554	6.61				
27	E	b 30°	0.342	5.76	4.46					
28	D	td 40°	0.179	2.13		1.69				

b = benzene; t = toluene; td = transdecaline; c = cyclohexane.

4. Finally, in the three last columns, we find quantities characterizing the polydispersion of the samples: ratios M_W/M_n and M_Z/M_W (different averages of the molecular mass) and polydispersity p. In Chapter 7, Section 3.2 we indicated how to determine the average M_W and M_n from the scattered intensity. We shall indicate in the next section how M_Z is determined.

The polydispersity p characterizes polydispersion if we assume that the distribution of the links obeys a Schulz and Zimm law (see Chapter 13, Section 2.4)

2. REFERENCE TO THE BROWNIAN STATE: THE ATACTIC POLYSTYRENE CASE

In order to determine the swelling of a polymer chain in a good solvent, we have to measure its size and to compare it with the size of the equivalent Brownian chain, i.e. the size that the chain would have in the Brownian state. This Brownian state is, in fact, only a concept; however, by setting the sample in an adequate environment, one can get a situation close to this concept.

2.1 Size of polystyrene in poor solvent and in the liquid state

Let us examine the measured values of averages ${}^0R^2_{G,Z}$ of the square radius of gyration, for several polystyrene samples (see Table 15.1) in cyclohexane, at temperatures 34.5 °C for team A, and 34.6 °C for teams C and E (T_F is 34 °C). The measured quantities can be written in the form

$$ {}^0R^2_{G,Z} = {}^0\Lambda_{Z,W} M_W \tag{15.2.1} $$

where M_W is the 'W' average of the molecular mass, which defines ${}^0\Lambda_{Z,W}$. If the conversion factor ${}^0\Lambda_{Z,W}$ does not depend upon M_W, we can assume that the chains are Brownian in this particular state. However, the quantities ${}^0\Lambda_{Z,W}$ do depend here upon polydispersion (and upon chemical features of the chain!). Results of Table 15.1 show that in the present case one can indeed consider ${}^0\Lambda_{Z,W}$ as independent of M_W for

$$ 10^6 \leqslant M_W \leqslant 50 \times 10^6. $$

We also noticed in Chapter 13, Section 2.6.3, that another physical situation exists in which the *effective* repulsive interactions between beads are zero: this is the case of a very concentrated polymer solution. Indeed, a concentration increase tends to screen the interaction: then each chain in the solution tends to a *Brownian* state.* Neutron scattering experiments[2] on mixtures of deuterated

* Let us note that it has not been formally proved that a polymer chain in the melt has an asymptotic strictly Brownian behaviour.

and non-deuterated polystyrene chains made is possible to confirm relation (15.2.1) in the melt state.

We can conclude that in a poor solvent solution at a temperature close to T_F, or in the melt, atactic polystyrene chains have a Brownian behaviour.

2.2 Brownian state and flexibility of the atactic polystyrene chain

Can we consider that a chain in solution at temperature T_F, or in the melt, is Brownian on a small scale? The answer is not simple, and the purpose of this section is to warn the reader against possible errors.

For any sequence of asymptotically Brownian chains, one has

$$\langle\langle |\vec{r}_i - \vec{r}_j|^2 \rangle\rangle = A|i-j|, \qquad A = \text{constant}$$

if $|i-j|$ is large enough, and this is approximatively true as long as

$$|i-j| > n_0.$$

One can try to determine the average value of the length

$$(\langle\langle |\vec{r}_{i+n_0} - \vec{r}_i|^2 \rangle\rangle)^{1/2}$$

which gives an order of magnitude of the persistence length l_p, related to the rigidity of the chain. It is, for instance, necessary to know l_p in order correctly to interpret experiments aimed at the determination of the *thermic* correlation length ξ_t (13.1.73) of a chain in good solvent.

To answer this question it is sufficient to set the polystyrene chain in the Brownian state and to measure its form function in the asymptotic domain $q^2 \gg 1$. Indeed, the form function $H(\vec{q})$, for a given q, is a revealer of correlations at distances of order q^{-1}. Observation of the behaviour of $H(\vec{q})$ when q increases should in principle show the crossover from Brownian to rigid conformation. However, the application of this method raises all sorts of practical difficulties which we are now going to examine.

When the measurement is made by light scattering, the result (see Fig. 15.2) can indicate a Brownian behaviour in all the observation window (15.1.22). However, in all cases, the smallest observable correlation length is of the order of 30 nm, which is still very great with respect to the persistence length. This method is therefore inadequate. On the other hand, it is more appropriate to use neutron radiation to perform these measurements [see (15.1.23) for the observation window]. There are many experimental results obtained in this way for atactic polystyrene in the Brownian state (polymer melt, etc.). An example is given in Fig. 15.4. This figure reproduces the result of measurements made by Tangari, King, and Summerfield[9] and the curve represents a plot of the quantity

$$\lim_{\rho \to 0} \frac{q^2 I(q)}{K\rho} = q^2 M_W H_Z(\vec{q})$$

against q^2.

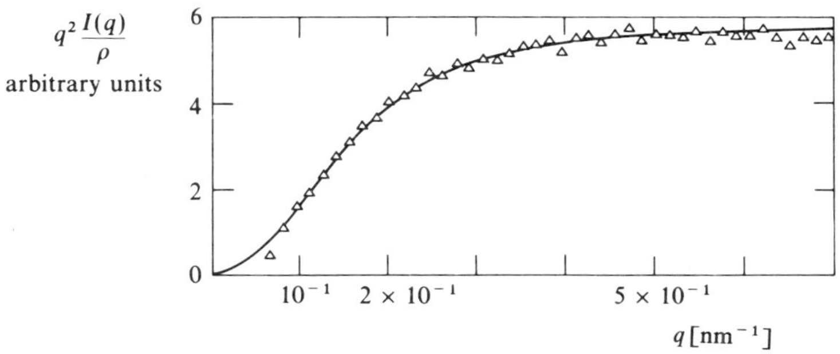

Fig. 15.4. Representation of experimental results of neutron scattering by blends of atactic polystyrenes, deuterated and non-deuterated. (From Tangari *et al.*[9].)
△ measured values
– curve obtained for the Brownian form function ${}^0H_Z(\vec{q})$ giving the best adjustment with the measured values.

At first sight, the result seems to indicate that the chain remains Brownian down to scale lengths as small as 0.15 nm! Indeed, the quantity defined above tends to a constant value when q increases, which is characteristic of the Brownian chain.

We can reproduce the result by calculation, assuming that the chain is Brownian and performing the adequate polydispersion corrections (with the dispersion law of Schulz and Zimm). In this way it is possible to obtain a function ${}^0H_Z(\vec{q})$ compatible with the experiment, and this leads to acceptable values of ${}^0R^2_{G,Z}$ and of p:

$$({}^0R^2_{G,Z})^{1/2} = 8.5 \text{ nm} \qquad \text{for } M_W = 1.14 \times 10^5 \text{ and } p = 0.1.$$

However, as we shall see, the results in the asymptotic domain should reveal a different property of polymer chains. Criticism came from Yoon and Flory,[10] disturbed by the absence of local chemical structure effects. These authors modelled the atactic polystyrene chain of N beads as realistically as they could (see Chapter 1). Thus, Yoon and Flory accounted for interactions between nearest neighbour monomers; in particular, these interactions are responsible for the fact that the three orientations of a bond j, relative to bonds $j-1, j-2$, are weighted differently in relation to the stereochemical composition. Moreover, we have seen in Chapter 1 that, for atactic polystyrene, pairs of successive benzene rings are slightly more frequent in syndiotactic than in isotactic positions.

Taking these facts into account, Yoon and Flory calculated the form function $H_{YF}(\vec{q})$ assuming that the only scattering centres are the carbons to which the benzene rings are attached. The form function is given by an expansion which involves averages of even powers of distances between centres. The chain can take only a finite number of configurations and is simulated by a Monte Carlo process. The allowed values of the valence angles and of the bond lengths ar predetermined; likewise, each configuration is associated with a well-defined weight. All these facts result from earlier work by Flory and co-workers. Self-intersections are allowed, and hence the chain is asymptotically Brownian. The results of Yoon and Flory are as follows:

1. For fixed N (for instance $N = 10^3$), the form function $H_{YF}(\vec{q})$ has an asymptotic behaviour characterized by a marked increase of the quantity $q^2 H_{YF}(\vec{q})$ in the interval $q \geqslant 3 \times 10^{-1}\,\text{nm}^{-1}$.

2. For fixed N, there exists, in continuous space, an equivalent chain made of N^* *independent* links of equal length. This number N^* is determined in the following way: consider the quantities $\langle\!\langle |\vec{r}_N - \vec{r}_0|^2 \rangle\!\rangle$ and $\langle\!\langle |\vec{r}_N - \vec{r}_0|^4 \rangle\!\rangle$ calculated for the initial chain model. From the characteristic ratio

$$\frac{\langle\!\langle |\vec{r}_N - \vec{r}_0|^4 \rangle\!\rangle}{(\langle\!\langle |\vec{r}_N - \vec{r}_0|^2 \rangle\!\rangle)^2}$$

we get the number N^* with the help of the following identity, valid for chains made of independent links with equal length, in continuous space

$$\frac{\langle\!\langle |\vec{r}_N - \vec{r}_0|^4 \rangle\!\rangle}{(\langle\!\langle |\vec{r}_N - \vec{r}_0|^2 \rangle\!\rangle)^2} = \frac{5}{3} - \frac{2}{3N^*} \qquad (d = 3).$$

We set

$$N^* = N/n$$

where n is the number of connected links in the model chain, necessary to form an independent sequence or a new link. This number is equal to 27 in the limit $N \to \infty$.

Result 1 contradicts the preceding result displayed in Fig. 15.4, where $q^2 H(\vec{q})$ tends towards a constant. The contradiction is only apparent, because the model chain of Yoon and Flory and the real polystyrene chain are different. The explanation has been found thanks to a very precise experiment carried out by Rawiso and Picot.[11] These experimentalists prepared their samples of atactic polystyrene in the following way: the chains were made of *partially* deuterated monomers, in which only the protons directly bound to the skeleton carbons are replaced by deuterium atoms. The labelled chains are immersed in a non-labelled polystyrene melt. Therefore, only the correlations between points close to the skeleton contribute to the form function. The result of the experiment (Fig. 15.5, curve a) is qualitatively in agreement with the predictions of Yoon and Flory, who in fact accounted only for the carbon skeleton in their calculation. The persistence length associated with the form function of Fig. 15.5, curve a, is

$$l_p = 2\,\text{nm}. \qquad (15.2.2)$$

When the real structure of the CH_2 and $CH\varphi$ groups are introduced in the calculation of the form function, we indeed find a behaviour analogous to the one represented in Fig. 15.5, curve b. The side-groups CH_2 and $CH\varphi$ (see

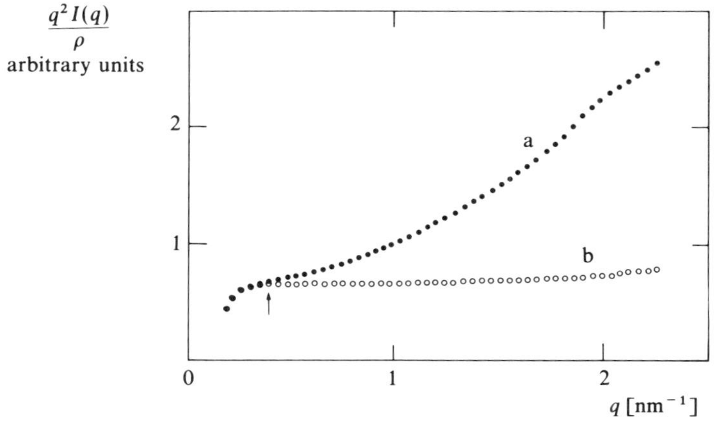

Fig. 15.5. Form functions of polystyrene in the melt; experimental result obtained by neutron scattering. (From Rawiso and Picot.[11])
o sample made of a blend of *totally* deuterated and non-deuterated chains
• sample made of *partially* deuterated $(CD_2 - CDC_6H_6)_N$ and non-deuterated chains.

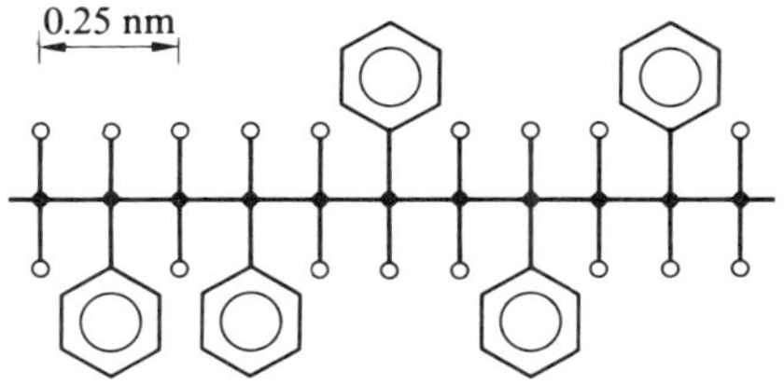

Fig. 15.6. Schematic representation of a sequence of a stretched polystyrene chain. The succession of points • displays the carbon skeleton. The hexagons represent benzene rings. Two consecutive rings lying on the same side of the skeleton are in isotactic position. Circles o represent hydrogens bound to the skeleton.

Fig. 15.6) contribute to the form function. The lateral extension R_l is about $R_l = 0.4$ nm and the weight of these contributions expresses itself with an additional factor

$$\exp(-q^2 R_l^2/2).$$

Multiplying the skeleton form function by this factor, we produce a decrease in the quantity $q^2 H(\vec{q})$. In this manner, the observed fact that $q^2 H(\vec{q})$ is constant in the asymptotic domain $q \geqslant 0.3\ \text{nm}^{-1}$ results from the compensation of two opposite effects.

We retain from this analysis that the atactic polystyrene chain can only be considered as Brownian on a length scale greater than 2 nm.

2.3 Area of the equivalent Brownian chain

The experiments described above indicate the existence of a Brownian behaviour for atactic polystyrenes in the two following physical situations: polymers in the melt or in dilute solutions at the Flory temperature T_F.

We can therefore evaluate experimentally the area of the Brownian chain which is equivalent to the polystyrene chain. This area gives us a geometrical intrinsic measure of the quantity of matter associated with a chain; its use allows us to interpret experimental results directly in terms of the continuous standard model.

For the experimentalist, the quantity of matter associated with a chain is given by its molecular mass. This quantity is not intrinsic, since it depends, for instance, on the percentage of isotopic substitution. Therefore, we shall relate the molecular mass to the area of the equivalent Brownian chain.

In the Brownian case, the mean square end-to-end distance ${}^0R^2$, the radius of gyration ${}^0R_G^2$, and the area S, are related by (in three-dimensional space)

$$ {}^0R_G^2 = {}^0R^2/6 = S/2. \tag{15.2.3}$$

The squared radius of gyration is quasi-proportional to the molecular mass. Let ${}^0\Lambda$ be the conversion factor

$$ {}^0R_G^2 = {}^0\Lambda\, M. \tag{15.2.4}$$

Therefore

$$ S = 2\ {}^0\Lambda\, M \tag{15.2.5}$$

which is the relation searched for.

We get the coefficient ${}^0\Lambda$ directly from the asymptotic behaviour of the mean form function $H_Z(\vec{q})$. Indeed, for a Brownian chain [see (7.3.37)]

$$ \lim_{\rho\to 0} \frac{q^4 I(q)}{K\rho} \simeq \frac{2}{{}^0\Lambda}\left(q^2 - \frac{1}{{}^0\Lambda\, M_n}\right) \qquad q^2\,{}^0R_{G,Z}^2 \gg 1. \tag{15.2.6}$$

The result is independent of polydispersion!

Light scattering experiments for very long chains in solution at the temperature $T \simeq T_F$ enable us to determine the value of ${}^0\Lambda$ with precision (see Fig. 15.2). Values of ${}^0\Lambda$ obtained for various samples are reported in Table 15.1, p.000. The average value is

$$ {}^0\Lambda = 8.3 \times 10^{-4}\ \text{nm}^2/\text{dalton}. \tag{15.2.7}$$

Moreover, we see from (15.2.6) that M_n can be obtained in this way, even for very high masses, when the measurement of osmotic pressures becomes unreliable.

Knowledge of the conversion factor ${}^0\Lambda$ allows us also to determine the average molecular mass M_Z from the 'Z' average of the square radius of gyration. Indeed, in the Brownian case, relation (15.2.4) can be written

$$ {}^0R_{G,Z}^2 = {}^0\Lambda\, M_Z. \tag{15.2.8}$$

The values M_Z which appear in Table 15.1 (column M_Z/M_W) are calculated by dividing ${}^0R_{G,Z}^2$ by ${}^\circ\Lambda$ [to obtain ${}^\circ R_{G,Z}^2$, see (15.1.11)].

It is of interest to determine the coefficient ${}^0\Lambda$ associated with atactic polystyrene in a less clever, but more convenient way. The idea is to measure the basic quantities M_W [to obtain M_W see (15.1.16)] and ${}^0R^2_{G,Z}$, for the greatest possible number of samples, in solution at a temperature T_F, or in the melt. For Brownian chains, the conversion coefficient ${}^0\Lambda$ is equal to

$$ {}^0\Lambda = \frac{{}^0R^2_{G,Z}}{M_Z} = \frac{{}^0R^2_{G,Z}}{M_W}\,\frac{M_W}{M_Z}. \tag{15.2.9}$$

We are left with the calculation of the ratio M_W/M_Z characterizing polydispersion. The result can be obtained in two ways.

1. The averages M_W and M_Z are taken directly from chromatograms obtained by size-exclusion chromatography (see Chapter 1).

2. The ratio M_W/M_Z is obtained indirectly from the ratio M_W/M_n determined by light scattering [see (15.1.16) and (15.2.1)]. Assuming that the distribution probability of the link numbers obeys the Schulz and Zimm law [see (13.2.22)], we can write

$$ M_W/M_n = 1 + p \tag{15.2.10}$$

$$ M_Z/M_W = 1 + 2p \tag{15.2.11}$$

where p is the polydispersion parameter. The value of p being calculated with the help of (15.2.10), we thus determine the ratio M_Z/M_W. The value of the coefficient ${}^0\Lambda$ obtained with this method by averaging over all samples A and C is ${}^0\Lambda = 8.41 \times 10^{-4}$ nm²/dalton.

Neutron scattering experiments made on shorter polystyrene chains $(({}^0R^2_{G,Z})^{1/2} \leqslant 20$ nm) in solution at temperature T_F, give a slightly smaller result

$$ {}^0\Lambda = 7.56 \times 10^{-4}\ \text{nm}^2/\text{dalton}. $$

This average value is also obtained for polystyrene chains in the melt.

Now, one value of ${}^0\Lambda$ must be adopted and we choose the value given by (15.2.7). Hence, for atactic polystyrene

$$ S = 16.6 \times 10^{-4} M\ \text{nm}^2. \tag{15.2.12}$$

In the next sections, we shall be able to characterize a chain by its area S. Note, however, that a polymer cannot be exactly represented by a Brownian chain: the area–mass correspondence is at best a good approximation.

3. POLYMER IN GOOD SOLVENT: DILUTE SOLUTION (POLYSTYRENE)

A dilute polymer solution is a *good* solution when the size of chain in solution is much greater than the size of the equivalent Brownian chain. On the other hand, we observe that in this case the osmotic pressure is higher than the Van't Hoff limit pressure. These two effects are due to the repulsive interaction between chain monomers. In this case, in order to interpret the experimental results, we can use the standard continuous model with simple two-body interaction. The only dimensionless parameter defining the interaction force is then the quantity z. In what follows, we shall try to see how this model accounts for the experimental properties of good solutions.

We shall start by establishing a correspondence between physical and theoretical parameters. Next, we shall study the intermediate regime (moderately long chains) and then the Kuhnian regime (very long chains)

3.1 Correspondence between basic quantities and theoretical parameters

The theoretical parameters which we need are: the area S of the equivalent Brownian chain, and the parameter z. For polystyrene, the area S is given as a function of mass by eqn (15.2.12). On the other hand, z will be determined, either from swelling (the best method) or from the second virial coefficient. Let us emphasize, however, that this correspondence is only a very good approximation.

3.1.1 Determination of z and b from gyration swelling

Theory gives us the end-to-end swelling $\mathfrak{X}_0(z) = R^2/{}^0R^2$; but the measured quantity is the 'Z' average of the *gyration* swelling

$$\mathfrak{X}_{\mathrm{G,Z}} = R^2_{\mathrm{G,Z}}/{}^0R^2_{\mathrm{G,Z}}.$$

For reasons of simplicity, we shall neglect polydispersion effects and write

$$\mathfrak{X}_{\mathrm{G,Z}}(z) = \mathfrak{X}_{\mathrm{G}}(z).$$

It is still necessary to determine the curve $\mathfrak{X}_{\mathrm{G}}(z)$ in order to associate a value of z with each experimental value of $\mathfrak{X}_{\mathrm{G}}$. For this, we shall write

$$\mathfrak{X}_{\mathrm{G}}(z) = \frac{R^2_{\mathrm{G}}(z)}{{}^0R^2_{\mathrm{G}}} = \frac{\mathrm{R}^2(z)}{{}^0R^2}\,\frac{{}^0R^2}{{}^0R^2_{\mathrm{G}}}\,\frac{R^2_{\mathrm{G}}(z)}{R^2(z)} = \mathfrak{X}_0(z)\,\aleph(z). \tag{15.3.1}$$

Note that, in this formula, the end-to-end swelling $\mathfrak{X}_0(z)$, the variation of which is strong, is well-known, whereas $\aleph(z)$, which is less well-known, does not vary much. We shall use the approximate expression (13.1.58)

$$\aleph(z) = \frac{1 + 1.1613z}{1 + 1.2185z}. \tag{15.3.2}$$

The function $\mathfrak{X}_0(z)$ has been determined by restructuration of the perturbation expansion for $d = 3$ [see (13.1.54)]. Figure 15.7 represents this function, as well as the function $\mathfrak{X}_{\mathrm{G}}(z)$ which we get from (15.3.1) and (15.3.2). Measured values of $\mathfrak{X}_{\mathrm{G,Z}}$ are given in Table 15.2, for different samples.

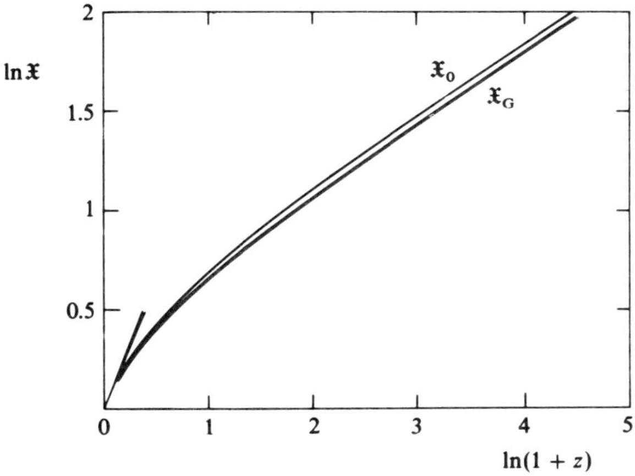

Fig. 15.7. End-to-end swelling (13.1.54) and gyration swelling in the (15.3.2) approximation, plotted against the interaction parameter z.

Table 15.2

Sample No.	Temperature [°C] (solvent = benzene)	$S_W^{1/2}$ [nm]	$\mathfrak{X}_{G,Z}$	z	b [nm^{-1}]
1	25°	305.7	4.93	30.5	1.57
2	”	252.8	4.59	25.2	1.57
3	”	228.7	4.47	23.2	1.60
4	”	197.1	4.20	19.4	1.55
6	”	158.3	3.84	15.12	1.50
7	30°	149.1	3.77	13.9	1.47
12	”	126.8	3.83	14.95	1.85
13	25°	120.47	3.49	11.2	1.46
15	30°	112.47	3.81	14.33	2.0
16	”	95.81	3.46	19.82	1.78
17	”	89	3.40	10.24	1.81
18	”	77.6	3.41	10.35	2.10
20	”	63.37	3.28	9.17	2.28
22	”	45.18	2.85	6.03	2.12

Inserting these values as ordinates in Fig. 15.7, we get values of z, which are found in Table 15.2. In this table we also find values of average areas S_W, calculated with the help of relation (15.2.12) from molecular masses M_W. The value of the interaction b is deduced from the formula

$$z = bS^{1/2}(2\pi)^{-3/2}, \tag{15.3.3}$$

and we shall write here

$$z \simeq b S_W^{1/2} (2\pi)^{-3/2}. \tag{15.3.4}$$

We check that for a given solvent, at a fixed temperature, the values of b obtained in this manner are practically independent of the areas S_W (Fig. 15.8).

Now we can study the variation of interaction b with temperature. Value of b will be calculated from the gyration swelling $\mathfrak{X}_{G,Z}$. We need a theoretical expression of $\mathfrak{X}_0(z)$, and for this purpose we shall use the formula obtained by restructuration of the perturbation expansion [see (13.1.54) and Fig. 15.7] and also, for comparison, Flory's formula and the formula derived from the ε-expansion [(13.1.43) and (13.1.36)]. With a given $\mathfrak{X}_{G,Z}$, we shall associate values of the parameter z which will be respectively written z, z_F, and z_ε (Table 15.3). From these values, we get respectively b, b_F, and b_ε for different temperatures T and for a fixed equivalent Brownian area S.

Let us plot these values against inverse temperature (Fig. 15.9). Since the interval in which temperature varies is very small, we expect the interaction to vary linearly with the inverse temperature. We have fitted straight lines through the values of b, b_F, and b_ε respectively, corresponding to different inverse temperatures; thus the intersection of these lines with the axis $b = 0$ must give $1/T_F$. We notice that the values of b give a better alignment than those of b_F and b_ε. This agrees with our predictions, and it shows that the standard continuous

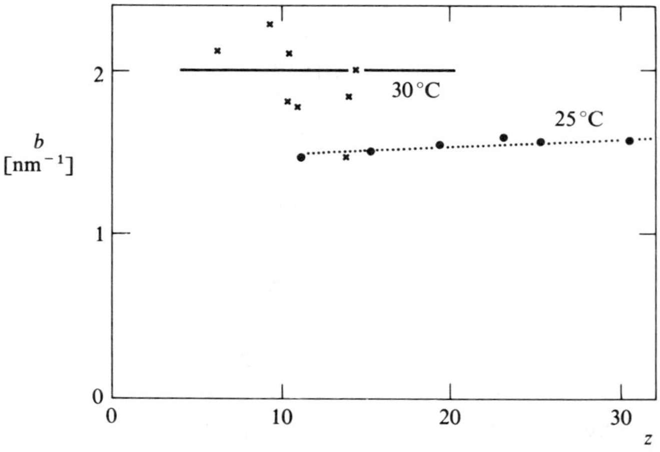

Fig. 15.8. Values of the two-body interaction b, derived from gyration swellings for two sets of polystyrene samples in benzene
● at 25 °C (from Miyaki *et al.*[3])
× at 30 °C (from Yamamoto *et al.*[8]).

Table 15.3

Sample No.	Solvent	Temperature [°C]	$(1/T) \times 10^3$ [K^{-1}]	$S_W^{1/2}$ [nm]	$\mathfrak{X}_{G,Z}$	z	b [nm^{-1}]	z_F	b_F [nm^{-1}]	z_ε	b_ε [nm^{-1}]
8	trans-decalin	40°	3.195	193.3	1.94	1.61	0.18	1.05	0.12	2.32	0.26
9	,,	30°	3.3	193.3	1.56	0.75	0.085	0.46	0.052	1.03	0.116
10	,,	25°	3.36	193.3	1.31	0.35	0.04	0.19	0.021	0.46	0.052
24	cyclohexane	60.1°	3.0	35.42	1.35	0.42	0.186	0.23	0.1	0.54	0.24
25	,,	50.1°	3.095	35.42	1.35	0.3	0.132	0.16	0.071	0.37	0.164
26	,,	45.4°	3.14	35.42	1.21	0.22	0.97	0.12	0.053	0.28	0.124

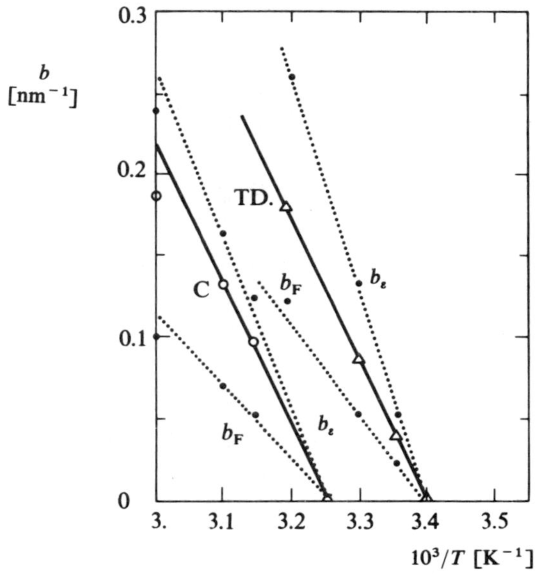

Fig. 15.9. Plot of the two-body interaction derived from the gyration swelling of polystyrene in cyclohexane (C) and in trans-decalin (TD), against inverse temperature:
o, △ values obtained with the standard continuous model;
● values obtained with Flory's formula or with the ε expansion.

model is a satisfactory representation of polymers. For polystyrene in cyclohexane, we shall adopt the relation

$$b = 2.918 - \frac{896}{T} \text{ nm}^{-1}, \tag{15.3.5}$$

which accounts for the known value of T_F (34 °C) which will be justified in Chapter 16.

Let us consider again the values of b derived from the swelling $\mathfrak{X}_{G,Z}(z)$ of polystyrene in benzene, at 25 °C and 30 °C (Table 15.2 and Fig. 15.10). For such a system, T_F is not known and it is believed that it is inferior to the fusion temperature of solid benzene (5.5 °C). Let us then represent the two values of b as a function of $1/T$ and let us extrapolate the straight line through these two points: the line intersects the $b = 0$ axis at $1/T = 3.59 \times 10^{-3}$. As a result, for this system, T_F seems to be equal to 279 K (6 °C).

3.1.2 Other determinations of b: osmotic two-body parameter

We can also determine the value of interaction b from the second virial coefficient A_2, a basic quantity associated with osmotic pressure as well as radiation scattering. Hence, for a monodisperse solute of molecular mass M, the

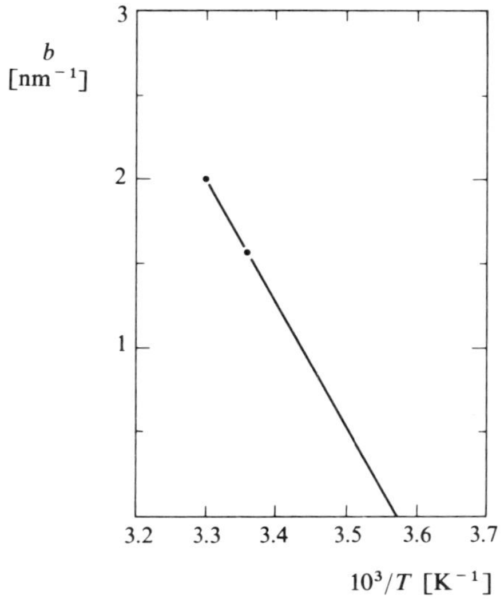

Fig. 15.10. Schematic plot of the two-body interaction b of polystyrene in benzene against inverse temperature. The inverse temperature of fusion of solid benzene is marked by a bar on the horizontal axis. (From Yamamoto *et al.*[8])

variation of the osmotic pressure with mass concentration ρ reads

$$\frac{\boldsymbol{\Pi}}{\boldsymbol{RT}} = \rho/M + A_2\rho^2 + \ldots .$$

When we express this pressure as a function of chain concentration $\boldsymbol{C}$ and of equivalent Brownian area S, we get, for $d = 3$ [see (10.8.8)],

$$\boldsymbol{\Pi\beta} = \boldsymbol{C} + \tfrac{1}{2}\boldsymbol{C}^2(2\pi S)^{3/2} z\, j(z) + \ldots .$$

Substituting $\rho = CM/$ Ⓐ in the first equation, we obtain A_2 by identification of the quadratic terms

$$A_2 = \frac{1}{2}(2\pi)^{3/2} \text{Ⓐ} \frac{S^{3/2}}{M^2} z j(z). \qquad (15.3.6)$$

Here $j(z)$ is a decreasing function of z and $j(0) = 1$; asymptotically $j(z) \propto z^{d(2\nu-1)-1}$. Introducing the interaction [see (15.3.3)]

$$b = (2\pi)^{3/2} S^{-1/2} z$$

and the conversion factor (15.2.5)

$${}^0\Lambda = S/2M$$

we have

$$A_2 = 2 \text{ Ⓐ } {}^0\Lambda^2\, \mathrm{b}\, j(z).$$

As a result

$$b = A_2/2 \text{Ⓐ}\, {}^0\!\Lambda^2 j(z).$$

In the simple-tree approximation, $j(z) \equiv 1$, and in this case

$$b = A_2/2 \text{Ⓐ}\, {}^0\!\Lambda^2. \tag{15.3.7}$$

For instance, for sample No. 3 we find

$$b = 0.13 \text{ nm}^{-1}$$

which is an order of magnitude smaller than the value derived from swelling ($b = 1.44 \text{ nm}^{-1}$). When z tends to infinity, the coefficient A_2 tends to zero (see Table 15.1 for the evolution of A_2 as a function of M). Therefore, in the simple-tree approximation, the value of b derived from A_2 tends to zero as the size of the chain tends to infinity! In fact, the function $j(z)$ is essential in eqn (15.3.6) and we see here how the simple-tree approximation fails.

Introducing an explicit expression of $j(z)$ in (15.3.6), we could determine the parameter z from A_2. We would then obtain a result analogous to the one reported in Table 15.2. However, it seems more interesting to proceed in the reverse way: let us determine $z\,j(z)$ from observed values of the second virial coefficient. We therefore write

$$z\,j(z) = \frac{2M^2 A_2}{(2\pi)^{3/2} \text{Ⓐ}\, S^{3/2}}. \tag{15.3.8}$$

The product $M^2 A_2$ is a volume, comparable to the effective volume occupied by Ⓐ chains. It is therefore natural to compare this expression to the 3/2 power of the swelling $\mathfrak{X}_0$. In such a manner, we introduce the two-body osmotic parameter $g(z)$ given by (10.8.10)

$$g(z) = z\,j(z)/[\mathfrak{X}_0(z)]^{3/2}. \tag{15.3.9}$$

Using (15.3.6), we can write $g(z)$ in the form

$$g(z) = \frac{2}{(2\pi)^{3/2}} \frac{M^2 A_2}{\text{Ⓐ}} \frac{1}{(S\mathfrak{X}_0(z))^{3/2}}. \tag{15.3.10}$$

Let us introduce the coefficient ${}^0\!\Lambda$ which converts areas into molecular masses [see (15.2.5)]. We get

$$g(z) = \frac{S^{1/2} A_2}{2(2\pi)^{3/2} \text{Ⓐ}\, {}^0\!\Lambda^2 [\mathfrak{X}_0(z)]^{3/2}}.$$

In order to derive $g(z)$ from the observed quantities S_{W}, A_2, and $\mathfrak{X}_{\mathrm{G,Z}}$, let us introduce the factor $\aleph(z)$ [see (15.3.1)]. We write

$$g_{\sim}(z) = \frac{S_{\mathrm{W}}^{1/2} A_{2,\sim}}{2(2\pi)^{3/2} \text{Ⓐ}\, {}^0\!\Lambda^2} \left(\frac{\aleph(z)}{\mathfrak{X}_{\mathrm{G,Z}}}\right)^{3/2}. \tag{15.3.11}$$

Values of $g_{\sim}(z)$ obtained in this way, are reported in Table 15.4 and Fig. 15.11.

Table 15.4

Sample No.	Solvent	Temperature [°C]	$S_W^{1/2}$ [nm]	$10^4 A_{2,\sim}$ [cm^3/dalton^2]	$\mathfrak{X}_{G,Z}$	$g_\sim(z)$
1	benzene	25°	305.7	1.09	4.93	0.218
2	”	25°	252.8	1.22	4.59	0.227
3	”	25°	228.7	1.28	4.47	0.222
4	”	25°	197.1	1.44	4.20	0.221
6	”	25°	158.3	1.66	3.84	0.250
7	”	30°	149.1	1.50	3.77	0.219
12	”	30°	126.8	1.60	3.83	0.194
13	”	25°	120.7	1.82	3.49	0.242
15	”	30°	112.47	1.73	3.81	0.186
16	”	30°	95.81	1.89	3.46	0.202
17	”	30°	89	2.03	3.40	0.206
18	”	30°	77.5	2.21	3.41	0.195
20	”	30°	63.37	2.33	3.28	0.178
22	”	30°	45.18	2.86	2.85	0.157
24	cyclohexane	60.1°	35.42	0.933	1.35	0.156
25	”	50.1°	35.42	0.677	1.27	0.12
26	”	45.4°	35.42	0.554	1.21	0.11

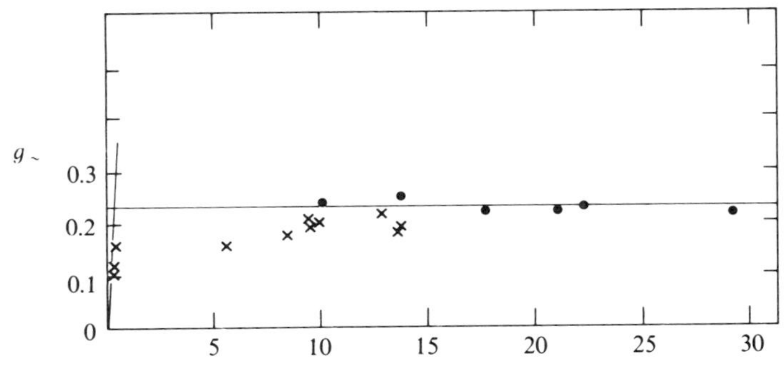

Fig. 15.11. Two-body osmotic parameter measured by light scattering, plotted against the interaction parameter z (ref. 14).

One can also determine the osmotic parameter from measurements of osmotic pressure. We will do this in Section 4.2.2 where we find a result in agreement with $g^* = g(z \to \infty) = 0.233$.

In conclusion, we can state that there is a good correspondence between physical and theoretical parameters. This correspondence will help us to study, in greater detail, the intermediate and asymptotic regimes of the polymer state in good solvent.

3.2 Intermediate regime

Concerning the chain swelling, a very sensitive test exists for the agreement between experiment and theory: it consists in determining the effective exponents associated with the increase in size of polymers, and in comparing them with the predictions of the standard continuous model.

Furthermore, by examining how an effective exponent varies with molecular mass and temperature, it is possible to determine whether one has to do with the intermediate or with the asymptotic regime of the polymer state in good solvent.

3.2.1 Effective exponents of gyration swelling

The effective exponent $\nu_G(z)$ for gyration swelling is by definition given by the relation*

$$2(2\nu_G(z) - 1) = \frac{d \ln \mathfrak{X}_G(z)}{d \ln z} \tag{15.3.12}$$

(see Section 3.1.1).

In the intermediate regime, this exponent varies slowly with z.

From experimental values of the gyration swelling of the chain $\mathfrak{X}_{G,Z}$, from values of the equivalent Brownian area S_W and from the temperature T of the solution, we can approximatively derive the variation of $\nu_G(z)$; the method consists in calculating, at a given value of z, the slope of the curve representing $\ln \mathfrak{X}_G$ versus $\ln z$. The equation for the tangent at a point of abscissa $\bar{z}$ is

$$\ln \mathfrak{X}_{G,Z} = 2(2\nu_G(\bar{z}) - 1)\ln z + A = (2\nu_G(\bar{z}) - 1)[\ln S + 2\ln b] + B \tag{15.3.13}$$

where A and B are constants.

In reality, the curve is defined by a succession of discrete points. The exponents $\nu_G(z)$ will be determined approximately from data reproduced in Table 15.2. We consider the temperature T as fixed.

Let us choose a succession of n samples ($j = 1, \ldots, n$) with areas S_{W_j} close to one another. The mean area is

$$\langle S_W \rangle = \frac{1}{n} \sum_j S_{W_j}.$$

We shall associate with it the value

$$\bar{z} = b(\langle S_W \rangle)^{1/2} (2\pi)^{-3/2}$$

and for the interaction b, we take $b = 1.57\ \text{nm}^{-1}$ (polystyrene-benzene at 25°C). Let us define

$$y_j = \ln \mathfrak{X}_{G,Z_j} \qquad \text{and} \qquad x_j = \ln S_{W_j}. \tag{15.3.14}$$

* Relation (15.3.12) is analogous to relation (15.3.16) which is obtained by setting $\sigma_0(z) = 2\nu(z) - 1$, in agreement with (12.3.69) and (12.3.66).

For this set of samples, we shall write (15.3.13) in the form

$$y = \sigma x + a \tag{15.3.15}$$

where $\sigma = 2\nu_G - 1$ and a are determined by minimization of the sum

$$\sum_j (y_j - \sigma x_j - a)^2.$$

Using notation (15.3.14), we find

$$\sigma = (2\nu_G(\bar{z}) - 1) = \frac{\langle (x - \langle x \rangle)(y - \langle y \rangle) \rangle}{\langle (x - \langle x \rangle)^2 \rangle}$$

Let us perform this operation for three sets of samples with different average areas $\langle S_W \rangle$. The result is given in Table 15.5, where we have also listed the mean square deviations

$$\delta S_W = [\langle (S_W - \langle S_W \rangle)^2 \rangle]^{1/2}.$$

Table 15.5 then gives the values of the effective exponent $\nu_G(\bar{z})$ for three values of $\bar{z}$. We now look at the variation of the swelling with temperature for a fixed equivalent Brownian area S (see Table 15.3, p. 735, and Fig. 15.10). A calculation, analogous to the preceding one, applied to samples 8, 9, 10, and then to 24, 25, and 26 of Table 15.1, gives $\nu_G(\bar{z}) = 0.546$ for $\bar{z} = 1$ and $\nu_G(\bar{z}) = 0.54$ for $\bar{z} = 0.6$.

Let us examine these results in the light of the standard continuous model. Our only theoretical reference is the information concerning the effective exponent for the end-to-end swelling

$$2(2\nu(z) - 1) = \frac{d \ln \mathfrak{X}_0}{d \ln z} \tag{15.3.16}$$

calculated by restructuration of the perturbation expansion (Chapter 13, Section 1.4.1). The variation of this quantity is plotted on Fig. 15.12, as a function of $\ln(1 + z)$

In order to interpret observations concerning gyration swelling in terms of $\mathfrak{X}(z)$, we must transform the square radii of gyration R_G^2, into square end-to-end distances R^2. We shall perform this change, using a perturbation expansion, and we shall discuss the result using the method of thermic sequences.

Table 15.5

Sample (see Table 15.1)	$\langle S_W \rangle$ [nm^2]	δS_W [nm^2]	$\bar{z}$	$\nu_G(\bar{z})$
1, 2, 3, 4, 6, 13	48134	26000	21.9	0.600
7, 12, 15, 16, 17 18, 20, 22	9985	6277	10	0.589
19, 21, 23, 27	2306	1603	4.8	0.584

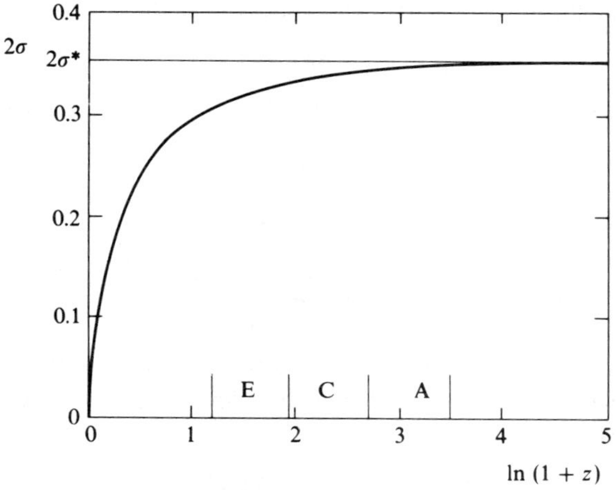

Fig. 15.12. The effective exponent $\sigma_0(z) = 2\nu(z) - 1$ is plotted against the interaction parameter z, calculated in Chapter 13, Section 1.3.1. The intervals in which z varies, associated respectively with samples A, C, and E, are marked by vertical lines.

3.2.1.1 *Calculation of the ratio $R_G^2(z)/R^2(z)$ by perturbation*

Let us express the gyration swelling $\mathfrak{X}_G$ as a function of the end-to-end swelling $\mathfrak{X}_0$. We have [see (15.3.1)]

$$\mathfrak{X}_G(z) = \aleph(z)\mathfrak{X}_0(z),$$

or

$$\ln \mathfrak{X}_G(z) = \ln \aleph(z) + \ln \mathfrak{X}_0(z).$$

Hence,

$$\frac{\mathrm{d} \ln \mathfrak{X}_G}{\mathrm{d} \ln z} = \frac{\mathrm{d} \ln \mathfrak{X}_0}{\mathrm{d} \ln z} + \frac{\mathrm{d} \ln \aleph}{\mathrm{d} \ln z}.$$

For $\ln \aleph(z)$ we assume the variation law (15.3.2)

$$\ln \aleph(z) = \ln \frac{(1 + 1.1613z)}{(1 + 1.2185z)} \tag{15.3.17}$$

If we neglect the polydispersion effect ($\mathfrak{X}_{G,Z} = \mathfrak{X}_G$), by combining (15.3.16) and (15.3.17) we obtain

$$\nu_G(z) = \nu(z) + \frac{1}{4} \frac{\mathrm{d} \ln \mathfrak{X}(z)}{\mathrm{d} \ln z}$$

$$= \nu(z) - \frac{0.0572}{4(1 + 1.1613z)(1 + 1.2185z)}. \tag{15.3.18}$$

In this approximation, $\nu_G(z) \leqslant \nu(z)$: the effective gyration swelling exponent converges more slowly towards its asymptotic value than the effective end-to-end swelling exponent.

3.2.1.2 *Note concerning the relative magnitudes of exponents $\nu_G(z)$ and $\nu(z)$*

The perturbation expansion shows that one always has $\nu_G(z) \leqslant \nu(z)$. Some authors have cast doubt on this result, and the question is of importance for the interpretation of experiments. We shall therefore show the validity of this inequality, by using the so-called thermic sequence method (see Chapter 13, Section 1.7.3). For this, we consider the radius of gyration of a discrete chain of $N-1$ links [see (7.3.25)]

$$R_G^2 = \frac{1}{2N^2} \sum_{ij}^{N} \langle\langle |\vec{r}_i - \vec{r}_j|^2 \rangle\rangle \,.$$

Neglecting inhomogeneity effects, we can write approximatively

$$R_G^2 \simeq \frac{1}{N} \sum_{n=0}^{N} (N-n) R_n^2 \tag{15.3.19}$$

where R_n^2 is the mean square end-to-end distance of a sequence of n links. The approximation consists in assuming that R_n^2 varies in a non-trivial way with n (whereby the position of the sequence in the chain is irrelevant). In particular, R_n^2 varies slightly with temperature, from which the name thermic sequence is derived. We shall write

$$R_n^2 = {}^0R_n^2\, \mathfrak{X}_{0,n} \tag{15.3.20}$$

where ${}^0R_n^2$ is the mean square end-to-end distance of the equivalent Brownian sequence, and $\mathfrak{X}_{0,n}$ the swelling of this sequence caused by the repulsive interaction. The same law will be assumed for the end-to-end swelling of a sequence, as for the end-to-end swelling of the total chain (this is only an approximation). Let z be the interaction parameter in the standard continuous model

$$z = b\, S^{1/2} (2\pi)^{-3/2}$$

where S is the equivalent Brownian area for the chain of $N-1$ links and b the interaction. Let s be the equivalent Brownian area for the sequence of n links and let

$$x = \frac{s}{S}.$$

We shall assume for the swelling law

$$\mathfrak{X}_{0,n} = \mathfrak{X}_0(z x^{1/2}).$$

We then calculate the square radius of gyration (15.3.19), replacing the sum on the number n of links by an integration on the variable x. From (15.3.20), we have

$$R_G^2 = \frac{{}^0R^2}{N} \sum_{n=0}^{N} (N-n)\frac{n}{N}\mathfrak{X}_{0,n} = {}^0R^2 \int_0^1 dx(1-x)x\,\mathfrak{X}_0(zx^{1/2}), \tag{15.3.21}$$

and for the gyration swelling

$$\mathfrak{X}_G = 6 \int_0^1 dx\,(1-x)x\,\mathfrak{X}_0(zx^{1/2}). \tag{15.3.22}$$

Let us now calculate the effective gyration swelling

$$2(2\nu_G(z) - 1) = \frac{d \ln \mathfrak{X}_G}{d \ln z}.$$

Taking (15.3.22) into account, we find

$$2(2\nu_G(z) - 1) = \frac{\int_0^1 dx\,(1-x)x\,\mathfrak{X}_0(zx^{1/2})2(2\nu(zx^{1/2}) - 1)}{\int_0^1 dx\,(1-x)x\,\mathfrak{X}_0(zx^{1/2})}. \tag{15.3.23}$$

Now, since according to Fig. 15.12, the function $\nu(z)$ is an increasing function, we have $\nu(zx^{1/2}) < \nu(z)$. Consequently, the equality (15.3.23) gives

$$\nu_G(z) \leqslant \nu(z) \tag{15.3.24}$$

and we shall admit that this inequality is also true for the standard continuous model with two-body interaction.

3.2.2 Asymptotic behaviour of the form function. Observation of the thermic sequence

The theory shows that a polymer chain, even though it is swollen in a good solvent, still has sequences (of links) which locally are Brownian-like. Of course, the number of links of these sequences depends on the repulsive interaction. Theoretically, for the standard continuous model, the area S_t corresponding to the crossover between Brownian and Kuhnian sequences varies as [see (13.1.173)]

$$S_t = \xi_t^2 \simeq 4b^{-2} \tag{15.3.25}$$

where ξ_t is the thermic correlation length and b the interaction. It will therefore be of interest to observe for a fixed parameter $z(z \geqslant z_0 \simeq 0.1)$ how a sequence with equivalent area s changes from Brownian to Kuhnian type when s increases.

Moreover, we shall examine how the crossover area S_t varies with the temperature on which the interaction b depends.

The most direct way to measure the swelling of a sequence characterized by an equivalent Brownian area s, would consist in labelling the ends of this sequence by isotopic substitution (deuteration) and in measuring the mean square end-to-end distance R^2 by neutron small-angle scattering.

There is, however, a simpler method, which consists in examining the asymptotic behaviour of the form function $H(\vec{q})$ of the entire chain. We know indeed [see (7.3.30) and Appendix E] that in the asymptotic domain defined by the condition

$$q^2 R^2 \gg 1 \qquad (15.3.26)$$

the function $H(\vec{q})$ results from intrachain correlations whose extension is of the order of q^{-1}. The variation of the form function in an appropriate interval of the wave-vector transfer q should reveal the crossover from Brownian to Kuhnian state. The asymptotic behaviour of the form function is expected to be of the Brownian type

$$H(\vec{q}) \simeq \frac{2}{q^2 \dfrac{{}^0R^2}{6}} \qquad \text{if } q^2 > \xi_t^{-2} \gg 1/R^2 \qquad (15.3.27)$$

and a behaviour of the Kuhnian type

$$H(\vec{q}) \simeq \frac{h_\infty}{\left(q^2 \dfrac{R^2}{6}\right)^{1/2\nu}} \qquad \text{if } 1/R^2 \ll q^2 < \xi_t^2. \qquad (15.3.28)$$

In order to make these observations in the best conditions, it is necessary to proceed in the following manner. First, the experiment should be carried out on very long chains, because the asymptotic domain must be as large as possible. The temperatures of the solution should be such that the corresponding thermic correlation lengths satisfy the conditions

$$R > \xi_t (= 2b^{-1}) > l_p \qquad (15.3.29)$$

where l_p is the persistence length associated with the local structure of the chain (see Section 2.2). Finally, the observation window (q_{min}, q_{max}) will be centred around ξ_t^{-1}:

$$q_{min} < \xi_t^{-1} < q_{max}. \qquad (15.3.30)$$

Farnoux *et al.*[12] carried out neutron scattering experiments on solutions of polystyrene ($M_W = 3.8 \times 10^6$) in cyclohexane, in order to test the predictions (15.3.27) and (15.3.28). Here, the area S has the value 6140.8 nm^2 [see (15.2.12)]. Values of the interaction b for polystyrene in cyclohexane are given as a function of temperature T by (15.3.5) and values of the interaction b corresponding to the different temperatures in the experiment of Farnoux *et al.* are given in Table 15.6.

The observation window here is the interval

$$q_{min} = 8 \times 10^{-2}\,\text{nm}^{-1}, \qquad q_{max} = 1\,\text{nm}^{-1}.$$

The square radius of the polystyrene chain of molecular mass $M_W = 3.8 \times 10^6$

Table 15.6

Temperature [°C]	b [nm^{-1}]	ξ_t [nm]	ξ_t^{-1} [nm^{-1}]	z	$\tau = S_t/S$
41.25°	0.06	33	3×10^{-2}	0.3	0.18
42.50°	0.067	29	3.3×10^{-2}	0.335	0.14
43.75°	0.08	25	4×10^{-2}	0.4	0.1
46.20°	0.103	19	5.1×10^{-2}	0.51	0.061
47.40°	0.112	17.8	5.5×10^{-2}	0.56	0.051
52.30°	0.150	13.3	7.5×10^{-2}	0.75	0.028
59.7°	0.21	9.52	1.05×10^{-2}	1.05	0.014
67.20°	0.268	7.46	1.34×10^{-2}	1.34	0.009

in cyclohexane is at least equal to 10^4 nm^2 [see Table 15.1, p. 724, and eqn (15.2.8)]. Condition (15.3.26) is therefore satisfied: it is really the asymptotic behaviour of the form function which we study here.

The thermic correlation lengths ξ_t derived from the interaction b are given in Table 15.6. Let us recall that the persistence length of atactic polystyrene is equal to 2 nm. Condition (15.3.29) is therefore satisfied.

Comparing the inverse length ξ_t^{-1} (see Table 15.6) to the quantities $q_{\min}$, $q_{\max}$ which define the observation window, we note that in the experiment of Farnoux *et al.*, only the samples heated to 59.7 °C and 67.2 °C satisfy condition (15.3.30)

$$q_{\min} < \xi_t^{-1} < q_{\max}.$$

It is at such temperatures that the variation of the form function with q should reveal a crossover between the Brownian and Kuhnian states. For the other temperatures, $\xi_t^{-1} < q_{\min}$; in other words, the experiment in this case should only reveal, in the whole window, the Brownian structure of the short sequences.

The results of the experiment by Farnoux *et al.* are shown in Fig. 15.13, the ratio $\rho/I(q)$ [where $I(q)$ is the scattered intensity] is represented as a function of q^2. We notice then that $\rho/I(q)$ varies linearly with q^2 in all cases where

$$\xi_t^{-1} < q_{\min} \text{ (temperatures smaller than 59.7 °C).}$$

On the other hand, when ξ_t falls within the observation window (59.7 °C and 67.2 °C) the function $\rho/I(q)$ varies linearly with q^2 only for the largest values of q^2. Moreover, in the vicinity of $q_{\min}$, the sign of the curvature is the one indicated by eqn (15.3.28).

Let then q_t be the boundary between the two regimes. We see that q_t increases with temperature. Now q_t must be proportional to $1/\xi_t$. Therefore, one observes that ξ_t decreases with temperature, and consequently with the interaction b, in agreement with the theory.

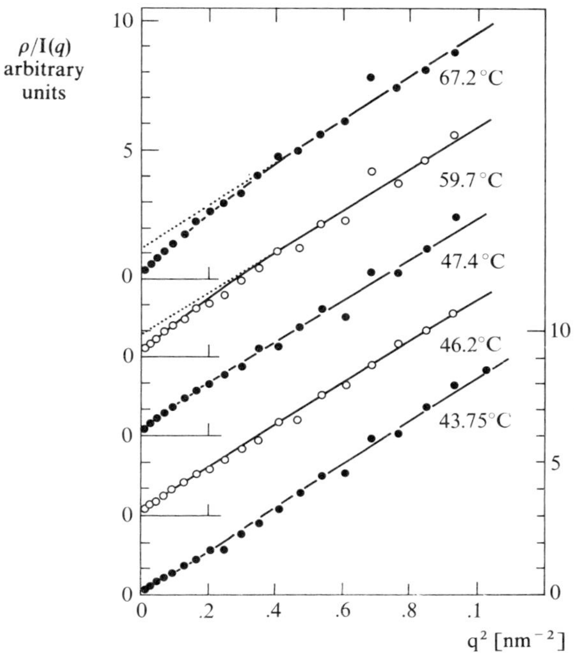

Fig. 15.13. Representation of the inverse form function $H(\vec{q})$ in the asymptotic domain $qR_G \gg 1$, at various temperatures of the solution:
o, ● values of $\rho/I(q)$ measured in arbitrary units. (From Farnoux *et al.*[12])
Values calculated with formula (13.1.174). The arrow indicates the abscissa corresponding to the calculated value of ξ_t^{-2}. The figure schematically shows the emergence, at small transfers q, of a Kuhnian behaviour when the solution is heated.

Farnoux *et al.* have interpreted their result with the thermic sequence model (see Chapter 13, Section 1.6.3). We recall that in this model, the long sequences have a Kuhnian behaviour. The form function associated with this model is calculated [see (13.1.174)] as a function of the parameters $x = q^2\,{}^0R^2/6$ and $\tau = S_t/S$ (see Table 15.6 for the values of τ). The expression of the form function is

$$H(x,\tau) = \frac{2\tau}{(x\tau^{2\nu})^2}\{x\tau^{2\nu} - \tau + [\tau - (1-\tau)x\tau^{2\nu}]\exp[-x\tau^{2\nu}]\}$$
$$+ \frac{1}{\nu}x^{-1/2\nu}[\Gamma(1/2\nu,x) - \Gamma(1/2\nu,x\tau^{2\nu})]$$
$$- \frac{1}{\nu}x^{-1/\nu}[\Gamma(1/\nu,x) - \Gamma(1/2\nu,x\tau^{2\nu})].$$

The adjustment of this form function to the values I/ρ measured in the window q_{min}, q_{max} is satisfactory. However, we cannot regard it as really significant, in particular if we take account of the warning given in Section 2.2 concerning persistence effects related to the chemical structure. Let us accept, however, that for a fixed value of z, the crossover from Brownian to Kuhnian state can be observed by letting the wave vector transfer q increase.

3.3 Kuhnian regime

The Kuhnian regime for an observable $\mathcal{O}$, function of the interaction parameter z, is often characterized by a power law in z. In this case, the effective exponent

$$\frac{\mathrm{d}\ln \mathcal{O}}{\mathrm{d}\ln z}$$

has a limit when $z \to \infty$. In practice the limit is reached when the exponent ceases to vary.

Experiments made in this limit are crucial, because they determine universal constants related to the polymer state in good solvent. The interval $z \geqslant z_a$ defining the asymptotic regime depends on the required precision. We shall be interested in the situation where the exponent $\mathrm{d}\ln\mathcal{O}/\mathrm{d}\ln z$ has 3 significant figures.

The interval $z \geqslant z_a$ also depends on the nature of the observable. Figure 15.11 shows, for instance, that the asymptotic behaviour of the osmotic parameter $g(z)$ is practically reached when $z \geqslant 10$. For the end-to-end swelling, it is necessary to wait until $z \geqslant 30$ (see Fig. 15.12). We shall now examine to what extent we can observe the asymptotic regimes of the gyration swelling, of the second virial coefficient (for $z \gg 1$) and of the form function (for $z \gg 1$ and $qR \gg 1$).

3.3.1 Measurement of the exponent ν

The exponent ν characterizes the swelling of a long polymer chain in very dilute solutions. In theory, it could be measured in several ways. However, can we trust results obtained by the simplest technique which consists in measuring the intrinsic viscosity? These measurements produce values for the exponent ν which are always lower than those obtained by light scattering measurements of the radius of gyration. It was necessary to explain this discrepancy in order to make a proper comparison between experimental and theoretical values of the exponent ν.

In fact, as shown by Weill and des Cloizeaux,[13] measurements related to dynamics should be avoided. Indeed, we pointed out in Chapter 13, Section 1.4.3, that the lengths which take part in the determination of dynamical physical quantities, include not only the mean square end-to-end distance or the radius of gyration of polymers, but also the hydrodynamic radius. Now, it is easy to show that the hydrodynamic radius reaches its asymptotic behaviour only very slowly, i.e. for a number of links much greater than that of the polymers used by experimentalists.* Hence, for the hydrodynamical radius, the asymptotic domain is practically unattainable.

Better experimental values of the exponent ν are obtained from measurements of the radius of gyration of polymers and of the second virial coefficient, in dilute solutions. Indeed, when the number of links increases, these quantities reach the asymptotic regime much more rapidly. However, the realization of a true asymptotic state for these quantities remains the main difficulty to overcome. Some progress has recently been made thanks to the elaboration of samples of high molecular mass: $M_W = 5 \times 10^7$.

The previous study of the intermediate regime shows that the boundary between this regime and the asymptotic regime occurs at about $z = 35$, which, in the most favourable cases, implies a molecular mass $M = 6 \times 10^7$. It seems therefore that static observation (radiation scattering) allows us a rather close approach to the asymptotic regime.

3.3.1.1 *Methods for the static measurement of the exponent* ν

When we increase the equivalent Brownian areas of a chain or of a sequence of a chain, the corresponding mean square radius of gyration increases asymptotically as s^ν. There are several methods of proving this power law by static measurements.

1. The first method is, in principle, the simplest one from the experimental point of view; it consists in measuring the intensity $I(q)$ scattered by the sample in the asymptotic domain $q^2 R^2_{G,Z} \gg 1$. The result we are looking for is then obtained by letting q vary in this domain. We have from (15.1.10)

$$\lim_{\rho \to 0} \frac{I(q)}{K\rho} = M_W H_Z(\vec{q})$$

where M_W is the 'W' average molecular mass and $H_Z(q)$ the 'Z' average form function of the chains. The law of asymptotic behaviour for a monodisperse system is given by the relation

$$H(\vec{q}) = \frac{\mathrm{h}_\infty}{\left(q^2 \dfrac{R^2}{6}\right)^{1/2\nu}} \qquad 1/R^2 \ll q^2 \ll 1/\xi_t^2$$

where ξ_t is the thermal correlation length [see (15.3.25)]. We get

$$\ln I(q) = -\frac{1}{\nu} \ln q + \text{constant}. \tag{15.3.31}$$

No polydispersion correction is necessary to interpret the experiment. However,

* The hydrodynamic radius has, however, the same asymptotic behaviour (characterized by the exponent ν) as the radius of gyration.

the form function has an asymptotic behaviour of type (15.3.31) only if the values of q belong to a certain interval. Now, this interval falls in between the observation windows (15.1.22) and (15.1.23) associated with light and neutron radiation respectively, and the overlap is only partial. The experiment is therefore difficult to work out.

2. The usual method consists in examining a set of samples of various molecular sizes, and in comparing the swellings associated with each of them.

The 'Z' average square radius of gyration measured by scattering [see (15.1.11)] is written as

$$R^2_{\mathrm{G,Z}} = \frac{\langle M^2 R^2_{\mathrm{G}} \rangle}{\langle M^2 \rangle} \tag{15.3.32}$$

where M is the molecular mass and R^2_{G} the corresponding square radius of gyration. The radius is to be compared to the average mass

$$M_{\mathrm{W}} = \frac{\langle M^2 \rangle}{\langle M \rangle} \tag{15.3.33}$$

also measured by scattering. The relation between R^2_{G} and M is the asymptotic swelling law

$$R^2_{\mathrm{G}} = \Lambda M^{2\nu} \tag{15.3.34}$$

where Λ is the conversion factor between molecular mass and Kuhnian area. This is the law to be tested by experiment, with the exponent ν as the unknown quantity.

The samples with great molecular mass are liable to be rather strongly polydisperse, and this polydispersion varies from one sample to another. It will be necessary to account for this. Now, the observable quantities $R^2_{\mathrm{G,Z}}$ and M_{W} are different averages of the molecular mass distribution in the sample. In order to relate these quantities to each other, let us choose as a reference system a set of monodisperse polymers with molecular mass equal to M_{W}. In this way (15.3.34) is written in the form

$$R^2_{\mathrm{G}}(M_{\mathrm{W}}) = \Lambda M_{\mathrm{W}}^{2\nu}.$$

Substituting (15.3.34) in (15.3.32), we obtain

$$R^2_{\mathrm{G,Z}} = \Lambda \frac{\langle M^{2(\nu+1)} \rangle}{\langle M^2 \rangle}, \tag{15.3.35}$$

and accounting for (15.3.33), we may write

$$R^2_{\mathrm{G,Z}} = \Lambda \left\{ \frac{\langle M^{2(\nu+1)} \rangle}{\langle M^2 \rangle} \left(\frac{\langle M \rangle}{\langle M^2 \rangle} \right)^{2\nu} \right\} M_{\mathrm{W}}^{2\nu}, \tag{15.3.36}$$

hence,

$$R^2_{\mathrm{G}}(M_{\mathrm{W}}) = R^2_{\mathrm{G,Z}} \Big/ \left\{ \frac{\langle M^{2(\nu+1)} \rangle}{\langle M^2 \rangle} \left(\frac{\langle M \rangle}{\langle M^2 \rangle} \right)^{2\nu} \right\}. \tag{15.3.37}$$

The unknowns ν and Λ are determined with the help of observed quantities and of relation (15.3.36).

Let us set

$$y = \ln R^2_{G,z}$$

$$x = \ln M_W$$

$$\lambda = \ln \Lambda$$

$$\epsilon(\nu) = \ln \left\{ \frac{\langle M^{2(\nu+1)} \rangle}{\langle M^2 \rangle} \left(\frac{\langle M \rangle}{\langle M^2 \rangle} \right)^{2\nu} \right\}. \qquad (15.3.38)$$

For a truly monodisperse system, $\varepsilon(\nu)$ is zero; for a physical system $\varepsilon(\nu)$ is small. This is the correction term for which the effective calculation will be given in the next section. Equation (15.3.36) reads

$$y = 2\nu x + \lambda + \varepsilon(\nu). \qquad (15.3.39)$$

The scattering experiment is done with the largest possible number of samples $j(j = 1, \ldots, n)$ of different molecular sizes, but always in the asymptotic range. The coefficients ν and λ are then determined by minimization of the sum

$$\sum_{j=1}^{n} [y_j - 2\nu x_j - \lambda - \varepsilon_j(\nu)]^2. \qquad (15.3.40)$$

3. A complementary method consists in determining the exponent ν by measuring the second virial coefficient A_2. This coefficient is related to the average volume R^3_G occupied by a chain. For a monodisperse solute

$$A_2 = B/M^2$$

where B has the dimension of a volume. In good solvent

$$B \propto X^3$$

where X represents the size of the chain ($R^2 = X^2 d$). As a consequence

$$A_2 = \Lambda_A^{3/2} M^{3\nu-2} \qquad (15.3.41)$$

where Λ_A is a conversion factor. Letting M vary, we can determine the exponent ν in a manner which is independent of the preceding one. However, it is more difficult in this case to account for polydispersion effects. Complex diagrammatic calculations would be needed for this purpose. In order to interpret the data from scattering experiments, we shall merely use the approximation

$$A_{2,\sim} = \Lambda_A^{3/2} M_W^{3\nu-2}$$

i.e,

$$\ln A_{2,\sim} = (3\nu - 2) \ln M_W + \tfrac{3}{2} \ln \Lambda_A. \qquad (15.3.42)$$

3.3.1.2 *Relation between molecular mass and square radius of gyration. Experimental value of the exponent ν*

For a long time, it was thought that experiment had to confirm the value of ν proposed by Flory, i.e. $\nu_F = 0.6$. This value was compared to the value $\nu_0 = 1/2$

associated with the Brownian chain. When the modern theory produced the value $\nu = 0.588$, it looked impossible to decide experimentally between this new value and ν_F.

Yet, in 1980, J.P. Cotton[14] tried to determine the unknown ν and A as precisely as possible from the best existing results with the help of relation (15.3.35).

$$R^2_{G,Z} = A \frac{\langle M^{2(\nu+1)} \rangle}{\langle M^2 \rangle} \tag{15.3.43}$$

Such a programme presents two difficulties.

The first one is related to the *systematic* error which might be caused by remoteness from the asymptotic state.

The second one is related to the experimental errors caused by polydispersion in the samples and insufficient resolution of the scattering apparatus.

Let us assume that the experiment is carried out with asymptotically swollen polymers; then polydispersion effects and a lack of resolution remain. To be sure, the samples are prepared with extreme care. The polystyrene chains are obtained by emulsion polymerization, which gives the highest possible molecular masses. Moreover, the samples are fractionated by precipitation and this fractionation is repeated several times. The final result is very homogeneous (see the characteristic values M_Z/M_W, M_W/M_n in Table 15.1, teams A and C). Nevertheless, there are fluctuations which increase the uncertainty on the unknown quantity ν, because of the corrective term $\epsilon(\nu)$ [see (15.3.38) and (15.3.36)].

Let us assume that, for the samples obtained by successive fractionations, the distribution in monomers N per chain obeys the Schulz and Zimm law [see (13.2.22)]

$$P(N) = \frac{1}{(-1 + 1/p)!} \frac{1}{\langle N \rangle p} \left(\frac{N}{\langle N \rangle p} \right)^{-1+1/p} \exp[-N/p\langle N \rangle] \tag{15.3.44}$$

where p characterizes the polydispersion [see Table 15.1 and (15.2.10)]. In this case, the corrective term $\varepsilon(\nu)$ [see (15.3.38)] is written as

$$\epsilon(\nu) = \ln \left\{ \frac{(2\nu + 1 + 1/p)!}{(1 + 1/p)!} \left(\frac{p}{1+p} \right)^{2\nu} \right\}. \tag{15.3.45}$$

Minimizing the quantity (15.3.40) iteratively,

$$\sum_{j=1}^{n} [y_j - 2\nu x_j - \lambda - \varepsilon_j(\nu)]^2$$

for the set of samples j of teams A and C (see Table 15.1), Cotton found the following values

$$\nu = 0.586$$

$$\lambda = \ln A = -8.566 \qquad (A = 1.904 \times 10^{-4}\ \text{nm}^2/\text{dalton}^{0.853}).$$

Consequently, for polystyrene-in-benzene, we shall write*

$$R_G^2(M_W) = 1.904 \times 10^{-4} M_W^{1.172}\ \text{nm}^2. \tag{15.3.46}$$

The method used to obtain these results is, however, not very convincing. It has not been proved that the samples of teams A and C are close enough to the Kuhnian state. Moreover, the temperatures of the solutions are different for teams A and C, and this introduces a systematic error in the results. Nevertheless, we shall adopt (15.3.46) as the best experimental expression for the asymptotic law relating the square radius of gyration to the molecular mass for polystyrene in benzene at 25 °C. On the other hand, for polystyrene in CS_2, we shall adopt the older result†

$$R_G^2(M_W) = 1.613 \times 10^{-4} M_W^{1.18}\ \text{nm}^2. \tag{15.3.47}$$

A simultaneous examination of (15.3.46) and (15.3.47) leads us to compare the interaction b for polystyrene in CS_2 and for polystyrene in C_6H_6. Using (15.3.12), we find at 25 °C,

$$\frac{b(CS_2)}{b(C_6H_6)} = \left(\frac{1.613}{1.904}\right)^{\frac{1}{2(2\nu - 1)}} = 0.62. \tag{15.3.48}$$

Let us finally mention a result obtained by comparing the 'volumes' $A_{2,\sim}$ and the masses M_W. From the comprehensive data A, C, and E (Table 15.1), we obtain (see Fig. 15.14)

$$A_{2,\sim} = 103 \times 10^{-4} M_W^{-0.254}\ \text{cm}^3/\text{dalton}^2 \tag{15.3.49}$$

The corresponding exponent ν is worth 0.582. This value is smaller than the value obtained earlier [see (15.3.46)], probably because samples E are not in a state of asymptotic swelling.

3.3.2 Study of the asymptotic behaviour of the form function

In the regime of asymptotic swelling ($z \gg 1$), the form function of the chain has a doubly asymptotic behaviour ($q^2 R^2 \gg 1$) given by the function

$$H(\vec{q}) \simeq \frac{\text{h}_\infty}{\left(q^2 \dfrac{R^2}{6}\right)^{1/2\nu}} \tag{15.3.50}$$

which is characteristic of fractal structure of dimension $D = 1/\nu$. The quantity h_∞ is a universal quantity. Des Cloizeaux and Duplantier calculated h_∞ to

* Or more generally for a solution of monodisperse polystyrene in benzene

$$R_G^2(M) = 1.904 \times 10^{-4} M^{1.172}\ \text{nm}^2.$$

† Or more generally for monodisperse polystyrene in CS_2

$$R_G^2(M) = 1.6113 \times 10^{-4} M^{1.18}\ \text{nm}^2.$$

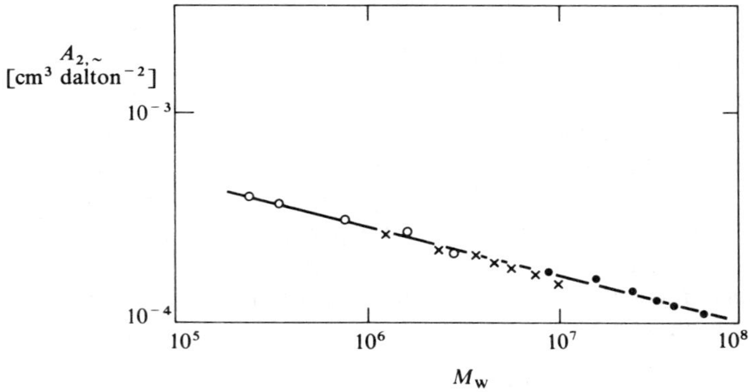

Fig. 15.14. The average $A_{2,\sim}$ of the second virial coefficient (observed by light scattering) is plotted against the average molecular mass M_W. (From Cotton.[14])

first-order in ε [see (13.1.162)] and obtained the (rather approximate) value

$$h_\infty = 1.04$$

(for a Brownian chain $h_\infty = 2$).

Let us examine how the law (15.3.50) is experimentally verified. We first mark the boundaries of reciprocal space q in which this law is valid. This interval is given by the inequality

$$1/R < q < 1/\xi_t \qquad (15.3.51)$$

where ξ_t is the thermal correlation length, which is of order 4 nm in good solvent. On the other hand, for $z = 30$, one has typically $R = 500$ nm. The experimentalist therefore disposes of two decades, in order to test (15.3.50). However, in order to explore the entire interval

$$2 \times 10^{-3} \leqslant q \leqslant 2 \times 10^{-1}\,\text{nm}^{-1} \qquad (15.3.52)$$

by radiation scattering, two types of radiation are necessary: indeed, the observation windows associated respectively with light [see (15.1.22)] and with neutrons [see (15.1.23)] do not separately cover the whole interval (15.3.52).

Experiments made so far have used only one or the other of these types of radiation. Hence the results obtained in this way, which should lead to the determination of ν and h_∞ [if (15.3.51) is verified], do not have all the precision which could be obtained from experiments by using successively the two types of radiation and the same sample.

Using light scattering in the doubly asymptotic regime, Loucheux, Weill, and Benoit[15] had already in 1953 shown the existence of an exponent different from 1/2 for polystyrene in benzene. In 1976, Farnoux *et al.*[16] tested the power law (15.3.50) by neutron scattering, plotting $\rho/I(q)$ against $q^{1/\nu}$ (where $\nu = 0.6$) (Fig. 15.15).

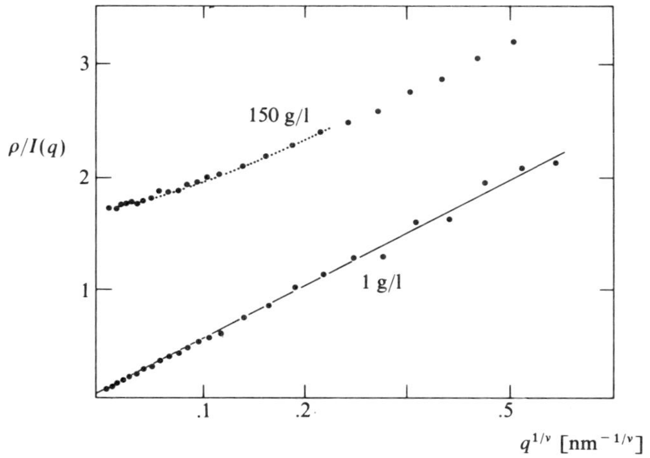

Fig. 15.15. Representation (in arbitrary unit) of the plot $\rho/I(q)$ against $q^{1/\nu}$, in the asymptotic domain $qR_G > 1$. Deuterated polystyrene ($M_W = 1.1 \times 10^6$) in CS_2. At the concentration $\rho = 1$ g/l, $K\rho/I(q) \simeq (M_W H(\vec{q}))^{-1}$ where $H(q)$ is the form function. At the concentration $\rho = 150$ g/l, $K\rho/I(q) \simeq (\rho \text{Ⓐ} H(\vec{q}))^{-1}$ where $H(\vec{q})$ is the structure function. Neutron scattering (after Farnoux *et al.*[12])

More recently, Rawiso, Duplessix, and Picot[17] made very precise *absolute* measurements of neutron scattering by long polystyrene chains in CS_2. The chains consisted of monomers labelled in the following manner

$$\text{+}(CD_2\text{—}CH\varphi\text{+})_N$$

where φ represents a benzene ring.

The incoherent contribution due to the proton spins was subtracted by the polarized flux method (Chapter 6, Section 5.3.3). The absolute measurement consists in measuring the (coherent) intensity $I(q)$ and the constant K related to the apparatus (see Chapter 7, Section 2.5.1, and Chapter 6, Section 5.4).

We represented in Fig. 15.16 the results of the above-mentioned authors, plotting $q^{1/\nu}I(q)/K\rho$ (where $\nu = 0.588$) against q.

Let us note here (see Fig. 15.16) that the observation window does not match condition (15.3.51): the measured values mostly reflect the local structure of the chain! However, extrapolating to $q \to 0$ the points plotted in this figure, one obtains a value

$$\left.\frac{q^{1/\nu}I}{K\rho}\right)_a = 2.06 \text{ nm}^{-1/\nu}\text{g}$$

Then eqn (15.1.10) leads to the result

$$q^{1/\nu}MH(\vec{q}) = 2.06 \text{ nm}^{-1/\nu}\text{g}.$$

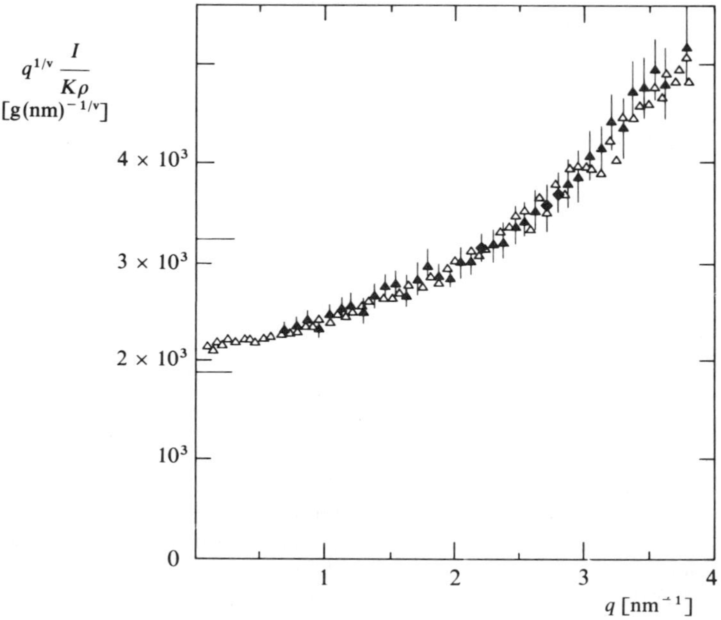

Fig. 15.16. Representation of the quantity $q^{1/\nu}I(q)/K\rho$ plotted against q, in the asymptotic domain $qR_G > 1$. The measurement of the scattered intensity is absolute here. Partially deuterated polystyrene $(CD_2{-}CHC_6H_6)_N$ of molecular mass $M_W = 1.2 \times 10^2$ in CS_2. The extrapolation at $q \to 0$ gives the quantity we are looking for. Neutron scattering (from Rawiso *et al.*[17]).

Let us replace $H(\vec{q})$ by its value (15.3.50). We get

$$\frac{\mathrm{h}_\infty}{(R^2/6M^{2\nu})^{1/2\nu}} = 2.06\ \mathrm{nm}^{-1/\nu}\mathrm{g}. \qquad (15.3.53)$$

Let us determine the value of the ratio $R^2/6M^{2\nu}$, which is a constant in the asymptotic state of swelling. The mean square end-to-end distance R^2 is deduced from the radius of gyration R_G^2 [see (15.2.3), (15.3.1), and (15.3.2)]

$$R^2 = 6R_G^2/\aleph = 6.29R_G^2.$$

Furthermore, from the conversion factor Λ between mass and Kuhnian area, we get

$$R_G^2 = \Lambda M^{2\nu}.$$

For polystyrene in CS_2, one has [see (15.3.47)]

$$\Lambda = 1.613 \times 10^{-4}\ \mathrm{nm}^2\,\mathrm{dalton}^{-1/2\nu}.$$

Substituting in eqn (15.3.53), we obtain

$$\mathrm{h}_\infty = 1.19. \qquad (15.3.54)$$

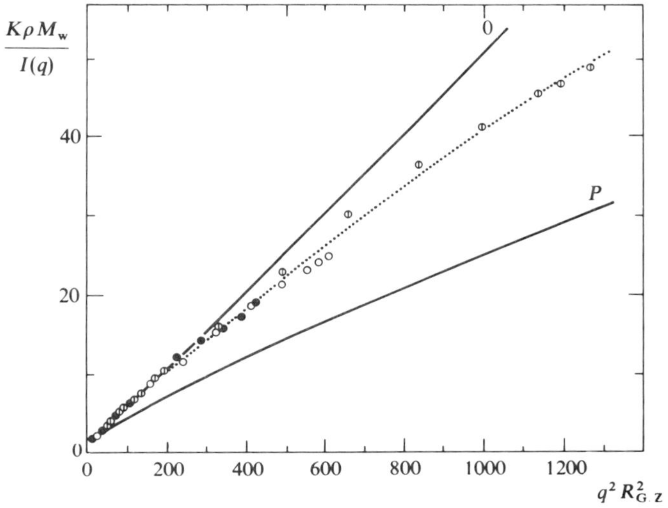

Fig. 15.17. The quantity $K\rho M_W/I(q)$ (pure number) is plotted against $q^2 R^2_{G,Z}$. Very dilute solutions of polystyrene in toluene; light scattering (from Noda *et al.*[5]):
● $M_W = 8.4 \times 10^6$
○ $M_W = 11 \times 10^6$
⦶ $M_W = 21 \times 10^6$
D: Debye function, P: function calculated by Peterlin (13.1.166).

This value is not very different from the calculated theoretical value. A confirmation of this result is given by the experiment of Noda *et al.*[5] (see Fig. 15.17 and samples B, Table 15.1). Here we are concerned with light scattering. In this case, the 'absolute' measurement of the scattered intensity, i.e. of the ratio I/K, is easily made: indeed, in the limit $q \to 0$, the quantity $I/K\rho$ is equal to the average molecular mass M_W [see (15.1.16)]. The asymptotic behaviour of the form function is displayed in Fig. 15.17, where $K\rho M_W/I = H^{-1}(\vec{q})$ is plotted against $q^2 R^2_{G,Z}$. The observation window is well set here with respect to condition (15.3.51). In the same figure, we plotted the curve representing the inverse form function calculated by Peterlin (see Chapter 13, Section 1.7.3) for which the value of h_∞ is $h_{\infty F} = 2(1/2\nu)! = 1.87$ [see (13.1.168)]. Comparing experimental and Peterlin curves, we see graphically that $h_\infty/h_{\infty F} = 1.62$, which leads to the value $h_\infty = 1.16$, close to (15.3.54).

3.4 Measurement of intrachain correlations

Measurements of intrachain correlations have been made by Matsushita *et al.*[18] in order to determine experimentally how the swelling of a portion of chain varies with its position within the chain.

Non-uniformity of swelling was predicted by Yamakawa and later by des Cloizeaux. This effect appears in the expansion of the mean square distance, as a function of the interaction b (see Chapter 10, Section 7.2). The inhomogeneity effect is related to a fundamental property of swelling. Let us then give an experimental evaluation of this phenomenon.

We are dealing here with distances greater than the persistence length l_p ($\simeq 2$ nm for polystyrene) related to a local rigidity of the chain. The experiment consists in the determination of gyration radii of chains or portions of chain having the same Brownian area S, but located in different places:

(a) the area S is the total Brownian area of the chain;

(b) the area S is the Brownian area of an extremal portion of the chain;

(c) the area S is the Brownian area of a central portion of the chain.

Matsushita *et al.* used three types of samples in order to determine the R_G^2 associated with the different areas S mentioned above:

(a) a deuterated polystyrene sample of molecular mass $M_W = 2.7 \times 10^4$ ($S = 44.82\ \text{nm}^2$);

(b) a two-block copolymer sample, made of a sequence of deuterated polystyrene with molecular mass $M_W = 2.7 \times 10^4$ and of a sequence of non-deuterated polystyrene, with molecular mass $M_W = 3.89 \times 10^5$;

(c) a three-block copolymer sample, made of a central deuterated polystyrene sequence with molecular mass $M_W = 3.1 \times 10^4$ and of two extreme sequences of non-deuterated polystyrene, with respective molecular masses $M_W = 1.4 \times 10^5$ and $M_W = 1.22 \times 10^5$.

The solvent used in all cases is CS_2, which is considered to be a fairly good solvent. The interaction b associated with polystyrene in CS_2 is worth $0.97\ \text{nm}^{-1}$ at 25 °C [see (15.3.48) and Fig. 15.8, p. 000]. The experiment being carried out at 20 °C, we shall take the value $b = 0.7\ \text{nm}^{-1}$ for the interaction. In this case, the thermic correlation length has the value $\xi_t = 2.85$ nm. We note that the latter is smaller than the end-to-end distance R of the sequences under consideration. ($R \geqslant \sqrt{3S} = 10$ nm); this point is not crucial, however.

The scattering experiment was done at the National Bureau of Standards Laboratory (Washington DC, USA). The polydispersion of the sequences has not been given by the authors. Consequently, we shall assume that these sequences are monodisperse, of molecular mass M_W, and in order to interpret the experiment we shall use the following expression (see Chapter 7, Sections 2.2.4 and 3.3.3.1)

$$I(q) = K_A V \sum_{\alpha\beta} \mathrm{b}_\alpha^{\circ\!\!\rightarrow} \mathrm{b}_\beta^{\circ\!\!\rightarrow} C_\alpha C_\beta H_{\alpha\beta}(\vec{q}). \tag{15.3.55}$$

where α is the isotopic index ($\alpha = 1$: deuterated; $\alpha = 2$: non-deuterated), $\mathrm{b}_\alpha^{\circ\!\!\rightarrow}$ is the contrast length of a monomer of type α with respect to the solvent, C_α is the concentration of monomer of type α and

$$\boldsymbol{H}_{\alpha\beta}(\vec{q}) = \frac{1}{C_\alpha C_\beta V} \sum_{a,b} \sum_{j\in\alpha} \sum_{l\in\beta} \langle\!\langle \exp(i\vec{q}\cdot(\vec{r}_{aj} - \vec{r}_{bl}))\rangle\!\rangle \tag{15.3.56}$$

is the structure function. The indices $\boldsymbol{a}$ and $\boldsymbol{b}$ mark the chains, the indices j and l mark the monomers, K_A is the instrument constant and V is the volume.

The contrast length of deuterated and non-deuterated monomers are, respectively,

$$\mathfrak{b}_1^{\circ\rightarrow} = 8.6 \times 10^{-5}\,\text{nm} \quad \text{and} \quad \mathfrak{b}_2^{\circ\rightarrow} = 0.3 \times 10^{-5}\,\text{nm}.$$

In practice, $\mathfrak{b}_2^{\circ\rightarrow}$ is negligible, and we set $\mathfrak{b}_2^{\circ\rightarrow} = 0$.

$$\begin{aligned} I(q) &\simeq K_A V(\mathfrak{b}_1^{\circ\rightarrow})^2 C_1 \boldsymbol{H}_{11}(\vec{q}) \\ &= K_A V(\mathfrak{b}_1^{\circ\rightarrow})^2 [C_1 N_1 H_1(\vec{q}) + C_1^2 \boldsymbol{H}_{11}^{\text{II}}(\vec{q})]. \end{aligned} \tag{15.3.57}$$

N_1 is the number of links, $H_1(\vec{q})$ the form function of the deuterated sequence and $\boldsymbol{H}_{11}^{\text{II}}(\vec{q})$ the intermolecular structure function.

For each sample (a, b, c), three solutions have been prepared with respective concentrations 2.5 per cent, 1.5 per cent, and 0.8 per cent. Extrapolating $I(q)$ at infinite dilution, the authors derived from $H_1(\vec{q})$ the square gyration radii of the deuterated blocks, R_G^2.

The values of R_G^2 are reproduced in Table 15.7 with those of the equivalent Brownian area S and of the gyration swelling $\mathfrak{X}_G = 2R_G^2/S$. There are two corrections to be made to these values.

1. Let us compare the swelling of the total chain (sample a), with the results of Section 3.1. For this, we bring $\ln \mathfrak{X}_G$ on the vertical axis of Fig. 15.7 and we measure the corresponding abscissa $\ln(1 + z)$: the result gives $z = 0.08$. From this value we get

$$b = zS^{-1/2}(2\pi)^{3/2} = 0.19\,\text{nm}^{-1}$$

in contradiction to the value indicated above, namely $b = 0.7\,\text{nm}^{-1}$. The corresponding interaction parameter has the value $z = bS^{1/2}(2\pi)^{-3/2} = 0.3$. On setting the corresponding value of $\ln(1 + z)$ on the horizontal axis of Fig. 15.7, we obtain the corrected swelling $\mathfrak{X}_G = 1.28$.

2. The Brownian area associated with sample c differs from the two others. In order to compare swellings between samples, let us reduce the swelling of this sample to that of a sequence of Brownian area $S = 44.82\,\text{nm}^2$. For this, we bring $\ln \mathfrak{X}_G)_c$ on the vertical axis of Fig. 15.7 which defines a point on the curve,* and measure the corresponding abscissa $\ln(1 + z)$, so obtaining z. Next, we

* In principle, Fig. 15.7 concerns only the total chain and not a sequence; however, it may be used to evaluate corrections.

Table 15.7

Sample	R_G^2 [nm²]	M_W	S [nm²]	$\mathfrak{X}_G$	$\mathfrak{X}_G$ corrected
a Total chain	24.6	2.7×10^4	44.82	1.10	1.28
b Small sequence of the two-block chain	31.24	2.7×10^4	44.82	1.39	1.39
c Central sequence of the three-block chain	40.7	3.1×10^4	51.46	1.58	
c′ Corrected central sequence of the three-block chain			44.82		1.53

calculate the corrected value $z' = z\left(\frac{S_b}{S_c}\right)^{1/2}$, and from the same curve we determine the value $\ln \mathfrak{X}_G$ corresponding to z'. This gives the corrected swelling, i.e. $\mathfrak{X}_G = 1.53$ (see line c′ of Table 15.7).

The results of Table 15.7 show that for equal Brownian surfaces, the swelling of the total chain is smaller than that of the extremal sequence, which itself is smaller than the swelling of the central sequence.

Let us compare the experimental result with the predictions of Chapter 10, Section 7.2. Strictly speaking, these apply only to small values of the interaction parameter z:

$$0 < z_\iota \ll 1 \qquad \iota = \text{a, b, c}' \tag{15.3.58}$$

but they must remain valid for all z_ι. The parameters z_ι are here associated respectively with the total chain (ι = a), with the small block of the two-block (ι = b) and with the central block of the three-block‡ (ι = c′). They are, in principle, equal to one another (provided that the radius of gyration of sample c is renormalized, as was the case).

Let us then calculate the parameters z_ι (ι = a, b, c′) using, on the one side the (corrected) values $\mathfrak{X}_G$ obtained by the experiment, and on the other side the theoretical results of Chapter 10. Equations (10.7.24), (10.7.26), and (10.7.28) give

$$\mathfrak{X}_{G\iota} = 1 + A_\iota z_\iota \tag{15.3.59}$$

where $A_a = 1.28$ (total chain), $A_b = 1.52$ (block of the two-block chain) and $A_{c'} = 2.44$ (central block of the three-block chain).

‡ Do not confuse $z_{c'}$ with z', the value of the latter being very approximate.

Table 15.8

Sample		'Gyration swelling' $\mathfrak{X}_{G,\iota}$		z_ι
		Experimental (corrected)	Calculated	
Total chain	a	1.28	$1 + 1.28 \times z_a$	0.22
Small block of the two-block chain	b	1.39	$1 + 1.52 \times z_b$	0.26
Central block of the three-block chain	c′	1.53	$1 + 2.44 \times z_{c'}$	0.22

Bringing the (corrected) experimental values of $\mathfrak{X}_{G\iota}$ and the theoretical values A_ι in eqns (15.3.59), we can determine the values of the interaction parameters z_ι, from a linear expansion in z. The result is reproduced in Table 15.8. Although the swellings $\mathfrak{X}_\iota$ are different, the interaction parameters z_ι must be equal. This is what we verify (but really at the price of the readjustment mentioned above!).

4. POLYMERS IN GOOD SOLVENT: SOLUTION WITH OVERLAP

Until recently (1970), the experimental effort has mostly been profitable in the field of dilute solutions, in the limit of zero concentration: in this limit, the basic quantities R_G^2 and A_2 are measured.

This situation changed not long ago: polymer solutions with overlap have aroused new interest. The use of neutron scattering and the technique of labelled molecules has contributed to this change. Furthermore, a decisive step was taken when it became obvious that experiments on solutions with overlap are easily interpretable in terms of the concentration ρ and of the basic quantities R_G^2 and A_2 (which are defined for $\rho \to 0$). In other words, it is necessary to account for swelling effects, which remain important.

4.1 Overlap factor and relative concentration

For the experimentalist, the quantity of monomers expresses itself naturally in terms of the mass-concentration ρ of solute. However, the important parameter is the rate of chain overlap, which can be defined as the ratio ρ/ρ^* where ρ^* represents the overlap concentration. More precisely, theory shows that overlap is easily expressed with the help of the product

$$CX^d$$

where C is the concentration in chains and where $X^2 = R^2/d$, R^2 being the mean square end-to-end distance in the limit $C \to 0$. We set

$$\rho/\rho^* = C/C^* = CX^d,$$

i.e.

$$\rho^* = MX^{-d}/Ⓐ$$

since

$$\rho = CM/Ⓐ.$$

The quantity CX^d is therefore a basic quantity, and it is important to determine it with great accuracy. However, it is not possible to propose a method which would be valid in all cases: for example, it is perfectly possible to use osmotic pressure to study solutions with overlap without being able to determine in this way the mean square distance $R^2 = X^2 d$.

There are, in fact, three methods of determining CX^d experimentally.

First method

In this case, it is necessary to know the basic quantities S, b, and ρ, whose values are determined mainly by light scattering (Section 1). From the area S and the interaction b, we get the parameter

$$z = bS^{1/2}(2\pi)^{-3/2},$$

and the swelling $\mathfrak{X}(z)$ is given with precision by theory (see Fig. 15.7, p. 000). We have, therefore,

$$X^2 = \mathfrak{X}_0(z)S.$$

Hence,

$$CX^3 = \frac{\rho Ⓐ}{M} S^{3/2}(\mathfrak{X}_0(z))^{3/2}$$

which is the result searched for.

Second method

In this case, the radius of gyration R_G^2 is measured in the limit of zero concentration. We set

$$\rho_G^* = \frac{M}{2^{3/2} R_G^3 Ⓐ} \tag{15.4.1}$$

in agreement with the definition of C_G^* given in Chapter 13, Section 2.1.

We get

$$\rho^* = \rho_G^*(6R_G^2/R^2)^{3/2} = \rho_G^*(\aleph(z))^{3/2}$$

where $\aleph(z)$ is a corrective factor which, for $d = 3$, is close to unity ($\aleph(0) = 1$; $\aleph(+\infty) = 0.952$ [see (13.1.63)].

We note that

$$CX^3 = \rho/\rho^* = (\rho/\rho_G^*)(\aleph(z))^{-3/2} \tag{15.4.2}$$

which is the result searched for.

Third method

This method is related to the measurement of the second virial coefficient. From the definitions given in Section 3.1.2

$$A_2\rho M = \tfrac{1}{2}(2\pi)^{3/2} g(z) CX^3(z) \tag{15.4.3}$$

where $g(z)$ is the osmotic parameter. In the regime of asymptotic swelling $g(z) \simeq 0.233$. In this case

$$A_2\rho M = 1.83\, CX^3(z)$$

which is the result searched for.

Of course, these definitions do not account for polydispersion. We shall introduce the necessary corrections when we discuss experimental results.

All this holds true only if the solution is not very concentrated. For the standard continuous model, the interval of variation of CX^3 is infinite, but in reality the fact that polymer chains cannot be made as long as desired is a limitation.

4.2 Swelling of the chain in solution with overlap: neutron scattering

In the limit of zero concentration, it may be assumed that the swelling of the total chain depends only on one parameter, which is the interaction parameter z.

In solution with overlap, the presence of other chains screens the interaction, which has the effect of decreasing the swelling. The latter depends then on chain-concentration as well as on z.

It is therefore of interest to observe the swelling of a chain as a function of concentration. Benoit and Picot tried in 1966[19] to derive the radius of gyration of long polymer chains in semi-dilute benzene solutions, from light scattering experiments. However, such experiments can only give an *apparent* radius [see (15.1.21)]. On the other hand, the *real* radius may be determined by neutron scattering, provided a properly labelled solute is used (see Chapter 7, Section 3.1.2). As a matter of fact, when neutron beams began to be considered as a tool to study polymer solutions, one of the first proposals was to investigate the variation of $R_G^2(\rho)$ with concentration ρ. The experiment was carried out in 1974, and we shall present its results.

4.2.1 Variation of the gyration swelling with concentration

In 1974, a team at Saclay[20] measured the radii of gyration $R_{G,Z}^2$ of polystyrene chains in solution in CS_2 by neutron scattering at room temperature, for different solute concentrations. Let us briefly describe the experiment. The

Table 15.9

	Type of polystyrene sample	M_W	M_W/M_n	p	$R^2_{G,Z}$ [nm^2]	$R^2_{G,W}$ [nm^2]	$\rho^*_{G,W}$ [g/l]	z	$\mathfrak{X}_0(z)$	$\nu(z)$
1	Deuterated	1.14×10^5	1.1	0.1	187	158	36(33)	0.87	1.65	0.565
	Non-deuterated	1.14×10^5	1.02	0.02						
2	Deuterated	5×10^5	1.14	0.14		836	12.1(11.5)	1.83	2.01	0.575
	Non-deuterated	5.3×10^5	1.1	0.1						

samples are mixtures of deuterated and non-deuterated polystyrene chains (see Table 15.9 for the basic quantities).

First, some comments on this table. The 'Z'-average of the square radius of gyration in the limit of zero concentration, $R^2_{G,Z}$, is measured by the usual scattering technique (Section 1.3.1): the sample of type 1 used here, is made exclusively of deuterated chains. This sample, being relatively polydisperse, we shall choose the monodisperse system of molecular mass M_W as a reference. The square radius of gyration of the reference chain is

$$R^2_G(M_W) = R^2_{G,Z} \exp[-\epsilon(\nu)] \tag{15.4.4}$$

where $\varepsilon(\nu)$ is given by (15.3.45), assuming that molecular mass distribution obeys the law of Schulz and Zimm.

The overlap concentration value [see (15.4.1)]

$$\rho^*_{G,W} = \frac{M_W}{2^{3/2}(R^2_{G,W})^{3/2} \text{Ⓐ}} \tag{15.4.5}$$

is given in Table 15.9. The value $\rho^*_{G,W}$, determined with the help of eqns (15.4.4) and (15.4.5), is given in parentheses in this table; the parameter z is calculated for an interaction $b = 0.97\ \text{nm}^{-1}$ [see (15.3.48) and Table 15.8].

The swelling $\mathfrak{X}_0(z)$ and the effective exponent $\nu(z)$ are determined with the help of Figs 15.7 (p. 733) and 15.12 (p. 742) respectively.

The basic quantity $R^2_{G,Z}$ for sample 2 could not be determined by neutron scattering: here, the observation window (15.1.23) is not adapted to the measurement of such a big radius. On the other hand, in semi-dilute solution, the swelling is less important and the quantities $R^2_{G,Z}(\rho)$ are accessible to experimentation.

In Table 15.10, we find the values of the square radius of gyration, $R_{G,Z}(\rho)$, for various overlap ratios ρ/ρ^*_G. A distinct decrease of $R^2_{G,Z}(\rho)$ is observed when this ratio increases. With which variation law, may we compare this result? A formula, analogous to (13.2.15)

$$R^2_G = A_G(CX^{1/\nu})^{-\frac{2\nu-1}{\nu d-1}} X^{1/\nu} = A_G(CX^d)^{-\frac{2\nu-1}{\nu d-1}} X^2 \tag{15.4.6}$$

Table 15.10

Polystyrene sample	ρ [g/l]	$\rho/\rho^*_{G,w}$	$R^2_{G,z}(\rho)$ [nm²]	A_W	$-\frac{Ⓐ\, H^{II}(\vec{0})}{2M^2_W} \times 10^4$ [cm³/g]
1.	30	0.94	144	0.45	3.54
	60	1.88	136.9	0.51	1.02
	100	3.2	123.2	0.52	0.44
	150	4.7	108.2	0.50	0.29
	200	6.4	102	0.50	0.15
	330	11	90.25	0.50	0.19
	500	16	82.8	0.50	
	1060	35	67.25	0.49	
2	60	5.22	585.6	0.51	
	200	17.4	511	0.59	

gives, in principle, the variation of the asymptotic radius of gyration. In this formula, the constant A_G is not known, but to a first approximation, for $CX^d = 1$, we may set

$$R^2_G \simeq A_G X^2 = 2A_G R^2_G/\aleph$$

which gives $A_G = \frac{\aleph}{2} = (\simeq 0.48)$. This corresponds to the fact that equality

$$CX^d = 1$$

defines a natural boundary between dilute and semi-dilute regimes.

Formula (15.4.6), validly describes the variations of R^2_G with respect to $CX^d\,(d = 3)$, in the asymptotic regime $z \gg 1$. However, the interaction parameter z here takes a value which is lower than the values corresponding to that regime (see Table 15.9). Also, since $\aleph(\infty) = 0.952$ [see (13.1.63)], the quantities CX^3 and X^2 are given by the relations

$$CX^3 = (\rho/\rho^*_G)(\aleph(\infty))^{-3/2} = 1.076\,\rho/\rho^*_G$$

$$X^2 \simeq X^2_W = \frac{2}{\aleph(\infty)} R^2_G(M_W) = 2.1 R^2_G(M_W)$$

where $R^2_G(M_W)$ given by (15.3.46) correspond to the limit $\rho \to 0$.

Taking the preceding remarks into account, we may examine whether the quantity

$$A_W = \frac{R^2_{G,z}(\rho)}{(1.076\,\rho/\rho^*_{G,w})^{-\frac{2\nu-1}{3\nu-1}} X^2_W} \qquad (15.4.7)$$

remains constant when the overlap varies.

The coefficient A_W has been calculated from experimental data, and we took for ν the asymptotic value $\nu = 0.588$. The results are displayed in Fig. 15.10. Note that A_W is nearly independent of swelling. Therefore, in semi-dilute solution, the square radius of gyration is certainly proportional to the molecular mass.

In the last column of Table 15.10, we find the values of the quantity

$$- Ⓐ \frac{H^{II}(\vec{0})}{2M_W^2} \tag{15.4.8}$$

where $H^{II}(\vec{q})$ is the inter molecular structure function [see (15.1.9)]. The quantity (15.4.8) is obtained by scattering (see Section 1). In the limit of zero concentration, this quantity is equal to the second virial coefficient $A_{2,\sim}$. In this way, by neutron scattering, we get

$$A_{2,\sim} = 6.21 \times 10^{-4}\,\text{cm}^3\,\text{g}^{-2},$$

in agreement with results obtained by light scattering and represented in Fig. 15.14.

For strong concentrations (see discussion, Chapter 7, Section 3.4)

$$H^{II}(\vec{0}) = -H^{I}(\vec{0})$$

where by definition

$$H^{I}(\vec{0}) = \frac{M}{\rho Ⓐ},$$

and consequently

$$-\frac{Ⓐ H^{II}(\vec{0})}{2M_W^2} \simeq \frac{1}{2\rho M_W}. \tag{15.4.9}$$

Therefore, when (15.4.9) is verified, the solution can no longer be considered as a semi-dilute solution of filaments without any thickness: the volume fraction occupied by the chains becomes an important factor. We observe from Tables 15.9 and 15.10 that relation (15.4.9) is closely verified when $\rho \geqslant 330$ g/l: in this case, we could interpret the swelling with the help of Edwards's formula [see (13.2.149)].

4.2.2 Variation of the form function with concentration

The scattering experiment described in the preceding section, indicates that the polymer chain, although still swollen, already has a characteristic property of Brownian chains: the square of its gyration radius is proportional to the molecular mass (see Table 15.10). However, the chain keeps a Kuhnian character in the sense that short sequences are swollen. Let S_c be the equivalent Brownian area of the sequence whose square end-to-end distance, R_c^2, is given by

$$R_c^2 = dS_c = \xi_k^2 = (CX^{1/\nu})^{-\frac{2\nu}{\nu d - 1}} = (CX^d)^{-\frac{2\nu}{\nu d - 1}} X^2 \tag{15.4.10}$$

where ξ_k is the Kuhnian overlap length (at concentration C). The theory predicts that sequences of area s

$$s > S_c$$

are Brownian-like, and that sequences of area s

$$S_t < s < S_c$$

are Kuhnian-like [the area S_t is associated with the thermal correlation length $\xi_t = 2b^{-1}$ (see (15.3.25)].

From the observation of the asymptotic behaviour of the form function $H(\vec{q})$ for a chain in semi-dilute solution, it is possible to verify this theory qualitatively. One expects for $H(q)$ a behaviour of the Kuhnian type

$$H(\vec{q}) \simeq \frac{\mathrm{h}_\infty}{\left(q^2 \dfrac{R^2}{6}\right)^{1/2\nu}} \qquad \text{if } \xi_k^{-2} < q^2 < \xi_t^{-2}, \tag{15.4.11}$$

and a behaviour of the Brownian type

$$H(\vec{q}) \simeq \frac{2}{q^2 \dfrac{R^2}{6}} \qquad \text{if } R^{-2} < q^2 < \xi_k^{-2}. \tag{15.4.12}$$

The situation here is at variance with the one described in Section 3.2.2, where the Brownian and Kuhnian types of behaviour appear in the reverse order. It is therefore interesting to test the predictions (15.4.11) and (15.4.12).

Farnoux *et al.*[12] carried out neutron scattering experiments with solutions of long polystyrene chains in order to measure the variation of the form function $H(\vec{q})$ when the overlap ratio increases. The solute was made of a mixture of deuterated and non-deuterated chains, from which it was possible to determine the form function $H(\vec{q})$ of a chain in solution with overlap (see Chapter 7, Section 3.2.1). The average molecular mass of the sample was $M_W = 1.1 \times 10^6$.

The radius of gyration, in the limit of zero concentration, for polystyrene in CS_2 with this molecular mass, has the value $R_G = 47$ nm [see (15.3.47)]. We verify that the observation window associated with neutron radiation

$$q_{\min} = 6 \times 10^{-2}\,\text{nm}^{-1} \qquad q_{\max} = 6\,\text{nm}^{-1} \tag{15.4.13}$$

[see (15.1.23)], allows us to observe the form function behaviour in the asymptotic domain $qR_{G,Z} \gg 1$.

The values calculated for the overlap ratios are given in Table 15.11. In this way, we see that ξ_k^{-1} falls within the observation window for all values of ρ/ρ_G^*. Figure 15.18 represents the inverse form functions plotted against $q^{1/\nu}$ (where $\nu = 0.6$), for various overlap ratios. For $\rho/\rho_G^* = 15.6$, $H(\vec{q})$ has a Kuhnian behaviour in the entire interval (15.4.13). But when this overlap ratio increases, the Kuhnian behaviour holds true only in the upper part of interval (15.4.13): in the vicinity of $q_{\min}$ a deviation becomes visible and we may interpret this as

Table 15.11

ρ [g/l]	$\frac{\rho}{\rho_G^*} = 0.928\, CX^d$	Calculated ξ_k [nm]	ξ_k^{-1} [nm^{-1}]
75	11.7	9.54	0.105
100	15.6	7.8	0.12
250	39	3.7	0.27
500	78.14	2.24	0.45

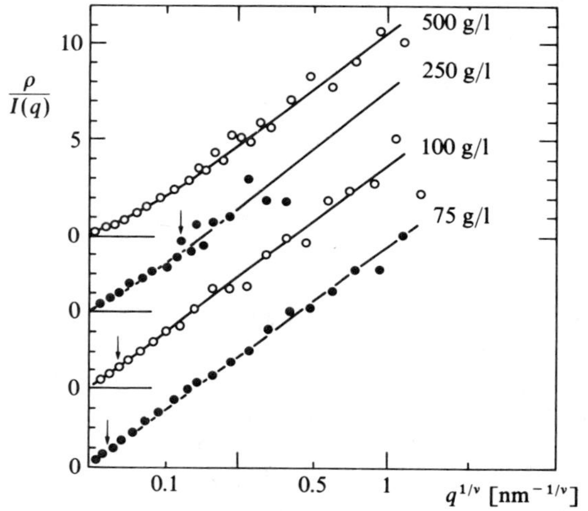

Fig. 15.18. Plots of $\rho/I(q)$ (in arbitrary units) for mixtures of deuterated and non-deuterated polystyrene in CS_2, at different total concentration in polystyrene. Neutron scattering (from Farnoux *et al.*[12]).

a behaviour of the Brownian type, in agreement with predictions (15.4.11) and (15.4.12).

Therefore, Fig. 15.18 indicates a change of behaviour which is opposite to that shown in Fig. 15.13, and this agrees with our expectations.

4.2.3 Variation of the gyration swelling with temperature

The square radius of gyration of a chain in semi-dilute solution ($CX^d \gg 1$) is written as [see (15.4.6)]

$$R_G^2 = A_G(CX^d)^{-\frac{2\nu-1}{\nu d-1}} X^2 \tag{15.4.14}$$

where $3X^2 = \mathfrak{X}_0\, {}^0R^2$ (let us recall that X defines the size of a chain in the limit of zero concentration ($CX^d \ll 1$)).

When we warm up the solution at fixed concentration C in the domain IV of the temperature–concentration diagram, the square radius $3X^2$ and the overlap ratio increase. The chain is then subject to two contradictory effects: first the solvent quality increases, second the screening of the interaction becomes stronger. Let us first recall that the average square distance $3X^2$ varies in the following manner (see Chapter 12, Section 3.3.2)

$$3X^2 = {}^0R^2\,\mathfrak{X}_0(z) \propto {}^0R^2 z^{2(2\nu-1)} \propto {}^0R^2 b^{2(2\nu-1)} S^{2\nu-1} \qquad (15.4.15)$$

where b is the interaction and S the equivalent Brownian area. Inserting this expression in (15.4.14), we obtain

$$R_G^2 \propto b^{\frac{2\nu-1}{\nu d-1}}, \qquad \text{for } CX^3 > 1. \qquad (15.4.16)$$

There is swelling here when the interaction (or the temperature) increases, but this swelling is less important than it is in dilute solution ($R_G^2 \propto b^{2(2\nu-1)}$, for $CX^3 \to 0$).

Cotton *et al.*[21] tried to verify relation (15.4.14). Unfortunately, the result of their experiment is not conclusive, essentially because the chain swellings derived from their observation of semi-dilute solutions (by neutron scattering) are greater, at a given temperature, than the swelling in the limit of zero concentration! The agreement that these authors found between the variations of measured gyration swellings and relation (15.4.15) is therefore misleading.

Richards *et al.*[22] did the same experiment and obtained the same kind of result. This British team had in view the verification of Edwards's variation law

$$R_G^2 = \frac{1}{2}\aleph S\left(1 + \frac{2}{3\pi}(b/CS)^{1/2}\right) \qquad (15.4.17)$$

[see (13.2.149)], by means of neutron scattering measurements. The radii determined in this way vary with temperature in a manner which, to this team, seemed to be in agreement with relation (15.4.17). But the swelling obtained in semi-dilute solutions is, at equal temperature, greater than the swelling in the limit of zero concentration!

After examination of the experimental reports, we can give the following explanation of these two paradoxical result. In both cases, the solute is made of a mixture of deuterated and non-deuterated polystyrene, the solvent being CS_2. Concentrations and molecular masses are, for instance:[21]

deuterated polystyrene	$\rho_D = 10$ g/1,	$M_W = 1.6 \times 10^5$
non-deuterated polystyrene	$\rho_H = 140$ g/1,	$M_W = 1.72 \times 10^5$

The quantity measured with the help of neutron scattering, by all the quoted authors, is the structure function $H_D(\vec{q})$ associated with the deuterated fraction of the solute. These authors then identified $H_D(\vec{q})$ with $\dfrac{M_W}{\rho_D} H_D(\vec{q})$ where $H_D(\vec{q})$ is

Table 15.12

Temperature [°C]	$R^2_{\mathrm{G\,app}}(M_{\mathrm{W}})\,[\mathrm{nm}^2]$	$\mathfrak{X}_{\mathrm{G\,app}}(\rho = 150\,[\mathrm{g/l}])$	$\mathfrak{X}_{\mathrm{G}}(\rho \to 0)$
32°	118	1.01	1.01
34°	121,4	1.039	1.03
36°	121.8	1.043	1.04
38°	124.3	1.064	1.06
40°	111.1	0.95	1.07
41°	127.2	1.089	1.08
43°	128.4	1.1	1.09
45°	132	1.13	1.1
51°	145.2	1.243	1.16
55°	150.8	1.29	1.18
61°	156.5	1.34	1.2
65°	165.1	1.41	1.22
68°	168.2	1.44	1.30

the form function, thereby neglecting the intermolecular contribution, $\boldsymbol{H}^{\mathrm{II}}_{\mathrm{D}}(\vec{q})$. In fact, the determination of $H_{\mathrm{D}}(\vec{q})$ requires complementary experiments to eliminate the contribution of $\boldsymbol{H}^{\mathrm{II}}_{\mathrm{D}}(\vec{q})$: hence, one should extrapolate at $\rho_{\mathrm{D}} \to 0$, with fixed $\rho = \rho_{\mathrm{H}} + \rho_{\mathrm{D}}$ (see Chapter 7, Section 3). The interpretation of the structure function $\boldsymbol{H}(\vec{q})$ as a form function, introduces errors already noticed in other circumstances by Benoit and Picot.[19] These errors may explain the results obtained by the authors quoted above.

4.3 Osmotic pressure and structure of the solution with overlap

For long polymer chains in dilute solutions, the osmotic pressure is always very weak, and the reasons for this are as follows. In order to keep the solution dilute when M increases, one must diminish the mass concentration ρ; when ρ decreases at fixed molecular mass, the pressure decreases; finally, when M increases at fixed ρ, the pressure decreases again. In conclusion, measurements of osmotic pressure are not adapted to the study of the asymptotic state of polymers in dilute solutions.

On the contrary, the measurement of the osmotic pressure for different overlap ratios provides essential information on the structure of semi-dilute solutions. Data have been available since the time when this technique was first used. However, this approach required a theoretical basis and we have a good theory only since 1975.

The essential difference between 'naïve' theories and theories which account for the critical aspects of the polymer solutions is clearly in the expression of the osmotic pressure. Let us consider a dilute solution. The simple-tree approximation, typical of older theories, gives

$$\frac{\boldsymbol{\Pi\beta}}{\boldsymbol{C}} = 1 + \frac{1}{2}(2\pi)^{3/2} z\, \boldsymbol{C} S^{3/2} \tag{15.4.18}$$

where S is the equivalent Brownian area.

On the contrary, in the asymptotic limit [see (12.3.74)], we have

$$\frac{\boldsymbol{\Pi\beta}}{\boldsymbol{C}} = 1 + \frac{1}{2}(2\pi)^{3/2} g^* \boldsymbol{C} X^3 + \dots . \tag{15.4.19}$$

In good solvents, the quantity $z\boldsymbol{C}S^{3/2} \propto S^2$ is much greater than $g^*\boldsymbol{C}X^3 \propto S^{3\nu}$. Hence, for $d = 3$, the theory which accounts for the chain swelling predicts for given ρ and S, a much lower pressure than the simple-tree approximation.

Very complete results have been obtained for the osmotic pressure ($d = 3$) and we shall discuss them. There are preliminary results for the surface pressure ($d = 2$). We mentioned them in Chapter 5 and we shall add nothing more.

4.3.1 Experimental verification of the universal law of osmotic pressure

In Chapter 5, Section 10, we discussed the result of osmotic pressure experiments, made at different concentrations ρ. In particular, we noted that, if we renormalize the concentration ρ by the volume R_G^3 associated with the polymer chain in the limit $\rho \to 0$, we can exactly superpose the results $\boldsymbol{\Pi} M/\boldsymbol{R}T\rho (= \boldsymbol{\Pi\beta}/\boldsymbol{C})$ obtained for two different molecular masses M_n of the solute and two different solvent qualities (see Fig. 5.10, Chapter 5). More generally, the osmotic pressure of a monodisperse solution is written in the following form (good solvent)

$$\frac{\boldsymbol{\Pi\beta}}{\boldsymbol{C}} = \mathrm{F}(\boldsymbol{C}X^d) \tag{15.4.20}$$

where $\boldsymbol{C}$ is the number of chains per unit volume and where $X^2 = R^2/d$. This is the form which accounts best for the structure of the polymer solution.

Let us express $\boldsymbol{C}X^3$ in (15.4.20), as a function of ρ/ρ_G^* [see (15.4.2)]. We have

$$\frac{\boldsymbol{\Pi\beta}}{\boldsymbol{C}} = \frac{\boldsymbol{\Pi} M}{\boldsymbol{R}T\rho} = \mathrm{G}(\rho/\rho_G^*) \tag{15.4.21}$$

where

$$\rho_G^* = \frac{M}{2^{3/2} R_G^3 \text{Ⓐ}} \tag{15.4.22}$$

and where (see 15.4.2) and (13.1.63)

$$\mathrm{F}(1.076x) = \mathrm{G}(x). \tag{15.4.23}$$

Table 15.13

Sample	$M_n \times 10^{-6}$	$M_W \times 10^{-6}$	$R^2_{G,Z}$ [nm^2]	$\rho^*_{G,ZW}$ [g/l]
1	0.0708		84	0.539
2	0.2		318	0.207
3	0.506		840	0.122
4	1.19	1.19	2300	0.0633
5		1.82	3970	0.0426
6		3.30	7330	0.0309
7		7.47	18700	0.0172

Noda, Kato, Kitano, and Nagasawa[23] have made systematic measurements of the osmotic pressure in order to verify (15.4.21). They used seven samples of poly(α-methyl)styrene of different molecular masses (see Table 15.13). Each sample was obtained by anionic polymerization, followed by a fractionation aimed at a reduction of polydispersity. The latter is characterized by the ratio $M_W/M_n \leqslant 1.01$. The solvent used was toluene. The interaction b for poly(α-methyl)styrene in toluene, experimentally determined, is about 1.5 nm^{-1}.

Relation (15.4.21) was tested by measuring:

1. The quantities M_n (or M_W) and $R^2_{G,Z}$ in the limit $\rho \to 0$ (the overlap concentrations

$$\rho^*_{G,ZW} = \frac{M_W}{2^{3/2}(R^2_{G,Z})^{3/2} \text{Ⓐ}} \tag{15.4.24}$$

calculated from measured quantities appear in Table 15.13).

2. The osmotic pressures Π associated with a broad enough interval of reduced concentrations $\rho/\rho^*_{G,Z}$

$$0.025 \leqslant \rho^*_{G,ZW} \leqslant 40.$$

In the dilute case, the osmotic pressure was measured at membrane equilibrium, with the help of an osmometer (Chapter 5, Section 1.9).

In the more concentrated case, the osmotic pressure was obtained by measurement of the solvent vapour-pressure-lowering (Chapter 5, Section 1.5).

The numbers $\Pi M_n/RT\rho$ are plotted on Fig. 15.19, against $\rho/\rho^*_{G,ZW}$. We see that the points represented on the figure are distributed along a well-defined curve, with slow variation.

We can therefore prove experimentally the existence of a universal relation between reduced osmotic pressure and overlap ratio (provided that we take into

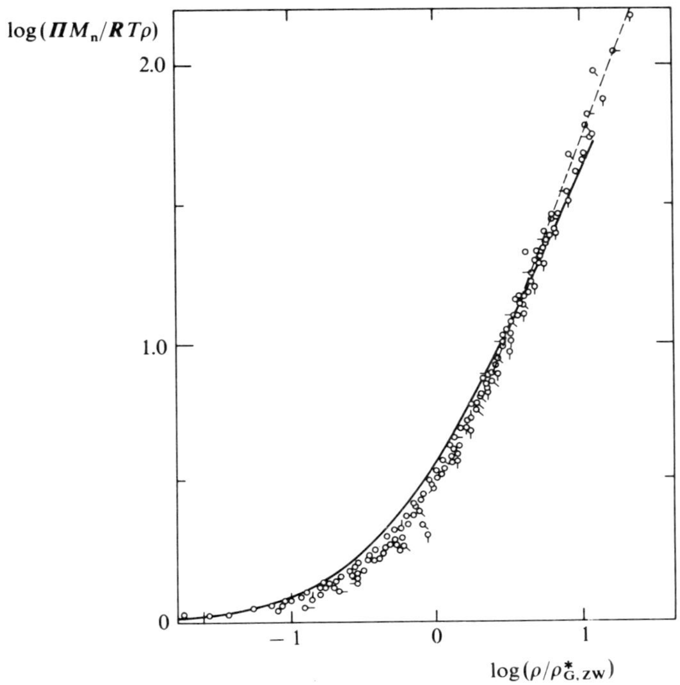

Fig. 15.19. Plot of measured osmotic pressures, against overlap ratio; poly(α-methyl)styrene in toluene (from Noda *et al.*[23]):

ο $M_n = 7.08 \times 10^4$ ϙ $M_w = 18.2 \times 10^5$
ό $M_n = 20 \times 10^4$ ⍴ $M_w = 3.3 \times 10^6$
ο- $M_n = 50.6 \times 10^4$ -ο $M_w = 7.47 \times 10^6$
ο $M_n = 119 \times 10^4$
—— value calculated from (13.2.76)
- - - - asymptotic value calculated from (13.2.82)

account polydispersion corrections). This relation is the more remarkable because it ceases to be true in the more concentrated solution regime, as we shall see in Chapter 16.

We wish to compare the experimental points with values derived from the universal function G(x) (15.4.21). For this, it is necessary to make polydispersion corrections. The authors claim that

$$M_W/M_n = 1.01,$$

i.e.

$$p = 0.01.$$

In order to make polydispersion corrections, we shall take as a reference system a monodisperse sample with concentration C and molecular mass $\overline{M}$, which we shall define as being equal to M_n. For this monodisperse system

$$\frac{\Pi\beta}{C} = \mathrm{F}(C\overline{X}^3) \simeq 1 + (g^*/2)(2\pi)^{3/2} C\overline{X}^3$$

where $\overline{X}$ corresponds to $\overline{M}$.

For the real system, the second virial coefficient is given by an average quantity. Such an average is difficult to compute. Nevertheless, it can easily by shown that the simple-tree approximation (see Chapter 10, Section 6) does not lead to any polydispersion correction. This brings up the hypothesis that, for $d = 3$, polydispersion corrections are negligible with respect to osmotic pressure, and we shall therefore assume to a first approximation that

$$\frac{\Pi\beta}{C} = \mathrm{F}(C\overline{X}^3) = \mathrm{G}(\rho/\bar{\rho}^*_{\mathrm{G}})$$

where

$$\overline{X} = X(\overline{M}) \qquad \text{and} \qquad \bar{\rho}^*_{\mathrm{G}} = \rho^*_{\mathrm{G}}(\overline{M}).$$

We must now express ρ^*_{G} as a function of the measured quantity $\rho^*_{\mathrm{G,zw}}$, defined by (15.4.24), and since $\overline{M} = \langle M \rangle$

$$\frac{\rho^*_{\mathrm{G,zw}}}{\bar{\rho}_{\mathrm{G}}} = \frac{M_{\mathrm{W}}}{\overline{M}} \left(\frac{R^2_{\mathrm{G}}(\overline{M})}{R^2_{\mathrm{G,Z}}}\right)^{3/2} = \frac{\langle M^2 \rangle}{(\langle M \rangle)^2} \left(\frac{\langle M \rangle^{2\nu+2}}{\langle M^{2\nu+2} \rangle}\right)^{3/2} \tag{15.4.25}$$

for $p \ll 1$, we have [see (13.2.23)]

$$\langle M^\alpha \rangle \simeq \langle M \rangle^\alpha \left(1 + \frac{\alpha(\alpha-1)}{2} p\right).$$

Hence,

$$\frac{\rho^*_{\mathrm{G,zw}}}{\bar{\rho}^*_{\mathrm{G}}} \simeq 1 + p\left(1 - \frac{3(\nu+1)(2\nu+1)}{2}\right) \tag{15.4.26}$$

and, as a consequence,

$$\frac{\rho}{\bar{\rho}^*_{\mathrm{G}}} = (\rho/\rho^*_{\mathrm{G,zw}})\left(1 + p\left(1 - \frac{3(\nu+1)(2\nu+1)}{2}\right)\right)$$
$$\simeq (\rho/\rho^*_{\mathrm{G,zw}})(1 - 4.2\, p).$$

Finally, the reduced osmotic pressure of a polydisperse system with $p = 0.01$, is approximatively

$$\frac{\Pi\beta}{C} = \mathrm{G}(0.958\rho/\rho^*_{\mathrm{G,zw}}) = \mathrm{F}(1.031\rho/\rho^*_{\mathrm{G,zw}})$$

since $\mathrm{F}(1.076\,x) = \mathrm{G}(x)$. In Fig. 15.19, we gave the correspondence between each overlap ratio $\rho/\rho^*_{G,zw}$ and two calculated values of the universal function $\Pi\beta/C = \Pi M_n/RT\rho$. The first one corresponds to expression (13.2.73); the second one, to the asymptotic expression (13.2.31). We note a good agreement, for $\rho/\rho^*_G \geqslant 10$, between the measured values and the asymptotic form.

Function (13.2.73) interpolates the measured points rather well. Let us recall that (13.2.73) is calculated only to first-order in $\varepsilon = 4 - d$. Besides, the polydispersion effects have been only crudely corrected.

4.3.2 Characteristic universal numbers

Characteristic numbers such as the osmotic parameter $g^*(\simeq 0.233)$ [*see* (13.1.16)], the ratio $A_3/A_2^2 M = 4h[g^*]/3(g^*)^2$ ($\simeq 0.517$) [see (13.1.28)] and the constant $F_\infty \simeq 2.46$ of the asymptotic form [see (13.2.81)] are associated with the universal law giving $\Pi\beta/C$. In principle, all these numbers are explicitly related to the curve plotted in Fig. 15.19. It is nevertheless interesting to test experimentally the value given to them theoretically.

We shall be interested here solely in the osmotic parameter g. Indeed, the experimental determination of the ratio $A_3/A_2^2 M$ from the data displayed in Fig. 15.19 is unfortunately not very precise: the contribution of this quantity to the function $\Pi\beta/C$ is difficult to separate from other virial coefficients. On the other hand, concerning the coefficient F_∞, there is obviously agreement between theory and experiment (see Fig. 15.19).

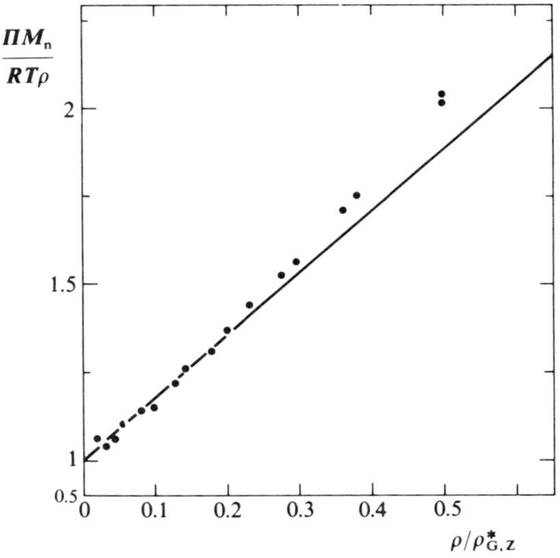

Fig. 15.20. .Representation of osmotic pressures measured for small overlap ratios (from Noda *et al.*[23]). Poly(α-methyl)styrene in toluene ($M_n = 7.08 \times 10^4$). Calculated value [eqn (15.4.28)].

The test concerning the value of the osmotic coefficient is made in the following way: the measured reduced pressures $\boldsymbol{\Pi} M_n/\boldsymbol{R}T\rho$ are plotted against $\rho/\rho^*_{G,zw}$ in Fig. 15.20, for $\rho/\rho^*_{G,zw} \leqslant 0.2$. On the same graph, we represent the equation [see (15.4.19)]

$$\frac{\boldsymbol{\Pi\beta}}{\boldsymbol{C}} = 1 + \frac{1}{2}(2\pi)^{3/2} g^* \boldsymbol{C} X^3 \tag{15.4.27}$$

For $g^* = 0.233$, and in agreement with the conversion (15.4.2) and the polydispersion correction (15.4.26), this equation is written as

$$\frac{\boldsymbol{\Pi\beta}}{\boldsymbol{C}} = \frac{\boldsymbol{\Pi} M_n}{\boldsymbol{R}T\rho} = 1 + 1.892(\rho/\rho^*_{G,zw}). \tag{15.4.28}$$

The straight line representing this equality on Fig. 15.20 is tangent to the measured curve, at the origin $\boldsymbol{C} = 0$, as predicted by the theory. The fit is very satisfactory.

4.4 Solute structure function: scattering experiments

The solute structure function $\boldsymbol{H}(\vec{q})$ is determined experimentally by radiation scattering. The interval q_{min}, q_{max} in which $\boldsymbol{H}(\vec{q})$ is measured depends upon the nature of the radiation used in the experiment [see (15.1.22) and (15.1.23)]. The choice of such a radiation is imposed by the choice of the interval which best suits the study of the observed structure.

A first measurement of the function $\boldsymbol{H}(\vec{q})$ for solutions with overlap was made in 1966 by Benoit and Picot,[19] using light scattering. Systematic measurements of the function $\boldsymbol{H}(\vec{q})$ for solutions with overlap have been made by the 'Strasacol'* group (Daoud *et al.* 1975[20]) using neutron scattering.

There is a relationship between the solute structure function $\boldsymbol{H}(\vec{q})$ at $\vec{q} = 0$ and the osmotic pressure. Nevertheless, $\boldsymbol{H}(\vec{q})$ reveals the structure of the solution in more detail than does the osmotic pressure, because $\vec{q}$ is an arbitrary vector.

4.4.1 General aspect of the structure function

Figures 15.21 and 15.22 show the results of scattering experiments on polystyrene solutions with overlap (light for Fig. 15.21, neutrons for Fig. 15.22).

The polymer in the experiment[19] reported in Fig. 15.21 is polystyrene of molecular mass $M_W = 7.5 \times 10^6$, at a concentration $\rho = 11.38$ g/l in benzene. In the limit of zero concentration, the radius of gyration measured by scattering has the value (see Table 15.1, p. 724)

$$R_{G,Z} = 150 \text{ nm}.$$

* Strasacol: Strasbourg, Saclay, College de France.

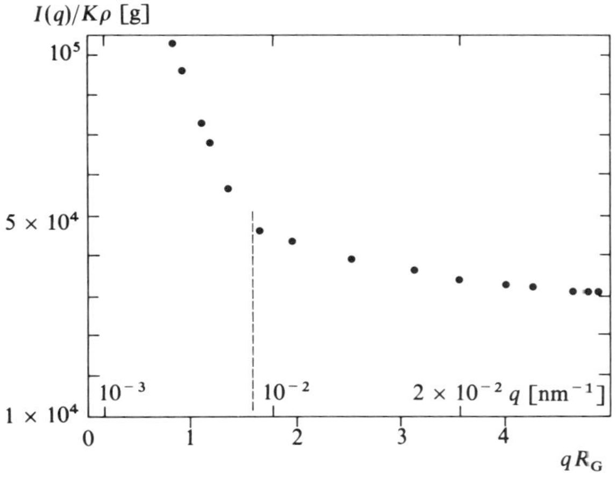

Fig. 15.21. Plot of the scattered intensity against transfer q and qR_G (from Benoit and Picot[19]). Polystyrene in benzene; concentration $\rho = 11.4$ g/l. Light scattering: absolute measurement. The vertical discontinuous line indicates the separation between 'abnormal' ($q \to 0$) and 'normal' behaviour.

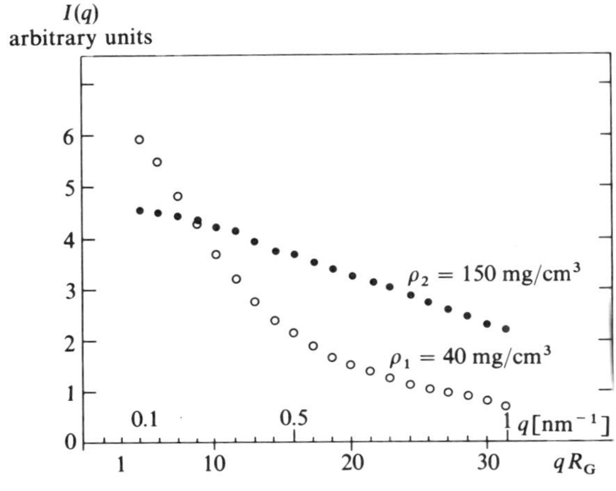

Fig. 15.22. Plot of the scattered intensity against transfer q and qR_G (from Farnoux[16]). Neutron scattering. Deuterated polystyrene, $M_W = 5 \times 10^5$ in CS_2.

Hence, the overlap concentration reads [see (15.2.4)]

$$\rho^*_{G,zw} = \frac{M_W}{2^{3/2}(R^2_{G,z})^{3/2} Ⓐ} = 2.1 \text{ g/l},$$

and, as a consequence, $\rho/\rho^*_G = CX^d \simeq 5.9$. The experiment is carried out with light scattering. The domain explored in reciprocal space is intermediate between the Guinier range $qR_G < 1$ and the asymptotic range $qR_G \gg 1$. The scattered intensity decreases regularly with wave vector transfer; but the authors have noticed an 'excess scattered intensity' in the interval $q \leqslant 10^{-2}\ \mathrm{nm}^{-1}$. This effect, generally called the Picot–Benoit effect,[19] has never really been explained: trapped dust particles are suggested, as well as inhomogeneities of the solute structure or a solvent effect.

Let us now consider the neutron scattering experiment,[20] the results of which are shown in Fig. 15.22. The corresponding samples are made of polystyrene of molecular mass $M_W = 5 \times 10^5$. The mass-concentrations are

$$\rho_1 = 40\ \mathrm{g/l} \qquad \text{and} \qquad \rho_2 = 150\ \mathrm{g/1}.$$

In the limit of zero concentration, the radius of gyration has the value (see Table 15.1)

$$R_{G,Z} = 25\ \mathrm{nm}.$$

Therefore,

$$\rho^*_{G,ZW} = 19\ \mathrm{g/1} \text{ and consequently } \rho_1/\rho^*_G \simeq 2.1,\ \ \rho_2/\rho^*_G \simeq 7.9.$$

The range of reciprocal space explored here with neutrons, is really in the asymptotic domain $qR_G \gg 1$; the reason for this is that the wave vector k_o is large with respect to $1/R_G$. This is the domain in which we are going to study the structure function, and in the semi-dilute regime, the latter is written as

$$H(\vec{q}) = H(\vec{0})\mathrm{J}(q\xi_k) \tag{15.4.29}$$

where ξ_k is the Kuhnian overlap length.

Note the screening effect in Fig. 15.22: when the solute concentration increases, the structure function drops at $q = 0$ and the curvature of $\mathrm{J}(x)$ decreases (in absolute value); besides, for large values of q, $I(q) \propto \rho^2 H(\vec{q})$ increases with ρ, which simply reflects the increase of substance. The drop of $I(0) \propto \rho^2 H(\vec{0})$ and the decrease of the curvature in $\mathrm{J}(q\xi_k)$ are related to a change of the analytical properties of $H(\vec{q})$, considered as a function of q. When ρ increases, the poles of $H(\vec{q})$ in the complex plane of q move away towards $\pm\infty$ along the imaginary axis.

4.4.2 Observation of the asymptotic form of $H(\vec{q})$

Let us examine the intensity $I(q)$ scattered by a solution with overlap [see (7.2.58)]

$$I(q) = K\rho^2 \,Ⓐ\, H(\vec{q}) \tag{15.4.30}$$

where $H(\vec{q})$ is the structure function of the solute. From the measured values of $I(q)$, we can get the asymptotic form of the structure function

$$H(\vec{q}) = H(\vec{0})\mathrm{J}(q\xi_k) \tag{15.4.31}$$

where $qX \gg 1$. This condition is certainly satisfied for the experiment of Fig. 15.23. Let us then examine the function $\mathrm{J}(x)$, as revealed by the experiment.

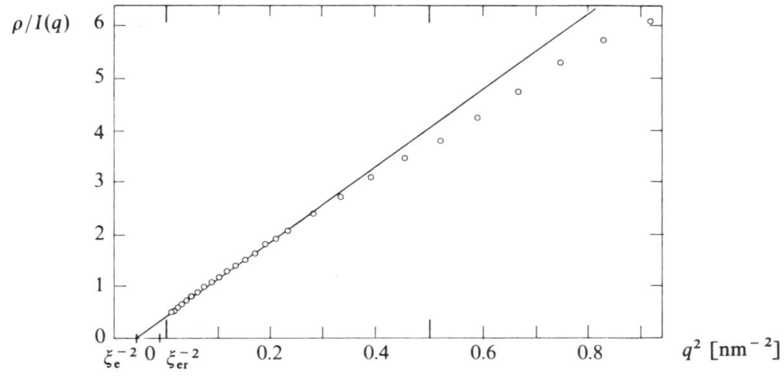

Fig. 15.23. Plot of $\rho/I(q)$ (in arbitrary units) against q^2(in nm^{-2}). Result obtained by neutron scattering. Semi-dilute solution of deuterated polystyrene ($M_W = 1.1 \times 10^6$) in CS_2 (from Farnoux[16]). Mass-concentration of the solute $\rho = 25$ g/l. The quantity ξ_e^{-2} is determined by linear extrapolation of the points which are near to $q^2 = 0$. We also marked ξ_{er}^{-2} calculated from eqn (13.2.122).

For rather small values of q (but for $qX \gg 1$), the behaviour of $\boldsymbol{H}(\vec{q})$ is dominated by the singularity located at $q^2 = -1/\xi_e^2$ where ξ_e is the screening length. This suggests representing $\mathrm{J}(x)$ in the form

$$\mathrm{J}(x) = \frac{\mathrm{J}_0}{1 + \Gamma^2 x^2} + \varphi(x) \qquad (15.4.32)$$

where Γ is a constant close to unity and where $\varphi(x)$ accounts for the more remote singularities in the complex plane. This representation is also in agreement with the definition [see (13.2.10)]

$$\xi_e = \Gamma\, \xi_k$$

where Γ is a universal constant.

Therefore the quantity $K\rho/I(q)$ given by the expression

$$\frac{K\rho}{I(q)} = \frac{1}{\rho \circledA \boldsymbol{H}(\vec{0})}(1 + \Gamma^2\xi_k^2 q^2)(\mathrm{J}_0 + \Gamma^2\xi_k^2 q^2\varphi(\xi_k q))^{-1} \qquad (15.4.33)$$

can be represented as a function of q^2 and Fig. 15.23 shows how convenient this representation is: indeed, there exists a complete domain of q^2 values in which this function is quasi-linear in q^2. It is then sufficient to extrapolate this function, for negative values of q^2, in order to determine, with precision, the negative value of q^2 for which $K\rho/I(q)$ becomes zero. From this value, we get the screening length ξ_e and here

$$\xi_e = 4.1 \text{ nm}$$

as shown in Fig. 15.23.

Now, as we have seen in Chapter 13 [eqn (13.2.9)], the Kuhnian overlap length is defined by

$$\xi_k = (CX^{1/\nu})^{-\frac{\nu}{\nu d - 1}}, \tag{15.4.34}$$

and we have here

$$\xi_k = 22 \text{ nm}.$$

The universal constant Γ which in good solvent for a semi-dilute solution is given by

$$\Gamma = \xi_e/\xi_k \tag{15.4.35}$$

is approximatively

$$\Gamma = 0.18.$$

This experimental value is close to the theoretical result given by (13.2.128). We see in Fig. 15.23 that the function $K\rho/I(q)$ bends downwards when $q^2 > \xi_e^{-2}$: this is a manifestation of the contributions of distant singularities in the complex q plane.

4.4.3 Similarity of structure functions associated with different overlaps, scaling law

In order to determine the screening length ξ_e of a solution with overlap, by extrapolation of measured values of the scattered intensity

$$I(q) = K\rho^2 \textcircled{A}\, \boldsymbol{H}(\vec{q}),$$

we assumed that, for $qR_G > 1$, the structure function has the universal behaviour

$$\boldsymbol{H}(\vec{q}) = \boldsymbol{H}(\vec{0})\mathrm{J}(q\xi_k)$$

where ξ_k is the Kuhnian overlap length. Let us check this scaling law by considering again the results of the scattering experiment[20] described earlier (Section 4.4.2), this time by comparing the results obtained for two different polystyrene samples:

$$\rho_1 = 25 \text{ g/l}; \qquad \rho_2 = 40 \text{ g/l}$$

$$M_1 = 1.1 \times 10^6; \qquad M_2 = 5 \times 10^5.$$

Let $I(0)$ then be the value obtained by extrapolating at $q = 0$ the measured scattered intensity. In Fig. 15.24, we plotted the quantities

$$I_1(0)/I_1(q) = 1/\mathrm{J}(q\xi_{k1})$$

and

$$I_2(0)/I_2(q) = 1/\mathrm{J}(q\xi_{k2})$$

against q.

If the function $I(0)/I(q)$ is universal the two results must superimpose after renormalization of the abscissa. We then draw three straight lines parallel to

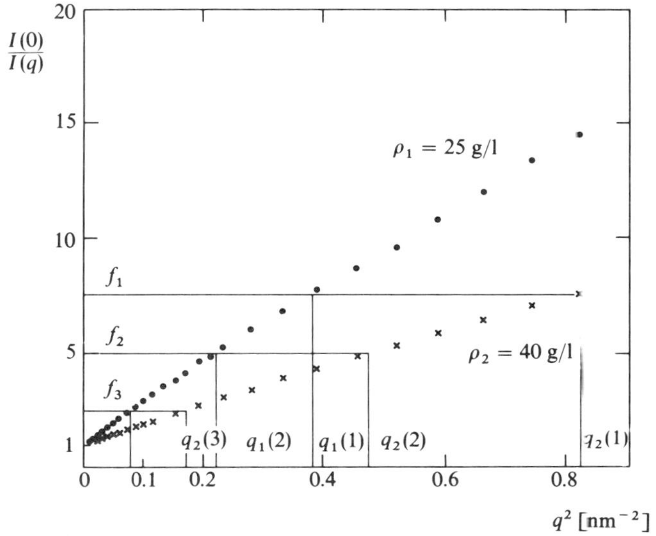

Fig. 15.24. Representation of the numbers $I(0)/I(q)$ obtained experimentally for two deuterated polystyrene solutions in CS_2:

• $M_W = 1.1 \times 10^6$, $\rho_1 = 25$ g/l

× $M_W = 5 \times 10^5$, $\rho_2 = 40$ g/l.

Neutron scattering (from Farnoux[16]). The lines parallel to the horizontal axis are used to show that the curves are affine.

the horizontal axis, and with arbitrary ordinates f_i $(i = 1, 2, 3)$. These lines intersect the function $I_1(0)/I_1(q)$ and $I_2(0)/I_2(q)$ at points of respective abscissa q_{1i} and q_{2i} $(i = 1, 2, 3)$: we now note that the quotients $q_{2i}/q_{1i}(\simeq 2.15)$ are practically independent of i. Therefore, the functions $I_1(0)/I_1(q)$ and $I_2(0)/I_2(q)$ are affine to each other. This remarkable result recalls the one we obtained when we studied the variation of osmotic pressure with solute concentration. Here the affinity ratio is, in principle, given by [see (15.4.34) and (15.1.1)]

$$\frac{\xi_{k2}}{\xi_{k1}} = \left(\frac{\rho_2}{\rho_1}\right)^{-\frac{\nu}{\nu d - 1}},$$

and in the present case the right-hand side is equal to 0.693 for $\nu = 0.588$. The left-hand side, derived from scattering experiments, has the value 0.727. There is a 3.5 per cent difference.

Let us establish in a more systematic way the existence and the characteristics of the scaling law $J(q\xi_k)$, by examining results of scattering experiments on four samples. The latter are defined by the following molecular masses and concentrations

Sample	1	2	3	4
M_W	1.1×10^6	5×10^5	5×10^5	5×10^5
ρ [g/l]	25	40	75	150

We represent all the results as a function only of the parameter $q\xi_e$. For this, we need the value of ξ_e corresponding to each sample; we get them from measured values of the intensity $I(q)$. Using the prescription given in Section 4.4.2, we obtain

Sample	1	2	3	4
ξ_e[nm]	4.1	2.98	1.83	1.05
ξ_k[nm]	22.2	15.8	9.54	5.6
Γ	0.18	0.19	0.19	0.19.

Bringing together all measured values $I(0)/I(q)$ on the same graph against $q\xi_e$ for the four samples, we obtain a single curve shown in Fig. 15.25. Therefore the experiment shows the existence of a universal scaling law $J(q\xi_e)$.

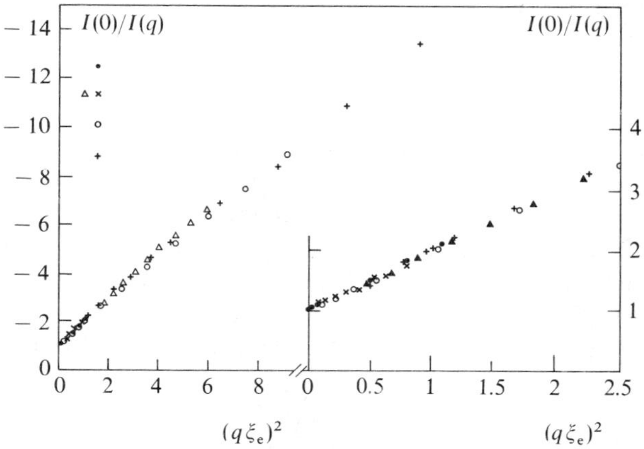

Fig. 15.25. Curve obtained by superposition of the numbers $I(0)/I(q)$ derived from experiments made at different concentrations and for different molecular masses Deuterated polystyrene in CS_2. The screening lengths ξ_e on the horizontal axis are determined experimentally (see Fig. 15.23)

$+$: $M_W = 1.1 \times 10^6$

o, ×, △ and ●: $M_W = 5.5 \times 10^5$.

4.4.4 Structure of the solution at distances smaller than the screening length

The repulsive interaction screens concentration correlations. The screening factor for two points placed at a distance r, is equal to $\exp(-r/\xi_e)$. Therefore, the screening effect is weak when $r < \xi_e$. In this case, the structure function $\boldsymbol{H}(\vec{q})$ is dominated by the contribution of the form function $H(\vec{q})$ and, in the q interval under consideration, it has the same behaviour as the form function $H(\vec{q})$ of the isolated chain. Now, for $qX \gg 1$ and $\boldsymbol{C} \to 0$, we have seen that

$$\boldsymbol{CH}(\vec{q}) \simeq \frac{\mathrm{h}_\infty}{\left(q^2 \dfrac{R^2}{6}\right)^{1/2\nu}}. \tag{15.4.36}$$

We expect to find a similar behaviour, when studying the structure function in the semi-dilute regime ($\boldsymbol{C}X^3 \gg 1$) and in the interval $q\xi_e \geqslant 1$.

Let us check this conjecture experimentally. For this, we consider Fig. 15.25, in which are displayed the points with ordinates $I(0)/I(q\xi_e)$ corresponding to four solutions at different concentrations. We multiply these ratios by $(q\xi_e)^{-1/\nu}$, which is proportional to the asymptotic form function of an isolated chain in good solvent. If there is an interval

$$q \geqslant q^* \propto 1/\xi_e$$

in which the structure function is not screened and remains proportional to the form function, then the quantities

$$[I(0)/I(q\xi_e)] \times (q\xi_e)^{-1/\nu} \tag{15.4.37}$$

are constant, independent of $q\xi_e$. Figure 15.26 shows us that such a domain exists. More precisely, we may determine a value q^*

$$q^* \simeq 2.6\ \xi_e^{-1}$$

such that for $q \geqslant q^*$

$$[I(0)/I(q\xi_e)] \times (q\xi_e)^{-1/\nu} \cong 1.51$$

[this quantity is to be compared to $\mathrm{h}_\infty^{-1} \simeq 0.96$ (15.3.50)].

Figure 15.26 suggests the following comment. In the interval $q \leqslant q^*$, the function $\boldsymbol{H}(\vec{q})$ is similar to a Lorentzian, in agreement with (15.4.31) and (15.4.32), and such behaviour is characteristic of the relatively homogeneous structure of semi-dilute solutions. However, in the complementary interval, $q > q^*$, the structure looks inhomogeneous, as in dilute solutions. Thus, we find that the boundary between dilute and semi-dilute depends on the value of q at which the observation is made.

4.4.5 Variation of the screening length with concentration

When the concentration ρ increases, the screening length decreases, and this is one of the most revealing facts of the structure of 'good' semi-dilute solutions.

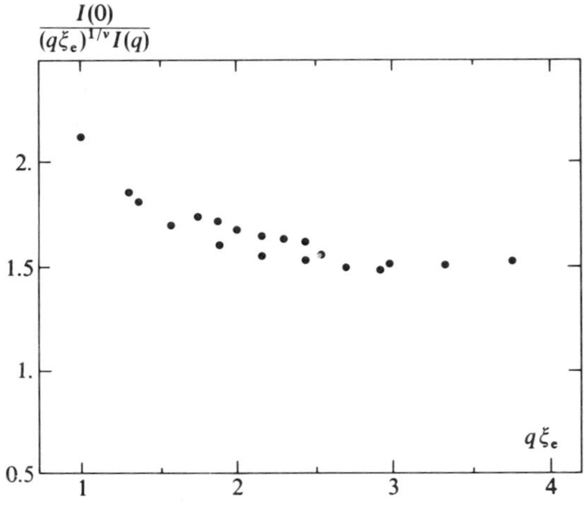

Fig. 15.26. Plot of the numbers $\frac{I(0)}{(q\xi_e)^{1/\nu}I(q)}$ against $q\xi_e$. Values taken from Fig. 15.25. For $q\xi_e \geqslant 2.6$, the structure function of the chains has about the same behaviour as the form function of a chain.

Theory defines the Kuhnian overlap length by the equality [see (15.4.34)]

$$\xi_k = X(CX^d)^{-\frac{\nu}{\nu d - 1}} \tag{15.4.38}$$

where $X^2 = R^2/3 = 2R_G^2/\aleph$ is determined in the limit $C \to 0$. Besides, [see (15.4.35)] in the asymptotic limit

$$\xi_e = \Gamma \xi_k$$

where Γ is a universal constant. Hence, the decrease of ξ_e with C is very different from the law $\xi_e \propto C^{-1/2}$ characteristic of the simple-tree approximation.

Let us examine the experimental results. Daoud *et al.*,[20] and thereafter Wiltzius, Haller, Cannell, and Schaeffer[24] made systematic measurements of ξ_e as a function of ρ for semi-dilute solutions of polystyrene (neutron and X-ray scattering). Figure 15.27 shows the results of Wiltzius *et al.* The quotients $\xi_e/R_{G,Z}$ are brought on the vertical axis, where $R_{G,Z}$ is the radius of gyration in the limit $\rho \to 0$. The overlap ratios are brought on the horizontal axis. These ratios are determined by the measurement of the second virial coefficient [see (15.4.3)]

$$CX^3 = \frac{A_2 \rho M}{1.83}.$$

In reality, the experiment gives the average $A_{2,\sim}$ and M_W, but we shall neglect polydispersion effects here. It is necessary to check that ξ_e has the required asymptotic behaviour. We note that this behaviour is already reached when $CX^3 \geqslant 1$.

Let us determine the value of the coefficient $\Gamma = \xi_e/\xi_k$. The definitions given above show that

$$\frac{\xi_e}{R_G} = \frac{\Gamma \xi_k}{X\sqrt{\aleph/2}} = \frac{\Gamma}{\sqrt{\aleph/2}}(CX^3)^{-\frac{\nu}{\nu d - 1}}. \tag{15.4.39}$$

The value taken from Fig. 15.27 is $\Gamma = 0.18$; the result is the same as in Section 4.4.3.

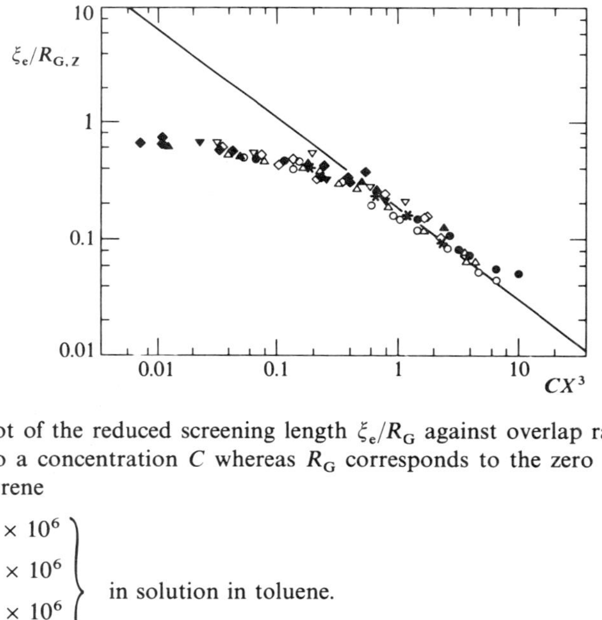

Fig. 15.27. Plot of the reduced screening length ξ_e/R_G against overlap ratio. (Here, ξ_e corresponds to a concentration C whereas R_G corresponds to the zero concentration limit.). Polystyrene

○ = $M_W = 26 \times 10^6$
△ = $M_W = 7.2 \times 10^6$
◇ = $M_W = 1.8 \times 10^6$
▽ = $M_W = 0.3 \times 10^6$
} in solution in toluene.

● = $M_W = 26 \times 10^6$
▲ = $M_W = 7.2 \times 10^6$
∗ = $M_W = 3.8 \times 10^6$
◆ = $M_W = 1.8 \times 10^6$
▼ = $M_W = 0.3 \times 10^6$
} in solution in methylethylcetone.

Results obtained by neutron or X-ray scattering (from Wiltzius *et al.*[24])
——asymptotic behaviour in good solvent.

4.4.6 Variation of the screening length with temperature

The study of the dependence of the screening length on temperature is also very significant. We know that in the semi-dilute regime, the screening length ξ_e is proportional to the Kuhnian overlap length. By definition

$$\xi_k = (CX^{1/\nu})^{-\frac{\nu}{\nu d - 1}}$$

where X defines the size of the chain in the limit $C \to 0$. Now X increases with temperature T; as a result, in the semi-dilute range, ξ_k decreases with T and the same holds true with ξ_e since then

$$\xi_e = \Gamma \xi_k = \Gamma (CX^{1/\nu})^{-\frac{\nu}{\nu d - 1}}. \tag{15.4.40}$$

Let us now examine the experimental results. Cotton *et al.*[21] measured the screening length ξ_e associated with polystyrene ($M_W = 1.4 \times 10^6$) solutions in deuterated cyclohexane, at fixed concentration ρ, for different temperatures. The result is seen in Table 15.14. We verify that ξ_e decreases as temperature increases.

Moreover, using the values of this table, we are going to verify that, in good solvent and in the overlap regime, the screening length obeys eqn (15.4.40). We shall observe that below a certain temperature this law ceases to be valid, as predicted from theory.

Table 15.14

	T[K]	ξ_e[nm]
	342.7	4.74
	340.05	4.82
	337.55	4.94
	335.25	5.06
	332.65	5.13
$\rho = 58$[g/l]	330.15	5.3
	327.6	5.42
	325.4	5.63
	322.9	5.98
	320.4	6.32
	319.25	6.45
	318.1	6.74
	315.35	7.35
	314.40	7.67
	313.25	8.16
	312.35	8.78
	311.35	9.13

For this, let us relate the screening length to the two-body interaction b. By definition, we have,

$$\xi_e = \lceil C^{-\frac{\nu}{\nu d - 1}} X^{-\frac{1}{\nu d - 1}}.$$

and [see (15.4.15)],

$$X^2 = \frac{R^2}{3} \propto b^{2(2\nu - 1)}.$$

Hence,

$$\xi_e \propto b^{-\frac{2\nu - 1}{\nu d - 1}}.$$

We have already verified that, in good solvent, interaction b varies linearly with $1/T$, for T close to T_F [see (15.3.5]. Therefore, for a good solvent and for an overlapping solute, we expect the quantity $\xi_e^{-\frac{\nu d - 1}{2\nu - 1}} \propto b$ to vary linearly with $1/T$.

In Fig. 15.28 we plotted the values of $\left(\frac{\xi_e}{{}^0R_G}\right)^{-\frac{\nu d - 1}{2\nu - 1}} \propto b$ against $1/T$. (Here ${}^0R_G$ does not depend on temperature.) Note the existence of an interval

$$1/T \leqslant 3.13 \times 10^{-3}\ \text{K}^{-1}$$

in which the points are aligned. Note also that the straight line interpolating these points intersects the horizontal axis at the point of abscissa

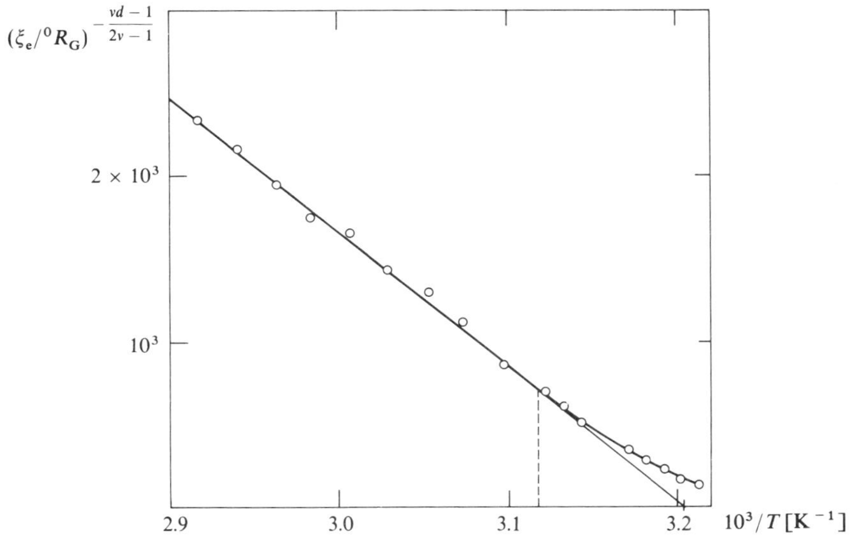

Fig. 15.28. Reduced screening lengths $\xi_e/{}^0R_G$ plotted against inverse temperature. Results obtained by neutron scattering (from Cotton *et al.*[21]). Polystyrene of mass $M_W = 1.4 \times 10^6$ in deuterated cyclohexane. Note that $1/T_F = 3.205 \times 10^{-3}$.

$1/T = 3.205 \times 10^{-3}\ K^{-1}$ (39 °C). This is the temperature T_F (39 °C) for non-deuterated polystyrene in deuterated cyclohexane, which is the system under investigation. There is therefore good agreement between experimental results and theoretical predictions.

Let us now examine the non-linear part of the curve represented in Fig. 15.28. The non-linearity appears at $1/T \geqslant 3.12 \times 10^{-3}\ K^{-1}$, i.e. below 48° C. We calculate the interaction parameter z corresponding to this temperature. By definition

$$z = bS^{1/2}(2\pi)^{-3/2}$$

where the interaction b is temperature-dependent.

For non-deuterated polystyrene in deuterated cyclohexane, which is studied here, the variation of b with T has not been determined experimentally. On the other hand, the dependence of b with respect to T is known for non-deuterated polystyrene in non-deuterated cyclohexane: it is given by eqn (15.3.5)

$$b = 2.918 - \frac{896}{T}\,\text{nm}^{-1}.$$

For the solutions studied in this section, we shall assume that the interaction has a dependence similar to that given above, the only difference being that the temperature T_F at which b vanishes is 39 °C (instead of 34 °C for the entirely non-deuterated system). In this manner, we have here

$$b = 2.872 - \frac{896}{T}\,\text{nm}^{-1}.$$

At 48 °C ($T = 321$ K) and for $S = 1800\ \text{nm}^2$, we find $z = 0.28$. This value of the parameter z should correspond to the crossover between *good* and *poor* solutions, for the mass-concentration in the experiment $\rho = 58$ g/l. Let us check this assertion on the Daoud and Jannink type diagram, represented on Fig. 13.26 (p. 000). For $\rho = 58$ g/l and $S = 1800\ \text{nm}^2$, one has

$$CS^{3/2} = 2.4.$$

On the diagram, the point of ordinate $z = 0.28$ and abscissa $CS^{3/2} = 2.4$ is close to the crossover line $C = C^{**}$, and, furthermore, formula (13.2.15) gives $C^{**}S^{3/2} = 2.2$ for $z = 0.28$, which is satisfactory.

4.4.7 Structure function at zero angle: osmotic compressibility

For a monodisperse solute, the structure function $H(\vec{0})$ is directly related to the osmotic compressibility

$$H(\vec{0}) = \frac{1}{\rho}\frac{\partial \rho}{\partial(\Pi\beta)} \tag{15.4.41}$$

and we may in this case test the law which relates compressibility and concentration, by radiation scattering in the limit of zero angle. Such an experiment somehow duplicates the one in which the osmotic pressure is measured mechanically (see Section 4.3). (Let us emphasize, however, that a sample is never monodisperse and that in this case $\boldsymbol{H}(\vec{0})$ does not remain strictly equal to the osmotic compressibility.) Equation (15.1.17) shows that the scattered intensity in the limit of zero angle is given by

$$I = \lim_{q \to 0} I(q) = K\rho^2 \text{Ⓐ} \boldsymbol{H}(\vec{0}).$$

We shall need to know the apparatus constant related to the instrument and to the sample. Extrapolation will be made from values measured in the asymptotic domain ($qR_G > 1$). In this manner, we avoid difficulties related to the Picot–Benoit effect (see Fig. 15.21, Section 4.4).

We shall examine the results in the representation

$$\frac{K\rho M}{I} = \frac{M}{RT}\frac{\partial \Pi}{\partial \rho}$$

as a function of the overlap ratio CX^d. Wiltzius *et al.*[24] made systematic measurements of the osmotic compressibility by neutron and X-ray scattering. Their results are shown in Fig. 15.29. The curve obtained in this way is very close to the one shown in Fig. 15.19, in which $\frac{M\Pi}{RT\rho}$ is plotted against CX^d.

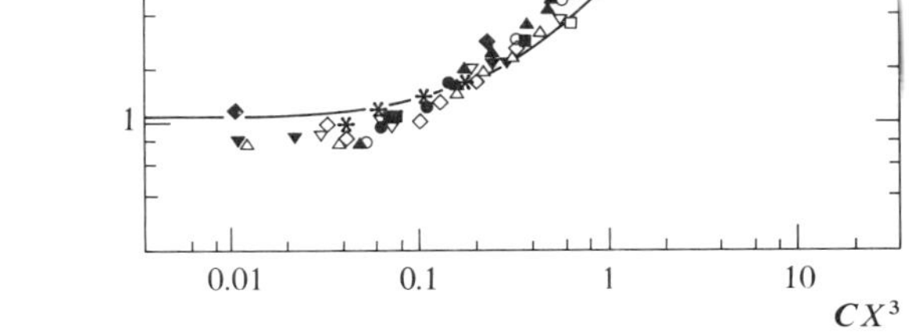

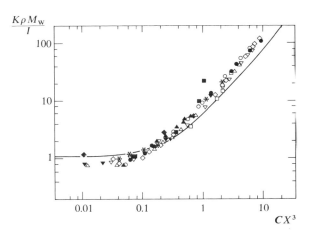

Fig. 15.29. Osmotic compressibility of polystyrene solutions, obtained by neutron and X-ray scattering, as a function of overlap ratio (from Wiltzius *et al.*[24]). Samples identical to those of Fig. 15.29.
—— calculated value
This figure is to be compared with Fig. 15.19.

It would be interesting to study the effect of polydispersion on the structure function and to compare them with those which show up in the osmotic pressure Π, but the calculation of these effects has not yet been made.

5. SOLUTIONS OF LONG CHAINS IN A MELT OF CHEMICALLY IDENTICAL BUT SHORTER CHAINS

For all experiments described in the preceding chapters, the polymer samples have been prepared in order to be as monodisperse as possible: the corresponding experimental results enable us to verify the variation laws predicted by theory.

However, there are other types of experimental tests of the theory of polymers in solution: these are experiments in which the chemically homogeneous solute is a mixture of two (or more) samples whose polymer chains have very different average sizes. However, we cannot test the theories of strongly polydisperse solutions as we did in the weakly polydisperse case i.e. by referring to a monodisperse solute of molecular mass equal to the average molecular mass of the sample. Polydispersion here becomes an essential parameter.

5.1 Structure function of a binary mixture of chemically identical chains with different lengths

The scattering cross-section per unit volume of an incompressible binary mixture is written as [see (15.2.37)]

$$\Xi(q) = (\mathfrak{b}_1 - \mathfrak{b}_0)^2 C_1^2 H_{11}(\vec{q}) \tag{15.5.1}$$

where the indices 0, 1 refer to the two species, and we note that

$$C_1^2 H_{11}(\vec{q}) = C_0^2 H_{00}(\vec{q})$$

for symmetry reasons [see (7.2.30)]

We then consider a mixture of rather short chains denoted by index zero and of longer chains denoted by index 1. The monomers constituting these chains are chemically identical but isotopically different. Hence, different collision lengths, $\mathfrak{b}_0$ and $\mathfrak{b}_1$ are associated with the two types of monomers. We note that in the limit $C_1 \to 0$ (see Chapter 7, Section 3.1.1)

$$\frac{(\mathfrak{b}_1 - \mathfrak{b}_0)^2}{\Xi(q)} \simeq \frac{1}{C_1^2 H_{11}^{\mathrm{I}}(\vec{q})} = \frac{1}{C_1 N_1 H_1(\vec{q})} \tag{15.5.2}$$

and the aim of the experiment is to determine the form function of a long chain isolated among shorter chains. In particular, from the experiment we can get the radius of gyration R_G of an isolated long chain.

We have shown, in Section 1.3.1, the manner in which the Zimm representation makes it possible to extrapolate experimental results, in the limit $C_1 \to 0$, by using the simple-tree approximation [see (13.2.115)]

$$\frac{(\mathrm{b}_1 - \mathrm{b}_0)^2}{\Xi(q)} = \frac{1}{C_1^2 \boldsymbol{H}^{\mathrm{I}}_{11}(\vec{q})} + \text{constant.} \tag{15.5.3}$$

However, there exists another representation of results which is better for studying blends of two types of chains. The approximate formula is written as

$$\frac{(\mathrm{b}_1 - \mathrm{b}_0)^2}{\Xi(q)} = \frac{1}{C_1^2 \boldsymbol{H}^{\mathrm{I}}_{11}(\vec{q})} + \frac{1}{C_0^2 \boldsymbol{H}^{\mathrm{I}}_{00}(\vec{q})}. \tag{15.5.4}$$

This simple formula can be justified in the following ways:

(1) It is symmetrical with respect to the indices zero and one.

(2) For $C_1 \to 0$, it again gives (15.5.3).

(3) When the chains are identical.

$$\boldsymbol{H}^{\mathrm{I}}_{11}(\vec{q}) = \frac{N}{C_1} H(\vec{q}) \qquad \boldsymbol{H}^{\mathrm{I}}_{00} = \frac{N}{C_0} H(\vec{q})$$

we find

$$\Xi(q) = (\mathrm{b}_1 - \mathrm{b}_0)^2 \frac{C_1 C_0}{C_1 + C_0} N\, H(\vec{q}) \tag{15.5.5}$$

in agreement with (7.2.30)

Formula (15.5.4) can also be derived by using the self-consistent approximation introduced by de Gennes.[25]

In the samples used for the scattering experiments the shorter chains strongly overlap; consequently, we shall assume that these chains are Brownian and that their form function is known. Hence, formula (15.5.4) contains only one unknown quantity $\boldsymbol{H}^{\mathrm{I}}_{11}(\vec{q})$ and it can be used to represent experimental results even for rather high concentrations of long chains.

5.2 Swelling of a long chain in a liquid of shorter chains

Let us consider the states of a chain made of N_1 monomers, in solution successively in its own monomer, in its dimer, in its trimer, etc. In the first case, the repulsive interaction between monomers of the chain produces a gyration swelling $\mathfrak{X}_G$ which obeys the laws already observed. In the other extreme case where the test chain finds itself in a melt of other identical chains, the repulsive interaction is screened and the test chain is in a quasi-Brownian state. In intermediate situations the swelling $\mathfrak{X}_0$ of the test chain of N_1 monomers varies with the number N of monomers (per chain) of the solvent chains. Observation of the variation law of $\mathfrak{X}_G$ with N clearly reveals the structure of polymer solutions. Such observation requires the labelling of the chain made of N_1 monomers: therefore it is appropriate to use neutron scattering. Kirste and Lehnen[2] made the experiment with blends of polydimethylsiloxane (PDMS)

Table 15.15

Sample	Solvent M_W	Solution $S_1^{1/2}/S$ [nm^{-1}]	Solute $R_{G,Z}^2$ [nm^2]	Solute $\mathfrak{X}_{G,Z}$	Solute $10\, A_{2,\sim}$ [cm^3/g^2]	Solute z_{eff}	Solution $z_{eff}S/S_1^{1/2}$ [nm]
1	180	70.2	361	2.88	4.6	6.24	0.089
2	3000	4.21	275.5	2.20	2.8	2.60	0.62
3	9600	1.31	182.25	1.45	0.87	0.57	0.435
4	5×10^4	0.25	169	1.35	0.54	0.42	1.67
5	2.5×10^5	0.05	130.9	1.043	-0.22		

chains, and they observed a variation of the swelling $\mathfrak{X}_G$, as predicted. The chain, whose radius of gyration is measured by scattering, is a polydimethylsiloxane chain of molecular mass $M_W = 2 \times 10^5$. The solvent chains are made of deuterated polydimethylsiloxane. The labelling is isotopic, and therefore the blend should remain homogeneous. The molecular masses of each sample used successively as a solvent, are reported in Table 15.15.

Kirste and Lehnen[2] determined the 'Z' average of the square radius of gyration, $R_{G,Z}^2$ of the long chain, in the limit of zero concentration, using the classical interpolation formula (15.5.3); see also Fig. 15.1. (The more adequate formula (15.5.4) was unknown to them.) In an earlier experiment, they also measured the 'Z' average of the square radius of gyration, ${}^0R_{G,Z}^2$, of the same chains in the quasi-Brownian state: this is the state in which polydimethylsiloxane chains are found when dispersed in dilute solution in bromocyclohexane at temperature T_F (29 °C). The authors quoted above obtained ${}^0R_{G,Z}^2 = 125.44\ nm^2$, i.e. an equivalent area $S_1 \simeq 2({}^0R_{G,Z}^2) = 250.88\ nm^2$. From this value, we get the conversion factor [see (15.2.1)]

$$ {}^0\mathit{\Lambda}_{Z,W} = {}^0R_{G,Z}^2/M_W = 6.24 \times 10^{-4}\ nm^2/dalton. $$

Hence, the equivalent Brownian area S (per chain) for the solvent chains may be calculated with the help of the relation $S \simeq 2\,{}^0\mathit{\Lambda}_{Z,W} M_W = 12.54 \times 10^{-4}\, M_W\, nm^2$.

Since the polydispersity of the sample has not been given by the authors, we have not been able to make the required corrections as in the case of polystyrene solution in benzene (Section 2.1).

The results of the experiment in itself, carried out at a temperature of 20 °C, are shown in Table 15.15; they are the quantities $R_{G,Z}^2$ and $A_{2,\sim}$ associated with the chain of equivalent area S_1, for various solvents of area S. From the radiation swelling

$$ \mathfrak{X}_{G,Z}(S_1, S) = R_{G,Z}^2/{}^0R_{G,Z}^2 $$

we derived the interaction parameter $z(S_1, S)$ with the help of the function plotted in Fig. 15.7. This parameter is denoted by z_{eff}. The theory (Chapter 13,

Section 2.10) has only been worked out for a long chain dissolved in a semi-dilute solution of short chains with an area-concentration $\mathscr{C}$. In this case [see (13.2.159)]

$$z_{\text{eff}} = \frac{1}{\mathscr{C}S} S_1^{1/2}(2\pi)^{-3/2}. \tag{15.5.6}$$

Nevertheless, we shall apply this formula to the present case, identifying polymer melt and semi-dilute solution. Let ρ_L then be the mass per unit volume of polymer melt. By definition

$$\rho_L = \frac{C_L M}{\text{Ⓐ}} = \frac{\mathscr{C}_L}{2\,{}^0\!\Lambda\,\text{Ⓐ}} \tag{15.5.7}$$

where ${}^0\!\Lambda$ is the molecular mass–area conversion factor (see above). Hence, setting $\mathscr{C} = \mathscr{C}_L$, we get from (15.5.6)

$$z_{\text{eff}} = \frac{1}{2\,{}^0\!\Lambda\text{Ⓐ}\,\rho_L}\,\frac{S_1^{1/2}}{S}\,(2\pi)^{-3/2}. \tag{15.5.8}$$

The quantity

$$z_{\text{eff}}\frac{S}{S_1^{1/2}} = \frac{1}{2\,{}^0\!\Lambda\,\text{Ⓐ}\,\rho_L}(2\pi)^{-3/2} = 0.084$$

should be independent of the equivalent Brownian area S of the solvent chains. The experiment made by Kirste and Lehnen[2] (see Table 15.15) indeed shows the existence of a range of values $S_1^{1/2}/S$, in which this quantity does not vary appreciably (as compared to S_1/S).

REFERENCES

1. Zimm, B.H. (1948). *J. Phys. Chem.* **16**, 1093.
2. Kirste, R.G. and Lehnen, B.R. (1976). *Makromol. Chem.* **177**, 1137.
3. Miyaki, Y., Einaga, Y., and Fujita, H. (1978). *Macromolecules* **11**, 1180.
4. Nierlich, M. and Cotton, J.P. (1980). Results presented at the 2nd CNRS–NSF meeting on macromolecules, Strasbourg.
5. Noda, I., Imai, M., Kitano, T., and Nagasawa, M. (1983). *Macromolecules* **16**, 425.
6. Fukuda, M., Fukutomi, M., Kato, Y., and Hashimoto, T. (1974). *J. Polym Sci. Polym Phys.* **12**, 871.
7. Nose, T. and Chu, B. (1979). *Macromolecules* **12**, 1122.
8. Yamamoto, A., Fuji, M., Tanaka, G., and Yamakawa, H. (1971). *Polym. J.* **2**, 799.
9. Tangari, C., King, J.S., and Summerfield, G.C. (1982). *Macromolecules* **15**, 132.
10. Yoon, D.Y. and Flory, P.J. (1976). *Macromolecules* **9**, 294.
11. Rawiso, M. and Picot, C. (1982). In *Static and dynamic properties of the polymer solid state* (eds. R. Pethrick and R. Richards), p.156. NATO Advanced Study Institute. Reidel Publishing Co.
12. Farnoux, B., Boué, F., Cotton, J.P., Daoud, M., Jannink, G., Nierlich, M., and de Gennes, P.G. (1978). *J. Physique* **39**, 77.
13. Weill, G. and des Cloizeaux, J. (1979). *J. Physique* **40**, 99.

14. Cotton, J.P. (1980). *J. Physique Lett.* **41**, L-231.
15. Loucheux, C., Weill, G., and Benoit, H. (1958). *J. Chim. Phys.* **55**, 540.
16. Farnoux, B. (1976). *Ann. Fr. Phys.* **1**, 73.
17. Rawiso, M., Duplessix, R., and Picot, C. (1987). *Macromolecules* **20**, 630.
18. Matsushita, Y., Noda, I., Nagasawa, M., Lodge, T.P., Amis, E.J., and Han, C.C. (1984). *Macromolecules* **17**, 1784.
19. Benoit, M., and Picot, C. (1966). *Pure Appl. Chem.* **12**, 545.
20. Daoud, M., Cotton, J.P., Farnoux, B., Jannink, G., Sarma, G., Benoit, H., Duplessix, R., Picot, C., and de Gennes, P.G. (1975). *Macromolecules* **8**, 804.
21. Cotton, J.P., Nierlich, M., Boué, F., Daoud, M., Farnoux, B., Jannink, G., Duplessix, R., and Picot, C. (1976). *J. Chem. Phys.* **65**, 1101.
22. Richards, R.W., Maconnachie, A., and Allen, G. (1981). *Polymer* **22**, 158.
23. Noda, I., Kato, N., Kitano, T., and Nagasawa, M. (1981). *Macromolecules* **14**, 668.
24. Wiltzius, P., Haller, H.R., Cannell, D.S., and Schaeffer, D.W. (1983). *Phys. Rev. Lett.* **51**, 1183.
25. de Gennes, P.G. (1979). *Scaling concepts in polymer physics*, Chapter IX, p. 258. Cornell University Press.

16

PARTIALLY ATTRACTIVE CHAINS: EXPERIMENTAL RESULTS

GENESIS

The behaviour of polymer solutions at various temperatures shows the existence of a demixtion phenomenon similar to that observed with binary mixtures of small molecules. Moreover, specific properties have been noted, characterizing the polymer state in poor solvent, and the progressive collapse of chains in dilute solutions when the temperature is lowered.

A systematic experimental study of demixtion was undertaken by Japanese physicists in the 1970s. They recognized the manifestations of a critical behaviour at the top of the demixtion curve and they verified that the critical exponents belong to the class $n = 1$, $d = 3$. Thus two different types of critical phenomena were experimentally studied for polymer solutions, at almost the same time; on one side, the French school (the 'Strasacol' team: Strasbourg, Saclay, Collège de France) observed the main phenomena pertaining to the class $n = 0$; on the other side, the Japanese school observed phenomena related to the class $n = 1$. The fact that for the same solution, there exist different situations in which each one of these critical phenomena appears independently without obvious contamination, is quite remarkable. However, all these results are related to one another and at the present time, physicists are accumulating experimental data which characterize the tricritical state defined as the confluence of two 'critical' lines ($1/N = 0$ and $T = T_c(N)$). Incidentally, let us recall that results of computer simulations must be included among these data.

The distinction between poor and good solvent was introduced in the 1950s by Fox and Flory after experimental studies of the intrinsic viscosity of polymer solutions. These authors recognized that the viscosity varies in relation to the dependence of the chain sizes on temperature; the poor solvent state is the state of a solution in which the chains have quasi-Brownian configurations. Systematic experiments have been made in this domain, for instance to determine the Flory temperature, but they have never given very precise results. Physicists are just now beginning to overcome the experimental and theoretical difficulties. Experiments have been made to show the existence of a collapse of the polymer chain, and certain authors have been prone to compare it with the coil–globule transition in proteins.

1. DETERMINATION OF THE BASIC PARAMETERS OF POLYSTYRENE IN CYCLOHEXANE AT A TEMPERATURE CLOSE TO T_F

Let us consider a solution of partially repulsive chains. The polystyrene–cyclohexane solutions in the vicinity of 34 °C (temperature T_F) are of this type, and have been studied a great deal. We shall interpret the observations with the help of the basic parameters of the standard continuous model, which are the two-body interaction b and the three-body interaction c. In the sequel (Section 2.2), we also use the osmotic parameters g and h which are more adapted to a study of the tricritical state.

Here, the two-body interaction b becomes attractive whereas the three-body interaction remains repulsive. In Chapter 15, Section 3.1.1, we determined how b varies with temperature, for the polystyrene–cyclohexane system. We obtained the result

$$b = 2.918 - 896/T \text{ nm}^{-1} \tag{16.1.1}$$

by measuring the T dependence of the swelling of a chain, in the zero concentration limit.

We shall assume that this expression remains valid when it is extrapolated to the poor solvent domain. Thus, $b = 0$ for $T = T_F$ (34.0 °C) and $b < 0$ for $T < T_F$. This does not really mean that the 'true' two-body interactions vanish for $T = T_F$ and that at $T = T_F$ there are only three-body interactions. Actually, the interaction b is a *bare* interaction but is nevertheless an interaction which is *additively renormalized* (see Chapter 14, Section 6) and its value depends partly on the 'true' three-body interactions. Incidentally, this remark also applies to good solutions; for this reason, if b is measured in good solvent, it is legitimate to extrapolate the result thus obtained to determine the value of b in poor solvent. Thus, in perturbation theory, when the diagram contributions are calculated by dimensional regularization, we may say that $b = 0$ for $T = T_F$!

The law (16.1.1) has been established by observing the swelling of the chain in good solvent and by interpreting the experiments in the framework of a one-parameter theory, this parameter being

$$z = bS^{1/2}(2\pi)^{-3/2}$$

where S is the equivalent Brownian area.

In a poor solvent, one parameter is no longer sufficient and now the three-body terms must be taken into account, not only implicitly but also explicitly. The second basic parameter required to study chains in the vicinity of T_F is the repulsive three-body interaction c. The comparison made in Chapter 14 between the Flory–Huggins theory and the continuous model suggests calculating c by the formula

$$c = v_r^2 \frac{N^3}{S^3}$$

where v_r is the volume of a monomer (volume per site on a lattice), a quantity which we shall identify with the partial volume v of solute per monomer.

The value of v_r for the polystyrene–cyclohexane system is given by Fig. 5.15 (Chapter 5)

$$v_r = 0.153 \text{ nm}^3.$$

On the other hand,

$$S = 2\,{}^0\Lambda M = 2\,{}^0\Lambda N m \qquad (16.1.2)$$

where m is the molecular mass of a monomer ($m = 104$ for styrene) and where ${}^0\Lambda$ is the area–molecular mass conversion-factor (see Chapter 14, Section 2.2). Thus,

$$c = \frac{v_r^2}{8\,{}^0\Lambda^3 m^3}$$

and therefore for polystyrene in cyclohexane

$$c = 6.8. \qquad (16.1.3)$$

Then, the corresponding parameter y is

$$y = \frac{c}{(2\pi)^3} = 0.027. \qquad (16.1.4)$$

This estimation of c is rather arbitrary. In the following section, the value of c will be estimated in a more realistic manner and a very different result will be found. To do this, we shall study how the top of the demixtion curve varies with temperature. Subsequently, using the value of c thus determined, we shall interpret experiments concerning the collapse of polymer chains in very dilute solutions.

2. DEMIXTION CURVES; GENERAL FEATURES

Demixtion is one of the clearest manifestations of the attractive forces between molecules of a solute, but it is sensitive to polydispersion effects. Therefore, demixtion is studied experimentally by using samples which are as monodisperse as possible.

The theory expounded in Chapter 14 (Sections 5 and 6) shows us that at a temperature lower than T_F, it is possible to find two mass concentrations $\rho^{|}(T)$ and $\rho^{||}(T)$, for which the chemical potentials of the solute molecules are equal,

$$\mu[\rho^{|}(T)] = \mu[\rho^{||}(T)],$$

and for which the osmotic pressures are also equal

$$\Pi[\rho^{|}(T)] = \Pi[\rho^{||}(T)].$$

The curve representing simultaneously the functions $\rho^{|}(T)$ and $\rho^{||}(T)$ which obey these equalities is by definition the demixtion curve.

Numerous experiments have been carried out to determine demixtion curves for various molecular masses of the solute. We shall begin with a description of the experimental methods. We shall then compare the results thus obtained with the predictions made in Chapter 14.

2.1 Experimental determination of a demixtion curve

The principle of the measurements is as follows. Let us consider a polymer solution containing polymers of average molecular mass M_{W}; the mass concentration of polymers is ρ and it is assumed that ρ is close to the overlap concentration

$$\rho^{*}_{\mathrm{G,WZ}} = \frac{M_{\mathrm{W}}}{(2\,{}^{0}R^{2}_{\mathrm{G,Z}})^{3/2}\,Ⓐ} \simeq \frac{M_{\mathrm{W}}}{S^{3/2}\,Ⓐ} \tag{16.2.1}$$

where ${}^{0}R^{2}_{\mathrm{G,Z}}$ is the 'Z' average of the square of the radius of gyration at temperature T_{F}, and where S is the area of the equivalent Brownian chain. We are in poor solvent; for instance, we use a solution of polystyrene in cyclohexane at a temperature $T_{\gamma\mathrm{F}}(S)$ defined by the condition

$$A_2(S) = 0$$

[see (14.2.12)].

The temperature of the solution is reduced to an arbitrary temperature T, at which demixtion may occur and is to be observed. Precautions have to be taken so that the system reaches the state of equilibrium at the end of the experiment. Thus, the cooling must be slow, and during this time, the solution must be kept continually stirred, until temperature T is reached. Gradually, a change in the appearance of the solution can be observed. It becomes milky, and finally after about 24 hours a meniscus appears, separating the solution into two different parts. This is a phase separation. The concentrations $\rho^{|}(T)$ and $\rho^{||}(T)$ ($\rho^{||} > \rho^{|}$) in the phases are now obtained by measuring the refraction indices of the two phases (for more information concerning the relation between the index of refraction and the mass-concentration, see Chapter 6, Section 4).

The whole process is repeated for various values of T. The set of points $\rho^{|}(T)$, $\rho^{||}(T)$ constitutes the demixtion curve. A characteristic example[1] of such a series of measurements is given by Fig. 16.1. Here, the solution consists of polystyrene of molecular mass $M_{\mathrm{W}} = 1.1 \times 10^5$ and of cyclohexane. The concentrations are expressed with the help of the mass fraction

$$\psi = \frac{\text{mass of solute}}{\text{mass of solution}}. \tag{16.2.2}$$

In Fig. 16.1, the middle of the points $\psi^{|}(T)$ and $\psi^{||}(T)$ was plotted for each value of T. These points define a curve which is nearly a straight line. This line cuts the demixtion curve at a point the coordinates ψ_{c}, T_{c} of which are the critical fraction and the critical temperature. The critical point is the top of the demixtion curve.

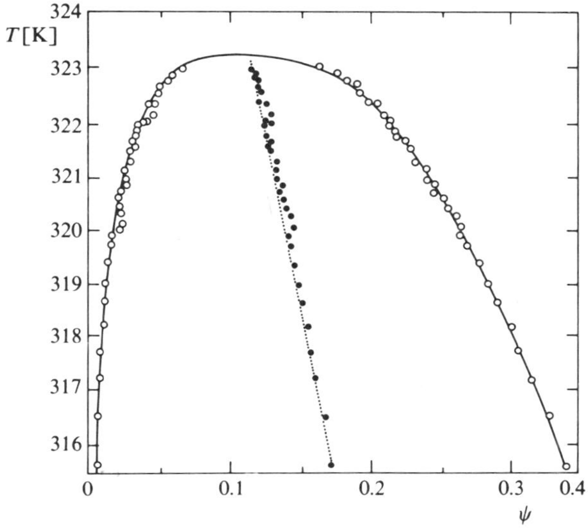

Fig. 16.1. Demixtion curve of a polystyrene solution ($M_W = 1.1 \times 10^5$) in cyclohexane. (From Shinozaki, Van Tan, Saito, and Nose.[1])
ψ is the mass fraction of solute;
○ measured point
● middle of the interval $\psi^{\text{II}} - \psi^{\text{I}}$
– visual guide
(the top of the demixtion curve should be on the prolongation of the dotted line; the drawing of the original figure is not perfect!)

The demixtion curves were measured for various molecular masses M_W. Figure 16.2 represents a result of experiments[2] made with polystyrene–cyclohexane solutions. Here the concentration is expressed with the help of the volume fraction

$$\varphi = \frac{\text{volume of solute}}{\text{volume of solution}}. \tag{16.2.3}$$

The volume of solute is the product of its mass $\mathscr{M}$ by the (average) partial volume per unit mass w_1 (see Chapter 5, Section 2.1). Thus,

$$\varphi = \rho w_1$$

where ρ is the mass-concentration.

We recognize in Fig. 16.2 characteristic features predicted theoretically in Chapter 14, Sections 5 and 6:

(1) the demixtion curve becomes more and more dissymmetrical when M_W increases;

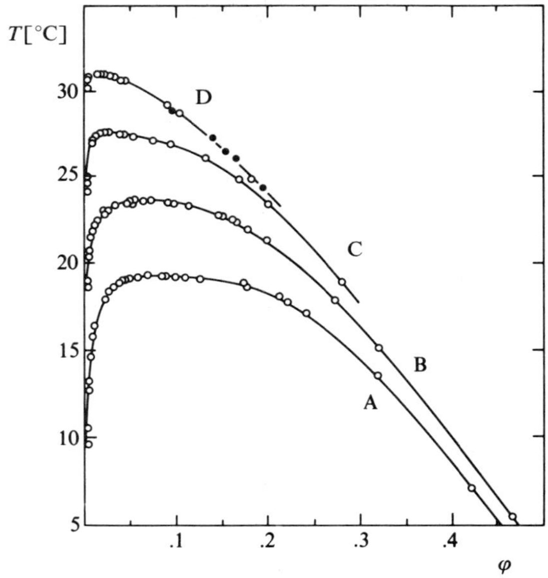

Fig. 16.2. Demixion curves for various solutions of polystyrene in cyclohexane. (From Perzynski[2].)
φ is the volume fraction of solute
A: $M_W = 4.36 \times 10^4$
B: $M_W = 8.9 \times 10^4$
C: $M_W = 2.5 \times 10^5$
D: $M_W = 1.27 \times 10^6$

(2) the curve exponentially approaches the vertical asymptote;

(3) the limit of the demixtion curves when $N \to \infty$ consists, on one side, of a half-line, and on the other side, of a curve which is nearly a half-line but still has a small downward curvature (see Fig. 14.21).

2.2 Demixtion curves and the Flory–Huggins approximation

The demixtion curves obtained experimentally (see Fig. 16.2) are in qualitative agreement with the theoretical predictions of Chapter 14.

Let us discuss these results in a more quantitative manner by using the Flory–Huggins approximation. We examine in particular how the coordinates T_c and φ_c of the top of the demixtion curve vary as functions of the equivalent Brownian area S of the chains. We can also replace the area S by the number N of links or the molecular mass M of the chain, all these quantities being proportional.

Let us recall that the solute is as monodisperse as possible. We suppose that the polydispersion is characterized by the ratio

$$\frac{M_W}{M_n} = 1.06 \tag{16.2.4}$$

where M_W, M_n are mean molecular masses defined in Chapters 1 and 15. Let m be the molecular mass of a monomer. We set

$$N = \left(\frac{M_W}{M_n}\right)\frac{M_n}{m} = 1.06\frac{M_n}{m} \tag{16.2.5}$$

and

$$N = \frac{S}{2\,{}^0\Lambda m} \tag{16.2.6}$$

where ${}^0\Lambda$ is the area–molecular-mass conversion-factor (see 15.2.4).

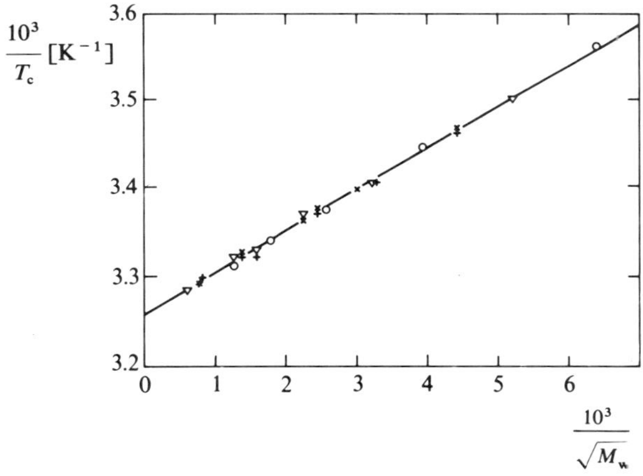

Fig. 16.3. The critical demixtion temperature is plotted against the molecular mass of the polymer, for polystyrene in cyclohexane (according to Perzynski[2]).

The straight line represents the equation

$$\frac{10^3}{T_c} = \frac{10^3}{T_F} + \frac{47}{\sqrt{M_W}} \qquad [\mathrm{K}^{-1}]$$

where $\frac{10^3}{T_F} = 3.256$ [K^{-1}] which gives

$$T_F = 307.13\ \mathrm{K} \qquad (34\,^\circ\mathrm{C}).$$

By converting M_W into a number of monomers, we obtain

$$\frac{10^3}{T_c} = \frac{10^3}{T_F} + \frac{4.57}{\sqrt{N}} \qquad [\mathrm{K}^{-1}]$$

The values of T_c deduced from Fig. 16.3 and the other experimental values of T_c concerning the polystyrene–cyclohexane system obey the empirical relation

$$\frac{10^3}{T_c} = 3.257 + \frac{4.57}{\sqrt{N}}. \tag{16.2.7}$$

The Flory–Huggins theory and the simple-tree approximation in the standard continuous model lead to formulae that are in both cases similar to (16.2.7). Let us consider the parameter $(\frac{1}{2} - \chi)$ of the Flory–Huggins theory (see Chapter 14, Section 5.3). The correspondence with the continuous model is given by (14.5.13)

$$\tfrac{1}{2} - \chi = \tfrac{1}{2} b S^{1/2} N^{-1/2} c^{-1/2}. \tag{16.2.8}$$

Thus, the parameter $(\frac{1}{2} - \chi)$ varies with temperature in the same manner as the parameter b. At T_c, we have $b = b_c$ and moreover [see (14.5.4)]

$$\frac{1}{2} - \chi_c = -\frac{1}{\sqrt{N}} - \frac{1}{2N}. \tag{16.2.9}$$

From the preceding expressions, we obtain

$$b_c = -2\left(1 + \frac{1}{2\sqrt{N}}\right) S^{-1/2} c^{1/2}. \tag{16.2.10}$$

On the other hand, the simple-tree approximation leads to the simpler formula

$$b_c = -2S^{-1/2} c^{1/2}.$$

Now, let us express b_c in terms of T_c with the help of (16.1.1)

$$b_c = 2.918 - 896/T_c. \tag{16.2.11}$$

Let us compare (16.2.10) and (16.2.11); taking (16.2.6) into account, we obtain

$$\begin{aligned} \frac{10^3}{T_c} &= 3.257 + \left(1 + \frac{1}{2\sqrt{N}}\right) \frac{2c^{1/2}}{0.896\, S^{1/2}} \\ &= 3.257 + \frac{1}{\sqrt{N}}\left(1 + \frac{1}{2\sqrt{N}}\right) \frac{2c^{1/2}}{0.896(2\ {}^0\!\Lambda m)^{1/2}}. \end{aligned}$$

Let us introduce the numerical values

$${}^0\!\Lambda = 8.3 \times 10^{-4}\ \mathrm{nm^2/dalton}$$

$$m = 104.$$

Thus, we obtain

$$\frac{10^3}{T_c} = 3.257 + \left(1 + \frac{1}{2\sqrt{N}}\right) \frac{5.36\, c^{1/2}}{\sqrt{N}} \tag{16.2.12}$$

in the framework of the Flory–Huggins theory and

$$\frac{10^3}{T_c} = 3.257 + \frac{5.36\, c^{1/2}}{\sqrt{N}}, \tag{16.2.13}$$

in the framework of the simple-tree approximation.

In these expressions, the interaction c has been left as an adjustable parameter. Equation (16.2.13) has the same form as eqn (16.2.7) which interpolates the experimental results. Now, let us identify the coefficients of $1/\sqrt{N}$ in the two equations: we obtain the following value of the interaction c

$$c = 0.73, \tag{16.2.14}$$

and this value is smaller, by one order of magnitude, than our first estimate (16.1.3). The value (16.2.14) of c results from a comparison between a theoretical prediction and observed values. It will be used to interpret the other experiments described in this chapter. The corresponding value of the three-body interaction parameter is

$$y = c/(2\pi)^3 = 2.9 \times 10^{-3}. \tag{16.2.15}$$

Thus, the Flory–Huggins theory and the simple-tree approximation explain correctly how T_c varies with N. The situation for the critical volume fraction is different. Equation (14.5.4) gives

$$\varphi_c = \frac{1}{1 + \sqrt{N}}.$$

As can be seen from Fig. 16.4, the experimental values do not decrease as fast.

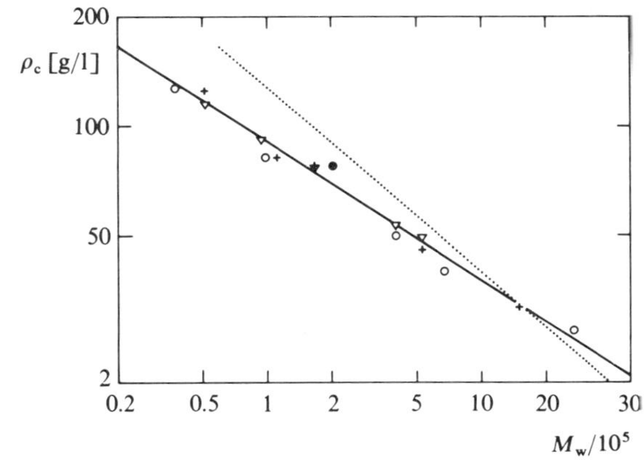

Fig. 16.4. Log–log plot of the critical concentration ρ_c as a function of the molecular mass M_W: polystyrene solutions in cyclohexane. (From Perzynski[2].)
—— equation $\rho_c = 6.8 \times M_W^{-0.38}$ (experimental result);
. . . . equation $\rho_c = 4.0 \times M_W^{-0.5}$ (Flory–Huggins theory).

Experimentalists represent the experimental results by a power law

$$\varphi_c \propto N^{-0.38}$$

but such a formulation has no theoretical basis.

Moreover, there is a discrepancy between the curvature at the top of the demixtion curve calculated in the framework of the Flory–Huggins theory (a locally parabolic curve; see Fig. 14.16) and the curvatures observed in Figs 16.1 and 16.2.

3. FLORY TEMPERATURE

The temperature T_F is related to the demixtion curve of polymer solutions, since

$$T_F = \lim_{S \to \infty} T_c(S) \qquad (16.3.1)$$

where $T_c(S)$ is the critical demixtion temperature associated with the top of the demixtion curve of a monodisperse solute whose equivalent Brownian area is S.

There are also special temperatures which are associated with a solvent–solute system of area S and which have the same limit T_F when $S \to \infty$. This fact reveals that the system is in a tricritical state.

3.1 Methods of measurement of T_F

1. T_F can be determined by measuring the temperature at which the two-body interaction b vanishes. We showed in Chapter 15 how values of b can be deduced from the swelling of a chain in *good* solvent. By letting the temperature T vary in good solvent, one obtains, in the zero concentration limit, a function $b(T)$ which is linear with respect to $1/T$ (see, for instance, Fig. 15.9). The extrapolation for $b \to 0$ gives a value of T_F defined by*

$$b(T_F) = 0.$$

2. The method which seems to give the best result consists in measuring the critical temperature $T_c(S)$ for various values of S (see, for instance, Fig. 16.2). The result T_F is obtained by extrapolation

$$T_F = \lim_{S \to \infty} T_c(S),$$

and for this purpose $1/T_c$ is plotted against $1/\sqrt{S}$ (see Fig. 16.3)

3. The most current method consists in determining the temperature $T_{\mathscr{V}F}(S)$ at which the second virial coefficient vanishes. This coefficient is obtained either by radiation scattering or by osmotic pressure measurements. The temperature

* The interaction $b(T)$ must be interpreted here as a renormalized interaction (see Section 1 and Chapter 14, Section 6).

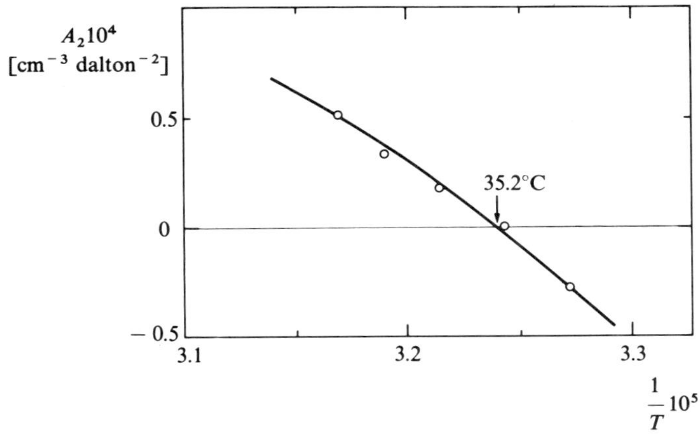

Fig. 16.5. Values of the second virial coefficient measured at various temperatures in the vicinity of $T_{\mathscr{V}F}(S)$. (From Strazielle and Benoit[4].)
Polystyrene in cyclohexane: $M_W = 1.28 \times 10^5$; $S = 200\ \text{nm}^2$.

$T_{\mathscr{V}F}$ for which $A_2(T_{\mathscr{V}F}) = 0$ is obtained by interpolation (see Fig. 16.5). Then, we have

$$T_F = \lim_{S\to\infty} T_{\mathscr{V}F}(S).$$

4. Another method consists in deducing T_F from measurements of the size of the chains at various temperatures and in the zero concentration limit. In principle, these sizes can be measured by light scattering. Actually, it turns out that it is more convenient to deduce them from measurements of the intrinsic viscosity

$$[\eta] = \lim_{\rho\to 0} \frac{\eta - \eta_0}{\rho\eta_0}$$

where η is the viscosity of the solution, η_0 the viscosity of the solvent and ρ the mass concentration (of the polymer). The intrinsic viscosity has the dimension of a volume per unit mass; it obeys the relation[3]

$$[\eta] \propto \frac{R_H R_G^2}{M}$$

where R_H is the hydrodynamical radius of an isolated chain and R_G its radius of gyration. The experiment is carried out as follows: using a series of samples with different molecular masses, an attempt is made to find a temperature T for which all the polymer solutions obey a relation of the form

$$[\eta] \propto M^{1/2}. \qquad (16.3.2)$$

This temperature is often identified with T_F.

Such an interpretation raises a problem. In fact, the theory shows that for each molecular mass M, there exists a temperature $T_{\mathscr{G}\,\infty\,\mathrm{F}}(S)$ (see Chapter 14, Section 6.9.6) for which an isolated chain is quasi-Brownian (here $S = 2\,{}^{0}\!\Lambda M$ is the equivalent Brownian area). Thus, as $T_{\mathscr{G}\,\infty\,\mathrm{F}}(S)$ depends on the area S, relation (16.3.2) cannot be satisfied for every M, at the same temperature. Actually, to define $T_{\mathscr{G}\,\infty\,\mathrm{F}}(S)$, we can replace (16.3.2) by the condition

$$\frac{1}{[\eta(M)]}\,\frac{\partial[\eta(M)]}{\partial M} = \frac{1}{2M}.$$

However, in practice, one tries to verify (16.3.2) for a certain number of samples and in this way an average temperature $\langle\!\langle T_{\mathscr{G}\,\infty\,\mathrm{F}}(S) \rangle\!\rangle$ can be obtained.

Let us also remark that the temperature $T_{\mathscr{G}\,\infty\,\mathrm{F}}$ was defined by the condition $\mathfrak{X}_0(S) = {}^{0}\!A_0(y)$ (see Chapter 14, Section 6.5.6) where $\mathfrak{X}_0(S)$ is the end-to-end swelling. Now, by viscosimetry, one measures a composite swelling which is nearly equal to

$$(R_G^2 R_H / {}^{0}R_G^2\, {}^{0}R_H)^{3/2}.$$

An expression of this swelling as a function of the parameters z, g, and h has not been calculated yet (in 1988), and for this reason we cannot precisely determine the value of z which, for the viscosity swelling, corresponds to the quasi-Brownian structure of chains of area S. Consequently, the temperature at which the chains have a quasi-Brownian behaviour cannot be determined rigorously by viscosimetry.

This discussion also applies to measurements of the gyration swelling by light scattering: the relation which enables us to test the quasi-Brownian state of a chain is

$${}^{0}R_G^2 = {}^{0}\!\Lambda M$$

where ${}^{0}\!\Lambda$ is the conversion-factor, and this relation can be criticized for the same reasons as (16.3.2).

In the following section, we give a few values of the average temperature $\langle\!\langle T_{\mathscr{G}\,\infty\,\mathrm{F}} \rangle\!\rangle$, measured by viscosimetry or by light scattering; however, these values cannot validly be compared with simple theoretical results.

3.2 Measured values of the temperatures $T_{\mathscr{V}\mathrm{F}}(S)$ and T_{F}

The temperatures T_{F} of the most current solvent–solute system were measured during the years 1960–70, and at that time the theories concerning polymer solutions in the vicinity of T_{F} were still inaccurate, if not erroneous. This explains why the experimentalists who determined T_{F} by using various methods did not obtain really concordant results. The renormalization theory applied to the tricritical state of polymer solutions explains the systematic discrepancies which the published results reveal. Let us, for instance, consider the following result obtained by Strazielle and Benoit in 1974[4] for different isotopes of polystyrene and of cyclohexane (see Table 16.1). Two remarkable facts appear in this table:

1. For S finite, the temperatures $T_{\gamma F}(S)$ are systematically *higher* than T_F.
2. The isotopic substitution of protons by deuterons in one constituent (but not in both of them simultaneously) modifies T_F appreciably.

The inequality

$$T_{\gamma F}(S) > T_F \tag{116.3.3}$$

is confirmed by older results collected in 1987 by Brandrup and Immergut[5] (see Table 16.2). However, this is contradictory to the results obtained in 1965 by Abe and Fujita[6] (see Fig. 16.6)

$$T_F\left(= \lim_{S \to \infty} T_{\gamma F}(S)\right) > T_{\gamma F}(S). \tag{16.3.4}$$

On the other hand, we note that the inequality (16.3.3) is in agreement with the tricritical theory of polymer solutions which applies in the vicinity of T_F [see (14.6.54)]. It would be good to have precise experimental results concerning these systems, which certainly have already been studied a great deal, but in a theoretical context which was inadequate. Concerning the sign of $T_{\gamma F}(S) - T_F$, we are tempted to trust the results contained in Tables 16.1 and 16.2 ($T_{\gamma F}(S) - T_F > 0$) because they are more recent than those which appear in Fig. 16.6. For $S = 200\ \text{nm}^2$, Table 16.1 gives

$$T_{\gamma F}(S) - T_F \simeq 0.8 \pm 0.8\ \text{K}. \tag{16.3.5}$$

Let us interpret this result in the context of the tricritical theory. We try to verify that this temperature difference is in agreement with the measured values of the parameters z and y (see Section 1).

For this purpose, let us calculate the temperature $T_{\gamma F}(S)$. The theory shows us [see (14.6.55)] that, at this temperature, the two-body interaction parameter

Table 16.1

System	$T_F = \lim_{S \to \infty} T_c(S)$ [°C]	$T_{\gamma F}$ for $S = 200\ \text{nm}^2$ [°C]
Polystyrene–cyclohexane	34.5°	35.2°
Deuterated polystyrene–cyclohexane	30°	30°
Deuterated polystyrene–deuterated cyclohexane	35.5°	36°
Polystyrene–deuterated cyclohexane	38.5°	40.2°

Table 16.2

System	Quantity whose measurement gives T_F	$T_F = \lim_{S\to\infty} T_c(S)$ [°C]	S[nm²]	$T_{\gamma F}(S)$ [°C]	Interval S[nm²]	$\langle\langle T_{\mathscr{G}\infty F}\rangle\rangle$ [°C]
Polystyrene–cyclohexane	A_2		300	34 °		
"	A_2		5300	34.4 °		
"	A_2			34.8 °		
"	A_2		2500	35 °		
"	T_c	34 °				
"	R_G				16000	34.5 °
"	[η]				5300	34.2 °
"	[η]					34.5 °
"	[η]					34.5 °
Polystyrene–methylcyclohexane	T_c	69.76 °				
Polydimethylsiloxane–bromobenzene	[η]					78.3 °
"	A_2			78.7 °		
Polydimethylsiloxane–cyclohexane	A_2			− 81 °		
"	[η]					− 68.3 °

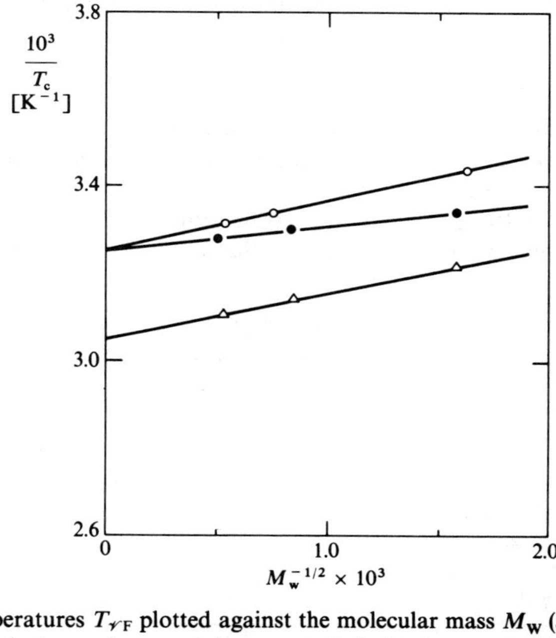

Fig. 16.6. Temperatures $T_{\gamma F}$ plotted against the molecular mass M_W (according to Abe and Fujita[6]). Polystyrene in: ● cyclohexane; ○ diethylmalonate; △ diethyloxalate.

obeys the relation

$$z_{\gamma F} = 8\, h^{7/11} y^{4/11} (A_4(y))^{-1} \tag{16.3.6}$$

where h is the three-body osmotic parameter and where $A_4(y)$ is a number close to unity. (We shall use the approximation $A_4(y) = 1$.) For $d = 3$, the osmotic parameter is a logarithmic function of the equivalent Brownian area S [see (14.6.36)]

$$h = \frac{1}{44\pi \ln (S/s_1)}. \tag{16.3.7}$$

We know that the ratio S/s_1 is proportional to the degree of polymerization $N = M/m$, and in the absence of more detailed information we shall assume that it is equal to N. For $S = 200\ \mathrm{nm}^2$, we have $N = 10^3$ (see Chapter 15, Section 2.2). In this way, we obtain

$$h = 1.05 \times 10^{-3}. \tag{16.3.8}$$

On the other hand, the value of the parameter y is 2.9×10^{-3}. Inserting these values in (16.3.6), we obtain

$$z_{\gamma F} = 0.015 = bS^{1/2}(2\pi)^{-3/2}$$

and, consequently, for $S = 200\ \mathrm{nm}^2$,

$$b = 0.017\ \mathrm{nm}^{-1}.$$

Finally, taking (16.1.1) into account, we find

$$T_{\gamma F}(S) - T_F = 1.5\ \mathrm{K} \tag{16.3.9}$$

which is the result we wanted. We observe that the evaluations deduced respectively from the experiment [see (16.3.5)] and from the theory are not contradictory. However, the result $z_{\gamma F} > 0$ must not be considered as obvious: thus, in the Flory theory, the state in which the value of the swelling is unity corresponds to a *negative* value $z = z_F < 0$ of the two-body interaction [see (14.3.7)].

Mean values $\langle\langle T_{\mathcal{G}\,\infty F} \rangle\rangle$ obtained by viscosity are listed in Table 16.2. We observe that these values obey the inequality

$$\langle\langle T_{\mathcal{G}\,\infty F} \rangle\rangle - T_F > 0, \tag{16.3.10}$$

and this fact suggests that the two-body interaction parameter z is positive when the chain is quasi-Brownian.

4. PROPERTIES OF A SOLUTION IN THE VICINITY OF THE CRITICAL DEMIXTION POINT

Figure 16.1 shows that the curvature at the top of the demixtion curve (φ_c, T_c) can be considered to be zero. This is a typical feature of binary systems. There the system has a critical behaviour which is revealed by the opalescence of the solution; in the vicinity of the point φ_c, T_c, the solution is able to scatter a radiation, and the effect becomes stronger as the critical point is approached. The Flory–Huggins theory, which is 'classical', cannot faithfully account for the observed critical phenomenon; for instance, it predicts that the demixtion curve has a finite curvature at the critical demixtion point T_c, φ_c. Numerous studies made with mixtures of simple liquids have enabled their authors to establish universal laws in agreement with the theoretical predictions. Solutions of monodisperse polymers are binary mixtures. The samples produced nowadays being nearly monodisperse, it is tempting to verify that the solutions of such samples really have this universal behaviour. This has been done by the Japanese school,[1, 7–10] and in particular by Kaneko and Kuwahara.

On the other hand, we note that a demixtion curve corresponds to each polymer length and that by changing the lengths, it is possible to study the variation of parameters associated with the opalescence phenomenon. Thus, in order to reveal the universal properties of the results, we must use scaling variables.

4.1 Theoretical considerations about the critical demixtion point

The theory concerning the top of the demixtion curve is well-understood in the case of binary mixtures of simple liquids. In general, to study the demixtion of polymer solutions, physicists contented themselves with a direct application of the theoretical results obtained for simple liquids. This classical approach will now be briefly described and subsequently a different approach will be expounded.

4.1.1 A recapitulatory list of scaling laws concerning critical demixtion: the case of simple binary liquids

Critical phenomena can be grouped into classes. It is admitted that in the vicinity of their respective critical points, a binary mixture, a liquid–vapour system and the three-dimensional Ising model belong to the same class. The space dimension ($d = 3$) and the dimension of the order parameter ($n = 1$) are numbers characterizing the class.

The order parameter associated with the demixtion phenomenon can be identified with the volume fraction $\varphi - \varphi_c$, but one could also choose the concentration $\boldsymbol{C} - \boldsymbol{C}_c$. The best candidate should be the parameter leading to the most symmetrical demixtion curve.

The *relevant* parameter is the reduced temperature

$$t = \frac{T - T_c}{T}.$$

For small values of t, the physical quantities associated with binary mixtures obey scaling laws, characterized by critical exponents. The calculated values[11] of the main exponents are [see (12.3.10)]

$$\mathrm{n} = 1, \quad d = 3 \qquad \beta = \frac{\nu d - \gamma}{2} = 0.325 \pm 0.002$$

$$\gamma = 1.2402 \pm 0.0009$$

$$\nu = 0.6300 \pm 0.0008$$

$$\eta = 2 - \gamma/\nu = 0.031 \pm 0.004.$$

We shall examine, for simple binary systems, two types of scaling laws:

1. Let $\varphi^{\mathrm{I}}(t)$, $\varphi^{\mathrm{II}}(t)$ be the volume fractions of solute corresponding, on the demixtion curve, to the reduced temperature t. For $|t| \ll 1$ we have

$$\frac{\varphi^{\mathrm{II}}(t) - \varphi^{\mathrm{I}}(t)}{\varphi_{\mathrm{c}}} = B_{\mathrm{a}}(-t)^{\beta}. \tag{16.4.1}$$

The parameter B_{a} is the critical amplitude associated with the singularity (subscript 'a' for atomic).

2. Let $\boldsymbol{H}(\vec{q})$ be the structure function of the solute and let $H(\vec{q})$ be the form factor of a solute molecule. In the case of simple liquids, the size of the molecules is small and we can admit that $\boldsymbol{H}(\vec{q}) = 1$ everywhere in the observation window. Consequently, for a solute of concentration C, we have [see (7.2.59)]

$$\boldsymbol{H}(\vec{q}) = \frac{1}{C} + \boldsymbol{H}^{\mathrm{II}}(\vec{q}) \tag{16.4.2}$$

where

$$\boldsymbol{H}^{\mathrm{II}}(\vec{q}) = \int \mathrm{d}^3 r\, \mathrm{e}^{i\vec{q}\cdot\vec{r}}[g(\vec{r}) - 1] \tag{16.4.3}$$

and where $g(\vec{r})$ is the pair correlation function of the solute [see (7.2.26)].

The theory[12] tells us that the following scaling law should be valid in the vicinity of the critical point

$$\boldsymbol{H}(\vec{q}) = \boldsymbol{H}(\vec{0})\, \mathrm{J_c}(q\xi) \tag{16.4.4}$$

where $\mathrm{J_c}(x)$ is a universal function ($\mathrm{J_c}(0) = 1$) and ξ a correlation length determined by the condition

$$\mathrm{J_c}(x) = 1/(1 + x^2 + \ldots) \qquad x \ll 1. \tag{16.4.5}$$

Moreover, we also have

$$\mathrm{J_c}(x) \simeq \mathrm{J}_{\mathrm{c}\infty}/x^{2-\eta} \qquad x \gg 1 \tag{16.4.6}$$

where η is an exponent of class $\mathrm{n} = 1$, $d = 3$.

The function $\mathrm{J_c}(x)$ is analytic with respect to $x = x' + ix''$ in the band $|x''| < 1/\Gamma_{\mathrm{c}}$ of the complex plane, and singular on the straight lines $|x''| = 1/\Gamma_{\mathrm{c}}$

where Γ_c is a constant close to unity. We may set

$$\xi_e = \Gamma_c \xi, \tag{16.4.7}$$

and we see immediately that ξ_e is the screening length associated with the concentration fluctuations; in other words, $[g(\vec{r}) - 1]$ decreases roughly as e^{-r/ξ_e} when $r \to \infty$. The screening length ξ_e and the length ξ diverge at the critical demixtion point

$$\xi/l_a = \Xi_a |t|^{-\nu} \tag{16.4.8}$$

where l_a is the atomic size associated with the two liquids and Ξ_a an amplitude. Let us recall that the exponent ν which appears here corresponds to the class $n = 1$, $d = 3$ (and not to the class $n = 0$, $d = 3$).

The volume $\boldsymbol{H}(\vec{0})$ is proportional to the osmotic compressibility since, according to (9.3.64),

$$\boldsymbol{H}(\vec{0}) = \frac{1}{\varphi} \frac{\partial \varphi}{\partial(\boldsymbol{\Pi \beta})}. \tag{16.4.9}$$

Let us fix the volume fraction φ at its critical value φ_c; the osmotic compressibility diverges when the reduced temperature t goes to zero. The variation law is given by the relation[13]

$$1/\boldsymbol{H}(\vec{0}) = \Psi_0 |t|^{\gamma} \tag{16.4.10}$$

where Ψ_∞ is the critical amplitude and where γ is an exponent of class $n = 1$, $d = 3$.

Now, we may let the volume fraction φ vary on the isotherm $t = 0$; then the osmotic compressibility diverges when $\varphi \to \varphi_c$ and the variation law is

$$1/\boldsymbol{H}(\vec{0}) = Y_\infty \left| \frac{\varphi - \varphi_c}{\varphi_c} \right|^{\gamma/\beta}. \tag{16.4.11}$$

More generally,[13] we have

$$1/\boldsymbol{H}(\vec{0}) = |t|^{\gamma}\, Y\left(|t|^{-\beta} \left| \frac{\varphi - \varphi_c}{\varphi_c} \right| \right) \tag{16.4.12}$$

where $Y(x)$ is a function such that $Y(0) = Y_0$ and $Y(x) \approx Y_\infty x^{\gamma/\beta}$, $x \gg 1$. We can verify that this formula is in agreement with (16.4.10) and (16.4.11).

4.1.2 Polymer solutions: coordinates of the demixtion critical point

We have already noted an essential feature of polymer solutions: the coordinates φ_c, T_c of the critical demixtion point are functions of the equivalent Brownian area S of the polymers. Let us recall here a few observed results.

1. Let T_F be the Flory temperature. The interval $T_F - T_c$ decreases when the equivalent Brownian area S of the polymers increases. For the polystyrene–cyclohexane system, we found a relation [see (16.2.7)] which can be written in

the form

$$\frac{1}{T_c} - \frac{1}{T_F} = \frac{4.57}{\sqrt{N}} \times 10^{-3} \tag{16.4.13}$$

where $N(\propto S)$ is the number of links

In a similar way, for the polystyrene–cyclohexane system, the interaction b is given by [see (16.1.1)]

$$b = 896\left(\frac{1}{T_F} - \frac{1}{T}\right) \tag{16.4.14}$$

In particular, for $T = T_c$, using (16.4.13), we get

$$b_c = -\frac{4.57}{\sqrt{N}}. \tag{16.4.15}$$

This result is in agreement with the predictions of the simple-tree approximation (and of the Flory–Huggins theory) which gives

$$b_c = -2S^{-1/2}c^{1/2}.$$

2. The critical fraction φ_c is a decreasing function of S, which can be represented by the empirical relation

$$\varphi_c \propto N^{-0.38}.$$

We observe that this relation is not in agreement with the simple-tree approximation which gives

$$\varphi_c \propto C_c = C_c S = S^{-1/2}c^{-1/2}.$$

4.1.3 Scaling laws for polymer solutions

In the case of polymer solutions, the solute molecules are flexible and may have a very large size in comparison to the solvent molecules. In such conditions, can we directly apply the scaling laws whose validity has been established for simple, liquid mixtures in the vicinity of the critical demixtion point? Anyway, this is what has been done until now.

When the correlation length ξ is much larger than the size ${}^0R$ of the chains, we may think that the scaling laws apply in a similar way, both for simple liquid mixtures and for polymer solutions. Actually, we can always manage so that, near the critical demixtion point, the condition $\xi/{}^0R \gg 1$ holds, and we verify experimentally that, in the vicinity of this point, polymer solutions have a behaviour which can be represented by standard scaling laws.

4.1.3.1 *The demixtion curve in the vicinity of the critical point*

Let us consider the scaling law (16.4.1)

$$\frac{\varphi^{\text{II}}(t) - \varphi^{\text{I}}(t)}{\varphi_c} = B_M(-t)^{\beta} \tag{16.4.16}$$

where $t = (T - T_c)/T$ and where B_M is the amplitude. The fact that here the solute is a polymer has three important consequences:

1. The amplitude B_M is a function of the molecular mass M.

2. The domain of values of t in which the scaling law is valid shrinks when M increases.

3. Instead of measuring the temperature by introducing the variable t which has no universal character, we can use the more intrinsic* variable τ which was defined in Chapter 14, Section 2.

$$\tau = \frac{\beta_c - \beta}{\beta_c - \beta_F} = \frac{T_F}{T_F - T_c} t. \tag{16.4.17}$$

Now, formula (16.4.16) reads

$$\frac{\varphi^{\text{II}}(\tau) - \varphi^{\text{I}}(\tau)}{\varphi_c} = B(-\tau)^{\beta} \tag{16.4.18}$$

where B could be a universal number. Let us write (16.4.18) in the simple-tree approximation. According to (14.6.7), we have,

$$\frac{\varphi^{\text{II}}(\tau) - \varphi^{\text{I}}(\tau)}{\varphi_c} \simeq \frac{C^{\text{II}} - C^{\text{I}}}{C^{\text{I}}} \simeq \left[24\left(\frac{b_c - b}{b_c}\right)\right]^{1/2} \tag{16.4.19}$$

Moreover, according to (16.1.1), the interaction b is given by an approximate expression of the following type

$$b = A\left(\frac{1}{T_F} - \frac{1}{T}\right) = A(\beta_F - \beta)$$

where A is a constant which is equal to 896 for the polystyrene–cyclohexane system. Combining the preceding equality and (16.4.17), we obtain

$$\frac{b_c - b}{b_c} = \tau.$$

Thus, in the simple-tree approximation

$$\frac{\varphi^{\text{II}}(\tau) - \varphi^{\text{I}}(\tau)}{\varphi_c} = \sqrt{24}(-\tau)^{1/2}. \tag{16.4.20}$$

In this approximation, the exponent β equals 1/2 and the value of the amplitude ${}^0B$ is

$${}^0B = \sqrt{24} = 4.9.$$

* The Japanese authors, quoted above[1, 7–10] studied demixtion by using (16.4.16), and they found that B_M does indeed depend on the molecular mass. However, using the *intrinsic* variable τ suppresses this dependence.

This value ${}^{0}B$ is to be compared with the measured value B (see Section 4.2.1 below).

4.1.3.2 *The structure function in the vicinity of the critical point*

Let us now consider the structure $\boldsymbol{H}(\vec{q})$. By definition [see (7.2.59) and (16.4.2)], we have

$$\boldsymbol{H}(\vec{q}) = \frac{H(\vec{q})}{\boldsymbol{C}} + \boldsymbol{H}^{\text{II}}(\vec{q})$$

where $H(\vec{q})$ is the form function of the chains, $\boldsymbol{C}$ their concentration (we have $M\boldsymbol{C} = \varphi/w_1$) and $\boldsymbol{H}^{\text{II}}(\vec{q})$ the intermolecular structure function.

Here, we shall be interested in the domain

$$q\,{}^{0}R_{\text{G}} \lesssim 1 \tag{16.4.21}$$

At temperature T_{F}, the effective two-body interaction b vanishes; the chains are quasi-Brownian and for $\varphi = \varphi_{\text{c}}$ ($\boldsymbol{C} = \boldsymbol{C}_{\text{c}}$), the intermolecular contribution to the structure function is negligible with respect to the intramolecular contribution. In this case

$$\boldsymbol{H}(\vec{q}) \simeq \frac{H(\vec{q})}{\boldsymbol{C}_{\text{c}}}.$$

Thus, in the domain (16.4.21)

$$\boldsymbol{H}(\vec{q}) \simeq \frac{1}{\boldsymbol{C}_{\text{c}}}(1 - q^2 R_{\text{G}}^2/3 + \ldots). \tag{16.4.22}$$

This expansion suggests the following approximation

$$\boldsymbol{H}(\vec{q}) \simeq \boldsymbol{H}(\vec{0})\frac{1}{1 + q^2 R_{\text{G}}^2/3} \tag{16.4.23}$$

where $\boldsymbol{H}(\vec{0}) = 1/\boldsymbol{C}_{\text{c}}$ is the osmotic compressibility. When the temperature of the solution diminishes and approaches T_{c}, the fluctuations of chain concentration become huge; in $\boldsymbol{H}(\vec{q})$, the intermolecular term $\boldsymbol{H}^{\text{II}}(\vec{q})$ becomes dominant. Then, in agreement with (16.4.4) and (16.4.5), we may write

$$\boldsymbol{H}(\vec{q}) \simeq \boldsymbol{H}(\vec{0})\frac{1}{1 + q^2\xi^2 + \ldots} \tag{16.4.24}$$

where ξ is the correlation length associated with the fluctuations of concentration.

To study the dependence of this length on temperature, we shall again use the parameter

$$\tau = \frac{\beta_{\text{c}} - \beta}{\beta_{\text{c}} - \beta_{\text{F}}} = \frac{T_{\text{F}}}{T_{\text{F}} - T_{\text{c}}}\,\frac{T - T_{\text{c}}}{T}.$$

In this case, the scaling law for the correlation length reads

$$\frac{\xi}{{}^{0}R_G} = \Xi|\tau|^{-\nu} \tag{16.4.25}$$

where the exponent ν is an exponent of class $n = 1$, $d = 3$; ${}^{0}R_G$ is the radius of gyration of the chain at temperature T_F and Ξ is an amplitude. This amplitude could be a universal number (?).* The formula (16.4.25) is valid only for $|\tau| \ll 1$. However, it can be extrapolated and, for $T = T_F$, this formula gives

$$\xi = \Xi\, {}^{0}R_G.$$

to a first approximation. On the other hand, (16.4.23) gives

$$\xi = {}^{0}R_G/\sqrt{3}.$$

Thus, by comparing the preceding equalities, we find the approximate result

$$\Xi \simeq 1/\sqrt{3}$$

which, in the following section will be compared to the experimental value determined by scattering.

4.2 An experimental study of the critical demixtion point

4.2.1 Experimental determination of the demixtion curve in the vicinity of the critical point

Dobashi, Nakata, and Kaneko[7] determined with precision the demixtion curves of polystyrene solutions in methylcyclohexane (the temperature T_F corresponds to 69.76 °C). These authors obtained data for a large number of good (nearly) monodisperse samples with various molecular masses. They interpreted their results by using the parameter $-t = \dfrac{T - T_c}{T}$ corresponding to each sample.

Let us try here to interpret the whole of their results by using only *one* parameter

$$\tau = \frac{\beta_c - \beta}{\beta_c - \beta_F} = \frac{T_F}{T_F - T_c}\, t.$$

For this purpose, we consider three samples *a*, *b*, *c*, well-studied by the authors quoted above. The experimental data concerning these samples are shown in Tables 16.3 and 16.4.

* Various expressions have been proposed for the scaling law concerning the correlation length. In 1959[14] Debye wrote the equation

$$\frac{\xi}{{}^{0}R_G} = \Xi_D|t|^{-\nu}$$

where $t = (T - T_c)/T$ and where Ξ_D is a non-universal constant. De Gennes showed in 1968,[15] that, in this form, the amplitude Ξ varies as $M^{-1/4}$. Finally, in 1979,[16] de Gennes introduced the formulation (16.4.25).

Table 16.3

Sample	M_W	T_c[K]	$T_F/(T_F - T_c)$	φ_c
a	1.73×10^4	296.13	7.35	0.1669
b	1.09×10^5	322.71	17.1	0.0840
c	7.10×10^6	334.82	43.14	0.0406

Table 16.4

Sample	T[K] ($\pm$ 0.003)	τ	$\dfrac{\varphi^{\text{II}} - \varphi^{\text{I}}}{\varphi_c}$	$\dfrac{\varphi^{\text{II}} + \varphi^{\text{I}}}{2\varphi_c}$
a	296.053	1.91×10^{-3}	0.484	1.005
	295.987	3.54×10^{-3}	0.597	1.0065
	295.898	5.76×10^{-3}	0.704	1.090
	295.676	1.13×10^{-2}	0.871	1.0185
	295.266	2.15×10^{-2}	1.095	1.0325
	294.475	4.13×10^{-2}	1.36	1.065
	292.865	8.28×10^{-2}	1.7	1.1215
b	322.695	7.93×10^{-4}	0.354	1.0035
	322.677	1.75×10^{-3}	0.496	1.005
	322.634	4.03×10^{-3}	0.638	1.0115
	322.548	8.58×10^{-3}	0.8154	1.024
	322.386	1.72×10^{-2}	1.027	1.0805
	322.00	3.91×10^{-2}	1.342	1.132
	321.41	6.91×10^{-2}	1.638	
c	334.803	2.19×10^{-3}	0.519	
	334.766	6.96×10^{-3}	0.71	
	334.702	1.52×10^{-2}	0.938	
	334.633	2.41×10^{-2}	1.076	
	334.513	3.96×10^{-2}	1.295	
	334.270	7.11×10^{-2}	1.584	

The values $(\varphi^{\text{II}} - \varphi^{\text{I}})/\varphi_c$ are plotted against τ in Fig. 16.7. We observe to our satisfaction that the measurements corresponding to different molecular masses superimpose exactly.

Moreover, we represented the function

$$\frac{\varphi^{\text{II}} - \varphi^{\text{I}}}{\varphi_c} = B(-\tau)^{\beta} \tag{16.4.26}$$

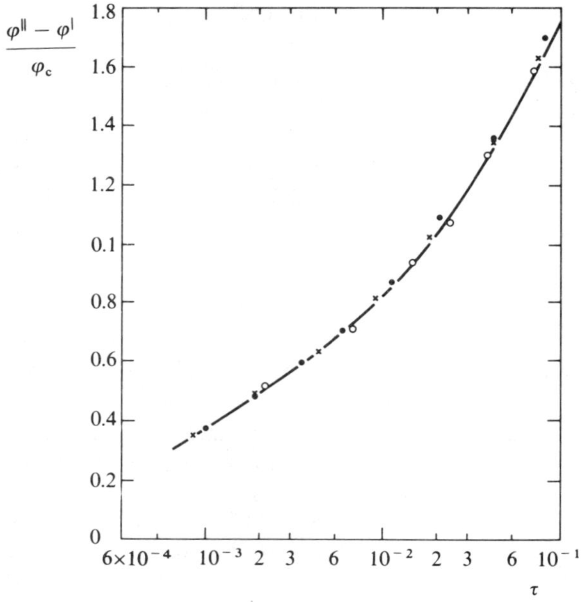

Fig. 16.7. The ratio $(\varphi^{\text{II}} - \varphi^{\text{I}})/\varphi_c$, plotted against the reduced variable

$$\tau = \frac{\beta_c - \beta}{\beta_c - \beta_F} = \frac{T_F}{T_F - T_c}\,\frac{T - T_c}{T}$$

for polystyrene in methylcyclohexane (from Dobashi *et al.*[7]):

● $M_W = 1.73 \times 10^4$
× $M_W = 1.09 \times 10^5$
o $M_W = 7.19 \times 10^6$

using for β the theoretical value $\beta = 0.325$ ($n = 1, d = 3$) and choosing for B the value which, for small τ, leads to the best fit with the experimental results of Fig. 16.6. This fit is excellent in the interval $\tau \leqslant 3 \times 10^{-2}$ and for B we obtain the value $B = 3.7$. This value is to be compared with the value ${}^0B = 4.90$ obtained with the simple-tree approximation. For $\tau > 3 \times 10^{-2}$, the experimental points do not remain on the theoretical curve defining the scaling law, but this is not surprising, since in this domain the behaviour is not critical.

4.2.2 Divergence of the structure function. Observations by light scattering

Numerous light-scattering experiments have been performed with polymer solutions in the vicinity of the critical demixtion point in order to determine the structure function $H(\vec{q})$.

The most conspicuous phenomenon is the opalescence of the irradiated sample. It results from the fact that, in the sample, the correlation length is of the same order of magnitude as the wavelength. This leads to strong scattering and in such conditions multiple scattering may occur. In this case, the intensity $I(q)$ of the scattered radiation falling on the detector is not given by the formula [see (7.2.58)]

$$I(q) = K\rho^2 \text{Ⓐ} H(\vec{q}) \qquad (16.4.27)$$

which applies only to single collisions. However, it is possible to prepare the sample so as to produce negligible multiple scattering effects, for instance by chosing a solvent–solute system with very close refraction indices, or more simply, by limiting the thickness of the sample to a size less than the *mean free length* of the radiation inside the target. Thus, when discussing the experimental result, we shall assume that precautions have been taken in order to ensure the validity of the simple scattering assumption and therefore of relation (16.4.27).

4.2.2.1 *Measurement of the structure function near T_c*

Experimental values of $1/I(q)$ obtained[8] for the same sample at various temperatures are plotted in Fig. 16.8 against q^2. The sample is a polystyrene ($M_W = 1.1 \times 10^5$) solution in cyclohexane. The volume fraction of polystyrene is $\varphi = 0.0813$

The critical fraction is $\varphi_c = 0.0825$. The demixtion temperature for $\varphi = 0.0813$ corresponds to 21.33 °C. We can see from Fig. 16.8 that the intensity $I(q)$ can depend on q^2 in three different ways:

1. For $T - T(\varphi) > 1$, the measured intensities are nearly independent of the transfer wave number q. This result can be understood as follows: the inverse scattering intensity (16.4.27) reads

$$I^{-1}(q) = \frac{1}{K\rho^2 \text{Ⓐ} H(\vec{q})}$$

Now for $T = T_F$ and $\varphi = \varphi_c$, the structure function can be represented by the approximate formula

$$H(\vec{q}) \simeq H(\vec{0}) \frac{1}{1 + q^2\,{}^0R_G^2/3} \quad (q^2\,{}^0R_G^2 < 1). \qquad (16.4.28)$$

The condition $q^2\ {}^0R_G^2 < 1$ is quite well verified in the experimental situation corresponding to Fig. 16.8. In fact, the value ${}^0R_G^2$ of the mean square radius of gyration of a Brownian chain of molecular mass $M_W = 1.1 \times 10^5$ is

$${}^0R_G^2 = {}^0\Lambda\, M_W$$

and for polystyrene [see (15.2.7)]

$${}^0R_G^2 = 91\ \text{nm}^2.$$

The observation window corresponding to the experiments of Fig. 16.8 is

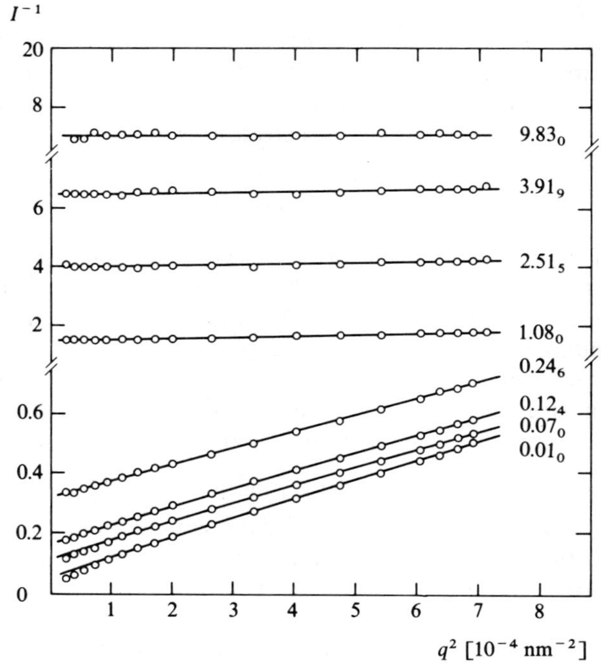

Fig. 16.8. Scattering of light by polystyrene ($M_W = 1.1 \times 10^5$) in cyclohexane: the inverse scattering intensities are plotted against q^2 (from Kojima, Kuwahara, and Kaneko[8]). The scattering experiments were performed at various temperatures along the isochore $\varphi = 0.0813$ (close to $\varphi_c = 0.0825$). For this volume fraction, the (nearly critical) temperature $T(\varphi)$ corresponds to 21.33 °C. The experimental temperature differences $T - T(\varphi)$ are indicated in the figure in front of the curves interpolating the measured points.

defined by the inequality

$$10^{-5} \leqslant q^2 \leqslant 10^{-3}\ \text{nm}^{-2}: \qquad (16.4.29)$$

We verify that in this interval

$$(q^2)\ {}^0R_G^2/3 < 0.03$$

in agreement with the above condition. The formula can thus be used to describe the experimental result. Moreover, the structure function $H(\vec{q})$ and consequently $I^{-1}(q)$ are practically constant in the interval (16.4.29). In the situation studied here, we have

$$\xi/{}^0R_G \leqslant 1.$$

2. For $0.1 < T - T(\varphi) < 1$, the function $I^{-1}(q)$ is a linear function of q^2 with an appreciable slope. In this temperature range, the function $I(q)$ can be

interpreted with the help of the expressions (16.4.24) and (16.4.27)

$$\frac{K\rho}{I(q)} = \frac{1}{\rho \, \text{Ⓐ} \, \boldsymbol{H}(\vec{q})} = \frac{1}{\rho \, \text{Ⓐ} \, \boldsymbol{H}(\vec{0})}(1 + q^2\xi^2 + \ldots). \tag{16.4.30}$$

The value $\boldsymbol{H}(\vec{0})$ is determined by linear extrapolation of the inverse intensity at $q = 0$, and ξ^2 by measuring the slope of the inverse intensity at $q = 0$. This method does not differ much from the method we used to determine the screening length ξ_e for good solutions of overlapping chains [see (15.4.3.3)]. In the present case, it is actually difficult to make a distinction between ξ_e and ξ, the inverse intensity being practically linear with respect to q^2 for small q^2. Let us also note that, in the situation described here, we have $\xi/{}^0R_G < 1$.

3. In the interval $T - T(\varphi) < 0.1$, the correlation length ξ is even larger, and the experimental points represented in Fig. 16.8 define a curve which has a downward concavity. We interpret this fact as resulting from the inequality

$$q^2\xi^2 \gg 1$$

for q^2 in the range (16.4.29).

In this case, the structure function is given by eqn (16.4.4) and we verify that this curve has a downward concavity. Figure 16.8 shows us that the osmotic compressibility

$$\frac{1}{\varphi}\frac{\partial\varphi}{\partial(\boldsymbol{\Pi\beta})} = \boldsymbol{H}(\vec{0})$$

and the correlation length ξ increase when $T \to T_c$. Actually, these quantities diverge, and we shall examine how.

4.2.2.2 *Compressibility and correlation length of the solution at the volume fraction φ_c*

In the preceding section, we showed how we can determine the osmotic compressibility $\boldsymbol{H}(\vec{0})$ and the correlation length ξ from results of scattering experiments. Let us now examine the experimental values of these quantities; of course, they depend on the fraction φ, the temperature T and the area S. Hamano, Kuwahara, and Kaneko,[9] and later on Shinozaki, Hamada, and Nose[10] measured $\boldsymbol{H}(\vec{0})$ and ξ for various temperatures. We shall discuss here the results obtained for the constant fraction $\varphi = \varphi_c$. The temperature dependence of the compressibility is in agreement with the scaling law (16.4.10)

$$1/\boldsymbol{H}(\vec{0}) = \Psi_0 |t|^\gamma$$

where γ is an exponent of class $n = 1$ $d = 3$, where $t = (T - T_c)/T$ and where Ψ_0 is an amplitude. The experimental value of the exponent γ is $\gamma = 1.22$, whereas its theoretical value is 1.241 ± 0.002.

We shall also examine with great interest how the correlation length ξ depends on temperature, for various Brownian areas S. The characteristics of the

Table 16.5

Sample	M_W	T_c [K]	$T_F/(T_F - T)$ ($T_F = 342.76$ K)	${}^0R_G$ [nm]
a	$9. \times 10^3$	281.20	5.56	2.733
b	1.1×10^5	323.24	17.56	9.55
c	1.26×10^6	336.97	59.2	32.3

Table 16.6

Sample	$(T - T_c)/T$	τ	$\xi(\simeq \xi_c)$ [nm]	$\xi/{}^0R_G$
a	8.55×10^{-4}	4.9×10^{-3}	328	120
	1.29×10^{-3}	7.1×10^{-3}	275	100
	2.20×10^{-3}	1.22×10^{-2}	201	73.7
	2.80×10^{-3}	1.55×10^{-2}	187.7	68.7
	3.50×10^{-3}	1.94×10^{-2}	152.8	56
	4.33×10^{-3}	2.41×10^{-2}	148	54.3
	5.17×10^{-3}	2.87×10^{-2}	117.3	42.9
	6.7×10^{-3}	3.74×10^{-2}	107.4	39.3
b	2.10×10^{-3}	3.75×10^{-2}	328	34.4
	2.60×10^{-3}	4.58×10^{-2}	322	33.78
	3.04×10^{-3}	5.34×10^{-2}	280	29.3
	3.63×10^{-3}	6.40×10^{-2}	237.6	24.88
	5.17×10^{-3}	9.10×10^{-2}	178.1	18.65
	6.16×10^{-3}	10.8×10^{-2}	157.4	16.5
	8.80×10^{-3}	15.4×10^{-2}	135.1	14.14
	9.80×10^{-3}	17.30×10^{-2}	117.3	12.3
	11.00×10^{-3}	19.50×10^{-2}	110.6	11.58
	12.00×10^{-3}	21.90×10^{-2}	111.2	11.6
c	1.82×10^{-3}	0.018	665	20.6
	3.00×10^{-3}	0.18	510	15.8
	5.3×10^{-3}	0.315	369	11.4
	7.36×10^{-3}	0.435	309.6	9.6
	8.4×10^{-3}	0.498	292	9.04
	10.2×10^{-3}	0.602	254.9	7.9
	13.1×10^{-3}	0.775	205.1	6.35

Table 16.7

T[K]	$10^2\, b$ [nm^{-1}]	z	$\mathfrak{X}$ Flory theory	$\mathfrak{X}_{0L}$ linear approximation	$\dfrac{\mathfrak{X}_0}{{}^0A_0(y)}$	$\mathfrak{X}_G$
308	0.89	0.026	1.08	1	1.01	1
307	0	0	1.04	0.96	0.99	0.97
306	− 1.05	− 0.0305	0.97	0.92	0.96	0.95
305	− 2.01	− 0.058	0.88	0.884	0.942	0.931
304	− 2.98	− 0.086	0.75	0.845	0.915	0.912
303	− 3.95	− 0.115	0.46	0.81	0.89	0.855
302	− 4.92	− 0.143	0.29	0.77	0.87	0.81
301	− 5.87	− 0.170	0.22	0.73	0.85	0.756

samples and the measured values of ξ are shown in Tables 16.6 and 16.7 respectively.

The ratio $\xi/{}^0R_G$ is plotted against $t = \dfrac{\beta_c - \beta}{\beta_c - \beta_F}$ on Fig. 16.9, for three samples with different molecular masses. We note, that with this representation, all

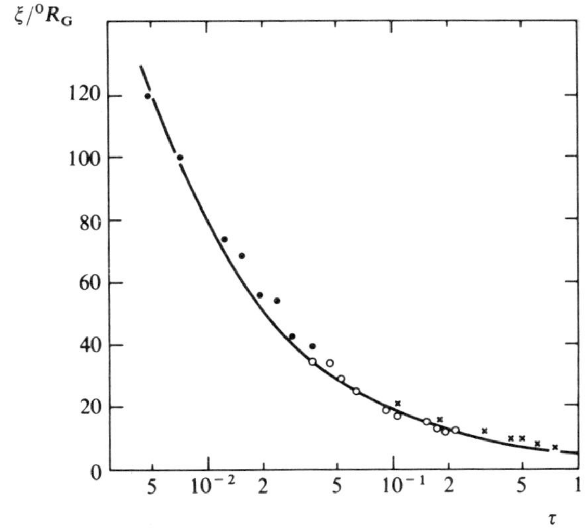

Fig. 16.9. The ratio $\xi/{}^0R_G$ plotted against the reduced temperature $\tau = \dfrac{\beta_c - \beta}{\beta_c - \beta_F} = \dfrac{T_F}{T_F - T_c}\,\dfrac{T - T_c}{T}$ (from Shinozaki *et al.*[10]).

Polystyrene in methylcyclohexane
● $M_W = 9 \times 10^3$
○ $M_W = 1.1 \times 10^5$
× $M_W = 1.26 \times 10^6$.

points are placed on a unique curve. We represented the function

$$\frac{\xi}{{}^{0}R_{G}} = \Xi\tau^{-\nu} \tag{16.4.31}$$

choosing for ν the theoretical value $\nu = 0.630$ (here the exponent ν belongs to the class $n = 1$, $d = 3$). The amplitude Ξ which leads to the best fit of the function (16.4.31) with the experimental data has the universal value

$$\Xi = 4.43. \tag{16.4.32}$$

We thus verify the fact that the correlation length ξ obeys a very general scaling law in the vicinity of the demixtion point. However, let us emphasize once more that the simplicity of the result comes from the fact that we represented the temperature of the solution by using the *intrinsic* variable τ (and not by using t; see Section 4.1.3.2).

5. COLLAPSE OF POLYMER CHAINS IN DILUTE SOLUTIONS

At temperature T_F, the swelling of the chains of equivalent Brownian area S is very close to unity ($\mathfrak{X}_0 \simeq 1$). When the temperature decreases, the two-body interaction b becomes negative. The consequence of this fact, for chains in sufficiently dilute solutions, is that the swelling $\mathfrak{X}_0$ becomes smaller than 1.

In this context, a chain in its Brownian configuration is sometimes called *coil* and in its more compact configuration it is called *globule*. The transformation corresponding to the fact that $\mathfrak{X}_0$ becomes smaller than one is called coil-globule transition, people also say that the chain collapses.

5.1 Experimental conditions required to observe the collapse phenomenon

The demixtion curves (Fig. 16.2) show us that there exists a range of volume fractions φ ($\varphi < \varphi_c$) and of temperatures $T (T < T_c)$ in which solutions are dilute and b is negative. Then for a fixed fraction φ_1, the predicted collapse (Chapter 14, Sections 3, 4, and 6) can be observed by lowering the temperature T down to the demixtion temperature $T(\varphi_1)$. The process is schematically shown in Fig. 16.10. We can see that the smaller φ_1 is, the larger the domain of temperature to be explored will be; however, the technical constraints of measurement require φ to be larger than a minimal value φ_{1m} below which observations are no longer possible.

We shall examine here results of scattering experiments which will give us the gyration swelling $\mathfrak{X}_G$, extrapolated t_0 concentration.

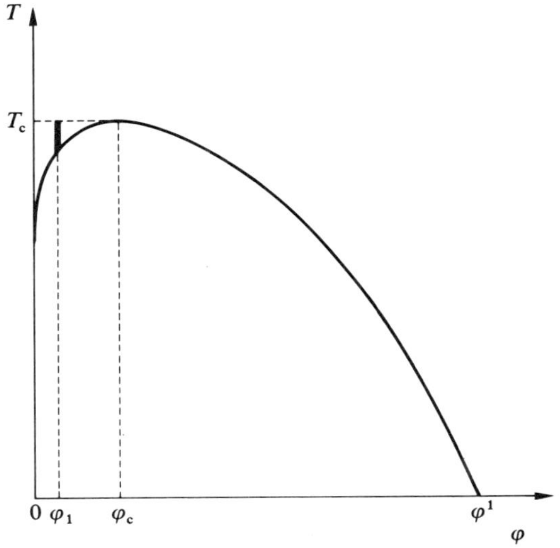

Fig. 16.10. Schematical representation of the demixtion curve and of a range of temperature, at a given φ ($\varphi = \varphi_1$; thick line), in which the collapse of polymer chains can be observed.

Now, the parameter S (the equivalent Brownian area of the polymer chains in the sample under study) has to be chosen so as to correspond to the best experimental conditions. In this connection, let us stress the following points:

1. The scattered intensity [see (7.2.60)]

$$I(q) = K\rho M H(\vec{q})$$

(where $H(\vec{q})$ is the form function, ρ the mass concentration, M the molecular mass and K the apparatus constant) is proportional to the product

$$\rho M \propto \varphi S.$$

The intensity increases proportionally to S. From this point of view, it is better to experiment with long chains.

2. When S increases, the demixtion curve comes exponentially nearer to the asymptote $\varphi = 0$. For a fixed fraction φ_1, the interval $T_c - T(\varphi_1)$ to be explored becomes larger when the demixtion curve moves away from the asymptote (see Fig. 16.2). Thus, from this point of view, it is better to experiment with short chains.

3. The shrinking obtained by lowering the temperature $T(\leqslant T_c)$ is a function of the two-body interaction z and of the three-body interaction y. For instance,

the Flory theory gives [see (14.3.7)]

$$\mathfrak{X}^{5/2} - \mathfrak{X}^{3/2} = \frac{1}{2}3^{3/2}z + 8 \times 3^{1/2}y\mathfrak{X}^{-3/2} \qquad (16.5.1)$$

and the tricritical linear theory [see (14.6.19)] leads to the approximation

$$\mathfrak{X}_0 \simeq \mathfrak{X}_{OL} = 1 - 4\pi y + \tfrac{4}{3}z + \ldots . \qquad (16.5.2)$$

Actually, the main effects come from variations of z, and, when z becomes negative, the chain goes from the coil state to the globule state. We now observe that this parameter is proportional to $S^{1/2}$, since by definition

$$z = bS^{1/2}(2\pi)^{-3/2}, \qquad (16.5.3)$$

and in this formula b is a function of temperature. On the other hand, the parameter y, associated with repulsive interactions, whose role is to ensure the stability* of the chain, remains, virtually constant, when T varies, and does not depend on S (for $d = 3$). For a fixed value of y, the swelling becomes larger when the area S increases. Thus, it seems better to experiment with long chains.

These conclusions are contradictory and actually it is difficult to determine explicitly an optimal value of S. Thus, experiments have been carried out with various areas S. However, the choice of S also depends on the nature of the radiation used in the experiment [see the observation windows (15.1.22) and (15.1.23)].

Nierlich, Cotton, and Farnoux[17] measured the radius of gyration of short chains ($S = 150\ \text{nm}^2$) by neutron scattering in a temperature range of 20 degrees below T_c. They observed a small shrinking.

Perzynski, Adam, and Delsanti[18] measured the hydrodynamic radius and subsequently the radius of gyration of very long polystyrene chains, in particular for $M_W = 1.26 \times 10^6$ ($S = 2092\ \text{nm}^2$), in a temperature range of 6 degrees below T_c. In the next section, we discuss the results concerning these radii of gyration.

5.2 Collapse observed by light scattering

Perzynski *et al.*[2] measured (by light scattering) the (mean) square $R_G^2(\varphi)$ of the radius of gyration of polystyrene chains of molecular mass $M_W = 1.26 \times 10^6$ ($S = 2092\ \text{nm}^2$) for various temperatures $T(\leqslant T_F)$ and for various volume fractions in the range $10^{-5} \leqslant \varphi \leqslant 10^{-4}$. The radii are extrapolated at zero concentration: for a given temperature T, we write

$$R_G^2 = \lim_{\varphi \to 0} R_G^2(\varphi).$$

* The stabilizing effect associated with the parameter y appears clearly in the expression of the third virial coefficient A_3 which is proportional to $h = y + \ldots$ [see (14.6.25)]. On the contrary, in the linear equation (16.5.2) it seems that the parameter y contributes to the shrinking of the chain, but the term of higher order gives a contribution with a positive sign. Let us also note that when the three-body interaction increases, not only does y grow but also z, owing to the existence of cut-off-dependent renormalization terms.

In principle, these results should be compared with the mean square radius of gyration of the same chains (of equivalent area S) measured at the temperature $T_{\mathscr{G}\infty\mathrm{F}}(S)$ (see Chapter 14, Section 6.9.6). However, as was seen in Section 3.1, it is in practice difficult to determine $T_{\mathscr{G}\infty\mathrm{F}}(S)$. This is why, with Perzynski, we define the experimental gyration swelling approximately, by choosing a reference temperature T_1 (35 °C) which is arbitrary but close to T_F (34 °C)

$$\mathfrak{X}_G = R_G^2(T)/R_G^2(T_1).$$

The vaues of $\mathfrak{X}_G$ obtained in the experiment for various temperatures inferior or equal to 35 °C are given in Table 16.7. We note that when the temperature decreases from 35 °C to 28 °C, the radius of gyration decreases by a factor 0.87.

Now let us interpret the results. For this purpose, we begin with the Flory equation (14.3.7), but we note that the swelling $\mathfrak{X}$ which appears in this equation is not very well defined (it is neither an end-to-end swelling nor a gyration swelling). Subsequently we shall use the tricritical theory which enabled us to calculate the end-to-end swelling. However, we note that in general this swelling differs from the gyration swelling. In good solvent, we have

$$\mathfrak{X}_G < \mathfrak{X}_0 \qquad \text{(see Chapters 13 and 15)}$$

in the Brownian state

$${}^0\mathfrak{X}_G = {}^0\mathfrak{X}_0 = 1$$

and in the globular state which is studied here

$$\mathfrak{X}_G > \mathfrak{X}_0.$$

Thus, let us consider a *collapsed* chain filling completely the volume of a sphere of radius R_0. We can admit that the end points are located at random in the sphere whose centre is assumed to coincide with the origin. Consequently, we have

$$R^2 = \langle\!\langle (\vec{r}_N - \vec{r}_0)^2 \rangle\!\rangle = \langle\!\langle r_N^2 \rangle\!\rangle + \langle\!\langle r_0^2 \rangle\!\rangle = 2R_G^2. \tag{16.5.4}$$

Accordingly, in this case, we find

$$\frac{\mathfrak{X}_G}{\mathfrak{X}_0} = \frac{R_G^2}{{}^0R_G^2} \bigg/ \frac{R^2}{{}^0R^2} = 6R_G^2/R^2 = 3. \tag{16.5.5}$$

In the intermediate cases, the swellings $\mathfrak{X}_0$ and $\mathfrak{X}_G$ obey the inequality

$$1 \leqslant \mathfrak{X}_G/\mathfrak{X}_0 \leqslant 3. \tag{16.5.6}$$

Let us return to the theoretical expressions of the (end-to-end) swelling. We denote by $\mathfrak{X}$ the swelling which is calculated with the help of eqn (14.3.7)

$$\mathfrak{X}^{5/2} - \mathfrak{X}^{3/2} = \frac{1}{2}3^{3/2}z + 8 \times 3^{1/2} y \mathfrak{X}^{-3/2}. \tag{16.5.7}$$

This swelling depends on the parameters y and z but only z depends on temperature [see (16.5.3) and (16.1.1)]. The parameter y is a constant whose numerical value $y = 0.0029$ was determined in Section 2.2 [demixtion curve: see (16.2.15)]. The value of y considered here is larger than the bound of the stability domain (see Chapter 14, Section 3.2)

$$y > 0.00164.$$

In this case, eqn (16.5.7) predicts an appreciable but continuous shrinking of the chain when temperature decreases, i.e. when z diminishes (see Table 16.7 and Fig. 16.11). Moreover, the same equation shows that the swelling is just equal to unity when $z = z_F = -(16/3)y = -0.015$. Comparing the calculated values of $\mathfrak{X}$ with the measured values of $\mathfrak{X}_G$, we observe that the Flory theory is not realistic.

We obtain a better agreement if we interpret the experimental results with the help of the linear expansion (16.5.2)

$$\mathfrak{X}_0 \simeq \mathfrak{X}_{0L} = 1 - 4\pi y + \tfrac{4}{3} z \tag{16.5.8}$$

where y has the value calculated above. In particular, with this approximation, the state in which the swelling is unity, corresponds to a positive value of z given

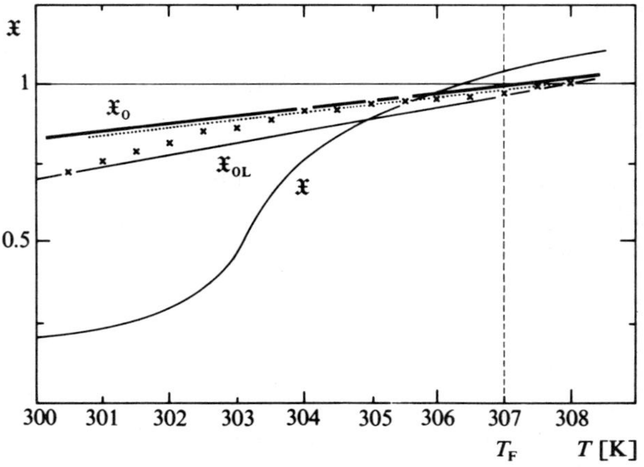

Fig. 16.11. The shrinking of the polymer chains when the temperature T decreases:
× the measured gyration swelling,
$\mathfrak{X}_G = R_G^2/{}^0R_G^2$ ($T = 308$ K) (from Perzynski[2])
—— the end-to-end swelling calculated by using various methods
$\mathfrak{X}$ eqn (16.5.7)
$\mathfrak{X}_{0L}$ eqn (16.5.8) (linear approximation)
$\mathfrak{X}_0/{}^0A_0(y)$ eqn (16.5.9) a value obtained by renormalization)
···· straight line resulting from a shift of $\mathfrak{X}_0/{}^0A(y)$, such that $\mathfrak{X}_0/{}^0A_0(y) = 1$ at $T = 308$ K.

by

$$z = z_{\mathrm{L}} = 3\pi y = 0.027.$$

However, let us observe that in the range $304 < T < T_{\mathrm{F}}$, the slope of the straight line representing $\mathfrak{X}_{0\mathrm{L}}$ is larger that the slope of the experimental curve corresponding to the observed values of the swelling.

Actually, as we showed in Chapter 15, Section 6, it is necessary to express the swelling, in the framework of the renormalized tricritical theory, in terms of the osmotic parameters g (two-body) and h (three-body). Thus, we can use the formula (14.6.54)

$$\mathfrak{X}_0 = {}^0A_0(y)\left[1 - \frac{148}{33}\pi h\right] + \frac{4}{3} z A_4(y)(h/y)^{4/11}$$

which is valid for values of T close to T_{F}. Using the reasonable approximation $A_4(y)/{}^0A_0(y) = 1$, we write

$$\mathfrak{X}_0/{}^0A_0(y) = \left(1 - \frac{148}{33}\pi h\right) + \frac{4}{3} z(h/y)^{4/11} \qquad (16.5.9)$$

where ${}^0A_0(y)$ can be considered as equal to unity and where, according to (14.6.52),

$$h \simeq \frac{1}{44\pi \ln (S/s_1)}.$$

For $M_{\mathrm{W}} = 1.26 \times 10^6$, we have $S/s_1 \simeq N = 1.21 \times 10^4$. Thus, $h = 7.7 \times 10^{-4}$. The most important correction term in the right-hand side of equation (16.5.9) is the term proportional to z. Then, we see that (16.5.9) and (16.5.8) are quite different, because of the factor $(h/y)^{4/11}$ the value of which here is 0.617, and this difference, a consequence of the tricritical renormalization, is observable. The calculated values of $\mathfrak{X}_0/{}^0A_0(y)$ are shown in Fig. 16.11 and in Table 16.7. We note that, when $\mathfrak{X}_0/{}^0A_0(y) = 1$, the two-body interaction parameter z equals 0.0133 and this value corresponds to a temperature $T_{\mathscr{G}\infty\mathrm{F}}(S)$ equal to 307.5 K for $S = 2092\ \mathrm{nm}^2$ [see eqns (16.5.3) and (16.1.1)].

This reference temperature defining the Brownian state differs from the one used by Perzynski[2] ($T = 308$ K) to determine the gyration swelling $\mathfrak{X}_{\mathrm{G}}$. Thus, to compare the calculated value of $\mathfrak{X}_0/{}^0A_0(y)$ with the measured value of $\mathfrak{X}_{\mathrm{G}}$, it is proper to shift the points of ordinate $\mathfrak{X}_0/{}^0A_0(y)$ by 0.5° to the right. The result of this operation appears in Fig. 16.11. The agreement between $\mathfrak{X}_{\mathrm{G}}$ and $\mathfrak{X}_0/{}^0A_0(y)$ is very satisfactory in the range $303 < T < T_{\mathrm{F}}$. The tricritical effects related to the attractive and repulsive interactions are now reasonably taken into account near T_{F}.

Finally, we see from Fig. 16.11 that a chain shrinks more strongly when the temperature becomes lower than 303 K: we say that the chain collapses.

6. A STUDY OF THE STRUCTURE OF THE SOLUTE AS A FUNCTION OF CONCENTRATION: OSMOTIC PRESSURE OF CONCENTRATED SOLUTIONS

For temperatures near T_F and overlapping chains (area III of the diagram represented on Fig. 13.26 (p. 642), the osmotic pressure reads [see (14.6.71)]

$$\Pi\beta A(y) = CA(y) + S^{3/2}\left[-4h + \frac{b}{2}S^{1/2}(2\pi)^{-3/2}A_4(y)(h/y)^{4/11}\right][CA(y)]^2 + S^3\frac{h}{3}[CA(y)]^3 \qquad (16.6.1)$$

where C is the chain-concentration. The osmotic coefficient h has a finite value when the overlap is finite. By definition, we have $A(y) = (2\pi)^{3/2}\,({}^0A_0(y))$ where ${}^0A_0(y)$ is probably close to unity.

The dominant term in this equation is the third term, which is proportional to C^3. Several authors[19, 20] tried to verify the existence of such a term by measuring the osmotic pressure and the compressibility of a solution at T_F, for various

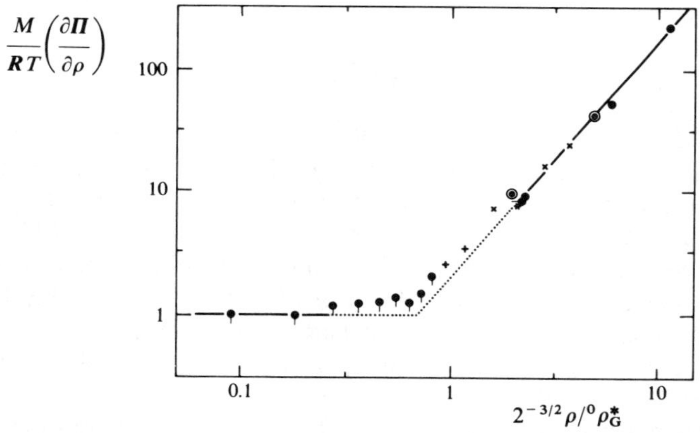

Fig. 16.12. The osmotic compressibility plotted against the overlap ratio at temperature T_F (from Stepanek, Perzynski, Delsanti, and Adam[19]). Polystyrene in cyclohexane Values measured by light scattering

+ $M_W = 4.22 \times 10^5$ ⊙ $M_W = 3.84 \times 10^6$
∗ $M_W = 1.26 \times 10^6$ × $M_W = 6.67 \times 10^6$

Values measured at the osmotic equilibrium (osmotic pressure)

- $M_n = 1.3 \times 10^5$
- $M_n = 7.2 \times 10^4$
- • $M_n = 20.6 \times 10^6$

concentrations with $C > S^{-3/2}$ where $3S = {}^0R^2$ is the mean square end-to-end distance of the equivalent Brownian chain. The result is given by Fig. 16.12 where

$$\frac{M}{RT}\left(\frac{\partial \Pi}{\partial \rho}\right)$$

is plotted against the ratio $\rho/{}^0\rho^*_{\mathrm{G,zw}}$, where

$${}^0\rho^*_{\mathrm{G,zw}} = \frac{M_\mathrm{W}}{2^{3/2}({}^0R^3_{\mathrm{G,z}}\text{Ⓐ})}. \tag{16.6.2}$$

If we neglect the polydispersion effects, we can write the overlap ratio as follows

$$CX^3 \simeq \rho/{}^0\rho^*_{\mathrm{G,zw}}.$$

Figure 16.12 shows that the reduced osmotic pressure $\Pi M/RT\rho$ is nearly proportional to $(\rho/{}^0\rho_\mathrm{G})^2$ which is what we expect. Let us calculate the proportionality factor. Keeping only the third term in formula (16.6.1), we get

$$\frac{M}{RT}\left(\frac{\partial \Pi}{\partial \rho}\right) \simeq h[A(y)]^2(\rho/{}^0\rho^*_\mathrm{G})^2. \tag{16.6.3}$$

With the help of the results shown in Fig. 16.12 we obtain the mean value

$$h[A(y)]^2 = 0.71,$$

and accordingly

$$h[{}^0A_0(y)]^2 = 2.8 \times 10^{-3}. \tag{16.6.4}$$

Nevertheless, the three-body osmotic parameter h is not really a constant: for a semi-dilute solution, it depends on the area S and on the concentration C

$$h \simeq \frac{1}{88\pi(\ln CSs_1^{1/2})} \qquad CSs_1^{1/2} < 1.$$

Thus, the slope h should slightly increase with $\mathscr{C} = CS$ but the effect cannot be detected from the results of Fig. 16.12.

Noda *et al.*[21] pursued this experimental study, but at higher concentrations. They measured the osmotic pressure of polymer chains in solution, for $T > T_\mathrm{F}$ and $\mathscr{C} > \mathscr{C}^{**}$ so as to remain always in the poor solvent state with a strong chain overlap (see Fig. 13.26, p. 642). In this physical situation, the volume fractions of polymer are high: for instance, it may occur that $\varphi > 0.3$. Thus, one gets out of the theoretical framework fixed in Chapters 13 and 14. To interpret the pressure measured by the authors quoted above, a theory for the liquid polymer state is needed. However, we present here their results without referring to any theory of this sort because, *per se*, these results manifest properties which are those of solutions of overlapping chains with $\varphi \ll 1$ (they were studied in Chapter 15 and at the beginning of this section).

When the condition $\varphi \ll 1$ is no longer satisfied, the osmotic pressure ceases from being a universal function of CX^3 (which represents the rate of overlap in semi-dilute solutions). Then, in such conditions, the volume fraction φ becomes the essential parameter which determines the pressure. In fact, this is just the property that Noda *et al.*[21] showed experimentally.

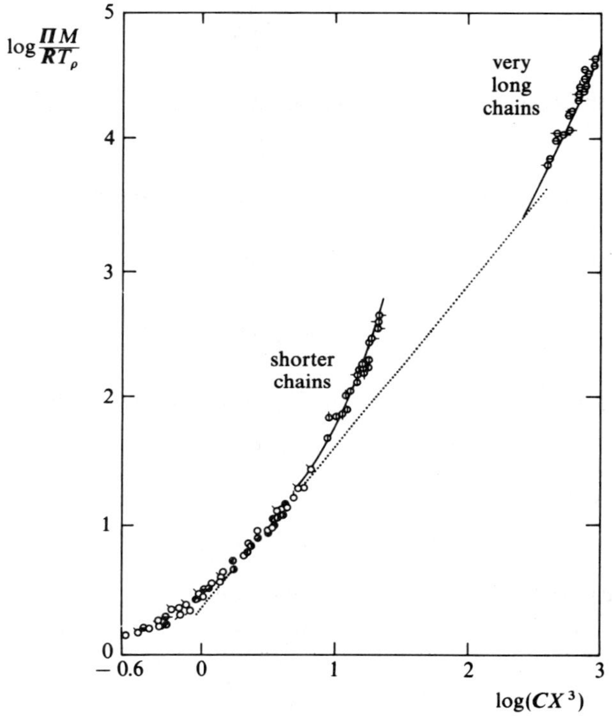

Fig. 16.13. The reduced osmotic pressure plotted against the rate of overlap (from Noda *et al.*[21]). Polystyrene in benzene.

o ȯ ƀ ρ	experimental points previously presented (see Fig. 15.19, p. 773)
o ɸ ȸ φ	sample $M_n = 6.3 \times 10^4$; osmotic pressure deduced from vapour pressures of solvent
ɵ -ɵ	sample $M_W = 9 \times 10^5$; osmotic pressure deduced from vapour pressures of solvent
. . . .	asymptotic form (13.2.29) of the universal function corresponding to overlapping polymers in good solvent; when the concentration goes beyond a certain limit, the measured pressures do not follow any universal law; this limit is smaller for short chains (Φ)then for long chains (⊖); these results do not depend on the temperature of the solution in the range 15–60 °C.
——	visual guide; it is possible to reproduce the results on the right-hand side of the figure with the help of the Flory–Huggins theory.

Let us consider the results they obtained for two different samples of polystyrene of molecular masses $M_n = 6.3 \times 10^4$ and $M_W = 9 \times 10^5$ respectively. The solvent is benzene. The solutions were prepared for various volume fractions up to the value $\varphi \simeq 0.8$.

The osmotic pressure is too high to be determined experimentally at the osmotic equilibrium. Thus, one measures the (solvent) vapour pressure $\boldsymbol{P}^{\mathrm{I}}$ of a solution containing a fraction φ of solute, and the vapour pressure $\boldsymbol{P}^{\mathrm{II}}$ of pure solvent. The osmotic pressure of the solution (with volume fraction φ) is deduced from the relation (see Chapter 5, Section 1.5)

$$\frac{\boldsymbol{P}^{\mathrm{I}}}{\boldsymbol{P}^{\mathrm{II}}} = \exp(-\,\boldsymbol{\beta\Pi}\, v_{0\mathrm{L}})$$

where $v_{0\mathrm{L}}$ is the partial volume of the solvent in the liquid state.

The measured pressures are shown in Fig. 16.13. The reduced osmotic pressure $\boldsymbol{\Pi M/RT}\rho$ is plotted against the rate of overlap [see (15.4.2)]

$$\boldsymbol{C}X^3 = (\rho/\rho_G^*)\aleph^{-3/2}$$

where $\boldsymbol{C}$ is the chain concentration, ρ the mass concentration and $\aleph = 6R_G^2/R^2$

Figure 16.13 shows, first of all, the existence of an interval in which the reduced osmotic pressure is a universal function of the rate of overlap. This result is a confirmation of those which were obtained in Chapter 15, Section 4. On the other hand, when the concentration becomes large (i.e. $\boldsymbol{C}X^d > 10$, in Fig. 16.13), the universal behaviour disappears.

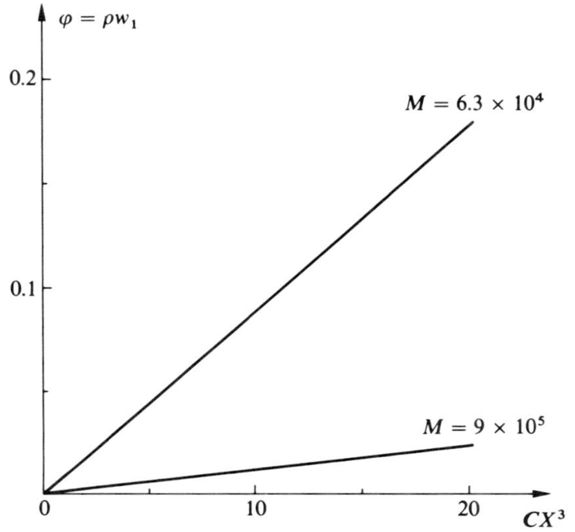

Fig. 16.14. Volume fraction φ of the solute plotted against the rate of overlap $\boldsymbol{C}X^3$, for various molecular masses M; polystyrene and benzene.

In particular, for the same rate of overlap, the deviation of the osmotic pressure of a solute with respect to the universal function is much larger if it is made of short chains than if it is made of long chains. Now we note that, for a fixed rate of overlap, the volume fraction φ of solute increases when the molecular mass of the chains decreases.* This fact appears in Fig. 16.14. Thus, when the concentration is large enough [see (8.2.14)], the volume fraction φ is the essential parameter which determines the osmotic pressure and this result looks very reasonable. Actually the authors quoted above[21] showed that under such conditions, the measured pressures can easily be interpreted with the help of the Flory–Huggins equation (14.5.4). Accordingly, the existence of a concentration domain where the reduced osmotic pressure is a universal function of the rate of overlap appears all the more remarkable.

7. DEMIXTION OF A LIQUID MADE OF DEUTERATED AND NON-DEUTERATED POLYMERS BELONGING TO THE SAME SPECIES

Mixtures of polymer chain belonging to the same chemical species but with different isotopic compositions (deuterated and non-deuterated) have been widely used for experimental studies of polymer structures, since good neutron beams became available. This technique, combining the preparation of adequate samples and neutron scattering experiments, enabled the experimentalists to determine the size of polymer chains (polystyrene or polydimethylsiloxane), in all kinds of polymer mixtures or concentrated polymer solutions. However, the technique relies on the fact that the deuterated and non-deuterated isotopic varieties of a same polymer are compatible with one another. It is admitted that under the experimental conditions described above, the mixture constitutes a unique phase. In fact, the mixing energy of deuterated and non-deuterated chains is probably very small. However, it is non-zero, in particular, because of differences in atomic volumes and polarizabilities. Thus, there is no doubt that demixtion may occur in mixtures of deuterated and undeuterated chains of very high molecular masses.

For polystyrene, Strazielle and Benoit[4] deduced from thermodynamical data that the chains used in current scattering experiments have a number of links which is much smaller than the critical number (for demixtion). However, the case of polystyrene is special. We note that demixtion can be observed only above the glass transition temperature and, for polystyrene, this temperature is rather high ($T_g = 380\,\mathrm{K}$).

However, for other polymers, the glass transition temperature is much lower: thus $T_g \simeq 188\,\mathrm{K}$ for polydimethylsiloxane (see Chapter 1, Section 4.5) and

* In fact, we have $\varphi = (\rho/\rho^*)\rho^* w_1 \propto (CX^3)M^{1-\nu d}$

$T_g = 174$ K for polybutadiene*.

$$-\!\!\left(\mathrm{CH_2{-}CH{=}CH{-}CH_2}\right)\!\!-_N$$

With such materials, it is possible to make experiments at lower temperatures and to observe demixtion for small degrees of polymerization.

For instance, neutron scattering experiments performed by Bates, Wignall, and Koehler[22] with an isotopic mixture of polybutadiene at temperature $T = 296$ K showed the existence of this effect (Oak Ridge, USA, 1985). The mixture was made of equal volumes of deuterated and non-deuterated polybutadiene ($N_W = 4600$ and $N_W/N_n = 1.10$ for the deuterated sample, $N_W = 4200$ and $N_W/N_n = 1.12$ for the undeuterated one). From the intensity $I(q)$ and the value K_A of the instrument constant, they deduced the scattering cross-section per unit volume [see (7.2.54)]

$$\Xi(q) = I(q)/K_A V$$

which is plotted against q in Fig. 16.15.

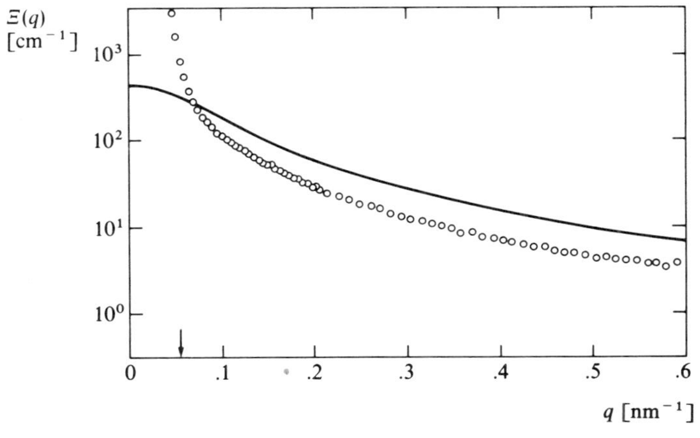

Fig. 16.15. The differential cross-section per unit volume $\Xi(q)$ plotted against the transfer wave number q. Scattering of neutrons by a mixture made of equal volumes of deuterated and undeuterated polybutadiene.

—— eqn (16.7.2); the value of S is $S = 2R_G^2 = 729.6\ \mathrm{nm}^2$ where R_G is the radius of gyration of the equivalent Brownian chain

o values measured by Bates *et al.*,[22] at $T = 296$ K. The arrow corresponds to the value $q = 1/R_G = 0.052\ \mathrm{nm}^{-1}$.

* This (synthetical) elastomer is analogous to the (natural) polyisoprene and can also be vulcanized to make rubber (see Chapter 1, Section 4.4).

For binary mixtures, we gave the general formula (7.2.29)

$$\Xi(q) = \hat{\mathfrak{b}}^2\hat{C}^2\hat{\boldsymbol{H}}(q) + \sum_{\mathcal{A}=1}^{2} (\hat{\mathfrak{b}}_{\mathcal{A}}^2 - \hat{\mathfrak{b}}^2)C_{\mathcal{A}}N_{\mathcal{A}}H_{\mathcal{A}}(\vec{q}) +$$

$$+ \sum_{\mathcal{A}=1}^{2}\sum_{\mathcal{B}=1}^{2} (\hat{\mathfrak{b}}_{\mathcal{A}} - \hat{\mathfrak{b}})(\hat{\mathfrak{b}}_{\mathcal{B}} + \hat{\mathfrak{b}})C_{\mathcal{A}}C_{\mathcal{B}}\boldsymbol{H}^{\mathrm{II}}_{\mathcal{A}\mathcal{B}}(\vec{q}), \qquad (16.7.1)$$

and we shall use it here. The mixture being incompressible, we set $\hat{\boldsymbol{H}}(\vec{q}) = 0$ for the values of $\vec{q}$ under consideration. We suppose that the chains have the same number N of links: consequently, in the homogeneous phase, they have the same form function $H(\vec{q})$. Moreover, we may assume that the chains are Brownian: in other words $H(\vec{q}) = {}^0H(\vec{q})$ [see (13.1.151)]. If the molecules belonging to the deuterated and non-deuterated isotopic varieties were uncorrelated, then $\boldsymbol{H}^{\mathrm{II}}_{\mathcal{A}\mathcal{B}}(\vec{q})$ would not depend on $\mathcal{A}$ and $\mathcal{B}$, and the last term in (16.7.1) would vanish (see Chapter 7, Section 2.3.1). In this way, one would get

$$\Xi(q) = (\mathfrak{b}_1 - \mathfrak{b}_2)^2 \frac{C_1C_2}{C} N\, {}^0H(\vec{q}) \qquad (16.7.2)$$

This expression is represented by a solid line in Fig. 16.15.

Actually, the scattering measurement shows that $\Xi(q)$ does not follow this simplified law. The contribution of the intermolecular structure functions $H^{\mathrm{II}}_{\mathcal{A}\mathcal{B}}(\vec{q})$ results in a strong forward scattering which shows us that the mixture is unhomogeneous and does not in fact constitute a unique phase. It is to be noted that demixtion in polymer melts is not quite the same as in polymer solutions. In particular, it seems that a simple-tree approximation is, in practice, enough to describe the coexistence curve in polymer melts.

REFERENCES

1. Shinozaki, K., Van Tan, T., Saito, Y., and Nose, T. (1982) *Polymer* **23**, 728.
2. Perzynski, R. (1984). Thesis. Paris VI.
3. Weill, G. and des Cloizeaux, J. (1979). *J. Physique* **40**, 99.
4. Strazielle, C. and Benoit, H. (1975). *Macromolecules* **8**, 203.
5. Brandrup, J. and Immergut, E.H. (eds.) (1975). *Polymer handbook.* John Wiley, New York.
6. Abe, M. and Fujita, H. (1965). *J. Phys. Chem.* **69**, 3263.
7. Dobashi, T., Nakata, M., Kaneko, M. (1980). *J. Chem. Phys.* **72**, 6685; (1980) **72**, 6692.
8. Kojima, J., Kuwahara, N., and Kaneko, M. (1975). *J. Chem. Phys.* **63**, 333.
9. Hamano, K., Kuwahara, N., and Kaneko, M. (1979). *Phys. Rev.* A **20**, 1135.
10. Shinozaki, K., Hamada, T., and Nose, T. (1982). *J. Chem. Phys.* **77**, 4734.
11. Le Guillou, J.C., and Zinn-Justin, J. (1977). *Phys. Rev. Lett.* **39**, 95.
12. Bray, A.J. (1976). *Phys. Rev. Lett.* **36**, 285.
13. Vincentini-Missoni, M. (1971). Contribution to *Proceedings of the International School of Physics 'Enrico Fermi', Varenna,* Course LI (ed. M.S. Green), p. 157. Academic Press.

14. Debye, P. (1989). *J. Chem. Phys.* **31**, 680.
15. de Gennes, P.G. (1968). *Phys. Lett.* **26A**, 313.
16. de Gennes, P.G. (1979). *Scaling concepts in polymer physics*, Chapter 7, p. 213. Cornell University Press, Ithaca, NY.
17. Nierlich, M., Cotton, J.P., and Farnoux, B. (1978). *J. Chem. Phys.* **69**, 1379.
18. Perzynski, R., Adam, M., and Delsanti; M. (1982). *J. Physique* **43**, 129.
19. Stepanek, P., Perzynski, R., Delsanti, M., and Adam, M. (1984). *Macromolecules* **17**, 2340.
20. Candau, F., Strazielle, C., and Benoit, H. (1976)). *Eur. Polym. J.* **12**, 95.
21. Noda, I., Higo, Y., Ueno, N., and Fujimoto, T. (1984). *Macromolecules* **17**, 1055.
22. Bates, F., Wignall, G.D., and Koehler, W.C. (1985). *Phys. Rev. Lett.* **55**, 2425.

APPENDIX A

GAUSSIAN RANDOM SETS, PROBABILITY OVER A SUB-SET, WICK'S THEOREM

1. Definition

Let us consider a set of random variables $x_1, \ldots, x_N$. We say that these variables constitute a Gaussian set if the probability (or the weight) which is associated with them has the form

$$P(x_1, \ldots, x_N) \propto \exp\left[-\frac{1}{2} \sum_{jl} A_{jl} x_j x_l \right] \tag{A.1}$$

To define the quadratic form, we can always choose symmetrical coefficients $A_{jl} = A_{lj}$ which must be considered as the matrix elements of a definite positive operator A. In other words, there always exists a positive number η having the property that for any non-zero set $x \equiv (x, \ldots, x_N)$,

$$\mathscr{U} = \sum_{jl} A_{jl} x_j x_l > \eta \sum_j x_j^2, \tag{A.2}$$

which can also be written in the form

$$\mathscr{U} = (x \cdot A \cdot x) > \eta x^2. \tag{A.3}$$

2. Generalizations

The concept of a Gaussian set can be extended to infinite sets of variables. These variables whose number is infinite, may be discrete and in this case $\mathscr{U}$ is expressed as an infinite sum; then the corresponding value of $\mathscr{U}$ is in general infinite and the postulate (A.1) becomes purely formal. However, the mean values associated with a finite subset of these variables may remain finite, and in this case they can easily be calculated.

The random variables may also be continuous functions of a parameter; then the sum is replaced by an integral.

3. Diagonalization

The quadratic form

$$\mathscr{U} = \sum_{j=1}^{N} \sum_{l=1}^{N} A_{jl} x_j x_l \tag{A.4}$$

where the A_{jl} are the elements of a positive-definite real-symmetric matrix, can be diagonalized with the help of a real orthogonal transformation $\mathscr{O}$

$$x_j = \sum_l \mathscr{O}_{jl} y_l$$

(here $\mathcal{O} = \mathcal{O}^*$, $\mathcal{O}\tilde{\mathcal{O}} = 1$, where $\mathcal{O}^*$ is the complex conjugate of $\mathcal{O}$ and $\tilde{\mathcal{O}}$ the transpose of $\mathcal{O}$).

The diagonalization of A with the help of $\mathcal{O}$ can be expressed by the equalities

$$x = \mathcal{O} \cdot y$$
$$\mathcal{U} = (x \cdot A \cdot x) = (y \cdot \tilde{\mathcal{O}} A \mathcal{O} \cdot y) = \sum_l A_l y_l^2 \tag{A.5}$$

where the A_l are the eigenvalues of A; the components of the eigenvector corresponding to A_l are constituted by the series $(\mathcal{O}_{1l}, \ldots, \mathcal{O}_{Nl})$.

4. Normalization

Let us consider

$$Z = \int \mathrm{d}x_1 \ldots \mathrm{d}x_N \exp\left[-\frac{1}{2} \sum_{jl} A_{jl} x_j x_l \right]. \tag{A.6}$$

By diagonalization, we obtain

$$Z = \int \mathrm{d}y_1 \ldots \mathrm{d}y_N \exp\left[-\frac{1}{2} \sum_{l=1}^{N} A_l y_l^2 \right]$$
$$Z = (2\pi)^{N/2} \left[\prod_1^N A_l \right]^{-1/2}$$

On the other hand, we can calculate the determinant associated with the quadratic form

$$\prod_1^N A_l = \det[\tilde{\mathcal{O}} A \mathcal{O}] = \det[\mathcal{O}^{-1} A \mathcal{O}] = \det A. \tag{A.8}$$

Consequently,

$$Z = (2\pi)^{N/2} [\det A]^{-1/2}. \tag{A.9}$$

Thus, the probability $P(x_1, \ldots, x_N)$ is given by

$$P(x_1, \ldots, x_N) = \frac{|\det A|^{1/2}}{(2\pi)^{N/2}} \exp\left[-\frac{1}{2}(x \cdot A \cdot x) \right]. \tag{A.10}$$

5. Characteristic function

The Fourier transform of the probability distribution is called the characteristic function; it is a function of N variables $k_1, \ldots, k_N$ and can be defined as follows

$$\tilde{P}(k_1, \ldots, k_N) = \langle\!\langle \exp[i(k_1 x_1 + \ldots + k_N x_N)] \rangle\!\rangle$$
$$= \int \mathrm{d}x_1 \ldots \mathrm{d}x_N \exp[i(k_1 x_1 + \ldots + k_N x_N)]\, P(x_1, \ldots, x_N). \tag{A.11}$$

Using the notation

$$k \cdot x = \sum_j k_j x_j$$

we have

$$\tilde{P}(k_1, \ldots, k_N) =$$
$$= \frac{1}{Z}\int dx_1 \ldots dx_N \exp\left[-\frac{1}{2}x\cdot A\cdot x + ik\cdot x\right]$$
$$= \frac{1}{Z}\int dx_1 \ldots dx_N \exp\left[-\frac{1}{2}(x - ik\cdot A^{-1})\cdot A\cdot(x - iA^{-1}\cdot k) - \frac{1}{2}k\cdot A^{-1}\cdot k\right]. \tag{A.12}$$

Let us now change variables

$$(x - iA^{-1}\cdot k) \to x$$

and let us use definition (A.6); we obtain

$$\tilde{P}(k_1, \ldots, k_N) = \exp[-\tfrac{1}{2}k\cdot A^{-1}\cdot k] \tag{A.13}$$

which can be written in the more explicit form

$$\tilde{P}(k_1, \ldots, k_N) = \exp\left[-\frac{1}{2}\sum_{jl} B_{jl}k_jk_l\right] \tag{A.14}$$

with

$$\sum_m B_{jm}A_{ml} = \delta_{jl} \qquad (B = A^{-1}).$$

Thus, the characteristic function of a Gaussian ensemble is Gaussian and conversely.

Consequently, a Gaussian ensemble can conveniently be defined by its characteristic function.

6. Probability over a subset

Any subset of variables obtained from linear combinations of Gaussian variables is Gaussian for the following reason. Let us consider independent linear combinations z_α of the x_j (with $\alpha = 1, \ldots, M$ and $M \leqslant N$)

$$z_\alpha = \sum_j L_{\alpha j}x_j \qquad (\text{we write: } z = L\cdot x). \tag{A.15}$$

Let us try to determine the probability law $Q(z_1, \ldots, z_M)$ associated with these variables. The corresponding characteristic function is

$$\tilde{Q}(q_1, \ldots, q_M) = \left\langle\!\!\left\langle \exp\left[i\sum_\alpha q_\alpha z_\alpha\right]\right\rangle\!\!\right\rangle \tag{A.16}$$

or in a more compact form

$$\tilde{Q}(q_1, \ldots, q_M) = \langle\!\langle\, e^{iq\cdot z}\rangle\!\rangle = \langle\!\langle\, e^{iq\cdot L\cdot x}\rangle\!\rangle = \exp[-\tfrac{1}{2}q\cdot LB\tilde{L}\cdot q]. \tag{A.17}$$

This new characteristic function is also Gaussian.

Let us now set

$$C = (LB\tilde{L})^{-1} = LA^{-1}\tilde{L} \tag{A.18}$$

where C is a square matrix. By using the general formulae (A.10) and (A.13), we obtain

$$\tilde{Q}(q_1, \ldots, q_M) = \exp\left[-\frac{1}{2}(q \cdot C \cdot q)\right]$$

$$Q(z_1, \ldots, z_M) = \frac{|\det C|^{-1/2}}{(2\pi)^{M/2}} \exp\left[-\frac{1}{2}(z \cdot C^{-1} \cdot z)\right]. \tag{A.19}$$

7. Mean values of products of random variables

The characteristic function is also the generating function of mean values. By expanding eqn (A.11), we find

$$\tilde{P}(k_1, \ldots, k_N) = \sum (i)^{n_1 + \ldots + n_N} \frac{k_1^{n_1} \ldots k_N^{n_N}}{n_1! \ldots n_N!} \langle\!\langle x_1^{n_1} \ldots x_N^{n_N} \rangle\!\rangle \tag{A.20}$$

This quantity is real because the mean values corresponding to an uneven number of variables vanish. In particular, by expanding (A.14) with respect to the k_j and by identifying the expansion with (A.20), we find

$$\begin{aligned} \langle\!\langle x_j \rangle\!\rangle &= 0 \\ \langle\!\langle x_j x_l \rangle\!\rangle &= B_{jl} \end{aligned} \tag{A.21}$$

Bringing these values into (A.14) and using definition (A.11), we obtain the classical result

$$\tilde{P}(k_1, \ldots, k_N) = \left\langle\!\!\left\langle \exp\left[i \sum_j k_j x_j\right] \right\rangle\!\!\right\rangle = \exp\left[-\frac{1}{2} \sum_{jl} \langle\!\langle x_j x_l \rangle\!\rangle k_j k_l\right]. \tag{A.22}$$

Thus, the set of the average values $\langle\!\langle x_j x_l \rangle\!\rangle$ completely determines the probability law of the Gaussian ensemble.

8. Wick's theorem

The mean values of products of variables x_j can directly be expressed in terms of the mean values $\langle\!\langle x_j x_l \rangle\!\rangle$. This property is a consequence of Wick's theorem, the importance of which is fundamental in perturbation theory.

Proposition

Let us consider the variables $X_1, \ldots, X_{2n}$, which may be different or not, and are such that each of them is a linear combination of Gaussian variables. Then, we have

$$\langle\!\langle X_1, \ldots, X_{2n} \rangle\!\rangle = \sum_j \langle\!\langle X_1 X_j \rangle\!\rangle \langle\!\langle X_2, \ldots, X_{j-1} X_{j+1} \ldots X_{2n} \rangle\!\rangle$$

which, by iteration, gives

$$\langle\langle X_1, \ldots, X_{2n} \rangle\rangle = \frac{1}{2^n n!} \sum_{\wp} \langle\langle X_{\wp 1} X_{\wp 2} \rangle\rangle \ldots \langle\langle X_{\wp(2n-1)} X_{\wp(2n)} \rangle\rangle \quad \text{(A.23)}$$

where $\wp$ is any permutation of the integers $(1, \ldots, 2n)$ and where $\wp j$ is the transform of j in this permutation. The formula means that, to calculate $\langle\langle X_1, \ldots, X_{2n} \rangle\rangle$, we have only to associate the factors x_j, two by two, in all possible ways. A product of n factors $\langle\langle X_j X_l \rangle\rangle$ corresponds to each association.

Proof

First, let us establish the result for one Gaussian variable. In this case, we have [see (A.22)]

$$\langle\langle e^{ikx} \rangle\rangle = \exp[-\tfrac{1}{2}k^2 \langle\langle x^2 \rangle\rangle]. \quad \text{(A.24)}$$

Expanding both sides, let us identify the coefficients of the powers of k. We obtain

$$\langle\langle x^{2n+1} \rangle\rangle = 0$$

$$\langle\langle x^{2n} \rangle\rangle = \frac{(2n)!}{2^n n!} [\langle\langle x^2 \rangle\rangle]^n = (2n-1)(2n-3) \ldots 1 \, [\langle\langle x^2 \rangle\rangle]^n. \quad \text{(A.25)}$$

As the number of permutations of $2n$ variables is $(2n)!$, we immediately see that in the case where there is only one variable, (A.23) coincides with (A.25). Let us now assume that each variable X_j in the product is equal to one of the independent variables y_l defined in Section 3. The Gaussian variables are independent if the quadratic form $\mathscr{U}$, which appears in the probability law, is diagonal [see (A.6)]. In this case, we can factorize

$$X_1 \ldots X_n = \prod_l (y_l)^{\alpha_l} \qquad (\alpha_l = \text{integer})$$

and, as the variables are independent,

$$\langle\langle X_1 \ldots X_n \rangle\rangle = \prod_l \langle\langle (y_l)^{\alpha_l} \rangle\rangle$$

Then, it is a straightforward matter to verify that Wick's theorem, which applies to each term separately, applies also the whole set, since we have here

$$\langle\langle y_l y_m \rangle\rangle = \langle\langle (y_l)^2 \rangle\rangle \, \delta_{lm}.$$

Now, we observe that Wick's theorem remains valid when the variables X undergo linear transformations. Accordingly, the theorem applies if each X is equal to a variable x_j which is a linear combination of the y_l; it also applies when the X are arbitrary linear combinations of Gaussian variables.

The theorem is thus proved in all cases.

APPENDIX B

NUMBER OF PARTITIONS OF AN INTEGER: HARDY–RAMANUJAN THEOREM

Let $\mathcal{N}(N)$ be the number of partitions of an integer N (Example: the set of numbers (1, 3, 3, 5) is a partition of 12).

This function $\mathcal{N}(N)$ can be defined by the characteristic function

$$Q(x) = \sum_{N=0}^{\infty} x^N \mathcal{N}(N) \qquad \text{(we set } \mathcal{N}(0) = 1\text{)}, \tag{B.1}$$

which is given by

$$Q(x) = \prod_{n=1}^{\infty} (1 - x^n)^{-1}, \tag{B.2}$$

as can be easily understood by expanding each term in the product. The function $Q(x)$ is analytic in a circle of modulus 1 whose edge is a natural boundary of $Q(x)$ and therefore cannot be crossed. The asymptotic value of $\mathcal{N}(N)$ for large values of n is given by the Hardy–Ramanujan theorem[1]

$$\mathcal{N}(N) \sim \exp[\pi(2N/3)^{1/2}] \equiv \kappa^{N^{1/2}} \tag{B.3}$$

To obtain this result, we can proceed as follows: the quantity $\mathcal{N}(N)$ is given by Cauchy's formula

$$\mathcal{N}(N) = \frac{1}{2\pi i} \int_C \frac{Q(x)}{x^{N+1}} \mathrm{d}x, \tag{B.4}$$

and the contour C is a closed curve around the origin. For large values of N, the integral can be evaluated by using the steepest descent method and we write

$$\mathcal{N}(N) = \frac{1}{2\pi i} \int_C \mathrm{e}^{\varphi(x)} \mathrm{d}x$$

where

$$\varphi(x) = \ln Q(x) - (N + 1)\ln x. \tag{B.5}$$

It is easy to convince oneself that the dominant term comes from a saddle point lying on the segment (0.1) very near to the point $x = 1$.

Let us evaluate $Q(x)$ in the vicinity of this point

$$\ln Q(x) = -\sum_{n=1}^{\infty} \ln(1 - x^n) = \sum_{n=1}^{\infty} \sum_{m=1}^{\infty} \frac{x^{mn}}{m} = \sum_{m=1}^{\infty} \frac{x^m}{(1 - x^m)m}.$$

Now, from the identity

$$\frac{1 - x}{1 - x^m} = \frac{1}{1 + x + \ldots + x^{m-1}} \tag{B.6}$$

we deduce that for $0 < x < 1$

$$\frac{1}{mx^{m-1}} > \frac{1-x}{1-x^m} > \frac{1}{m}.$$

Multiplying these inequalities by $x^m/m(1-x)$ and summing, we obtain the bounds

$$\frac{1}{1-x}\sum_{m=1}^{\infty}\frac{x}{m^2} > \ln Q(x) > \frac{1}{1-x}\sum_{m=1}^{\infty}\frac{x^m}{m^2}. \tag{B.7}$$

Moreover, we have

$$\sum_{m=1}^{\infty}\frac{1}{m^2} = \frac{\pi^2}{6}. \tag{B.8}$$

Consequently, for values of x close to unity

$$\ln Q(x) \simeq \frac{\pi^2}{6(1-x)} \tag{B.9}$$

The value x_c of x at the saddle point is given by $\partial\varphi(x)/\partial x = 0$ which leads to the result

$$(1-x_c)^2 \simeq \frac{\pi^2}{6N}. \tag{B.10}$$

Let us bring this value of x_c in $\varphi(x)$

$$\varphi(x_c) = Q(x_c) - (N+1)\ln x_c \simeq \frac{\pi^2}{6(1-x_c)} + N(1-x_c) = \pi(2N/3)^{1/2}. \tag{B.11}$$

As a consequence, we find the predicted result

$$\mathcal{N}(N) \sim \exp[\varphi(x_c)] = \exp[\pi(2N/3)^{1/2}].$$

The validity of this approach has been more rigorously established by G.H. Hardy[1] who discussed it in the book which he devoted to Ramanujan.[2] In particular, he shows in this book that for large values of N, $\mathcal{N}(N)$ is given more precisely by the formula

$$\mathcal{N}(N) \simeq \frac{1}{2\pi(2)^{1/2}}\frac{\mathrm{d}}{\mathrm{d}N}\exp\left[\pi\left(\frac{2N}{3}-\frac{1}{36}\right)^{1/2}\right]. \tag{B.12}$$

References

1. Hardy, G.H. and Ramanujan, S. (1917). *Proc. London Math. Soc.* **16**(2), 112, and (1918) **17**(2).
2. Hardy, G.H. (1940). *Ramanujan*, Chapter VIII: Asymptotic theory of partitions. Chelsea Publ. Co., New York.
   Also see Andrews, G.E. (1976). The theory of partitions. Addison-Wesley, London.

APPENDIX C

FREE ENERGY OF AN IDEAL INCOMPRESSIBLE SOLUTION AND CHEMICAL POTENTIALS

Let us consider an ideal solution made of m components in a volume V. Each chemical species is denoted by an index $\mathscr{A}$ (with $\mathscr{A} = 1, \ldots, m$). The numbers of molecules belonging to each species in the solution will be represented by the symbols $N_{\mathscr{A}}$ and the concentrations by $C_{\mathscr{A}}$

$$C_{\mathscr{A}} = N_{\mathscr{A}}/V$$

$$N = \sum_{\mathscr{A}=1}^{m} N_{\mathscr{A}}. \tag{C.1}$$

Let us try to calculate the Gibbs free energy of the solution; it reads

$$\mathbb{G} = \mathbb{F} + PV \tag{C.2}$$

where $\mathbb{F}$ is the Helmholtz free energy and P the pressure

$$\mathbb{F} = \mathbb{E} - \beta^{-1}\,\mathbb{S}. \tag{C.3}$$

Here, $\mathbb{E}$ is the energy and $\mathbb{S}$ the entropy. For an ideal solution, we may write

$$\mathbb{E} = \sum_{\mathscr{A}=1}^{m} C_{\mathscr{A}} E_{\mathscr{A}} \tag{C.4}$$

where the energies $E_{\mathscr{A}}$ are considered as constants. We have now to calculate the entropy which, in principle, depends only on the molecular volumes of the chemical species.

When the molecular volumes of the various chemical species are equal, the problem is simple. In fact, in this case, we can admit that the molecules occupy the N sites of a lattice. Then the number of different configurations is

$$Z = \frac{N!}{\prod_{\mathscr{A}} N_{\mathscr{A}}!} \tag{C.5}$$

The entropy can thus be written in the form

$$\mathbb{S} = \ln Z + \sum_{\mathscr{A}=1}^{m} S_{\mathscr{A}} N_{\mathscr{A}} \tag{C.6}$$

where the internal entropies $S_{\mathscr{A}}$ are considered as constants.

Applying Moivre's formula

$$n! \simeq (n/\mathrm{e})^n$$

we find

$$\mathbb{S} = N \ln N - \sum_{\mathscr{A}=1}^{m} N_{\mathscr{A}} \ln N_{\mathscr{A}} + \sum_{\mathscr{A}=1}^{m} S_{\mathscr{A}} N_{\mathscr{A}}. \tag{C.7}$$

Consequently, the Helmholtz free energy is given by

$$\mathbb{F} = \mathscr{F}\{N_{\mathscr{A}}\} \equiv \sum_{\mathscr{A}=1}^{m} (E_{\mathscr{A}} - \beta^{-1} S_{\mathscr{A}}) N_{\mathscr{A}} + \beta^{-1} \sum_{\mathscr{A}=1}^{m} N_{\mathscr{A}} \ln (N_{\mathscr{A}}/N) \tag{C.8}$$

Note that this quantity is calculated only for a volume $V = Nv$ depending explicitly on the $N_{\mathscr{A}}$. It is impossible to let the $N_{\mathscr{A}}$ vary without simultaneously changing the volume.

Consequently, such an expression does not allow us to calculate the chemical potentials directly

$$\mu_{\mathscr{A}} = (\partial \mathbb{F}/\partial N_{\mathscr{A}})_V$$

We need the Gibbs potential (Gibbs free energy). The Gibbs potential is given by (C.2) and (C.8); thus, we find

$$\mathbb{G} = \sum_{\mathscr{A}=1}^{m} (E_{\mathscr{A}} - \beta^{-1} S_{\mathscr{A}}) N_{\mathscr{A}} + \beta^{-1} \sum_{\mathscr{A}=1}^{m} N_{\mathscr{A}} \ln(N_{\mathscr{A}}/N) + Pv \sum_{\mathscr{A}=1}^{m} N_{\mathscr{A}}. \tag{C.9}$$

We can now extract the value of the chemical potentials from the preceding formula since they can also be defined by the expression

$$\mu_{\mathscr{A}} = (\partial \mathbb{G}/\partial N_{\mathscr{A}})_P \tag{C.10}$$

In fact, when P varies, the volume remains constant and given by the equality $V = Nv$. Thus, setting

$${}^0\mu_{\mathscr{A}} = E_{\mathscr{A}} - \beta^{-1} S_{\mathscr{A}} \tag{C.11}$$

we obtain

$$\mu_{\mathscr{A}} = {}^0\mu_{\mathscr{A}} + \beta^{-1} \ln(N_{\mathscr{A}}/N) + Pv. \tag{C.12}$$

These relations can also be written in the form

$$\mathbb{G}/V = \sum_{\mathscr{A}=1}^{m} ({}^0\mu_{\mathscr{A}} + Pv) C_{\mathscr{A}} + \beta^{-1} \sum_{\mathscr{A}=1}^{m} C_{\mathscr{A}} \ln (C_{\mathscr{A}} v)$$

$$\mu_{\mathscr{A}} = {}^0\mu_{\mathscr{A}} + \beta^{-1} \ln(C_{\mathscr{A}} v) + Pv \tag{C.13}$$

When the volumes $v_{\mathscr{A}}$ are not equal, the generalization of the preceding formulae is far from being obvious. However, we observe that in a one-dimensional space, the preceding calculation of Z remains valid. Consequently, in this case, eqn (C.8) also remains valid. Moreover, eqn (C.9) can be written as follows

$$\mathbb{G} = \sum_{\mathscr{A}=1}^{m} (E_{\mathscr{A}} - \beta^{-1} S_{\mathscr{A}}) N_{\mathscr{A}} + \beta^{-1} \sum_{\mathscr{A}=1}^{m} N_{\mathscr{A}} \ln(N_{\mathscr{A}}/N) + P \sum_{\mathscr{A}=1}^{m} N_{\mathscr{A}} v_{\mathscr{A}} \tag{C.14}$$

and (C.12) becomes

$$\mu_{\mathscr{A}} = {}^{0}\mu_{\mathscr{A}} + \beta^{-1}\ln(N_{\mathscr{A}}/N) + Pv_{\mathscr{A}}. \tag{C.15}$$

In the same way, (C.13) can be generalized provided we set

$$v = \frac{V}{N} = \frac{\Sigma N_{\mathscr{A}} v_{\mathscr{A}}}{\Sigma N_{\mathscr{A}}} \tag{C.16}$$

and in this way, we find

$$\mathbb{G}/V = \sum_{\mathscr{A}=1}^{m} ({}^{0}\mu_{\mathscr{A}} + Pv_{\mathscr{A}})C_{\mathscr{A}} + \beta^{-1} \sum_{\mathscr{A}=1}^{m} C_{\mathscr{A}} \ln(C_{\mathscr{A}} v)$$

$$\mu_{\mathscr{A}} = {}^{0}\mu_{\mathscr{A}} + \beta^{-1}\ln(C_{\mathscr{A}} v) + Pv_{\mathscr{A}}. \tag{C.17}$$

In three dimensions, the situation is more complex and volume effects certainly occur, but it seems that they have not been taken into account until now in a reasonable manner; in this case, again, we must content ourselves with the approximation described above.

When the volumes are not equal, certain authors, instead of using (C.8), prefer the expression

$$\mathbb{S} = -\sum_{\mathscr{A}=1}^{m} N_{\mathscr{A}} \ln\left(\frac{N_{\mathscr{A}} v_{\mathscr{A}}}{\sum_{\mathscr{B}=1}^{m} N_{\mathscr{B}} v_{\mathscr{B}}}\right) + \sum_{\mathscr{A}=1}^{m} S_{\mathscr{A}} N_{\mathscr{A}}. \tag{C.18}$$

However, the reasons they give for this are not really convincing, and it is difficult to see why such an expression, which is wrong in one-dimension, could be valid in three dimensions.

APPENDIX D

ELECTROMAGNETIC FIELD PRODUCED BY AN OSCILLATING ELECTRICAL DIPOLE

Let $\vec{D}(\mathrm{t}) = \vec{D}\,\mathrm{e}^{-i\omega\mathrm{t}}$ be the electrical moment of an oscillating dipole located at the origin of the co-ordinates. For instance, this moment could be created by applying an external field $\vec{E}_0(\mathrm{t})$ to a small dielectric object of polarizability α, located in 0

$$\vec{D}(\mathrm{t}) \equiv \vec{D}\mathrm{e}^{-i\omega\mathrm{t}} = \alpha\vec{E}_0(\mathrm{t}). \tag{D.1}$$

We want to show that the electric field $\vec{E}$ and the magnetic field $\vec{B}$ produced at a point of position vector $\vec{r}$ are given, for large values of r, by the following expressions

$$\vec{E}(\vec{r},\mathrm{t}) = -\left(\frac{\omega}{\mathrm{c}}\right)^2 \frac{\vec{r}\wedge(\vec{r}\wedge\vec{D})}{r^3}\,\mathrm{e}^{-i\omega(\mathrm{t}-r/\mathrm{c})} = \left(\frac{\omega}{\mathrm{c}}\right)^2 \left[\frac{\vec{D}}{r} - \frac{\vec{r}(\vec{r}\cdot\vec{D})}{r^3}\right]\mathrm{e}^{-i\omega(\mathrm{t}-r/\mathrm{c})}$$

$$\vec{B}(\vec{r},\mathrm{t}) = \left(\frac{\omega}{\mathrm{c}}\right)^2 \frac{\vec{r}\wedge\vec{D}}{r^2}\,\mathrm{e}^{-i\omega(\mathrm{t}-r/\mathrm{c})}. \tag{D.2}$$

To find this result, we have only to calculate the vector potential $\vec{A}(\vec{r},\mathrm{t})$ and the potential $V(\vec{r},\mathrm{t})$ corresponding to these fields, since

$$\vec{E}(\vec{r},\mathrm{t}) = -\frac{\partial}{\mathrm{c}\partial\mathrm{t}}\vec{A}(\vec{r},\mathrm{t}) - \vec{\nabla}V(\vec{r},\mathrm{t})$$

$$\vec{B}(\vec{r},\mathrm{t}) = \vec{\nabla}\wedge\vec{A}(\vec{r},\mathrm{t}). \tag{D.3}$$

The potential and the vector potential are related to the current $\vec{j}(\vec{r},\mathrm{t})$ and to the charge density $\rho(\vec{r},\mathrm{t})$ by the equations

$$\triangle V - \frac{1}{\mathrm{c}^2}\frac{\partial^2 V}{\partial\mathrm{t}^2} = -4\pi\rho(\vec{\mathrm{r}},\mathrm{t})$$

$$\triangle\vec{A} - \frac{1}{\mathrm{c}^2}\frac{\partial^2\vec{A}}{\partial\mathrm{t}^2} = -\frac{4\pi}{\mathrm{c}}\vec{j}(\vec{r},\mathrm{t})$$

$$\vec{\nabla}\cdot\vec{A} + \frac{\partial V}{\partial\mathrm{t}} = 0. \tag{D.4}$$

The current and the charge density obey the conservation equation

$$\frac{\partial\rho}{\partial\mathrm{t}} + \vec{\nabla}\cdot\vec{j} = 0$$

and are given by

$$\vec{j}(\vec{r},t) = \delta(\vec{r})\frac{\partial \vec{D}(t)}{\partial t}$$

$$\rho(\vec{r},t) = -\vec{D}(t)\cdot\vec{\nabla}\,\delta(\vec{r}). \tag{D.5}$$

Thus, the vector potential and the potential can be written in the form

$$\vec{A}(\vec{r},t) = \frac{\partial}{c\,\partial t}\vec{G}(\vec{r},t)$$

$$V(\vec{r},t) = -\vec{\nabla}\cdot\vec{G}(\vec{r},t) \tag{D.6}$$

where $\vec{G}(\vec{r},t)$ obeys the equation

$$\left(\triangle - \frac{1}{c^2}\frac{\partial^2}{\partial t^2}\right)\vec{G}(\vec{r},t) = -4\pi\,\vec{D}(t)\,\delta(\vec{r}). \tag{D.7}$$

Solving this equation, we find

$$\vec{G}(\vec{r},t) = \frac{\vec{D}}{r}e^{-i\omega(t-r/c)}, \tag{D.8}$$

as can easily be verified.

Now, (D.3) and (D.4) give

$$\vec{E}(\vec{r},t) = -\frac{\partial^2}{c^2\partial t^2}\vec{G}(\vec{r},t) + \vec{\nabla}[\vec{\nabla}\cdot\vec{G}(\vec{r},t)]$$

$$\vec{B}(\vec{r},t) = \frac{\partial}{c\partial t}[\vec{\nabla}\wedge\vec{G}(\vec{r},t)]. \tag{D.9}$$

Let us replace $\vec{G}(\vec{r},t)$ by its value (D.8). Taking into account the fact that $\vec{G}(\vec{r},t)$ depends on $\vec{r}$ only through its modulus r, we can write

$$\vec{E}(\vec{r},t) = \left(\frac{\omega}{c}\right)^2\vec{G}(\vec{r},t) + \frac{1}{r}\frac{\partial}{\partial r}\vec{G}(\vec{r},t) + \frac{\vec{r}}{r}\left[\vec{r}\frac{\partial}{\partial r}\left(\frac{1}{r}\frac{\partial}{\partial r}\vec{G}(\vec{r},t)\right)\right],$$

$$\vec{B}(\vec{r},t) = \left(-\frac{i\omega}{c}\right)\frac{\vec{r}}{r}\wedge\frac{\partial}{\partial r}\vec{G}(\vec{r},t), \tag{D.10}$$

or more explicitly

$$\vec{E}(\vec{r},t)e^{i\omega(t-r/c)} = \frac{\vec{D}}{r}\left[\left(\frac{\omega}{c}\right)^2 + i\left(\frac{\omega}{c}\right)\frac{1}{r} - \frac{1}{r^2}\right] + $$

$$+\frac{\vec{r}(\vec{D}\cdot\vec{r})}{r^3}\left[-\left(\frac{\omega}{c}\right)^2 - 3i\left(\frac{\omega}{c}\right)\frac{1}{r} + \frac{3}{r^2}\right]$$

$$\vec{B}(\vec{r},t)e^{i\omega(t-r/c)} = \left[\left(\frac{\omega}{c}\right)^2 + i\left(\frac{\omega}{c}\right)\frac{1}{r}\right]\frac{\vec{r}}{r^2}\wedge\vec{D}. \tag{D.11}$$

For $r \gg c/\omega$, these equalities lead to the announced result (D.2)

APPENDIX E

FORM FUNCTION OF A FEW CHARACTERISTIC STRUCTURES; BEHAVIOUR FOR LARGE WAVE VECTOR TRANSFER

1. Form function

The form function associated with a spatial structure, which may be random and which is embedded in d-dimensional space, can be defined by

$$H(\vec{q}) = \left\langle\!\!\left\langle \left| \frac{\tilde{A}(\vec{q})}{\tilde{A}(\vec{0})} \right|^2 \right\rangle\!\!\right\rangle \tag{E.1}$$

where

$$\tilde{A}(\vec{q}) = \int \mathrm{d}^d r\, A(\vec{r}) \mathrm{e}^{\mathrm{i}\vec{q}\cdot\vec{r}} \tag{E.2}$$

is the amplitude and where $A(\vec{r})$ is a density (a function or a distribution).

It is convenient to define the form function by starting from the density correlation

$$p(\vec{r}) = \int \mathrm{d}^d r'\, A(\vec{r}\,')\, A(\vec{r}\,' + \vec{r}) \tag{E.3}$$

because $H(\vec{q})$ can also be written in the form

$$H(\vec{q}) = \frac{\displaystyle\int \mathrm{d}^d r\, \mathrm{e}^{-\mathrm{i}\vec{q}\cdot\vec{r}}\, p(\vec{r})}{\displaystyle\int \mathrm{d}^d r\, p(\vec{r})} \tag{E.4}$$

In particular, we have

$$H(\vec{0}) = 1$$

$$\int \mathrm{d}^d q\, H(\vec{q}) = (2\pi)^d \frac{p(\vec{0})}{\displaystyle\int \mathrm{d}^d r\, p(\vec{r})}. \tag{E.5}$$

When the structure under consideration is a well-defined solid-body of volume v in a d-dimensional space, we set $A(\vec{r}) = 1$ if $\vec{r}$ corresponds to a point belonging to the body, and $A(\vec{r}) = 0$ otherwise. In this case, (E.3) gives

$$p(\vec{0}) = v$$
$$\int \mathrm{d}^d r\, p(\vec{r}) = v^2, \tag{E.6}$$

and consequently

$$\int \mathrm{d}^d q\, H(\vec{q}) = \frac{(2\pi)^d}{v}. \tag{E.7}$$

Moreover, we note that the density correlation $p(\vec{r})$ [see (E.3)] is just equal to the volume of the space domain belonging simultaneously to the solid body and to its replica, obtained by the translation of vector $\vec{r}$.

2. Asymptotic properties of $H(\vec{q})$

The properties of $H(\vec{q})$ for large q are directly related to the singularities of $p(\vec{r})$ and this fact is a consequence of (E.4). In particular, for a random structure, or a structure with a random position, the asymptotic behaviour of $H(\vec{q})$ depends mainly on the singularities of $p(\vec{r})$ at the origin.

A singularity of the form

$$\frac{p(\vec{r})}{\int \mathrm{d}^d r'\, p(\vec{r}\,')} = A r^{-\alpha}$$

leads to the following asymptotic behaviour

$$H(\vec{q}) \simeq A J(\alpha, d) q^{\alpha - d} \qquad \text{(for large } q\text{)} \tag{E.9}$$

where

$$J(\alpha, d) = \pi^{d/2} 2^{d-\alpha} \frac{\Gamma\left(\dfrac{d}{2} - \dfrac{\alpha}{2}\right)}{\Gamma\left(\dfrac{\alpha}{2}\right)} \tag{E.10}$$

[see Appendix M, eqns (M.2) and (M.3)].

Here, we aim at showing how the asymptotic behaviour of $H(\vec{q})$ depends, for large q, on the properties of the corresponding structure.

In particular, it will be shown that a solid D-dimensional body with a random orientation in a d-dimensional space ($d \geqslant D$) has a form function whose asymptotic behaviour is not at all the same when $d > D$ and when $d = D$.

First, we shall illustrate this remark by studying the form function of a rigid rod in various cases; then, afterwards, we shall consider the form factors associated with convex solid bodies of dimension D. Finally, we shall determine the asymptotic behaviour of the form factor of critical continuous polymer chains (Brownian chains and Kuhnian chains).

3. Form function of a rigid rod

3.1 *A rod in one-dimensional space*

Let us consider a rod of length L in a one-dimensional space. Here, we label the calculated quantities with the subscript 1. We have

$$\tilde{A}_1(\vec{q}) = \int_{-L/2}^{L/2} \mathrm{d}x\, \mathrm{e}^{iqx} = \frac{2 \sin q L/2}{q}.$$

Therefore, according to (E.1),

$$H_1(\vec{q}) = \frac{4 \sin^2 q L/2}{q^2 L^2} = \frac{2}{q^2 L^2} - \frac{2 \cos q L}{q^2 L^2}. \tag{E.11}$$

Moreover, according to (E.3)

$$\begin{aligned} p_1(\vec{x}) &= L(1 - |x|/L) \qquad |x| \leqslant L \\ &= 0 \qquad\qquad\qquad\quad |x| > L. \end{aligned} \tag{E.12}$$

The function $p_1(\vec{x})$ has singularities for $x = 0$ and $x = \pm L$. The singularity at the origin gives, in $H_1(\vec{q})$, the behaviour $2/q^2 L^2$ for large qL; the other singularity gives the oscillating term.

3.2 *Rigid rod with a random orientation in a d-dimensional space*

The rod is assumed to be homogeneous; it has a length L and zero volume. We associate with it a distribution probability $A(\vec{r})$ with $\int \mathrm{d}^d r\, A(\vec{r}) = L$. The function $p(\vec{r})$, corresponding to the fact that the configurations of the rod are random in a d-dimensional space, can be written as follows

$$p(\vec{r}) = \frac{2}{\Omega(d) r^{d-1}}\, p_1(\vec{r})$$

where $p_1(\vec{r})$ is given by (E.12) and where $\Omega(d)$ is the 'area' of a ball of unit radius in d dimensions [see Appendix F, eqn (F.1)].

$$\Omega(d) = \frac{2\pi^{d/2}}{\Gamma(d/2)}. \tag{E.13}$$

Thus, we have

$$\int \mathrm{d}^d r\, p(\vec{r}) = 2 \int_0^{\infty} \mathrm{d}r\, p_1(\vec{r}) = L^2.$$

The strongest singularity of $p(\vec{r})$ is at the origin

$$\frac{p(\vec{r})}{\displaystyle\int \mathrm{d}^d r'\, p(\vec{r}\,')} \simeq \frac{2 r^{-(d-1)}}{\Omega(d) L}$$

and, taking into account (E.9), (E.10) on one side, and (E.12), (E.13) on the other side, we deduce the asymptotic value of $H(\vec{q})$ from the preceding result

$$H(\vec{q}) \simeq 2\frac{J(d-1,d)}{\Omega(d)\,qL} = 2\pi^{1/2}\frac{\Gamma(d/2)}{\Gamma(d/2-1/2)}\frac{1}{qL}. \tag{E.14}$$

4. Convex *D*-dimensional solid body

4.1 *D-dimensional solid body embedded in a D-dimensional space*

As was previously pointed out, $p(\vec{r})$ is the common volume of the solid body and of its replica obtained by the translation of vector $\vec{r}$. For small values of r, compared to the radii of curvature of the 'surface' of the solid body, we have (see Fig. E.1)

$$p(\vec{r}) = v - \frac{1}{2}\int \mathrm{d}\sigma\,[\hat{n}\cdot\vec{r}] + \ldots \tag{E.15}$$

where v is the volume of the body, $\mathrm{d}\sigma$ an element of its 'area' in $D-1$ dimensions and $\hat{n}$ a unit vector perpendicular to the 'surface'. Let us now average over all orientations; we can easily see that

$$\langle\!\langle |\hat{n}.\vec{r}| \rangle\!\rangle = r\pi^{-1/2}\frac{\Gamma(D/2)}{\Gamma(D/2+1/2)}. \tag{E.16}$$

Thus, the correlation function corresponding to a D-dimensional solid body with a random orientation in a D-dimensional space, is given for small r by the

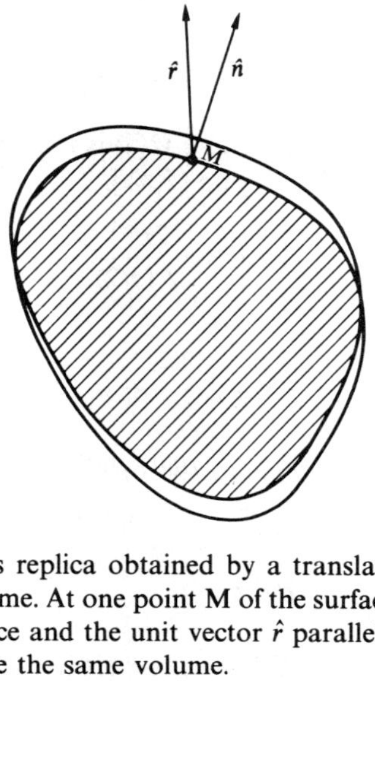

Fig. E.1. Solid body and its replica obtained by a translation of vector $\vec{r}$. Hachuring indicates their common volume. At one point M of the surface, we represented the vector $\hat{n}$ perpendicular to the surface and the unit vector $\hat{r}$ parallel to $\vec{r}$. We note that the two non-hachured crescents have the same volume.

expression

$$p_D(\vec{r}) = v - rs\pi^{-1/2}\frac{\Gamma(D/2)}{2\Gamma(D/2 + 1/2)} \tag{E.17}$$

when s is the 'area' of the $(D - 1)$-dimensional 'surface'. Moreover, we have

$$\int \mathrm{d}^d r\, p_D(\vec{r}) = v^2.$$

It can be shown that, when the solid body is not spherical, the dominant singularity is at the origin. Thus, in this case, for large q, (E.8) and (E.9) give

$$H(\vec{q}) \simeq -\frac{s}{v^2 q^{D+1}} J(-1, D)\pi^{-1/2}\frac{\Gamma(D/2)}{2\Gamma(D/2 + 1/2)},$$

or with the help of (E.10).

$$H(\vec{q}) \simeq \frac{s}{v^2 q^{D+1}} 2^{D-1}\pi^{-1+D/2}\Gamma(D/2). \tag{E.18}$$

Incidentally, we note that for $D = 1$, this formula gives the non-oscillating term of (E.11).

4.2 *D-dimensional solid body in a d-dimensional space*

Now let us put the D-dimensional solid body in a d-dimensional space (with $d > D$) with a random orientation. The corresponding function $p(\vec{r})$ is related to $p_D(\vec{r})$ by the equality

$$p(\vec{r}) = \frac{\Omega(D)}{\Omega(d) r^{d-D}} p_D(\vec{r}), \tag{E.19}$$

and consequently

$$\int \mathrm{d}^d r\, p(\vec{r}) = \int \mathrm{d}^D r\, p_D(\vec{r}) = v^2.$$

The dominant singularity of $p(\vec{r})$ is always at the origin. Then, for small r, formula (E.17) gives

$$p(\vec{r}) \simeq \frac{\Omega(D)}{\Omega(d)}\frac{v}{r^{d-D}}.$$

Thus, for large q, taking into account (E.8), (E.9) and (E.10), (E.13), we find,

$$H(\vec{q}) \simeq \frac{1}{vq^D}\frac{\Omega(D)}{\Omega(d)} J(d - D, d) = \frac{1}{vq^D}(4\pi)^{D/2}\frac{\Gamma(d/2)}{\Gamma(d/2 - D/2)}. \tag{E.20}$$

Incidentally, we observe that, by putting $D = 1$ in this formula, we recover (E.14).

Now, we note that (E.20) and (E.18) correspond to very different types of behaviour.

5. Polymer chains in a d-dimensional space

Let us consider a continuous polymer chain and let $\vec{r}(n)$ be the position vector of a point on the chain. The coordinate n of the point along the chain varies from $-N/2$ to $+N/2$, and it will be assumed that this quantity is proportional to the number of links. The form factor is then defined by

$$H(\vec{q}) = \int_{-N/2}^{N/2} \frac{\mathrm{d}n}{N} \int_{-N/2}^{N/2} \frac{\mathrm{d}n'}{N} \langle\!\langle \mathrm{e}^{\mathrm{i}\vec{q}\cdot(\vec{r}(n)) - \vec{r}(n'))} \rangle\!\rangle .$$

For large values of q, this expression simplifies and, neglecting the contributions of the chain ends, we may write

$$H(\vec{q}) = \int_{-\infty}^{+\infty} \frac{\mathrm{d}n}{N} \langle\!\langle \mathrm{e}^{\mathrm{i}\vec{q}\cdot(\vec{r}(n) - \vec{r}(0))} \rangle\!\rangle .$$

On the other hand, we know that for small $\vec{q}$

$$H(\vec{q}) = 1 - q^2 X_G^2 + \ldots$$

where X_G is related to the radius of gyration R_G by

$$R_G^2 = d X_G^2.$$

A *critical* chain, like the Brownian chain or the Kuhnian chain, is characterized by a size exponent ν corresponding to the dimensionality $D = 1/\nu$. In this case, the order of magnitude of the distances $|\vec{r}(n) - \vec{r}(0)|$ contributing to $H(\vec{q})$ is given by the length x defined by

$$\frac{x}{X_G} = \left(\frac{n}{N}\right)^{\nu} \qquad \nu = 1/D \tag{E.21}$$

and, we can also write $H(\vec{q})$ in the form

$$H(\vec{q}) = \int_{-\infty}^{+\infty} \frac{\mathrm{d}n}{N} f(qx).$$

Let us now choose another integration variable. From (E.21), we get

$$\frac{\mathrm{d}n}{N} = D \frac{x^{D-1}}{(X_G)^D} \mathrm{d}x$$

and consequently

$$H(\vec{q}) = D \int_{-\infty}^{+\infty} \mathrm{d}x \frac{x^{D-1}}{(X_G)^D} f(qx),$$

or more precisely

$$H(\vec{q}) = \frac{h_G}{(q X_G)^D}$$

where

$$h_G = D \int_{-\infty}^{+\infty} \mathrm{d}t \, t^{D-1} f(t).$$

Now we can see that (E.22) is very analogous to (E.20): the form function of a random chain of dimension D, embedded in a d-dimensional space, has the same asymptotic dependence on to q, as a solid body of dimension D, oriented at random in a d-dimensional space (with $d > D$).

APPENDIX F

AREA OF A BALL IN A SPACE OF DIMENSION d

The area of a ball of unit radius in a space of dimension d, is given by the formula

$$\Omega(d) = \frac{2\pi^{d/2}}{\Gamma(d/2)} \tag{F.1}$$

where

$$\Gamma(x+1) = x!$$

In particular, we have

$\Omega(1) = 2$ (a segment has two end points)
$\Omega(2) = 2\pi$ (length of a circle)
$\Omega(3) = 4\pi$ (area of a sphere)
$\Omega(4) = 2\pi^2$ (volume limiting an hypersphere in 4 dimensions)

Proof

Let us consider the integral

$$I = \int_{-\infty}^{+\infty} e^{-x^2}\, dx. \tag{F.2}$$

We can also write

$$I^d = \left[\int_{-\infty}^{+\infty} e^{-x^2}\,dx\right]^d = \int d^d r\, e^{-r^2} = \frac{\Omega(d)}{2}\int_0^{\infty} dx\, x^{d/2-1} e^{-x} \tag{F.3}$$

or

$$I^d = \frac{\Omega(d)}{2}\left(\frac{d}{2}-1\right)! = \frac{\Omega(d)}{2}\Gamma(d/2). \tag{F.4}$$

In particular, for $d = 2$, we have

$$I^2 = \frac{\Omega(2)}{2} = \pi.$$

By bringing the value of I in eqn (F.4), we obtain (F.1).

APPENDIX G

ANALYTIC CONTINUATION OF CERTAIN INTEGRALS

1. Analytic continuation of certain one-variable integrals and subtractions

1.1 *Problem and result*

Let us consider the integral

$$I(\alpha) = \int_0^\infty \mathrm{d}x\, x^{-\alpha} f(x) \tag{G.1}$$

where $f(x)$ is a function which has a Taylor expansion at the origin and which decreases at infinity fast enough to insure the convergence of the integral for $\alpha_0 < \alpha < 1$.

We want to extend the definition of $I(\alpha)$, by analytic continuation to all (non-integral) values of α larger than α_0.

By definition, we represent this analytic continuation by the same symbol $I(\alpha)$. It will be shown that $I(\alpha)$ is a meromorphic function which has poles for integer values of $\alpha (\alpha = 1, 2, \ldots)$ and that, for $n + 1 < \alpha < n + 2$, $I(\alpha)$ is given by the integral

$$I(\alpha) = \int_0^\infty \mathrm{d}x\, x^{-\alpha}\left[f(x) - f(0) - \ldots - f^{(n)}(0)\frac{x^r}{n!}\right] \tag{G.2}$$

or by the following expression obtained by integration by parts (under conditions of convergence)

$$I(\alpha) = \frac{\Gamma(\alpha - n - 1)}{\Gamma(\alpha)} \int_0^\infty \mathrm{d}x\, x^{-\alpha+n+1} f^{(n+1)}(x).$$

1.2 *Proof*

Let us first assume that we have $\alpha_0 < \alpha < 1$. We can write $I(\alpha)$ in the form

$$I(\alpha) = \int_0^1 \mathrm{d}x\, x^{-\alpha}[f(x) - f(0)] + \frac{f(0)}{1-\alpha} + \int_1^\infty \mathrm{d}x\, x^{-\alpha} f(x) \tag{G.3}$$

since for $\alpha_0 < \alpha < 1$, we have

$$\int_0^1 \mathrm{d}x\, x^{-\alpha} = \frac{1}{1-\alpha}.$$

The expression which appears in the right-hand side of (G.3) is equivalent to (G.1) in the domain $\alpha_0 < \alpha < 1$ but it remains meaningful when one lets α vary in the complex plane so as to bring it on the segment $1 < \alpha < 2$ by turning around the pole located at $\alpha = 1$; thus, the preceding expression defines $I(\alpha)$ in the whole interval $\alpha_0 < \alpha < 2$.

Now, let us assume that we have $1 < \alpha < 2$. In this case, we can rewrite $I(\alpha)$, given by (G.3), in the form

$$I(\alpha) = \int_0^\infty \mathrm{d}x\, x^{-\alpha}[f(x) - f(0)], \tag{G.4}$$

by combining the last two terms of (G.3) because, for $1 < \alpha < 2$, we have

$$\int_1^\infty \mathrm{d}x\, x^{-\alpha} = \frac{1}{\alpha - 1}.$$

The process can be iterated without difficulty and a simple recurrence argument leads to the general formula (G.2).

1.3 *Remarks*

The symbol $\mathfrak{T}^n$ is commonly used to represent the operation which, applied to a function $f(x)$, consists in extracting from it, its Taylor series upto order n

$$\mathfrak{T}^n f(x) = \sum_{p=0}^{n} f^{(p)}(0)\frac{x^p}{p!}$$

Then, with this notation, we can rewrite $I(\alpha)$ in the more compact form

$$I(\alpha) = \int_0^\infty \mathrm{d}x\, x^{-\alpha}(1 - \mathfrak{T}^n)f(x)$$

$$(n + 1 < \alpha < n + 2).$$

1.4 *Example*

We know that for $\alpha > 0$, $\beta > 0$

$$I(\alpha,\beta) \equiv \int_0^1 \mathrm{d}x\, x^{\alpha-1}(1 - x)^{\beta-1} = \frac{\Gamma(\alpha)\Gamma(\beta)}{\Gamma(\alpha + \beta)}$$

Let us look for an analytic continuation of this integral for $-1 < \alpha < 0$.

For $\alpha > 0$, $\beta > 0$, we can write

$$I(\alpha,\beta) = \int_0^\infty \mathrm{d}x\, x^{\alpha-1}(1 - x)^{\beta-1}\,\ominus(1 - x)$$

where $\ominus(x)$ is the step function. Assuming now that $-1 < \alpha < 0$, $\beta < 0$, we can apply the preceding rule; in this case, by analytic continuation, we find [see (G.4)]

$$I(\alpha,\beta) = \int_0^\infty \mathrm{d}x\, x^{\alpha-1}[(1 - x)^{\beta-1}\,\ominus(1 - x) - 1]$$

$$= \int_0^1 \mathrm{d}x\, x^{\alpha-1}[(1 - x)^{\beta-1} - 1] + \frac{1}{\alpha}.$$

On the other hand, we know that for all values of α and β, the analytic continuation of $I(\alpha, \beta)$ is

$$I(\alpha, \beta) = \frac{\Gamma(\alpha)\Gamma(\beta)}{\Gamma(\alpha + \beta)}.$$

2. Analytic continuation of certain multiple integrals

2.1 *Problem*

Let us now consider an integral of the form

$$I(\beta) = \int_0^\infty \mathrm{d}x_1 \ldots \int_0^\infty \mathrm{d}x_n \frac{f(x_1, \ldots, x_n)}{[\triangle(x_1, \ldots, x_n)]^\beta} \tag{G.5}$$

Here $\triangle(x_1, \ldots, x_n)$ is a homogeneous polynomial of degree p with respect to the variables $x_1, \ldots, x_n$, having the property that, the condition $x_j > 0$ for all j entails the inequality $\triangle(x_1, \ldots, x_n) > 0$. Moreover, in this integral $f(x_1, \ldots, x_n)$ is a function of $x_1, \ldots, x_n$, which has a Taylor expansion at the origin and which decreases at infinity fast enough to insure, for certain values of β, the universal convergence of the integral. For instance, we could assume that, when one x_j (or several x_j) becomes infinite, $f(x_1, \ldots, x_n)$ decreases faster than any power law. This implies the convergence of (G.5) for $\mathrm{Re}(\beta n) < 1$ and the problem is to find an analytic continuation of $I(\beta)$ for larger values of $\mathrm{Re}(\beta n)$.

2.2 *Continuation method*

To find an analytic continuation of $I(\beta)$, it is possible to generalize to the case of several variables, the approach described in Section 1, by introducing subtractions rather similar to those appearing in formula (G.2). The interest of this method, in physics, comes from the fact that, in general, the substracted terms have a physical interpretation. The applicability condition for this method is the following: in (G.5), the integrand must possess *the fundamental property of simultaneous Taylor expansion* the definition of which is given in Appendix H. It is then possible to show that the analytic continuation of $I(\beta)$ is a meromorphic function of β. Moreover, the subtractions which have to be made must be interpreted as renormalizations, in polymer theory as in field theory; this is why this subtractive method is so interesting from a theoretical point of view (see the definition of the operator $\mathfrak{R}$ in Appendix H).

Reference

1. Bergère, M.C. and David, F. (1979). *J. Math. Phys.* **20**, 1244.

APPENDIX H

DIMENSIONAL REGULARIZATION AND RENORMALIZATION OF THE SHORT-RANGE CUT-OFF-DEPENDENT TERMS

1. Generalities

In this appendix, we deal with short-range divergences which, for $d < 4$, occur in the calculation of the diagrams contributing to the connected partition functions $\mathscr{Z}(\ldots; S_1, \ldots, S_N)$. The question is rather simple when there are only two-body interactions, and this case has been treated in Chapter 10, Sections 4.1.6 and 4.2.4. Then, to eliminate the divergences, we only have to renormalize the partition functions, and when we calculate the mean value of a physical quantity the renormalization factors cancel out.

The situation is more complex when two-body forces and three-body forces (or p-body forces with $p > 2$) act on the polymer; in particular, in this case, the three-body forces can generate cut-off-dependent two-body forces. A simple renormalization of the partition functions is not sufficient to eliminate the divergent terms which depend on the cut-off; the bare interactions are also to be renormalized. The aim of this appendix is to show briefly how it is possible to proceed, but the question deserves a very careful study. Thus the appendix presents a programme rather than results, and anyone wishing to carry out this programme may encounter difficulties.

2. Form of the contributions associated with the diagrams

The various terms contributing to the function $^{+}\mathscr{Z}(\ldots; S_1, \ldots, S_N)$ are represented by diagrams. After summing over the internal wave vectors, we can write the contribution of a diagram as a function of the areas s_j of polymer segments whose end points on the diagram are either interaction points or correlation points (or end points of polymers). The contribution $\mathscr{D}\{s\}$, which is a function of the s_j, also depends on vectors $\vec{k}_j$; these vectors are born by the segments j and directly related to the external wave vectors (see Chapter 10, Sections 4.1.5 and 4.2.6). In what follows, these vectors $\vec{k}_j$ will be considered as constants. As was previously shown [see (10.4.57)], $\mathscr{D}\{s\}$ is of the form

$$\mathscr{D}\{s\} = \mathscr{D}_0 \, |\det \mathscr{S}|^{-d/2} \exp\left[-\sum_j s_j k_j^2 + \frac{1}{2} \sum_{\alpha,\beta}^{\circlearrowleft} \vec{k}_\alpha (\mathscr{S}^{-1})_{\alpha\beta} \vec{k}_\beta \right] \tag{H.1}$$

where $\circlearrowleft$ is the number of loops, $\mathscr{S}$ a matrix with elements $\mathscr{S}_{\alpha\beta}$, and $\vec{k}_\alpha$ a vector in a space of dimension d (with $\alpha = 1, \ldots, \circlearrowleft$ and $\beta = 1, \ldots, \circlearrowleft$). Moreover, according to (10.4.56), we have

$$\mathscr{S}_{\alpha\beta} = \sum_j s_j \ominus_{j\alpha} \ominus_{j\beta}$$

$$\vec{k}_\alpha = \sum_j s_j \ominus_{j\alpha} \vec{k}_j.$$

The contribution $\mathscr{D}$ of the diagram is obtained by integrating over the variables s_j. The only term depending on the space dimension d is the factor $|\det \mathscr{S}|^{-d/2}$. We thus see that in the integral giving $\mathscr{D}$, this factor $|\det \mathscr{S}|^{-d/2}$ is rather similar to the factor $|\triangle(x_1, \ldots, x_n)|^{-\beta}$ which appears in eqn (G.4) of Appendix G.

The areas s_j, are subjected to constraints. First, the sum of the areas of the segments constituting a polymer $\boldsymbol{a}$ is equal to the area $S_{\boldsymbol{a}}$ of this polymer. Secondly, if there are correlation points on a polymer, these correlations points determine polymer sections having well-defined areas, and one must write that the sum of the areas of the segments constituting a polymer section is equal to the area of the section. To simplify, in what follows we consider only diagrams which have no correlation points in the middle of a chain, and we admit that the principles developed in this simple case are actually quite general.

By definition, we have J polymer segments on the diagram, each segment being represented by an index j ($j = 1, \ldots, J$). Then, let us consider a polymer $\boldsymbol{a}$ and let $\mathscr{I}_{\boldsymbol{a}}$ be the set of indices corresponding to this polymer. For each $\boldsymbol{a}$, we have

$$\sum_{i \in \mathscr{I}_{\boldsymbol{a}}} s_i = S_{\boldsymbol{a}} \tag{H.2}$$

and, moreover, we have

$$J = \sum_{\boldsymbol{a}} \sum_{i \in \mathscr{I}_{\boldsymbol{a}}} 1.$$

There are J areas on the diagram but only I of them (with $I < J$) appear in the matrix elements of $\mathscr{S}$ and therefore in (det $\mathscr{S}$). These *internal* areas are the areas of segments bound by interaction points belonging either to the same interaction or to overlapping interactions. With them, we associate, the indices $j = 1, \ldots, I$. It is because a few of these areas, or all of these areas, vanish that divergences occur. In order to avoid such divergences, we introduce a cut-off area s_0 and we postulate that the internal areas s_j (with $j = 1, \ldots, J$) obey the regularization conditions

$$s_j > s_0 \qquad (j = 1, \ldots, I), \tag{H.3}$$

whereas, for the other areas, we have only

$$s_j > 0 \qquad (j = I + 1, \ldots, J).$$

Finally, in agreement with (H.2) and (H.3), the contribution $\mathscr{D}(s_0, \{S\})$ of the diagram can be expressed as the integral

$$\mathscr{D}(s_0, \{S\}) =$$
$$\int_{s_0}^{\infty} ds_1 \ldots \int_{s_0}^{\infty} ds_I \int_{s_0}^{\infty} ds_{I+1} \ldots \int_{s_0}^{\infty} ds_J \left[\prod_a \delta\left(S_a - \sum_{j \in \mathscr{I}_a} s_j\right)\right] \mathscr{D}\{s\}. \quad \text{(H.4)}$$

Now, we can write $\mathscr{D}\{s\}$ in the form

$$\mathscr{D}\{s\} = \exp\left[-\frac{1}{2}\sum_{j=I+1}^{J} k_j^2 s_j\right] \mathscr{D}^{\subset}\{s\}$$

where $\mathscr{D}^{\subset}\{s\}$ depends only on the internal areas $s_1, \ldots, s_I$ [see (H.1)]. Then, let us set

$$\ominus\{S,s\} = \int_0^{\infty} ds_{I+1} \ldots \int_0^{\infty} ds_J \prod_a \delta\left(S_a - \sum_{j \in \mathscr{I}_a} s_j\right) \exp\left[-\frac{1}{2}\sum_{j=I+1}^{J} k_j^2 s_j\right]. \quad \text{(H.5)}$$

Of course, the integral is perfectly convergent, and $\ominus\{S,s\}$ appears as a product of step functions and of entire functions of s_j. Incidentally, these functions are just polynomials in the s_j when the $\vec{k}_j$ vanish. Thus, we can rewrite (H.4) in the form

$$\mathscr{D}(s_0, \{S\}) = \int^{\infty} ds_1 \ldots \int^{\infty} ds_I \ominus\{S,s\} \mathscr{D}^{\subset}\{s\}. \quad \text{(H.6)}$$

The integral (H.6) which gives $\mathscr{D}(s_0, \{S\})$ is finite for $d < 4$, but is a function of s_0. However, we feel that the properties of long chains should not depend on the details of the chemical microstructure. Consequently, they must be independent of s_0, or, in special cases, depend on them through constants which diverge when $s_0 \to 0$. Thus, we shall try to rewrite $\mathscr{D}(s_0, \{S\})$ in the form of a finite sum of the type

$$\mathscr{D}(s_0, \{S\}) = \mathscr{D}_0(s_0, \{S\}) + \sum_m \frac{1}{(s_0)^{\gamma_m}} \mathscr{D}_m\{S\} \quad \text{(H.7)}$$

where $\mathscr{D}_0(s_0, \{S\})$ has a finite limit when $s_0 \to 0$

$$\lim_{s_0 \to 0} \mathscr{D}_0(s_0, \{S\}) = \mathscr{D}_0\{S\}.$$

When this is possible, we can always admit that the γ_m are positive and this condition determines $\mathscr{D}_0(s_0, \{S\})$. In particular, we note that, for $d < 2$, the integral in (H.6) converges in the limit $s_0 \to 0$. Thus, in this case

$$\mathscr{D}(s_0, \{S\}) = \mathscr{D}_0(s_0, \{S\}).$$

Moreover, we remark that the exponents γ_m are functions of the space dimension d. Thus, when d increases, new divergent terms (with respect to s_0)

appear for particular (fractional) values of d. Of course, for these values of d, the sum (H.7) does not exist.

We can also remark that for $d < 2$

$$\mathscr{D}_0\{S\} = \mathscr{D}(0, \{S\}) = \int_0^\infty ds_1 \ldots \int_0^\infty ds_I \ominus \{S,s\}\, \mathscr{D}^{\subset}\{s\}. \tag{H.8}$$

The quantity $\mathscr{D}(0, \{S\})$ is defined only for $d < 2$ but $\mathscr{D}_0\{S\}$ is the analytic continuation of $\mathscr{D}(0, \{S\})$ for $d > 2$. However, the function $\mathscr{D}\{S\}$ has poles for particular values of d and dimension three is one of them.

3. A study of certain functions of several variables with singularities at the origin

3.1 *The simultaneous Taylor series property*

In this section, we consider a function $F\{s\}$ of a set of variables s_j and for this function we define the *simultaneous Taylor series property* (in brief STSP).

Let us denote by $\mathscr{E}$ a subset of the set of the variables s_j. A collection $\mathscr{E}_1, \ldots, \mathscr{E}_A$ of subsets is called *a nest* if the condition $\sigma > \tau$ implies $\mathscr{E}_\sigma \supset \mathscr{E}_\tau$. Thus, considering a nest, we can associate a supplementary variable $\rho_{\mathscr{E}}$ with each subset $\mathscr{E}$ contained in this nest. We set

$$s_j\{\rho\} = s_j \prod_{\mathscr{E}(s_j \in \mathscr{E})} \rho_{\mathscr{E}}$$

where the product is performed over all the subsets $\mathscr{E}$ for which s_j belongs to $\mathscr{E}$.

Let us denote by $F_{\mathscr{N}}\{s,\rho\}$ the quantity obtained by replacing the s_j by the $s_j\{\rho\}$ in $F\{s\}$. With M. Bergère,[1] we say that $F\{s\}$ has *the simultaneous Taylor series property with respect to the nest* $\mathscr{N}$ if there exist real numbers (which may be negative) such that the function

$$\prod_{\mathscr{E} \in \mathscr{N}} (\rho_{\mathscr{E}})^{-\mu_{\mathscr{E}}} F_{\mathscr{N}}\{s,\rho\} \tag{H.9}$$

has a finite limit when all the $\rho_{\mathscr{E}}$ vanish and, at the origin, has a Taylor series in terms of all these variables simultaneously.

Moreover, if for each $\mathscr{N}$, $F\{s\}$ has the preceding property, we say that $F\{s\}$ has *the simultaneous Taylor series property* (STSP). For instance the function $(s_1 + s_2)^{-1}$ has this fundamental property. Moreover we note that if $F\{s\}$ has the STSP, any product of $F\{s\}$ by a polynomial has the same property.

3.2 *Superficial degree of divergence of a function of several variables and convergence of an integral of this function*

Consider a function $F\{s\}$ of a set of variables $s_j (s_j > 0)$. If this function has the STSP, the superficial degree of divergence $\omega_{\mathscr{E}}$ associated with a subset $\mathscr{E}$ of these variables, made of $m_{\mathscr{E}}$ variables, is defined by

$$\omega_{\mathscr{E}} = -\mu_{\mathscr{E}} - m_{\mathscr{E}}$$

where $\mu_{\mathscr{E}}$ is given by (H.9). Then, one can prove the following theorem.

Theorem

If $F\{s\}$ has the Taylor series property (STSP), if for any $\mathscr{E}$, we have $\omega_{\mathscr{E}} < 0$, and if $F\{s\}$ decreases exponentially at infinity, the integral

$$I = \int_0^\infty \mathrm{d}s_1 \ldots \int_0^\infty \mathrm{d}s_m \, F\{s\}$$

is absolutely convergent.

3.3 *The operator* $\mathfrak{R}$

Consider a function $F\{s\}$ having the STSP. The operator $\mathfrak{R}$ is introduced to calculate the finite part of a divergent integral of the form

$$I = \int \mathrm{d}s_1 \ldots \mathrm{d}s_m \, F\{s\}$$

in the case where certain superficial degrees of divergence $\omega_{\mathscr{E}}$ are positive.

Definition of $\mathfrak{T}^n(x)$

If $f(x)$ is such that $g(x) = x^{-\mu} f(x)$ has a Taylor expansion for $x = 0$, then we can define the operator $\mathfrak{T}^n(x)$ by the equality

$$\mathfrak{T}^n(x) f(x) = x^\mu \mathfrak{T}^n(x) g(x) = x^\mu \left[\sum_{p=0}^{n} g^{(p)}(0) \frac{x^p}{p!} \right]$$

Thus, $\mathfrak{T}^n(x)$ extracts $(n+1)$ terms from the Taylor expansion

Definition of the operator $\mathfrak{R}$

The function $\mathfrak{R} F\{s\}$ is defined by the equality

$$\mathfrak{R} F\{s\} = F\{s\} + \sum_{\mathscr{N}} \prod_{\mathscr{E} \varepsilon \mathscr{N}} (-\mathfrak{T}^{[\omega_{\mathscr{E}}]}(\rho_{\mathscr{E}})) F\{s, \rho\} |_{\rho_{\mathscr{E}} = 1 \text{ for all } \mathscr{E}}$$

where the sum is over all the nests of the subsets $\mathscr{E}$ and where $[\omega_{\mathscr{E}}]$ is the integral part of the superficial degree of divergence $\omega_{\mathscr{E}}$.

We also write in a symbolic form

$$\mathfrak{R} = 1 + \sum_{\mathscr{N}} \prod_{\mathscr{E} \in \mathscr{N}} (-\mathfrak{T}_{\mathscr{E}}^{[\omega_{\mathscr{E}}]}) \tag{H.10}$$

where $\mathfrak{T}_{\mathscr{E}}$ denotes a subtraction related to the subset $\mathscr{E}$ of the set of variables s_j ($\mathscr{E}$ may coincide with the whole set).

Proposition

If $F\{s\}$ has the STSP and if the integral $\int \mathrm{d}s_1 \ldots \int \mathrm{d}s_m \; \mathfrak{R} F\{s\}$ converges uniformly at infinity, then the integral

$$I = \int_0^\infty \mathrm{d}s_1 \ldots \int_0^\infty \mathrm{d}s_m \, \mathfrak{R} F\{s\}$$

is absolutely convergent.

This statement is just a variant of well-known results applying to field theory, and in particular of a theorem established by Bergère and Lam.[1] Let us note that, concerning our problems, this convergence condition at infinity depends on the space dimension d and that for special values of d it is not effective.

Another expression of the operator $\mathfrak{R}$

Let $\mathfrak{R}_{\mathscr{E}}$ be the operator $\mathfrak{R}$ corresponding to a subset $\mathscr{E}$ of variables s_j; with $\mathfrak{R}_{\mathscr{E}}$, we can associate the operator $\bar{\mathfrak{R}}_{\mathscr{E}}$ defined by

$$\mathfrak{R}_{\mathscr{E}} = (1 - \mathfrak{T}_{\mathscr{E}}^{[\omega_{\mathscr{E}}]})\bar{\mathfrak{R}}_{\mathscr{E}}, \tag{H.11}$$

and, in the same way for the whole set, we write

$$\mathfrak{R} = (1 - \mathfrak{T}^{[\omega]})\bar{\mathfrak{R}} \tag{H.12}$$

where ω is the superficial degree of divergence of the whole set.

Now, let $\mathscr{S}$ be a set of sub-sets $\mathscr{E}$, such that two sub-sets have always a zero intersection, and that $\mathscr{S}$ does not contain the whole set. Then, the operator $\mathfrak{R}$ can be written in the form

$$\bar{\mathfrak{R}} = 1 - \sum_{\mathscr{S}} \prod_{\mathscr{E} \in \mathscr{S}} \mathfrak{T}_{\mathscr{E}}^{[\omega_{\mathscr{E}}]} \bar{\mathfrak{R}}_{\mathscr{E}} \tag{H.13}$$

and $\bar{\mathfrak{R}}_{\mathscr{E}}$ can be expressed in a very similar manner in terms of the $\bar{\mathfrak{R}}_{\mathscr{E}'}$ with $\mathscr{E}' \subset \mathscr{E}$.

Therefore, $\bar{\mathfrak{R}}$ is an operator whose action is defined step by step. Thus, from (H.12) and (H.13), we obtain

$$\mathfrak{R} = 1 - \mathfrak{T}_{\mathscr{E}}^{[\omega]} \bar{\mathfrak{R}} - \sum_{\mathscr{S}} \prod_{\mathscr{E} \in \mathscr{S}} \mathfrak{T}_{\mathscr{E}}^{[\omega_{\mathscr{E}}]} \bar{\mathfrak{R}}_{\mathscr{E}}. \tag{H.14}$$

Remark

It is easy to see that a subtraction term like $\mathfrak{T}_{\mathscr{E}} F(s)$ can be written in the form

$$\mathfrak{T}_{\mathscr{E}} F(s) = \sum_i A_i\{s \in \mathscr{E}\} \, B_i\{s \notin \mathscr{E}\}. \tag{H.15}$$

When the operator $\mathfrak{R}$ is applied to diagram contributions, the essentially divergent terms $A_i\{s \in \mathscr{E}\}$ must be interpreted as renormalization terms or counter-terms; they are functions of s_0, which become infinite when $s_0 \to 0$, and are associated with the complete set of areas s_j constituting $\mathscr{E}$. On the contrary, $B_i\{s \notin \mathscr{E}\}$ appears as the non-divergent contribution of diagrams which are derived from the initial diagrams and which contain counter-terms acting as interactions. These secondary diagrams represented by B_j are obtained by contracting the segments corresponding to the areas s_j of the set $\mathscr{E}$; the contraction of these lines creates new vertices, i.e. additional interaction terms represented by A_j.

4. Dimensional regularization of the contributions of polymer diagrams

The functions $\mathscr{D}\{s\}$ defined above (Section 2) appear not only in polymer theory but also in field theory where the Schwinger method is used to calculate diagrams (see Appendix I): in fact, this, method relies on the transformation

$$\frac{1}{k^2 + a} = \int_0^\infty \mathrm{d}\alpha\, \mathrm{e}^{-\alpha(a+k^2)} \qquad (a = m^2)$$

which can be considered as a Laplace–de Gennes transformation, and we note that the areas s_j are generally represented in field theory by the symbols α_j. Of course, the two formalisms are somewhat different. It happens, in field theory, that the study of the divergences of the Green's functions can be reduced to that of the divergences of the vertex functions: such a reduction is not possible in polymer theory; we must then study the divergences of the partition functions which are homologous to the Green's functions. Finally, let us note that in (H.1), the external wave vectors appear through the vectors $\vec{k}_j$ and $\vec{k}_\alpha$ and that, consequently, the dimension of the matrix $\mathscr{S}$ which appears in the exponential is equal to the number of loops. In field theory, the dependence with respect to the external vectors is generally expressed in a more direct way and, instead of $\mathscr{S}$, one often introduces a matrix whose dimension is not the number of loops but the number of external legs or of polymer segments.[1,2] However, these differences are rather superficial. The divergences coming from the s_j (with $s_j > s_0$ for $j = 1, \ldots, T$) when the cut-off s_0 goes to zero, are the same in polymer theory and in field theory.

Now, it has been shown, in field theory, that the functions of the variables α_j which represent the contributions of the diagrams have *the simultaneous Taylor series property*. This indicates that the functions $\mathscr{D}\{s\}$ in polymer theory also have the same property. Moreover, when the space dimension is small enough, the integral which appears in (H.4) converges even for $s_0 = 0$. In this case [see (H.8)]

$$\mathscr{D}(s_0, \{S\}) \simeq \mathscr{D}_0\{S\} = \int_0^\infty \mathrm{d}s_1 \ldots \int_0^\infty \mathrm{d}s_I \ominus \{S,s\}\, \mathscr{D}^{\subset}\{s\}.$$

It is then possible to show that the operation $\mathfrak{R}$ makes it possible to get an analytic continuation of $\mathscr{D}_0\{S\}$ with respect to d, for all values of d that are not poles. In this case, the analytic continuation of $\mathscr{D}_0\{S\}$ is given by

$$\mathscr{D}_0\{S_p\} = \int_0^\infty ds_1 \ldots \int_0^\infty ds_J\, \mathfrak{R}[\ominus\{S,s\}\, \mathscr{D}^{\subset}\{s\}].$$

This analytic continuation is generally called dimensional regularization

5. Renormalization of the contribution of a diagram

We can now rewrite $\mathscr{D}(s_0,\{S_p\})$ given by (H.6) so as to make the divergent parts appear more clearly. For this purpose, we use the expression (H.14) of $\mathfrak{R}$.

$$\mathscr{D}(s_0,\{S_p\}) =$$

$$\int_{s_0}^\infty ds_1 \ldots \int_{s_0}^\infty ds_I \left[\mathfrak{R} + \mathfrak{T}^{[\omega]}\bar{\mathfrak{R}} + \sum_{\mathscr{S}} \prod_{\mathscr{E}\in\mathscr{S}} \mathfrak{T}_{\mathscr{E}}^{[\omega_{\mathscr{E}}]}\bar{\mathfrak{R}}_{\mathscr{E}}\right][\ominus\{S,s\}\,\mathscr{D}^{\subset}\{s\}]$$

and, as we can take the limit $s_0 \to 0$ in the convergent terms, we can also write

$$\mathscr{D}(s_0,\{S_p\} =$$

$$\int_0^\infty ds_1 \ldots \int_0^\infty ds_J\, \mathfrak{R}[\ominus\{S,s\}\,\mathscr{D}^{\subset}\{s\}] +$$

$$+ \int_{s_0}^\infty ds_1 \ldots \int_{s_0}^\infty ds_I \left[\mathfrak{T}^{[\omega]}\bar{\mathfrak{R}} + \sum_{\mathscr{S}} \prod_{\mathscr{E}\in\mathscr{S}} \mathfrak{T}_{\mathscr{E}}^{[\omega_{\mathfrak{E}}]}\bar{\mathfrak{R}}_{\mathscr{E}}\right][\ominus\{S,s\}\,\mathscr{D}^{\subset}\{s\}]. \tag{H.16}$$

On the right-hand side, the first term is finite: this is the term given by dimensional regularization; the other terms are divergent when $s_0 \to 0$. In the same way, the operations $\bar{\mathfrak{R}}_{\mathscr{E}}$ suppress all the divergences which occur when a certain number of variables s_j, constituting a subset of $\mathscr{E}$, distinct from $\mathscr{E}$, simultaneously vanish.

Let us now examine more in detail the structure of the renormalization terms. In (H.6) or (H.16), the subsets $\mathscr{E}$ which give rise to subtractions are related to divergences coming from $\mathscr{D}^{\subset}\{s\}$ and not from $\ominus\{S,s\}$. Then, let us set

$$F\{s\} = \mathfrak{R}_{\mathscr{E}}\ \ominus\{S,s\}\,\mathscr{D}^{\subset}\{s\}. \tag{H.17}$$

in (H.15). It is not difficult to see that the factors $B_i\{s\notin\mathscr{E}\}$ are functions of the areas S_p of the polymers whereas these areas do not appear in the factors $A_i\{s\in\mathscr{E}\}$. Thus, by integration, the $B_j\{s\notin\mathscr{E}\}$ lead to terms which have limits when $s_0 \to 0$ but depend on the constraints of the systems (the areas of the polymers, for instance), whereas the integration of the $A_i\{s\in\mathscr{E}\}$ give terms which strongly depend on s_0 (divergent terms) but are independent of the external constraints. This fact and the manner in which the renormalization terms appear in (H.16) enable us better to understand how these subtraction terms can be absorbed with the help of a proper redefinition of the renormalization factors and of the bare interactions. Moreover, we expect the new interaction constants to appear in our formalism, as quantities which do not depend on the areas (the lengths) of the interacting polymers. Of course, this is exactly what we want and it is in order to obtain this result that we regularized the integral (H.6) by introducing the function $\mathfrak{R} \ominus \{S,s\} \mathscr{D}^{\subset}\{s\}$ in (H.16) instead of using the function $\ominus \{S,s\} \mathfrak{R} \mathscr{D}^{\subset}\{s\}$.

Nevertheless, it must be noted that in this way, we do not obtain a completely renormalized theory for all values of the space dimension d. In the regularized theory, the contributions of the diagrams remain meromorphic functions of d. Moreover, poles appear for physical values of d, in particular for $d = 3$.

Thus, in order to study the marginal behaviour of chains for $d = 4$ and $d = 3$, it is convenient to study the nature of the diagram expansions for $d = 4 - \varepsilon$ and $d = 3 - {}'\varepsilon$, as was done in the present book.

In field theory, the dimensional regularization can also be used but, in general, one prefers other renormalization methods which always give finite results for all values of d. These methods are equivalent to a direct introduction of $\mathfrak{R} \mathscr{D}\{s\}$; then, the renormalization constants depend on the masses, but this is not a drawback, since in field theory, the masses are considered as constant.

In brief, the preceding discussion leads to the following result. The dimensional regularization of the diagram integrals, which are considered as function of d, is an analytic continuation of these functions. The dimensional regularization is equivalent to a renormalization of the bare interaction constants and of the partition functions. The partition functions obtained by dimensional regularization can be directly used to study the properties of long chains.

References

1. Bergère, M.C. and Lam, Y.M.P. (1976). *J. Math. Phys.* **17**, 1546.
2. Bergère, M.C. and Lam, Y.M.P. Unpublished course on renormalization techniques, given at Saclay (January 1975).
3. Bergère, M.C. and David, F. (1979). *J. Math. Phys.* **20**, 1244.

APPENDIX I

LEGENDRE TRANFORMATION AND TREE DIAGRAMS

1. Legendre transformation: definition

The Legendre transformation plays a fundamental role in Thermodynamics, in Statistical Mechanics, and in Field Theory.

In its elementary form, it defines a correspondence between an arbitrary function $A(x)$ and a function $B(y)$ which is deduced from $A(x)$ in the following way.

With each value of x, we associate a value of y by setting

$$y = A'(x). \tag{I.1}$$

We admit that y is well-defined and that this relation can be inverted. This means that (I.1) defines x as a function of y and therefore $A(x)$ as an implicit function of y. Now, taking (I.1) into account, we can define $B(y)$ by the equality

$$A(x) + B(y) - xy = 0. \tag{I.2}$$

Differentiating this equation with respect to y, we obtain

$$B'(y) - x + \frac{\mathrm{d}x}{\mathrm{d}y}[A'(x) - y] = 0,$$

or according to (I.1)

$$x = B'(y). \tag{I.3}$$

In brief, the Legendre transformation, which is reciprocal, transforms any couple x, $A(x)$ into a couple y, $B(y)$ and this correspondence is defined by the set of equations

$$\begin{aligned} A(x) + B(y) - xy &= 0, \\ y &= A'(x), \\ x &= B'(y). \end{aligned} \tag{I.4}$$

2. Generalization

The Legendre transformation can immediately be extended to functions of several variables. Actually, let us consider a function $A(x_1, \ldots, x_n)$ and the associated variables

$$y_j = \frac{\partial A(x_1, \ldots, x_n)}{\partial x_j}. \tag{I.5}$$

The Legendre transform $B(y_1, \ldots, y_n)$ of $A(x_1, \ldots, x_n)$ is defined by setting

$$A(x_1, \ldots, x_n) + B(y_1, \ldots, y_n) - \sum_1^n x_j y_j = 0,$$

and from this relation we immediately deduce the relations

$$x_j = \frac{\partial B(y_1, \ldots, y_n)}{\partial y_j}, \tag{I.6}$$

which show that the transformation is reciprocal.

In the same way, if $A\{x(\vec{r})\}$ is a functional of the function $x(\vec{r})$ defined in a d-dimensional space, we can, by Legendre transformation, associate with it a functional $B\{y(\vec{r})\}$ of the function $y(\vec{r})$ defined in the same space.

3. Generating functions of tree graphs and Legendre transformation

The tree graphs which we consider here are connected graphs made of m-leg vertices ($m > 0$) (see Fig. I.1) and of interaction lines which connect these vertices. Two vertices are connected at most by one interaction line, and consequently the graph does not contain any cycle. In general, a graph has external legs (see Fig. I.2). The graph consisting of only one interaction line without vertex is considered as acceptable. We construct in this way so-called *rootless tree graphs* which have a certain number m of external legs. These graphs will be represented by the symbol ${}^\alpha\mathfrak{G}_m$ where α defines the type of the graph ($m \geqslant 0$). The symmetry number of the graph is denoted by ${}^\alpha\mathfrak{S}_m$.

In the same way, by marking one external leg of a rootless graph ${}^\alpha\mathfrak{G}_m$ ($m > 0$), we define a one-root graph ${}^\beta\mathfrak{G}_m$, where β defines the type of the graph (see Fig. I.3). The symmetry number of the graph is denoted by ${}^\beta\mathfrak{S}'_m$.

We want to define the generating functions of these graphs. For this purpose, we associate a contribution with each graph. The contribution of a graph ${}^\alpha\mathfrak{G}_m$ is ${}^\alpha C_m$, the contribution of a graph ${}^\beta\mathfrak{G}'_m$ is ${}^\beta C'_m$. The contributions ${}^\alpha C_m$ and ${}^\beta C'_m$ are calculated in the same manner.

We associate a factor v_m with each m-leg vertex and a factor $-u$ with any internal or external interaction line (an external leg, for instance). The contribution of a graph is then defined as the product of all the factors associated with

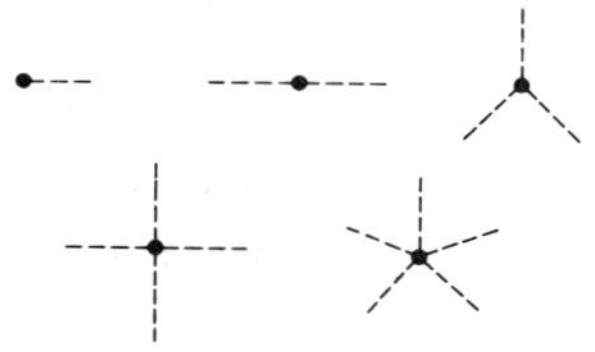

Fig. I.1. The figure represents vertices of order 1, 2, 3, 4, 5. A coefficient v_m is associated with each m-leg vertex.

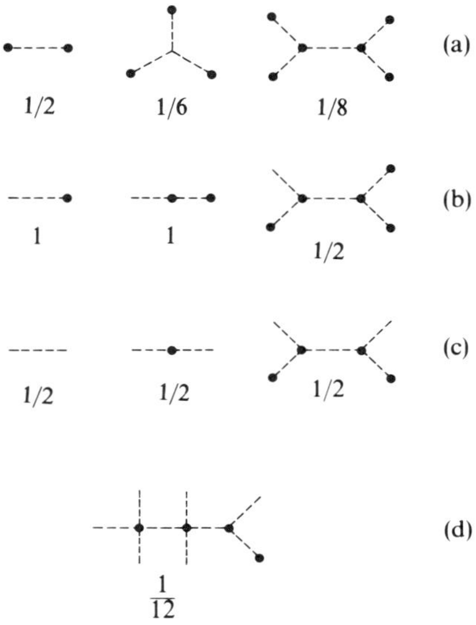

Fig. I.2. Examples of rootless tree graphs and their symmetry factors: (a) trees without legs; (b) trees with one leg; (c) trees with two legs; (d) a tree with six legs.

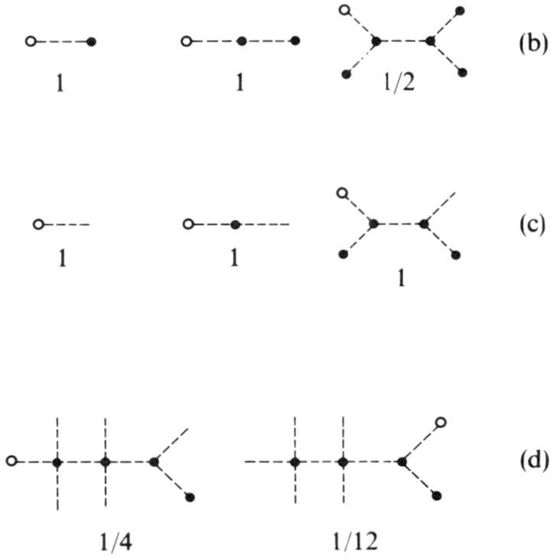

Fig. I.3. Examples of one-root graphs with their symmetry factors. The root is symbolized by a small open circle. The one-root graphs which are represented here, derive from the rootless graphs represented in Fig. I.2.

this graph. Thus, we attribute the contribution $-u$ to the graph having one interaction line without vertex.

The generating function of the rootless graphs with m legs is given by

$$A(x) \equiv \sum_{m=0}^{\infty} A_m x^m = \sum_{m=0}^{\infty} x^m \sum_{\alpha} \frac{{}^{\alpha}C_m}{{}^{\alpha}\mathcal{S}_m}. \tag{I.7}$$

In the same way, the generating function of the one-root graphs with $(m-1)$ legs is given by

$$y = A'(x) = \sum_{m=1}^{\infty} m A_m x^{m-1} = \sum_{m=1}^{\infty} x^{m-1} \sum_{\beta} \frac{{}^{\beta}C'_m}{{}^{\beta}\mathcal{S}'_m}. \tag{I.8}$$

We shall now show that the Legendre transform $B(y)$ of $A(x)$ has a much simpler structure than $A(x)$.

By looking at the graphs, we can see that the series $y = A'(x)$ can easily be constructed by iteration. The iterative process is contained in the equation

$$y = -ux - u \sum_{m=1}^{\infty} \frac{v_m}{(m-1)!} y^{m-1}, \tag{I.9}$$

and from it we get

$$x = B'(y) = -u^{-1} y - \sum_{m=1}^{\infty} \frac{v_m}{(m-1)!} y^{m-1}. \tag{I.10}$$

The value y_0 of y which corresponds to the value $x = 0$ is given by the equation

$$B'(y_0) = 0. \tag{I.11}$$

and, moreover, we have

$$B(y_0) = -A(0) = -A_0. \tag{I.12}$$

We can now integrate (I.10). We obtain

$$B(y) = -A_0 - \frac{1}{2} u^{-1} (y^2 - y_0^2) - \sum_{m=1}^{\infty} \frac{v_m}{m!} (y^m - y_0^m) \tag{I.13}$$

where

$$A_0 = \sum_{\alpha} \frac{{}^{\alpha}C_0}{{}^{\alpha}\mathcal{S}_0} \tag{I.14}$$

and y_0 is a root of the equation

$$y_0 = -u \sum_{m=1}^{\infty} \frac{v_m}{(m-1)!} y_0^{m-1}. \tag{I.15}$$

Thus, the function $A(x)$, which is the generating function of the tree graphs, has a Legendre transform $B(y)$ whose structure is much simpler. In fact, we have

$$B(y) - B(0) = -\frac{1}{2} u^{-1} y^2 - \sum_{m=1}^{\infty} \frac{v_m}{m!} y^m. \tag{I.16}$$

The result simplifies when $v_1 = 0$. In this case, on one side $A_0 = 0$ because it is impossible to construct graphs without external legs, and on the other side the value $y_0 = 0$ is a solution of (I.15). Then we obtain

$$B(y) = -\frac{1}{2}u^{-1}y^2 - \sum_{m=2}^{\infty} \frac{v_m}{m!} y^m. \tag{I.17}$$

The case $v_1 \neq 0$ is more complicated, but it is still possible to calculate A_0 by reducing the problem to the case $v_1 = 0$, and by considering the vertices v_1 as plain leg ends.

APPENDIX J

SCHWINGER'S AND FEYNMAN'S IDENTITIES: APPLICATION TO THE CALCULATION OF DIAGRAMS IN FIELD THEORY

The calculation of the diagrams representing the various terms of a perturbation series, usually requires multiple integrations which are simplified with the help of one of the two following identities, which can be used indifferently.

1. Schwinger's and Feynman's identities

From the identity

$$\frac{\Gamma(\alpha)}{a^{\alpha}} = \int_0^{\infty} \mathrm{d}t\, t^{\alpha-1}\,\mathrm{e}^{-at} \tag{J.1}$$

we immediately deduce *Schwinger's identity*

$$\frac{\Gamma(\alpha_1)\ldots\Gamma(\alpha_n)}{(a_1)^{\alpha_1}\ldots(a_n)^{\alpha_n}} = \int_{t_j>0} \mathrm{d}t_1\ldots\mathrm{d}t_n\, t_1^{\alpha_1-1}\ldots t_n^{\alpha_n-1}\exp(-(a_1t_1+\ldots+a_nt_n)) \tag{J.2}$$

We can also write it in the form

$$\frac{\Gamma(\alpha_1)\ldots\Gamma(\alpha_n)}{(a_1)^{\alpha_1}\ldots(a_n)^{\alpha_n}} =$$

$$= \int_0^{\infty}\mathrm{d}T\int_{t_j>0}\mathrm{d}t_1\ldots\mathrm{d}t_n\,\delta(t_1+\ldots+t_n-T)\,t_1^{\alpha_1-1}\ldots t_n^{\alpha_n-1}$$

$$\times\exp(-(a_1t_1+\ldots+a_nt_n)) \tag{J.3}$$

Let us set $t_n = Ts_n$; we obtain

$$\frac{\Gamma(\alpha_1)\ldots\Gamma(\alpha_n)}{(a_1)^{\alpha_1}\ldots(a_n)^{\alpha_n}} = \int_{s_j>0}\mathrm{d}s_1\ldots\mathrm{d}s_n\,\delta(s_1+\ldots+s_n-1)x_1^{\alpha_1-1}\ldots x_n^{\alpha_n-1}\times$$

$$\times\int_0^{\infty}\mathrm{d}T\,T^{\alpha_1+\ldots+\alpha_n-1}\exp(-T(a_1s_1+\ldots+a_ns_n))$$

By integrating with respect to T, *we find Feynman's identity*

$$\frac{1}{(a_1)^{\alpha_1}\ldots(a_n)^{\alpha_n}} =$$

$$= \frac{\Gamma(\alpha_1+\ldots+\alpha_n)}{\Gamma(\alpha_1)\ldots\Gamma(\alpha_n)}\int_{s_j>0}\mathrm{d}s_1\ldots ds_n\,\frac{\delta(s_1+\ldots+s_n-1)s_1^{\alpha_1-1}\ldots s_n^{\alpha_n-1}}{[s_1a_1+\ldots+s_na_n]^{\alpha_1+\ldots+\alpha_n}} \tag{J.4}$$

and, in particular, in the case $n = 2$

$$\frac{1}{a^\alpha b^\beta} = \frac{\Gamma(\alpha)\Gamma(\beta)}{\Gamma(\alpha+\beta)} \int_0^1 \mathrm{d}s \frac{s^{\alpha-1}[1-s]^{\beta-1}}{[sa+(1-s)b]^{\alpha+\beta}}. \tag{J.5}$$

2. The use of Schwinger's and Feynman's identities for the calculation of certain integrals

Let us now assume that $a_1, \ldots, a_n$ are polynomials of second-order with respect to m variables $p_1, \ldots, p_m$ and let us try to calculate the integral

$$I = \int_{-\infty}^{+\infty} \mathrm{d}p_1 \ldots \int_{-\infty}^{+\infty} \mathrm{d}p_m \frac{1}{(a_1)^{\alpha_1} \ldots (a_n)^{\alpha_n}} \tag{J.6}$$

which can also be written as follows [see (J.2)]

$$I = \frac{1}{\Gamma(\alpha_1)\ldots\Gamma(\alpha_n)} \int \mathrm{d}p_1 \ldots \mathrm{d}p_m \int_{t_j>0} \mathrm{d}t_1 \ldots \mathrm{d}t_n\, t_1^{\alpha_1-1} \ldots t_n^{\alpha_n-1}$$
$$\times \exp[-(a_1 t_1 + \ldots + a_n t_n)].$$

To calculate I, we shall use the Schwinger method and the Feynman method successively, but first let us define the problem more precisely. Consider the expression

$$G\{s\} = s_1 a_1 + \ldots + s_n a_n \tag{J.7}$$

where $s_1, \ldots, s_n$ are auxiliary variables. We can write $G\{s\}$ by ordering the terms according to their degree in $p_1, \ldots, p_m$, quantities which can be considered as the components of a vector p

$$G\{s\} = (p \cdot A\{s\} \cdot p) + L\{s\} \cdot p + B\{s\}. \tag{J.8}$$

Here, $A\{s\}$ is a matrix, $L\{s\}$ is a vector, and $B\{s\}$ is a constant. We can also write

$$G\{s\} = (p + \tfrac{1}{2}L\{s\} \cdot A^{-1}\{s\}) \cdot A\{s\} \cdot (p + \tfrac{1}{2}A^{-1}\{s\} \cdot L\{s\}) +$$
$$+ B\{s\} - \tfrac{1}{4}L\{s\} \cdot A^{-1}\{s\} \cdot L\{s\}.$$

The variables $p_1, \ldots, p_m$ can be expressed in terms of new variables $q_1, \ldots, q_m$ defined by

$$q = p + \tfrac{1}{2}A\{s\} \cdot L\{s\}, \tag{J.9}$$

and it is also convenient to set

$$C(s\} = B\{s\} - \tfrac{1}{4}L\{s\} \cdot A^{-1}\{s\} \cdot L\{s\}.$$

Thus, we can write

$$G\{s\} = (q \cdot A\{s\} \cdot q) + C\{s\}, \tag{J.10}$$

and we note that $A(s)$ and $C\{s\}$ are homogeneous of order one with respect to s

$$A\{\lambda s\} = \lambda A\{s\} \qquad C\{\lambda s\} = \lambda C\{s\}.$$

Let us now try to calculate I by using Schwinger's identity, which enables us to write I [see (J.6)] in the form

$$I = \frac{1}{\Gamma(\alpha_1)\ldots\Gamma(\alpha_n)} \int \mathrm{d}p_1 \ldots \mathrm{d}p_m \int_{t_j > 0} \mathrm{d}t_1 \ldots \mathrm{d}t_n \, t_1^{\alpha_1 - 1} \ldots t_n^{\alpha_n - 1} \times \exp[-(a_1 t_1 + \ldots + a_n t_n)].$$

In other words, according to (J.7),

$$I = \frac{1}{\Gamma(\alpha_1)\ldots\Gamma(\alpha_n)} \int_{t_j > 0} \mathrm{d}t_1 \ldots \mathrm{d}t_n \, t_1^{\alpha_1 - 1} \ldots t_n^{\alpha_n - 1} J_S\{t\} \tag{J.11}$$

where (S for Schwinger)

$$J_S\{t\} = \int \mathrm{d}p_1 \ldots \mathrm{d}p_m \, \mathrm{e}^{-G\{t\}}, \tag{J.12}$$

or more explicitly

$$J_S\{t\} = \mathrm{e}^{-C\{t\}} \int \mathrm{d}q_1 \ldots \mathrm{d}q_m \, \mathrm{e}^{-(q \cdot A\{t\} \cdot q)}.$$

This expression can easily be integrated after diagonalization of the matrix and we find

$$J_S\{t\} = \mathrm{e}^{-C\{t\}} \frac{\pi^{m/2}}{|\det A\{t\}|^{1/2}}.$$

Let us bring this result in (J.11); we obtain

$$I = \frac{\pi^{m/2}}{\Gamma(\alpha_1)\ldots\Gamma(\alpha_n)} \int_{t_j > 0} \mathrm{d}t_1 \ldots \mathrm{d}t_n \, t_1^{\alpha_1 - 1} \ldots t_n^{\alpha_n - 1} \frac{\mathrm{e}^{-C\{t\}}}{|\det A\{t\}|^{1/2}}. \tag{J.13}$$

Now the homogeneity properties of $A\{t\}$ and $C\{t\}$ enable us to eliminate an integration variable. In fact, let us set $t_j = Ts_j$; we have

$$A\{t\} = TA\{s\} \qquad C\{t\} = TC\{s\}$$

$A = (m \times m)$ matrix
Using this result, we can transform I as follows

$$I = \frac{\pi^{m/2}}{\Gamma(\alpha_1)\ldots\Gamma(\alpha_n)} \int_0^\infty \mathrm{d}T \int_{t_j > 0} \mathrm{d}t_1 \ldots \mathrm{d}t_n \, \delta(t_1 + \ldots + t_n - T) \, t_1^{\alpha_1 - 1} \ldots t_n^{\alpha_n - 1} \times \frac{\mathrm{e}^{-C\{t\}}}{|\det A\{t\}|^{1/2}}$$

$$I = \frac{\pi^{m/2}}{\Gamma(\alpha_1)\ldots\Gamma(\alpha_n)} \int_{s_j > 0} \mathrm{d}s_1 \ldots \mathrm{d}s_n \, \delta(s_1 + \ldots + s_n - 1) \frac{s_1^{\alpha_1 - 1} \ldots s_n^{\alpha_n - 1}}{|\det A\{s\}|^{1/2}} \times \int_0^\infty \mathrm{d}T \, T^{(\alpha_1 + \ldots + \alpha_n - m/2)} \, \mathrm{e}^{-TC\{s\}}.$$

Thus, we obtain the *final result*

$$I = \frac{\pi^{m/2}\,\Gamma(\alpha_1 + \ldots + \alpha_n - m/2)}{\Gamma(\alpha_1)\ldots\Gamma(\alpha_n)} \int_{s_j > 0} \mathrm{d}s_1 \ldots \mathrm{d}s_n$$
$$\times \frac{s_1^{\alpha_1 - 1} \ldots s_n^{\alpha_n - 1}\,\delta(s_1 + \ldots + s_n - 1)}{|\det A\{s\}|^{1/2}\,|C\{s\}|^{\alpha_1 + \ldots + \alpha_n - m/2}}. \quad \text{(J.14)}$$

Let us now use Feynman's method to calculate I. Starting from the definition (J.6) and combining (J.4) with (J.7), we get

$$I = \frac{\Gamma(\alpha_1 + \ldots + \alpha_n)}{\Gamma(\alpha_1)\ldots\Gamma(\alpha_n)} \int_{s_j > 0} \mathrm{d}s_1 \ldots \mathrm{d}s_n\, \delta(s_1 + \ldots + s_n - 1) s_1^{\alpha_1 - 1} \ldots s_n^{\alpha_n - 1} J_{\mathrm{F}}\{s\} \quad \text{(J.15)}$$

where (F for Feynman)

$$J_{\mathrm{F}}\{s\} = \int \mathrm{d}p_1 \ldots \mathrm{d}p_m \frac{1}{[G\{s\}]^{\alpha_1 + \ldots + \alpha_n}} \quad \text{(J.16)}$$

which according to (J.10) can also be written as follows

$$J_{\mathrm{F}}\{s\} = \int \mathrm{d}q_1 \ldots \mathrm{d}q_m \frac{1}{[(q \cdot A\{s\} \cdot q) + C\{s\}]^{\alpha_1 + \ldots + \alpha_n}}.$$

Let us perform a linear transformation of the q variables in order to diagonalize the quadratic form $(q \cdot A \cdot q)$; we obtain

$$J_{\mathrm{F}}\{s\} = \frac{1}{|\det A\{s\}|^{m/2}} \int \frac{\mathrm{d}^m p}{[p^2 + C\{s\}]^{\alpha_1 + \ldots + \alpha_n}}$$

where $\vec{p}$ is now an m-dimensional vector. In this way, we find

$$J_{\mathrm{F}}\{s\} = \frac{1}{|\det A\{s\}|^{m/2}\,[C\{s\}]^{\alpha_1 + \ldots + \alpha_n}} \int \frac{\mathrm{d}^m p}{[p^2 + 1]^{\alpha_1 + \ldots + \alpha_n}}. \quad \text{(J.17)}$$

The integral on the right-hand side is a number which can be easily calculated. In fact, let us recall that the 'area' $\Omega(d)$ of the surface of a ball of unit radius in a d-dimensional space is given by (F.1)

$$\Omega(d) = \frac{2\pi^{d/2}}{\Gamma(d/2)}.$$

Consequently, we have

$$\int \frac{\mathrm{d}^m p}{(p^2 + 1)^{\alpha_1 + \ldots + \alpha_n}} = \Omega(m) \int_0^\infty \frac{p^{m-1}\,\mathrm{d}p}{(p^2 + 1)^{\alpha_1 + \ldots + \alpha_n}}$$
$$= \frac{\pi^{m/2}}{\Gamma(m/2)} \int_0^\infty \frac{t^{-1 + m/2}}{(t + 1)^{\alpha_1 + \ldots + \alpha_n}} = \frac{\pi^{m/2}\,\Gamma(\alpha_1 + \ldots + \alpha_n - m/2)}{\Gamma(\alpha_1 + \ldots + \alpha_n)}. \quad \text{(J.18)}$$

By combining (J.15), (J.17), and (J.18), we recover the result (J.14). Both methods lead to an identical result. Schwinger's method is slightly simpler and has the advantage of corresponding to the Laplace–de Gennes transformation.

3. Generalization and application to the calculation of diagrams

Let us now assume that the quantities $a_1, \ldots, a_n$ are polynomials of order two with respect to m vectors $\vec{p}_1, \ldots, \vec{p}_m$ of a d-dimensional space. The component of order i of the vector $\vec{p}_j$ will be denoted by p_j^i. Our problem is to calculate

$$I_d = \int \frac{\mathrm{d}^d p_1 \ldots \mathrm{d}^d p_m}{(a_1)^{\alpha_1} \ldots (a_n)^{\alpha_n}}. \tag{J.19}$$

Let us set as before [see (J.7)]

$$G\{s\} = s_1 a_1 + \ldots + s_n a_n.$$

We admit that $G\{s\}$ can be written in the form

$$G\{s\} = \sum_i (p^i \cdot A\{s\} \cdot p^i) + \sum_i L^i\{s\} \cdot p^i + B\{s\}. \tag{J.20}$$

Here, the index i labels a component and p^i represents the set of the projections of the vectors $\vec{p}_1, \ldots, \vec{p}_m$ on the axis of index i. Now, we introduce

$$q^i = p^i + \tfrac{1}{2} A\{s\} \cdot L^i\{s\}$$

and, in this way, we can write

$$G\{s\} = \sum_i q^i \cdot A\{s\} \cdot q^i + C\{s\} \tag{J.21}$$

where

$$C\{s\} = B\{s\} - \frac{1}{4} \sum_i L^i\{s\} \cdot A\{s\} \cdot L^i\{s\}.$$

The new expression of I_d is obtained by generalizing (J.14): for that, we only have to replace m by md and $|\det A|$ by $|\det A|^d$ in this equation. We thus obtain the *final result*

$$I_d = \frac{\pi^{md/2}\,\Gamma(\alpha_1 + \ldots + \alpha_n - md/2)}{\Gamma(\alpha_1) \ldots \Gamma(\alpha_n)}$$

$$\int_{s_j > 0} \mathrm{d}s_1 \ldots \mathrm{d}s_n \frac{s_1^{\alpha_1 - 1} \ldots s_n^{\alpha_n - 1} \delta(s_1 + \ldots + s_n - 1)}{|\det A\{s\}|^{d/2} \, |C\{s\}|^{\alpha_1 + \ldots + \alpha_n - md/2}}. \tag{J.22}$$

The initial integral was over md variables, the above integral is over n variables. Thus, for $n < md$, the transformation is useful.

APPENDIX K

PROBABILITY DISTRIBUTION OF THE END-TO-END VECTEUR OF A LONG SELF-AVOIDING POLYMER: FIRST-ORDER CALCULATION IN $\varepsilon = 4 - d$

We want to calculate the probability distribution

$$P(\vec{r}) = \langle\!\langle \delta(\vec{r} - \vec{r}(s) + \vec{r}(0)) \rangle\!\rangle$$

for an isolated chain in the framework of the standard continuous model for $z \gg 1$. Its Fourier transform $\tilde{P}(\vec{k})$ is given by

$$\tilde{P}(\vec{k}) = \langle\!\langle \exp[\mathrm{i}\vec{k}\cdot(\vec{r}(s) - \vec{r}(0))] \rangle\!\rangle = \frac{\mathscr{Z}(k,S)}{\mathscr{Z}(S)}$$

To obtain $\tilde{P}(\vec{k})$ to first-order in $\varepsilon = 4 - d$, we must start by calculating $\tilde{P}(\vec{k})$ to first-order in z. We have

$$\tilde{P}(\vec{k}) = \frac{\exp(-k^2 S/2) + {}^1\mathscr{Z}(k,S)}{1 + {}^1\mathscr{Z}(S)}$$

where ${}^1\mathscr{Z}(k,S)$ is given by (10.7.4) and ${}^1\mathscr{Z}(S)$ by (10.7.6)

$$\begin{aligned}{}^1\mathscr{Z}(k,S) = {} & -z\exp(-k^2 S/2)\frac{1}{(1 - d/2)(2 - d/2)} \\ & - z\int_0^1 \mathrm{d}x(1 - x)x^{-d/2}[\exp(-(1 - x)k^2 S/2) - \exp(-k^2 S/2)]\end{aligned}$$

$${}^1\mathscr{Z}(S) = -z\frac{1}{(1 - d/2)(2 - d/2)}$$

From these equalities, we obtain

$$\tilde{P}(\vec{k}) = \exp(-k^2 S/2)\left[1 - z\int_0^1 \mathrm{d}x(1 - x)x^{-d/2}(\exp(xk^2 S/2) - 1)\right].$$

Before going to the asymptotic limit, we must re-express $\tilde{P}(\vec{k})$ in terms of the real size of a polymer. Now, according to (10.7.6), we have

$$S = X^2\left(1 - \frac{z}{(2 - d/2)(3 - d/2)}\right).$$

Moreover, we can choose the osmotic parameter g as the expansion parameter, instead of z, and we know that to first-order $z = g + \dots$. In this way, we obtain an expression which converges when $z \to \infty$, namely

$$\tilde{P}(\vec{k}) = \exp(-k^2X^2/2)\left\{1 + \frac{g}{(2-d/2)(3-d/2)}k^2X^2/2 - g\int_0^1 \mathrm{d}x(1-x)x^{-d/2}\left[\exp(-xk^2X^2/2) - 1\right]\right\}$$

or

$$\tilde{P}(\vec{k}) = \exp(-k^2X^2/2)\left\{1 - g\int_0^1 \mathrm{d}x(1-x)x^{-d/2}(\exp(xk^2X^2/2) - 1 - xk^2X^2/2)\right\}.$$

The integral which appears in the preceding equality converges when $d \to 4$. Now, we have only to put $d = 4$ in this expression to obtain an expansion of $\tilde{P}(\vec{k})$ which is valid in the asymptotic limit $g \to g^* = \frac{\varepsilon}{8} + \ldots$ [see (12.3.102)]

$$\tilde{P}(\vec{k}) = \exp(-k^2X^2/2) \times \left\{1 - g\int_0^1 \mathrm{d}x(1-x)x^{-2}(\exp(xk^2X^2/2) - 1 - xk^2X^2/2)\right\}. \quad \text{(K.1)}$$

Now let us calculate

$$P(\vec{r}) = \int \frac{\mathrm{d}^d k}{(2\pi)^d}\exp(i\vec{k}\cdot\vec{r})\tilde{P}(\vec{k})$$

from (K.1). Using the equality

$$\int_{-\infty}^{+\infty} \frac{\mathrm{d}^d k}{(2\pi)^d} e^{i\vec{k}\cdot\vec{r}}\exp(-ak^2/2) = \frac{1}{(2\pi a)^{d/2}}\exp(-r^2/2a)$$

we obtain

$$\tilde{P}(\vec{r})(2\pi)^{d/2}X^d = \exp(-r^2/2X^2) - g\int_0^1 \mathrm{d}x(1-x)x^{-2} \times \left\{\frac{\exp(-r^2/2X^2(1-x))}{(1-x)^{d/2}} - \exp(-r^2/2X^2)\left[1 + x\left(\frac{d}{2} - \frac{r^2}{2X^2}\right)\right]\right\}.$$

As we deal with an expansion of order ε, and as we know that g^* is of order ε, we can put $d = 4$ in the right-hand side of the preceding equation. Thus, let us change the integration variable by setting

$$\frac{1}{1-x} = 1 + t.$$

We obtain

$$\tilde{P}(\vec{r})(2\pi)^{d/2}X^d = \exp(-r^2/2X^2)\left[1 - g\psi\left(\frac{r^2}{2X^2}\right)\right]$$

where

$$\psi(y) = \int_0^\infty \frac{\mathrm{d}t}{t^2}\left[(1+t)e^{-yt} - \frac{1}{1+t} - \frac{t}{(1+t)^2}(2-y)\right].$$

Calculating this integral is elementary and gives

$$\psi(y) = -\ln y - \mathcal{E} + 2 + y(\ln y + \mathcal{E} - 2)$$

where $\mathcal{E}$ is Euler's constant

$$\mathcal{E} = -\int_0^\infty \mathrm{d}t \ln t\, \mathrm{e}^{-t} \simeq 0.577.$$

Finally, we obtain the result

$$\tilde{P}(\vec{r})(2\pi)^{d/2} X^d = \mathrm{e}^{-y}\{1 + g(\ln y + \mathcal{E} - 2) - gy(\ln y + \mathcal{E} - 2)\}$$

$$y = r^2/2X^2. \qquad \text{(K.2)}$$

APPENDIX L

MAXWELL'S RULES

For a solution which separates into two phases characterized by concentrations $\boldsymbol{C}^{\mathrm{I}}$ and $\boldsymbol{C}^{\mathrm{II}}$, the equilibrium conditions can be expressed as follows

$$\boldsymbol{\Pi}(\boldsymbol{C}^{\mathrm{I}}) = \boldsymbol{\Pi}(\boldsymbol{C}^{\mathrm{II}}) \qquad (= \bar{\boldsymbol{\Pi}}) \tag{L.1}$$

$$\boldsymbol{\mu}(\boldsymbol{C}^{\mathrm{I}}) = \boldsymbol{\mu}(\boldsymbol{C}^{\mathrm{II}}) \qquad (= \bar{\boldsymbol{\mu}}) \tag{L.2}$$

where $\boldsymbol{\Pi}(\boldsymbol{C})$ is the osmotic pressure and $\boldsymbol{\mu}(\boldsymbol{C})$ the chemical potential at the concentration $\boldsymbol{C} = \boldsymbol{N}/\boldsymbol{V}$. The symbols $\bar{\boldsymbol{\Pi}}$ and $\bar{\boldsymbol{C}}$ represent the osmotic pressure and the concentration at equilibrium.

These equilibrium conditions can be written in terms of $\boldsymbol{\Pi}(\boldsymbol{C})$ alone or of $\boldsymbol{\mu}(\boldsymbol{C})$ alone, by introducing Maxwell's rules, which will be now established. For this purpose, we assume that the free energy $\mathbb{F}$ can be written in the form

$$\mathbb{F} = \boldsymbol{V} f(\boldsymbol{N}/\boldsymbol{V})$$

and consequently

$$\boldsymbol{\mu}(\boldsymbol{C}) = \left.\frac{\partial \mathbb{F}}{\partial \boldsymbol{N}}\right)_{\boldsymbol{V}} = f'(\boldsymbol{C})$$

$$\boldsymbol{\beta}\boldsymbol{\Pi}(\boldsymbol{C}) = -\left.\frac{\partial \mathbb{F}}{\partial \boldsymbol{V}}\right)_{\boldsymbol{N}} = \boldsymbol{C} f'(\boldsymbol{C}) - f(\boldsymbol{C}). \tag{L.3}$$

The *Maxwell rule* for pressure reads

$$\int_{\boldsymbol{C}^{\mathrm{I}}}^{\boldsymbol{C}^{\mathrm{II}}} \frac{\mathrm{d}\boldsymbol{C}}{\boldsymbol{C}^2} (\boldsymbol{\Pi}(\boldsymbol{C}) - \bar{\boldsymbol{\Pi}}) = 0 \tag{L.4}$$

where

$$\bar{\boldsymbol{\Pi}} = \boldsymbol{\Pi}(\boldsymbol{C}^{\mathrm{I}}) = \boldsymbol{\Pi}(\boldsymbol{C}^{\mathrm{II}}).$$

It can be obtained as follows: from (L.3) we derive

$$\int_{\boldsymbol{C}^{\mathrm{I}}}^{\boldsymbol{C}^{\mathrm{II}}} \frac{\mathrm{d}\boldsymbol{C}}{\boldsymbol{C}^2} \boldsymbol{\beta}\boldsymbol{\Pi}(\boldsymbol{C}) = \int_{\boldsymbol{C}^{\mathrm{I}}}^{\boldsymbol{C}^{\mathrm{II}}} \mathrm{d}\boldsymbol{C} \left[\frac{f'(\boldsymbol{C})}{\boldsymbol{C}} - \frac{f(\boldsymbol{C})}{\boldsymbol{C}^2}\right] = \frac{f(\boldsymbol{C}^{\mathrm{II}})}{\boldsymbol{C}^{\mathrm{II}}} - \frac{f(\boldsymbol{C}^{\mathrm{I}})}{\boldsymbol{C}^{\mathrm{I}}}$$

$$= \boldsymbol{\beta}\frac{\boldsymbol{\Pi}(\boldsymbol{C}^{\mathrm{I}})}{\boldsymbol{C}^{\mathrm{I}}} - \boldsymbol{\beta}\frac{\boldsymbol{\Pi}(\boldsymbol{C}^{\mathrm{II}})}{\boldsymbol{C}^{\mathrm{II}}} + \boldsymbol{\mu}(\boldsymbol{C}^{\mathrm{II}}) - \boldsymbol{\mu}(\boldsymbol{C}^{\mathrm{I}})$$

and to establish (L.4) we have only to apply the equilibrium conditions (L.1) and (L.2).

The *Maxwell rule* for the chemical potential reads

$$\int_{\boldsymbol{C}^{\mathrm{I}}}^{\boldsymbol{C}^{\mathrm{II}}} \mathrm{d}\boldsymbol{C}[\boldsymbol{\mu}(\boldsymbol{C}) - \bar{\boldsymbol{\mu}}] = 0 \tag{L.5}$$

where

$$\bar{\mu} = \mu(C^{|}) = \mu(C^{||}).$$

It can be obtained as follows: from (L.3), we deduce

$$\int_{C^{|}}^{C^{||}} \mathrm{d}C\, \mu(C) = f(C^{|}) - f(C^{||})$$

$$= C^{|}\mu(C^{|}) - C^{||}\mu(C^{||}) - \beta\Pi(C^{||}) + \beta\Pi(C^{|}),$$

and to establish (L.5) we have only to apply the equilibrium conditions (L.1) and (L.2)

Thus, the rule (L.4) and the condition (L.1) entail (L.2); in the same way, the rule (L.5) and the condition (L.2) entail (L.1).

APPENDIX M

FOURIER TRANSFORM OF $r^{-\alpha}$

The integral $\int \mathrm{d}^d r\, r^{-\alpha}\, \mathrm{e}^{i\vec{q}\cdot\vec{r}}$ is convergent for

$$0 < \alpha < d. \tag{M.1}$$

In this case,

$$\int \mathrm{d}^d r\, r^{-\alpha}\, \mathrm{e}^{i\vec{q}\cdot\vec{r}} = J(\alpha,d) q^{\alpha-d} \tag{M.2}$$

where $J(\alpha,d)$ is a function which, as we shall see, has an analytic continuation for all values of d. Thus, formula (M.2) can be used to determine the asymptotic large q behaviour, of the d-dimensional Fourier transforms of the functions $F(\vec{r})$ which at the origin have a singularity of the form $r^{-\alpha}$.

Let us calculate $J(\alpha,d)$. For this purpose, we apply the operation

$$\int \mathrm{d}^d q \exp(-q^2/2)\ldots$$

to both sides of (M.2). Thus, we obtain (see Appendix A)

$$(2\pi)^{d/2} \int \mathrm{d}^d r\, r^{-\alpha} \exp(-r^2/2) = J(\alpha,d) \int \mathrm{d}^d q\, q^{\alpha-d} \exp(-q^2/2),$$

or more explicitly

$$(2\pi)^{d/2} \int_0^\infty \mathrm{d}r\, r^{d-\alpha-1} \exp(-r^2/2) = J(\alpha,d) \int_0^\infty \mathrm{d}q\, q^{\alpha-1} \exp(-q^2/2).$$

Let us set $r = 2x$ and $q^2 = 2y$. We obtain

$$(2\pi)^{d/2} \int_0^\infty \mathrm{d}x\, (2x)^{\frac{d}{2}-\frac{\alpha}{2}-1}\, \mathrm{e}^{-x} = J(\alpha,d) \int_0^\infty \mathrm{d}y\, (2y)^{\frac{\alpha}{2}-1}\, \mathrm{e}^{-y},$$

and this equality leads to the result

$$J(\alpha,d) = \pi^{d/2}\, 2^{d-\alpha} \frac{\Gamma(d/2-\alpha/2)}{\Gamma(\alpha/2)} \tag{M.3}$$

where $\Gamma(x)$ is the gamma function ($\Gamma(x+1) = x!$).

AUTHOR INDEX

Each number refers to the page where the corresponding author is quoted. A number with an exponent indicates a page containing a list of references; the exponent gives the order in the list.

SUBJECT INDEX

Important definitions are indicated by bold face numbers; a list of important symbols is also given at the beginning of the book.